THE
RISING
SEA

THE RISING SEA

Foundations of Algebraic Geometry

Ravi Vakil

PRINCETON UNIVERSITY PRESS

PRINCETON AND OXFORD

Published by Princeton University Press
41 William Street, Princeton, New Jersey 08540
99 Banbury Road, Oxford OX2 6JX

press.princeton.edu

All Rights Reserved

ISBN 978-0-691-26866-8
ISBN (pbk.) 978-0-691-26867-5
ISBN (e-book) 978-0-691-26868-2

Library of Congress Control Number: 2025938020

British Library Cataloging-in-Publication Data is available

Editorial: Diana Gillooly and Whitney Rauenhorst
Production Editorial: Kathleen Cioffi
Text and Cover Design: Wanda España
Production: Erin Suydam
Publicity: William Pagdatoon
Copyeditor: Bhisham Bherwani

Cover image: Hokusai's *The Great Wave at Kanagawa* (1760–1849) vintage Japanese Ukiyo-e woodcut print. Original public domain / Rawpixel

This book has been composed in Palatino

Printed in the United States of America

10 9 8 7 6 5 4 3 2 1

*This book is dedicated to
Alice, Benjamin, and Jacob.*

Je pourrais illustrer la . . . approche, en gardant l'image de la noix qu'il s'agit d'ouvrir. La première parabole qui m'est venue à l'esprit tantôt, c'est qu'on plonge la noix dans un liquide émollient, de l'eau simplement pourquoi pas, de temps en temps on frotte pour qu'elle pénètre mieux, pour le reste on laisse faire le temps. La coque s'assouplit au fil des semaines et des mois—quand le temps est mûr, une pression de la main suffit, la coque s'ouvre comme celle d'un avocat mûr à point! . . .

L'image qui m'était venue il y a quelques semaines était différente encore, la chose inconnue qu'il s'agit de connaître m'apparaissait comme quelque étendue de terre ou de marnes compactes, réticente à se laisser pénétrer. . . . La mer s'avance insensiblement et sans bruit, rien ne semble se casser rien ne bouge l'eau est si loin on l'entend à peine. . . . Pourtant elle finit par entourer la substance rétive.

I can illustrate the . . . approach with the . . . image of a nut to be opened. The first analogy that came to my mind is of immersing the nut in some softening liquid, and why not simply water? From time to time you rub so the liquid penetrates better, and otherwise you let time pass. The shell becomes more flexible through weeks and months—when the time is ripe, hand pressure is enough, the shell opens like a perfectly ripened avocado! . . .

A different image came to me a few weeks ago. The unknown thing to be known appeared to me as some stretch of earth or hard marl, resisting penetration. . . . The sea advances insensibly in silence, nothing seems to happen, nothing moves, the water is so far off you hardly hear it. . . . Yet finally it surrounds the resistant substance.

—A. Grothendieck [Gr5, p. 552-3], translation by C. McLarty [Mc, p. 1]

Contents

PART I

Preliminaries

PART II

Schemes

PART III
Morphisms of Schemes

PART IV
"Geometric" Properties of Schemes

PART V

Quasicoherent Sheaves on Schemes, and Their Uses

PART VI
More Cohomological Tools

Preface

This book is intended to give a serious and reasonably complete introduction to algebraic geometry, and is not just for (future) experts in the field. The exposition serves a narrow set of goals (see §0.4), and necessarily takes a particular point of view on the subject.

It has now been many decades since David Mumford wrote that algebraic geometry "seems to have acquired the reputation of being esoteric, exclusive, and very abstract, with adherents who are secretly plotting to take over all the rest of mathematics! In one respect this last point is accurate" ([Mu4, preface] and [Mu7, p. 227]). The revolution has now fully come to pass, and has fundamentally changed how we think about many fields of pure mathematics. A remarkable number of celebrated advances rely in some way on the insights and ideas first articulated by Alexander Grothendieck, Jean-Pierre Serre, and others.

For a number of reasons, algebraic geometry has earned a reputation of being inaccessible. The power of the subject comes from rather abstract heavy machinery, and it is easy to lose sight of the intuitive nature of the objects and methods. Many in nearby fields have only a vague sense of the fundamental ideas of the subject. Algebraic geometry itself has fractured into many parts, and even within algebraic geometry, new researchers are often unaware of the basic ideas in subfields removed from their own.

But there is another more optimistic perspective to be taken. The ideas that allow algebraic geometry to connect several parts of mathematics are fundamental, and well-motivated. Many people in nearby fields would find it useful to develop a working knowledge of the foundations of the subject, and not just at a superficial level. Within algebraic geometry itself, there is a canon (at least for those approaching the subject from this particular direction), which everyone in the field can and should be familiar with. The rough edges of scheme theory have been sanded down over the past half century, although there remains an inescapable need to understand the subject on its own terms.

0.0.1. *The importance of exercises.* This book has a lot of exercises. I have found that unless I have some problems I can think through, ideas don't get fixed in my mind. Some exercises are trivial— some experts find this offensive, but I find this desirable. A very few necessary ones may be hard, but the reader should have been given the background to deal with them—the difficult exercises are not just an excuse to push hard material out of the text. The exercises are interspersed with the exposition; they are not left for the end. Most have been extensively field-tested. The point of view here is one I explored with Kedlaya and Poonen in [KPV], a book that was ostensibly about problems, but secretly a case for how one should learn and think about and do mathematics. Most people learn by doing, rather than just passively reading. Judiciously chosen problems can be the best way of guiding the learner toward enlightenment.

0.0.2. *Structure.* You will quickly notice that everything is labeled x.y.z, where x is the chapter; y is the section; and z is a number, except that exercises are indicated by letters (and are sprinkled throughout the text, rather than at the end of sections). Individual paragraphs often get labels for ease of reference, or to indicate a new topic. Definitions are in bold, and are sometimes given in passing.

0.1   For the Reader

This book is intended to be a collection of communal wisdom, necessarily distilled through an imperfect filter. I wish to say a few words on how you might use it, although it is not clear to me whether you should follow this advice.

Before discussing details, I want to say clearly at the outset: the wonderful machine of modern algebraic geometry was created to understand basic and naive questions about geometry (broadly construed). The purpose of this book is to give you a thorough foundation in these powerful ideas. *Do not be seduced by the lotus-eaters into infatuation with untethered abstraction.* Hold tight to your geometric motivation as you learn the formal structures which have proved to be so effective in studying fundamental questions. When introduced to a new idea, always ask why you should care. Do not expect an answer right away, but demand an answer eventually. Try at least to apply any new abstraction to some concrete example you can understand well. See if you can make a rough picture to capture the essence of the idea. (I deliberately asked an uncoordinated and confused three-year-old to make most of the figures in the book in order to show that even quick sketches can enlighten and clarify.)

Understanding algebraic geometry is often thought to be hard because it consists of large complicated pieces of machinery. In fact the opposite is true; to switch metaphors, rather than being narrow and deep, algebraic geometry is shallow but extremely broad. It is built out of a large number of very small parts, in keeping with Grothendieck's vision of mathematics. It is a challenge to hold the entire organic structure, with its messy interconnections, in your head.

A reasonable place to start is with the idea of "affine complex varieties": subsets of $\mathbb{C}^n$ cut out by some polynomial equations. Your geometric intuition can immediately come into play—you may already have some ideas or questions about dimension, or smoothness, or solutions over subfields such as $\mathbb{R}$ or $\mathbb{Q}$. Wiser heads would counsel spending time understanding complex varieties in some detail before learning about schemes. Instead, I encourage you to learn about schemes immediately, learning about affine complex varieties as the central (but not exclusive) example. This is not ideal, but can save time, and is surprisingly workable. An alternative is to learn about varieties elsewhere, and then come back to this book later.

The intuition for schemes can be built on the intuition for affine complex varieties. Allen Knutson and Terry Tao have pointed out that this involves three different simultaneous generalizations, which can be interpreted as three large themes in mathematics. (i) We allow nilpotents in the ring of functions, which is basically *analysis* (looking at near-solutions of equations instead of exact solutions). (ii) We glue these affine schemes together, which is what we do in *differential geometry* (looking at manifolds instead of coordinate patches). (iii) Instead of working over $\mathbb{C}$ (or another algebraically closed field), we work more generally over a ring that isn't an algebraically closed field, or even a field at all, which is basically *number theory* (solving equations over number fields, rings of integers, etc.).

Because our goal is to be comprehensive, and to understand everything one should know after a first course, it will necessarily take longer to get to interesting sample applications. You may be misled into thinking that one has to work this hard to get to these applications—it is not true! You should deliberately keep an eye out for examples you would have cared about before. This will take some time and patience.

As you learn algebraic geometry, you should pay attention to crucial stepping stones. Of course, the steps get bigger the farther you go.

Chapter 1. Category theory is only language, but it is language with an embedded logic. Category theory is much easier once you realize that it is designed to formalize and abstract things you already know. The initial chapter on category theory prepares you to think cleanly. For example, when someone names something a "cokernel" or a "product," you should want to know why it deserves that name, and what the name really should mean. The conceptual advantages of thinking this way will gradually become apparent over time. Yoneda's Lemma—and more generally, the idea of understanding an object through the maps to it—will play an important role.

Chapter 2. The theory of sheaves again abstracts something you already understand well (see the motivating example of §2.1), and what is difficult is understanding how one best packages and works with the information of a sheaf (stalks, sheafification, sheaves on a base, etc.).

Chapters 1 and 2 are a risky gamble, and they attempt a delicate balance. Attempts to explain algebraic geometry often leave such background to the reader, refer to other sources the reader won't read, or punt it to a telegraphic appendix. Instead, this book attempts to explain everything necessary, but as little as possible, and tries to get across how you should think about (and work with) these fundamental ideas, and why they are more grounded than you might fear.

Chapters 3–5. Armed with this background, you will be able to think cleanly about various sorts of "spaces" studied in different parts of geometry (including real manifolds, topological spaces, and complex manifolds), as ringed spaces that locally are of a certain form. A scheme is just another kind of "geometric space," and we are then ready to transport lots of intuition from "classical geometry" to this new setting. (This also will set you up to later think about other geometric kinds of spaces in algebraic geometry, such as complex analytic spaces, algebraic spaces, orbifolds, stacks, rigid analytic spaces, and formal schemes.) The ways in which schemes differ from your geometric intuition can be internalized, and your intuition can be expanded to accommodate them. There are many properties you will realize you will want, as well as other properties that will later prove important. These all deserve names. Take your time becoming familiar with them.

Chapters 7–11. Thinking categorically will lead you to ask about morphisms of schemes (and other spaces in geometry). One of Grothendieck's fundamental lessons is that the morphisms are central. Important geometric properties should really be understood as properties of morphisms. There are many classes of morphisms with special names, and in each case you should think through why that class deserves a name.

Chapters 12–13. You will then be in a good position to think about fundamental geometric properties of schemes: dimension and smoothness. You may be surprised that these are subtle ideas, but you should keep in mind that they are subtle everywhere in mathematics.

Chapters 14–22. Vector bundles are ubiquitous tools in geometry, and algebraic geometry is no exception. They lead us to the more general notion of quasicoherent sheaves, much as free modules over a ring lead us to modules more generally. We study their properties next, including cohomology. Chapter 19, applying these ideas to study curves, may help make clear how useful they are.

Chapters 23–29. With this in hand, you are ready to learn more advanced tools widely used in the subject. Many examples of what you can do are given, and the classical story of the 27 lines on a smooth cubic surface (Chapter 27) is a good opportunity to see many ideas come together.

The rough logical dependencies among the chapters are shown in Figure 0.1. (Caution: This should be taken with a grain of salt. For example, you can avoid using much of Chapter 19 on curves in later chapters, but it is a crucial source of examples, and a great way to consolidate your understanding.)

In general, I prefer to have as few hypotheses as possible. Certainly a hypothesis that isn't necessary to the proof is a red herring. But if a reasonable hypothesis can make the proof cleaner and more memorable, I am willing to include it.

In particular, Noetherian hypotheses are handy when necessary, but are otherwise misleading. Even Noetherian-minded readers (normal human beings) are better off having the right hypotheses, as they will make clearer why things are true.

We often state results particular to varieties, especially when there are techniques unique to this situation that one should know. But restricting to algebraically closed fields is useful surprisingly rarely. Geometers needn't be afraid of arithmetic examples or of algebraic examples; a central insight of algebraic geometry is that the same formalism applies without change.

Pathological examples are useful to know. On mountain highways, there are tall sticks on the sides of the road designed for bad weather. In winter, you cannot see the road clearly, and the sticks serve as warning signs: if you cross this line, you will die! Pathologies and (counter)examples serve a similar goal. They also serve as a reality check when confronting a new statement, theorem, or conjecture whose veracity you may doubt. (See, for example, §4.1.8.)

When working through a book in algebraic geometry, it is particularly helpful to have other algebraic geometry books at hand, to see different approaches and to have alternative expositions when things become difficult. This book may serve as a good secondary book. If it is your primary source, then two other excellent books with what I consider a similar philosophy are [Liu] and [GW]. De Jong's encyclopedic online reference [Stacks] is peerless. There are many other outstanding sources out there, perhaps one for each approach to the subject; you should browse around and find one you find sympathetic. If you are looking for a correct or complete history of the subject, you have come to the wrong place. This book is not intended to be a complete guide to the literature, and many important sources are ignored or left out, due to my own ignorance and laziness.

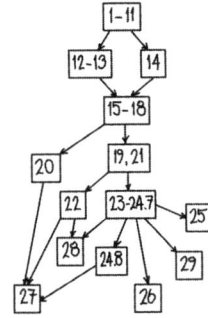

Figure 0.1 *Important logical dependencies among chapters (or, more precisely, a directed graph showing which chapter should be read before which other chapter).*

Finally, if you attempt to read this without working through a significant number of exercises (see §0.0.1), I will come to your house and pummel you with [Gr-EGA] until you beg for mercy. It is important to not just have a vague sense of what is true, but to be able to actually get your hands dirty. To quote Mark Kisin: "You can wave your hands all you want, but it still won't make you fly." Note: The hints may help you, but sometimes they may not.

0.2   For the Expert

If you use this book for a course, you should of course adapt it to your own point of view and your own interests. In particular, you should think about an application or theorem you want to reach at the end of the course (which may well not be in this book), and then work toward it. You should feel no compulsion to sprint to the end; I advise instead taking more time, and ending at the right place for your students. (Figure 0.1, showing large-scale dependencies among the chapters, may help you map out a course.) I have found that the theory of curves (Chapter 19) and the 27 lines on the cubic surface (Chapter 27) have served this purpose well at the end of winter and spring quarters. This has been true even if some of the needed background has not been covered, and has had to be taken by students as some sort of black box. For the first quarter, the goal is to build a common language for many kinds of geometry and geometric spaces (not just algebraic geometry, but also manifolds, complex geometry, some differential geometry, and more)—a sort of archetypal form of a geometric space.

Faithfulness to the goals of §0.4 required a brutal triage, and I have made a number of decisions you may wish to reverse. I will briefly describe some choices made that may be controversial.

Decisions on how to describe things were made for the sake of the learners. If there were two approaches, and one was "correct" from an advanced point of view, and one was direct and natural from a naive point of view, I went with the latter.

On the other hand, the theory of varieties (over an algebraically closed field, say) was *not* done first and separately. This choice brought me close to tears, but in the end I am convinced that it can work well, if done in the right spirit.

Instead of spending the first part of the course on varieties, I spent the time in a different way. It is tempting to assume that students will either arrive with great comfort and experience with category theory and sheaf theory, or that they will pick up these ideas on their own time. I would love to live in that world. I encourage you to not skimp on these foundational issues. I have found that although these first lectures felt painfully slow to me, they were revelatory to a number of the students, and those with more experience were not bored and did not waste their time. This investment paid off in spades when I was able to rely on their ability to think cleanly and to use these tools in practice. Furthermore, if they left the course with nothing more than hands-on experience with these ideas, the world was still better off for it.

For the most part, we will state results in the maximal generality that the proof justifies, but we will not give a much harder proof if the generality of the stronger result will not be used. There are a few cases where we work harder to prove a somewhat more general result that many readers may not appreciate. For example, we prove a number of theorems for proper morphisms, not just projective morphisms. But in such cases, readers are invited or encouraged to ignore the subtleties required for the greater generality.

I consider line bundles (and maps to projective space) more fundamental than divisors. General Cartier divisors are not seriously discussed (§15.6.4), although *effective* Cartier divisors play an essential role.

Cohomology is done first using the Čech approach (as Serre first did), and derived functor cohomology is introduced only later. I am well aware that Grothendieck thinks that the agreement of Čech cohomology with derived functor cohomology "should be considered as an accidental phenomenon," and that "it is important for technical reasons not to take as *definition* of cohomology the Čech cohomology," [Gr3, p. 108]. But I am convinced that this is the right way for most people to see this kind of cohomology *for the first time*. (It is certainly true that many topics in algebraic geometry are best understood in the language of derived functors. But this is a view from the mountaintop, looking down, and not the best way to explore the forests. In order to appreciate derived functors appropriately, one must understand the homological algebra behind it, and not just take it as a black box.)

We restrict to the Noetherian case only when it is necessary, or (rarely) when it really saves effort. In this way, non-Noetherian people will clearly see where they should be careful, and Noetherian people will realize that non-Noetherian things are not so terrible. Moreover, even if you are interested primarily in Noetherian schemes, it helps to see "Noetherian" in the hypotheses of theorems only when necessary, as it will help you remember how and when this property gets used.

There are some cases where Noetherian readers will suffer a little more than they would otherwise. As an inflammatory example, instead of using Noetherian hypotheses, we invoke the notion of quasiseparatedness early and often. The cost is that one extra word has to be remembered, on top of an overwhelming number of other words. But once that is done, it is not hard to remember that essentially every scheme anyone cares about is quasiseparated. Furthermore, whenever the hypotheses "quasicompact and quasiseparated" turn up, the reader will immediately guess a key idea of the proof. As another example, coherent sheaves and finite type (quasicoherent) sheaves are the same in the Noetherian situation, but are still worth distinguishing in statements of the theorems and exercises, for the same reason: to be clearer on what is used in the proof.

Many important topics are not discussed. Valuative criteria are not proved (see §13.7), and their statement is relegated to an optional section. Completely omitted: dévissage, formal schemes, and cohomology with supports. Sorry!

0.3   Background and Conventions

"Should you just be an algebraist or a geometer?" is like saying, "Would you rather be deaf or blind?"

—M. Atiyah, [At2, p. 659]

All rings are assumed to be commutative unless explicitly stated otherwise. All rings are assumed to contain a unit, denoted by 1. Maps of rings must send 1 to 1. We don't require that $0 \neq 1$; in other words, the "0-ring" (with one element) is a ring. (There is a ring map from any ring to the 0-ring; the 0-ring only maps to itself. The 0-ring is the final object in the category of rings.) The definition of "integral domain" includes $1 \neq 0$, so the 0-ring is not an integral domain. We accept the Axiom of Choice, usually in the guise of Zorn's Lemma. In particular, any proper ideal in a ring is contained in a maximal ideal. (The Axiom of Choice also arises in the argument that the category of A-modules has enough injectives; see Exercise 23.2.G.)

The reader should be familiar with some basic notions in commutative ring theory, in particular, the notion of ideals (including prime and maximal ideals), various types of rings (including integral domains, principal ideal domains, unique factorization domains, and local rings), localization, and modules. Tensor products and exact sequences of A-modules will be important. We will use the notation $(A, \mathfrak{m})$ or $(A, \mathfrak{m}, k)$ for local rings (rings with a unique maximal ideal)—A is the ring, $\mathfrak{m}$ its maximal ideal, and $k = A/\mathfrak{m}$ its residue field. We will use the structure theorem for finitely generated modules over a principal ideal domain A: any such module can be written as the direct sum of principal modules $A/(a)$. Some experience with field theory will be important from time to time.

Manifolds will be brought up periodically as examples, but for the most part, they are meant for motivation, so we will often not specify whether they are topological (real) manifolds, differentiable (C^∞, i.e., smooth) real manifolds, analytic (real) manifolds, or complex (holomorphic) manifolds. Nonetheless, we will define all four (Definition 4.3.9).

0.3.1. *Caution about foundational issues.* We will not concern ourselves with subtle foundational issues (set-theoretic issues, universes, etc.). It is true that some people should be careful about these issues. But is that really how you want to live your life? (If you are one of these rare people, a good start is [KS2, §1.1].)

0.3.2. *Further background.* It may be helpful to have books on other subjects at hand that you can dip into for specific facts, rather than reading them in advance. In commutative algebra, [E] is good for this. Other popular choices are [AtM] and [Mat2]. The book [Al] takes a point of view useful to algebraic geometry. For homological algebra, [Weib] is simultaneously detailed and readable.

Background from other parts of mathematics (topology, geometry, complex analysis, number theory, . . .) will of course be helpful for intuition and grounding. Some previous exposure to topology is certainly essential.

0.3.3. *Nonmathematical conventions.* "Unimportant" means "unimportant for the current exposition," *not* necessarily unimportant in the larger scheme of things. Other words may be used idiosyncratically as well.

There are optional sections of topics worth knowing on a second or third (but not first) reading. They are marked with a star: ∗. Starred sections are not necessarily harder, but merely unimportant. You should not read double-starred sections (∗∗) unless you really really want to, but you should be aware of their existence. (It may be strange to have parts of a book that should *not* be read!)

Let's now find out if you are taking my advice about double-starred sections.

0.4∗∗   The Goals of This Book

There are a number of possible introductions to the field of algebraic geometry: Riemann surfaces; complex geometry; the theory of varieties; a nonrigorous examples-based introduction; algebraic geometry for number theorists; an abstract functorial approach; and more. All have their place. Different approaches suit different students (and different advisors). This book takes only one route.

Our intent is to cover a canon completely and rigorously, with enough examples and calculations to help develop intuition for the machinery. This is often the content of a "second course" in algebraic geometry, and in an ideal world, people would learn this material over many years, after having background courses in commutative algebra, algebraic topology, differential geometry, complex analysis, homological algebra, number theory, and French literature. We do not live in an ideal world. For this reason, the book is written as a first introduction, but a challenging one.

This book seeks to do a very few things, but tries to do them well. Our goals and premises are as follows.

The core of the material should be digestible over a single year. After a year of blood, sweat, and tears, readers should have a broad familiarity with the foundations of the subject, and be ready to attend seminars, and learn more advanced material. They should not just have a vague intuitive understanding of the ideas of the subject; they should know interesting examples, know why they are interesting, and be able to work through their details. Readers in other fields of mathematics should know enough to understand the algebro-geometric ideas that arise in their area of interest.

This means that this book is not encyclopedic, and even beyond that, hard choices have to be made. (In particular, analytic aspects are essentially ignored, and are at best dealt with in passing, without proof. This is a book about *algebraic* algebraic geometry.)

This book is usable (and has been used) for a course, but the course should (as always) take on the personality of the instructor. With a good course, people should be able to leave early and still get something useful from the experience. With this book, it is possible to leave without regret after learning about category theory, or about sheaves, or about geometric spaces, having become a better person.

The book is also usable (and has been used) for learning on one's own. But most mortals cannot learn algebraic geometry fully on their own; ideally, you should read in a group, and even if not, you should have someone you can ask questions to (both stupid and smart questions).

There is certainly more than a year's material here, but I have tried to make clear which topics are essential, and which are not. Those teaching a class will choose which "inessential" things are important for the point they wish to get across, and use them.

There is a canon (at least for this particular approach to algebraic geometry). I have been repeatedly surprised at how much people in different parts of algebraic geometry agree on what every civilized algebraic geometer should know after a first (serious) year. (There are of course different canons for different parts of the subject, e.g., complex algebraic geometry, combinatorial algebraic geometry, computational algebraic geometry, etc.) There are extra bells and whistles that different instructors might add on, to prepare students for their particular part of the field or their own point of view, but the core of the subject remains unified, despite the diversity and richness of the subject. There are some serious and painful compromises to be made to reconcile this goal with the previous one.

Algebraic geometry is for everyone (with the appropriate definition of "everyone"). Algebraic geometry courses tend to require a lot of background, which makes them inaccessible to all but those who know they will go deeply into the subject. Algebraic geometry is too important for that; it is essential that many of those in nearby fields develop some serious familiarity with the foundational ideas and tools of the subject, and not just at a superficial level. (Similarly, algebraic geometers uninterested in any nearby field are necessarily arid, narrow thinkers. Do not be such a person!)

For this reason, this book attempts to require as little background as possible. The background required will, in a technical sense, be surprisingly minimal—ideally just some commutative ring theory and point-set topology, and call for some comfort with things like prime ideals and localization. This is misleading, of course—the more you know, the better. And the less background you have, the harder you will have to work—this is not a light read. On a related note . . .

The book is intended to be as self-contained as possible. I have tried to follow the motto: "If you use it, you must prove it." I have noticed that most students are human beings: if you tell them that some algebraic fact is in some late chapter of a book on commutative algebra, they will not immediately go and read it. Surprisingly often, what we need can be developed quickly from scratch, and even if people do not read it, they can see what is involved. The cost is that the book is much denser, and that significant sophistication and maturity is demanded of the reader. The benefit is that more people can follow it; they are less likely to reach a point where they get thrown. On the other hand, people who already have some familiarity with algebraic geometry, but want to understand the foundations more completely, should not be bored, and can focus on more subtle issues.

This goal is important because one should not just know what is true, but also know why things are true, and what is hard, and what is not hard. Also, this helps the previous goal, by reducing the number of prerequisites.

The book is intended to build intuition for the formidable machinery of algebraic geometry. The exercises are central for this (see §0.0.1). Informal language can sometimes be helpful. Many examples are given. (If you do not have pictures in your head which provide you with insight into why things are true, and how to prove things, then you cannot really say that you are "thinking geometrically.") Learning how to think cleanly (and, in particular, categorically) is essential. The advantages of *appropriate* generality should be made clear by example, and not through intimidation. The motivation is more local than global. For example, there is no introductory chapter explaining why one might be interested in algebraic geometry, and instead there is an introductory chapter explaining why you should want to think categorically (and how to actually do this).

Balancing the above goals is already impossible. We must thus give up any hope of achieving any other desiderata. **There are no other goals.**

0.4.1. Inadequate acknowledgments.
This entire project consists of communal wisdom passed from person to person. I have tried to collect it and distill it and curate it, but I can in no way begin to give correct credit for the ideas. I cannot even begin to correctly credit the people I personally learned from. There are far too many to list, and any list I have tried to make has had too many painful omissions. Instead, I will try to describe the broad classes of people who have influenced this work.

My life began in Toronto, and I had an early glimpse of algebraic geometry at the University of Toronto, through Bierstone, Milman, Murty, Arthur, and many others. In my graduate and postdoctoral years at Harvard, Princeton, and MIT, I fell in love with algebraic geometry and developed into the mathematician that I am, and if you know whom I met during those years—older role models and teachers, slightly older mentors, many peers, and brilliant younger students—you will see their personalities in these pages. In some important algebro-geometric sense, Joe Harris gave me a heart, Brendan Hassett gave me a brain, and Johan de Jong gave me a spine. But the entire ethos of algebraic geometry at the time—where subfields were rapidly developing, yet people learned from and talked to each other across the entire field, with generosity and friendship—is what attracted me into the subject. Perhaps most of the famous algebraic geometers of today have their ideas in these pages (although they may not realize it), coming from long conversations at conferences, or gleaned from talks.

It may surprise the next group how much I learned from them: my students and postdocs (both official ones and those who merely passed through). My main contribution as a mathematician has been in thinking about how to develop talent, and I have have taken great joy in watching extraordinary talent develop.

I (and soon, you) also owe an extreme debt to those have contributed to these notes over the years. Many people have individually given hundreds of detailed useful comments, and the resulting conversations have been particularly rewarding. At one point I kept a list of those in this group who merited particular thanks, but the length and arbitrariness of that list led me to decide to leave the evidence in public on the website where these notes came to life, math216.wordpress.com. (Many people preferred to give their comments by email, so their contributions may be less visible.)

The phrase "the rising sea" is due to Grothendieck [Gr5, pp. 552–3], with this particular translation by McLarty [Mc, p. 1]. The phrase was popularized as the title of Daniel Murfet's excellent blog [Mur], and I want to particularly thank Daniel for his generosity in sharing this title. The cover is an ominous variation on Hokusai's famous woodblock print, *The Great Wave off Kanagawa*. Mike Stay is the author of Jokes 1.2.12 and 21.5.2. Many particulars of the design of the prepublication version of the book are due to a number of people, but in particular to Sándor Kovács. I am also indebted to Diana Gillooly for her care and wisdom in shepherding this book to publication.

I am grateful for the financial support this project has indirectly received over the years, from Stanford, the National Science Foundation, and the Simons Foundation. The generous support of fundamental research and basic science has been an important engine of human development over centuries, so it is a pleasure to acknowledge it here.

Finally, at the most basic level, I am grateful to my family, and for my family.

PART I

Preliminaries

I

Chapter 1

Just Enough Category Theory to Be Dangerous

Was mich nicht umbringt, macht mich stärker. That which does not kill me, makes me stronger.
—F. Nietzsche [N, aphorism number 8]

Before we get to any interesting geometry, we need to develop a language to discuss things cleanly and effectively. This is best done in the language of categories. There is not much to know about categories to get started; it is just a very useful language. Like all mathematical languages, category theory comes with an embedded logic, which allows us to abstract intuitions in settings we know well to far more general situations.

Our motivation is as follows. We will be creating some new mathematical objects (such as schemes, and certain kinds of sheaves), and we expect them to act like objects we have seen before. We could try to nail down precisely what we mean by "act like," and what minimal set of things we have to check in order to verify that they act the way we expect. Fortunately, we don't have to—other people have done this before us, by defining key notions, such as *abelian categories*, which behave like modules over a ring.

Our general approach will be as follows. I will try to tell you what you need to know, and no more. (This I promise: If I use the word "topoi," you can shoot me.) I will begin by telling you things you already know, and describing what is essential about the examples, in a way that we can abstract a more general definition. We will then see this definition in less familiar settings, and get comfortable with using it to solve problems and prove theorems.

1.0.1. *Example: product.* For example, we will define the notion of *product* of schemes. We could just give a definition of product, but then you should want to know why this precise definition deserves the name of "product." As a motivation, we revisit the notion of product in a situation we know well: (the category of) sets. One way to define the product of sets U and V is as the set of ordered pairs $\{(u, v) : u \in U, v \in V\}$. But someone from a different mathematical culture might reasonably define it as the set of symbols $\{{}^u_v : u \in U, v \in V\}$. These notions are "obviously the same." Better: There is "an obvious bijection between the two."

This can be made precise by giving a better definition of product, in terms of a *universal property*. Given two sets M and N, a product is a set P, along with maps $\mu: P \to M$ and $\nu: P \to N$, such that for *any set P' with maps $\mu': P' \to M$ and $\nu': P' \to N$*, these maps factor *uniquely* through P:

(1.0.1.1)

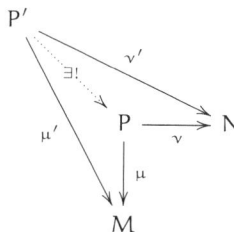

(The symbol $\exists$ means "there exists," and the symbol ! means "unique.") Thus a **product** is a *diagram*

$$P \xrightarrow{\nu} N$$
$$\mu \downarrow$$
$$M \qquad ,$$

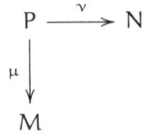

and not just a set P, although the maps μ and ν are often left implicit.

This definition agrees with the traditional definition, with one twist: there isn't just a single product; but any two products come with a *unique* isomorphism between them. In other words, the product is unique up to unique isomorphism. Here is why: If you have a product

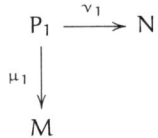

$$P_1 \xrightarrow{\nu_1} N$$
$$\mu_1 \downarrow$$
$$M$$

and I have a product

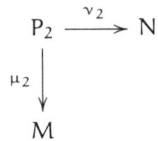

$$P_2 \xrightarrow{\nu_2} N$$
$$\mu_2 \downarrow$$
$$M$$

then by the universal property of my product (letting (P_2, μ_2, ν_2) play the role of (P, μ, ν) and (P_1, μ_1, ν_1) play the role of (P', μ', ν') in (1.0.1.1)), there is a unique map $f\colon P_1 \to P_2$ making the appropriate diagram commute (i.e., $\mu_1 = \mu_2 \circ f$ and $\nu_1 = \nu_2 \circ f$). Similarly, by the universal property of your product, there is a unique map $g\colon P_2 \to P_1$ making the appropriate diagram commute. Now consider the universal property of my product, this time letting (P_2, μ_2, ν_2) play the role of both (P, μ, ν) and (P', μ', ν') in (1.0.1.1). There is a unique map $h\colon P_2 \to P_2$ such that

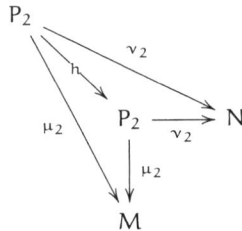

$$P_2$$
$$\mu_2 \quad h \quad \nu_2$$
$$P_2 \xrightarrow{\nu_2} N$$
$$\mu_2 \downarrow$$
$$M$$

commutes. However, I can name two such maps: the identity map id_{P_2} and $f \circ g$. Thus $f \circ g = \mathrm{id}_{P_2}$. Similarly, $g \circ f = \mathrm{id}_{P_1}$. Thus the maps f and g arising from the universal property are bijections. In short, there is a unique bijection between P_1 and P_2 preserving the "product structure" (the maps to M and N). This gives us the right to name any such product $M \times N$, since any two such products are uniquely identified.

This definition has the advantage that it works in many circumstances, and once we define categories, we will soon see that the above argument applies verbatim in any category to show that products, if they exist, are unique up to unique isomorphism. Even if you haven't seen the definition of category before, you can verify that this agrees with your notion of product in some category that you have seen before (such as the category of vector spaces, or the category of manifolds).

This is handy even in cases that you understand. For example, one way of defining the product of two manifolds M and N is to cut them both up into charts, then take products of charts, then glue them together. But if I cut up the manifolds M and N in one way, and you cut them up in another, how do we know our resulting product manifolds are the "same"? We could wave our hands, or make an annoying argument about refining covers, but instead, we should just show that they are "categorical products" and hence canonically the "same" (i.e., isomorphic). We will formalize this argument in §1.2.

Another set of notions we will abstract is that of categories that "behave like modules." We will want to define kernels and cokernels for new notions, and we should make sure that these notions behave the way we expect them to. This leads us to the definition of *abelian categories*, first defined by Grothendieck in his Tôhoku paper [Gr1].

In this chapter, we will give an informal introduction to these and related notions, in the hope of our developing just enough familiarity to comfortably use them in practice.

1.1 Categories and Functors

The introduction of the digit 0 or the group concept was general nonsense too, and mathematics was more or less stagnating for thousands of years because nobody was around to take such childish steps.

— A. Grothendieck, [BroP, pp. 4–5]

Before functoriality, people lived in caves.

— B. Conrad

We begin with an informal definition of categories and functors.

1.1.1. Categories.

A **category** consists of a collection of **objects**, and for each pair of objects, a set of **morphisms** (or **arrows**) between them. (For experts: technically, this is the definition of a *locally small category*. In the correct definition, the morphisms need only form a class, not necessarily a set, but see Caution 0.3.1.) Morphisms are often informally called **maps**. The collection of objects of a category $\mathscr{C}$ is often denoted by $\mathrm{obj}(\mathscr{C})$, but we will usually denote the collection also by $\mathscr{C}$. If $A, B \in \mathscr{C}$, then the set of morphisms from A to B is denoted by $\mathrm{Mor}(A, B)$. A morphism is often written $f: A \to B$, and A is said to be the **source** of f, and B the **target** of f. (Of course, $\mathrm{Mor}(A, B)$ is taken to be disjoint from $\mathrm{Mor}(A', B')$ unless $A = A'$ and $B = B'$.)

Morphisms compose as expected: there is a composition $\mathrm{Mor}(B, C) \times \mathrm{Mor}(A, B) \to \mathrm{Mor}(A, C)$, and if $f \in \mathrm{Mor}(A, B)$ and $g \in \mathrm{Mor}(B, C)$, then their composition is denoted by $g \circ f$. Composition is associative: if $f \in \mathrm{Mor}(A, B)$, $g \in \mathrm{Mor}(B, C)$, and $h \in \mathrm{Mor}(C, D)$, then $h \circ (g \circ f) = (h \circ g) \circ f$. For each object $A \in \mathscr{C}$, there is always an **identity morphism** $\mathrm{id}_A : A \to A$, such that when you (left- or right-) compose a morphism with the identity, you get the same morphism. More precisely, for any morphisms $f: A \to B$ and $g: B \to C$, $\mathrm{id}_B \circ f = f$ and $g \circ \mathrm{id}_B = g$. (If you wish, you may check that "identity morphisms are unique": there is only one morphism deserving the name id_A.) This ends the definition of a category.

We have a notion of **isomorphism** between two objects of a category (a morphism $f: A \to B$ such that there exists some—necessarily unique—morphism $g: B \to A$, where $f \circ g$ and $g \circ f$ are the identities on B and A respectively).

1.1.2. *Example.* The prototypical example to keep in mind is the category of sets, denoted by *Sets*. The objects are sets, and the morphisms are maps of sets. (Because Russell's paradox shows that there is no set of all sets, we did not say earlier that there is a set of all objects. But as stated in §0.3, we are deliberately omitting all set-theoretic issues.)

1.1.3. *Example.* Another good example is the category Vec_k of vector spaces over a given field k. The objects are k-vector spaces, and the morphisms are linear transformations. (What are the isomorphisms?)

1.1.A. UNIMPORTANT EXERCISE. A category in which each morphism is an isomorphism is called a **groupoid**. (This notion is not important in what we will discuss. The point of this exercise is to give you some practice with categories, by relating them to an object you know well.)

(a) A perverse definition of a *group* is: a groupoid with one object. Make sense of this. (Similarly, in case you care, a perverse definition of a **monoid** is: a category with one object.)

(b) Describe a groupoid that is not a group.

1.1.B. EXERCISE. If A is an object in a category $\mathscr{C}$, show that the invertible elements of $\mathrm{Mor}(A, A)$ form a group (called the **automorphism group of** A, denoted by $\mathrm{Aut}(A)$). What are the automorphism groups of the objects in Examples 1.1.2 and 1.1.3? Show that two isomorphic objects have isomorphic automorphism groups. (For readers with a topological background: if X is a topological space, then the fundamental groupoid is the category where the objects are points of X, and the morphisms $x \to y$ are paths from x to y, up to homotopy. Then the automorphism group of x_0 is the (pointed) fundamental group $\pi_1(X, x_0)$. In the case where X is connected, and $\pi_1(X)$ is not abelian, this illustrates the fact that for a connected groupoid—whose definition you can guess—the automorphism groups of the objects are all isomorphic, but not canonically isomorphic.)

1.1.4. *Example: Abelian groups.* The abelian groups, along with group homomorphisms, form a category *Ab*.

1.1.5. *Important Example: Modules over a ring.* If A is a ring, then the A-modules form a category Mod_A. (This category has additional structure; it will be the prototypical example of an *abelian category*; see §1.5.) Taking $A = k$, we obtain Example 1.1.3; taking $A = \mathbb{Z}$, we obtain Example 1.1.4.

1.1.6. *Example: Rings.* There is a category *Rings*, where the objects are rings, and the morphisms are maps of rings in the usual sense (maps of sets that respect addition and multiplication, and that send 1 to 1 by our conventions; see §0.3).

1.1.7. *Example: Topological spaces.* The topological spaces, along with continuous maps, form a category *Top*. The isomorphisms are homeomorphisms.

In all of the above examples, the objects of the categories were in obvious ways sets with additional structure (a **concrete category**, although we won't use this terminology). This needn't be the case, as the next example shows.

1.1.8. *Example: Partially ordered sets.* A **partially ordered set** (or **poset**) is a set S along with a binary relation $\geq$ on S satisfying:

(i) $x \geq x$ (reflexivity),
(ii) $x \geq y$ and $y \geq z$ imply $x \geq z$ (transitivity), and
(iii) if $x \geq y$ and $y \geq x$ then $x = y$ (antisymmetry).

A partially ordered set $(S, \geq)$ can be interpreted as a category whose objects are the elements of S, and with a single morphism from x to y if and only if $x \geq y$ (and no morphism otherwise).

A trivial example is $(S, \geq)$ where $x \geq y$ if and only if $x = y$. Another example is

(1.1.8.1)

Here there are three objects. The identity morphisms are omitted for convenience, and the two non-identity morphisms are depicted. A third example is

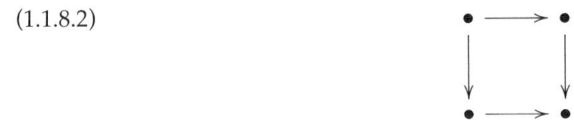

(1.1.8.2)

Here the "obvious" morphisms are again omitted: the identity morphisms, and the morphism from the upper left to the lower right. Similarly,

$$\cdots \longrightarrow \bullet \longrightarrow \bullet \longrightarrow \bullet$$

depicts a partially ordered set, where, again, only the "generating morphisms" are depicted.

1.1.9. *Example: The category of subsets of a set, and the category of open subsets of a topological space.* If X is a set, then the subsets form a partially ordered set, where arrows are given by inclusion. (Be careful: you may be expecting the arrows to go the other way, because of Example 1.1.8.) Informally, if $U \subset V$, then we have exactly one morphism $U \to V$ in the category (and otherwise none). Similarly, if X is a topological space, then the *open* sets form a partially ordered set, where the maps are given by inclusions.

1.1.10. *Definition.* A **subcategory** $\mathscr{A}$ of a category $\mathscr{B}$ has as its objects some of the objects of $\mathscr{B}$, and some of the morphisms of $\mathscr{B}$, such that the objects of $\mathscr{A}$ include the sources and targets of the morphisms of $\mathscr{A}$, and the morphisms of $\mathscr{A}$ include the identity morphisms of the objects of $\mathscr{A}$, and are preserved by composition. (For example, (1.1.8.1) is in an obvious way a subcategory of (1.1.8.2). Also, we have an obvious "inclusion" $i: \mathscr{A} \to \mathscr{B}$, which will soon be an example of a functor.)

1.1.11. Functors.
A **covariant functor** F from a category $\mathscr{A}$ to a category $\mathscr{B}$, denoted by $F: \mathscr{A} \to \mathscr{B}$, is the following data. It is a map of objects $F: \mathrm{obj}(\mathscr{A}) \to \mathrm{obj}(\mathscr{B})$, and for each $A_1, A_2 \in \mathscr{A}$, and morphism $m: A_1 \to A_2$, a morphism $F(m): F(A_1) \to F(A_2)$ in $\mathscr{B}$. We require that F preserve identity morphisms (for $A \in \mathscr{A}$, $F(\mathrm{id}_A) = \mathrm{id}_{F(A)}$), and that F preserve composition ($F(m_2 \circ m_1) = F(m_2) \circ F(m_1)$). (You may wish to verify that covariant functors send isomorphisms to isomorphisms.) A trivial example is the **identity functor** $\mathrm{id}: \mathscr{A} \to \mathscr{A}$, whose definition you can guess. Here are some less trivial examples.

1.1.12. *Example: A forgetful functor.* Consider the functor from the category of vector spaces (over a field k) Vec_k to *Sets* that associates to each vector space its underlying set. The functor sends a linear transformation to its underlying map of sets. This is an example of a **forgetful functor**, where some additional structure is forgotten. Another example of a forgetful functor is $Mod_A \to Ab$ from A-modules to abelian groups, which remembers only the abelian group structure of the A-module.

1.1.13. *Topological examples.* Examples of covariant functors include the fundamental group functor π_1, which sends a topological space X with choice of a point $x_0 \in X$ to a group $\pi_1(X, x_0)$ (what are the objects and morphisms of the source category?), and the ith homology functor $Top \to Ab$, which sends a topological space X to its ith homology group $H_i(X, \mathbb{Z})$. The covariance corresponds to the fact that a (continuous) morphism of pointed topological spaces $\phi: X \to Y$ with $\phi(x_0) = y_0$ induces a map of fundamental groups $\pi_1(X, x_0) \to \pi_1(Y, y_0)$, and similarly for homology groups.

1.1.14. *Example.* Suppose A is an object in a category $\mathscr{C}$. Then there is a functor $h^A: \mathscr{C} \to Sets$ sending $B \in \mathscr{C}$ to $\mathrm{Mor}(A, B)$ and sending $f: B_1 \to B_2$ to $\mathrm{Mor}(A, B_1) \to \mathrm{Mor}(A, B_2)$, described by

$$[g: A \to B_1] \longmapsto [f \circ g: A \to B_1 \to B_2].$$

This seemingly silly functor ends up surprisingly being an important concept.

1.1.15. Definitions. If $F: \mathscr{A} \to \mathscr{B}$ and $G: \mathscr{B} \to \mathscr{C}$ are covariant functors, then we define a functor $G \circ F: \mathscr{A} \to \mathscr{C}$ (the **composition** of G and F) in the obvious way. Composition of functors is associative in an evident sense.

A covariant functor $F: \mathscr{A} \to \mathscr{B}$ is **faithful** if for all $A, A' \in \mathscr{A}$, the map $\mathrm{Mor}_{\mathscr{A}}(A, A') \to \mathrm{Mor}_{\mathscr{B}}(F(A), F(A'))$ is injective, and **full** if it is surjective. A functor that is full and faithful is **fully faithful**. (For various philosophical reasons, the notion of "full" functor on its own is unimportant; "fully faithful" is the useful notion.) A subcategory $i: \mathscr{A} \to \mathscr{B}$ is a **full subcategory** if i is full.

(Inclusions are always faithful, so there is no need for the phrase "faithful subcategory.") Thus a subcategory $\mathscr{A}'$ of $\mathscr{A}$ is full if and only if for all A, B $\in$ obj($\mathscr{A}'$), $\mathrm{Mor}_{\mathscr{A}'}(A, B) = \mathrm{Mor}_{\mathscr{A}}(A, B)$. For example, the forgetful functor $Vec_k \to Sets$ is faithful, but not full; and if A is a ring, the category of finitely generated A-modules is a full subcategory of the category Mod_A of A-modules.

1.1.16. Definition. A **contravariant functor** is defined in the same way as a covariant functor, except the arrows switch directions: in the above language, $F(A_1 \to A_2)$ is now an arrow from $F(A_2)$ to $F(A_1)$. (Thus $F(m_2 \circ m_1) = F(m_1) \circ F(m_2)$, not $F(m_2) \circ F(m_1)$.)

It is wise to state whether a functor is covariant or contravariant, unless the context makes it very clear. If it is not stated (and the context does not make it clear), the functor is often assumed to be covariant.

Sometimes people describe a contravariant functor $\mathscr{C} \to \mathscr{D}$ as a covariant functor $\mathscr{C}^{\mathrm{opp}} \to \mathscr{D}$, where $\mathscr{C}^{\mathrm{opp}}$ is the same category as $\mathscr{C}$ except that the arrows go in the opposite direction. Here $\mathscr{C}^{\mathrm{opp}}$ is said to be the **opposite category** to $\mathscr{C}$.

One can define fullness, etc. for contravariant functors, and you should do so.

1.1.17. *Linear algebra example.* If Vec_k is the category of k-vector spaces (introduced in Example 1.1.3), then taking duals gives a contravariant functor $(\cdot)^\vee \colon Vec_k \to Vec_k$. Indeed, to each linear transformation $f \colon V \to W$, we have a dual transformation $f^\vee \colon W^\vee \to V^\vee$, and $(f \circ g)^\vee = g^\vee \circ f^\vee$.

1.1.18. *Topological example (cf. Example 1.1.13) for those who have seen cohomology.* The ith cohomology functor $H^i(\cdot, \mathbb{Z}) \colon Top \to Ab$ is a contravariant functor.

1.1.19. *Example.* There is a contravariant functor $Top \to Rings$ taking a topological space X to the ring of real-valued continuous functions on X. A morphism of topological spaces $X \to Y$ (a continuous map) induces the pullback map from functions on Y to functions on X.

1.1.20. *Example (the functor of points; cf. Example 1.1.14).* Suppose A is an object of a category $\mathscr{C}$. Then there is a contravariant functor $h_A \colon \mathscr{C} \to Sets$ sending $B \in \mathscr{C}$ to $\mathrm{Mor}(B, A)$, and sending the morphism $f \colon B_1 \to B_2$ to the morphism $\mathrm{Mor}(B_2, A) \to \mathrm{Mor}(B_1, A)$ via

$$[g \colon B_2 \to A] \longmapsto [g \circ f \colon B_1 \to B_2 \to A].$$

This example initially looks weird and different, but Examples 1.1.17 and 1.1.19 may be interpreted as special cases; do you see how? What is A in each case? This functor might reasonably be called the *functor of maps* (to A), but is actually known as the **functor of points**. We will meet this functor again in §1.2.11 and (in the category of schemes) in Definition 7.3.10.

1.1.21.* Natural transformations (and natural isomorphisms) of covariant functors, and equivalences of categories.
(This notion won't come up in an essential way until at least Chapter 7, so you shouldn't read this section until then.) Suppose F and G are two covariant functors from $\mathscr{A}$ to $\mathscr{B}$. A **natural transformation of covariant functors** $F \to G$ is the data of a morphism $m_A \colon F(A) \to G(A)$ for each $A \in \mathscr{A}$ such that for each $f \colon A \to A'$ in $\mathscr{A}$, the diagram

$$
\begin{array}{ccc}
F(A) & \xrightarrow{\ F(f)\ } & F(A') \\
{\scriptstyle m_A}\downarrow & & \downarrow{\scriptstyle m_{A'}} \\
G(A) & \xrightarrow[\ G(f)\ ]{} & G(A')
\end{array}
$$

commutes. A **natural isomorphism** of functors is a natural transformation such that each m_A is an isomorphism. (We make analogous definitions when F and G are both contravariant.)

The data of functors $F \colon \mathscr{A} \to \mathscr{B}$ and $F' \colon \mathscr{B} \to \mathscr{A}$ such that $F \circ F'$ is naturally isomorphic to the identity functor $\mathrm{id}_{\mathscr{B}}$ on $\mathscr{B}$ and $F' \circ F$ is naturally isomorphic to $\mathrm{id}_{\mathscr{A}}$ is said to be an **equivalence of categories**. The right notion of when two categories are "essentially the same" is not *isomorphism* (a functor giving bijections of objects and morphisms) but *equivalence*. Exercises 1.1.C and 1.1.D might give you some vague sense of this. Later exercises (for example, to show that "rings" and "affine schemes" are essentially the same once arrows are reversed, Exercise 7.3.E) may help, too.

Two examples might make this strange concept more comprehensible. The double dual of a finite-dimensional vector space V is *not* V, but we learn early to say that it is canonically isomorphic to V. We can make that precise as follows. Let $f.d.Vec_k$ be the category of finite-dimensional vector spaces over k. Note that this category contains oodles of vector spaces of each dimension.

1.1.C. EXERCISE. Let $(\cdot)^{\vee\vee}: f.d.Vec_k \to f.d.Vec_k$ be the double dual functor from the category of finite-dimensional vector spaces over k to itself. Show that $(\cdot)^{\vee\vee}$ is naturally isomorphic to the identity functor on $f.d.Vec_k$. (Without the finite-dimensionality hypothesis, we only get a natural transformation of functors from id to $(\cdot)^{\vee\vee}$.)

Let $\mathscr{V}$ be the category whose objects are the k-vector spaces k^n for each $n \geq 0$ (there is one vector space for each n), and whose morphisms are linear transformations. The objects of $\mathscr{V}$ can be thought of as vector spaces with bases, and the morphisms as matrices. There is an obvious functor $\mathscr{V} \to f.d.Vec_k$, as each k^n *is* a finite-dimensional vector space.

1.1.D. EXERCISE. Show that $\mathscr{V} \to f.d.Vec_k$ gives an equivalence of categories, by describing an "inverse" functor. (Recall that we are being cavalier about set-theoretic assumptions—see Caution 0.3.1—so feel free to simultaneously choose bases for each vector space in $f.d.Vec_k$. To make this precise, you will need to use Gödel-Bernays set theory or else replace $f.d.Vec_k$ with a very similar small category, but we won't worry about this.)

1.1.22.** *Aside for experts.* Your argument for Exercise 1.1.D will show that (modulo set-theoretic issues) this definition of equivalence of categories is the same as another one commonly given: a covariant functor $F: \mathscr{A} \to \mathscr{B}$ is an equivalence of categories if it is fully faithful and every object of $\mathscr{B}$ is isomorphic to an object of the form $F(A)$ for some $A \in \mathscr{A}$ (F is **essentially surjective**, a term we will not need).

1.2   Universal Properties Determine an Object up to Unique Isomorphism

Given some category that we come up with, we often will have ways of producing new objects from old. In good circumstances, definitions will be usefully made using the notion of a *universal property*. Informally, we wish that there were an object with some property. We first show that if it exists, then it is essentially unique, or more precisely, is unique up to unique isomorphism. Then we go about constructing an example of such an object to show existence.

Explicit constructions are sometimes easier to work with than universal properties, but with a little practice, universal properties are useful in proving things quickly and slickly. Indeed, when learning the subject, people often find explicit constructions more appealing, and use them more often in proofs, but as they become more experienced, they find universal property arguments more elegant and insightful.

1.2.1. Products were defined by a universal property. We have seen one important example of a universal property argument already in §1.0.1: products. You should go back and verify that our discussion there gives a notion of product in any category, and shows that products, *if they exist*, are unique up to unique isomorphism.

1.2.2. Initial, final, and zero objects. Here are some simple but useful concepts that will give you practice with universal property arguments. An object of a category $\mathscr{C}$ is an **initial object** if it has precisely one map to every object. It is a **final object** if it has precisely one map from every object. It is a **zero object** if it is both an initial object and a final object.

1.2.A. EXERCISE. Show that any two initial objects are uniquely isomorphic. Show that any two final objects are uniquely isomorphic.

In other words, *if* an initial object exists, it is unique up to unique isomorphism, and similarly for final objects. This (partially) justifies the phrase "*the* initial object" rather than "*an* initial object," and similarly for "*the* final object" and "*the* zero object." (Convention: We often say "the," not "a," for anything defined up to unique isomorphism.)

1.2.B. EXERCISE. What are the initial and final objects in *Sets*, *Rings*, and *Top* (if they exist)? How about in the two examples of §1.1.9?

1.2.3. Localization of rings and modules. Another important example of a definition by universal property is the notion of *localization* of a ring. We first review a constructive definition, and then reinterpret the notion in terms of universal property. A **multiplicative subset** S of a ring A is a subset closed under multiplication containing 1. We define a ring $S^{-1}A$. The elements of $S^{-1}A$ are of the form a/s where $a \in A$ and $s \in S$, and where $a_1/s_1 = a_2/s_2$ if (and only if) *for some* $s \in S$, $s(s_2 a_1 - s_1 a_2) = 0$. We define $(a_1/s_1) + (a_2/s_2) = (s_2 a_1 + s_1 a_2)/(s_1 s_2)$, and $(a_1/s_1) \times (a_2/s_2) = (a_1 a_2)/(s_1 s_2)$. (If you wish, you may check that this equality of fractions really is an equivalence relation and the two binary operations on fractions are well-defined on equivalence classes and make $S^{-1}A$ into a ring.) We have a canonical ring map

$$(1.2.3.1) \qquad\qquad A \longrightarrow S^{-1}A$$

given by $a \mapsto a/1$. Note that if $0 \in S$, $S^{-1}A$ is the 0-ring.

There are two particularly important flavors of multiplicative subsets. The first is $\{1, f, f^2, \dots\}$, where $f \in A$. This localization is denoted by A_f. (Can you describe an isomorphism $A_f \longleftrightarrow A[t]/(tf-1)$?) The second is $A \setminus \mathfrak{p}$, where $\mathfrak{p}$ is a prime ideal. This localization $S^{-1}A$ is denoted by $A_{\mathfrak{p}}$. (Notational warning: If $\mathfrak{p}$ is a prime ideal, then $A_{\mathfrak{p}}$ means you're allowed to divide by elements not in $\mathfrak{p}$. However, if $f \in A$, A_f means you're allowed to divide by f. This can be confusing. For example, if (f) is a prime ideal, then $A_f \neq A_{(f)}$.)

Warning: Sometimes localization is first introduced in the special case where A is an integral domain and $0 \notin S$. In that case, $A \hookrightarrow S^{-1}A$, but this isn't always true, as shown by the following exercise. (But we will see that noninjective localizations needn't be pathological, and we can sometimes understand them geometrically; see Exercise 3.2.L.)

1.2.C. EXERCISE. Show that $A \to S^{-1}A$ is injective if and only if S contains no zerodivisors. (A **zerodivisor** of a ring A is an element a such that there is a nonzero element b with $ab = 0$. The other elements of A are called **non-zerodivisors**. For example, an invertible element is never a zerodivisor. Counterintuitively, 0 is a zerodivisor in every ring but the 0-ring. More generally, if M is an A-module, then $a \in A$ is a **zerodivisor for** M if there is a nonzero $m \in M$ with $am = 0$. The other elements of A are called **non-zerodivisors for** M. Equivalently, and *very* usefully, $a \in A$ is a non-zerodivisor for M if and only if $\times a : M \to M$ is an injection, or equivalently in the language of §1.4.4, if

$$0 \longrightarrow M \xrightarrow{\times a} M$$

is exact.)

If A is an integral domain and $S = A \setminus \{0\}$, then $S^{-1}A$ is called the **fraction field** of A, which we denote by $K(A)$. The previous exercise shows that A is a subring of its fraction field $K(A)$. We now return to the case where A is a general (commutative) ring.

1.2.D. EXERCISE. Verify that $A \to S^{-1}A$ satisfies the following universal property: $S^{-1}A$ is initial among A-algebras B where every element of S is sent to an invertible element in B. (Recall: The data of "an A-algebra B" and "a ring map $A \to B$" are the same.) Translation: Any map $A \to B$ where every element of S is sent to an invertible element factors uniquely through $A \to S^{-1}A$. Another translation: a ring map out of $S^{-1}A$ is the same thing as a ring map from A that sends every element of S to an invertible element. Furthermore, an $S^{-1}A$-module is the same thing as an A-module for which $s \times \cdot : M \to M$ is an A-module isomorphism for all $s \in S$.

In fact, it is cleaner to *define* $A \to S^{-1}A$ by the universal property, and to show that it exists, and to use the universal property to check various properties $S^{-1}A$ has. Let's get some practice with this by *defining* localizations of modules by the universal property. Suppose M is an A-module. We define the A-module map $\phi \colon M \to S^{-1}M$ as being initial among A-module maps $M \to N$ such that elements of S are invertible in N ($s \times \cdot \colon N \to N$ is an isomorphism for all $s \in S$). More precisely, any such map $\alpha \colon M \to N$ factors uniquely through ϕ:

$$
\begin{array}{ccc}
M & \xrightarrow{\ \phi\ } & S^{-1}M \\
& \alpha \searrow & \downarrow \exists! \\
& & N
\end{array}
$$

(Translation: $M \to S^{-1}M$ is universal (initial) among A-module maps from M to modules that are actually $S^{-1}A$-modules. Can you make this precise by defining clearly the objects and morphisms in this category?)

Notice: (i) This determines $\phi \colon M \to S^{-1}M$ up to unique isomorphism (you should think through what this means); (ii) we are defining not only $S^{-1}M$, but also the map ϕ at the same time; and (iii) essentially by definition the A-module structure on $S^{-1}M$ extends to an $S^{-1}A$-module structure.

1.2.E. EXERCISE. Show that $\phi \colon M \to S^{-1}M$ exists, by constructing something satisfying the universal property. Hint: Define elements of $S^{-1}M$ to be of the form m/s where $m \in M$ and $s \in S$, and $m_1/s_1 = m_2/s_2$ if and only if for some $s \in S$, $s(s_2 m_1 - s_1 m_2) = 0$. Define the additive structure by $(m_1/s_1) + (m_2/s_2) = (s_2 m_1 + s_1 m_2)/(s_1 s_2)$, and the $S^{-1}A$-module structure (and hence the A-module structure) as given by $(a_1/s_1) \cdot (m_2/s_2) = (a_1 m_2)/(s_1 s_2)$.

1.2.F. EXERCISE.

(a) Show that localization commutes with finite products, or equivalently, with finite direct sums. In other words, if $M_1, \ldots, M_n$ are A-modules, describe an isomorphism (of A-modules, and of $S^{-1}A$-modules) $S^{-1}(M_1 \times \cdots \times M_n) \to S^{-1}M_1 \times \cdots \times S^{-1}M_n$.

(b) Show that localization commutes with *arbitrary* direct sums.

(c) Show that "localization does not necessarily commute with infinite products": the obvious map $S^{-1}(\prod_i M_i) \to \prod_i S^{-1}M_i$ induced by the universal property of localization is not always an isomorphism. (Hint: $(1, 1/2, 1/3, 1/4, \ldots) \in \mathbb{Q} \times \mathbb{Q} \times \cdots$.)

1.2.4. *Remark.* Localization does not always commute with Hom; see Example 1.5.10. But Exercise 1.5.H will show that in good situations (if the first argument of Hom is *finitely presented*), localization *does* commute with Hom.

1.2.5. Tensor products. Another important example of a universal property construction is the notion of a **tensor product** of A-modules:

$$
\begin{array}{ccc}
\otimes_A \colon & \mathrm{obj}(Mod_A) \times \mathrm{obj}(Mod_A) & \longrightarrow & \mathrm{obj}(Mod_A) \\
& (M, N) & \longmapsto & M \otimes_A N
\end{array}
$$

The subscript A is often suppressed when it is clear from context. The tensor product is often defined as follows. Suppose you have two A-modules M and N. Then elements of the tensor product $M \otimes_A N$ are finite A-linear combinations of symbols $m \otimes n$ ($m \in M$, $n \in N$), subject to relations $(m_1 + m_2) \otimes n = m_1 \otimes n + m_2 \otimes n$, $m \otimes (n_1 + n_2) = m \otimes n_1 + m \otimes n_2$, $a(m \otimes n) = (am) \otimes n = m \otimes (an)$ (where $a \in A$, $m_1, m_2 \in M$, $n_1, n_2 \in N$). More formally, $M \otimes_A N$ is the free A-module generated by $M \times N$, quotiented by the submodule generated by $(m_1 + m_2, n) - (m_1, n) - (m_2, n)$, $(m, n_1 + n_2) - (m, n_1) - (m, n_2)$, $a(m, n) - (am, n)$, and $a(m, n) - (m, an)$ for $a \in A$, $m, m_1, m_2 \in M$, $n, n_1, n_2 \in N$. The image of (m, n) in this quotient is $m \otimes n$.

If A is a field k, we recover the tensor product of vector spaces.

1.2.G. EXERCISE (IF YOU HAVEN'T SEEN TENSOR PRODUCTS BEFORE). Show that $\mathbb{Z}/(10)$ $\otimes_{\mathbb{Z}} \mathbb{Z}/(12) \cong \mathbb{Z}/(2)$. (This exercise is intended to give some hands-on practice with tensor products.)

1.2.H. IMPORTANT EXERCISE: RIGHT-EXACTNESS OF $(\cdot) \otimes_A N$. Show that $(\cdot) \otimes_A N$ gives a covariant functor $Mod_A \to Mod_A$. Show that $(\cdot) \otimes_A N$ is a **right-exact functor**, i.e., if

$$M' \to M \to M'' \to 0$$

is an exact sequence of A-modules (which means $f : M \to M''$ is surjective, and M' surjects onto the kernel of f; see §1.4.4), then the induced sequence

$$M' \otimes_A N \to M \otimes_A N \to M'' \otimes_A N \to 0$$

is also exact. This exercise is repeated in Exercise 1.5.G, but you may get a lot out of doing it now. (You will be reminded of the definition of right-exactness in §1.5.6.)

In contrast, you can quickly check that the tensor product is not left-exact: tensor the exact sequence of $\mathbb{Z}$-modules

$$0 \longrightarrow \mathbb{Z} \xrightarrow{\times 2} \mathbb{Z} \longrightarrow \mathbb{Z}/(2) \longrightarrow 0$$

with $\mathbb{Z}/(2)$.

The constructive definition of $\otimes$ is a weird definition, and really the "wrong" definition. To motivate a better one: notice that there is a natural A-bilinear map $M \times N \to M \otimes_A N$. (If $M, N, P \in Mod_A$, a map $f : M \times N \to P$ is **A-bilinear** if $f(m_1 + m_2, n) = f(m_1, n) + f(m_2, n)$, $f(m, n_1 + n_2) = f(m, n_1) + f(m, n_2)$, and $f(am, n) = f(m, an) = af(m, n)$.) *Any* A-bilinear map $M \times N \to P$ factors through the tensor product uniquely: $M \times N \to M \otimes_A N \to P$. (Think this through!)

We can take this as the *definition* of the tensor product as follows. It is an A-module T along with an A-bilinear map $t : M \times N \to T$, such that given any A-bilinear map $t' : M \times N \to T'$, there is a unique A-linear map $f : T \to T'$ such that $t' = f \circ t$.

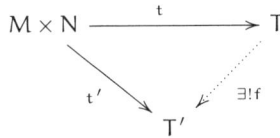

1.2.I. EXERCISE. Show that $(T, t : M \times N \to T)$ is unique up to unique isomorphism. Hint: First figure out what "unique up to unique isomorphism" means for such pairs, using a category of pairs (T, t). Then follow the analogous argument for the product.

In short, given M and N, there is an A-bilinear map $t : M \times N \to M \otimes_A N$, unique up to unique isomorphism, defined by the following universal property: for any A-bilinear map $t' : M \times N \to T'$ there is a unique A-linear map $f : M \otimes_A N \to T'$ such that $t' = f \circ t$.

As with all universal property arguments, this argument shows uniqueness *assuming existence*. To show existence, we need an explicit construction.

1.2.J. EXERCISE. Show that the construction of §1.2.5 satisfies the universal property of tensor product.

The three exercises below are useful facts about tensor products with which you should be familiar.

1.2.K. IMPORTANT EXERCISE.

(a) If M is an A-module and $A \to B$ is a morphism of rings, give $B \otimes_A M$ the structure of a B-module (this is part of the exercise). Show that this describes a functor $Mod_A \to Mod_B$.

(b) **(tensor product of rings)** If further $A \to C$ is another morphism of rings, show that $B \otimes_A C$ has a natural structure of a ring. Hint: Multiplication will be given by $(b_1 \otimes c_1)(b_2 \otimes c_2) = (b_1 b_2) \otimes (c_1 c_2)$. (Exercise 1.2.U will interpret this construction as a fibered coproduct.)

1.2.L. IMPORTANT EXERCISE. If S is a multiplicative subset of A and M is an A-module, describe a natural isomorphism $(S^{-1}A) \otimes_A M \xrightarrow{\sim} S^{-1}M$ (as $S^{-1}A$-modules *and* as A-modules).

1.2.M. EXERCISE ($\otimes$ COMMUTES WITH $\oplus$). Show that tensor products commute with arbitrary direct sums: if M and $\{N_i\}_{i\in I}$ are all A-modules, describe an isomorphism

$$M \otimes (\oplus_{i\in I} N_i) \xrightarrow{\sim} \oplus_{i\in I} (M \otimes N_i).$$

1.2.6. Essential Example: Fibered products. Suppose we have morphisms $\alpha: X \to Z$ and $\beta: Y \to Z$ (in *any* category). Then the **fibered product** (or *fibred product*) is an object $X \times_Z Y$ along with morphisms $\mathrm{pr}_X: X \times_Z Y \to X$ and $\mathrm{pr}_Y: X \times_Z Y \to Y$, where the two compositions $\alpha \circ \mathrm{pr}_X, \beta \circ \mathrm{pr}_Y: X \times_Z Y \to Z$ agree, such that given any object W with maps to X and Y (whose compositions to Z agree), these maps factor through some unique $W \to X \times_Z Y$:

(1.2.6.1)

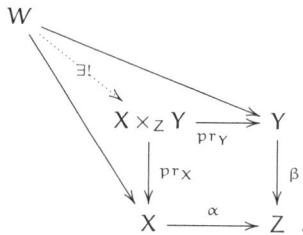

(Warning: The definition of the fibered product depends on α and β, even though they are omitted from the notation $X \times_Z Y$.)

By the usual universal property argument, if it exists, it is unique up to unique isomorphism. (You should think this through until it is clear to you.) Thus the use of the phrase "the fibered product" (rather than "a fibered product") is reasonable, and we should reasonably be allowed to give it the name $X \times_Z Y$. We know what maps to it are: they are precisely maps to X and maps to Y that agree as maps to Z.

1.2.7. *Definition.* As an example, if $\pi: X \to Y$ is a morphism, and the fibered product $X \times_Y X$ exists, then this determines a **diagonal morphism** $\delta_\pi: X \to X \times_Y X$. The diagonal morphism will turn out to be a very useful notion.

Depending on your religion, the diagram

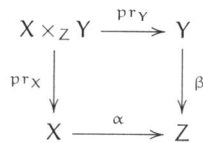

is called a **fibered/pullback/Cartesian diagram/square** (six possibilities—and even more are possible if you prefer "fibred" to "fibered").

The right way to interpret the notion of fibered product is first to think about what it means in the category of sets.

1.2.N. EXERCISE (FIBERED PRODUCTS OF SETS). Show that in *Sets*,

$$X \times_Z Y = \{(x, y) \in X \times Y : \alpha(x) = \beta(y)\}.$$

More precisely, show that the right side, equipped with its evident maps to X and Y, satisfies the universal property of the fibered product. (This will help you build intuition for fibered products.)

1.2.O. EXERCISE. If X is a topological space, show that fibered products always exist in the category of open sets of X, by describing what a fibered product is. (Hint: It has a one-word description.)

1.2.P. EXERCISE. If Z is the final object in a category $\mathscr{C}$, and $X, Y \in \mathscr{C}$, show that "$X \times_Z Y = X \times Y$": "the" fibered product over Z is uniquely isomorphic to "the" product. Assume all relevant (fibered) products exist. (This is an exercise about unwinding the definition.)

1.2.Q. USEFUL EXERCISE: TOWERS OF CARTESIAN DIAGRAMS ARE CARTESIAN DIAGRAMS. If the two squares in the following commutative diagram are Cartesian diagrams, show that the "outside rectangle" (involving U, V, Y, and Z) is also a Cartesian diagram.

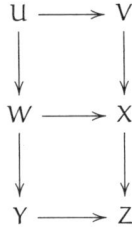

$$
\begin{array}{ccc}
U & \longrightarrow & V \\
\downarrow & & \downarrow \\
W & \longrightarrow & X \\
\downarrow & & \downarrow \\
Y & \longrightarrow & Z
\end{array}
$$

1.2.R. EXERCISE. Given morphisms $X_1 \to Y$, $X_2 \to Y$, and $Y \to Z$, show that there is a natural morphism $X_1 \times_Y X_2 \to X_1 \times_Z X_2$, assuming that both fibered products exist. (This is trivial once you figure out what it is saying. The point of this exercise is to see why it is trivial.)

1.2.S. IMPORTANT EXERCISE: THE DIAGONAL-BASE-CHANGE DIAGRAM. Suppose we are given morphisms $X_1, X_2 \to Y$ and $Y \to Z$. Show that the following diagram is a Cartesian square.

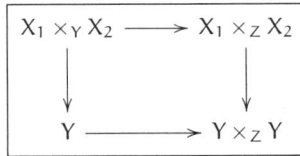

$$
\begin{array}{ccc}
X_1 \times_Y X_2 & \longrightarrow & X_1 \times_Z X_2 \\
\downarrow & & \downarrow \\
Y & \longrightarrow & Y \times_Z Y
\end{array}
$$

Assume all relevant (fibered) products exist. (If this exercise is too hard now, you can try it again at Exercise 1.3.B.) You will appreciate how useful this diagram is when you repeatedly use the diagonal morphism in proofs and constructions.

If you liked this problem, you may enjoy Exercise 11.1.C.

1.2.8. *Coproducts.* Define **coproduct** in a category by reversing all the arrows in the definition of product. Define **fibered coproduct** in a category by reversing all the arrows in the definition of fibered product. Coproduct is denoted by $\coprod$.

1.2.T. EXERCISE. Show that coproduct for *Sets* is disjoint union. This is why we use the notation $\coprod$ for disjoint union.

1.2.U. EXERCISE. Suppose $A \to B$ and $A \to C$ are two ring morphisms, so in particular B and C are A-modules. Recall (Exercise 1.2.K) that $B \otimes_A C$ has a ring structure. Show that there is a natural morphism $B \to B \otimes_A C$ given by $b \mapsto b \otimes 1$. (This is not necessarily an inclusion; see Exercise 1.2.G.) Similarly, there is a natural morphism $C \to B \otimes_A C$. Show that this gives a fibered coproduct on rings, i.e., that

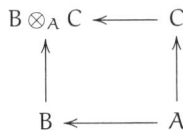

$$
\begin{array}{ccc}
B \otimes_A C & \longleftarrow & C \\
\uparrow & & \uparrow \\
B & \longleftarrow & A
\end{array}
$$

satisfies the universal property of fibered coproduct.

1.2.9. Monomorphisms and epimorphisms.

1.2.10. Definition. A morphism $\pi\colon X \to Y$ is a **monomorphism** if any two morphisms $\mu_1\colon Z \to X$ and $\mu_2\colon Z \to X$ such that $\pi \circ \mu_1 = \pi \circ \mu_2$ must satisfy $\mu_1 = \mu_2$. In other words, there is at most one way of filling in the dotted arrow so that the diagram

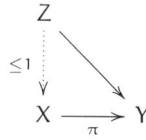

$$
\begin{array}{ccc}
 & Z & \\
{\scriptstyle \leq 1}\downarrow & & \searrow \\
X & \xrightarrow[\pi]{} & Y
\end{array}
$$

commutes—for any object Z, the natural map $\mathrm{Mor}(Z, X) \to \mathrm{Mor}(Z, Y)$ is an injection. Intuitively, it is the categorical version of an injective map, and indeed this notion generalizes the familiar notion of injective maps of sets. (The reason we don't use the word "injective" is that in some contexts, "injective" will have an intuitive meaning that may not agree with "monomorphism." One example: in the category of divisible groups, the map $\mathbb{Q} \to \mathbb{Q}/\mathbb{Z}$ is a monomorphism but not injective. This is also the case with "epimorphism"—to be defined shortly—in contract with "surjective.")

1.2.V. EXERCISE. Show that the composition of two monomorphisms is a monomorphism.

1.2.W. EXERCISE. Prove that a morphism $\pi\colon X \to Y$ is a monomorphism if and only if the fibered product $X \times_Y X$ exists, and the induced diagonal morphism $\delta_\pi\colon X \to X \times_Y X$ (Definition 1.2.7) is an isomorphism. We may then take this as the definition of monomorphism. (Monomorphisms aren't central to future discussions, although they will come up again. This exercise is just good practice.)

1.2.X. EASY EXERCISE. We use the notation of Exercise 1.2.R. Show that if $Y \to Z$ is a monomorphism, then the morphism $X_1 \times_Y X_2 \to X_1 \times_Z X_2$ you described in Exercise 1.2.R is an isomorphism. (Hint: For any object V, give a natural bijection between maps from V to the first and maps from V to the second. It is also possible to use the Diagonal-Base-Change diagram, Exercise 1.2.S.)

The notion of **epimorphism** is "dual" to the definition of monomorphism, where all the arrows are reversed. This concept will not be central for us, although it turns up in the definition of an abelian category. Intuitively, it is the categorical version of a surjective map. (But be careful when working with categories of objects that are sets with additional structure, as epimorphisms need not be surjective. Example: In the category *Rings*, $\mathbb{Z} \to \mathbb{Q}$ is an epimorphism, but obviously not surjective.)

1.2.11. Representable functors and Yoneda's Lemma. Much of our discussion about universal properties can be cleanly expressed in terms of representable functors, under the rubric of "Yoneda's Lemma." Yoneda's lemma is an easy fact stated in a complicated way. Informally speaking, you can essentially recover an object in a category by knowing the maps into it. For example, we have seen that the data of maps to $X \times Y$ are naturally (canonically) the data of maps to X and to Y. Indeed, we have now taken this as the *definition* of $X \times Y$.

Recall Example 1.1.20. Suppose A is an object of category $\mathscr{C}$. For any object $C \in \mathscr{C}$, we have a set of morphisms $\mathrm{Mor}(C, A)$. If we have a morphism $f\colon B \to C$, we get a map of sets

(1.2.11.1) $\mathrm{Mor}(C, A) \longrightarrow \mathrm{Mor}(B, A)$

by composition: given a map from C to A, we get a map from B to A by precomposing with $f\colon B \to C$. Hence this gives a contravariant functor $h_A\colon \mathscr{C} \to Sets$. Yoneda's Lemma states that the functor h_A determines A up to unique isomorphism. More precisely:

1.2.Y. IMPORTANT EXERCISE THAT YOU SHOULD DO ONCE IN YOUR LIFE (YONEDA'S LEMMA).

(a) Suppose you have two objects A and A' in a category $\mathscr{C}$, and maps

(1.2.11.2) $i_C\colon \mathrm{Mor}(C, A) \longrightarrow \mathrm{Mor}(C, A')$

that commute with the maps (1.2.11.1). Show that the i_C (as C ranges over the objects of $\mathscr{C}$) are induced from a unique morphism $g\colon A \to A'$. More precisely, show that there is a unique morphism $g\colon A \to A'$ such that for all $C \in \mathscr{C}$, i_C is $u \mapsto g \circ u$.

(b) If furthermore the i_C are all bijections, show that the resulting g is an isomorphism. (Hint for both: This is much easier than it looks. This statement is so general that there are really only a couple of things that you could possibly try. For example, if you're hoping to find a morphism $A \to A'$, where will you find it? Well, you are looking for an element $\mathrm{Mor}(A, A')$. So just plug $C = A$ into (1.2.11.2), and see where the identity goes.)

There is an analogous statement with the arrows reversed, where instead of maps into A, you think of maps *from* A. The role of the contravariant functor h_A of Example 1.1.20 is played by the covariant functor h^A of Example 1.1.14. Because the proof is the same (with the arrows reversed), you needn't think it through.

The phrase "Yoneda's Lemma" properly refers to a more general statement. Although it looks more complicated, it is no harder to prove.

1.2.Z.* EXERCISE.

(a) Suppose A and B are objects in a category $\mathscr{C}$. Give a bijection between the natural transformations $h^A \to h^B$ of covariant functors $\mathscr{C} \to Sets$ (see Example 1.1.14 for the definition) and the morphisms $B \to A$.

(b) State and prove the corresponding fact for contravariant functors h_A (see Example 1.1.20).
     Remark: A contravariant functor F from $\mathscr{C}$ to *Sets* is said to be **representable** if there is a natural isomorphism

$$\xi \colon F \xrightarrow{\ \sim\ } h_A.$$

Thus the representing object A is determined up to unique isomorphism by the pair (F, ξ). There is a similar definition for covariant functors. (We will revisit this in §7.6, and this problem will appear again as Exercise 7.6.C. The element $\xi^{-1}(\mathrm{id}_A) \in F(A)$ is often called the "universal object"; do you see why?)

(c) **Yoneda's Lemma.** Suppose F is a covariant functor $\mathscr{C} \to Sets$, and $A \in \mathscr{C}$. Give a bijection between the natural transformations $h^A \to F$ and $F(A)$. (The corresponding fact for contravariant functors is essentially Exercise 10.1.B.)

In fancy terms, Yoneda's lemma states the following. Given a category $\mathscr{C}$, we can produce a new category, called the **functor category** of $\mathscr{C}$, where the objects are contravariant functors $\mathscr{C} \to Sets$, and the morphisms are natural transformations of such functors. We have a functor (which we can usefully call h) from $\mathscr{C}$ to its functor category, which sends A to h_A. Yoneda's Lemma states that this is a fully faithful functor, called the *Yoneda embedding*. (Fully faithful functors were defined in §1.1.15.)

1.2.12. *Joke.* The Yoda embedding, contravariant it is.

1.3   Limits and Colimits

Two important definitions, those of limits and colimits, are determined by universal properties. They generalize a number of familiar constructions. I will give the definitions first, and then show you why they are familiar. For example, fractions will be motivating examples of colimits (Exercise 1.3.D(a)), and the p-adic integers (Example 1.3.4) will be motivating examples of limits.

1.3.1. Limits. We say that a category is a **small category** if the objects form a set and the morphisms form a set. (This is a technical condition intended only for experts.) Suppose $\mathscr{I}$ is any small category, and $\mathscr{C}$ is any category. Then a functor $F \colon \mathscr{I} \to \mathscr{C}$ (i.e., with an object $A_i \in \mathscr{C}$ for each element $i \in \mathscr{I}$, and appropriate commuting morphisms dictated by $\mathscr{I}$) is said to be a **diagram indexed by** $\mathscr{I}$. We call $\mathscr{I}$ an **index category**. Our index categories will usually be partially ordered sets (Example 1.1.8), in which, in particular, there is at most one morphism between any two objects. (But other examples are sometimes useful.) For example, if $\square$ is the category

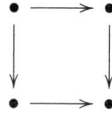

and $\mathscr{A}$ is a category, then a functor $\square \to \mathscr{A}$ is precisely the data of a commuting square in $\mathscr{A}$.

Then the **limit of the diagram** is an object $\varprojlim_{\mathscr{I}} A_i$ (or $\varprojlim_{\mathscr{I}} A_i$) of $\mathscr{C}$ along with morphisms $f_j \colon \varprojlim_{\mathscr{I}} A_i \to A_j$ for each $j \in \mathscr{I}$, such that if $m \colon j \to k$ is a morphism in $\mathscr{I}$, then

(1.3.1.1)

$$
\begin{array}{ccc}
 & \varprojlim_{\mathscr{I}} A_i & \\
f_j \downarrow & & \searrow f_k \\
A_j & \xrightarrow{\ F(m)\ } & A_k
\end{array}
$$

commutes, and the object and the maps to each A_i are universal (final) with respect to this property. More precisely, given any other object W along with maps $g_i \colon W \to A_i$ commuting with the $F(m)$ (if $m \colon j \to k$ is a morphism in $\mathscr{I}$, then $g_k = F(m) \circ g_j$), there is a unique map

$$ g \colon W \to \varprojlim_{\mathscr{I}} A_i $$

so that $g_i = f_i \circ g$ for all i. (In some cases, the limit is sometimes called the **inverse limit** or **projective limit**. We won't use this language.) By the usual universal property argument, if the limit exists, it is unique up to unique isomorphism.

1.3.2. *Examples: Products.* For example, if $\mathscr{I}$ is the partially ordered set

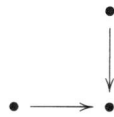

we obtain the fibered product.

If $\mathscr{I}$ is

we obtain the product.

If $\mathscr{I}$ is a set (i.e., its only morphisms are the identity maps), then the limit is called the **product** of the A_i, and is denoted by $\prod_i A_i$. The special case where $\mathscr{I}$ has two elements is the example of the previous paragraph.

1.3.A. EXERCISE (REALITY CHECK). Suppose that the partially ordered set $\mathscr{I}$ has an initial object e. Show that the limit of any diagram indexed by $\mathscr{I}$ exists.

1.3.B. EXERCISE: THE DIAGONAL-BASE-CHANGE DIAGRAM, AGAIN. Solve 1.2.S again by identifying both $X_1 \times_Y X_2$ and $Y \times_{(Y \times_Z Y)} (X_1 \times_Z X_2)$ as the limit of the diagram

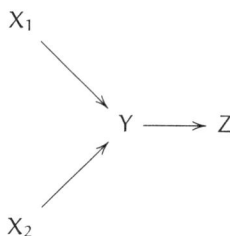

1.3.3. *Example: Formal power series.* For a ring A, the **(formal) power series**, $A[[x]]$, are often described informally (and somewhat unnaturally) as being the ring

$$A[[x]] = \{a_0 + a_1 x + a_2 x^2 + \cdots \}$$

(where $a_i \in A$, and the ring operations are the "obvious" ones). It is an example of a limit in the category of rings:

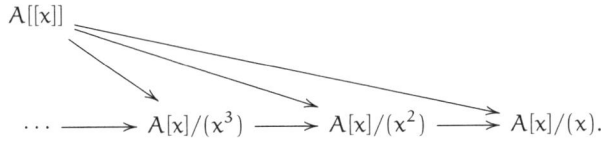

The universal property of limits yields a natural ring morphism $A[x] \to A[[x]]$. If $A = \mathbb{R}$ or $\mathbb{C}$, this map factors through the ring of *convergent power series*.

1.3.4. *Example: The p-adic integers.* For a prime number p, the **p-adic integers** (or more informally, p-**adics**), $\mathbb{Z}_p$, are often described informally (and somewhat unnaturally) as being of the form

$$a_0 + a_1 p + a_2 p^2 + \cdots$$

(where $0 \le a_i < p$). They are an example of a limit in the category of rings:

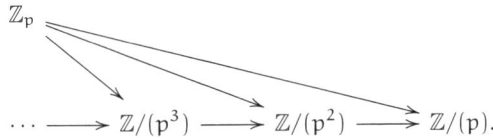

(Warning: $\mathbb{Z}_p$ is sometimes used to denote the integers modulo p, but $\mathbb{Z}/(p)$ or $\mathbb{Z}/p\mathbb{Z}$ is better to use for this, to avoid confusion. Worse: by §1.2.3, $\mathbb{Z}_p$ also denotes those rationals whose denominators are a power of p. Hopefully the meaning of $\mathbb{Z}_p$ will be clear from the context.)

The similarity of Examples 1.3.3 and 1.3.4 is no coincidence. Formal power series and the p-adic integers are examples of *completions*, the topic of Chapter 28.

Limits do not always exist for any index category $\mathscr{I}$. However, you can often easily check that limits exist if the objects of your category can be interpreted as sets with additional structure, and arbitrary products exist (respecting the set-like structure).

1.3.C. IMPORTANT EXERCISE. Show that in the category *Sets*,

$$\left\{ (a_i)_{i \in \mathscr{I}} \in \prod_i A_i : F(m)(a_j) = a_k \text{ for all } m \in \mathrm{Mor}_\mathscr{I}(j,k) \subset \mathrm{Mor}(\mathscr{I}) \right\},$$

along with the obvious projection maps to each A_i, is the limit $\varprojlim_\mathscr{I} A_i$.

This clearly also works in the category *Mod*$_A$ of A-modules (in particular, *Vec*$_k$ and *Ab*), as well as *Rings*.

From this point of view, $2 + 3p + 2p^2 + \cdots \in \mathbb{Z}_p$ can be understood as the sequence $(2, 2 + 3p, 2 + 3p + 2p^2, \dots)$.

1.3.5. Colimits. More immediately relevant for us will be the dual (arrow-reversed version) of the notion of limit (or inverse limit). We just flip the arrows f_i in (1.3.1.1), and get the notion of a **colimit**, which is denoted by $\mathrm{colim}_\mathscr{I} A_i$ (or $\varinjlim_\mathscr{I} A_i$). (You should draw the corresponding diagram.) Again, if it exists, it is unique up to unique isomorphism. (In some cases, the colimit is sometimes called the **direct limit**, **inductive limit**, or **injective limit**. We won't use this language. I prefer using limit/colimit in analogy with kernel/cokernel and product/coproduct. This is more than analogy, as kernels and products may be interpreted as limits, and similarly with cokernels

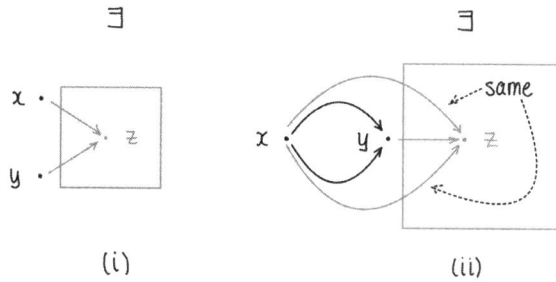

Figure 1.1 *A filtered category (pictorial definition).*

and coproducts. Also, I remember that kernels "map to," and cokernels are "mapped to," which reminds me that a limit maps *to* all the objects in the big commutative diagram indexed by $\mathscr{I}$; and a colimit has a map *from* all the objects.)

1.3.6. *Joke.* A comathematician is a device for turning cotheorems into ffee.

Even though we have just flipped the arrows, colimits behave quite differently from limits.

1.3.7. *Example.* The abelian group $5^{-\infty}\mathbb{Z}$ of rational numbers whose denominators are powers of 5 is a colimit $\operatorname{colim}_{i\in\mathbb{Z}^+} 5^{-i}\mathbb{Z}$. More precisely, $5^{-\infty}\mathbb{Z}$ is the colimit of the diagram

$$\mathbb{Z} \longrightarrow 5^{-1}\mathbb{Z} \longrightarrow 5^{-2}\mathbb{Z} \longrightarrow \cdots$$

in the category of abelian groups.

The colimit over an index *set* I is called the **coproduct**, denoted by $\coprod_i A_i$, and is the dual (arrow-reversed) notion to the product.

1.3.D. EXERCISE.

(a) Interpret the statement "$\mathbb{Q} = \operatorname{colim} \frac{1}{n}\mathbb{Z}$."
(b) Interpret the union of some subsets of a given set as a colimit. (Dually, the intersection can be interpreted as a limit.) The objects of the category in question are the subsets of the given set.

Colimits do not always exist, but there are two useful large classes of examples for which they do.

1.3.8. Definition. A nonempty partially ordered set $(S, \geq)$ is **filtered** (or is said to be a **filtered set**) if for each $x, y \in S$, there is a z such that $x \geq z$ and $y \geq z$. More generally (see Figure 1.1), a nonempty category $\mathscr{I}$ is **filtered** if:

(i)   for each $x, y \in \mathscr{I}$, there is a $z \in \mathscr{I}$ and arrows $x \to z$ and $y \to z$, and
(ii) for every two arrows $u \colon x \to y$ and $v \colon x \to y$, there is an arrow $w \colon y \to z$ such that $w \circ u = w \circ v$.

(Other terminologies are also commonly used, such as "directed partially ordered set" and "filtered index category," respectively.)

1.3.E. EXERCISE. Suppose $\mathscr{I}$ is filtered. (We will almost exclusively use the case where $\mathscr{I}$ is a filtered set.) Recall the symbol $\coprod$ for disjoint union of sets. Show that any diagram in *Sets* indexed by $\mathscr{I}$ has the following, with the obvious maps to it, as a colimit:

$$\left\{ (a_i, i) \in \coprod_{i \in \mathscr{I}} A_i \right\} \Big/ \left(\begin{array}{l} (a_i, i) \sim (a_j, j) \text{ if and only if there are } f \colon A_i \to A_k \text{ and} \\ g \colon A_j \to A_k \text{ in the diagram for which } f(a_i) = g(a_j) \text{ in } A_k \end{array} \right)$$

(You will see that the "filtered" hypothesis is there is to ensure that $\sim$ is an equivalence relation.)

For example, in Example 1.3.7, each element of the colimit is an element of something upstairs, but you can't say in advance what it is an element of. For instance, $17/125$ is an element of $5^{-3}\mathbb{Z}$ (or $5^{-4}\mathbb{Z}$, or later ones), but not $5^{-2}\mathbb{Z}$.

This idea applies to many categories whose objects can be interpreted as sets with additional structure (such as abelian groups, A-modules, groups, etc.). For example, the colimit $\operatorname{colim} M_i$ in the category of A-modules Mod_A can be described as follows. The set underlying $\operatorname{colim} M_i$ is defined as in Exercise 1.3.E. To add the elements $m_i \in M_i$ and $m_j \in M_j$, choose an $\ell \in \mathscr{I}$ with arrows $u \colon i \to \ell$ and $v \colon j \to \ell$, and then define the sum of m_i and m_j to be $F(u)(m_i) + F(v)(m_j) \in M_\ell$. The element $m_i \in M_i$ is 0 if and only if there is some arrow $u \colon i \to k$ for which $F(u)(m_i) = 0$, i.e., if and only if it becomes 0 "later in the diagram." Last, multiplication by an element of A is defined in the obvious way.

1.3.F. EXERCISE. Verify that the A-module described above is indeed the colimit. (Make sure you verify that addition is well-defined, i.e., is independent of the choice of representatives m_i and m_j, the choice of ℓ, and the choice of arrows u and v. Similarly, make sure that scalar multiplication is well-defined.)

1.3.G. USEFUL EXERCISE (LOCALIZATION AS A COLIMIT). Generalize Exercise 1.3.D(a) to interpret localization of an integral domain as a colimit over a filtered set: suppose S is a multiplicative set of A, and interpret $S^{-1}A = \operatorname{colim} \frac{1}{s}A$ where the colimit is over $s \in S$, and in the category of A-modules. (Aside: Can you make some version of this work even if A isn't an integral domain, e.g., $S^{-1}A = \operatorname{colim} A_s$? This will work in the category of A-algebras.)

A variant of this construction works without the filtered condition if you have another means of "connecting elements in different objects of your diagram." For example:

1.3.H. EXERCISE: COLIMITS OF A-MODULES WITHOUT THE FILTERED CONDITION. Suppose you are given a diagram of A-modules indexed by $\mathscr{I}$: $F \colon \mathscr{I} \to Mod_A$, where we let $M_i := F(i)$. Show that the colimit is $\oplus_{i \in \mathscr{I}} M_i$ modulo the relations $m_i - F(n)(m_i)$ for every $n \colon i \to j$ in $\mathscr{I}$ (i.e., for every arrow in the diagram). (Somewhat more precisely: "modulo" means "quotiented by the submodule generated by.")

1.3.9. Summary. One useful thing to informally keep in mind is the following. In a category where the objects are "set-like," an element of a limit can be thought of as a family of elements of each object in the diagram that are "compatible" (Exercise 1.3.C). And an element of a colimit can be thought of as ("has a representative that is") an element of a single object in the diagram (Exercise 1.3.E). Even though the definitions of limit and colimit are the same, just with arrows reversed, these interpretations are quite different.

1.3.10. *Small remark.* In fact, colimits exist in the category of sets for all reasonable ("small") index categories (see for example [E, Thm. A6.1]), but that won't matter to us.

1.3.11. *Joke.* What do you call someone who reads a paper on category theory? Answer: A coauthor.

1.4   Adjoints

We next come to a very useful notion closely related to universal properties. Just as a universal property "essentially" (up to unique isomorphism) determines an object in a category (assuming such an object exists), "adjoints" essentially determine a functor (again, assuming it exists). Two *covariant* functors $F \colon \mathscr{A} \to \mathscr{B}$ and $G \colon \mathscr{B} \to \mathscr{A}$ are **adjoint** if there is a natural bijection for all $A \in \mathscr{A}$ and $B \in \mathscr{B}$,

$$(1.4.0.1) \qquad\qquad \tau_{AB} \colon \operatorname{Mor}_{\mathscr{B}}(F(A), B) \to \operatorname{Mor}_{\mathscr{A}}(A, G(B)).$$

We say that (F, G) form an **adjoint pair**, and that F is **left-adjoint** to G (and G is **right-adjoint** to F). We say F is a **left adjoint** (and G is a **right adjoint**). By "natural" we mean the following. For all f: $A \to A'$ in $\mathscr{A}$, we require

(1.4.0.2)
$$
\begin{array}{ccc}
\mathrm{Mor}_{\mathscr{B}}(F(A'), B) & \xrightarrow{\ Ff^* \ } & \mathrm{Mor}_{\mathscr{B}}(F(A), B) \\
\downarrow{\scriptstyle \tau_{A'B}} & & \downarrow{\scriptstyle \tau_{AB}} \\
\mathrm{Mor}_{\mathscr{A}}(A', G(B)) & \xrightarrow{\ f^* \ } & \mathrm{Mor}_{\mathscr{A}}(A, G(B))
\end{array}
$$

to commute, and for all $g: B \to B'$ in $\mathscr{B}$ we want a similar commutative diagram to commute. (Here f^* is the map induced by $f: A \to A'$, and Ff^* is the map induced by $Ff: F(A) \to F(A')$.)

1.4.A. EXERCISE. Write down what this diagram should be.

1.4.B. EXERCISE. Show that the map τ_{AB} (1.4.0.1) has the following properties. For each A there is a map $\eta_A: A \to GF(A)$ so that for any $g: F(A) \to B$, the corresponding $\tau_{AB}(g): A \to G(B)$ is given by the composition

$$
A \xrightarrow{\ \eta_A \ } GF(A) \xrightarrow{\ Gg \ } G(B).
$$

Similarly, there is a map $\epsilon_B: FG(B) \to B$ for each B so that for any $f: A \to G(B)$, the corresponding map $\tau_{AB}^{-1}(f): F(A) \to B$ is given by the composition

$$
F(A) \xrightarrow{\ Ff \ } FG(B) \xrightarrow{\ \epsilon_B \ } B.
$$

Here is a key example of an adjoint pair.

1.4.C. EXERCISE. Suppose M, N, and P are A-modules (where A is a ring). Describe a bijection $\mathrm{Hom}_A(M \otimes_A N, P) \leftrightarrow \mathrm{Hom}_A(M, \mathrm{Hom}_A(N, P))$. (Hint: Try to use the universal property of $\otimes$.)

1.4.D. EXERCISE (TENSOR-HOM ADJUNCTION). Show that $(\cdot) \otimes_A N$ and $\mathrm{Hom}_A(N, \cdot)$ are adjoint functors.

1.4.E. EXERCISE. Suppose $B \to A$ is a morphism of rings. If M is an A-module, you can create a B-module M_B by considering it as a B-module. This gives a functor $\cdot_B: Mod_A \to Mod_B$. Show that this functor is right-adjoint to $\cdot \otimes_B A$. In other words, describe a bijection

$$
\mathrm{Hom}_A(N \otimes_B A, M) \cong \mathrm{Hom}_B(N, M_B)
$$

functorial in both arguments. (This adjoint pair is very important.)

1.4.1.* *Fancier remarks we won't use.* You can check that the left adjoint determines the right adjoint up to natural isomorphism, and vice versa. The maps η_A and ϵ_B of Exercise 1.4.B are called the *unit* and *counit* of the adjunction. This leads to a different characterization of adjunction. Suppose functors $F: \mathscr{A} \to \mathscr{B}$ and $G: \mathscr{B} \to \mathscr{A}$ are given, along with natural transformations $\eta: \mathrm{id}_{\mathscr{A}} \to GF$ and $\epsilon: FG \to \mathrm{id}_{\mathscr{B}}$ with the property that $G\epsilon \circ \eta G = \mathrm{id}_G$ (for each $B \in \mathscr{B}$, the composition of $\eta_{G(B)}: G(B) \to GFG(B)$ and $G(\epsilon_B): GFG(B) \to G(B)$ is the identity) and $\epsilon F \circ F\eta = \mathrm{id}_F$. Then you can check that F is left-adjoint to G. These facts aren't hard to check, so if you want to use them, you should verify everything for yourself.

1.4.2. *Examples from other fields.* For those familiar with representation theory: Frobenius reciprocity may be understood in terms of adjoints. Suppose V is a finite-dimensional representation of a finite group G, and W is a representation of a subgroup $H < G$. Then induction and restriction are an adjoint pair $(\mathrm{Ind}_H^G, \mathrm{Res}_H^G)$ between the category of G-modules and the category of H-modules.

Topologists' favorite adjoint pair may be the suspension functor and the loop space functor.

1.4.3. Example: Groupification of abelian semigroups. Here is another motivating example: getting an abelian group from an abelian semigroup. (An **abelian semigroup** is just like an abelian group, except we don't require an identity or an inverse. Morphisms of abelian semigroups are maps of sets preserving the binary operation. One example is the nonnegative integers $\mathbb{Z}^{\geq 0} = \{0, 1, 2, \dots\}$ under addition. Another is the positive integers $\mathbb{Z}^+ = \{1, 2, 3, \dots\}$ under addition. Yet another is the positive integers $\mathbb{Z}^+$ under multiplication. You may enjoy groupifying all three.) From an abelian semigroup, you can create an abelian group. In our examples, from the nonnegative integers under addition $(\mathbb{Z}^{\geq 0}, +)$, we create the integers $(\mathbb{Z}, +)$, and from the positive integers under multiplication $(\mathbb{Z}^+, \times)$, we create the positive rationals $(\mathbb{Q}^+, \times)$. Here is a formalization of that notion. A **groupification** of an abelian semigroup S is a map of abelian semigroups $\pi \colon S \to G$ such that G is an abelian group, and any map of abelian semigroups from S to an abelian *group* G′ factors *uniquely* through G:

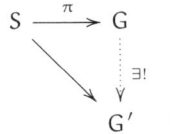

$$
\begin{array}{ccc}
S & \xrightarrow{\ \pi\ } & G \\
 & \searrow & \downarrow {\scriptstyle \exists!} \\
 & & G'
\end{array} \quad .
$$

(Perhaps "abelian groupification" would be more precise than "groupification.")

1.4.F. EXERCISE (AN ABELIAN GROUP IS GROUPIFIED BY ITSELF). Show that if an abelian semigroup is *already* a group then the identity morphism is the groupification. (More correct: the identity morphism is *a* groupification.) Note that you don't need to construct groupification (or even know that it exists in general) to solve this exercise.

1.4.G. EXERCISE. Construct the "groupification functor" H from the category of nonempty abelian semigroups to the category of abelian groups. (One possible construction: given an abelian semigroup S, the elements of its groupification $H(S)$ are ordered pairs $(a, b) \in S \times S$, which you may think of as $a - b$, with the equivalence that $(a, b) \sim (c, d)$ if $a + d + e = b + c + e$ for some $e \in S$. Describe addition in this group, and show that it satisfies the properties of an abelian group. Describe the abelian semigroup map $S \to H(S)$.) Let F be the forgetful functor from the category of abelian groups *Ab* to the category of abelian semigroups. Show that H is left-adjoint to F.

(Here is the general idea for experts: We have a full subcategory of a category. We want to "project" from the category to the subcategory. We have

$$
\mathrm{Mor}_{\mathrm{category}}(S, H) = \mathrm{Mor}_{\mathrm{subcategory}}(G, H)
$$

automatically; thus we are describing the left adjoint to the forgetful functor. How the argument worked: we constructed something which was in the smaller category, which automatically satisfies the universal property.)

1.4.H. EXERCISE (CF. EXERCISE 1.4.E). The purpose of this exercise is to give you more practice with "left adjoints of forgetful functors," the means by which we get abelian groups from abelian semigroups, and sheaves from presheaves. Suppose A is a ring, and S is a multiplicative subset. Then $S^{-1}A$-modules are a full subcategory (§1.1.15) of the category of A-modules (via the obvious inclusion $Mod_{S^{-1}A} \hookrightarrow Mod_A$). Then $Mod_A \to Mod_{S^{-1}A}$ can be interpreted as an adjoint to the forgetful functor $Mod_{S^{-1}A} \to Mod_A$. State and prove the correct statements.

(Here is the larger story. Every $S^{-1}A$-module is an A-module, and this is an injective map, so we have a forgetful functor F: $Mod_{S^{-1}A} \to Mod_A$. In fact this is a fully faithful functor: it is injective on objects, and the morphisms between any two $S^{-1}A$-modules *as A-modules* are just the same when they are considered as $S^{-1}A$-modules. Then there is a functor G: $Mod_A \to Mod_{S^{-1}A}$, which might reasonably be called "localization with respect to S," which is left-adjoint to the forgetful functor. Translation: If M is an A-module, and N is an $S^{-1}A$-module, then Mor(GM, N) (morphisms as $S^{-1}A$-modules, which are the same as morphisms as A-modules) are in natural bijection with Mor(M, FN) (morphisms as A-modules).)

situation	category $\mathscr{A}$	category $\mathscr{B}$	left adjoint $F\colon \mathscr{A}\to\mathscr{B}$	right adjoint $G\colon \mathscr{B}\to\mathscr{A}$
A-modules (Ex. 1.4.D)	Mod_A	Mod_A	$(\cdot)\otimes_A N$	$\mathrm{Hom}_A(N,\cdot)$
ring maps $B\to A$ (Ex. 1.4.E)	Mod_B	Mod_A	$(\cdot)\otimes_B A$ (extension of scalars)	$M\mapsto M_B$ (restriction of scalars)
(pre)sheaves on a topological space X (Ex. 2.4.K)	presheaves on X	sheaves on X	sheafification	forgetful
(semi)groups (§1.4.3)	semigroups	groups	groupification	forgetful
sheaves, $\pi\colon X\to Y$ (Ex. 2.7.B)	sheaves on Y	sheaves on X	π^{-1}	π_*
sheaves of abelian groups or $\mathscr{O}$-modules, open embeddings $\pi\colon U\hookrightarrow Y$ (Ex. 23.4.G)	sheaves on U	sheaves on Y	$\pi_!$	π^{-1}
quasicoherent sheaves, $\pi\colon X\to Y$ (Prop. 14.5.7)	$QCoh_Y$	$QCoh_X$	π^*	π_*
ring maps $B\to A$ (Ex. 17.1.J)	Mod_A	Mod_B	$M\mapsto M_B$ (restriction of scalars)	$N\mapsto \mathrm{Hom}_B(A,N)$
quasicoherent sheaves, affine $\pi\colon X\to Y$ (Ex. 17.1.K(b))	$QCoh_X$	$QCoh_Y$	π_*	$\pi^{!?}$

Table 1.1 *Some important adjoint pairs.*

Table 1.1 gives most of the adjoints that will come up for us. Other examples will also come up, such as the adjoint pair $(\sim, \Gamma_\bullet)$ between graded modules over a graded ring, and quasicoherent sheaves on the corresponding projective scheme (§15.7).

1.4.4. *Last comments only for people who have seen adjoints before.* If (F, G) is an adjoint pair of functors, then F commutes with colimits, and G commutes with limits. Also, limits commute with limits and colimits commute with colimits. We will prove these facts (and a little more) in §1.5.14.

1.5   An Introduction to Abelian Categories

> *Ton papier sur l'Algèbre homologique a été lu soigneusement, et a converti tout le monde (même Dieudonné, qui semble complètement fonctorisé!) à ton point de vue.*
>
> *Your paper on homological algebra was read carefully and converted everyone (even Dieudonné, who seems to be completely functorised!) to your point of view.*
>
> —J.-P. Serre, letter to A. Grothendieck, July 13, 1955 [GrS, pp. 17–18]

Since learning linear algebra, you have been familiar with the notions and behaviors of kernels, cokernels, etc. Later in your life you saw them in the category of abelian groups, and later still in the category of A-modules.

We will soon define some new categories (certain sheaves) that will have familiar-looking behavior, reminiscent of that of modules over a ring. The notions of kernels, cokernels, images, and more will make sense, and they will behave "the way we expect" from our experience with modules. This can be made precise through the notion of an *abelian category*. Abelian categories are the right general setting in which one can do "homological algebra," in which notions of kernel, cokernel, and so on are used, and one can work with complexes and exact sequences.

We will see enough to motivate the definitions that we will see in general: monomorphism (and subobject), epimorphism, kernel, cokernel, and image. But in this book we will avoid showing that they behave "the way we expect" in a general abelian category because the examples we will see are directly interpretable in terms of modules over rings. In particular, it is not worth memorizing the definition of abelian category.

Two central examples of an abelian category are the category *Ab* of abelian groups, and the category *Mod*$_A$ of A-modules. The first is a special case of the second (just take A = $\mathbb{Z}$). As we give the definitions, you should verify that *Mod*$_A$ is an abelian category.

We first define the notion of *additive category*. We will use it only as a stepping stone to the notion of an abelian category. Two examples you can keep in mind while reading the definition: the category of free A-modules (where A is a ring), and real (or complex) Banach spaces.

1.5.1. Definition. A category $\mathscr{C}$ is said to be **additive** if it satisfies the following properties.

Ad1.   For each A, B $\in \mathscr{C}$, Mor(A, B) is an abelian group, such that composition of morphisms distributes over addition. (You should think about what this means—it translates to two distinct statements.)

Ad2. $\mathscr{C}$ has a zero object, denoted by 0. (This is an object that is simultaneously an initial object and a final object, Definition 1.2.2.)

Ad3.   It has products of two objects (a product A × B for any pair of objects), and hence, by induction, products of any finite number of objects.

In an additive category, the morphisms are often called **homomorphisms**, and Mor is denoted by Hom. In fact, the notation Hom is a good indication that you're working in an additive category. A functor between additive categories preserving the additive structure of Hom is called an **additive functor**.

1.5.2. *Remarks.* It is a consequence of the definition of additive category that finite products are also finite coproducts (i.e., sums)—the details don't matter to us. The symbol $\oplus$ is used for this notion. Also, it is quick to show that additive functors send zero objects to zero objects (show that Z is a 0-object if and only if $\mathrm{id}_Z = 0_Z$; additive functors preserve both id and 0), and preserve products.

One motivation for the name 0-object is that the 0-morphism in the abelian group Hom(A, B) is the composition A $\to$ 0 $\to$ B. (We also remark that the notion of 0-morphism thus makes sense in any category with a 0-object.)

(A cleaner axiomatization of additive categories that makes clear that the abelian group structure of Mor(A, B) is intrinsic to the category itself is the following [Lur, pp. 21–22]. *A0. $\mathscr{C}$ has a zero object. A1. $\mathscr{C}$ has products of any two objects, and coproducts of any two objects.* By the universal property of product and coproduct, we have natural morphisms $\phi_{AB} : A \coprod B \to A \times B$. *A2. ϕ_{AB} is an isomorphism.* This allows us to define a binary operation on Mor(A, B), with f + g (for f, g $\in$ Mor(A, B)) defined by the composition

$$A \xrightarrow{\ (f,g)\ } B \times B \xrightarrow{\ \phi_{BB}^{-1}\ } B \coprod B \longrightarrow B,$$

where the last map is the "codiagonal" defined by the universal property of coproduct. A little work shows that this endows Mor(A, B) with the structure of a **commutative monoid**, i.e., an abelian semigroup with identity. The identity is the composition A $\to$ 0 $\to$ B. *A3. This commutative monoid* Mor(A, B) *is an abelian group.*)

1.5.3. The category of A-modules Mod_A is clearly an additive category, but it has even more structure. We now formalize some essential aspects of this structure in the notion of *abelian category*.

1.5.4. Definition. Let $\mathscr{C}$ be a category with a 0-object (and thus 0-morphisms). A **kernel** of a morphism $f\colon B \to C$ is *defined to be* a map $i\colon A \to B$ such that $f \circ i = 0$, and that is universal with respect to this property. Diagramatically:

(1.5.4.1)

(Note that the kernel is not just an object; it is a morphism of an object to B. In practice, the term is often applied to just the object, and the intended interpretation is clear from the context.) Hence it is unique up to unique isomorphism by universal property nonsense. The kernel is written $\ker f \to B$. A **cokernel** (denoted by $\operatorname{coker} f$) is defined dually by reversing the arrows—do this yourself. The kernel of $f\colon B \to C$ is the limit (§1.3) of the diagram

(1.5.4.2)

$$
\begin{array}{ccc}
 & & 0 \\
 & & \downarrow \\
B & \xrightarrow{\ f\ } & C
\end{array}
$$

and similarly the cokernel is a colimit (see (2.6.0.1)).

If $i\colon A \to B$ is a monomorphism, then we say that A is a **subobject** of B, where the map i is implicit. There is also the notion of **quotient object**, defined dually to subobject.

An **abelian category** is an additive category satisfying three additional properties.

(1) Every map has a kernel and cokernel.
(2) Every monomorphism is the kernel of its cokernel.
(3) Every epimorphism is the cokernel of its kernel.

It is a nonobvious (and imprecisely stated) fact that every property you want to be true about kernels, cokernels, etc. follows from these three. (Warning: In part of the literature, additional hypotheses are imposed as part of the definition.)

The **image** of a morphism $f\colon A \to B$ is defined as $\operatorname{im}(f) = \ker(\operatorname{coker} f)$ whenever it exists (e.g., in every abelian category). The morphism $f\colon A \to B$ factors uniquely through $\operatorname{im} f \to B$ whenever $\operatorname{im} f$ exists, and $A \to \operatorname{im} f$ is an epimorphism and a cokernel of $\ker f \to A$ in every abelian category. The reader may want to verify this as a (hard!) exercise.

The cokernel of a monomorphism is called the **quotient**. The quotient of a monomorphism $A \to B$ is often denoted by B/A (with the map from B implicit).

We will leave the foundations of abelian categories untouched. The key thing to remember is that if you understand kernels, cokernels, images and so on in the category of modules over a given ring, you can manipulate objects in any abelian category. This is made precise by the Freyd–Mitchell Embedding Theorem (Remark 1.5.5).

However, the abelian categories we will come across will obviously be related to modules, and our intuition will clearly carry over, so we needn't invoke a theorem whose proof we haven't read. For example, we will show that sheaves of abelian groups on a topological space X form an abelian category (§2.6), and the interpretation in terms of "compatible germs" will connect notions of kernels, cokernels etc. of sheaves of abelian groups to the corresponding notions of abelian groups.

1.5.5. *Small remark on chasing diagrams.* It is useful to prove facts (and solve exercises) about abelian categories by chasing elements. Unfortunately, some commonly used abelian categories, such as the category of complexes (to be defined in Exercise 1.5.C), do not have "elements"—they are not naturally "sets with additional structure" in any obvious way. Nonetheless, proof by element-chasing can be justified by the Freyd–Mitchell Embedding Theorem: If $\mathscr{C}$ is an abelian category whose objects form a set, then there is a ring A and an exact, fully faithful functor from $\mathscr{C}$ into Mod_A, which embeds $\mathscr{C}$ as a full subcategory. (Unfortunately, the ring A need not be commutative.) A proof is sketched in [Weib, §1.6], and references to a complete proof are given there. A proof is also given in [KS2, §9.6]. The upshot is that to prove something about a diagram in some abelian category, we may assume that it is a diagram of modules over some ring, and we may then "diagram-chase" elements. Moreover, any fact about kernels, cokernels, and so on that holds in Mod_A holds in any abelian category.

If invoking a theorem whose proof you haven't read bothers you, a short alternative is Mac Lane's "elementary rules for chasing diagrams," [Mac, Thm. 3, p. 204]; [Mac, Lem. 4, p. 205] gives a proof of the Five Lemma (Exercise 1.6.6) as an example.

But in any case, do what you need to do to put your mind at ease, so you can move forward. Do as little as your conscience will allow.

1.5.6. Complexes, exactness, and homology.

(In this entire discussion, we assume we are working in an abelian category.) We say a sequence

$$(1.5.6.1) \qquad \cdots \longrightarrow A \xrightarrow{\ f\ } B \xrightarrow{\ g\ } C \longrightarrow \cdots$$

is a **complex at** B if $g \circ f = 0$, and is **exact at** B if $\ker g = \operatorname{im} f$. (More specifically, g has a kernel that is an image of f. Exactness at B implies being a complex at B—do you see why?) A sequence is a **complex** (resp., **exact**) if it is a complex (resp., exact) at each (internal) term. A **short exact sequence** is an exact sequence with five terms, the first and last of which are zeros—in other words, an exact sequence of the form

$$0 \longrightarrow A \longrightarrow B \longrightarrow C \longrightarrow 0.$$

For example, $0 \longrightarrow A \longrightarrow 0$ is exact if and only if $A = 0$;

$$0 \longrightarrow A \xrightarrow{\ f\ } B$$

is exact if and only if f is a monomorphism (with a similar statement for $A \xrightarrow{\ f\ } B \longrightarrow 0$);

$$0 \longrightarrow A \xrightarrow{\ f\ } B \longrightarrow 0$$

is exact if and only if f is an isomorphism; and

$$0 \longrightarrow A \xrightarrow{\ f\ } B \xrightarrow{\ g\ } C$$

is exact if and only if f is a kernel of g (with a similar statement for $A \xrightarrow{\ f\ } B \xrightarrow{\ g\ } C \longrightarrow 0$). To show some of these facts it may be helpful to prove that (1.5.6.1) is exact at B if and only if the cokernel of f is a cokernel of the kernel of g.

If you would like practice in playing with these notions before thinking about homology, you can prove the Snake Lemma (stated in Example 1.6.5, with a stronger version in Exercise 1.6.B), or the Five Lemma (stated in Example 1.6.6, with a stronger version in Exercise 1.6.C). (I would do this in the category of A-modules, but see [KS2, Lem. 12.1.1, Lem. 8.3.13] for proofs in general.)

If (1.5.6.1) is a complex at B, then its **homology at** B (often denoted by H) is $\ker g / \operatorname{im} f$. (More precisely, there is some monomorphism $\operatorname{im} f \hookrightarrow \ker g$, and H is the cokernel of this monomorphism.) Therefore, (1.5.6.1) is exact at B if and only if its homology at B is 0. We say that elements of $\ker g$ (assuming the objects of the category are sets with some additional structure) are the **cycles**, and elements of $\operatorname{im} f$ are the **boundaries** (so homology is "cycles mod boundaries"). If the complex is indexed in decreasing order, the indices are often written as subscripts, and H_i is the homology at $A_{i+1} \to A_i \to A_{i-1}$. If the complex is indexed in increasing order, the indices are often written as superscripts, and the homology H^i at $A^{i-1} \to A^i \to A^{i+1}$ is often called **cohomology**.

An exact sequence

$$(1.5.6.2) \qquad A^\bullet: \qquad \cdots \longrightarrow A^{i-1} \xrightarrow{f^{i-1}} A^i \xrightarrow{f^i} A^{i+1} \xrightarrow{f^{i+1}} \cdots$$

can be "factored" into short exact sequences

$$0 \longrightarrow \ker f^i \longrightarrow A^i \longrightarrow \ker f^{i+1} \longrightarrow 0,$$

which is helpful in proving facts about long exact sequences by reducing them to facts about short exact sequences.

More generally, if (1.5.6.2) is assumed only to be a complex, then it can be "factored" into short exact sequences

$$(1.5.6.3) \qquad 0 \longrightarrow \ker f^i \longrightarrow A^i \longrightarrow \operatorname{im} f^i \longrightarrow 0,$$

$$0 \longrightarrow \operatorname{im} f^{i-1} \longrightarrow \ker f^i \longrightarrow H^i(A^\bullet) \longrightarrow 0$$

1.5.A. EXERCISE. Describe exact sequences

$$(1.5.6.4) \qquad 0 \longrightarrow \operatorname{im} f^i \longrightarrow A^{i+1} \longrightarrow \operatorname{coker} f^i \longrightarrow 0,$$

$$0 \longrightarrow H^i(A^\bullet) \longrightarrow \operatorname{coker} f^{i-1} \longrightarrow \operatorname{im} f^i \longrightarrow 0$$

(These are somehow dual to (1.5.6.3). In fact, in some mirror universe this might have been given as the standard definition of homology.) Assume the category is that of modules over a fixed ring for convenience, but be aware that the result is true for any abelian category.

1.5.B. EXERCISE AND IMPORTANT DEFINITION. Suppose

$$0 \xrightarrow{d^0} A^1 \xrightarrow{d^1} \cdots \xrightarrow{d^{n-1}} A^n \xrightarrow{d^n} 0$$

is a complex of finite-dimensional k-vector spaces (often called $A^\bullet$ for short). Define $h^i(A^\bullet) := \dim H^i(A^\bullet)$. Show that $\sum (-1)^i \dim A^i = \sum (-1)^i h^i(A^\bullet)$. In particular, if $A^\bullet$ is exact, then $\sum (-1)^i \dim A^i = 0$. (If you haven't dealt much with cohomology, this will give you some practice.)

1.5.C. IMPORTANT EXERCISE. Suppose $\mathscr{C}$ is an abelian category. Define the **category $Com_\mathscr{C}$ of complexes**) as follows. The objects are infinite complexes

$$A^\bullet: \qquad \cdots \longrightarrow A^{i-1} \xrightarrow{f^{i-1}} A^i \xrightarrow{f^i} A^{i+1} \xrightarrow{f^{i+1}} \cdots$$

in $\mathscr{C}$, and the morphisms $A^\bullet \to B^\bullet$ are commuting diagrams

$$(1.5.6.5)$$

$$
\begin{array}{ccccccccc}
\cdots \longrightarrow & A^{i-1} & \xrightarrow{f^{i-1}} & A^i & \xrightarrow{f^i} & A^{i+1} & \xrightarrow{f^{i+1}} & \cdots \\
& \downarrow & & \downarrow & & \downarrow & & \\
\cdots \longrightarrow & B^{i-1} & \xrightarrow{g^{i-1}} & B^i & \xrightarrow{g^i} & B^{i+1} & \xrightarrow{g^{i+1}} & \cdots
\end{array}
$$

Show that $Com_\mathscr{C}$ is an abelian category.

Feel free to deal with the special case of modules over a fixed ring. (Remark for experts: Essentially the same argument shows that $\mathscr{C}^\mathscr{I}$ is an abelian category for any small category $\mathscr{I}$ and any abelian category $\mathscr{C}$. This immediately implies that the category of presheaves on a topological space X with values in an abelian category $\mathscr{C}$ is automatically an abelian category; cf. §2.3.5.)

1.5.D. IMPORTANT EXERCISE. Show that (1.5.6.5) induces a map of homology $H^i(A^\bullet) \to H^i(B^\bullet)$. Show furthermore that H^i is a covariant functor $Com_\mathscr{C} \to \mathscr{C}$. (Again, feel free to deal with the special case Mod_A.)

1.5.7. *Homotopic maps induce the same maps on homology.* We say two maps of complexes $f\colon C^\bullet \to D^\bullet$ and $g\colon C^\bullet \to D^\bullet$ are **homotopic** if there is a sequence of maps $w\colon C^i \to D^{i-1}$ such that $f - g = dw + wd$.

1.5.E. EXERCISE. Show that two homotopic maps give the same map on homology.

1.5.8. Theorem (long exact sequences) — *A short exact sequence of complexes*

induces a **long exact sequence in cohomology**

$$\cdots \longrightarrow H^{i-1}(C^\bullet) \longrightarrow$$
$$H^i(A^\bullet) \longrightarrow H^i(B^\bullet) \longrightarrow H^i(C^\bullet) \longrightarrow$$
$$H^{i+1}(A^\bullet) \longrightarrow \cdots$$

(This requires a definition of the **connecting homomorphism** $H^{i-1}(C^\bullet) \to H^i(A^\bullet)$, which is "natural" in an appropriate sense.) In the category of modules over a ring, Theorem 1.5.8 will come out of our discussion of spectral sequences—see Exercise 1.6.F—but this is a somewhat perverse way of proving it. For a proof in general, see [KS2, Theorem 12.3.3]. You may want to prove it yourself, by first proving a weaker version of the Snake Lemma (Example 1.6.5), where in the hypotheses (1.6.5.1), the 0's in the bottom left and top right are removed, and in the conclusion (1.6.5.2), the first and last 0's are removed.

1.5.9. *Exactness of functors.* If $F\colon \mathscr{A} \to \mathscr{B}$ is an additive covariant functor from one abelian category to another, we say that F is **right-exact** if the exactness of

$$A' \longrightarrow A \longrightarrow A'' \longrightarrow 0$$

in $\mathscr{A}$ implies that

$$F(A') \longrightarrow F(A) \longrightarrow F(A'') \longrightarrow 0$$

is also exact. Dually, we say that F is **left-exact** if the exactness of

$$0 \longrightarrow A' \longrightarrow A \longrightarrow A'' \qquad \text{implies}$$

$$0 \longrightarrow F(A') \longrightarrow F(A) \longrightarrow F(A'') \qquad \text{is exact.}$$

An additive contravariant functor is **left-exact** if the exactness of

$$A' \longrightarrow A \longrightarrow A'' \longrightarrow 0 \qquad \text{implies}$$

$$0 \longrightarrow F(A'') \longrightarrow F(A) \longrightarrow F(A') \qquad \text{is exact.}$$

The reader should be able to deduce what it means for a contravariant functor to be **right-exact**. An additive covariant or contravariant functor is **exact** if it is both left-exact and right-exact.

1.5.F. EXERCISE. Suppose F is an exact functor. Show that applying F to an exact sequence preserves exactness. For example, if F is covariant, and $A' \to A \to A''$ is exact, then $FA' \to FA \to FA''$ is exact. (This will be generalized in Exercise 1.5.I(c).)

1.5.G. EXERCISE. Suppose A is a ring, $S \subset A$ is a multiplicative subset, and M is an A-module.

(a) Show that localization of A-modules $Mod_A \to Mod_{S^{-1}A}$ is an exact covariant functor.
(b) Show that $(\cdot) \otimes_A M$ is a right-exact covariant functor $Mod_A \to Mod_A$. (This is a repeat of Exercise 1.2.H.)
(c) Show that $Hom(M, \cdot)$ is a left-exact covariant functor $Mod_A \to Mod_A$. If $\mathscr{C}$ is any abelian category, and $C \in \mathscr{C}$, show that $Hom(C, \cdot)$ is a left-exact covariant functor $\mathscr{C} \to Ab$.
(d) Show that $Hom(\cdot, M)$ is a left-exact contravariant functor $Mod_A \to Mod_A$. If $\mathscr{C}$ is any abelian category, and $C \in \mathscr{C}$, show that $Hom(\cdot, C)$ is a left-exact contravariant functor $\mathscr{C} \to Ab$.

1.5.H. EXERCISE. Suppose M is a **finitely presented A-module**: M has a finite number of generators, and with these generators it has a finite number of relations; or, equivalently, and usefully fits in an exact sequence

$$(1.5.9.1) \qquad A^{\oplus q} \longrightarrow A^{\oplus p} \longrightarrow M \longrightarrow 0.$$

Use (1.5.9.1) and the left-exactness of Hom to describe an isomorphism

$$S^{-1} Hom_A(M, N) \xleftarrow{\sim} Hom_{S^{-1}A}(S^{-1}M, S^{-1}N).$$

(You might be able to interpret this in light of a variant of Exercise 1.5.I below, for left-exact contravariant functors rather than right-exact covariant functors.)

1.5.10. *Example:* Hom *doesn't always commute with localization.* In the language of Exercise 1.5.H, take $A = N = \mathbb{Z}$, $M = \mathbb{Q}$, and $S = \mathbb{Z} \setminus \{0\}$.

1.5.11.* Two useful facts in homological algebra.
We now come to two (sets of) facts I wish I had learned as a child, as they would have saved me lots of grief. They encapsulate what is best and worst of abstract nonsense. The statements are so general as to be nonintuitive. The proofs are very short. They generalize some specific behavior that is easy to prove on an ad hoc basis. Once they are second nature to you, many subtle facts will become obvious to you as special cases. And you will see that they will get used (implicitly or explicitly) repeatedly.

1.5.12.* *Interaction of homology and (right/left-)exact functors.*
You might wait to prove this until you learn about cohomology in Chapter 18, when it will first be used in a serious way.

1.5.I. IMPORTANT EXERCISE (THE FHHF THEOREM). This result can take you far, and perhaps for that reason it has sometimes been called the Fernbahnhof (FernbaHnHoF) Theorem, notably in [Vak1, Exer. 1.5.I]. Suppose F: $\mathscr{A} \to \mathscr{B}$ is a covariant functor of abelian categories, and $C^\bullet$ is a complex in $\mathscr{A}$.

(a) (F *right-exact yields* $FH^\bullet \longrightarrow H^\bullet F$) If F is right-exact, describe a natural morphism $FH^\bullet \to H^\bullet F$. (More precisely, for each i, the left side is F applied to the cohomology at piece i of $C^\bullet$, while the right side is the cohomology at piece i of $FC^\bullet$.)

(b) (F *left-exact yields* $FH^\bullet \longleftarrow H^\bullet F$) If F is left-exact, describe a natural morphism $H^\bullet F \to FH^\bullet$.

(c) (F *exact yields* $FH^\bullet \overset{\sim}{\longleftrightarrow} H^\bullet F$) If F is exact, show that the morphisms of (a) and (b) are inverses and thus isomorphisms.

Hint for (a): Use $C^i \overset{d^i}{\longrightarrow} C^{i+1} \longrightarrow \operatorname{coker} d^i \longrightarrow 0$ to give an isomorphism F coker $d^i \overset{\sim}{\longleftrightarrow}$ coker Fd^i. Then use the first line of (1.5.6.4) to give an epimorphism F im $d^i \twoheadrightarrow$ im Fd^i. Then use the second line of (1.5.6.4) to give the desired map $FH^i C^\bullet \to H^i FC^\bullet$. While you are at it, you may as well describe a map for the fourth member of the quartet {coker, im, H, ker}: F ker $d^i \to$ ker Fd^i.

1.5.13. If this makes your head spin, you may prefer to think of it in the following specific case, where both $\mathscr{A}$ and $\mathscr{B}$ are the category of A-modules, and F is $(\cdot) \otimes_A N$ for some fixed A-module N. Your argument in this case will translate without change to yield a solution to Exercise 1.5.I(a) and (c) in general. If $\otimes N$ is exact, then N is called a **flat** A-module. (The notion of flatness will turn out to be very important, and is discussed in detail in Chapter 24.)

For example, localization is exact (Exercise 1.5.G(a)), so $S^{-1}A$ is a *flat* A-algebra for all multiplicative sets S. Thus taking cohomology of a complex of A-modules commutes with localization—something you could verify directly.

1.5.14. *Interaction of adjoints, (co)limits, and (left- and right-) exactness.*
A surprising number of arguments boil down to the following statement:

Limits commute with limits and right adjoints. In particular, in an abelian category, because kernels are limits, both limits and right adjoints are left-exact.

And to its dual:

Colimits commute with colimits and left adjoints. In particular, because cokernels are colimits, both colimits and left adjoints are right-exact.

These statements were promised in §1.4.4, and will be proved below. The latter has a useful extension:

In Mod_A, colimits over filtered index categories are exact. "Filtered" was defined in §1.3.8.

1.5.15. ** *Caution.* It is not true that in abelian categories in general, colimits over filtered index categories are exact. (Grothendieck realized the desirability of such colimits being exact, and formalized this as his "AB5" axiom; see, for example, [Stacks, tag 079A].) Here is a counterexample. Because the axioms of abelian categories are self-dual, it suffices to give an example in which a *cofiltered limit* fails to be exact (where **cofiltered** has the obvious dual definition to *filtered*), and we do this. Fix a prime p. In the category *Ab* of abelian groups, for each positive integer n, we have an exact sequence $\mathbb{Z} \to \mathbb{Z}/(p^n) \to 0$. Taking the limit over all n in the obvious way, we obtain $\mathbb{Z} \to \mathbb{Z}_p \to 0$, which is certainly not exact.)

Unimportant Remark 1.5.18 will dash another hope you may have.

1.5.16. If you want to use these statements (for example, later in this book), you will have to prove them. Let's now make them precise.

1.5.J. EXERCISE (KERNELS COMMUTE WITH LIMITS). Suppose $\mathscr{C}$ is an abelian category, and $a\colon \mathscr{I} \to \mathscr{C}$ and $b\colon \mathscr{I} \to \mathscr{C}$ are two diagrams in $\mathscr{C}$ indexed by $\mathscr{I}$. For convenience, let $A_i = a(i)$ and $B_i = b(i)$ be the objects in those two diagrams. Let $h_i\colon A_i \to B_i$ be maps commuting with the maps in the diagrams. (Translation: h is a natural transformation of functors $a \to b$; see §1.1.21.) Then the ker h_i form another diagram in $\mathscr{C}$ indexed by $\mathscr{I}$. Describe a canonical isomorphism lim ker $h_i \overset{\sim}{\longleftrightarrow}$ ker(lim $A_i \to$ lim B_i), assuming the limits exist.

Implicit in the previous exercise is the idea that limits should somehow be understood as functors.

1.5.K. EXERCISE. Make sense of the statement that "limits commute with limits" in a general category, and prove it. (Hint: Recall that kernels are limits. The previous exercise should be a corollary of this one.)

1.5.17. Proposition (right adjoints commute with limits) — *Suppose* $(F \colon \mathscr{C} \to \mathscr{D}, G \colon \mathscr{D} \to \mathscr{C})$ *is a pair of adjoint functors. If* $A = \lim A_i$ *is a limit in* $\mathscr{D}$ *of a diagram indexed by* $\mathscr{I}$, *then* $GA = \lim GA_i$ *(with the corresponding maps* $GA \to GA_i$*) is a limit in* $\mathscr{C}$.

Proof. We must show that $GA \to GA_i$ satisfies the universal property of limits. Suppose we have maps $W \to GA_i$ commuting with the maps of $\mathscr{I}$. We wish to show that there exists a unique $W \to GA$ extending the $W \to GA_i$. By adjointness of F and G, we can restate this as: Suppose we have maps $FW \to A_i$ commuting with the maps of $\mathscr{I}$. We wish to show that there exists a unique $FW \to A$ extending the $FW \to A_i$. But this is precisely the universal property of the limit. $\qquad\square$

Of course, the dual statements to Exercise 1.5.K and Proposition 1.5.17 hold by the dual arguments.

If F and G are additive functors between abelian categories, and (F, G) is an adjoint pair, then (as kernels are limits and cokernels are colimits) G is left-exact and F is right-exact.

1.5.L. EXERCISE. Show that in Mod_A, colimits over filtered index categories are exact. (Your argument will apply without change to any abelian category whose objects can be interpreted as "sets with additional structure.") Right-exactness follows from the above discussion, so the issue is left-exactness. (Possible hint: After you show that localization is exact, Exercise 1.5.G(a), or stalkification is exact, Exercise 2.6.E, in a hands-on way, you will be easily able to prove this. Conversely, if you do this exercise, those two will be easy.)

1.5.M. EXERCISE. Show that filtered colimits commute with homology in Mod_A. Hint: Use the FHHF Theorem (Exercise 1.5.I), and the previous exercise.

In light of Exercise 1.5.M, you may want to think about how limits (and colimits) commute with homology in general, and which way maps go. The statement of the FHHF Theorem should suggest the answer. (Are limits analogous to left-exact functors, or right-exact functors?) We won't directly use this insight, but see §18.1 **(vii)** for an example.

Just as colimits are exact (not just right-exact) in especially good circumstances, limits are exact (not just left-exact), too. The following will be used twice in Chapter 28.

1.5.N. EXERCISE. Suppose

is an inverse system of exact sequences of modules over a ring, such that the maps $A_{n+1} \to A_n$ are surjective. (We say: "transition maps of the left term are surjective.") Show that the limit

$$(1.5.17.1) \qquad 0 \longrightarrow \lim A_n \longrightarrow \lim B_n \longrightarrow \lim C_n \longrightarrow 0$$

is also exact. (You will need to define the maps in (1.5.17.1).)

1.5.18. *Unimportant remark.* Based on these ideas, you may suspect that right-exact functors always commute with colimits. The fact that the tensor product commutes with infinite direct sums (Exercise 1.2.M) may reinforce this idea. Unfortunately, it is not true—"double dual" $\cdot^{\vee\vee}$: $Vec_k \to Vec_k$ is covariant and right exact (in fact, exact), but does not commute with infinite direct sums, as $\oplus_{i=1}^{\infty}(k^{\vee\vee})$ is not isomorphic to $(\oplus_{i=1}^{\infty}k)^{\vee\vee}$.

1.5.19.* Dreaming of derived functors. When you see a left-exact functor, you should always dream that you are seeing the end of a long exact sequence. If

$$0 \longrightarrow M' \longrightarrow M \longrightarrow M'' \longrightarrow 0$$

is an exact sequence in abelian category $\mathscr{A}$, and $F \colon \mathscr{A} \to \mathscr{B}$ is a left-exact functor, then

$$0 \longrightarrow FM' \longrightarrow FM \longrightarrow FM''$$

is exact, and you should always dream that it continues in some natural way. For example, the next term should depend only on M'—call it R^1FM'—and if it is zero, then $FM \to FM''$ is an epimorphism. This remark holds true for left-exact and contravariant functors too. In good cases, such a continuation exists, and is incredibly useful. We will discuss this in Chapter 23.

1.6*   Spectral Sequences

> *Je suis quelque peu affolé par ce déluge de cohomologie, mais j'ai courageusement tenu le coup. Ta suite spectrale me paraît raisonnable (je croyais, sur un cas particulier, l'avoir mise en défaut, mais je m'étais trompé, et cela marche au contraire admirablement bien).*
>
> *I am a bit panic-stricken by this flood of cohomology, but have borne up courageously. Your spectral sequence seems reasonable to me (I thought I had shown that it was wrong in a special case, but I was mistaken, on the contrary it works remarkably well).*
>
> —J.-P. Serre, letter to A. Grothendieck, March 14, 1956 [GrS, p. 38]

Spectral sequences are a powerful bookkeeping tool for proving things involving complicated commutative diagrams. They were introduced by Leray in the 1940s at the same time as he introduced sheaves. They have a reputation for being abstruse and difficult. It has been suggested that the name 'spectral' was given because, like specters, spectral sequences are terrifying, evil, and dangerous. I have heard no one disagree with this interpretation, which is perhaps not surprising since I just made it up.

Nonetheless, the goal of this section is to tell you enough that you can use spectral sequences without hesitation or fear, and why you shouldn't be frightened when they come up in a seminar. What is perhaps different in this presentation is that we will use spectral sequences to prove things that you may have already seen, and that you can prove easily in other ways. This will allow you to get some hands-on experience in how to use them. We will also see them only in the special case of double complexes (the version by far the most often used in algebraic geometry), and not in the general form usually presented (filtered complexes, exact couples, etc.). See [Weib, Ch. 5] for more detailed information if you wish.

You should *not* read this section when you are reading the rest of Chapter 1. Instead, you should read it just before you need it for the first time. When you finally *do* read this section, you *must* do the exercises up to Exercise 1.6.F.

For concreteness, we work in the category Mod_A of module over a ring A. However, everything we say will apply in any abelian category. (And if it helps you feel secure, work instead in the category Vec_k of vector spaces over a field k.)

1.6.1. Double complexes.

A **double complex** is a collection of A-modules $E^{p,q}$ (p, q $\in \mathbb{Z}$), and "rightward" morphisms $d_\rightarrow^{p,q}: E^{p,q} \rightarrow E^{p+1,q}$ and "upward" morphisms $d_\uparrow^{p,q}: E^{p,q} \rightarrow E^{p,q+1}$. In the superscript, the first entry denotes the column number (the "x-coordinate"), and the second entry denotes the row number (the "y-coordinate"). (Warning: This is opposite to the convention for matrices.) The subscript is meant to suggest the direction of the arrows. We will always write these as $d_\rightarrow$ and $d_\uparrow$ and ignore the superscripts. We require that $d_\rightarrow$ and $d_\uparrow$ satisfy (a) $d_\rightarrow^2 = 0$, (b) $d_\uparrow^2 = 0$, and one more condition: (c) either $d_\rightarrow d_\uparrow = d_\uparrow d_\rightarrow$ (all the squares commute) or $d_\rightarrow d_\uparrow + d_\uparrow d_\rightarrow = 0$ (they all anticommute). Both come up in nature, and you can switch from one to the other by replacing $d_\uparrow^{p,q}$ with $(-1)^p d_\uparrow^{p,q}$. So I will assume that all the squares anticommute, and that you know how to turn the commuting case into this one. (You will see that there is no difference in the recipe, basically because the image and kernel of a homomorphism f equal the image and kernel respectively of $-f$.)

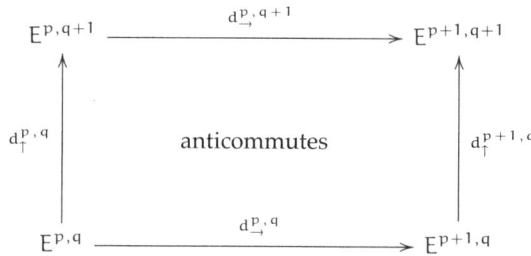

$$
\begin{array}{ccc}
E^{p,q+1} & \xrightarrow{d_\rightarrow^{p,q+1}} & E^{p+1,q+1} \\
\Big\uparrow{\scriptstyle d_\uparrow^{p,q}} & \text{anticommutes} & \Big\uparrow{\scriptstyle d_\uparrow^{p+1,q}} \\
E^{p,q} & \xrightarrow{d_\rightarrow^{p,q}} & E^{p+1,q}
\end{array}
$$

There are variations on this definition, where, for example, the vertical arrows go downward, or some subset of the $E^{p,q}$ is required to be zero.

From the double complex we construct a corresponding (single) complex $E^\bullet$ with $E^k = \oplus_i E^{i,k-i}$, with $d = d_\rightarrow + d_\uparrow$. In other words, when there is a *single* superscript k, we mean a sum of the kth antidiagonal of the double complex. The single complex is sometimes called the **total complex**. Note that $d^2 = (d_\rightarrow + d_\uparrow)^2 = d_\rightarrow^2 + (d_\rightarrow d_\uparrow + d_\uparrow d_\rightarrow) + d_\uparrow^2 = 0$, so $E^\bullet$ is indeed a complex.

The cohomology of the single complex is sometimes called the **hypercohomology** of the double complex. We will instead use the phrase **cohomology of the double complex**.

Our initial goal will be to find the cohomology of the double complex. You will see later that we secretly also have other goals.

A spectral sequence is a recipe for computing some information about the cohomology of the double complex. I won't yet give the full recipe. Surprisingly, this fragmentary bit of information is sufficient to prove lots of things.

1.6.2. Approximate definition. A **spectral sequence** with **rightward orientation** is a sequence of tables or **pages** $\rightarrow E_0^{p,q}$, $\rightarrow E_1^{p,q}$, $\rightarrow E_2^{p,q}$, ...(p, q $\in \mathbb{Z}$), where $\rightarrow E_0^{p,q} = E^{p,q}$, along with a differential

$$
\rightarrow d_r^{p,q}: \ \rightarrow E_r^{p,q} \longrightarrow \ \rightarrow E_r^{p-r+1,q+r}
$$

(r $\in \mathbb{Z}^{\geq 0}$) with $\rightarrow d_r^{p,q} \circ \rightarrow d_r^{p+r-1,q-r} = 0$, and with an isomorphism of the cohomology of $\rightarrow d_r$ at $\rightarrow E_r^{p,q}$ (i.e., ker $\rightarrow d_r^{p,q}$ / im $\rightarrow d_r^{p+r-1,q-r}$) with $\rightarrow E_{r+1}^{p,q}$.

The orientation indicates that our 0th differential is the rightward one: $d_0 = d_\rightarrow$. The left subscript "$\rightarrow$" is usually omitted.

The order of the morphisms is best understood visually:

(1.6.2.1)

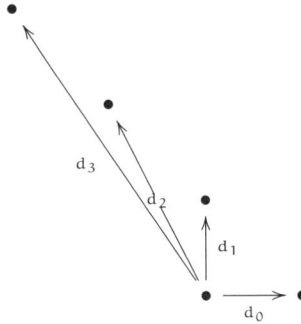

(the morphisms each apply to different pages). Notice that the map always is "degree 1" in terms of the grading of the single complex $E^\bullet$. (You should figure out what this informal statement really means.)

The actual definition describes what $E_r^{\bullet,\bullet}$ and $d_r^{\bullet,\bullet}$ really are, in terms of $E^{\bullet,\bullet}$. We will describe d_0, d_1, and d_2 below, and you should for now take on faith that this sequence continues in some natural way.

Note that $E_r^{p,q}$ is always a subquotient of the corresponding term on the ith page $E_i^{p,q}$ for all $i < r$. In particular, if $E^{p,q} = 0$, then $E_r^{p,q} = 0$ for all r.

Suppose now that $E^{\bullet,\bullet}$ is a **first quadrant double complex**, i.e., $E^{p,q} = 0$ for $p < 0$ or $q < 0$ (so $E_r^{p,q} = 0$ for all r unless $p, q \in \mathbb{Z}^{\geq 0}$). Then for any fixed p, q, once r is sufficiently large, $E_{r+1}^{p,q}$ is computed from $(E_r^{\bullet,\bullet}, d_r)$ using the complex

(1.6.2.2)

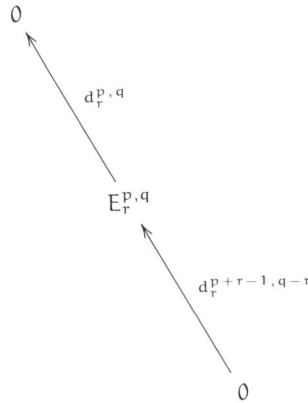

and thus we have canonical isomorphisms

$$E_r^{p,q} \cong E_{r+1}^{p,q} \cong E_{r+2}^{p,q} \cong \cdots .$$

We denote this module by $E_\infty^{p,q}$. The same idea works in other circumstances, for example, when the double complex is only nonzero in a finite number of rows—$E^{p,q} = 0$ unless $q_0 < q < q_1$. This will come up for example in the mapping cone and long exact sequence discussion (Exercises 1.6.F and 1.6.E below).

We now describe the first few pages of the spectral sequence explicitly. As stated above, the differential d_0 on $E_0^{\bullet,\bullet} = E^{\bullet,\bullet}$ is defined to be $d_\rightarrow$. The rows are complexes.

The 0th page E_0:

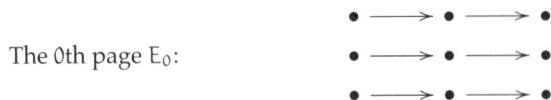

So E_1 is just the table of cohomologies of the rows. You should check that there are now vertical maps $d_1^{p,q}: E_1^{p,q} \to E_1^{p,q+1}$ of the row cohomology groups, induced by $d_{\uparrow}$, and that these make the columns into complexes. (This is essentially the fact that a map of complexes induces a map on homology.) We have "used up the horizontal morphisms," but "the vertical differentials live on."

The 1st page E_1:

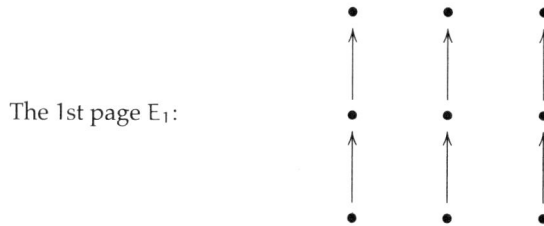

We take cohomology of d_1 on E_1, giving us a new table, $E_2^{p,q}$. It turns out that there are natural morphisms from each entry to the entry two above and one to the left, and that the composition of these two is 0. (It is a very worthwhile exercise to work out how this natural morphism d_2 should be defined. Your argument may be reminiscent of the connecting homomorphism in the Snake Lemma 1.6.5 or in the long exact sequence in cohomology arising from a short exact sequence of complexes, Theorem 1.5.8. This is no coincidence.)

The 2nd page E_2:

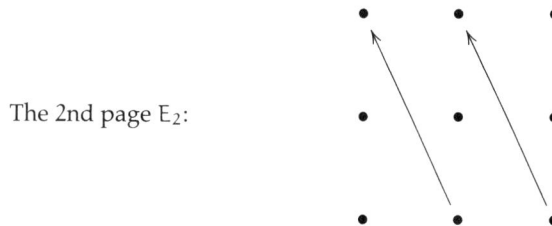

This is the beginning of a pattern.

Then it is a theorem that there is a filtration of $H^k(E^\bullet)$ by $E_\infty^{p,q}$ where $p + q = k$. (We can't yet state it as an official **Theorem** because we haven't precisely defined the pages and differentials in the spectral sequence.) More precisely, there is a filtration

$$(1.6.2.3) \qquad E_\infty^{0,k} \xrightarrow{E_\infty^{1,k-1}} ? \xrightarrow{E_\infty^{2,k-2}} \cdots \xrightarrow{E_\infty^{k,0}} H^k(E^\bullet)$$

where the quotients are displayed above each inclusion. (Here is a tip for remembering which way the quotients are supposed to go. The differentials on later and later pages point deeper and deeper into the filtration. Thus the entries in the direction of the later arrowheads are the subobjects, and the entries in the direction of the later "arrowtails" are quotients. This tip has the advantage of being independent of the details of the spectral sequence, e.g., the "quadrant" or the orientation.)

We say that the spectral sequence $_\to E_\bullet^{\bullet,\bullet}$ **converges** to $H^\bullet(E^\bullet)$. We often say that $_\to E_2^{\bullet,\bullet}$ (or any other page) **abuts** to $H^\bullet(E^\bullet)$.

Although the filtration gives only partial information about $H^\bullet(E^\bullet)$, sometimes one can find $H^\bullet(E^\bullet)$ precisely. One example is if all $E_\infty^{i,k-i}$ are zero, or if all but one of them are zero (e.g., if $E_r^{\bullet,\bullet}$ has precisely one nonzero row or column, in which case one says that the spectral sequence **collapses** at the rth step, although we will not use this term). Another example is in the category of vector spaces over a field, in which case we can find the dimension of $H^k(E^\bullet)$. Also, in lucky circumstances, E_2 (or some other small page) already equals E_∞.

1.6.A. EXERCISE: INFORMATION FROM THE SECOND PAGE. Show that $H^0(E^\bullet) = E_\infty^{0,0} = E_2^{0,0}$ and

$$0 \longrightarrow E_2^{0,1} \longrightarrow H^1(E^\bullet) \longrightarrow E_2^{1,0} \xrightarrow{d_2^{1,0}} E_2^{0,2} \longrightarrow H^2(E^\bullet)$$

is exact.

1.6.3. *The other orientation.*
You may have observed that we could as well have done everything in the opposite direction, i.e., reversing the roles of horizontal and vertical morphisms. Then the sequences of arrows giving the spectral sequence would look like this (compare to (1.6.2.1)):

(1.6.3.1)

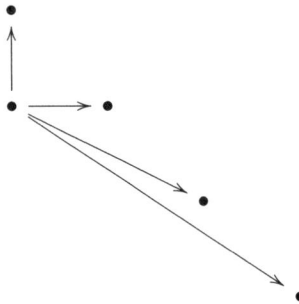

This spectral sequence is denoted by $_\uparrow E_\bullet^{\bullet,\bullet}$ ("with the **upward orientation**"). Then we would again get pieces of a filtration of $H^\bullet(E^\bullet)$ (where we have to be a bit careful with the order with which $_\uparrow E_\infty^{p,q}$ corresponds to the subquotients—it is the opposite order to that of (1.6.2.3) for $_\to E_\infty^{p,q}$). Warning: in general there is no isomorphism between $_\to E_\infty^{p,q}$ and $_\uparrow E_\infty^{p,q}$.

In fact, the observation that we can start with either the horizontal or vertical maps was our secret goal all along. Both algorithms compute information about the same thing ($H^\bullet(E^\bullet)$), and usually we don't care about the final answer—we often care about the answer we get in one way, and we get at it by doing the spectral sequence in the *other* way.

1.6.4. Examples.
We are now ready to see how this is useful. The moral of these examples is the following. In the past, you may have proved various facts involving various sorts of diagrams, by chasing elements around. Now, you will just plug them into a spectral sequence, and let the spectral sequence machinery do the chasing for you.

1.6.5. *Example: Proving the Snake Lemma.* Consider the diagram

(1.6.5.1)
$$0 \longrightarrow D \longrightarrow E \longrightarrow F \longrightarrow 0$$
$$\alpha \uparrow \qquad \beta \uparrow \qquad \gamma \uparrow$$
$$0 \longrightarrow A \longrightarrow B \longrightarrow C \longrightarrow 0$$

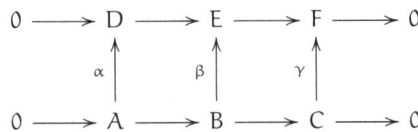

where the rows are exact in the middle (at A, B, C, D, E, F) and the squares commute. (Normally the Snake Lemma is described with the vertical arrows pointing downward, but I want to fit this into my spectral sequence conventions.) We wish to show that there is an exact sequence

(1.6.5.2) $0 \to \ker\alpha \to \ker\beta \to \ker\gamma \to \operatorname{coker}\alpha \to \operatorname{coker}\beta \to \operatorname{coker}\gamma \to 0.$

We plug this into our spectral sequence machinery. We first compute the cohomology using the rightward orientation, i.e., using the order (1.6.2.1). Then because the rows are exact, $E_1^{p,q}=0$, so the spectral sequence has already converged: $E_\infty^{p,q}=0$.

We next compute this "0" in another way, by computing the spectral sequence using the upward orientation. Then $_\uparrow E_1^{\bullet,\bullet}$ (with its differentials) is:

$$0 \longrightarrow \operatorname{coker}\alpha \longrightarrow \operatorname{coker}\beta \longrightarrow \operatorname{coker}\gamma \longrightarrow 0$$
$$0 \longrightarrow \ker\alpha \longrightarrow \ker\beta \longrightarrow \ker\gamma \longrightarrow 0.$$

Then $_\uparrow E_2^{\bullet,\bullet}$ is of the form:

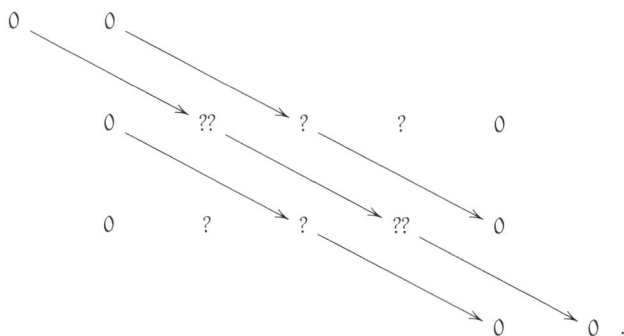

We see that after $_\uparrow E_2$, all the terms will stabilize except for the double question marks—all maps to and from the single question marks are to and from 0-entries. And after $_\uparrow E_3$, even the two double-question-mark terms will stabilize. But in the end our complex must be the 0 complex. This means that in $_\uparrow E_2$, all the entries must be zero, except for the two double question marks, and these two must be isomorphic. This means that $0 \to \ker \alpha \to \ker \beta \to \ker \gamma$ and $\operatorname{coker} \alpha \to \operatorname{coker} \beta \to \operatorname{coker} \gamma \to 0$ are both exact (which comes from the vanishing of the single question marks), and

$$\operatorname{coker}(\ker \beta \to \ker \gamma) \cong \ker(\operatorname{coker} \alpha \to \operatorname{coker} \beta)$$

is an isomorphism (which comes from the equality of the double question marks). Taken together, we have proved the exactness of (1.6.5.2), and hence the Snake Lemma! (Notice: In the end we didn't really care about the double complex. We just used it as a prop to prove the Snake Lemma.)

Spectral sequences make it easy to see how to generalize results further. For example, if $A \to B$ is no longer assumed to be injective, how would the conclusion change?

1.6.B. UNIMPORTANT EXERCISE (GRAFTING EXACT SEQUENCES, A VARIANT OF THE SNAKE LEMMA). Extend the Snake Lemma as follows. Suppose we have a commuting diagram

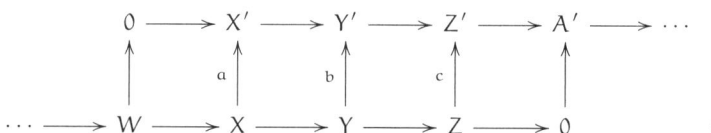

where the top and bottom rows are exact. Show that the top and bottom rows can be "grafted together" to an exact sequence

1.6.6. *Example: The Five Lemma.* Suppose

(1.6.6.1)

where the rows are exact and the squares commute.

Suppose α, β, δ, ϵ are isomorphisms. We will show that γ is an isomorphism.

We first compute the cohomology of the total complex using the rightward orientation (1.6.2.1). We choose this because we see that we will get lots of zeros. Then $_{\rightarrow}E_1^{\bullet,\bullet}$ looks like this:

$$
\begin{array}{ccccc}
? & 0 & 0 & 0 & ? \\
\uparrow & \uparrow & \uparrow & \uparrow & \uparrow \\
\\
? & 0 & 0 & 0 & ?
\end{array}
$$

Then $_{\rightarrow}E_2$ looks similar, and the sequence will converge by E_2, as we will never get any arrows between two nonzero entries in a table thereafter. We can't conclude that the cohomology of the total complex vanishes, but we can note that it vanishes in all but four degrees—and most importantly, it vanishes in the two degrees corresponding to the entries C and H (the source and target of γ).

We next compute this using the upward orientation (1.6.3.1). Then $_{\uparrow}E_1$ looks like

$$
\begin{array}{ccccc}
0 \longrightarrow & 0 \longrightarrow & ? \longrightarrow & 0 \longrightarrow & 0 \\
0 \longrightarrow & 0 \longrightarrow & ? \longrightarrow & 0 \longrightarrow & 0,
\end{array}
$$

and the spectral sequence converges at this step. We wish to show that those two question marks are zero. But they are precisely the cohomology groups of the total complex that we just showed *were* zero—so we are done!

The best way to become comfortable with this sort of argument is to try it out yourself several times, and realize that it really is easy. So you should do the following exercises! Many can readily be done directly, but you should deliberately try to use this spectral sequence machinery in order to get practice and develop confidence.

1.6.C. EXERCISE: A SUBTLER FIVE LEMMA. By looking at the spectral sequence proof of the Five Lemma above, prove a subtler version of it, where one of the isomorphisms can instead be required to be just an injection, and another can instead be required to be just a surjection. (I am deliberately not telling you which ones, so you can see how the spectral sequence is telling you how to improve the result.)

1.6.D. EXERCISE: ANOTHER SUBTLE VERSION OF THE FIVE LEMMA. If β and δ (in (1.6.6.1)) are injective, and α is surjective, show that γ is injective. Give the dual statement (whose proof is of course essentially the same).

The next two exercises no longer involve first quadrant double complexes. You will have to think a little to realize why there is no reason for confusion or alarm.

1.6.E. IMPORTANT EXERCISE (THE MAPPING CONE). Suppose $\mu\colon A^{\bullet} \to B^{\bullet}$ is a morphism of complexes. Suppose $C^{\bullet}$ is the single complex associated to the double complex $A^{\bullet} \to B^{\bullet}$. ($C^{\bullet}$ is called the **mapping cone** of μ.) Show that there is a long exact sequence of complexes:

$$\cdots \to H^{i-1}(C^{\bullet}) \to H^i(A^{\bullet}) \to H^i(B^{\bullet}) \to H^i(C^{\bullet}) \to H^{i+1}(A^{\bullet}) \to \cdots.$$

(There is a slight notational ambiguity here; depending on how you index your double complex, your long exact sequence might look slightly different.) In particular, we will use the fact that μ induces an isomorphism on cohomology if and only if the mapping cone is exact. (We won't use it until the proof of Theorem 18.2.4.)

1.6.F. EXERCISE. Use spectral sequences to show that a short exact sequence of complexes gives a long exact sequence in cohomology (Theorem 1.5.8). (This is a generalization of Exercise 1.6.E.)

The Grothendieck composition-of-functors spectral sequence (Theorem 23.3.5) will be an important example of a spectral sequence that specializes in a number of useful ways.

You are now ready to go out into the world and use spectral sequences to your heart's content!

1.6.7. Complete definition of spectral sequences, and proof.

You should most definitely not read the precise definition of a spectral sequence, and the proof that they work as advertised, any time soon after reading the introduction to spectral sequences above. But after a suitable interval, you should at least flip through a construction and proof to convince yourself that nothing fancy is involved. The idea is not as bad as you might think; see [Vak2].

It is useful to notice that the proof implies that spectral sequences are functorial in the 0th page: the spectral sequence formalism has good functorial properties in the double complex. Unfortunately, Grothendieck's terminology, "spectral functor," [Gr1, §2.4] did not catch on.

Chapter 2

Sheaves

It is perhaps surprising that geometric spaces are often best understood in terms of (nice) functions on them. For example, a differentiable manifold that is a subset of $\mathbb{R}^n$ can be studied in terms of its smooth (C^∞) functions. Because "geometric spaces" can have few (everywhere-defined) functions, a more precise version of this insight is that the structure of the space can be well understood by considering all functions on all open subsets of the space. This information is encoded and organized in something called a *sheaf*. Sheaves were introduced by Leray in the 1940s, and Serre introduced them to algebraic geometry. (The reason for the name will be somewhat explained in Remark 2.4.3.) We will define sheaves and describe useful facts about them. We will begin with a motivating example to convince you that the notion is not so foreign.

One reason sheaves are slippery to work with is that they keep track of a huge amount of information, and there are some subtle local-to-global issues. There are also three different ways of getting a hold of them:

- in terms of open sets (the definition §2.2)—intuitive but in some ways the least helpful;
- in terms of stalks (see §2.4.1); and
- in terms of a base of a topology (§2.5).

(Some people strongly prefer the espace étalé interpretation, §2.2.11, as well.) Knowing which to use requires experience, so it is essential to do a number of exercises on different aspects of sheaves in order to truly understand the concept.

2.1   Motivating Example: The Sheaf of Smooth Functions

Consider smooth (C^∞) functions on the topological space $X = \mathbb{R}^n$ (or more generally on a manifold X). The sheaf of smooth functions on X is the data of all smooth functions on all open subsets on X. We will see how to manage these data, and observe some of their properties. On each open set $U \subset X$, we have a ring of smooth functions. We denote this ring of functions by $\mathscr{O}(U)$.

Given a smooth function on an open set, you can restrict it to a smaller open set, obtaining a smooth function there. In other words, if $U \subset V$ is an inclusion of open sets, we have a "restriction map" $\text{res}_{V,U} \colon \mathscr{O}(V) \to \mathscr{O}(U)$.

Take a smooth function on a big open set, and restrict it to a medium open set, and then restrict that to a small open set. The result is the same as if you restrict the smooth function on the big open set directly to the small open set. In other words, if $U \hookrightarrow V \hookrightarrow W$, then the following diagram commutes:

$$\mathscr{O}(W) \xrightarrow{\text{res}_{W,V}} \mathscr{O}(V)$$
$$\mathscr{O}(W) \xrightarrow{\text{res}_{W,U}} \mathscr{O}(U) \xleftarrow{\text{res}_{V,U}} \mathscr{O}(V)$$

Next take two smooth functions f_1 and f_2 on a big open set U, and an open cover of U by some collection of open subsets $\{U_i\}$. (We say $\{U_i\}$ **covers** U, or is an **open cover of** U, if $U = \cup U_i$.) Suppose that f_1 and f_2 agree on each of these U_i. Then they must have been the same function to begin with. In other words, if $\{U_i\}_{i \in I}$ is a cover of U, and $f_1, f_2 \in \mathscr{O}(U)$, and $\text{res}_{U,U_i} f_1 = \text{res}_{U,U_i} f_2$, then $f_1 = f_2$. Thus we can *identify* functions on an open set by looking at them on a covering by small open sets.

Finally, suppose you are given the same U and cover $\{U_i\}$; take a smooth function on each of the U_i—a function f_1 on U_1, a function f_2 on U_2, and so on—and assume they agree on the pairwise overlaps. Then they can be "glued together" to make one smooth function on all of U. In other words, given $f_i \in \mathscr{O}(U_i)$ for all i such that $\mathrm{res}_{U_i, U_i \cap U_j} f_i = \mathrm{res}_{U_j, U_i \cap U_j} f_j$ for all i and j, there is some $f \in \mathscr{O}(U)$ such that $\mathrm{res}_{U, U_i} f = f_i$ for all i.

The entire example above would have worked just as well with continuous functions, or real-analytic functions, or just plain real-valued functions. Thus all of these classes of "nice" functions share some common properties. We will soon formalize these properties in the notion of a sheaf.

2.1.1. The germ of a smooth function. Before we do, we first give another definition, that of the germ of a smooth function at a point $p \in X$. Intuitively, it is a "shred" of a smooth function at p. Germs are objects of the form

$$(f, \text{open set } U) \quad \text{such that} \quad p \in U, f \in \mathscr{O}(U)$$

modulo the relation that $(f, U) \sim (g, V)$ if there is some open set $W \subset U, V$ containing p where $f|_W = g|_W$ (i.e., $\mathrm{res}_{U,W} f = \mathrm{res}_{V,W} g$). In other words, two functions that are the same in an open neighborhood of p (but may differ elsewhere) have the same germ. We call this set of germs the stalk at p, and denote it by $\mathscr{O}_p$. Notice that the stalk is a ring: you can add two germs, and get another germ: if you have a function f defined on U, and a function g defined on V, then $f + g$ is defined on $U \cap V$. Moreover, $f + g$ is well-defined: if $\tilde{f}$ has the same germ as f, meaning that there is some open set W containing p on which they agree, and $\tilde{g}$ has the same germ as g, meaning they agree on some open W' containing p, then $\tilde{f} + \tilde{g}$ is the same function as $f + g$ on $U \cap V \cap W \cap W'$.

Notice also that if $p \in U$, you get a map $\mathscr{O}(U) \to \mathscr{O}_p$. Experts may already see that we are talking about germs as colimits.

We can see that $\mathscr{O}_p$ is a local ring as follows. Consider those germs vanishing at p, which we denote by $\mathfrak{m}_p \subset \mathscr{O}_p$. They certainly form an ideal: $\mathfrak{m}_p$ is closed under addition, and when you multiply something vanishing at p by any function, the result also vanishes at p. We check that this ideal is maximal by showing that the quotient ring is a field:

$$(2.1.1.1) \qquad 0 \longrightarrow \mathfrak{m}_p := \text{ideal of germs vanishing at } p \longrightarrow \mathscr{O}_p \xrightarrow{f \mapsto f(p)} \mathbb{R} \longrightarrow 0.$$

2.1.A. EXERCISE. Show that this is the only maximal ideal of $\mathscr{O}_p$. (Hint: show that every element of $\mathscr{O}_p \setminus \mathfrak{m}_p$ is invertible.)

Note that we can interpret the value of a function at a point, or the value of a germ at a point, as an element of the local ring modulo the maximal ideal. (We will see that this doesn't work for more general sheaves, but *does* work for things behaving like sheaves of functions. This will be formalized in the notion of a *locally ringed space*, which we will see, briefly, in §7.3.)

2.1.2. *Aside.* Notice that $\mathfrak{m}_p/\mathfrak{m}_p^2$ is a module over $\mathscr{O}_p/\mathfrak{m}_p \cong \mathbb{R}$, i.e., it is a real vector space. It turns out to be naturally (whatever that means) the cotangent space to the differentiable or analytic manifold at p. This insight will prove handy later, when we define tangent and cotangent spaces of schemes.

2.1.B.* EXERCISE FOR THOSE WITH DIFFERENTIAL GEOMETRIC BACKGROUND. Prove this. (Rhetorical question for experts: What goes wrong if the sheaf of continuous functions is substituted for the sheaf of smooth functions? What goes wrong if you use the sheaf of C^1 functions?)

2.2   Definition of Sheaf and Presheaf

We now formalize these notions, by defining presheaves and sheaves. Presheaves are simpler to define, and notions such as kernel and cokernel are straightforward. Sheaves are more complicated to define, and some notions such as cokernel require more thought. But sheaves are more useful because they are in some vague sense more geometric; you can get information about a sheaf locally.

2.2.1. Definition of sheaf and presheaf on a topological space X.

To be concrete, we will define sheaves of sets. However, in the definition the category *Sets* can be replaced by any reasonable category, and other important examples are abelian groups *Ab*, k-vector spaces *Vec*$_k$, rings *Rings*, modules over a ring *Mod*$_A$, and more. (You may have to think more when dealing with a category of objects that isn't "sets with additional structure," but there aren't any surprising complications. In any case, this won't be relevant for us, although people who want to do this should start by solving Exercise 2.2.C.) Sheaves (and presheaves) are often written in calligraphic font. The fact that $\mathscr{F}$ is a sheaf on a topological space X is often written as

$$\mathscr{F}$$
$$\vert$$
$$X.$$

2.2.2. Definition: Presheaf.
A **presheaf of sets** $\mathscr{F}$ on a topological space X is the following data.

- To each open set $U \subset X$, we have a set $\mathscr{F}(U)$ (e.g., the set of differentiable functions in our motivating example). (Notational warning: Several notations are in use, for various good reasons; such as $\mathscr{F}(U) = \Gamma(U, \mathscr{F}) = H^0(U, \mathscr{F})$. We will use them all.) The elements of $\mathscr{F}(U)$ are called **sections of** $\mathscr{F}$ **over** U. (§2.2.11 combined with Exercise 2.2.G gives a motivation for this terminology, although this isn't so important for us.)

  By convention, if the "U" is omitted, it is implicitly taken to be X: "**sections of** $\mathscr{F}$" means "sections of $\mathscr{F}$ over X." These are also called **global sections**.
- For each inclusion $U \hookrightarrow V$ of open sets, we have a **restriction map**

$$\mathrm{res}_{V,U} : \mathscr{F}(V) \to \mathscr{F}(U)$$

(just as we did for differentiable functions). If $f \in \mathscr{F}(V)$, we often write

$$f\vert_U$$

for $\mathrm{res}_{V,U}(f)$.

  The data is required to satisfy the following two conditions.
- The map $\mathrm{res}_{U,U}$ is the identity: $\mathrm{res}_{U,U} = \mathrm{id}_{\mathscr{F}(U)}$.
- If $U \hookrightarrow V \hookrightarrow W$ are inclusions of open sets, then the restriction maps compose as you would expect, i.e., the diagram

$$\mathscr{F}(W) \xrightarrow{\ \mathrm{res}_{W,V}\ } \mathscr{F}(V)$$
$$\mathrm{res}_{W,U} \searrow \qquad \swarrow \mathrm{res}_{V,U}$$
$$\mathscr{F}(U)$$

commutes.

2.2.A. EXERCISE FOR CATEGORY-LOVERS: "A PRESHEAF IS THE SAME AS A CONTRAVARI-ANT FUNCTOR." Given any topological space X, we have a "category of open sets" (Example 1.1.9), where the objects are the open sets and the morphisms are inclusions. Verify that the data of a presheaf is precisely the data of a contravariant functor from the category of open sets of X to the category of sets. (This interpretation is surprisingly useful.)

2.2.3. Definition: Stalks and germs.
We define the stalk of a presheaf at a point in two equivalent ways. One will be hands-on, and the other will be as a colimit.

2.2.4. Define the **stalk** of a presheaf $\mathscr{F}$ at a point p to be the set of **germs** of $\mathscr{F}$ at p, denoted by $\mathscr{F}_p$, as in the example of §2.1.1. Germs correspond to sections over some open set containing p, and two of these sections are considered the same if they agree on some smaller open set containing p. More precisely: the stalk is

$$\{(f, \text{open } U) \ : \ p \in U, f \in \mathscr{F}(U)\}$$

modulo the relation that $(f, U) \sim (g, V)$ if there is some open set $W \subset U, V$ where $p \in W$ and $\mathrm{res}_{U,W} f = \mathrm{res}_{V,W} g$. (To explain the agricultural terminology: the French name "germe" is meant to suggest a tiny shoot sprouting from a seed, cf. "germinate".)

2.2.5. A useful equivalent definition of a stalk is as a colimit of all $\mathscr{F}(U)$ over all open sets U containing p:

$$\mathscr{F}_p = \mathrm{colim} \, \mathscr{F}(U).$$

The index category is a filtered set (given any two such open sets, there is a third such set contained in both), so these two definitions are the same by Exercise 1.3.E. Hence we can define stalks for sheaves of sets, groups, rings, and other things for which colimits exist for directed sets. It is very helpful to keep both definitions of stalk in mind at the same time.

If $p \in U$, and $f \in \mathscr{F}(U)$, then the image of f in $\mathscr{F}_p$ is called the **germ of** f at p. (Warning: Unlike in the example of §2.1.1, in general, the value of a section at a point doesn't make sense.)

2.2.6. Definition: Sheaf. A presheaf is a **sheaf** if it satisfies two more axioms. Notice that these axioms use the additional information of when some open sets cover another.

Identity axiom. For any open set U, if $\{U_i\}_{i \in I}$ is an open cover of U, and $f_1, f_2 \in \mathscr{F}(U)$, and $f_1|_{U_i} = f_2|_{U_i}$ for all i, then $f_1 = f_2$.

(A presheaf satisfying the identity axiom is called a **separated presheaf**, but we will not use that notation in any essential way.)

Gluability axiom. If $\{U_i\}_{i \in I}$ is an open cover of U, and we are given $f_i \in \mathscr{F}(U_i)$ for all i, such that $f_i|_{U_i \cap U_j} = f_j|_{U_i \cap U_j}$ for all i, j, then there is some $f \in \mathscr{F}(U)$ such that $\mathrm{res}_{U,U_i} f = f_i$ for all i.

In mathematics, definitions often come paired: "at most one" and "at least one." In this case, identity means there is at most one way to glue, and gluability means that there is at least one way to glue.

(For experts and scholars of the empty set only: an additional axiom sometimes included is that $\mathscr{F}(\varnothing)$ is a one-element set, and in general, for a sheaf with values in a category, $\mathscr{F}(\varnothing)$ is required to be the final object in the category. This actually follows from the above definitions, assuming that the empty product is appropriately defined as the final object.)

Example. If U and V are disjoint, then $\mathscr{F}(U \cup V) = \mathscr{F}(U) \times \mathscr{F}(V)$. Here we use the fact that $\mathscr{F}(\varnothing)$ is the final object.

The **stalk of a sheaf** at a point is just its stalk as a presheaf—the same definition applies—and similarly for the **germs of a section of a sheaf**.

2.2.B. UNIMPORTANT EXERCISE: PRESHEAVES THAT ARE NOT SHEAVES. Show that the following are presheaves on $\mathbb{C}$ (with the classical topology), but not sheaves: (a) bounded functions, (b) holomorphic functions admitting a holomorphic square root.

Both of the presheaves in the previous exercise satisfy the identity axiom. A "natural" example failing even the identity axiom is implicit in Remark 2.5.5.

We now make a couple of points intended only for category-lovers.

2.2.7. *Interpretation in terms of the equalizer exact sequence.* The two axioms for a presheaf to be a sheaf can be interpreted as "exactness" of the "equalizer exact sequence":

$\cdot \longrightarrow \mathscr{F}(U) \longrightarrow \prod \mathscr{F}(U_i) \rightrightarrows \prod \mathscr{F}(U_i \cap U_j)$. Identity is exactness at $\mathscr{F}(U)$, and gluability is exactness at $\prod \mathscr{F}(U_i)$. I won't make this precise, or even explain what the double right arrow means. (What is an exact sequence of sets?!) But you may be able to figure it out from the context.

2.2.C. EXERCISE. The identity and gluability axioms may be interpreted as saying that $\mathscr{F}(\cup_{i\in I}U_i)$ is a certain limit. What is that limit?

Here are a number of examples of sheaves.

2.2.D. EXERCISE.

(a) Verify that the examples of §2.1 are indeed sheaves (of smooth functions, or continuous functions, or real-analytic functions, or plain real-valued functions, on a manifold or $\mathbb{R}^n$).
(b) Show that real-valued continuous functions on (open sets of) a topological space X form a sheaf.

2.2.8. *Important example: Restriction of a sheaf.* Suppose $\mathscr{F}$ is a sheaf on X, and U is an open subset of X. Define the **restriction of $\mathscr{F}$ to** U, denoted by $\mathscr{F}|_U$, to be the collection $\mathscr{F}|_U(V) = \mathscr{F}(V)$ for all open subsets $V \subset U$. Clearly this is a sheaf on U. (Unimportant but fun fact: §2.7 will tell us how to restrict sheaves to *arbitrary* subsets.)

2.2.9. *Important example: The skyscraper sheaf.* Suppose X is a topological space, with $p \in X$, and S is a set. Let $i_p : p \to X$ be the inclusion. Then $i_{p,*}S$ defined by

$$i_{p,*}S(U) = \begin{cases} S & \text{if } p \in U, \text{ and} \\ \{e\} & \text{if } p \notin U \end{cases}$$

forms a sheaf. Here $\{e\}$ is any one-element set. (Check this if it isn't clear to you—what are the restriction maps?) This is called a **skyscraper sheaf** supported at p, because the informal picture of it looks like a skyscraper at p. (Mild caution: This informal picture suggests that the only nontrivial stalk of a skyscraper sheaf is at p, which isn't the case. Exercise 6.1.B(b) gives an example, although it certainly isn't the simplest one.) There is an analogous definition for sheaves of abelian groups, except $i_{p,*}S(U) = \{0\}$ if $p \notin U$; and for sheaves with values in a category more generally, $i_{p,*}S(U)$ should be a final object.

(This notation is admittedly hideous, and the alternative $(i_p)_*S$ is equally bad. In §2.2.12 we explain this notation.)

2.2.10. *Constant presheaves and constant sheaves.* Let X be a topological space, and S a set. Define $\underline{S}_{\text{pre}}(U) = S$ for all open sets U. You will readily verify that $\underline{S}_{\text{pre}}$ forms a presheaf (with restriction maps the identity). This is called the **constant presheaf associated to** S. This isn't (in general) a sheaf. (It may be distracting to say why. Lovers of the empty set will insist that the sheaf axioms force the sections over the empty set to be the final object in the category, i.e., a one-element set. But even if we patch the definition by setting $\underline{S}_{\text{pre}}(\varnothing) = \{e\}$, if S has more than one element, and X is the two-point space with the **discrete topology**, i.e., where every subset is open, you can check that $\underline{S}_{\text{pre}}$ fails gluability.)

2.2.E. EXERCISE (CONSTANT SHEAVES). Now let $\mathscr{F}(U)$ be the set of maps $U \to S$ that are *locally constant*, i.e., for any point p in U, there is an open neighborhood of p where the function is constant. Show that this is a *sheaf*. (A better description is this: endow S with the discrete topology, and let $\mathscr{F}(U)$ be the continuous maps $U \to S$.) This is called the **constant sheaf** (with values in S); do not confuse it with the constant presheaf. (I would prefer the name "*locally* constant sheaf," but it is too late in history for this change.) We denote this sheaf by $\underline{S}$.

2.2.F. EXERCISE ("MORPHISMS GLUE"). Suppose Y is a topological space. Show that "continuous maps to Y" form a sheaf of sets on X. More precisely, to each open set U of X, we associate the set of continuous maps of U to Y. Show that this forms a sheaf. (Exercise 2.2.D(b), with $Y = \mathbb{R}$, and Exercise 2.2.E, with $Y = S$ with the discrete topology, are both special cases.)

2.2.G. EXERCISE. This is a fancier version of the previous exercise.

(a) (*Sheaf of sections of a map.*) Suppose we are given a continuous map $\mu\colon Y\to X$. Show that "sections of μ" form a sheaf. More precisely, to each open set U of X, associate the set of continuous maps $s\colon U\to Y$ such that $\mu\circ s=\mathrm{id}|_U$. Show that this forms a sheaf. (For those who have heard of vector bundles, these are a good example.) This is motivation for the phrase "section of a sheaf."

(b) (This exercise is for those who know what a topological group is. If you don't know what a topological group is, you might be able to guess.) Suppose that Y is a topological group. Show that continuous maps to Y form a sheaf of *groups*.

2.2.11.* *The space of sections (espace étalé) of a (pre)sheaf.* Depending on your background, you may prefer the following perspective on sheaves. Suppose $\mathscr{F}$ is a presheaf of sets (e.g., a sheaf) on a topological space X. Construct a topological space F along with a continuous map $\pi\colon F\to X$ as follows: as a set, F is the disjoint union of all the stalks of $\mathscr{F}$. This naturally gives a map of sets $\pi\colon F\to X$. Topologize F as follows. Each s in $\mathscr{F}(U)$ determines a subset $\{(x,s_x)\;:\;x\in U\}$ of F. The topology on F is the weakest topology such that these subsets are open. (These subsets form a base of the topology. For each $y\in F$, there is an open neighborhood V of y and an open neighborhood U of $\pi(y)$ such that $\pi|_V$ is a homeomorphism from V to U. Do you see why these facts are true?) The topological space F could be thought of as the **space of sections** of $\mathscr{F}$ (in French called the **espace étalé** of $\mathscr{F}$). We will not discuss this construction at any length, but it can have some advantages: (a) It is always better to know as many ways as possible of thinking about a concept. (b) "Inverse image" (informally, "pullback") has a natural interpretation in this language (mentioned briefly in Exercise 2.7.C). (c) Sheafification has a natural interpretation in this language (see Remark 2.4.7).

2.2.H. IMPORTANT EXERCISE/DEFINITION: THE PUSHFORWARD SHEAF OR DIRECT IMAGE SHEAF. Suppose $\pi\colon X\to Y$ is a continuous map, and $\mathscr{F}$ is a presheaf on X. Then define a presheaf $\pi_*\mathscr{F}$ on Y by $\pi_*\mathscr{F}(V)=\mathscr{F}(\pi^{-1}(V))$, where V is an open subset of Y. Show that $\pi_*\mathscr{F}$ is a presheaf on Y, and is a sheaf if $\mathscr{F}$ is. This is called the **pushforward** (or **direct image**) of $\mathscr{F}$. More precisely, $\pi_*\mathscr{F}$ is called the **pushforward of $\mathscr{F}$ by π**.

2.2.12. As the notation suggests, the skyscraper sheaf (Example 2.2.9) can be interpreted as the pushforward of the constant sheaf $\underline{S}$ on a one-point space p, under the inclusion morphism $i_p\colon\{p\}\to X$.

   Once we endow sheaves with the structure of a category, we will see that the pushforward is a functor from sheaves on X to sheaves on Y (Exercise 2.3.B).

2.2.I. EXERCISE (PUSHFORWARD INDUCES MAPS OF STALKS). Suppose $\pi\colon X\to Y$ is a continuous map, and $\mathscr{F}$ is a sheaf of sets (or rings or A-modules) on X. If $\pi(p)=q$, describe the natural morphism of stalks $(\pi_*\mathscr{F})_q\to\mathscr{F}_p$. (You can use the explicit definition of stalk using representatives, §2.2.4, or the universal property, §2.2.5. If you prefer one way, you should try the other.)

2.2.13. Important example: Ringed spaces, and $\mathscr{O}_X$-modules. Suppose $\mathscr{O}_X$ is a sheaf of rings on a topological space X (i.e., a sheaf on X with values in the category of *Rings*). Then $(X,\mathscr{O}_X)$ is called a **ringed space**. The sheaf of rings is often denoted by $\mathscr{O}_X$, pronounced "oh-X." This sheaf is called the **structure sheaf** of the ringed space. Sections of the structure sheaf $\mathscr{O}_X$ over an open subset U are called **functions on** U. Functions on X are called **global functions**, or just **functions**. (*Caution: what we call "functions," others sometimes call "regular functions." Furthermore, we will later define "rational functions" on schemes in §5.2.I and §6.6.36, which are not precisely functions in this sense; they are a particular type of "partially defined function.*")

   The symbol $\mathscr{O}_X$ will always refer to the structure sheaf of a ringed space X. The restriction $\mathscr{O}_X|_U$ of $\mathscr{O}_X$ to an open subset $U\subset X$ is denoted by $\mathscr{O}_U$. (We will later call $(U,\mathscr{O}_U)\to(X,\mathscr{O}_X)$ an *open embedding* of ringed spaces; see Definition 7.2.1.) The stalk of $\mathscr{O}_X$ at a point p is written "$\mathscr{O}_{X,p}$," because this looks less hideous than "$\mathscr{O}_{X_p}$."

Just as we have modules over a ring, we have $\mathscr{O}_X$-modules over a sheaf of rings $\mathscr{O}_X$. There is only one possible definition that could go with the name $\mathscr{O}_X$-**module** (or often $\mathscr{O}$-**module**)— a sheaf of abelian groups $\mathscr{F}$ with the following additional structure. For each U, $\mathscr{F}(U)$ is an $\mathscr{O}_X(U)$-module. Furthermore, this structure should behave well with respect to restriction maps: if $U \subset V$, then

(2.2.13.1)

$$
\begin{array}{ccc}
\mathscr{O}_X(V) \times \mathscr{F}(V) & \xrightarrow{\text{action}} & \mathscr{F}(V) \\
{\scriptstyle \text{res}_{V,U} \times \text{res}_{V,U}} \downarrow & & \downarrow {\scriptstyle \text{res}_{V,U}} \\
\mathscr{O}_X(U) \times \mathscr{F}(U) & \xrightarrow{\text{action}} & \mathscr{F}(U)
\end{array}
$$

commutes. (You should convince yourself that I haven't forgotten anything.)

Recall that the notion of A-module generalizes the notion of abelian group, because an abelian group is the same thing as a $\mathbb{Z}$-module. Similarly, the notion of $\mathscr{O}_X$-module generalizes the notion of sheaf of abelian groups, because the latter is the same thing as a $\underline{\mathbb{Z}}$-module. Hence when we are proving things about $\mathscr{O}_X$-modules, we are also proving things about sheaves of abelian groups.

2.2.J. EXERCISE. If $(X, \mathscr{O}_X)$ is a ringed space, and $\mathscr{F}$ is an $\mathscr{O}_X$-module, describe how for each $p \in X$, $\mathscr{F}_p$ is an $\mathscr{O}_{X,p}$-module.

2.2.14. *For those who know about vector bundles.* The motivating example of $\mathscr{O}_X$-modules is the sheaf of sections of a vector bundle. If $(X, \mathscr{O}_X)$ is a differentiable manifold (so $\mathscr{O}_X$ is the sheaf of smooth functions), and $\pi\colon V \to X$ is a vector bundle over X, then the sheaf of smooth sections $\sigma\colon X \to V$ is an $\mathscr{O}_X$-module. Indeed, given a section s of π over an open subset $U \subset X$, and a function f on U, we can multiply s by f to get a new section fs of π over U. Moreover, if U' is a smaller subset, then we could multiply f by s and then restrict to U', or we could restrict both f and s to U' and then multiply, and we would get the same answer. That is precisely the commutativity of (2.2.13.1).

2.3   Morphisms of Presheaves and Sheaves

2.3.1. *Definitions.* Whenever one defines a new mathematical object, category theory teaches us to try to understand maps from one such object to another. We now define morphisms of presheaves, and similarly for sheaves. In other words, we will describe the *category of presheaves* (of sets, abelian groups, etc.) and the *category of sheaves*.

A **morphism of presheaves** of sets (or indeed of presheaves with values in any category) on X, $\phi\colon \mathscr{F} \to \mathscr{G}$, is the data of maps $\phi(U)\colon \mathscr{F}(U) \to \mathscr{G}(U)$ for all U behaving well with respect to restriction: if $U \hookrightarrow V$ then

$$
\begin{array}{ccc}
\mathscr{F}(V) & \xrightarrow{\phi(V)} & \mathscr{G}(V) \\
{\scriptstyle \text{res}_{V,U}} \downarrow & & \downarrow {\scriptstyle \text{res}_{V,U}} \\
\mathscr{F}(U) & \xrightarrow{\phi(U)} & \mathscr{G}(U)
\end{array}
$$

commutes. (Notice: The underlying space of both $\mathscr{F}$ and $\mathscr{G}$ is X.)

Morphisms of sheaves are defined identically: the morphisms from a sheaf $\mathscr{F}$ to a sheaf $\mathscr{G}$ are precisely the morphisms from $\mathscr{F}$ to $\mathscr{G}$ as presheaves. (Translation: The category of sheaves on X is a full subcategory of the category of presheaves on X.) If $(X, \mathscr{O}_X)$ is a ringed space, then morphisms of $\mathscr{O}_X$-modules have the obvious definition. (Can you write it down?)

An example of a morphism of sheaves is the map from the sheaf of smooth functions on $\mathbb{R}$ to the sheaf of continuous functions. This is a "forgetful map": we are forgetting that these functions are differentiable, and remembering only that they are continuous.

2.3.2. *Notation.* We may as well set some notation: let $Sets_X$, Ab_X, etc. denote the category of sheaves of sets, abelian groups, etc. on a topological space X. Let $Mod_{\mathscr{O}_X}$ denote the category of $\mathscr{O}_X$-modules on a ringed space $(X, \mathscr{O}_X)$. Let $Sets_X^{pre}$, etc. denote the category of presheaves of sets, etc. on X.

2.3.3. *Aside for category-lovers.* If you interpret a presheaf on X as a contravariant functor (from the category of open sets), a morphism of presheaves on X is a natural transformation of functors (§1.1.21).

2.3.A. EXERCISE: MORPHISMS OF (PRE)SHEAVES INDUCE MORPHISMS OF STALKS. If $\phi\colon \mathscr{F} \to \mathscr{G}$ is a morphism of presheaves on X, and $p \in X$, describe an induced morphism of stalks $\phi_p\colon \mathscr{F}_p \to \mathscr{G}_p$. Translation: Taking the stalk at p induces a **stalkification functor** $Sets_X \to Sets$. (Your proof will extend in obvious ways. For example, if ϕ is a morphism of $\mathscr{O}_X$-modules, then ϕ_p is a map of $\mathscr{O}_{X,p}$-modules.)

2.3.B. EXERCISE. Suppose $\pi\colon X \to Y$ is a continuous map of topological spaces (i.e., a morphism in the category of topological spaces). Show that pushforward gives a functor $\pi_*\colon Sets_X \to Sets_Y$. Here *Sets* can be replaced by other categories. (Watch out for some possible confusion: a presheaf is a functor, and presheaves form a category. It may be best to forget that presheaves are functors for now.)

2.3.C. IMPORTANT EXERCISE AND DEFINITION: "SHEAF Hom." Suppose $\mathscr{F}$ and $\mathscr{G}$ are two sheaves of sets on X. (In fact, it will suffice that $\mathscr{F}$ is a presheaf.) Let $\mathcal{H}om(\mathscr{F}, \mathscr{G})$ be the collection of data

$$\mathcal{H}om(\mathscr{F}, \mathscr{G})(U) := \mathrm{Mor}(\mathscr{F}|_U, \mathscr{G}|_U).$$

(Recall the notation $\mathscr{F}|_U$, the restriction of the sheaf to the open set U, Example 2.2.8.) Show that this is a sheaf of sets on X. (To avoid a common confusion: the right side does *not* say $\mathrm{Mor}(\mathscr{F}(U), \mathscr{G}(U))$.) This sheaf is called "sheaf $\mathcal{H}om$." (Strictly speaking, we should reserve Hom for when we are in an additive category, so this should possibly be called "sheaf Mor." But the terminology "sheaf Hom" is too established to uproot.) It will be clear from your construction that, like Hom, $\mathcal{H}om$ is a contravariant functor in its first argument and a covariant functor in its second argument.

Warning: $\mathcal{H}om$ does not commute with taking stalks. More precisely: it is not true that $\mathcal{H}om(\mathscr{F}, \mathscr{G})_p$ is isomorphic to $\mathrm{Hom}(\mathscr{F}_p, \mathscr{G}_p)$. (Can you think of a counterexample? There is at least a map from one of these to the other—in which direction?)

2.3.4. We will use many variants of the definition of $\mathcal{H}om$. For example, if $\mathscr{F}$ and $\mathscr{G}$ are sheaves of abelian groups on X, then $\mathcal{H}om_{Ab_X}(\mathscr{F}, \mathscr{G})$ is defined by taking $\mathcal{H}om_{Ab_X}(\mathscr{F}, \mathscr{G})(U)$ to be the maps *as sheaves of abelian groups* $\mathscr{F}|_U \to \mathscr{G}|_U$. (Note that $\mathcal{H}om_{Ab_X}(\mathscr{F}, \mathscr{G})$ has the structure of a sheaf of abelian groups in a natural way.) Similarly, if $\mathscr{F}$ and $\mathscr{G}$ are $\mathscr{O}_X$-modules, we define $\mathcal{H}om_{Mod_{\mathscr{O}_X}}(\mathscr{F}, \mathscr{G})$ in the analogous way (and it is an $\mathscr{O}_X$-module). Obnoxiously, the subscripts Ab_X and $Mod_{\mathscr{O}_X}$ are often dropped (here and in the literature), so be careful which category you are working in! We call $\mathcal{H}om_{Mod_{\mathscr{O}_X}}(\mathscr{F}, \mathscr{O}_X)$ the **dual** of the $\mathscr{O}_X$-module $\mathscr{F}$, and denote it by $\mathscr{F}^\vee$.

2.3.D. UNIMPORTANT EXERCISE (REALITY CHECK).

(a) If $\mathscr{F}$ is a sheaf of sets on X, then show that $\mathcal{H}om(\underline{\{p\}}, \mathscr{F}) \cong \mathscr{F}$, where $\{p\}$ is the constant sheaf "with values in the one element set $\{p\}$."

(b) If $\mathscr{F}$ is a sheaf of abelian groups on X, then show that $\mathcal{H}om_{Ab_X}(\underline{\mathbb{Z}}, \mathscr{F}) \cong \mathscr{F}$ (an isomorphism of sheaves of abelian groups).

(c) If $\mathscr{F}$ is an $\mathscr{O}_X$-module, then show that $\mathcal{H}om_{Mod_{\mathscr{O}_X}}(\mathscr{O}_X, \mathscr{F}) \cong \mathscr{F}$ (an isomorphism of $\mathscr{O}_X$-modules).

A key idea in (b) and (c) is that 1 "generates" (in some sense) $\mathbb{Z}$ (in (b)) and $\mathscr{O}_X$ (in (c)).

2.3.5. Presheaves of abelian groups (and even "presheaf $\mathscr{O}_X$-modules") form an abelian category.

We can make module-like constructions using presheaves of abelian groups on a topological space X. (Throughout this section, all (pre)sheaves are of abelian groups.) For example, we can clearly add maps of presheaves and get another map of presheaves: if $\phi, \psi: \mathscr{F} \to \mathscr{G}$, then we define the map $\phi + \psi$ by $(\phi + \psi)(V) = \phi(V) + \psi(V)$. (There is something small to check here: that the result is indeed a map of presheaves.) In this way, presheaves of abelian groups form an additive category (Definition 1.5.1: the morphisms between any two presheaves of abelian groups form an abelian group; there is a 0-object; and one can take finite products). For exactly the same reasons, sheaves of abelian groups also form an additive category.

If $\phi: \mathscr{F} \to \mathscr{G}$ is a morphism of presheaves, define the **presheaf kernel** $\ker_{\mathrm{pre}} \phi$ by $(\ker_{\mathrm{pre}} \phi)(U) := \ker \phi(U)$.

2.3.E. EXERCISE. Show that $\ker_{\mathrm{pre}} \phi$ is a presheaf. (Hint: If $U \hookrightarrow V$, define the restriction map by chasing the following diagram:

$$
\begin{array}{ccccccc}
0 & \longrightarrow & \ker_{\mathrm{pre}} \phi(V) & \longrightarrow & \mathscr{F}(V) & \longrightarrow & \mathscr{G}(V) \\
& & \big\downarrow {\scriptstyle \exists!} & & \big\downarrow {\scriptstyle \mathrm{res}_{V,U}} & & \big\downarrow {\scriptstyle \mathrm{res}_{V,U}} \\
0 & \longrightarrow & \ker_{\mathrm{pre}} \phi(U) & \longrightarrow & \mathscr{F}(U) & \longrightarrow & \mathscr{G}(U).
\end{array}
$$

You should check that the restriction maps compose as desired.)

Define the **presheaf cokernel** $\mathrm{coker}_{\mathrm{pre}} \phi$ similarly. It is a presheaf by essentially the same (dual) argument.

2.3.F. EXERCISE: THE COKERNEL DESERVES ITS NAME. Show that the presheaf cokernel satisfies the universal property of cokernels (Definition 1.5.4) in the category of presheaves.

Similarly, $\ker_{\mathrm{pre}} \phi \to \mathscr{F}$ satisfies the universal property for kernels in the category of presheaves.

It is not too tedious to verify that presheaves of abelian groups form an abelian category, and the reader is free to do so. The key idea is that all abelian-categorical notions may be defined and verified "open set by open set." We needn't worry about restriction maps—they "come along for the ride." Hence we can define terms such as **subpresheaf**, **image presheaf** (or **presheaf image**), and **quotient presheaf** (or **presheaf quotient**), and they behave as you would expect. You construct kernels, quotients, cokernels, and images open set by open set. Homological algebra (exact sequences and so forth) works, and also "works open set by open set." In particular:

2.3.G. EASY EXERCISE. Show (or observe) that for a topological space X with open set U, $\mathscr{F} \mapsto \mathscr{F}(U)$ gives a functor from presheaves of abelian groups on X, Ab_X^{pre}, to abelian groups, Ab. Then show that this functor is exact.

2.3.H. EXERCISE. Show that a sequence of presheaves $0 \to \mathscr{F}_1 \to \mathscr{F}_2 \to \cdots \to \mathscr{F}_n \to 0$ is exact if and only if $0 \to \mathscr{F}_1(U) \to \mathscr{F}_2(U) \to \cdots \to \mathscr{F}_n(U) \to 0$ is exact for all U.

The above discussion essentially carries over without change to presheaves with values in any abelian category. (Think this through if you wish.)

However, we are interested in more geometric objects, sheaves, where things can be understood in terms of their local behavior, thanks to the identity and gluing axioms. We will soon see that sheaves of abelian groups also form an abelian category, but a complication will arise that will force the notion of *sheafification* on us. Sheafification will be the answer to many of our prayers. We just haven't yet realized what we should be praying for.

To begin with, sheaves Ab_X form an additive category, as described in the first paragraph of §2.3.5.

Kernels work just as with presheaves:

2.3.I. IMPORTANT EXERCISE. Suppose $\phi\colon \mathscr{F}\to\mathscr{G}$ is a morphism of *sheaves*. Show that the presheaf kernel $\ker_{\text{pre}}\phi$ is in fact a sheaf. Show that it satisfies the universal property of kernels (Definition 1.5.4). (Hint: The second question follows immediately from the fact that $\ker_{\text{pre}}\phi$ satisfies the universal property in the category of *presheaves*.)

Thus if ϕ is a morphism of sheaves, we define

$$\ker\phi := \ker_{\text{pre}}\phi.$$

The problem arises with the cokernel.

2.3.J. IMPORTANT EXERCISE. Let X be $\mathbb{C}$ with the classical topology, let $\mathscr{O}_X$ be the sheaf of holomorphic functions, and let $\mathscr{F}$ be the *presheaf* of functions admitting a holomorphic logarithm. Describe an exact sequence of presheaves on X:

$$0 \longrightarrow \mathbb{Z} \longrightarrow \mathscr{O}_X \longrightarrow \mathscr{F} \longrightarrow 0,$$

where $\mathbb{Z}\to\mathscr{O}_X$ is the natural inclusion and $\mathscr{O}_X\to\mathscr{F}$ is given by $f\mapsto\exp(2\pi i f)$. (Be sure to verify exactness.) Show that $\mathscr{F}$ is *not* a sheaf. (Hint: $\mathscr{F}$ does not satisfy the gluability axiom. The problem is that there are functions that don't have a logarithm but locally have a logarithm.) This will come up again in Example 2.4.9.

We will have to put our hopes for understanding cokernels of sheaves on hold for a while. We will first learn to understand sheaves using stalks.

2.4   Properties Determined at the Level of Stalks, and Sheafification

2.4.1. Properties determined by stalks. We now come to the second way of understanding sheaves mentioned at the start of the chapter. In this section, we will see that lots of facts about sheaves can be checked "at the level of stalks." We call any property determined at the level of stalks **stalk-local**. This isn't true for presheaves, and reflects the local nature of sheaves. We will see that sections and morphisms are determined "by their stalks," and the property of a morphism being an isomorphism may be checked at stalks. (The last one is the trickiest.)

2.4.A. IMPORTANT EASY EXERCISE **(SECTIONS ARE DETERMINED BY GERMS)**. Prove that a section of a sheaf of sets is determined by its germs, i.e., the natural map

(2.4.1.1) $$\mathscr{F}(U) \longrightarrow \prod_{p\in U}\mathscr{F}_p$$

is injective. Hint 1: you won't use the gluability axiom, so this is true for separated presheaves. Hint 2: it is false for presheaves in general—see Exercise 2.4.E—so you *will* use the identity axiom. (Your proof will also apply to sheaves of groups, rings, etc.—to categories of "sets with additional structure." The same is true for many exercises in this section.)

Exercise 2.4.A suggests a question: Which elements of the right side of (2.4.1.1) are in the image of the left side?

2.4.2. Important definition. We say that an element $(s_p)_{p\in U}$ of the right side $\prod_{p\in U}\mathscr{F}_p$ of (2.4.1.1) consists of **compatible germs** if for all $p\in U$, there is some representative

$$(\tilde{s}_p\in\mathscr{F}(U_p), U_p \text{ open in } U)$$

for s_p (where $p\in U_p\subset U$) such that the germ of $\tilde{s}_p$ at all $q\in U_p$ is s_q. Equivalently, there is an open cover $\{U_i\}$ of U, and sections $f_i\in\mathscr{F}(U_i)$, such that if $p\in U_i$, then s_p is the germ of f_i at p. Clearly any section s of $\mathscr{F}$ over U gives a choice of compatible germs for U.

2.4.B. IMPORTANT EXERCISE. Prove that any choice of compatible germs for a sheaf of sets $\mathscr{F}$ over U is the image of a section of $\mathscr{F}$ over U. (Hint: You will use gluability.)

We have thus completely described the image of (2.4.1.1), in a way that will prove useful.

2.4.3. *Remark.* This perspective motivates the agricultural terminology "sheaf": a sheaf is (the data of) a bunch of stalks, bundled together appropriately.

Now we throw morphisms into the mix. Recall Exercise 2.3.A: morphisms of (pre)sheaves induce morphisms of stalks.

2.4.C. EXERCISE (MORPHISMS ARE DETERMINED BY STALKS). If ϕ_1 and ϕ_2 are morphisms from a presheaf of sets $\mathscr{F}$ to a sheaf of sets $\mathscr{G}$ that induce the same maps on each stalk, show that $\phi_1 = \phi_2$. Hint: consider the following diagram.

(2.4.3.1)

$$
\begin{array}{ccc}
\mathscr{F}(U) & \longrightarrow & \mathscr{G}(U) \\
\downarrow & & \uparrow \\
\prod_{p \in U} \mathscr{F}_p & \longrightarrow & \prod_{p \in U} \mathscr{G}_p
\end{array}
$$

2.4.D. TRICKY EXERCISE (ISOMORPHISMS ARE DETERMINED BY STALKS). Show that a morphism of sheaves of sets is an isomorphism if and only if it induces an isomorphism of all stalks. Hint: Use (2.4.3.1). Once you have injectivity, show surjectivity, perhaps using Exercise 2.4.B, or gluability in some other way; this is more subtle. Warning: This exercise does *not* say that if two sheaves have isomorphic stalks, then they are isomorphic.

2.4.E. EXERCISE.

(a) Show that Exercise 2.4.A is false for general presheaves.
(b) Show that Exercise 2.4.C is false for general presheaves.
(c) Show that Exercise 2.4.D is false for general presheaves.
     (General hint for finding counterexamples of this sort: Consider a 2-point space with the discrete topology.)

2.4.4. Sheafification.
Every sheaf is a presheaf (and indeed by definition sheaves on X form a full subcategory of the category of presheaves on X). Just as groupification (§1.4.3) gives an abelian group that best approximates an abelian semigroup, sheafification gives the sheaf that best approximates a presheaf, with an analogous universal property. (One possible example to keep in mind is the sheafification of the presheaf of holomorphic functions admitting a square root on $\mathbb{C}$ with the classical topology. See also the exponential exact sequence, Example 2.4.9.)

2.4.5. *Definition.* If $\mathscr{F}$ is a presheaf on X, then a morphism of presheaves sh: $\mathscr{F} \to \mathscr{F}^{\mathrm{sh}}$ on X is a **sheafification of** $\mathscr{F}$ if $\mathscr{F}^{\mathrm{sh}}$ is a sheaf, and for any sheaf $\mathscr{G}$, and any presheaf morphism g: $\mathscr{F} \to \mathscr{G}$, there *exists* a *unique* morphism of sheaves f: $\mathscr{F}^{\mathrm{sh}} \to \mathscr{G}$ making the diagram

$$
\begin{array}{ccc}
\mathscr{F} & \xrightarrow{\ \mathrm{sh}\ } & \mathscr{F}^{\mathrm{sh}} \\
 & {\scriptstyle g} \searrow & \downarrow {\scriptstyle f} \\
 & & \mathscr{G}
\end{array}
$$

commute.

We still have to show that it exists. The following two exercises require existence (which we will show shortly), but not the details of the construction.

2.4.F. EXERCISE. Show that sheafification is unique up to unique isomorphism, assuming it exists. Show that if $\mathscr{F}$ is a sheaf, then the sheafification is $\mathrm{id}: \mathscr{F} \longrightarrow \mathscr{F}$. (This should be second nature by now.)

2.4.G. EASY EXERCISE (SHEAFIFICATION IS A FUNCTOR). Assume for now that sheafification exists. Use the universal property to show that for any morphism of presheaves $\phi: \mathscr{F} \to \mathscr{G}$, we get a natural induced morphism of sheaves $\phi^{\mathrm{sh}}: \mathscr{F}^{\mathrm{sh}} \to \mathscr{G}^{\mathrm{sh}}$. Show that sheafification is a functor from presheaves on X to sheaves on X.

2.4.6. *Construction.* We next show that any presheaf of sets (or groups, rings, etc.) has a sheafification. Suppose $\mathscr{F}$ is a *presheaf*. Define $\mathscr{F}^{\mathrm{sh}}$ by defining $\mathscr{F}^{\mathrm{sh}}(U)$ as the set of "compatible (families of) germs" of the presheaf $\mathscr{F}$ over U. Explicitly:

$$\mathscr{F}^{\mathrm{sh}}(U) := \{(f_p \in \mathscr{F}_p)_{p \in U} : \text{for all } p \in U, \text{ there exists an open neighborhood } V$$
$$\text{of } p, \text{ contained in } U, \text{ and } s \in \mathscr{F}(V) \text{ with } s_q = f_q \text{ for all } q \in V\}.$$

Here s_q means the germ of s at q—the image of s in the stalk $\mathscr{F}_q$.

2.4.H. EASY EXERCISE. Show that $\mathscr{F}^{\mathrm{sh}}$ (using the tautological restriction maps) forms a sheaf.

2.4.I. EASY EXERCISE. Describe a natural map of presheaves $\mathrm{sh}: \mathscr{F} \to \mathscr{F}^{\mathrm{sh}}$.

2.4.J. EXERCISE. Show that the map sh satisfies the universal property of sheafification (Definition 2.4.5). (This is easier than you might fear.)

2.4.K. USEFUL EXERCISE, NOT JUST FOR CATEGORY-LOVERS. Show that the sheafification functor is left-adjoint to the forgetful functor from sheaves on X to presheaves on X. This is not difficult—it is largely a restatement of the universal property. But it lets you use results from §1.5.14, and can "explain" why you don't need to sheafify when taking kernel (why the presheaf kernel is already the sheaf kernel), and why you need to sheafify when taking cokernel and (soon, in Exercise 2.6.K) tensor product.

2.4.L. EXERCISE. Show $\mathscr{F} \to \mathscr{F}^{\mathrm{sh}}$ induces an isomorphism of stalks. (Possible hint: Use the concrete description of the stalks. Another possibility once you read Remark 2.7.3: judicious use of adjoints.)

As a reality check, you may want to verify that "the sheafification of a constant presheaf is the corresponding constant sheaf" (see §2.2.10): If X is a topological space and S is a set, then $(\underline{S}_{\mathrm{pre}})^{\mathrm{sh}}$ may be naturally identified with $\underline{S}$.

2.4.7.* *Remark.* The "space of sections" (or "espace étalé") construction (§2.2.11) yields a different-sounding description of sheafification that may be preferred by some readers. The main idea is identical: if $\mathscr{F}$ is a presheaf, let F be the space of sections (or espace étalé) of $\mathscr{F}$. You may wish to show that if $\mathscr{F}$ is a presheaf, the sheaf of sections of $F \to X$ (defined in Exercise 2.2.G(a)) is the sheafification of $\mathscr{F}$. Exercise 2.2.E may be interpreted as an example of this construction. The "space of sections" construction of the sheafification is essentially the same as Construction 2.4.6.

2.4.8. Subsheaves and quotient sheaves.
We now discuss subsheaves and quotient sheaves from the perspective of stalks.

2.4.M. EXERCISE. Suppose $\phi: \mathscr{F} \to \mathscr{G}$ is a morphism of sheaves of sets on a topological space X. Show that the following are equivalent.

(a) ϕ is a monomorphism in the category of sheaves.
(b) ϕ is injective on the level of stalks: $\phi_p: \mathscr{F}_p \to \mathscr{G}_p$ is injective for all $p \in X$.
(c) ϕ is injective on the level of open sets: $\phi(U): \mathscr{F}(U) \to \mathscr{G}(U)$ is injective for all open $U \subset X$.

(Possible hints: For (b) implies (a), recall that morphisms are determined by stalks, Exercise 2.4.C. For (a) implies (c), use the "indicator sheaf" with one section over every open set contained in U, and no section over any other open set.) If these conditions hold, we say that $\mathscr{F}$ is a **subsheaf** of $\mathscr{G}$ (where the "inclusion" ϕ is sometimes left implicit).

(You may later wish to extend your solution to Exercise 2.4.M to show that for any morphism of *presheaves*, if all maps of sections are injective, then all stalk maps are injective. And furthermore, if $\phi\colon \mathscr{F}\to\mathscr{G}$ is a morphism from a separated presheaf to an arbitrary presheaf, then injectivity on the level of stalks implies that ϕ is a monomorphism in the category of presheaves. This is useful in some approaches to Exercise 2.6.C.)

2.4.N. EXERCISE. Continuing the notation of the previous exercise, show that the following are equivalent.

(a) ϕ is an epimorphism in the category of sheaves.
(b) ϕ is surjective on the level of stalks: $\phi_p\colon \mathscr{F}_p\to\mathscr{G}_p$ is surjective for all $p\in X$.

(Possible hint: Use a skyscraper sheaf.)

If these conditions hold, we say that $\mathscr{G}$ is a **quotient sheaf** of $\mathscr{F}$.

Thus *monomorphisms and epimorphisms — subsheafiness and quotient sheafiness — can be checked at the level of stalks.*

Both exercises generalize readily to sheaves with values in any reasonable category, where "injective" is replaced by "monomorphism" and "surjective" is replaced by "epimorphism."

Notice that there was no part (c) to Exercise 2.4.N, and Example 2.4.9 shows why. (But there is a version of (c) that *implies* (a) and (b): surjectivity on all open sets in *any* base of a topology implies that the corresponding map of sheaves is an epimorphism, Exercise 2.5.D.)

2.4.9. *Example (cf. Exercise 2.3.J).* Let $X=\mathbb{C}$ with the classical topology, and define $\mathscr{O}_X$ to be the sheaf of holomorphic functions, and $\mathscr{O}_X^*$ to be the sheaf of invertible (nowhere zero) holomorphic functions. This is a sheaf of abelian groups under multiplication. We have maps of sheaves

$$(2.4.9.1) \qquad 0 \longrightarrow \underline{\mathbb{Z}} \xrightarrow{\times 2\pi i} \mathscr{O}_X \xrightarrow{\exp} \mathscr{O}_X^* \longrightarrow 1.$$

(You can figure out what the sheaves 0 and 1 mean; they are isomorphic, and are written in this way for reasons that may be clear.) We will soon interpret this as an exact sequence of sheaves of abelian groups (the **exponential exact sequence**; see Exercise 2.6.F), although we don't yet have the language to do so.

2.4.O. ENLIGHTENING EXERCISE. Show that $\exp\colon \mathscr{O}_X \longrightarrow \mathscr{O}_X^*$ describes $\mathscr{O}_X^*$ as a quotient sheaf of $\mathscr{O}_X$. Find an open set on which this map is not surjective.

This is a great example to get a sense of what "surjectivity" means for sheaves: nowhere vanishing holomorphic functions (such as the function x away from the origin) have logarithms locally, but they need not have logarithms globally.

2.5   Recovering Sheaves from a "Sheaf on a Base"

Sheaves are natural things to want to think about, but hard to get our hands on. We like the identity and gluability axioms, but they make proving things trickier than for presheaves. We have discussed how we can understand sheaves using stalks (using "compatible germs"). We now introduce a second way of getting a hold of sheaves, by introducing the notion of a *sheaf on a base*. Warning: This way of understanding an entire sheaf from limited information is confusing. You may find it helpful to focus on the central insight that this partial information suffices to determine the germs, and that they are compatible (with nearby germs).

First, we define a **base of a topology**. Suppose we have a topological space X, i.e., we know which subsets U_i of X are open. Then a base of a topology is a subcollection of the open sets $\{B_j\}\subset\{U_i\}$, such that each U_i is a union of some of the B_j. Here is one example that you have seen early in your mathematical life. Suppose $X = \mathbb{R}^n$. Then the way the classical topology is often first defined is by defining *open balls* $B_r(x) = \{y \in \mathbb{R}^n : |y - x| < r\}$, and declaring that any union of open balls is open. So the balls form a base of the classical topology—we say they *generate* the classical topology. As an application of how we use them, to check continuity of some map $\pi: X \to \mathbb{R}^n$, you need only think about the pullback of balls on $\mathbb{R}^n$—part of the traditional δ-ϵ definition of continuity.

Now suppose we have a sheaf $\mathscr{F}$ on a topological space X, and a base $\{B_i\}$ of open sets on X. Then consider the information

$$(\{\mathscr{F}(B_i)\}, \{\mathrm{res}_{B_i, B_j} : \mathscr{F}(B_i) \to \mathscr{F}(B_j)\}),$$

which is a subset of the information contained in the sheaf—we are paying attention only to the information involving elements of the base, not to all open sets.

We can recover the entire sheaf from this information. This is because we can determine the stalks from this information, and we can determine when germs are compatible.

2.5.A. IMPORTANT EXERCISE. Make this precise. How can you recover a sheaf $\mathscr{F}$ from this partial information?

This suggests a notion, called a **sheaf on a base**. A sheaf of sets (or abelian groups, rings, . . .) on a base $\{B_i\}$ is the following. For each B_i in the base, we have a set $F(B_i)$. If $B_i \subset B_j$, we have maps $\mathrm{res}_{B_j, B_i} : F(B_j) \to F(B_i)$, with $\mathrm{res}_{B_i, B_i} = \mathrm{id}_{F(B_i)}$. (Things called "B" are always assumed to be in the base.) If $B_i \subset B_j \subset B_k$, then $\mathrm{res}_{B_k, B_i} = \mathrm{res}_{B_j, B_i} \circ \mathrm{res}_{B_k, B_j}$. So far we have defined a **presheaf on a base** $\{B_i\}$.

We also require the **base identity** axiom: If $B = \cup B_i$, then if $f, g \in F(B)$ are such that $\mathrm{res}_{B, B_i} f = \mathrm{res}_{B, B_i} g$ for all i, then $f = g$.

We require the **base gluability** axiom, too: If $B = \cup B_i$, and we have $f_i \in F(B_i)$ such that f_i agrees with f_j on any basic open set contained in $B_i \cap B_j$ (i.e., $\mathrm{res}_{B_i, B_k} f_i = \mathrm{res}_{B_j, B_k} f_j$ for all $B_k \subset B_i \cap B_j$), then there exists $f \in F(B)$ such that $\mathrm{res}_{B, B_i} f = f_i$ for all i.

2.5.1. Theorem — *Suppose $\{B_i\}$ is a base on X, and F is a sheaf of sets on this base. Then there is a sheaf $\mathscr{F}$ extending F (with isomorphisms $\mathscr{F}(B_i) \longleftrightarrow F(B_i)$ agreeing with the restriction maps). This sheaf $\mathscr{F}$ is unique up to unique isomorphism.*

Proof. We will define $\mathscr{F}$ as the sheaf of (families of) compatible germs of F.

Define the **stalk** of a presheaf F on a base at $p \in X$ by

$$F_p = \mathrm{colim}\, F(B_i),$$

where the colimit is over all B_i (in the base) containing p.

We will say a family of germs in an open set U is compatible near p if there is a section s of F over some B_i containing p such that the germs over B_i are precisely the germs of s. More formally, define

$$\mathscr{F}(U) := \{(f_p \in F_p)_{p \in U} : \text{for all } p \in U, \text{ there exists } B \text{ with } p \in B \subset U, s \in F(B),$$
$$\text{with } s_q = f_q \text{ for all } q \in B\},$$

where each B is in our base.

This is a sheaf (for the same reasons that the sheaf of compatible germs was; cf. Exercise 2.4.H).

I next claim that if B is in our base, the natural map $F(B) \to \mathscr{F}(B)$ is an isomorphism.

2.5.B. EXERCISE. Verify that $F(B) \to \mathscr{F}(B)$ is an isomorphism, likely by showing that it is injective and surjective (or else by describing the inverse map and verifying that it is indeed inverse). Possible hint: Elements of $\mathscr{F}(B)$ are determined by stalks, as are elements of $F(B)$.

It will be clear from your solution to Exercise 2.5.B that the restriction maps for F are the same as the restriction maps of $\mathscr{F}$ (for elements of the base).

Finally, you should verify to your satisfaction that $\mathscr{F}$ is indeed unique up to unique isomorphism. (First be sure that you understand what this means!) □

Theorem 2.5.1 shows that sheaves on X can be recovered from their "restriction to a base." It is clear from the argument (and in particular the solution to the Exercise 2.5.B) that if $\mathscr{F}$ is a sheaf and F is the corresponding sheaf on the base B, then for any $p \in X$, $\mathscr{F}_p$ is naturally isomorphic to F_p.

Theorem 2.5.1 is a statement about *objects* in a category, so we should hope for a similar statement about *morphisms*.

2.5.C. IMPORTANT EXERCISE: MORPHISMS OF SHEAVES CORRESPOND TO MORPHISMS OF SHEAVES ON A BASE. Suppose $\{B_i\}$ is a base for the topology of X. A morphism $F \to G$ of sheaves on the base is a collection of maps $F(B_k) \to G(B_k)$ such that the diagram

$$
\begin{array}{ccc}
F(B_i) & \longrightarrow & G(B_i) \\
{\scriptstyle \mathrm{res}_{B_i, B_j}} \downarrow & & \downarrow {\scriptstyle \mathrm{res}_{B_i, B_j}} \\
F(B_j) & \longrightarrow & G(B_j)
\end{array}
$$

commutes for all $B_j \hookrightarrow B_i$.

(a) Verify that a morphism of sheaves is determined by the induced morphism of sheaves on the base.
(b) Show that a morphism of sheaves on the base gives a morphism of the induced sheaves. (Possible hint: Compatible stalks.)

2.5.2. *Remark*. The above constructions and arguments describe an equivalence of categories (§1.1.21) between sheaves on X and sheaves on a given base of X. There is no new content to this statement, but you may wish to think through what it means. What are the functors in each direction? Why aren't their compositions the identity?

2.5.3. *Remark*. It will be useful to extend these notions to $\mathscr{O}_X$-modules (see, for example, Exercise 6.2.C). You will readily be able to verify that there is a correspondence (really, equivalence of categories) between $\mathscr{O}_X$-modules and "$\mathscr{O}_X$-modules on a base." Rather than working out the details, you should just informally think through the main points: What is an "$\mathscr{O}_X$-module on a base"? Given an $\mathscr{O}_X$-module on a base, why is the corresponding sheaf naturally an $\mathscr{O}_X$-module? Later, if you are forced at gunpoint to fill in details, you will be able to.

2.5.D. UNIMPORTANT EXERCISE. Suppose a morphism of sheaves of sets $F \to G$ on a base $\{B_i\}$ is surjective for all B_i (i.e., $F(B_i) \to G(B_i)$ is surjective for all i). Show that the corresponding morphism of sheaves (*not* on the base) is surjective (or, more precisely, is an epimorphism). The converse is not true, unlike the case for injectivity. This gives a useful sufficient criterion for "surjectivity": a morphism of sheaves is an epimorphism ("surjective") if it is surjective for sections on a base. You may enjoy trying this out with Example 2.4.9 (dealing with holomorphic functions in the classical topology on $X = \mathbb{C}$), showing that the exponential map $\exp: \mathscr{O}_X \to \mathscr{O}_X^*$ is surjective, using the base of contractible open sets.

2.5.4. Gluing sheaves.
We will repeatedly see the theme of constructing some object by gluing, in many different contexts. Keep an eye out for it! In each case, we carefully consider what information we need in order to glue.

2.5.E. IMPORTANT EXERCISE. Suppose $X = \cup U_i$ is an open cover of X, and we have sheaves $\mathscr{F}_i$ on U_i along with isomorphisms

$$
\phi_{ij} \colon \mathscr{F}_i|_{U_i \cap U_j} \xrightarrow{\sim} \mathscr{F}_j|_{U_i \cap U_j}
$$

(with ϕ_{ii} the identity) that agree on triple overlaps, i.e.,

$$(2.5.4.1) \qquad\qquad\qquad \phi_{jk} \circ \phi_{ij} = \phi_{ik}$$

on $U_i \cap U_j \cap U_k$ (this is called the **cocycle condition**, for reasons we ignore). Show that these sheaves can be glued together into a sheaf $\mathscr{F}$ on X (unique up to unique isomorphism), with isomorphisms $\mathscr{F}|_{U_i} \xrightarrow{\sim} \mathscr{F}_i$, and the isomorphisms over $U_i \cap U_j$ are the obvious ones. Warning: We are not assuming this is a finite cover, so you cannot use induction. For this reason this exercise can be perplexing. (You can use the ideas of this section to solve this problem, but you don't necessarily need to. Hint: As the base, take those open sets contained in *some* U_i.)

Thus we can "glue sheaves together," using limited patching information. Small observation: The hypothesis ϕ_{ii} is the identity is extraneous, as it follows from the cocycle condition.

2.5.5. *Remark for experts.* Exercise 2.5.E almost says that the "set" of sheaves forms a sheaf itself, but not quite. Making this precise leads one to the notion of a *stack*.

2.6   Sheaves of Abelian Groups, and $\mathcal{O}_X$-Modules, Form Abelian Categories

We are now ready to see that sheaves of abelian groups, and their cousins, $\mathcal{O}_X$-modules, form abelian categories. In other words, we may treat them similarly to vector spaces, and modules over a ring. In the process of doing this, we will see that this is much stronger than an analogy; kernels, cokernels, exactness, and so forth can be understood at the level of stalks (which are just abelian groups), and the compatibility of the germs will come for free.

The category of sheaves of abelian groups on a topological space X is clearly an additive category (Definition 1.5.1). In order to show that it is an abelian category, we must begin by showing that any morphism $\phi \colon \mathscr{F} \to \mathscr{G}$ has a kernel and a cokernel. We have already seen that ϕ has a kernel (Exercise 2.3.I): the presheaf kernel is a sheaf, and is a kernel in the category of sheaves.

2.6.A. EXERCISE. Show that the stalk of the kernel is the kernel of the stalks: for all $p \in X$, there is a natural isomorphism

$$(\ker(\mathscr{F} \to \mathscr{G}))_p \xleftarrow{\sim} \ker(\mathscr{F}_p \to \mathscr{G}_p).$$

We next address the issue of the cokernel. Now $\phi \colon \mathscr{F} \to \mathscr{G}$ has a cokernel in the category of presheaves; call it $\mathscr{H}_{pre}$ (where the subscript is meant to remind us that this is a presheaf). Let $sh \colon \mathscr{H}_{pre} \to \mathscr{H}$ be its sheafification. Recall that the cokernel is defined using a universal property: it is the colimit of the diagram

$$(2.6.0.1) \qquad\qquad\qquad \begin{array}{ccc} \mathscr{F} & \xrightarrow{\phi} & \mathscr{G} \\ \downarrow & & \\ 0 & & \end{array}$$

in the category of presheaves (cf. (1.5.4.2) and the comment thereafter).

2.6.1. Proposition — *The composition $\mathscr{G} \to \mathscr{H}$ is the cokernel of ϕ in the category of sheaves.*

Proof. We show that it satisfies the universal property. Given any sheaf $\mathscr{E}$ and a commutative diagram

$$\begin{array}{ccc} \mathscr{F} & \xrightarrow{\phi} & \mathscr{G} \\ \downarrow & & \downarrow \\ 0 & \longrightarrow & \mathscr{E} \end{array}$$

we construct

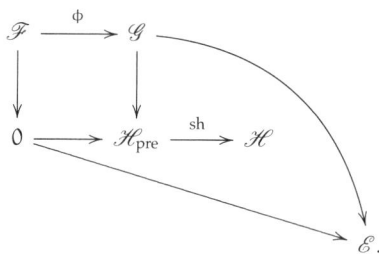

We show that there is a unique morphism $\mathscr{H} \to \mathscr{E}$ making the diagram commute. As $\mathscr{H}_{\mathrm{pre}}$ is the cokernel in the category of presheaves, there is a unique morphism of presheaves $\mathscr{H}_{\mathrm{pre}} \to \mathscr{E}$ making the diagram commute. But then by the universal property of sheafification (Definition 2.4.5), there is a unique morphism of *sheaves* $\mathscr{H} \to \mathscr{E}$ making the diagram commute.   □

2.6.B. EXERCISE. Show that the stalk of the cokernel is naturally isomorphic to the cokernel of the stalk.

We have now defined the notions of kernel and cokernel, and verified that they may be checked at the level of stalks. We have also verified that the properties of a morphism being a monomorphism or an epimorphism are also determined at the level of stalks (Exercises 2.4.M and 2.4.N). Hence we have proved the following:

2.6.2. Theorem — *Sheaves of abelian groups on a topological space X form an abelian category.*

That's all there is to it—what needs to be proved has been shifted to the stalks, where everything works because stalks are abelian groups!

And we see more: all structures coming from the abelian nature of this category may be checked at the level of stalks. For example:

2.6.C. EXERCISE. Suppose $\phi \colon \mathscr{F} \to \mathscr{G}$ is a morphism of sheaves of abelian groups. Show that the image sheaf im ϕ is the sheafification of the image presheaf. (You must use the definition of image in an abelian category. In fact, this gives the accepted definition of image sheaf for a morphism of sheaves of sets.) Show that the stalk of the image is the image of the stalk.

As a consequence, **exactness of a sequence of sheaves may be checked at the level of stalks**. If you are not sure about this, you should do the following exercise.

2.6.D. EXERCISE. Suppose $\alpha \colon \mathscr{F} \to \mathscr{G}$ and $\beta \colon \mathscr{G} \to \mathscr{H}$ are two morphisms of sheaves of abelian groups on X. Show that

$$\mathscr{F} \xrightarrow{\ \alpha\ } \mathscr{G} \xrightarrow{\ \beta\ } \mathscr{H}$$

is exact (at $\mathscr{G}$) if and only if for all $p \in X$,

$$\mathscr{F}_p \xrightarrow{\ \alpha_p\ } \mathscr{G}_p \xrightarrow{\ \beta_p\ } \mathscr{H}_p$$

is exact.

In particular:

2.6.E. IMPORTANT EXERCISE (CF. EXERCISE 2.3.A). Show that taking the stalk of a sheaf of abelian groups is an exact functor. More precisely, if X is a topological space and $p \in X$ is a point, show that taking the stalk at p defines an exact functor $Ab_X \to Ab$.

2.6.F. EXERCISE. Check that the exponential exact sequence (2.4.9.1) is indeed an exact sequence of sheaves of abelian groups.

2.6.G. EXERCISE: LEFT-EXACTNESS OF THE FUNCTOR OF "SECTIONS OVER U." Suppose $U \subset X$ is an open set, and $0 \to \mathscr{F} \to \mathscr{G} \to \mathscr{H}$ is an exact sequence of sheaves of abelian groups. Show that

$$0 \longrightarrow \mathscr{F}(U) \longrightarrow \mathscr{G}(U) \longrightarrow \mathscr{H}(U)$$

is exact. (You should do this "by hand," even if you realize there is a very fast proof using the left-exactness of the "forgetful" right adjoint to the sheafification functor.) Show that the section functor need not be exact: show that if $0 \to \mathscr{F} \to \mathscr{G} \to \mathscr{H} \to 0$ is an exact sequence of sheaves of abelian groups, then

$$0 \longrightarrow \mathscr{F}(U) \longrightarrow \mathscr{G}(U) \longrightarrow \mathscr{H}(U) \longrightarrow 0$$

need not be exact. (Hint: Recall the exponential exact sequence (2.4.9.1). But feel free to make up a different example.)

2.6.H. EXERCISE: LEFT-EXACTNESS OF PUSHFORWARD. Suppose $0 \to \mathscr{F} \to \mathscr{G} \to \mathscr{H}$ is an exact sequence of sheaves of abelian groups on X. If $\pi \colon X \to Y$ is a continuous map, show that

$$0 \longrightarrow \pi_* \mathscr{F} \longrightarrow \pi_* \mathscr{G} \longrightarrow \pi_* \mathscr{H}$$

is exact. (The previous exercise, dealing with the left-exactness of the global section functor can be interpreted as a special case of this, the case where Y is a point.)

2.6.I. EXERCISE: LEFT-EXACTNESS OF $\mathcal{H}om$ (CF. EXERCISE 1.5.G(C) AND (D)). Suppose $\mathscr{F}$ is a sheaf of abelian groups on a topological space X. Show that $\mathcal{H}om(\mathscr{F}, \cdot)$ is a left-exact covariant functor $Ab_X \to Ab_X$. Show that $\mathcal{H}om(\cdot, \mathscr{F})$ is a left-exact contravariant functor $Ab_X \to Ab_X$.

2.6.3. $\mathscr{O}_X$-modules.

2.6.J. EXERCISE. Show that if $(X, \mathscr{O}_X)$ is a ringed space, then $\mathscr{O}_X$-modules form an abelian category. (There is a fair bit to check, but there aren't many new ideas.)

2.6.4. Many facts about sheaves of abelian groups carry over to $\mathscr{O}_X$-modules without change, because a sequence of $\mathscr{O}_X$-modules is exact if and only if the underlying sequence of sheaves of abelian groups is exact. You should be able to easily check that all of the statements of the earlier exercises in §2.6 also hold for $\mathscr{O}_X$-modules, when modified appropriately. For example (Exercise 2.6.I), $\mathcal{H}om_{\mathscr{O}_X}(\cdot, \cdot)$ is a left-exact contravariant functor in its first argument and a left-exact covariant functor in its second argument.

We end with a useful construction using some of the ideas in this section.

2.6.K. IMPORTANT EXERCISE: TENSOR PRODUCTS OF $\mathscr{O}_X$-MODULES.

(a) Suppose $\mathscr{O}_X$ is a sheaf of rings on X. Define (categorically) what we should mean by **tensor product of two $\mathscr{O}_X$-modules**. Give an explicit construction, and show that it satisfies your categorical definition. Hint: Take the "presheaf tensor product"—which needs to be defined— and sheafify. Note: $\otimes_{\mathscr{O}_X}$ is often written $\otimes$ when the subscript is clear from the context. (An example showing sheafification is necessary will arise in Example 15.1.1.)
(b) Show that the tensor product of stalks is the stalk of the tensor product. (If you can show this, you may be able to make sense of the phrase "colimits commute with tensor products.")

2.6.5. Conclusion. Just as presheaves of abelian groups on a topological space form an abelian category because all abelian-categorical notions make sense open set by open set, sheaves of abelian groups on a topological space form an abelian category because all abelian-categorical notions make sense stalk by stalk.

2.7   The Inverse Image Sheaf

We next describe a notion that is fundamental, but rather intricate. We will not need it for some time, so this may be best left for a second reading. Suppose we have a continuous map $\pi \colon X \to Y$.

If $\mathscr{F}$ is a sheaf on X, we have defined the pushforward or direct image sheaf $\pi_*\mathscr{F}$, which is a sheaf on Y. There is also a notion of inverse image sheaf. (We will not call it the pullback sheaf, reserving that name for a later construction for quasicoherent sheaves, §14.5.) This is a covariant functor π^{-1} from sheaves on Y to sheaves on X. If the sheaves on Y have some additional structure (e.g., group or ring), then this structure is respected by π^{-1}.

2.7.1. *Definition by adjoint: Elegant but abstract.* We define the inverse image π^{-1} as the left adjoint to π_*.

This isn't really a definition; we need a construction to show that the adjoint exists. Note that we then get canonical maps $\pi^{-1}\pi_*\mathscr{F} \to \mathscr{F}$ (associated to the identity in $\mathrm{Mor}_Y(\pi_*\mathscr{F}, \pi_*\mathscr{F})$) and $\mathscr{G} \to \pi_*\pi^{-1}\mathscr{G}$ (associated to the identity in $\mathrm{Mor}_X(\pi^{-1}\mathscr{G}, \pi^{-1}\mathscr{G})$).

2.7.2. *Construction: Concrete but ugly.* Define the temporary notation

$$\pi^{-1}_{\mathrm{pre}}\mathscr{G}(U) = \mathrm{colim}_{V \supset \pi(U)} \mathscr{G}(V).$$

(Recall the explicit description of colimit: Sections of π^{-1}_{pre} over U are sections on open sets containing $\pi(U)$, with an equivalence relation. Note that $\pi(U)$ won't be an open set in general.)

2.7.A. EXERCISE. Show that this defines a presheaf on X. Show that it needn't form a sheaf. (Hint: Map two points to one point.)

Now define the **inverse image of** $\mathscr{G}$ by $\pi^{-1}\mathscr{G} := (\pi^{-1}_{\mathrm{pre}}\mathscr{G})^{\mathrm{sh}}$. Note that π^{-1} is a functor from sheaves on Y to sheaves on X. The next exercise shows that π^{-1} is indeed left-adjoint to π_*. But you may wish to try the later exercises first, and come back to Exercise 2.7.B at another time. (For the later exercises, try to give two proofs, one using the universal property, and the other using the explicit description.)

2.7.B. IMPORTANT TRICKY EXERCISE ((π^{-1}, π_*) ARE ADJOINT). If $\pi\colon X \to Y$ is a continuous map, and $\mathscr{F}$ is a sheaf on X and $\mathscr{G}$ is a sheaf on Y, describe a bijection

$$\mathrm{Mor}_X(\pi^{-1}\mathscr{G}, \mathscr{F}) \xleftarrow{\;\sim\;} \mathrm{Mor}_Y(\mathscr{G}, \pi_*\mathscr{F}).$$

Observe that your bijection is "natural" in the sense of the definition of adjoints (i.e., it is functorial in both $\mathscr{F}$ and $\mathscr{G}$). Thus Construction 2.7.2 satisfies the universal property of Definition 2.7.1. Possible hint: Show that both sides agree with the following third construction, which we denote by $\mathrm{Mor}_{YX}(\mathscr{G}, \mathscr{F})$. A collection of maps $\phi_{VU}\colon \mathscr{G}(V) \to \mathscr{F}(U)$ (as U runs through all open sets of X, and V runs through all open sets of Y containing $\pi(U)$) is said to be *compatible* if for all open $U' \subset U \subset X$ and all open $V' \subset V \subset Y$ with $\pi(U) \subset V$, $\pi(U') \subset V'$, the diagram

(2.7.2.1)
$$
\begin{array}{ccc}
\mathscr{G}(V) & \xrightarrow{\;\phi_{VU}\;} & \mathscr{F}(U) \\
\mathrm{res}_{V,V'} \downarrow & & \downarrow \mathrm{res}_{U,U'} \\
\mathscr{G}(V') & \xrightarrow{\;\phi_{V'U'}\;} & \mathscr{F}(U')
\end{array}
$$

commutes. Define $\mathrm{Mor}_{YX}(\mathscr{G}, \mathscr{F})$ to be the set of all compatible collections $\phi = \{\phi_{VU}\}$.

2.7.3. *Remark ("stalk and skyscraper are an adjoint pair").* As a special case, if X is a point $p \in Y$, we see that $\pi^{-1}\mathscr{G}$ is the stalk $\mathscr{G}_p$ of $\mathscr{G}$, and maps from the stalk $\mathscr{G}_p$ to a set S are the same as maps of sheaves on Y from $\mathscr{G}$ to the skyscraper sheaf with set S supported at p. You may prefer to prove this special case by hand directly before solving Exercise 2.7.B, as it is enlightening. (It can also be useful—can you use it to solve Exercises 2.4.L and 2.4.N?)

2.7.C. EXERCISE. Show that the stalks of $\pi^{-1}\mathscr{G}$ are the same as the stalks of $\mathscr{G}$. More precisely, if $\pi(p) = q$, describe a natural isomorphism $\mathscr{G}_q \xrightarrow{\sim} (\pi^{-1}\mathscr{G})_p$. (Possible hint: use the concrete description of the stalk, as a colimit. Recall that stalks are preserved by sheafification, Exercise 2.4.L. Alternatively, use adjointness.)

Exercise 2.7.C, along with the notion of compatible germs, may give you a simple way of thinking about (and perhaps visualizing) inverse image sheaves. Closely related: you can think of sections of the inverse image sheaf as, locally, inverse images of sections on the target. (Those preferring the "espace étalé" or "space of sections" perspective, §2.2.11, can check that the "inverse image of the space of sections" is the "space of sections" of the inverse image.)

2.7.D. EXERCISE (EASY BUT USEFUL). If U is an open subset of Y, $i: U \to Y$ is the inclusion, and $\mathscr{G}$ is a sheaf on Y, show that $i^{-1}\mathscr{G}$ is naturally isomorphic to $\mathscr{G}|_U$ (the restriction of $\mathscr{G}$ to U, §2.2.8).

2.7.E. EXERCISE. Show that π^{-1} is an exact functor from sheaves of abelian groups on Y to sheaves of abelian groups on X (cf. Exercise 2.6.E). (Hint: exactness can be checked on stalks, and by Exercise 2.7.C, the stalks are the same.) Essentially the same argument will show that π^{-1} is an exact functor from $\mathscr{O}_Y$-modules (on Y) to $(\pi^{-1}\mathscr{O}_Y)$-modules (on X), but don't bother writing that down. (Remark for experts: π^{-1} is a left adjoint, hence right-exact by abstract nonsense, as discussed in §1.5.14. Left-exactness holds because colimits of abelian groups over filtered index sets are exact, Exercise 1.5.L.)

2.7.4. Definition: The push-pull map.
Suppose

(2.7.4.1)

$$\begin{array}{ccc} W & \xrightarrow{\;\beta'\;} & X \\ {\scriptstyle \alpha'}\big\downarrow & & \big\downarrow{\scriptstyle \alpha} \\ Y & \xrightarrow{\;\beta\;} & Z \end{array}$$

is a commutative (not necessarily Cartesian!) diagram, and $\mathscr{F}$ is a sheaf on X. Define the **push-pull map**

(2.7.4.2)
$$\beta^{-1}\alpha_*\mathscr{F} \longrightarrow \alpha'_*(\beta')^{-1}\mathscr{F}$$

of sheaves on Y as follows. Start with the identity $(\beta')^{-1}\mathscr{F} \xrightarrow{\sim} (\beta')^{-1}\mathscr{F}$ on W. By adjointness of $((\beta')^{-1}, \beta'_*)$, this is the same as the data of a morphism $\mathscr{F} \to (\beta'_*)(\beta')^{-1}\mathscr{F}$ on X. Apply α_* to get a map $\alpha_*\mathscr{F} \to \alpha_*(\beta'_*)(\beta')^{-1}\mathscr{F}$ on Z. By the commutativity of (2.7.4.1), this is the map $\alpha_*\mathscr{F} \to \beta_*(\alpha'_*)(\beta')^{-1}\mathscr{F}$ on Z. By adjointness of (β^{-1}, β_*), this yields a map (2.7.4.2).

We observe that this entire construction is functorial in $\mathscr{F}$ (i.e., given a map $\mathscr{F} \to \mathscr{G}$ of sheaves on X, we get a certain commutative diagram of sheaves on Y—what is it?). (We will later extend this to $\mathscr{O}$-modules, quasicoherent sheaves, and cohomology; see Exercises 7.2.D(f), 14.5.K, and 18.7.B.)

2.7.F. ** SURPRISINGLY HARD EXERCISE. We could have defined the push-pull map in a "dual way" starting with the identity $\alpha_*\mathscr{F} \to \alpha_*\mathscr{F}$ on Z, then using adjointness of (α^{-1}, α_*), and continuing from there. Why does this give the *same* definition of the push-pull map?

2.7.5. The support of a sheaf, and the support of a section of a sheaf.
Exercise 2.7.H below gives us an excuse to introduce the notion of *support*, which we use repeatedly later.

2.7.6. *Definition.* Suppose $\mathscr{F}$ is a sheaf (or, indeed, a separated presheaf) of abelian groups on X, and s is a global section of $\mathscr{F}$. Define the **support of the section** s, denoted by Supp s, to be the set of points p of X where s has a nonzero germ:

$$\text{Supp } s := \{p \in X \;:\; s_p \neq 0 \text{ in } \mathscr{F}_p\}.$$

We think of this as the subset of X where "the section s lives"—the complement is the locus where s is the 0-section. (Unimportant: We could define this even if $\mathscr{F}$ is a presheaf, but without the inclusion $\mathscr{F}(U) \hookrightarrow \prod_{p \in U} \mathscr{F}_p$ of Exercise 2.4.A, we could have the strange situation where we have a nonzero section that "lives nowhere," because it is 0 "near every point," i.e., is 0 in every stalk.)

2.7.G. EXERCISE (THE SUPPORT OF A SECTION IS CLOSED). Show that Supp s is a closed subset of X.

2.7.7. *Caution: The locus where a continuous function is nonzero is open; the locus where the germ of a function is nonzero is closed.* Basically by the definition of continuity, the locus where the *value* of a continuous function is nonzero is *open*. (More generally, the locus where the value of a function on a locally ringed space is nonzero is open; see Exercise 4.3.F(a).) In contrast, Exercise 2.7.G shows that the locus where the *germ* of a function is nonzero is *closed*. We will try to avoid misunderstanding by using phrases like "f is 0 at p" (the value of f is zero, i.e., $f(p) = 0$) and "f is 0 near p" (the germ of f is zero, i.e., $f = 0$ in $\mathscr{O}_{X,p}$, or equivalently, f is zero in some neighborhood of p).

2.7.8. *Definition.* Define the **support of a sheaf** of groups $\mathscr{G}$, denoted by Supp $\mathscr{G}$, as the locus where the stalks are nontrivial:

$$\text{Supp}\,\mathscr{G} := \{p \in X \ : \ |\mathscr{G}_p| \neq 1\}.$$

Equivalently, Supp $\mathscr{G}$ is the union of supports of sections over all open sets. Clearly support is a "stalk-local notion," and hence "commutes" with restriction to open sets. (Irrelevant for us: more generally, if the sheaf has values in some category, such as the category of sets, the support can be defined as the points where the stalk is not the terminal object.)

2.7.H. EXERCISE.

(a) Suppose $Z \subset Y$ is a closed subset, and $i \colon Z \hookrightarrow Y$ is the inclusion. If $\mathscr{F}$ is a sheaf of groups on Z, then show that the stalk $(i_* \mathscr{F})_q$ is the one-element group if $q \notin Z$, and $\mathscr{F}_q$ if $q \in Z$.

(b) Suppose Supp $\mathscr{G} \subset Z$, where Z is closed. Show that the natural map $\mathscr{G} \to i_* i^{-1} \mathscr{G}$ is an isomorphism. Thus a sheaf supported on a closed subset can be considered a sheaf on that closed subset.

2.7.9. Extension by zero, an occasional *left adjoint* to the inverse image functor. In addition to always being a left adjoint, π^{-1} can also be a right adjoint when π is an inclusion of an open subset. We discuss this when we need it, in §23.4.7.

PART II

Schemes

II

L'idée même de schéma est d'une simplicité enfantine—si simple, si humble, que personne avant moi n'avait songé à se pencher si bas. Si "bébête" même, pour tout dire, que pendant des années encore et en dépit de l'évidence, pour beaucoup de mes savants collègues, ça faisait vraiment "pas sérieux"!

The very idea of scheme is of infantile simplicity—so simple, so humble, that no one before me thought of stooping so low. So childish, in short, that for years, despite all the evidence, for many of my erudite colleagues, it was really "not serious"!

—A. Grothendieck [Gr5, p. P32], translated by C. McLarty [Mc, p. 313]

Chapter 3

Toward Affine Schemes: The Underlying Set, and Topological Space

There is no serious historical question of how Grothendieck found his definition of schemes. It was in the air. Serre has well said that no one invented schemes... The question is, what made Grothendieck believe he should use this definition to simplify an 80 page paper by Serre into some 1,000 pages of Éléments de Géométrie Algébrique?

— C. McLarty [Mc, p. 313]

We are now ready to consider the notion of a *scheme*, which is the type of geometric space central to algebraic geometry. We should first think through what we mean by "geometric space." You have likely seen the notion of a manifold, and we wish to abstract this notion so that it can be generalized to other settings, notably so that we can deal with non-smooth and arithmetic objects.

3.1 Toward Schemes

The key insight behind this generalization from the notion of something like a manifold to a more versatile notion of a "geometric space" is the following: we can understand a geometric space (such as a manifold) well by understanding the functions on this space. More precisely, we will understand it through the sheaf of functions on the space. If we are interested in differentiable manifolds, we will consider smooth functions; if we are interested in analytic manifolds, we will consider real analytic functions; and so on.

Thus we will define a scheme to be the following data

- *The set:* the points of the scheme
- *The topology:* the open sets of the scheme
- *The structure sheaf:* the sheaf of "algebraic functions" (a sheaf of rings) on the scheme.

Recall that a topological space with a sheaf of rings is called a *ringed space* (§2.2.13).

We will try to draw pictures throughout. Pictures can help develop geometric intuition, which can guide the algebraic development (and, eventually, vice versa). Some people find pictures very helpful, while others are repulsed or confused.

We will try to make all three notions as intuitive as possible. For the set, in the key example of complex (affine) varieties (roughly, things cut out in $\mathbb{C}^n$ by polynomials), we will see that the points are the "traditional points" (n-tuples of complex numbers), plus some extra points that will be handy to have around. For the topology, we will require that "the subset where an algebraic function vanishes must be closed," and require nothing else. For the sheaf of algebraic functions (the structure sheaf), we will expect that in the complex plane $\mathbb{C}^2$, $(3x^2 + y^2)/(2x + 4xy + 1)$ should be an algebraic function on the open set consisting of points where the denominator doesn't vanish, and this will largely motivate our definition.

3.1.1. Example: Differentiable manifolds. As motivation, we return to our example of differentiable manifolds, reinterpreting them from this perspective. We will be quite informal in this discussion. Suppose X is a differentiable manifold. It is a topological space, and has a *sheaf of smooth* (C^∞) *functions* $\mathcal{O}_X$ (see §2.1). This gives X the structure of a ringed space. We have observed that evaluation at a point $p \in X$ gives a surjective map from the stalk to $\mathbb{R}$

$$\mathcal{O}_{X,p} \longrightarrow \mathbb{R},$$

so the kernel, the (germs of) functions vanishing at p, is a maximal ideal $\mathfrak{m}_{X,p}$ (see §2.1.1).

We could *define* a differentiable real manifold as a topological space X with a sheaf of rings (see Definition 4.3.9). We would require that there is a cover of X by open sets such that on each open set the ringed space is isomorphic to a ball around the origin in $\mathbb{R}^n$ (with the sheaf of smooth functions on that ball). With this definition, the ball is the basic patch, and a general manifold is obtained by gluing these patches together. (Admittedly, a great deal of geometry comes from how one chooses to patch the balls together!) In the algebraic setting, the basic patch is an *affine scheme*, which we will discuss soon. (In the definition of manifold, there is an additional requirement that the topological space be Hausdorff and second-countable, to avoid pathologies. Schemes are often required to be "separated" to avoid essentially the same pathologies. Separatedness will be discussed in Chapter 11.)

Functions are determined by their values at points. This is an obvious statement, but won't be true for schemes in general. We will see an example in Exercise 3.2.A(a), and discuss this behavior further in §3.2.12.

Morphisms of manifolds. How can we describe maps of differentiable manifolds $\pi\colon X \to Y$? They are certainly continuous maps—but which ones? We can pull back functions along continuous maps. Smooth functions pull back to smooth functions. More formally, we have a map $\pi^{-1}\mathcal{O}_Y \to \mathcal{O}_X$. (The inverse image sheaf π^{-1} was defined in §2.7.) Inverse image is left-adjoint to pushforward, so we also get a map $\pi^\sharp\colon \mathcal{O}_Y \to \pi_* \mathcal{O}_X$.

Certainly, given a map of differentiable manifolds, smooth functions pull back to smooth functions. It is less obvious that *this is a sufficient condition for a continuous map to be smooth.*

3.1.A. IMPORTANT EXERCISE FOR THOSE WITH A LITTLE EXPERIENCE WITH MANIFOLDS. Suppose that $\pi\colon X \to Y$ is a continuous map of differentiable manifolds (as topological spaces—not a priori smooth). Show that π is smooth if smooth functions pull back to smooth functions, i.e., if pullback by π gives a map $\mathcal{O}_Y \to \pi_* \mathcal{O}_X$. (Hint: Check this on small patches. Once you figure out what you are trying to show, you will realize that the result is immediate.)

3.1.B. EXERCISE. Show that a morphism of differentiable manifolds $\pi\colon X \to Y$ with $\pi(p) = q$ induces a morphism of stalks $\pi^\sharp\colon \mathcal{O}_{Y,q} \to \mathcal{O}_{X,p}$. Show that $\pi^\sharp(\mathfrak{m}_{Y,q}) \subset \mathfrak{m}_{X,p}$. In other words, if you pull back a function that vanishes at q, you get a function that vanishes at p—not a huge surprise. (In §7.3, we formalize this by saying that maps of manifolds are maps of locally ringed spaces.)

3.1.2. *Aside.* Here is a little more for experts: Notice that π induces a map on tangent spaces (see Aside 2.1.2),

$$(\mathfrak{m}_{X,p}/\mathfrak{m}_{X,p}^2)^\vee \longrightarrow (\mathfrak{m}_{Y,q}/\mathfrak{m}_{Y,q}^2)^\vee.$$

This is the tangent map you would geometrically expect. Again, it is interesting that the cotangent map $\mathfrak{m}_{Y,q}/\mathfrak{m}_{Y,q}^2 \to \mathfrak{m}_{X,p}/\mathfrak{m}_{X,p}^2$ is algebraically more natural than the tangent map (there are no "duals").

Experts are now free to try to interpret other differential-geometric information using only the map of topological spaces and the map of sheaves. For example: How can one check if π is a submersion of manifolds? How can one check if f is an immersion? (We will see that the algebro-geometric version of these notions are *smooth morphism* and *unramified morphism*; see Definition 13.6.2 and §21.7, respectively.)

3.1.3. *Side Remark.* Manifolds are covered by disks that are all isomorphic. This isn't true for schemes (even for "smooth complex varieties"). There are examples of two "smooth complex curves" (the algebraic version of Riemann surfaces) X and Y so that no nonempty open subset of X is isomorphic to a nonempty open subset of Y (see Exercise 7.5.L). And there is a Riemann surface X such that no two open subsets of X are isomorphic (see Exercise 19.7.E). Informally, this is because in the Zariski topology on schemes, all nonempty open sets are "huge" and have more "structure."

3.1.4. Other examples. If you are interested in differential geometry, you might be interested in differentiable manifolds, on which the functions under consideration are smooth functions. Similarly, if you are interested in topology, you will be interested in topological spaces, on which you will consider the continuous functions. If you are interested in complex geometry, you will be interested in complex manifolds (or possibly "complex analytic varieties"), on which the functions are holomorphic functions. In each of these cases of interesting "geometric spaces," the topological space and sheaf of functions are clear. The notion of scheme fits naturally into this family.

3.2   The Underlying Set of an Affine Scheme

For any ring A, we are going to define something called $\operatorname{Spec} A$, the **spectrum of** A. In this section, we will define it as a set, but we will soon endow it with a topology, and later we will define a sheaf of rings on it (the structure sheaf). Such an object is called an *affine scheme*. Later $\operatorname{Spec} A$ will denote the set along with the topology, and a sheaf of functions. But for now, as there is no possibility of confusion, $\operatorname{Spec} A$ will just be the set.

3.2.1. The set $\operatorname{Spec} A$ is the set of prime ideals of A. The prime ideal $\mathfrak{p}$ of A when considered as an element of $\operatorname{Spec} A$ will be denoted by $[\mathfrak{p}]$, to avoid confusion. Elements $a \in A$ will be called **functions on** $\operatorname{Spec} A$, and their **value** at the point $[\mathfrak{p}]$ will be $a \pmod{\mathfrak{p}}$. *This is weird: a function can take values in different rings at different points — the function 5 on* $\operatorname{Spec} \mathbb{Z}$ *takes the value* 1 (mod 2) *at* $[(2)]$ *and* 2 (mod 3) *at* $[(3)]$.

"*An element a of the ring lying in a prime ideal* $\mathfrak{p}$" translates to "*a function a that is 0 at the point* $[\mathfrak{p}]$" or "*a function a vanishing at the point* $[\mathfrak{p}]$," and we will use these phrases interchangeably. Notice that if you add or multiply two functions, you add or multiply their values at all points; this is a translation of the fact that $A \to A/\mathfrak{p}$ is a ring morphism. These translations are important—make sure you are very comfortable with them! They should become second nature.

If A is generated over a base field (or base ring) by elements $x_1, \ldots, x_r$, the elements $x_1, \ldots, x_r$ are often called **coordinates**, because we will later be able to reinterpret them as restrictions of "coordinates on r-space," via the idea of §3.2.10, made precise in Exercise 7.2.F.

3.2.2. *Beginning of a grand dictionary between algebra and geometry.* We are building up the beginning of a grand dictionary; see the first part of §3.7.2. At some point you will get the sense that we are slowly decoding some timeless Rosetta Stone whose etchings we struggle to make out.

3.2.3. *Glimpses of the future.* In §4.1.3: we will interpret functions on $\operatorname{Spec} A$ as global sections of the "structure sheaf," i.e., as functions on a ringed space, in the sense of §2.2.13. We repeat a caution from §2.2.13: what we will call "functions," others may call "regular functions." And we will later define "rational functions" (§6.6.36), which are not precisely functions in this sense; they are a particular type of "partially-defined function."

The notion of "value of a function" will be later interpreted as a value of a function on a particular locally ringed space; see Definition 4.3.7.

3.2.4. We now give some examples.

Example 1 (the complex affine line): $\mathbb{A}^1_{\mathbb{C}} := \operatorname{Spec} \mathbb{C}[x]$. Let's find the prime ideals of $\mathbb{C}[x]$. As $\mathbb{C}[x]$ is an integral domain, 0 is prime. Also, $(x - a)$ is prime, for any $a \in \mathbb{C}$; it is even a maximal ideal, as the quotient by this ideal is a field:

$$0 \longrightarrow (x - a) \longrightarrow \mathbb{C}[x] \xrightarrow{f \mapsto f(a)} \mathbb{C} \longrightarrow 0.$$

(This exact sequence may remind you of (2.1.1.1) in our motivating example of manifolds.)

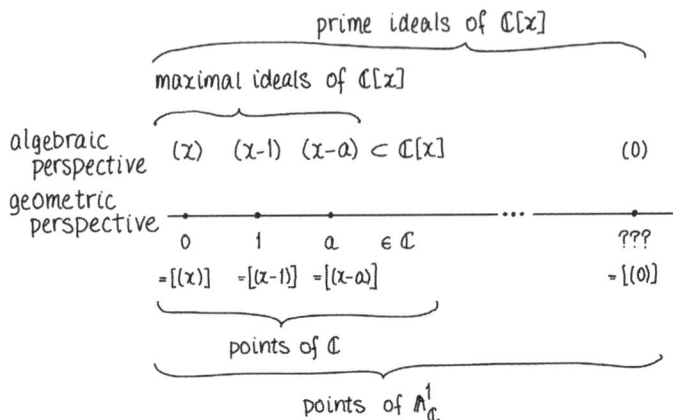

Figure 3.1 *A picture of $\mathbb{A}^1_\mathbb{C} = \operatorname{Spec} \mathbb{C}[x]$.*

We now show that there are no other prime ideals. We use the fact that $\mathbb{C}[x]$ has a division algorithm, and is a unique factorization domain. Suppose $\mathfrak{p}$ is a prime ideal. If $\mathfrak{p} \neq (0)$, then suppose $f(x) \in \mathfrak{p}$ is a nonzero element of smallest degree. It is not constant, as prime ideals can't contain 1. If $f(x)$ is not linear, then factor $f(x) = g(x)h(x)$, where $g(x)$ and $h(x)$ have positive degree. (Here we use the fact that $\mathbb{C}$ is algebraically closed.) Then $g(x) \in \mathfrak{p}$ or $h(x) \in \mathfrak{p}$, contradicting the minimality of the degree of f. Hence there is a linear element $x - a$ of $\mathfrak{p}$. Then I claim that $\mathfrak{p} = (x - a)$. Suppose $f(x) \in \mathfrak{p}$. Then the division algorithm would give $f(x) = g(x)(x - a) + m$, where $m \in \mathbb{C}$. Then $m = f(x) - g(x)(x - a) \in \mathfrak{p}$. If $m \neq 0$, then $1 \in \mathfrak{p}$, giving a contradiction.

Thus we can and should (and must!) make a picture of $\mathbb{A}^1_\mathbb{C} = \operatorname{Spec} \mathbb{C}[x]$ (see Figure 3.1). This is just the first illustration of a point of view of Sophie Germain [Ge]: "L'algèbre n'est qu'une géométrie écrite; la géométrie n'est qu'une algèbre figurée." (Algebra is but written geometry; geometry is but drawn algebra.)

There is one "traditional" point for each complex number, plus one extra ("bonus") point $[(0)]$. We can mostly picture $\mathbb{A}^1_\mathbb{C}$ as $\mathbb{C}$: the point $[(x - a)]$ we will reasonably associate to $a \in \mathbb{C}$. Where should we picture the point $[(0)]$? The best way of thinking about it is somewhat zen. It is somewhere on the complex line, but nowhere in particular. Because (0) is contained in all of these prime ideals, we will somehow associate it with this line passing through all the other points. This new point $[(0)]$ is called the "generic point" of the line. (We will formally define "generic point" in §3.6.) It is "generically on the line" but you can't pin it down any further than that. It is not at any particular place on the line. (This is misleading too—we will see in Easy Exercise 3.6.N that it is "near" every point. So it is near everything, but located nowhere precisely.) We will place it far to the right for lack of anywhere better to put it. You will notice that we sketch $\mathbb{A}^1_\mathbb{C}$ as one-(real-)dimensional (even though we picture it as an enhanced version of $\mathbb{C}$); this is to later remind ourselves that this will be a one-dimensional space, where dimensions are defined in an algebraic (or complex-geometric) sense. (Dimension will be defined in Chapter 12.)

To give you some feeling for this space, we make some statements that are currently unde-fined, but suggestive. The functions on $\mathbb{A}^1_\mathbb{C}$ are the polynomials. So $f(x) = x^2 - 3x + 1$ is a function. What is its value at $[(x - 1)]$, which we think of as the point $1 \in \mathbb{C}$? Answer: $f(1)$! Or equivalently, we can evaluate $f(x)$ modulo $x - 1$—this is the same thing by the division algorithm. (What is its value at $[(0)]$? It is $f(x) \pmod 0$, which is just $f(x)$.)

Here is a more complicated example: $g(x) = (x - 3)^3/(x - 2)$ is a "rational function." It is defined everywhere but $x = 2$. (When we know what the structure sheaf is, we will be able to say that it is an element of the structure sheaf on the open set $\mathbb{A}^1_\mathbb{C} \setminus \{2\}$.) We want to say that $g(x)$ has a triple zero at 3, and a single pole at 2, and we will be able to after §13.5.

Example 2 (the affine line over $k = \bar{k}$): $\mathbb{A}^1_k := \operatorname{Spec} k[x]$ **where k is an algebraically closed field.** This is called the affine line over k. All of our discussion in the previous example carries

Figure 3.2 *A "picture" of* Spec ℤ, *which looks suspiciously like Figure 3.1.*

over without change. We will use the same picture, which is after all intended to just be a metaphor.

Example 3: Spec ℤ. An amazing fact is that from our perspective, this will look a lot like the affine line $\mathbb{A}^1_{\bar{k}}$. The integers, like $\bar{k}[x]$, form a unique factorization domain, with a division algorithm. The prime ideals are: (0) and (p), where p is prime. Thus everything from Example 1 carries over without change, even the picture. Our picture of Spec ℤ is shown in Figure 3.2.

Let's blithely carry over our discussion of functions to this space. 100 is a function on Spec ℤ. Its value at (3) is "1 (mod 3)." Its value at (2) is "0 (mod 2)," and in fact it has a double zero. $27/4$ is a "rational function" on Spec ℤ, defined away from (2). We want to say that it has a double pole at (2), and a triple zero at (3). Its value at (5) is

$$27 \times 4^{-1} \equiv 2 \times (-1) \equiv 3 \quad (\text{mod } 5).$$

(We will gradually make this discussion precise over time.)

Example 4: Silly but important examples, and the German word for bacon. The set Spec k, where k is any field, is boring: one point. Spec 0, where 0 is the zero-ring, is the empty set, as 0 has no prime ideals.

3.2.A. A SMALL EXERCISE ABOUT SMALL SCHEMES.

(a) Describe the set Spec $k[\epsilon]/(\epsilon^2)$. The ring $k[\epsilon]/(\epsilon^2)$ is called the ring of **dual numbers**, and will turn out to be quite useful. You should think of ϵ as a very small number, so small that its square is 0 (although it itself is not 0). It is a nonzero function whose value at all points is zero, thus giving our first example of functions not being determined by their values at points. We will discuss this phenomenon further in §3.2.12.

(b) Describe the set Spec $k[x]_{(x)}$ (see §1.2.3 for a discussion of localization). We will see this scheme again repeatedly, starting with §3.2.9 and Exercise 3.4.K. You might later think of it as a shred of a particularly nice "smooth curve."

In Example 2, we restricted to the case of algebraically closed fields for a reason: things are more subtle if the field is not algebraically closed.

Example 5 (the affine line over ℝ): $\mathbb{A}^1_{\mathbb{R}} = \text{Spec } \mathbb{R}[x]$. Using the fact that $\mathbb{R}[x]$ is a Euclidean domain, similar arguments to those of Examples 1–3 show that the prime ideals are (0), $(x-a)$, where $a \in \mathbb{R}$, and $(x^2 + ax + b)$, where $x^2 + ax + b$ is an irreducible quadratic. The latter two are maximal ideals, i.e., their quotients are fields. For example: $\mathbb{R}[x]/(x-3) \cong \mathbb{R}$, $\mathbb{R}[x]/(x^2+1) \cong \mathbb{C}$.

3.2.B. UNIMPORTANT EXERCISE. Show that for the last type of prime, of the form $(x^2 + ax + b)$, the quotient is *always* isomorphic to $\mathbb{C}$.

So we have the points that we would normally expect to see on the real line, corresponding to real numbers; the generic point 0; and new points which we may interpret as *conjugate pairs* of complex numbers (the roots of the quadratic). This last type of point should be seen as more akin to the real numbers than to the generic point. You can picture $\mathbb{A}^1_{\mathbb{R}}$ as the complex plane, folded along the real axis. But the key point is that Galois-conjugate points (such as i and $-i$) are considered glued.

Let's explore functions on this space. Consider the function $f(x) = x^3 - 1$. Its value at the point $[(x-2)]$ is 7, or perhaps better, "7 (mod $x-2$)." How about at (x^2+1)? We get

$$x^3 - 1 \equiv -x - 1 \quad (\text{mod } x^2 + 1),$$

which may be profitably interpreted as $-i - 1$.

One moral of this example is that we can work over a non-algebraically closed field if we wish. It is more complicated, but we can recover much of the information we care about.

3.2.C. IMPORTANT EXERCISE.   Describe the set $\mathbb{A}^1_{\mathbb{Q}}$. (This is harder to picture in a way analogous to $\mathbb{A}^1_{\mathbb{R}}$. But the rough cartoon of points on a line, as in Figure 3.1, remains a reasonable sketch.)

Example 6 (the affine line over $\mathbb{F}_p$): $\mathbb{A}^1_{\mathbb{F}_p} = \operatorname{Spec} \mathbb{F}_p[x]$. As in the previous examples, $\mathbb{F}_p[x]$ is a Euclidean domain, so the prime ideals are of the form (0) or $(f(x))$, where $f(x) \in \mathbb{F}_p[x]$ is an irreducible polynomial, which can be of any degree. Irreducible polynomials correspond to sets of Galois conjugates in $\overline{\mathbb{F}}_p$.

Note that $\operatorname{Spec} \mathbb{F}_p[x]$ has p points corresponding to the elements of $\mathbb{F}_p$, but also many more (infinitely more, see Exercise 3.2.D). This makes this space much richer than simply p points. For example, a polynomial $f(x)$ is not determined by its values at the p elements of $\mathbb{F}_p$, but it *is* determined by its values at the points of $\operatorname{Spec} \mathbb{F}_p[x]$. (As we have mentioned before, this is not true for all schemes.)

You should think about this, even if you are a geometric person—this intuition will later turn up in geometric situations. Even if you think you are interested only in working over an algebraically closed field (such as $\mathbb{C}$), you will have non-algebraically closed fields (such as $\mathbb{C}(x)$) forced upon you.

3.2.D. EXERCISE. If k is a field, show that $\operatorname{Spec} k[x]$ has infinitely many points. (Hint: Euclid's proof of the infinitude of primes of $\mathbb{Z}$.)

Example 7 (the complex affine plane): $\mathbb{A}^2_{\mathbb{C}} = \operatorname{Spec} \mathbb{C}[x, y]$. (As with Examples 1 and 2, our discussion will apply with $\mathbb{C}$ replaced by *any* algebraically closed field.) Sadly, $\mathbb{C}[x, y]$ is not a principal ideal domain: (x, y) is not a principal ideal. We can quickly name *some* prime ideals. One is (0), which has the same flavor as the (0) ideals in the previous examples. $(x - 2, y - 3)$ is prime, and indeed maximal, because $\mathbb{C}[x, y]/(x - 2, y - 3) \cong \mathbb{C}$, where this isomorphism is via $f(x, y) \mapsto f(2, 3)$. More generally, $(x - a, y - b)$ is prime for any $(a, b) \in \mathbb{C}^2$. Also, if $f(x, y)$ is an irreducible polynomial (e.g., $y - x^2$ or $y^2 - x^3$), then $(f(x, y))$ is prime.

3.2.E. EXERCISE. Show that we have identified all the prime ideals of $\mathbb{C}[x, y]$. Hint: Suppose $\mathfrak{p}$ is a prime ideal that is not principal. Show you can find $f(x, y), g(x, y) \in \mathfrak{p}$ with no common factor. By considering the Euclidean algorithm in the Euclidean domain $\mathbb{C}(x)[y]$, show that you can find a nonzero $h(x) \in (f(x, y), g(x, y)) \subset \mathfrak{p}$. Using primality, show that one of the linear factors of $h(x)$, say $(x - a)$, is in $\mathfrak{p}$. Similarly show there is some $(y - b) \in \mathfrak{p}$.

We now attempt to draw a picture of $\mathbb{A}^2_{\mathbb{C}}$ (see Figure 3.3). The maximal prime ideals of $\mathbb{C}[x, y]$ correspond to the traditional points in $\mathbb{C}^2$: $[(x - a, y - b)]$ corresponds to $(a, b) \in \mathbb{C}^2$. We now have to visualize the "bonus points." $[(0)]$ somehow lives behind all of the traditional points; it is somewhere on the affine plane, but nowhere in particular. So, for example, it does not lie on the parabola $y = x^2$. The point $[(y - x^2)]$ lies on the parabola $y = x^2$, but nowhere in particular on it. (Figure 3.3 is a bit misleading. For example, the point $[(0)]$ isn't in the fourth quadrant; it is somehow near every other point, which is why it is depicted as a somewhat diffuse large dot.) You can see from this picture that we already are implicitly thinking about "dimension." The prime ideals $(x - a, y - b)$ are somehow of dimension 0, the prime ideals $(f(x, y))$ are of dimension 1, and (0) is of dimension 2. (All of our dimensions here are *complex* or *algebraic* dimensions. The complex plane $\mathbb{C}^2$ has real dimension 4, but complex dimension 2. Complex dimensions are in general half of real dimensions.) We won't define dimension precisely until Chapter 12, but you should feel free to keep it in mind before then.

Note, too, that maximal ideals correspond to the "smallest" points. Smaller ideals correspond to "bigger" points. "One prime ideal contains another" means that the points "have the opposite containment." All of this will be made precise once we have a topology. This order-reversal is a little confusing, and will remain so even once we have made the notions precise.

We now come to the obvious generalization of Example 7.

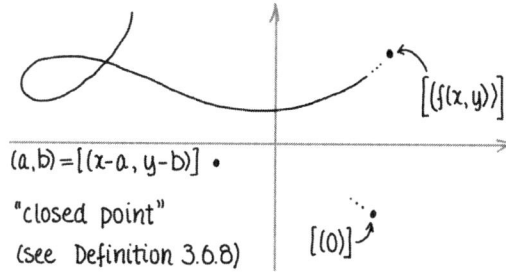

Figure 3.3 *Picturing* $\mathbb{A}^2_{\mathbb{C}} = \mathrm{Spec}\,\mathbb{C}[x, y]$.

Example 8 (complex affine n-space—important!): Let $\mathbb{A}^n_{\mathbb{C}} := \mathrm{Spec}\,\mathbb{C}[x_1, \ldots, x_n]$. (Important definition: More generally, $\mathbb{A}^n_A$ is defined to be $\mathrm{Spec}\,A[x_1, \ldots, x_n]$, where A is an arbitrary ring. When the base ring is clear from context, the subscript A is often omitted. For pedants: the notation $\mathbb{A}^n_A$ implicitly includes the data of the n **coordinate functions** $x_1, \ldots, x_n$.) For concreteness, let's consider $n = 3$. We now have an interesting question in what at first appears to be pure algebra: What are the prime ideals of $\mathbb{C}[x, y, z]$?

Analogously to before, $(x - a, y - b, z - c)$ is a prime ideal. This is a maximal ideal, because its residue ring is a field ($\mathbb{C}$); we think of these as "zero-dimensional points." We will often write (a, b, c) for $[(x - a, y - b, z - c)]$ because of our geometric interpretation of these ideals. There are no more maximal ideals, by Hilbert's weak Nullstellensatz.

3.2.5. Hilbert's weak Nullstellensatz — *If k is an algebraically closed field, then the maximal ideals of $k[x_1, \ldots, x_n]$ are precisely those ideals of the form $(x_1 - a_1, \ldots, x_n - a_n)$, where $a_i \in k$.*

We may as well state a slightly stronger version now.

3.2.6. Hilbert's Nullstellensatz — *If k is any field, every maximal ideal of $k[x_1, \ldots, x_n]$ has residue field a finite extension of k. Translation: any field extension of k that is finitely generated as a k-algebra is necessarily also finitely generated as a k-vector space (i.e., is a finite extension of fields).*

This statement is also often called *Zariski's Lemma*.

3.2.F. EXERCISE. Show that the Nullstellensatz 3.2.6 implies the weak Nullstellensatz 3.2.5.

We will prove the Nullstellensatz in §8.4.3, and again in Exercise 12.2.B.

The following fact is a useful accompaniment to the Nullstellensatz.

3.2.G. EXERCISE (NOT REQUIRING THE NULLSTELLENSATZ). Any integral domain A which is a finite k-algebra (i.e., a k-algebra that is a finite-dimensional vector space over k) must be a field. Hint: For any nonzero $x \in A$, show $\times x \colon A \to A$ is an isomorphism. (Thus, in combination with the Nullstellensatz 3.2.6, we see that *prime ideals of $k[x_1, \ldots, x_n]$ with finite-dimensional residue ring are the same as maximal ideals of $k[x_1, \ldots, x_n]$*. This is worth remembering.)

There are other prime ideals of $\mathbb{C}[x, y, z]$, too. We have (0), which corresponds to a "three-dimensional point." We have $(f(x, y, z))$, where f is irreducible. To this we associate the "hypersurface" $f = 0$, so this is "two-dimensional" in nature. But we have not found them all! One clue: we have prime ideals of "dimension" 0, 2, and 3—we are missing "dimension 1." Here is one such prime ideal: (x, y). We picture this as the locus where $x = y = 0$, which is the z-axis. This is a prime ideal, as the corresponding quotient $\mathbb{C}[x, y, z]/(x, y) \cong \mathbb{C}[z]$ is an integral domain (and should be interpreted as the functions on the z-axis). There are lots of "one-dimensional prime ideals," and it is not possible to classify them in a reasonable way. It will turn out that they correspond to things that we think of as irreducible curves. Thus remarkably the answer to the purely algebraic question ("what are the prime ideals of $\mathbb{C}[x, y, z]$") is fundamentally geometric!

The fact that the points of $\mathbb{A}^1_{\mathbb{Q}}$ corresponding to maximal ideals of the ring $\mathbb{Q}[x]$ (what we will soon call "closed points"; see Definition 3.6.8) can be interpreted as points of $\overline{\mathbb{Q}}$ where Galois-conjugates are glued together (Exercise 3.2.C) extends to $\mathbb{A}^n_{\mathbb{Q}}$. For example, in $\mathbb{A}^2_{\mathbb{Q}}$, $(\sqrt{2}, \sqrt{2})$ is glued to $(-\sqrt{2}, -\sqrt{2})$ but not to $(\sqrt{2}, -\sqrt{2})$. The following exercise will give you some idea of how this works.

3.2.H. EXERCISE. Describe the maximal ideal of $\mathbb{Q}[x, y]$ corresponding to $(\sqrt{2}, \sqrt{2})$ and $(-\sqrt{2}, -\sqrt{2})$. Describe the maximal ideal of $\mathbb{Q}[x, y]$ corresponding to $(\sqrt{2}, -\sqrt{2})$ and $(-\sqrt{2}, \sqrt{2})$. What are the residue fields in each case?

The description of "closed points" of $\mathbb{A}^2_{\mathbb{Q}}$ (those points corresponding to maximal ideals of the ring $\mathbb{Q}[x, y]$) as Galois-orbits of points in $\overline{\mathbb{Q}}^2$ can even be extended to other "non-closed" points, as follows.

3.2.I. UNIMPORTANT AND TRICKY BUT FUN EXERCISE. Consider the map of sets $\phi: \mathbb{C}^2 \to \mathbb{A}^2_{\mathbb{Q}}$ defined as follows. (z_1, z_2) is sent to the prime ideal of $\mathbb{Q}[x, y]$ consisting of polynomials vanishing at (z_1, z_2).

(a) What is the image of (π, π^2)?
(b)* Show that ϕ is surjective. (Warning: You will need some ideas we haven't discussed in order to solve this. Once we define the Zariski topology on $\mathbb{A}^2_{\mathbb{Q}}$, you will be able to check that ϕ is continuous, where we give $\mathbb{C}^2$ the classical topology. This example generalizes. For example, you may later be able to generalize this to arbitrary dimension.)

3.2.7. Quotients and localizations. Two natural ways of getting new rings from old ones—quotients and localizations—have interpretations in terms of spectra.

3.2.8. *Quotients:* $\operatorname{Spec} A/I$ *as a subset of* $\operatorname{Spec} A$. It is an important fact that the prime ideals of A/I are in bijection with the prime ideals of A containing I.

3.2.J. ESSENTIAL ALGEBRA EXERCISE (MANDATORY IF YOU HAVEN'T SEEN IT BEFORE). Suppose A is a ring, and I an ideal of A. Let $\phi: A \to A/I$. Show that ϕ^{-1} gives an inclusion-preserving bijection between prime ideals of A/I and prime ideals of A containing I. Thus we can picture $\operatorname{Spec} A/I$ as a subset of $\operatorname{Spec} A$.

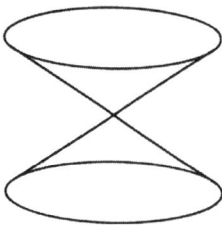

Figure 3.4 A "picture" of $\operatorname{Spec} \mathbb{C}[x, y, z]/(x^2 + y^2 - z^2)$.

As an important motivational special case, you now have a picture of *affine complex varieties*. Suppose A is a finitely generated $\mathbb{C}$-algebra, generated by $x_1, \ldots, x_n$, with relations $f_1(x_1, \ldots, x_n) = \cdots = f_r(x_1, \ldots, x_n) = 0$. Then this description in terms of generators and relations naturally gives us an interpretation of $\operatorname{Spec} A$ as a subset of $\mathbb{A}^n_{\mathbb{C}}$, which we think of as "traditional points" (n-tuples of complex numbers) along with some "bonus" points we haven't yet fully described. To see which of the traditional points are in $\operatorname{Spec} A$, we simply solve the equations $f_1 = \cdots = f_r = 0$. For example, $\operatorname{Spec} \mathbb{C}[x, y, z]/(x^2 + y^2 - z^2)$ may be pictured as shown in Figure 3.4. (Admittedly this is just a "sketch of the $\mathbb{R}$-points," but we will still find it helpful later.) This entire picture carries over (along with the Nullstellensatz) with $\mathbb{C}$ replaced by any algebraically closed field. Indeed, the picture of Figure 3.4 can be said to depict $\operatorname{Spec} k[x, y, z]/(x^2 + y^2 - z^2)$ for most algebraically closed fields k (although it is misleading in characteristic 2, because of the coincidence $x^2 + y^2 - z^2 = (x + y + z)^2$).

3.2.9. *Localizations:* $\operatorname{Spec} S^{-1}A$ *as a subset of* $\operatorname{Spec} A$. The following exercise shows how prime ideals behave under localization.

3.2.K. ESSENTIAL ALGEBRA EXERCISE (MANDATORY IF YOU HAVEN'T SEEN IT BEFORE). Suppose S is a multiplicative subset of A. Describe an order-preserving bijection of the prime ideals of $S^{-1}A$ with the prime ideals of A that *don't meet* the multiplicative set S.

Recall from §1.2.3 that there are two important flavors of localization. The first is $A_f = \{1, f, f^2, \dots\}^{-1}A$ where $f \in A$. A motivating example is $A = \mathbb{C}[x, y]$, $f = y - x^2$. The second is $A_{\mathfrak{p}} = (A \setminus \mathfrak{p})^{-1}A$, where $\mathfrak{p}$ is a prime ideal. A motivating example is $A = \mathbb{C}[x, y]$, $S = A \setminus (x, y)$.

If $S = \{1, f, f^2, \dots\}$, the prime ideals of $S^{-1}A$ are just those prime ideals not containing f—the points where "f doesn't vanish." (In §3.5, we will call this a *distinguished open set*, once we know what open sets are.) So to picture $\operatorname{Spec}\mathbb{C}[x, y]_{y-x^2}$, we picture the affine plane, and throw out those points on the parabola $y - x^2 = 0$—the points (a, a^2) for $a \in \mathbb{C}$ (by which we mean $[(x - a, y - a^2)]$), as well as the "new kind of point" $[(y - x^2)]$.

It can be initially confusing to think about localization in the case where zerodivisors are inverted, because localization $A \to S^{-1}A$ is not injective (Exercise 1.2.C). Geometric intuition can help. Consider the case $A = \mathbb{C}[x, y]/(xy)$ and $f = x$. What is the localization A_f? The space $\operatorname{Spec}\mathbb{C}[x, y]/(xy)$ "is" the union of the two axes in the affine plane. Localizing means throwing out the locus where x vanishes. So we are left with the x-axis, minus the origin, so we expect $\operatorname{Spec}\mathbb{C}[x]_x$. So there should be some natural isomorphism

$$(\mathbb{C}[x, y]/(xy))_x \xrightarrow{\sim} \mathbb{C}[x]_x.$$

3.2.L. EXERCISE. Show that these two rings are isomorphic. (You will see that y on the left goes to 0 on the right.)

If $S = A \setminus \mathfrak{p}$, the prime ideals of $S^{-1}A$ are just the prime ideals of A contained in $\mathfrak{p}$. In our example $A = \mathbb{C}[x, y]$, $\mathfrak{p} = (x, y)$, we keep all those points corresponding to "things through the origin," i.e., the zero-dimensional point (x, y), the two-dimensional point (0), and those one-dimensional points $(f(x, y))$ where $f(0, 0) = 0$, i.e., those "irreducible curves through the origin." You can think of this being a shred of the plane near the origin; anything not actually "visible" at the origin is discarded (see Figure 3.5).

Another example is when $A = k[x]$, and $\mathfrak{p} = (x)$ (or more generally when $\mathfrak{p}$ is any maximal ideal). Then $A_{\mathfrak{p}}$ has only two prime ideals (Exercise 3.2.A(b)). You should see this as the germ of a "smooth curve," where one point is the "classical

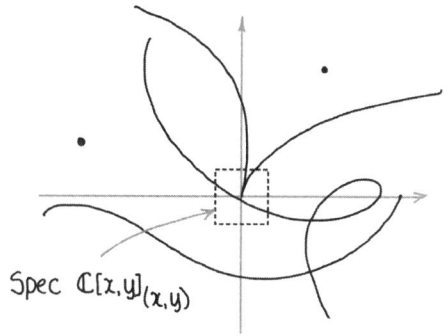

$\operatorname{Spec} \mathbb{C}[x,y]_{(x,y)}$

Figure 3.5 *Picturing* $\operatorname{Spec}\mathbb{C}[x, y]_{(x,y)}$ *as a "shred of $\mathbb{A}^2_{\mathbb{C}}$." Only those points near the origin remain.*

point," and the other is the "generic point of the curve." This is an example of a discrete valuation ring, and indeed all discrete valuation rings should be visualized in such a way. We will discuss discrete valuation rings in §13.5. By then we will have justified the use of the words "smooth" and "curve." (Reality check: try to picture Spec of $\mathbb{Z}$ localized at (2) and at (0). How do the two pictures differ?)

3.2.10. Important fact: Maps of rings induce maps of spectra (as sets). We now make an observation that will later grow to be the notion of morphisms of schemes.

3.2.M. IMPORTANT EASY EXERCISE. If $\phi\colon B \to A$ is a map of rings, and $\mathfrak{p}$ is a prime ideal of A, show that $\phi^{-1}(\mathfrak{p})$ is a prime ideal of B.

Hence a map of rings $\phi\colon B \to A$ induces a map of sets $\operatorname{Spec}A \to \operatorname{Spec}B$ "in the opposite direction." This gives a contravariant functor from the category of rings to the category of sets: the composition of two maps of rings induces the composition of the corresponding maps of spectra.

3.2.N. EASY EXERCISE (REALITY CHECK). Let B be a ring.

(a) Suppose $I \subset B$ is an ideal. Show that the map $\operatorname{Spec} B/I \to \operatorname{Spec} B$ is the inclusion of §3.2.8.
(b) Suppose $S \subset B$ is a multiplicative set. Show that the map $\operatorname{Spec} S^{-1}B \to \operatorname{Spec} B$ is the inclusion of §3.2.9.

3.2.11. *An explicit example.* In the case of "affine complex varieties" (or indeed affine varieties over any algebraically closed field), the translation between maps given by explicit formulas and maps of rings is quite direct. For example, consider a map from the parabola in $\mathbb{C}^2$ (with coordinates a and b) given by $b = a^2$, to the "curve" in $\mathbb{C}^3$ (with coordinates x, y, and z) cut out by the equations $y = x^2$ and $z = y^2$. Suppose the map sends the point $(a, b) \in \mathbb{C}^2$ to the point $(a, b, b^2) \in \mathbb{C}^3$. In our new language, we have a map

$$\operatorname{Spec} \mathbb{C}[a, b]/(b - a^2) \longrightarrow \operatorname{Spec} \mathbb{C}[x, y, z]/(y - x^2, z - y^2)$$

given by

$$\mathbb{C}[a, b]/(b - a^2) \longleftarrow \mathbb{C}[x, y, z]/(y - x^2, z - y^2),$$

$$(a, b, b^2) \longleftarrow (x, y, z),$$

i.e., $x \mapsto a$, $y \mapsto b$, and $z \mapsto b^2$. If the idea is not yet clear, the following two exercises are very much worth doing—they can be very confusing the first time you see them, and very enlightening (and trivial) when you finally figure them out.

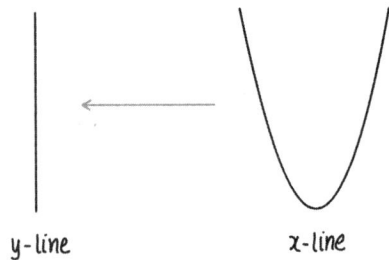

y-line   x-line

Figure 3.6 *The map $\mathbb{C} \to \mathbb{C}$ given by $x \mapsto y = x^2$.*

3.2.O. IMPORTANT EXERCISE (SPECIAL CASE). Consider the map of complex manifolds sending $\mathbb{C} \to \mathbb{C}$ via $x \mapsto y = x^2$. We interpret the "source" $\mathbb{C}$ as the "x-line," and the "target" $\mathbb{C}$ the "y-line." You can picture it as the projection of the parabola $y = x^2$ in the xy-plane to the y-axis (see Figure 3.6). Interpret the corresponding map of rings as given by $\mathbb{C}[y] \to \mathbb{C}[x]$ by $y \mapsto x^2$. Verify that the preimage (the fiber) above the point $a \in \mathbb{C}$ is the point(s) $\pm\sqrt{a} \in \mathbb{C}$, using the definition given above (identifying a with $[(y - a)]$, and $\sqrt{a}$ with $[(x - \sqrt{a})]$). (A more sophisticated version of this example appears in Example 10.3.4. Warning: The roles of x and y are swapped there, in order to picture double covers in a certain way.)

3.2.P. IMPORTANT EXERCISE (GENERALIZING EXAMPLE 3.2.11). Suppose k is a field, and $f_1, \ldots, f_n \in k[x_1, \ldots, x_m]$ are given. Let $\phi \colon k[y_1, \ldots, y_n] \to k[x_1, \ldots, x_m]$ be the morphism of k-algebras defined by $y_i \mapsto f_i$.

(a) Show that ϕ induces a map of sets $\operatorname{Spec} k[x_1, \ldots, x_m]/I \to \operatorname{Spec} k[y_1, \ldots, y_n]/J$ for any ideals $I \subset k[x_1, \ldots, x_m]$ and $J \subset k[y_1, \ldots, y_n]$ such that $\phi(J) \subset I$. (You may wish to consider the case $I = 0$ and $J = 0$ first. In fact, part (a) has nothing to do with k-algebras; you may wish to prove the statement when the rings $k[x_1, \ldots, x_m]$ and $k[y_1, \ldots, y_n]$ are replaced by general rings A and B.)
(b) Show that the map of part (a) sends the point $(a_1, \ldots, a_m) \in k^m$ (or more precisely, the point $[(x_1 - a_1, \ldots, x_m - a_m)] \in \operatorname{Spec} k[x_1, \ldots, x_m]$) to

$$(f_1(a_1, \ldots, a_m), \ldots, f_n(a_1, \ldots, a_m)) \in k^n.$$

3.2.Q. EXERCISE: PICTURING $\mathbb{A}^N_{\mathbb{Z}}$. Consider the map of sets $\pi\colon \mathbb{A}^n_{\mathbb{Z}} \to \operatorname{Spec}\mathbb{Z}$, given by the ring map $\mathbb{Z} \to \mathbb{Z}[x_1,\ldots,x_n]$. If p is prime, describe a bijection between the fiber $\pi^{-1}([(p)])$ and $\mathbb{A}^n_{\mathbb{F}_p}$. (You won't need to describe either set! Which is good because you can't.) This exercise may give you a sense of how to picture maps (see Figure 3.7), and in particular why you can think of $\mathbb{A}^n_{\mathbb{Z}}$ as an "$\mathbb{A}^n$-bundle" over $\operatorname{Spec}\mathbb{Z}$. (Can you interpret the fiber over $[(0)]$ as $\mathbb{A}^n_k$ for some field k?)

3.2.12. Functions are not determined by their values at points: the fault of nilpotents. We conclude this section by describing some strange behavior. We are developing machinery that will let us bring our geometric intuition to algebra. There is one serious serious point where your intuition will be false, so you should know now, and adjust your intuition appropriately. As noted by Mumford ([Mu2, p. 12]), "it is this aspect of schemes which was most scandalous when Grothendieck defined them."

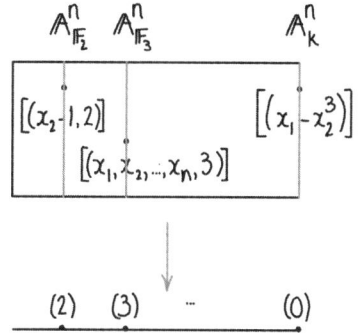

Figure 3.7 *A picture of $\mathbb{A}^n_{\mathbb{Z}} \to \operatorname{Spec}\mathbb{Z}$ as a "family of $\mathbb{A}^n$'s," or an "$\mathbb{A}^n$-bundle over $\operatorname{Spec}\mathbb{Z}$." What is the field k? How should you "geometrically" think of the three points indicated?*

Suppose we have a function (ring element) vanishing at all points. Then it is not necessarily the zero function! The translation of this question is: is the intersection of all prime ideals necessarily just 0? The answer is no, as is shown by the example of the ring of dual numbers $k[\epsilon]/(\epsilon^2)\colon \epsilon \neq 0$, but $\epsilon^2 = 0$. (We saw this ring in Exercise 3.2.A(a).) Any function whose power is zero certainly lies in the intersection of all prime ideals.

3.2.R. EXERCISE. Ring elements that have a power that is 0 are called **nilpotents**.

(a) Show that if I is an ideal of nilpotents, then the inclusion $\operatorname{Spec}B/I \to \operatorname{Spec}B$ of Exercise 3.2.J is a bijection. Thus nilpotents don't affect the underlying set. (We will soon see in §3.4.7 that they won't affect the topology either—the difference will be in the structure sheaf.)

(b) Show that the nilpotents of a ring B form an ideal. This ideal is called the **nilradical**, and is denoted by $\mathfrak{N} = \mathfrak{N}(B)$.

Thus the nilradical is contained in the intersection of all the prime ideals. The converse is also true:

3.2.13. Theorem — *The nilradical $\mathfrak{N}(A)$ is the intersection of all the prime ideals of A. Geometrically: a function on $\operatorname{Spec}A$ vanishes at every point if and only if it is nilpotent.*

3.2.S. EXERCISE. If you don't know this theorem, then look it up, or better yet, prove it yourself. (Hint: Use the fact that any proper ideal of A is contained in a maximal ideal, which requires Zorn's Lemma. Possible further hint: Suppose $x \notin \mathfrak{N}(A)$. We wish to show that there is a prime ideal not containing x. Show that A_x is not the 0-ring, by showing that $1 \neq 0$.)

3.2.14. In particular, although it is upsetting that functions are not determined by their values at points, we have precisely specified what the failure of this intuition is: two functions have the same values at points if and only if they differ by a nilpotent. You should think of this geometrically: a function vanishes at every point of the spectrum of a ring if and only if it has a power that is zero. And if there are no nonzero nilpotents—if $\mathfrak{N} = (0)$—then functions *are* determined by their values at points. If a ring has no nonzero nilpotents, we say that it is **reduced**.

3.2.T. FUN UNIMPORTANT EXERCISE: DERIVATIVES WITHOUT DELTAS AND EPSILONS (OR AT LEAST WITHOUT DELTAS). Suppose we have a polynomial $f(x) \in k[x]$. Instead, we work in $k[x,\epsilon]/(\epsilon^2)$. What then is $f(x+\epsilon)$? (Do a couple of examples, then prove the pattern you observe.) This is a hint that nilpotents will be important in defining differential information (Chapter 21).

3.3   Visualizing Schemes: Generic Points

A heavy warning used to be given that pictures are not rigorous; this has never had its bluff called and has permanently frightened its victims into playing for safety. Some pictures, of course, are not rigorous, but I should say most are (and I use them whenever possible myself).

— J. E. Littlewood [Lit, p. 54]

We all know that Art is not truth. Art is a lie that makes us realize truth, at least the truth that is given us to understand. The artist must know the manner whereby to convince others of the truthfulness of his lies.

—P. Picasso [Pi, p. 315]

For years, you have been able to picture $x^2 + y^2 = 1$ in the plane, and you now have an idea of how to picture $\operatorname{Spec}\mathbb{Z}$. If we are claiming to understand rings as geometric objects (through the Spec functor), then we should wish to develop geometric insight into them. To develop geometric intuition about schemes, it is helpful to have pictures in your mind, extending your intuition about geometric spaces you are already familiar with. As we go along, we will empirically develop some idea of what schemes should look like. This section summarizes what we have gleaned so far.

Some mathematicians prefer to think completely algebraically, and never think in terms of pictures. Others will be disturbed by the fact that this is an art, not a science. And finally, this hand-waving will necessarily never be used in the rigorous development of the theory. For these reasons, you may wish to skip these sections. However, having the right picture in your mind can greatly help in understanding what facts should be true, and how to prove them. Fitzgerald's exhortation on the importance of stretching one's mind is essential advice to a mathematician.

The test of a first-rate intelligence is the ability to hold two opposed ideas in the mind at the same time, and still retain the ability to function.

— F. Scott Fitzgerald, *The Crack-Up* [Fi, p. 41]

Our starting point is the example of "affine complex varieties" (things cut out by equations involving a finite number of variables over $\mathbb{C}$), and, more generally, similar examples over arbitrary algebraically closed fields. We begin with notions that are intuitive ("traditional" points behaving the way you expect them to), and then add in the two features that are new and disturbing, generic points and nonreduced behavior. You can then extend this notion to seemingly different spaces, such as $\operatorname{Spec}\mathbb{Z}$.

Hilbert's weak Nullstellensatz 3.2.5 shows that the "traditional points" are present as points of the scheme, and this carries over to any algebraically closed field. If the field is not algebraically closed, the traditional points are glued together into clumps by Galois conjugation, as in Examples 5 (the real affine line) and 6 (the affine line over $\mathbb{F}_p$) in §3.2. This is a geometric interpretation of Hilbert's Nullstellensatz 3.2.6.

But we have some additional points to add to the picture. You should remember that they "correspond" to "irreducible" "closed" (algebraic) subsets. As motivation, consider the case of the complex affine plane (Example 7): we had one for each irreducible polynomial, plus one corresponding to the entire plane. We will make "closed" precise when we define the Zariski topology (in the next section). You may already have an idea of what "irreducible" should mean; we make that precise at the start of §3.6. By "correspond" we mean that each closed irreducible subset has a corresponding point sitting on it, called its *generic point* (defined in §3.6). It is a new point, distinct from all the other points in the subset. (The correspondence is described in Exercise 3.7.F for

Spec A, and in Exercise 5.1.B for schemes in general.) We don't know precisely where to draw the generic point, so we may stick it arbitrarily anywhere, but you should think of it as being "almost everywhere," and, in particular, near every other point in the subset.

In §3.2.8, we saw how the points of Spec A/I should be interpreted as a subset of Spec A. So, for example, when you see Spec $\mathbb{C}[x,y]/(x+y)$, you should picture this not just as a line, but as a line in the xy-plane; the choice of generators x and y of the algebra $\mathbb{C}[x,y]$ implies an inclusion into affine space.

In §3.2.9, we saw how the points of Spec $S^{-1}A$ should be interpreted as subsets of Spec A. The two most important cases were discussed. The points of Spec A_f correspond to the points of Spec A where f doesn't vanish; we will later (§3.5) interpret this as a distinguished open set.

If $\mathfrak{p}$ is a prime ideal, then Spec $A_{\mathfrak{p}}$ should be seen as a "shred of the space Spec A near the subset corresponding to $\mathfrak{p}$." The simplest nontrivial case of this is $A = k[x]$ and $\mathfrak{p} = (x) \subset A$ (see Exercise 3.2.A, which we discuss again in Exercise 3.4.K).

> *"If any one of them can explain it," said Alice, (she had grown so large in the last few minutes that she wasn't a bit afraid of interrupting him), "I'll give him sixpence. I don't believe there's an atom of meaning in it."*
>
> *"If there's no meaning in it," said the King, "that saves a world of trouble, you know, as we needn't try to find any."*
>
> — Lewis Carroll [Carr, Ch. XII]

3.4   The Underlying Topological Space of an Affine Scheme

We next introduce the *Zariski topology* on the spectrum of a ring. When you first hear the definition, it seems odd, but with a little experience it becomes reasonable. As motivation, consider $\mathbb{A}^2_\mathbb{C} =$ Spec $\mathbb{C}[x,y]$, the complex (affine) plane (with a few extra points). In algebraic geometry, we will only be allowed to consider algebraic functions, i.e., polynomials in x and y. The locus where a polynomial vanishes should reasonably be a closed set, and the Zariski topology is defined by saying that the only sets we should consider closed should be these sets, and other sets forced to be closed by these. In other words, it is the coarsest topology where these sets are closed.

3.4.1. In particular, although topologies are often described using open subsets, it will be more convenient for us to define this topology in terms of closed subsets. If S is a subset of a ring A, define the **vanishing set** of S by

$$V(S) := \{[\mathfrak{p}] \in \text{Spec } A \ : \ S \subset \mathfrak{p}\}.$$

It is the set of points on which all elements of S are zero. (It should now be second nature to equate "vanishing at a point" with "contained in a prime.") We declare that these—and no others—are the closed subsets.

For example, consider $V(xy, yz) \subset \mathbb{A}^3_\mathbb{C} = \text{Spec } \mathbb{C}[x,y,z]$. Which points are contained in this locus? We think of this as solving $xy = yz = 0$. Of the "traditional" points (interpreted as ordered triples of complex numbers, thanks to the Hilbert's Nullstellensatz 3.2.5), we have the points where $y = 0$ or $x = z = 0$: the xz-plane and the y-axis respectively. Of the "new" points, we have the generic point of the xz-plane (also known as the point $[(y)]$), and the generic point of the y-axis (also known as the point $[(x,z)]$). You might imagine that we also have a number of "one-dimensional" points contained in the xz-plane.

3.4.A. EASY EXERCISE. Check that the x-axis is contained in $V(xy, yz)$. (The x-axis is defined by $y = z = 0$, and the y-axis and z-axis are defined analogously.)

Let's return to the general situation. The following exercise lets us restrict attention to vanishing sets of *ideals*.

3.4.B. EASY EXERCISE. Show that if (S) is the ideal generated by S, then $V(S) = V((S))$.

3.4.2. *Definition.* We define the **Zariski topology** by declaring that $V(S)$ is closed for all S. (We may as well state here that **Zariski closure** means closure in the Zariski topology.) Let's check that the Zariski topology is indeed a topology:

3.4.C. EXERCISE.

(a)   Show that $\varnothing$ and $\operatorname{Spec} A$ are both open subsets of $\operatorname{Spec} A$.

(b)   If I_i is a collection of ideals (as i runs over some index set), show that $\cap_i V(I_i) = V(\sum_i I_i)$. Hence the union of any collection of open sets is open.

(c)   Show that $V(I_1) \cup V(I_2) = V(I_1 I_2)$. (The **product of two ideals** I_1 and I_2 of A are finite A-linear combinations of products of elements of I_1 and I_2, i.e., elements of the form $\sum_{j=1}^{n} i_{1,j} i_{2,j}$, where $i_{k,j} \in I_k$. Equivalently, it is the ideal generated by products of elements of I_1 and I_2. You should quickly check that this is an ideal, and that products are associative, i.e., $(I_1 I_2) I_3 = I_1 (I_2 I_3)$.) Hence the intersection of any finite number of open sets is open.

3.4.3. Properties of the "vanishing set" function $V(\cdot)$. The function $V(\cdot)$ is obviously inclusion-reversing: If $S_1 \subset S_2$, then $V(S_2) \subset V(S_1)$. Warning: We could have equality in the second inclusion without equality in the first, as the next exercise shows.

3.4.D. EXERCISE/DEFINITION. If $I \subset A$ is an ideal, then define its **radical** by

$$\sqrt{I} := \{r \in A \ : \ r^n \in I \text{ for some } n \in \mathbb{Z}^{>0}\}.$$

For example, the nilradical $\mathfrak{N}$ (§3.2.R) is $\sqrt{(0)}$. Show that $\sqrt{I}$ is an ideal (cf. Exercise 3.2.R(b)). Show that $V(\sqrt{I}) = V(I)$. We say *an ideal is* **radical** if it equals its own radical. Show that $\sqrt{\sqrt{I}} = \sqrt{I}$, and that prime ideals are radical.

Here are two useful consequences. As $(I \cap J)^2 \subset IJ \subset I \cap J$ (products of ideals were defined in Exercise 3.4.C), we have that $V(IJ) = V(I \cap J)$ $(= V(I) \cup V(J)$ by Exercise 3.4.C(c)). Also, combining this with Exercise 3.4.B, we see $V(S) = V((S)) = V(\sqrt{(S)})$.

3.4.E. EXERCISE (RADICALS COMMUTE WITH FINITE INTERSECTIONS). If $I_1, \ldots, I_n$ are ideals of a ring A, show that $\sqrt{\cap_{i=1}^{n} I_i} = \cap_{i=1}^{n} \sqrt{I_i}$. We will use this property repeatedly without referring back to this exercise.

3.4.F. EXERCISE FOR LATER USE. Show that $\sqrt{I}$ is the intersection of all the prime ideals containing I. (Hint: Use Theorem 3.2.13 on an appropriate ring.)

3.4.4. Examples. Let's see how this meshes with our examples from the previous section.

Recall that $\mathbb{A}^1_{\mathbb{C}}$, as a set, was just the "traditional" points (corresponding to maximal ideals, in bijection with $a \in \mathbb{C}$), and one "new" point $[(0)]$. The Zariski topology on $\mathbb{A}^1_{\mathbb{C}}$ is not that exciting: the open sets are the empty set, and $\mathbb{A}^1_{\mathbb{C}}$ minus a finite number of maximal ideals. (It "almost" has the cofinite topology. Notice that the open sets are determined by their intersections with the "traditional points". The "new" point $[(0)]$ comes along for the ride, which is a good sign that it is harmless. Ignoring the "new" point, observe that the topology on $\mathbb{A}^1_{\mathbb{C}}$ is a coarser topology than the classical topology on $\mathbb{C}$.)

3.4.G. EXERCISE. Describe the topological space $\mathbb{A}^1_k$ (cf. Exercise 3.2.D). (Notice that the strange new point $[(0)]$ is "near every other point"—every neighborhood of every point contains $[(0)]$. This is typical of these new points; see Easy Exercise 3.6.N.)

The case of $\operatorname{Spec} \mathbb{Z}$ is similar. The topology is "almost" the cofinite topology in the same way. The open sets are the empty set, and $\operatorname{Spec} \mathbb{Z}$ minus a finite number of "ordinary" primes $((p)$, where p is prime).

3.4.5. *Closed subsets of* $\mathbb{A}^2_{\mathbb{C}}$. The case $\mathbb{A}^2_{\mathbb{C}}$ is more interesting. You should think through where the "one-dimensional prime ideals" fit into the picture. In Exercise 3.2.E, we identified all the prime ideals of $\mathbb{C}[x, y]$ (i.e., the points of $\mathbb{A}^2_\mathbb{C}$) as the maximal ideals $[(x - a, y - b)]$ (where $a, b \in \mathbb{C}$—"zero-dimensional points"), the "one-dimensional points" $[(f(x, y))]$ (where $f(x, y)$ is irreducible), and the "two-dimensional point" $[(0)]$.

Then the closed subsets are of the following form:

(a) the entire space (the closure of the "two-dimensional point" $[(0)]$),
(b) a finite number (possibly none) of "curves" (each the closure of a "one-dimensional point"—the "one-dimensional point" along with the "zero-dimensional points" "lying on it") and a finite number (possibly none) of "zero-dimensional" points (what we will soon call "closed points"; see Definition 3.6.8).

We will soon know enough to verify this using general theory, but you can prove it yourself now, using ideas in Exercise 3.2.E. (The key idea: if $f(x, y)$ and $g(x, y)$ are irreducible polynomials that are not multiples of each other, why do their zero sets intersect at a finite number of points?)

3.4.6. Important fact: Maps of rings induce continuous maps of topological spaces. We saw in §3.2.10 that a map of rings $\phi: B \to A$ induces a map of sets $\pi: \text{Spec } A \to \text{Spec } B$.

3.4.H. IMPORTANT EASY EXERCISE. By showing that closed sets pull back to closed sets, show that π is a *continuous* map. Interpret Spec as a contravariant functor *Rings* → *Top*.

Not all continuous maps arise in this way. Consider for example the continuous map on $\mathbb{A}^1_\mathbb{C}$ that is the identity except 0 and 1 (i.e., $[(x)]$ and $[(x - 1)]$) are swapped; no polynomial can manage this marvelous feat.

In §3.2.10, we saw that Spec B/I and Spec $S^{-1}B$ are naturally *subsets* of Spec B. It is natural to ask if the Zariski topology behaves well with respect to these inclusions, and indeed it does.

3.4.I. IMPORTANT EXERCISE (CF. EXERCISE 3.2.N). Suppose that $I, S \subset B$ are an ideal and a multiplicative subset respectively.

(a) Show that Spec B/I is naturally a *closed* subset of Spec B. If $S = \{1, f, f^2, \dots\}$ ($f \in B$), show that Spec $S^{-1}B$ is naturally an *open* subset of Spec B. Show that for arbitrary S, Spec $S^{-1}B$ need not be open or closed. (Hint: Spec $\mathbb{Q} \subset \text{Spec } \mathbb{Z}$, or possibly Figure 3.5.)
(b) Show that the Zariski topology on Spec B/I (resp., Spec $S^{-1}B$) is the subspace topology induced by inclusion in Spec B. (Hint: Compare closed subsets.)

3.4.7. In particular, if $I \subset \mathfrak{N}$ is an ideal of nilpotents, the bijection Spec $B/I \to \text{Spec } B$ (Exercise 3.2.R) is a homeomorphism. Thus nilpotents don't affect the topological space. (The difference will be in the structure sheaf.)

3.4.J. USEFUL EXERCISE FOR LATER. Suppose $I \subset B$ is an ideal. Show that f vanishes on $V(I)$ if and only if $f \in \sqrt{I}$ (i.e., $f^n \in I$ for some $n \geq 1$). (Hint: Exercise 3.4.F. If you are stuck, you will get another hint when you see Exercise 3.5.E.)

3.4.K. EASY EXERCISE (CF. EXERCISE 3.2.A). Describe the topological space Spec $k[x]_{(x)}$.

3.5   A Base of the Zariski Topology on Spec A: Distinguished Open Sets

If $f \in A$, define the **distinguished open set**

$$D(f) := \{[\mathfrak{p}] \in \text{Spec } A \,:\, f \notin \mathfrak{p}\}$$
$$= \{[\mathfrak{p}] \in \text{Spec } A \,:\, f([\mathfrak{p}]) \neq 0\}.$$

It is the locus where f doesn't vanish. (I often privately write this as $D(f \neq 0)$ to remind myself of this. I also privately call this the "**Doesn't-vanish set**" of f in analogy with $V(f)$ being the **V**anishing set of f.) We have already seen this set when discussing Spec A_f as a subset of Spec A. For example, we have observed that the Zariski topology on the distinguished open set $D(f) \subset \operatorname{Spec} A$ coincides with the Zariski topology on $\operatorname{Spec} A_f$ (Exercise 3.4.I).

The reason these sets are important is that they form a particularly nice base for the (Zariski) topology:

3.5.A. EASY EXERCISE. Show that the distinguished open sets form a base for the (Zariski) topology. (Hint: Given a subset $S \subset A$, show that the complement of $V(S)$ is $\cup_{f \in S} D(f)$.)

Here are some important but not difficult exercises to give you a feel for this concept.

3.5.B. NIFTY EXERCISE. Suppose $f_i \in A$ as i runs over some index set J. Show that $\cup_{i \in J} D(f_i) = \operatorname{Spec} A$ if and only if $(\{f_i\}_{i \in J}) = A$, or equivalently and very usefully, if there are a_i $(i \in J)$, all but finitely many 0, such that $\sum_{i \in J} a_i f_i = 1$. (One of the directions will use the fact that any proper ideal of A is contained in some maximal ideal.)

3.5.C. EXERCISE. Show that if $\operatorname{Spec} A$ is an infinite union of distinguished open sets $\cup_{j \in J} D(f_j)$, then in fact it is a union of a finite number of these, i.e., there is a finite subset J' so that $\operatorname{Spec} A = \cup_{j \in J'} D(f_j)$. (Hint: Exercise 3.5.B.)

3.5.D. EASY EXERCISE. Show that $D(f) \cap D(g) = D(fg)$.

3.5.E. IMPORTANT EXERCISE (CF. EXERCISE 3.4.J). Show that $D(f) \subset D(g)$ if and only if $f^n \in (g)$ for some $n \geq 1$, if and only if g is an invertible element of A_f.

We will use Exercise 3.5.E often. You can solve it thinking purely algebraically, but the following geometric interpretation may be helpful. (You should try to draw your own picture to go with this discussion.) Inside $\operatorname{Spec} A$, we have the closed subset $V(g) = \operatorname{Spec} A/(g)$, where g vanishes, and its complement $D(g)$, where g doesn't vanish. Then f is a function on this closed subset $V(g)$ (or more precisely, on $\operatorname{Spec} A/(g)$), and by assumption it vanishes at all points of the closed subset. Now any function vanishing at every point of the spectrum of a ring must be nilpotent (Theorem 3.2.13). In other words, there is some n such that $f^n = 0$ in $A/(g)$, i.e., $f^n \equiv 0 \pmod{g}$ in A, i.e., $f^n \in (g)$.

3.5.F. EASY EXERCISE. Show that $D(f) = \varnothing$ if and only if $f \in \mathfrak{N}$.

3.5.G. UNIMPORTANT EXERCISE (INJECTIVE MAPS OF RINGS INDUCE MAPS OF SPECTRA WITH DENSE IMAGE). Show that if $B \hookrightarrow A$, then the induced map of topological spaces $\operatorname{Spec} A \to \operatorname{Spec} B$ has dense image.

3.5.1. *Remark/caution about notation.* There will be two variations on the notation $D(f)$: the projective distinguished open set $D_+(f)$ (the locus on a projective scheme where a "form f of positive degree" doesn't vanish, §4.5.8), and X_f (the locus on a scheme X where a function f doesn't vanish, Definition 6.2.8). These are *almost* the same thing as $D(f)$, but not quite, so we keep the notation separate for reasons of hygiene.

3.6   Topological (and Noetherian) Properties

Many topological notions are useful when applied to the topological space $\operatorname{Spec} A$ and, later, to schemes.

3.6.1. Possible topological attributes of $\operatorname{Spec} A$: Connectedness, irreducibility, quasicompactness.

3.6.2. *Connectedness.*

A topological space X is **connected** if it cannot be written as the disjoint union of two nonempty open sets. Exercise 3.6.A below gives an example of a nonconnected $\operatorname{Spec} A$, and the subsequent remark explains that all examples are of this form.

3.6.A. EXERCISE. If $A = A_1 \times A_2 \times \cdots \times A_n$, describe a homeomorphism $\operatorname{Spec} A_1 \coprod \operatorname{Spec} A_2 \coprod \cdots \coprod \operatorname{Spec} A_n \to \operatorname{Spec} A$ for which each $\operatorname{Spec} A_i$ is mapped onto a distinguished open subset $D(f_i)$ of $\operatorname{Spec} A$. Thus $\operatorname{Spec} \prod_{i=1}^n A_i = \coprod_{i=1}^n \operatorname{Spec} A_i$ as topological spaces. (Hint: Reduce to $n = 2$ for convenience. Let $f_1 = (1, 0)$ and $f_2 = (0, 1)$.)

3.6.3. *Remark: The idempotent-connectedness package.* An extension of Exercise 3.6.A is that $\operatorname{Spec} A$ is not connected if and only if A is isomorphic to the product of nonzero rings A_1 and A_2. The key idea is to show that both conditions are equivalent to there existing nonzero $a_1, a_2 \in A$ for which $a_1^2 = a_1$, $a_2^2 = a_2$, $a_1 + a_2 = 1$, and hence $a_1 a_2 = 0$. An element $a \in A$ satisfying $a^2 = a$ is called an **idempotent**. This will appear as Exercise 10.5.L.

3.6.4. *Irreducibility.*

A topological space is said to be **irreducible** if it is nonempty, and it is not the union of two proper closed subsets. In other words, a nonempty topological space X is irreducible if whenever $X = Y \cup Z$ with Y and Z closed in X, we have $Y = X$ or $Z = X$. Equivalently (and helpfully): any two nonempty open subsets of X intersect. This is a less useful notion in classical geometry—$\mathbb{C}^2$ is *reducible* (i.e., not irreducible), but we will see that $\mathbb{A}_{\mathbb{C}}^2$ is irreducible (Exercise 3.6.C).

3.6.B. EASY EXERCISE.

(a) Show that in an irreducible topological space, any nonempty open set is dense. (For this reason, you will see that unlike in the classical topology, in the Zariski topology, nonempty open sets are all "huge.")

(b) If X is a topological space, and Z is a subset (with the subspace topology), show that Z is irreducible if and only if the closure $\overline{Z}$ in X is irreducible.

3.6.C. EASY EXERCISE. If A is an integral domain, show that $\operatorname{Spec} A$ is irreducible. (Hint: Pay attention to the generic point $[(0)]$.) We will generalize this in Exercise 3.7.G.

3.6.D. EXERCISE. Show that an irreducible topological space is connected.

3.6.E. EXERCISE. Give (with proof!) an example of a ring A where $\operatorname{Spec} A$ is connected but reducible. (Possible hint: A picture may help. The symbol "$\times$" has two "pieces" yet is connected.)

3.6.F. TRICKY EXERCISE.

(a) Suppose $I = (wz - xy, wy - x^2, xz - y^2) \subset k[w, x, y, z]$. Show that $\operatorname{Spec} k[w, x, y, z]/I$ is irreducible, by showing that $k[w, x, y, z]/I$ is an integral domain. (This is hard, so here is one of several possible hints: Show that $k[w, x, y, z]/I$ is isomorphic to the k-subalgebra of $k[a, b]$ generated by monomials of degree divisible by 3. There are other approaches as well, some of which we will see later. This is an example of a hard question: how do you tell if an ideal is prime?) We will later see this as the cone over the *twisted cubic curve* (the twisted cubic curve is defined in Exercise 9.3.A, and is a special case of a *Veronese embedding*, §9.3.6).

(b)* Note that the generators of the ideal of part (a) may be rewritten as the equations ensuring that

$$\operatorname{rank} \begin{pmatrix} w & x & y \\ x & y & z \end{pmatrix} \leq 1,$$

i.e., as the determinants of the 2×2 submatrices. Generalize part (a) to the ideal of rank one $2 \times n$ matrices. This notion will correspond to the cone (§3.3.11) over the *degree n rational normal curve* (Exercise 9.3.I).

3.6.5. *Quasicompactness.*

A topological space X is **quasicompact** if given any cover $X = \cup_{i \in I} U_i$ by open sets, there is a finite subset S of the index set I such that $X = \cup_{i \in S} U_i$. Informally: every open cover has a finite subcover. We will like this condition, because we are afraid of infinity. Depending on your definition of "compactness," this is the definition of compactness, minus possibly a Hausdorff condition. However, this isn't really the algebro-geometric analog of "compact" (we certainly wouldn't want $\mathbb{A}^1_{\mathbb{C}}$ to be compact)—the right analog is "properness" (§11.4).

3.6.G. EXERCISE.

(a) Show that Spec A is quasicompact. (Hint: Exercise 3.5.C.)
(b)* (Less important) Show that in general Spec A can have nonquasicompact open sets. Possible hint: Let $A = k[x_1, x_2, x_3, \dots]$ and $\mathfrak{m} = (x_1, x_2, \dots) \subset A$, and consider the complement of $V(\mathfrak{m})$. This example will be useful for constructing other "counterexamples" later, e.g., Exercises 8.1.F and 5.1.K. In Exercise 3.6.U, we will see that such weird behavior doesn't happen for "suitably nice" (Noetherian) rings.

3.6.H. EXERCISE.

(a) If X is a topological space that is a finite union of quasicompact spaces, show that X is quasicompact.
(b) Show that every closed subset of a quasicompact topological space is quasicompact.

3.6.6. ** *Fun but irrelevant remark.* Exercise 3.6.A shows that $\coprod_{i=1}^{n} \operatorname{Spec} A_i \cong \operatorname{Spec} \prod_{i=1}^{n} A_i$, but this *never* holds if "n is infinite" and all A_i are nonzero, as Spec of any ring is quasicompact (Exercise 3.6.G(a)). This leads to an interesting phenomenon. We show that $\operatorname{Spec} \prod_{i=1}^{\infty} A_i$ is "strictly bigger" than $\coprod_{i=1}^{\infty} \operatorname{Spec} A_i$, where each A_i is isomorphic to the field k. First, we have an inclusion of sets $\coprod_{i=1}^{\infty} \operatorname{Spec} A_i \hookrightarrow \operatorname{Spec} \prod_{i=1}^{\infty} A_i$, as there is a maximal ideal of $\prod A_i$ corresponding to each i (precisely, those elements that are 0 in the ith component.) But there are other maximal ideals of $\prod A_i$. Hint: Describe a proper ideal not contained in any of these maximal ideals. (One idea: consider elements $\prod a_i$ that are "eventually zero," i.e., $a_i = 0$ for $i \gg 0$.) This leads to the notion of *ultrafilters*, which are very useful, but irrelevant to our current discussion.

3.6.7. Possible topological properties of points of $\operatorname{Spec} A$.

3.6.8. *Definition.* A point of a topological space $p \in X$ is said to be a **closed point** if $\{p\}$ is a closed subset. In the classical topology on $\mathbb{C}^n$, all points are closed. In $\operatorname{Spec} \mathbb{Z}$ and $\operatorname{Spec} k[t]$, all the points are closed except for $[(0)]$.

3.6.I. EXERCISE. Show that the closed points of $\operatorname{Spec} A$ correspond to the maximal ideals. (In particular, nonempty affine schemes *have* closed points, as nonzero rings have maximal ideals, §0.3.)

3.6.9. *Connection to the classical theory of varieties.* Hilbert's Nullstellensatz lets us interpret the closed points of $\mathbb{A}^n_{\mathbb{C}}$ as the n-tuples of complex numbers. More generally, the closed points of $\operatorname{Spec} \overline{k}[x_1, \dots, x_n]/(f_1, \dots, f_r)$ (where $\overline{k}$ is an algebraically closed field) are naturally interpreted as those points in $\overline{k}^n$ satisfying the equations $f_1 = \cdots = f_r = 0$ (see Exercises 3.2.J and 3.2.N(a), for example). Hence from now on we will say "closed point" instead of "traditional point" and "non-closed point" instead of "bonus" point when discussing subsets of $\mathbb{A}^n_{\overline{k}}$. The following exercise is the reason that on algebraic varieties, the set of closed points is dense (cf. Exercise 5.3.F and Definition 11.2.9).

3.6.J. EXERCISE.

(a) Suppose that k is a field, and A is a finitely generated k-algebra. Show that the set of closed points of $\operatorname{Spec} A$ is dense, by showing that if $f \in A$, and $D(f)$ is a nonempty (distinguished) open subset of $\operatorname{Spec} A$, then $D(f)$ contains a closed point of $\operatorname{Spec} A$. Hint: Note that A_f is *also* a finitely generated k-algebra. Use the Nullstellensatz 3.2.6 to recognize closed points of Spec of a finitely generated k-algebra B as those for which the residue field (the quotient of A by the corresponding maximal ideal) is a finite extension of k. Apply this to both $B = A$ and $B = A_f$.

(b) Show that if A is a k-algebra that is not finitely generated, the set of closed points need not be dense. (Hint: Exercise 3.4.K.)

3.6.K. EXERCISE. Suppose k is an algebraically closed field, and $A = k[x_1, \ldots, x_n]/I$ is a finitely generated k-algebra with $\mathfrak{N}(A) = \{0\}$ (so the discussion of §3.2.14 applies). Consider the set $X = \operatorname{Spec} A$ as a subset of $\mathbb{A}_k^n$. The space $\mathbb{A}_k^n$ contains the "classical" points k^n. Show that functions on X are determined by their values on the closed points (by the weak Nullstellensatz 3.2.5, the "classical" points $k^n \cap \operatorname{Spec} A$ of $\operatorname{Spec} A$). Hint: If f and g are different functions on X, then $f - g$ is nowhere zero on an open subset of X. Use Exercise 3.6.J(a).

Once we know what a variety is (Definition 11.2.**9**), this will immediately imply that *a function on a variety over an algebraically closed field is determined by its values on the "classical points."* (Before the advent of scheme theory, functions on varieties—over algebraically closed fields—were thought of as functions on "classical" points, and Exercise 3.6.K basically shows that there is no harm in thinking of "traditional" varieties as a particular flavor of schemes.)

3.6.10. *Specialization and generization.* Given two points x, y of a topological space X, we say that x is a **specialization** of y, and y is a **generization** of x, if $x \in \overline{\{y\}}$. This (and Exercise 3.6.L) now makes precise our hand-waving about "one point containing another." It is of course nonsense for a point to contain another. But it is not nonsense to say that the closure of a point contains another. For example, in $\mathbb{A}_{\mathbb{C}}^2 = \operatorname{Spec} \mathbb{C}[x, y]$, $[(y - x^2)]$ is a generization of $[(x - 2, y - 4)] = (2, 4) \in \mathbb{C}^2$, and $(2, 4)$ is a specialization of $[(y - x^2)]$ (see Figure 3.8).

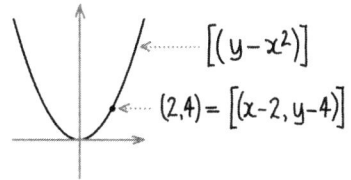

Figure 3.8 $(2, 4) = [(x - 2, y - 4)]$ *is a specialization of* $[(y - x^2)]$. $[(y - x^2)]$ *is a generization of* $(2, 4)$.

3.6.L. EXERCISE. If $X = \operatorname{Spec} A$, show that $[\mathfrak{q}]$ is a specialization of $[\mathfrak{p}]$ if and only if $\mathfrak{p} \subset \mathfrak{q}$. Hence show that $V(\mathfrak{p}) = \overline{\{[\mathfrak{p}]\}}$.

3.6.11. *Definition.* We say that a point $p \in X$ is a **generic point** for a closed subset K if $\overline{\{p\}} = K$.

This important notion predates Grothendieck. The early twentieth-century Italian algebraic geometers had a notion of "generic points" of a variety, by which they meant points with no special properties, so that anything proved of "a generic point" was true of "almost all" the points on that variety. The modern "generic point" has the same intuitive meaning. If something is "generically" or "mostly" true for the points of an irreducible subset, in the sense of being true for a dense open subset (for "almost all points"), then it is true for the generic point, and vice versa. (This is a statement of principle, not of fact. An interesting case is "reducedness," for which this principle does not hold in general, but *does* hold for "reasonable" schemes such as varieties; see Remark 5.2.2.) For example, a function has value zero at the generic point of an irreducible scheme if and only if it has the value zero at all points. You should keep an eye out for other examples of this.

The phrase **general point** does not have the same meaning. The phrase "the general point of X satisfies such-and-such a property" means "every point of some dense open subset of X satisfies such-and-such a property." Be careful not to confuse "general" and "generic." But be warned that accepted terminology does not always follow this convention; witness "generic smoothness" (§21.6.3, for example) and "generic flatness" (Theorem 24.5.13).

3.6.M. EXERCISE. Verify that $[(y - x^2)] \in \mathbb{A}_{\mathbb{C}}^2$ is a generic point for $V(y - x^2)$.

As more motivation for the terminology "generic": we think of $[(y - x^2)]$ as being some non-specific point on the parabola (with the closed points $(a, a^2) \in \mathbb{C}^2$, i.e., $(x - a, y - a^2)$ for $a \in \mathbb{C}$, being "specific points"); it is "generic" in the conventional sense of the word. We might "special-ize it" to a specific point of the parabola; hence, for example, $(2,4)$ is a specialization of $[(y-x^2)]$. (Again, see Figure 3.8.) To make this somewhat more precise:

3.6.N. EASY EXERCISE. Suppose p is a generic point for the closed subset K. Show that it is "near every point q of K" (every neighborhood of q contains p), and "not near any point r not in K" (there is a neighborhood of r not containing p). (This idea was mentioned in §3.2.4 Example 1 and in Exercise 3.4.G.)

We will soon see (Exercise 3.7.F) that there is a natural bijection between points of $\operatorname{Spec} A$ and irreducible closed subsets of $\operatorname{Spec} A$, sending each point to its closure, and each irreducible closed subset to its (unique) generic point. You can prove this now, but we will wait until we have developed some convenient terminology.

3.6.12. Irreducible and connected components, and Noetherian conditions.
An **irreducible component** of a topological space is a maximal irreducible subset (an irreducible subset not contained in any larger irreducible subset). Irreducible components are closed (as the closure of irreducible subsets are irreducible, Exercise 3.6.B(b)), and it can be helpful to think of irreducible components of a topological space X as maximal among the irreducible *closed* subsets of X. We think of these as the "pieces of X" (see Figure 3.9).

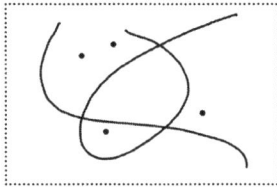

Figure 3.9 *This closed subset of $\mathbb{A}^2_{\mathbb{C}}$ has six irreducible components.*

Similarly, a subset Y of a topological space X is a **connected component** if it is a maximal connected subset (a connected subset not contained in any larger connected subset).

3.6.O. EXERCISE (EVERY TOPOLOGICAL SPACE IS THE UNION OF IRREDUCIBLE COMPONENTS). Show that every point p of a topological space X is contained in an irreducible component of X. Hint: Zorn's Lemma. More precisely, consider the partially ordered set $\mathscr{S}$ of irreducible subsets of X containing p. Show that there exists a maximal totally ordered subset $\{Z_\alpha\}$ of $\mathscr{S}$. Show that $\cup Z_\alpha$ is irreducible.

3.6.13. *Remark.* Every point is contained in a connected component, and connected components are always closed. You can prove this now, but we deliberately postpone this until we need it, in an optional starred section (Exercise 10.5.J). On the other hand, connected components need not be open; see [Stacks, tag 004T]. An example of an affine scheme with connected components that are not open is $\operatorname{Spec}(\prod_1^\infty \mathbb{F}_2)$.

3.6.14. In the examples we have considered, the spaces have naturally broken up into a finite number of irreducible components. For example, the locus $xy=0$ in $\mathbb{A}^2_{\mathbb{C}}$ we think of as having two "pieces"—the two axes. The reason for this is that their underlying topological spaces (as we shall soon establish) are *Noetherian*. A topological space X is called **Noetherian** if it satisfies the **descending chain condition** for closed subsets: any sequence $Z_1 \supset Z_2 \supset \cdots \supset Z_n \supset \cdots$ of closed subsets eventually stabilizes—there is an r such that $Z_r = Z_{r+1} = \cdots$. Here is a first example (which you should work out explicitly, not using Noetherian rings).

3.6.P. EXERCISE. Show that $\mathbb{A}^2_{\mathbb{C}}$ is a Noetherian topological space: any decreasing sequence of closed subsets of $\mathbb{A}^2_{\mathbb{C}} = \operatorname{Spec}\mathbb{C}[x,y]$ must eventually stabilize. Note that it can take arbitrarily long to stabilize. (The closed subsets of $\mathbb{A}^2_{\mathbb{C}}$ were described in §3.4.5.) Show that $\mathbb{C}^2$ with the classical topology is *not* a Noetherian topological space.

3.6.15. Proposition — *Suppose X is a Noetherian topological space. Then every nonempty closed subset Z can be expressed uniquely as a finite union $Z = Z_1 \cup \cdots \cup Z_n$ of irreducible closed subsets, none contained in any other.*

Translation: Any closed subset Z has a finite number of "pieces."

Proof. The following technique is called **Noetherian induction**, for reasons that will be clear. We will use it again, many times.

Consider the collection of nonempty closed subsets of X that *cannot* be expressed as a finite union of irreducible closed subsets. We will show that this collection is empty. Otherwise, let Y_1 be one such. If Y_1 properly contains another such, then choose one, and call it Y_2. If Y_2 properly contains another such, then choose one, and call it Y_3, and so on. By the descending chain condition, this must eventually stop, and we must have some Y_r that cannot be written as a finite union of irreducible closed subsets, but every closed subset properly contained in it can be so written. But then Y_r is not itself irreducible, so we can write $Y_r = Y' \cup Y''$, where Y' and Y'' are both proper closed subsets. Both of these by hypothesis can be written as the union of a finite number of irreducible closed subsets, and hence so can Y_r, yielding a contradiction. Thus each closed subset can be written as a finite union of irreducible closed subsets. We can assume that none of these irreducible closed subsets contain any others, by discarding some of them.

We now show uniqueness. Suppose

$$Z = Z_1 \cup Z_2 \cup \cdots \cup Z_r = Z_1' \cup Z_2' \cup \cdots \cup Z_s'$$

are two such representations. Then $Z_1' \subset Z_1 \cup Z_2 \cup \cdots \cup Z_r$, so $Z_1' = (Z_1 \cap Z_1') \cup \cdots \cup (Z_r \cap Z_1')$. Now Z_1' is irreducible, so one of these is Z_1' itself, say (without loss of generality) $Z_1 \cap Z_1'$. Thus $Z_1' \subset Z_1$. Similarly, $Z_1 \subset Z_a'$ for some a; but because $Z_1' \subset Z_1 \subset Z_a'$, and Z_1' is contained in no other Z_i', we must have $a = 1$, and $Z_1' = Z_1$. Thus each element of the list of Z's is in the list of Z''s, and vice versa, so they must be the same list. □

3.6.Q. EXERCISE. Show that every connected component of a topological space X is the union of irreducible components of X. Show that any subset of X that is simultaneously open and closed must be the union of some of the connected components of X. If X is a *Noetherian* topological space, show that the union of any subset of the connected components of X is always open and closed in X. (In particular, connected components of Noetherian topological spaces are always open, which is not true for more general topological spaces; see Remark 3.6.13.)

3.6.16. Noetherian rings. It turns out that all of the spectra we have considered (except in starred Exercise 3.6.G(b)) are Noetherian topological spaces, but that isn't true of the spectra of all rings. The key characteristic all of our examples have had in common is that the rings were *Noetherian*. A ring is **Noetherian** if every ascending sequence $I_1 \subset I_2 \subset \cdots$ of ideals eventually stabilizes: there is an r such that $I_r = I_{r+1} = \cdots$. (This is called the **ascending chain condition** on ideals.)

Here are some quick facts about Noetherian rings. You should be able to prove them all.

- Fields are Noetherian. $\mathbb{Z}$ is Noetherian.
- (*Noetherianness is preserved by quotients.*) If A is Noetherian, and I is any ideal of A, then A/I is Noetherian.
- (*Noetherianness is preserved by localization.*) If A is Noetherian, and S is any multiplicative set, then $S^{-1}A$ is Noetherian.

3.6.R. EXERCISE. Show that principal ideal domains are Noetherian rings.

3.6.S. IMPORTANT EXERCISE. Show that a ring A is Noetherian if and only if every ideal of A is finitely generated.

The next fact is nontrivial.

3.6.17. The Hilbert Basis Theorem — *If A is Noetherian, then so is A[x].*

Hilbert proved this in the epochal paper [Hil] where he also proved the Hilbert Syzygy Theorem (§16.1.2), and defined Hilbert functions and showed that they are eventually polynomial (§18.6).

3.6.18. By the results described above, any polynomial ring in finitely many variables over any field, or over the integers, is Noetherian—and also any quotient or localization thereof. Hence, for example, any finitely generated algebra over k or $\mathbb{Z}$, or any localization thereof, is Noetherian. Most "nice" rings are Noetherian, but not all rings are Noetherian: $k[x_1, x_2, \ldots]$ is not, because $(x_1) \subset (x_1, x_2) \subset (x_1, x_2, x_3) \subset \cdots$ is a strictly ascending chain of ideals (cf. Exercise 3.6.G(b)).

3.6.19. *Proof of the Hilbert Basis Theorem 3.6.17.* We show that any ideal $I \subset A[x]$ is finitely generated. We inductively produce a set of generators $f_1, \ldots$, as follows. For $n \geq 0$, if $I \neq (f_1, \ldots, f_n)$, let f_{n+1} be any nonzero element of $I - (f_1, \ldots, f_n)$ of lowest degree. Thus f_1 is any element of I of lowest degree, assuming $I \neq (0)$. If this procedure terminates, we are done. Otherwise, let $a_n \in A$ be the initial coefficient of f_n for all $n > 0$. As A is Noetherian, $(a_1, a_2, \ldots) = (a_1, \ldots, a_N)$ for some N. Say $a_{N+1} = \sum_{i=1}^N b_i a_i$. Then

$$f_{N+1} - \sum_{i=1}^N b_i f_i x^{\deg f_{N+1} - \deg f_i}$$

is an element of I that is not in $(f_1, \ldots, f_N)$ (as $f_{N+1} \notin (f_1, \ldots, f_N)$), and of lower degree than f_{N+1}, contradicting how we were supposed to have chosen f_{N+1}. $\qquad\square$

We now connect Noetherian rings and Noetherian topological spaces.

3.6.T. EXERCISE. If A is Noetherian, show that Spec A is a Noetherian topological space. Describe a ring A such that Spec A is not a Noetherian topological space. (Aside: If Spec A is a Noetherian topological space, A need not be Noetherian. One example is $A = k[x_1, x_2, x_3, \ldots]/(x_1, x_2^2, x_3^3, \ldots)$. Then Spec A has one point, so is Noetherian. But A is not Noetherian as $(x_1) \subsetneq (x_1, x_2) \subsetneq (x_1, x_2, x_3) \subsetneq \cdots$ in A.)

3.6.U. EXERCISE (PROMISED IN EXERCISE 3.6.G(B)). Show that every open subset of a Noetherian topological space is quasicompact. Hence if A is Noetherian, every open subset of Spec A is quasicompact. (If you prefer, show the result for *any* subset, with the induced topology.)

3.6.20. For future use: Noetherian conditions for modules. An important related notion is that of a Noetherian *module*. Although we won't use this notion for some time (§10.7.3), we will develop their most important properties, while Noetherian ideas are still fresh in your mind. If A is any ring, not necessarily Noetherian, we say an A-**module is Noetherian** if it satisfies the ascending chain condition for submodules. Thus, for example, a ring A is Noetherian if and only if it is a Noetherian A-module.

3.6.V. EXERCISE (CF. IMPORTANT EXERCISE 3.6.S). Show that if M is a Noetherian A-module, then any submodule of M is a finitely generated A-module. (In fact the converse is true, and you are welcome to prove that too.)

3.6.W. EXERCISE. If $0 \to M' \to M \to M'' \to 0$ is exact, show that M' and M'' are Noetherian if and only if M is Noetherian. (Hint: Given an ascending chain in M, we get two simultaneous ascending chains in M' and M''. Possible further hint: Prove that if

$$M' \longrightarrow M \xrightarrow{\phi} M''$$

is exact, and $N \subset N' \subset M$, and $N \cap M' = N' \cap M'$ and $\phi(N) = \phi(N')$, then $N = N'$.)

3.6.X. EXERCISE. Show that if A is a Noetherian ring, then $A^{\oplus n}$ is a Noetherian A-module.

3.6.Y. EXERCISE. Show that if A is a Noetherian ring and M is a finitely generated A-module, then M is a Noetherian module. Hence by Exercise 3.6.V, any submodule of a finitely generated module over a Noetherian ring is finitely generated.

3.6.21. Why you should not worry about Noetherian hypotheses. Should you work hard to eliminate Noetherian hypotheses? Should you worry about Noetherian hypotheses? Should you stay up at night thinking about non-Noetherian rings? For the most part, the answer to these questions is "no." Most people will never need to worry about non-Noetherian rings, but there are reasons to be open to them. First, they can actually come up. For example, fibered products of Noetherian schemes over Noetherian schemes (and even fibered products of Noetherian points over Noetherian points!) can be non-Noetherian (Warning 10.1.5), and the normalization of Noetherian rings can be non-Noetherian (Warning 10.7.4). You can either work hard to show that the rings or schemes you care about don't have this pathology, or you can just relax and not worry about it. Second, there is often no harm in working with schemes in general. Knowing when Noetherian conditions are needed will help you remember why results are true, because you will have some sense of where Noetherian conditions enter into arguments. Finally, for some people, non-Noetherian rings naturally come up. For example, adeles are not Noetherian. And many valuation rings that naturally arise in arithmetic and tropical geometry are not Noetherian.

3.7   The Function I(·), Taking Subsets of Spec A to Ideals of A

We now introduce a notion that is in some sense "inverse" to the vanishing set function V(·). Given a subset $S \subset \operatorname{Spec} A$, I(S) is the set of functions vanishing on S. In other words, $I(S) = \bigcap_{[\mathfrak{p}] \in S} \mathfrak{p} \subset A$ (at least when S is nonempty).

We make three quick observations. (Do you see why they are true?)

- I(S) is clearly an ideal of A.
- I(·) is inclusion-reversing: if $S_1 \subset S_2$, then $I(S_2) \subset I(S_1)$.
- $I(\overline{S}) = I(S)$.

3.7.A. EXERCISE. Let $A = k[x, y]$. If $S = \{[(y)], [(x, y - 1)]\}$ (see Figure 3.10), then I(S) consists of those polynomials vanishing on the x-axis, and at the point $(0, 1)$. Give generators for this ideal.

3.7.B. EXERCISE. Suppose $S \subset \mathbb{A}_{\mathbb{C}}^3$ is the union of the three axes. Give generators for the ideal I(S). Be sure to prove it! We will see in Exercise 13.1.F that this ideal is not generated by less than three elements.

$\bullet\ (0,1) = [(x, y-1)]$

3.7.C. EXERCISE. Show that $V(I(S)) = \overline{S}$. Hence $V(I(S)) = S$ for a closed set S.

$[(y)]$

Note that I(S) is always a radical ideal—if $f \in \sqrt{I(S)}$, then f^n vanishes on S for some $n > 0$, so then f vanishes on S, so $f \in I(S)$.

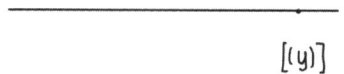

Figure 3.10 *The set S of Exercise 3.7.A, pictured as a subset of $\mathbb{A}^2$.*

3.7.D. EASY EXERCISE. Prove that if $J \subset A$ is an ideal, then $I(V(J)) = \sqrt{J}$. (Huge hint: Exercise 3.4.J.)

Exercises 3.7.C and 3.7.D show that V and I are "almost" inverse. More precisely:

3.7.1. Theorem — $V(·)$ and $I(·)$ give an inclusion-reversing bijection between closed subsets of Spec A and radical ideals of A (where a closed subset gives a radical ideal by $I(·)$, and a radical ideal gives a closed subset by $V(·)$).

Theorem 3.7.1 is sometimes called Hilbert's Nullstellensatz, but we reserve that name for Theorem 3.2.6.

3.7.E. EXERCISE. Let $J = (x^2 + y^2 - 1, y - 1)$. Find, with proof, an element of $I(V(J)) \setminus J$.

3.7.F. IMPORTANT EXERCISE (CF. EXERCISE 3.7.G). Show that $V(·)$ and $I(·)$ give a bijection between *irreducible closed subsets* of Spec A and *prime* ideals of A. From this conclude that in Spec A there is a bijection between points of Spec A and irreducible closed subsets of Spec A (where a point determines an irreducible closed subset by taking the closure). Hence *each irreducible closed subset* *of* Spec A *has precisely one generic point*—any irreducible closed subset Z can be written uniquely as $\overline{\{z\}}$.

3.7.G. EXERCISE/DEFINITION. A prime ideal of a ring A is a **minimal prime ideal** (or, more simply, **minimal prime**) if it is minimal with respect to inclusion. (For example, the only minimal prime ideal of $k[x, y]$ is (0).) If A is any ring, show that the irreducible components of Spec A are in bijection with the minimal prime ideals of A. In particular, Spec A is irreducible if and only if A has only one minimal prime ideal; this generalizes Exercise 3.6.C.

Proposition 3.6.15, Exercise 3.6.T, and Exercise 3.7.G imply that every Noetherian ring has a finite number of minimal prime ideals. An algebraic fact is now revealed to be really a "geometric" fact!

3.7.H. EXERCISE (REALITY CHECK). In $\mathbb{A}_k^n = \operatorname{Spec} k[x_1, \dots, x_n]$, the subset cut out by $f(x_1, \dots, x_n) \in k[x_1, \dots, x_n]$ should certainly have irreducible components corresponding to the distinct irreducible factors of f. Prove this. (What property of the ring $k[x_1, \dots, x_n]$ makes this work?)

3.7.I. EXERCISE. What are the minimal prime ideals of $k[x, y]/(xy)$ (where k is a field)?

3.7.2. Beginning of a grand dictionary between algebra and geometry. We are now well on our away to building a grand dictionary between algebra and geometry. You should add more entries to the table as your understanding deepens.

algebra	geometry	reference
ring A	affine scheme Spec A	§3.2.1
$\mathfrak{p} = I(p)$ prime ideal of A	$p = [\mathfrak{p}]$ point of Spec A	§3.2.1, §3.7
element $f \in A$	function f on Spec A	§3.2.1
$f \pmod{\mathfrak{p}}$	$f(p)$; value of the function f at p	§3.2.1
$f \in \mathfrak{p}$	the function f vanishes (is zero) at p	§3.2.1
nilradical ideal $\mathfrak{N} = \sqrt{(0)} = \cap \mathfrak{p}$ of A	functions vanishing at every point	Thm. 3.2.13
maximal ideal of A	irreducible closed subset of Spec A that is a point (= closed point of Spec A)	Ex. 3.6.I
$\mathfrak{p} = I(Z)$ prime ideal of A (a repeat!)	irreducible closed subset $Z = V(\mathfrak{p})$ of Spec A	Ex. 3.7.F
minimal prime ideal of A	irreducible component of Spec A	Ex. 3.7.G
radical ideal $I = \sqrt{I} = I(S)$ of A	closed subset $S = V(I)$ of Spec A	Thm. 3.7.1

Chapter 4

The Structure Sheaf, and the Definition of Schemes in General

The final ingredient in the definition of an affine scheme is the *structure sheaf* $\mathcal{O}_{\mathrm{Spec}\,A}$, which we think of as the "sheaf of algebraic functions." You should keep in your mind the example of "algebraic functions" on $\mathbb{C}^n$, which you understand well. For example, for the plane $\mathbb{A}^2$, we expect that on the open set $D(xy)$ (away from the two axes), $(3x^4 + y + 4)/x^7 y^3$ should be an algebraic function.

These functions will have values at points, but won't be determined by their values at points. But like all sections of sheaves, they will be determined by their germs (see §4.3.6).

4.1   The Structure Sheaf of an Affine Scheme

We want now to give a ring of functions on *any* open set, but we know that it suffices to describe the functions on a base of the topology. We will see that the base of distinguished open sets will make things much more tractable than you might fear. So we now describe the structure sheaf as a sheaf (of rings) on the base of distinguished open sets (Theorem 2.5.1 and Exercise 3.5.A).

4.1.1. *Definition.* Define $\mathcal{O}_{\mathrm{Spec}\,A}(D(f))$ to be the localization of A at the multiplicative set of all functions that do not vanish outside of $V(f)$ (i.e., those $g \in A$ such that $V(g) \subset V(f)$, or equivalently $D(f) \subset D(g)$; cf. Exercise 3.5.E). This depends only on $D(f)$, and not on f itself. (Scholars of the empty set might notice that by Exercise 3.5.F, we have that $\mathcal{O}_{\mathrm{Spec}\,A}(\varnothing) = \{0\}$.)

4.1.A. GREAT EXERCISE. Show that the natural map $A_f \to \mathcal{O}_{\mathrm{Spec}\,A}(D(f))$ is an isomorphism. (Possible hint: Exercise 3.5.E.)

If $D(f') \subset D(f)$, define the restriction map

$$\mathrm{res}_{D(f),D(f')} \colon \mathcal{O}_{\mathrm{Spec}\,A}(D(f)) \longrightarrow \mathcal{O}_{\mathrm{Spec}\,A}(D(f'))$$

in the obvious way: the latter ring is a further localization of the former ring. The restriction maps obviously compose: this is a "presheaf on the distinguished base."

4.1.2. Theorem — *The data just described give a sheaf on the distinguished base, and hence determine a sheaf on the topological space* $\mathrm{Spec}\,A$.

4.1.3. This sheaf is called the **structure sheaf**, and will be denoted by $\mathcal{O}_{\mathrm{Spec}\,A}$, or sometimes $\mathcal{O}$ if the subscript is clear from the context. The notation $\mathrm{Spec}\,A$ will hereafter denote the data of a topological space with a structure sheaf. Such a topological space, with a structure sheaf, will be called an **affine scheme** (Definition 4.3.1). We continue to use the language of ringed spaces (§2.2.13): functions on open sets, global functions, and so forth. An important lesson of Theorem 4.1.2 is not just that $\mathcal{O}_{\mathrm{Spec}\,A}$ is a sheaf, but also that the distinguished base provides a good way of working with $\mathcal{O}_{\mathrm{Spec}\,A}$. Notice also that we have justified interpreting elements of A as functions on $\mathrm{Spec}\,A$.

Proof. We must show that the base identity and base gluability axioms hold (§2.5). We show that they both hold for the open set that is the entire space $\mathrm{Spec}\,A$, and leave to you the trick that extends them to arbitrary distinguished open sets (Exercises 4.1.B and 4.1.C). Suppose $\mathrm{Spec}\,A = \cup_{i \in I} D(f_i)$, or, equivalently (Exercise 3.5.B), the ideal generated by the f_i is the entire ring A.

(Aside: Experts familiar with the equalizer exact sequence of §2.2.7 will realize that we are showing exactness of

$$(4.1.3.1) \qquad 0 \longrightarrow A \longrightarrow \prod_{i \in I} A_{f_i} \longrightarrow \prod_{i \neq j \in I} A_{f_i f_j},$$

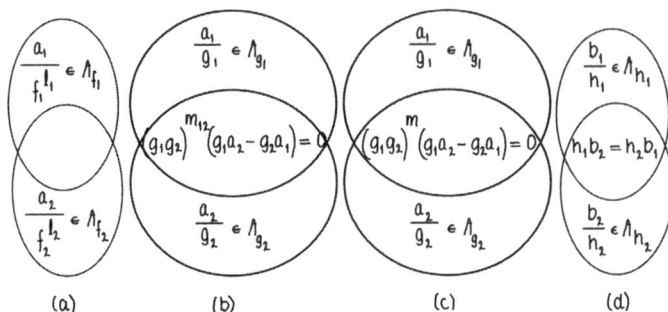

Figure 4.1 *Base gluability of the structure sheaf.*

where $\{f_i\}_{i \in I}$ is a set of functions with $(f_i)_{i \in I} = A$. Signs are involved in the right-hand map: the map $A_{f_i} \to A_{f_i f_j}$ is the localization map, and the map $A_{f_j} \to A_{f_i f_j}$ is the *negative* of the localization map. Base identity corresponds to injectivity at A, and gluability corresponds to exactness at $\prod_i A_{f_i}$.)

We check identity on the base. Suppose that $\mathrm{Spec}\, A = \cup_{i \in I} D(f_i)$, where i runs over some index set I. Then there is some finite subset of I, which we name $\{1, \ldots, n\}$, such that $\mathrm{Spec}\, A = \cup_{i=1}^n D(f_i)$, i.e., $(f_1, \ldots, f_n) = A$ (Exercise 3.5.C, a special case of quasicompactness of $\mathrm{Spec}\, A$, Exercise 3.6.G(a)). Suppose we are given $s \in A$ such that $\mathrm{res}_{\mathrm{Spec}\, A, D(f_i)}\, s = 0$ in A_{f_i} for all i. We wish to show that $s = 0$. The fact that $\mathrm{res}_{\mathrm{Spec}\, A, D(f_i)}\, s = 0$ in A_{f_i} implies that there is some m such that for each $i \in \{1, \ldots, n\}$, $f_i^m s = 0$. Now $(f_1^m, \ldots, f_n^m) = A$ (for example, from $\mathrm{Spec}\, A = \cup D(f_i) = \cup D(f_i^m)$), so there are $r_i \in A$ with $\sum_{i=1}^n r_i f_i^m = 1$ in A, from which

$$s = \left(\sum r_i f_i^m \right) s = \sum r_i (f_i^m s) = 0.$$

Thus we have checked the "base identity" axiom for $\mathrm{Spec}\, A$.

4.1.B. EXERCISE. Make tiny changes to the above argument to show that base identity holds for any distinguished open $D(f)$. (Hint: Judiciously replace A by A_f in the above argument.)

We next show base gluability. (Serre has described this as a "partition of unity" argument, and if you look at it in the right way, his insight is very enlightening.) Suppose again $\cup_{i \in I} D(f_i) = \mathrm{Spec}\, A$, where I is an index set (possibly horribly infinite). Suppose we are given elements in each A_{f_i} that agree on the overlaps $A_{f_i f_j}$. Note that intersections of distinguished open sets are also distinguished open sets.

Assume first that I is finite, say, $I = \{1, \ldots, n\}$. We have elements $a_i / f_i^{l_i} \in A_{f_i}$ agreeing on overlaps $A_{f_i f_j}$ (see Figure 4.1(a)). Letting $g_i = f_i^{l_i}$, using $D(f_i) = D(g_i)$, we can simplify notation by considering our elements as of the form $a_i / g_i \in A_{g_i}$ (Figure 4.1(b)).

The fact that a_i / g_i and a_j / g_j "agree on the overlap" (i.e., in $A_{g_i g_j}$) means that for some m_{ij},

$$(g_i g_j)^{m_{ij}} (g_j a_i - g_i a_j) = 0$$

in A. By taking $m = \max m_{ij}$ (here we use the finiteness of I), we can simplify notation:

$$(g_i g_j)^m (g_j a_i - g_i a_j) = 0$$

for all i, j (Figure 4.1(c)). Let $b_i = a_i g_i^m$ for all i, and $h_i = g_i^{m+1}$ (so $D(h_i) = D(g_i)$). Then we can simplify notation even more (Figure 4.1(d)): on each $D(h_i)$, we have a function b_i / h_i, and the overlap condition is

(4.1.3.2) $$h_j b_i = h_i b_j.$$

Now $\cup_i D(h_i) = \mathrm{Spec}\, A$, implying that $1 = \sum_{i=1}^n r_i h_i$ for some $r_i \in A$. Define

(4.1.3.3) $$r = \sum r_i b_i.$$

This will be the element of A that restricts to each b_j/h_j. Indeed, from the overlap condition (4.1.3.2),

$$rh_j = \sum_i r_i b_i h_j = \sum_i r_i h_i b_j = b_j.$$

We next deal with the case where I is infinite. Choose a finite subset $\{1, \ldots, n\} \subset I$ with $(f_1, \ldots, f_n) = A$ (or equivalently, use quasicompactness of Spec A to choose a finite subcover by $D(f_i)$). Construct r as above, using (4.1.3.3). We will show that for any $z \in I - \{1, \ldots, n\}$, r restricts to the desired element $a_z/f_z^{l_z}$ of A_{f_z}. Repeat the entire process above with $\{1, \ldots, n, z\}$ in place of $\{1, \ldots, n\}$, to obtain $r' \in A$ that restricts to $a_i/f_i^{l_i}$ for $i \in \{1, \ldots, n, z\}$. Then by the base identity axiom, $r' = r$. (Note that we use base identity to *prove* base gluability. This is an example of how the identity axiom is somehow "prior" to the gluability axiom.) Hence r restricts to the desired element $a_z/f_z^{l_z}$ of A_{f_z}.

4.1.C. EXERCISE. Alter this argument appropriately to show base gluability for any distinguished open $D(f)$.

We have now completed the proof of Theorem 4.1.2. $\square$

4.1.4. ** *Remark.* Definition 4.1.1 and Theorem 4.1.2 suggest a potentially slick way of describing sections of $\mathscr{O}_{\text{Spec} A}$ over *any* open subset: perhaps $\mathscr{O}_{\text{Spec} A}(U)$ is the localization of A at the multiplicative set of all functions that do not vanish at any point of U. This is not true. A counterexample (that you will later be able to make precise): Let Spec A be two copies of $\mathbb{A}_k^2$ glued together at their origins (see (24.3.8.1) for explicit equations) and let U be the complement of the origin(s). Then the function which is 1 on the first copy of $\mathbb{A}_k^2 \setminus \{(0,0)\}$ and 0 on the second copy of $\mathbb{A}_k^2 \setminus \{(0,0)\}$ is not of this form. (Follow-up question: Why would this discussion not work for two copies of $\mathbb{A}_k^1$ glued at a point?)

4.1.5. Important notion for future use: $\mathscr{O}_{\text{Spec} A}$-modules coming from A-modules. The following generalization of Theorem 4.1.2 will be essential in the definition of a *quasicoherent sheaf* in Chapter 6. You can leave these exercises for then, but they may be easiest to solve now, while the ideas are fresh in your mind.

4.1.D. IMPORTANT EXERCISE/DEFINITION. Suppose M is an A-module. Show that the following construction describes a sheaf $\widetilde{M}$ on the distinguished base. Define $\widetilde{M}(D(f))$ to be the localization of M at the multiplicative set of all functions that do not vanish outside of $V(f)$. Define restriction maps $\text{res}_{D(f), D(g)}$ in the analogous way to $\mathscr{O}_{\text{Spec} A}$. Show that this defines a sheaf on the distinguished base, and hence a sheaf on Spec A. Then show that this is an $\mathscr{O}_{\text{Spec} A}$-module.

4.1.6. *Important remark.* In the course of answering the previous exercise, you will show that if $(f_i)_{i \in I} = A$,

(4.1.6.1) $$0 \longrightarrow M \longrightarrow \prod_{i \in I} M_{f_i} \longrightarrow \prod_{i \neq j \in I} M_{f_i f_j}$$

(cf. (4.1.3.1)) is exact. In particular, M can be identified with a specific submodule of $M_{f_1} \times \cdots \times M_{f_r}$. Even though $M \to M_{f_i}$ may not be an inclusion for any f_i, $M \to M_{f_1} \times \cdots \times M_{f_r}$ *is* an inclusion. This will be useful later: we will want to show that if M has some nice property, then M_f does too, which will be easy. We will also want to show that if $(f_1, \ldots, f_n) = A$, and the M_{f_i} have this property, then M does too. This idea will be made precise in the Affine Communication Lemma 5.3.2.

4.1.E. EXERCISE. Suppose $\mathfrak{p}$ is a prime ideal of A. Describe a canonical isomorphism $\widetilde{M}_{[\mathfrak{p}]} \xrightarrow{\sim} M_{\mathfrak{p}}$ between the stalk of $\widetilde{M}$ and the localization of M. In particular, the stalk of $\mathscr{O}_{\text{Spec} A}$ at the point $[\mathfrak{p}]$ is the local ring $A_{\mathfrak{p}}$.

The following exercise shows how an important result can be understood quite differently from algebraic and geometric perspectives.

4.1.F. EXERCISE. Suppose M is an A-module. Show that the natural map

$$M \hookrightarrow \prod_{\mathfrak{p} \in \text{Spec } A} M_{\mathfrak{p}}$$

is an injection in two ways: (a) by considering the kernel of the map, and showing that any element of it must be 0, and (b) by applying Exercise 2.4.A (sections of a sheaf are determined by germs) to the sheaf $\widetilde{M}$. (Thus an A-module is zero if and only if all its localizations at prime ideals are zero, and we see this as simultaneously (a) an algebraic and (b) a geometric fact.)

4.1.G. EXERCISE. Describe a bijection between maps of A-modules $M \to N$, and maps of $\mathcal{O}_{\text{Spec } A}$-modules $\widetilde{M} \to \widetilde{N}$. (Fancy translation: you will have described Mod_A as a full subcategory of $Mod_{\mathcal{O}_{\text{Spec } A}}$.)

4.1.7. *Definition: Support of a module.* Motivated by Exercise 4.1.E, and the notion of support of a section of a sheaf (Definition 2.7.6), define the **support of** $m \in M$ by

$$\text{Supp } m := \{[\mathfrak{p}] \in \text{Spec } A \ : \ m_{\mathfrak{p}} \neq 0\} \subset \text{Spec } A,$$

which is the support of m considered as a section of $\widetilde{M}$. By Exercise 2.7.G, Supp m is a closed subset of Spec A.

Similarly (following Definition 2.7.8), define the **support of** M by

$$\text{Supp } M := \text{Supp } \widetilde{M} = \{[\mathfrak{p}] \in \text{Spec } A \ : \ M_{\mathfrak{p}} \neq 0\} \subset \text{Spec } A.$$

These notions will come up repeatedly. We will discuss support in more detail in §6.6.3.

4.1.8. Recurring (counter)examples. The example of two planes meeting at a point (§4.1.4) will appear many times as an example or a counterexample. It is important to have such examples (or counterexamples) at the back of your mind, as "boundary markers" reminding you of what is reasonable, and what to watch out for when you hear a new statement. The appearances of the following are listed in the index (under "recurring (counter)examples"), so you can see how often they come up, and where. You do not have the language to understand many of these yet, but you can come back to this list later.

- two planes meeting at a point
- affine space minus the origin, and its inclusion in affine space
- affine space with doubled origin
- (the cone over) the quadric surface, $A = k[w, x, y, z]/(wz - xy)$
- an embedded point on a line, $k[x, y]/(y^2, xy)$
- $x^2 = 0$ in the projective plane
- infinite disjoint unions of schemes (especially $\coprod \text{Spec } k[x]/(x^n)$)
- the morphism $\text{Spec } \overline{\mathbb{Q}} \to \text{Spec } \mathbb{Q}$
- $k(x) \otimes_k k(y)$
- $\prod^{\infty} \mathbb{F}_2$

4.2   Visualizing Schemes: Nilpotents

> *The price of metaphor is eternal vigilance.*
>
> —A. Rosenbluth and N. Wiener (attribution by [Lew, p. 4])

In §3.3, we discussed how to visualize the underlying set of schemes, adding in generic points to our previous intuition of "classical" (or closed) points. Our later discussion of the Zariski topology fit well with that picture. In our definition of the "affine scheme" $(\operatorname{Spec} A, \mathcal{O}_{\operatorname{Spec} A})$, we have the additional information of nilpotents, which are invisible on the level of points (§3.2.12), so now we figure out how to picture them. We will then readily be able to glue them together to picture schemes in general, once we have made the appropriate definitions. As we are building intuition, we cannot be rigorous or precise.

As motivation, note that we have inclusion-reversing bijections

$$\text{maximal ideals of } A \longleftrightarrow \text{closed points of Spec } A, \qquad \text{(Exercise 3.6.I)}$$

$$\text{prime ideals of } A \longleftrightarrow \text{irreducible closed subsets of Spec } A, \qquad \text{(Exercise 3.7.F)}$$

$$\text{radical ideals of } A \longleftrightarrow \text{closed subsets of Spec } A. \qquad \text{(Theorem 3.7.1)}$$

If we take the things on the right as "pictures," our goal is to figure out how to picture ideals that are not radical:

$$\text{ideals of } A \longleftrightarrow \text{???}$$

(We will later fill this in rigorously in a different way with the notion of a *closed subscheme*, the scheme-theoretic version of closed subsets, §9.1. But our goal now is to create a picture.)

As motivation, when we see the expression $\operatorname{Spec} \mathbb{C}[x]/(x(x-1)(x-2))$, we immediately interpret it as a closed subset of $\mathbb{A}^1_{\mathbb{C}}$, namely $\{0, 1, 2\}$. In particular, the map $\mathbb{C}[x] \to \mathbb{C}[x]/(x(x-1)(x-2))$ can be interpreted (via the Chinese Remainder Theorem) as: take a function on $\mathbb{A}^1$, and restrict it to the three points 0, 1, and 2.

This will guide us in how to visualize a nonradical ideal. The simplest example to consider is $\operatorname{Spec} \mathbb{C}[x]/(x^2)$ (Exercise 3.2.A(a)). As a subset of $\mathbb{A}^1$, it is just the origin $0 = [(x)]$, which we are used to thinking of as $\operatorname{Spec} \mathbb{C}[x]/(x)$ (i.e., corresponding to the ideal (x), not (x^2)). We want to enrich this picture in some way. We should picture $\mathbb{C}[x]/(x^2)$ in terms of the information the quotient remembers. The image of a polynomial $f(x)$ is the information of its value at 0, and its derivative (cf. Exercise 3.2.T). We thus picture this as being the point, plus a little bit more—a little bit of infinitesimal "fuzz" on the point (see Figure 4.2). The sequence of restrictions $\mathbb{C}[x] \to \mathbb{C}[x]/(x^2) \to \mathbb{C}[x]/(x)$ should be interpreted as nested pictures:

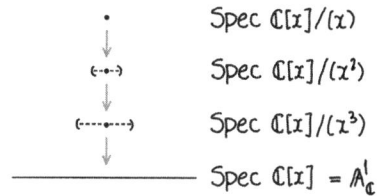

Figure 4.2 *Picturing quotients of $\mathbb{C}[x]$.*

$$\mathbb{C}[x] \longrightarrow\!\!\!\!\rightarrow \mathbb{C}[x]/(x^2) \longrightarrow\!\!\!\!\rightarrow \mathbb{C}[x]/(x),$$

$$f(x) \longmapsto f(0).$$

Similarly, $\mathbb{C}[x]/(x^3)$ remembers even more information—the second derivative as well. Thus we picture this as the point 0 with even more fuzz.

More subtleties arise in two dimensions (see Figure 4.3). Consider

$$\operatorname{Spec} \mathbb{C}[x, y]/(x, y)^2,$$

which is sandwiched between two rings we know well:

$$\mathbb{C}[x, y] \longrightarrow\!\!\!\!\rightarrow \mathbb{C}[x, y]/(x, y)^2 \longrightarrow\!\!\!\!\rightarrow \mathbb{C}[x, y]/(x, y),$$

$$f(x, y) \longmapsto f(0, 0).$$

Again, the quotient by $(x, y)^2$ remembers the first derivative "in all directions." We picture this as fuzz around the point, in the shape of a circle (no direction is privileged). Similarly, $(x, y)^3$ remembers the second derivative "in all directions"—bigger circular fuzz.

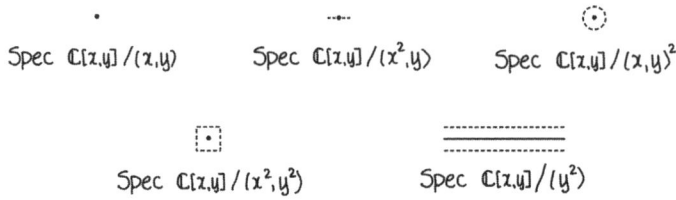

Figure 4.3 *Picturing quotients of $\mathbb{C}[x, y]$.*

Consider instead the ideal (x^2, y). What it remembers is the derivative only in the x direction—given a polynomial, we remember its value at 0, and the coefficient of x. We remember this by picturing the fuzz only in the x direction.

This gives us some handle on picturing more things of this sort, but now it becomes more an art than a science. For example, $\operatorname{Spec} \mathbb{C}[x, y]/(x^2, y^2)$ we might picture as a fuzzy square around the origin. (Could you believe that this square is circumscribed by the circular fuzz $\operatorname{Spec} \mathbb{C}[x, y]/(x, y)^3$, and inscribed by the circular fuzz $\operatorname{Spec} \mathbb{C}[x, y]/(x, y)^2$?) One feature of this example is that given two ideals I and J of a ring A (such as $\mathbb{C}[x, y]$), your fuzzy picture of $\operatorname{Spec} A/(I, J)$ should be the "intersection" of your picture of $\operatorname{Spec} A/I$ and $\operatorname{Spec} A/J$ in $\operatorname{Spec} A$. (You will make this precise in Exercise 9.1.I(a).) For example, $\operatorname{Spec} \mathbb{C}[x, y]/(x^2, y^2)$ should be the intersection of two thickened lines. (How would you picture $\operatorname{Spec} \mathbb{C}[x, y]/(x^5, y^3)$? $\operatorname{Spec} \mathbb{C}[x, y, z]/(x^3, y^4, z^5, (x + y + z)^2)$? $\operatorname{Spec} \mathbb{C}[x, y]/((x, y)^5, y^3)$?)

Figure 4.4 *A picture of the scheme $\operatorname{Spec} k[x, y]/(y^2, xy)$. The fuzz at the origin indicates where "there is nonreducedness."*

One final example that will motivate us in §6.6 is $\operatorname{Spec} \mathbb{C}[x, y]/(y^2, xy)$. Knowing what a polynomial in $\mathbb{C}[x, y]$ is modulo (y^2, xy) is the same as knowing its value on the x-axis, as well as first-order differential information around the origin. This is worth thinking through carefully: do you see how this information is captured (however imperfectly) in Figure 4.4?

Our pictures capture useful information that you already have some intuition for. For example, consider the intersection of the parabola $y = x^2$ and the x-axis (in the xy-plane); see Figure 4.5. You already have a sense that the intersection has multiplicity 2. In terms of this visualization, we interpret this as intersecting (in $\operatorname{Spec} \mathbb{C}[x, y]$):

$$\operatorname{Spec} \mathbb{C}[x, y]/(y - x^2) \cap \operatorname{Spec} \mathbb{C}[x, y]/(y) = \operatorname{Spec} \mathbb{C}[x, y]/(y - x^2, y)$$
$$= \operatorname{Spec} \mathbb{C}[x, y]/(y, x^2),$$

which we interpret as the fact that the parabola and line not just meet with multiplicity 2, but that the "multiplicity 2" part is in the direction of the x-axis. You will make this example precise in Exercise 9.1.I(b).

4.2.1. We will later make the location of the fuzz more precise when we discuss associated points. We will see that, in reasonable circumstances, the fuzz is concentrated on closed subsets (Exercise 6.6.N).

> *On a bien souvent répété que la Géométrie est l'art de bien raisonner sur des figures mal faites; encore ces figures, pour ne pas nous tromper, doivent-elles satisfaire à certaines conditions; les proportions peuvent être grossièrement altérées, mais les positions relatives des diverses parties ne doivent pas être bouleversées.*
>
> *It is often said that geometry is the art of reasoning well from badly made figures; however, these figures, if they are not to deceive us, must satisfy certain conditions; the proportions may be grossly altered, but the relative positions of the different parts must not be upset.*
>
> —H. Poincaré [Po1, p. 2] (see J. Stillwell's translation [Po2, p. ix])

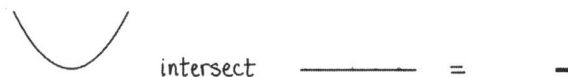

Figure 4.5 *The "scheme-theoretic" intersection of the parabola $y = x^2$ and the x-axis is a nonreduced scheme (with fuzz in the x-direction).*

4.3   Definition of Schemes

4.3.1. *Definitions.* We can now define *scheme* in general. First, define an **isomorphism of ringed spaces** $(X, \mathscr{O}_X) \to (Y, \mathscr{O}_Y)$ as (i) a homeomorphism $\pi \colon X \to Y$, and (ii) an isomorphism of sheaves $\mathscr{O}_X$ and $\mathscr{O}_Y$, considered to be on the same space via π. (Part (ii), more precisely, is an isomorphism $\mathscr{O}_Y \xrightarrow{\sim} \pi_* \mathscr{O}_X$ of sheaves on Y, or equivalently, an isomorphism $\pi^{-1} \mathscr{O}_Y \xrightarrow{\sim} \mathscr{O}_X$ of sheaves on X.) In other words, we have a "correspondence" of sets, topologies, and structure sheaves. Every isomorphism $\alpha \colon (X, \mathscr{O}_X) \to (Y, \mathscr{O}_Y)$ clearly has an "inverse isomorphism" $\alpha^{-1} \colon (Y, \mathscr{O}_Y) \to (X, \mathscr{O}_X)$.

An **affine scheme** is a ringed space that is isomorphic to $(\operatorname{Spec} A, \mathscr{O}_{\operatorname{Spec} A})$ for some A. A **scheme** $(X, \mathscr{O}_X)$ is a ringed space such that any point of X has an open neighborhood U such that $(U, \mathscr{O}_X|_U)$ is an affine scheme. The topology on a scheme is called the **Zariski topology**. The scheme can be denoted by $(X, \mathscr{O}_X)$, although it is often denoted by X, with the structure sheaf $\mathscr{O}_X$ left implicit.

An **isomorphism of schemes** $(X, \mathscr{O}_X) \to (Y, \mathscr{O}_Y)$ is an isomorphism as ringed spaces. Recall the definition of $\Gamma(\cdot, \cdot)$ in §2.2.2. If $U \subset X$ is an open subset, then the elements of $\Gamma(U, \mathscr{O}_X)$ are said to be the **functions on** U; this generalizes the definition of functions on an affine scheme, §3.2.1.

4.3.2. *Remark.* From the definition of the structure sheaf on an affine scheme, several things are clear. First of all, if we are told that $(X, \mathscr{O}_X)$ is an affine scheme, we may recover its ring (i.e., find the ring A such that $\operatorname{Spec} A = X$) by taking the ring of global sections, as $X = D(1)$, so:

$$\Gamma(X, \mathscr{O}_X) = \Gamma(D(1), \mathscr{O}_{\operatorname{Spec} A}) \quad \text{as } D(1) = \operatorname{Spec} A$$
$$= A.$$

(You can verify that we get more, and can "recognize X as the scheme $\operatorname{Spec} A$": we get an isomorphism $\pi \colon (\operatorname{Spec} \Gamma(X, \mathscr{O}_X), \mathscr{O}_{\operatorname{Spec} \Gamma(X, \mathscr{O}_X)}) \xrightarrow{\sim} (X, \mathscr{O}_X)$. For example, if $\mathfrak{m}$ is a maximal ideal of $\Gamma(X, \mathscr{O}_X)$, then $\{\pi([\mathfrak{m}])\} = V(\mathfrak{m})$.) The following exercise will make these ideas rigorous—they are subtler than they first appear.

4.3.3. *Caution.* It is not part of the definition that the overlap of two affine open subsets is also affine (see Exercise 4.4.C for an example), although this is true in most cases of interest (see Proposition 11.2.13).

4.3.A. ENLIGHTENING EXERCISE (WHICH CAN BE STRANGELY CONFUSING). Describe a bijection between the isomorphisms $\operatorname{Spec} A \xrightarrow{\sim} \operatorname{Spec} A'$ (of ringed spaces) and the ring isomorphisms $A' \xrightarrow{\sim} A$. Hint: The hardest part is to show that if an isomorphism $\pi \colon \operatorname{Spec} A \xrightarrow{\sim} \operatorname{Spec} A'$ induces an isomorphism $\pi^\sharp \colon A' \xrightarrow{\sim} A$, which in turn induces an isomorphism $\rho \colon \operatorname{Spec} A \xrightarrow{\sim} \operatorname{Spec} A'$, then $\pi = \rho$. First show this on the level of points; this is (surprisingly) the trickiest part. Then show $\pi = \rho$ as maps of topological spaces. Finally, to show $\pi = \rho$ on the level of structure sheaves, use the distinguished base. Feel free to use insights from later in this section, but be careful to avoid circular arguments. Even struggling with this exercise and failing (until reading later sections) will be helpful.

More generally, given $f \in A$, $\Gamma(D(f), \mathscr{O}_{\operatorname{Spec} A}) \cong A_f$. Thus under the natural inclusion of sets $\operatorname{Spec} A_f \hookrightarrow \operatorname{Spec} A$, the Zariski topology on $\operatorname{Spec} A$ restricts to give the Zariski topology on $\operatorname{Spec} A_f$ (Exercise 3.4.I), and the structure sheaf of $\operatorname{Spec} A$ restricts to the structure sheaf of $\operatorname{Spec} A_f$, as the next exercise shows.

4.3.B. IMPORTANT BUT EASY EXERCISE. Suppose $f \in A$. Show that under the identification of $D(f)$ in $\operatorname{Spec} A$ with $\operatorname{Spec} A_f$ (§3.5), there is a natural isomorphism of ringed spaces $(D(f), \mathscr{O}_{\operatorname{Spec} A}|_{D(f)}) \xleftrightarrow{\sim} (\operatorname{Spec} A_f, \mathscr{O}_{\operatorname{Spec} A_f})$. Hint: Notice that distinguished open sets of $\operatorname{Spec} A_f$ are already distinguished open sets in $\operatorname{Spec} A$.

4.3.C. EASY EXERCISE. If X is a scheme, and U is *any* open subset, prove that $(U, \mathscr{O}_X|_U)$ is also a scheme.

4.3.4. *Definitions.* We say $(U, \mathscr{O}_X|_U)$ is an **open subscheme of** X. If U is also an affine scheme, we often say U is an **affine open subset**, or an **affine open subscheme**, or sometimes, informally, just an **affine open**. For example, $D(f)$ is an affine open subscheme of $\operatorname{Spec} A$. (Fun fact: It is not true that every affine open subscheme of $\operatorname{Spec} A$ is of the form $D(f)$; see §19.11.10 for an example.)

4.3.D. EASY EXERCISE. Show that if X is a scheme, then the affine open sets form a base for the Zariski topology.

4.3.E. EASY EXERCISE. The **disjoint union of schemes** is defined as you would expect: it is the disjoint union of sets, with the expected topology (thus it is the disjoint union of topological spaces), with the expected sheaf. We use the symbol $X \coprod Y$ for the disjoint union of X and Y, because once we know what morphisms of schemes are, this construction will turn out to be the coproduct in the category of schemes (Exercise 7.3.K).

(a) Show that the disjoint union of a *finite* number of affine schemes is also an affine scheme. (Hint: Exercise 3.6.A.)
(b) (*A first example of a non-affine scheme*) Show that an infinite disjoint union of (nonempty) affine schemes is not an affine scheme. (Hint: Affine schemes are quasicompact, Exercise 3.6.G(a). This is basically answered in Remark 3.6.6.)

4.3.5. *Remark: A first glimpse of closed subschemes.* Open subsets of a scheme come with a natural scheme structure (Definition 4.3.4). For comparison, closed subsets can have many "natural" scheme structures. We will discuss this later (in §9.1), but for now, it suffices for you to know that a closed subscheme of X is, informally, a particular kind of scheme structure on a closed subset of X. As an example: if $I \subset A$ is an ideal, then $\operatorname{Spec} A/I$ endows the closed subset $V(I) \subset \operatorname{Spec} A$ with a scheme structure; but note that there can be different ideals with the same vanishing set (for example, (x) and (x^2) in $k[x]$).

4.3.6. Stalks of the structure sheaf: Germs, values at a point, and the residue field of a point. Like every sheaf, the structure sheaf has stalks, and, unsurprisingly, they are interesting from an algebraic point of view. In fact, we have seen them before.

4.3.7. *Definition.* We say a ringed space is a **locally ringed space** if its stalks are local rings. By Exercise 4.1.E, the stalk of $\mathscr{O}_{\operatorname{Spec} A}$ at the point $[\mathfrak{p}]$ is the local ring $A_\mathfrak{p}$. Hence schemes are locally ringed spaces. Manifolds are another example of locally ringed spaces; see §2.1.1. In both cases, taking the quotient by the maximal ideal may be interpreted as evaluating at the point. The maximal ideal of the local ring $\mathscr{O}_{X,\mathfrak{p}}$ is denoted by $\mathfrak{m}_{X,\mathfrak{p}}$ or $\mathfrak{m}_\mathfrak{p}$, and the **residue field** $\mathscr{O}_{X,\mathfrak{p}}/\mathfrak{m}_\mathfrak{p}$ is denoted by $\kappa(\mathfrak{p})$. Each function f on an open subset U of a locally ringed space X has a value at each point $\mathfrak{p}$ of U: the **value of** f **at** $\mathfrak{p}$ (denoted by $f(\mathfrak{p})$) is the image of the germ of f at $\mathfrak{p}$ under the canonical map $\mathscr{O}_{X,\mathfrak{p}} \to \kappa(\mathfrak{p})$. We say that a function **vanishes** at a point $\mathfrak{p}$ if its value at $\mathfrak{p}$ is 0. (This generalizes our notion of the value of a function on $\operatorname{Spec} A$, defined in §3.2.1.) Notice that we can't even make sense of the phrase "function vanishing" on ringed spaces in general.

4.3.F. USEFUL EXERCISE.

(a) If f is a function on a locally ringed space X, show that the subset of X where f vanishes is closed. Hint: If f is a function on a *ringed space* X, show that the subset of X where the germ of f is invertible is open. (Thus the locus on a locally ringed space where a function has nonzero *value* is open. In contrast, the locus it has nonzero *germ* is *closed*, as mentioned in Caution 2.7.7.)

(b) Show that if f is a function on a locally ringed space that vanishes nowhere, then f is invertible.

Consider a point $[\mathfrak{p}]$ of an affine scheme Spec A. (Of course, any point of a scheme can be interpreted in this way, as each point has an affine open neighborhood.) The residue field at $[\mathfrak{p}]$ is $A_\mathfrak{p}/\mathfrak{p}A_\mathfrak{p}$, which is isomorphic to $K(A/\mathfrak{p})$, the fraction field of the quotient. It is useful to note that localizing at $\mathfrak{p}$ and taking the quotient by $\mathfrak{p}$ "commute," i.e., the following diagram commutes.

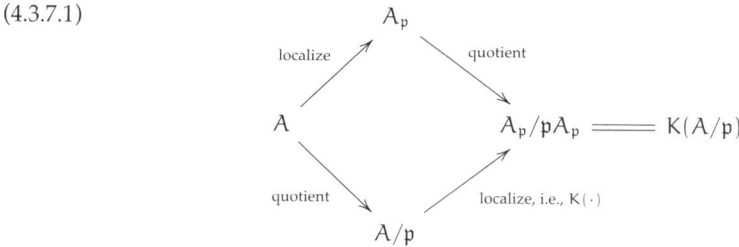

(4.3.7.1)

$$
\begin{array}{ccc}
& A_\mathfrak{p} & \\
\nearrow^{\text{localize}} & & \searrow^{\text{quotient}} \\
A & & A_\mathfrak{p}/\mathfrak{p}A_\mathfrak{p} = K(A/\mathfrak{p}) \\
\searrow_{\text{quotient}} & & \nearrow_{\text{localize, i.e., } K(\cdot)} \\
& A/\mathfrak{p} &
\end{array}
$$

For example, consider the scheme $\mathbb{A}^2_k = \operatorname{Spec} k[x, y]$, where k is a field of characteristic not 2. Then $(x^2 + y^2)/x(y^2 - x^5)$ is a function away from the y-axis and the curve $y^2 - x^5$. Its value at $(2, 4)$ (by which we mean $[(x - 2, y - 4)]$) is $(2^2 + 4^2)/(2(4^2 - 2^5))$, as

$$
\frac{x^2 + y^2}{x(y^2 - x^5)} \equiv \frac{2^2 + 4^2}{2(4^2 - 2^5)}
$$

in the residue field—check this if it seems mysterious. And its value at $[(y)]$, the generic point of the x-axis, is $\frac{x^2}{-x^6} = -1/x^4$, which we see by setting y to 0. This is indeed an element of the fraction field of $k[x, y]/(y)$, i.e., $k(x)$. (If you think you care only about algebraically closed fields, let this example be a first warning: $A_\mathfrak{p}/\mathfrak{p}A_\mathfrak{p}$ won't be algebraically closed in general, even if A is a finitely generated $\mathbb{C}$-algebra!)

If anything makes you nervous, you should make up an example to make you feel better. Here is one: 27/4 is a function on $\operatorname{Spec} \mathbb{Z} \setminus \{[(2)], [(7)]\}$ or indeed on an even bigger open set. What is its value at $[(5)]$? Answer: $2/(-1) \equiv -2 \pmod 5$. What is its value at the generic point $[(0)]$? Answer: 27/4. Where does it vanish? At $[(3)]$.

4.3.8. *Definition: The fiber and rank of an $\mathcal{O}_X$-module at a point.* If $\mathscr{F}$ is an $\mathcal{O}_X$-module on a scheme X (or, more generally, a locally ringed space), define the **fiber** (or *fibre*) **of $\mathscr{F}$ at a point** $p \in X$ by

$$
\mathscr{F}|_p := \mathscr{F}_p \otimes_{\mathcal{O}_{X,p}} \kappa(p).
$$

The fiber is a vector space over $\kappa(p)$. The dimension of this vector space $\dim_{\kappa(p)} \mathscr{F}|_p$ is called the **rank of $\mathscr{F}$ at** p, and is denoted by $\operatorname{rank}_p \mathscr{F}$. For example, $\mathcal{O}_X|_p$ is $\kappa(p)$, and $\operatorname{rank}_p \mathcal{O}_X = 1$.

4.3.9.* **Definition: Manifolds of various sorts.** In the same way, we can cleanly and concisely define (various notions of) manifolds. A **topological** (respectively, **differentiable**, C^∞ = **smooth**) (real) **manifold** is a ringed space $(X, \mathcal{O}_X)$ whose underlying topological space X is Hausdorff and second countable, such that any point of X has an open neighborhood U such that $(U, \mathcal{O}_X|_U)$ is isomorphic to an open ball in $\mathbb{R}^n$ with the sheaf of continuous (respectively, differentiable, C^∞ = smooth) real-valued functions on it. The phrase **manifold** (without adjective) often means **smooth manifold**. A **complex manifold** is a ringed space $(X, \mathcal{O}_X)$ whose underlying topological space X is Hausdorff and second countable, such that any point of X has an open neighborhood U such that $(U, \mathcal{O}_X|_U)$ is isomorphic to an open ball in $\mathbb{C}^n$ with the sheaf of holomorphic functions on it.

4.4 Three Examples

We now give three extended examples. Our short-term goal is to see that we can really work with the structure sheaf, and can compute the ring of sections of interesting open sets that aren't just distinguished open sets of affine schemes. Our long-term goal is to meet interesting examples that will come up repeatedly in the future.

4.4.1. First example: The (affine) plane minus the origin. This example will show you that the distinguished base is something that you can work with. Let $A = k[x, y]$, so $\operatorname{Spec} A = \mathbb{A}_k^2$. Let's work out the space of functions on the open set $U = \mathbb{A}^2 \setminus \{(0, 0)\} = \mathbb{A}^2 \setminus \{[(x, y)]\}$.

It is not immediately obvious whether this is a distinguished open set. (In fact it is not—you may be able to figure out why within a few paragraphs, if you can't right now. It is not enough to show that (x, y) is not a principal ideal.) But in any case, we can describe it as the union of two things that *are* distinguished open sets: $U = D(x) \cup D(y)$. We will find the functions on U by gluing together functions on $D(x)$ and $D(y)$.

The functions on $D(x)$ are, by Definition 4.1.1, $A_x = k[x, y, 1/x]$. The functions on $D(y)$ are $A_y = k[x, y, 1/y]$. Note that A injects into its localizations (if 0 is not inverted), as it is an integral domain (Exercise 1.2.C), so A injects into both A_x and A_y, and both inject into A_{xy} (and indeed into $k(x, y) = K(A)$). So we are looking for functions on $D(x)$ and $D(y)$ that agree on $D(x) \cap D(y) = D(xy)$, i.e., we are interpreting $A_x \cap A_y$ in A_{xy} (or in $k(x, y)$). Clearly those rational functions with only powers of x in the denominator, and also with only powers of y in the denominator, are the polynomials. Translation: $A_x \cap A_y = A$. Thus we conclude:

(4.4.1.1) $$\Gamma(U, \mathscr{O}_{\mathbb{A}^2}) \equiv k[x, y].$$

In other words, we get no extra functions by removing the origin. Notice how easy that was to calculate!

4.4.2. *Aside.* Notice that any function on $\mathbb{A}^2 \setminus \{(0, 0)\}$ extends over all of $\mathbb{A}^2$. This is an analog of *Hartogs's Lemma* in complex geometry: you can extend a holomorphic function defined on the complement of a set of codimension at least 2 on a complex manifold over the missing set. This will work more generally in the algebraic setting: you can extend over points in codimension at least 2 not only if they are "smooth," but also if they are mildly singular—what we will call *normal*. We will make this precise in §13.5.19. This fact will be very useful for us.

4.4.3. We now show an interesting fact: $(U, \mathscr{O}_{\mathbb{A}^2}|_U)$ is a scheme, but it is not an affine scheme. (This is confusing, so you will have to pay attention.) Here's why: otherwise, if $(U, \mathscr{O}_{\mathbb{A}^2}|_U) \cong (\operatorname{Spec} A', \mathscr{O}_{\operatorname{Spec} A'})$, then we can recover A' by taking global sections:

$$A' = \Gamma(U, \mathscr{O}_{\mathbb{A}^2}|_U),$$

which we have already identified in (4.4.1.1) as $k[x, y]$. So if U is affine, then we have the isomorphism

$$U \xleftarrow{\sim} \operatorname{Spec} A' = \operatorname{Spec} k[x, y] = \mathbb{A}_k^2.$$

But this bijection between prime ideals in a ring and points of the spectrum is more constructive than that: *given the prime ideal I, you can recover the point as the generic point of the closed subset cut out by I*, i.e., $V(I)$, *and given the point p, you can recover the ideal as those functions vanishing at p*, i.e., $I(p)$. In particular, the prime ideal (x, y) of A' should cut out a point of $\operatorname{Spec} A'$. But on U, $V(x) \cap V(y) = \varnothing$. Conclusion: U is *not* an affine scheme. (If you are ever looking for a counterexample to something, and you are expecting one involving a non-affine scheme, keep this example in mind!)

4.4.4. Gluing two copies of $\mathbb{A}^1$ together in two different ways. We have now seen two examples of non-affine schemes: an infinite disjoint union of nonempty schemes (Exercise 4.3.E) and $\mathbb{A}^2 \setminus \{(0, 0)\}$. I want to give you two more examples. They are important because they are the first examples of fundamental behavior, the first pathological, and the second central.

First, I need to tell you how to glue two schemes together. Before that, you should review how to glue topological spaces together along isomorphic open sets. Given two topological spaces X and Y, and open subsets $U \subset X$ and $V \subset Y$ along with a homeomorphism $U \xleftarrow{\sim} V$, we can create a new topological space W, which we think of as gluing X and Y together along $U \overset{\sim}{\longleftrightarrow} V$. It is the quotient of the disjoint union $X \coprod Y$ by the equivalence relation $U \sim V$, where the quotient is given the quotient topology. Then X and Y are naturally (identified with) open subsets of W, and, indeed, cover W. Can you restate this cleanly with an arbitrary (not necessarily finite) number of topological spaces?

Now that we have discussed gluing topological spaces, let's glue schemes together. (This applies without change more generally to ringed spaces.) Suppose you have two schemes $(X, \mathscr{O}_X)$ and $(Y, \mathscr{O}_Y)$, and open subsets $U \subset X$ and $V \subset Y$, along with a homeomorphism $\pi\colon U \longrightarrow V$, and an isomorphism of structure sheaves $\mathscr{O}_V \longrightarrow \pi_* \mathscr{O}_U$ (i.e., an isomorphism *of schemes* $(U, \mathscr{O}_X|_U) \overset{\sim}{\longleftrightarrow} (V, \mathscr{O}_Y|_V)$). Then we can glue these together to get a single scheme. Reason: Let W be X and Y glued together using the isomorphism $U \overset{\sim}{\longleftrightarrow} V$. Then Exercise 2.5.E shows that the structure sheaves can be glued together to get a sheaf of rings. Note that this is indeed a scheme: any point has an open neighborhood that is an affine scheme. (Do you see why?)

4.4.A. ESSENTIAL EXERCISE (CF. EXERCISE 2.5.E). Show that you can glue an arbitrary collection of schemes together. Suppose we are given:

- schemes X_i (as i runs over some index set I, not necessarily finite),
- open subschemes $X_{ij} \subset X_i$ with $X_{ii} = X_i$,
- isomorphisms $f_{ij}\colon X_{ij} \overset{\sim}{\longrightarrow} X_{ji}$ with f_{ii} the identity

such that

- (the cocycle condition) the isomorphisms "agree on triple intersections," i.e., $f_{ik}|_{X_{ij} \cap X_{ik}} = f_{jk}|_{X_{ji} \cap X_{jk}} \circ f_{ij}|_{X_{ij} \cap X_{ik}}$ (so, implicitly, to make sense of the right side, $f_{ij}(X_{ik} \cap X_{ij}) \subset X_{jk}$).

(The cocycle condition ensures that f_{ij} and f_{ji} are inverses. In fact, the hypothesis that f_{ii} is the identity also follows from the cocycle condition.) Show that there is a unique scheme X (up to unique isomorphism) along with open subschemes isomorphic to the X_i respecting this gluing data in the obvious sense. (Hint: What is X as a set? What is the topology on this set? In terms of your description of the open sets of X, what are the sections of this sheaf over each open set?) You may wish to show uniqueness (up to unique isomorphism) by describing a universal property satisfied by X.

I will now give you two non-affine schemes. Both are handy to know. In both cases, I will glue together two copies of the affine line $\mathbb{A}^1_k$. Let $X = \operatorname{Spec} k[t]$ and $Y = \operatorname{Spec} k[u]$. Let $U = D(t) = \operatorname{Spec} k[t, 1/t] \subset X$ and $V = D(u) = \operatorname{Spec} k[u, 1/u] \subset Y$. We will get both examples by gluing X and Y together along U and V. The difference will be in how we glue.

4.4.5. Second example: The affine line with the doubled origin. Consider the isomorphism $U \overset{\sim}{\longleftrightarrow} V$ via the isomorphism $k[t, 1/t] \overset{\sim}{\longleftrightarrow} k[u, 1/u]$ given by $t \longleftrightarrow u$ (cf. Exercise 4.3.A). The resulting scheme is called the **affine line with doubled origin**. Figure 4.6 is a picture of it.

Figure 4.6 *The affine line with doubled origin.*

As the picture suggests, intuitively this is an analog of a failure of Hausdorffness. Now $\mathbb{A}^1$ itself is not Hausdorff, so we can't say that it is a failure of Hausdorffness. We see this as weird and bad, so we will want to make a definition that will prevent this from happening. This will be the notion of *separatedness* (to be discussed in Chapter 11). (In fact, in some sense, separatedness is the *right* definition of Hausdorfness; see Exercise 11.2.B!) This will answer other of our prayers as well. For example, on a separated scheme, the "affine base of the Zariski topology" is nice—the intersection of two affine open sets will be affine (Proposition 11.2.13).

4.4.B. EXERCISE. Show that the affine line with doubled origin is not affine. Hint: Calculate the ring of global sections, and look back at the argument for $\mathbb{A}^2 \setminus \{(0,0)\}$.

4.4.C. EASY EXERCISE. Do the same construction with $\mathbb{A}^1$ replaced by $\mathbb{A}^2$. You will have defined the **affine plane with doubled origin**. Describe two affine open subsets of this scheme whose intersection is not an affine open subset. (This example was promised in Caution 4.3.3. Also, an "infinite-dimensional" version comes up in Exercise 5.1.K.)

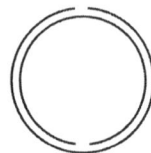

Figure 4.7 *Gluing two affine lines together to get* $\mathbb{P}^1$.

4.4.6. Third example: The projective line. Consider the isomorphism $U \cong V$ via the isomorphism $k[t, 1/t] \cong k[u, 1/u]$ given by $t \leftrightarrow 1/u$. Figure 4.7 is a suggestive picture of this gluing. The resulting scheme is called the **projective line over the field** k, and is denoted by $\mathbb{P}^1_k$.

Notice how the points glue. Let me assume that k is algebraically closed for convenience. (You can think about how this changes otherwise.) On the first affine line, we have the closed ("traditional") points $[(t - a)]$, which we think of as "a on the t-line," and we have the generic point $[(0)]$. On the second affine line, we have closed points that are "b on the u-line," and the generic point. Then a on the t-line is glued to $1/a$ on the u-line (if $a \neq 0$, of course), and the generic point is glued to the generic point (the ideal (0) of $k[t]$ becomes the ideal (0) of $k[t, 1/t]$ upon localization, and the ideal (0) of $k[u]$ becomes the ideal (0) of $k[u, 1/u]$. And (0) in $k[t, 1/t]$ is (0) in $k[u, 1/u]$ under the isomorphism $t \xleftarrow{\sim} 1/u$).

4.4.7. If k is algebraically closed, we can interpret the closed points of $\mathbb{P}^1_k$ in the following way, which may make this sound closer to the way you have seen projective space defined earlier. The points are of the form $[a, b]$, where a and b are not both zero, and $[a, b]$ is identified with $[ac, bc]$, where $c \in k^\times$. Then if $b \neq 0$, this is identified with a/b on the t-line, and if $a \neq 0$, this is identified with b/a on the u-line.

4.4.8. Proposition — $\mathbb{P}^1_k$ *is not affine.*

Proof. We do this by calculating the ring of global sections. The global sections correspond to sections over X and sections over Y that agree on the overlap. A section on X is a polynomial $f(t)$. A section on Y is a polynomial $g(u)$. If we restrict $f(t)$ to the overlap, we get something we can still call $f(t)$; and similarly for $g(u)$. Now we want them to be equal: $f(t) = g(1/t)$. But the only polynomials in t that are at the same time polynomials in $1/t$ are the constants k. Thus $\Gamma(\mathbb{P}^1, \mathscr{O}_{\mathbb{P}^1}) = k$. If $\mathbb{P}^1$ were affine, then it would be $\operatorname{Spec} \Gamma(\mathbb{P}^1, \mathscr{O}_{\mathbb{P}^1}) = \operatorname{Spec} k$, i.e., one point. But it isn't—it has lots of points. $\square$

We have proved an analog of an important theorem: the only holomorphic functions on $\mathbb{CP}^1$ are the constants! (See §11.4.7 for a serious yet easy generalization.)

4.4.9. Important example: Projective space. We now make a preliminary definition of **projective n-space over a field** k, denoted by $\mathbb{P}^n_k$, by gluing together $n + 1$ open sets each isomorphic to $\mathbb{A}^n_k$. Judicious choice of notation for these open sets will make our life easier. Our motivation is as follows. In the construction of $\mathbb{P}^1$ above, we thought of points of projective space as $[x_0, x_1]$, where (x_0, x_1) are only determined up to scalars, i.e., (x_0, x_1) is considered the same as $(\lambda x_0, \lambda x_1)$. Then the first patch can be interpreted by taking the locus where $x_0 \neq 0$, and then we consider the points $[1, t]$, and we think of t as x_1/x_0; even though x_0 and x_1 are not well-defined, x_1/x_0 *is*. The second corresponds to where $x_1 \neq 0$, and we consider the points $[u, 1]$, and we think of u as x_0/x_1. It will be useful to instead use the notation $x_{1/0}$ for t and $x_{0/1}$ for u.

For $\mathbb{P}^n$, we glue together $n + 1$ open sets, one for each of $i = 0, \ldots, n$. The ith open set U_i will have coordinates $x_{0/i}, \ldots, x_{(i-1)/i}, x_{(i+1)/i}, \ldots, x_{n/i}$. It will be convenient to write this as

(4.4.9.1) $$\operatorname{Spec} k[x_{0/i}, x_{1/i}, \ldots, x_{n/i}]/(x_{i/i} - 1)$$

(so we have introduced a "dummy variable" $x_{i/i}$, which we immediately set to 1). We glue the distinguished open set $D(x_{j/i})$ of U_i to the distinguished open set $D(x_{i/j})$ of U_j, by identifying these two schemes by describing the identification of rings

$$k[x_{0/i}, x_{1/i}, \ldots, x_{n/i}, 1/x_{j/i}]/(x_{i/i} - 1) \cong$$

$$k[x_{0/j}, x_{1/j}, \ldots, x_{n/j}, 1/x_{i/j}]/(x_{j/j} - 1)$$

via $x_{k/i} = x_{k/j}/x_{i/j}$ and $x_{k/j} = x_{k/i}/x_{j/i}$ (which implies $x_{i/j}x_{j/i} = 1$). We need to check that this gluing information agrees over triple overlaps.

4.4.D. EXERCISE. Check this, as painlessly as possible. (Possible hint: The triple intersection is affine; describe the corresponding ring.)

4.4.10. Definition: $\mathbb{P}^n_A$. Note that our definition does not use the fact that k is a field. Hence we may as well define $\mathbb{P}^n_A$ for any *ring* A. This will be useful later.

4.4.E. EXERCISE. Show that the only functions on $\mathbb{P}^n_k$ are constants ($\Gamma(\mathbb{P}^n_k, \mathscr{O}) \cong k$), and hence that $\mathbb{P}^n_k$ is not affine if $n > 0$. Hint: You might fear that you will need some delicate interplay among all of your affine open sets, but you will only need two of your open sets to see this. There is even some geometric intuition behind this: the complement of the union of two open sets has codimension 2. But "Algebraic Hartogs's Lemma" (discussed informally in §4.4.2, and to be stated rigorously in Theorem 13.5.19) says that any function defined on this union extends to a function on all of projective space. Because we are expecting to see only constants as functions on all of projective space, we should already see this for this union of our two affine open sets.

4.4.F. EXERCISE (GENERALIZING §4.4.7). Show that if k is algebraically closed, the closed points of $\mathbb{P}^n_k$ may be interpreted in the traditional way: the points are of the form $[a_0, \ldots, a_n]$, where the a_i are not all zero, and $[a_0, \ldots, a_n]$ is identified with $[\lambda a_0, \ldots, \lambda a_n]$, where $\lambda \in k^\times$.

Helpful translation: We think of the closed points of $\mathbb{P}^n_k$ (where $k = \bar{k}$) as $[x_0, x_1, \ldots, x_n]$, with $x_i \in k$ (not all x_i zero), and we identify $[x_0, x_1, \ldots, x_n]$ with $[\lambda x_0, \lambda x_1, \ldots, \lambda x_n]$ for all $\lambda \in k^\times$. Then $[x_0, x_1, \ldots, x_n]$ corresponds to a line through the origin, and $(x_{0/i}, x_{1/i}, \ldots, x_{n/i})$ is where this line meets the hyperplane $x_i = 1$.

We will later give other definitions of projective space (Definition 4.5.9, §15.2.3), and the x_i will be called *projective coordinates* there. Our first definition here will often be handy for computing things. But there is something unnatural about it—projective space is highly symmetric, and that isn't clear from our current definition. Furthermore, as noted by Herman Weyl [Wey, p. 90]: "The introduction of numbers as coordinates is an act of violence."

4.4.11. Fun aside: The Chinese Remainder Theorem is a *geometric* fact. The Chinese Remainder Theorem is embedded in what we have done. We will see this in a single example, but you should then figure out the general statement. The Chinese Remainder Theorem says that knowing an integer modulo 60 is the same as knowing an integer modulo 3, 4, and 5. Here is how to see this in the language of schemes. What is $\operatorname{Spec} \mathbb{Z}/(60)$? What are the prime ideals of this ring? Answer: those prime ideals of $\mathbb{Z}$ containing (60), i.e., those primes dividing 60, i.e., (2), (3), and (5).

Figure 4.8 is a sketch of $\operatorname{Spec} \mathbb{Z}/(60)$. They are all closed points, as these are all maximal ideals, so the topology is the discrete topology. What are the stalks? You can check that they are $\mathbb{Z}/(4)$, $\mathbb{Z}/(3)$, and $\mathbb{Z}/(5)$. The nilpotents "at (2)" are indicated by the "fuzz" on that point. (We discussed visualizing nilpotents with "infinitesimal fuzz" in §4.2.) So what are global sections on this scheme? They are

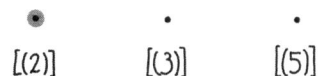

[(2)]	[(3)]	[(5)]

Figure 4.8 *A picture of the scheme* $\operatorname{Spec} \mathbb{Z}/(60)$.

sections on this open set (2), this other open set (3), and this third open set (5). In other words, we have a natural isomorphism of rings

$$\mathbb{Z}/(60) \xrightarrow{\;\sim\;} \mathbb{Z}/(2^2) \times \mathbb{Z}/(3) \times \mathbb{Z}/(5).$$

4.4.12. ** *Example.* Here is an example of a function on an open subset of a scheme with some surprising behavior. On $X = \operatorname{Spec} k[w, x, y, z]/(wz - xy)$, consider the open subset $D(y) \cup D(w)$. Clearly the function z/y on $D(y)$ agrees with x/w on $D(w)$ on the overlap $D(y) \cap D(w)$. Hence they glue together to give a section. You may have seen this before when thinking about analytic continuation in complex geometry—we have a "holomorphic" function that has the description z/y on an open set, and this description breaks down elsewhere, but you can still "analytically continue" it by giving the function a different definition on different parts of the space.

  Follow-up for curious experts. This function has no "single description" as a well-defined expression in terms of w, x, y, z! There is a lot of interesting geometry here, and this scheme will be a constant source of (counter)examples for us (look in the index under "recurring (counter)examples"). Here is a glimpse, in terms of words we have not yet defined. The space $\operatorname{Spec} k[w, x, y, z]$ is $\mathbb{A}^4$, and is, not surprisingly, four-dimensional. We are working with the subset X, which is a hypersurface, and is three-dimensional. It is a cone over a smooth quadric surface in $\mathbb{P}^3$ (flip to Figure 9.2). The open subset $D(y) \subset X$ is X minus some hypersurface, so we are throwing away a codimension 1 locus. The open subset $D(w)$ involves throwing away another codimension 1 locus. You might think that the intersection of these two discarded loci is then codimension 2, and that failure of extending this weird function to a global polynomial comes because of a failure of our Hartogs's Lemma-type theorem. But that's not true—$V(y) \cap V(w)$ is in fact codimension 1. Here is what is actually going on. The space $V(y)$ is obtained by throwing away the (cone over the) union of two lines, ℓ and m_1, one in each "ruling" of the surface, and $V(w)$ also involves throwing away the (cone over the) union of two lines, ℓ and m_2. The intersection is the (cone over the) line ℓ, which is a codimension 1 set. Remarkably, despite being "pure codimension 1" the cone over ℓ is not cut out even set-theoretically by a single equation (see Exercise 15.5.N). This means that any expression $f(w, x, y, z)/g(w, x, y, z)$ for our function cannot correctly describe our function on $D(y) \cup D(w)$—at some point of $D(y) \cup D(w)$ it must be $0/0$. Here's why. Our function can't be defined on $V(y) \cap V(w)$, so g must vanish here. But g can't vanish just on the cone over ℓ—it must vanish elsewhere too.

4.5   Projective Schemes, and the Proj Construction

Projective schemes are important for a number of reasons. Here are a few. Schemes that were of "classical interest" in geometry—and those that you would have cared about before knowing about schemes—are all projective or an open subset thereof (basically, *quasiprojective*; see §4.5.10). Moreover, most "schemes of interest" tend to be projective or quasiprojective. In fact, it is very hard to even give an example of a scheme satisfying basic properties—for example, finite type and "Hausdorff" ("separated") over a field—that is provably not quasiprojective. For complex geometers: it is hard to find a compact complex variety that is provably not projective (see Remark 11.4.6), and it is quite hard to come up with a complex variety that is provably not an open subset of a projective variety. So projective schemes are really ubiquitous. Also, the notion of "projective k-scheme" is a good approximation of the algebro-geometric version of compactness ("properness"; see §11.4).

  Finally, although projective schemes may be obtained by gluing together affine schemes, and we know that keeping track of gluing can be annoying, there is a simple means of dealing with them without worrying about gluing. Just as there is a rough dictionary between rings and affine schemes, we will have a (slightly looser) analogous dictionary between graded rings and projective schemes. Just as one can work with affine schemes by instead working with rings, one can work with projective schemes by instead working with graded rings.

4.5.1. Motivation from classical geometry.

For geometric intuition, we recall how one thinks of projective space "classically" (in the classical topology, over the real numbers). $\mathbb{P}^n$ can be interpreted as the lines through the origin in

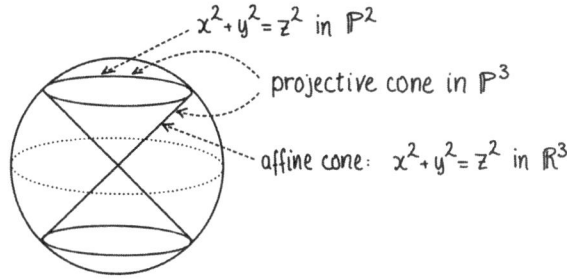

Figure 4.9 *The affine and projective cone of $x^2 + y^2 = z^2$ in classical geometry.*

$\mathbb{R}^{n+1}$. Thus subsets of $\mathbb{P}^n$ correspond to unions of lines through the origin of $\mathbb{R}^{n+1}$, and closed subsets correspond to such unions that are closed. (The same is not true with "closed" replaced by "open"!)

One often pictures $\mathbb{P}^n$ as being the "points at infinite distance" in $\mathbb{R}^{n+1}$, where the points infinitely far in one direction are associated with the points infinitely far in the opposite direction. We can make this more precise using the decomposition

$$\mathbb{P}^{n+1} = \mathbb{R}^{n+1} \coprod \mathbb{P}^n,$$

by which we mean that there is an open subset in $\mathbb{P}^{n+1}$ identified with $\mathbb{R}^{n+1}$ (the points with last "projective coordinate" nonzero), and the complementary closed subset identified with $\mathbb{P}^n$ (the points with last "projective coordinate" zero). (The phrase "projective coordinate" will be formally defined in §4.5.9, but we will use it even before then, in Exercises 4.5.B and 4.5.E.)

Then, for example, any equation cutting out some set V of points in $\mathbb{P}^n$ will also cut out some set of points in $\mathbb{R}^{n+1}$ that will be a closed union of lines. We call this the *affine cone* of V. These equations will cut out some union of $\mathbb{P}^1$s in $\mathbb{P}^{n+1}$, and we call this the *projective cone* of V. The projective cone is the disjoint union of the affine cone and V. For example, the affine cone over $x^2 + y^2 = z^2$ in $\mathbb{P}^2$ is just the "classical" picture of a cone in $\mathbb{R}^3$; see Figure 4.9. We will make this analogy precise in our algebraic setting in §9.3.11.

4.5.2. Projective schemes, a first description.
We now describe a construction of projective schemes, which will help motivate the Proj construction. We begin by giving an algebraic interpretation of the cone just described, and more generally, getting some practice with transforming between projective coordinates, and coordinates on the standard affine open subsets. We switch coordinates from x, y, z to x_0, x_1, x_2 in order to use the notation of §4.4.9. For the next questions, in $\mathbb{P}^2_k$, we take projective coordinates x_0, x_1, x_2. (The terminology "projective coordinate" will not be formally defined until §4.5.9, but you should be able to solve these exercises anyway.) The big open set (or "coordinate chart") $U_0 = \{[x_0, x_1, x_2] : x_0 \neq 0\}$ has coordinates $x_{1/0}$ and $x_{2/0}$, which we interpret as x_1/x_0 and x_2/x_0. We have similar definitions for U_1 and U_2. It will be convenient to define $x_{0/0}$ as 1.

4.5.A. EXERCISE. Describe $(x_{0/1}, x_{2/1})$ in terms of $(x_{1/0}, x_{2/0})$.

4.5.B. EXERCISE. Explain how to define a scheme that should be interpreted as $x_0^2 + x_1^2 - x_2^2 = 0$ "in $\mathbb{P}^2_k$." Hint: In the affine open subset corresponding to $x_2 \neq 0$, it should (in the language of 4.4.9) be cut out by $x_{0/2}^2 + x_{1/2}^2 - 1 = 0$, i.e., it should "be" the scheme $\operatorname{Spec} k[x_{0/2}, x_{1/2}]/(x_{0/2}^2 + x_{1/2}^2 - 1)$. You can similarly guess what it should be on the other two standard open sets, and show that the three schemes glue together.

4.5.C. EXERCISE. (a) Describe *all* homogeneous polynomials in x_0, x_1, x_2 that can be "dehomogenized" to $x_{1/0}^2 + x_{2/0}^5 = 1$. Explain which one is "best." (b) In what points do the vanishing sets of these polynomials meet the "line at infinity" $x_0 = 0$ in $\mathbb{P}^2_k$?

4.5.D. EXERCISE. Consider the parabola $x_{2/0} = x_{1/0}^2$ (or, if you prefer, $y = x^2$). How does it meet the line at infinity? (What does this mean?) What is its description in the different coordinate patches U_1 and U_2?

4.5.3. *Remark: Degree* d *hypersurfaces in* $\mathbb{P}^n$. The degree d homogeneous polynomials in $n + 1$ variables over a field form a vector space of dimension $\binom{n+d}{d}$. (This is essentially the content of Exercises 9.3.J and 15.1.C.) It is almost true that two polynomials cut out the same subset of $\mathbb{P}_k^n$ if and only if one is a nonzero scalar multiple of the other. Unfortunately, the examples of $x^2y = 0$ and $xy^2 = 0$ show that the "only if" doesn't always hold. You will later be able to check (for example, using Exercise 9.5.C) that two polynomials cut out the same *closed subscheme* (whatever that means) if *and only if* one is a nonzero scalar multiple of the other. (This is some evidence that the notion of a "closed subscheme" is better than that of a "closed subset.") The zero polynomial doesn't really cut out a hypersurface in any reasonable sense of the word. Thus we informally imagine that "degree d hypersurfaces in $\mathbb{P}^n$ are parametrized by $\mathbb{P}^{\binom{n+d}{d}-1}$." This intuition will come up repeatedly (in special cases), and we will give it precise meaning in §25.3.5. (We will properly define *hypersurfaces* in §9.3.2, once we have the language of closed subschemes. At that time we will also define line, hyperplane, quadric hypersurface, conic curve, and other wondrous notions.)

4.5.E. EXERCISE (GENERALIZING EXERCISE 4.5.B). Consider $\mathbb{P}_A^n$, with projective coordinates $x_0, \ldots, x_n$. Given a collection of homogeneous polynomials $f_i \in A[x_0, \ldots, x_n]$, make sense of the scheme "cut out in $\mathbb{P}_A^n$ by the f_i." (This will later be made precise as an example of a "vanishing scheme"; see Exercise 4.5.S.)

This could be taken as the definition of a *projective A-scheme*, but we will wait until §4.5.10 to state it a little better.

4.5.4. Preliminaries on graded rings.
The Proj construction produces a scheme out of a graded ring. We now give some background on graded rings.

4.5.5. *Definition:* $\mathbb{Z}$-*graded rings.* A $\mathbb{Z}$-**graded ring** is a ring $S_\bullet = \oplus_{n \in \mathbb{Z}} S_n$ (the subscript is called the **grading**) where multiplication respects the grading, i.e., sends $S_m \times S_n$ to S_{m+n}. Clearly S_0 is a subring, each S_n is an S_0-module, and $S_\bullet$ is an S_0-algebra. Suppose for the remainder of §4.5.5 that $S_\bullet$ is a $\mathbb{Z}$-graded ring. Elements of some individual S_n are called **homogeneous elements** of $S_\bullet$. An ideal I of $S_\bullet$ is a **homogeneous ideal** (or a **graded ideal**, although we won't use this terminology) if it is generated by homogeneous elements. Nonzero homogeneous elements have an obvious **degree**. An element of S_d is called a **form of degree** d.

4.5.F. EXERCISE.

(a) Show that an ideal I is homogeneous if and only if it contains the degree n piece of each of its elements for each n. (Hence I can be decomposed into homogeneous pieces, $I = \oplus I_n$, and $S_\bullet/I$ has a natural $\mathbb{Z}$-grading. This is the reason for the name *homogeneous ideal.*)
(b) Show that the set of homogeneous ideals of a given $\mathbb{Z}$-graded ring $S_\bullet$ is closed under sum, product, intersection, and radical.
(c) Show that a homogeneous ideal $I \subset S_\bullet$ is prime if $I \neq S_\bullet$, and if for any *homogeneous* $a, b \in S_\bullet$, if $ab \in I$, then $a \in I$ or $b \in I$.

   If T is a multiplicative subset of $S_\bullet$ containing only homogeneous elements, then $T^{-1}S_\bullet$ has a natural structure as a $\mathbb{Z}$-graded ring.

4.5.6. $\mathbb{Z}^{\geq 0}$-*graded rings, graded ring over* A, *and finitely generated graded rings.* A $\mathbb{Z}^{\geq 0}$-**graded ring** is a $\mathbb{Z}$-graded ring with no elements of negative degree.

For the remainder of the book, *graded ring* will refer to a $\mathbb{Z}^{\geq 0}$-graded ring. **Warning: This convention is nonstandard (for good reason).**

From now on, unless otherwise stated, $S_\bullet$ is assumed to be a graded ring. Fix a ring A, which we call the **base ring**. If $S_0 = A$, we say that $S_\bullet$ is a **graded ring over** A. A key example is $A[x_0, \ldots, x_n]$, or more generally $A[x_0, \ldots, x_n]/I$, where I is a homogeneous ideal with $I \cap S_0 = 0$ (cf. Exercise 4.5.E). Here we take the conventional grading on $A[x_0, \ldots, x_n]$, where each x_i has weight 1.

The subset $S_+ := \oplus_{i>0} S_i \subset S_\bullet$ is an ideal, called the **irrelevant ideal**. The reason for the name "irrelevant" will be clearer in a few paragraphs. If the irrelevant ideal S_+ is a finitely generated ideal, we say that $S_\bullet$ is a **finitely generated graded ring over** A. If $S_\bullet$ is generated by S_1 as an A-algebra, we say that $S_\bullet$ is **generated in degree** 1. (We will later find it useful to interpret "$S_\bullet$ is generated in degree 1" as "the natural map $\text{Sym}^\bullet S_1 \to S_\bullet$ is a surjection." The *symmetric algebra* construction will be briefly discussed in §14.2.6.)

4.5.G. EXERCISE.

(a) Show that a graded ring $S_\bullet$ over A is a finitely generated graded ring (over A) if and only if $S_\bullet$ is a finitely generated graded A-algebra, i.e., generated over $A = S_0$ by a finite number of homogeneous elements of positive degree. (Hint for the forward implication: Show that the generators of S_+ as an ideal are also generators of $S_\bullet$ as an algebra.)

(b) Show that a graded ring $S_\bullet$ over A is Noetherian if and only if $A = S_0$ is Noetherian and $S_\bullet$ is a finitely generated graded ring.

4.5.7. The Proj **construction.**

We now define a scheme $\text{Proj} \, S_\bullet$, where $S_\bullet$ is a ($\mathbb{Z}^{\geq 0}$-)graded ring. Here are two examples, to provide a light at the end of the tunnel. If $S_\bullet = A[x_0, \ldots, x_n]$, we will recover $\mathbb{P}^n_A$; and if $S_\bullet = A[x_0, \ldots, x_n]/(f(x_0, \ldots, x_n))$ where f is homogeneous, we will construct something "cut out in $\mathbb{P}^n_A$ by the equation $f = 0$" (cf. Exercise 4.5.E).

As we did with Spec of a ring, we will build $\text{Proj} \, S_\bullet$ first as a set, then as a topological space, and finally as a ringed space. In our preliminary definition of $\mathbb{P}^n_A$, we glued together $n + 1$ well-chosen affine pieces, but we don't want to make any choices, so we do this by simultaneously considering "all possible" affine open sets. Our affine building blocks will be as follows. For each homogeneous $f \in S_+$, note that the localization $(S_\bullet)_f$ is naturally a $\mathbb{Z}$-graded ring, where $\deg(1/f) = -\deg f$. Consider

$$(4.5.7.1) \qquad\qquad \text{Spec}((S_\bullet)_f)_0,$$

where $((S_\bullet)_f)_0$ means the 0-graded piece of the graded ring $(S_\bullet)_f$. (These $\text{Spec}((S_\bullet)_f)_0$ will be our affine building blocks, as f varies over the homogeneous elements of S_+.) The notation $((S_\bullet)_f)_0$ is admittedly horrible—the first and third subscripts refer to the grading, and the second refers to localization. As motivation for considering this construction: applying this to $S_\bullet = k[x_0, \ldots, x_n]$, with $f = x_i$, we obtain the ring appearing in (4.4.9.1),

$$k[x_{0/i}, x_{1/i}, \ldots, x_{n/i}]/(x_{i/i} - 1).$$

(Before we begin the construction: another possible way of defining $\text{Proj} \, S_\bullet$ is by gluing together affines of this form, by jumping straight to Exercises 4.5.N and 4.5.O. If you prefer that, by all means do so.)

The *points* of $\text{Proj} \, S_\bullet$ are the homogeneous prime ideals of $S_\bullet$ not containing the irrelevant ideal S_+ (the "relevant homogeneous prime ideals").

4.5.H. IMPORTANT AND TRICKY EXERCISE. Suppose $f \in S_+$ is homogeneous.

(a) Give a bijection between the prime ideals of $((S_\bullet)_f)_0$ and the homogeneous prime ideals of $(S_\bullet)_f$. Hint: Avoid notational confusion by proving instead that if A is a $\mathbb{Z}$-graded ring with a homogeneous invertible element f in positive degree, then there is a bijection between prime ideals of A_0 and homogeneous prime ideals of A. Using the ring map $A_0 \to A$, from each homogeneous prime ideal P of $A_\bullet$ we find a prime ideal $P_0 := P \cap A_0$ of A_0. For the other direction, for each prime ideal P_0 of A_0, show that $P := \sqrt{P_0 A_\bullet}$ is a prime ideal of $A_\bullet$. Be sure to show that these maps indeed give a bijection!

(b) Interpret the set of prime ideals of $((S_\bullet)_f)_0$ as a subset of $\operatorname{Proj} S_\bullet$.

The correspondence of the points of $\operatorname{Proj} S_\bullet$ with homogeneous prime ideals helps us picture $\operatorname{Proj} S_\bullet$. For example, if $S_\bullet = k[x, y, z]$ with the usual grading, then we picture the homogeneous prime ideal $(z^2 - x^2 - y^2)$ first as a subset of $\operatorname{Spec} S_\bullet$; it is a cone (see Figure 4.9). As in §4.5.1, we picture $\mathbb{P}^2_k$ as the "plane at infinity." Thus we picture this equation as cutting out a conic "at infinity" (in $\operatorname{Proj} S_\bullet$). We will make this intuition somewhat more precise in §9.3.11.

4.5.8. Motivated by the affine case, if T is a set of homogeneous elements of $S_\bullet$, define the (projective) **vanishing set of** T, $V(T) \subset \operatorname{Proj} S_\bullet$, to be those homogeneous prime ideals containing T but not S_+. If I is a homogeneous ideal, define $V(I)$ in the same way; notice that it is the same as $V(T)$ where T is any homogeneous set of generators of I.

Define $V(f)$ if f is a homogeneous element of *positive* degree in the obvious way. Let $D_+(f) :=$ $\operatorname{Proj} S_\bullet \setminus V(f)$ (the **projective distinguished open set**) be the complement of $V(f)$. (The subscript is intended to distinguish this notation from the similar $D(f)$.) Once we define a scheme structure on $\operatorname{Proj} S_\bullet$, we will (without comment) use $D_+(f)$ to refer to the open *subscheme*, not just the open subset. (Although the definition of $D_+(f)$ makes sense even if f has degree 0, we deliberately allow only f of positive degree. For example, we will want the $D_+(f)$ to form an affine cover, and if f has degree 0, then $D_+(f)$ needn't be affine.)

4.5.I. EXERCISE. Show that $D_+(f)$ "is" (or more precisely, "corresponds to") the subset $\operatorname{Spec}((S_\bullet)_f)_0$ you described in Exercise 4.5.H(b). For example, in §4.4.9, the $D_+(x_i)$ are the standard open sets covering projective space.

As in the affine case, the $V(I)$'s satisfy the axioms of the closed sets of a topology, and we call this the **Zariski topology** on $\operatorname{Proj} S_\bullet$. (Other definitions given in the literature may look superficially different, but can be easily shown to be the same.) Many statements about the Zariski topology on Spec of a ring carry over to this situation with little extra work. Clearly $D_+(f) \cap D_+(g) = D_+(fg)$, by the same immediate argument as in the affine case (Exercise 3.5.D).

4.5.J. EASY EXERCISE. Verify that the projective distinguished open sets $D_+(f)$ (as f runs through the homogeneous elements of S_+) form a base of the Zariski topology.

4.5.K. EXERCISE. Fix a graded ring $S_\bullet$.

(a) Suppose I is any homogeneous ideal of $S_\bullet$ contained in S_+, and f is a homogeneous element of positive degree. Show that f vanishes on $V(I)$ (i.e., $V(I) \subset V(f)$) if and only if $f^n \in I$ for some n. (Hint: Mimic the affine case; see Exercise 3.4.J.) In particular, as in the affine case (Exercise 3.5.E), if $D_+(f) \subset D_+(g)$, then $f^n \in (g)$ for some n, and vice versa. (Here g is also homogeneous of positive degree.)

(b) If $Z \subset \operatorname{Proj} S_\bullet$, define $I(Z) \subset S_+$. Show that it is a homogeneous ideal of $S_\bullet$. For any two subsets, show that $I(Z_1 \cup Z_2) = I(Z_1) \cap I(Z_2)$.

(c) For any subset $Z \subset \operatorname{Proj} S_\bullet$, show that $V(I(Z)) = \overline{Z}$.

4.5.L. EXERCISE (CF. EXERCISE 3.5.B). Fix a graded ring $S_\bullet$, and a homogeneous ideal $I \subseteq S_+$. Show that the following are equivalent.

(a) $V(I) = \varnothing$.

(b) For any homogeneous f_i (as i runs through some index set) generating I, $\cup D_+(f_i) = \operatorname{Proj} S_\bullet$.

(c) $\sqrt{I} \supset S_+$.

This is more motivation for the ideal S_+ being "irrelevant:" any ideal whose radical contains it is "geometrically irrelevant."

We now construct $\operatorname{Proj} S_\bullet$ as a *scheme*.

4.5.M. EXERCISE. Suppose some homogeneous $f \in S_+$ is given. Via the inclusion

$$D_+(f) = \operatorname{Spec}((S_\bullet)_f)_0 \hookrightarrow \operatorname{Proj} S_\bullet$$

of Exercise 4.5.I, show that the Zariski topology on $\operatorname{Proj} S_\bullet$ restricts to the Zariski topology on $\operatorname{Spec}((S_\bullet)_f)_0$.

Now that we have defined $\operatorname{Proj} S_\bullet$ as a topological space, we are ready to define the structure sheaf. On $D_+(f)$, we wish it to be the structure sheaf of $\operatorname{Spec}((S_\bullet)_f)_0$. We will glue these sheaves together using Exercise 2.5.E on gluing sheaves.

4.5.N. EXERCISE. If $f, g \in S_+$ are homogeneous and nonzero, describe an isomorphism between $\operatorname{Spec}((S_\bullet)_{fg})_0$ and the distinguished open subset $D(g^{\deg f}/f^{\deg g})$ of $\operatorname{Spec}((S_\bullet)_f)_0$.

Similarly, $\operatorname{Spec}((S_\bullet)_{fg})_0$ is identified with a distinguished open subset of $\operatorname{Spec}((S_\bullet)_g)_0$. We then glue the various $\operatorname{Spec}((S_\bullet)_f)_0$ (as f varies) together, using these pairwise gluings.

4.5.O. EXERCISE. By checking that these gluings behave well on triple overlaps (see Exercise 2.5.E), finish the definition of the scheme $\operatorname{Proj} S_\bullet$.

4.5.P. EXERCISE. (Some will find this essential, while others will prefer to ignore it.) (Re)interpret the structure sheaf of $\operatorname{Proj} S_\bullet$ in terms of compatible germs.

4.5.9. *Definition.* We (re)define **projective space** (over a ring A) by

$$\mathbb{P}_A^n := \operatorname{Proj} A[x_0, \dots, x_n].$$

This definition involves no messy gluing, or special choice of patches. Note that the variables $x_0, \dots, x_n$, which we call the **projective coordinates** on $\mathbb{P}_A^n$, are part of the definition. (They may have other names than x's, depending on the context.)

4.5.Q. EXERCISE. Check that this agrees with our earlier construction of $\mathbb{P}_A^n$ (§4.4.9). (How do you know that the $D_+(x_i)$ cover $\operatorname{Proj} A[x_0, \dots, x_n]$?)

Notice that with our old definition of projective space, it would have been a nontrivial exercise to show that $D_+(x^2 + y^2 - z^2) \subset \mathbb{P}_k^2$ (the complement of a plane conic) is affine; with our new perspective, it is immediate—it is $\operatorname{Spec}(k[x, y, z]_{(x^2+y^2-z^2)})_0$.

4.5.R. EXERCISE. Suppose that k is an algebraically closed field. We know from Exercise 4.4.F that the closed points of $\mathbb{P}_k^n$, as defined in §4.4.9, are in bijection with the points of "classical" projective space. By Exercise 4.5.Q, the scheme $\mathbb{P}_k^n$ as defined in §4.4.9 is isomorphic to $\operatorname{Proj} k[x_0, \dots, x_n]$. Therefore, each point $[a_0, \dots, a_n]$ of classical projective space corresponds to a homogeneous prime ideal of $k[x_0, \dots, x_n]$. Which homogeneous prime ideal is it?

We now figure out the "right definition" of the vanishing scheme, in analogy with the vanishing *set* $V(\cdot)$ (§3.4.1). You will be defining a *closed subscheme* (mentioned in Remark 4.3.5, and to be properly defined in §9.1).

4.5.S. EXERCISE. If $S_\bullet$ is generated in degree 1, and $f \in S_+$ is homogeneous, explain how to define $V(f)$ "in" $\operatorname{Proj} S_\bullet$, the **vanishing *scheme*** of f. This is the same notation as that of the "vanishing set" $V(\cdot)$, but now means something richer than a mere set. (Exercise 9.1.K is similar. Warning: In general, f isn't a function on $\operatorname{Proj} S_\bullet$. We will later interpret it as something close: a section of a line bundle; see, for example, §15.1.2.) Hence define $V(I)$ for any homogeneous ideal I of S_+. (Another solution in more general circumstances will be given in Exercise 14.2.D.)

4.5.10. Projective and quasiprojective schemes.

We call a scheme of the form (i.e., isomorphic to) Proj $S_\bullet$, where $S_\bullet$ is a *finitely generated* graded ring over A, a **projective scheme over** A, or a **projective A-scheme**. A **quasiprojective A-scheme** is a quasicompact open subscheme of a projective A-scheme. The "A" is omitted if it is clear from the context; often A is a field.

4.5.11. *Unimportant remarks.*

(i) Note that Proj $S_\bullet$ makes sense even when $S_\bullet$ is not finitely generated. This can be useful. For example, you will later be able to do Exercise 7.4.D without worrying about Exercise 7.4.H.

(ii) The quasicompact requirement in the definition of quasiprojectivity is of course redundant in the Noetherian case (cf. Exercise 3.6.U), which is all that matters to most.

4.5.12. *Silly example.* Note that $\mathbb{P}_A^0 = \operatorname{Proj} A[T] \cong \operatorname{Spec} A$. Thus "Spec A is a projective A-scheme."

4.5.13. Example: Projectivization of a vector space $\mathbb{P}V$.

We can make this definition of projective space even more choice-free as follows. Let V be an $(n + 1)$-dimensional vector space over k. (Here k can be replaced by any ring A and V by a free module, as usual.) Define

$$\operatorname{Sym}^\bullet V^\vee = k \oplus V^\vee \oplus \operatorname{Sym}^2 V^\vee \oplus \cdots .$$

(The reason for the dual is explained by the next exercise.) If, for example, V is the dual of the vector space with basis associated to $x_0, \ldots, x_n$, we would have $\operatorname{Sym}^\bullet V^\vee = k[x_0, \ldots, x_n]$. Then we can define $\mathbb{P}V := \operatorname{Proj}(\operatorname{Sym}^\bullet V^\vee)$. In this language, we have an interpretation for $x_0, \ldots, x_n$: they are linear functionals on the underlying vector space V. (Warning: Some authors use the definition $\mathbb{P}V = \operatorname{Proj}(\operatorname{Sym}^\bullet V)$, so be cautious.)

4.5.T. UNIMPORTANT EXERCISE. Suppose k is algebraically closed. Describe a natural bijection between one-dimensional subspaces of V and the closed points of $\mathbb{P}V$. Thus this construction canonically (in a basis-free manner) describes the one-dimensional subspaces of the vector space V.

Unimportant remark: You may be surprised at the appearance of the dual in the definition of $\mathbb{P}V$. This is partially explained by the previous exercise. Most normal (traditional) people define the projectivization of a vector space V to be the space of one-dimensional subspaces of V. Grothendieck considered the projectivization to be the space of one-dimensional *quotients*. One motivation for this is that it gets rid of the annoying dual in the definition above. There are better reasons, which we won't go into here. In a nutshell, quotients tend to be better behaved than subobjects for coherent sheaves, which generalize the notion of vector bundle. (Coherent sheaves are discussed in Chapter 14.)

On another note related to Exercise 4.5.T: you can also describe a natural bijection between points of V and the closed points of $\operatorname{Spec}(\operatorname{Sym}^\bullet V^\vee)$. This construction respects the affine/projective cone picture of §9.3.11.

4.5.14. *The Grassmannian.*

At this point, we could describe the fundamental geometric object known as the *Grassmannian*, and give the "wrong" (but correct) definition of it. We will instead wait until §7.7 to give the wrong definition, when we will know enough to sense that something is amiss. The right definition will be given in §16.4.

Chapter 5

Some Properties of Schemes

Now that we have defined the notion of a scheme, we can define some useful properties of schemes. As you see each definition, you should try it out in specific examples of your choice, such as your favorite schemes of the form $\operatorname{Spec} \mathbb{C}[x_1 \ldots, x_n]/(f_1, \ldots, f_r)$.

5.1 Topological Properties

The definitions of *connected, connected component, (ir)reducible, irreducible component, quasicompact, generization, specialization, generic point, Noetherian topological space,* and *closed point* were given in §3.6. You should have pictures in your mind of each of these notions.

Exercise 3.6.C shows that $\mathbb{A}_k^n$ is irreducible (it was easy). This argument "behaves well under gluing," yielding:

5.1.A. EASY EXERCISE. Show that $\mathbb{P}_k^n$ is irreducible.

5.1.B. EXERCISE. Exercise 3.7.F showed that there is a bijection between irreducible closed subsets and points for affine schemes (the map sending a point p to the closed subset $\overline{\{p\}}$ is a bijection). Show that this is true of schemes in general.

5.1.C. EASY EXERCISE. Prove that if X is a scheme that has a finite cover $X = \cup_{i=1}^n \operatorname{Spec} A_i$ where A_i is Noetherian, then X is a Noetherian topological space (§3.6.14). (We will soon call a scheme with such a cover a *Noetherian scheme*, §5.3.4.) Hint: Show that a topological space that is a finite union of Noetherian subspaces is itself Noetherian.

Thus $\mathbb{P}_k^n$ and $\mathbb{P}_\mathbb{Z}^n$ are Noetherian topological spaces: we built them by gluing together a finite number of spectra of Noetherian rings.

5.1.D. EASY EXERCISE. Show that a scheme X is quasicompact if and only if it can be written as a finite union of affine open subschemes. (Hence $\mathbb{P}_A^n$ is quasicompact for any ring A.)

5.1.E. EXERCISE: QUASICOMPACT SCHEMES HAVE CLOSED POINTS. Show that if X is a quasicompact scheme, then every point has a closed point in its closure. Show that every nonempty closed subset of X contains a closed point of X. In particular, every nonempty quasicompact scheme has a closed point. (Warning: There exist nonempty schemes with no closed points, such as in Exercise 15.2.P, so your argument had better use the quasicompactness hypothesis!)

This exercise will often be used in the following way. If there is some property P of points of a scheme that is "open" (if a point p has P, then there is some open neighborhood U of p such that all the points in U have P), then to check if *all* points of a quasicompact scheme have P, it suffices to check only the closed points. (A first example of this philosophy is Exercise 5.2.E.) This provides a connection between schemes and the classical theory of varieties—the points of traditional varieties are the *closed* points of the corresponding schemes (essentially by the Nullstellensatz; see §3.6.9 and Exercise 5.3.F). In many good situations, the closed points are dense (such as for varieties; see §3.6.9 and Exercise 5.3.F again), but this is not true in some fundamental cases (see Exercise 3.6.J(b)).

5.1.1. Quasiseparated schemes. Quasiseparatedness is a weird notion that comes in handy for certain people. (Warning: We will later realize that this is really a property of *morphisms*, not of schemes, §8.3.1.) Most people, however, can ignore this notion, as the schemes they will encounter in real life will all have this property. A topological space is **quasiseparated** if the intersection of any two quasicompact open subsets is quasicompact. (The motivation for the "separatedness" in the name is that it is a weakened version of *separated*, for which the intersection of any two *affine* open sets is affine; see Proposition 11.2.13.)

5.1.F. SHORT EXERCISE. Show that a scheme is quasiseparated if and only if the intersection of any two affine open subsets is a finite union of affine open subsets.

5.1.G. EXERCISE. Show that affine schemes are quasiseparated. (Possible hint: Exercise 5.1.F.)

5.1.H. EXERCISE. Show that a scheme X is quasiseparated if and only if there *exists* a cover of X by affine open subsets, any two of which have intersection also covered by a *finite* number of affine open subsets.

   We will see that quasiseparatedness will be a useful hypothesis in theorems (in conjunction with quasicompactness), and that various interesting kinds of schemes (affine, locally Noetherian, and separated—see Exercises 5.1.G, 5.3.A, and 11.2.D, respectively) are quasiseparated, and this will allow us to state theorems more succinctly (e.g., "if X is quasicompact and quasiseparated" rather than "if X is quasicompact, and either this or that or the other thing holds").
   "Quasicompact and quasiseparated" means something concrete:

5.1.I. EXERCISE. Show that a scheme X is quasicompact and quasiseparated if and only if X can be covered by a finite number of affine open subsets, any two of which have intersection also covered by a finite number of affine open subsets.

5.1.2. *Qcqs.* When you see "quasicompact and quasiseparated" as hypotheses in a theorem, you should take this as a clue that you will use this interpretation, and that finiteness will be used in an essential way. This combination of hypotheses is so common that people often describe it as **qcqs** for short (see, for example, the Qcqs Lemma 6.2.9).

5.1.J. EASY EXERCISE. Show that all projective A-schemes are quasicompact and quasiseparated. (Hint: Use the fact that the graded ring in the definition is finitely generated—those finite number of generators will lead you to a covering set.)

5.1.K. EXERCISE (A NONQUASISEPARATED SCHEME). Let $X = \operatorname{Spec} k[x_1, x_2, \dots]$, and let U be $X \setminus [\mathfrak{m}]$, where $\mathfrak{m}$ is the maximal ideal $(x_1, x_2, \dots)$. Take two copies of X, glued along U ("affine ∞-space with a doubled origin"; see Example 4.4.5 and Exercise 4.4.C for "finite-dimensional" versions). Show that the result is not quasiseparated. Hint: This "open embedding" $U \subset X$ came up earlier in Exercise 3.6.G(b) as an example of a nonquasicompact open subset of an affine scheme.

5.1.3. Dimension. One very important topological notion is *dimension*. (It is amazing that this is a *topological* idea.) But despite being intuitively fundamental, it is more difficult, so we postpone it until Chapter 12.

5.2 Reducedness and Integrality

Recall that one of the alarming things about schemes is that functions are not determined by their values at points, and that is because of the presence of nilpotents (§3.2.12).

5.2.1. *Definition.* We say that a scheme X is **reduced** if its stalks $\mathcal{O}_{X,p}$ have no nonzero nilpotents (for all p). Reducedness is by definition a stalk-local property (§2.4.1).

5.2.A. EXERCISE. Show that if f and g are two functions (global sections of $\mathcal{O}_X$) on a reduced scheme that agree at all points, then f = g. (Two hints: $\mathcal{O}_X(U) \hookrightarrow \prod_{p \in U} \mathcal{O}_{X,p}$, from Exercise 2.4.A, and the nilradical is the intersection of all prime ideals, from Theorem 3.2.13.)

5.2.B. EXERCISE. Recall that a ring is said to be *reduced* if it has no nonzero nilpotents (§3.2.14). Show that A is a reduced ring if and only if Spec A is reduced. Show that a scheme X is reduced if $\mathcal{O}_X(U)$ is reduced for every open set U of X.

5.2.C. EASY EXERCISE. Show that $\mathbb{A}_k^n$ and $\mathbb{P}_k^n$ are reduced.

The scheme $\operatorname{Spec} k[x,y]/(y^2, xy)$ is nonreduced. When we sketched it in Figure 4.4, we indicated that the fuzz represented nonreducedness at the origin. The following exercise is a first stab at making this precise.

5.2.D. EXERCISE. Show that $\left(k[x,y]/(y^2,xy)\right)_x$ has no nonzero nilpotent elements. (Possible hint: Show that it is isomorphic to another ring, by considering the geometric picture. Exercise 3.2.L may give another hint.) Show that the only point of $\operatorname{Spec} k[x,y]/(y^2,xy)$ with a nonreduced stalk is the origin.

5.2.E. UNIMPORTANT EXERCISE. If X is a quasicompact scheme, show that it suffices to check reducedness at closed points. Hint: Do not try to show that reducedness is an open condition (see Remark 5.2.2). Instead, show that any nonreduced point has a nonreduced closed point in its closure, using Exercise 5.1.E. (This result is interesting, but we won't need it.)

5.2.2. *Remark.* We will see in Exercise 6.6.N that reducedness is an open condition for locally Noetherian schemes. But in general, reducedness is not an open condition. You may be able to identify the underlying topological space of

$$X = \operatorname{Spec} \mathbb{C}[x, y_1, y_2, \ldots]/(y_1^2, y_2^2, y_3^2, \ldots, (x-1)y_1, (x-2)y_2, (x-3)y_3, \cdots)$$

with that of $\operatorname{Spec} \mathbb{C}[x]$, and then to show that the nonreduced points of X are precisely the closed points corresponding to the positive integers. The complement of this set is not Zariski open.

5.2.3.* *Warning for experts.* If a scheme X is reduced, then (from Exercise 5.2.B) its ring of global sections is reduced. However, the converse is not true. You already know enough to verify that if X is the scheme cut out by $x^2 = 0$ in $\mathbb{P}_k^2$, then $\Gamma(X, \mathscr{O}_X) \cong k$, and that X is nonreduced. (This example will come up again in §11.4.7 and §18.6.U.)

5.2.F. EXERCISE. Suppose X is quasicompact, and f is a function that vanishes at all points of X. Show that there is some n such that $f^n = 0$. Show that this may fail if X is not quasicompact. (This exercise is less important, but shows why we like quasicompactness, and gives a standard pathology when quasicompactness doesn't hold.) Hint: Take an infinite disjoint union of $\operatorname{Spec} A_n$ with $A_n := k[\epsilon]/(\epsilon^n)$. This scheme arises again in §9.4.2 (see Figure 9.5 for a picture) and in Caution/Example 9.4.11.

5.2.4. *Definition.* A scheme X is **integral** if it is nonempty, and $\mathscr{O}_X(U)$ is an integral domain for every nonempty open subset U of X.

5.2.G. IMPORTANT EXERCISE. Show that a scheme X is integral if and only if it is irreducible and reduced. (Thus we picture integral schemes as: "one piece, no fuzz." Possible hint: Exercise 4.3.F.)

5.2.H. EXERCISE. Show that an affine scheme $\operatorname{Spec} A$ is integral if and only if A is an integral domain.

5.2.I. EXERCISE. Suppose X is an integral scheme. Then X (being irreducible) has a generic point η. Suppose $\operatorname{Spec} A$ is any nonempty affine open subset of X. Show that $\mathscr{O}_{X,\eta}$ (the stalk of $\mathscr{O}_X$ at η) is naturally identified with $K(A)$, the fraction field of A. This is called the **function field** $K(X)$ of X. It can be computed on any nonempty open set of X, as any such open set contains the generic point. The elements of $K(X)$ are called **rational functions on** X (to be generalized further in Definition 6.6.36).

5.2.J. EXERCISE. Suppose X is an integral scheme. Show that the restriction maps $\operatorname{res}_{U,V} : \mathscr{O}_X(U) \to \mathscr{O}_X(V)$ are inclusions so long as $V \neq \varnothing$. Suppose $\operatorname{Spec} A$ is any nonempty affine open subset of X (so A is an integral domain). Show that the natural map $\mathscr{O}_X(U) \to \mathscr{O}_{X,\eta} = K(A)$ (where U is any nonempty open subset and η is the generic point) is an inclusion.

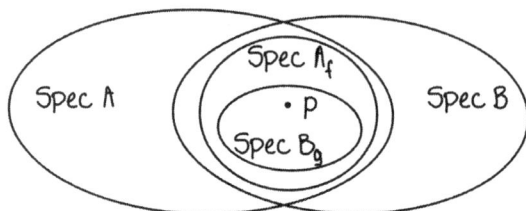

Figure 5.1 *A trick to show that the intersection of two affine open sets may be covered by open sets that are simultaneously distinguished in both affine open sets.*

Thus irreducible varieties (an important example of integral schemes defined later) have the convenient property that sections over different open sets can be considered subsets of the same ring. In particular, restriction maps (except to the empty set) are always inclusions, and gluing is easy: functions f_i on a cover U_i of U (as i runs over an index set) glue if and only if they are the same element of $K(X)$. This is one reason why (irreducible) varieties are usually introduced before schemes.

5.2.5. *Caution.* Integrality is not stalk-local (the disjoint union of two integral schemes is not integral, as $\operatorname{Spec} A \coprod \operatorname{Spec} B = \operatorname{Spec}(A \times B)$ by Exercise 3.6.A), but it almost is; see Exercise 5.3.C.

5.3 The Affine Communication Lemma, and Properties of Schemes That Can Be Checked "Affine-Locally"

This section is intended to address something tricky in the definition of schemes. We have defined a scheme as a topological space with a sheaf of rings that can be covered by affine schemes. Hence we have all of the affine open sets in the cover, but we don't know how to communicate between any two of them. Somewhat more explicitly, if I have an affine cover, and you have an affine cover, and we want to compare them, and I calculate something on my cover, there should be some way of us getting together, and figuring out how to translate my calculation over to your cover. The Affine Communication Lemma 5.3.2 will provide a convenient machine for doing this.

Thanks to this lemma, we can define a host of important properties of schemes. All of these are "affine-local," in that they can be checked on any affine cover, i.e., a covering by open affine sets. (We state this more formally after the statement of the Affine Communication Lemma.) We like such properties because we can check them using any affine cover we like. If the scheme in question is quasicompact, then we need only check a finite number of affine open sets.

5.3.1. Proposition — *Suppose* $\operatorname{Spec} A$ *and* $\operatorname{Spec} B$ *are affine open subschemes of a scheme* X*. Then* $\operatorname{Spec} A \cap \operatorname{Spec} B$ *is the union of open sets that are simultaneously distinguished open subschemes of* $\operatorname{Spec} A$ *and* $\operatorname{Spec} B$.

Proof. (See Figure 5.1.) Given any point $p \in \operatorname{Spec} A \cap \operatorname{Spec} B$, we produce an open neighborhood of p in $\operatorname{Spec} A \cap \operatorname{Spec} B$ that is simultaneously distinguished in both $\operatorname{Spec} A$ and $\operatorname{Spec} B$. Let $\operatorname{Spec} A_f$ be a distinguished open subset of $\operatorname{Spec} A$ contained in $\operatorname{Spec} A \cap \operatorname{Spec} B$ and containing p. Let $\operatorname{Spec} B_g$ be a distinguished open subset of $\operatorname{Spec} B$ contained in $\operatorname{Spec} A_f$ and containing p. Then $g \in \Gamma(\operatorname{Spec} B, \mathcal{O}_X)$ restricts to an element $g' \in \Gamma(\operatorname{Spec} A_f, \mathcal{O}_X) = A_f$. The points where g and g' vanish on $\operatorname{Spec} A_f$ are the same, so

$$\operatorname{Spec} B_g = \operatorname{Spec} A_f \setminus \{[\mathfrak{p}] \ : \ g' \in \mathfrak{p}\}$$
$$= \operatorname{Spec}(A_f)_{g'}.$$

If $g' = g''/f^n$ ($g'' \in A$), then $\operatorname{Spec}(A_f)_{g'} = \operatorname{Spec} A_{fg''}$, and we are done. $\square$

The following easy result will be crucial for us.

5.3.2. Affine Communication Lemma — *Let* P *be some property enjoyed by some affine open subsets of a scheme* X, *such that for any affine open subset* $\operatorname{Spec} A \hookrightarrow X$,

(i)   *if* $\operatorname{Spec} A \hookrightarrow X$ *has property* P, *then for any* $f \in A$, $\operatorname{Spec} A_f \hookrightarrow X$ *does too;*
(ii) *if* $(f_1, \dots, f_n) = A$ *(i.e., the* $\operatorname{Spec} A_{f_i}$ *cover* $\operatorname{Spec} A$; *see Exercise 3.5.B), and* $\operatorname{Spec} A_{f_i} \hookrightarrow X$ *has* P *for all* i, *then so does* $\operatorname{Spec} A \hookrightarrow X$.

Suppose that $X = \cup_{i \in I} \operatorname{Spec} A_i$, *where* $\operatorname{Spec} A_i$ *has property* P. *Then every affine open subset of* X *has* P, *too.*

A property satisfying (i) and (ii) is said to be **affine-local**. One motivation for this name is that, by this lemma, we can determine if *every* affine open subscheme has the property by checking on *any* affine cover.

As an example, reducedness is an affine-local property. More generally, any property that is stalk-local (a scheme has property P if and only if all its stalks have property Q) is necessarily affine-local (a scheme has property P if and only if all of its affine open sets have property R, where an affine scheme has property R if and only if all its stalks have property Q). But it is sometimes not so obvious what the right definition of Q is; see, for example, the discussion of normality in the next section.

Proof. Let $\operatorname{Spec} A$ be an affine subscheme of X. Cover $\operatorname{Spec} A$ with a finite number of distinguished open sets $\operatorname{Spec} A_{g_j}$, each of which is distinguished in some $\operatorname{Spec} A_i$. This is possible by Proposition 5.3.1 and the quasicompactness of $\operatorname{Spec} A$ (Exercise 3.6.G(a)). By (i), each $\operatorname{Spec} A_{g_j}$ has P. By (ii), $\operatorname{Spec} A$ has P. $\qquad\square$

By choosing the property P appropriately, we define some important properties of schemes.

5.3.3. Proposition — *Suppose* A *is a ring, and* $(f_1, \dots, f_n) = A$.

(a) *If* A *is a Noetherian ring, then so is* A_{f_i}. *If each* A_{f_i} *is Noetherian, then so is* A.
(b) *Suppose* B *is a ring, and* A *is a* B*-algebra. (Hence* A_g *is a* B*-algebra for all* $g \in A$.) *If* A *is a finitely generated* B*-algebra, then so is* A_{f_i}. *If each* A_{f_i} *is a finitely generated* B*-algebra, then so is* A.

We will prove these shortly (§5.3.9). But let's first motivate you to read the proof by giving some interesting definitions and results *assuming* Proposition 5.3.3 is true.

5.3.4. *Important definition.* Suppose X is a scheme. If X can be covered by affine open sets $\operatorname{Spec} A$ where A is Noetherian, we say that X is a **locally Noetherian scheme**. If in addition X is quasicompact, or equivalently can be covered by finitely many such affine open sets, we say that X is a **Noetherian scheme**. (We will see a number of definitions of the form "if X has this property, we say that it is locally Q; if further X is quasicompact, we say that it is Q.") By Exercise 5.1.C, the underlying topological space of a Noetherian scheme is Noetherian. Hence by Exercise 3.6.U, all open subsets of a Noetherian scheme are quasicompact.

5.3.A. EXERCISE. Show that locally Noetherian schemes are quasiseparated.

5.3.B. EXERCISE. Show that a Noetherian scheme has a finite number of irreducible components. (Hint: Proposition 3.6.15.) Show that a Noetherian scheme has a finite number of connected components, each a finite union of irreducible components.

5.3.C. EXERCISE. Show that a Noetherian scheme X is integral if and only if X is nonempty and connected and all stalks $\mathscr{O}_{X,p}$ are integral domains. Thus, in "good situations," integrality is the union of local (stalks are integral domains) and global (connected) conditions. Hint: Recall that integral = irreducible + reduced (Exercise 5.2.G). If a scheme's stalks are integral domains, then it is reduced (reducedness is a stalk-local condition, Definition 5.2.1). If a scheme X has underlying topological space that is Noetherian, then X has finitely many irreducible components (by the previous exercise); if two of them meet at a point p, then $\mathscr{O}_{X,p}$ is not an integral domain. (You can readily extend this from Noetherian schemes to locally Noetherian schemes, by showing that a connected scheme is irreducible if and only if it is nonempty and has a cover by open irreducible subsets. But some Noetherian hypotheses are necessary; see [Stacks, tag 0568].)

5.3.5. *Unimportant caution.* The ring of global sections of a Noetherian scheme need not be Noetherian; see Exercise 19.11.H.

5.3.6. Schemes over a given field k**, or more generally over a given ring** A **(A-schemes).** You may be particularly interested in working over a particular field, such as $\mathbb{C}$ or $\mathbb{Q}$, or over a ring such as $\mathbb{Z}$. Motivated by this, we define the notion of A-**scheme**, or **scheme over** A, where A is a ring, as a scheme where all the rings of sections of the structure sheaf (over all open sets) are A-algebras, and all restriction maps are maps of A-algebras. (Like some earlier notions such as quasiseparatedness, this will later, in Exercise 7.3.I, be properly understood as a "relative notion"; it is the data of a morphism $X \to \operatorname{Spec} A$.) Suppose now X is an A-scheme. If X can be covered by affine open sets $\operatorname{Spec} B_i$ where each B_i is a *finitely generated* A-algebra, we say that X is **locally of finite type over** A, or that it is a **locally finite type** A**-scheme**. (This is admittedly cumbersome terminology; it will make more sense later, once we know about morphisms in §8.3.9.) If furthermore X is quasicompact, X is (of) **finite type over** A, or a **finite type** A**-scheme**. Note that a scheme locally of finite type over k or $\mathbb{Z}$ (or indeed any Noetherian ring) is locally Noetherian, and similarly a scheme of finite type over any Noetherian ring is Noetherian. As our key "geometric" examples: (i) $\operatorname{Spec} \mathbb{C}[x_1, \dots, x_n]/I$ is a finite type $\mathbb{C}$-scheme; and (ii) $\mathbb{P}^n_{\mathbb{C}}$ is a finite type $\mathbb{C}$-scheme. (The field $\mathbb{C}$ may be replaced by an arbitrary ring A.)

5.3.D. EXERCISE.

(a) *(Quasiprojective implies finite type)* If X is a quasiprojective A-scheme (Definition 4.5.10), show that X is of finite type over A. If A is furthermore assumed to be Noetherian, show that X is a Noetherian scheme, and hence has a finite number of irreducible components.

(b) Suppose U is an open subscheme of a projective A-scheme. Show that U is locally of finite type over A. If A is Noetherian, show that U is quasicompact, and hence quasiprojective over A, and hence by (a) of finite type over A. Show this need not be true if A is not Noetherian. Better: Give an example of an open subscheme of a projective A-scheme that is not quasicompact, necessarily for some non-Noetherian A. (Hint: Silly example 4.5.12.)

5.3.7. Varieties. We now make a connection to the classical language of varieties. An affine scheme that is reduced and of finite type over k is said to be an **affine variety (over** k**)**, or an **affine** k**-variety**. A reduced (quasi)projective k-scheme is a **(quasi)projective variety (over** k**)**, or a **(quasi)projective** k**-variety**. (Warning: In the literature, it is sometimes also assumed in the definition of variety that the scheme is irreducible, or that k is algebraically closed.)

5.3.E. EXERCISE.

(a) Show that $\operatorname{Spec} k[x_1, \dots, x_n]/I$ is an affine k-variety if and only if $I \subset k[x_1, \dots, x_n]$ is a radical ideal.

(b) Suppose $I \subset k[x_0, \dots, x_n]$ is a radical graded ideal. Show that $\operatorname{Proj} k[x_0, \dots, x_n]/I$ is a projective k-variety. (Caution: The example of $I = (x_0^2, x_0 x_1, \dots, x_0 x_n)$ shows that $\operatorname{Proj} k[x_0, \dots, x_n]/I$ can be a projective k-variety without I being radical.)

We will not define varieties in general until §11.2.9; we will need the notion of separatedness first, to exclude abominations like the line with the doubled origin (Example 4.4.5). But many of the statements we will make in this section about affine k-varieties will automatically apply more generally to k-varieties.

5.3.F. EXERCISE. Show that a point of a locally finite type k-scheme is a closed point if and only if the residue field of the stalk of the structure sheaf at that point is a finite extension of k. Show that on a locally finite type k-scheme, the closed points are dense. Hint: §3.6.9. (Warning: Closed points need not be dense even on quite reasonable schemes; see Exercise 3.6.J(b).)

5.3.8. *Definition.* The **degree of a closed point** p of a locally finite type k-scheme (e.g., a variety over k) is the degree of the field extension $\kappa(p)/k$. For example, in $\mathbb{A}_k^1 = \operatorname{Spec} k[t]$, the point $[(p(t))]$ ($p(t) \in k[t]$ irreducible) is $\deg p(t)$. If k is algebraically closed, the degree of every closed point is 1.

5.3.9. *Proof of Proposition 5.3.3.* We divide each part into (i) and (ii) following the statement of the Affine Communication Lemma 5.3.2.

Proof of Proposition 5.3.3(a). (i) Let $\phi : A \to A_f$ be the localization map. If $I_1 \subsetneq I_2 \subsetneq I_3 \subsetneq \cdots$ is a strictly increasing chain of ideals of A_f, then we can verify that $J_1 \subsetneq J_2 \subsetneq J_3 \subsetneq \cdots$ is a strictly increasing chain of ideals of A, where $J_j = \phi^{-1}(I_j)$. (We think of J_j as $I_j \cap A$, except that in general A needn't inject into A_{f_i}.) Clearly, J_j is an ideal of A. If $x/f^n \in I_{j+1} \setminus I_j$ where $x \in A$, then $x \in J_{j+1}$, and $x \notin J_j$ (or else $x(1/f)^n \in I_j$ as well).

(ii) Suppose $I_1 \subsetneq I_2 \subsetneq I_3 \subsetneq \cdots$ is a strictly increasing chain of ideals of A. Then for each $1 \le i \le n$,

$$I_{i,1} \subset I_{i,2} \subset I_{i,3} \subset \cdots$$

is an increasing chain of ideals in A_{f_i}, where $I_{i,j} = I_j \otimes_A A_{f_i}$. It remains to show that for each j, $I_{i,j} \subsetneq I_{i,j+1}$ for some i; the result will then follow.

5.3.G. EXERCISE. Finish the proof of Proposition 5.3.3(a). (Hint: If $M := I_j/I_{j-1} \ne 0$, then from $M \hookrightarrow \prod M_{f_i}$ (4.1.6.1), one of the M_{f_i} is nonzero.)

Proof of Proposition 5.3.3(b). Part (i) is clear: if A is generated over B by $r_1, \ldots, r_n$, then A_f is generated over B by $r_1, \ldots, r_n, 1/f$.

(ii) Here is the idea. As the f_i generate A, we can write $1 = \sum c_i f_i$ for $c_i \in A$. We have generators of A_{f_i}: $r_{ij}/f_i^{k_j}$, where $r_{ij} \in A$. I claim that $\{f_i\}_i \cup \{c_i\} \cup \{r_{ij}\}_{ij}$ generate A as a B-algebra. Here is why. Suppose you have any $r \in A$. Then in A_{f_i}, we can write r as some polynomial in the r_{ij}'s and f_i, divided by some huge power of f_i. So "in each A_{f_i}, we have described r in the desired way," except for this annoying denominator. Now use a "partition of unity"-type argument as in the proof of Theorem 4.1.2 to combine all of these into a single expression, killing the denominator. Show that the resulting expression you build still agrees with r in each of the A_{f_i}. Thus it is indeed r (by the identity axiom for the structure sheaf).

5.3.H. EXERCISE. Make this argument precise.

This concludes the proof of Proposition 5.3.3. □

5.4   Normality and Factoriality

5.4.1. Normality.

We can now define a property of schemes that says that they are "not too far from smooth," called *normality*, which will come in very handy. We will see later that "locally Noetherian normal schemes satisfy Hartogs's Lemma" (Algebraic Hartogs's Lemma 13.5.19 for Noetherian normal schemes): functions defined away from a set of codimension 2 extend over that set. (We saw a first glimpse of this in §4.4.2.) As a consequence, rational functions (defined in Exercise 5.2.I) that have no poles (certain sets of codimension 1 where the function isn't defined) are defined everywhere. We need definitions of dimension and poles to make this precise. See §13.8.7 and §26.3.5 for the fact that "smoothness" (really, "regularity") implies normality.

Recall that an integral domain A is **integrally closed** if the only zeros in $K(A)$ to any monic polynomial in $A[x]$ must lie in A itself. The basic example is $\mathbb{Z}$ (see Exercise 5.4.F for a reason). We say a scheme X is **normal** if all of its stalks $\mathcal{O}_{X,p}$ are normal, i.e., are integral domains, and integrally closed in their fraction fields. (So, by definition, normality is a *stalk-local* property.) As reducedness is a stalk-local property (Definition 5.2.1), normal schemes are reduced. (One might say that a ring A is a **normal ring** if Spec A is normal, thereby extending the notion of "integral closure" to rings that are not integral domains.)

5.4.A. EXERCISE. Show that integrally closed domains behave well under localization: if A is an integrally closed domain, and S is a multiplicative subset not containing 0, show that $S^{-1}A$ is an integrally closed domain. (Hint: Assume that $x^n + a_{n-1}x^{n-1} + \cdots + a_0 = 0$, where $a_i \in S^{-1}A$ has a root in the fraction field. Turn this into another equation in $A[x]$ that also has a root in the fraction field.)

It is no fun checking normality at every single point of a scheme. Thanks to this exercise, we know that if A is an integrally closed domain, then Spec A is normal. Also, for schemes that are quasicompact or locally of finite type over a field, normality can be checked at closed points, thanks to this exercise, and the fact that for such schemes, any point is a generization of a closed point (Exercises 5.1.E and 5.3.F).

It is not true that normal schemes are integral. For example, the disjoint union of two normal schemes is normal. Thus $\operatorname{Spec} k \coprod \operatorname{Spec} k \cong \operatorname{Spec}(k \times k) \cong \operatorname{Spec} k[x]/(x(x-1))$ is normal, but its ring of global sections is not an integral domain.

5.4.B. EXERCISE. Show that a Noetherian scheme is normal if and only if it is the finite disjoint union of integral Noetherian normal schemes. (Hint: Exercise 5.3.C.)

We are close to proving a useful result in commutative algebra, so we may as well go all the way.

5.4.2. Proposition — *If A is an integral domain, then the following are equivalent.*

(i) *A is integrally closed.*
(ii) *$A_\mathfrak{p}$ is integrally closed for all prime ideals $\mathfrak{p} \subset A$.*
(iii) *$A_\mathfrak{m}$ is integrally closed for all maximal ideals $\mathfrak{m} \subset A$.*

Proof. Exercise 5.4.A shows that integral closedness is preserved by localization, so (i) implies (ii). Clearly (ii) implies (iii).

It remains to show that (iii) implies (i). This argument involves a pretty construction that we will use again. Suppose A is not integrally closed. We show that there is some $\mathfrak{m}$ such that $A_\mathfrak{m}$ is also not integrally closed. Suppose

$$(5.4.2.1) \qquad\qquad x^n + a_{n-1}x^{n-1} + \cdots + a_0 = 0$$

(with $a_i \in A$) has a solution s in $K(A) \setminus A$. Let I be the **ideal of denominators of** s:

$$(5.4.2.2) \qquad\qquad I := \{r \in A \; : \; rs \in A\}.$$

(Note that I is clearly an ideal of A.) Now $I \neq A$, as $1 \notin I$. Thus there is some maximal ideal $\mathfrak{m}$ containing I. Then $s \notin A_\mathfrak{m}$, so equation (5.4.2.1) in $A_\mathfrak{m}[x]$ shows that $A_\mathfrak{m}$ is not integrally closed as well, as desired. $\qquad\square$

5.4.C. UNIMPORTANT EXERCISE. If A is an integral domain, show that $A = \cap A_\mathfrak{m}$, where the intersection runs over all maximal ideals of A. (We won't use this exercise, but it gives good practice with the ideal of denominators.)

5.4.D. UNIMPORTANT EXERCISE RELATING TO THE IDEAL OF DENOMINATORS. One might naively hope from experience with unique factorization domains that the ideal of denominators is principal. This is not true. As a counterexample, consider our new friend $A = k[w, x, y, z]/(wz - xy)$ (which we first met in Example 4.4.12, and which we will later recognize as the cone over the quadric surface), and $w/y = x/z \in K(A)$. Show that the ideal of denominators of this element of $K(A)$ is (y, z).

We will see that the I in the above exercise is not principal (Exercise 13.1.D—you may be able to show it directly, using the fact that I is a homogeneous ideal of a graded ring). But we will also see that in good situations (Noetherian, normal), the ideal of denominators is "pure codimension 1"—this is the content of Algebraic Hartogs's Lemma 13.5.19. In its proof, §13.5.21, we give a geometric interpretation of the ideal of denominators.

5.4.3. Factoriality.

If all the stalks of a scheme X are unique factorization domains, we say that X is **factorial**. (Unimportant remark: The locus of points on an affine variety over an algebraically closed field that are factorial is an open subset, [BGS, p. 1].)

5.4.E. EXERCISE. Show that any nonzero localization of a unique factorization domain is a unique factorization domain.

5.4.4. Thus if A is a unique factorization domain, then Spec A is factorial. The converse need not hold; see Exercise 5.4.N. In fact, we will see that elliptic curves are factorial, yet *no* affine open set is the Spec of a unique factorization domain, §19.11.1. Hence one can show factoriality by finding an appropriate affine cover, but there need not *be* such a cover of a factorial scheme. (One might reasonably call a ring A such that Spec A is factorial, a *factorial ring*; this is a strictly weaker notion than unique factorization domain. We won't need this terminology.)

5.4.5. *Remark: How to check if a ring is a unique factorization domain.* There are very few means of checking that a Noetherian integral domain is a unique factorization domain. Some useful ones are: (0) elementary means: rings with a Euclidean algorithm such as $\mathbb{Z}$, $k[t]$, and $\mathbb{Z}[i]$; polynomial rings over a unique factorization domain, by Gauss's Lemma (see, e.g., [Lan, IV.2.3]); (1) the localization of a unique factorization domain is also a unique factorization domain (Exercise 5.4.E); (2) height 1 prime ideals are principal (Proposition 12.3.7); (3) normal and Cl = 0 (Exercise 15.5.P); (4) Nagata's Lemma (Exercise 15.5.Q). (Caution: Even if A is a unique factorization domain, A[[t]] need not be; see [Mat2, p. 165]. The first example, due to P. Salmon, was $A = k(u)[[x, y, z]]/(z^2 + x^3 + uz^6)$; see [Sa, Dan].)

5.4.6. *Factoriality implies normality.* One reason we like factoriality is that it implies normality.

5.4.F. IMPORTANT EXERCISE. Show that unique factorization domains are integrally closed. Hence factorial schemes are normal, and if A is a unique factorization domain, then Spec A is normal. (However, rings can be integrally closed without being unique factorization domains, as we will see in Exercise 5.4.L. Another example is given without proof in Exercise 5.4.N; in that example, Spec of the ring is factorial. A variation on Exercise 5.4.L will show that schemes can be normal without being factorial; see Exercise 13.1.E.)

5.4.7. *Examples.*

5.4.G. EASY EXERCISE. Show that the following schemes are normal: $\mathbb{A}_k^n$, $\mathbb{P}_k^n$, Spec $\mathbb{Z}$. (As usual, k is a field. Although it is true that if A is integrally closed then A[x] is as well—see [Bo, Ch. 5, §1, no. 3, Cor. 2] or [E, Ex. 4.18]—this is not an easy fact, so do not use it here.)

5.4.H. HANDY EXERCISE (YIELDING MANY ENLIGHTENING EXAMPLES LATER). Suppose A is a unique factorization domain with 2 invertible, and $z^2 - f$ is irreducible in A[z].

(a) Show that if $f \in A$ has no repeated prime factors, then Spec $A[z]/(z^2 - f)$ is normal. Hint: $B := A[z]/(z^2 - f)$ is an integral domain, as $(z^2 - f)$ is prime in A[z]. Suppose we have monic $F(T) \in B[T]$ so that $F(T) = 0$ has a solution α in $K(B) \setminus K(A)$. Then by replacing $F(T)$ by $\bar{F}(T)F(T)$, we can assume $F(T) \in A[T]$. Also, $\alpha = g + hz$, where $g, h \in K(A)$. Now α is the solution of $Q(T) = 0$ for monic $Q(T) = T^2 - 2gT + (g^2 - h^2f) \in K(A)[T]$, so we can factor $F(T) = P(T)Q(T)$ in K(A)[T]. By Gauss's lemma, $2g$, $g^2 - h^2f \in A$. Say, $g = r/2$, $h = s/t$ (s and t have no common factors, $r, s, t \in A$). Then $g^2 - h^2f = (r^2t^2 - 4s^2f)/4t^2$. Then t is invertible.

(b) Show that if $f \in A$ has repeated prime factors, then Spec $A[z]/(z^2 - f)$ is *not* normal.

5.4.I. EXERCISE. Show that the following schemes are normal:

(a) $\operatorname{Spec}\mathbb{Z}[x]/(x^2 - n)$, where n is a square-free integer congruent to 3 modulo 4. Caution: The hypotheses of Exercise 5.4.H do not apply, so you will have to do this directly. (Your argument may also show the result when 3 is replaced by 2. A similar argument shows that the ring $\mathbb{Z}[(1 + \sqrt{n})/2]$ is integrally closed if $n \equiv 1 \pmod{4}$ is square-free.)

(b) $\operatorname{Spec}k[x_1, \ldots, x_n]/(x_1^2 + x_2^2 + \cdots + x_m^2)$, where $\operatorname{char}k \neq 2$, $n \geq m \geq 3$.

(c) $\operatorname{Spec}k[w, x, y, z]/(wz - xy)$, where $\operatorname{char}k \neq 2$. This is our cone over a quadric surface example from Example 4.4.12 and Exercise 5.4.D. Hint: Exercise 5.4.J may help. (The result also holds for $\operatorname{char}k = 2$, but don't worry about this.)

5.4.J. EXERCISE (DIAGONALIZING QUADRATIC FORMS). Suppose k is an algebraically closed field of characteristic not 2. (The hypothesis that k is algebraically closed is not necessary, so feel free to deal with this more general case. Part (a) should then be rewritten slightly: "sum of at most n squares" should be "k-linear combination of at most n squares.")

(a) Show that any quadratic form in n variables can be "diagonalized" by changing coordinates to be a sum of at most n squares (e.g., $uw - v^2 = ((u+w)/2)^2 + (i(u-w)/2)^2 + (iv)^2)$, *where the linear forms appearing in the squares are linearly independent*. (Hint: Use induction on the number of variables, by "completing the square" at each step.)

(b) Show that the number of squares appearing depends only on the quadratic form. For example, $x^2 + y^2 + z^2$ cannot be written as a sum of two squares. (Possible approach: Given a basis $x_1, \ldots, x_n$ of the linear forms, write the quadratic form as

$$\begin{pmatrix} x_1 & \cdots & x_n \end{pmatrix} M \begin{pmatrix} x_1 \\ \vdots \\ x_n \end{pmatrix},$$

where M is a symmetric matrix. Determine how M transforms under a change of basis, and show that the rank of M is independent of the choice of basis.)

The **rank** of the quadratic form is the number of ("linearly independent") squares needed. If the number of squares equals the number of variables, the quadratic form is said to be **full rank** or (of) **maximal rank**.

5.4.K. EASY EXERCISE (RINGS CAN BE INTEGRALLY CLOSED BUT NOT UNIQUE FACTORIZATION DOMAINS, ARITHMETIC VERSION). Show that $\mathbb{Z}[\sqrt{-5}]$ is integrally closed but not a unique factorization domain. (Hints: Exercise 5.4.I(a) and $2 \times 3 = (1 + \sqrt{-5})(1 - \sqrt{-5})$.)

5.4.L. EASY EXERCISE (RINGS CAN BE INTEGRALLY CLOSED BUT NOT UNIQUE FACTORIZATION DOMAINS, GEOMETRIC VERSION). Suppose $\operatorname{char}k \neq 2$. Let $A = k[w, x, y, z]/(wz - xy)$, so $\operatorname{Spec}A$ is the cone over the smooth quadric surface (cf. Exercise 4.4.12 and §5.4.D).

(a) Show that A is integrally closed. (Hint: Exercises 5.4.I(c) and 5.4.J.)

(b) Show that A is not a unique factorization domain. (Clearly, $wz = xy$. But why are w, x, y, and z irreducible? Hint: A is a graded integral domain. Show that if a homogeneous element factors, the factors must be homogeneous.)

The previous two exercises look similar, but there is a difference. The cone over the quadric surface is normal (by Exercise 5.4.L) but not factorial; see Exercise 13.1.E. On the other hand, $\operatorname{Spec}\mathbb{Z}[\sqrt{-5}]$ *is* factorial—all of its stalks are unique factorization domains. (You will later be able to show this by showing that $\mathbb{Z}[\sqrt{-5}]$ is a Dedekind domain, §13.5.11, whose stalks are necessarily unique factorization domains by Theorem 13.5.8(f).)

5.4.M. EXERCISE. Suppose A is a k-algebra, and l/k is a finite extension of fields. (Your proof might not use finiteness; this hypothesis is included to avoid distraction by infinite-dimensional vector spaces.) Show that if $A \otimes_k l$ is a normal integral domain, then A is a normal integral domain as well. (Although we won't need this, a version of the converse is true if l/k is separable, [Gr-EGA, IV$_2$.6.14.2].) Hint: Fix a k-basis for l, $b_1 = 1, \ldots, b_d$. Explain why $1 \otimes b_1, \ldots, 1 \otimes b_d$ forms a free A-basis for $A \otimes_k l$. Explain why we have injections

$$
\begin{array}{ccc}
A & \longrightarrow & K(A) \\
\downarrow & & \downarrow \\
A \otimes_k l & \longrightarrow & K(A) \otimes_k l.
\end{array}
$$

Show that $K(A) \otimes_k l = K(A \otimes_k l)$. (Idea: $A \otimes_k l \subset K(A) \otimes_k l \subset K(A \otimes_k l)$. Why is $K(A) \otimes_k l$ a field?) Show that $(A \otimes_k l) \cap K(A) = A$. Now assume $P(T) \in A[T]$ is monic and has a root $\alpha \in K(A)$, and proceed from there.

5.4.N. EXERCISE (UFD-NESS IS NOT AFFINE-LOCAL). Let $A = (\mathbb{Q}[x,y]_{x^2+y^2})_0$ denote the homogeneous degree 0 part of the ring $\mathbb{Q}[x,y]_{x^2+y^2}$. In other words, it consists of quotients $f(x,y)/(x^2+y^2)^n$, where f has pure degree $2n$. Show that the distinguished open sets $D(\frac{x^2}{x^2+y^2})$ and $D(\frac{y^2}{x^2+y^2})$ cover Spec A. (Hint: The sum of those two fractions is 1.) Show that $A_{\frac{x^2}{x^2+y^2}}$ and $A_{\frac{y^2}{x^2+y^2}}$ are unique factorization domains. (Hint: Show that both rings are isomorphic to $\mathbb{Q}[t]_{t^2+1}$; this is a localization of the unique factorization domain $\mathbb{Q}[t]$.) Finally, show that A is not a unique factorization domain. Possible hint:

$$
\left(\frac{xy}{x^2+y^2} \right)^2 = \left(\frac{x^2}{x^2+y^2} \right) \left(\frac{y^2}{x^2+y^2} \right).
$$

(This is generalized in Exercise 15.5.H . It is also related to Exercise 15.5.M.)

Number theorists may prefer the example of Exercise 5.4.K: $\mathbb{Z}[\sqrt{-5}]$ is not a unique factorization domain, but it turns out that you can cover it with two affine open subsets $D(2)$ and $D(3)$, each corresponding to unique factorization domains. (For number theorists: to show that $\mathbb{Z}[\sqrt{-5}]_2$ and $\mathbb{Z}[\sqrt{-5}]_3$ are unique factorization domains, first show that the class group of $\mathbb{Z}[\sqrt{-5}]$ is $\mathbb{Z}/2$ using the geometry of numbers, as in [Ar3, Ch. 11, Thm. 7.9]. Then show that the ideals $(1 + \sqrt{-5}, 2)$ and $(1 + \sqrt{-5}, 3)$ are not principal, using the usual norm in $\mathbb{C}$.) The ring $\mathbb{Z}[\sqrt{-5}]$ is an example of a Dedekind domain, as we will discuss in §13.5.11.

5.4.8. *Remark.* For an example of k-algebra A that is not a unique factorization domain, but becomes one after a particular field extension; see Exercise 15.5.J.

Chapter 6

Rings Are to Modules as Schemes Are to . . .

When considering the notion of rings for the first time, one is quickly led to defining the notion of module. For example, any ring morphism $R \to S$ expresses S as an R-module. An ideal I of a ring R is also an R-module, as is the quotient R/I; and the exact sequence $0 \to I \to R \to R/I \to 0$ is a first example of the utility of the abelian category structure of Mod_R.

With a scheme X, you may think that we already have the right analog of modules: the category of $\mathscr{O}_X$-modules $Mod_{\mathscr{O}_X}$, which also forms an abelian category (§2.6). But the right analog for a scheme X turns out to be a better-behaved subset (subcategory!) of $Mod_{\mathscr{O}_X}$, called *quasicoherent* $\mathscr{O}_X$-*modules*, or *quasicoherent sheaves*.

6.1 Quasicoherent Sheaves

Given an A-module M, we defined an $\mathscr{O}$-module $\widetilde{M}$ on Spec A in §4.1.5—the sections over $D(f)$ were M_f. (Now is a good time to complete the exercises in §4.1.5 if you have not done so already.) These are our "local models" for quasicoherent sheaves. In the same way that a scheme is defined by "gluing together rings," a quasicoherent sheaf over that scheme is obtained by "gluing together modules over those rings."

6.1.1. *Definition: The category of quasicoherent sheaves.* If X is a scheme, then an $\mathscr{O}_X$-module $\mathscr{F}$ is a **quasicoherent sheaf** if for every affine open subset $\operatorname{Spec} A \subset X$, $\mathscr{F}|_{\operatorname{Spec} A} \cong \widetilde{M}$ for some A-module M. We make quasicoherent sheaves on a scheme X into category $QCoh_X$ by taking the morphisms to be morphisms as $\mathscr{O}_X$-modules.

6.1.2. Theorem — *Let X be a scheme, and $\mathscr{F}$ an $\mathscr{O}_X$-module. Suppose P is the property of affine open subschemes $\operatorname{Spec} A \subset X$ that $\mathscr{F}|_{\operatorname{Spec} A} \cong \widetilde{M}$ for some A-module M. Then P satisfies the two hypotheses of the Affine Communication Lemma 5.3.2.*

Thus to check quasicoherence of an $\mathscr{O}$-module, it suffices to check a collection of affine open sets covering X. For example, $\widetilde{M}$ is a quasicoherent sheaf on Spec A. Exercise 4.1.G shows that the category Mod_A and the category $QCoh_{\operatorname{Spec} A}$ are essentially the same (we have described an equivalence $Mod_A \overset{\sim}{\longleftrightarrow} QCoh_{\operatorname{Spec} A}$).

Proof. As usual, the first hypothesis of the Affine Communication Lemma 5.3.2 is easier: if Spec A has property P, then so does the distinguished open $\operatorname{Spec} A_f$: if M is an A-module, then $\widetilde{M}|_{\operatorname{Spec} A_f} \cong \widetilde{M_f}$ as sheaves of $\mathscr{O}_{\operatorname{Spec} A_f}$-modules (both sides agree on the level of distinguished open sets and their restriction maps).

We next show the second hypothesis of the Affine Communication Lemma 5.3.2. We are given:

- an $\mathscr{O}_{\operatorname{Spec} A}$-module $\mathscr{F}$,
- elements $f_1, \ldots, f_n$ generating A,
- A_{f_i}-modules M_i, and
- isomorphisms $\phi_i : \mathscr{F}|_{D(f_i)} \overset{\sim}{\longrightarrow} \widetilde{M_i}$.

The isomorphisms

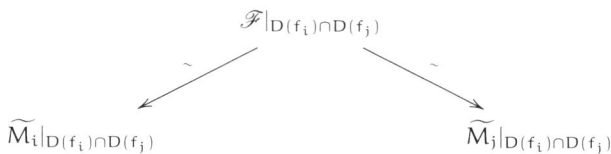

$$\mathscr{F}|_{D(f_i) \cap D(f_j)}$$

$$\widetilde{M_i}|_{D(f_i) \cap D(f_j)} \qquad\qquad \widetilde{M_j}|_{D(f_i) \cap D(f_j)}$$

yield (by taking sections over $D(f_i) \cap D(f_j)$) isomorphisms $\phi_{ij} \colon (M_i)_{f_j} \xrightarrow{\sim} (M_j)_{f_i}$ of $A_{f_i f_j}$-modules, satisfying the cocycle condition (2.5.4.1). As suggestive notation, let $M_{ij} := (M_i)_{f_j}$ (identified with $(M_j)_{f_i}$ via ϕ_{ij}).

We seek an A-module M with an isomorphism $\widetilde{M} \xleftrightarrow{\sim} \mathscr{F}$, and (4.1.6.1) suggests that we should define M by

$$(6.1.2.1) \qquad 0 \longrightarrow M \longrightarrow M_1 \times \cdots \times M_n \xrightarrow{\gamma} M_{12} \times M_{13} \times \cdots \times M_{(n-1)n}.$$

(The discussion at (4.1.6.1) suggests how the map γ should be defined, with appropriate signs.) Translation: $M = \ker(\gamma)$. If we can describe an isomorphism $\widetilde{M}|_{D(f_i)} \xrightarrow{\sim} \widetilde{M_i}$ on each $D(f_i)$ (or equivalently, an isomorphism $M_{f_i} \xrightarrow{\sim} M_i$ of A_{f_i}-modules) such that the triangle

of sheaves on $\operatorname{Spec} A_{f_i f_j}$ commutes (for all i, j), then Exercise 2.5.E (on gluing sheaves) gives our desired isomorphism $\widetilde{M} \xleftrightarrow{\sim} \mathscr{F}$.

By Exercise 4.1.G, we need only describe an isomorphism $M_{f_i} \xrightarrow{\sim} M_i$ such that the triangles

$$(6.1.2.2)$$

commute.

For notational convenience, we assume $i = 1$. Because localization is exact, from (6.1.2.1) we have an exact sequence

$$0 \longrightarrow M_{f_1} \longrightarrow (M_1)_{f_1} \times \cdots \times (M_n)_{f_1} \xrightarrow{\gamma_{f_1}} (M_{12})_{f_1} \times (M_{13})_{f_1} \times \cdots \times (M_{(n-1)n})_{f_1}.$$

We will show that

$$(6.1.2.3)$$

$$0 \longrightarrow M_1 \xrightarrow{\beta} (M_1)_{f_1} \times \cdots \times (M_n)_{f_1} \xrightarrow{\gamma_{f_1}} (M_{12})_{f_1} \times (M_{13})_{f_1} \times \cdots \times (M_{(n-1)n})_{f_1}$$

is an exact sequence (do you see how β should be defined?), yielding our desired isomorphism $M_{f_1} \xrightarrow{\sim} M_1$. We rewrite (6.1.2.3) as

$$(6.1.2.4) \qquad 0 \longrightarrow M_1 \longrightarrow (M_1)_{f_1} \times (M_1)_{f_2} \times \cdots \times (M_1)_{f_n}$$

$$\xrightarrow{\alpha} (M_1)_{f_1 f_2} \times \cdots \times (M_1)_{f_1 f_n} \times (M_1)_{f_2 f_3} \times \cdots \times (M_1)_{f_{n-1} f_n}.$$

But this is precisely the exact sequence (4.1.6.1), except the ring A is replaced by A_{f_1}, and the module M is replaced by M_1.

6.1.A. EXERCISE. Tie up the last loose end: Why does the triangle (6.1.2.2) commute? □

Remark for experts: The proof of Theorem 6.1.2, phrased slightly more carefully, shows that quasicoherent sheaves satisfy faithfully flat descent.

6.1.3. Stray concluding topics.

6.1.B. UNIMPORTANT EXERCISE (NOT EVERY $\mathscr{O}_X$-MODULE IS A QUASICOHERENT SHEAF).

(a) Suppose $X = \operatorname{Spec} k[t]$. Let $\mathscr{F}$ be the skyscraper sheaf supported at the origin $[(t)]$, with group $k(t)$ and the usual $k[t]$-module structure. Show that this is an $\mathscr{O}_X$-module that is not a quasicoherent sheaf. (More generally, if X is an integral scheme, and $p \in X$ is not the generic point, we could take the skyscraper sheaf at p with group the function field of X. Except in silly circumstances, this sheaf won't be quasicoherent.) See Exercises 9.1.G and 6.2.H for more (pathological) examples of $\mathscr{O}_X$-modules that are not quasicoherent.

(b) Suppose $X = \operatorname{Spec} k[t]$. Let $\mathscr{F}$ be the skyscraper sheaf supported at the generic point $[(0)]$, with group $k(t)$. Give this the structure of an $\mathscr{O}_X$-module. Show that this *is* a quasicoherent sheaf. Describe all the restriction maps in the distinguished topology of X. (Remark: Your argument will apply more generally, for example, when X is an integral scheme with generic point η, and $\mathscr{F}$ is the skyscraper sheaf $i_{\eta,*} K(X)$.)

6.1.4. Torsion-free sheaves (a stalk-local condition) and torsion sheaves.
An A-module M is said to be **torsion-free** if $am = 0$ implies that either a is a zerodivisor in A or $m = 0$.

In the case where A is an integral domain, which is basically the only context in which we will use this concept, the definition of torsion-freeness can be restated as $am = 0$ only if $a = 0$ or $m = 0$. In this case, the **torsion submodule** of M, denoted by M_{tors}, consists of those elements of M annihilated by some nonzero element of A. (If A is not an integral domain, this construction needn't yield an A-module.) Clearly M is torsion-free if and only if $M_{tors} = 0$. We say a module M over an integral domain A is **torsion** if $M = M_{tors}$; this is equivalent to $M \otimes_A K(A) = 0$.

If X is a scheme, then an $\mathscr{O}_X$-module $\mathscr{F}$ is said to be **torsion-free** if $\mathscr{F}_p$ is a torsion-free $\mathscr{O}_{X,p}$-module for all p.

6.1.5. *Definition: Torsion quasicoherent sheaves on reduced schemes.*
Motivated by the definition of M_{tors} above, we say that a quasicoherent sheaf on a *reduced* scheme is **torsion** if its stalk at the generic point of every irreducible component is 0. We will mainly use this for coherent sheaves on regular curves, where this notion is very simple indeed (see Exercise 14.3.F(b)), but in the literature it comes up in more general situations.

6.2   Characterizing Quasicoherence Using the Distinguished Affine Base

Because quasicoherent sheaves are locally of a very special form, in order to "know" a quasicoherent sheaf, we need only know what the sections are over every affine open set, and how to restrict sections from an affine open set U to a *distinguished* affine open subset of U. We make this precise by defining what we will call the *distinguished affine base* of the Zariski topology—not a base in the usual sense. This is a refinement of the notion of a *sheaf on a base*. The point of this discussion is to give a useful characterization of quasicoherence, but you may wish to just jump to §6.2.3.

The open sets of the distinguished affine base are the affine open subsets of X. We have already observed that this forms a base. But forget that fact. We like distinguished open sets $\operatorname{Spec} A_f \hookrightarrow \operatorname{Spec} A$, and we don't really understand open embeddings of one random affine open subset in another. So we just remember the "nice" inclusions.

6.2.1. Definition.
The **distinguished affine base** of a scheme X is the data of the affine open sets and the distinguished inclusions.

In other words, we remember only some of the open sets (the affine open sets), and *only some of the morphisms between them* (the distinguished morphisms). For experts: if you think of a topology as a category (the category of open sets), we have described a subcategory.

We can define a sheaf on the distinguished affine base in the obvious way: we have a set (or abelian group, or ring) for each affine open set, and we know how to restrict to distinguished open sets. (You should think through the statement of the identity and gluability axioms yourself.)

Given a sheaf $\mathscr{F}$ on X, we get a sheaf on the distinguished affine base. You can guess where we are going: we will show that all the information of the sheaf is contained in the information of the sheaf on the distinguished affine base.

As a warm-up, we can recover stalks as follows. (We will be implicitly using only the following fact. We have a collection of open subsets, and *some* inclusions among these subsets, such that if we have any $p \in U, V$ where U and V are in our collection of open sets, there is some W containing p, and contained in U and V such that $W \hookrightarrow U$ and $W \hookrightarrow V$ are both in our collection of inclusions. In the case we are considering here, this is the key Proposition 5.3.1 that, given any two affine open sets Spec A, Spec B in X, Spec $A \cap$ Spec B could be covered by affine open sets that were simultaneously distinguished in Spec A and Spec B. In fancy language: the category of affine open sets, and distinguished inclusions, forms a filtered set.)

The stalk $\mathscr{F}_p$ is the colimit $\mathrm{colim}(f \in \mathscr{F}(U))$ where the colimit is over all open sets containing p. We compare this to $\mathrm{colim}(f \in \mathscr{F}(U))$ where the colimit is over all affine open sets containing p, and all distinguished inclusions. You can check that the elements of one correspond to elements of the other. (Think carefully about this!)

6.2.A. EXERCISE. Show that a section of a sheaf on the distinguished affine base is determined by the section's germs.

6.2.2. Theorem —

(a) *A sheaf on the distinguished affine base $\mathscr{F}^b$ determines a unique (up to unique isomorphism) sheaf $\mathscr{F}$ that, when restricted to the affine base, is $\mathscr{F}^b$. (Hence if you start with a sheaf, and take the sheaf on the distinguished affine base, and then take the induced sheaf, you get the sheaf you started with.)*

(b) *A morphism of sheaves on a distinguished affine base uniquely determines a morphism of sheaves.*

(c) *An $\mathscr{O}_X$-module "on the distinguished affine base" yields an $\mathscr{O}_X$-module.*

This proof is identical to our argument of §2.5 showing that sheaves are (essentially) the same as sheaves on a base, using the "sheaf of compatible germs" construction. The main reason for repeating it is to let you see that all that is needed is for the open sets to form a filtered set (or, in the current case, that the category of open sets and distinguished inclusions is filtered).

For experts: (a) and (b) are describing an equivalence of categories between sheaves on the Zariski topology of X and sheaves on the distinguished affine base of X.

Proof. (a) Suppose $\mathscr{F}^b$ is a sheaf on the distinguished affine base. Then we can define stalks.

For *any* open set U of X, define the sheaf of compatible germs

$$\mathscr{F}(U) := \{(f_p \in \mathscr{F}^b_p)_{p \in U} : \text{ for all } p \in U, \text{ there exists } U_p \text{ with } p \in U_p \subset U,$$
$$s \in \mathscr{F}^b(U_p) \text{ such that } s_q = f_q \text{ for all } q \in U_p\},$$

where each U_p is in our base, and s_q means "the germ of s at q." (As usual, those who want to worry about the empty set are welcome to.)

This really is a sheaf: convince yourself that we have restriction maps, identity, and gluability, really quite easily.

I next claim that if U is in our base, then $\mathscr{F}(U) = \mathscr{F}^b(U)$. We clearly have a map $\mathscr{F}^b(U) \to \mathscr{F}(U)$. This is an isomorphism on stalks, and hence an isomorphism by Exercise 2.4.D.

6.2.B. EXERCISE. Prove (b) (cf. Exercise 2.5.C).

6.2.C. EXERCISE. Prove (c) (cf. Remark 2.5.3).

The proof of Theorem 6.2.2 is now complete.   □

6.2.3. A characterization of quasicoherent sheaves in terms of distinguished inclusions. We use this perspective to give a useful characterization of quasicoherent sheaves among $\mathscr{O}_X$-modules. Suppose $\mathscr{F}$ is an $\mathscr{O}_X$-module, and $\operatorname{Spec} A_f \hookrightarrow \operatorname{Spec} A \subset X$ is a distinguished open subscheme of an affine open subscheme of X. Let $\phi \colon \Gamma(\operatorname{Spec} A, \mathscr{F}) \to \Gamma(\operatorname{Spec} A_f, \mathscr{F})$ be the restriction map. The source of ϕ is an A-module, and the target is an A_f-module, so by the universal property of localization (Exercise 1.2.D), ϕ naturally factors as:

$$
\begin{array}{ccc}
\Gamma(\operatorname{Spec} A, \mathscr{F}) & \xrightarrow{\ \ \phi\ \ } & \Gamma(\operatorname{Spec} A_f, \mathscr{F}) \\
& \searrow_{\otimes_A A_f} \qquad \nearrow_{\alpha} & \\
& \Gamma(\operatorname{Spec} A, \mathscr{F})_f. &
\end{array}
$$

6.2.D. VERY IMPORTANT EXERCISE. Show that an $\mathscr{O}_X$-module $\mathscr{F}$ is quasicoherent if and only if for each such distinguished $\operatorname{Spec} A_f \hookrightarrow \operatorname{Spec} A$, α is an isomorphism.

Thus a quasicoherent sheaf is (equivalent to) the data of one module for each affine open subset (a module over the corresponding ring), such that the module over a distinguished open set $\operatorname{Spec} A_f$ is given by localizing the module over $\operatorname{Spec} A$.

6.2.E. EXERCISE (GOOD PRACTICE: THE SHEAF OF NILPOTENTS). If A is a ring, and $f \in A$, show that $\mathfrak{N}(A_f) \cong \mathfrak{N}(A)_f$. Use this to define/construct the quasicoherent **sheaf of nilpotents** on any scheme X. This is an example of an ideal sheaf (of $\mathscr{O}_X$).

6.2.4. *Tensor products.* Tensor product in the category of quasicoherent sheaves on X can also be cleanly described in terms of affine open sets.

6.2.F. EXERCISE. If $\mathscr{F}$ and $\mathscr{G}$ are quasicoherent sheaves on a scheme X, show that $\mathscr{F} \otimes \mathscr{G}$ is a quasicoherent sheaf described by the following information: If $\operatorname{Spec} A \subset X$ is an affine open subset, and $\Gamma(\operatorname{Spec} A, \mathscr{F}) = M$ and $\Gamma(\operatorname{Spec} A, \mathscr{G}) = N$, then $\Gamma(\operatorname{Spec} A, \mathscr{F} \otimes \mathscr{G}) = M \otimes_A N$, and the restriction map $\Gamma(\operatorname{Spec} A, \mathscr{F} \otimes \mathscr{G}) \to \Gamma(\operatorname{Spec} A_f, \mathscr{F} \otimes \mathscr{G})$ is precisely the localization map $M \otimes_A N \to (M \otimes_A N)_f \cong M_f \otimes_{A_f} N_f$. (We are using the algebraic fact that $(M \otimes_A N)_f \cong M_f \otimes_{A_f} N_f$. You can prove this by universal property if you want, or by using the explicit construction.)

Note that, thanks to the machinery behind the distinguished affine base, sheafification is taken care of. This is a feature we will use often: constructions involving quasicoherent sheaves that involve sheafification for general sheaves don't require sheafification when considered on the distinguished affine base. Along with the fact that injectivity, surjectivity, kernels, tensor products, and so on may be computed on affine opens, this is the reason that it is particularly convenient to think about quasicoherent sheaves in terms of affine open sets. (There is a slight caveat in the case of *Hom*; see Exercise 14.3.A.)

6.2.5.** *Elegant side remark, but perhaps too fancy for now.* In Exercise 6.2.F, the tensor product is in the category of quasicoherent sheaves. But in fact this is even the tensor product in the category of $\mathscr{O}_X$-modules. To show that "$\mathscr{F} \otimes_{QCoh_X} \mathscr{G}$ satisfies the universal property of $\mathscr{F} \otimes_{\mathscr{O}_X} \mathscr{G}$," first show the following. Let $\mathscr{M}$ be a quasicoherent sheaf on X, and $\mathscr{N}$ any $\mathscr{O}_X$-module. Let $\operatorname{Spec} A \subset X$ be an affine open subset, and $M = \mathscr{M}(\operatorname{Spec} A)$. Then the "global sections" map

$$
\operatorname{Hom}_{\operatorname{Spec} A}(\mathscr{M}|_{\operatorname{Spec} A}, \mathscr{N}|_{\operatorname{Spec} A}) \longrightarrow \operatorname{Hom}_A(M, \mathscr{N}(\operatorname{Spec} A))
$$

is an isomorphism. Our slogan: On an affine scheme, maps from a quasicoherent sheaf to *any* $\mathscr{O}$-module are determined by global sections. You will hear an echo of this slogan in the parenthetical comment in Exercise 7.3.G.

6.2.6. *Definition.* Given a section s of $\mathscr{F}$ and a section t of $\mathscr{G}$, we have a section $s \otimes t$ of $\mathscr{F} \otimes \mathscr{G}$.

6.2.7. X_f and the Qcqs lemma. The next result will be useful in the future in showing that certain constructions yield quasicoherent sheaves.

6.2.8. *Definition: X_f.* If X is a scheme, and $f \in \Gamma(X, \mathscr{O}_X)$ is a function, let $X_f \subset X$ be the subset on which f doesn't vanish. By Exercise 4.3.F(a), X_f is open. (We avoid the notation $D(f)$ to avoid any suggestion that X is affine.)

6.2.9. The Qcqs Lemma — *Suppose X is a quasicompact and quasiseparated scheme, $\mathscr{F}$ is a quasicoherent sheaf on X, and $f \in \Gamma(X, \mathscr{O}_X)$ is a function on X. Then the restriction map*

$$\mathrm{res}_{X_f \subset X} \colon \Gamma(X, \mathscr{F}) \longrightarrow \Gamma(X_f, \mathscr{F})$$

is precisely localization: there is an isomorphism $\Gamma(X, \mathscr{F})_f \overset{\sim}{\longrightarrow} \Gamma(X_f, \mathscr{F})$ making the following diagram commute.

$$
\begin{array}{ccc}
\Gamma(X, \mathscr{F}) & \xrightarrow{\ \mathrm{res}_{X_f \subset X}\ } & \Gamma(X_f, \mathscr{F}) \\
{\scriptstyle \otimes_{\Gamma(X,\mathscr{O}_X)}(\Gamma(X,\mathscr{O}_X)_f)} \searrow & & \nearrow {\scriptstyle \sim} \\
& \Gamma(X, \mathscr{F})_f &
\end{array}
$$

6.2.G. IMPORTANT EXERCISE. Prove the Qcqs Lemma 6.2.9. Hint: The "quasicompact quasiseparated" hypothesis translates precisely to "covered by a finite number of affine open sets, the pairwise intersection of which is also covered by a finite number of affine open sets." Apply the exact functor $\otimes_{\Gamma(X,\mathscr{O}_X)}\Gamma(X, \mathscr{O}_X)_f$ to the exact sequence

$$0 \longrightarrow \Gamma(X, \mathscr{F}) \longrightarrow \oplus_i \Gamma(U_i, \mathscr{F}) \longrightarrow \oplus \Gamma(U_{ijk}, \mathscr{F}),$$

where the U_i form a finite affine cover of X and the U_{ijk} form a finite affine cover of $U_i \cap U_j$.

6.2.H. LESS IMPORTANT EXERCISE. Give a counterexample to show that the Qcqs Lemma 6.2.9 is false without the quasicompactness hypothesis. (Possible hint: Take an infinite disjoint union of affine schemes. The key idea is that infinite products do not commute with localization.)

6.2.10. ** **Grothendieck topologies.** The distinguished affine base isn't a topology in the usual sense—the union of two affine sets isn't necessarily affine, for example. It is, however, a first new example of a generalization of a topology—the notion of a **site** or a **Grothendieck topology**. We give the definition to satisfy the curious, but we certainly won't use this notion. (For a clean statement, see [Stacks, tag 00VH]; our discussion here is intended only as motivation.) The idea is that we should abstract away only those notions we need to define sheaves. We need the notion of open set, but it turns out that we won't even need an underlying set, i.e., we won't even need the notion of points! Let's think through how little we need. For our discussion of sheaves to work, we needed to know what the open sets were, and what the (allowed) inclusions were, and that these should "behave well," and, in particular, that the data of the open sets and inclusions should form a category. (For example, the composition of an allowed inclusion with another allowed inclusion should be an allowed inclusion—in the distinguished affine base, a distinguished open set of a distinguished open set is a distinguished open set.) So we just require the data of *this category*. At this point, we can already define presheaf (as just a contravariant functor from this category of "open sets"). We saw this idea earlier in Exercise 2.2.A.

   In order to extend this definition to that of a sheaf, we need more information. We want two open subsets of an open set to intersect in an open set, so *we want the category to be closed under fibered products* (cf. Exercise 1.2.O). For the identity and gluability axioms, we need to know *when some open sets cover another*, so we also include this as part of the data of a Grothendieck topology. The data of the coverings satisfy some obvious properties. Every open set covers itself (i.e., *the identity map in the category of open sets is a covering*). Coverings pull back: *if we have a map $Y \to X$, then any cover of X pulls back to a cover of Y.* Finally, *a cover of a cover should be a cover.* Such data (satisfying these axioms) is called a *Grothendieck topology* or a *site*. (There are useful variants of this definition in the literature. Again, we are following [Stacks].) We can define the notion of a sheaf on a Grothendieck topology in the usual way, with no change. A **topos** is a scary name for a category of sheaves of sets on a Grothendieck topology.

Grothendieck topologies are used in a wide variety of contexts in and near algebraic geometry. Étale cohomology (using the étale topology), a generalization of Galois cohomology, is a central tool, as are more general flat topologies, such as the smooth topology. The definitions of Deligne-Mumford and Artin stack use the étale and smooth topology, respectively. Tate developed a good theory of non-Archimedean analytic geometry over totally disconnected ground fields such as $\mathbb{Q}_p$ using a suitable Grothendieck topology. Work in K-theory (related, for example, to Voevodsky's work) uses exotic topologies.

6.3 Quasicoherent Sheaves Form an Abelian Category

Morphisms from one quasicoherent sheaf on a scheme X to another are defined to be just morphisms as $\mathcal{O}_X$-modules. In this way, the quasicoherent sheaves on a scheme X form a category, denoted by $QCoh_X$. (By definition, it is a full subcategory of $Mod_{\mathcal{O}_X}$.) We now show that quasicoherent sheaves on X form an *abelian* category.

When you show that something is an abelian category, you have to check many things, because the definition has many parts. However, if the objects you are considering lie in some ambient abelian category, then it is much easier. You have seen this idea before: there are several things you have to do to check that something is a group. But if you have a subset of group elements, it is much easier to check that it forms a subgroup.

You can look back at the definition of an abelian category, and you will see that in order to check that a subcategory is an abelian subcategory, it suffices to check only the following:

(i) 0 is in the subcategory,
(ii) the subcategory is closed under finite sums, and
(iii) the subcategory is closed under kernels and cokernels.

In our case of $QCoh_X \subset Mod_{\mathcal{O}_X}$, the first two are cheap: 0 is certainly quasicoherent, and the subcategory is closed under finite sums: if $\mathscr{F}$ and $\mathscr{G}$ are sheaves on X, and over Spec A, $\mathscr{F} \cong \widetilde{M}$ and $\mathscr{G} \cong \widetilde{N}$, then $\mathscr{F} \oplus \mathscr{G} \cong \widetilde{M \oplus N}$ (Do you see why?), so $\mathscr{F} \oplus \mathscr{G}$ is a quasicoherent sheaf.

We now check (iii), using the characterization of Important Exercise 6.2.D. Suppose $\alpha \colon \mathscr{F} \to \mathscr{G}$ is a morphism of quasicoherent sheaves. Then on any affine open set U, where the morphism is given by $\beta \colon M \to N$, define $(\ker \alpha)(U) = \ker \beta$ and $(\operatorname{coker} \alpha)(U) = \operatorname{coker} \beta$. Then these behave well under inversion of a single element: if

$$0 \longrightarrow K \longrightarrow M \longrightarrow N \longrightarrow P \longrightarrow 0$$

is exact, then so is

$$0 \longrightarrow K_f \longrightarrow M_f \longrightarrow N_f \longrightarrow P_f \longrightarrow 0,$$

from which $(\ker \beta)_f \cong \ker(\beta_f)$ and $(\operatorname{coker} \beta)_f \cong \operatorname{coker}(\beta_f)$. Thus both of these define quasicoherent sheaves. Moreover, by checking stalks, they are indeed the kernel and cokernel of α (exactness can be checked stalk-locally). Thus the quasicoherent sheaves indeed form an abelian category.

6.3.A. EXERCISE. Show that a sequence of quasicoherent sheaves $\mathscr{F} \to \mathscr{G} \to \mathscr{H}$ on X is exact if and only if it is exact on every open set in any given affine cover of X. (In particular, "taking sections over an affine open subset Spec A" is an exact functor from the category of quasicoherent sheaves on X to the category of A-modules. Recall that taking sections is only left-exact in general; see §2.6.G.) Thus we may check injectivity or surjectivity of a morphism of quasicoherent sheaves by checking on an affine cover of our choice.

Caution: If $0 \to \mathscr{F} \to \mathscr{G} \to \mathscr{H} \to 0$ is an exact sequence of quasicoherent sheaves, then, for any open set,

$$0 \longrightarrow \mathscr{F}(U) \longrightarrow \mathscr{G}(U) \longrightarrow \mathscr{H}(U)$$

is exact, and exactness on the right is guaranteed to hold only if U is affine. (To set you up for cohomology: whenever you see left-exactness, you expect to eventually interpret this as a start of a long exact sequence. So we are expecting H^1's on the right, and now we expect that $H^1(\operatorname{Spec} A, \mathscr{F}) = 0$. This will indeed be the case.)

6.3.B. LESS IMPORTANT EXERCISE (CONNECTION TO ANOTHER DEFINITION, AND QUASICOHERENT SHEAVES ON RINGED SPACES IN GENERAL). Show that an $\mathscr{O}_X$-module $\mathscr{F}$ on a scheme X is quasicoherent if and only if there exists an open cover by U_i such that on each U_i, $\mathscr{F}|_{U_i}$ is isomorphic to the cokernel of a map of two "free sheaves":

$$\mathscr{O}_{U_i}^{\oplus I} \longrightarrow \mathscr{O}_{U_i}^{\oplus J} \longrightarrow \mathscr{F}|_{U_i} \longrightarrow 0$$

is exact. This is the definition of a quasicoherent sheaf on a ringed space in general. It is useful in many circumstances, for example, in complex analytic geometry.

6.4   Finite Type Quasicoherent, Finitely Presented, and Coherent Sheaves

Here are three natural finiteness conditions on an A-module M. If A is a Noetherian ring, which is the case that almost all of you will ever care about, they are all the same.

The first is the simplest: a module could be **finitely generated**. In other words, there is a surjection $A^{\oplus p} \to M \to 0$.

The second is reasonable too. It could be finitely presented— it could have a finite number of generators with a finite number of relations. Translation: There exists a **finite presentation**, i.e., an exact sequence

$$A^{\oplus q} \longrightarrow A^{\oplus p} \longrightarrow M \longrightarrow 0.$$

6.4.1. The third notion is frankly a bit surprising. We say that an A-module M is **coherent** if (i) it is finitely generated, and (ii) for any map $A^{\oplus p} \to M$ (not necessarily surjective!), the kernel is finitely generated.

6.4.2. Proposition — *If A is Noetherian, then these three definitions are the same.*

Proof. Clearly, coherent implies finitely presented, which in turn implies finitely generated. So suppose M is finitely generated. Take any $\alpha \colon A^{\oplus p} \to M$. Then $\ker \alpha$ is a submodule of a finitely generated module over A, and is thus finitely generated by Exercise 3.6.Y. Thus M is coherent. $\square$

Hence most people can think of these three notions as the same.

6.4.3. Proposition — *The coherent A-modules form an abelian subcategory of the category of A-modules.*

The proof in general is a series of short exercises in §6.7. You should try them only if you are particularly curious.

Proof if the ring A is Noetherian. Recall from our discussion at the start of §6.3 that we must check three things:

 (i)   The 0-module is coherent.
 (ii) The category of coherent modules is closed under finite sums.
 (iii) The category of coherent modules is closed under kernels and cokernels.

The first two are clear. For (iii), suppose that $f \colon M \to N$ is a map of finitely generated modules. Then $\operatorname{coker} f$ is finitely generated (it is the image of N), and $\ker f$ is too (it is a submodule of a finitely generated module over a Noetherian ring, Exercise 3.6.Y). $\square$

6.4.4. *Important remark: Finitely generated modules over a principal ideal domain.* We record for future reference that finitely generated modules over a principal ideal domain are finite direct sums of cyclic modules (see, for example, [DF, §12.1, Thm. 5]). We only mention it here because it needs mentioning *somewhere*.

6.4.5. We now extend the definitions of these three classes of modules to quasicoherent sheaves.

6.4.A. EXERCISE (FINITE GENERATION SATISFIES THE HYPOTHESES OF THE AFFINE COMMUNICATION LEMMA 5.3.2).

(a) If $f \in A$, and M is a finitely generated A-module, show that M_f is a finitely generated A_f-module.
(b) If $(f_1, \dots, f_n) = A$, and M_{f_i} is a finitely generated A_{f_i}-module for all i, show that M is a finitely generated A-module.

For the finite presentation case, the following result is handy, and versions apply in other situations (see Lemma 8.3.R for the "algebra" version). For the longest time I didn't even suspect such a thing could be true.

6.4.6. Lemma: Finitely presented implies always finitely presented (module version) — *Suppose M is a finitely presented A-module (for **some** surjection from a finite free module, the kernel is finitely generated), and $\phi \colon A^{\oplus p} \to M$ is **any** surjection. Then $\ker \phi$ is finitely generated.*

Proof. Choose a finite presentation of M:

$$\oplus_{k=1}^r A f_k \xrightarrow{\ \beta\ } \oplus_{i=1}^n A x_i \xrightarrow{\ \alpha\ } M \longrightarrow 0,$$

so $\alpha(x_1), \dots, \alpha(x_n)$ generate M. Write the surjection of the hypothesis as $\phi \colon \oplus_{j=1}^p A y_j \to M$. Choose lifts $g_j \in \oplus A x_i$ of $\phi(y_j) \in M$. Then we can write

$$M = \oplus A x_i \,/\, (\beta(f_1), \dots, \beta(f_r))$$

$$= (\oplus A x_i \oplus A y_j) \,/\, (\beta(f_1), \dots, \beta(f_r), y_1 - g_1, \dots, y_p - g_p).$$

As the $\{\phi(y_j)\}$ generate M, for each $i = 1, \dots, n$ we can choose $h_i \in \oplus_{j=1}^p A y_j$ (i.e., a map $h \colon \oplus_{i=1}^n A x_i \to \oplus_{j=1}^p A y_j$) so that $\alpha(x_i) = \phi(h_i)$. Thus

$$M = (\oplus A x_i \oplus A y_j) \,/\, (\beta(f_1), \dots, \beta(f_r), y_1 - g_1, \dots, y_p - g_p, x_1 - h_1, \dots, x_n - h_n)$$

$$= \oplus A y_j \,/\, (h(\beta(f_1)), \dots, h(\beta(f_r)), y_1 - h(g_1), \dots, y_p - h(g_p)). \qquad \square$$

6.4.B. EXERCISE (FINITE PRESENTATION SATISFIES THE HYPOTHESES OF THE AFFINE COMMUNICATION LEMMA 5.3.2).

(a) If $f \in A$, and if M is a finitely presented A-module, show that M_f is a finitely presented A_f-module.
(b) If $(f_1, \dots, f_n) = A$, and M_{f_i} is a finitely presented A_{f_i}-module for all i, show that M is a finitely presented A-module. (Hint: Use Exercise 6.4.A(b) to show first that M is finitely generated, so we may write $M = \operatorname{coker}(\alpha \colon N \hookrightarrow A^{\oplus n})$ for some submodule $N \subset A^{\oplus n}$. We wish to show that N is finitely generated. Localize at the f_i, use Lemma 6.4.6 ("finitely presented implies always finitely presented") to show N_{f_i} is a finitely generated A_{f_i} module, then apply Exercise 6.4.A(b) to N.)

6.4.C. EXERCISE (COHERENCE SATISFIES THE HYPOTHESES OF THE AFFINE COMMUNICATION LEMMA 5.3.2).

(a) If $f \in A$, and M is a coherent A-module, show that M_f is a coherent A_f-module.

(b) If $(f_1, \dots, f_n) = A$, and M_{f_i} is a coherent A_{f_i}-module for all i, show that M is a coherent A-module. (Hint: If $\phi : A^{\oplus p} \to M$, then $(\ker \phi)_{f_i} = \ker(\phi_{f_i})$, which is finitely generated for all i. Then apply Exercise 6.4.A(b).)

6.4.7. Definition. A quasicoherent sheaf $\mathscr{F}$ is **finite type** (resp., **finitely presented**, **coherent**) if for every affine open Spec A, $\Gamma(\mathrm{Spec}\, A, \mathscr{F})$ is a finitely generated (resp., finitely presented, coherent) A-module. Note that coherent sheaves are always finite type, and that on a locally Noetherian scheme, all three notions are the same (by Proposition 6.4.2). Proposition 6.4.3 implies that the coherent sheaves on X form an abelian category, which we denote by Coh_X.

By the Affine Communication Lemma 5.3.2, and Exercises 6.4.A, 6.4.B, and 6.4.C, it suffices to check "finite typeness" (resp., finite presentation, coherence) on the open sets in a single affine cover.

6.4.8. *Warning.* It is not uncommon in the later literature to incorrectly define coherent as finitely generated. Please only use the correct definition, as the wrong definition causes confusion. Besides your doing this for reasons of honesty, it will also help you see what hypotheses are actually necessary to prove things. And that always helps you remember what the proofs are—and hence why things are true.

6.4.9. *Why coherence?* Proposition 6.4.3 is a good motivation for the definition of coherence: it gives a small (in a nontechnical sense) abelian category in which we can think about vector bundles.

There are two sorts of people who should care about the details of this definition, rather than living in a Noetherian world where coherent means finite type. Complex geometers should care. They consider complex-analytic spaces with the classical topology. One can define the notion of coherent $\mathscr{O}_X$-module in a way analogous to this (see [Se1, Def. 2]). Then Oka's Theorem states that the structure sheaf of $\mathbb{C}^n$ (hence of any complex manifold) is coherent, and this is very hard (see [GR, §2.5] or [Rem, §7.2]).

The second sort of people who should care are the sort of arithmetic people who may need to work with non-Noetherian rings—see §3.6.21—or work in non-Archimedean analytic geometry.

6.4.10. *Remark: Quasicoherent and coherent sheaves on ringed spaces.* We will discuss quasicoherent and coherent sheaves on schemes, but they can be defined more generally (see Exercise 6.3.B for quasicoherent sheaves, and [Se1, Def. 2] for coherent sheaves). Many of the results we state will hold in greater generality, but because the proofs look slightly different, we restrict ourselves to schemes to avoid distraction.

6.4.11.** *Coherence is not a good notion in smooth geometry.* The following example from B. Conrad shows that in quite reasonable (but less "rigid") situations, the structure sheaf is not coherent over itself. Consider the ring $\mathscr{O}_0$ of germs of smooth (C^∞) functions at $0 \in \mathbb{R}$, with coordinate x. Now $\mathscr{O}_0$ is a local ring. Its maximal ideal $\mathfrak{m}$ is generated by x. (Key idea: Suppose $f \in \mathfrak{m}$, and suppose f has a representative defined on $(-\epsilon, \epsilon)$. Then for $t \in (-\epsilon, \epsilon)$, $f(t) = \int_0^t f'(u)\, du = t \int_0^1 f'(tv)\, dv$. By "differentiating under the integral sign" repeatedly, we may check that $\int_0^1 f'(tv)\, dv$ is smooth.)

Let $\phi \in \mathscr{O}_0$ be the germ of a smooth function that is 0 for $x \le 0$, and positive for $x > 0$ (such as $\phi(x) = e^{-1/x^2}$ for $x > 0$). Consider the map $\times\phi : \mathscr{O}_0 \to \mathscr{O}_0$. The kernel is the ideal I_ϕ of functions vanishing for $x \ge 0$. Clearly, I_ϕ is nonzero (for example, $\phi(-x) \in I_\phi$), but as $\mathfrak{m} = (x)$, $I_\phi = xI_\phi$, I_ϕ cannot be finitely generated, or else Nakayama's Lemma 8.2.9 would be contradicted. (Essentially the same argument shows that the sheaf of smooth functions on $\mathbb{R}$ is not coherent.) This is why coherence has no useful meaning for smooth manifolds.

6.5   Algebraic Interlude: The Jordan–Hölder Package

The Jordan–Hölder Theorem in group theory is part of a more fundamental and somehow simpler story. The Jordan–Hölder "yoga" is why you can often factor some sort of algebraic object into

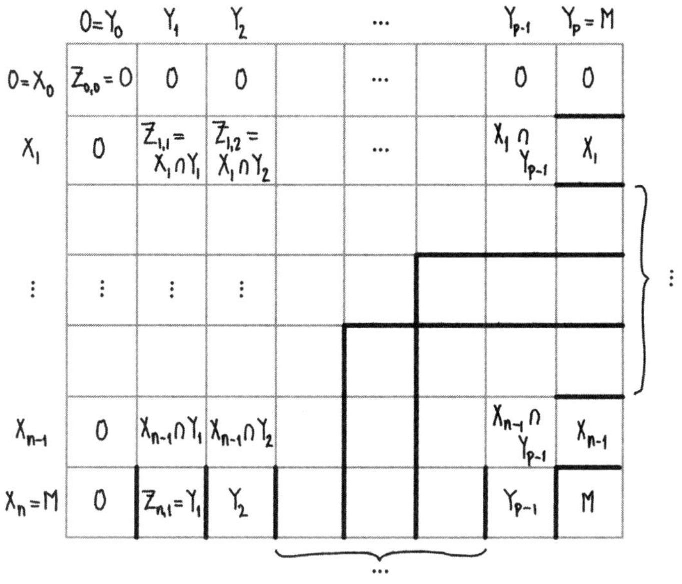

Figure 6.1 *The rectangular array in the proof of the Jordan–Hölder Theorem.*

primes or irreducibles, uniquely (in an appropriate sense), where each prime/irreducible appears the same number of times no matter how you factor. From this point of view, it generalizes unique factorization of integers; well-definedness of dimension of vector spaces; classification of finitely generated abelian groups; unique factorization of ideals in a Dedekind domain; and the traditional Jordan–Hölder Theorem in group theory.

We will be mostly interested in modules over a ring, but there is no harm in working in a general abelian category $\mathscr{C}$. (This can be readily generalized further, as in Exercise 6.5.G.)

We say an object $M \in \mathscr{C}$ is **simple** (or **irreducible**) if its only subobjects are 0 and itself. A **composition series** for M is a (finite) filtration

$$(6.5.0.1) \qquad 0 = X_0 \subsetneq X_1 \subsetneq \cdots \subsetneq X_{n-1} \subsetneq X_n = M$$

such that the quotients X_{i+1}/X_i are all simple. If M has a composition series, we say that M has **finite length**.

6.5.1. The Jordan–Hölder Theorem — *If M has a finite composition series, then all composition series for M have the same length, and the quotients are all the same, possibly rearranged.*

6.5.2. *Definition.* We call the length of any of the composition series for M the **length** of M, denoted by $\ell(M)$. This notion is well-defined by the Jordan–Hölder Theorem. But we even have a refined notion: we have the multiplicity with which each simple object appears in any composition series for M.

If M is not of finite length, we say $\ell(M) = \infty$.

6.5.3. *Example.* In the category of abelian groups, the finite-length objects are the *finite* abelian groups. The Jordan–Hölder Theorem in this case, applied to $\mathbb{Z}/n\mathbb{Z}$, can be used to give the unique factorization of n.

6.5.4. *Proof of the Jordan–Hölder Theorem 6.5.1.*
Suppose we have two finite composition series, (6.5.0.1) and

$$0 = Y_0 \subsetneq Y_1 \subsetneq \cdots \subsetneq Y_{p-1} \subsetneq Y_p = M,$$

of one object $M \in \mathscr{C}$. Make a rectangular array with entries $Z_{i,j} := X_i \cap Y_j$, as in Figure 6.1. Figure 6.2 shows this construction applied to two

(12)	(12)	(12)	(12)
(12)	(12)	(6)	(6)
(12)	(4)	(2)	(2)
(12)	(4)	(2)	(1)

Figure 6.2 *A sample "Jordan–Hölder table" for two composition series for $\mathbb{Z}/(12)$.*

composition series for the $\mathbb{Z}$-module $\mathbb{Z}/(12)$,

$$(12) \subsetneq (6) \subsetneq (2) \subsetneq (1) \quad \text{and} \quad (12) \subsetneq (4) \subsetneq (2) \subsetneq (1).$$

6.5.5. Observe that

- $Z_{i,j} \subset Z_{i',j'}$ if $i \leq i'$ and $j \leq j'$,
- $Z_{n,j} = Y_j$ and $Z_{i,p} = X_i$,
- $Z_{0,j} = Z_{i,0} = 0$, and
- $Z_{i,j} \cap Z_{i',j'} = Z_{\min(i,i'),\min(j,j')}$.

6.5.A. EXERCISE. Show (by descending induction on i) that $Z_{i,j}/Z_{i,j-1}$ is 0 or isomorphic to the simple element Y_j/Y_{j-1}.

Similarly, we have the analogous statement for $Z_{i,j}/Z_{i-1,j}$. In the rectangular array, draw a *thick line* between these two (horizontally or vertically adjacent) entries if the quotient is nonzero, and label that line with the (isomorphism class of) the simple group (again, see Figure 6.1).

Consider any 2×2 subsquare of the array:

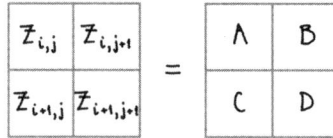

$$\begin{array}{|c|c|}\hline Z_{i,j} & Z_{i,j+1} \\ \hline Z_{i+1,j} & Z_{i+1,j+1} \\ \hline \end{array} = \begin{array}{|c|c|}\hline A & B \\ \hline C & D \\ \hline \end{array}$$

We will see that the thick lines inside the square form one of the following five patterns (each of which appears in Figure 6.2).

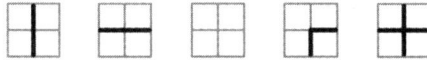

From $A = B \cap C$, we see that (i) if $D = B$, then $A = C$, and (ii) if $D = C$, then $A = B$. That takes care of the first three cases.

Suppose next that D/B and D/C are *both* nonzero (hence simple). If $B = C$, we are in the fourth ("elbow") case.

Finally, otherwise, we will see that $D/B \cong C/A$ (and similarly, $D/C \cong B/A$), and we are in the fifth case. Consider the map $C \to D/B$. Then A is precisely the kernel, from $A = B \cap C$ (§6.5.5). Thus we have an injection $C/A \hookrightarrow D/B$. By simplicity of D/B, either C/A is zero, or this injection is actually an isomorphism $C/A \xrightarrow{\sim} D/B$. $\qquad\qquad\square$

6.5.B. EXERCISE. Prove the Jordan–Hölder Theorem. Hint: Notice that the thickened lines can be interpreted as paths from the right side of the table to the bottom of the table, with one left turn. This will give a bijection between the simple quotients of one filtration and the simple quotients of the other filtration.

6.5.C. EXERCISE. Show that every subquotient of a finite-length object M is finite length. Possible approach: Suppose the subquotient is M''/M', where $M' \subseteq M'' \subseteq M$. Choose a composition series $M_\bullet$ for M. Make a new rectangular table in a similar way, using the composition series $M_\bullet$, and the filtration $0 \subseteq M' \subseteq M'' \subseteq M$. Think suitably about paths, similarly to the approach to the proof of the Jordan–Hölder Theorem.

6.5.D. EXERCISE. Show that length is additive in exact sequences: if $0 \to M' \to M \to M'' \to 0$ is an exact sequence, then $\ell(M) = \ell(M') + \ell(M'')$. (Do *not* assume these objects have finite length.)

6.5.E. EXERCISE. Show that any filtration of a finite length module can be refined into a composition series.

6.5.F. UNIMPORTANT EXERCISE. Show that the finite length objects in $\mathscr{C}$ form a full subcategory of $\mathscr{C}$.

The category of groups does not form an abelian category, so Theorem 6.5.1 can't immediately imply the traditional Jordan–Hölder Theorem for groups. However, the same proof applies without change, with only one additional input.

6.5.G. EXERCISE (THAT WE WON'T USE). Prove the Jordan–Hölder Theorem for groups. You will need the second isomorphism theorem: if N_1 and N_2 are normal subgroups of a group G, then $N_1 N_2$ forms a *normal* subgroup of G, and $N_2/(N_1 \cap N_2) \cong (N_1 N_2)/N_1$.

6.5.6. Additional facts particular to modules over a ring.
We now apply these concepts specifically to the category Mod_A.

6.5.H. EXERCISE. Show that the simple objects of Mod_A are precisely the objects of the form $A/\mathfrak{m}$, where $\mathfrak{m}$ is a maximal ideal of A.

6.5.I. EXERCISE. Suppose M is a finite length A-module, and (6.5.0.1) is a composition series for M, with $M_i/M_{i-1} \cong A/\mathfrak{m}_i$ (where the $\mathfrak{m}_i$ are maximal ideals, not necessarily distinct). Show that M is annihilated by $\mathfrak{m}_1 \cdots \mathfrak{m}_n$. Equivalently, M is an $A/(\mathfrak{m}_1 \cdots \mathfrak{m}_n)$-module.

Suppose now that the list $(\mathfrak{m}_1, \ldots, \mathfrak{m}_n)$ consists of the distinct maximal ideals $\mathfrak{n}_1, \ldots, \mathfrak{n}_s$, appearing with multiplicity $\ell_1, \ldots, \ell_s$. (These are the "refined" lengths mentioned in Definition 6.5.2.)

By the Chinese Remainder Theorem,

$$A/\mathfrak{m}_1 \ldots \mathfrak{m}_n \cong A/\mathfrak{n}_1^{\ell_1} \cdots \mathfrak{n}_s^{\ell_s} \cong A/\mathfrak{n}_1^{\ell_1} \times \cdots A/\mathfrak{n}_s^{\ell_s}.$$

For $1 \le i \le s$, let $e_i \in A$ be an element of A so that $e_i \equiv 1 \pmod{\mathfrak{n}_i^{\ell_i}}$ and $e_i \equiv 0 \pmod{\mathfrak{n}_j^{\ell_j}}$ for $i \ne j$. The e_i exist by the Chinese Remainder Theorem.

6.5.J. EXERCISE. Show that $M \cong e_1 M \times \cdots \times e_s M$, and $e_i M$ is a finite-length module where all the simple quotients are $A/\mathfrak{n}_i$. Thus M is a product of pieces, each with composition series with only one type of "simple factor."

6.5.K. EXERCISE. Suppose M is a finite length A-module. Show that M is finitely generated, and Supp M consists of finitely many points of Spec A, all closed. (A converse under Noetherian hypotheses will be proved in Exercise 6.6.X(a).) We thus have a notion of the "length of M at each of these closed points."

6.5.7. *Definition.* If A has finite length as a module over itself, we say A is an **Artinian ring**. If A is furthermore a local ring, we say A is an **Artin** or **Artinian local ring**.

6.5.8. *Applying this language to schemes.* We next consider the category $QCoh_X$ of quasicoherent sheaves on a scheme X. We have the notion of the length $\ell(\mathscr{F})$ of a finite-length quasicoherent sheaf on X.

6.5.L. EXERCISE. Show that the simple objects of $QCoh_X$ are those isomorphic to the structure sheaves of closed points.

6.5.M. EXERCISE. Suppose that $\mathscr{F}$ is a finite length element of $QCoh_X$. Show that $\mathscr{F}$ is finite type, and Supp $\mathscr{F}$ consists of finitely many points of X, all closed. (A converse under Noetherian hypotheses will be proved in Exercise 6.6.X(b).) Explain how to define the length of a $\mathscr{F}$ at one of the points of Supp $\mathscr{F}$.

6.5.9. *Definition.* The **length of a scheme** X is the length of the structure sheaf $\mathscr{O}_X$ (in $QCoh_X$). A scheme X is **finite length** or **Artinian** if $\mathscr{O}_X$ is finite length.

6.6   Visualizing Schemes: Associated Points and Zerodivisors

The theory of *associated points* of a module refines the notion of support (§4.1.7). Associated points will help us understand and visualize nilpotents, and generalize the notion of "rational functions" to nonintegral schemes. They are useful in ways we won't use, for example, through their connection to primary decomposition. They might be most useful for us in helping us understand and visualize (non-)zerodivisors, which will come up repeatedly, through effective Cartier divisors and line bundles, regular sequences, depth and Cohen–Macaulayness, and more.

There is no particular reason to discuss associated points now, and this section can be read independently, at your leisure. But it is a good opportunity to practice visualizing geometry, and to learn some useful algebra.

6.6.1. Motivation. Figure 6.3 is a sketch of a scheme X. We see two connected components, and three irreducible components. The irreducible components of X have dimensions 2, 1, and 1, although we won't be able to make sense of "dimension" until Chapter 12. Both connected components are nonreduced.

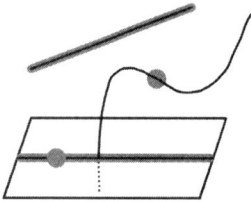

Figure 6.3 *This scheme has six associated points, of which three are embedded points. A function is a zerodivisor if it vanishes at any of these six points.*

We see a little more in this picture, which we will make precise in this section, in terms of "associated points." The reducible connected component seems to have different amounts of nonreduced behavior on different loci. The scheme X has six associated points, which are the generic points of the irreducible subsets "visible" in the picture. A function on X is a zerodivisor if its zero locus contains any of these six irreducible subvarieties.

Suppose M is a finitely generated module over a Noetherian ring A. For example, M could be A itself. Then there are some special points of Spec A that are particularly crucial to understanding M. These are the *associated points* of M (or equivalently, the *associated prime ideals* of M—we will use these terms interchangeably). As motivation, we give a zillion properties of associated points, and leave it to you to verify them from the theory developed in the rest of this section.

As you read this section, you may wish to keep in mind

$$M = A = k[x, y]/(y^2, xy)$$

(Figure 4.4) as a running example.

6.6.2. A zillion properties of associated points. Here are some of the properties of associated points that we will prove.

There are finitely many associated points $\mathrm{Ass}_A M \subset \mathrm{Spec}\, A$.

The support of M is the closure of the associated points of M: $\mathrm{Supp}\, M = \overline{\mathrm{Ass}_A M}$. The support of any submodule of M is the closure of some subset of the associated points of M. The support of any element of M is the closure of some subset of the associated points.

The associated points of M are precisely the generic points of irreducible components of $\mathrm{Supp}\, m$ for all $m \in M$. The associated points of M are precisely the generic points of those $\mathrm{Supp}\, m$ that are irreducible (as m runs over the elements of M). The associated primes of M are precisely those prime ideals that are annihilators of some element of M.

Taking "associated points" commutes with localization. Hence this notion is "geometric in nature," which will (in §6.6.2) allow us to extend the notion to coherent sheaves on locally Noetherian schemes.

Associated points behave *fairly* well in exact sequences. For example, the associated points of a submodule are a subset of the associated points of the module.

If $I \subset A$ is an ideal, the associated primes $\mathfrak{p}$ of A/I are precisely those $\mathfrak{p}$ such that a $\mathfrak{p}$-primary ideal appears in the primary decomposition of I.

We will repeatedly use the fact that *an element of A is a zerodivisor if and only if it vanishes at an associated point.*

An element of A is nilpotent if and only if it vanishes at every associated point. The locus of points $[\mathfrak{p}]$ of $\operatorname{Spec} A$ where the stalk $A_\mathfrak{p}$ is nonreduced is the closure of some subset of the associated points.

An associated point that is in the closure of another associated point is said to be an *embedded point*. If A is reduced, then $\operatorname{Spec} A$ has no embedded points. Hypersurfaces in $\mathbb{A}_k^n$ have no embedded points. We will later see that complete intersections have no embedded points (§26.2.7).

Elements of M are determined by their localization at the associated points. Sections of the corresponding sheaf $\widetilde{M}$ (Exercise/Definition 4.1.D) are determined by their germs at the associated points.

This discussion immediately implies a notion of **associated point** for a coherent sheaf on a locally Noetherian scheme, with all the good properties described here. The phrase **associated point of a locally Noetherian scheme** X (without explicit mention of a coherent sheaf) means "associated point of $\mathscr{O}_X$," and similarly for **embedded points**.

We now establish these zillion facts.

6.6.3. More on the notion of support.

The notion of *associated points* of an A-module M refines the notion of *support* (in the case where M is finitely generated over a Noetherian ring A). (In what follows, we make no assumptions that A is Noetherian or that M is finitely generated until we need to.) To set this up, recall (§4.1.7) that the *support of* $m \in M$,

$$\operatorname{Supp} m = \{[\mathfrak{p}] \in \operatorname{Spec} A \; : \; m_\mathfrak{p} \neq 0\},$$

is a closed subset, and thus of the form $V(I)$ for some I. Exercise 6.6.A gives the "best such" I. Define the **annihilator ideal** $\operatorname{Ann}_A m \subset A$ of an element m of an A-module M by:

$$\operatorname{Ann}_A m := \{a \in A \; : \; am = 0\} = \ker(A \xrightarrow{\times m} M).$$

The subscript A is omitted if it is clear from context.

6.6.A. EASY IMPORTANT EXERCISE. Show that $\operatorname{Supp} m = V(\operatorname{Ann} m)$.

Recall (Definition 2.7.6) that

$$\operatorname{Supp} \widetilde{M} = \{\mathfrak{p} \in \operatorname{Spec} A : \widetilde{M}_\mathfrak{p} \neq 0\},$$

and the analogous Definition 4.1.7 of the *support of the module* M,

(6.6.3.1) $$\operatorname{Supp} M := \{\mathfrak{p} \in \operatorname{Spec} A \; : \; M_\mathfrak{p} \neq 0\},$$

so $\operatorname{Supp} M = \operatorname{Supp} \widetilde{M}$. If M is a principal module generated by $m \in M$, then

$$\operatorname{Supp} M = \operatorname{Supp} Am = \operatorname{Supp} m = V(\operatorname{Ann} m).$$

The notions of support and associated points behave well in exact sequences, and under localization. We begin to explain this now.

6.6.4. *The notion of support behaves well in exact sequences.*

6.6.B. EXERCISE. Suppose that $0 \to M' \to M \to M'' \to 0$ is a short exact sequence of A-modules.

(a) Show that $\operatorname{Supp} M = \operatorname{Supp} M' \cup \operatorname{Supp} M''$.
(b) Show that if M is a finitely generated module, then $\operatorname{Supp} M$ is a closed subset of $\operatorname{Spec} A$. (Hint: Induction on the number of generators.)

Warning: $\operatorname{Supp} M$ need not be closed in general; consider $A = \mathbb{Z}$ and $M = \oplus_{\mathfrak{p} \text{ prime}} \mathbb{Z}/(\mathfrak{p})$.

6.6.C. EXERCISE. Suppose M is a finitely generated A-module, and $x \in A$ has value 0 at all the points of $\operatorname{Supp} M$, i.e., x is contained in all of the primes where M is supported. Show that some power x^n of x annihilates every element of M. (Hint: Annihilate a generating set.)

6.6.D. EXERCISE (FOR USE IN §18.6). Suppose M is a finitely generated A-module, and $x \in A$. Show that $\text{Supp}(M/xM) = (\text{Supp } M) \cap V(x)$. Here M/xM is defined by the exact sequence

$$M \xrightarrow{\times x} M \longrightarrow M/xM \longrightarrow 0.$$

6.6.5. Definition: Associated points and associated primes.

Define the **associated prime ideals** of an A-module M to be the *prime* ideals of A of the form $\text{Ann}_A(m)$ for some $m \in M$. Define the **associated points** of M to be the corresponding points of Spec A; we use the terminology "associated points" and "associated primes" interchangeably. The set of associated points is denoted by $\text{Ass}_A M \subset \text{Spec } A$. The subscript A is dropped if it is clear from the context. (To help remember the definition and the notation, some call these the *assassins*, as they are the primes that can ruthlessly annihilate elements of the module, without remorse. But we will not use this term.)

6.6.6. The **associated primes of a ring** A are the associated primes of A considered as an A-module (i.e., $M = A$).

6.6.E. EASY EXERCISE (ASSOCIATED POINTS OF INTEGRAL DOMAINS). If A is an integral domain, show that $\text{Ass } A = \{[(0)]\}$—the zero ideal is the only associated prime.

6.6.F. EXERCISE (ASSOCIATED POINTS OF HYPERSURFACES). Given $f \in k[x_1, \ldots, x_n]$, show that the associated primes of $k[x_1, \ldots, x_n]/(f)$ are those principal ideals generated by the prime factors of f. (Your argument will apply more generally to any $f \in A$, where A is a unique factorization domain.)

6.6.7. *The observation that* $[\mathfrak{p}] \in \text{Ass}_A(M)$ *if and only if there is an injection* $A/\mathfrak{p} \hookrightarrow M$ *of A-modules will be essential.* The next exercise might drive this home.

6.6.G. EXERCISE.

(a) Suppose $M' \subset M$. Show that $\text{Ass } M' \subset \text{Ass } M$. (The corresponding statement for "support" is implicit in Exercise 6.6.B(a).)
(b) Show that $\text{Ass } M \subset \text{Supp } M$.

If M is finitely generated, then Supp M is closed (Exercise 6.6.B), so $\overline{\text{Ass } M} \subset \text{Supp } M$. Equality will be shown in Proposition 6.6.24, when A is Noetherian.

6.6.8. Nonzero modules over Noetherian rings have associated points.

Suppose m is a nonzero element of an A-module M. Observe that for any nonzero multiple xm of m, $\text{Ann } m \subseteq \text{Ann } xm \subsetneq A$.

6.6.H. EXERCISE. Suppose A is Noetherian. Show that there is some multiple $n = xm$ such that *any* nonzero multiple $yn \neq 0$ of n satisfies $\text{Ann } yn = \text{Ann } n$.

6.6.9. Proposition — *Continuing the notation of the previous exercise, we have that* $\text{Ann } n$ *is a prime ideal.*

Proof. Suppose $ab \in \text{Ann } n$, so $abn = 0$. Then either $bn = 0$ (in which case $b \in \text{Ann } n$), or else $a \in \text{Ann } bn = \text{Ann } n$. $\square$

We have thus proved the following.

6.6.10. Proposition (nonzero modules over Noetherian rings have associated primes)

— *If M is a nonzero module over a Noetherian ring A, then* $\text{Ass}_A M$ *is nonempty. More precisely, for any* $m \neq 0$ *in M, there is an associated prime* $\mathfrak{p}$ *containing* $\text{Ann } m$, *and* $\mathfrak{p} = \text{Ann } xm$ *for some* $x \in A$.

6.6.11. Localizations at the associated primes.

Recall the useful fact that $M \to \prod_{\mathfrak{p} \in \text{Spec } A} M_{\mathfrak{p}}$ is an injection (Exercise 4.1.F). Our current situation is much better: we can take the product over only the localization at *associated* primes.

6.6.I. EXERCISE. Suppose M is a module over a Noetherian ring A. Show that the natural map

$$(6.6.11.1) \qquad\qquad M \longrightarrow \prod_{\mathfrak{p} \in \text{Ass } M} M_{\mathfrak{p}}$$

is an injection. Hint: If the kernel K is nonzero, then K has an associated prime $\mathfrak{p}$, which is the annihilator of some $m \in K \subset M$, and m is nonzero in $M_{\mathfrak{p}}$.

Clearly we need only the maximal among the associated primes in (6.6.11.1).

6.6.12. Zerodivisors = elements of associated primes.

6.6.13. Proposition — *Suppose* $f \in A$, *with* A *Noetherian. Then* f *is a zerodivisor on* M *if and only if* f *vanishes at an associated point of* M. *Translation: The set of zerodivisors is the union of the associated prime ideals.*

Again, we need only the maximal among the associated primes. For example, if $(A, \mathfrak{m})$ is a local ring, then $\mathfrak{m}$ is an associated prime if and only if every element of $\mathfrak{m}$ is a zerodivisor.

Proof. Suppose f vanishes at an associated point $[\mathfrak{p}]$ of M. Choose m with Ann $m = \mathfrak{p}$. Then $fm = 0$ while $m \neq 0$, so f is a zerodivisor.

Conversely, if f vanishes at no associated point, consider the commuting diagram

$$
\begin{array}{ccc}
M & \longhookrightarrow & \prod_{\mathfrak{p} \in \text{Ass } M} M_{\mathfrak{p}} \\
\scriptstyle \times f \downarrow & & \downarrow \scriptstyle \times f \\
M & \longhookrightarrow & \prod_{\mathfrak{p} \in \text{Ass } M} M_{\mathfrak{p}},
\end{array}
$$

where the rows are the maps of (6.6.11.1). The vertical arrow on the right (multiplication by f) is an injection by hypothesis, so the vertical arrow on the left must be an injection as well. □

6.6.14. Associated points behave *fairly* well in exact sequences.

6.6.15. Proposition — *Suppose*

$$(6.6.15.1) \qquad\qquad 0 \longrightarrow M' \longrightarrow M \longrightarrow M'' \longrightarrow 0$$

is a short exact sequence of A-*modules. Then*

$$(6.6.15.2) \qquad\qquad \text{Ass } M' \subset \text{Ass } M \subset \text{Ass } M' \cup \text{Ass } M''.$$

We come to our first complicated proof of the section.

Proof. The first inclusion of (6.6.15.2) was shown in Exercise 6.6.G(a).

Suppose next that $[\mathfrak{p}] \in \text{Ass } M$, so there is some $m \in M$ with Ann $m = \mathfrak{p}$. We wish to find a submodule of M' or M'' isomorphic to $A/\mathfrak{p}$. If this proposition were true, we would expect to find such a submodule in the "part of (6.6.15.1) spanned by m." So we consider instead the exact sequence

$$0 \longrightarrow Am \cap M' \longrightarrow Am \longrightarrow Am/(Am \cap M') \longrightarrow 0,$$

noting that the three modules appearing here are submodules of the corresponding modules in (6.6.15.1). So, by Exercise 6.6.G(a), it suffices to prove the result in this "special case," which can be rewritten as

$$0 \longrightarrow I/\mathfrak{p} \longrightarrow A/\mathfrak{p} \longrightarrow A/I \longrightarrow 0,$$

where I is the annihilator of m considered as an element of the module $Am/(Am \cap M')$. For convenience, let $B = A/\mathfrak{p}$ (an integral domain), so we rewrite the exact sequence further as the top row of

$$
\begin{array}{ccccccccc}
0 & \longrightarrow & J & \longrightarrow & B & \longrightarrow & B/J & \longrightarrow & 0 \\
& & \big\uparrow & & \big\uparrow & & \big\uparrow & & \\
0 & \longrightarrow & M' & \longrightarrow & M & \longrightarrow & M'' & \longrightarrow & 0.
\end{array}
$$

Now localize the top row of B-modules at $(0) \subset B$, so it becomes an exact sequence of vector spaces over the fraction field $K(B)$, and the central element is one-dimensional:

$$
0 \longrightarrow J \otimes K(B) \longrightarrow K(B) \longrightarrow (B/J) \otimes K(B) \longrightarrow 0.
$$

Thus one of the outside terms $J \otimes K(B)$ and $(B/J) \otimes K(B)$ has a nonzero element, which (tracing our argument backward) gives an element of M' or M'' whose annihilator is precisely $\mathfrak{p}$. $\qquad \square$

6.6.16. *Cautionary example.* The short exact sequence of $\mathbb{Z}$-modules

$$
0 \longrightarrow \mathbb{Z} \xrightarrow{\ \times 2\ } \mathbb{Z} \longrightarrow \mathbb{Z}/2 \longrightarrow 0
$$

(and Easy Exercise 6.6.E) shows it is not always true that $\operatorname{Ass} M = \operatorname{Ass} M' \cup \operatorname{Ass} M''$. However, sometimes we can still ensure some associated primes of M'' lift to associated primes of M, as we shall see in §6.6.20.

6.6.17. Finitely generated modules over Noetherian rings have finitely many associated points/primes.

6.6.J. IMPORTANT EXERCISE. Suppose that M is a finitely generated module over a Noetherian ring A. Show that M has a (finite) filtration

(6.6.17.1) $\qquad 0 = M_0 \subset M_1 \subset \cdots \subset M_n = M, \quad$ where $M_{i+1}/M_i \cong A/\mathfrak{p}_i$ for some prime $\mathfrak{p}_i$.

Hint: Build (6.6.17.1) inductively from left to right, using Proposition 6.6.10, and show the process terminates.

6.6.K. EXERCISE. Suppose an A-module M has a finite filtration (6.6.17.1), with no assumptions of finite generation or Noetherianity. Show that every associated prime of M appears as one of the $\mathfrak{p}_i$. In particular, under this hypothesis (for example, if M is finitely generated over a Noetherian ring) M *has finitely many associated points/primes*. (Hint: Exercise 6.6.E and Proposition 6.6.15.)

6.6.18. *Caution: Nonassociated prime ideals may unavoidably appear among the quotients in (6.6.17.1).* Example 6.6.16 shows that the nonassociated prime ideals may be among the quotients in (6.6.17.1), although in that case it is because the filtration was chosen unwisely. But better choices will not always remedy the problem:

6.6.L. EXERCISE. Consider the module $M = (x, y) \subset A = k[x, y]$ over the ring A. Show that *any* filtration (6.6.17.1) of M contains a quotient $A/\mathfrak{p}_i$, where $\mathfrak{p}_i$ is *not* an associated prime.

6.6.19. *Remark.* However, not all is lost: Exercise 6.6.P will show that for any quotient $A/\mathfrak{p}_i$ in any filtration (6.6.17.1) of M, $\mathfrak{p}_i$ must *contain* an associated prime of M.

6.6.20. Associated points behave *fairly* well in exact sequences, continued.

6.6.21. Proposition — *Suppose A is a Noetherian ring. We continue to consider the short exact sequence*

$$
0 \longrightarrow M' \longrightarrow M \longrightarrow M'' \longrightarrow 0
$$

of A-modules. Suppose $\mathfrak{p} \in \operatorname{Ass} M''$, *but* $\mathfrak{p} \notin \operatorname{Supp} M'$ *(a stronger hypothesis than* $\mathfrak{p} \notin \operatorname{Ass} M'$*). Then* $\mathfrak{p} \in \operatorname{Ass} M$.

We come to our second complicated proof of the section.

Proof. Choose $m'' \in M''$ with $\mathfrak{p} = \operatorname{Ann} m''$ in M''. Choose a lift of $m \in M$ of $m'' \in M''$. We apply a strategy similar to that of our proof of Proposition 6.6.15. Consider the "inclusion of short exact sequences"

$$
\begin{array}{ccccccccc}
0 & \longrightarrow & \ker(Am \to Am'') & \longrightarrow & Am & \longrightarrow & Am'' & \longrightarrow & 0 \\
 & & \uparrow & & \uparrow & & \uparrow & & \\
0 & \longrightarrow & M' & \longrightarrow & M & \longrightarrow & M'' & \longrightarrow & 0.
\end{array}
$$

As $\operatorname{Supp}(\ker(Am \to Am'')) \subset \operatorname{Supp} M'$, $\operatorname{Ass}(Am) \subset \operatorname{Ass} M$, and $\operatorname{Ass}(Am'') \subset \operatorname{Ass} M''$ (Exercises 6.6.B(a) and 6.6.G(a)), we have reduced to considering the top row instead of the bottom row. The top row can be conveniently rewritten as

$$
0 \longrightarrow \mathfrak{p}/I \longrightarrow A/I \longrightarrow A/\mathfrak{p} \longrightarrow 0
$$

(here $I = \operatorname{Ann}(m)$), where our hypothesis translates to $[\mathfrak{p}] \notin \operatorname{Supp}(\mathfrak{p}/I)$. For convenience, let $B = A/I$, so we may now consider the sequence

$$
0 \longrightarrow \mathfrak{q} \longrightarrow B \longrightarrow B/\mathfrak{q} \longrightarrow 0,
$$

where $\mathfrak{q}$ is prime, with the confusion-inducing hypothesis $[\mathfrak{q}] \notin \operatorname{Supp} \mathfrak{q}$.

From the confusing hypothesis, there is an element b of B that vanishes on $\operatorname{Supp} \mathfrak{q}$ but doesn't vanish at $[\mathfrak{q}]$. Translation: (i) b lies in all the primes of $\operatorname{Supp} \mathfrak{q}$, but (ii) $b \notin \mathfrak{q}$. Then from (i) there is some power b^n of b that annihilates all elements of $\mathfrak{q}$ (Exercise 6.6.C, which applies because $\mathfrak{q}$ is finitely generated, which in turn follows from Noetherianity of A). From (ii), $b^n \notin \mathfrak{q}$.

Then $\operatorname{Ann}(b^n)$ contains $\mathfrak{q}$ from (i), but does not contain any element of $B \setminus \mathfrak{q}$ from (ii), so $\operatorname{Ann}(b^n) = \mathfrak{q}$. Hence $\mathfrak{q}$ is an associated prime of B, which (unwinding our argument) shows that $\mathfrak{p}$ is an associated prime of M. $\square$

6.6.22. Minimal primes are associated.

6.6.M. EXERCISE: MINIMAL PRIMES ("IRREDUCIBLE COMPONENTS") ARE ASSOCIATED. Suppose M is a finitely generated module over Noetherian A, and $\mathfrak{p} \subset A$ is a prime ideal corresponding to an irreducible component of $\operatorname{Supp} M \subset \operatorname{Spec} A$. Show that $[\mathfrak{p}] \in \operatorname{Ass} M$. Hint: Exercise 6.6.J and Proposition 6.6.21.

6.6.23. *Non-Noetherian Remark.* By combining Proposition 6.6.13 with Exercise 6.6.M, we see that if A is a Noetherian ring, then any element of any minimal prime $\mathfrak{p}$ is a zerodivisor. This is true without Noetherian hypotheses: suppose $s \in \mathfrak{p}$; then by minimality of $\mathfrak{p}$, $\mathfrak{p}A_{\mathfrak{p}}$ is the unique prime ideal in $A_{\mathfrak{p}}$, so the element $s/1$ of $A_{\mathfrak{p}}$ is nilpotent (because it is contained in all prime ideals of $A_{\mathfrak{p}}$, Theorem 3.2.13). Thus for some $t \in A \setminus \mathfrak{p}$, $ts^n = 0$, so s is a zerodivisor in A. We will use this in Exercise 12.1.H.

6.6.24. Proposition — *Suppose M is a finitely generated module over a Noetherian ring A. Then* $\operatorname{Supp} M = \overline{\operatorname{Ass} M}$.

Proof. Combine Exercises 6.6.G(b) and 6.6.M. $\square$

6.6.N. EXERCISE. Show that the locus of points $[\mathfrak{p}]$ where $A_{\mathfrak{p}}$ is nonreduced is the support of the nilradical $\operatorname{Supp} \mathfrak{N}$. Hence show that the "reduced locus" of a locally Noetherian scheme is open.

The following justifies a simple way of thinking about associated primes of a ring.

6.6.O. EXERCISE. Show that a prime ideal $\mathfrak{p} \subset A$ is an associated prime of a Noetherian ring A if and only if there is $f \in A$ such that $\operatorname{Supp} f = V(\mathfrak{p}) = \overline{[\mathfrak{p}]}$.

6.6.P. EXERCISE, PROMISED IN REMARK 6.6.19. Show that for each quotient in the filtration (6.6.17.1) of M, $\operatorname{Supp} A/\mathfrak{p}_i = \overline{[\mathfrak{p}_i]} \subset \operatorname{Supp} M$, and that every $\mathfrak{p}_i$ contains a minimal prime, and hence an associated prime.

6.6.25. "Support" and "associated points" commute with localization.
Suppose S is a multiplicative subset of A, and $\mathfrak{p} \subset A$ is a prime ideal not meeting S, so (abusing notation slightly) $[\mathfrak{p}] \in \operatorname{Spec} S^{-1}A \subset \operatorname{Spec} A$ (§3.2.9).

6.6.Q. EXERCISE (Supp COMMUTES WITH LOCALIZATION). Show that for any A-module M, $\operatorname{Supp}_A M \cap \operatorname{Spec} S^{-1}A = \operatorname{Supp}_{S^{-1}A} S^{-1}M$.

6.6.26. Proposition (Ass commutes with localization) — *Let M be a finitely generated module over a Noetherian ring A. Then $\operatorname{Ass}_A M \cap \operatorname{Spec} S^{-1}A = \operatorname{Ass}_{S^{-1}A} S^{-1}M$.*

Proof. We first show that $\operatorname{Ass}_A M \cap \operatorname{Spec} S^{-1}A \subset \operatorname{Ass}_{S^{-1}A} S^{-1}M$. If $\mathfrak{p} \in \operatorname{Ass}_A M$, then we have an injection $A/\mathfrak{p} \hookrightarrow M$. Localizing by S (which preserves injectivity), we have $(S^{-1}A)/(S^{-1}\mathfrak{p}) \hookrightarrow M$. (Where did we use $\mathfrak{p} \in \operatorname{Spec} S^{-1}A$?)

We next show that $\operatorname{Ass}_{S^{-1}A} S^{-1}M \subset \operatorname{Ass}_A M \cap \operatorname{Spec} S^{-1}A$. Suppose $\mathfrak{q} := S^{-1}\mathfrak{p} \in \operatorname{Ass}_{S^{-1}A} S^{-1}M$, so $\mathfrak{q} = \operatorname{Ann}_{S^{-1}A}(m/s)$, for some $s \in S$, and $m \in M$. Since the elements of S are units in $S^{-1}A$, we have that $\mathfrak{q} = \operatorname{Ann}_{S^{-1}A} m$. As support commutes with localization, $V(\mathfrak{p})$ must be an irreducible component of $\operatorname{Supp} m$ (as $\operatorname{Supp} m \cap \operatorname{Spec} S^{-1}M$ contains $[\mathfrak{q}]$, but no generizations of $[\mathfrak{q}]$). Then by Exercise 6.6.M, $\mathfrak{p}$ is an associated prime of M. $\square$

Unimportant remark: We can remove the finite generation hypothesis in this Proposition by replacing the last sentence of the proof with the following. "Equivalently, $\mathfrak{p}$ is minimal over $A/\operatorname{Ann}_A m$. Then by Exercise 6.6.M, $\mathfrak{p}$ is the annihilator of some $a \in A/\operatorname{Ann}_A m$, and thus P is the annihilator of am in M."

6.6.27. Embedded points/primes.

6.6.28. *Definition.* The associated points that are *not* the generic points of irreducible components of $\operatorname{Supp} M$ are called **embedded points**, and their corresponding primes are called **embedded primes**. The motivation for this language is their geometric incarnation as embedded points (§6.6.32 shortly). For example, the scheme of Figure 6.3 has three embedded points (the corresponding ring has three embedded primes).

6.6.29. *Remark.* Exercise 6.6.F translates to "hypersurfaces in $\mathbb{A}^n_k$ have no embedded points." More generally, if A is a unique factorization domain, then $\operatorname{Spec} A/(f)$ has no embedded points for any $f \in A$. Generalizing in a different direction, we will see that "complete intersections have no embedded points" in §26.2.7.

6.6.R. EXERCISE. Suppose A is a reduced ring (i.e., A has no nonzero nilpotents). Show that $\operatorname{Spec} A$ has no embedded primes. (Hints: If $\mathfrak{p} = \operatorname{Ann} a$ is an embedded prime, and $\mathfrak{q}$ is a minimal prime, choose $b \in \mathfrak{p} \setminus \mathfrak{q}$. From $ab = 0$, show that $a \in \mathfrak{q}$. Hence a is contained in every minimal prime. From $ab = 0$, show that a is contained in all the minimal primes. Show that a vanishes at all points of $\operatorname{Spec} A$, and hence by Theorem 3.2.13 is nilpotent.)

Thus *reduced rings have no embedded primes.* Even better: *The only elements of a ring that an embedded prime can annihilate are nilpotents.*

6.6.30. *Remark.* The converse to Exercise 6.6.R is false. Rings without embedded primes can still have nilpotents—witness $k[x]/(x^2)$.

6.6.S. EXERCISE. Show that if $\mathfrak{p}$ is an embedded prime of a ring A, then $A_\mathfrak{p}$ is nonreduced.

6.6.31. Get your hands dirty: Explicit algebraic exercises.

6.6.T. EXERCISE. (See Figure 4.4.)

(a) Suppose f is a function on $\operatorname{Spec} k[x,y]/(y^2, xy)$, i.e., $f \in k[x,y]/(y^2, xy)$. Show that $\operatorname{Supp} f$ is either the empty set, or the origin, or the entire space. Hence find the associated points of $\operatorname{Spec} k[x,y]/(y^2, xy)$.

(b) Show explicitly by hand that $f \in k[x,y]/(y^2, xy)$ is a zerodivisor if and only if $f(0,0)=0$.

6.6.U. EXERCISE (PRACTICE WITH FUZZY PICTURES). Suppose $X = \operatorname{Spec} \mathbb{C}[x,y]/I$, and that the associated points of X are $[(y - x^2)]$, $[(x - 1, y - 1)]$, and $[(x - 2, y - 2)]$. (Exercise 6.6.U will verify that such an X actually exists.)

(a) Sketch X as a subset of $\mathbb{A}^2_{\mathbb{C}} = \operatorname{Spec} \mathbb{C}[x,y]$, including fuzz.
(b) Do you have enough information to know if X is reduced?
(c) Do you have enough information to know if $x + y - 2$ is a zerodivisor? How about $x + y - 3$? How about $y - x^2$?
(d) Let $I = (y - x^2)^3 \cap (x - 1, y - 1)^{15} \cap (x - 2, y - 2)$. Show that $X = \operatorname{Spec} \mathbb{C}[x,y]/I$ satisfies the hypotheses of this exercise. (Rhetorical question: Is there a "smaller" example? Is there a "smallest"?)

6.6.32. Geometric definitions.

6.6.V. EXERCISE/DEFINITION. Define the **associated points** and **embedded points of a coherent sheaf on a locally Noetherian scheme**. (Idea/hint: Do it affine-locally.)

6.6.W. EXERCISE. Suppose X is a locally Noetherian scheme, and $U \subset X$ is an open subscheme. Show that the natural map

$$(6.6.32.1) \qquad\qquad \Gamma(U, \mathscr{O}_X) \longrightarrow \prod_{p \in \operatorname{Ass} X \cap U} \mathscr{O}_{X,p}$$

(cf. (6.6.11.1)) is an injection.

6.6.33. *Scheme-theoretic density (and the adjective "scheme-theoretic" more generally).* An open subscheme U of a scheme X is said to be **scheme-theoretically dense** if any function on any open set V is 0 if it restricts to 0 on $U \cap V$. This is stronger than density in the usual (topological) sense. For example, if X is locally Noetherian, then you can use the injection (6.6.32.1) (with $M = A$) to show that an open subscheme $U \subset X$ is scheme-theoretically dense if and only if it contains all the associated points of X.

  More generally, the adjective **scheme-theoretic** typically indicates an ideal-theoretic definition, enriching and refining a more naive set-theoretic definition. Examples will include:

- scheme-theoretically dense (here),
- scheme-theoretic intersection (Figure 4.5, Exercise 9.1.I, Exercise 10.4.B),
- scheme-theoretic support (Definition 9.1.4),
- scheme-theoretic image (§9.4.1),
- scheme-theoretic closure (§9.4.8),
- scheme-theoretic preimage (§10.3.1), and
- scheme-theoretic fiber (Definition 10.3.3).

6.6.34. Revisiting the notion of length.

6.6.X. EXERCISE.

(a) *(Noetherian converse to Exercise 6.5.K)* Suppose A is a Noetherian ring. Show that any finitely generated module whose support consists of finitely many points of $\operatorname{Spec} A$, all closed, has finite length.

(b)   (*Noetherian converse to Exercise 6.5.M*) Suppose $\mathscr{F}$ is a finite type quasicoherent sheaf on a locally Noetherian scheme X, supported at finitely many points of X, all closed. Show that $\mathscr{F}$ has finite length.

6.6.Y. EXERCISE.

(a)   Show that each finite length scheme (Definition 6.5.9) has a finite number of points, with the discrete topology.
(b)   If X is locally finite type over $\overline{k}$, of finite length ℓ, show that $\dim_{\overline{k}} \Gamma(X, \mathscr{O}_X) = \ell$.

6.6.35. Generalizing the fraction field: The total fraction ring.

6.6.36. *Definition: Rational function on a locally Noetherian scheme.* A **rational function** on a locally Noetherian scheme is an element of the image of $\Gamma(U, \mathscr{O}_U)$ in (6.6.32.1) for some U containing all the associated points. Equivalently, the set of rational functions is the colimit of $\mathscr{O}_X(U)$ over all open sets containing the associated points.

For example, on $\operatorname{Spec} k[x, y]/(y^2, xy)$ (Figure 4.4), $\frac{x-2}{(x-1)(x-3)}$ is a rational function, but $\frac{x-2}{x(x-1)}$ is not.

A rational function has a maximal **domain of definition**, because any two actual functions on an open set (i.e., sections of the structure sheaf over that open set) that agree as "rational functions" (i.e., on small enough open sets containing associated points) must be the same function, by the injectivity of (6.6.32.1). We say that a rational function f is **regular at a point** p if p is contained in this maximal domain of definition (or, equivalently, if there is some open set containing p where f is defined). For example, on $\operatorname{Spec} k[x, y]/(y^2, xy)$, the rational function $\frac{x-2}{(x-1)(x-3)}$ has domain of definition consisting of everything but 1 and 3 (i.e., $[(x-1)]$ and $[(x-3)]$), and is regular away from those two points. A rational function is **regular** if it is regular at all points. (Unfortunately, "regular" is a regularly overused word in mathematics, and in algebraic geometry in particular.)

The complement of the domain of definition of a rational function f is called the **indeterminacy locus of** f (a phrase we'll see again in §11.3.3).

The rational functions form a ring, called the **total fraction ring** or **total quotient ring** of X. If $X = \operatorname{Spec} A$ is affine, then this ring is called the **total fraction (or quotient) ring** of A. If X is integral, the total fraction ring is the function field $K(X)$—the stalk at the generic point—so this extends our earlier Definition 5.2.I of $K(\cdot)$.

6.6.Z. EXERCISE. Show that the ring of rational functions on a locally Noetherian scheme is the colimit of the functions over all open sets containing the associated points:

$$\operatorname{colim}_{\{U \,:\, \operatorname{Ass} X \subset U\}} \mathscr{O}(U).$$

Slightly better (but slightly different): show that a rational function is the data of a function f defined on an open set U containing all associated points, where two such data (U, f) and (U', f') define the same rational function if and only if $f|_{U \cap U'} = f'|_{U \cap U'}$ (cf. §1.3.9). If X is reduced, show that this is the same as requiring that they are defined on an open set of each of the irreducible components.

6.6.37. *Remark: Associated points of integral schemes.* In order for some of our discussion elsewhere to make sense in non-Noetherian settings, we note that the notion of associated points for integral schemes works perfectly, because it works for integral domains—only the generic point is associated. In particular, the definition above of rational functions on an integral scheme X agrees with Definition 5.2.I, as precisely the elements of the function field $K(X)$.

6.6.38.** Aside: Primary ideals and associated primes. Primary decomposition was introduced by the world chess champion E. Lasker, and axiomatized by world math champion E. Noether. We won't need it, but here is the beginning of the story, for the curious reader. An ideal $I \subset A$ in a ring is **primary** if $I \neq A$, and $xy \in I$ implies either $x \in I$ or $y^n \in I$ for some $n > 0$. In this case, $\sqrt{I}$ is prime, and I is said to be $\mathfrak{p}$-primary. Equivalently, if $I \subset A$ then I is $\mathfrak{p}$-primary if and only if A/I has only one associated prime $\mathfrak{p}$. If I is an ideal of a Noetherian ring A, then the associated prime ideals A/I turn out to be precisely the radicals of ideals in a primary decomposition. See [E, §3.3], for example, for more of this important story.

6.7**   Coherent Modules over Non-Noetherian Rings

This section is intended for people who might work with non-Noetherian rings, or who otherwise might want to understand coherent sheaves in a more general setting. Even these people should read this much later. Read this only if you really want to!

Suppose A is a ring. Recall the definition of when an A-module M is finitely generated, finitely presented, and coherent (§6.4). The reason we like coherence is that coherent modules form an abelian category. Here are some accessible exercises working out why these notions behave well. Some repeat earlier discussion in order to keep this section self-contained.

The notion of coherence of a module is only interesting in the case that a ring is coherent over itself. Similarly, coherent sheaves on a scheme X will be interesting only when $\mathscr{O}_X$ is coherent ("over itself"). In this case, coherence is clearly the same as finite presentation. An example where non-Noetherian coherence comes up is the ring $R\langle x_1, \ldots, x_n \rangle$ of "restricted power series" over a valuation ring R of a non-discretely valued K (for example, a completion of the algebraic closure of $\mathbb{Q}_p$). This is relevant to Tate's theory of non-Archimedean analytic geometry over K (which you can read about in [BCDKT], for example). The importance of the coherence of the structure sheaf underlines the importance of Oka's Theorem in complex geometry (stated in §6.4.8).

6.7.A. EXERCISE. Show that coherent implies finitely presented implies finitely generated. (This was discussed at the start of §6.4.)

6.7.B. EXERCISE. Show that 0 is coherent.

Suppose for Exercises 6.7.C–6.7.I that

$$(6.7.0.1) \qquad 0 \longrightarrow M \longrightarrow N \longrightarrow P \longrightarrow 0$$

is an exact sequence of A-modules. In this series of exercises, we will show that if two of $\{M, N, P\}$ are coherent, the third is as well, which will prove very useful.

6.7.1. Hint. The following hint applies to several of the exercises: try to write

$$
\begin{array}{ccccccccc}
0 & \longrightarrow & A^{\oplus p} & \longrightarrow & A^{\oplus(p+q)} & \longrightarrow & A^{\oplus q} & \longrightarrow & 0 \\
 & & \downarrow & & \downarrow & & \downarrow & & \\
0 & \longrightarrow & M & \longrightarrow & N & \longrightarrow & P & \longrightarrow & 0,
\end{array}
$$

and possibly use the Snake Lemma 1.6.5.

6.7.C. EXERCISE. Show that N finitely generated imply P finitely generated. (You will only need right-exactness of (6.7.0.1).)

6.7.D. EXERCISE. Show that M, P finitely generated implies N finitely generated. Possible hint: §6.7.1. (You will only need right-exactness of (6.7.0.1).)

6.7.E. EXERCISE. Show that N, P finitely generated need not imply M finitely generated. (Hint: If I is an ideal, we have $0 \to I \to A \to A/I \to 0$.)

6.7.F. EXERCISE. Show that N coherent, M finitely generated imply M coherent. (You will only need left-exactness of (6.7.0.1).)

6.7.G. EXERCISE. Show that N, P coherent imply M coherent. Hint for (i) in the definition of coherence (§6.4.1):

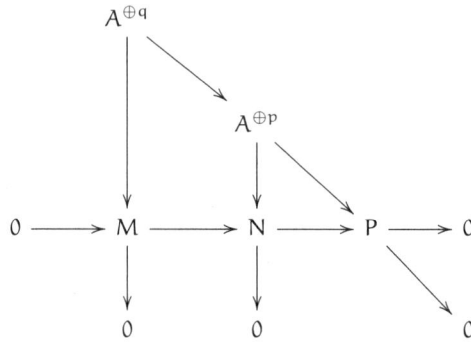

$$
\begin{array}{ccccccccc}
 & & A^{\oplus q} & & & & & & \\
 & & \downarrow & \searrow & & & & & \\
 & & & & A^{\oplus p} & & & & \\
 & & \downarrow & & \downarrow & \searrow & & & \\
0 & \longrightarrow & M & \longrightarrow & N & \longrightarrow & P & \longrightarrow & 0 \\
 & & \downarrow & & \downarrow & & \searrow & & \\
 & & 0 & & 0 & & & & 0
\end{array}
$$

(You will only need left-exactness of (6.7.0.1).)

6.7.H. EXERCISE. Show that M finitely generated and N coherent imply P coherent. Hint for (ii) in the definition of coherence (§6.4.1): §6.7.1.

6.7.I. EXERCISE. Show that M, P coherent imply N coherent. (Hint: §6.7.1.)

6.7.J. EXERCISE. Show that a finite direct sum of coherent modules is coherent.

6.7.K. EXERCISE. Suppose M is finitely generated and N is coherent. If $\phi\colon M \to N$ is any map, show that Im ϕ is coherent.

6.7.L. EXERCISE. Show that the kernel and cokernel of maps of coherent modules are coherent.

   At this point, we have verified that coherent A-modules form an abelian subcategory of the category of A-modules. (Things you have to check: 0 should be in this set; it should be closed under finite sums; and it should be closed under taking kernels and cokernels.)

6.7.M. EXERCISE. Suppose M and N are coherent submodules of the coherent module P. Show that $M + N$ and $M \cap N$ are coherent. (Hint: Consider the right map $M \oplus N \to P$.)

6.7.N. EXERCISE. Show that if A is coherent (as an A-module), then finitely presented modules are coherent. (Of course, if finitely presented modules are coherent, then A is coherent, as A is finitely presented!)

6.7.O. EXERCISE. If M is finitely presented and N is coherent, show that $\mathrm{Hom}(M, N)$ is coherent. (Hint: Hom is left-exact in its first argument.)

6.7.P. EXERCISE. If M is finitely presented, and N is coherent, show that $M \otimes_A N$ is coherent.

PART III

Morphisms of Schemes

III

Chapter 7

Morphisms of Schemes

We now describe the morphisms between schemes. We will define some easy-to-state properties of morphisms, but leave more subtle properties for later.

Recall that a scheme is (i) a set, (ii) with a topology, (iii) and a (structure) sheaf of rings, and that it is sometimes helpful to think of the definition as having three steps. In the same way, the notion of morphism of schemes $X \to Y$ may be defined (i) as a map of sets, (ii) that is continuous, and (iii) with some further information involving the sheaves of functions. In the case of affine schemes, we have already seen the map of sets (§3.2.10), and later seen that this map is continuous (Exercise 3.4.H).

7.1 Motivations for the "Right" Definition of Morphism of Schemes

Here are two motivations for how morphisms should behave. The first is algebraic, and the second is geometric.

7.1.1. *Algebraic motivation.* We will want morphisms of affine schemes $\operatorname{Spec} A \to \operatorname{Spec} B$ to be precisely the ring maps $B \to A$. We have already seen that ring maps $B \to A$ induce maps of topological spaces in the opposite direction (Exercise 3.4.H); the main new ingredient will be the structure sheaf of functions to add into the mix. Then a morphism of schemes should be something that "on the level of affine open sets, looks like this."

7.1.2. *Geometric motivation.* Motivated by the theory of manifolds (§3.1.1), which, like schemes, are ringed spaces, we want morphisms of schemes at the very least to be morphisms of ringed spaces; we now motivate what these are. (We will formalize this in the next section.) Notice that if $\pi \colon X \to Y$ is a map of differentiable manifolds, then a smooth function on Y pulls back to a smooth function on X. More precisely, given an open subset $U \subset Y$, there is a natural map $\Gamma(U, \mathscr{O}_Y) \to \Gamma(\pi^{-1}(U), \mathscr{O}_X)$. This behaves well with respect to restriction (restricting a function to a smaller open set and pulling back yields the same result as pulling back and then restricting), so in fact we have a map of sheaves on Y: $\mathscr{O}_Y \to \pi_* \mathscr{O}_X$. Similarly a morphism of schemes $\pi \colon X \to Y$ should induce a map $\mathscr{O}_Y \to \pi_* \mathscr{O}_X$. But in fact in the category of differentiable manifolds a continuous map $\pi \colon X \to Y$ is a map of differentiable manifolds precisely when smooth functions on Y pull back to smooth functions on X (i.e., the pullback map from smooth functions on Y to *functions* on X lies in the subset of *smooth functions*, i.e., the continuous map π induces a pullback of smooth functions, which can be interpreted as a map $\mathscr{O}_Y \to \pi_* \mathscr{O}_X$), so this map of sheaves *characterizes* morphisms in the differentiable category. So we could use this as the *definition* of morphism in the differentiable category (see Exercise 3.1.A).

But how do we apply this to the category of schemes? In the category of differentiable manifolds, a continuous map $\pi \colon X \to Y$ *induces* a pullback of (the sheaf of) functions, and we can ask when this induces a pullback of *smooth* functions. However, functions are odder on schemes, and we can't recover the pullback map just from the map of topological spaces. The right patch is to hardwire this into the definition of morphism, i.e., to have a continuous map $\pi \colon X \to Y$, along with a pullback map $\pi^\sharp \colon \mathscr{O}_Y \to \pi_* \mathscr{O}_X$. This leads to the definition of the *category* of ringed spaces.

One might hope to define morphisms of schemes as morphisms of ringed spaces. This isn't quite right, as then Motivation 7.1.1 isn't satisfied: as desired, for each morphism $A \to B$ there is a morphism $\operatorname{Spec} B \to \operatorname{Spec} A$, but there can be additional morphisms of ringed spaces $\operatorname{Spec} B \to \operatorname{Spec} A$ not arising in this way (see Exercise 7.2.G). A revised definition as morphisms of ringed spaces that locally look of this form will work, but to avoid the awkwardness of sorting out the details, we take a different approach, using locally ringed spaces, which corresponds to asking not just that functions pull back, but also that *values of functions* pull back. However, we will check that our eventual definition actually is equivalent to this (Exercise 7.3.D).

We begin by formally defining morphisms of ringed spaces.

7.2   Morphisms of Ringed Spaces

7.2.1. Definition. A **morphism of ringed spaces** $\pi \colon X \to Y$ is a continuous map of topological spaces (which we unfortunately also call π) along with a map $\mathscr{O}_Y \to \pi_* \mathscr{O}_X$, which we think of as a "pullback map." By adjointness (§2.7.1), this is the same as a map $\pi^{-1} \mathscr{O}_Y \to \mathscr{O}_X$. (It can be convenient to package this information as in the diagram (2.7.2.1).) There is an obvious notion of composition of morphisms, so ringed spaces form a category. Hence we have notion of automorphisms and isomorphisms. You can easily verify that an isomorphism of ringed spaces means the same thing as it did before (Definition 4.3.1).

If $U \subset Y$ is an open subset, then there is a natural morphism of ringed spaces $(U, \mathscr{O}_Y|_U) \to (Y, \mathscr{O}_Y)$ (which implicitly appeared earlier in §2.2.13). More precisely, if $U \to Y$ is an isomorphism of U with an open subset V of Y, and we are given an isomorphism $(U, \mathscr{O}_U) \xrightarrow{\sim} (V, \mathscr{O}_Y|_V)$ (via the isomorphism $U \xrightarrow{\sim} V$), then the resulting map of ringed spaces is called an **open embedding** (or **open immersion**) of ringed spaces, and the morphism $U \to Y$ is often written $U \hookrightarrow Y$.

7.2.A. EXERCISE (MORPHISMS OF RINGED SPACES GLUE). Suppose $(X, \mathscr{O}_X)$ and $(Y, \mathscr{O}_Y)$ are ringed spaces, $X = \cup_i U_i$ is an open cover of X, and we have morphisms of ringed spaces $\pi_i \colon U_i \to Y$ that "agree on the overlaps," i.e., $\pi_i|_{U_i \cap U_j} = \pi_j|_{U_i \cap U_j}$. Show that there is a unique morphism of ringed spaces $\pi \colon X \to Y$ such that $\pi|_{U_i} = \pi_i$. (Exercise 2.2.F essentially showed this for topological spaces.)

7.2.B. EXERCISE. Show that open embeddings of ringed spaces are monomorphisms (in the category of ringed spaces).

7.2.C. EASY IMPORTANT EXERCISE: PUSHING FORWARD $\mathscr{O}$-MODULES. Given a morphism of ringed spaces $\pi \colon X \to Y$, show that sheaf pushforward induces a functor $\pi_* \colon Mod_{\mathscr{O}_X} \to Mod_{\mathscr{O}_Y}$.

7.2.D. IMPORTANT EXERCISE FOR LATER: PULLING BACK $\mathscr{O}$-MODULES. (You may wish to ignore this exercise until when you need it, because you may not appreciate it until then.) Suppose $\pi \colon (X, \mathscr{O}_X) \to (Y, \mathscr{O}_Y)$ is a morphism of ringed spaces. A slight variation on the inverse image allows us to **pull back an $\mathscr{O}_Y$-module** $\mathscr{G}$ to get an $\mathscr{O}_X$-module $\pi^* \mathscr{G}$.

(a) Show that $(X, \pi^{-1} \mathscr{O}_Y)$ is a ringed space, and that the morphism of ringed spaces π factors into $(X, \mathscr{O}_X) \to (X, \pi^{-1} \mathscr{O}_Y) \to (Y, \mathscr{O}_Y)$.

(b) Show that $\pi^{-1} \mathscr{G}$ is a $\pi^{-1} \mathscr{O}_Y$-module, and describe how $\pi^{-1} \mathscr{G} \otimes_{\pi^{-1} \mathscr{O}_Y} \mathscr{O}_X$ is an $\mathscr{O}_X$-module, which we call $\pi^* \mathscr{G}$, the **pullback** of $\mathscr{G}$.

(c) Show that π^* is a covariant functor $Mod_{\mathscr{O}_Y} \to Mod_{\mathscr{O}_X}$.

(d) If furthermore $\rho \colon (W, \mathscr{O}_W) \to (X, \mathscr{O}_X)$ is another morphism of ringed spaces, show that "$\rho^* \pi^* = (\pi \circ \rho)^*$" (or, more precisely, describe a natural isomorphism of functors $\rho^* \pi^* \xrightarrow{\sim} (\pi \circ \rho)^*$).

(e) (Cf. Exercise 2.7.B.) Show that π^* is left-adjoint to π_*. More precisely, for any $\mathscr{F} \in Mod_{\mathscr{O}_X}$, describe a natural bijection

$$\operatorname{Hom}_{(X, \mathscr{O}_X)}(\pi^* \mathscr{G}, \mathscr{F}) \longleftrightarrow \operatorname{Hom}_{(Y, \mathscr{O}_Y)}(\mathscr{G}, \pi_* \mathscr{F})$$

that is functorial in both $\mathscr{F} \in Mod_{\mathscr{O}_X}$ and $\mathscr{G} \in Mod_{\mathscr{O}_Y}$.

(f) *(the push-pull map for $\mathscr{O}$-modules)* Suppose

$$
\begin{array}{ccc}
W & \xrightarrow{\ \beta'\ } & X \\
{\scriptstyle \alpha'}\downarrow & & \downarrow{\scriptstyle \alpha} \\
Y & \xrightarrow{\ \beta\ } & Z
\end{array}
$$

is a commutative (not necessarily Cartesian) diagram of *ringed spaces*, and $\mathscr{F}$ is an $\mathscr{O}_X$-module. Explain how to define the **push-pull map**

$$\beta^* \alpha_* \mathscr{F} \longrightarrow \alpha'_* (\beta')^* \mathscr{F}$$

(cf. (2.7.4.2)) of $\mathscr{O}_Y$-modules, by following the discussion of §2.7.4 for the push-pull map for sheaves.

7.2.E. EASY IMPORTANT EXERCISE. Given a morphism of ringed spaces $\pi\colon X \to Y$ with $\pi(p) = q$, show that there is a map of stalks $(\mathscr{O}_Y)_q \to (\mathscr{O}_X)_p$.

7.2.F. KEY EXERCISE. Suppose $\pi^\sharp\colon B \to A$ is a morphism of rings. Define a morphism of ringed spaces $\pi\colon \operatorname{Spec} A \to \operatorname{Spec} B$ as follows. The map of topological spaces was given in Exercise 3.4.H. To describe a morphism of sheaves $\mathscr{O}_{\operatorname{Spec} B} \to \pi_* \mathscr{O}_{\operatorname{Spec} A}$ on $\operatorname{Spec} B$, it suffices to describe a morphism of sheaves on the distinguished base of $\operatorname{Spec} B$. On $D(g) \subset \operatorname{Spec} B$, we define

$$\mathscr{O}_{\operatorname{Spec} B}(D(g)) \longrightarrow \mathscr{O}_{\operatorname{Spec} A}(\pi^{-1}D(g)) = \mathscr{O}_{\operatorname{Spec} A}(D(\pi^\sharp g))$$

by $B_g \to A_{\pi^\sharp g}$. Verify that this makes sense (e.g., is independent of g), and that this describes a morphism of sheaves on the distinguished base. (This is the third in a series of exercises. We saw that a morphism of rings induces a map of sets in §3.2.10, a map of topological spaces in Exercise 3.4.H, and now a map of ringed spaces here.)

The map of ringed spaces of Key Exercise 7.2.F is really not complicated. Here is an example. Consider the ring map $\mathbb{C}[y] \to \mathbb{C}[x]$ given by $y \mapsto x^2$ (see Figure 3.6). We are mapping the affine line with coordinate x to the affine line with coordinate y. The map is (on closed points) $a \mapsto a^2$. For example, where does $[(x-3)]$ go to? Answer: $[(y-9)]$, i.e., $3 \mapsto 9$. What is the preimage of $[(y-4)]$? Answer: Those prime ideals in $\mathbb{C}[x]$ containing $[(x^2-4)]$, i.e., $[(x-2)]$ and $[(x+2)]$, so the preimage of 4 is indeed ± 2. This is just about the map of sets, which is old news (§3.2.10), so let's now think about functions pulling back. What is the pullback of the function $3/(y-4)$ on $D(y-4) = \mathbb{A}^1 - \{4\}$? Of course it is $3/(x^2-4)$ on $\mathbb{A}^1 \setminus \{-2, 2\}$.

The construction of Key Exercise 7.2.F will soon be an example of a morphism of schemes. In fact, we could make that definition right now. Before we do, we point out (via the next exercise) that not every morphism of ringed spaces between affine schemes is of the form of Key Exercise 7.2.F. (In the language of §7.3, this morphism of ringed spaces is not a morphism of locally ringed spaces.) We won't use this exercise, but it answers a question that everyone will have.

7.2.G. EXERCISE. Recall (Exercise 3.4.K) that $\operatorname{Spec} k[y]_{(y)}$ has two points, $[(0)]$ and $[(y)]$, where the second point is closed, and the first is not. Describe a map *of ringed spaces* $\operatorname{Spec} k(x) \to \operatorname{Spec} k[y]_{(y)}$ sending the unique point of $\operatorname{Spec} k(x)$ to the closed point $[(y)]$, where the pullback map on global sections sends k to k by the identity, and sends y to x. Show that this map of ringed spaces is not of the form described in Key Exercise 7.2.F.

7.2.2. Tentative definition we won't use (cf. Motivation 7.1.1). A morphism of schemes $\pi\colon (X, \mathscr{O}_X) \to (Y, \mathscr{O}_Y)$ is a morphism of ringed spaces that "locally looks like" the maps of affine schemes described in Key Exercise 7.2.F. Precisely, for each choice of affine open sets $\operatorname{Spec} A \subset X$, $\operatorname{Spec} B \subset Y$, such that $\pi(\operatorname{Spec} A) \subset \operatorname{Spec} B$, the induced map of ringed spaces should be of the form shown in Key Exercise 7.2.F.

We would like this definition to be checkable on any affine cover, and we might hope to use the Affine Communication Lemma to develop the theory in this way. This works, but it will be more convenient to use a clever trick: in the next section, we will use the notion of locally ringed spaces, and then once we have used it, we will discard it like yesterday's garbage.

7.3 From Locally Ringed Spaces to Morphisms of Schemes

In order to prove that morphisms behave in a way we hope, we will use the notion of a *locally ringed space*. It will not be used later, although it is useful elsewhere in geometry. The notion of locally ringed spaces (and maps between them) is inspired by what we know about manifolds (see Exercise 3.1.B). If $\pi\colon X \to Y$ is a morphism of manifolds, with $\pi(p) = q$, and f is a function on Y vanishing at q, then the pulled back function $\pi^\sharp(f)$ on X should vanish on p. Put differently: germs of functions (at $q \in Y$) vanishing at q should pull back to germs of functions (at $p \in X$) vanishing at p.

7.3.1. *Definition.* Recall (Definition 4.3.7) that a *locally ringed space* is a ringed space $(X, \mathscr{O}_X)$ such that the stalks $\mathscr{O}_{X,p}$ are all local rings. A **morphism of locally ringed spaces** $\pi\colon X \to Y$ is a morphism of ringed spaces such that the induced map of stalks $\pi^\sharp\colon \mathscr{O}_{Y,q} \to \mathscr{O}_{X,p}$ (Exercise 7.2.E) sends the maximal ideal $\mathfrak{m}_{Y,q}$ of $\mathscr{O}_{Y,q}$ into the maximal ideal $\mathfrak{m}_{X,p}$ of $\mathscr{O}_{X,p}$ (a **"morphism of local rings"**). (Note, in particular, that locally ringed spaces form a category.) This means something rather concrete and intuitive: "if $p \mapsto q$, and g is a function vanishing at q, then it will pull back to a function vanishing at p." You would also want: "if $p \mapsto q$, and g is a function *not* vanishing at q, then it will pull back to a function *not* vanishing at p." This is a consequence of the following exercise.

7.3.A. EXERCISE (VALUES OF FUNCTIONS ON LOCALLY RINGED SPACES PULL BACK). If $\pi\colon X \to Y$ is a morphism of locally ringed spaces, and $\pi(p) = q$, show that π induces an inclusion $\kappa(q) \hookrightarrow \kappa(p)$ of residue fields. This captures the fact that the pullback sends the locus where a function is zero (respectively, nonzero) to the locus where the pulled back function is zero (respectively, nonzero). In particular, by Exercise 4.3.F(b), the pullback of invertible functions are invertible functions.

To summarize: we use the notion of locally ringed space only to define morphisms of schemes, and to show that morphisms have reasonable properties. The main things you need to remember about locally ringed spaces are (i) that the functions have values at points, and (ii) that given a map of locally ringed spaces, values of functions "pull back," and, in particular, zeros of functions "pull back."

7.3.B. EXERCISE. Show that morphisms of locally ringed spaces glue (cf. Exercise 7.2.A). (Hint: Your solution to Exercise 7.2.A may work without change.)

7.3.C. EASY IMPORTANT EXERCISE. Recall (§4.3.7) that Spec A is a locally ringed space. Show that the morphism of ringed spaces $\pi\colon \operatorname{Spec} A \to \operatorname{Spec} B$ defined by a ring morphism $\pi^\sharp\colon B \to A$ (Exercise 7.2.F) is a morphism of locally ringed spaces.

7.3.2. Key proposition — *If* $\pi\colon \operatorname{Spec} A \to \operatorname{Spec} B$ *is a morphism of locally ringed spaces, then it is the morphism of locally ringed spaces induced by the map* $\pi^\sharp\colon B = \Gamma(\operatorname{Spec} B, \mathscr{O}_{\operatorname{Spec} B}) \to \Gamma(\operatorname{Spec} A, \mathscr{O}_{\operatorname{Spec} A}) = A$, *as in Exercise 7.3.C.*

(Aside: Exercise 4.3.A is a special case of Key Proposition 7.3.2. You should look back at your solution to Exercise 4.3.A, and see where you implicitly used ideas about locally ringed spaces.)

Proof. Suppose $\pi\colon \operatorname{Spec} A \to \operatorname{Spec} B$ is a morphism of locally ringed spaces. We wish to show that it is determined by its map on global sections $\pi^\sharp\colon B \to A$. We first need to check that the map of points is determined by global sections. Now a point p of Spec A can be identified with the prime ideal of global functions vanishing on it. By Exercise 7.3.A, $\pi(p)$ as an ideal is the kernel of the natural map $B \to \kappa(p)$. (Here we use the fact that π is a map of locally ringed spaces.) This is precisely the way in which the map of sets Spec $A \to$ Spec B induced by a ring map $B \to A$ was defined (§3.2.10).

Note in particular that if $g \in B$, $\pi^{-1}(D(g)) = D(\pi^\sharp g)$, again using the hypothesis that π is a morphism of locally ringed spaces.

It remains to show that $\pi^\sharp \colon \mathscr{O}_{\text{Spec } B} \to \pi_* \mathscr{O}_{\text{Spec } A}$ is the morphism of sheaves given by Exercise 7.2.F (cf. Exercise 7.3.C). It suffices to check this on the distinguished base (Exercise 2.5.C(a)). We now want to check that for any map of locally ringed spaces inducing the map of sheaves $\mathscr{O}_{\text{Spec } B} \to \pi_* \mathscr{O}_{\text{Spec } A}$, the map of sections on any distinguished open set $D(g) \subset$ Spec B $(g \in B)$ is determined by the map of global sections $B \to A$.

Consider the commutative diagram

$$
\begin{array}{ccccc}
B = \Gamma(\text{Spec } B, \mathscr{O}_{\text{Spec } B}) & \xrightarrow{\ \pi^\sharp_{\text{Spec } B}\ } & \Gamma(\text{Spec } A, \mathscr{O}_{\text{Spec } A}) & = & A \\[2mm]
\Big\downarrow{\scriptstyle \text{res}_{\text{Spec } B, D(g)}} & & \Big\downarrow{\scriptstyle \text{res}_{\text{Spec } A, D(\pi^\sharp g)}} & & \\[2mm]
B_g = \Gamma(D(g), \mathscr{O}_{\text{Spec } B}) & \xrightarrow{\ \pi^\sharp_{D(g)}\ } & \Gamma(D(\pi^\sharp g), \mathscr{O}_{\text{Spec } A}) & = & A_{\pi^\sharp g} = A \otimes_B B_g.
\end{array}
$$

The vertical arrows (restrictions to distinguished open sets) are localizations by g, so the lower horizontal map $\pi^\sharp_{D(g)}$ is determined by the upper map (it is just localization by g). $\qquad\square$

We are ready for our definition.

7.3.3. Definition. If X and Y are schemes, then a morphism $\pi \colon X \to Y$ as locally ringed spaces is called a **morphism of schemes**. We have thus defined the **category of schemes**, which we denote by *Sch*. (We then have notions of **isomorphism**—the same as before, §4.3.1—and **automorphism**. The *target* Y of π is sometimes called the **base scheme** or the **base**, when we are interpreting π as a "family of schemes parametrized by Y"—this may become clearer once we have defined the fibers of morphisms in §10.3.2.)

The definition in terms of locally ringed spaces easily implies Tentative Definition 7.2.2:

7.3.D. IMPORTANT EXERCISE. Show that a morphism of schemes $\pi \colon X \to Y$ is a morphism of ringed spaces that looks locally like morphisms of affine schemes. Precisely, if Spec A is an affine open subset of X and Spec B is an affine open subset of Y, and $\pi(\text{Spec } A) \subset$ Spec B, then the induced morphism of ringed spaces is a morphism of affine schemes. (In case it helps, note: If $W \subset X$ and $Y \subset Z$ are both open embeddings of ringed spaces, then any morphism of ringed spaces $X \to Y$ induces a morphism of ringed spaces $W \to Z$, by composition $W \to X \to Y \to Z$.) Show that it suffices to check on a set $(\text{Spec } A_i, \text{Spec } B_i)$ where the Spec A_i form an open cover of X and the Spec B_i form an open cover of Y.

In practice, we will use the affine cover interpretation, and forget completely about locally ringed spaces. In particular, put imprecisely, the category of affine schemes is the category of rings with the arrows reversed. More precisely:

7.3.E. EXERCISE. Show that the category of rings and the opposite category of affine schemes are equivalent (see §1.1.16 and §1.1.21 to read about opposite and equivalent categories, respectively).

In particular, there can be many different maps from one point to another! For example, here are two different maps from the point Spec $\mathbb{C}$ to the point Spec $\mathbb{C}$: the identity (corresponding to the identity $\mathbb{C} \to \mathbb{C}$), and complex conjugation. (There are even more such maps!)

It is clear (from the corresponding facts about locally ringed spaces) that *morphisms of schemes glue* (Exercise 7.3.B), and the composition of two morphisms is a morphism. Isomorphisms in this category are precisely what we defined them to be earlier (§4.3.7).

7.3.F. ENLIGHTENING EXERCISE. (This exercise can give you some practice with understanding morphisms of schemes by cutting them up into affine open sets.) Make sense of the following sentence: "$\mathbb{A}_k^{n+1} \setminus \{\vec{0}\} \to \mathbb{P}_k^n$ given by

$$(x_0, x_1, \ldots, x_n) \longmapsto [x_0, x_1, \ldots, x_n]$$

is a morphism of schemes." Caution: You can't just say where points go; you have to say where functions go. So you may have to divide $\mathbb{A}_k^{n+1} \setminus \{\vec{0}\}$ up into affines, and describe the maps, and check that they glue. (Can you generalize to the case where k is replaced by a general ring B? See Exercise 7.3.O for an answer.)

7.3.4. Morphisms to affine schemes.

The following result shows that it is easy to describe morphisms to an affine scheme without working hard to cover the source with affine open sets.

7.3.G. ESSENTIAL EXERCISE. Show that morphisms $X \to \operatorname{Spec} A$ are in natural bijection with ring morphisms $A \to \Gamma(X, \mathscr{O}_X)$. Hint: Show that this is true when X is affine. Use the fact that morphisms glue, Exercise 7.3.B. (This is even true in the category of locally ringed spaces. You are free to prove it in this generality, but it is easier in the category of schemes. You might find in this more general statement an echo of the slogan of double-starred Remark 6.2.5.)

In particular, there is a canonical morphism from a scheme to the Spec of its ring of global sections. (Warning: Even if X is a finite type k-scheme, the ring of global sections might be nasty! In particular, it might not be finitely generated; see §19.11.13.) The canonical morphism $X \to \operatorname{Spec} \Gamma(X, \mathscr{O}_X)$ is an isomorphism if and only if X is affine (i.e., isomorphic to Spec A for some ring A), and in this case it is the isomorphism hinted at in Remark 4.3.2.

7.3.H. EASY EXERCISE. If $S_\bullet$ is a finitely generated graded A-algebra, describe a natural "structure morphism" $\operatorname{Proj} S_\bullet \to \operatorname{Spec} A$.

7.3.5. *Remark.* From Essential Exercise 7.3.G, it is one small step to show that some products of schemes exist: if A and B are rings, then $\operatorname{Spec} A \times \operatorname{Spec} B = \operatorname{Spec}(A \otimes_{\mathbb{Z}} B)$; and if A and B are C-algebras, then $\operatorname{Spec} A \times_{\operatorname{Spec} C} \operatorname{Spec} B = \operatorname{Spec}(A \otimes_C B)$. But we are in no hurry, so we wait until Exercise 10.1.A to discuss this properly.

7.3.6.** *Side fact for experts: Γ and Spec are adjoints.* We have a contravariant functor Spec from rings to locally ringed spaces, and a contravariant functor Γ from locally ringed spaces to rings. In fact $(\Gamma, \operatorname{Spec})$ is an adjoint pair! (Caution: We have only discussed adjoints for covariant functors; if you care, you will have to figure out how to define adjoints for contravariant functors.) Thus we could have defined Spec by requiring it to be right-adjoint to Γ. (Fun but irrelevant side question: If you used ringed spaces rather than locally ringed spaces, Γ again has a right adjoint. What is it?)

Our ability to easily describe morphisms to affine schemes will allow us to revisit our earlier discussions of schemes over a given field or ring (§5.3.6).

7.3.7. The category of complex schemes (or more generally the category of k-schemes where k is a field, or more generally the category of A-schemes where A is a ring, or more generally the category of S-schemes where S is a scheme).
The category of S-schemes Sch_S (where S is a scheme) is defined as follows. The objects (S-schemes) are morphisms of the form

$$\begin{array}{c} X \\ \downarrow \\ S. \end{array}$$

(The morphism to S is called the **structure morphism**. A motivation for this terminology is the fact that if $S = \mathrm{Spec}\, A$, the structure morphism gives the functions on each open set of X the structure of an A-algebra; cf. §5.3.6.) The morphisms in the category of S-schemes are defined to be the commutative diagrams

$$
\begin{array}{ccc}
X & \longrightarrow & Y \\
\downarrow & & \downarrow \\
S & \overset{=}{\longrightarrow} & S,
\end{array}
$$

which are more conveniently written as

$$
\begin{array}{ccc}
X & \longrightarrow & Y \\
& \searrow \quad \swarrow & \\
& S. &
\end{array}
$$

When there is no confusion (if the base scheme is clear), simply the top row of the diagram is given. In the case where $S = \mathrm{Spec}\, A$, where A is a ring, we get the notion of an A-scheme, which is the same as the same definition in §5.3.6 (Exercise 7.3.I), but in a more satisfactory form. For example, complex geometers may consider the category of $\mathbb{C}$-schemes.

7.3.I. EASY EXERCISE. Show that this definition of A-scheme given in §7.3.7 agrees with the earlier definition of §5.3.6.

7.3.J. EASY EXERCISE. Show that $\mathrm{Spec}\,\mathbb{Z}$ is the final object in the category of schemes. In other words, if X is any scheme, there exists a unique morphism to $\mathrm{Spec}\,\mathbb{Z}$. (Hence the category of schemes is isomorphic to the category of $\mathbb{Z}$-schemes.) If A is any ring (for example, a field k), show that $\mathrm{Spec}\, A$ is the final object in the category of A-schemes.

7.3.8. Coproducts of schemes.

7.3.K. EASY EXERCISE. Suppose X and Y are schemes. Show that their disjoint union $X \coprod Y$ (defined in Exercise 4.3.E) is the coproduct of X and Y in the category of schemes, justifying the use of the coproduct symbol $\coprod$. (See §3.6.3 for a surprising connection to idempotents.)

7.3.9. Morphisms from (some) affine schemes.

Morphisms *from* affine schemes are not quite as simple as morphisms *to* affine schemes, but some cases are worth pointing out.

7.3.L. EXERCISE.

(a) Suppose p is a point of a scheme X. Describe a canonical (choice-free) morphism $\mathrm{Spec}\,\mathscr{O}_{X,p} \to$ X. (Hint: Do this for affine X first. But then for general X be sure to show that your morphism is independent of choice.)

(b) Define a canonical morphism $\mathrm{Spec}\,\kappa(p) \to X$. (This is often written $p \to X$; one gives p the obvious interpretation as a scheme, $\mathrm{Spec}\,\kappa(p)$.)

7.3.M. EXERCISE (MORPHISMS FROM Spec OF A LOCAL RING TO X). Suppose X is a scheme, and $(A, \mathfrak{m})$ is a local ring. Suppose we have a scheme morphism $\pi\colon \mathrm{Spec}\, A \to X$ sending $[\mathfrak{m}]$ to p. Show that any open set containing p contains the image of π. Show that there is a bijection between $\mathrm{Mor}(\mathrm{Spec}\, A, X)$ and
$$\{(p \in X, \text{some local homomorphism } \mathscr{O}_{X,p} \to A)\}.$$

(Possible hint: Exercise 7.3.L(a).)

These exercises lead us to the notion of field-valued points, and more generally ring-valued points, and more generally still, scheme-valued points.

7.3.10. Definition: The functor of points, and scheme-valued points (and ring-valued points, and field-valued points) of a scheme. If Z is a scheme, then Z-**valued points** of a scheme X, denoted by X(Z), are defined to be maps Z → X. If A is a ring, then A-**valued points** of a scheme X, denoted by X(A), are defined to be the (Spec A)-valued points of the scheme. (The most common case of this is when A is a field.)

If you are working over a base scheme B—for example, complex algebraic geometers will consider only schemes and morphisms over B = Spec ℂ—then in the above definition, there is an implicit structure map Z → B (or Spec A → B in the case of X(A)). For example, for a complex geometer, if X is a scheme over ℂ, the ℂ(t)-valued points of X correspond to commutative diagrams of the form

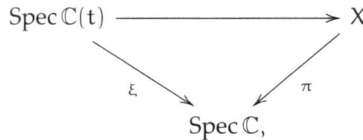

$$\text{Spec}\,\mathbb{C}(t) \longrightarrow X$$

$$\xi \searrow \qquad \swarrow \pi$$

$$\text{Spec}\,\mathbb{C},$$

where π: X → Spec ℂ is the structure map for X, and ξ corresponds to the obvious inclusion of rings ℂ → ℂ(t). (Warning: A k-valued point of a k-scheme X is sometimes called a "rational point" of X, which is dangerous, as for most of the world, "rational" refers to ℚ. We will use the safer phrase "k-valued point" of X. Another safe choice is "k-rational point" of X or k-point.)

The terminology "Z-valued point" (and A-valued point) is unfortunate, because we earlier defined the notion of points of a scheme, and Z-valued points (and A-valued points) are not (necessarily) points! But these usages are well established in the literature. (Look in the index under "point" to see even more inconsistent use of adjectives that modify this word.)

7.3.N. EXERCISE.

(a) (easy) Show that a morphism of schemes X → Y induces a map of Z-valued points X(Z) → Y(Z).

(b) Note that morphisms of schemes X → Y are not determined by their "underlying" map of points. (What is an example?) Show that they *are* determined by their induced maps of Z-valued points, as Z varies over all schemes. (Hint: Pick Z = X. In the course of doing this exercise, you will largely prove Yoneda's Lemma in the guise of Exercise 10.1.B.)

7.3.11. Furthermore, we will see that "products of Z-valued points" behave as you might hope (§10.1.3). A related reason this language is suggestive: The notation X(Z) suggests the interpretation of X as a (contravariant) functor h_X from schemes to sets—the **functor of** (scheme-valued) **points** of the scheme X (cf. Example 1.1.20).

Here is a low-brow reason A-valued points are a useful notion: *The A-valued points of an affine scheme* Spec $\mathbb{Z}[x_1, \ldots, x_n]/(f_1, \ldots, f_r)$ *(where $f_i \in \mathbb{Z}[x_1, \ldots, x_n]$ are relations) are precisely the solutions to the equations*

$$f_1(x_1, \ldots, x_n) = \cdots = f_r(x_1, \ldots, x_n) = 0$$

in the ring A. For example, the rational solutions to $x^2 + y^2 = 16$ are precisely the ℚ-valued points of Spec $\mathbb{Z}[x, y]/(x^2 + y^2 - 16)$. The integral solutions are precisely the ℤ-valued points. So A-valued points of an affine scheme (finite type over ℤ) can be interpreted simply. In the special case where A is local, A-valued points of a general scheme have a good interpretation too (Exercises 7.3.L and 7.3.M).

On the other hand, Z-valued points of projective space can be subtle. There are some maps we can write down easily, as shown by applying the next exercise in the case X = Spec A, where A is a B-algebra.

7.3.O. EASY (BUT SURPRISINGLY ENLIGHTENING) EXERCISE (CF. EXERCISE 7.3.F).

(a) Suppose B is a ring. If X is a B-scheme, and $f_0, \ldots, f_n$ are $n + 1$ functions on X with no common zeros, then show that $[f_0, \ldots, f_n]$ gives a morphism of B-schemes X → $\mathbb{P}^n_B$.

(b) Suppose g is a nowhere vanishing function on X, and f_i are as in part (a). Show that the morphisms $[f_0, \ldots, f_n]$ and $[gf_0, \ldots, gf_n]$ to $\mathbb{P}_B^n$ are the same.

7.3.12. *Example (Exercise 7.3.F revisited).* Consider the $n+1$ functions $x_0, \ldots, x_n$ on $\mathbb{A}^{n+1}$ (otherwise known as $n+1$ sections of the trivial bundle). They have no common zeros on $\mathbb{A}^{n+1} \setminus \{\vec{0}\}$. Hence they determine a morphism $\mathbb{A}^{n+1} \setminus \{\vec{0}\} \to \mathbb{P}^n$.

7.3.13. You might hope that Exercise 7.3.O(a) gives all morphisms to projective space (over B). But this isn't the case. Indeed, even the identity morphism $X = \mathbb{P}_k^1 \to \mathbb{P}_k^1$ isn't of this form, as the source $\mathbb{P}^1$ has no nonconstant global functions with which to build this map. (There are similar examples with an affine source.) However, there is a correct generalization (characterizing *all* maps from schemes to projective schemes) in Theorem 15.2.2. This result roughly states that this idea works, so long as the f_i are not quite functions, but sections of a line bundle. Our desire to understand maps to projective schemes in a clean way will be one important motivation for understanding line bundles.

We will see more ways to describe maps to projective space in the next section. A different description directly generalizing Exercise 7.3.O(a) will be given in Exercise 15.2.A, which will turn out (in Theorem 15.2.2) to be a "universal" description.

Incidentally, before Grothendieck, it was considered a real problem to figure out the right way to interpret points of projective space with "coordinates" in a ring. These difficulties were due to a lack of functorial reasoning. And the clues to the right answer already existed (the same problems arise for maps from a manifold to $\mathbb{RP}^n$)—if you ask such a geometric question (for projective space is geometric), the answer is necessarily geometric, not purely algebraic!

7.3.14. *Visualizing morphisms: Picturing maps of schemes when nilpotents are present.* You now know how to visualize the points of schemes (§3.3), and nilpotents (§4.2 and §6.6). The following imprecise exercise will give you some sense of how to visualize maps of schemes when nilpotents are involved. Suppose $a \in \mathbb{C}$. Consider the map of rings $\mathbb{C}[x] \to \mathbb{C}[\epsilon]/(\epsilon^2)$ given by $x \mapsto a\epsilon$. Recall that $\operatorname{Spec} \mathbb{C}[\epsilon]/(\epsilon^2)$ may be pictured as a point with a tangent vector (§4.2). How would you picture this map if $a \neq 0$? How does your picture change if $a = 0$? (The tangent vector should be "crushed" in this case.)

Exercise 13.1.G will extend this considerably. You may enjoy reading its statement now.

7.3.15. ** For readers with appropriate background: Analytification of complex algebraic varieties.** Warning: Any discussion of analytification is only for readers who are familiar with the notion of complex analytic varieties, or willing to develop it on their own in parallel with our development of schemes.

7.3.P. EXERCISE (ANALYTIFYING COMPLEX ALGEBRAIC VARIETIES). Suppose X is a reduced, finite type $\mathbb{C}$-scheme. Define the corresponding complex analytic prevariety X_{an}. (The definition of an analytic prevariety is the same as the definition of a variety without the Hausdorff condition.) Caution: Your definition should not depend on a choice of an affine cover of X. (Hint: First explain how to analytify reduced finite type affine $\mathbb{C}$-schemes. Then glue.) Give a bijection between the closed points of X and the points of X_{an}, using the weak Nullstellensatz 3.2.5. (In fact one may construct a continuous map of sets $X_{an} \to X$, generalizing Exercise 3.2.I.)

7.3.Q. EXERCISE (THE ANALYTIFICATION FUNCTOR). Recall the analytification construction of Exercise 7.3.16. For each morphism of reduced finite type $\mathbb{C}$-schemes $\pi \colon X \to Y$ (over $\mathbb{C}$), define a morphism of complex analytic prevarieties $\pi_{an} \colon X_{an} \to Y_{an}$ (the **analytification of** π). Show that analytification gives a functor from the category of reduced finite type $\mathbb{C}$-schemes to the category of complex analytic prevarieties.

7.3.16. *Remark.* Two nonisomorphic varieties can have isomorphic analytifications. For example, Serre described two different algebraic structures on the complex manifold $\mathbb{C}^* \times \mathbb{C}^*$ (see [Ha2, p. 232] and [MO68421]) one is "the obvious one," and the other is a $\mathbb{P}^1$-bundle over an elliptic curve, with a section removed. For an example of a smooth complex surface with infinitely many algebraic structures, see §19.11.3. On the other hand, a compact complex variety can have only one algebraic structure (see [Se3, §19]).

7.3.17. *Further facts about analytification.* For more on analytification, see Exercises 11.2.K (Hausdorffness), 14.1.J, 18.4.H, 18.4.U (line bundles), §18.6.5, §20.1.9 (degree), and Exercise 21.2.W (manifolds). For background on complex analytic spaces, see [GR].

7.4 Maps of Graded Rings and Maps of Projective Schemes

As maps of rings correspond to maps of affine schemes in the opposite direction, maps of graded rings (over a base ring A) sometimes give maps of projective schemes in the opposite direction. This is an imperfect generalization: not every map of graded rings gives a map of projective schemes (§7.4.2); not every map of projective schemes comes from a map of graded rings (§19.11.9); and different maps of graded rings can yield the same map of schemes (Exercise 7.4.C).

You may find it helpful to think through Examples 7.4.1 and 7.4.2 while working through the following exercise.

7.4.A. ESSENTIAL EXERCISE. Suppose that $\varphi \colon S_\bullet \longrightarrow R_\bullet$ is a morphism of ($\mathbb{Z}^{\geq 0}$-)graded rings. (By **map of graded rings**, we mean a map of rings that preserves the grading as a map of "graded abelian semigroups" (or "graded commutative monoids"). In other words, there is a $d > 0$ such that S_n maps to R_{dn} for all n.) Show that this induces a morphism of schemes

$$(\operatorname{Proj} R_\bullet) \setminus V(\varphi(S_+)) \to \operatorname{Proj} S_\bullet.$$

(Hint: Suppose f is a homogeneous element of S_+. Define a map $D_+(\varphi(f)) \to D_+(f)$. Show that they glue together (as f runs over all homogeneous elements of S_+). Show that this defines a map from all of $\operatorname{Proj} R_\bullet \setminus V(\varphi(S_+))$.) In particular, if

(7.4.0.1) $$V(\varphi(S_+)) = \varnothing,$$

then we have a morphism $\operatorname{Proj} R_\bullet \to \operatorname{Proj} S_\bullet$. From your solution, it will be clear that if φ is furthermore a morphism of A-algebras, then the induced morphism $\operatorname{Proj} R_\bullet \setminus V(\varphi(S_+)) \to \operatorname{Proj} S_\bullet$ is a morphism of A-schemes.

7.4.1. *Example.* Let's see Exercise 7.4.A in action. We will scheme-theoretically interpret the map of complex projective manifolds $\mathbb{CP}^1$ to $\mathbb{CP}^2$ given by

$$\mathbb{CP}^1 \longrightarrow \mathbb{CP}^2$$
$$[s, t] \longmapsto [s^{20}, s^9 t^{11}, t^{20}].$$

Notice first that this is well-defined: $[\lambda s, \lambda t]$ is sent to the same point of $\mathbb{CP}^2$ as $[s, t]$. The reason for its being well-defined is that the three polynomials s^{20}, $s^9 t^{11}$, and t^{20} are all homogeneous of degree 20.

Algebraically, this corresponds to a map of graded rings in the opposite direction,

$$\mathbb{C}[x, y, z] \longrightarrow \mathbb{C}[s, t],$$

given by $x \mapsto s^{20}$, $y \mapsto s^9 t^{11}$, $z \mapsto t^{20}$. You should interpret this in light of your solution to Exercise 7.4.A, and compare this to the affine example of §3.2.11.

7.4.2. *Example.* Notice that there is no map of complex manifolds $\mathbb{CP}^2 \to \mathbb{CP}^1$ given by $[x, y, z] \mapsto [x, y]$, because the map is not defined when $x = y = 0$. This corresponds to the fact that the map of graded rings $\mathbb{C}[s, t] \to \mathbb{C}[x, y, z]$ given by $s \mapsto x$ and $t \mapsto y$ doesn't satisfy hypothesis (7.4.0.1).

7.4.B. EXERCISE. Show that if $\phi \colon S_\bullet \to R_\bullet$ satisfies $\sqrt{(\phi(S_+))} = R_+$, then hypothesis (7.4.0.1) is satisfied. (Hint: Exercise 4.5.L.) This algebraic formulation of the more geometric hypothesis can sometimes be easier to verify.

7.4.C. UNIMPORTANT EXERCISE. This exercise shows that different maps of graded rings can give the same map of schemes. Let $R_\bullet = k[x, y, z]/(xz, yz, z^2)$ and $S_\bullet = k[a, b, c]/(ac, bc, c^2)$, where every variable has degree 1. Show that $\operatorname{Proj} R_\bullet \cong \operatorname{Proj} S_\bullet \cong \mathbb{P}_k^1$. Show that the maps $S_\bullet \to R_\bullet$ given by $(a, b, c) \mapsto (x, y, z)$ and $(a, b, c) \mapsto (x, y, 0)$ give the same (iso)morphism $\operatorname{Proj} R_\bullet \to \operatorname{Proj} S_\bullet$. (The real reason is that all of these constructions are insensitive to what happens in a finite number of degrees. This will be made precise in a number of ways later, most immediately in Exercise 7.4.F.)

7.4.3. *Unimportant remark.* Exercise 15.2.H shows that not every morphism of schemes $\operatorname{Proj} R_\bullet \to \operatorname{Proj} S_\bullet$ comes from a map of graded rings $S_\bullet \to R_\bullet$, even in quite reasonable circumstances.

7.4.4. Veronese subrings.
Here is a useful construction. Suppose $S_\bullet$ is a finitely generated graded ring. Define the n**th Veronese subring** of $S_\bullet$ by $S_{n\bullet} = \oplus_{j=0}^{\infty} S_{nj}$. (The "old degree" n is "new degree" 1.) The geometric interpretation of this construction is the important *Veronese embedding*, discussed in §9.3.6.

7.4.D. EXERCISE. Show that the map of graded rings $S_{n\bullet} \hookrightarrow S_\bullet$ induces an *isomorphism* $\operatorname{Proj} S_\bullet \xrightarrow{\sim} \operatorname{Proj} S_{n\bullet}$. (Hint: If $f \in S_+$ is homogeneous of degree divisible by n, identify $D(f)$ on $\operatorname{Proj} S_\bullet$ with $D(f)$ on $\operatorname{Proj} S_{n\bullet}$. Why do such distinguished open sets cover $\operatorname{Proj} S_\bullet$?)

7.4.E. EXERCISE. If $S_\bullet$ is generated in degree 1, show that $S_{n\bullet}$ is also generated in degree 1. (You may want to consider the case of the polynomial ring first.)

7.4.F. EXERCISE. Show that if $R_\bullet$ and $S_\bullet$ are the same finitely generated graded rings except in a finite number of nonzero degrees (make this precise!), then $\operatorname{Proj} R_\bullet \cong \operatorname{Proj} S_\bullet$.

7.4.G. EXERCISE. Suppose $S_\bullet$ is generated over S_0 by $f_1, \ldots, f_n$. Find a d such that $S_{d\bullet}$ is finitely generated in "new" degree 1 (= "old" degree d). (This is surprisingly tricky, so here is a hint. Suppose there are generators $x_1, \ldots, x_n$ of degrees $d_1, \ldots, d_n$, respectively. Show that any monomial $x_1^{a_1} \cdots x_n^{a_n}$ of degree at least $nd_1 \ldots d_n$ has $a_i \geq (\prod_j d_j)/d_i$ for some i. Show that the $(nd_1 \cdots d_n)$th Veronese subring is generated by elements in "new" degree 1.)

Exercise 7.4.G, in combination with Exercise 7.4.D, shows that there is little harm in assuming that finitely generated graded rings are generated in degree 1, as after a regrading (or more precisely, keeping only terms of degree a multiple of d, then dividing the degree by d), this is indeed the case. This is handy, as it means that, using Exercise 7.4.D, we can assume that any finitely generated graded ring is generated in degree 1. Exercise 9.3.F will later imply as a consequence that we can embed every Proj in some projective space.

7.4.H. LESS IMPORTANT EXERCISE. Suppose $S_\bullet$ is a finitely generated graded ring. Show that $S_{n\bullet}$ is a finitely generated graded ring. (Possible approach: Use the previous exercise, or something similar, to show there is some N such that $S_{nN\bullet}$ is generated in degree 1, so the graded ring $S_{nN\bullet}$ is finitely generated. Then show that for each $0 < j < N$, $S_{nN\bullet+nj}$ is a finitely generated module over $S_{nN\bullet}$.)

7.5   Rational Maps from Reduced Schemes

Informally speaking, a "rational map" is "a morphism defined almost everywhere," much as a rational function (Definition 6.6.36, Exercise 5.2.I) is a name for a function defined almost everywhere. We will later see that in good situations, just as with rational functions, where a rational map is defined, it is uniquely defined (the Reduced-to-Separated Theorem 11.3.2), and has a largest "*domain of definition*" (§11.3.3). For this section only, *we assume X to be reduced*, although we will try to still be careful and state these hypotheses in all theorems. A key example will be irreducible varieties (§7.5.7), and the language of rational maps is most often used in this case.

7.5.1. *Definition.* A **rational map** π from a reduced scheme X to a scheme Y, denoted by $\pi: X \dashrightarrow Y$, is the data of a morphism $\alpha: U \to Y$ from a dense open set $U \subset X$, with the equivalence relation $(\alpha: U \to Y) \sim (\beta: V \to Y)$ if there is a dense open set $Z \subset U \cap V$ such that $\alpha|_Z = \beta|_Z$. (In §11.3.3, we will improve this to: if $\alpha|_{U \cap V} = \beta|_{U \cap V}$ in good circumstances—when Y is separated.) People often use the word "map" for "morphism," which is quite reasonable, except that a rational map need not be a map. So to avoid confusion, when one means "rational map," one should never just say "map."

We will also discuss rational maps of S-schemes for a scheme S. The definition is the same, except now X and Y are S-schemes, and $\alpha: U \to Y$ is a morphism of S-schemes.

7.5.2.* *Rational maps more generally.* Just as with rational functions, Definition 7.5.1 can be extended to where X is not reduced, as is (using the same name, "rational map"), or in a version that imposes some control over what happens over the nonreduced locus (*pseudo-morphisms*, [Stacks, tag 01RX]). We will see in §11.3 that rational maps from reduced schemes to separated schemes behave particularly well, which is why they are usually considered in this context. The reason for the definition of pseudo-morphisms is to extend these results to when X is nonreduced. We will not use the notion of pseudo-morphism.

7.5.3. An obvious example of a rational map is a morphism. Another important example is the projection $\mathbb{P}_A^n \dashrightarrow \mathbb{P}_A^{n-1}$ given by $[x_0, \cdots, x_n] \mapsto [x_0, \cdots, x_{n-1}]$. (How precisely is this a rational map in the sense of Definition 7.5.1?)

7.5.A. EXERCISE. Suppose X is a reduced scheme. Describe how rational maps $X \dashrightarrow \mathbb{A}^1$ should be identified with rational functions on X.

A rational map $\pi: X \dashrightarrow Y$ is **dominant** (or, in some sources, *dominating*) if for some (and hence every) representative $U \to Y$, the image is dense in Y. A morphism is a **dominant morphism** (or *dominating morphism*) if it is dominant as a rational map.

7.5.B. EXERCISE. Show that a rational map $\pi: X \dashrightarrow Y$ of integral schemes is dominant if and only if π sends the generic point of X to the generic point of Y.

7.5.4. *Composition of rational maps.* A little thought will convince you that you can compose (in a well-defined way) a dominant map $\pi: X \dashrightarrow Y$ from a reduced scheme X to a reduced scheme Y with a rational map $\rho: Y \dashrightarrow Z$. Furthermore, the composition $\rho \circ \pi$ will be dominant if ρ is dominant. Integral schemes and dominant rational maps between them form a category that is geometrically interesting.

7.5.C. EASY EXERCISE. Show that dominant rational maps of integral schemes give morphisms of function fields in the opposite direction.

In "suitably classical situations" (integral finite type k-schemes — and, in particular, irreducible varieties, to be defined in §11.2.9), this is reversible: dominant rational maps correspond to inclusions of function fields in the opposite direction. We make this precise in §7.5.7 below. But it is not true that morphisms of function fields *always* give dominant rational maps, or even rational maps. For example, Spec $k[x]$ and Spec $k(x)$ have the same function field $k(x)$, but there is no corresponding rational map Spec $k[x] \dashrightarrow$ Spec $k(x)$ of k-schemes. Reason: Such a rational map would correspond to a morphism from an open subset U of Spec $k[x]$, say, Spec $k[x, 1/f(x)]$, to Spec $k(x)$. But there is no map of rings $k(x) \to k[x, 1/f(x)]$ (sending k identically to k and x to x) for any one $f(x)$.

(If you want more evidence that the topologically defined notion of dominance is simultaneously algebraic, you can show that if $\phi: A \to B$ is a ring morphism, then the corresponding morphism Spec $B \to$ Spec A is dominant if and only if ϕ has kernel contained in the nilradical of A.)

f	$g \circ f$	$g \circ f$ $= \mathrm{id}$ as:			g	$f \circ g$	$f \circ g$ $= \mathrm{id}$ as:
rat'l map	rat'l map	rat'l map	$X \xleftarrow{\ \sim\ } Y$		rat'l map	rat'l map	rat'l map
morphism	"	"	$X_1 \qquad Y_1$		morphism	"	"
"	morphism	"	X_2		"	morphism	"
"	"	morphism	$X_3 \longleftarrow Y_2$		"	morphism	"
"	"	"	$X_4 \xleftrightarrow{\ \sim\ } Y_3$		"	"	morphism

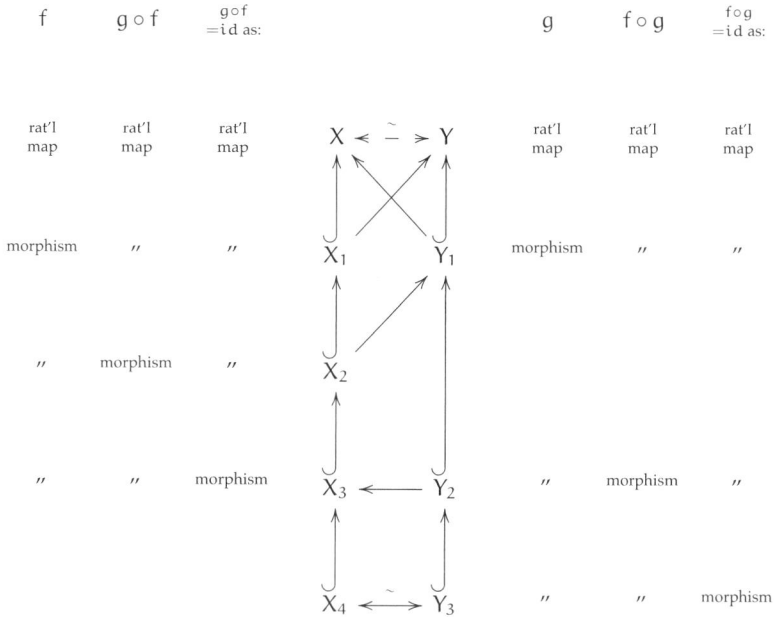

Figure 7.1 *Outline of the proof of Proposition 7.5.6.*

7.5.5. *Definition.* A rational map $\pi: X \dashrightarrow Y$ of integral schemes is said to be **birational** if it is dominant, and there is another rational map (a "rational inverse") ψ that is also dominant, such that $\pi \circ \psi$ is (in the same equivalence class as) the identity on Y, and $\psi \circ \pi$ is (in the same equivalence class as) the identity on X. This is the notion of isomorphism in the category of integral schemes and dominant rational maps. (In the differentiable category, this is not particularly interesting.) A *morphism* is a **birational morphism** if it is birational as a rational map. (Note that the "inverse" to a birational morphism may be only a rational map, not a morphism.)

We say X and Y are **birational** (to each other) if there exists a birational map $X \dashrightarrow Y$. If X and Y are irreducible, then birational maps induce isomorphisms of function fields. The fact that maps of function fields correspond to rational maps in the opposite direction for integral finite type k-schemes, to be proved in Proposition 7.5.8, shows that a map between integral finite type k-schemes that induces an isomorphism of function fields is birational. An integral finite type k-scheme is said to be **rational** if it is birational to $\mathbb{A}_k^n$ for some n.

7.5.6. Proposition — *Suppose X and Y are irreducible and reduced (i.e., integral, by Exercise 5.2.G) schemes. Then X and Y are birational if and only if there is a dense open subscheme U of X and a dense open subscheme V of Y such that $U \cong V$.*

(We have "integral" hypotheses for simplicity. The hypotheses can be dropped with sufficient care and modified definitions.)

Proposition 7.5.6 tells you how to think of birational maps. Just as a rational map is a "mostly defined function," two birational reduced schemes are "mostly isomorphic." For example, a reduced finite type k-scheme (such as an affine k-variety) is rational if and only if it has a dense open subscheme isomorphic to an open subscheme of $\mathbb{A}_k^n$.

Proof. The "if" direction is immediate, so we prove the "only if" direction. We basically follow our nose, and use what we are given (see Figure 7.1).

We have inverse rational maps $F: X \dashrightarrow Y$ and $G: Y \dashrightarrow X$. Choose representative *morphisms* $f: X_1 \to Y$ (where $X_1 \subset X$) and $g: Y_1 \to X$ (where $Y_1 \subset Y$) for the rational maps F and G, respectively. If $X_2 = (f|_{X_1})^{-1}(Y_1) \subset X_1$, then $g \circ f$ is a *morphism* from X_2 to X that is the identity *as a rational map*.

Thus there is a dense open subset $X_3 \subset X_2$ such that the morphism $g \circ f : X_3 \to X$ is the identity morphism onto X_3 (or, more precisely, it is the open embedding $X_3 \hookrightarrow X$).

Similarly, let $Y_2 = (g|_{Y_1})^{-1}(X_3) \subset Y_1$, so $f \circ g$ is a morphism from Y_2 to Y that is the identity as a rational map. Then let $Y_3 \subset Y_2$ be a dense open subset such that $f \circ g : Y_3 \to Y$ is the inclusion (the "identity").

Finally, if $X_4 = (f|_{X_3})^{-1}(Y_3) \subset X_3$, then $(g \circ f)|_{X_4}$ is the identity morphism on X_4 (by way of Y_3), and $(f \circ g)|_{Y_3}$ is the identity morphism on Y_3 (by way of X_4), so we have found our isomorphism of open sets that is a "representative" for our birational map. $\square$

7.5.7. Rational maps of irreducible varieties.

7.5.8. Proposition — *Suppose X is an integral k-scheme and Y is an integral finite type k-scheme, and we are given an extension of function fields $\phi^\sharp : K(Y) \hookrightarrow K(X)$ preserving k. Then there exists a dominant rational map of k-schemes $\phi : X \dashrightarrow Y$ inducing $\phi^\sharp$.*

Proof. By replacing Y with an open subset, we may assume that Y is affine, say, $\operatorname{Spec} B$, where B is generated over k by finitely many elements $y_1, \ldots, y_n$. Since we need to define ϕ only on an open subset of X, we may similarly assume that $X = \operatorname{Spec} A$ is affine. Then $\phi^\sharp$ gives an inclusion $\phi^\sharp : B \hookrightarrow K(A)$. Write $\phi^\sharp(y_i)$ as f_i/g_i ($f_i, g_i \in A$), and let $g := \prod g_i$. Then $\phi^\sharp$ further induces an inclusion $B \hookrightarrow A_g$. Therefore $\phi : \operatorname{Spec} A_g \to \operatorname{Spec} B$ induces $\phi^\sharp$. The morphism ϕ is dominant because the inverse image of the zero ideal under the inclusion $B \hookrightarrow A_g$ is the zero ideal, so ϕ takes the generic point of X to the generic point of Y. $\square$

7.5.D. EXERCISE. Let K be a finitely generated field extension of k. (Recall that a field extension K over k is **finitely generated** if there is a finite "generating set" $\{x_1, \ldots, x_n\}$ in K such that every element of K can be written as a rational function in $x_1, \ldots, x_n$ with coefficients in k.) Show that there exists an irreducible affine k-variety with function field K. (Hint: Consider the map $k[t_1, \ldots, t_n] \to K$ given by $t_i \mapsto x_i$, and show that the kernel is a prime ideal $\mathfrak{p}$, and that $k[t_1, \ldots, t_n]/\mathfrak{p}$ has fraction field K. Interpreted geometrically: consider the map $\operatorname{Spec} K \to \operatorname{Spec} k[t_1, \ldots, t_n]$ given by the ring map $t_i \mapsto x_i$, and take the closure of the one-point image.)

7.5.E. EXERCISE. Describe equivalences of categories among the following.

(a) the category with objects "integral *affine* k-varieties," and morphisms "dominant rational maps defined over k"; and

(b) the opposite ("arrows-reversed") category with objects "finitely generated field extensions of k," and morphisms "inclusions extending the identity on k."

(Once we define varieties in general in §11.2.9, you can add in: (c) the category with objects "integral k-varieties," and morphisms "dominant rational maps defined over k.")

In particular, an integral affine k-variety X is rational if its function field $K(X)$ is a purely transcendental extension of k, i.e., $K(X) \cong k(x_1, \ldots, x_n)$ for some n. (This needs to be said more precisely: the map $k \hookrightarrow K(X)$ induced by $X \to \operatorname{Spec} k$ should agree with the "obvious" map $k \hookrightarrow k(x_1, \ldots, x_n)$ under this isomorphism.)

7.5.9. More examples of rational maps.

A recurring theme in these examples is that domains of definition of rational maps to projective schemes extend over regular codimension 1 points. We will make this precise in the Curve-to-Projective Extension Theorem 15.3.1, when we discuss curves.

The first example is the classical formula for Pythagorean triples, and its derivation by "stereographic projection." Suppose you are looking for rational points on the circle C given by $x^2 + y^2 = 1$ (Figure 7.2). One rational point is $p = (1, 0)$. If q is another rational point, then pq is a line of rational (non-infinite) slope. This gives a rational map from the conic C (now interpreted as $\operatorname{Spec} \mathbb{Q}[x, y]/(x^2 + y^2 - 1)$) to $\mathbb{A}^1_{\mathbb{Q}}$, given by $(x, y) \mapsto y/(x - 1)$. (Something subtle just happened: we were talking about $\mathbb{Q}$-points on a circle, and ended up with a rational map of schemes.)

Conversely, given a line of slope m through p, where m is rational, we can recover q by solving the equations $y = m(x-1)$, $x^2 + y^2 = 1$. We substitute the first equation into the second, to get a quadratic equation in x. We know that we will have a solution $x = 1$ (because the line meets the circle at $(x, y) = (1, 0)$), so we expect to be able to factor this out, and find the other factor. This indeed works:

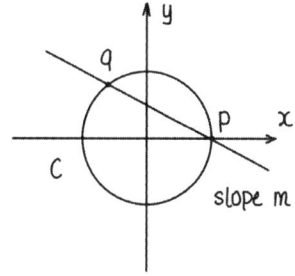

$$x^2 + (m(x-1))^2 = 1$$
$$\implies (m^2 + 1)x^2 + (-2m^2)x + (m^2 - 1) = 0$$
$$\implies (x-1)((m^2 + 1)x - (m^2 - 1)) = 0.$$

Figure 7.2 *Finding primitive Pythagorean triples using geometry.*

The other solution is $x = (m^2 - 1)/(m^2 + 1)$, which gives $y = -2m/(m^2 + 1)$. Thus we get a birational map between the conic C and $\mathbb{A}^1$ with coordinate m, given by $f: (x, y) \mapsto y/(x-1)$ (defined for $x \neq 1$), and with inverse rational map given by $m \mapsto ((m^2 - 1)/(m^2 + 1), -2m/(m^2 + 1))$ (defined away from $m^2 + 1 = 0$).

We can extend this to a rational map $C \dashrightarrow \mathbb{P}^1_{\mathbb{Q}}$ via the "inclusion" $\mathbb{A}^1_{\mathbb{Q}} \to \mathbb{P}^1_{\mathbb{Q}}$ (which we later call an open embedding). Then f is given by $(x, y) \mapsto [y, x-1]$. We then have an interesting question: what is the domain of definition of f? It appears to be defined everywhere except for where $y = x - 1 = 0$, i.e., everywhere but p. But in fact it can be extended over p! Note that $(x, y) \mapsto [x+1, -y]$ (where $(x, y) \neq (-1, 0)$) agrees with f on the common domain of definition, as $[x+1, -y] = [y, x-1]$. Hence this rational map can be extended further than we at first thought. This will be a special case of the Curve-to-Projective Extension Theorem 15.3.1.

7.5.F. EXERCISE. Use the above to find a "formula" yielding all Pythagorean triples.

7.5.G. EXERCISE. Show that the conic $x^2 + y^2 = z^2$ in $\mathbb{P}^2_k$ is isomorphic to $\mathbb{P}^1_k$ for any field k of characteristic not 2. (See Exercise 4.5.S for the definition of "$x^2 + y^2 = z^2$ in $\mathbb{P}^2_k$" if you need it.)

7.5.10. In fact, any conic in $\mathbb{P}^2_k$ with a k-valued point (i.e., a point with residue field k) of rank 3 (after base change to $\overline{k}$, so "rank" makes sense; see Exercise 5.4.J) is isomorphic to $\mathbb{P}^1_k$. (The hypothesis of having a k-valued point is certainly necessary: $x^2 + y^2 + z^2 = 0$ over $k = \mathbb{R}$ is a conic that is not isomorphic to $\mathbb{P}^1_k$.)

7.5.H. EXERCISE. Find all rational solutions to $y^2 = x^3 + x^2$, by finding a birational map to $\mathbb{A}^1_{\mathbb{Q}}$, mimicking what worked with the conic. Hint: What point should you project from? (In Exercise 19.10.I, we will see that these points basically form a group, and that this is a degenerate elliptic curve.)

You will obtain a rational map to $\mathbb{P}^1_{\mathbb{Q}}$ that is not defined over the "node" $x = y = 0$, and *cannot* be extended over this codimension 1 set. This is an example of the limits of our future result, the Curve-to-Projective Extension Theorem 15.3.1, showing how to extend rational maps to projective space over codimension 1 sets: the codimension 1 sets have to be regular.

7.5.I. EXERCISE. Use a similar idea to find a birational map from the quadric surface $Q = \{x^2 + y^2 = w^2 + z^2\} \subset \mathbb{P}^3_{\mathbb{Q}}$ to $\mathbb{P}^2_{\mathbb{Q}}$. Use this to find all rational points on Q. (This illustrates a good way of solving Diophantine equations. You will find a dense open subset of Q that is isomorphic to a dense open subset of $\mathbb{P}^2$, where you can easily find all the rational points. There will be a closed subset of Q where the rational map is not defined, or is not an isomorphism, but you can deal with this subset in an ad hoc fashion.)

7.5.J. EXERCISE (THE CREMONA TRANSFORMATION, A USEFUL CLASSICAL CONSTRU-CTION). Consider the rational map $\mathbb{P}^2_k \dashrightarrow \mathbb{P}^2_k$, given by $[x, y, z] \to [1/x, 1/y, 1/z]$. What is the domain of definition? (It is bigger than the locus where $xyz \neq 0$!) You will observe that you can extend it over "codimension 1 sets" (ignoring the fact that we don't yet know what codimension means). This again foreshadows the Curve-to-Projective Extension Theorem 15.3.1.

7.5.11.* Complex curves that are not rational (fun but inessential).

We now describe two examples of curves C that do not admit a nonconstant rational map from $\mathbb{P}^1_{\mathbb{C}}$. (Admittedly, we do not yet know what "curve" means, but no matter.) Both proofs are by Fermat's method of *infinite descent*. These results can be interpreted (as you will later be able to check using Theorem 16.3.3) as the fact that these curves have no "nontrivial" $\mathbb{C}(t)$-valued points, by which we mean that any $\mathbb{C}(t)$-valued point is secretly a $\mathbb{C}$-valued point. You may notice that if you consider the same examples with $\mathbb{C}(t)$ replaced by $\mathbb{Q}$ (and where C is a curve over $\mathbb{Q}$ rather than $\mathbb{C}$), you get two fundamental questions in number theory and geometry. The analog of Exercise 7.5.M is the question of rational points on elliptic curves, and you may realize that the analog of Exercise 7.5.K is even more famous. Also, the arithmetic analog of Exercise 7.5.M(a) is the "Four Squares Theorem" (there are not four integer squares in arithmetic progression), first stated by Fermat. These examples will give you a glimpse of how and why facts over number fields are often paralleled by facts over function fields of curves. This parallelism is a recurring deep theme in the subject.

7.5.K. EXERCISE. If $n > 2$, show that $\mathbb{P}^1_{\mathbb{C}}$ has no dominant rational maps to the "Fermat curve" $x^n + y^n = z^n$ in $\mathbb{P}^2_{\mathbb{C}}$. Hint: Reduce this to showing that there is no "nonconstant" solution $(f(t), g(t), h(t))$ to $f(t)^n + g(t)^n = h(t)^n$, where $f(t)$, $g(t)$, and $h(t)$ are rational functions in t (that is, elements of $\mathbb{C}(t)$). By clearing denominators, reduce this to showing that there is no nonconstant solution where $f(t)$, $g(t)$, and $h(t)$ are relatively prime polynomials. For this, assume there is a solution, and consider one of the lowest positive degree. Then use the fact that $\mathbb{C}[t]$ is a unique factorization domain, and $h(t)^n - g(t)^n = \prod_{i=1}^{n}(h(t) - \zeta^i g(t))$, where ζ is a primitive nth root of unity. Argue that each $h(t) - \zeta^i g(t)$ is an nth power. Then use

$$(h(t) - g(t)) + \alpha\,(h(t) - \zeta g(t)) = \beta\left(h(t) - \zeta^2 g(t)\right)$$

for suitably chosen α and β to get a solution of smaller degree. (How does this argument fail for $n = 2$?)

7.5.L. EXERCISE. Give two smooth complex curves X and Y so that no nonempty open subset of X is isomorphic to a nonempty open subset of Y. (Try not to be bothered by the fact that we have not yet defined "smoothness.") Hint: Exercise 7.5.K.

7.5.M. EXERCISE. Suppose a, b, and c are distinct complex numbers. By the following steps, show that if $x(t)$ and $y(t)$ are two rational functions of t such that

(7.5.11.1) $$y(t)^2 = (x(t) - a)(x(t) - b)(x(t) - c),$$

then $x(t)$ and $y(t)$ are constants ($x(t), y(t) \in \mathbb{C}$). (Here $\mathbb{C}$ may be replaced by any field K of characteristic not 2; slightly extra care is needed if K is not algebraically closed.)

(a) Suppose $P, Q \in \mathbb{C}[t]$ are relatively prime polynomials such that four linear combinations of them are perfect squares, no two of which are constant multiples of each other. Show that P and Q are constant (i.e., $P, Q \in \mathbb{C}$). Hint: By renaming P and Q, show that you may assume that the perfect squares are $P, Q, P - Q, P - \lambda Q$ (for some $\lambda \in \mathbb{C}$). Define u and v to be square roots of P and Q, respectively. Show that $u - v, u + v, u - \sqrt{\lambda}v, u + \sqrt{\lambda}v$ are perfect squares, and that u and v are relatively prime. If P and Q are not both constant, note that $0 < \max(\deg u, \deg v) < \max(\deg P, \deg Q)$. Assume from the start that P and Q were chosen as a counterexample with minimal $\max(\deg P, \deg Q)$ to obtain a contradiction. (Aside: It is possible to have *three* distinct linear combinations that are perfect squares. Such examples essentially correspond to primitive Pythagorean triples in $\mathbb{C}(t)$—can you see how?)

(b) Suppose $(x, y) = (p/q, r/s)$ is a solution to (7.5.11.1), where $p, q, r, s \in \mathbb{C}[t]$, and p/q and r/s are in lowest terms. Clear denominators to show that $r^2 q^3 = s^2(p - aq)(p - bq)(p - cq)$. Show that $s^2 | q^3$ and $q^3 | s^2$, hence that $s^2 = \delta q^3$ for some $\delta \in \mathbb{C}$. From $r^2 = \delta(p - aq)(p - bq)(p - cq)$, show that $(p - aq), (p - bq), (p - cq)$ are perfect squares. Show that q is also a perfect square, and then apply part (a).

A much better geometric approach to Exercises 7.5.K and 7.5.M is given in Exercise 21.4.H.

7.6* Representable Functors and Group Schemes

7.6.1. Maps to $\mathbb{A}^1$ correspond to functions. If X is a scheme, there is a bijection between the maps $X \to \mathbb{A}^1$ and global sections of the structure sheaf: by Exercise 7.3.G, maps $\pi: X \to \mathbb{A}^1_{\mathbb{Z}}$ correspond to maps of rings $\pi^\sharp: \mathbb{Z}[t] \to \Gamma(X, \mathscr{O}_X)$, and $\pi^\sharp(t)$ is a function on X; this is reversible.

This map is very natural in an informal sense: you can even picture this map to $\mathbb{A}^1$ as being *given* by the function. (By analogy, a function on a manifold is a map to $\mathbb{R}$.) But it is natural in a more precise sense: this bijection is functorial in X. We will ponder this example at length, and see that it leads us to two important sophisticated notions: representable functors and group schemes.

7.6.A. EASY EXERCISE. Suppose X is a $\mathbb{C}$-scheme. Verify that there is a natural bijection between maps $X \to \mathbb{A}^1_{\mathbb{C}}$ in the category of $\mathbb{C}$-schemes and functions on X. (Here the base ring $\mathbb{C}$ can be replaced by any ring A.)

This interpretation can be extended to rational maps, as follows.

7.6.B. UNIMPORTANT EXERCISE. Interpret rational functions on an integral scheme (Exercise 5.2.I)) as rational maps to $\mathbb{A}^1_{\mathbb{Z}}$.

(If you wish, you can extend your argument to rational maps on locally Noetherian schemes; see Definition 6.6.36.)

7.6.2. Representable functors. We restate the bijection of §7.6.1 as follows. We have two different contravariant functors from *Sch* to *Sets*: maps to $\mathbb{A}^1$ (i.e., H: $X \mapsto \mathrm{Mor}(X, \mathbb{A}^1_{\mathbb{Z}})$), and functions on X (F: $X \mapsto \Gamma(X, \mathscr{O}_X)$). The "naturality" of the bijection—the functoriality in X—is precisely the statement that the bijection gives a natural isomorphism of functors (§1.1.21): given any $\pi: X \to X'$, the diagram

$$
\begin{array}{ccc}
H(X') & \xrightarrow{H(\pi)} & H(X) \\
\downarrow & & \downarrow \\
F(X') & \xrightarrow{F(\pi)} & F(X)
\end{array}
$$

(where the vertical maps are the bijections given in §7.6.1) commutes.

More generally, if Y is an element of a category $\mathscr{C}$ (we care about the special case $\mathscr{C} = Sch$), recall the contravariant functor $h_Y: \mathscr{C} \to Sets$ defined by $h_Y(X) = \mathrm{Mor}(X, Y)$ (Example 1.1.20). We say a contravariant functor from $\mathscr{C}$ to *Sets* is **represented by** Y if it is naturally isomorphic to the functor h_Y. We say it is **representable** if it is represented by *some* Y.

The bijection of §7.6.1 may now be restated as: *the global section functor is represented by* $\mathbb{A}^1$.

7.6.C. IMPORTANT EASY EXERCISE (REPRESENTING OBJECTS ARE UNIQUE UP TO UNIQUE ISOMORPHISM). Show that if a contravariant functor F is represented by Y and by Z, then we have a unique isomorphism $Y \xrightarrow{\sim} Z$ induced by the natural isomorphism of functors $h_Y \xrightarrow{\sim} h_Z$. Hint: This is a version of the universal property arguments of §1.2: once again, we are recognizing an object (up to unique isomorphism) by maps to that object. This exercise is essentially Exercise 1.2.Z(b). (This extends readily to Yoneda's Lemma in this setting, Exercise 10.1.B. You are welcome to try that now.)

You have implicitly seen this notion before: you can interpret the existence of products and fibered products in a category as examples of representable functors. (You may wish to work out how a natural isomorphism $h_{Y \times Z} \xrightarrow{\sim} h_Y \times h_Z$ induces the projection maps $Y \times Z \to Y$ and $Y \times Z \to Z$.)

7.6.D. EXERCISE (WARM-UP). Suppose F is the contravariant functor $Sch \to Sets$ defined by $F(X) = \{\text{Grothendieck}\}$ for all schemes X. Show that F is representable. (What is it representable by?)

7.6.E. EXERCISE. In this exercise, $\mathbb{Z}$ may be replaced by any ring.

(a) *(Affine n-space represents the functor of n functions.)* Show that the contravariant functor from ($\mathbb{Z}$-)schemes to *Sets*

$$X \longmapsto \{(f_1, \dots, f_n) : f_i \in \Gamma(X, \mathscr{O}_X)\}$$

is represented by $\mathbb{A}_{\mathbb{Z}}^n$. Show that $\mathbb{A}_{\mathbb{Z}}^1 \times_{\mathbb{Z}} \mathbb{A}_{\mathbb{Z}}^1 \cong \mathbb{A}_{\mathbb{Z}}^2$, i.e., that $\mathbb{A}^2$ satisfies the functorial description of $\mathbb{A}^1 \times \mathbb{A}^1$. (You will undoubtedly be able to immediately show that $\prod \mathbb{A}_{\mathbb{Z}}^{m_i} \xrightarrow{\sim} \mathbb{A}_{\mathbb{Z}}^{\sum m_i}$.)

(b) *(The functor of invertible functions is representable.)* Show that the contravariant functor from ($\mathbb{Z}$-)schemes to *Sets* taking X to invertible functions on X is representable by $\operatorname{Spec} \mathbb{Z}[t, t^{-1}]$.

7.6.3. Definition. The scheme defined in Exercise 7.6.E(b) is called the **multiplicative group** (or multiplicative group scheme) $\mathbb{G}_m$. "$\mathbb{G}_m$ over a field k" ("the multiplicative group over k") means $\operatorname{Spec} k[t, t^{-1}]$, with the same group operations. Better: it represents the group of invertible functions in the category of k-schemes. We can similarly define $\mathbb{G}_m$ over an arbitrary ring or even arbitrary scheme.

7.6.F. LESS IMPORTANT EXERCISE. Fix a ring A. Consider the functor H from the category of locally ringed spaces to *Sets* given by $H(X) = \{A \to \Gamma(X, \mathscr{O}_X)\}$. Show that this functor is representable (by $\operatorname{Spec} A$). This gives another (admittedly odd) motivation for the definition of $\operatorname{Spec} A$, closely related to that of §7.3.6.

7.6.4.** Group schemes (or, more generally, group objects in a category).
(The rest of §7.6 should be read only for entertainment.) We return again to Example 7.6.1. Functions on X are better than a set: they form a group. (Indeed they even form a ring, but we will worry about this later.) Given a morphism $X \to Y$, pullback of functions $\Gamma(Y, \mathscr{O}_Y) \to \Gamma(X, \mathscr{O}_X)$ is a group homomorphism. So we should expect $\mathbb{A}^1$ to have some group-like structure. This leads us to the notion of *group scheme*, or more generally a *group object* in a category, which we now define.

Suppose $\mathscr{C}$ is a category with a final object Z and with products. (We know that *Sch* has a final object $Z = \operatorname{Spec} \mathbb{Z}$, by Exercise 7.3.J. We will later see that it has products, §10.1. But in Exercise 7.6.K we will give an alternative characterization of group objects that applies in any category, so we won't worry about this.)

A **group object** in $\mathscr{C}$ is an object X along with three morphisms:

- *multiplication:* $m: X \times X \to X$,
- *inverse:* $i: X \to X$, and
- *identity element:* $e: Z \to X$ (*not* the identity map).

These morphisms are required to satisfy several conditions.

(i) Associativity axiom:

$$
\begin{array}{ccc}
X \times X \times X & \xrightarrow{\ m \times \mathrm{id}\ } & X \times X \\
{\scriptstyle \mathrm{id} \times m} \downarrow & & \downarrow {\scriptstyle m} \\
X \times X & \xrightarrow{\ \ \ m\ \ \ } & X
\end{array}
$$

commutes. (Here id means the equality $X \to X$.)

(ii) Identity axiom:

$$X \xrightarrow{\sim} Z \times X \xrightarrow{\ e \times \mathrm{id}\ } X \times X \xrightarrow{\ m\ } X$$

and

$$X \xrightarrow{\sim} X \times Z \xrightarrow{\ \mathrm{id} \times e\ } X \times X \xrightarrow{\ m\ } X$$

are both the identity map $X \to X$. (This corresponds to the group axiom: "multiplication by the identity element is the identity map.")

(iii) Inverse axiom: $X \xrightarrow{i \times \mathrm{id}} X \times X \xrightarrow{m} X$ and $X \xrightarrow{\mathrm{id} \times i} X \times X \xrightarrow{m} X$ are both the map that is the composition $X \longrightarrow Z \xrightarrow{e} X$.

As motivation, you can check that a group object in the category of sets is in fact the same thing as a group. (This is symptomatic of how you take some notion and make it categorical. You write down its axioms in a categorical way, and if all goes well, if you specialize to the category of sets, you get your original notion. You can apply this to the notion of "rings" in an exercise below.)

A **group scheme** is defined to be a group object in the category of schemes. A **group scheme** over a ring A (or a scheme S) is defined to be a group object in the category of A-schemes (or S-schemes).

7.6.G. EXERCISE. Give $\mathbb{A}^1_{\mathbb{Z}}$ the structure of a group scheme, by describing the three structural morphisms, and showing that they satisfy the axioms. (Hint: The morphisms should not be surprising. For example, inverse is given by $t \mapsto -t$. Note that we know that the product $\mathbb{A}^1_{\mathbb{Z}} \times \mathbb{A}^1_{\mathbb{Z}}$ exists, by Exercise 7.6.E(a).)

7.6.H. EXERCISE. Show that if G is a group object in a category $\mathscr{C}$, then for any $X \in \mathscr{C}$, $\mathrm{Mor}(X, G)$ has the structure of a group, and the group structure is preserved by pullback (i.e., $\mathrm{Mor}(\cdot, G)$ is a contravariant functor to the category of groups, *Groups*).

7.6.I. EXERCISE. Show that the group structure described by the previous exercise translates the group scheme structure on $\mathbb{A}^1_{\mathbb{Z}}$ to the group structure on $\Gamma(X, \mathscr{O}_X)$, via the bijection of §7.6.1.

7.6.J. EXERCISE. Define the notion of **abelian group scheme**, and **ring scheme**. (You will undoubtedly at the same time figure out how to define the notion of abelian group object and ring object in any category $\mathscr{C}$. You may discover a more efficient approach to such questions after reading §7.6.5.)

7.6.5. *Group schemes, more functorially.* There was something unsatisfactory about our discussion of the "group-respecting" nature of the bijection in §7.6.1: we observed that the right side (functions on X) formed a group, then we developed the axioms of a group scheme, then we cleverly figured out the maps that made $\mathbb{A}^1_{\mathbb{Z}}$ into a group scheme, then we showed that this induced a group structure on the left side of the bijection $(\mathrm{Mor}(X, \mathbb{A}^1))$ that precisely corresponded to the group structure on the right side (functions on X).

The picture is more cleanly explained as follows. The language of scheme-valued points (Definition 7.3.10) has the following advantage: notice that the *points* of a group scheme need not themselves form a group (consider $\mathbb{A}^1_{\mathbb{Z}}$). But Exercise 7.6.H shows that the Z-*valued points* of a group scheme (where Z is any given scheme) indeed form a group.

7.6.K. EXERCISE. Suppose we have a contravariant functor F from *Sch* (or indeed any category) to *Groups*. Suppose further that F composed with the forgetful functor *Groups* $\to$ *Sets* is represented by an object Y. Show that the group operations on F(X) (as X varies through *Sch*) uniquely determine $m: Y \times Y \to Y$, $i: Y \to Y$, $e: Z \to Y$ satisfying the axioms defining a group scheme such that the group operation on $\mathrm{Mor}(X, Y)$ is the same as that on F(X).

In particular, the definition of a group object in a category was forced upon us by the definition of group. More generally, you should expect that any class of objects that can be interpreted as sets with additional structure should fit into this picture.

You should apply this exercise to $\mathbb{A}^1_{\mathbb{Z}}$, and see how the explicit formulas you found in Exercise 7.6.G are forced on you.

7.6.L. EXERCISE. Work out the maps m, i, and e in the group schemes of Exercise 7.6.E.

7.6.M. EXERCISE. Explain why the product of group objects in a category can be naturally interpreted as a group object in that category.

7.6.N. EXERCISE.

(a) Define **morphism of group schemes**.
(b) Recall that if A is a ring, then $GL_n(A)$ (the **general linear group over** A) is the group of invertible $n \times n$ matrices with entries in the ring A. Figure out the right definition of **the group scheme** GL_n **(over a ring** A**)**, and describe the **determinant map** $\det : GL_n \to \mathbb{G}_m$.
(c) Make sense of the statement: "$(\cdot^n) : \mathbb{G}_m \to \mathbb{G}_m$ given by $t \mapsto t^n$ is a morphism of group schemes."

The language of Exercise 7.6.N(a) suggests that group schemes form a category; feel free to prove this if you want. In fact, the category of group schemes has a zero object. What is it?

7.6.O. EXERCISE (KERNELS OF MAPS OF GROUP SCHEMES). Suppose $F : G_1 \to G_2$ is a morphism of group schemes. Consider the contravariant functor $Sch \to Groups$ given by $X \mapsto \ker(\mathrm{Mor}(X, G_1) \to \mathrm{Mor}(X, G_2))$. If this is representable, by a group scheme G_0, say, show that $G_0 \to G_1$ is the kernel of F in the category of group schemes.

7.6.P. EXERCISE. Show that the kernel of $(\cdot^n)$ (Exercise 7.6.N) is representable. If $n > 0$, show that over a field k of characteristic p dividing n, this group scheme is nonreduced. (This group scheme, denoted by μ_n, is important, although we will not use it.)

7.6.Q. EXERCISE. Show that the kernel of $\det : GL_n \to \mathbb{G}_m$ is representable. This is the group scheme SL_n. (You can do this over $\mathbb{Z}$, or over a field k, or even over an arbitrary ring A; the algebra is the same.)

7.6.R. EXERCISE. Show (as easily as possible) that $\mathbb{A}^1_k$ is a ring k-scheme. (Here k can be replaced by any ring.)

7.6.S. EXERCISE.

(a) Define the notion of a (left) **group scheme action** (of a group scheme on a scheme).
(b) Suppose A is a ring. Show that specifying an integer-valued grading on A is equivalent to specifying an action of $\mathbb{G}_m$ on $\operatorname{Spec} A$. (This interpretation of a grading is surprisingly enlightening. Caution: There are two possible choices of the integer-valued grading, and there are reasons for both. Both are used in the literature.)

7.6.6. *Aside: Hopf algebras.* Here is a notion that we won't use, but it is easy enough to define now. Suppose $G = \operatorname{Spec} A$ is an affine group scheme, i.e., a group scheme that is an affine scheme. The categorical definition of group scheme can be restated in terms of the ring A. (This requires thinking through Remark 7.3.5; see Exercise 10.1.A.) Then these axioms define a **Hopf algebra**. For example, we have a "comultiplication map" $A \to A \otimes A$.

7.6.T. EXERCISE. As $\mathbb{A}^1_{\mathbb{Z}}$ is a group scheme, $\mathbb{Z}[t]$ has a Hopf algebra structure. Describe the comultiplication map $\mathbb{Z}[t] \to \mathbb{Z}[t] \otimes_{\mathbb{Z}} \mathbb{Z}[t]$.

7.7** The Grassmannian: First Construction

The Grassmannian is a useful geometric construction that is "the geometric object underlying linear algebra." In (classical) geometry over a field $K = \mathbb{R}$ or $\mathbb{C}$, just as projective space parametrizes one-dimensional subspaces of a given n-dimensional vector space, the Grassmannian parametrizes k-dimensional subspaces of n-dimensional space. The Grassmannian $G(k, n)$ is a manifold of dimension $k(n - k)$ (over the field). The manifold structure is given as follows.

Given a basis $(v_1, \ldots, v_n)$ of n-space, "most" k-planes can be described as the span of the k vectors,

(7.7.0.1)
$$\left\langle v_1 + \sum_{i=k+1}^{n} a_{1i}v_i, v_2 + \sum_{i=k+1}^{n} a_{2i}v_i, \ldots, v_k + \sum_{i=k+1}^{n} a_{ki}v_i \right\rangle.$$

(Can you describe which k-planes are *not* of this form? Hint: Row-reduced echelon form. Aside: The stratification of $G(k, n)$ by normal form is the decomposition of the Grassmannian into *Schubert cells*. You may be able to show using the normal form that each Schubert cell is isomorphic to an affine space.) Any k-plane of this form can be described in such a way uniquely. We use this to identify the k-planes of this form with the manifold $K^{k(n-k)}$ (with coordinates a_{ji}). This is a large affine patch on the Grassmannian (called the "open Schubert cell" with respect to this basis). As the v_i vary, these patches cover the Grassmannian (why?), and the manifold structures agree (a harder fact).

We now *define* the Grassmannian in algebraic geometry, over a ring A. Suppose $v = (v_1, \ldots, v_n)$ is a basis for $A^{\oplus n}$. More precisely: $v_i \in A^{\oplus n}$, and the map $A^{\oplus n} \to A^{\oplus n}$ given by $(a_1, \ldots, a_n) \mapsto a_1 v_1 + \cdots + a_n v_n$ is an isomorphism.

7.7.A. EXERCISE. Show that any two bases are related by an invertible $n \times n$ matrix over A—a matrix with entries in A whose determinant is an invertible element of A.

For each such basis v, consider the scheme $U_v \cong \mathbb{A}_A^{k(n-k)}$, with coordinates a_{ji} ($k+1 \leq i \leq n$, $1 \leq j \leq k$), which we imagine as corresponding to the k-plane spanned by the vectors (7.7.0.1).

7.7.B. EXERCISE. Given two bases v and w, explain how to glue U_v to U_w along appropriate open sets. You may find it convenient to work with coordinates a_{ji} where i runs from 1 to n, not just $k+1$ to n, but imposing $a_{ji} = \delta_{ji}$ (i.e., 1 when $i = j$ and 0 otherwise) when $i \leq k$. This convention is analogous to coordinates $x_{i/j}$ on the patches of projective space (§4.4.9). Hint: The relevant open subset of U_v will be where a certain determinant doesn't vanish.

7.7.C. EXERCISE/DEFINITION. By checking triple intersections, verify that these patches (over all possible bases) glue together to a single scheme (Exercise 4.4.A). This is the **Grassmannian** $G(k, n)$ over the ring A. Because it can be interpreted as a space of linear "$\mathbb{P}_A^{k-1}$'s" in $\mathbb{P}_A^{n-1}$, it is often also written $\mathbb{G}(k-1, n-1)$. (You will see that this is wise notation, in Exercise 12.2.N, for example.)

Although this definition is pleasantly explicit (it is immediate that the Grassmannian is covered by $\mathbb{A}^{k(n-k)}$'s), and perhaps more "natural" than our original definition of projective space in §4.4.9 (we aren't making a choice of basis; we use *all* bases), there are several things unsatisfactory about it. In fact the Grassmannian is projective; this isn't obvious with this definition. Furthermore, the Grassmannian comes with a natural morphism to $\mathbb{P}^{\binom{n}{k}-1}$ (which is an example of a *closed embedding*, a notion we will meet in Definition 9.1.1), called the **Plücker embedding**. Finally, there is an action of GL_n on the space of k-planes in n-space, so we should be able to see this action in our algebraic incarnation. We will address these issues in §16.4, by giving a better description, as a moduli space.

7.7.1. *(Partial) flag varieties.* Just as the Grassmannian "parametrizes" k-planes in n-space, the **flag variety** parametrizes "flags": nested sequences of subspaces of n-space

$$F_0 \subset F_1 \subset \cdots \subset F_n,$$

where $\dim F_i = i$. Generalizing both of these is the notion of a **partial flag variety** associated to data $0 \leq a_1 < \cdots < a_\ell \leq n$, parametrizing nested subspaces of n-space

$$F_{a_1} \subset \cdots \subset F_{a_\ell},$$

where $\dim F_{a_i} = a_i$. You should be able to generalize all of the discussion in §7.7 to this setting.

Chapter 8

Useful Classes of Morphisms of Schemes

We now define an excessive number of types of morphisms. Some properties (often finiteness properties) are useful because *most* "reasonable" morphisms have such properties, and they will be used in proofs in obvious ways. Others correspond to geometrically meaningful behavior, and you should have a picture of what each means.

8.0.1. *Change of perspective:* **Morphisms** *are more fundamental than* **objects**. One of Grothendieck's lessons is that things that we often think of as properties of *objects* are better understood as properties of *morphisms*. One way of turning properties of objects into properties of morphisms is as follows. If P is a property of schemes, we often (but not always) say that a *morphism* $\pi\colon X \to Y$ has P if for every affine open subset $U \subset Y$, $\pi^{-1}(U)$ has P. We will see this for P = quasicompact, quasiseparated, affine, and more. (As you might hope, in good circumstances, P will satisfy the hypotheses of the Affine Communication Lemma 5.3.2, so we don't have to check *every* affine open subset.) Informally, you can think of such a morphism as one where all the fibers have P. (You can quickly define the fiber of a morphism as a topological space, but once we define fibered product, we will define the *scheme-theoretic* fiber, and then this discussion will make sense.) But it means more than that: it means that "being P" is really not just fiber-by-fiber, but behaves well as the fiber varies. (For comparison, "submersion of manifolds" means more than that the fibers are smooth.)

8.1 "Reasonable" Classes of Morphisms (Such as Open Embeddings)

8.1.1. You will notice that almost all classes of morphisms that are useful have many properties in common, which follow from the following three basic properties. We call any class of morphisms satisfying these properties a **"reasonable" class of morphisms** of schemes. (To avoid any stupidities, we assume that the class includes all isomorphisms; but this requirement doesn't deserve a number.)

(i)   The class is **preserved by composition**: if $\pi\colon X \to Y$ and $\rho\colon Y \to Z$ are both in this class, then so is $\rho \circ \pi$.

(ii) The class is **preserved by "base change"** (or "pullback" or "fibered product"). Precisely: if $\pi\colon X \to Y$ is in this class, then for any $Y' \to Y$, the induced map $X \times_Y Y' \to Y'$ is also in this class. (Implicit in this statement is that the fibered product $X \times_Y Y'$ exists; but we will soon see that all fibered products of schemes exist, §10.1.)

(iii) The class is **local on the target**. In other words, (a) if $\pi\colon X \to Y$ is in the class, then for any open subset V of Y, the restricted morphism $\pi^{-1}(V) \to V$ is in the class; and (b) for a morphism $\pi\colon X \to Y$, if there is an open cover $\{V_i\}$ of Y for which each restricted morphism $\pi^{-1}(V_i) \to V_i$ is in the class, then π is in the class. In particular, as schemes are built out of affine schemes, properties are often easy to verify on any affine cover (the properties are "affine-local" on the target), as described in §8.0.1. (Stalk-local properties are automatically local on the target.)

Properties (i) and (ii) imply a useful additional property—(iv) "reasonable" classes of morphisms are **preserved by product**:

8.1.A. EXERCISE. Suppose P is a property of morphisms preserved by composition and base change ((i), (ii) above), and $X \to Y$ and $X' \to Y'$ are two morphisms of S-schemes with property P.

Assume that $X \times_S X'$ and $Y \times_S Y'$ exist (which is indeed true, §10.1). Show that $X \times_S X' \to Y \times_S Y'$ has property P as well. Hint:

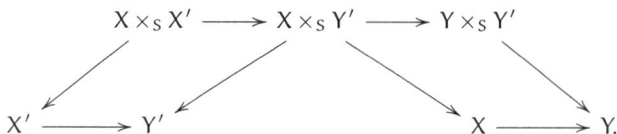

$$X \times_S X' \longrightarrow X \times_S Y' \longrightarrow Y \times_S Y'$$

$$X' \longrightarrow Y' \qquad X \longrightarrow Y.$$

(v) Another extremely important consequence of (i) and (ii) is the *Cancellation Theorem* for such morphisms, Theorem 11.1.1.

When you learn of a new class of morphisms, you should immediately ask whether the properties (i)–(iii) (and hence (iv) and (v)) hold. The answer will almost always be "yes"; and if it is "no," then you should exercise caution, and possibly figure out how to make the definition better (see, for example, §10.5).

To prepare to think about these properties for schemes, you may want to verify that properties analogous to (i)–(iii) hold in the category of topological spaces with the classes "injections," "surjections," "open embeddings" (open embedding of topological spaces = isomorphism with an open subset), and "closed embeddings" (closed embedding of topological spaces = isomorphism with a closed subset).

As a first example in the category of schemes:

8.1.B. EASY EXERCISE. Show that "isomorphisms" of schemes are "reasonable" in this sense, i.e., that they satisfy the three properties (i)–(iii) above.

We are now ready to introduce a fundamental class of morphisms of schemes.

8.1.2. *Definition.* A morphism $\pi \colon X \to Y$ of schemes is an **open embedding** (or **open immersion**) if it is an open embedding as ringed spaces (§7.2.1). In other words, a morphism $\pi \colon (X, \mathscr{O}_X) \to (Y, \mathscr{O}_Y)$ of schemes is an open embedding if π factors as

$$(X, \mathscr{O}_X) \xrightarrow[\sim]{\rho} (U, \mathscr{O}_Y|_U) \overset{\tau}{\hookrightarrow} (Y, \mathscr{O}_Y),$$

where ρ is an isomorphism, and $\tau \colon U \hookrightarrow Y$ is an inclusion of an open set. It is immediate that isomorphisms are open embeddings. The symbol $\hookrightarrow$ is often used to indicate that a morphism is an open embedding (or more generally, a locally closed embedding, see §9.2).

If X is actually a subset of Y (and π is the inclusion, i.e., ρ is the identity), then we say $(X, \mathscr{O}_X)$ is an **open subscheme** of $(Y, \mathscr{O}_Y)$. The difference between open embeddings and open subschemes is a bit confusing, and not too important: at the level of sets, open subschemes *are* subsets, while open embeddings are *bijections onto* subsets.

"Open subschemes" are scheme-theoretic analogs of open subsets. ("Closed subschemes" are scheme-theoretic analogs of closed subsets, but they have a surprisingly different flavor, as we will see in §9.1.)

The next two exercises verify that the class of open embeddings is "reasonable" in the sense described above.

8.1.C. EXERCISE. Verify that the class of open embeddings satisfies properties (i) and (iii) of §8.1.

Property (ii) is the content of the next exercise.

8.1.D. IMPORTANT BUT EASY EXERCISE: FIBERED PRODUCTS WITH OPEN EMBEDDINGS EXIST. Verify that the class of open embeddings satisfies property (ii) of §8.1. More specifically: suppose $i \colon U \to Z$ is an open embedding, and $\rho \colon Y \to Z$ is any morphism. Show that $U \times_Z Y$ exists and $U \times_Z Y \to Y$ is an open embedding. (Hint: I'll even tell you what $U \times_Z Y$ is: $(\rho^{-1}(U), \mathscr{O}_Y|_{\rho^{-1}(U)})$.) In particular, if $U \hookrightarrow Z$ and $V \hookrightarrow Z$ are open embeddings, $U \times_Z V \cong U \cap V$: "intersection of open embeddings is a fiber product."

8.1.E. EASY EXERCISE. Show that open embeddings of schemes are monomorphisms (in the category of schemes). Hint: Exercise 7.2.B.

8.1.F. EASY EXERCISE. Suppose $\pi\colon X \to Y$ is an open embedding. Show that if Y is locally Noetherian, then X is too. Show that if Y is Noetherian, then X is too. However, show that if Y is quasicompact, X need not be. (Hint: Let Y be affine but not Noetherian; see Exercise 3.6.G(b).)

8.1.3. *Definition.* In analogy with "local on the target" (§8.1.1), we define what it means for a property P of morphisms to be **local on the source**: if $\pi\colon X \to Y$ has P, and $i\colon U \hookrightarrow X$ is an open subset, then $\pi \circ i\colon U \to Y$ has P; and to check if a morphism $\pi\colon X \to Y$ has P, it suffices to check on any open cover $\{U_i\}$ of X. We then define **affine-local on the source** (and **affine-local on the target**) in the obvious way. For example, to check if $\pi\colon X \to Y$ has P it suffices to check on any affine open cover of the source (resp., target). The use of "affine-local" rather than "local" is to emphasize that the criterion on affine schemes is simple to describe. (Clearly the Affine Communication Lemma 5.3.2 will be very handy.)

8.1.G. EXERCISE (PRACTICE WITH THE CONCEPT). Show that the notion of "open embedding" is not local on the source.

8.2   Another Algebraic Interlude: Lying Over and Nakayama

> Algebra is the offer made by the devil to the mathematician. The devil says: I will give you this powerful machine, it will answer any question you like. All you need to do is give me your soul: give up geometry and you will have this marvelous machine.
>
> —M. Atiyah, [At2, p. 659]; but see the Atiyah quote at the start of §0.3

To set up our discussion in the next section on integral morphisms, we develop some algebraic preliminaries. A clever trick we use can also be used to show Nakayama's Lemma, so we discuss this as well.

Suppose $\phi\colon B \to A$ is a ring morphism. We say $a \in A$ is **integral** over B if a satisfies some monic polynomial

$$a^n + ?a^{n-1} + \cdots + ? = 0$$

where the coefficients lie in $\phi(B)$. A ring *morphism* $\phi\colon B \to A$ is **integral** if every element of A is integral over $\phi(B)$. An integral ring morphism ϕ is an **integral extension** if ϕ is an *inclusion* of rings. You should think of integral morphisms and integral extensions as ring-theoretic generalizations of the notion of algebraic extensions of fields.

8.2.A. EXERCISE. Show that if $\phi\colon B \to A$ is a ring morphism, $(b_1, \ldots, b_n) = 1$ in B, and $B_{b_i} \to A_{\phi(b_i)}$ is integral for all i, then ϕ is integral. Hint: Replace B by $\phi(B)$ to reduce to the case where B is a subring of A. Suppose $a \in A$. Show that there is some t and m such that $b_i^t a^m \in B + Ba + Ba^2 + \cdots + Ba^{m-1}$ for some t and m independent of i. Use a "partition of unity" argument as in the proof of Theorem 4.1.2 to show that $a^m \in B + Ba + Ba^2 + \cdots + Ba^{m-1}$.

8.2.B. EXERCISE.

(a) Show that the property of a *morphism* $\phi\colon B \to A$ being integral is always preserved by localization and quotient of B, and quotient of A, but not localization of A. More precisely: suppose ϕ is integral. Show that the induced maps $T^{-1}B \to \phi(T)^{-1}A$, $B/J \to A/\phi(J)A$, and $B \to A/I$ are integral (where T is a multiplicative subset of B, J is an ideal of B, and I is an ideal of A), but $B \to S^{-1}A$ need not be integral (where S is a multiplicative subset of A). (Hint for the latter: Show that $k[t] \to k[t]$ is an integral ring morphism, but $k[t] \to k[t]_{(t)}$ is not.)

(b) Show that the property of ϕ being an integral *extension* is preserved by localization of B, but not localization or quotient of A. (Hint for the latter: $k[t] \to k[t]$ is an integral extension, but $k[t] \to k[t]/(t)$ is not.)

(c) In fact, the property of ϕ being an integral *extension* (as opposed to integral *morphism*) is not preserved by taking quotients of B either. (Let $B = k[x, y]/(y^2)$ and $A = k[x, y, z]/(z^2, xz - y)$. Then B injects into A, but $B/(x)$ doesn't inject into $A/(x)$.) But it is preserved in some cases. Suppose $\phi \colon B \to A$ is an integral extension, and $J \subset B$ is the restriction of an ideal $I \subset A$. (Side remark: You can show that this holds if J is prime.) Show that the induced map $B/J \to A/JA$ is an integral extension. (Hint: Show that the composition $B/J \to A/JA \to A/I$ is an injection.)

The following lemma uses a useful but sneaky trick.

8.2.1. Lemma — *Suppose $\phi \colon B \to A$ is a ring morphism. Then $a \in A$ is integral over B if and only if it is contained in a subalgebra of A that is a finitely generated B-module.*

Proof. If a satisfies a monic polynomial equation of degree n, then the B-submodule of A generated by $1, a, \ldots, a^{n-1}$ is closed under multiplication, and hence a subalgebra of A.

Assume conversely that a is contained in a subalgebra A' of A that is a finitely generated B-module. Choose a finite generating set $m_1, \ldots, m_n$ of A' (as a B-module). Then $am_i = \sum b_{ij} m_j$, for some $b_{ij} \in B$. Thus

$$(8.2.1.1) \qquad \left(a \operatorname{Id}_{n \times n} - [b_{ij}]_{ij} \right) \begin{pmatrix} m_1 \\ \vdots \\ m_n \end{pmatrix} = \begin{pmatrix} 0 \\ \vdots \\ 0 \end{pmatrix},$$

where Id_n is the $n \times n$ identity matrix in A. We can't invert the matrix $(a \operatorname{Id}_{n \times n} - [b_{ij}]_{ij})$, but we almost can. Recall that an $n \times n$ matrix M has an *adjugate matrix* $\operatorname{adj}(M)$ such that $\operatorname{adj}(M) M = \det(M) \operatorname{Id}_n$. (The (i, j)th entry of $\operatorname{adj}(M)$ is the determinant of the matrix obtained from M by deleting the ith column and jth row, times $(-1)^{i+j}$. You have likely seen this in the form of a formula for M^{-1} when there *is* an inverse; see, for example, [DF, p. 440].) The coefficients of $\operatorname{adj}(M)$ are polynomials in the coefficients of M. Multiplying (8.2.1.1) by $\operatorname{adj}(a \operatorname{Id}_{n \times n} - [b_{ij}]_{ij})$, we get

$$\det(a \operatorname{Id}_{n \times n} - [b_{ij}]_{ij}) \begin{pmatrix} m_1 \\ \vdots \\ m_n \end{pmatrix} = \begin{pmatrix} 0 \\ \vdots \\ 0 \end{pmatrix}.$$

So $\det(aI - [b_{ij}])$ annihilates the generating elements m_i, and hence every element of A', i.e., $\det(aI - [b_{ij}]) = 0$. But expanding the determinant yields an integral equation for a with coefficients in B. $\qquad \square$

8.2.2. Corollary (finite implies integral) — *If A is a **finite** B-algebra (a finitely generated B-module), then ϕ is an integral ring morphism.*

The converse is false: integral does not imply finite, as $\mathbb{Q} \hookrightarrow \overline{\mathbb{Q}}$ is an integral ring morphism, but $\overline{\mathbb{Q}}$ is not a finite $\mathbb{Q}$-module. (A field extension is integral if it is algebraic.)

8.2.C. EXERCISE. Show that if $C \to B$ and $B \to A$ are both integral ring morphisms, then so is their composition.

8.2.D. EXERCISE. Suppose $\phi \colon B \to A$ is a ring morphism. Show that the elements of A integral over B form a subalgebra of A.

8.2.3. *Remark: Transcendence theory.* These ideas lead to the main facts about transcendence theory we will need for a discussion of dimension of varieties; see Exercise/Definition 12.2.A.

8.2.4. The Lying Over and Going-Up Theorems. The Lying Over Theorem is a useful property of integral extensions.

8.2.5. The Lying Over Theorem — *Suppose* $\phi\colon B \to A$ *is an* integral *extension. Then for any prime ideal* $\mathfrak{q} \subset B$, *there is a prime ideal* $\mathfrak{p} \subset A$ *such that* $\mathfrak{p} \cap B = \mathfrak{q}$.

To be clear on how weak the hypotheses are: B need not be Noetherian, and A need not be finitely generated over B.

8.2.6. *Geometric translation: If* $B \to A$ *is an integral extension, then* $\operatorname{Spec} A \to \operatorname{Spec} B$ *is surjective.* Although the Lying Over Theorem 8.2.5 is a theorem in algebra, the name can be interpreted geometrically: the theorem asserts that the corresponding morphism of schemes is surjective. (A map of schemes is **surjective** if the underlying map of sets is surjective.) Translation: "Above" every prime $\mathfrak{q}$ "downstairs," there is a prime $\mathfrak{p}$ "upstairs"; see Figure 8.1. (For this reason, it is often said that $\mathfrak{p}$ "lies over" $\mathfrak{q}$ if $\mathfrak{p} \cap B = \mathfrak{q}$.)

The following exercise sets up the proof.

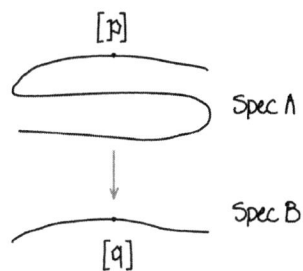

Figure 8.1 *A picture of the Lying Over Theorem 8.2.5: If* $\phi\colon B \to A$ *is an integral extension, then* $\operatorname{Spec} A \to \operatorname{Spec} B$ *is surjective.*

8.2.E.* EXERCISE. Show that the special case where A is a field translates to: if $B \subset A$ is a subring with A integral over B, then B is a field. Prove this. (Hint: You must show that all nonzero elements in B have inverses in B. Here is the start: If $b \in B$, then $1/b \in A$, and this satisfies some integral equation over B.)

Proof of the Lying Over Theorem 8.2.5. We first make a reduction: by localizing at $\mathfrak{q}$ (preserving integrality by Exercise 8.2.B(b)), we can assume that $(B, \mathfrak{q})$ is a local ring. Then let $\mathfrak{p}$ be any *maximal* ideal of A. Consider the following diagram.

$$
\begin{array}{ccc}
A & \longrightarrow\!\!\!\!\!\rightarrow & A/\mathfrak{p} \qquad\qquad \text{field} \\
\big\uparrow & & \big\uparrow \\
B & \longrightarrow\!\!\!\!\!\rightarrow & B/(\mathfrak{p} \cap B)
\end{array}
$$

The right vertical arrow is an integral extension by Exercise 8.2.B(c). By Exercise 8.2.E, $B/(\mathfrak{p} \cap B)$ is a field too, so $\mathfrak{p} \cap B$ is a maximal ideal, hence it is $\mathfrak{q}$. $\qquad\square$

8.2.F. IMPORTANT EXERCISE (THE **Going-Up Theorem**).

(a) Suppose $\phi\colon B \to A$ is an integral ring *morphism* (not necessarily an integral extension). Show that if $\mathfrak{q}_1 \subset \mathfrak{q}_2 \subset \cdots \subset \mathfrak{q}_n$ is a chain of prime ideals of B, and $\mathfrak{p}_1 \subset \cdots \subset \mathfrak{p}_m$ is a chain of prime ideals of A such that $\mathfrak{p}_i$ "lies over" $\mathfrak{q}_i$ (and $1 \le m < n$), then the second chain can be extended to $\mathfrak{p}_1 \subset \cdots \subset \mathfrak{p}_n$ so that this remains true. (Hint: Reduce to the case $m = 1, n = 2$; reduce to the case where $\mathfrak{q}_1 = (0)$ and $\mathfrak{p}_1 = (0)$; use the Lying Over Theorem 8.2.5.)
(b) Draw a picture of this theorem (akin to Figure 8.1).

There are analogous "Going-Down" results (requiring quite different hypotheses); see, for example, Theorem 12.2.12 and Exercise 24.5.E.

8.2.7. Nakayama's Lemma.
The trick in the proof of Lemma 8.2.1 can be used to quickly prove Nakayama's Lemma, which we will use repeatedly in the future. This name is used for several different but related results, which we discuss here. (A geometric interpretation will be given in Exercise 14.3.D.) We may as well prove it while the trick is fresh in our minds.

8.2.8. Nakayama's Lemma version 1 — *Suppose A is a ring, I is an ideal of A, and M is a finitely generated A-module, such that $M = IM$. Then there exists an $a \in A$ with $a \equiv 1 \pmod{I}$ with $aM = 0$.*

(Equivalently, there is some $i \in I$ for which multiplication by i induces the identity on M: $im = m$ for all $m \in M$.)

Proof. Say M is generated by $m_1, \ldots, m_n$. Then as $M = IM$, we have $m_i = \sum_j a_{ij} m_j$ for some $a_{ij} \in I$. Thus

$$(8.2.8.1) \qquad (\mathrm{Id}_n - Z) \begin{pmatrix} m_1 \\ \vdots \\ m_n \end{pmatrix} = 0,$$

where $Z = (a_{ij})$. Multiplying both sides of (8.2.8.1) on the left by $\mathrm{adj}(\mathrm{Id}_n - Z)$, we obtain

$$\det(\mathrm{Id}_n - Z) \begin{pmatrix} m_1 \\ \vdots \\ m_n \end{pmatrix} = 0.$$

But when you expand $\det(\mathrm{Id}_n - Z)$, as Z has entries in I, you get something that is 1 (mod I). $\square$

Here is why you care. Suppose I is contained in all maximal ideals of A. (The intersection of all the maximal ideals is called the **Jacobson radical**, but we won't use this phrase. For comparison, recall that the nilradical was the intersection of the *prime ideals* of A.) Then any $a \equiv 1$ (mod I) is invertible. (We are not using Nakayama yet!) Reason: Otherwise $(a) \neq A$, so the ideal (a) is contained in some maximal ideal m—but $a \equiv 1$ (mod m), contradiction. As a is invertible, we have the following.

8.2.9. Nakayama's Lemma version 2 — *Suppose A is a ring, I is an ideal of A contained in all maximal ideals, and M is a finitely generated A-module. (The most interesting case is when A is a local ring, and I is the maximal ideal.) Suppose $M = IM$. Then $M = 0$.*

8.2.G. EXERCISE (NAKAYAMA'S LEMMA VERSION 3). Suppose A is a ring, and I is an ideal of A contained in all maximal ideals. Suppose M is a *finitely generated* A-module, and $N \subset M$ is a submodule. If $N/IN \to M/IM$ is surjective, then $M = N$.

8.2.H. IMPORTANT EXERCISE (NAKAYAMA'S LEMMA VERSION 4: GENERATORS OF M/mM LIFT TO GENERATORS OF M). Suppose (A, m) is a local ring. Suppose M is a finitely generated A-module, and $f_1, \ldots, f_n \in M$, with (the images of) $f_1, \ldots, f_n$ generating M/mM. Then $f_1, \ldots, f_n$ generate M. (In particular, taking $M = m$, if we have generators of m/m^2, they also generate m.)

8.2.I. IMPORTANT EXERCISE GENERALIZING LEMMA 8.2.1. Recall that a B-module N is said to be **faithful** if the only element of B acting on N by the identity is 1 (or equivalently, if the only element of B acting as the 0-map on N is 0). Suppose S is a subring of a ring A, and $r \in A$. Suppose there is a faithful $S[r]$-module M that is finitely generated as an S-module. Show that r is integral over S. (Hint: Change a few words in the proof of version 1 of Nakayama, Lemma 8.2.8.)

8.2.J. EXERCISE. Suppose A is an integral domain, and $\widetilde{A}$ is the **integral closure** of A in $K(A)$, i.e., those elements of $K(A)$ integral over A, which form a subalgebra by Exercise 8.2.D. Show that $\widetilde{A}$ is integrally closed in $K(\widetilde{A}) = K(A)$.

8.3 A Gazillion Finiteness Conditions on Morphisms

By the end of this section, you will have seen the following types of morphisms: quasicompact, quasiseparated, affine, finite, integral, closed, (locally) of finite type, quasifinite—and possibly, (locally) of finite presentation.

8.3.1. Quasicompact and quasiseparated morphisms.

A morphism $\pi \colon X \to Y$ of schemes is **quasicompact** if for every open affine subset U of Y, $\pi^{-1}(U)$ is quasicompact. (Equivalently, the preimage of any quasicompact open subset is quasicompact. This is the right definition in other parts of geometry.)

We will like this notion because (i) finite sets have advantages over infinite sets (e.g., a finite set of integers has a maximum; also, things can be proved inductively), and (ii) most reasonable schemes will be quasicompact.

Along with quasicompactness comes the weird notion of quasiseparatedness. A morphism $\pi \colon X \to Y$ is **quasiseparated** if for every affine open subset U of Y, $\pi^{-1}(U)$ is a quasiseparated scheme (§5.1.1). (Equivalently, the preimage of any quasiseparated open subset is quasiseparated, although we won't worry about proving this. This is the definition that extends to other parts of geometry.) This will be a useful hypothesis in theorems, usually in conjunction with quasicompactness. (For this reason, "quasicompact and quasiseparated" is often abbreviated as **qcqs**; cf. §5.1.2.) Various interesting kinds of morphisms (locally Noetherian source, affine, separated—see Exercises 8.3.B(b), 8.3.D, and 11.2.D, resp.) are quasiseparated, and having the word "quasiseparated" will allow us to state theorems more succinctly. *Important remark:* We will give an equivalent definition of quasiseparatedness in §11.1.4 ("quasiseparated = quasicompact diagonal"), which will be much simpler to use.

8.3.A. EASY EXERCISE. Show that the composition of two quasicompact morphisms is quasicompact. (It is also true—but not easy—that the composition of two quasiseparated morphisms is quasiseparated. You are free to show this directly, but it will in any case follow easily once we understand it in a more sophisticated way; see §11.1.5.)

Following Grothendieck's philosophy of thinking that the important notions are properties of morphisms, not of objects (§8.0.1), we can restate the definition of a quasicompact (resp., quasiseparated) scheme as a scheme that is quasicompact (resp., quasiseparated) over the final object $\operatorname{Spec} \mathbb{Z}$ in the category of schemes (Exercise 7.3.J).

8.3.B. EASY EXERCISE.

(a) Show that any morphism from a Noetherian scheme is quasicompact.
(b) Show that any morphism from a quasiseparated scheme is quasiseparated. Thus by Exercise 5.3.A, any morphism from a locally Noetherian scheme is quasiseparated. Thus readers working only with locally Noetherian schemes may take quasiseparatedness as a standing hypothesis.

8.3.2. *Caution.* The two parts of Exercise 8.3.B may lead you to suspect that any morphism $\pi \colon X \to Y$ with quasicompact source and target is necessarily quasicompact. This is false, and you may verify that the following is a counterexample. Let Z be the nonquasiseparated scheme constructed in Exercise 5.1.K, and let $X = \operatorname{Spec} k[x_1, x_2, \dots]$ as in Exercise 5.1.K. The obvious open embedding $\pi \colon X \to Z$ (identifying X with one of the two pieces glued together to get Z) is not quasicompact. (But once you see the Cancellation Theorem 11.1.1, you will quickly see that any morphism from a quasicompact source to a *quasiseparated* target is necessarily quasicompact.)

8.3.C. EXERCISE. (Obvious hint for both parts: The Affine Communication Lemma 5.3.2.)

(a) *(Quasicompactness is affine-local on the target.)* Show that a morphism $\pi \colon X \to Y$ is quasicompact if there is a cover of Y by affine open sets U_i such that $\pi^{-1}(U_i)$ is quasicompact.
(b) *(Quasiseparatedness is affine-local on the target.)* Show that a morphism $\pi \colon X \to Y$ is quasiseparated if there is a cover of Y by affine open sets U_i such that $\pi^{-1}(U_i)$ is quasi-separated.

8.3.3. Affine morphisms.

A morphism $\pi \colon X \to Y$ is **affine** if for every affine open set U of Y, $\pi^{-1}(U)$ (interpreted as an open subscheme of X) is an affine scheme. Trivially, the composition of two affine morphisms is affine.

8.3.D. FAST EXERCISE. Show that affine morphisms are quasicompact and quasiseparated. (Hint for the latter: Exercise 5.1.G.)

8.3.4. Proposition (the property of "affineness" of a morphism is affine-local on the target) — *A morphism $\pi\colon X \to Y$ is affine if there is a cover of Y by affine open sets U such that $\pi^{-1}(U)$ is affine.*

Proof. We apply the Affine Communication Lemma 5.3.2 to the condition "π is affine over." We check our two criteria. First, suppose $\pi\colon X \to Y$ is affine over $\operatorname{Spec} B$, i.e., $\pi^{-1}(\operatorname{Spec} B) = \operatorname{Spec} A$. Then for any $s \in B$, $\pi^{-1}(\operatorname{Spec} B_s) = \operatorname{Spec} A_{\pi^\sharp s}$. (Do you see why?)

Second, suppose we are given $\pi\colon X \to \operatorname{Spec} B$ and $(s_1, \ldots, s_n) = B$ with $X_{\pi^\sharp s_i}$ affine ($\operatorname{Spec} A_i$, say). (Recall from Definition 6.2.8 that $X_{\pi^\sharp s_i}$ is the open subset of X where $\pi^\sharp s_i$ doesn't vanish.) We wish to show that X is affine, too. Let $A = \Gamma(X, \mathscr{O}_X)$. Then $X \to \operatorname{Spec} B$ factors through the tautological map $\alpha\colon X \to \operatorname{Spec} A$ (arising from the (iso)morphism $A \to \Gamma(X, \mathscr{O}_X)$, Exercise 7.3.G).

$$\cup_i X_{\pi^\sharp s_i} = X \xrightarrow{\;\;\alpha\;\;} \operatorname{Spec} A$$
$$\pi \searrow \qquad \swarrow \beta$$
$$\cup_i D(s_i) = \operatorname{Spec} B$$

We want to show that α is an isomorphism. Now $\beta^{-1}(D(s_i)) = D(\beta^\sharp s_i) \cong \operatorname{Spec} A_{\beta^\sharp s_i}$ (the preimage of a distinguished open set is a distinguished open set), and $\pi^{-1}(D(s_i)) = \operatorname{Spec} A_i$. Furthermore, X is quasicompact and quasiseparated by the affine-locality of these notions (Exercise 8.3.C), so the hypotheses of the Qcqs Lemma 6.2.9 are satisfied. Hence by the Qcqs Lemma 6.2.9, we have an induced isomorphism of $A_{\beta^\sharp s_i} = \Gamma(X, \mathscr{O}_X)_{\beta^\sharp s_i} \cong \Gamma(X_{\beta^\sharp s_i}, \mathscr{O}_X) = A_i$. Thus α induces an isomorphism $\operatorname{Spec} A_i \xrightarrow{\sim} \operatorname{Spec} A_{\beta^\sharp s_i}$ (an isomorphism of rings induces an isomorphism of affine schemes, Exercise 4.3.A). Thus α is an isomorphism over all $\operatorname{Spec} A_{\beta^\sharp s_i}$, which cover $\operatorname{Spec} A$, and hence α is an isomorphism. Therefore $X \to \operatorname{Spec} A$ is an isomorphism, so X is affine as desired. $\qquad\square$

The affine-locality of affine morphisms (Proposition 8.3.4) has some nonobvious consequences, as shown in the next exercise.

8.3.E. USEFUL EXERCISE. Suppose Z is a closed subset of an affine scheme $\operatorname{Spec} A$ locally cut out by one equation. (In other words, $\operatorname{Spec} A$ can be covered by smaller open sets, and on each such set Z is cut out by one equation.) Show that the complement Y of Z is affine. (This is clear if Z is globally cut out by one equation f, even set-theoretically; then $Y = \operatorname{Spec} A_f$. However, Z is not always of this form; see §19.11.10.)

8.3.5. Finite and integral morphisms.
Before defining finite and integral morphisms, we give an example to keep in mind. If L/K is a field extension, then $\operatorname{Spec} L \to \operatorname{Spec} K$ (i) is always affine; (ii) is integral if L/K is algebraic; and (iii) is finite if L/K is finite.

Recall that if we have a ring morphism $B \to A$ such that A is a finitely generated B-*module*, then we say that A is a **finite** B-algebra. This is stronger than its being a finitely generated B-*algebra*. (The similarity of the terminology "finite" and "finitely generated" B-algebra is unfortunate.) We say a morphism $\pi\colon X \to Y$ is **finite** if for every affine open set $\operatorname{Spec} B$ of Y, $\pi^{-1}(\operatorname{Spec} B)$ is the spectrum of a *finite* B-algebra. By definition, finite morphisms are affine.

8.3.F. EXERCISE (THE PROPERTY OF FINITENESS IS AFFINE-LOCAL ON THE TARGET). Show that a morphism $\pi\colon X \to Y$ is finite if there is a cover of Y by affine open sets $\operatorname{Spec} A$ such that $\pi^{-1}(\operatorname{Spec} A)$ is the spectrum of a finite A-algebra.

The following four examples will give you some feeling for finite morphisms. In each example, you will notice two things. In each case, the maps are always finite-to-one (as maps of sets). We will verify this in general in Exercise 8.3.J. You will also notice that the morphisms are **closed** as maps of topological spaces, i.e., the images of closed sets are closed. We will show

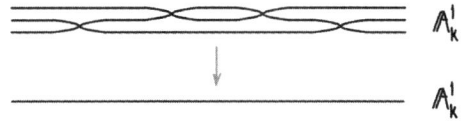

Figure 8.2 *The "branched cover"* $\mathbb{A}^1_k \to \mathbb{A}^1_k$ *of the "u-line" by the "t-line" given by* $u \mapsto p(t)$ *is finite.*

that finite morphisms are always closed in Exercise 8.3.L (and give a second proof in §9.3.5). Intuitively, you should think of finite as being closed plus finite fibers, although this isn't quite true. We will make this precise in Theorem 28.5.2.

Example 1: Branched covers. Consider the morphism $\operatorname{Spec} k[t] \to \operatorname{Spec} k[u]$ given by $u \mapsto p(t)$, where $p(t) \in k[t]$ is a degree n polynomial (see Figure 8.2). This is finite: $k[t]$ is generated as a $k[u]$-module by $1, t, t^2, \ldots, t^{n-1}$.

Example 2: Closed embeddings (to be defined soon, in §9.1.1). If I is an ideal of a ring A, consider the morphism $\operatorname{Spec} A/I \to \operatorname{Spec} A$ given by the obvious map $A \to A/I$ (see Figure 8.3 for an example, with $A = k[t]$, $I = (t)$). This is a finite morphism (A/I is generated as a A-module by the element $1 \in A/I$).

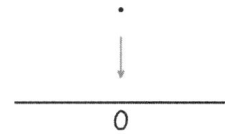

Figure 8.3 *The "closed embedding"* $\operatorname{Spec} k \to \operatorname{Spec} k[t]$ *given by* $t \mapsto 0$ *is finite.*

Example 3: Normalization (to be defined in §10.7). Consider the morphism $\operatorname{Spec} k[t] \to \operatorname{Spec} k[x, y]/(y^2 - x^2 - x^3)$ corresponding to $k[x, y]/(y^2 - x^2 - x^3) \to k[t]$ given by $x \mapsto t^2 - 1$, $y \mapsto t^3 - t$ (check that this is a well-defined ring map!); see Figure 8.4. This is a finite morphism, as $k[t]$ is generated as a $(k[x, y]/(y^2 - x^2 - x^3))$-module by 1 and t. (The figure suggests that this is an isomorphism away from the "node" of the target. You can verify this, by checking that it induces an isomorphism between $D(t^2 - 1)$ in the source and $D(x)$ in the target. We will meet this example again repeatedly!)

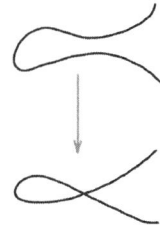

Figure 8.4 *The "normalization" of the nodal cubic* $\operatorname{Spec} k[t] \to \operatorname{Spec} k[x, y]/(y^2 - x^2 - x^3)$ *given by* $(x, y) \mapsto (t^2 - 1, t^3 - t)$ *is finite.*

8.3.G. IMPORTANT EXERCISE (EXAMPLE 4, FINITE MORPHISMS TO $\operatorname{Spec} k$). Show that if $X \to \operatorname{Spec} k$ is a finite morphism, then X is a finite union of points with the discrete topology, each point with residue field a finite extension of k; see Figure 8.5. (An example is $\operatorname{Spec}(\mathbb{F}_8 \times \mathbb{F}_4[x, y]/(x^2, y^4) \times \mathbb{F}_4[t]/(t^9) \times \mathbb{F}_2) \to \operatorname{Spec} \mathbb{F}_2$.) Do *not* just quote some fancy theorem! Possible approach: By Exercise 3.2.G, any integral domain that is a finite k-algebra must be a field. If $X = \operatorname{Spec} A$, show that every prime $\mathfrak{p}$ of A is maximal. Show that the irreducible components of $\operatorname{Spec} A$ are closed points. Show $\operatorname{Spec} A$ is finite, then discrete. Show that the residue fields $A/\mathfrak{m}$ of A are finite extensions of k. (See Exercise 8.4.D for an extension to quasifinite morphisms.)

Figure 8.5 *A picture of a finite morphism to* $\operatorname{Spec} k$. *Bigger fields are depicted as bigger points.*

8.3.H. EASY EXERCISE (CF. EXERCISE 8.2.C). Show that the composition of two finite morphisms is also finite.

8.3.I. EXERCISE ("FINITE MORPHISMS TO $\operatorname{Spec} A$ ARE PROJECTIVE"). If R is an A-algebra, define a graded ring $S_\bullet$ by $S_0 = A$, and $S_n = R$ for $n > 0$. (What is the multiplicative structure? Hint: You know how to multiply elements of R together, and how to multiply elements of A with elements of R.) Describe an isomorphism $\operatorname{Proj} S_\bullet \xrightarrow{\sim} \operatorname{Spec} R$. Show that if R is a *finite* A-algebra (finitely generated as an A-module), then $S_\bullet$ is a finitely generated graded ring over A, and that hence Spec R is a projective A-scheme (§4.5.10).

Hence finite morphisms are affine (by definition) and projective. The converse is also true; see Corollary 18.1.5.

8.3.J. IMPORTANT EXERCISE. Show that finite morphisms have finite fibers. (This is a useful exercise, because you will have to figure out how to get at points in a fiber of a morphism: given $\pi: X \to Y$, and $q \in Y$, what are the points of $\pi^{-1}(q)$? This will be easier to do once we discuss fibers in greater detail—see Remark 10.3.5—but it will be enlightening to do it now.) Hint: If $X = \operatorname{Spec} A$ and $Y = \operatorname{Spec} B$ are both affine, and $q = [\mathfrak{q}]$, then we can throw out everything in B outside $\overline{\mathfrak{q}}$ by modding out by $\mathfrak{q}$; show that the preimage is $\operatorname{Spec}(A/\pi^\sharp \mathfrak{q} A)$. Then you have reduced to the case where Y is the Spec of an integral domain B, and $[\mathfrak{q}] = [(0)]$ is the generic point. We can throw out the rest of the points of B by localizing at (0). Show that the preimage is Spec of A localized at $\pi^\sharp(B \setminus \{0\})$. Show that the condition of finiteness is preserved by the constructions you have done, and thus reduce the problem to Exercise 8.3.G.

The previous two exercises show that finite morphisms are projective and have finite fibers. The converse is also true; see Theorem 18.1.6.

There is more to finiteness than finite fibers, and three examples to keep in mind are Example 8.3.6, Exercise 8.3.K, and Figure 19.1 (a variant of Figure 8.4 showing a morphism that is affine, closed, one-to-one on points, and even a monomorphism, but not finite).

8.3.6. *Example.* The open embedding $\mathbb{A}^2_k \setminus \{(0,0)\} \to \mathbb{A}^2_k$ has finite fibers, but is not affine (as $\mathbb{A}^2_k - \{(0,0)\}$ isn't affine, §4.4.1) and hence not finite.

8.3.K. EASY EXERCISE. Show that the open embedding $\mathbb{A}^1_\mathbb{C} \setminus \{0\} \to \mathbb{A}^1_\mathbb{C}$ has finite fibers and is affine, but is not finite.

8.3.7. *Definition.* A morphism $\pi: X \to Y$ of schemes is **integral** if π is affine, and for every affine open subset $\operatorname{Spec} B \subset Y$, with $\pi^{-1}(\operatorname{Spec} B) = \operatorname{Spec} A$, the induced map $B \to A$ is an integral ring morphism. This is an affine-local condition on the target by Exercises 8.2.A and 8.2.B, and the Affine Communication Lemma 5.3.2. It is preserved by composition by Exercise 8.2.C. Integral morphisms are mostly useful because finite morphisms are integral by Corollary 8.2.2. Note that the converse implication doesn't hold (witness $\operatorname{Spec} \overline{\mathbb{Q}} \to \operatorname{Spec} \mathbb{Q}$, as discussed after the statement of Corollary 8.2.2).

8.3.L. EXERCISE. Prove that integral morphisms are closed, i.e., that the image of closed subsets is closed. (Hence finite morphisms are closed. A second proof will be given in §9.3.5.) Hint: Reduce to the affine case. If $\pi^\sharp: B \to A$ is a ring map, inducing the integral morphism $\pi: \operatorname{Spec} A \to \operatorname{Spec} B$, then suppose $I \subset A$ cuts out a closed set of $\operatorname{Spec} A$, and $J = (\pi^\sharp)^{-1}(I)$, then note that $B/J \subset A/I$, and apply the Lying Over Theorem 8.2.5.

8.3.M. UNIMPORTANT EXERCISE. Suppose $B \to A$ is integral. Show that for any ring morphism $B \to C$, the induced map $C \to A \otimes_B C$ is integral. (Hint: We wish to show that any $\sum_{i=1}^n a_i \otimes c_i \in A \otimes_B C$ is integral over C. Use the fact that each of the finitely many a_i are integral over B, and then Exercise 8.2.D.) Once we know what "base change" is, this will imply that the property of integrality of a morphism is preserved by base change, Exercise 10.4.D(e).

8.3.8. *Fibers of integral morphisms.* Unlike finite morphisms (Exercise 8.3.J), integral morphisms don't always have finite fibers. (Can you think of an example?) However, once we make sense of fibers as topological spaces (or even schemes) in §10.3.2, you can check (Exercise 12.1.E) that the fibers have the property that no point is in the closure of any other point.

8.3.9. Morphisms (locally) of finite type.

A morphism $\pi: X \to Y$ is **locally of finite type** if for every affine open set $\operatorname{Spec} B$ of Y, and every affine open subset $\operatorname{Spec} A$ of $\pi^{-1}(\operatorname{Spec} B)$, the induced morphism $B \to A$ expresses A as a finitely generated B-algebra. By the affine-locality of finite typeness of B-schemes (Proposition 5.3.3(b)), this is equivalent to: $\pi^{-1}(\operatorname{Spec} B)$ can be covered by affine open subsets $\operatorname{Spec} A_i$ so that each A_i is a finitely generated B-algebra.

A morphism π is **of finite type** if it is locally of finite type and quasicompact. Translation: For every affine open set $\operatorname{Spec} B$ of Y, $\pi^{-1}(\operatorname{Spec} B)$ can be covered with *a finite number of* open sets $\operatorname{Spec} A_i$ so that the induced morphism $B \to A_i$ expresses A_i as a finitely generated B-algebra.

8.3.10. *Linguistic curiosity.* It is a common practice to name properties as follows: P equals locally P plus quasicompact. Two exceptions are "ringed space" (§7.3) and "finite presentation" (§8.3.13).

Example: The "structure morphism" $\mathbb{P}^n_A \to \operatorname{Spec} A$ is of finite type, as $\mathbb{P}^n_A$ is covered by $n+1$ open sets of the form $\operatorname{Spec} A[x_1, \ldots, x_n]$.

Our earlier definition of schemes of "finite type over k" (or "finite type k-schemes") from §5.3.6 is now a special case of this more general notion: the phrase "a scheme X is of finite type over k" means that we are given a morphism $X \to \operatorname{Spec} k$ (the "structure morphism") that is of finite type.

Here are some properties enjoyed by morphisms of finite type.

8.3.N. EXERCISE (THE NOTIONS "LOCALLY FINITE TYPE" AND "FINITE TYPE" ARE AFFINE-LOCAL ON THE TARGET). Show that a morphism $\pi\colon X \to Y$ is locally of finite type if there is a cover of Y by affine open sets $\operatorname{Spec} B_i$ such that $\pi^{-1}(\operatorname{Spec} B_i)$ is locally finite type over B_i. Hence the notions of "locally finite type" and "finite type" are both affine-local on the target.

8.3.O. EXERCISE (FINITE = INTEGRAL + FINITE TYPE).

(a) (easier) Show that finite morphisms are of finite type.
(b) Show that a morphism is finite if and only if it is integral and of finite type.

8.3.P. EXERCISES (NOT HARD, BUT IMPORTANT).

(a) Show that every open embedding is locally of finite type, and hence that every quasicompact open embedding is of finite type. Show that every open embedding into a locally Noetherian scheme is of finite type.
(b) Show that the composition of two morphisms locally of finite type is locally of finite type. (Hence as the composition of two quasicompact morphisms is quasicompact, Easy Exercise 8.3.A, the composition of two morphisms of finite type is of finite type.)
(c) Suppose $\pi\colon X \to Y$ is locally of finite type, and Y is locally Noetherian. Show that X is also locally Noetherian. If $\pi\colon X \to Y$ is a morphism of finite type, and Y is Noetherian, show that X is Noetherian.

8.3.Q. EXERCISE. Suppose X is a scheme over $\mathbb{F}_p$. Explain how to define (without choice) an endomorphism $F\colon X \to X$ such that for each affine open subset $\operatorname{Spec} A \subset X$, F corresponds to the map $A \to A$ given by $f \mapsto f^p$ for all $f \in A$. (The morphism F is called the **absolute Frobenius morphism**.) Prove that if X is locally of finite type over $\mathbb{F}_p$, then F is finite.

8.3.11. *Less important definition.* A morphism $\pi\colon X \to Y$ is **quasifinite** if it is of finite type, and for all $q \in Y$, $\pi^{-1}(q)$ is a finite set. The main point of this definition is the "finite fiber" part; the "finite type" hypothesis will ensure that this notion is "preserved by fibered product," Exercise 10.4.F.

Combining Exercise 8.3.J with Exercise 8.3.O(a), we see that finite morphisms are quasifinite. There are quasifinite morphisms that are not finite, such as $\mathbb{A}^2_k \setminus \{(0,0)\} \to \mathbb{A}^2_k$ (Example 8.3.6). However, we will soon see that quasifinite morphisms to $\operatorname{Spec} k$ are finite (Exercise 8.4.D). A key example of a morphism with finite fibers that is not quasifinite is $\operatorname{Spec} \mathbb{C}(t) \to \operatorname{Spec} \mathbb{C}$. Another is $\operatorname{Spec} \overline{\mathbb{Q}} \to \operatorname{Spec} \mathbb{Q}$. (For interesting behavior caused by the fact that $\operatorname{Spec} \overline{\mathbb{Q}} \to \operatorname{Spec} \mathbb{Q}$ is not of finite type, see Warning 10.1.5.)

8.3.12. *How to picture quasifinite morphisms.* If $\pi\colon X \to Y$ is a finite morphism of locally Noetherian schemes, then for any quasicompact open subset $U \subset X$, the induced morphism $\pi|_U\colon U \to Y$ is quasifinite. In fact *every* reasonable quasifinite morphism arises in this way. (This simple-sounding statement is in fact a deep and important result—a form of Zariski's Main Theorem; see Exercise 28.5.F.) Thus the right way to visualize quasifiniteness is as a finite map with some (closed locus of) points removed.

8.3.13.** (Locally) finitely presented morphisms. The following variant of "locally of finite type" is useful in non-Noetherian situations.

8.3.14. *Definition.* A ring A is a **finitely presented** B-algebra (or $B \to A$ is **finitely presented**) if "A has a finite number of generators and a finite number of relations over B":

$$A \cong B[x_1,\ldots,x_n]/(f_1(x_1,\ldots,x_n),\ldots,f_r(x_1,\ldots,x_n)).$$

If B is Noetherian, then finitely presented is the same as finite type ("finite number of relations" comes for free, Exercise 3.6.S), so normal people will not care.

8.3.R. EXERCISE: FINITELY PRESENTED IMPLIES ALWAYS FINITELY PRESENTED (ALGEBRA VERSION). Suppose $A = B[x_1,\ldots,x_n]/(f_1,\ldots,f_r)$, where $f_i \in B[x_1,\ldots,x_n]$. Suppose also that $Y_1,\ldots,Y_m$ generate A as an B-algebra, so we have a surjection $\phi\colon B[y_1,\ldots,y_m] \twoheadrightarrow A$, where $y_j \mapsto Y_j$. Show that the ideal $\ker(\phi) \subset B[y_1,\ldots,y_m]$ is finitely generated. Hint: Follow the proof of Lemma 6.4.6.

8.3.15. Lemma (finite presentation of A over B is an affine-local property in A) — *The condition of a B-algebra being finitely presented satisfies the hypotheses of the Affine Communication Lemma 5.3.2.*

Proof. (i) The first hypothesis of the Affine Communication Lemma is the content of the following Exercise.

8.3.S. EXERCISE. Suppose $A = B[x_1,\ldots,x_n]/(f_1,\ldots,f_r)$, and $g \in A$. Show that A_g is a finitely presented B-algebra. (Hint: It has one more generator and one more relation.)

(ii) (This argument is the hardest part of our discussion on finite presentation.) We suppose that A is a B-algebra, and $g_1,\ldots,g_m \in A$ with $(g_1,\ldots,g_m) = A$, and A_{g_i} finitely presented over B. We wish to show that A is finitely presented over B. Now the A_{g_i} are finitely generated over B, so by Proposition 5.3.3(b), A is finitely generated over B, say $\phi\colon B[x_1,\ldots,x_n] \twoheadrightarrow A$. We will show that $I := \ker(\phi)$ is a finitely generated ideal of $B[x_1,\ldots,x_n]$.

As $(g_1,\ldots,g_m) = A$, we may choose $h_j \in A$ so that $1 = h_1 g_1 + \cdots + h_m g_m$. Choose lifts H_j and G_j of the h_j and g_j, respectively, to $B[x_1,\ldots,x_n]$. Define $G_0 = 1 - H_1 G_1 - \cdots - H_m G_m$, so $G_0 \in I$, and $(G_0, G_1,\ldots,G_m) = 1$ in $B[x_1,\ldots,x_n]$.

8.3.T. EXERCISE. Show that I_{G_j} is a finitely generated module over $B[x_1,\ldots,x_n]_{G_j} = B[x_1,\ldots,x_n,y_j]/(y_j G_j - 1)$. (The case $j = 0$ is different but easy—$I_{G_0} = (1)$!) Hint: Lemma 8.3.R.

Then by Exercise 6.4.A ("finite generation of modules is an affine-local condition"), I is a finitely generated $B[x_1,\ldots,x_n]$-module, i.e., a finitely generated ideal of $B[x_1,\ldots,x_n]$, as desired. $\qquad\square$

8.3.16. *Definition.* We say a B-scheme $X \to \operatorname{Spec} B$ is **locally finitely presented over** B if there exists an affine cover $X = \cup \operatorname{Spec} A_i$ such that A_i is finitely presented over B. This is equivalent (by Lemma 8.3.15) to A being finitely presented over B for *all* affine open subsets $\operatorname{Spec} A$ of X.

8.3.U. EXERCISE (THE NOTION OF "LOCALLY FINITELY PRESENTED MORPHISM" IS AFFINE-LOCAL IN THE TARGET). Suppose $\pi\colon X \to Y$ is a morphism of schemes. Show that as

Spec B ⊂ Y runs through the affine open subsets of Y, the condition that $\pi\colon \pi^{-1}(\operatorname{Spec} B) \to \operatorname{Spec} B$ is locally finitely presented over Spec B satisfies the hypotheses of the Affine Communication Lemma. (This is notably easier to show than Lemma 8.3.15.)

8.3.17. *Definition.* A morphism $\pi\colon X \to Y$ is **locally finitely presented** (or **locally of finite presentation**) if for every affine open set Spec B of Y, $\pi^{-1}(\operatorname{Spec} B) \to \operatorname{Spec} B$ is locally finitely presented over Spec B. By Exercise 8.3.U, it suffices to check a single affine cover of Y. (Hence this subsumes Definition 8.3.16.)

8.3.V. EXERCISE. Suppose $\pi\colon X \to Y$ is a morphism of schemes.

(a) Show that if π is locally of finite presentation, then for *any* open subscheme Spec B of Y and *any* affine open subscheme Spec A of X with $\pi(\operatorname{Spec} A) \subset \operatorname{Spec} B$, A is a finitely presented B-algebra.

(b) Suppose $\cup \operatorname{Spec} A_i = X$ is an affine open cover of X, and Spec B_i are affine open subsets of Y with $\pi(\operatorname{Spec} A_i) \subset \operatorname{Spec} B_i$. Suppose A_i is finitely presented over B_i. Show that π is locally finitely presented.

8.3.W. EXERCISE. Show that open embeddings are locally finitely presented.

8.3.X. EXERCISE. Show that the composition of two locally finitely presented morphisms is locally finitely presented.

8.3.18.** *Remark.* A morphism $\pi\colon X \to Y$ is locally finitely presented if and only if for every projective system of Y-schemes $\{S_\lambda\}_{\lambda \in I}$ with each S_λ an affine scheme, the natural map

$$\operatorname{colim}_\lambda \operatorname{Hom}_Y(S_\lambda, X) \longrightarrow \operatorname{Hom}_Y(\lim_\lambda S_\lambda, X)$$

is a bijection (see [Gr-EGA, IV$_3$.8.14.2]). This characterization of locally finitely presented morphisms as "limit-preserving" can be useful.

8.3.19. *Definition.* A morphism is **finitely presented** (or **of finite presentation**) if it is locally finitely presented, quasicompact, and quasiseparated. In particular, the notion of finite presented morphisms is affine-local on the target.

This definition violates the principle that erasing "locally" is the same as adding "quasicompact and" (Remark 8.3.10). But it is well motivated: finite presentation means "finite in all possible ways" (the algebra corresponding to each affine open set has a finite number of generators and a finite number of relations, and a finite number of such affine open sets cover, and their intersections are also covered by a finite number affine open sets)—it is all you would hope for in a scheme without it actually being Noetherian. Exercise 10.3.H makes this precise, and explains how this notion often arises in practice.

If Y is locally Noetherian, then locally of finite presentation is the same as locally of finite type, and finite presentation is the same as finite type. So if you are a Noetherian person, you don't need to worry about this notion.

8.4 Images of Morphisms: Chevalley's Theorem and Elimination Theory

In this section, we will answer a question that you may have wondered about long before hearing the phrase "algebraic geometry." If you have a number of polynomial equations in a number of variables with indeterminate coefficients, you would reasonably ask what conditions there are on the coefficients for a (common) solution to exist. Given the algebraic nature of the problem, you might hope that the answer should be purely algebraic in nature—it shouldn't be "random," or involve bizarre functions like exponentials or cosines. You should expect the answer to be given by "algebraic conditions." This is indeed the case, and it can be profitably interpreted as a question about images of maps of varieties or schemes, in which guise it is answered by Chevalley's

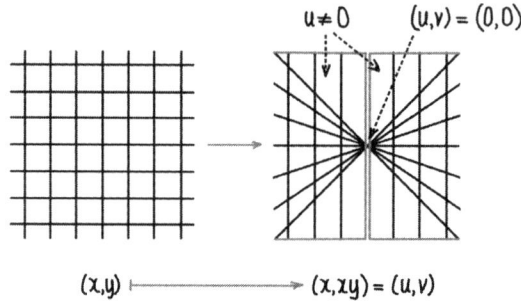

Figure 8.6 *The image of* $(x, y) \mapsto (x, xy) = (u, v)$.

Theorem 8.4.2 (see §8.4.8 for a more precise proof). As a consequence, we get an immediate proof of the Nullstellensatz 3.2.6 (§8.4.3).

In certain cases, the image is nicer still. For example, the image of a finite morphism is always closed (Exercise 8.3.L). We will prove a classical result, the Fundamental Theorem of Elimination Theory 8.4.10, which essentially generalizes this (as explained in §9.3.5) to maps from projective space.

In a different direction, in the distant future we will see that in certain good circumstances ("flat" plus a bit more; see Exercise 24.5.G), morphisms are open (the image of open subsets is open); one example (which you can try to show directly) is $\mathbb{A}_B^n \to \operatorname{Spec} B$.

8.4.1. Chevalley's Theorem.

If $\pi\colon X \to Y$ is a morphism of schemes, the notion of the image of π as *sets* is clear: we just take the points in Y that are the image of points in X. We know that the image can be open (open embeddings), and we have seen examples where it is closed, and more generally, locally closed. But it can be weirder still: consider the morphism $\mathbb{A}_k^2 \to \mathbb{A}_k^2$ given by $(x, y) \mapsto (x, xy)$. The image is the plane, with the y-axis removed, but the origin put back in (Figure 8.6). We make a definition to capture this phenomenon. A **constructible subset** of a Noetherian topological space is a subset that belongs to the smallest family of subsets such that (i) every open set is in the family, (ii) a finite intersection of family members is in the family, and (iii) the complement of a family member is also in the family. For example, the image of $(x, y) \mapsto (x, xy)$ is constructible.

8.4.A. EXERCISE: CONSTRUCTIBLE SUBSETS ARE FINITE DISJOINT UNIONS OF LOCALLY CLOSED SUBSETS. Recall that a subset of a topological space X is **locally closed** if it is the intersection of an open subset and a closed subset. (Equivalently, it is an open subset of a closed subset, or a closed subset of an open subset. We will later have trouble extending this to open and closed and locally closed subschemes; see Exercise 9.2.C.) Show that a subset of a Noetherian topological space X is constructible if and only if it is the finite disjoint union of locally closed subsets.

As a consequence, if $X \to Y$ is a continuous map of Noetherian topological spaces, then the preimage of a constructible set is a constructible set. (Important remark: The only reason for the hypothesis of the topological space in question being Noetherian is that this is the only setting in which we have defined constructible sets. An extension of the notion of constructibility to more general topological spaces is mentioned in Exercise 10.3.I.)

8.4.B. EASY EXERCISE (USED IN §8.4.3). Show that the generic point of $\mathbb{A}_k^1$ does not form a constructible subset of $\mathbb{A}_k^1$ (where k is a field).

8.4.C. EXERCISE (USED IN EXERCISE 24.5.G).

(a) Show that a constructible subset of a Noetherian scheme is closed if and only if it is "stable under specialization." More precisely, if Z is a constructible subset of a Noetherian scheme X, then Z is closed if and only if for every pair of points y_1 and y_2 with $y_1 \in \overline{y_2}$, if $y_2 \in Z$, then $y_1 \in Z$. Hint for the "if" implication: Show that Z can be written as $\coprod_{i=1}^{n} U_i \cap Z_i$, where $U_i \subset X$ is open and $Z_i \subset X$ is closed. Show that Z can be written as $\coprod_{i=1}^{n} U_i \cap Z_i$, (with possibly different n, U_i, Z_i), where each Z_i is irreducible and meets U_i. Now use "stability under specialization" and the generic point of Z_i to show that $Z_i \subset Z$ for all i, so $Z = \cup Z_i$.

(b) Show that a constructible subset of a Noetherian scheme is open if and only if it is "stable under generization." (Hint: This follows in one line from (a).)

The image of a morphism of schemes can be stranger than a constructible set. Indeed, if S is *any* subset of a scheme Y, it can be the image of a morphism: let X be the disjoint union of spectra of the residue fields of all the points of S, and let $\pi\colon X \to Y$ be the natural map. This is quite pathological, but in any reasonable situation, the image is essentially no worse than what arose in the previous example of $(x, y) \mapsto (x, xy)$. This is made precise by Chevalley's Theorem.

8.4.2. Chevalley's Theorem — *If $\pi\colon X \to Y$ is a finite type morphism of Noetherian schemes, the image of any constructible set is constructible. In particular, the image of π is constructible.*

(For the minority who might care: see §10.3.8 for an extension to locally finitely presented morphisms.) We discuss the proof after giving some important consequences that may seem surprising, in that they are algebraic corollaries of a seemingly quite geometric and topological theorem. The first is a proof of the Nullstellensatz.

8.4.3. *Proof of the Nullstellensatz 3.2.6.* We wish to show that if K is a field extension of k that is finitely generated as a k-algebra, say, by $x_1, \ldots, x_n$, then it is a finite extension of fields. It suffices to show that each x_i is algebraic over k. But if any x_i is not algebraic over k, then we have an inclusion of rings $k[x_i] \to K$, corresponding to a dominant morphism $\pi\colon \operatorname{Spec} K \to \mathbb{A}_k^1$ of finite type k-schemes. Of course, $\operatorname{Spec} K$ is a single point, so the image of π is one point. By Chevalley's Theorem 8.4.2 and Exercise 8.4.B, the image of π is not the generic point of $\mathbb{A}_k^1$, so $\operatorname{im}(\pi)$ is a closed point of $\mathbb{A}_k^1$, and thus π is not dominant. $\qquad \square$

A similar idea can be used in the following exercise.

8.4.D. EXERCISE (QUASIFINITE MORPHISMS TO A FIELD ARE FINITE). Suppose $\pi\colon X \to \operatorname{Spec} k$ is a quasifinite morphism. Show that π is finite. (Hint: Deal first with the affine case, $X = \operatorname{Spec} A$, where A is finitely generated over k. Suppose A contains an element x that is not algebraic over k, i.e., we have an inclusion $k[x] \hookrightarrow A$. Exercise 8.3.G may help.)

8.4.E. EXERCISE. Suppose X is a scheme of finite type over $\operatorname{Spec} \mathbb{Z}$, and p is a closed point of X. Show that the residue field $\kappa(p)$ is a finite field.

8.4.F. EXERCISE (FOR MAPS OF VARIETIES, SURJECTIVITY CAN BE CHECKED ON CLOSED POINTS). Assume Chevalley's Theorem 8.4.2. Show that a morphism of affine k-varieties $\pi\colon X \to Y$ is surjective if and only if it is surjective on closed points (i.e., if every closed point of Y is the image of a closed point of X). (Once we define varieties in general, in Definition 11.2.9, you will see that your argument works without change with the adjective "affine" removed.)

8.4.4. Proof of Chevalley's Theorem 8.4.2.

8.4.G. EXERCISE.

(a) Reduce the proof of Chevalley's Theorem to the case where Y is affine, say, $Y = \operatorname{Spec} B$.
(b) Reduce further to the case where X is affine, say, $X = \operatorname{Spec} B[t_1, \ldots, t_n]/I$.
(c) Reduce further to the case where $X = \mathbb{A}_B^n = \operatorname{Spec} B[t_1, \ldots, t_n]$.
(d) Reduce further to the case where $X = \mathbb{A}_B^1 = \operatorname{Spec} B[t]$.

We have simplified the task to showing that given $\pi\colon X = \operatorname{Spec} B[t] \to \operatorname{Spec} B = Y$, the image of any constructible subset of X is constructible in Y. Because constructible sets are finite unions of locally closed subsets (Exercise 8.4.A—we are in the Noetherian situation), we need only show that the image of any *locally closed* subset Z of X is constructible in Y.

8.4.5. *An important special case.* We begin with the case when Z is a *closed* subset of $X = \mathbb{A}^1_B$, say,

$$Z = V(f_1(t), \dots, f_r(t)), \quad \text{where} \quad f_i(t) = a_{id_i} t^{d_i} + \dots + a_{i1} t + a_{i0},$$

with $a_{ij} \in B$, so $f_i(t) \in B[t]$ and $d_i = \deg f_i(t)$.

Now $Y = \operatorname{Spec} B$ is the finite disjoint union of locally closed subsets $Y_{\vec{e}}$ indexed by $\vec{e} = (e_1, \dots, e_r)$, where $e_i \in \mathbb{Z}$, $-1 \le e_i \le d_i$, and $Y_{\vec{e}}$ is the locus where $f_1(t), \dots, f_r(t)$ have "true degrees" $e_1, \dots, e_r$, respectively: the coefficient of t^{e_i} in $f_i(t)$ is nonzero, and the "higher" coefficients are zero (at the points of $Y_{\vec{e}}$). More precisely: $Y_{\vec{e}} \subset Y$ is the locus where $a_{ie_i} \ne 0$, and $a_{ij} = 0$ for $j > e_i$. (The zero polynomial has "true degree" -1 under this convention.) More precisely still, we define $Y_{\vec{e}} := \operatorname{Spec} B_{\vec{e}}$, where

$$B_{\vec{e}} := \frac{B_{\prod_i a_{ie_i}}}{(a_{ij})_{j > e_i}}, \quad \text{so} \quad Y_{\vec{e}} = (\cap_i D(a_{ie_i})) \cap \left(\bigcap_{\substack{i \\ j > e_i}} V(a_{ij}) \right).$$

If $\vec{e} = (-1, \dots, -1)$, then

(8.4.5.1) $$\pi^{-1}(Y_{\vec{e}}) \xrightarrow{\ \pi\ } Y_{\vec{e}}$$

is surjective (the map (8.4.5.1) is $\operatorname{Spec} B_{(-1,\dots,-1)}[t] \to \operatorname{Spec} B_{(-1,\dots,-1)}$), and if $\vec{e} \ne (-1, \dots, -1)$, (8.4.5.1) is a finite morphism corresponding to the map of rings

$$B_{\vec{e}} \longrightarrow B_{\vec{e}}[t] / \left(t^{e_1} + \frac{a_{1(e_1-1)}}{a_{1e_1}} t^{e_1-1} + \dots + \frac{a_{10}}{a_{1e_1}}, \dots, t^{e_r} + \frac{a_{r(e_r-1)}}{a_{re_r}} t^{e_r-1} + \dots + \frac{a_{r0}}{a_{re_r}} \right).$$

(The ring on the right is generated as a $B_{\vec{e}}$-module by $1, t, \dots, t^{\max(e_i)}$.) Because the images of finite morphisms are closed (Exercise 8.3.L), $\pi(Z) \cap Y_{\vec{e}}$ (the image of (8.4.5.1)) is a locally closed subset of Y, so $\pi(Z)$ is indeed constructible when Z is closed. We have completed our important special case.

8.4.6. Note that the locus of points $q \in Y$ where the entire fiber $\pi^{-1}(q)$ is contained in Z, $\{q \in Y : \pi^{-1}(q) \subset Z\}$, is a *closed* subset of Y: it is $\operatorname{Spec} B_{(-1,\dots,-1)} = \operatorname{Spec} B/(a_{ij})$. We call this $\operatorname{FL}(Z) \subset Y$, the "fibral locus for Z," only for the rest of this proof.

8.4.7. Finally, if Z is *locally closed*, then $\overline{Z} \subset X$ is closed, so $\pi(\overline{Z})$ is a constructible subset of $Y = \operatorname{Spec} B$. Define $\delta Z := \overline{Z} \setminus Z$, a closed subset of X; $\operatorname{FL}(\overline{Z})$ and $\operatorname{FL}(\delta Z)$ are both closed subsets of Y by §8.4.6. Since $\operatorname{FL}(\overline{Z}) \setminus \operatorname{FL}(\delta Z)$ is contained in $\pi(Z)$, and $\operatorname{FL}(\delta Z)$ is contained in the complement of $\pi(Z)$, it suffices to prove the result for the open subset $Y' := Y \setminus (\operatorname{FL}(\overline{Z}) \cup \operatorname{FL}(\delta Z)) \subset Y$. Let $\pi' : X' = \mathbb{A}^1_{Y'} \to Y'$ be the restriction of π above Y', and let $Z' = Z \cap \mathbb{A}^1_{Y'}$ (which is quasifinite above Y'), and $\delta Z' = \overline{Z'} \setminus Z' \subset \mathbb{A}^1_{Y'}$ (where $\overline{Z'}$ is the closure of Z' in $\mathbb{A}^1_{Y'}$).

8.4.H. EXERCISE. Show that the closed subset $\delta Z' \subset \mathbb{A}^1_{Y'}$ does not meet any generic fiber of π', i.e., the preimage of any generic point of Y'. Hence show that there is a dense open subset of Y' not intersecting the constructible set $\pi'(\delta Z')$.

We have now proved the result above a dense open subset of $\operatorname{Spec} B$.

8.4.I. EXERCISE. Finish the proof of Chevalley's Theorem 8.4.2 by invoking Noetherian induction (in a suitably rigorous manner). $\square$

8.4.8.* **Elimination of quantifiers.** A basic sort of question that arises in any number of contexts is when a system of equations has a solution. Suppose, for example, you have some polynomials in variables $x_1, \dots, x_n$ over an algebraically closed field $\overline{k}$, some of which you set to be zero, and some of which you set to be nonzero. (This question is of fundamental interest even before you know any scheme theory!) Then there is an algebraic condition on the coefficients that will tell you if there is a solution. Define the **Zariski topology on** $\overline{k}^n$ in the obvious way: Closed subsets are cut out by equations. (A mild generalization of this appears in Exercise 12.2.L.)

8.4.J. EXERCISE (ELIMINATION OF QUANTIFIERS, OVER AN ALGEBRAICALLY CLOSED FIELD). Fix an algebraically closed field $\overline{k}$. Suppose

$$f_1, \ldots, f_p, g_1, \ldots, g_q \in \overline{k}[W_1, \ldots, W_m, X_1, \ldots X_n]$$

are given. Show that there is a (Zariski-)constructible subset Y of $\overline{k}^m$ such that

$$(8.4.8.1) \qquad f_1(w_1, \ldots, w_m, X_1, \ldots, X_n) = \cdots = f_p(w_1, \ldots, w_m, X_1, \ldots, X_n) = 0$$

and

$$(8.4.8.2) \qquad g_1(w_1, \ldots, w_m, X_1, \ldots, X_n) \neq 0 \quad \cdots \quad g_q(w_1, \ldots, w_m, X_1, \ldots, X_n) \neq 0$$

has a solution $(X_1, \ldots, X_n) = (x_1, \ldots, x_n) \in \overline{k}^n$ if and only if $(w_1, \ldots, w_m) \in Y$. Hints: If Z is a finite type scheme over $\overline{k}$, and the closed points are denoted by Z^{cl} ("cl" is for either "closed" or "classical"), then under the inclusion of topological spaces $Z^{cl} \hookrightarrow Z$, the Zariski topology on Z induces the Zariski topology on Z^{cl}. Note that we can identify $(\mathbb{A}_{\overline{k}}^p)^{cl}$ with $\overline{k}^p$ by the Nullstellensatz (Exercise 5.3.F). If X is the locally closed subset of $\mathbb{A}^{m+n}$ cut out by the equalities and inequalities (8.4.8.1) and (8.4.8.2), we have the diagram

$$(8.4.8.3)$$

$$
\begin{array}{ccc}
X^{cl} \ \hookrightarrow & X \xrightarrow{\;\text{loc. cl.}\;} & \mathbb{A}^{m+n} \\
{\scriptstyle \pi^{cl}} \downarrow & {\scriptstyle \pi} \downarrow & \swarrow \\
\overline{k}^m \ \hookrightarrow & \mathbb{A}^m &
\end{array}
$$

where $Y = \operatorname{im} \pi^{cl}$. By Chevalley's Theorem 8.4.2, $\operatorname{im} \pi$ is constructible, and hence so is $(\operatorname{im} \pi) \cap \overline{k}^m$. It remains to show that $(\operatorname{im} \pi) \cap \overline{k}^m = Y \,(= \operatorname{im} \pi^{cl})$. You might use the Nullstellensatz.

This is called "elimination of quantifiers" because it gets rid of the quantifier "there exists a solution." The analogous statement for real numbers, where inequalities are also allowed, is a special case of Tarski's celebrated theorem of elimination of quantifiers for real closed fields (see, for example, [Ta]).

8.4.9. The Fundamental Theorem of Elimination Theory.

In the case of projective space (and later, projective morphisms), one can do better than Chevalley (at least in describing images of *closed* subsets).

8.4.10. Theorem (Fundamental Theorem of Elimination Theory) — *The morphism* $\pi \colon \mathbb{P}_A^n \to \operatorname{Spec} A$ *is* **closed** *(sends closed sets to closed sets).*

Note that *no* Noetherian hypotheses are needed.

A great deal of classical algebra and geometry is contained in this theorem as special cases. Here are some examples.

First, let $A = k[a, b, c, \ldots, i]$, and consider the closed subset of $\mathbb{P}_A^2$ (taken with coordinates x, y, z) corresponding to $ax + by + cz = 0$, $dx + ey + fz = 0$, $gx + hy + iz = 0$. Then we are looking for the locus in $\operatorname{Spec} A$ where these equations have a nontrivial solution. This indeed corresponds to a Zariski-closed set—where

$$\det \begin{pmatrix} a & b & c \\ d & e & f \\ g & h & i \end{pmatrix} = 0.$$

Thus the idea of the determinant is embedded in elimination theory.

As a second example, let $A = k[a_0, a_1, \ldots, a_m, b_0, b_1, \ldots, b_n]$. Now consider the closed subset of $\mathbb{P}_A^1$ (taken with coordinates x and y) corresponding to $a_0 x^m + a_1 x^{m-1} y + \cdots + a_m y^m = 0$ and $b_0 x^n + b_1 x^{n-1} y + \cdots + b_n y^n = 0$. Then there is a polynomial in the coefficients $a_0, \ldots, b_n$ (an element of A) that vanishes if and only if these two polynomials have a common nonzero root—this polynomial is called the *resultant*.

More generally, given a number of homogeneous equations in $n + 1$ variables with indeterminate coefficients, Theorem 8.4.10 implies that one can write down equations in the coefficients that precisely determine when the equations have a nontrivial solution.

8.4.11. *Proof of the Fundamental Theorem of Elimination Theory 8.4.10.* Suppose $Z \hookrightarrow \mathbb{P}^n_A$ is a closed subset. We wish to show that $\pi(Z)$ is closed.

By the definition of the Zariski topology on $\text{Proj}\, A[x_0, \ldots, x_n]$ (§4.5.7), Z is cut out (set-theoretically) by some homogeneous elements $f_1, f_2, \ldots \in A[x_0, \ldots, x_n]$. We wish to show that the points $p \in \text{Spec}\, A$ that are in $\pi(Z)$ form a closed subset. Equivalently, we want to show that those p for which $f_1, f_2, \ldots$ have a common zero in $\text{Proj}\, \kappa(p)[x_0, \ldots, x_n]$ form a closed subset of $\text{Spec}\, A$.

To motivate our argument, we consider a related question. Suppose that $S_\bullet := k[x_0, \ldots, x_n]$, and that $g_1, g_2, \ldots \in k[x_0, \ldots, x_n]$ are homogeneous polynomials. How can we tell if $g_1, g_2, \ldots$ have a common zero in $\text{Proj}\, S_\bullet = \mathbb{P}^n_k$?

They would have a common zero if and only if in

$$\mathbb{A}^{n+1}_k = \text{Spec}\, S_\bullet,$$

they cut out (set-theoretically) more than the origin, i.e., if set-theoretically

$$V(g_1, g_2, \ldots) \not\subset V(x_0, \ldots, x_n).$$

By the inclusion-reversing bijection between closed subsets and radical ideals (Theorem 3.7.1), this is true if and only if

$$\sqrt{(g_1, g_2, \ldots)} \not\supset \sqrt{(x_0, \ldots, x_n)}.$$

But $(x_0, \ldots, x_n)$ is radical, so this is true if and only if

$$\sqrt{(g_1, g_2, \ldots)} \not\supset (x_0, \ldots, x_n).$$

This is true if and only if not all of the generators $x_0, \ldots, x_n$ of the ideal $(x_0, \ldots, x_n)$ are in $\sqrt{(g_1, g_2, \ldots)}$, which is true if only if there is some i such that no power of x_i is in $(g_1, g_2, \ldots)$. This is true if and only if

$$(x_0, \ldots, x_n)^N \not\subset (g_1, g_2, \ldots) \quad \text{for all N.}$$

(Do you see why this implication goes both ways?) This is equivalent to

$$S_N \not\subset (g_1, g_2, \ldots) \quad \text{for all N,}$$

which may be rewritten as

(8.4.11.1) $$S_N \not\subset g_1 S_{N - \deg g_1} \oplus g_2 S_{N - \deg g_2} \oplus \cdots$$

for all N. In other words, this is equivalent to the statement that the k-linear map

$$S_{N - \deg g_1} \oplus S_{N - \deg g_2} \oplus \cdots \longrightarrow S_N$$

is not surjective. This map is given by a matrix with $\dim S_N$ rows. (It may have an infinite number of columns, but this will not bother us.) To check that this linear map is not surjective, we need only check that all the "maximal" ($\dim S_N \times \dim S_N$) determinants are zero. (Of course, we need to check this for all N.) Thus the condition that $g_1, g_2, \ldots$ have no common zeros in $\mathbb{P}^n_k$ is the same as checking some (admittedly infinite) number of equations. In other words, it is a Zariski-closed condition on the coefficients of the polynomials $g_1, g_2, \ldots$.

8.4.K. EXERCISE. Complete the proof of the Fundamental Theorem of Elimination Theory 8.4.10. (Hint: Follow precisely the same argument, with k replaced by A, and the g_i replaced by the f_i. How and why does this prove the theorem? If the previous sentence is mysterious, you may want to wait until you reach §10.3.2, or perhaps glance there now.) □

Notice that projectivity was crucial to the proof: we used graded rings in an essential way. Notice also that the proof is essentially just linear algebra.

Chapter 9

Closed Embeddings and Related Notions

The scheme-theoretic analog of closed subsets has a surprisingly different flavor from the analog of open sets (open subschemes/embeddings). However, just as open subschemes (the scheme-theoretic version of open set) are locally modeled on open sets $U \subset Y$, the analog of closed subsets also has a local model. This was foreshadowed by our understanding of closed subsets of Spec B as roughly corresponding to ideals. If $I \subset B$ is an ideal, then Spec $B/I \hookrightarrow$ Spec B is a morphism of schemes, and we have checked that on the level of topological spaces, this describes Spec B/I as a closed subset of Spec B, with the subspace topology (Exercise 3.4.I). This morphism is our "local model" of a closed subscheme/embedding.

9.1 Closed Embeddings and Closed Subschemes

9.1.1. *Definition*. A morphism $\pi\colon X \to Y$ is a **closed embedding** (or **closed immersion**) if it is an affine morphism, and for every affine open subset Spec $B \subset Y$, with $\pi^{-1}(\text{Spec }B) \cong \text{Spec }A$, the map $B \to A$ is surjective (i.e., of the form $B \to B/I$, our desired local model). ("Embedding" is preferable to "immersion," because the differential geometric notion of immersion is closer to what algebraic geometers call unramified, which we will define in §21.7.) The symbol $\hookrightarrow$ often is used to indicate that a morphism is a closed embedding (or, more generally, a locally closed embedding, §9.2).

If X is a *subset* of Y (and π on the level of sets is the inclusion), we say that X is a **closed subscheme** of Y. A closed embedding is the same thing as an isomorphism with a closed subscheme.

9.1.A. EXERCISE. Show that a closed embedding identifies the topological space of X with a closed subset of the topological space of Y. (Caution: The closed embeddings Spec $k[x]/(x) \hookrightarrow$ Spec $k[x]$ and Spec $k[x]/(x^2) \hookrightarrow$ Spec $k[x]$ show that the closed subset does not determine the closed subscheme. The "infinitesimal" information, or "fuzz," is lost.)

9.1.B. EASY EXERCISE. Show that closed embeddings are finite morphisms, hence of finite type.

9.1.C. EASY EXERCISE. Show that the composition of two closed embeddings is a closed embedding.

9.1.D. EXERCISE. Show that the property of being a closed embedding is affine-local on the target.

In particular, if $B \to A$ is a surjection of rings, then the induced morphism Spec $A \to$ Spec B is a closed embedding.

9.1.2. Important: Closed subschemes correspond to quasicoherent sheaves of ideals. We call a sub-$\mathscr{O}_Y$-module of $\mathscr{O}_Y$ an **ideal sheaf** (or *sheaf of ideals*) on Y. A closed subscheme $\pi\colon X \hookrightarrow Y$ gives a surjection of $\mathscr{O}_Y$-modules $\mathscr{O}_Y \to \pi_* \mathscr{O}_X$. The kernel $\mathscr{I}_{X/Y}$ of this map is then a sheaf of ideals on Y. On each open subset U, it gives an ideal $\mathscr{I}_{X/Y}(U)$ of the ring $\mathscr{O}_Y(U)$; on the affine open subset Spec B, it gives the ideal I of B determining the surjection $B \to A = B/I$ in the definition of "closed subscheme."

9.1.E. EXERCISE. Show that $\pi_* \mathscr{O}_X$ is a quasicoherent sheaf on Y. (One way of proceeding is to use Exercise 6.2.D.)

Quasicoherent sheaves form an abelian category, so $\mathscr{I}_{X/Y} = \ker(\mathscr{O}_Y \to \pi_* \mathscr{O}_X)$ is also a quasi-coherent sheaf on Y. Thus every closed subscheme of Y yields a *quasicoherent* sheaf of ideals on Y. We call the exact sequence (of quasicoherent sheaves on Y)

(9.1.2.1) $$0 \longrightarrow \mathscr{I}_{X/Y} \longrightarrow \mathscr{O}_Y \longrightarrow \pi_* \mathscr{O}_X \longrightarrow 0$$

the **closed subscheme exact sequence** for $\pi : X \hookrightarrow Y$.

We recover X (and its map π to Y) from the ideal sheaf $\mathscr{I}_{X/Y}$ via (9.1.2.1): for each affine open subset $\operatorname{Spec} B \subset Y$, the sections over $\operatorname{Spec} B$ give us an exact sequence

$$0 \longrightarrow I \longrightarrow B \longrightarrow B/I \longrightarrow 0,$$

and $\operatorname{Spec} B/I$ is $\pi^{-1}(\operatorname{Spec} B)$.

The converse holds:

9.1.F. IMPORTANT EXERCISE. Prove that every quasicoherent sheaf of ideals on Y produces a closed subscheme on Y in this way.

This is a hard exercise, so as a hint, here are two different ways of proceeding; some combination of them may work for you. *Approach 1.* For each affine open $\operatorname{Spec} B$, we have a closed subscheme $\operatorname{Spec} B/I(B) \hookrightarrow \operatorname{Spec} B$. (i) For any two affine open subschemes $\operatorname{Spec} A$ and $\operatorname{Spec} B$, show that the two closed subschemes $\operatorname{Spec} A/I(A) \hookrightarrow \operatorname{Spec} A$ and $\operatorname{Spec} B/I(B) \hookrightarrow \operatorname{Spec} B$ restrict to the *same* closed subscheme of their intersection. (Hint: Cover their intersection with open sets simultaneously distinguished in both affine open sets, Proposition 5.3.1.) Thus, for example, we can glue these two closed subschemes together to get a closed subscheme of $\operatorname{Spec} A \cup \operatorname{Spec} B$. (ii) Use Exercise 4.4.A on gluing schemes (or the ideas therein) to glue together the closed embeddings in all affine open subschemes simultaneously. You will only need to worry about triple intersections. *Approach 2.* (i) Describe X first as a subset of Y. (ii) Check that X is closed. (iii) Define the sheaf of functions $\mathscr{O}_X$ on this subset, perhaps using compatible germs. (iv) Check that this resulting ringed space is indeed locally the closed subscheme given by $\operatorname{Spec} B/I \hookrightarrow \operatorname{Spec} B$.

9.1.G. UNIMPORTANT EXERCISE: A SHEAF OF IDEALS (AS $\mathscr{O}$-MODULES) THAT IS NOT QUASICOHERENT. Let $X = \operatorname{Spec} k[x]_{(x)}$, the "germ of the affine line at the origin," which has two points, the closed point and the generic point η. Define $\mathscr{I}(X) = \{0\} \subset \mathscr{O}_X(X) = k[x]_{(x)}$ and $\mathscr{I}(\eta) = k(x) = \mathscr{O}_X(\eta)$. Show that this sheaf of ideals does not correspond to a closed subscheme.

9.1.H. ** HARD (AND FRANKLY UNIMPORTANT) EXERCISE (NOT USED LATER). In the literature, the usual definition of a closed embedding is a morphism $\pi : X \to Y$ such that π induces a homeomorphism of the underlying topological space of X onto a closed subset of the topological space of Y, and the induced map $\pi^\sharp : \mathscr{O}_Y \to \pi_* \mathscr{O}_X$ of sheaves on Y is surjective. (By "surjective" we mean that the ring morphism on stalks is surjective.) Show that this definition agrees with the one given above. (To show that our definition involving surjectivity on the level of affine open sets implies this definition, you can use the fact that surjectivity of a morphism of sheaves can be checked on a suitably chosen base, Exercise 2.5.D.)

We have now defined the analog of open subsets and closed subsets in the land of schemes. Their definition is slightly less "symmetric" than in the classical topological setting: the "complement" of a closed subscheme is a unique open subscheme, but there are many "complementary" closed subschemes to a given open subscheme in general. (We will soon define one that is "best," that has a reduced structure, §9.4.9.)

9.1.3. Examples and properties.

9.1.I. IMPORTANT EXERCISE.

(a) In analogy with closed subsets, define the notion of a **finite (scheme-theoretic) union of closed subschemes** of X, and an arbitrary (not necessarily finite) **(scheme-theoretic)**

intersection of closed subschemes of X. (Exercise 9.1.F may help.) Hint: If X is affine, then you might expect that the union of closed subschemes corresponding to I_1 and I_2 would be the closed subscheme corresponding to either $I_1 \cap I_2$ or $I_1 I_2$ — but which one? We would want the union of a closed subscheme with itself to *be* itself, so the right choice is $I_1 \cap I_2$.

(b) Describe the scheme-theoretic intersection of $V(y - x^2)$ and $V(y)$ in $\mathbb{A}^2$. See Figure 4.5 for a picture. (For example, explain informally how this corresponds to two curves meeting at a single point with multiplicity 2—notice how the 2 is visible in your answer. Alternatively, what is the nonreducedness telling you about the intersection—both its "size" and its "direction"?) Describe their scheme-theoretic union.

(c) Show that the underlying set of a finite union of closed subschemes is the finite union of the underlying sets, and similarly for arbitrary intersections.

(d) Describe the scheme-theoretic intersection of $V(y^2 - x^2)$ and $V(y)$ in $\mathbb{A}^2$. Draw a picture. (Did you expect the intersection to have multiplicity 1 or multiplicity 2?) Hence show that if X, Y, and Z are closed subschemes of W, then $(X \cap Z) \cup (Y \cap Z) \neq (X \cup Y) \cap Z$ in general. In particular, not all properties of intersection and union carry over from sets to schemes.

9.1.J. EXERCISE. Show that closed embeddings (like open embeddings, Easy Exercise 8.1.E) are monomorphisms.

In many cases, loci that are a priori closed sets often have enriched versions as closed subschemes. The remaining exercises are examples of this. (As described in §6.6.33, we often indicate that we are considering scheme-theoretic enrichments by adding the adjective *scheme-theoretic*—for example, we say "scheme-theoretic intersection," "scheme-theoretic union," and so forth.)

9.1.K. IMPORTANT EXERCISE/DEFINITION: THE VANISHING SCHEME (CF. EXERCISE 4.5.S).

(a) Suppose Y is a scheme, and $s \in \Gamma(Y, \mathscr{O}_Y)$. Define the closed subscheme **cut out by** s. We call this the **vanishing scheme** $V(s)$ of s, as it is the scheme-theoretic version of our earlier (set-theoretical) version of $V(s)$ (§3.4.1). (Hint: On affine open Spec B, we just take Spec $B/(s_B)$, where s_B is the restriction of s to Spec B. Use Exercise 9.1.F to show that this yields a well-defined closed subscheme.)

(b) If u is an invertible function, show that $V(s) = V(su)$.

(c) If S is a set of functions, define $V(S)$.

9.1.4. *Definition: Scheme-theoretic support.* If M is an A-module, and $m \in M$, we call the closed subscheme $\text{Spec}(A/\text{Ann } m)$ of Spec A the **scheme-theoretic support** of m. (This is often described as the "smallest closed subscheme on which m is defined." Can you make this precise?)

9.1.L. EXERCISE. Show that the underlying set of the scheme-theoretic support of m is the usual "set-theoretic" support, Supp m.

9.1.M. EXERCISE. Define the **scheme-theoretic support of a finitely generated A-module** M to be the scheme-theoretic intersection of all closed subschemes Spec $A/I \hookrightarrow$ Spec A for which M is an (A/I)-module. Show that it is the scheme-theoretic union of the scheme-theoretic supports of any finite generating set of M.

9.1.N. EXERCISE (GLOBALIZING THE PREVIOUS EXERCISE). Suppose X is a scheme, and $\mathscr{F}$ is a finite type quasicoherent sheaf on X. How will you define the **scheme-theoretic support of $\mathscr{F}$**?

9.2 Locally Closed Embeddings and Locally Closed Subschemes

Now that we have defined analogs of open and closed subsets, it is natural to define the analog of locally closed subsets. Recall that locally closed subsets are intersections of open subsets and closed subsets. Hence they are closed subsets of open subsets, or equivalently open subsets of closed subsets. The analog of these equivalences will be a little problematic in the land of schemes. (Similarly, the notions of subgroup and quotient group are not quite complementary. Rhetorical question: Should "subquotient" of a group G be defined as a subgroup of a quotient of G, or a quotient of a subgroup of G, or are these the same? We will confront the same issue when defining locally closed embeddings.)

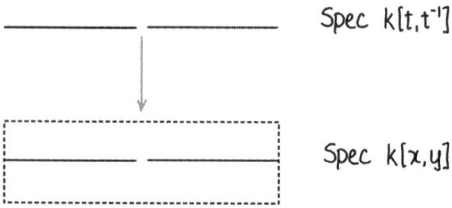

Figure 9.1 *The locally closed embedding* $\mathrm{Spec}\, k[t, t^{-1}] \to \mathrm{Spec}\, k[x, y]$ $(t \mapsto (t, 0) = (x, y),$ *i.e.,* $(x, y) \to (t, 0))$.

We say a morphism $\pi\colon X \to Y$ is a **locally closed embedding** (or **locally closed immersion**) if π can be factored into

$$X \overset{\rho}{\underset{\text{closed}}{\hookrightarrow}} Z \overset{\tau}{\underset{\text{open}}{\hookrightarrow}} Y,$$

where (as indicated) ρ is a closed embedding and τ is an open embedding. For example, the morphism $\mathrm{Spec}\, k[t, t^{-1}] \to \mathrm{Spec}\, k[x, y]$ given by $(x, y) \mapsto (t, 0)$ is a locally closed embedding (Figure 9.1). The symbol $\hookrightarrow$ is often used to indicate that a morphism is a locally closed embedding. Because closed embeddings and open embeddings are monomorphisms (Exercises 9.1.J and 8.1.E, respectively), and monomorphisms are preserved by composition, locally closed embeddings are also monomorphisms.

(Warning: The term *immersion* is often used instead of *locally closed embedding* or *locally closed immersion*, but this is unwise terminology, for reasons that already arose for closed embeddings in Definition 9.1.1: The differential geometric notion of immersion is closer to what algebraic geometers call *unramified*, which we will define in §21.7. Also, the naked term *embedding* should be avoided, because it is needlessly imprecise.)

9.2.1. If X is a subset of Y (and π on the level of sets is the inclusion), we say X is a **locally closed subscheme** of Y. (Warning: The naked term **subscheme** is often used to mean *locally closed subscheme*, but we will avoid this unhappy usage.)

9.2.A. EASY EXERCISE. Show that locally closed embeddings are locally of finite type.

9.2.B. EXERCISE. Suppose $\pi\colon X \to Y$ is a locally closed embedding whose image is a closed subset of Y. Show that π is a closed embedding. Possible hint: The property of being a closed embedding is affine-local on the target, Exercise 9.1.D. (This is philosophically useful; see, for example, §11.2.2.)

At this point, you could define the intersection of two locally closed embeddings in a scheme X (which will also be a locally closed embedding in X). But it would be awkward, as you would have to show that your construction is independent of the factorizations of each locally closed embedding into a closed embedding and an open embedding. Instead, we wait until Exercise 10.2.C, when recognizing the intersection as a fibered product will make this easier.

Clearly an open subscheme U of a closed subscheme V of X can be interpreted as a closed subscheme of an open subscheme: as the topology on V is induced from the topology on X, the underlying set of U is the intersection of some open subset U$'$ on X with V. We can take V$' = V \cap U'$, and then V$' \to U'$ is a closed embedding, and U$' \to X$ is an open embedding.

It is not clear that a closed subscheme V$'$ of an open subscheme U$'$ can be expressed as an open subscheme of a closed subscheme V. In the category of topological spaces, we would take V as the closure of V$'$, so we are now motivated to define the analogous construction, which will give us an excuse to introduce several related ideas, in §9.4. We will then resolve this issue in good cases (e.g., if X is Noetherian) in Exercise 9.4.C.

We formalize our discussion in an exercise.

9.2.C. EXERCISE. Suppose $V \to X$ is a morphism. Consider three conditions:

(i) V is the intersection of an open subscheme of X and a closed subscheme of X (you will have to define the meaning of "intersection" here, see Exercise 8.1.D, or else see the hint below).

(ii)   V is an open subscheme of a closed subscheme of X, i.e., it factors into an open embedding followed by a closed embedding.

(iii)   V is a closed subscheme of an open subscheme of X, i.e., V is a locally closed embedding.

Show that (i) and (ii) are equivalent, and both imply (iii). (Remark: (iii) does *not* always imply (i) and (ii), see the pathological example [Stacks, tag 01QW].) Hint: It may be helpful to think of the problem as follows. You might hope to think of a locally closed embedding as a Cartesian diagram.

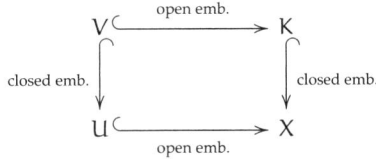

$$\begin{array}{ccc} V & \xrightarrow{\text{open emb.}} & K \\ {\scriptstyle\text{closed emb.}}\downarrow & & \downarrow{\scriptstyle\text{closed emb.}} \\ U & \xrightarrow[\text{open emb.}]{} & X \end{array}$$

(We already know that fibered products with open embeddings exist, by Exercise 8.1.D.) Interpret (i) as the existence of the diagram. Interpret (ii) as this diagram minus the lower left corner. Interpret (iii) as the diagram minus the upper right corner.

9.2.D. EXERCISE. Show that the composition of two locally closed embeddings is a locally closed embedding. (Hint: You might use (ii) implies (iii) in the previous exercise.)

9.2.2. *Unimportant but fun remark.* It may feel odd that in the definition of a locally closed embedding, we had to make a choice (of a composition of a closed embedding followed by an open embedding, rather than vice versa); but this type of issue came up earlier: a subquotient of a group can be defined as the quotient of a subgroup, or a subgroup of a quotient. Which is the right definition? Or are they the same? (Hint: The subquotient of a subquotient should certainly be a subquotient, cf. Exercise 9.2.D.)

9.3   Important Examples from Projective Geometry

We now interpret closed embeddings in terms of graded rings. Don't worry; most of the annoying foundational discussion of graded rings is complete, and we now just take advantage of our earlier work.

9.3.1. Example: Closed embeddings in projective space $\mathbb{P}_A^n$. Recall the definition of projective space $\mathbb{P}_A^n$ given in §4.4.10 (and the terminology defined there). Any *homogeneous* polynomial f in $x_0, \ldots, x_n$ defines a closed subscheme. (Thus even if f doesn't make sense as a function, its vanishing scheme still makes sense.) On the open set U_i, the closed subscheme is $V(f(x_{0/i}, \ldots, x_{n/i}))$, which we privately think of as $V(f(x_0, \ldots, x_n)/x_i^{\deg f})$. On the overlap

$$U_i \cap U_j = \operatorname{Spec} A[x_{0/i}, \ldots, x_{n/i}, x_{j/i}^{-1}]/(x_{i/i} - 1),$$

these functions on U_i and U_j don't exactly agree, but they agree up to a nonvanishing scalar, and hence cut out the same closed subscheme of $U_i \cap U_j$ (Exercise 9.1.K(b)):

$$f(x_{0/i}, \ldots, x_{n/i}) = x_{j/i}^{\deg f} f(x_{0/j}, \ldots, x_{n/j}).$$

Similarly, a collection of homogeneous polynomials in $A[x_0, \ldots, x_n]$ cuts out a closed subscheme of $\mathbb{P}_A^n$. (Exercise 15.7.I will show that all closed subschemes of $\mathbb{P}_A^n$ are of this form.)

9.3.2. *Definition.* As usual, let k be a field. A closed subscheme of $\mathbb{P}_k^n$ cut out by a single (nonzero, homogeneous) equation is called a **hypersurface** (in $\mathbb{P}_k^n$). (Be careful: the word "hypersurface" can be used to mean slightly different things in algebraic geometry, for entirely reasonable reasons; see, for example, §12.1.5.) Of course, a hypersurface is not cut out by a single global function on $\mathbb{P}^n_k$: there *are* no nonconstant global functions (Exercise 4.4.E). The **degree of a hypersurface** is the degree of the polynomial. (Implicit in this is that this notion can be determined from the subscheme itself. You may have the tools to prove this now, but we won't formally prove it until Exercise 18.6.G.) A hypersurface of degree 1 (resp., degree $2, 3, \dots$) is called a **hyperplane** (resp., **quadric, cubic, quartic, quintic, sextic, septic, octic**, $\dots$, **hypersurface**). If $n = 2$, hypersurfaces are called **curves**; a degree 1 hypersurface is called a **line**, and a degree 2 hypersurface is called a **conic curve**, or a **conic** for short. If $n = 3$, a hypersurface is called a **surface**. (In Chapter 12, we will justify the terms *curve* and *surface*.)

9.3.A. EXERCISE.

(a) Show that $wz = xy, x^2 = wy, y^2 = xz$ describes an irreducible closed subscheme in $\mathbb{P}^3_k$. In fact it is a curve, a notion we will define once we know what dimension is. This curve is called the **twisted cubic**. (The twisted cubic is a good nontrivial example of many things, so you should make friends with it as soon as possible. It implicitly appeared earlier in Exercise 3.6.F.)

(b) Show that the twisted cubic is isomorphic to $\mathbb{P}^1_k$.

We now extend this discussion to projective schemes in general.

9.3.B. EXERCISE. Suppose that $S_\bullet \longrightarrow R_\bullet$ is a surjection of graded rings. Show that the domain of the induced morphism (Exercise 7.4.A) is $\operatorname{Proj} R_\bullet$, and that the induced morphism $\operatorname{Proj} R_\bullet \to \operatorname{Proj} S_\bullet$ is a closed embedding.

Exercise 15.7.I will show the converse to Exercise 9.3.B: every closed embedding $X \hookrightarrow \operatorname{Proj} S_\bullet$ into a projective A-scheme ($S_\bullet$ is a finitely generated graded A-algebra) is of the form $\operatorname{Proj}(S_\bullet/I)$, where I is a homogeneous ideal, of "projective functions" vanishing on X. (You may be able to prove this now.)

9.3.C. EXERCISE. Show that an injective linear map of k-vector spaces $V \hookrightarrow W$ induces a closed embedding $\mathbb{P}V \hookrightarrow \mathbb{P}W$. (This is another justification for the definition of $\mathbb{P}V$ in Example 4.5.13 in terms of the *dual* of V.)

9.3.3. *Definition*. The closed subscheme defined in Exercise 9.3.C is called a **linear space**. Once we know about dimension, we will call this closed subscheme a linear space of dimension $\dim V - 1 = \dim \mathbb{P}V$. More explicitly, a linear space of dimension n in $\mathbb{P}^N$ is any closed subscheme cut out by $N - n$ k-linearly independent homogeneous linear polynomials in $x_0, \dots, x_N$. A linear space of dimension 1 (resp., 2, n, $\dim \mathbb{P}V - 1$) is called a **line** (resp., **plane**, **n-plane**, **hyperplane**). (If the linear map in the previous exercise is not injective, then the hypothesis (7.4.0.1) of Exercise 7.4.A fails.)

9.3.D. EXERCISE (A SPECIAL CASE OF BÉZOUT'S THEOREM). Suppose $X \subset \mathbb{P}^n_k$ is a degree d hypersurface cut out by $f = 0$, and ℓ is a line not contained in X. A very special case of Bézout's Theorem (Exercise 18.6.J) implies that X and ℓ meet with multiplicity d, "counted correctly." Make sense of this, by restricting the homogeneous degree d polynomial f to the line ℓ, and using the fact that a degree d polynomial in $k[x]$ has d roots, counted properly. (If it makes you feel better, assume $k = \bar{k}$.)

9.3.E. EXERCISE. Show that the map of graded rings $k[w, x, y, z] \to k[s, t]$ given by $(w, x, y, z) \mapsto (s^3, s^2t, st^2, t^3)$ induces a closed embedding $\mathbb{P}^1_k \hookrightarrow \mathbb{P}^3_k$, which yields an isomorphism of $\mathbb{P}^1_k$ with the twisted cubic (defined in Exercise 9.3.A — in fact, this will solve Exercise 9.3.A(b)). Doing this in a hands-on way will set you up well for the general Veronese construction of §9.3.6; see Exercise 9.3.I for a generalization.

9.3.4. A particularly nice case: When $S_\bullet$ is generated in degree 1.

Suppose $S_\bullet$ is a finitely generated graded ring generated in degree 1. Then S_1 is a finitely generated S_0-module, and the irrelevant ideal S_+ is generated in degree 1 (cf. Exercise 4.5.G(a)).

9.3.F. EXERCISE. Show that if $S_\bullet$ is generated (as an A-algebra) in degree 1 by $n+1$ elements $x_0, \ldots, x_n$, then $\operatorname{Proj} S_\bullet$ may be described as a closed subscheme of $\mathbb{P}_A^n$ as follows. Consider $A^{\oplus(n+1)}$ as a free module with generators $t_0, \ldots, t_n$ associated to $x_0, \ldots, x_n$. The surjection of

$$\operatorname{Sym}^\bullet\left(A^{\oplus(n+1)}\right) = A[t_0, t_1, \ldots, t_n] \twoheadrightarrow S_\bullet,$$

$$t_i \longmapsto x_i,$$

implies $S_\bullet = A[t_0, t_1, \ldots t_n]/I$, where I is a homogeneous ideal. (In particular, $\operatorname{Proj} S_\bullet$ can always be interpreted as a closed subscheme of some $\mathbb{P}_A^n$ if $S_\bullet$ is finitely generated in degree 1. Then using Exercises 7.4.D and 7.4.G, you can remove the hypothesis of generation in degree 1.)

This is analogous to the fact that if R is a finitely generated A-algebra, then choosing n generators of R as an algebra is the same as describing $\operatorname{Spec} R$ as a closed subscheme of $\mathbb{A}_A^n$. In the affine case, this is "choosing coordinates"; in the projective case, this is "choosing projective coordinates."

Recall (Exercise 4.4.F) that if k is algebraically closed, then we can interpret the closed points of $\mathbb{P}^n$ as the lines through the origin in $(n+1)$-space. The following exercise states this more generally.

9.3.G. EXERCISE. Suppose $S_\bullet$ is a finitely generated graded ring over an algebraically closed field k, generated in degree 1 by $x_0, \ldots, x_n$, inducing closed embeddings $\operatorname{Proj} S_\bullet \hookrightarrow \mathbb{P}^n$ and $\operatorname{Spec} S_\bullet \hookrightarrow \mathbb{A}^{n+1}$. Give a bijection between the closed points of $\operatorname{Proj} S_\bullet$ and the "lines through the origin" in $\operatorname{Spec} S_\bullet \subset \mathbb{A}^{n+1}$.

9.3.5. *A second proof that finite morphisms are closed.*

This interpretation of $\operatorname{Proj} S_\bullet$ as a closed subscheme of projective space (when it is generated in degree 1) yields the following second proof of the fact (shown in Exercise 8.3.L) that finite morphisms are closed. Suppose $\pi: X \to Y$ is a finite morphism. The question is local on the target, so it suffices to consider the affine case $Y = \operatorname{Spec} B$. It suffices to show that $\pi(X)$ is closed. Then by Exercise 8.3.I, X is a projective B-scheme, and hence by the Fundamental Theorem of Elimination Theory 8.4.10, its image is closed.

9.3.6. Important classical construction: The Veronese embedding.

Suppose $S_\bullet = k[x, y]$, so $\operatorname{Proj} S_\bullet = \mathbb{P}_k^1$. Then $S_{2\bullet} = k[x^2, xy, y^2] \subset k[x, y]$ (see §7.4.4 on the Veronese subring). We identify this subring as follows.

9.3.H. EXERCISE. Let $u = x^2$, $v = xy$, $w = y^2$. Show that $S_{2\bullet} \cong k[u, v, w]/(uw - v^2)$, by mapping u, v, w to x^2, xy, y^2, respectively.

We have a graded ring generated by three elements in degree 1. Thus we think of it as sitting "in" $\mathbb{P}^2$, via the construction of §9.3.F. This can be interpreted as "$\mathbb{P}^1$ as a conic in $\mathbb{P}^2$."

9.3.7. Thus if k is algebraically closed of characteristic not 2, using the fact that we can diagonalize quadratics (Exercise 5.4.J), the conics in $\mathbb{P}^2$, up to change of coordinates, come in only a few flavors: sums of three (and no fewer) squares (e.g., our conic of the previous exercise), sums of two (and no fewer) squares (e.g., $y^2 - x^2 = 0$, the union of two lines), a single (nonzero) square (e.g., $x^2 = 0$, which looks set-theoretically like a line, and is nonreduced), and 0 (perhaps not a conic at all). Thus we have proved: any plane conic (over an algebraically closed field of characteristic not 2) that can be written as the sum of three nonzero squares (and no fewer) is isomorphic to $\mathbb{P}^1$. (See Exercise 7.5.G for a closely related fact.)

We now soup up this example.

9.3.I. EXERCISE. We continue to take $S_\bullet = k[x, y]$. Show that $\operatorname{Proj} S_{d\bullet}$ is given by the equations that

$$\begin{pmatrix} y_0 & y_1 & \cdots & y_{d-1} \\ y_1 & y_2 & \cdots & y_d \end{pmatrix}$$

is rank 1 (i.e., that all the 2×2 minors vanish). This is called the **degree d rational normal curve** "in" $\mathbb{P}^d$. You did the *twisted cubic* case $d = 3$ in Exercises 9.3.A and 9.3.E.

9.3.8. *Definition.* More generally, if $S_\bullet = k[x_0, \ldots, x_n]$, then $\nu_d : \operatorname{Proj} S_{d\bullet} \to \mathbb{P}^{N-1}$ (where N is the dimension of the vector space of homogeneous degree d polynomials in $x_0, \ldots, x_n$) is called the d-**uple embedding** or the (d-uple) **Veronese embedding**. (Combining Exercise 7.4.E with Exercise 9.3.F shows that ν_d is a closed embedding.)

9.3.J. COMBINATORIAL EXERCISE (CF. REMARK 4.5.3). Show that $N = \binom{n+d}{d}$.

9.3.K. UNIMPORTANT EXERCISE. Find six linearly independent quadratic equations vanishing on the **Veronese surface** $\operatorname{Proj} S_{2\bullet}$, where $S_\bullet = k[x_0, x_1, x_2]$, which sits naturally in $\mathbb{P}^5$. (You needn't show that these equations generate all the equations cutting out the Veronese surface, although this is in fact true.) Possible hint: Use the identity

$$\det \begin{pmatrix} x_0 x_0 & x_0 x_1 & x_0 x_2 \\ x_1 x_0 & x_1 x_1 & x_1 x_2 \\ x_2 x_0 & x_2 x_1 & x_2 x_2 \end{pmatrix} = 0.$$

9.3.9. Rulings on the quadric surface. We return to rulings on the quadric surface, which first appeared in the optional (starred) section §4.4.12.

9.3.L. USEFUL GEOMETRIC EXERCISE: THE RULINGS ON THE QUADRIC SURFACE $WZ = XY$. This exercise is about the lines on the quadric surface X given by $wz - xy = 0$ in $\mathbb{P}_k^3$ (where the projective coordinates on $\mathbb{P}_k^3$ are ordered w, x, y, z). This construction arises all over the place in nature.

(a) Suppose a_0 and b_0 are given elements of k, not both zero. Make sense of the statement: as $[c, d]$ varies in $\mathbb{P}^1$, $[a_0 c, b_0 c, a_0 d, b_0 d]$ is a line in the quadric surface. (This describes "a family of lines parametrized by $\mathbb{P}^1$," although we can't yet make this precise.) Find another family of lines. These are the two **rulings** of the smooth quadric surface.

(b) Show that through every k-valued point of the quadric surface X, there passes one line from each ruling.

(c) Show there are no other lines. (There are many ways of proceeding. At risk of predisposing you to one approach, here is a germ of an idea. Suppose L is a line on the quadric surface, and $[1, x, y, z]$ and $[1, x', y', z']$ are distinct points on it. Because they are both on the quadric, $z = xy$ and $z' = x'y'$. Because all of L is on the quadric, $(1 + t)(z + tz') - (x + tx')(y + ty') = 0$ for all t. After some algebraic manipulation, this translates into $(x - x')(y - y') = 0$. How can this be made watertight? Another possible approach uses Bézout's Theorem, in the form of Exercise 9.3.D.)

Hence by Exercise 5.4.J, if we are working over an algebraically closed field of characteristic not 2, we have shown that all rank 4 quadric surfaces have two rulings of lines. (In Example 10.6.2, we will recognize this quadric as $\mathbb{P}^1 \times \mathbb{P}^1$.)

9.3.10. *Side Remark: Rulings on quadric hypersurfaces.* The existence of these two rulings is the first chapter of a number of important and beautiful stories. The second chapter is often the following. If k is an algebraically closed field, then a maximal rank (Exercise 5.4.J) quadric hypersurface X of dimension m contains no linear spaces of dimension greater than $m/2$. (We will see in Exercise 13.3.D that the maximal rank quadric hypersurfaces are the "smooth" quadrics.) If $m = 2a + 1$, then X contains an irreducible $\binom{a+2}{2}$-dimensional family of a-planes. If $m = 2a$, then X contains two irreducible $\binom{a+1}{2}$-dimensional families of a-planes, and furthermore two a-planes Λ and Λ' are in the same family if and only if $\dim(\Lambda \cap \Lambda') \equiv a \pmod 2$. These families of linear spaces are also called **rulings**. (For more information, see [GH1, §6.1, p. 735, Prop.]. You already know enough to think through the examples of $m = 0$, 1, and 2. The case $m = 2$, rulings on a quadric surface, turns up in §4.4.12, Figure 9.2, Exercise 9.3.L, §10.6.2, §12.2.19, §13.2.12, Exercise 15.5.Q, and Exercise 19.10.M. The case $m = 3$ is discussed in Exercise 16.4.K.)

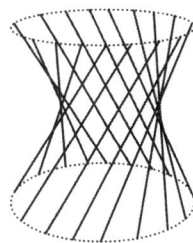

Figure 9.2 *One of the two rulings on the quadric surface* $S := V(wz - xy) \subset \mathbb{P}^3$*. One ruling contains the line* $V(w, x)$*, and the other contains the line* $V(w, y)$*. (The quadric surface sketched is actually* $a^2 + b^2 = c^2 + 1$*. Can you see why this an affine chart for* S*? What are* a*, *b*, and *c *in terms of* w*, *x*, *y*, and *z*? Hint: Exercise 5.4.J.)*

9.3.11. Affine and projective cones (Figure 9.3).

If $S_\bullet$ is a finitely generated graded ring, then the **affine cone** of $\operatorname{Proj} S_\bullet$ is $\operatorname{Spec} S_\bullet$. Caution: This terminology is not ideal, as this construction depends on $S_\bullet$, not just on $\operatorname{Proj} S_\bullet$. As motivation, consider the graded ring $S_\bullet = \mathbb{C}[x, y, z]/(z^2 - x^2 - y^2)$. Figure 9.4 is a sketch of $\operatorname{Spec} S_\bullet$. (Here we draw the "real picture" of $z^2 = x^2 + y^2$ in $\mathbb{R}^3$.) It is a cone in the traditional sense; the origin $(0, 0, 0)$ is the "cone point."

This gives a useful way of picturing Proj (even over arbitrary rings, not just $\mathbb{C}$). Intuitively, you could imagine that if you discarded the origin, you would get something that would project onto $\operatorname{Proj} S_\bullet$. The following exercise makes that precise.

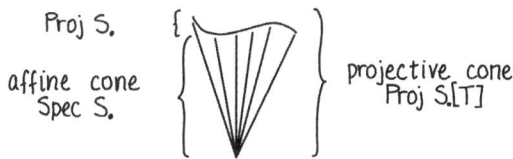

Figure 9.3 *The affine and projective cones of* $\operatorname{Proj} S_\bullet$*.*

9.3.M. EXERCISE. If $\operatorname{Proj} S_\bullet$ is a projective scheme over a field k, describe a natural morphism $\operatorname{Spec} S_\bullet \setminus V(S_+) \to \operatorname{Proj} S_\bullet$. (Can you see why $V(S_+)$ is a single point, and should reasonably be called the origin?)

This readily generalizes to the following exercise, which again motivates the terminology "irrelevant."

9.3.N. EASY EXERCISE. If $S_\bullet$ is a finitely generated graded ring over a base ring A, describe a natural morphism $\operatorname{Spec} S_\bullet \setminus V(S_+) \to \operatorname{Proj} S_\bullet$.

In fact, it can be made precise that $\operatorname{Proj} S_\bullet$ is the quotient (by the multiplicative group of scalars) of the affine cone minus the origin.

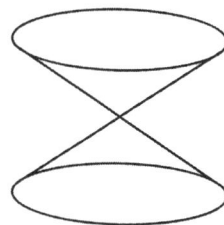

Figure 9.4 *The cone* $x^2 + y^2 = z^2$ *in* $\mathbb{A}^3$*.*

9.3.12. *Definition.* The **projective cone** of $\operatorname{Proj} S_\bullet$ is $\operatorname{Proj} S_\bullet[T]$, where T is a new variable of degree 1. For example, the cone corresponding to the conic $\operatorname{Proj} k[x, y, z]/(z^2 - x^2 - y^2)$ is $\operatorname{Proj} k[x, y, z, T]/(z^2 - x^2 - y^2)$. The projective cone is sometimes called the **projective completion** of $\operatorname{Spec} S_\bullet$. (Note: This depends on $S_\bullet$, not just on $\operatorname{Proj} S_\bullet$.)

9.3.O. LESS IMPORTANT EXERCISE (CF. §4.5.1). Show that the "projective cone" $\operatorname{Proj} S_\bullet[T]$ of $\operatorname{Proj} S_\bullet$ has a closed subscheme isomorphic to $\operatorname{Proj} S_\bullet$ (informally, corresponding to $T = 0$), whose complement (the distinguished open set $D(T)$) is isomorphic to the affine cone $\operatorname{Spec} S_\bullet$.

This construction can be usefully pictured as the affine cone union some points "at infinity," and the points at infinity form the Proj. (The reader may wish to start with Figure 9.4, and try to visualize the conic curve "at infinity," and then compare this visualization to Figure 4.9.)

We have thus completely described the algebraic analog of the classical picture of §4.5.1.

9.3.13. *Definition.* If we put a nonstandard weighting on the variables of $k[x_0, \ldots, x_n]$ — say, we give x_i degree d_i — then $\text{Proj}\, k[x_0, \ldots, x_n]$ is called **weighted projective space** $\mathbb{P}(d_0, d_1, \ldots, d_n)$.

9.3.P. EXERCISE. (a) Show that $\mathbb{P}(m, n)$ is isomorphic to $\mathbb{P}^1$. (b) Show that $\mathbb{P}(1, 1, 2) \cong \text{Proj}\, k[u, v, w, z]/(uw - v^2)$. Hint: Do this by looking at the even-graded parts of $k[x_0, x_1, x_2]$; cf. Exercise 7.4.D. (This is a projective cone over a conic curve. Over a field of characteristic not 2, it is isomorphic to the traditional cone $x^2 + y^2 = z^2$ in $\mathbb{P}^3$; see Figure 9.4.)

9.3.Q. EXERCISE (COMBINING A NUMBER OF THE ABOVE CONCEPTS). Show that $\mathbb{P}(1, 1, n)$ is isomorphic to the projective cone over the degree n rational normal curve.

9.4 The (Closed Sub)scheme-Theoretic Image

We now define a series of notions that are all of the form, "the smallest closed subscheme such that something is true." One example will be the notion of scheme-theoretic closure of a locally closed embedding, which will allow us to interpret locally closed embeddings in three equivalent ways (open subscheme intersect closed subscheme; open subscheme of closed subscheme; and closed subscheme of open subscheme—cf. Exercise 9.2.C).

9.4.1. Scheme-theoretic image.
We start with the notion of scheme-theoretic image. Set-theoretic images are badly behaved in general (§8.4.1), and even with reasonable hypotheses, such as those in Chevalley's Theorem 8.4.2, things can be confusing. For example, there is no reasonable way to impose a scheme structure on the image of $\mathbb{A}_k^2 \to \mathbb{A}_k^2$ given by $(x, y) \mapsto (x, xy)$. It will be useful (e.g., Exercise 9.4.C) to define a notion of a closed subscheme of the target that "best approximates" the image. This will incorporate the notion that the image of something with nonreduced structure ("fuzz") can also have nonreduced structure. As usual, we will need to impose reasonable hypotheses to make this notion behave well (see Theorem 9.4.4 and Corollary 9.4.5).

9.4.2. *Definition.* Suppose $i: Z \hookrightarrow Y$ is a closed subscheme, giving an exact sequence $0 \to \mathscr{I}_{Z/Y} \to \mathscr{O}_Y \to i_* \mathscr{O}_Z \to 0$. We say that **the image of** $\pi: X \to Y$ **lies in** Z if the composition $\mathscr{I}_{Z/Y} \to \mathscr{O}_Y \to \pi_* \mathscr{O}_X$ is zero. Informally, locally, functions vanishing on Z pull back to the zero function on X. If the image of π lies in some subschemes Z_j (as j runs over some index set), it clearly lies in their intersection (cf. Exercise 9.1.I(a) on intersections of closed subschemes). We then define the **scheme-theoretic image of** π, a closed subscheme of Y, as the "smallest closed subscheme containing the image," i.e., the intersection of all closed subschemes containing the image. In particular (and in our first examples), if Y is affine, the scheme-theoretic image is cut out by functions on Y that are 0 when pulled back to X.

Other reasonable names for this concept are *closed subscheme-theoretic image*, or *image closed subscheme*. They have the advantage that they remind the reader that the scheme-theoretic image is a *closed* subscheme, but we won't use these phrases.

Example 1. Consider $\pi: \text{Spec}\, k[\epsilon]/(\epsilon^2) \to \text{Spec}\, k[x] = \mathbb{A}_k^1$, given by $x \mapsto \epsilon$. Then the scheme-theoretic image of π is given by $\text{Spec}\, k[x]/(x^2)$ (the polynomials pulling back to 0 are precisely multiples of x^2). Thus the image of the fuzzy point still has some fuzz.

Example 2. Consider $\pi: \text{Spec}\, k[\epsilon]/(\epsilon^2) \to \text{Spec}\, k[x] = \mathbb{A}_k^1$, given by $x \mapsto 0$. Then the scheme-theoretic image is given by $\text{Spec}\, k[x]/(x)$: the image is reduced. In this picture, the fuzz is "collapsed" by π.

Example 3. Consider $\pi: \text{Spec}\, k[t, t^{-1}] = \mathbb{A}^1 - \{0\} \to \mathbb{A}^1 = \text{Spec}\, k[u]$, given by $u \mapsto t$. Any function $g(u)$ that pulls back to 0 as a function of t must be the zero-function. Thus the scheme-theoretic image is everything. The set-theoretic image, on the other hand, is the distinguished open set $\mathbb{A}^1 \setminus \{0\}$. Thus in not-too-pathological cases, the underlying set of the scheme-theoretic image is not the set-theoretic image. But the situation isn't terrible: the underlying set of the scheme-theoretic image must be closed, and indeed it is the closure of the set-theoretic image. We might imagine that in reasonable cases this will be true, and in even nicer cases, the underlying set of the scheme-theoretic image will be the set-theoretic image. We will later see that this is indeed the case (§9.4.6).

But sadly pathologies can sometimes happen in, well, pathological situations.

Example 4 (Figure 9.5). Let $X = \coprod \operatorname{Spec} k[\epsilon_n]/((\epsilon_n)^n)$ (a scheme that appeared in the hint to Exercise 5.2.F) and $Y = \operatorname{Spec} k[x]$, and define $X \to Y$ by $x \to \epsilon_n$ on the nth component of X. If a function $g(x)$ on Y pulls back to 0 on X, then its Taylor expansion is 0 to order n (by examining the pullback to the nth component of X) for all n, so $g(x)$ must be 0. (This argument will be vastly generalized in Exercise 13.9.A(b).) Thus the scheme-theoretic image is $V(0)$ on Y, i.e., Y itself, while the set-theoretic image is easily seen to be just the origin (the closed point 0). (This morphism implicitly arises in Caution/Example 9.4.11.)

9.4.3. *Criteria for computing scheme-theoretic images affine-locally.* Example 4 clearly is weird though, and we can show that in "reasonable circumstances" such pathology doesn't occur.

In the special case where the target Y is affine (pay attention to how this argument works in Examples 1 through 3), say $Y = \operatorname{Spec} B$, almost by definition the scheme-theoretic image of $\pi\colon X \to Y$ is cut out by the ideal $I \subset B$ of functions on Y that pull back to zero on X.

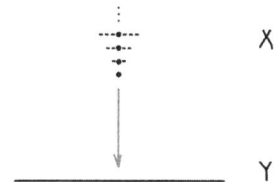

Figure 9.5 *Yuck (Example 4 of Definition 9.4.2).*

It would be great to use this to compute the scheme-theoretic image affine by affine (affine-locally). On the affine open set $\operatorname{Spec} B \subset Y$, define the ideal $I(B) \subset B$ of functions that pull back to 0 on X. Formally, $I(B) := \ker(B \to \Gamma(\operatorname{Spec} B, \pi_*(\mathscr{O}_X)))$. If for each such B, and each $g \in B$, the map $\phi\colon I(B)_g \to I(B_g)$ is an isomorphism, then we will have defined the scheme-theoretic image as a closed subscheme (see Exercise 9.1.F). Injectivity of ϕ is straightforward: if $r \in B$ and r/g^n pulls back to 0 on $\pi^{-1}(\operatorname{Spec} B_g)$, then as g is nonvanishing at all points of $\operatorname{Spec} B_g = D(g)$, we have that $\pi^\sharp(r)$ is 0 on $\pi^{-1}(\operatorname{Spec} B_g)$ as well.

So the question is the surjectivity of ϕ, which translates to: given a function r/g^n on $D(g)$ that pulls back to zero on $\pi^{-1}(D(g))$, is it true that for some m, $rg^m = 0$ when pulled back to $\pi^{-1}(\operatorname{Spec} B)$?

(i)   We first show that the answer is yes if $\pi^{-1}(\operatorname{Spec} B)$ is reduced. In this case we can take advantage of Exercise 5.2.A, that a function on a reduced scheme is zero if it has value zero at every point. We may then take $m = 1$ (as $\pi^\sharp(r)$ vanishes on $\pi^{-1}(D(g))$ and g vanishes on $V(g)$, so $\pi^\sharp(rg)$ vanishes on $\pi^{-1}(\operatorname{Spec} B) = \pi^{-1}(D(g)) \cup \pi^{-1}(V(g))$.)

(ii)   The answer is also yes if $\pi^{-1}(\operatorname{Spec} B)$ is affine, say, $\operatorname{Spec} A$: if $r' = \pi^\sharp(r)$ and $g' = \pi^\sharp(g)$ in A, then if $r' = 0$ on $D(g')$, there is an m such that $r'(g')^m = 0$ (as the statement $r' = 0$ in $D(g')$ means precisely this — the functions on $D(g')$ are $A_{g'}$).

(ii)′   More generally, the answer is yes if $\pi^{-1}(\operatorname{Spec} B)$ is quasicompact: cover $\pi^{-1}(\operatorname{Spec} B)$ with finitely many affine open sets. For each one there will be some m_i so that $rg^{m_i} = 0$ when pulled back to this open set. Then let $m = \max(m_i)$. (We see again that quasicompactness is our friend!)

In conclusion, we have proved the following (subtle) theorem.

9.4.4. Theorem — *Suppose $\pi\colon X \to Y$ is a morphism of schemes. If X is reduced or π is quasicompact, then the scheme-theoretic image of π may be computed affine-locally: on $\operatorname{Spec} A \subset Y$, it is cut out by the functions (elements of A) that pull back to the function 0 (on $\pi^{-1}(\operatorname{Spec} A)$).*

9.4.5. Corollary — *Under the hypotheses of Theorem 9.4.4, the closure of the set-theoretic image of π is the underlying set of the scheme-theoretic image.*

(Example 4 above shows how we need these hypotheses.)

9.4.6. In particular, if the set-theoretic image is closed (e.g., if π is finite or projective), the set-theoretic image is the underlying set of the scheme-theoretic image, as promised in Example 3 above.

9.4.7. *Proof of Corollary 9.4.5.* The set-theoretic image is in the underlying set of the scheme-theoretic image. (Check this!) The underlying set of the scheme-theoretic image is closed, so the closure of the set-theoretic image is contained in the underlying set of the scheme-theoretic image. On the other hand, if U is the complement of the closure of the set-theoretic image, $\pi^{-1}(U) = \varnothing$. Under these hypotheses, the scheme-theoretic image can be computed locally, so the scheme-theoretic image is the empty set on U. $\qquad\square$

We conclude with a few stray remarks.

9.4.A. EASY EXERCISE. If X is reduced, show that the scheme-theoretic image of $\pi: X \to Y$ is also reduced.

More generally, you might expect there to be no unnecessary nonreduced structure on the image not forced by nonreduced structure on the source. The following makes this precise in the locally Noetherian case (when we can talk about associated points).

9.4.B. * UNIMPORTANT EXERCISE. If $\pi: X \to Y$ is a *quasicompact* morphism of locally Noetherian schemes, show that the associated points of the image subscheme are a subset of the image of the associated points of X. (The example of $\coprod_{a \in \mathbb{C}} \operatorname{Spec} \mathbb{C}[t]/(t - a) \to \operatorname{Spec} \mathbb{C}[t]$ shows what can go wrong if you give up quasicompactness — note that reducedness of the source doesn't help.) Hint: Reduce to the case where X and Y are affine. (Can you develop your geometric intuition so that this becomes plausible to you?)

9.4.8. Scheme-theoretic closure of a locally closed subscheme.

We define the **scheme-theoretic closure** of a locally closed embedding $\pi: X \hookrightarrow Y$ as the scheme-theoretic image of π. (A shorter phrase for this is *schematic closure*, although this more elegant nomenclature has not caught on.)

9.4.C. EXERCISE. If a locally closed embedding $V \hookrightarrow X$ is quasicompact (e.g., if V is Noetherian, Exercise 8.3.B(a)), or if V is reduced, show that (iii) implies (i) and (ii) in Exercise 9.2.C. Thus in this fortunate situation, a locally closed embedding can be thought of in three different ways, whichever is convenient. (Hint: Corollary 9.4.5.)

9.4.D. UNIMPORTANT EXERCISE, USEFUL FOR INTUITION. If $\pi: X \hookrightarrow Y$ is a locally closed embedding into a locally Noetherian scheme (so X is also locally Noetherian), then the associated points of the scheme-theoretic closure are (naturally in bijection with) the associated points of X. (Hint: Exercise 9.4.B.) Informally, the only nonreduced structure on the scheme-theoretic closure is that "forced by" the nonreduced structure on X.

9.4.9. The (reduced) subscheme structure on a closed subset.

Suppose X^{set} is a closed sub*set* of a scheme Y. Then we can define a canonical scheme structure X on X^{set} that is reduced. We could describe it as being cut out by those functions whose values are zero at all the points of X^{set}. On the affine open set $\operatorname{Spec} B$ of Y, if the set X^{set} corresponds to the radical ideal $I = I(X^{\mathrm{set}} \cap \operatorname{Spec} B)$ (recall the $I(\cdot)$ function from §3.7), the scheme X corresponds to $\operatorname{Spec} B/I$. You can quickly check that this behaves well with respect to any distinguished inclusion $\operatorname{Spec} B_f \hookrightarrow \operatorname{Spec} B$. We could also consider this construction as an example of a scheme-theoretic image in the following crazy way: let W be the scheme that is a disjoint union of all the points of X^{set}, where the point corresponding to p in X^{set} is Spec of the residue field of $\mathscr{O}_{Y,p}$. Let $\rho: W \to Y$ be the "canonical" map sending "p to p," and giving an isomorphism on residue fields. Then the scheme structure on X is the scheme-theoretic image of ρ. A third definition: It is the smallest closed subscheme whose underlying set contains X^{set}.

9.4.E. EXERCISE. Show that all three definitions are the same.

This construction is called the (induced) **reduced subscheme structure** on the closed subset X^{set}. (Vague exercise: Make a definition of the reduced subscheme structure precise and rigorous to your satisfaction.)

9.4.F. EXERCISE. Show that the underlying set of the induced reduced subscheme $X \hookrightarrow Y$ is indeed the closed subset X^{set}. Show that X is reduced.

9.4.10. Reduced version of a scheme.
In the main interesting case where X^{set} is all of Y, we obtain a *reduced closed subscheme* $Y^{red} \rightarrow Y$, called the **reduction** of Y. On the affine open subset $\operatorname{Spec} B \hookrightarrow Y$, $Y^{red} \hookrightarrow Y$ corresponds to (the quotient by) the nilradical $\mathfrak{N}(B)$ of B. The reduction of a scheme is the "reduced version" of the scheme, and informally corresponds to "shearing off the fuzz."

An alternative equivalent definition of the reduction: On the affine open subset $\operatorname{Spec} B \hookrightarrow Y$, the reduction of Y corresponds to the nilradical $\mathfrak{N}(B) \subset B$. We easily check that for any $f \in B$, $\mathfrak{N}(B)_f = \mathfrak{N}(B_f)$. By Exercise 9.1.F this defines a closed subscheme, which we call the reduction.

9.4.11.* *Caution/example.* It is not true that for *every* open subset $U \subset Y$, $\Gamma(U, \mathscr{O}_{Y^{red}})$ is $\Gamma(U, \mathscr{O}_Y)$ modulo its nilpotents. For example, on $Y = \coprod \operatorname{Spec} k[x]/(x^n)$, the function x is not nilpotent, but is 0 on Y^{red}, as it is "locally nilpotent." This may remind you of Example 4 after Definition 9.4.2.

9.5   Slicing by Effective Cartier Divisors, Regular Sequences and Regular Embeddings

We now introduce regular embeddings, an important class of locally closed embeddings. Locally closed embeddings of regular schemes in regular schemes are one important example of regular embeddings (Exercise 13.2.M(b)). Effective Cartier divisors, basically the codimension 1 case, will turn out to be repeatedly useful as well, with repeated "slicing by effective Cartier divisors" providing the inductive step in many arguments. (When we say **slicing** in many future contexts, we will mean slicing by effective Cartier divisors.)

9.5.1. Locally principal closed subschemes, and effective Cartier divisors.
A closed subscheme is **locally principal** if on each open set in a small enough open cover it is cut out by a single equation (i.e., by a principal ideal, hence the terminology). More specifically, a locally principal closed subscheme is a closed subscheme $\pi\colon X \hookrightarrow Y$ for which there is an open cover $\{U_i\}$ of Y for which for each i, $\pi^{-1}(U_i) \hookrightarrow U_i$ is isomorphic to the closed subscheme $V(s_i) \hookrightarrow U_i$ for some $s_i \in \Gamma(U_i, \mathscr{O}_Y)$. (This open cover can clearly be chosen to be by affine open subsets.)

For example, hypersurfaces in $\mathbb{P}^n_k$ (Definition 9.3.2) are locally principal: each homogeneous polynomial in $n + 1$ variables defines a locally principal closed subscheme of $\mathbb{P}^n_A$. (Warnings: Unlike "local principality," "principality" is not an affine-local condition; see §19.11.10! Also, the example of a projective hypersurface, §9.3.1, shows that a locally principal closed subscheme need not be cut out by a globally defined function. Finally, we unfortunately use the phrase "locally principal" in a different way in §15.4.8.)

If the ideal sheaf is locally generated near every point by a function that is not a zerodivisor, we call the closed subscheme an *effective Cartier divisor*. More precisely: if $\pi\colon X \rightarrow Y$ is a closed embedding, and there is a cover Y by *affine* open subsets $\operatorname{Spec} A_i \subset Y$, and there exist non-zerodivisors $t_i \in A_i$ with $V(t_i) = X|_{\operatorname{Spec} A_i}$ (scheme-theoretically — i.e., the ideal sheaf of X over $\operatorname{Spec} A_i$ is generated by t_i), then we say that X is an **effective Cartier divisor** on Y. (We will not explain the origin of the phrase, as it is not relevant for this point of view.) We will often use the evocative phrase **slicing by an effective Cartier divisor** when we mean "restricting to an effective Cartier divisor."

The notion of an effective Cartier divisor is central. The notion of a locally principal closed subscheme is much less so.

9.5.A. EXERCISE. Suppose $t \in A$ is a non-zerodivisor. Show that t is a non-zerodivisor in $A_\mathfrak{p}$ for each prime $\mathfrak{p}$.

9.5.2. *Caution.* If D is an effective Cartier divisor on an affine scheme $\operatorname{Spec} A$, it is not necessarily true that $D = V(t)$ for some $t \in A$ (see Exercise 15.5.I—§19.11.10 gives a different flavor of example). In other words, the condition of a closed subscheme being an effective Cartier divisor can be verified on *an* affine cover, but cannot be checked on an *arbitrary* affine cover—it is not an affine-local condition in this obvious a way.

9.5.B. EXERCISE. Suppose X is a locally Noetherian scheme. If $t \in \Gamma(X, \mathscr{O}_X)$ is a function on X, show that t (or more precisely the closed subscheme $V(t)$) is an effective Cartier divisor if and only if it doesn't vanish on any associated point of X. Show that a locally principal closed subscheme of X is an effective Cartier divisor if and only if it doesn't contain any associated points of X.

9.5.C. EXERCISE. Suppose $V(t) = V(t') \hookrightarrow \operatorname{Spec} A$ is an effective Cartier divisor, with t and t' non-zerodivisors in A. Show that t is an invertible function times t'.

The idea of effective Cartier divisors leads us to the notion of regular sequences. (We will close the loop in Exercise 9.5.H, where we will interpret effective Cartier divisors on any reasonable scheme as regular embeddings of codimension 1.)

9.5.3. Regular sequences.
The notion of "regular sequence," like so much else, is due to Serre, [Se2]. The definition of regular sequence is the algebraic version of the following geometric idea: locally, we take an effective Cartier divisor (a non-zerodivisor); then an effective Cartier divisor on that; then an effective Cartier divisor on that; and so on, a finite number of times. A little care is necessary; for example, we might want this to be independent of the order of the equations imposed, and this is true only when we say this in the right way.

We make the definition of regular sequence for a ring A, and more generally for an A-module M.

9.5.4. *Definition.* If M is an A-module, a sequence $x_1, \ldots, x_r \in A$ is called an M-**regular sequence** (or a **regular sequence for** M) if the following two conditions are satisfied.

(i) For each i, x_i is not a zerodivisor for $M/(x_1, \ldots, x_{i-1})M$. (The case $i = 1$ should be interpreted as: "x_1 is not a zerodivisor of M." Recall that zerodivisors were defined in Exercise 1.2.C.)

(ii) We have a proper inclusion $(x_1, \ldots, x_r)M \subsetneq M$.

In the case most relevant to us, when $M = A$, this should be seen as a reasonable approximation of a "complete intersection," and indeed we will use this as the definition (§9.5.7).

We say r is the **length of the regular sequence** $x_1, \ldots, x_r \in A$. An A-regular sequence is just called a **regular sequence**.

9.5.D. EXERCISE. If M is an A-module, show that an M-regular sequence continues to satisfy condition (i) of the Definition 9.5.4 of regular sequence upon any localization. (Once you know what flatness is, you will see that your argument shows that condition (i) is preserved by any flat base change.)

9.5.E. EXERCISE. If x, y is an M-regular sequence, show that x^N, y is an M-regular sequence. Hint: The difficult part is showing that y is not a zerodivisor of $M/(x^N M)$. Show this by induction on N. If y *is* a zerodivisor of $M/(x^N M)$, then $ym \equiv 0 \pmod{x^N}$ for some $m \in M \setminus x^N M$. Hence $ym = x^N k$ for some $k \in M$. Use the fact that x, y is a regular sequence to show that m is a multiple of x. (Your argument will easily extend to show more generally that if $x_1, \ldots, x_n$ is a regular sequence, and $a_1, \ldots, a_n \in \mathbb{Z}^+$, then $x_1^{a_1}, \ldots, x_n^{a_n}$ is a regular sequence.)

9.5.5. *The regularity of a sequence depends on its order.* We now give an example ([E, Example 17.3]) showing that the order of a regular sequence matters. Suppose $A = k[x, y, z]/(x - 1)z$, so $X = \operatorname{Spec} A$ is the union of the $z = 0$ plane and the $x = 1$ plane—X is reduced and has two components (see Figure 9.6). You can readily verify that x is a non-zerodivisor of A ($x = 0$ misses one component of X, and doesn't vanish entirely on the other), and that the corresponding effective Cartier divisor $X' = \operatorname{Spec} k[x, y, z]/(x, z)$ (on X) is integral. Then $(x - 1)y$ gives an effective Cartier divisor on X' (it doesn't vanish entirely on X'), so x, $(x - 1)y$ is a regular sequence for A. However, $(x - 1)y$ is *not* a non-zerodivisor of

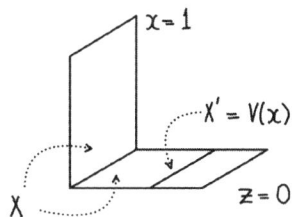

Figure 9.6 *Order matters in a regular sequence (in the "non-local" situation), for unsurprising reasons.*

A, as it *does* vanish entirely on one of the two components. Thus $(x - 1)y$, x is *not* a regular sequence. The reason that reordering the regular sequence x, $(x - 1)y$ ruins regularity is clear: there is a locus on which $(x - 1)y$ isn't effective Cartier, but it disappears if we enforce $x = 0$ first. The problem is one of "nonlocality"—"near" $x = y = z = 0$ there is no problem. This may motivate the fact that in the (Noetherian) local situation, this problem disappears. We now make this precise.

9.5.6. Theorem — *Suppose* $(A, \mathfrak{m})$ *is a Noetherian local ring, and M is a finitely generated A-module. Then any M-regular sequence* $(x_1, \ldots, x_r)$ *in* $\mathfrak{m}$ *remains a regular sequence upon any reordering.*

(In [Di], Dieudonné gives a half-page example showing that Noetherian hypotheses are necessary in Theorem 9.5.6. Also, in Exercise 24.6.C, we give a different, somehow simpler proof in the case where A contains a field.)

Before proving Theorem 9.5.6 (in Exercise 9.5.F), we prove the first nontrivial case, when $r = 2$. This discussion is secretly a baby case of the Koszul complex (which is briefly mentioned in §24.4.1).

Suppose $x, y \in \mathfrak{m}$, and x, y is an M-regular sequence. Translation: $x \in \mathfrak{m}$ is a non-zerodivisor on M, and $y \in \mathfrak{m}$ is a non-zerodivisor on M/xM.

Consider the double complex

$$(9.5.6.1) \qquad \begin{array}{ccc} M & \xrightarrow{\times(-x)} & M \\ {\scriptstyle \times y}\big\uparrow & & \big\uparrow{\scriptstyle \times y} \\ M & \xrightarrow{\times x} & M \end{array} ,$$

where the bottom left is considered to be in position $(0, 0)$. (The only reason for the minus sign in the top row is our arbitrary preference for anti-commuting rather than commuting squares in §1.6.1, but it really doesn't matter.)

We compute the cohomology of the total complex using a (simple) spectral sequence, beginning with the rightward orientation. (The gratuitous use of spectral sequences here, as in many of our other applications, is overkill; we do this partially in order to get practice with the machine.) On the first page, we have

$$(9.5.6.2) \qquad \begin{array}{ccc} \ker(M \xrightarrow{\times x} M) & & M/xM \\ {\scriptstyle \times y}\big\uparrow & & \big\uparrow{\scriptstyle \times y} \\ \ker(M \xrightarrow{\times x} M) & & M/xM . \end{array}$$

The entries $\ker(M \xrightarrow{\times x} M)$ in the first column are 0, as x is a non-zerodivisor on M. Taking homology in the vertical direction to obtain the second page, we find

(9.5.6.3) $\qquad\qquad\qquad\qquad$ 0 $\qquad$ $M/(x,y)M$

$$0 \qquad\qquad 0 \qquad ,$$

using the fact that y is a non-zerodivisor on M/xM. The sequence clearly converges here. Thus the original double complex (9.5.6.1) only has nonzero cohomology in degree 2, where it is $M/(x,y)M$.

Now we run the spectral sequence on (9.5.6.1) using the upward orientation. The first page of the sequence is:

$$M/yM \xrightarrow{\times(-x)} M/yM$$

$$\ker(M \xrightarrow{\times y} M) \xrightarrow{\times x} \ker(M \xrightarrow{\times y} M).$$

The sequence must converge to (9.5.6.3) after the next step. From the top row, we see that multiplication by x must be injective on M/yM, so x is a non-zerodivisor on M/yM. From the bottom row, multiplication by x gives an isomorphism of $\ker(M \xrightarrow{\times y} M)$ with itself. As $x \in \mathfrak{m}$, by version 2 of Nakayama's Lemma (Lemma 8.2.9), this implies that $\ker(M \xrightarrow{\times y} M) = 0$, so y is a non-zerodivisor on M. Thus we have shown that y, x is a regular sequence on M—the $r = 2$ case of Theorem 9.5.6.

9.5.F. EASY EXERCISE. Prove Theorem 9.5.6. (Hint: Show it first in the case of a reordering where only two adjacent x_i are swapped, using the $r = 2$ case just discussed.) Where are the Noetherian hypotheses used?

9.5.7. Regular embeddings.

Suppose $\pi\colon X \hookrightarrow Y$ is a locally closed embedding. We say that π is a **regular embedding (of codimension r) at a point** $p \in X$ if in the local ring $\mathscr{O}_{Y,p}$, the ideal of X is generated by a regular sequence (of length r). We say that π is a **regular embedding (of codimension r)** if it is a regular embedding (of codimension r) at all $p \in X$. (Another reasonable name for a regular embedding might be "local complete intersection." Unfortunately, "local complete intersection morphism," or "lci morphism," is already used for a related notion; see [Stacks, tag 068E].)

Our terminology uses the word "codimension," which we have not defined as a word on its own. The reason for using this word will become clearer once you meet Krull's Principal Ideal Theorem 12.3.3 and Krull's Height Theorem 12.3.10.

9.5.G. EXERCISE (THE CONDITION OF A LOCALLY CLOSED EMBEDDING BEING A REGULAR EMBEDDING IS OPEN). Show that if a locally closed embedding $\pi\colon X \hookrightarrow Y$ of locally Noetherian schemes is a regular embedding at p, then it is a regular embedding in some open neighborhood of p in X. Hint: Reduce to the case where π is a closed embedding, and then where Y (hence X) is affine—say $Y = \operatorname{Spec} B$, $X = \operatorname{Spec} B/I$, and $p = [\mathfrak{p}]$—and there are $f_1, \ldots, f_r$ such that in $\mathscr{O}_{Y,p}$, the images of the f_i are a regular sequence generating $I_{\mathfrak{p}}$. We wish to show that $(f_1, \ldots, f_r) = I$ "in an open neighborhood of p." Prove the following fact in algebra: if I and J are ideals of a Noetherian ring A, and $\mathfrak{p} \subset A$ is a prime ideal such that $I_{\mathfrak{p}} = J_{\mathfrak{p}}$, show that there exists $a \in A \setminus \mathfrak{p}$ such that $I_a = J_a$ in A_a. To do this, show that it suffices to consider the special case $I \subset J$, by considering $I \cap J$ and J instead of I and J. To show this special case, let $K = J/I$, a finitely generated module, and show that if $K_{\mathfrak{p}} = 0$ then $K_a = 0$ for some $a \in A \setminus \mathfrak{p}$.

Hence if X is locally of finite type over a field (e.g., a variety), then to check that a closed embedding π is a regular embedding it suffices to check at closed points of X (Exercise 5.3.F).

Exercise 13.1.F(b) will show that not all closed embeddings are regular embeddings.

9.5.H. EXERCISE. Show that a closed embedding $X \hookrightarrow Y$ of locally Noetherian schemes is a regular embedding of codimension 1 if and only if X is an effective Cartier divisor on Y. Unimportant remark: The Noetherian hypotheses can be replaced by requiring $\mathscr{O}_Y$ to be coherent, and essentially the same argument applies. It is interesting to note that "effective Cartier divisor" implies "regular embedding of codimension 1" always, but that the converse argument requires Noetherian(-like) assumptions. (See [MO129242] for a counterexample to the converse.)

9.5.8. *Definition.* A **codimension** r **complete intersection** in a scheme Y is a closed subscheme X that can be written as the scheme-theoretic intersection of r effective Cartier divisors $D_1, \ldots, D_r$, such that at every point $p \in X$, the equations corresponding to $D_1, \ldots, D_r$ form a regular sequence. The phrase **complete intersection** means "codimension r complete intersection for some r.'

Chapter 10

Fibered Products of Schemes, and Base Change

Fibered products have an unexpectedly central place in algebraic geometry. Experience will gradually teach you why they play such an outsized role. Until you have acquired that experience, you should pay closer attention to them than you think they might deserve.

10.1   They Exist

Before we get to products, we note that coproducts exist in the category of schemes: just as with the category of sets (Exercise 1.2.T), coproduct is disjoint union.

We will now construct the fibered product in the category of schemes.

10.1.1. Theorem: Fibered products exist — *Suppose* $\alpha\colon X \to Z$ *and* $\beta\colon Y \to Z$ *are morphisms of schemes. Then the fibered product*

$$
\begin{array}{ccc}
X \times_Z Y & \xrightarrow{\ \alpha'\ } & Y \\
{\scriptstyle \beta'}\big\downarrow & & \big\downarrow{\scriptstyle \beta} \\
X & \xrightarrow{\ \alpha\ } & Z
\end{array}
$$

exists in the category of schemes.

Note: If A is a ring, people often sloppily write $\times_A$ for $\times_{\operatorname{Spec} A}$. If B is an A-algebra, and X is an A-scheme, people often write X_B or $X \times_A B$ for $X \times_{\operatorname{Spec} A} \operatorname{Spec} B$.

10.1.2. *Warning: Products of schemes aren't products of sets.* Before showing existence, here is a warning: The product of schemes isn't a product of sets (and more generally for fibered products). We have made a big deal about schemes being *sets*, endowed with a *topology*, upon which we have a *structure sheaf*. So you might think that we will construct the product in this order: take the product set, endow it with the product topology, and then figure out what the structure sheaf must be. But we won't, because scheme products behave oddly on the level of sets. You may have checked (Exercise 7.6.E(a)) that the product of two affine lines over your favorite algebraically closed field $\bar{k}$ is the affine plane: $\mathbb{A}^1_{\bar{k}} \times_{\bar{k}} \mathbb{A}^1_{\bar{k}} \cong \mathbb{A}^2_{\bar{k}}$. But the underlying set of the latter is *not* the underlying set of the former—we get additional points, corresponding to curves in $\mathbb{A}^2$ that are not lines parallel to the axes!

10.1.3. On the other hand, W-valued points (where W is a scheme, Definition 7.3.10) *do* behave well under (fibered) products (as mentioned in §7.3.11). This is just the universal property *definition* of fibered product: a W-valued point of a scheme X is defined as an element of $\operatorname{Hom}(W, X)$, and the fibered product is defined (via Yoneda; see §1.2.1) by

$$(10.1.3.1) \qquad \operatorname{Hom}(W, X \times_Z Y) = \operatorname{Hom}(W, X) \times_{\operatorname{Hom}(W, Z)} \operatorname{Hom}(W, Y).$$

This is one justification for making the definition of scheme-valued point. For this reason, those classical people preferring to think only about varieties over an algebraically closed field $\bar{k}$ (or, more generally, finite type schemes over $\bar{k}$), and preferring to understand them through their closed points—or, equivalently, the $\bar{k}$-valued points, by the Nullstellensatz (Exercise 5.3.F)—needn't worry: the closed points of the product of two finite type $\bar{k}$-schemes over $\bar{k}$ are (naturally identified with) the product of the closed points of the factors. This will follow from the fact that the product is also finite type over $\bar{k}$, which we verify in Exercise 10.2.E. This is one of the reasons that varieties over algebraically closed fields can be easier to work with. But over a nonalgebraically closed field, things become even more interesting; Example 10.2.3 is a first glimpse.

10.1.4. *Fancy aside.* You may feel that (i) "products of topological spaces are products on the underlying sets" is natural, while (ii) "products of schemes are not necessarily products on the underlying sets" is weird. But really (i) is the lucky consequence of the fact that the underlying set of a topological space can be interpreted as set of p-valued points, where p is a point, so it is best seen as a consequence of paragraph 10.1.3, which is the "more correct"—i.e., more general—fact.

10.1.5. *Warning on Noetherianness.* The fibered product of Noetherian schemes need not be Noetherian. You will later be able to verify that Exercise 10.2.F gives an example, i.e., that $A := \overline{\mathbb{Q}} \otimes_{\mathbb{Q}} \overline{\mathbb{Q}}$ is not Noetherian, as follows. By Exercise 12.1.H(a), $\dim A = 0$. A Noetherian dimension 0 scheme has a finite number of points (Exercise 12.1.C). But by Exercise 10.2.F, $\operatorname{Spec} A$ has an infinite number of points.

On the other hand, the fibered product of finite type k-schemes over finite type k-schemes is a finite type k-scheme (Exercise 10.2.E), so this pathology does not arise for varieties.

10.1.6. Philosophy behind the proof of Theorem 10.1.1. The proof of Theorem 10.1.1 can be confusing. The following comments may help a little.

We already basically know existence of fibered products in two cases: the case where X, Y, and Z are affine (stated explicitly below), and the case where $\beta \colon Y \to Z$ is an open embedding (Exercise 8.1.D).

10.1.A. EXERCISE (PROMISED IN REMARK 7.3.5). Use Exercise 7.3.G ($\operatorname{Hom}_{Sch}(W, \operatorname{Spec} A) = \operatorname{Hom}_{Rings}(A, \Gamma(W, \mathscr{O}_W))$) to show that given ring maps $C \to A$ and $C \to B$,

$$\operatorname{Spec}(A \otimes_C B) \cong \operatorname{Spec} A \times_{\operatorname{Spec} C} \operatorname{Spec} B.$$

(Interpret tensor product as the "fibered coproduct" in the category of rings.) Hence the fibered product of affine schemes exists (in the category of schemes). (This generalizes the fact that the product of affine lines exist, Exercise 7.6.E(a).)

The main theme of the proof of Theorem 10.1.1 is that because schemes are built by gluing affine schemes along open subsets, these two special cases will be all that we need. The argument will repeatedly use the same ideas—roughly, that schemes glue (Exercise 4.4.A), and that morphisms of schemes glue (Exercise 7.3.B). This is a sign that something more structural is going on; §10.1.7 describes this for experts.

Proof of Theorem 10.1.1. The key idea is this: we cut everything up into affine open sets, do fibered products there, and show that everything glues nicely. The conceptually difficult part of the proof comes from the gluing, and the realization that we have to check almost nothing. We divide the proof up into a number of bite-sized pieces.

Step 1: Fibered products of affine with "almost-affine" over affine. We begin by combining the affine case with the open embedding case as follows. Suppose X and Z are affine, and $\beta \colon Y \to Z$ factors as $Y \overset{\iota}{\hookrightarrow} Y' \longrightarrow Z$, where ι is an open embedding and Y' is affine. Then $X \times_Z Y$ exists. This is because if the two small squares of are Cartesian diagrams, then the "outside rectangle" is also a Cartesian diagram. (This was Exercise 1.2.Q, although you should be able to see this on the spot.) It will be important to remember (from Important Exercise 8.1.D) that "open embeddings" are "preserved by fibered product": the fact that $Y \to Y'$ is an open embedding implies that $W \to W'$ is an open embedding.

Key step 2: Fibered product of affine with arbitrary over affine exists. We now come to the key part of the argument: if X and Z are affine, and Y is arbitrary. This is confusing when you first see it, so we first deal with a special case, when Y is the union of two affine open sets, $Y_1 \cup Y_2$. Let $Y_{12} = Y_1 \cap Y_2$.

For $i = 1$ and 2, $X \times_Z Y_i$ exists by the affine case, Exercise 10.1.A. Call this W_i. Also, $X \times_Z Y_{12}$ exists by Step 1 (call it W_{12}), and comes with *canonical* open embeddings into W_1 and W_2 (by construction of fibered products with open embeddings; see the last sentence of Step 1). Thus we can glue W_1 to W_2 along W_{12}; call this resulting scheme W.

We check that the result is the fibered product by verifying that it satisfies the universal property. Suppose we have maps $\alpha'' \colon V \to X$ and $\beta'' \colon V \to Y$ that compose (with α and β, respectively) to the same map $V \to Z$. We need to construct a unique map $\gamma \colon V \to W$, so that $\alpha' \circ \gamma = \beta''$ and $\beta' \circ \gamma = \alpha''$.

(10.1.6.1)

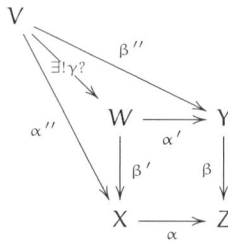

For $i = 1$ and 2, define $V_i := (\beta'')^{-1}(Y_i)$. Define $V_{12} := (\beta'')^{-1}(Y_{12}) = V_1 \cap V_2$. Then there is a unique map $V_i \to W_i$ such that the composed maps $V_i \to X$ and $V_i \to Y_i$ are as desired (by the universal property of the fibered product $X \times_Z Y_i = W_i$), hence a unique map $\gamma_i \colon V_i \to W$. Similarly, there is a unique map $\gamma_{12} \colon V_{12} \to W$ such that the composed maps $V_{12} \to X$ and $V_{12} \to Y$ are as desired. But the restriction of γ_i to V_{12} is one such map, so it must be γ_{12}. Thus the maps γ_1 and γ_2 agree on V_{12}, and glue together to a unique map $\gamma \colon V \to W$. We have shown existence and uniqueness of the desired γ.

We have thus shown that if Y is the union of two affine open sets, and X and Z are affine, then $X \times_Z Y$ exists.

We now tackle the general case. (You may prefer to first think through the case where "two" is replaced by "three.") We now cover Y with open sets Y_i, as i runs over some index set (not necessarily finite!). As before, we define W_i and W_{ij}. We can glue these together to produce a scheme W along with open sets we identify with W_i (Exercise 4.4.A—you should check the triple intersection cocycle condition).

As in the two-affine case, we show that W is the fibered product by showing that it satisfies the universal property. Suppose we have maps $\alpha'' \colon V \to X$, $\beta'' \colon V \to Y$ that compose to the same map $V \to Z$. We construct a unique map $\gamma \colon V \to W$, so that $\alpha' \circ \gamma = \beta''$ and $\beta' \circ \gamma = \alpha''$. Define $V_i = (\beta'')^{-1}(Y_i)$ and $V_{ij} := (\beta'')^{-1}(Y_{ij}) = V_i \cap V_j$. Then there is a unique map $V_i \to W_i$ such that the composed maps $V_i \to X$ and $V_i \to Y_i$ are as desired, hence a unique map $\gamma_i \colon V_i \to W$. Similarly, there is a unique map $\gamma_{ij} \colon V_{ij} \to W$ such that the composed maps $V_{ij} \to X$ and $V_{ij} \to Y$ are as desired. But the restriction of γ_i to V_{ij} is one such map, so it must be γ_{ij}. Thus the maps γ_i and γ_j agree on V_{ij}. Thus the γ_i glue together to a unique map $\gamma \colon V \to W$. We have shown existence and uniqueness of the desired γ, completing this step.

Step 3: Z affine, X and Y arbitrary. We next show that if Z is affine, and X and Y are arbitrary schemes, then $X \times_Z Y$ exists. We just follow Step 2, with the roles of X and Y reversed, using the fact that, by the previous step, we can assume that the fibered product of an affine scheme with an arbitrary scheme over an affine scheme exists.

Step 4: Z "almost-affine," X and Y arbitrary. This is akin to Step 1. Let $Z \hookrightarrow Z'$ be an open embedding into an affine scheme. Then $X \times_{Z'} Y$ satisfies the universal property of $X \times_Z Y$. (If you don't see this directly, one argument is to use Easy Exercise 1.2.X and the fact that open embeddings are monomorphisms, Exercise 8.1.E.)

Step 5: The general case. We employ the same trick yet again. Suppose $\alpha\colon X \to Z$, $\beta\colon Y \to Z$ are two morphisms of schemes. Cover Z with affine open subschemes Z_i, and let $X_i = \alpha^{-1}(Z_i)$ and $Y_i = \beta^{-1}(Z_i)$. Define $Z_{ij} := Z_i \cap Z_j$, $X_{ij} := \alpha^{-1}(Z_{ij})$, and $Y_{ij} := \beta^{-1}(Z_{ij})$. Then $W_i := X_i \times_{Z_i} Y_i$ exists for all i (Step 3), and $W_{ij} := X_{ij} \times_{Z_{ij}} Y_{ij}$ exists for all i, j (Step 4), and for each i and j, W_{ij} comes with a canonical open embedding into both W_i and W_j (see the last sentence in Step 1). As W_i satisfies the universal property of $X \times_Z Y_i$ (do you see why?), we may canonically identify W_i (which we know to exist by Step 3) with $X \times_Z Y_i$. Similarly, we identify W_{ij} with $X \times_Z Y_{ij}$.

We then proceed exactly as in Step 2: The W_i's can be glued together along the W_{ij} (the cocycle condition can be readily checked to be satisfied), and W can be checked to satisfy the universal property of $X \times_Z Y$ (again, exactly as in Step 2). $\qquad\square$

10.1.7.** Describing the existence of fibered products using the fancy language of representable functors.

The proof above can be described more cleanly in the language of representable functors (§7.6). This will be enlightening only after you have absorbed the above argument and meditated on it for a long time. It may be most useful to shed light on representable functors, rather than on the existence of the fibered product.

Until the end of §10.1 only, by functor *we mean contravariant functor from the category Sch of schemes to the category of Sets.* For each scheme X, we have a functor h_X, taking a scheme Y to the set $\mathrm{Mor}(Y, X)$ (§1.1.20). Recall (§1.2.11, §7.6) that a functor is *representable* if it is naturally isomorphic to some h_X. If a functor is representable, then the representing scheme is unique up to unique isomorphism (Exercise 7.6.C). This can be usefully extended as follows:

10.1.B. EXERCISE (YONEDA'S LEMMA).

(a) If X and Y are schemes, describe a bijection between morphisms of schemes $X \to Y$ and natural transformations of functors $h_X \to h_Y$. Hence show that the category of schemes is a full subcategory (defined in §1.1.15) of the "functor category" (all contravariant functors $Sch \to Sets$).

(b) If X is a scheme and F is a contravariant functor $Sch \to Sets$, give a bijection between $F(X)$ and the natural transformation of functors $h_X \to F$. (You are free to check that your bijection is "functorial" in both F and X.)

Hint: This question has nothing to do with schemes; your argument will work in any category. This is the contravariant version of Exercise 1.2.Z(c).

One of Grothendieck's insights is that we should try to treat such functors as "geometric spaces," without worrying about representability. Many notions carry over to this more general setting without change, and some notions are easier. For example, fibered products of functors always exist: $h \times_{h''} h'$ may be defined by

$$(h \times_{h''} h')(W) = h(W) \times_{h''(W)} h'(W),$$

where the fibered product on the right is a fibered product of sets, which always exists. (This isn't quite enough to define a functor; we have only described where objects go. You should work out where morphisms go, too.) We didn't use anything about schemes; this works with Sch replaced by any category.

Then "$X \times_Z Y$ exists" translates to "$h_X \times_{h_Z} h_Y$ is representable."

10.1.8. *Representable functors are Zariski sheaves.*

Because "morphisms to schemes glue" (Exercise 7.3.B), we have a necessary condition for a functor to be representable. We know that if $\{U_i\}$ is an open cover of Y, a morphism $Y \to X$ is determined by its restrictions $U_i \to X$, and given morphisms $U_i \to X$ that agree on the overlap $U_i \cap U_j \to X$, we can glue them together to get a morphism $Y \to X$. In the language of equalizer exact sequences (§2.2.7),

$$\cdot \longrightarrow h_X(Y) \longrightarrow \prod h_X(U_i) \rightrightarrows \prod h_X(U_i \cap U_j)$$

is exact. Thus morphisms to X (i.e., the functor h_X) form a sheaf on every scheme Y. If this holds, we say that the functor is a **Zariski sheaf**. (You can impress your friends by telling them that this is a *sheaf on the big Zariski site*.) We can repeat this discussion with *Sch* replaced by the category *Sch_S* of schemes over a given base scheme S. We have proved (or observed) that *in order for a functor to be representable, it is necessary for it to be a Zariski sheaf.*

The fibered product passes this test:

10.1.C. EXERCISE. If $X, Y \to Z$ are schemes, show that $h_X \times_{h_Z} h_Y$ is a Zariski sheaf. (Do not use the fact that $h_X \times_{h_Z} h_Y$ is representable! The point of this section is to recover representability from a more sophisticated perspective.)

We can make some other definitions that extend notions from schemes to functors. We say that a map (i.e., natural transformation) of functors $h' \to h$ expresses h' as a **subfunctor** of h if for all schemes X, the induced map of sets $h'(X) \to h(X)$ is an inclusion. We say that a subfunctor $h' \hookrightarrow h$ is an **open subfunctor** if for all schemes X and maps $h_X \to h$ (i.e., all elements of $h(X)$), the fibered product $h_X \times_h h'$ is representable, by the scheme U, say, and $h_U \to h_X$ corresponds to an open embedding of schemes $U \to X$. The following fibered diagram may help.

$$
\begin{array}{ccc}
h_U & \xrightarrow{\ \text{open}\ } & h_X \\
\downarrow & & \downarrow \\
h' & \longrightarrow & h
\end{array}
$$

10.1.D. EXERCISE. Show that a map of representable functors $h_W \to h_Z$ is an open subfunctor if and only if $W \to Z$ is an open embedding, so this indeed extends the notion of open embedding to (contravariant) functors (*Sch* $\to$ *Sets*).

10.1.E. EXERCISE (THE GEOMETRIC NATURE OF THE NOTION OF "OPEN SUBFUNCTOR").

(a) Show that an open subfunctor of an open subfunctor is also an open subfunctor.
(b) Suppose $h' \to h$ and $h'' \to h$ are two open subfunctors of h. Define the intersection of these two open subfunctors, which should also be an open subfunctor of h.
(c) Suppose U and V are two open subschemes of a scheme X, so $h_U \to h_X$ and $h_V \to h_X$ are open subfunctors. Show that the intersection of these two open subfunctors is, as you would expect, $h_{U \cap V}$.

10.1.F. EXERCISE. Suppose $\alpha \colon X \to Z$ and $\beta \colon Y \to Z$ are morphisms of schemes, and $U \subset X$, $V \subset Y$, $W \subset Z$ are open embeddings, where U and V map to W. Interpret $h_U \times_{h_W} h_V$ as an open subfunctor of $h_X \times_{h_Z} h_Y$. (Hint: Given a map $h_T \to h_{X \times_Z Y}$, what open subset of T should correspond to $U \times_W V$?)

A collection h_i of open subfunctors of h is said to **cover** h if for *every* map $h_X \to h$ from a representable subfunctor, the corresponding open subsets $U_i \hookrightarrow X$ cover X.

Given that functors do not have an obvious underlying set (let alone a topology), it is rather amazing that we are talking about when one is an "open subset" of another, or when some functors "cover" another!

10.1.G. EXERCISE. Suppose $\{Z_i\}_i$ is an affine cover of Z, $\{X_{ij}\}_j$ is an affine cover of the preimage of Z_i in X, and $\{Y_{ik}\}_k$ is an affine cover of the preimage of Z_i in Y. Show that $\{h_{X_{ij}} \times_{h_{Z_i}} h_{Y_{ik}}\}_{ijk}$ is an open cover of the functor $h_X \times_{h_Z} h_Y$. (Hint: Consider a map $h_T \to h_X \times_{h_Z} h_Y$, and extend your solution to Exercise 10.1.F.)

We now come to a key point: a Zariski sheaf that is "locally representable" must be representable.

10.1.H. KEY EXERCISE. If a functor h is a Zariski sheaf that has an open cover by representable functors ("is covered by schemes"), then h is representable. (Hint: Use Exercise 4.4.A to glue together the schemes representing the open subfunctors.)

This immediately leads to the existence of fibered products as follows. Exercise 10.1.C shows that $h_X \times_{h_Z} h_Y$ is a Zariski sheaf. But $h_{X_{ij}} \times_{h_{Z_i}} h_{Y_{ik}}$ is representable for each i, j, k (fibered products of affines over an affine exist, Exercise 10.1.A), and these functors are an open cover of $h_X \times_{h_Z} h_Y$ by Exercise 10.1.G, so by Key Exercise 10.1.H we are done.

10.2 Computing Fibered Products in Practice

Before giving some examples, we first see how to compute fibered products in practice. There are four types of morphisms, **(1)–(4)**, that it is particularly easy to take fibered products with, and all morphisms can be built from these atomic components. More precisely, **(1)** will imply that we can compute fibered products locally on the source and target. Thus to understand fibered products in general, it suffices to understand them on the level of affine sets, i.e., to be able to compute $A \otimes_B C$ given ring maps $B \to A$ and $B \to C$. Any map $B \to A$ (and similarly $B \to C$) may be expressed as $B \to B[t_1, \dots]/I$, so if we know how to base change by "adding variables" **(2)** and "taking quotients" **(3)**, we can "compute" any fibered product (at least in theory). The fourth type of morphism **(4)**, corresponding to localization, is useful to understand explicitly as well.

(1) Base change by open embeddings.

We have already done this (Exercise 8.1.D), and we used it in the proof that fibered products of schemes exist.

(2) Adding an extra variable.

10.2.A. EASY ALGEBRA EXERCISE. Show that $A \otimes_B B[t] \cong A[t]$, so the following is a Cartesian diagram. (Your argument might naturally extend to allow the addition of infinitely many variables, but we won't need this generality.) Hint: Show that $A[t]$ satisfies an appropriate universal property.

$$
\begin{array}{ccc}
\operatorname{Spec} A[t] & \longrightarrow & \operatorname{Spec} B[t] \\
\downarrow & & \downarrow \\
\operatorname{Spec} A & \longrightarrow & \operatorname{Spec} B
\end{array}
$$

10.2.1. *Definition (affine space over an arbitrary scheme).* If X is any scheme, we define $\mathbb{A}_X^n$ as $X \times_{\mathbb{Z}} \operatorname{Spec} \mathbb{Z}[x_1, \dots, x_n]$. Clearly, $\mathbb{A}_{\operatorname{Spec} A}^n$ is canonically the same as $\mathbb{A}_A^n$ as defined in Example 8 of §3.2.4.

(3) Base change by closed embeddings.

10.2.B. EXERCISE. Suppose $\phi \colon B \to A$ is a ring morphism, and $I \subset B$ is an ideal. Let $I^e := \langle \phi(i) \rangle_{i \in I} \subset A$ be the **extension of** I **to** A. Describe a natural isomorphism $A/I^e \overset{\sim}{\longleftrightarrow} A \otimes_B (B/I)$. (Hint: Consider $I \to B \to B/I \to 0$, and use the right-exactness of $\otimes_B A$, Exercise 1.2.H.)

10.2.2. As an immediate consequence: the fibered product with a closed subscheme is a closed subscheme of the fibered product in the obvious way. We say that "closed embeddings are preserved by base change."

10.2.C. EXERCISE.

(a) Interpret the intersection of two closed embeddings into X (cf. Exercise 9.1.I) as their fibered product over X.

(b) Show that "locally closed embeddings" are preserved by base change.

(c) Define the **intersection of n locally closed embeddings** $X_i \hookrightarrow Z$ ($1 \leq i \leq n$) by the fibered product of the X_i over Z (mapping to Z). Show that the intersection of (a finite number of) locally closed embeddings is also a locally closed embedding.

10.2.D. EXERCISE. Suppose $I \subset A[x_1, \ldots, x_m]$ and $J \subset A[y_1, \ldots, y_n]$ are ideals. Prove the isomorphism

$$A[x_1, \ldots, x_m]/I \otimes_A A[y_1, \ldots, y_n]/J \xleftarrow{\sim} A[x_1, \ldots, x_m, y_1, \ldots, y_n]/(I, J).$$

Hint: Exercises 10.2.A and 10.2.B.

As an application, we can compute tensor products of finitely generated k-algebras over k. For example, we have

$$k[x_1, x_2]/(x_1^2 - x_2) \otimes_k k[y_1, y_2]/(y_1^3 + y_2^3) \cong k[x_1, x_2, y_1, y_2]/(x_1^2 - x_2, y_1^3 + y_2^3).$$

10.2.E. EXERCISE. Suppose X and Y are locally of finite type A-schemes. Show that $X \times_A Y$ is also locally of finite type over A. Prove the same thing with "locally" removed from both the hypothesis and the conclusion.

10.2.3. *Example.* We can use these ideas to compute $\mathbb{C} \otimes_\mathbb{R} \mathbb{C}$:

$$
\begin{aligned}
\mathbb{C} \otimes_\mathbb{R} \mathbb{C} &\cong \mathbb{C} \otimes_\mathbb{R} (\mathbb{R}[x]/(x^2 + 1)) \\
&\cong (\mathbb{C} \otimes_\mathbb{R} \mathbb{R}[x])/(x^2 + 1) && \text{by §10.2(3)} \\
&\cong \mathbb{C}[x]/(x^2 + 1) && \text{by §10.2(2)} \\
&\cong \mathbb{C}[x]/((x - i)(x + i)) \\
&\cong \mathbb{C}[x]/(x - i) \times \mathbb{C}[x]/(x + i) && \text{by the Chinese Remainder Theorem} \\
&\cong \mathbb{C} \times \mathbb{C}
\end{aligned}
$$

Thus $\operatorname{Spec} \mathbb{C} \times_\mathbb{R} \operatorname{Spec} \mathbb{C} \cong \operatorname{Spec} \mathbb{C} \coprod \operatorname{Spec} \mathbb{C}$. This example is the first example of many different behaviors. Notice, for example, that two points somehow correspond to the Galois group of $\mathbb{C}$ over $\mathbb{R}$; for one of them, x (the "i" in one of the copies of $\mathbb{C}$) equals i (the "i" in the other copy of $\mathbb{C}$), and in the other, $x = -i$.

10.2.4.* *Remark.* Here is a clue that there is something deep going on behind Example 10.2.3. If L/K is a (finite) Galois extension with Galois group G, then $L \otimes_K L$ is isomorphic to L^G (the product of $|G|$ copies of L). This turns out to be a restatement of the classical form of linear independence of characters! In the language of schemes, $\operatorname{Spec} L \times_K \operatorname{Spec} L$ is a union of a number of copies of $\operatorname{Spec} L$ that naturally form a torsor over the Galois group G; but we will not define torsor here.

10.2.F.* HARD BUT FASCINATING EXERCISE FOR THOSE FAMILIAR WITH $\operatorname{Gal}(\overline{\mathbb{Q}}/\mathbb{Q})$. Show that the points of $\operatorname{Spec} \overline{\mathbb{Q}} \otimes_\mathbb{Q} \overline{\mathbb{Q}}$ are in natural bijection with $\operatorname{Gal}(\overline{\mathbb{Q}}/\mathbb{Q})$, and the Zariski topology on the former agrees with the profinite topology on the latter. (Some hints: First do the case of finite Galois extensions. Relate the topology on Spec of a direct limit of rings to the inverse limit of Specs. Can you see which point corresponds to the identity of the Galois group?)

At this point, we can compute any $A \otimes_B C$ (where A and C are B-algebras): any map of rings $\phi \colon B \to A$ can be interpreted by adding variables (perhaps infinitely many) to B, and then imposing relations. But, in practice, **(4)** is useful, as we will see in examples.

(4) Base change of affine schemes by localization.

10.2.G. EXERCISE. Suppose $\phi \colon B \to A$ is a ring morphism, and $S \subset B$ is a multiplicative subset of B, which implies that $\phi(S)$ is a multiplicative subset of A. Describe a natural isomorphism $\phi(S)^{-1} A \xleftarrow{\sim} A \otimes_B (S^{-1}B)$.

Informal translation: "The fibered product with a localization is the localization of the fibered product in the obvious way." We say that "localizations are preserved by base change." This is handy if the localization is of the form $B \hookrightarrow B_f$ (corresponding to taking distinguished open sets) or (if B is an integral domain) $B \hookrightarrow K(B)$ (corresponding to taking the generic point), or various things in between.

10.2.5. Examples. These four facts let you calculate lots of things in practice, and we will use them freely.

10.2.H. EXERCISE: THE THREE IMPORTANT TYPES OF MONOMORPHISMS OF SCHEMES. Show that the following are monomorphisms (Definition 1.2.10): open embeddings, closed embeddings, and localization of affine schemes. As monomorphisms are preserved by composition (Exercise 1.2.V), compositions of the above are also monomorphisms—for example, locally closed embeddings, or maps from "Spec of stalks at points of X" to X. (Caution: If $\mathfrak{p}$ is a point of a scheme X, the natural morphism $\operatorname{Spec} \mathscr{O}_{X,\mathfrak{p}} \to X$—cf. Exercise 7.3.M—is a monomorphism but is not in general an open embedding.)

10.2.I. EXERCISE. Recall that $\mathbb{A}^n_A \cong \mathbb{A}^n_\mathbb{Z} \times_{\operatorname{Spec} \mathbb{Z}} \operatorname{Spec} A$ (§10.2.1). Prove similarly that $\mathbb{P}^n_A \cong \mathbb{P}^n_\mathbb{Z} \times_{\operatorname{Spec} \mathbb{Z}} \operatorname{Spec} A$. Thus affine space and projective space are pulled back from their "universal manifestation" over the final object $\operatorname{Spec} \mathbb{Z}$.

10.2.6. *Extending the base field.* One special case of base change is called **extending the base field**: if X is a k-scheme, and ℓ is a field extension (often ℓ is the algebraic closure of k), then $X \times_{\operatorname{Spec} k} \operatorname{Spec} \ell$ (sometimes informally written $X \times_k \ell$ or X_ℓ) is an ℓ-scheme. Often properties of X can be checked by verifying them instead on X_ℓ. This is the subject of *descent*—certain properties "descend" from X_ℓ to X. We have already seen that the property of being the Spec of a normal integral domain descends in this way (Exercise 5.4.M). Exercises 10.2.J and 10.2.K give other examples of properties which descend: the property of two morphisms being equal, and the property of a(n affine) morphism being a closed embedding, both descend in this way. Those interested in schemes over non-algebraically closed fields will use this philosophy repeatedly, to reduce results to the algebraically closed case.

10.2.J. EXERCISE. Suppose $\pi\colon X \to Y$ and $\rho\colon X \to Y$ are morphisms of k-schemes, ℓ/k is a field extension, and $\pi_\ell\colon X \times_{\operatorname{Spec} k} \operatorname{Spec} \ell \to Y \times_{\operatorname{Spec} k} \operatorname{Spec} \ell$ and $\rho_\ell\colon X \times_{\operatorname{Spec} k} \operatorname{Spec} \ell \to Y \times_{\operatorname{Spec} k} \operatorname{Spec} \ell$ are the induced maps of ℓ-schemes. (Be sure you understand what this means!) Show that if $\pi_\ell = \rho_\ell$ then $\pi = \rho$. (Hint: Show that π and ρ are the same on the level of sets. To do this, you may use that $X \times_{\operatorname{Spec} k} \operatorname{Spec} \ell \to X$ is surjective, which we will soon prove in Exercise 10.4.G. Then reduce to the case where X and Y are affine.)

10.2.K. EASY EXERCISE. Suppose $\pi\colon X \to Y$ is an affine morphism over k, and ℓ/k is a field extension. Show that π is a closed embedding if and only if $\pi \times_k \ell\colon X \times_k \ell \to Y \times_k \ell$ is. (The affine hypothesis is not necessary for this result, but it makes the proof easier, and this is the situation in which we will most need it.)

10.2.L. EXERCISE (SEEMINGLY PATHOLOGICAL BEHAVIOR IN NONPATHOLOGICAL CIRCUMSTANCES). Suppose k is a field, and $A = k(x) \otimes_k k(y)$. Show that A is a (nonzero) localization of $k[x, y]$, hence an integral domain. Thus (0) is the unique minimal prime ideal of A. Show that the remaining prime ideals of A correspond to ideals $(f(x, y)) \subset k[x, y]$, where $f(x, y)$ is an irreducible polynomial in $k[x, y]$ containing *both* the variables x and y. (Hint: **(2)** and **(4)** above.)

This example will come up again in Remarks 12.2.17 and 24.5.9. The idea in the solution to Exercise 10.2.L also yields the following.

10.2.M. UNIMPORTANT BUT FUN EXERCISE. Show that $\operatorname{Spec} \mathbb{Q}(t) \otimes_\mathbb{Q} \mathbb{C}$ has closed points in natural correspondence with the transcendental complex numbers. This scheme doesn't come up in nature, but it is certainly neat!

10.3 Interpretations: Pulling Back Families, and Fibers of Morphisms

10.3.1. Pulling back families.

Before making any definitions, we give a motivating informal example. Consider the "family of curves"

$$y^2 = x^3 + tx$$

in the xy-plane parametrized by t. Translation: Consider

$$\operatorname{Spec} k[x, y, t]/(y^2 - x^3 - tx) \longrightarrow \operatorname{Spec} k[t].$$

If we "pull back this family" to the uv-plane via $uv = t$, we get the family

$$y^2 = x^3 + uvx.$$

If instead we set t to 3, we get the curve

$$y^2 = x^3 + 3x,$$

which we interpret as the fiber of the original family above $t = 3$. You may have noticed the fiber products underlying in these constructions:

$$
\begin{array}{ccc}
\operatorname{Spec} k[x, y, u, v]/(y^2 - x^3 - uvx) & \longrightarrow & \operatorname{Spec} k[x, y, t]/(y^2 - x^3 - tx) \\
\downarrow & & \downarrow \\
\operatorname{Spec} k[u, v] & \xrightarrow{\;\;uv \mapsfrom t\;\;} & \operatorname{Spec} k[t]
\end{array}
$$

and

(10.1.1)
$$
\begin{array}{ccc}
\operatorname{Spec} k[x, y]/(y^2 - x^3 - 3x) & \longrightarrow & \operatorname{Spec} k[x, y, t]/(y^2 - x^3 - tx) \\
\downarrow & & \downarrow \\
\operatorname{Spec} k & \xrightarrow{\;\;3 \mapsfrom t\;\;} & \operatorname{Spec} k[t]
\end{array}
$$
.

We now formalize this.

Suppose $Y \to Z$ is a morphism. We interpret this as a "family of schemes parametrized by a **base scheme** (or just plain **base**) Z." Then if we have another morphism $\psi \colon X \to Z$, we interpret the induced map $X \times_Z Y \to X$ as the "pulled back family" (see Figure 10.1).

$$
\begin{array}{ccc}
X \times_Z Y & \longrightarrow & Y \\
{\scriptstyle\text{pulled back family}} \downarrow & & \downarrow {\scriptstyle\text{family}} \\
X & \xrightarrow{\;\;\psi\;\;} & Z
\end{array}
$$

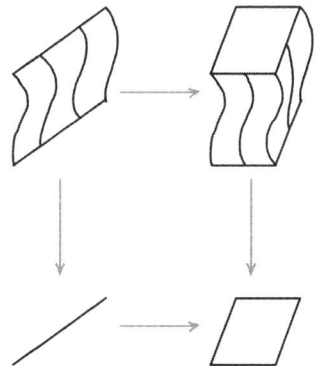

Figure 10.1 *A picture of a pulled back family.*

We sometimes say that $X \times_Z Y$ is the **scheme-theoretic preimage** (or *scheme-theoretic pullback*, or *scheme-theoretic inverse image*, or *inverse image scheme*) of Y. (Our forthcoming discussion of fibers may give some motivation for this.) For this reason, fibered product is often called **base change** or **change of base** or **pullback**. In addition to the various names for a Cartesian diagram given in §1.2.7, in algebraic geometry it is often called a **base change diagram** or a **pullback diagram**, and $X \times_Z Y \to X$ is called the **pullback** of $Y \to Z$ by ψ, and $X \times_Z Y$ is called the **pullback** of Y by ψ. One often uses the phrase "over X" or "above X" when discussing $X \times_Z Y$, especially if X is a locally closed subscheme of Z. (Random side remark: Scheme-theoretic preimage always makes sense, while the notion of scheme-theoretic image is somehow problematic, as discussed in §9.4.1.)

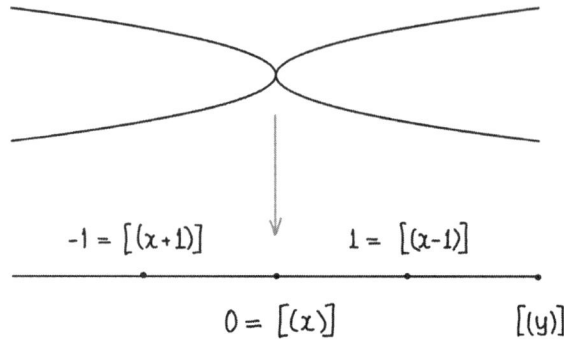

Figure 10.2 *The map* $\mathbb{C} \to \mathbb{C}$ *given by* $y \mapsto y^2$ *(cf. Figure 3.6).*

10.3.2. Fibers of morphisms.

(If you did Exercise 8.3.J, that finite morphisms have finite fibers, you will not find the following discussion surprising.) A special case of pullback is the notion of a fiber of a morphism. We motivate this with the notion of fiber in the category of topological spaces.

10.3.A. EXERCISE. Show that if $Y \to Z$ is a continuous map of topological spaces, and X is a point p of Z, then the fiber of Y over p (the set-theoretic fiber, with the induced topology) is naturally identified with $X \times_Z Y$.

More generally, for any $\pi\colon X \to Z$, the fiber of $X \times_Z Y \to X$ over a point p of X is naturally identified with the fiber of $Y \to Z$ over $\pi(p)$.

10.3.3. *Definition.* Motivated by topology, we return to the category of schemes. Suppose $p \to Z$ is the inclusion of a point (not necessarily closed). More precisely, if p is a point with residue field K, consider the map $\operatorname{Spec} K \to Z$ sending $\operatorname{Spec} K$ to p, with the natural isomorphism of residue fields. Then if $g\colon Y \to Z$ is any morphism, the scheme-theoretic preimage of p is called the **scheme-theoretic fiber** (or *fibre*) **of** g **above** p, and is denoted by $g^{-1}(p)$. If Z is irreducible, the scheme-theoretic fiber above the generic point of Z is called the **generic fiber** (of g). In an affine open subscheme $\operatorname{Spec} A$ containing p, p corresponds to some prime ideal $\mathfrak{p}$, and the morphism $\operatorname{Spec} K \to Z$ corresponds to the ring map $A \to A_{\mathfrak{p}}/\mathfrak{p}A_{\mathfrak{p}}$. This is the composition of localization and closed embedding, and thus can be computed by the tricks above. (Note that $p \to Z$ is a monomorphism, by Exercise 10.2.H.)

10.3.B. EXERCISE. Show that the underlying topological space of the scheme-theoretic fiber of $X \to Y$ above a point p is naturally identified with the topological fiber of $X \to Y$ above p.

10.3.C. EXERCISE (ANALOG OF EXERCISE 10.3.A). Suppose that $\pi\colon Y \to Z$ and $\tau\colon X \to Z$ are morphisms, and $p \in X$ is a point. Show that the fiber of $X \times_Z Y \to X$ over p is (isomorphic to) the base change to p of the fiber of $\pi\colon Y \to Z$ over $\tau(p)$.

10.3.4. *Example (enlightening in several ways).* Consider the projection of the parabola $y^2 = x$ to the x-axis over $\mathbb{Q}$, corresponding to the map of rings $\mathbb{Q}[x] \to \mathbb{Q}[y]$, with $x \mapsto y^2$. If $\mathbb{Q}$ alarms you, replace it with your favorite field and see what happens. (You should look at Figure 10.2, which is a flipped version of the parabola of Figure 3.6, and figure out how to edit it to reflect what we glean here.) Writing $\mathbb{Q}[y]$ as $\mathbb{Q}[x, y]/(y^2 - x)$ helps us interpret the morphism conveniently.

(i)   Then the preimage of 1 is two points:

$$\operatorname{Spec} \mathbb{Q}[x, y]/(y^2 - x) \otimes_{\mathbb{Q}[x]} \mathbb{Q}[x]/(x - 1)$$
$$\cong \operatorname{Spec} \mathbb{Q}[x, y]/(y^2 - x, x - 1)$$

$$\cong \mathrm{Spec}\, \mathbb{Q}[y]/(y^2-1)$$
$$\cong \mathrm{Spec}\, \mathbb{Q}[y]/(y-1) \coprod \mathrm{Spec}\, \mathbb{Q}[y]/(y+1).$$

(ii) The preimage of 0 is one nonreduced point:

$$\mathrm{Spec}\, \mathbb{Q}[x,y]/(y^2-x,x) \cong \mathrm{Spec}\, \mathbb{Q}[y]/(y^2).$$

(iii) The preimage of -1 is one reduced point, but of "size 2 over the base field."

$$\mathrm{Spec}\, \mathbb{Q}[x,y]/(y^2-x,x+1) \cong \mathrm{Spec}\, \mathbb{Q}[y]/(y^2+1)$$
$$\cong \mathrm{Spec}\, \mathbb{Q}[i]$$
$$= \mathrm{Spec}\, \mathbb{Q}(i).$$

(iv) The preimage of the generic point is again one reduced point, but of "size 2 over the residue field," as we verify now.

$$\mathrm{Spec}\, \mathbb{Q}[x,y]/(y^2-x) \otimes_{\mathbb{Q}[x]} \mathbb{Q}(x) \cong \mathrm{Spec}\, \mathbb{Q}[y] \otimes_{\mathbb{Q}[y^2]} \mathbb{Q}(y^2),$$

i.e., (informally) the Spec of the ring of polynomials in y divided by polynomials in y^2. A little thought shows you that in this ring you may invert *any* polynomial in y, as if $f(y)$ is any polynomial in y, then

$$\frac{1}{f(y)} = \frac{f(-y)}{f(y)f(-y)},$$

and the latter denominator is a polynomial in y^2. Thus

$$\mathbb{Q}[x,y]/(y^2-x) \otimes_{\mathbb{Q}[x]} \mathbb{Q}(x) \cong \mathbb{Q}(y),$$

which is a degree 2 field extension of $\mathbb{Q}(x)$ (note that $\mathbb{Q}(x) = \mathbb{Q}(y^2)$).

(You might want to work through the preimages of other points of $\mathbb{A}^1_{\mathbb{Q}}$, such as $[(x^3-4)]$ and $[(x^2-2)]$.)

Notice the following interesting fact. In each of the four cases, the number of preimages can be interpreted as 2, where you count to two in several ways: you can count points (as in the case of the preimage of 1); you can get nonreduced behavior (as in the case of the preimage of 0); or you can have a field extension of degree 2 (as in the case of the preimage of -1 or the generic point). In each case, the fiber is an affine scheme whose dimension as a vector space over the residue field of the point is 2. Number-theoretic readers may have seen this behavior before. We will discuss this example again in §16.3.10. This is going to be symptomatic of a very important kind of morphism (a finite flat morphism; see Remark 24.3.8 and §24.3.12).

Try to draw a picture of this morphism if you can, so you can develop a pictorial shorthand for what is going on. A good first approximation is the parabola of Figure 10.2, but you will want to somehow depict the peculiarities of (iii) and (iv).

10.3.5. *Remark: Finite morphisms have finite fibers.* If you haven't done Exercise 8.3.J, that finite morphisms have finite fibers, now would be a good time to do it, as you will find it more straightforward given what you know now.

10.3.D. EXERCISE (IMPORTANT FOR THOSE WITH MORE ARITHMETIC BACKGROUND). What is the scheme-theoretic fiber of $\mathrm{Spec}\, \mathbb{Z}[i] \to \mathrm{Spec}\, \mathbb{Z}$ over the prime (p)? Your answer will depend on p, and there are four cases, corresponding to the four cases of Example 10.3.4. (Can you draw a picture?)

10.3.E. EXERCISE. (This exercise will give you practice in computing a fibered product over something that is not a field.) Consider the morphism of schemes $X = \mathrm{Spec}\, k[t] \to Y = \mathrm{Spec}\, k[u]$ corresponding to $k[u] \to k[t]$, $u \mapsto t^2$, where char $k \neq 2$. Show that $X \times_Y X$ has two irreducible components. (What happens if char $k = 2$? See Exercise 10.5.A for a clue.)

10.3.6. *A first view of a blow-up.*

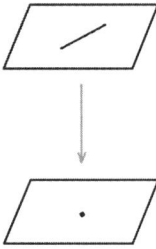

Figure 10.3 *A first example of a blow-up.*

10.3.F. IMPORTANT CONCRETE EXERCISE. (The discussion here immediately generalizes to $\mathbb{A}^n_A$.) Define a closed subscheme $\text{Bl}_{(0,0)} \mathbb{A}^2_k$ of $\mathbb{A}^2_k \times_k \mathbb{P}^1_k$ as follows (see Figure 10.3). If the coordinates on $\mathbb{A}^2_k$ are x, y, and the projective coordinates on $\mathbb{P}^1_k$ are u, v, this subscheme is cut out in $\mathbb{A}^2_k \times_k \mathbb{P}^1_k$ by the single equation $xv = yu$. (You may wish to interpret $\text{Bl}_{(0,0)} \mathbb{A}^2_k$ as follows. The $\mathbb{P}^1_k$ parametrizes lines through the origin. The blow-up corresponds to ordered pairs of (point p, line ℓ) such that $(0,0)$ and p both lie on ℓ.) Describe the fiber of the morphism $\text{Bl}_{(0,0)} \mathbb{A}^2_k \to \mathbb{P}^1_k$ over each closed point of $\mathbb{P}^1_k$. Show that the morphism $\text{Bl}_{(0,0)} \mathbb{A}^2_k \to \mathbb{A}^2_k$ is an isomorphism away from $(0,0) \in \mathbb{A}^2_k$. Show that the fiber over $(0,0)$ is an effective Cartier divisor (§9.5.1, a closed subscheme that is locally cut out by a single equation that is not a zerodivisor). It is called the **exceptional divisor**. We will discuss blow-ups in Chapter 22. This particular example will come up in the motivating example of §22.1, and in Exercise 20.2.D.

We haven't yet discussed regularity, but here is a hand-waving argument suggesting that the $\text{Bl}_{(0,0)} \mathbb{A}^2_k$ is "smooth": the preimage above either standard open set $U_i \subset \mathbb{P}^1$ is isomorphic to $\mathbb{A}^2$. Thus "the blow-up is a surgery that takes the smooth surface $\mathbb{A}^2_k$, cuts out a point, and glues back in a $\mathbb{P}^1$, in such a way that the outcome is another smooth surface."

10.3.7. General fibers, generic fibers, generically finite morphisms.
The phrases "generic fiber" and "general fiber" parallel the phrases "generic point" and "general point" (Definition 3.6.11). Suppose $\pi\colon X \to Y$ is a morphism of schemes. When one says the **general fiber** (or *a* general fiber) of π has a certain property, this means that there exists a dense open subset $U \subset Y$ such that the fibers above any point in U have that property.

When one says *the generic fiber* of $\pi\colon X \to Y$ (Definition 10.3.3), this implicitly means that Y is irreducible, and the phrase refers to the fiber over the generic point. *General fiber* and *generic fiber* are not the same thing! Clearly, if something holds for the general fiber, then it holds for the generic fiber, but the converse is not always true. However, in good circumstances, it can be—properties of the generic fiber extend to an honest open neighborhood. For example, if Y is irreducible and Noetherian, and π is finite type, then if the generic fiber of π is empty (resp., nonempty), then the general fiber is empty (resp., nonempty), by Chevalley's Theorem 8.4.2.

If $\pi\colon X \to Y$ is finite type, we say π is **generically finite** if π is finite after base change to the generic point of each irreducible component (or, equivalently, by Exercise 8.4.D, if the preimage of the generic point of each irreducible component of Y is a finite set). (The notion of generic finiteness can be defined in more general circumstances; see [Stacks, tag 073A].)

10.3.G. EXERCISE ("GENERICALLY FINITE" USUALLY MEANS "GENERALLY FINITE"). Suppose $\pi\colon X \to Y$ is an affine, finite-type, generically finite morphism of locally Noetherian schemes, and Y is reduced. Show that there is an open neighborhood of each generic point of Y over which π is actually finite. (The hypotheses can be weakened considerably; see [Stacks, tag 02NW].) Hint: Reduce to the case where Y is $\text{Spec}\,B$, where B is an integral domain. Then X is affine, say, $X = \text{Spec}\,A$. Write $A = B[x_1, \dots, x_n]/I$. Now $A \otimes_B K(B)$ is a finite $K(B)$-module (finite-dimensional vector space) by hypothesis, so there are monic polynomials $f_i(t) \in K(B)[t]$ such that $f_i(x_i) = 0$ in $A \otimes_B K(B)$. Let b be the product of the (finite number of) denominators appearing in the coefficients in the $f_i(x)$. By replacing B by B_b, argue that you can assume that $f_i(t) \in B[t]$. Then $f_i(x_i) = 0$ in $A \otimes_B K(B)$, meaning that $f_i(x_i)$ is annihilated by some nonzero element of B. By replacing B by its localization at the product of these n nonzero elements ("shrinking $\text{Spec}\,B$ further"), argue that $f_i(x_i) = 0$ in A. Then conclude.

10.3.8.** Finitely presented families (morphisms) are locally pullbacks of particularly nice families.
If you are macho and are embarrassed by Noetherian rings, the following exercise

can be used to extend results from the Noetherian case to locally finitely presented situations. Exercise 10.3.I, an extension of Chevalley's Theorem 8.4.2, is a good example.

10.3.H. EXERCISE. Suppose $\pi\colon X \to \operatorname{Spec} B$ is a finitely presented morphism. Show that there exists a base change diagram of the form

$$
\begin{array}{ccc}
X & \longrightarrow & X' \\
\pi \downarrow & & \downarrow \pi' \\
\operatorname{Spec} B & \xrightarrow{\ \rho\ } & \operatorname{Spec} \mathbb{Z}[x_1, \ldots, x_N],
\end{array}
$$

where N is some integer, and π' is finitely presented (= finite type, as the target is Noetherian; see §8.3.13). Thus each finitely presented morphism is locally (on the base) a pullback of a finite type morphism to a Noetherian scheme. Hence any result proved for Noetherian schemes and stable under base change is automatically proved for locally finitely presented morphisms to arbitrary schemes. Hint: Think about the case where X is affine first. If $X = \operatorname{Spec} A$, then $A = B[y_1, \ldots, y_n]/(f_1, \ldots, f_r)$. Choose one variable x_i for each coefficient of $f_i \in B[y_1, \ldots, y_n]$. What is X' in this case? Then consider the case where X is the union of two affine open sets that intersect in an affine open set. Then consider more general cases until you solve the full problem. You will need to use every part of the definition of finite presentation. (Exercise 25.2.B extends this result.)

10.3.I. EXERCISE (CHEVALLEY'S THEOREM FOR LOCALLY FINITELY PRESENTED MORPHISMS).

(a) Suppose that A is a finitely presented B-algebra (B not necessarily Noetherian), so $A = B[x_1, \ldots, x_n]/(f_1, \ldots, f_r)$. Show that the image of $\operatorname{Spec} A \to \operatorname{Spec} B$ is a finite union of locally closed subsets of $\operatorname{Spec} B$. Hint: Exercise 10.3.H (the simpler affine case).

(b) Show that if $\pi\colon X \to Y$ is a quasicompact locally finitely presented morphism, and Y is quasicompact, then $\pi(X)$ is a finite union of locally closed subsets. (For hardened experts only: [Gr-EGA, $0_{\mathrm{III}}.9.1$] gives a definition of *local constructibility*, and of constructibility in more generality. The general form of Chevalley's Constructibility Theorem [Gr-EGA, $\mathrm{IV}_1.1.8.4$] is that the image of a locally constructible subset, under a finitely presented map, is also locally constructible.)

10.4 Properties Preserved by Base Change

All "reasonable" properties of morphisms are preserved under base change (cf. §8.1(ii)). We discuss this, and in §10.5.1 we will explain how to fix those that don't fit this pattern.

We have already shown that the notion of "open embedding" is preserved by base change (Exercise 8.1.D). We did this by explicitly describing what the fibered product of an open embedding is: if $Y \hookrightarrow Z$ is an open embedding, and $\psi\colon X \to Z$ is any morphism, then we checked that the open subscheme $\psi^{-1}(Y)$ of X satisfies the universal property of fibered products.

We have also shown that the notion of "closed embedding" is preserved by base change (§10.2 (3)). In other words, given a Cartesian diagram

$$
\begin{array}{ccc}
W & \longrightarrow & Y \\
\downarrow & & \downarrow{\scriptstyle \text{cl. emb.}} \\
X & \longrightarrow & Z
\end{array}
$$

where $Y \hookrightarrow Z$ is a closed embedding, $W \to X$ is as well.

10.4.A. EASY EXERCISE. Show that the closed embeddings and locally closed embeddings are both "reasonable" classes of morphisms in the sense of §8.1.

10.4.B. EXERCISE. Suppose $X \to Z$ and $Y \to Z$ are both locally closed embeddings.

(a) Show that $X \times_Z Y \to Z$ is a locally closed embedding.
(b) In this way we define the **(scheme-theoretic) intersection** of any two locally closed subschemes of Z. Verify that this is in agreement with the previous definition of intersections of two open subschemes of Z, and the previous definition of the intersection of two closed subschemes of Z.

10.4.C. EASY EXERCISE. Show that locally principal closed subschemes (Definition 9.5.1) pull back to locally principal closed subschemes.

Similarly, other important properties are preserved by base change.

10.4.D. EXERCISE. Show that the following properties of morphisms are preserved by base change.

(a) quasicompact
(b) affine morphism
(c) finite
(d) integral
(e) locally of finite type
(f) finite type
(g)* locally of finite presentation

10.4.E. EXERCISE. Show that the properties of morphisms of the previous exercise are all "reasonable" in the sense of §8.1.

10.4.F. EXERCISE.

(a)* Show that the notion of "quasifinite morphism" (finite type + finite fibers, Definition 8.3.11) is preserved by base change. (Warning: The notion of "finite fibers" is not preserved by base change. $\operatorname{Spec} \overline{\mathbb{Q}} \to \operatorname{Spec} \mathbb{Q}$ has finite fibers, but $\operatorname{Spec} \overline{\mathbb{Q}} \otimes_{\mathbb{Q}} \overline{\mathbb{Q}} \to \operatorname{Spec} \overline{\mathbb{Q}}$ has one point for each element of $\operatorname{Gal}(\overline{\mathbb{Q}}/\mathbb{Q})$; see Exercise 10.2.F.) Hint: Reduce to the case $\operatorname{Spec} A \to \operatorname{Spec} B$. Reduce to the case $\phi: \operatorname{Spec} A \to \operatorname{Spec} k$. By Exercise 8.4.D, such ϕ are actually finite, and finiteness is preserved by base change.
(b) Show that the notion of "quasifinite morphism" is "reasonable" in the sense of §8.1.

10.4.G. EXERCISE.

(a) Show that surjectivity is preserved by base change. (**Surjectivity** has its usual meaning: surjective as a map of sets.) You may end up showing that for any fields k_1 and k_2 containing k_3, $k_1 \otimes_{k_3} k_2$ is nonzero, and using Zorn's Lemma to find a maximal ideal in $k_1 \otimes_{k_3} k_2$.
(b) Show that surjective morphisms form a "reasonable" class (§8.1).

10.4.1. On the other hand, injectivity is not preserved by base change—witness the bijection $\operatorname{Spec} \mathbb{C} \to \operatorname{Spec} \mathbb{R}$, which loses injectivity upon base change by $\operatorname{Spec} \mathbb{C} \to \operatorname{Spec} \mathbb{R}$ (see Example 10.2.3). This can be rectified (see §10.5.I).

10.4.H. EXERCISE (CF. EXERCISE 10.2.E). Suppose X and Y are integral finite type $\overline{k}$-schemes. Show that $X \times_{\overline{k}} Y$ is an integral finite type $\overline{k}$-scheme. (Once we define "variety," this will become the important fact that the product of irreducible varieties over an algebraically closed field is an irreducible variety; cf. Proposition 11.2.11. The fact that the base field $\overline{k}$ is algebraically closed is important; see §10.5.1. See Exercises 10.5.O and 10.5.R for improvements.) Hint: Reduce to the case where X and Y are both affine, say, $X = \operatorname{Spec} A$ and $Y = \operatorname{Spec} B$, with A and B integral domains. You might flip ahead to Easy Exercise 10.5.N to see how to do this. Suppose $(\sum a_i \otimes b_i)(\sum a'_j \otimes b'_j) = 0$ in $A \otimes_{\overline{k}} B$, with $a_i, a'_j \in A$, $b_i, b'_j \in B$, where both $\{b_i\}$ and $\{b'_j\}$ are linearly independent over $\overline{k}$, and a_1 and a_1' are nonzero. Show that $D(a_1 a_1') \subset \operatorname{Spec} A$ is nonempty. By the weak Nullstellensatz 3.2.5, there is a maximal $\mathfrak{m} \subset A$ in $D(a_1 a_1')$ with $A/\mathfrak{m} = \overline{k}$. By reducing modulo $\mathfrak{m}$, deduce $\left(\sum \overline{a}_i \otimes b_i \right) \left(\sum \overline{a'}_j \otimes b_j' \right) = 0$ in B, where the overline indicates residue modulo $\mathfrak{m}$. Show that this contradicts the fact that B is an integral domain.

10.5*   Properties Not Preserved by Base Change, and How to Fix Them

We saw in the previous section that many useful properties of morphisms are preserved by base change. Indeed, "preserved by base change" is one of the properties any "reasonable" class of morphisms should have (§8.1). But some seemingly reasonable notions are not preserved by base change, and in this section we discuss how to modify them appropriately so that they are. We do this not for our own amusement, but because the resulting notions come up in nature, and perhaps in retrospect are the notions we should have started with.

The "universal" patch to this problem is as follows. If P is some property of schemes, then a morphism of schemes is said to be **universally** P if it remains P under any base change. Then the class of universally P morphisms is preserved by base change. (Do you see why?)

An important example (§11.4) is the notion of *universally closed morphisms*. Another example is that of *universally injective morphisms*, which turns out to generalize (and "geometrize") purely inseparable field extensions (§10.5.4).

One problem with "universally P" morphisms is that it is a priori hard to determine whether a given morphism is universally P—you seem to have to check every single base change of the morphism. Finding other equivalent criteria is thus essential.

10.5.1. Geometric fibers.

There are some notions that you should reasonably expect to be preserved by pullback based on your geometric intuition. Given a family in the topological category, fibers pull back in reasonable ways. So for example, any pullback of a family in which all the fibers are irreducible will also have this property; ditto for connected. Unfortunately, both of these fail in algebraic geometry, as Example 10.2.3 shows:

(10.5.1.1)
$$
\begin{array}{ccc}
\operatorname{Spec} \mathbb{C} \coprod \operatorname{Spec} \mathbb{C} & \longrightarrow & \operatorname{Spec} \mathbb{C} \\
\downarrow & & \downarrow \\
\operatorname{Spec} \mathbb{C} & \longrightarrow & \operatorname{Spec} \mathbb{R}.
\end{array}
$$

The family on the right (the vertical map) has irreducible and connected fibers, and the one on the left doesn't. The same example shows that the notion of "integral fibers" also doesn't behave well under pullback. And we used it in §10.4.1 to show that injectivity isn't preserved by base change.

10.5.A. EXERCISE. Suppose k is a field of characteristic p, so $k(u)/k(u^p)$ is a purely inseparable extension. By considering $k(u) \otimes_{k(u^p)} k(u)$, show that the notion of "reduced fibers" does not necessarily behave well under pullback. (We will soon see that this happens only in characteristic p, in the presence of inseparability.)

We rectify this problem as follows.

10.5.2. A **geometric point** of a scheme X is defined to be a morphism $\operatorname{Spec} k \to X$, where k is an algebraically closed field. Awkwardly, this is now the third kind of "point" of a scheme! There are just plain points, which are elements of the underlying set; there are Z-valued points (Z a scheme), which are maps $Z \to X$, §7.3.10; and there are geometric points. Geometric points are clearly a flavor of a scheme-valued point, but they are also an enriched version of a (plain) point: they are the data of a point with an inclusion of the residue field of the point in an algebraically closed field.

A **geometric fiber** of a morphism $X \to Y$ is defined to be the fiber over a geometric point of Y, i.e., the fibered product with the geometric point $\operatorname{Spec} k \to Y$. A morphism has **connected** (resp., **irreducible**, **integral**, **reduced**) **geometric fibers** if all its geometric fibers are connected (resp., irreducible, integral, reduced). One usually says that the morphism has **geometrically connected** (resp., **geometrically irreducible**, **geometrically integral**, **geometrically reduced**) fibers. A k-scheme X is **geometrically connected** (resp., **geometrically irreducible**, **geometrically integral**, **geometrically reduced**) if the structure morphism $X \to \operatorname{Spec} k$ has geometrically connected (resp., irreducible, integral, reduced) fibers. We will soon see that to check any of these conditions, we need only base change to $\overline{k}$.

(Warning: In some sources, in the definition of "geometric point," "algebraically closed" is replaced by "separably closed.")

10.5.B. EXERCISE. Show that the notion of "connected (resp., irreducible, integral, reduced) geometric fibers" behaves well under base change.

10.5.C. EXERCISE FOR THE ARITHMETICALLY-MINDED. Show that for the morphism $\operatorname{Spec} \mathbb{C} \to \operatorname{Spec} \mathbb{R}$, all geometric fibers consist of two reduced points. (Cf. Example 10.2.3.) Thus $\operatorname{Spec} \mathbb{C}$ is a geometrically reduced but not a geometrically irreducible $\mathbb{R}$-scheme.

10.5.D. EASY EXERCISE. Give examples of k-schemes that:

(a) are reduced but not geometrically reduced;
(b) are connected but not geometrically connected;
(c) are integral but not geometrically integral.

10.5.E. EXERCISE (TO CONVINCE GEOMETERS WHY GEOMETRIC FIBERS ARE MEANINGFUL). Recall Example 10.3.4, the projection of the parabola $y^2 = x$ to the x-axis, corresponding to the map of rings $\mathbb{Q}[x] \to \mathbb{Q}[y]$, with $x \mapsto y^2$. Show that the geometric fibers of this map are always two points, except for those geometric fibers "over $0 = [(x)]$." (Note that $\operatorname{Spec} \mathbb{C} \to \operatorname{Spec} \mathbb{Q}[x]$ and $\operatorname{Spec} \overline{\mathbb{Q}} \to \operatorname{Spec} \mathbb{Q}[x]$, both corresponding to ring maps with $x \mapsto 0$, are both geometric points "above 0.")

Checking whether a k-scheme is geometrically connected, etc., seems annoying: you need to check every single algebraically closed field containing k. However, in each of these four cases, the failure of nice behavior of geometric fibers can already be detected after a finite extension of fields. For example, $\operatorname{Spec} \mathbb{Q}(i) \to \operatorname{Spec} \mathbb{Q}$ is not geometrically connected, and in fact you only need to base change by $\operatorname{Spec} \mathbb{Q}(i)$ to see this. We make this precise as follows.

Suppose X is a k-scheme. If K/k is a field extension, define $X_K = X \times_k \operatorname{Spec} K$. Consider the following twelve statements of the form "X_K is [property] for all fields [condition]":

- (C_K), $(C_{K=\overline{K}})$, $(C_{\overline{K}})$, (C_{k^s}),
- (I_K), $(I_{K=\overline{K}})$, $(I_{\overline{K}})$, (I_{k^s}),
- (R_K), $(R_{K=\overline{K}})$, $(R_{\overline{K}})$, (R_{k^p}).

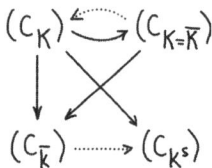

Figure 10.4 *Connectedness conditions that will later turn out to be equivalent (solid arrows are trivial, dotted arrows are Exercise 10.5.F).*

Here C means "connected," I means "irreducible," and R means "reduced." Also, k^s is the separable closure of k, and k^p is the perfect closure of k. If $\operatorname{char} k = 0$, then $\overline{k} = k^s$ and $k = k^p$, so life is simpler.

Thus (C_K) means "X_K is connected for all fields K," $(I_{\overline{K}})$ means "$X_{\overline{K}}$ is irreducible," and (R_{k^p}) means "X_{k^p} is reduced." Geometrically connected (resp., geometrically irreducible, geometrically reduced) translates to $(C_{K=\overline{K}})$ (resp., $(I_{K=\overline{K}})$, $(R_{K=\overline{K}})$).

Trivially (C_K) implies $(C_{K=\overline{K}})$ which implies $(C_{\overline{K}})$, and (C_K) implies (C_{k^s}), and similarly (with appropriate changes) with "connected" replaced by "irreducible" and "reduced." (See Figure 10.4.)

10.5.F. EXERCISE.

(a) Suppose that E/F is a field extension, and A is an F-algebra. Show that A is a subalgebra of $A \otimes_F E$. (Hint: Think of these as vector spaces over F.)

(b) Show that: $(C_{K=\overline{K}})$ implies (C_K) and $(C_{\overline{K}})$ implies (C_{K^s}).

(c) Show that: $(I_{K=\overline{K}})$ implies (I_K) and $(I_{\overline{K}})$ implies (I_{K^s}).

(d) Show that: $(R_{K=\overline{K}})$ implies (R_K) and $(R_{\overline{K}})$ implies (R_{K^p}).

Thus, for example, a k-scheme is geometrically integral if and only if it remains integral under any field extension. Hence "geometrically integral" means "universally, integral fibers."

10.5.3. Harder fact. In fact, (C_K) through (C_{K^s}) are all equivalent, and similarly for the other two properties (irreducibility and reducedness instead of connectedness, with k^p replacing k^s in the case of reducedness). We defer the explanation to the end of this section, in a double-starred discussion (§10.5.5). On a first reading, you may want to read only Corollary 10.5.12 on connectedness, Proposition 10.5.15 on irreducibility, Proposition 10.5.22 on reducedness, and Proposition 11.2.11 on varieties, and then to use them to solve Exercise 10.5.Q. You can later come back and read the proofs, which include some useful tricks turning questions about general schemes over a field to questions about finite type schemes.

The following exercise may help even the geometrically minded reader appreciate the utility of these notions. (There is nothing important about the dimension 2 and the degree 4 in this exercise!)

10.5.G.* EXERCISE. Recall from Remark 4.5.3 that the quartic curves in $\mathbb{P}^2_k$ are parametrized by a $\mathbb{P}^{14}_k$. (This will be made much more precise in §25.3.5.) Show that the points of $\mathbb{P}^{14}_k$ corresponding to geometrically integral curves form an open subset. Explain the necessity of the modifier "geometrically" (even if k is algebraically closed).

10.5.4. Universally injective (= radical) morphisms.

As remarked in §10.4.1, injectivity is not preserved by base change. A better notion is that of **universally injective** morphisms: morphisms that are injections of sets after any base change. In keeping with the traditional agricultural terminology (sheaves, germs, ...; cf. Remark 2.4.3), these morphisms were named **radical** after radishes. As a first example: all locally closed embeddings are universally injective (as they are injective, and remain locally closed embeddings upon any base change). If you wish, you can show more generally that all monomorphisms are universally injective. (Hint: Show that monomorphisms are injective, and that the class of monomorphisms is preserved by base change.)

Suppose you want to determine whether a given morphism is universally injective. A map of sets is injective if and only if each fiber contains at most one point. Now "the fiber of a base change is the base change of the fiber" (Exercise 10.3.C). Also, the underlying set of the scheme-theoretic fiber is the fiber of the map on underlying sets. Hence the question of universal injectivity turns into one about field theory. We make this precise in Exercise 10.5.I, after first making the field theory question precise in Exercise 10.5.H. En route, we will see why universal injectivity is the algebro-geometric generalization of the notion of purely inseparable extensions of fields.

10.5.H. EXERCISE. Suppose E/F is an extension of fields, inducing $\phi : \operatorname{Spec} E \to \operatorname{Spec} F$.

(a) If E contains an element x transcendental over F, show that ϕ is not universally injective. Hint: Exercise 10.2.L (on $k(x) \otimes_k k(y)$) and the Cartesian diagrams

$$
\begin{array}{ccccc}
? & \longrightarrow & \operatorname{Spec} F(x) \otimes_F F(y) & \longrightarrow & \operatorname{Spec} F(y) \\
\downarrow & & \downarrow & & \downarrow \\
\operatorname{Spec} E & \longrightarrow & \operatorname{Spec} F(x) & \longrightarrow & \operatorname{Spec} F.
\end{array}
$$

Now $\operatorname{Spec} E \to \operatorname{Spec} F(x)$ (a map from a point to a point) is surjective, and surjectivity is preserved by base change (Exercise 10.4.G(a)).

(b) If $E \setminus F$ contains an element x algebraic over F (with minimal polynomial $m_x(t) \in F[t]$, say), and separable over F, show that ϕ is not universally injective. Hint: The Cartesian diagrams

$$
\begin{array}{ccccc}
? & \longrightarrow & \operatorname{Spec} F[t]/(m_x(t)) \otimes_F \overline{F} & \longrightarrow & \operatorname{Spec} \overline{F} \\
\downarrow & & \downarrow & & \downarrow \\
\operatorname{Spec} E & \longrightarrow & \operatorname{Spec} F[t]/(m_x(t)) & \longrightarrow & \operatorname{Spec} F.
\end{array}
$$

(c) Suppose E contains no elements of the above forms, i.e., E/F is a **purely inseparable** extension (all elements of E are algebraic over F, and have some p^Nth power in F). If E'/F is any other field extension, show that $\operatorname{Spec} E \otimes_F E'$ contains precisely one point.

10.5.I. EXERCISE. Suppose $\pi \colon X \to Y$ is a morphism of schemes. Show that the following are equivalent.

(i) The morphism π is universally injective.
(ii) The morphism π is injective, and for each $p \in X$, the field extension $\kappa(p)/\kappa(\pi(p))$ is purely inseparable.

You may already see that the class of universally injective morphisms is also preserved by composition, and local on the target, and hence forms a "reasonable" class (§8.1). In any case, we will see this in a different way in §11.2.16.

10.5.5.** Proof of Harder Fact 10.5.3.
We will use one fact that we will not prove until much later. Recall that a map of topological spaces is an **open map** if the image of every open set is an open set.

10.5.6. Fact (to be proved in Theorem 24.5.11) — *Suppose X is a k-scheme. Then $X \to \operatorname{Spec} k$ is* **universally open**, *i.e., remains open after any base change.*

We could prove Fact 10.5.6 now without *too* much trouble, but it will follow much more easily with the magic of flatness.

10.5.7. Lemma — *Suppose the field extension E/F is purely inseparable. Suppose X is any F-scheme. Then $\phi \colon X_E \to X$ is a homeomorphism.*

Proof. The morphism ϕ is a bijection, so we may identify the points of X and X_E. (Reason: For any point $p \in X$, the scheme-theoretic fiber $\phi^{-1}(p)$ is a single point, by the definition of pure inseparability.) The morphism ϕ is continuous (so open sets in X are also open in X_E), and by Fact 10.5.6, ϕ is open (so open sets in X_E are also open in X). $\qquad \square$

10.5.8. Connectedness.
Recall that a connected component of a topological space X is a maximal connected subset of X (§3.6.12).

10.5.J. EXERCISE (PROMISED IN REMARK 3.6.13). Show that every point is contained in a connected component, and that connected components are closed. (Hint: See the hint for Exercise 3.6.O.)

10.5.K. EASY TOPOLOGICAL EXERCISE. Suppose $\phi \colon X \to Y$ is open, and has nonempty connected fibers. Show that ϕ induces a bijection of connected components.

10.5.9. Lemma — *Suppose X is geometrically connected over k. Then for any scheme Y/k, $X \times_k Y \to Y$ induces a bijection of connected components.*

Proof. Combine Fact 10.5.6 and Exercise 10.5.K. $\qquad \square$

We come next to a repeatedly useful algebra-geometry link, promised in Remark 3.6.3: idempotent functions on X on the algebraic side correspond to a decomposition of X into two open and closed sets on the geometric side.

10.5.L. EXERCISE (THE IDEMPOTENT-CONNECTEDNESS PACKAGE). Show that a scheme X is disconnected if and only if there exists a function $e \in \Gamma(X, \mathscr{O}_X)$ that is an idempotent ($e^2 = e$) distinct from 0 and 1. (Hint: If X is the disjoint union of two open sets X_0 and X_1, let e be the function that is 0 on X_0 and 1 on X_1. Conversely, given such an idempotent, define $X_0 = V(e)$ and $X_1 = V(1-e)$.)

10.5.10. Proposition — *Suppose k is separably closed, and A is a k-algebra with $\operatorname{Spec} A$ connected. Then $\operatorname{Spec} A$ is geometrically connected over k. More generally, for any field extension K/k, $\operatorname{Spec} A \otimes_k K$ is connected.*

Proof. It suffices to assume that K is algebraically closed (as $\operatorname{Spec} A \otimes_k \overline{K} \to \operatorname{Spec} A \otimes_k K$ is surjective). By choosing an embedding $\overline{k} \hookrightarrow K$ and considering the diagram

(10.5.10.1)
$$
\begin{array}{ccccc}
\operatorname{Spec} A \otimes_k K & \longrightarrow & \operatorname{Spec} A \otimes_k \overline{k} & \xrightarrow[\text{by Lem. 10.5.7}]{\text{homeo.}} & \operatorname{Spec} A \\
\downarrow & & \downarrow & & \downarrow \\
\operatorname{Spec} K & \longrightarrow & \operatorname{Spec} \overline{k} & \longrightarrow & \operatorname{Spec} k \quad ,
\end{array}
$$

it suffices to assume k is algebraically closed.

If $\operatorname{Spec} A \otimes_k K$ is disconnected, then $A \otimes_k K$ contains an idempotent $e \neq 0, 1$ (by Exercise 10.5.L). Write $e = \sum_{i=1}^{N} a_i \otimes \ell_i$, where $a_i \in A$ and $\ell_i \in K$. Define B as the subalgebra of K generated over k by the ℓ_i (as an algebra), so B is an integral domain of finite type over k. Then $K(B)$ is a finitely generated field extension of k.

$$
\begin{array}{ccccc}
\operatorname{Spec} A \otimes_k K & \longrightarrow & \operatorname{Spec} A \otimes_k B & \longrightarrow & \operatorname{Spec} A \\
\downarrow & & \downarrow & & \downarrow {\scriptstyle \text{universally open}} \\
\operatorname{Spec} K & \longrightarrow & \operatorname{Spec} B & \longrightarrow & \operatorname{Spec} k
\end{array}
$$

Also, $e \in A \otimes_k B$, and we recall that e is an idempotent distinct from 0 and 1 (even in this ring). By the idempotent-connectedness package (Exercise 10.5.L) applied to $A \otimes_k B$, $\operatorname{Spec} A \otimes_k B$ is disconnected, say, with nonempty open subsets U and V with $U \coprod V = \operatorname{Spec} A \otimes_k B$.

Now $\phi \colon \operatorname{Spec} A \otimes_k B \to \operatorname{Spec} B$ is an open map (because $\operatorname{Spec} A \to \operatorname{Spec} k$ is universally open, Fact 10.5.6), so $\phi(U)$ and $\phi(V)$ are nonempty open sets of $\operatorname{Spec} B$. As $\operatorname{Spec} B$ is connected, the intersection $\phi(U) \cap \phi(V)$ is a nonempty open set, which has a closed point p (with residue field k, as $k = \overline{k}$). But then $\phi^{-1}(p) \cong \operatorname{Spec} A$, and we have covered $\operatorname{Spec} A$ with two disjoint open sets, yielding a contradiction. $\qquad \square$

10.5.11. *Sneaky "tensor-finiteness" trick that was just used.* This argument used a particularly clever trick that we will use again: we conjured a finitely generated algebra B out of nowhere, and then used classical "geometry over algebraically closed fields" to prove something far removed from classical algebraic geometry. What let us do that was that any single element of a tensor product (such as e in the argument above) is a *finite sum* of "elementary tensors," and so we worked instead in a subring generated by the "parts" of those "elementary tensors." We will call this the **tensor-finiteness trick** in order to help direct your attention to where it can be used again.

10.5.12. Corollary — *If k is separably closed, and Y is a connected k-scheme, then Y is geometrically connected.*

Proof. We wish to show that for any field extension K/k, Y_K is connected. By Proposition 10.5.10, $\operatorname{Spec} K$ is geometrically connected over k. Apply Lemma 10.5.9 with $X = \operatorname{Spec} K$. □

10.5.13. Irreducibility.

10.5.14. Proposition — *Suppose k is separably closed, A is a k-algebra with $\operatorname{Spec} A$ irreducible, and K/k is a field extension. Then $\operatorname{Spec} A \otimes_k K$ is irreducible.*

Proof. We follow the philosophy of the proof of Proposition 10.5.10. As in the first paragraph of that proof, it suffices to assume that k is algebraically closed.

If $A \otimes_k K$ is not irreducible, then we can find x and y with $V(x), V(y) \neq \operatorname{Spec} A \otimes_k K$ and $V(x) \cup V(y) = \operatorname{Spec} A \otimes_k K$. As in the second paragraph of the proof of Proposition 10.5.10, we use the "tensor-finiteness trick" (§10.5.11). We choose expressions $x = \sum a_i \otimes \ell_i$ and $y = \sum a_j' \otimes \ell_j'$ ($a_i, a_j' \in A$, $\ell_i, \ell_j' \in K$), and then define B to be the subring of K generated by the ℓ_i and ℓ_j'. Thus B is an integral domain finitely generated over $k = \bar{k}$, and $x, y \in A \otimes_k B$. Then $D(x)$ and $D(y)$ are nonempty open subsets of $\operatorname{Spec} A \otimes_k B$, whose images in $\operatorname{Spec} B$ are nonempty open sets (by universal openness of $\operatorname{Spec} A \to \operatorname{Spec} k$, Fact 10.5.6), and thus their intersection is nonempty and contains a closed point p (with residue field $k = \bar{k}$). But then $\phi^{-1}(p) \cong \operatorname{Spec} A$, and we have covered $\operatorname{Spec} A$ with two proper closed sets (the restrictions of $V(x)$ and $V(y)$), yielding a contradiction. □

10.5.M. EXERCISE. Suppose k is separably closed, and A and B are k-algebras, both with irreducible Spec (i.e., with one minimal prime). Show that $A \otimes_k B$ has irreducible Spec too. (Hint: Reduce to the case where A and B are finite type over k. Extend the proof of the previous proposition.)

10.5.N. EASY EXERCISE. Show that a scheme X is irreducible if and only if there exists an open cover $X = \cup U_i$ with U_i irreducible for all i, and $U_i \cap U_j \neq \varnothing$ for all i, j.

10.5.15. Proposition — *Suppose K/k is a field extension of a separably closed field and X_k is irreducible. Then X_K is irreducible.*

Proof. Take an open cover $X_k = \cup U_i$ by pairwise intersecting irreducible affine open subsets. The base change of each U_i to K is irreducible by Proposition 10.5.14, and they pairwise intersect. The result then follows from Exercise 10.5.N. □

10.5.O. EXERCISE (GEOMETRICALLY INTEGRAL × INTEGRAL = INTEGRAL). Suppose B is a k-algebra such that $B \otimes_k \bar{k}$ is an integral domain ($\operatorname{Spec} B$ is geometrically integral), and A is a k-algebra that is an integral domain ($\operatorname{Spec} A$ is integral). Show that $A \otimes_k B$ is an integral domain ($\operatorname{Spec} A \otimes_k B$ is integral). (Once we define "variety," this will imply that the product of a geometrically integral variety with an integral variety is an integral variety.) Hint: Revisit the proof of Exercise 10.4.H.

10.5.16. Reducedness.

We recall a statement from transcendence theory (the basics of which are developed in Exercise 12.2.A). Because this is a starred section, we content ourselves with a reference rather than a proof.

10.5.17. Theorem: Finitely generated extensions of perfect fields are separably generated — *Any finitely generated extension of a perfect field E/F can be factored into a finite separable part and a purely transcendental part:*

$$E$$

$$\bigg| \quad \textit{finite separable}$$

$$F(t_1, \ldots, t_n)$$

$$\bigg| \quad \textit{purely transcendental}$$

$$F.$$

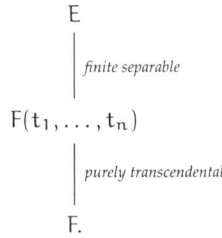

See [E, Cor. 16.17(b)] or [van, §19.7] for a proof. If you wish to prove it yourself, see Exercise 10.5.P.

10.5.18. Finitely generated extensions that can be factored in this way are said to be **separably generated**. The transcendence basis $\{t_1, \ldots, t_n\}$ is said to be a **separating transcendence basis**.

10.5.P. EXERCISE (PROOF OF THEOREM 10.5.17). Suppose E/F is a finitely generated extension of a perfect field of characteristic p. Fix a transcendence basis $\{x_1, \ldots, x_d\}$ of E/F.

(a) Suppose the inseparable degree $n := [E : F(x_1, \ldots, x_d)]_i > 1$. Show there is $a \in E$ *not* separable over $F(x_1, \ldots, x_d)$ such that its minimal polynomial $f \in F(x_1, \ldots, x_d)[T]$ lies in $F[x_1, \ldots, x_d, T]$, and prove f is irreducible in $F[x_1, \ldots, x_d, T]$ as well as in $F(x_1, \ldots, x_{d-1}, T)[x_d]$.

(b) For a and f as in (a), prove $\frac{\partial f}{\partial x_j} \neq 0$ for some j. Relabel the x_i's so $j = d$. Prove that $\{x_1, \ldots, x_{d-1}, a\}$ is a transcendence basis of E over F, and use multiplicativity of inseparable degree to show $[E : F(x_1, \ldots, x_{d-1}, a)]_i < n$.

(c) Deduce the existence of a separating transcendence basis of E/F.

10.5.19. Proposition (geometrically reduced $\times$ reduced = reduced) — *Suppose B is a geometrically reduced k-algebra, and A is a reduced k-algebra. Then $A \otimes_k B$ is reduced.*

(Compare this to Exercise 10.5.O.)

Proof. Reduce to the case where A is finitely generated over k using the tensor-finiteness trick (§10.5.11). (Suppose we have $x \in A \otimes_k B$ with $x^n = 0$. Suppose $x = \sum a_i \otimes b_i$. Let A' be the finitely generated subring of A generated by the a_i. Then $A' \otimes_k B$ is a subring of $A \otimes_k B$. Replace A by A'.) Then A is a subring of the product $\prod K_i$ of the function fields of its irreducible components. (One can see this as follows: the kernel of $A \to \prod A/\mathfrak{p}_i$ is the intersection of the primes $\mathfrak{p}_i$, which is the ideal of nilpotents $\mathfrak{N}$ by Theorem 3.2.13; but $\mathfrak{n} = 0$ as A is reduced. And $A/\mathfrak{p}_i \hookrightarrow K(A/\mathfrak{p}_i) = K_i$, clearly. Alternatively, you may find this clear from our discussion of associated points (§6.6).) So it suffices to prove the result for A a product of fields. Then it suffices to prove the result when A is a field. But then we are done, by the definition of geometric reducedness. $\square$

10.5.20. Proposition — *Suppose A is a reduced k-algebra. Then:*

(a) $A \otimes_k k(t)$ *is reduced.*

(b) *If E/k is a finite separable extension, then $A \otimes_k E$ is reduced.*

Proof. (a) Clearly $A \otimes k[t]$ is reduced, and localization preserves reducedness (as reducedness is stalk-local, Exercise 5.2.A).

(b) Working inductively, we can assume E is generated by a single element, with minimal polynomial $p(t)$. By the tensor-finiteness trick, we can assume A is finitely generated over k. (Why can we do this? What are the finitely many elementary tensors in this argument?) Then by the same trick as in the proof of Proposition 10.5.19, we can replace A by the product of the function fields of its components, and then we can assume A is a field. But then $A[t]/p(t)$ is reduced by the definition of separability of p. $\square$

10.5.21. Lemma — *Suppose E/k is a field extension of a perfect field, and A is a reduced k-algebra. Then $A \otimes_k E$ is reduced.*

Proof. By the tensor-finiteness trick, we may assume E is finitely generated over k. (Be sure you see how!) By Theorem 10.5.17, we can factor E/k into extensions of the forms of Proposition 10.5.20 (a) and (b). We then apply Proposition 10.5.20. □

10.5.22. Proposition — *Suppose E/k is an extension of a perfect field, and X is a reduced k-scheme. Then X_E is reduced.*

Proof. Reduce to the case where X is affine. Use Lemma 10.5.21. □

10.5.23. Corollary — *Suppose k is perfect, and A and B are reduced k-algebras. Then $A \otimes_k B$ is reduced.*

Proof. By Lemma 10.5.21, A is a geometrically reduced k-algebra. Then apply Lemma 10.5.19. □

10.5.Q. EXERCISE (COMPLETING HARD FACT 10.5.3). Show that (R_{k^p}) implies (R_K), (I_{k^s}) implies (I_K), and (C_{k^s}) implies (C_K).

10.5.R. EXERCISE. Suppose that A and B are two integral domains that are $\overline{k}$-algebras. Show that $A \otimes_{\overline{k}} B$ is an integral domain. (Compare this to Exercise 10.4.H, which had finite type hypotheses.)

10.6   Products of Projective Schemes: The Segre Embedding

We next describe products of projective A-schemes over A. (The case of greatest initial interest is $A = k$.) To do this, we need only describe $\mathbb{P}^m_A \times_A \mathbb{P}^n_A$, because any projective A-scheme has a closed embedding in some $\mathbb{P}^m_A$, and closed embeddings behave well under base change, so if $X \hookrightarrow \mathbb{P}^m_A$ and $Y \hookrightarrow \mathbb{P}^n_A$ are closed embeddings, then $X \times_A Y \hookrightarrow \mathbb{P}^m_A \times_A \mathbb{P}^n_A$ is also a closed embedding, cut out by the equations of X and Y (§10.2(**3**)). We will describe $\mathbb{P}^m_A \times_A \mathbb{P}^n_A$, and see that it, too, is a projective A-scheme. (Hence if X and Y are projective A-schemes, then their product $X \times_A Y$ over A is also a projective A-scheme.)

Before we do this, we will get some motivation from classical projective spaces (nonzero vectors modulo nonzero scalars, Exercise 4.4.F) in a special case. Our map will send $[x_0, x_1, x_2] \times [y_0, y_1]$ to a point in $\mathbb{P}^5$, whose coordinates we think of as being entries in the "multiplication table"

$$\begin{bmatrix} x_0 y_0, & x_1 y_0, & x_2 y_0, \\ x_0 y_1, & x_1 y_1, & x_2 y_1 \end{bmatrix}.$$

This is indeed a well-defined map of sets. Notice that the resulting matrix is rank 1, and from the matrix, we can read off $[x_0, x_1, x_2]$ and $[y_0, y_1]$ up to multiplication by nonzero scalars. For example, to read off the point $[x_0, x_1, x_2] \in \mathbb{P}^2$, we take the first row, unless it is all zero, in which case we take the second row. (They can't both be all zero.) In conclusion: In classical projective geometry, given a point of $\mathbb{P}^m$ and $\mathbb{P}^n$, we have produced a point in $\mathbb{P}^{mn+m+n}$, and from this point in $\mathbb{P}^{mn+m+n}$, we can recover the points of $\mathbb{P}^m$ and $\mathbb{P}^n$.

Suitably motivated, we return to algebraic geometry. We define a map

$$\mathbb{P}^m_A \times_A \mathbb{P}^n_A \longrightarrow \mathbb{P}^{mn+m+n}_A$$

by

$$([x_0, \ldots, x_m], [y_0, \ldots, y_n]) \longmapsto [z_{00}, z_{01}, \ldots, z_{ij}, \ldots, z_{mn}]$$

$$= [x_0 y_0, x_0 y_1, \ldots, x_i y_j, \ldots, x_m y_n].$$

More explicitly, we consider the map from the affine open set $U_i \times V_j$ (where $U_i = D(x_i)$ and $V_j = D(y_j)$) to the affine open set $W_{ij} = D(z_{ij})$ by

$$(x_{0/i}, \ldots, x_{m/i}, y_{0/j}, \ldots, y_{n/j}) \mapsto (x_{0/i}y_{0/j}, \ldots, x_{i/i}y_{j/j}, \ldots, x_{m/i}y_{n/j})$$

or, in terms of algebras, $z_{ab/ij} \mapsto x_{a/i}y_{b/j}$.

10.6.A. EXERCISE. Check that these maps glue to give a well-defined morphism $\mathbb{P}_A^m \times_A \mathbb{P}_A^n \to \mathbb{P}_A^{mn+m+n}$.

10.6.1. We next show that this morphism is a closed embedding. We can check this on an open cover of the target (the notion of being a closed embedding is affine-local, Exercise 9.1.D). Let's check this on the open set where $z_{ij} \neq 0$. The preimage of this open set in $\mathbb{P}_A^m \times \mathbb{P}_A^n$ is the locus where $x_i \neq 0$ and $y_j \neq 0$, i.e., $U_i \times V_j$. As described above, the map of rings is given by $z_{ab/ij} \mapsto x_{a/i}y_{b/j}$; this is clearly a surjection, as $z_{aj/ij} \mapsto x_{a/i}$ and $z_{ib/ij} \mapsto y_{b/j}$. (A generalization of this ad hoc description will be given in Exercise 15.2.E.)

  This map is called the **Segre morphism** or the **Segre embedding**. If A is a field, the image is called the **Segre variety**.

10.6.B. EXERCISE. Show that the Segre scheme (the image of the Segre embedding) is cut out (scheme-theoretically) by the equations corresponding to

$$\text{rank} \begin{pmatrix} z_{00} & \cdots & z_{0n} \\ \vdots & \ddots & \vdots \\ z_{m0} & \cdots & z_{mn} \end{pmatrix} = 1,$$

i.e., that all 2×2 minors vanish. Hint: Suppose you have a polynomial in the z_{ij} that becomes zero upon the substitution $z_{ij} = x_iy_j$. Give a recipe for subtracting polynomials of the form "monomial times 2×2 minor" so that the end result is 0. (The analogous question for the Veronese embedding in special cases is the content of Exercises 9.3.I and 9.3.K.)

10.6.2. *Important example.* Let's consider the first nontrivial example, when $m = n = 1$. We get $\mathbb{P}_k^1 \times_k \mathbb{P}_k^1 \hookrightarrow \mathbb{P}_k^3$ (where k is a field). We get a single equation

$$\text{rank} \begin{pmatrix} z_{00} & z_{01} \\ z_{10} & z_{11} \end{pmatrix} = 1,$$

i.e., $z_{00}z_{11} - z_{01}z_{10} = 0$. We again meet our old friend, the quadric surface (§9.3.9)! Hence: the smooth quadric surface $wz - xy = 0$ (Figure 9.2) is isomorphic to $\mathbb{P}_k^1 \times_k \mathbb{P}_k^1$. Recall from Exercise 9.3.L that the quadric surface has two families (rulings) of lines. You may wish to check that one family of lines corresponds to the image of $\{x\} \times_k \mathbb{P}_k^1$ as x varies, and the other corresponds to the image $\mathbb{P}_k^1 \times_k \{y\}$ as y varies.

  If k is an algebraically closed field of characteristic not 2, then by diagonalizability of quadratics (Exercise 5.4.J), all rank 4 ("full rank") quadratics in four variables are isomorphic, so all rank 4 quadric surfaces over an algebraically closed field of characteristic not 2 are isomorphic to $\mathbb{P}_k^1 \times_k \mathbb{P}_k^1$.

  Note that this is not true over a field that is not algebraically closed. For example, over $\mathbb{R}$, $w^2 + x^2 + y^2 + z^2 = 0$ (in $\mathbb{P}_\mathbb{R}^3$) is not isomorphic to $\mathbb{P}_\mathbb{R}^1 \times_\mathbb{R} \mathbb{P}_\mathbb{R}^1$. Reason: The former has no real points, while the latter has lots of real points.

  You may do the next two exercises in either order. The second can be used to show the first, but the first may give you insight into the second.

10.6.C. EXERCISE: A COORDINATE-FREE DESCRIPTION OF THE SEGRE EMBEDDING. Show that the Segre embedding can be interpreted as $\mathbb{P}V \times \mathbb{P}W \to \mathbb{P}(V \otimes W)$

via the surjective map of graded rings

$$\mathrm{Sym}^\bullet(V^\vee \otimes W^\vee) \longrightarrow \oplus_{i=0}^\infty \left(\mathrm{Sym}^i V^\vee\right) \otimes \left(\mathrm{Sym}^i W^\vee\right)$$

"in the opposite direction."

10.6.D. EXERCISE: A COORDINATE-FREE DESCRIPTION OF PRODUCTS OF PROJECTIVE A-SCHEMES IN GENERAL. Suppose that $S_\bullet$ and $T_\bullet$ are finitely generated graded rings over A. Describe an isomorphism

$$(\mathrm{Proj}\, S_\bullet) \times_A (\mathrm{Proj}\, T_\bullet) \overset{\sim}{\longleftrightarrow} \mathrm{Proj} \oplus_{n=0}^\infty (S_n \otimes_A T_n)$$

(where hopefully the definition of multiplication in the graded ring $\oplus_{n=0}^\infty (S_n \otimes_A T_n)$ is clear).

10.7 Normalization

We discuss normalization now only because the central construction gives practice with the central idea behind the construction in §10.1 of the fibered product (see Exercises 10.7.B and 10.7.I).

Normalization is a means of turning a *reduced* scheme into a normal scheme. (Unimportant remark: Reducedness is not a necessary hypothesis, and is included only to avoid distraction. If you care, you can follow through the construction, and realize that the normalization of a scheme factors through the reduction, and is the normalization of the reduction.) A *normalization* of a reduced scheme X is a morphism $\nu\colon \widetilde{X} \to X$ from a normal scheme, where ν induces a bijection of irreducible components of $\widetilde{X}$ and X, and in reasonable cases (cf. Exercise 10.7.N) ν gives a birational morphism on each of the irreducible components. It will satisfy a universal property, and hence it will be unique up to unique isomorphism. Figure 8.4 is an example of a normalization.

We begin with the case where X is irreducible, and hence (by Exercise 5.2.G) integral. (We will then deal with a more general case, and also discuss normalization in a function field extension.) In this case of irreducible X, the **normalization** $\nu\colon \widetilde{X} \to X$ is a dominant morphism (not just a dominant rational map!) from an irreducible normal scheme to X, such that any other such morphism factors through ν:

(10.7.0.1)

Thus if the normalization exists, then it is unique up to unique isomorphism. We now have to show that it exists, and we do this in a way that will look familiar. We deal first with the case where X is affine, say, $X = \mathrm{Spec}\, A$, where A is an integral domain. Then let $\widetilde{A}$ be the integral closure of A in its fraction field $K(A)$.

10.7.A. EXERCISE. (Recall that A is an integral domain.) Show that $\nu\colon \mathrm{Spec}\, \widetilde{A} \to \mathrm{Spec}\, A$ satisfies the universal property of normalization. (En route, you might show that the global sections of an irreducible normal scheme are also "normal," i.e., integrally closed.)

10.7.B. IMPORTANT (BUT SURPRISINGLY EASY) EXERCISE. Show that normalizations of integral schemes exist in general. (Hint: Ideas from the existence of fibered products, §10.1, may help.)

10.7.C. EASY EXERCISE. Show that normalizations are integral and surjective. (Hint for surjectivity: The Lying Over Theorem; see §8.2.6.)

We will soon see that normalization of integral finite type k-schemes is always a birational morphism, in Exercise 10.7.N.

10.7.D. EXERCISE. Explain (by defining a universal property) how to extend the notion of normalization to the case where X is a reduced scheme, with possibly more than one component, but under the hypothesis that every affine open subset of X has finitely many irreducible components. Note that this includes all locally Noetherian schemes. (If you wish, you can show that the normalization exists in this case. See [Stacks, tag 035Q] for more.)

Here are some examples.

10.7.E. EXERCISE (NORMALIZATION OF THE NODAL CUBIC). Suppose that char $k \neq 2$ (only so we can use the word "nodal"). Show that $\operatorname{Spec} k[t] \to \operatorname{Spec} k[x, y]/(y^2 - x^2(x + 1))$ given by $(x, y) \mapsto (t^2 - 1, t(t^2 - 1))$ (see Figure 8.4) is a normalization. The target curve is called the **nodal cubic curve**. (Hint: Show that the domains $k[t]$ and $k[x, y]/(y^2 - x^2(x + 1))$ have the same fraction field. Show that $k[t]$ is integrally closed. Show that $k[t]$ is contained in the integral closure of $k[x, y]/(y^2 - x^2(x + 1))$.)

You will see from the previous exercise that once we guess what the normalization is, it isn't hard to verify that it is indeed the normalization. Perhaps a few words are in order as to where the polynomials $t^2 - 1$ and $t(t^2 - 1)$ arose in the previous exercise. The key idea is to guess $t = y/x$. (Then $t^2 = x + 1$ and $y = xt$ quickly.) This idea comes from three possible places. We begin by sketching the curve, and noticing the "node" at the origin. ("Node" will be formally defined in §28.2.1.)

(a) The function y/x is well-defined away from the node, and at the node, the two branches have "values" $y/x = 1$ and $y/x = -1$.
(b) We can also note that if $t = y/x$, then t^2 is a polynomial, so we will need to adjoin t in order to obtain the normalization.
(c) The curve is cubic, so we expect a general line to meet the cubic in three points, counted with multiplicity. (We will make this precise when we discuss Bézout's Theorem, Exercise 18.6.J, but in this case we have already gotten a hint of this in Exercise 7.5.H.) There is a $\mathbb{P}^1$ parametrizing lines through the origin (with coordinate equal to the slope of the line, y/x), and most such lines meet the curve with multiplicity 2 at the origin, and hence meet the curve at precisely one other point of the curve. So this "coordinatizes" most of the curve, and we try adding in this coordinate.

10.7.F. EXERCISE. Find the normalization of the *cusp* $y^2 = x^3$ (see Figure 10.5). ("Cusp" will be formally defined in Definition 28.2.2.)

10.7.G. EXERCISE. Suppose char $k \neq 2$. Find the normalization of the *tacnode* $y^2 = x^4$, and draw a picture analogous to Figure 10.5. ("Tacnode" will be formally defined in Definition 28.2.2.)

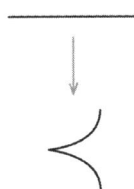

Figure 10.5 *Normalization of a cusp.*

(Although we haven't defined "singularity," "smooth," "curve," or "dimension," you should still read this.) Notice that in the previous examples, normalization "resolves" the singularities ("non-smooth" points) of the curve. In general, it will do so in dimension 1 (in reasonable Noetherian circumstances, as normal local Noetherian integral domains of dimension 1 are all discrete valuation rings, §13.5), but won't do so in higher dimension (the cone $z^2 = x^2 + y^2$ over a field k of characteristic not 2 is normal, Exercise 5.4.I(b)).

10.7.H. EXERCISE. Suppose $X = \operatorname{Spec} \mathbb{Z}[15i]$. Describe the normalization $\tilde{X} \to X$. (Hint: $\mathbb{Z}[i]$ is a unique factorization domain, §5.4.5(0), and hence is integrally closed by Exercise 5.4.F.) Over what points of X is the normalization not an isomorphism?

Another exercise in a similar vein is the normalization of the "pinched plane," Exercise 13.5.H.

10.7.I. EXERCISE (NORMALIZATION IN A FUNCTION FIELD EXTENSION, AN IMPORTANT GENERALIZATION). Suppose X is an integral scheme. The **normalization** $\nu\colon \widetilde{X} \to X$ **of** X **in a given algebraic field extension** L **of the function field** K(X) **of** X is a dominant morphism from a normal integral scheme $\widetilde{X}$ with function field L, such that ν induces the inclusion $K(X) \hookrightarrow L$, which is universal with respect to this property.

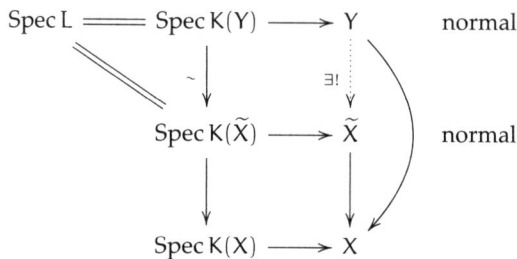

Show that the normalization in a (given) extension of fields exists.

The following two examples, one arithmetic and one geometric, show that this is an interesting construction.

10.7.J. EXERCISE. Suppose $X = \operatorname{Spec} \mathbb{Z}$ (with function field $\mathbb{Q}$). Find its integral closure in the field extension $\mathbb{Q}(i)$. (There is no "geometric" way to do this; it is purely an algebraic problem, although the answer should be understood geometrically.)

10.7.1. *Remark: Rings of integers in number fields.* A finite extension K of $\mathbb{Q}$ is called a **number field**, and the integral closure of $\mathbb{Z}$ in K the **ring of integers in** K, denoted by $\mathscr{O}_K$. (This notation is awkward given our other use of the symbol $\mathscr{O}$.)

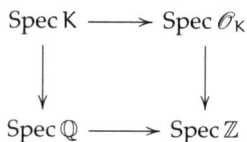

By the previous exercises, $\mathscr{O}_K$ is a normal integral domain, and we will see soon (Theorem 10.7.3(a)) that it is Noetherian, and later (Exercise 12.1.G) that it has "dimension 1." This is an example of a *Dedekind domain*; see §13.5.11. We will think of it as a "smooth" curve as soon as we define what "smooth" (really, regular) and "curve" mean.

10.7.K. EXERCISE. Find the ring of integers in $\mathbb{Q}(\sqrt{n})$, where n is square-free and $n \equiv 3 \pmod 4$. (Hint: Exercise 5.4.I(a), where you will also be able to figure out the answer for square-free n in general.)

10.7.L. EXERCISE. Suppose char $k \neq 2$ for convenience (although it isn't necessary).

(a) Suppose $X = \operatorname{Spec} k[x]$ (with function field $k(x)$). Find its normalization in the field extension $k(x)(y)$, where $y^2 = x^2 + x$. (Again, we get a Dedekind domain.) Hint: This can be done without too much pain. Show that $\operatorname{Spec} k[x,y]/(x^2 + x - y^2)$ is normal, possibly by identifying it as an open subset of $\mathbb{P}^1_k$, or possibly using Exercise 5.4.H.

(b) Suppose $X = \mathbb{P}^1$, with distinguished open subscheme $\operatorname{Spec} k[x]$. Find its normalization in the field extension $k(x, y)$, where $y^2 = x^2 + x$. (Part (a) involves computing the normalization over one affine open set; now figure out what happens over the "other" affine open set, and how to glue.)

10.7.2. Fancy fact: Finiteness of integral closure.

The following fact is useful.

10.7.3. Theorem (finiteness of integral closure) — *Suppose A is a Noetherian integral domain,* $K = K(A)$, *L/K is a finite extension of fields, and B is the integral closure of A in L ("the integral closure of A in the field extension L/K," i.e., those elements of L integral over A).*

(a) *If A is integrally closed and L/K is separable, then B is a finitely generated A-module.*
(b) *(E. Noether) If A is a finitely generated k-algebra, then B is a finitely generated A-module.*

Eisenbud gives a proof in a page and a half: (a) is [E, Prop. 13.14] and (b) is [E, Cor. 13.13]. A sketch is given in §10.7.5.

10.7.4. *Warning.* Part (b) does *not* hold for Noetherian A in general—the integral closure of a Noetherian ring need not be Noetherian. (See [Rei2, §9.4] or [E, §13.3] for some discussion.) This is alarming. The existence of such an example is a sign that Theorem 10.7.3 is not easy.

In the following four exercises, you can replace "integral finite type k-scheme" by "integral variety" once you know what a variety is.

10.7.M. EXERCISE (FINITENESS OF NORMALIZATION).

(a) Show that if X is an integral, locally finite type k-scheme, then its normalization $\nu \colon \widetilde{X} \to X$ is a finite morphism.
(b) Suppose X is a locally Noetherian integral scheme. Show that if X is normal, then the normalization in a *finite* separable field extension is a finite morphism. Show that if X is an integral finite type k-scheme, then the normalization in a finite extension of fields is a finite morphism. In particular, once we define "variety" (Definition 11.2.9), you will see that this implies that the normalization of a variety (including in a finite extension of the function field) is a variety.

10.7.N. EXERCISE. Show that if X is an integral finite type k-scheme, then the normalization morphism is birational. (Hint: Proposition 7.5.8; or solve Exercise 10.7.O first.)

10.7.O. EXERCISE. Suppose that X is an integral finite type k-scheme. Show that the normalization map of X is an isomorphism on an open dense subset of X. (Hint: Proposition 7.5.6.) Show that the normal points of X form a dense open subset. (By Exercise 10.7.D, the "integral" hypothesis can be relaxed to "reduced.")

10.7.P. EXERCISE. Suppose $\rho \colon Z \to X$ is a finite birational morphism from an integral finite type k-scheme to a *normal* integral finite type k-scheme. Show that ρ is an isomorphism.

10.7.5.** *Sketch of proof of finiteness of integral closure, Theorem 10.7.3.* Here is a sketch to show the structure of the argument. It uses commutative algebra ideas from Chapter 12, so you should only glance at this to see that nothing fancy is going on. *Part (a)*: reduce to the case where L/K is Galois, with group $\{\sigma_1, \ldots, \sigma_n\}$. Choose $b_1, \ldots, b_n \in B$ forming a K-vector space basis of L. Let M be the matrix (familiar from Galois theory) with ijth entry $\sigma_i b_j$, and let $d = \det M$. Show that the entries of M lie in B, and that $d^2 \in K$ (as d^2 is Galois-fixed). Show that $d \neq 0$ using linear independence of characters. Then complete the proof by showing that $B \subset d^{-2}(Ab_1 + \cdots + Ab_n)$ (submodules of finitely generated modules over Noetherian rings are also Noetherian, Exercise 3.6.Y) as follows. Suppose $b \in B$, and write $b = \sum c_i b_i$ ($c_i \in K$). If c is the column vector with entries c_i, show that the ith entry of the column vector Mc is $\sigma_i b \in B$. Multiplying Mc on the left by $\operatorname{adj} M$ (see the trick of the proof of Lemma 8.2.1), show that $dc_i \in B$. Thus $d^2 c_i \in B \cap K = A$ (as A is integrally closed), as desired.

For *(b)*, use the Noether Normalization Lemma 12.2.4 to reduce to the case $A = k[x_1, \ldots, x_n]$. Reduce to the case where L is normally closed over K. Let L' be the subextension of L/K so that L/L' is Galois and L'/K is purely inseparable. Use part (a) to reduce to the case $L = L'$. If $L' \neq K$, then for some q, L' is generated over K by the qth root of a finite set of rational functions. Reduce to the case $L' = k'(x_1^{1/q}, \ldots, x_n^{1/q})$, where k'/k is a finite purely inseparable extension. In this case, show that $B = k'[x_1^{1/q}, \ldots, x_n^{1/q}]$, which is indeed finite over $k[x_1, \ldots, x_n]$. $\square$

10.7.6. Reduction of schemes, revisited. It may be enlightening to revisit the "reduction" of a scheme using this point of view. The argument is clean and simple, and by working it out for yourself, you will better appreciate the various moving parts. We now define "reduction" from scratch, and then show that it agrees with the definition of §9.4.10.

10.7.7. *Definition.* Suppose X is a scheme. In analogy with (10.7.0.1), we say $\rho : X^{red} \to X$ is "the" **reduction of** X if any morphism $W \to X$ from a reduced scheme W factors uniquely through ρ:

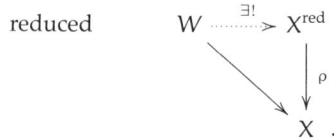

$$
\text{reduced} \qquad W \xdashrightarrow{\ \exists!\ } X^{red}
$$
$$
\searrow \quad \downarrow \rho
$$
$$
X \ .
$$

The "the" is in quotes because this is a slight abuse of notation. In analogy with fibered products and normalization, we now check that such an X^{red} exists:

10.7.Q. EXERCISE.

(a) In analogy with Exercise 10.7.A, show that $\rho : \operatorname{Spec} A/\mathfrak{N}(A) \to \operatorname{Spec} A$ satisfies the universal property of reduction.
(b) In analogy with Exercise 10.7.B, show that the reduction of schemes exists in general.
(c) To make sure we haven't contradicted our earlier selves: show that the construction of §9.4.10 satisfies this universal property.

10.7.8. *Definition.* By the universal property of reduction, any morphism of schemes $\pi : X \to Y$ induces a morphism $X^{red} \to Y^{red}$, the **reduction of** π which we quite reasonably name π^{red}. (You likely see that "reduction" gives a functor from the category of schemes to the category of reduced schemes.)

Chapter 11

Separated and Proper Morphisms, and (Finally!) Varieties

Separatedness is a fundamental notion. It is the analog of the Hausdorff condition for manifolds (see Exercise 11.2.B), and as with Hausdorffness, this geometrically intuitive notion ends up being just the right hypothesis to make theorems work. Although the definition initially looks odd, in retrospect it is just perfect.

Let's review why we like Hausdorffness. Recall that a topological space is **Hausdorff** if for every two points x and y, there are disjoint open neighborhoods of x and y. The real line is Hausdorff, but the "real line with doubled origin" (of which Figure 4.6 may be taken as a sketch) is not. Many proofs and results about manifolds use Hausdorffness in an essential way. For example, the classification of compact one-dimensional manifolds is very simple, but if the Hausdorff condition were removed, we would have a very wild set.

So once armed with this definition, we can cheerfully exclude the line with doubled origin from civilized discussion, and we can (finally) define the notion of a *variety*, in a way that corresponds to the classical definition.

With our motivation from manifolds, we shouldn't be surprised that all of our affine and projective schemes are separated. Subsets of Hausdorff topological spaces are automatically Hausdorff. Hence if Y is a manifold, and X is a subset that satisfies all the hypotheses of a manifold except possibly Hausdorffness, X is a manifold. Similarly, we will see that locally closed embeddings in something separated are also separated (combine Exercise 11.2.C and Proposition 11.2.3).

Grothendieck taught us that one should try to define properties of morphisms, not of objects; then we can say that an object has that property if its morphism to the final object has that property. We discussed this briefly at the start of Chapter 8. In this spirit, separatedness will be a property of morphisms, not schemes.

As an unexpected bonus, a separated morphism to an affine scheme has the property that the intersection of two affine open sets in the source is affine (Proposition 11.2.13). This will make Čech cohomology work very easily on (quasicompact) schemes (Chapter 18). You might consider this an analog of the fact that, in $\mathbb{R}^n$, the intersection of two convex sets is also convex. As affine schemes are trivial from the point of view of quasicoherent cohomology, just as convex sets in $\mathbb{R}^n$ have no cohomology, this metaphor is apt.

A lesson arising from the construction is the importance of the diagonal morphism (Definition 1.2.7). More precisely, given a morphism $\pi \colon X \to Y$, good consequences can be leveraged from good behavior of the **diagonal morphism** $\delta_\pi \colon X \to X \times_Y X$ (the product of the identity morphism $X \to X$ with itself), usually through fun diagram chases. This lesson applies across many fields of geometry. (Another nice gift of the diagonal morphism: it will give us a good algebraic definition of differentials, in Chapter 21.)

So before we define separatedness, we make some observations about diagonal morphisms.

11.1 Fun with Diagonal Morphisms, and Quasiseparatedness Made Easy

Recall that most classes of morphisms are "reasonable" in the sense of §8.1: they are (i) preserved by composition, (ii) presesrved by base change, (iii) local on the target, and, as a consequence, (iv) preserved by product (Exercise 8.1.A).

Another useful consequence is that (v) "reasonable" classes of morphisms have a "cancellation property." This involves morphisms whose *diagonal* lies in P, i.e., those $\pi \colon X \to Y$ such that $\delta_\pi \colon X \to X \times_Y X$ lies in P. We call this auxiliary class of morphisms Pδ ("P-diagonal").

The Cancellation Theorem is a strange-looking, but very useful, result. Like the Diagonal-Base-Change diagram 1.2.S, this result is unexpectedly ubiquitous.

11.1.1. The Cancellation Theorem for a property P of morphisms — *Let P be a class of morphisms that is preserved by base change and composition. (Every "reasonable" class of morphisms satisfies this, §8.1.) Suppose*

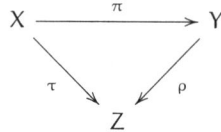

is a commuting diagram of schemes. Suppose ρ is in Pδ and τ is in P. Then π is in P.

The following diagram summarizes this important theorem:

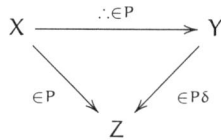

When you plug in different P, you get very different-looking (and nonobvious) consequences, the first of which is given in Exercise 11.2.E. (Here are some facts you can prove easily, but which can be interpreted as applications of the Cancellation Theorem in *Sets*, and which may thus shed light on how the Cancellation Theorem works. If $\pi\colon X \to Y$ and $\rho\colon Y \to Z$ are maps of sets, and $\rho \circ \pi$ is injective, then so is π; and if $\rho \circ \pi$ is surjective and ρ is injective, then π is surjective.)

11.1.A. EXERCISE. Prove the Cancellation Theorem 11.1.1. Possible hint: Interpret and make use of the following diagram.

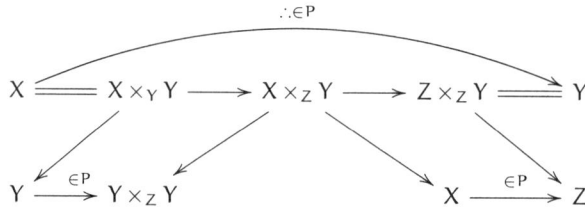

(vi) If P is a "reasonable" class, so is Pδ:

11.1.2. Theorem — *If P is a "reasonable" class of morphisms, then the class Pδ of morphisms whose diagonal is in P is also a "reasonable" class.*

Proof. We verify that Pδ satisfies the "three axioms of reasonableness."

11.1.B. EXERCISE. Prove that "Pδ is preserved by composition." Hint: Interpret and make use of the following diagram.

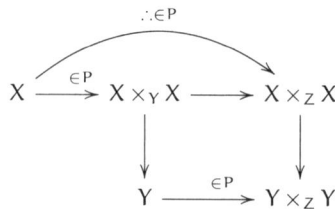

11.1.C. EXERCISE. Prove that "Pδ is preserved by base change." Hint: Suppose (a) below is Cartesian. Show that (b) is also Cartesian.

(a)

$$\begin{CD} V @>>> W \\ @VVV @VVV \\ X @>>> Y \end{CD}$$

(b)

$$\begin{CD} V @>>> V \times_W V \\ @VVV @VVV \\ X @>>> X \times_Y X \end{CD}$$

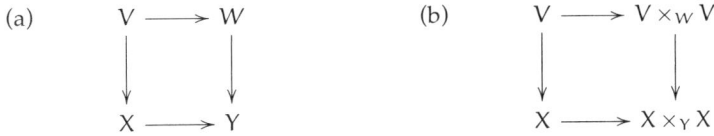

11.1.D. EXERCISE. Prove that "Pδ is local on the target." Hint: Suppose we have a cover $\{U_i \to Y\}$ of Y so that we have a Cartesian diagram (a). Then we have Cartesian diagram (b).

(a)

$$\begin{CD} V_i @>{\in P\delta}>> U_i \\ @AAA @AAA \\ X @>>> Y \end{CD}$$

(b)

$$\begin{CD} V_i @>{\therefore \in P}>> V_i \times_{U_i} V_i \\ @AAA @VVV \\ X @>>> X \times_Y X \end{CD}$$

$\square$

We continue our list of useful properties of "reasonable" classes of morphisms.

(vii) Finally, if you have two morphisms from X that are of certain form, the next exercise shows what more you need to ensure that the map to the product is also of that form. This time you can find the "proof by diagram" by yourself.

11.1.E. EXERCISE. Suppose P is a "reasonable" class of morphisms. Suppose $\rho : X \to Y$ and $\sigma : X \to Z$ are morphisms of S-schemes in class P, and $X \to S$ is in Pδ. Show that $(\rho, \sigma) : X \to Y \times_S Z$ is in class P as well. (See Exercise 11.2.A for an application.)

11.1.F.* EXERCISE. Show that finitely presented morphisms form a "reasonable" class of morphisms.

11.1.3. Example: Monomorphisms. Recall that a morphism $\pi : X \to Y$ is a monomorphism if and only if the diagonal $\delta_\pi : X \to X \times_Y X$ is an isomorphism (Exercise 1.2.W). Because isomorphisms form a "reasonable" class of morphisms, we have immediately shown that monomorphisms form a "reasonable" class as well (while accidentally solving Exercise 1.2.V, and proving that monomorphisms are preserved by base change).

11.1.4. Redefinition: Quasiseparated morphisms.
We (now) say a morphism $\pi : X \to Y$ is **quasiseparated** if the diagonal morphism $\delta_\pi : X \to X \times_Y X$ is quasicompact.

11.1.G. EXERCISE. Show that this agrees with our earlier definition of quasiseparated (§8.3.1): show that $\pi : X \to Y$ is quasiseparated if and only if for any affine open Spec A of Y, and two affine open subsets U and V of X mapping to Spec A, $U \cap V$ is a *finite* union of affine open sets. (Possible hint: Compare this to Proposition 11.2.13. Another possible hint: The Diagonal-Base-Change diagram, Exercise 1.2.S.)

11.1.5. Because quasicompact morphisms are a "reasonable" class (§10.4.E), Theorem 11.1.2 implies that quasiseparated morphisms are a reasonable class! In particular, quasiseparated morphisms are preserved by composition and base change, and the notion is local on the target.

11.1.6. All of this discussion was remarkably removed from geometry. To bring in the geometry, we now look more closely at the diagonal, which isn't just any old kind of morphism.

11.2   Separatedness, and Varieties

11.2.1. Proposition — *Let $\pi : X \to Y$ be a morphism of schemes.*

(a) *If X and Y are affine, then the diagonal morphism $\delta_\pi : X \to X \times_Y X$ is a closed embedding.*
(b) *In general, the diagonal morphism $\delta_\pi : X \to X \times_Y X$ is a locally closed embedding.*

The corresponding locally closed subscheme of $X \times_Y X$—the **diagonal (locally closed) subscheme**—will be denoted by Δ.

Proof. (a) Suppose $X = \operatorname{Spec} A$ and $Y = \operatorname{Spec} B$. Then the diagonal morphism corresponds to the natural ring map $A \otimes_B A \to A$ given by $a_1 \otimes a_2 \mapsto a_1 a_2$, which is obviously surjective.

(b) We will describe a union of open subsets of $X \times_Y X$ covering the image of X, such that the image of X is a closed embedding in this union.

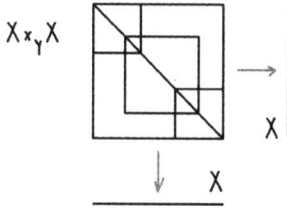

Figure 11.1 *An open neighborhood of the diagonal is covered by $U_{ij} \times_{V_i} U_{ij}$.*

Say Y is covered with affine open sets V_i and X is covered with affine open sets U_{ij}, with $\pi\colon U_{ij} \to V_i$. Note that $U_{ij} \times_{V_i} U_{ij}$ is an affine open subscheme of the product $X \times_Y X$ (basically this is how we constructed the product, by gluing together affine building blocks). Then the diagonal is covered by these affine open subsets $U_{ij} \times_{V_i} U_{ij}$. (Any point $p \in X$ lies in some U_{ij}; then $\delta(p) \in U_{ij} \times_{V_i} U_{ij}$. Figure 11.1 may be helpful.) Note that $\delta^{-1}(U_{ij} \times_{V_i} U_{ij}) = U_{ij}$: clearly, $U_{ij} \subset \delta^{-1}(U_{ij} \times_{V_i} U_{ij})$, and because $\operatorname{pr}_1 \circ \delta = \operatorname{id}_X$ (where pr_1 is the first projection), $\delta^{-1}(U_{ij} \times_{V_i} U_{ij}) \subset U_{ij}$. Finally, by part (a), $U_{ij} \to U_{ij} \times_{V_i} U_{ij}$ is a closed embedding. $\square$

The open subsets we described may not cover $X \times_Y X$, so we have not shown that δ is a closed embedding.

11.2.A. EASY EXERCISE. Suppose $X \to Y$ and $X \to Z$ are both locally closed embeddings. Prove that $X \to Y \times Z$ is a locally closed embedding. Hint: Exercise 11.1.E.

11.2.2. *Definition.* A morphism $\pi\colon X \to Y$ is **separated** if the diagonal morphism $\delta_\pi\colon X \to X \times_Y X$ is a closed embedding. An A-scheme X is said to be **separated over** A if the structure morphism $X \to \operatorname{Spec} A$ is separated. When people say that a scheme (rather than a morphism) X is separated, they mean implicitly that some "structure morphism" is separated. For example, if they are talking about A-schemes, they mean that X is separated over A.

Thanks to Proposition 11.2.1(b) (and the fact that a locally closed embedding whose image is closed is actually a closed embedding, Exercise 9.2.B), a morphism is separated if and only if the diagonal Δ is a closed subset—a purely topological condition on the diagonal. This is reminiscent of a definition of Hausdorff, as the next exercise shows.

11.2.B. EXERCISE (ONLY FOR THOSE SEEKING TOPOLOGICAL MOTIVATION). Show that a topological space X is Hausdorff if and only if the diagonal is a closed subset of $X \times X$. (The reason separatedness of schemes doesn't give Hausdorffness—i.e., for any two open points x and y there aren't necessarily disjoint open neighborhoods—is that in the category of schemes, the topological space $X \times X$ is not in general the product of the topological space X with itself; see §10.1.2.)

11.2.3. Proposition — *Separated morphisms form a "reasonable" class in the sense of §8.1—they are by preserved by composition and base change, and the notion is local on the target.*

Proof. The class of closed embeddings is "reasonable" in this sense (Exercise 10.4.A), so by Theorem 11.1.2 the same is true of separated morphisms. $\square$

11.2.C. EXERCISE. Show that monomorphisms are separated. Hence locally closed embeddings (and, in particular, open and closed embeddings) are separated.

11.2.D. EXERCISE. Show that separated morphisms are quasiseparated. (Hint: Closed embeddings are affine, hence quasicompact.)

11.2.4. Proposition — *Affine morphisms are separated. In particular, finite morphisms are separated.*

Proof. By Proposition 11.2.1(a), every morphism of affine schemes is separated. The notion of separatedness is local on the target, so we are done. $\square$

Many of the following exercises will drive home the importance of the Cancellation Theorem 11.1.1.

11.2.E. EXERCISE. Show that an A-scheme is separated (over A) if and only if it is separated over $\mathbb{Z}$. In particular, a complex scheme is separated over $\mathbb{C}$ if and only if it is separated over $\mathbb{Z}$, so complex geometers and arithmetic geometers can discuss separated schemes with each other without confusion.

11.2.F. EASY EXERCISE. Suppose we have morphisms $X \xrightarrow{\pi} Y \xrightarrow{\rho} Z$.

(a) Show that if $\rho \circ \pi$ is a locally closed embedding (resp., locally of finite type, separated), then so is π.
(b) If $\rho \circ \pi$ is quasicompact, and Y is Noetherian, show that π is quasicompact. Hint: Exercise 8.3.B(a).
(c) If $\rho \circ \pi$ is quasiseparated, show that π is quasiseparated. Hint: Exercise 8.3.B(b).

11.2.G. EXERCISE. Suppose $\mu\colon Z \to X$ is a morphism, and $\sigma\colon X \to Z$ is a section of μ, i.e., $\mu \circ \sigma$ is the identity on X.

$$Z$$
$$\mu \big\downarrow\big\uparrow \sigma$$
$$X$$

Show that σ is a locally closed embedding. Show that if μ is separated, then σ is a closed embedding. Show that if μ is quasiseparated, then σ is quasicompact. Give an example to show that σ need not be a closed embedding if μ is not separated.

11.2.5. *The graph morphism.* Suppose $\pi\colon X \to Y$ is a morphism of Z-schemes. The morphism $\gamma_\pi\colon X \to X \times_Z Y$, given by $\gamma_\pi = (\mathrm{id}, \pi)$, is called the **graph morphism** (see Figure 11.2). (It secretly appeared in Exercise 11.1.A.) Then π factors as $\mathrm{pr}_2 \circ \gamma_\pi$, where pr_2 is the second projection. The diagram of Figure 11.2 is often called the **graph of a morphism** π. (We will discuss graphs of rational maps in §11.3.4.) Exercise 11.2.G applied to $X \times_Z Y \to X$ yields the following.

11.2.6. Proposition — *The graph morphism γ_π is always a locally closed embedding. If Y is a separated Z-scheme (i.e., $Y \to Z$ is separated), then γ_π is a closed embedding. If Y is a quasiseparated Z-scheme, then γ_π is quasicompact.*

The image of γ_π (a locally closed subscheme of $X \times_Z Y$) is often also called the **graph of** π; we might denote it by Γ_π.

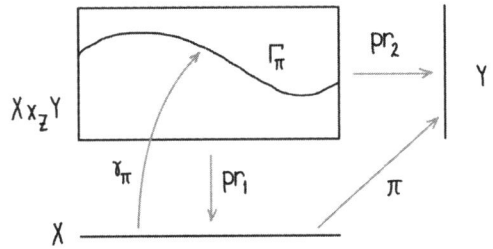

Figure 11.2 *The graph morphism.*

11.2.H. EXERCISE WE WON'T USE. Suppose P is a class of morphisms such that closed embeddings are in P, and P is preserved by fibered product and composition. Recall the Definition 10.7.8 of the "reduction" of a morphism. Show that if $\pi\colon X \to Y$ is in P then $\pi^{\mathrm{red}}\colon X^{\mathrm{red}} \to Y^{\mathrm{red}}$ is in P. (Two examples are the classes of separated morphisms and quasiseparated morphisms.) Possible hint:

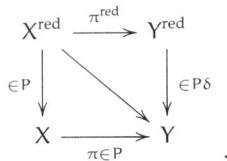

$$X^{\mathrm{red}} \xrightarrow{\pi^{\mathrm{red}}} Y^{\mathrm{red}}$$
$$\in P \downarrow \qquad \downarrow \in P\delta$$
$$X \xrightarrow[\pi \in P]{} Y$$

11.2.7. Important examples.

11.2.I. EXERCISE. Show that the line with doubled origin (Example 4.4.5) is not separated, by verifying that the image of the diagonal morphism is not closed. (Another argument is given below, in Exercise 11.3.C. A fancy argument is given in Exercise 13.7.C.)

We next come to our first example of something separated but not affine. The following single calculation will imply that all quasiprojective A-schemes are separated (once we know that the composition of separated morphisms are separated, Proposition 11.2.3).

11.2.8. Proposition — *The morphism* $\mathbb{P}^n_A \to \operatorname{Spec} A$ *is separated.*

We give two proofs. The first is by direct calculation. The second requires no calculation, and just requires that you remember some classical constructions described earlier.

Proof 1: Direct calculation. We cover $\mathbb{P}^n_A \times_A \mathbb{P}^n_A$ with open sets of the form $U_i \times_A U_j$, where $U_0, \ldots, U_n$ form the "usual" affine open cover. The case $i = j$ was taken care of before, in the proof of Proposition 11.2.1(a). If $i \neq j$ then

$$U_i \times_A U_j \cong \operatorname{Spec} A[x_{0/i}, \ldots, x_{n/i}, y_{0/j}, \ldots, y_{n/j}]/(x_{i/i} - 1, y_{j/j} - 1).$$

Now $\delta^{-1}(\operatorname{pr}_1^{-1}(U_i))$ is U_i (as the diagonal morphism composed with projection to the first factor is the identity), and similarly $U_j = \delta^{-1}(\operatorname{pr}_2^{-1}(U_j))$. Thus $\delta^{-1}(U_i \times_A U_j) = U_i \cap U_j$, so the diagonal morphism over $U_i \times_A U_j$ is $U_i \cap U_j \to U_i \times_A U_j$. This is a closed embedding, as the corresponding map of rings

$$A[x_{0/i}, \ldots, x_{n/i}, y_{0/j}, \ldots, y_{n/j}] \longrightarrow A[x_{0/i}, \ldots, x_{n/i}, x_{j/i}^{-1}]/(x_{i/i} - 1)$$

(given by $x_{k/i} \mapsto x_{k/i}$, $y_{k/j} \mapsto x_{k/i}/x_{j/i}$) is clearly a surjection (as each generator of the ring on the right is clearly in the image—note that $x_{j/i}^{-1}$ is the image of $y_{i/j}$). $\qquad\square$

Proof 2: Classical geometry. Note that the diagonal morphism $\delta: \mathbb{P}^n_A \to \mathbb{P}^n_A \times_A \mathbb{P}^n_A$ followed by the Segre embedding $S: \mathbb{P}^n_A \times_A \mathbb{P}^n_A \to \mathbb{P}^{n^2+2n}$ (§10.6, a closed embedding) can also be factored as the second Veronese embedding $\nu_2: \mathbb{P}^n_A \to \mathbb{P}^{\binom{n+2}{2}-1}$ (§9.3.6) followed by a linear map $L: \mathbb{P}^{\binom{n+2}{2}-1} \to \mathbb{P}^{n^2+2n}$, both of which are closed embeddings (the linear map L by Exercise 9.3.C).

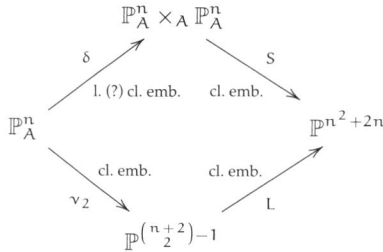

Informally, in coordinates:

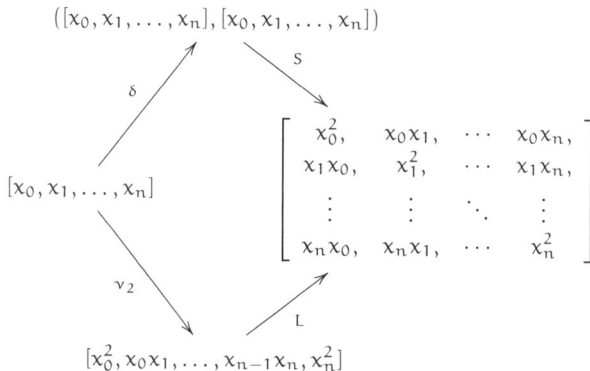

The composed map $\mathbb{P}_A^n \to \mathbb{P}_A^{n^2+2n}$ may be written as

$$[x_0, \ldots, x_n] \mapsto [x_0^2, x_0x_1, x_0x_2, \ldots, x_n^2],$$

where the subscripts on the right run over all ordered pairs (i, j) where $0 \leq i, j \leq n$.) This forces δ to send closed sets to closed sets (or else $S \circ \delta$ won't, but $L \circ \nu_2$ will). □

11.2.9. Important definition (finally!). A **variety** over a field k, or k-**variety**, is a reduced, separated scheme of finite type over k. For example, a reduced finite type affine k-scheme is a variety. (Do you see why separatedness holds?) The notion of variety generalizes our earlier notion of affine variety (§5.3.7) and projective variety (§5.3.7; see Corollary 11.2.15). By Exercise 5.3.F, the closed points of a variety are dense. (Notational caution: In some sources, the additional condition of irreducibility is imposed. Also, it is often assumed that k is algebraically closed.)

Varieties over k form a category: morphisms of k-varieties are just morphisms as k-schemes. (Of course, the category of varieties over an algebraically closed field k long predates modern scheme theory.)

11.2.J. EASY EXERCISE. Show that morphisms of k-varieties are finite type and separated.

A **subvariety** of a variety X is a *reduced* locally closed subscheme of X (which you can quickly check is a variety itself). An open subvariety of X is an open subscheme of X. (Reducedness is automatic in this case.) A closed subvariety of X is a reduced closed subscheme of X.

11.2.K.** EXERCISE (FOR THOSE WITH APPROPRIATE BACKGROUND: COMPLEX ALGEBRAIC VARIETIES YIELD COMPLEX ANALYTIC VARIETIES). Show that the analytification (Exercise 7.3.P) of a complex algebraic variety is a complex analytic variety.

11.2.10. *Products of varieties.*

11.2.11. Proposition —

(a) *If k is perfect, the product of k-varieties (the fibered product over* Spec k*) is a k-variety.*
(b) *If k is algebraically closed, the product of irreducible k-varieties is an irreducible k-variety.*
(c) *If k is algebraically closed, the product of connected k-varieties is a connected k-variety.*

Proof. (a) The finite type and separated statements are straightforward, as both properties are preserved by base change and composition. For reducedness, reduce to the affine case, then use Corollary 10.5.23.

(b) It only remains to show irreducibility. Reduce to the affine case using Exercise 10.5.N (as in the proof of Proposition 10.5.15). Then use Exercise 10.5.M, which requires separable closure. (As an exercise, you can avoid the complications of §10.5 by instead using Exercise 10.4.H.)

11.2.L. EXERCISE. Prove (c). □

11.2.12. Back to separatedness. Here is a very handy consequence of separatedness.

11.2.13. Proposition — *Suppose* $X \to$ Spec A *is a separated morphism to an affine scheme, and* U *and* V *are affine open subsets of* X. *Then* $U \cap V$ *is an affine open subset of* X.

Before proving this, we give an unexpected consequence. Taking $X =$ Spec A, we see that the intersection of any two affine open subsets of Spec A is also an affine open subset. This is certainly not obvious! We know the intersection of two distinguished affine open subsets is affine (from $D(f) \cap D(g) = D(fg)$), but we have little handle on affine open subsets in general.

Warning: This property does not characterize separatedness. (The correct property is "affine-diagonal," which we will not discuss, but which is useful in the theory of stacks.) For example, if A is a field k and X is the line with doubled origin over k, then X also has this property.

In the proof of Proposition 11.2.13, we will use the following nifty result.

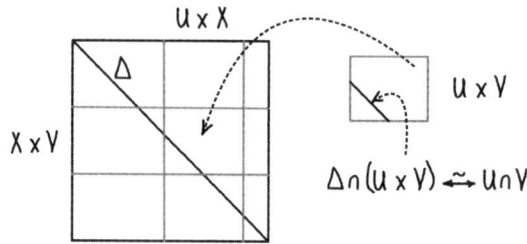

Figure 11.3 $\Delta \cap (U \times V) \cong U \cap V$ *(Exercise 11.2.M).*

11.2.M. EASY EXERCISE. Suppose U and V are open subsets of an A-scheme X. Show that $\Delta \cap (U \times_A V) \cong U \cap V$. (The intersection of two locally closed subschemes was defined in Exercise 10.2.C.) Hints: (i) This was basically proved in the course of Proof 1 of Proposition 11.2.8. (ii) See Figure 11.3. (iii) You can interpret this statement using the Diagonal-Base-Change diagram, Exercise 1.2.S:

$$
\begin{array}{ccc}
U \cap V & \longrightarrow & U \times_A V \\
\downarrow & & \downarrow \\
X & \xrightarrow{\ \delta\ } & X \times_A X.
\end{array}
$$

11.2.14. *Proof of Proposition 11.2.13.* By Exercise 11.2.M, $(U \times_A V) \cap \Delta \cong U \cap V$, where Δ is the diagonal. But $U \times_A V$ is affine (the fibered product of two affine schemes over an affine scheme is affine, Step 1 of our construction of fibered products, Theorem 10.1.1), and $(U \times_A V) \cap \Delta$ is a closed subscheme of the affine scheme $U \times_A V$, and hence $U \cap V$ is affine. $\qquad \square$

11.2.15. Corollary — *Every quasiprojective A-scheme is separated over A. In particular, every reduced quasiprojective k-scheme is a k-variety.*

Proof. Suppose $X \to \operatorname{Spec} A$ is a quasiprojective A-scheme. The structure morphism can be factored into an open embedding composed with a closed embedding followed by $\mathbb{P}^n_A \to \operatorname{Spec} A$. Open embeddings and closed embeddings are separated (Exercise 11.2.C), and $\mathbb{P}^n_A \to \operatorname{Spec} A$ is separated (Proposition 11.2.8). Compositions of separated morphisms are separated (Proposition 11.2.3), so we are done. $\qquad \square$

Here is an exercise we won't use, but you may like. It includes a converse to Proposition 11.2.13.

11.2.N. EXERCISE. Show that $X \to \operatorname{Spec} A$ is separated if and only if, for all affine open subsets U and V of X, (i) the intersection $U \cap V$ is affine, and (ii) the map $\mathcal{O}(U) \otimes_A \mathcal{O}(V) \to \mathcal{O}(U \cap V)$ is surjective. Show that it is enough to check that this holds as U and V range over the sets in any affine cover $X = \cup U_i$. (Hint: We largely did this in Proof 1 of Proposition 11.2.8.)

11.2.16. ** **Universally injective morphisms and the diagonal.**

11.2.O. EXERCISE. Show that $\pi \colon X \to Y$ is universally injective if and only if the diagonal morphism $\delta_\pi \colon X \to X \times_Y X$ is surjective. (Recall that δ_π is *always* injective, by Proposition 11.2.1(b).)

Because surjective morphisms form a "reasonable" class (Exercise 10.4.G), we see that universally injective morphisms also form a "reasonable" class.

11.2.P. EASY EXERCISE. If $\pi \colon X \to Y$ and $\rho \colon Y \to Z$ are morphisms, and $\rho \circ \pi$ is universally injective, show that π is universally injective.

11.2.Q. EXERCISE.

(a) Show that universally injective morphisms are separated.

(b) Show that a map between finite type schemes over an algebraically closed field $\overline{k}$ is universally injective if and only if it is injective on closed points.

11.3 The Locus where Two Morphisms from X to Y Agree, and the "Reduced-to-Separated" Theorem

When we introduced rational maps in §7.5, we promised that in good circumstances, a rational map has a "largest domain of definition." We are now ready to make precise what "good circumstances" means, in the Reduced-to-Separated Theorem 11.3.2. We first introduce an important result making sense of the locus where two morphisms with the same source and target "agree."

11.3.A. USEFUL EXERCISE: THE LOCUS WHERE TWO MORPHISMS AGREE. Suppose $\pi\colon X \to Y$ and $\pi'\colon X \to Y$ are two morphisms over some scheme Z.

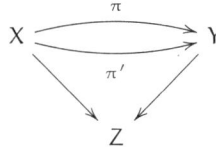

We can now give meaning to the phrase "the locus where π and π' agree," and to the fact that, in particular, there is a largest locally closed subscheme where they agree—which is closed if Y is separated over Z. Suppose $\mu\colon W \to X$ is some morphism (not assumed to be a locally closed embedding). We say that π and π' agree on μ if $\pi \circ \mu = \pi' \circ \mu$. Show that there is a locally closed subscheme $i\colon V \hookrightarrow X$ on which π and π' agree, such that any morphism $\mu\colon W \to X$ on which π and π' agree factors uniquely through i, i.e., there is a unique $j\colon W \to V$ such that $\mu = i \circ j$. Show further that if $Y \to Z$ is separated, then $i\colon V \hookrightarrow X$ is a closed embedding. Hint: Define V to be the following fibered product:

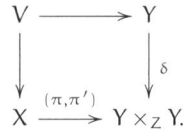

$$
\begin{array}{ccc}
V & \longrightarrow & Y \\
\downarrow & & \downarrow{\scriptstyle \delta} \\
X & \xrightarrow{(\pi,\pi')} & Y \times_Z Y.
\end{array}
$$

As δ is a locally closed embedding, $V \to X$ is, too. Then if $\mu\colon W \to X$ is any morphism such that $\pi \circ \mu = \pi' \circ \mu$, then μ factors through V.

The fact that the locus where two maps agree can be nonreduced should not come as a surprise: consider two maps from $\mathbb{A}^1_k$ to itself, $\pi(x) = 0$ and $\pi'(x) = x^2$. They agree when $x = 0$, but the situation is (epsilonically) better than that—they should agree even on $\operatorname{Spec} k[x]/(x^2)$.

11.3.1. *Minor remarks.*

(i) In the previous exercise, we describe $V \hookrightarrow X$ by a universal property. Taking this as the definition, it is not a priori clear that V is a locally closed subscheme of X, or even that it exists.

(ii) Warning: Consider two maps from $\operatorname{Spec} \mathbb{C}$ to itself over $\operatorname{Spec} \mathbb{R}$, the identity and complex conjugation. These are both maps from a point to a point, yet they do not agree despite agreeing as maps of sets. (If you are not convinced that they disagree as morphisms, this might help: after base change $\operatorname{Spec} \mathbb{C} \to \operatorname{Spec} \mathbb{R}$, they do not agree even as maps of sets.)

(iii) More generally, the (set-theoretic) locus where π and π' agree can be interpreted as follows: π and π' agree at p if $\pi(p) = \pi'(p)$ and the two maps of residue fields are the same. (You may enjoying thinking this through as an exercise.)

11.3.B. EXERCISE: MAPS OF $\overline{k}$-VARIETIES ARE DETERMINED BY THE MAPS ON CLOSED POINTS. Suppose $\pi\colon X \to Y$ and $\pi'\colon X \to Y$ are two morphisms of $\overline{k}$-varieties that are the same at the level of closed points (i.e., for each closed point $p \in X$, $\pi(p) = \pi'(p)$). Show that $\pi = \pi'$. (This implies that the functor from the category of "classical varieties over $\overline{k}$", which we won't define here, to the category of $\overline{k}$-schemes, is fully faithful. Can you generalize this appropriately to non-algebraically closed fields?)

11.3.C. LESS IMPORTANT EXERCISE. Show that the line with doubled origin X (Example 4.4.5) is not separated, by finding two morphisms $\pi\colon W \to X$, $\pi'\colon W \to X$ whose domain of agreement is not a closed subscheme (cf. Proposition 11.2.1(b)). (Another argument was given above, in Exercise 11.2.I. A fancy argument will be given in Exercise 13.7.C.)

We now come to the central result of this section.

11.3.2. Reduced-to-Separated Theorem — *Two S-morphisms $\pi\colon U \to Z$, $\pi'\colon U \to Z$ from a reduced scheme to a separated S-scheme agreeing on a dense open subset of U are the same.*

Proof. Let V be the locus where π and π' agree. It is a closed subscheme of U (by Exercise 11.3.A) that contains a dense open set. But the only closed subscheme of a reduced scheme U whose underlying set is dense is all of U. (Do you see why this last statement is true? Can you give a counterexample if the reducedness hypothesis is removed?) □

11.3.3. *Consequence 1.* Hence (as X is reduced and Y is separated) if we have two morphisms from open subsets of X to Y, say, $\pi\colon U \to Y$ and $\pi'\colon V \to Y$, and they agree on a dense open subset Z of $U \cap V$, then they necessarily agree on $U \cap V$.

Consequence 2. A rational map has a largest **domain of definition** on which $\pi\colon U \dashrightarrow Y$ is a morphism, which is the union of all the domains of definition. In particular, a rational function on a reduced scheme has a largest domain of definition. For example, the domain of definition of $\mathbb{A}_k^2 \dashrightarrow \mathbb{P}_k^1$ given by $(x,y) \mapsto [x,y]$ has domain of definition $\mathbb{A}_k^2 \setminus \{(0,0)\}$ (cf. §7.5.3). This partially extends the definition of the domain of a rational function on a locally Noetherian scheme (Definition 6.6.36). The complement of the domain of definition is called the **indeterminacy locus** (a phrase used earlier in §6.6.36), and its points are sometimes called **fundamental points** of the rational map, although we won't use these phrases. (We will see in Exercise 22.4.L that a rational map to a projective scheme can be upgraded to an honest morphism by "blowing up" a scheme-theoretic version of the locus of indeterminacy.)

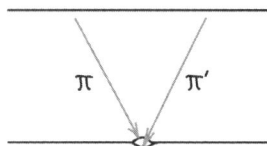

Figure 11.4 *Two different maps to a nonseparated scheme agreeing on a dense open set (see Exercise 11.3.D).*

11.3.D. EXERCISE. Show that the Reduced-to-Separated Theorem 11.3.2 is false if we give up reducedness of the source or separatedness of the target. Here are some possibilities. For the first, consider the two maps from $\operatorname{Spec} k[x,y]/(y^2,xy)$ to $\operatorname{Spec} k[t]$, where we take π given by $t \mapsto x$ and π' given by $t \mapsto x+y$; π and π' agree on the closed subscheme cut out by $y = 0$. For the second, consider the two maps from $\operatorname{Spec} k[t]$ to the line with the doubled origin, one of which maps to the "upper half," and one of which maps to the "lower half." These two morphisms agree on the dense open set $D(t)$; see Figure 11.4.

11.3.4. *Graphs of rational maps.* (Graphs of *morphisms* were defined in §11.2.5.) If X is reduced and Y is separated, define the **graph Γ_π of a rational map** $\pi\colon X \dashrightarrow Y$ as follows. Let (U,π') be any representative of this rational map (so $\pi'\colon U \to Y$ is a morphism). Let Γ_π be the scheme-theoretic closure of $\Gamma_{\pi'} \hookrightarrow U \times Y \hookrightarrow X \times Y$, where the first map is a closed embedding (Proposition 11.2.6), and the second is an open embedding. The product here should be taken in the category you are working in. For example, if you are working with k-schemes, the fibered product should be taken over k.

11.3.E. EXERCISE. Show that the graph of a rational map $\pi\colon X \dashrightarrow Y$ is independent of the choice of representative of π. Hint: Suppose $\xi'\colon U \to Y$ and $\xi\colon V \to Y$ are two representatives of π. Reduce to the case where V is the domain of definition of π (§11.3.3), and $\xi' = \xi|_U$. Reduce to the case $V = X$. Show an isomorphism $\Gamma_\pi \cong X$, and $\Gamma_{\xi|_U} \cong U$. (Remark: The separatedness of Y is not necessary.)

In analogy with graphs of morphisms, the following diagram of a graph of a rational map π can be useful (cf. Figure 11.2).

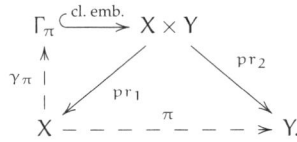

$$
\begin{array}{ccc}
\Gamma_\pi & \xrightarrow{\text{cl. emb.}} & X \times Y \\
\uparrow{\scriptstyle\gamma_\pi} & \swarrow{\scriptstyle\mathrm{pr}_1} & \downarrow{\scriptstyle\mathrm{pr}_2} \\
X & \dashrightarrow{\scriptstyle\pi} & Y.
\end{array}
$$

11.3.F. EXERCISE (THE BLOW-UP OF THE AFFINE PLANE AS THE GRAPH OF A RATIONAL MAP). Consider the rational map $\mathbb{A}_k^2 \dashrightarrow \mathbb{P}_k^1$ given by $(x, y) \mapsto [x, y]$. Show that this rational map cannot be extended over the origin. (A similar argument arises in Exercise 7.5.J on the Cremona transformation.) Show that the graph of the rational map is the morphism (the blow-up) described in Exercise 10.3.F. (When we define blow-ups in general, we will see that they are often graphs of rational maps; see Exercise 22.4.M.)

11.3.5. Variations.
Variations of the short proof of Theorem 11.3.2 yield other useful results. Exercise 11.3.B is one example. Here is another.

11.3.G. EXERCISE (MAPS TO A SEPARATED SCHEME CAN BE EXTENDED OVER AN EFFECTIVE CARTIER DIVISOR IN AT MOST ONE WAY). Suppose X is a Z-scheme (not necessarily reduced!), and Y is a separated Z-scheme. Suppose further that D is an effective Cartier divisor on X. Show that any Z-morphism $X \setminus D \to Y$ can be extended in at most one way to a Z-morphism $X \to Y$. (Hint: Reduce to the case where $X = \operatorname{Spec} A$, and D is the vanishing scheme of $t \in A$. Reduce to showing that the scheme-theoretic image of $D(t)$ in X is all of X. Show this by showing that $A \to A_t$ is an inclusion.)

As noted in §7.5.2, rational maps can be defined from any X that has associated points to any Y. The Reduced-to-Separated Theorem 11.3.2 can be extended to this setting, as follows.

11.3.H. EXERCISE (THE "ASSOCIATED-TO-SEPARATED THEOREM"). Prove that two S-morphisms π and π' from a locally Noetherian scheme U to a separated S-scheme Z, agreeing on an open subset containing the associated points of U, are the same.

11.4   Proper Morphisms

A map of topological spaces (also known as a continuous map!) is said to be *proper* if the preimage of any compact set is compact. *Properness* of morphisms is an analogous property. For example, a variety over $\mathbb{C}$ will be proper if it is compact in the classical topology (see [Se3, §7]). Alternatively, we will see that projective A-schemes are proper over A—so this is a nice property satisfied by projective schemes, which also is convenient to work with.

Recall (§8.3.5) that a (continuous) map of topological spaces $\pi\colon X \to Y$ is *closed* if for each closed subset $S \subset X$, $\pi(S)$ is also closed. A morphism of schemes is closed if the underlying continuous map is closed. We say that a morphism of schemes $\pi\colon X \to Y$ is **universally closed** if for every morphism $Z \to Y$, the induced morphism $Z \times_Y X \to Z$ is closed. In other words, a morphism is universally closed if it remains closed under any base change. By now you will automatically expect the following exercise.

11.4.A. EASY EXERCISE. Show that the class of universally closed morphisms is "reasonable" in the sense of §8.1.

To motivate the definition of properness for schemes, we remark that a continuous map $\pi\colon X \to Y$ of locally compact Hausdorff spaces that have countable bases for their topologies is universally closed if and only if it is proper (i.e., preimages of compact subsets are compact). You are welcome to prove this as an exercise.

11.4.1. Definition. *A morphism* $\pi\colon X \to Y$ *is* **proper** *if it is separated, finite type, and universally closed.* A scheme X is often said to be proper if some implicit structure morphism is proper. For example, a k-scheme X is often described as proper if $X \to \operatorname{Spec} k$ is proper. (A k-scheme is often said to be **complete** if it is proper. We will not use this terminology.) If A is a ring, one often says that an A-scheme is **proper over** A if it is proper over Spec A.

Let's try this idea out in practice. We expect that $\mathbb{A}^1_{\mathbb{C}} \to \operatorname{Spec} \mathbb{C}$ is not proper, because the complex manifold corresponding to $\mathbb{A}^1_{\mathbb{C}}$ is not compact. However, note that this map is separated (it is a map of affine schemes), finite type, and (trivially) closed. So the "universally" is what matters here.

11.4.B. EASY EXERCISE. Show that $\pi\colon \mathbb{A}^1_{\mathbb{C}} \to \operatorname{Spec} \mathbb{C}$ is a closed map, but not universally closed. Hence π is not proper. Possible hint: Consider a well-chosen map such as $\mathbb{A}^1_{\mathbb{C}} \times_{\mathbb{C}} \mathbb{P}^1_{\mathbb{C}} \to \mathbb{P}^1_{\mathbb{C}}$. (See Figure 19.1 for another finite type, separated, closed morphism that is not proper. Showing *that* morphism is not proper requires more creativity.)

11.4.2. *Example.* As a first example: Closed embeddings are proper. They are clearly separated, as affine morphisms are separated, Proposition 11.2.4. They are finite type. After base change, they remain closed embeddings (§10.2.2), and closed embeddings are always closed. This easily extends further as follows.

11.4.3. Proposition — *Finite morphisms are proper.*

Proof. Finite morphisms are separated (as they are affine by definition, and affine morphisms are separated, Proposition 11.2.4), and finite type (basically because finite modules over a ring are automatically finitely generated). To show that finite morphisms are closed after any base change, we note that they remain finite after any base change (finiteness is preserved by base change, Exercise 10.4.D(d)), and finite morphisms are closed (Exercise 8.3.L). $\qquad\square$

11.4.4. Proposition —

(a) *The notion of "proper morphism" is stable under base change.*

(b) *The notion of "proper morphism" is local on the target (i.e., $\pi\colon X \to Y$ is proper if and only if for any affine open cover $U_i \to Y$, $\pi^{-1}(U_i) \to U_i$ is proper). Note that the "only if" direction follows from (a) — consider base change by $U_i \hookrightarrow Y$.*

(c) *The notion of "proper morphism" is preserved by composition.*

(d) *The product of two proper morphisms is proper: if $\pi\colon X \to Y$ and $\pi'\colon X' \to Y'$ are proper, where all morphisms are morphisms of Z-schemes, then $\pi \times \pi'\colon X \times_Z X' \to Y \times_Z Y'$ is proper.*

(e) *Suppose*

(11.4.4.1)

$$
\begin{array}{ccc}
X & \xrightarrow{\ \pi\ } & Y \\
 & {\scriptstyle \tau}\searrow \quad \swarrow {\scriptstyle \rho} & \\
 & Z &
\end{array}
$$

is a commutative diagram, τ is proper, and ρ is separated. Then π is proper.

A sample application of (e): A morphism (over Spec k) from a proper k-scheme to a separated k-scheme is always proper.

Proof. Parts (a) through (c) say that proper morphisms form a "reasonable" class of morphisms in the sense of §8.1. So they follow because separated morphisms, finite type morphisms, and universal closed morphisms are all reasonable classes. Then (d) follows from (a) and (c) by Exercise 8.1.A. Closed embeddings are proper (Example 11.4.2), so (e) follows from the Cancellation Theorem 11.1.1 for proper morphisms. $\qquad\square$

11.4.C. EXERCISE ("IMAGES OF PROPER SCHEMES ARE PROPER"). Suppose in the diagram

$$X \xrightarrow{\;\;\pi\;\;} Y$$
$$\tau \searrow \qquad \swarrow \rho$$
$$Z$$

that τ is proper and ρ is separated and finite type. Show that the scheme-theoretic image of X under π is a proper Z-scheme. (Thus properness in algebraic geometry behaves like properness in topology.)

We now come to the most important example of proper morphisms.

11.4.5. Theorem — *Projective A-schemes are proper over A.*

(As finite morphisms to Spec A are projective A-schemes, Exercise 8.3.I, Theorem 11.4.5 can be used to give a second proof that finite morphisms are proper, Proposition 11.4.3.)

Proof. The structure morphism of a projective A-scheme $X \to \operatorname{Spec} A$ factors as a closed embedding followed by $\mathbb{P}_A^n \to \operatorname{Spec} A$. Closed embeddings are proper (Example 11.4.2), and compositions of proper morphisms are proper (Proposition 11.4.4), so it suffices to show that $\mathbb{P}_A^n \to \operatorname{Spec} A$ is proper. We have already seen that this morphism is finite type (Easy Exercise 5.3.D) and separated (Proposition 11.2.8), so it suffices to show that $\mathbb{P}_A^n \to \operatorname{Spec} A$ is universally closed. As $\mathbb{P}_A^n = \mathbb{P}_{\mathbb{Z}}^n \times_{\mathbb{Z}} \operatorname{Spec} A$, it suffices to show that $\mathbb{P}_X^n := \mathbb{P}_{\mathbb{Z}}^n \times_{\mathbb{Z}} X \to X$ is closed for any scheme X. But the property of being closed is local on the target on X, so by covering X with affine open subsets, it suffices to show that $\mathbb{P}_B^n \to \operatorname{Spec} B$ is closed for all rings B. This is the Fundamental Theorem of Elimination Theory (Theorem 8.4.10). $\qquad \square$

11.4.6. *Remark: "Reasonable" proper schemes are projective.* It is not easy to come up with an example of an A-scheme that is proper but not projective! Over a field, all proper curves are projective (see Remark 19.2.I), and all smooth surfaces are projective (see [Ba, Thm. 1.28] for a proof of this theorem of Zariski; smoothness of course is not yet defined). We will meet a first example of a proper but not projective variety (a singular threefold) in §15.2.11. We will later see an example of a proper nonprojective surface in §19.11.11, and a simpler one in Exercise 20.2.G. Once we know about flatness, we will see Hironaka's example of a proper nonprojective irreducible smooth threefold over $\mathbb{C}$ (§24.7.6).

11.4.7. Functions on connected reduced proper $\overline{k}$-schemes must be constant.

As an enlightening application of these ideas, we show that if X is a connected reduced proper k-scheme where $k = \overline{k}$, then $\Gamma(X, \mathscr{O}_X) = k$. The analogous fact for connected compact complex manifolds uses the maximum principle. We saw this in the special case $X = \mathbb{P}^n$ in Exercise 4.4.E; it will be vastly generalized by Grothendieck's Coherence Theorem 18.9.1.

Suppose $f \in \Gamma(X, \mathscr{O}_X)$ (f is a function on X). This is the same as a map $\pi \colon X \to \mathbb{A}_k^1$ (Exercise 7.3.G, discussed further in §7.6.1). Let π' be the composition of π with the open embedding $\mathbb{A}^1 \hookrightarrow \mathbb{P}^1$. By Proposition 11.4.4(e), π' is proper, and, in particular, closed. As X is connected, the image of π' is as well. Thus the set-theoretic image of π' must be either a closed point, or all of $\mathbb{P}^1$. But the set-theoretic image of π' lies in $\mathbb{A}^1$, so it must be a closed point p (which we identify with an element of k).

By Corollary 9.4.5, the support of the *scheme-theoretic* image of π is the closed point p. By Exercise 9.4.A, the scheme-theoretic image is precisely p (with the reduced structure). Thus π can be interpreted as the structure map to Spec k, followed by a closed embedding to $\mathbb{A}^1$ identifying Spec k with p. You should be able to verify that this is the map to $\mathbb{A}^1$ corresponding to the constant function $f = p$.

(What are counterexamples if different hypotheses are relaxed? The condition $k = \bar{k}$ cannot be ignored—if $k = \mathbb{R}$, then $X = \operatorname{Spec} \mathbb{C}$ is a connected reduced proper projective k-scheme, with $\Gamma(X, \mathcal{O}_X) = k^2$. Interesting fact: $x^2 = 0$ in $\mathbb{P}^2_k$ shows that even nonreduced proper k-schemes can have only constants as functions; see §5.2.3 and Exercise 18.6.U.)

11.4.8. Facts (not yet proved) that may help you correctly think about finiteness.
The following facts may shed some light on the notion of finiteness. We will prove them later.

A morphism of locally Noetherian schemes is finite if and only if it is proper and affine, if and only if it is proper and quasifinite (Theorem 28.5.2). We have proved parts of this statement, but we will only finish the proof once we know Zariski's Main Theorem; cf. §8.3.12).

As an application: quasifinite morphisms from proper schemes to separated schemes are finite. Here is why: Suppose $\pi\colon X \to Y$ is a quasifinite morphism over Z, where X is proper over Z. Then by the Cancellation Theorem 11.1.1 for proper morphisms, $X \to Y$ is proper. Hence as π is quasifinite and proper, π is finite.

As an explicit example, consider the map $\pi\colon \mathbb{P}^1_k \to \mathbb{P}^1_k$ given by $[x, y] \mapsto [f(x, y), g(x, y)]$, where f and g are homogeneous polynomials of the same degree with no common roots in $\mathbb{P}^1$. The fibers are finite, and π is proper (from the Cancellation Theorem 11.1.1 for proper morphisms, as discussed after the statement of Theorem 11.4.4), so π is finite. This could be checked directly as well, but now we can save ourselves the annoyance.

11.4.9.** Group varieties. The rest of §11.4 is double-starred, and we will also throughout work in the category of *varieties over a field* k. We briefly discuss group varieties, mainly because we can, not because we have anything profound to say. We discuss them right now only because properness gives an unexpected consequence, that proper group varieties are abelian (thanks to the surprisingly simple Rigidity Lemma 11.4.12).

As discussed in §11.2.9, now that we have the category of varieties over k, we immediately have the notion of a **group variety** (thanks to §7.6.4 on group objects in any category). An **algebraic group** (over k) is a smooth group variety. Examples include GL_n (Exercise 7.6.N(b)), which includes $\mathbb{G}_m$ as a special case, and SL_n (Exercise 7.6.Q).

11.4.10. *Side Remarks (that we won't prove or use).* Group varieties are automatically smooth (hence algebraic groups) in characteristic 0. Group varieties are not necessarily smooth in positive characteristic. (Let k be an imperfect field of characteristic p, and choose $t \in k$ without a pth root in k. Consider the "closed sub-group scheme" $x^p = ty^p$ of the "additive group" $\mathbb{G}^2_a = \operatorname{Spec} k[x, y]$. This group scheme is reduced, but not geometrically reduced; and not smooth, or even regular.) Algebraic groups are automatically quasiprojective. An algebraic group G is affine (as a scheme) if and only if it admits a closed immersion into GL_n for some $n \geq 0$, such that $G \to GL_n$ is a homomorphism of group schemes (Exercise 7.6.N(a)). For further discussion on these and other issues, see [P2, §5].

11.4.11. *Definition: Abelian varieties.* We can now define one of the most important classes of varieties: abelian varieties. The bad news is that we have no real example yet. We will later see elliptic curves as an example (§19.10), but we will not meet anything more exotic than that. Still, it is pleasant to know that we can make the definition this early.

An **abelian variety** over a field k is an algebraic group that is geometrically integral and projective (over k). It turns out that the analytification of any abelian variety over $\mathbb{C}$ is a complex torus, i.e., $\mathbb{C}^n/\Lambda$, where Λ is a lattice (of rank 2n). (The key idea: Connected compact complex Lie groups are commutative; see Remark 11.4.14; the universal cover is a simply connected commutative complex Lie group, and thus $\mathbb{C}^n$. See [BL, Lem. 1.1.1].)

We now show that abelian varieties are abelian, i.e., abelian varieties are commutative algebraic group varieties. (This is less tautological than it sounds! The adjective "abelian" is used in two different senses; what they have in common is that they are derived from the name Niels Henrik

Abel.) As algebraic groups need not be commutative (witness GL_n), it is somehow surprising that commutativity could be forced by "compactness."

The key result is the following fact, beautiful in its own right.

11.4.12. Rigidity Lemma — *Let* X, Y, *and* Z *be varieties, where* X *is proper and geometrically integral, with a* k-*valued point* p, *and* Y *is irreducible. Let* $\mathrm{pr}\colon X \times Y \to Y$ *be the projection, and let* $\alpha\colon X \times Y \to Z$ *be any morphism. Suppose* $q \in Y$ *is such that* α *is constant on* $X \times \{q\}$; *say* $\alpha(X \times \{q\}) = r \in Z$.

(a) Then there is a morphism $\psi\colon Y \to Z$ *such that* $\alpha = \psi \circ \mathrm{pr}$. *(In particular, for every* $q' \in Y$, α *is constant on* $X \times \{q'\}$.)

(b) If α *is also constant on* $\{p\} \times Y$, *then* α *is a constant function on all of* $X \times Y$.

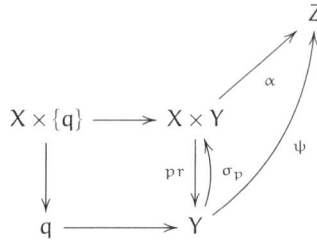

The hypotheses can be relaxed in a number of ways, but this version suffices for our purposes. If you have not read §10.5.1, where geometric integrality is discussed, you need not worry too much; this is automatic if $k = \bar{k}$ and X is integral (Hard Fact 10.5.3). Even if you *have* read §10.5.1, it is helpful to first read this proof under the assumption that $k = \bar{k}$, to avoid being distracted from the main idea by geometric points. But of course all of §11.4.9 is double-starred, so you shouldn't be reading this anyway.

11.4.D. EXERCISE. Show that the properness hypothesis on X is necessary. (Can you make this accord with your intuition?)

Proof of the Rigidity Lemma 11.4.12. Define $\beta\colon X \times Y \to Z$ by $\beta(x, y) = \alpha(p, y)$. (More precisely, if σ_p is the section of $X \times Y \to Y$ pulled back from $\operatorname{Spec} k \mapsto p$, then $\beta = \alpha \circ \sigma_p \circ \mathrm{pr}$.) We will show that $\beta(x, y) = \alpha(x, y)$. This will imply (a) (take $\psi = \alpha \circ \sigma_p$), which in turn immediately implies (b).

Proposition 10.5.19 (geometrically reduced times reduced is reduced) implies that $X \times Y$ is reduced. Exercise 10.5.O (geometrically irreducible times irreducible is irreducible) implies that $X \times Y$ is irreducible. As Z is separated, by Exercise 11.3.A, it suffices to show that $\alpha = \beta$ on a nonempty (hence dense) open subset of $X \times Y$.

Let $U \subset Z$ be an affine open neighborhood of r. Then $\alpha^{-1}(Z \setminus U)$ is a closed subset of $X \times Y$. As the projection pr is proper (using properness of X, and preservation of properness under base change, Proposition 11.4.4(a)), we have $\mathrm{pr}(\alpha^{-1}(Z \setminus U))$ is closed. Its complement is open, and contains q; let $V \subset Y$ be an open neighborhood of q disjoint from $\mathrm{pr}(\alpha^{-1}(Z \setminus U))$.

11.4.E. EXERCISE. Suppose $\gamma\colon \operatorname{Spec} \overline{K} \to V$ is a geometric point of V. Show that the fiber $X_{\overline{K}}$ over $\operatorname{Spec} \overline{K}$ (or, more precisely, over γ) is mapped to a point in Z by the restriction $\alpha_{\overline{K}}$ of α to $X_{\overline{K}}$. Hint: Review §11.4.7.

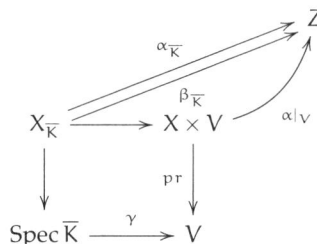

By Exercise 11.3.B (maps of varieties over an algebraically closed field are determined by the map of closed points), $\alpha_{\overline{K}} = \beta_{\overline{K}}$.

11.4.F. EXERCISE. Use Exercise 11.3.A (the fact that the locus where $\alpha = \beta$ is a closed subscheme—and represents a certain functor) to show that $\alpha = \beta$ on $\mathrm{pr}^{-1}(V)$.

$\square$

11.4.13. Corollary — *Suppose A is an abelian variety, G is a group variety, and $\phi\colon A \to G$ is any morphism of varieties. Then ϕ is the composition of a translation and a homomorphism.*

Proof. Let e_A and e_G be the identity points of A and G, respectively. Composing ϕ with a translation, we may assume that $\phi(e_A) = e_G$. Consider the morphism $\alpha\colon A \times A \to G$ given by $\alpha(a_1, a_2) = \phi(a_1 a_2)\phi(a_2)^{-1}\phi(a_1)^{-1}$. Then $\alpha(\{e_A\} \times A) = \alpha(A \times \{e_A\}) = \{e_G\}$, so by the Rigidity Lemma 11.4.12, α is a constant, and sends $A \times A$ to e_G. Thus $\phi(a_1 a_2) = \phi(a_1)\phi(a_2)$, so ϕ is a homomorphism. $\square$

11.4.G. EXERCISE (ABELIAN VARIETIES ARE ABELIAN). Show that an abelian variety is an abelian group variety. Hint: Apply Corollary 11.4.13 to the inversion morphism $i\colon A \to A$.

11.4.14. *Remark.* A similar idea is used to show that connected compact complex Lie groups are abelian; see, for example, [BL, Lem. 1.1.1].

PART IV

"Geometric" Properties of Schemes

IV

Chapter 12

Dimension

Everyone knows what a curve is, until he has studied enough mathematics to become confused.

—F. Klein, [RS, p. 90]

At this point, you know a fair bit about schemes, but there are some fundamental notions you cannot yet define. In particular, you cannot use the phrase "smooth surface," as it involves the notions of dimension and of smoothness. You may be surprised that we have gotten so far without using these ideas. You may also be disturbed to find that these notions can be subtle, but you should keep in mind that they are subtle in all parts of mathematics.

In this chapter, we will address the first notion, that of dimension of schemes. This should agree with, and generalize, our geometric intuition. Although we think of dimension as a basic notion in geometry, it is a slippery concept, as it is throughout mathematics. Even in linear algebra, the definition of dimension of a vector space is surprising the first time you see it, even though it quickly becomes second nature. The definition of dimension for manifolds is equally nontrivial. For example, how do we know that there isn't an isomorphism between some two-dimensional manifold and some three-dimensional manifold? Your answer will likely use topology, and hence you should not be surprised that the notion of dimension is often quite topological in nature.

A caution for those thinking over the complex numbers: our dimensions will be algebraic, and hence half that of the "real" picture. For example, we will see very shortly that $\mathbb{A}^1_{\mathbb{C}}$, which you may picture as the complex numbers (plus one generic point), has dimension 1.

12.1   Dimension and Codimension

12.1.1. *Definition(s): Dimension.* Surprisingly, the right definition is purely topological—it depends just on the topological space, and not on the structure sheaf. We define the **dimension** of a topological space X (denoted by $\dim X$) as the supremum of the lengths of the chains of closed irreducible sets, starting the indexing of the closed irreducible sets with 0. (The dimension may be infinite.) Scholars of the empty set can take the dimension of the empty set to be $-\infty$. (An analogy from linear algebra: The dimension of a vector space is the supremum of the lengths of the chains of subspaces.) Define the **dimension** of a ring as the supremum of the lengths of the chains of nested prime ideals (where indexing starts at zero). These two definitions of dimension are sometimes called **Krull dimension**. (You might think a Noetherian ring has finite dimension because all chains of prime ideals are finite, but this isn't necessarily true—see Exercise 12.1.M.)

12.1.A. EASY EXERCISE. Show that $\dim \operatorname{Spec} A = \dim A$. (Hint: Exercise 3.7.F gives a bijection between irreducible closed subsets of $\operatorname{Spec} A$ and prime ideals of A. It is "inclusion-reversing.")

The homeomorphism between $\operatorname{Spec} A$ and $\operatorname{Spec} A/\mathfrak{N}(A)$ (§3.4.7: the Zariski topology disregards nilpotents) implies that $\dim \operatorname{Spec} A = \dim \operatorname{Spec} A/\mathfrak{N}(A)$.

12.1.2. *Examples.* We have identified all the prime ideals of $k[t]$ (they are 0, and $(f(t))$ for irreducible polynomials $f(t)$), $\mathbb{Z}$ ((0) and (p)), k (only (0)), and $k[x]/(x^2)$ (only (x)), so we can quickly check that $\dim \mathbb{A}^1_k = \dim \operatorname{Spec} \mathbb{Z} = 1$, $\dim \operatorname{Spec} k = 0$, $\dim \operatorname{Spec} k[x]/(x^2) = 0$.

12.1.3. We must be careful with the notion of dimension for reducible spaces. If Z is the union of two closed subsets X and Y, then $\dim Z = \max(\dim X, \dim Y)$. Thus dimension is not a "local" characteristic of a space. This sometimes bothers us, so we try to only talk about dimensions of irreducible topological spaces. We say a topological space is **pure dimensional** or **equidimensional** (resp., **pure dimension** n or **equidimensional of dimension** n) if each of its irreducible components has the same dimension (resp., they are all of dimension n). A pure dimension 1 (resp., 2, n) topological space is said to be a **curve** (resp., **surface**, **n-fold**).

12.1.B. USEFUL EXERCISE. Show that a scheme has dimension n if and only if it admits an open cover by affine open subsets of dimension at most n, where equality is achieved for some affine open subset. Hint: You may find it helpful, here and later, to show the following. For any topological space X and open subset $U \subset X$, there is a bijection between irreducible closed subsets of U, and irreducible closed subsets of X that meet U.

12.1.C. EASY EXERCISE. Show that a Noetherian scheme of dimension 0 has a finite number of points.

12.1.D. EASY EXERCISE. Show that a surjection of integral domains of the same finite dimension must be an isomorphism.

12.1.E. EXERCISE (FIBERS OF INTEGRAL MORPHISMS, PROMISED IN §8.3.8). Suppose $\pi: X \to Y$ is an integral morphism. Show that every (nonempty) fiber of π has dimension 0. Hint: As integral morphisms are preserved by base change (Exercise 10.4.D(d)), we assume that $Y = \operatorname{Spec} k$. Hence we must show that if $\phi: k \to A$ is an integral extension, then $\dim A = 0$. Outline of proof: Suppose $\mathfrak{p} \subset \mathfrak{m}$ are two prime ideals of A. Mod out by $\mathfrak{p}$, so we can assume that A is an integral domain. I claim that any nonzero element is invertible: Say $x \in A$, and $x \neq 0$. Then the minimal monic polynomial for x has nonzero constant term. But then x is invertible—recall the coefficients are in a field.

12.1.F. IMPORTANT EXERCISE. Show that if $\pi: \operatorname{Spec} A \to \operatorname{Spec} B$ corresponds to an integral *extension* of rings, then $\dim \operatorname{Spec} A = \dim \operatorname{Spec} B$. Hint: Show that a chain of prime ideals downstairs gives a chain upstairs of the same length, by the Going-Up Theorem (Exercise 8.2.F). Conversely, a chain upstairs gives a chain downstairs. Use Exercise 12.1.E to show that no two elements of the chain upstairs go to the same element $[\mathfrak{q}] \in \operatorname{Spec} B$ of the chain downstairs.

12.1.G. EXERCISE. Show that if $\nu: \tilde{X} \to X$ is the normalization of a scheme (possibly in a finite extension of the function field of X), then $\dim \tilde{X} = \dim X$. Feel free to assume that X is integral for convenience.

12.1.H. USEFUL EXERCISE. Suppose X is an affine k-scheme, and K/k is an *algebraic* field extension.

(a)   Suppose X has pure dimension n. Show that $X_K := X \times_k K$ has pure dimension n. (See Exercise 24.5.F for a generalization, which, for example, removes the affine hypothesis. Also, see Exercise 12.2.M and Remark 12.2.17 for the fate of possible generalizations to arbitrary field extensions.) Hint: If $X = \operatorname{Spec} A$, reduce to the case where A is an integral domain. An irreducible component of X_K corresponds to a minimal prime $\mathfrak{p}$ of $A_K := A \otimes_k K$. Suppose $a \in \ker(A \to A_K/\mathfrak{p})$. Show that $a = 0$, using the fact that a lies in a minimal prime $\mathfrak{p}$ of A_K (and is hence a zerodivisor, by §6.6.23), and A_K is a free A-module (so multiplication in A_K by $a \in A$ is injective if a is nonzero). Thus $A \to A_K/\mathfrak{p}$ is injective. Then use Exercise 12.1.F.

(b)   Prove the converse to (a): Show that if X_K has pure dimension n, then X has pure dimension n.

12.1.I. EXERCISE. Show that $\dim \mathbb{Z}[x] = 2$. (Hint: The prime ideals of $\mathbb{Z}[x]$ were implicitly determined in Exercise 3.2.Q.)

12.1.4. Codimension. Because dimension behaves oddly for disjoint unions, we need some care when defining codimension, and in using the phrase. For example, if Y is a closed subset of X, we might define the codimension to be $\dim X - \dim Y$, but this behaves badly. For example, if X is the disjoint union of a point Y and a curve Z, then $\dim X - \dim Y = 1$, but this has nothing to do with the local behavior of X near Y.

A better definition is as follows. In order to avoid excessive pathology, we define the codimension of Y in X *only when Y is irreducible*. (Use extreme caution when using this word in any other setting.) Define the **codimension of an irreducible subset** $Y \subset X$ of a topological space as the supremum of the lengths of *increasing* chains of irreducible closed subsets starting with $\overline{Y}$ (where indexing starts at 0—recall that the closure of an irreducible set is irreducible, Exercise 3.6.B(b)). In particular, the **codimension of a point** is the codimension of its closure. The codimension of Y in X is denoted by $\mathrm{codim}_X Y$.

We say that a prime ideal $\mathfrak{p}$ in a ring has **codimension** equal to the supremum of the lengths of the chains of decreasing prime ideals starting at $\mathfrak{p}$, with indexing starting at 0. Thus in an integral domain, the ideal (0) has codimension 0; and in $\mathbb{Z}$, the ideal (23) has codimension 1. Note that the codimension of the prime ideal $\mathfrak{p}$ in A is $\dim A_{\mathfrak{p}}$ (see §3.2.9). (This notion is often called **height**.) Thus the codimension of $\mathfrak{p}$ in A is the codimension of $[\mathfrak{p}]$ in $\operatorname{Spec} A$.

(Continuing an analogy with linear algebra: the codimension of a vector subspace $Y \subset X$ is the supremum of the lengths of increasing chains of subspaces starting with Y. This is a better definition than $\dim X - \dim Y$, because it works even when $\dim X = \infty$. You might prefer to define $\mathrm{codim}_X Y$ as $\dim(X/Y)$; that is analogous to defining the codimension of $\mathfrak{p}$ in A as the dimension of $A_{\mathfrak{p}}$—see the previous paragraph.)

12.1.J. EXERCISE. Show that if Y is an irreducible closed subset of a scheme X, and η is the generic point of Y, then the codimension of Y is the dimension of the local ring $\mathscr{O}_{X,\eta}$.

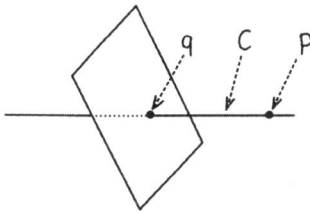

Figure 12.1 *Behavior of codimension.*

12.1.5. Notice that Y is codimension 0 in X if it is an irreducible component of X. Similarly, Y is codimension 1 if it is not an irreducible component, and for every irreducible component Y' it is contained in, there is no irreducible closed subset strictly between Y and Y'. (See Figure 12.1 for examples.) A closed subset all of whose irreducible components are codimension 1 in some ambient space X is said to be a **hypersurface** in X. (Be careful: The word "hypersurface" can be used to mean slightly different things in algebraic geometry, for entirely reasonable reasons; see, for example, Definition 9.3.2.)

12.1.K. EASY EXERCISE. If Y is an irreducible closed subset of a scheme X, show that

$$(12.1.5.1) \qquad\qquad \mathrm{codim}_X Y + \dim Y \leq \dim X.$$

We will soon see that equality always holds if X and Y are irreducible varieties (Theorem 12.2.9), but equality doesn't hold in general (§12.3.13).

Warning. The notion of codimension still can behave slightly oddly. For example, consider Figure 12.1. (You should think of this as an intuitive sketch.) Here the total space X has dimension 2, but point p is dimension 0, and codimension 1. We also have an example of a codimension 2 subset q contained in a codimension 0 subset C with no codimension 1 subset "in between."

Worse things can happen; we will soon see an example of a closed point in an *irreducible* surface that is nonetheless codimension 1, not 2, in §12.3.13. However, for irreducible *varieties* this can't happen, and inequality (12.1.5.1) must be an equality (Theorem 12.2.9).

12.1.6. In unique factorization domains, codimension 1 prime ideals are principal. We make a short observation, for future reference.

12.1.7. Lemma — *In a unique factorization domain A, all codimension 1 prime ideals are principal.*

This is a first glimpse of the fact that codimension 1 is rather special—this theme will continue in §12.3. We will see that the converse of Lemma 12.1.7 holds as well (when A is a Noetherian integral domain, Proposition 12.3.7).

Proof. Suppose $\mathfrak{p}$ is a codimension 1 prime. Choose any $f \neq 0$ in $\mathfrak{p}$, and let g be any irreducible/prime factor of f that is in $\mathfrak{p}$ (there is at least one). Then (g) is a nonzero prime ideal contained in $\mathfrak{p}$, so $(0) \subset (g) \subset \mathfrak{p}$. As $\mathfrak{p}$ is codimension 1, we must have $\mathfrak{p} = (g)$, and thus $\mathfrak{p}$ is principal. $\qquad\square$

12.1.8. Local rings of dimension 0.
Dimension 0 objects aren't just points; they have further information, which might be thought of as some sort of infinitesimal "shape." We caught a glimpse of this earlier in §6.5 and Exercise 8.3.G.

12.1.9. Useful proposition — *Suppose $(A, \mathfrak{m})$ is a Noetherian local ring. Then the following are equivalent.*

 (i) $\dim A = 0$;
 (ii) $\mathfrak{m} = \mathfrak{N}(A)$;
 (iii) *for some* $n > 1$, $\mathfrak{m}^n = 0$;
 (iv) *A has finite length;*
 (v)   *every descending sequence of ideals of A is eventually constant.*

Such a ring is called an *Artinian* or *finite length* local ring (Definition 6.5.7).

12.1.L. EXERCISE. Prove Proposition 12.1.9.

12.1.10. A fun but unimportant counterexample. We end this introductory section with a fun pathology. As a Noetherian ring has no infinite chain of prime ideals, you may think that Noetherian rings must have finite dimension. Nagata, the master of counterexamples, shows you otherwise with the following example.

12.1.M. ** EXERCISE: AN INFINITE-DIMENSIONAL NOETHERIAN RING. Let $A = k[x_1, x_2, \ldots]$. Choose an increasing sequence of positive integers $m_1, m_2, \ldots$ whose *differences* are also increasing ($m_{i+1} - m_i > m_i - m_{i-1}$). Let $\mathfrak{p}_i = (x_{m_i+1}, \ldots, x_{m_{i+1}})$ and $S = A \setminus \cup_i \mathfrak{p}_i$.

(a) Show that S is a multiplicative set.
(b) Show that $S^{-1}\mathfrak{p}_i$ in $S^{-1}A$ is the largest prime ideal in a chain of prime ideals of length $m_{i+1} - m_i$. Hence conclude that $\dim S^{-1}A = \infty$.
(c) Suppose B is a ring such that (i) for every maximal ideal $\mathfrak{m}$, $B_\mathfrak{m}$ is Noetherian, and (ii) every nonzero $b \in B$ is contained in finitely many maximal ideals. Show that B is Noetherian. (One possible approach: Show that for any $x_1, x_2, \ldots$, $(x_1, x_2, \ldots)$ is finitely generated.)
(d) Use (c) to show that $S^{-1}A$ is Noetherian.

12.1.11. *Remark: Noetherian local rings have finite dimension.* However, we shall see in Exercise 12.3.K(a) that Noetherian *local* rings always have finite dimension. (This requires a surprisingly hard fact, Krull's Height Theorem 12.3.10.) Thus points of locally Noetherian schemes always have finite codimension.

12.2   Dimension, Transcendence Degree, and Noether Normalization

We now give a powerful alternative interpretation for dimension for irreducible varieties, in terms of transcendence degree. The proof will involve a classical construction, *Noether normalization*, which will be useful in other ways as well. In case you haven't seen transcendence theory, here is a lightning introduction.

12.2.A. EXERCISE/DEFINITION. Recall that an *element* of a field extension E/F is *algebraic over* F if it is integral over F. Recall also that a *field extension* E/F is an *algebraic extension* if it is an integral extension (if all elements of E are algebraic over F; cf. §8.2). The composition of two algebraic extensions is algebraic, by Exercise 8.2.C. If E/F is a field extension, and F' and F'' are two intermediate field extensions, then we write $F' \sim F''$ if $F'F''$ is algebraic over both F' and F''. Here $F'F''$ is the **compositum** of F' and F'', the smallest field extension in E containing F' and F''.]

(a) *Show that $\sim$ is an equivalence relation on subextensions of* E/F. A **transcendence basis** of E/F is a set of elements $\{x_i\}$ that are algebraically independent over F (there is no nontrivial polynomial relation among the x_i with coefficients in F) such that $F(\{x_i\}) \sim E$.
(b) *Show that if E/F has two transcendence bases, and one has cardinality n, then both have cardinality n.* (Hint: Show that you can substitute elements from the one basis into the other one at a time.) The size of any transcendence basis is called the **transcendence degree** (which may be ∞), and is denoted by trdeg. Any finitely generated field extension necessarily has finite transcendence degree. (Remark: See Theorem 10.5.17 for a related result.)

12.2.1. Theorem (dimension = transcendence degree) — *Suppose A is a* **finitely generated domain over a field** k *(i.e., a finitely generated k-algebra that is an integral domain). Then* $\dim \operatorname{Spec} A =$ $\operatorname{trdeg} K(A)/k$. *Hence if X is an irreducible k-variety, then $\dim X = \operatorname{trdeg} K(X)/k$.*

We will prove Theorem 12.2.1 shortly (§12.2.7). We first show that it is useful by giving some immediate consequences. We seem to have immediately $\dim \mathbb{A}_k^n = n$. However, our proof of Theorem 12.2.1 will go *through* this fact, so it isn't really a consequence.

A more substantive consequence is the following. If X is an irreducible k-variety, then $\dim X$ is the transcendence degree of the function field (the stalk at the generic point) $\mathscr{O}_{X,\eta}$ over k. Thus (as the generic point lies in all nonempty open sets) the dimension can be computed in any open set of X. (Warning: This is false without the finite type hypothesis, even in quite reasonable circumstances—let X be the two-point space $\operatorname{Spec} k[x]_{(x)}$, and let U consist of only the generic point; see Exercise 3.4.K.)

Another consequence is a second proof of the Nullstellensatz 3.2.6.

12.2.B. EXERCISE: THE NULLSTELLENSATZ FROM DIMENSION THEORY. Suppose $A = k[x_1, \ldots, x_n]/I$. Show that the residue field of any maximal ideal of A is a finite extension of k. (Hint: The maximal ideals correspond to dimension 0 points, which correspond to transcendence degree 0 finitely generated extensions of k, i.e., finite extensions of k.)

Yet another consequence is geometrically believable.

12.2.C. EXERCISE. If $\pi \colon X \to Y$ is a dominant morphism of irreducible k-varieties, then $\dim X \geq \dim Y$. (This is false more generally: consider the inclusion of the generic point into an irreducible curve.)

12.2.D. EXERCISE. Suppose $f_1(x_1, \ldots, x_n), \ldots, f_m(x_1, \ldots, x_n) \in k[x_1, \ldots, x_n]$ are m polynomials in n variables over a field, with $m > n$. Show that the polynomials $f_1, \ldots, f_m$ are algebraically dependent. (Remark: If $m = n$, how can you tell if the f_i are algebraically dependent? There is a remarkably simple answer in characteristic 0: they are algebraically dependent if and only if the determinant of the Jacobian matrix is the zero polynomial; see Exercise 21.6.C! This is false in characteristic p: if $f_i(x_1, \ldots, x_n) = x_i^p$, then the f_i are algebraically independent, yet the Jacobian matrix is the zero matrix.)

12.2.E. EXERCISE (PRACTICE WITH THE CONCEPT). Show that the three equations

$$wz - xy = 0, \quad wy - x^2 = 0, \quad xz - y^2 = 0$$

cut out an integral *surface* S in $\mathbb{A}_k^4$. (You may recognize these equations from Exercises 3.6.F and 9.3.A.) You might expect S to be a curve, because it is cut out by three equations in four-space. One of many ways to proceed: cut S into pieces. For example, show that $D(w) \cong \operatorname{Spec} k[x, w]_w$. (You may recognize S as the affine cone over the twisted cubic. The twisted cubic was defined in

Exercise 9.3.A.) It turns out that you need three equations to cut out this surface. The first equation cuts out a threefold in $\mathbb{A}_k^4$ (by Krull's Principal Ideal Theorem 12.3.3, which we will meet soon). The second equation cuts out a surface: our surface, along with another surface. The third equation cuts out our surface, and removes the "extraneous component." One last aside: Notice once again that the cone over the quadric surface $k[w, x, y, z]/(wz - xy)$ makes an appearance.

12.2.2. *Definition: Degree of a rational map of irreducible varieties.* If $\pi \colon X \dashrightarrow Y$ is a dominant rational map of irreducible (hence integral) k-varieties of the same dimension, the degree of the field extension is called the **degree** of the rational map. This readily extends if X is reducible: we add up the degrees on each of the components of X. If π is a rational map of integral affine k-varieties of the same dimension that is *not* dominant, we say the degree is 0. We will interpret this degree in terms of counting preimages of general points of Y in §24.5.14. Note that degree is multiplicative under composition: If $\rho \colon Y \dashrightarrow Z$ is a rational map of integral k-varieties of the same dimension, then $\deg(\rho \circ \pi) = \deg(\rho) \deg(\pi)$, as degrees of field extensions are multiplicative in towers.

12.2.3. Noether normalization.
Our proof of Theorem 12.2.1 will use another important classical notion, Noether normalization.

12.2.4. Noether Normalization Lemma — *Suppose A is an integral domain, finitely generated over a field k. If* $\operatorname{trdeg} K(A)/k = n$*, then there are elements* $x_1, \ldots, x_n \in A$*, algebraically independent over k, such that A is a finite (hence integral by Corollary 8.2.2) extension of* $k[x_1, \ldots, x_n]$*.*

The geometric content behind this result is that given any integral finite type affine k-scheme X, we can find a surjective finite morphism $X \to \mathbb{A}_k^n$, where n is the transcendence degree of the function field of X (over k). Surjectivity follows from the Lying Over Theorem 8.2.5, in particular, Exercise 12.1.F. This interpretation is sometimes called *geometric Noether normalization*. (See Remark 29.3.7 for a projective version of Noether normalization.)

12.2.5. *Nagata's proof of the Noether Normalization Lemma 12.2.4.* Suppose we can write $A = k[y_1, \ldots, y_m]/\mathfrak{p}$, i.e., that A can be chosen to have m generators. Note that $m \geq n$. We show the result by induction on m. The base case $m = n$ is immediate.

Assume now that $m > n$, and that we have proved the result for smaller m. We will find $m - 1$ elements $z_1, \ldots, z_{m-1}$ of A such that A is finite over $k[z_1, \ldots, z_{m-1}]/\mathfrak{q}$ (i.e., the subring of A generated by $z_1, \ldots, z_{m-1}$, where the z_i satisfy the relations given by the ideal $\mathfrak{q}$). Then by the inductive hypothesis, $k[z_1, \ldots, z_{m-1}]/\mathfrak{q}$ is finite over some $k[x_1, \ldots, x_n]$, and A is finite over $k[z_1, \ldots, z_{m-1}]$, so by Exercise 8.3.H, A is finite over $k[x_1, \ldots, x_n]$.

$$A = k[y_1, \ldots, y_m]/\mathfrak{p}$$

$$\Big| \text{ finite}$$

$$k[z_1, \ldots, z_{m-1}]/\mathfrak{q}$$

$$\Big| \text{ finite}$$

$$k[x_1, \ldots, x_n]$$

As $y_1, \ldots, y_m$ are algebraically dependent in A, there is some nonzero algebraic relation $f(y_1, \ldots, y_m) = 0$ among them (where f is a polynomial in m variables over k).

Let $z_1 = y_1 - y_m^{r_1}$, $z_2 = y_2 - y_m^{r_2}, \ldots, z_{m-1} = y_{m-1} - y_m^{r_{m-1}}$, where $r_1, \ldots, r_{m-1}$ are positive integers to be chosen shortly. Then

$$f(z_1 + y_m^{r_1}, z_2 + y_m^{r_2}, \ldots, z_{m-1} + y_m^{r_{m-1}}, y_m) = 0.$$

Then upon expanding this out, each monomial in f (as a polynomial in m variables) will yield a single term that is a constant times a power of y_m (with no z_i factors). By choosing the r_i so that

$0 \ll r_1 \ll r_2 \ll \cdots \ll r_{m-1}$, we can ensure that the powers of y_m appearing are all distinct, so that, in particular, there is a leading term y_m^N, and all other terms (including those with factors of z_i) are of smaller degree in y_m. Thus we have described an integral dependence of y_m on $z_1, \ldots, z_{m-1}$, as desired. $\qquad\square$

12.2.6. *The geometry behind Nagata's proof.* Here is the geometric intuition behind Nagata's argument. Suppose we have an n-dimensional variety in $\mathbb{A}_k^m$ with $n < m$, for example, $xy = 1$ in $\mathbb{A}^2$. One approach is to hope the projection to a hyperplane is a finite morphism. In the case of $xy = 1$, if we projected to the x-axis, it wouldn't be finite, roughly speaking, because the asymptote $x = 0$ prevents the map from being closed (cf. Exercise 8.3.K). If we instead projected to a random line, we might hope that we would get rid of this problem, and indeed we usually can: this problem arises for only a finite number of directions. But we might have a problem if the field were finite: perhaps the finite number of directions in which to project each has a problem. (You can show that if k is an infinite field, then the substitution in the above proof $z_i = y_i - y_m^{r_i}$ can be replaced by the linear substitution $z_i = y_i - a_i y_m$, where $a_i \in k$, and that for a nonempty Zariski-open choice of a_i, we indeed obtain a finite morphism.) Nagata's trick in general is to "jiggle" the variables in a nonlinear way, and this jiggling kills the nonfiniteness of the map.

12.2.F. EXERCISE (DIMENSION IS ADDITIVE FOR PRODUCTS OF FINITE TYPE K-SCHEMES). If X and Y are irreducible k-varieties, show that $\dim X \times_k Y = \dim X + \dim Y$. Hint: If we had surjective finite morphisms $X \to \mathbb{A}_k^{\dim X}$ and $Y \to \mathbb{A}_k^{\dim Y}$, we could construct a surjective finite morphism $X \times_k Y \to \mathbb{A}_k^{\dim X + \dim Y}$. (Warning: This can be false without the finite type assumption; see Remark 12.2.17.)

In particular, we have $\dim A[x] = \dim A + 1$ when A is finitely generated over a field. (This is true more generally when A is Noetherian, although we will not prove this.)

12.2.G. LESS IMPORTANT (BUT FUN) EXERCISE.

(a) Suppose $X \subset \mathbb{P}^n$ is an irreducible projective k-variety. Show that the affine cone and projective cone over X both have dimension $\dim X + 1$.
(b) Show that a prime ideal of $k[x_0, \ldots, x_n]$ that is minimal over a homogeneous ideal is also homogeneous.
(c) Suppose $\mathfrak{p} \subset \mathfrak{q}$ are two homogeneous prime ideals of $k[x_0, \ldots, x_n]$ such that there is no homogeneous prime ideal strictly between the two. Show that there is no prime ideal (not necessarily homogeneous) strictly between the two.

12.2.7. *Proof of Theorem 12.2.1 on dimension and transcendence degree.* Suppose X is an integral affine k-scheme of finite type. We show that $\dim X$ equals the transcendence degree n of its function field, by induction on n. (The idea is that we reduce from X to $\mathbb{A}^n$ to a hypersurface in $\mathbb{A}^n$ to $\mathbb{A}^{n-1}$.) Assume the result is known for all transcendence degrees less than n.

By the Noether Normalization Lemma 12.2.4, there exists a surjective finite morphism $X \to \mathbb{A}_k^n$. By Exercise 12.1.F, $\dim X = \dim \mathbb{A}_k^n$. If $n = 0$, we are done, as $\dim \mathbb{A}_k^0 = 0$.

We now show that $\dim \mathbb{A}_k^n = n$ for $n > 0$, by induction. Clearly, $\dim \mathbb{A}_k^n \geq n$, as we can describe a chain of irreducible subsets of length n: if $x_1, \ldots, x_n$ are coordinates on $\mathbb{A}^n$, consider the chain of ideals

$$(0) \subset (x_1) \subset \cdots \subset (x_1, \ldots, x_n)$$

in $k[x_1, \ldots, x_n]$. Suppose we have a chain of prime ideals of length at least $n + 1$:

$$(0) = \mathfrak{p}_0 \subset \cdots \subset \mathfrak{p}_m.$$

Choose any nonzero element g of $\mathfrak{p}_1$, and let f be any irreducible factor of g lying in $\mathfrak{p}_1$. Then replace $\mathfrak{p}_1$ by (f). (Of course, $\mathfrak{p}_1$ may have been (f) to begin with.) Then $K(k[x_1, \ldots, x_n]/(f(x_1, \ldots, x_n)))$ has transcendence degree $n-1$, so by induction,

$$\dim k[x_1,\ldots,x_n]/(f) = n-1,$$

so $\dim k[x_1,\ldots,x_n] \leq n$, completing the proof. $\qquad\square$

12.2.8. Codimension is the difference of dimensions for irreducible varieties.
Noether normalization will help us show that codimension is the difference of dimensions for irreducible varieties, i.e., that the inequality (12.1.5.1) is always an equality.

12.2.9. Theorem — *Suppose X is a pure-dimensional k-scheme locally of finite type (for example, an irreducible k-variety), Y is an irreducible closed subset, and η is the generic point of Y. Then $\dim Y + \dim \mathcal{O}_{X,\eta} = \dim X$. Hence by Exercise 12.1.J, $\dim Y + \operatorname{codim}_X Y = \dim X$—inequality (12.1.5.1) is always an equality.*

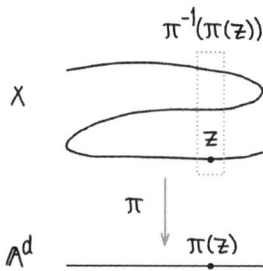

Figure 12.2 *Key step in proof of Theorem 12.2.9.*

Proving this will give us an excuse to introduce some useful notions, such as the Going-Down Theorem for finite extensions of integrally closed domains (Theorem 12.2.12). Before we begin the proof, we give an algebraic translation.

12.2.H. EXERCISE. A ring A is called **catenary** if for every nested pair of prime ideals $\mathfrak{p} \subset \mathfrak{q} \subset A$, every strictly increasing chain of prime ideals "from $\mathfrak{p}$ to $\mathfrak{q}$" is contained in a maximal such chain, and all maximal chains have the same (finite) length. (We will not use this term in any serious way later.) Show that if A is a localization of a finitely generated ring over a field k, then A is catenary.

12.2.10. *Remark.* Most rings arising naturally in algebraic geometry are catenary. Important examples include: localizations of finitely generated $\mathbb{Z}$-algebras; complete Noetherian local rings; Dedekind domains (§13.5.11); and Cohen–Macaulay rings (see §26.2.13). It is hard to give an example of a noncatenary ring; see, for example, [Stacks, tag 02JE] or [He].

12.2.11. *Proof of Theorem 12.2.9.*

12.2.I. EXERCISE. Reduce the proof of Theorem 12.2.9 to the following problem. If X is an irreducible affine k-variety and Z is a closed irreducible subset maximal among those smaller than X (the only larger closed irreducible subset is X), then $\dim Z = \dim X - 1$.

Let $d = \dim X$ for convenience. By Noether normalization 12.2.4, we have a finite morphism $\pi\colon X \to \mathbb{A}^d$ corresponding to a finite extension of rings. Then $\pi(Z)$ is an irreducible closed subset of $\mathbb{A}^d$ (finite morphisms are closed, Exercise 8.3.L).

12.2.J. EXERCISE. Show that it suffices to show that $\pi(Z)$ is a hypersurface. (Hint: The dimension of any hypersurface is $d-1$ by Theorem 12.2.1 on dimension and transcendence degree. Exercise 12.1.F implies that $\dim \pi^{-1}(\pi(Z)) = \dim \pi(Z)$. But be careful: Z is not $\pi^{-1}(\pi(Z))$ in general.)

Now if $\pi(Z)$ is not a hypersurface, then it is properly contained in an irreducible hypersurface H, so by the Going-Down Theorem 12.2.12 for finite extensions of integrally closed domains (which we shall now prove), there is some closed irreducible subset Z' of X properly containing Z, contradicting the maximality of Z. $\qquad\square$

12.2.12. Theorem (Going-Down Theorem for finite extensions of integrally closed domains) — *Suppose $\phi\colon B \hookrightarrow A$ is a finite extension of rings (so A is a finite B-module), B is an integrally closed domain, and A is an integral domain. Then, given nested prime ideals $\mathfrak{q} \subset \mathfrak{q}'$ of B, and a prime $\mathfrak{p}'$ of A lying over $\mathfrak{q}'$ (i.e., $\mathfrak{p}' \cap B = \mathfrak{q}'$), there exists a prime $\mathfrak{p}$ of A contained in $\mathfrak{p}'$, lying over $\mathfrak{q}$.*

As usual, you should sketch a geometric picture of the statement. This theorem is usually stated in the context of extending a chain of ideals, in the same way as the Going-Up Theorem (Exercise 8.2.F), and you may want to think this through. (Another Going-Down Theorem, for flat morphisms, will be given in Exercise 24.5.E.)

This theorem is true more generally with "finite" replaced by "integral"; see [E, p. 291] ("Completion of the proof of 13.9") for the extension of Theorem 12.2.12, or else see [AtM, Thm. 5.16] or [Mat1, Thm. 5(v)] for completely different proofs. See [E, Fig. 10.4] for an example (in the form of a picture) of why the "integrally closed" hypothesis on B cannot be removed.

In the course of the proof, we will use the following easy fact. We could prove it now, but we leave it until Exercise 12.3.E, because the proof uses a trick arising in Exercise 12.3.D.

12.2.13. Proposition (prime avoidance) — *Suppose $\mathfrak{p}_1, \ldots, \mathfrak{p}_n$ are prime ideals of a ring A, and I is another ideal of A not contained in any $\mathfrak{p}_i$. Then I is not contained in $\cup \mathfrak{p}_i$: there is an element $f \in I$ not in any of the $\mathfrak{p}_i$.*

(Can you give a geometric interpretation of this result? Can you figure out why it is called "prime avoidance"?)

12.2.14. *Proof of Theorem 12.2.12 (Going-Down Theorem for finite extensions of integrally closed domains).* The proof uses Galois theory. Let L be the normal closure of $K(A)/K(B)$ (the smallest subfield of $\overline{K(B)}$ containing $K(A)$ that is mapped to itself by any automorphism over $\overline{K(B)}/K(B)$). Let C be the integral closure of B in L (discussed in Exercise 10.7.I). Because $A \hookrightarrow C$ is an integral extension, there is a prime Q of C lying over $\mathfrak{q} \subset B$ (by the Lying Over Theorem 8.2.5), and a prime Q' of C containing Q lying over $\mathfrak{q}'$ (by the Going-Up Theorem, Exercise 8.2.F). Similarly, there is a prime P' of C lying over $\mathfrak{p}' \subset A$ (and thus over $\mathfrak{q}' \subset B$). We would be done if $P' = Q'$ (just take $\mathfrak{p} = Q \cap A$), but this needn't be the case. However, Lemma 12.2.15 below shows there is an automorphism σ of C over B that sends Q' to P', and then the image of $\sigma(Q)$ in A does the trick, completing the proof. The following diagram, in geometric terms, may help.

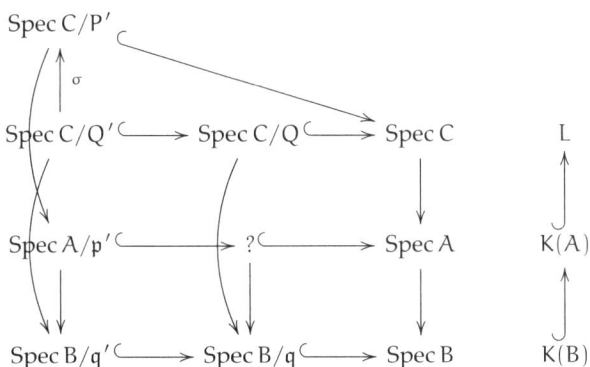

12.2.15. Lemma — *Suppose B is an integrally closed domain, $L/K(B)$ is a finite normal field extension, and C is the integral closure of B in L. If $\mathfrak{q}'$ is a prime ideal of B, then automorphisms of $L/K(B)$ act transitively on the prime ideals of C lying over $\mathfrak{q}'$.*

This result is often first seen in number theory, with $B = \mathbb{Z}$ and L a Galois extension of $\mathbb{Q}$.

Proof. Let P and Q_1 be two prime ideals of C lying over $\mathfrak{q}'$, and let $Q_2, \ldots, Q_n$ be the prime ideals of C conjugate to Q_1 (the image of Q_1 under $\mathrm{Aut}(L/K(B))$). If P is not one of the Q_i, then P is not contained in any of the Q_i. (Do you see why? Hint: Exercise 12.1.E or 12.1.F.) Hence by prime avoidance (Proposition 12.2.13), P is not contained in their union, so there is some $a \in P$ not contained in any Q_i. Thus no conjugate of a can be contained in Q_1, so the norm $N_{L/K(B)}(a) \in B$ is not contained in $Q_1 \cap B = \mathfrak{q}'$. (Recall: If $L/K(B)$ is separable, the norm is just the product of the conjugates. But even if $L/K(B)$ is not separable, the norm is the product of conjugates to the appropriate power, because we can factor $L/K(B)$ as a separable extension followed by a purely inseparable extension, or because we can take an explicit basis for the extension and calculate the norm of a. See [Lan, §VI.5] for more on norms.) But since $a \in P$, its norm lies in P, but also in B, and hence in $P \cap B = \mathfrak{q}'$, yielding a contradiction. $\qquad\square$

We end with some exercises for your practice and enlightenment.

12.2.K. EXERCISE/DEFINITION. Suppose X is a finite-dimensional Noetherian scheme. Define a function $\dim_X(\cdot) : X \to \mathbb{Z}^{\geq 0}$ from the underlying set of X, by taking $\dim_X p$ to be the dimension of the largest irreducible component of X containing p. We call this **the dimension of X at p**. Caution: This definition differs from [Stacks, tag 0055]. Show that $\dim_X(\cdot)$ is an upper semicontinuous function. (Recall that a function f from a topological space X to $\mathbb{R}$ is **upper semicontinuous** if for each $x \in \mathbb{R}$, $f^{-1}((-\infty, x))$ is open. Informally: the function can jump "up" upon taking limits. Our upper semicontinuous functions will map to $\mathbb{Z}$, so informally functions jump "up" on closed subsets. Similarly, f is **lower semicontinuous** if for each $x \in \mathbb{R}$, $f^{-1}((x, \infty))$ is open.)

12.2.16. *Recurring Theme: Semicontinuities.* Semicontinuity is an important notion in algebraic geometry. Here are some examples.

 (i)   The rank of a matrix of functions (because rank drops on closed subsets, where various discriminants vanish),
 (ii) the dimension of a variety at a point (Exercise 12.2.K above),
 (iii) fiber dimension (Theorem 12.4.3),
 (iv) the rank of a finite type quasicoherent sheaf (Exercise 14.3.J),
 (v)   the degree of a finite morphism, as a function of the target (§14.3.4),
 (vi) the dimension of tangent space at closed points of a variety over an algebraically closed field (Exercise 21.2.G), and
 (vii) the rank of cohomology groups of coherent sheaves, in proper flat families (the Semicontinuity Theorem 25.1.1).

All but (i) are upper semicontinuous; (i) is a lower semicontinuous function.

12.2.L. EXERCISE. Suppose p is a closed point of a locally finite type k-scheme X. Show that the following three integers are the same: (i) $\dim_X p$ (Exercise 12.2.K); (ii) $\dim \mathcal{O}_{X,p}$; and (iii) $\operatorname{codim}_X p$. (We define the **Zariski topology on the closed points of a finite type k-scheme** in the obvious way: Closed subsets are cut out by equations, as in §8.4.8.)

12.2.M. EXERCISE (THE DIMENSION OF A LOCALLY FINITE TYPE K-SCHEME IS PRE-SERVED BY any FIELD EXTENSION; CF. EXERCISE 12.1.H(A)). Suppose X is a locally finite type k-scheme of pure dimension n, and K/k is a field extension (not necessarily algebraic). Show that X_K has pure dimension n. Hint: Reduce to the case where X is affine, so, say, $X = \operatorname{Spec} A$. Reduce to the case where A is an integral domain. Show (using the axiom of choice) that K/k can be written as an algebraic extension of a purely transcendental extension. Hence by Exercise 12.1.H(a), it suffices to deal with the case where K/k is purely transcendental, say, with transcendence basis $\{e_i\}_{i \in I}$ (possibly infinite). Show that $A_K := A \otimes_k K$ is an integral domain, by interpreting it as a certain localization of the domain $A[\{e_i\}]$. If $t_1, \ldots, t_d$ is a transcendence basis for $K(A)/k$, show that $\{e_i\} \cup \{t_j\}$ is a transcendence basis for $K(A_K)/k$. Show that $\{t_j\}$ is a transcendence basis for $K(A_K)/K$.

Exercise 12.2.M is conceptually important. For example, if X is described by some equations with $\mathbb{Q}$-coefficients, the dimension of X doesn't depend on whether we consider it as a $\mathbb{Q}$-scheme or as a $\mathbb{C}$-scheme.

12.2.17. *Remark.* Unlike Exercise 12.1.H, Exercise 12.2.M has finite type hypotheses on X. It is not true that if X is an arbitrary k-scheme of pure dimension n, and K/k is an arbitrary extension, then

X_K necessarily has pure dimension n. For example, Exercise 10.2.L shows that if k is a field, then $\dim k(x) \otimes_k k(y) = 1$.

12.2.18.* **Lines on hypersurfaces, part 1.** Notice: Although dimension theory is not central to the following statement, it is essential to the proof.

12.2.N. ENLIGHTENING STRENUOUS EXERCISE: MOST SURFACES IN THREE-SPACE OF DEGREE $d > 3$ HAVE NO LINES. In this exercise, we work over an algebraically closed field $\bar{k}$. For any $d > 3$, show that most degree d surfaces in $\mathbb{P}^3$ contain no lines. Here, "most" means "all closed points of a Zariski-open subset of the parameter space for degree d homogeneous polynomials in four variables, up to scalars." As there are $\binom{d+3}{3}$ such monomials, the degree d hypersurfaces are parametrized by $\mathbb{P}^{\binom{d+3}{3}-1}$ (see Remark 4.5.3). Hint: Construct an "incidence correspondence"

$$X = \{(\ell, H) \; : \; [\ell] \in \mathbb{G}(1,3), [H] \in \mathbb{P}^{\binom{d+3}{3}-1}, \ell \subset H\},$$

parametrizing lines in $\mathbb{P}^3$ contained in a hypersurface: define a closed subscheme X of $\mathbb{P}^{\binom{d+3}{3}-1} \times \mathbb{G}(1,3)$ that makes this notion precise. (Recall that $\mathbb{G}(1,3)$ is a Grassmannian.) Show that X is a $\mathbb{P}^{\binom{d+3}{3}-1-(d+1)}$-bundle over $\mathbb{G}(1,3)$. (Possible hint for this: How many degree d hypersurfaces contain the line $x = y = 0$?) Show that $\dim \mathbb{G}(1,3) = 4$ (see §7.7: $\mathbb{G}(1,3)$ has an open cover by $\mathbb{A}^4$'s). Show that $\dim X = \binom{d+3}{3} - 1 - (d+1) + 4$. Show that the image of the projection $X \to \mathbb{P}^{\binom{d+3}{3}-1}$ must lie in a proper closed subset. The following diagram may help.

$$\dim \binom{d+3}{3} - 1 - (d+1) + 4$$

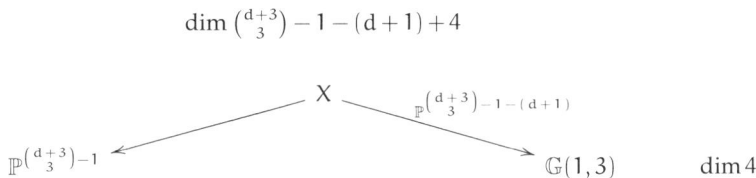

(The argument readily generalizes to show that if $d > 2n - 3$, then "most" degree d hypersurfaces in $\mathbb{P}^n$ have no lines. The case $n = 1$ and $n = 2$ are trivial but worth thinking through.)

12.2.19. *Side remark.* If you do the previous exercise, your dimension count will suggest the true facts that degree 1 hypersurfaces—i.e., hyperplanes—have two-dimensional families of lines, and that most degree 2 hypersurfaces have one-dimensional families (rulings) of lines, as shown in Exercise 9.3.L. They will also suggest that most degree 3 hypersurfaces contain a finite number of lines, which reflects the celebrated fact that regular cubic surfaces over an algebraically closed field always contain 27 lines (Theorem 27.0.1), and we will use this "incidence correspondence" or "incidence variety" to prove it (§27.3). The statement about quartic surfaces generalizes to the Noether–Lefschetz Theorem, implying that a very general surface of degree d at least 4 contains no curves that are not the intersection of the surface with a hypersurface; see [Lef, GH2]. The term **"very general"** means that in the parameter space (in this case, the projective space parametrizing surfaces of degree d), the statement is true away from a countable union of proper Zariski-closed subsets. Like "general" (which was defined in §10.3.7), "very general" is a weaker version of the phrase "almost every."

12.3 Krull's Theorems

We next introduce a number of results that we will repeatedly use, relating functions and the codimension 1 points where they vanish. The key result is Krull's Principal Ideal Theorem, which will be geometrically believable, and whose proof is short yet somehow nonintuitive and algebraic. This section will require Noetherian hypotheses.

12.3.1. Krull's Principal Ideal Theorem. In a vector space, a single linear equation always cuts out a subspace of codimension 0 or 1 (and codimension 0 occurs only when the equation is 0). The Principal Ideal Theorem, due to W. Krull, generalizes this linear algebra fact.

12.3.2. Krull's Principal Ideal Theorem (geometric version) — *Suppose X is a locally Noetherian scheme, and f is a function. The irreducible components of $V(f)$ are codimension 0 or 1.*

This is clearly a consequence of the following algebraic statement.

12.3.3. Krull's Principal Ideal Theorem (algebraic version) — *Suppose A is a Noetherian ring, and $f \in A$. Then every prime $\mathfrak{p}$ minimal among those containing f has codimension at most 1. If furthermore f is not a zerodivisor, then every such prime $\mathfrak{p}$ containing f has codimension precisely 1.*

For example, locally principal closed subschemes have "codimension 0 or 1," and effective Cartier divisors have "pure codimension 1." Here is another example (which you could certainly prove directly, without the Principal Ideal Theorem).

12.3.A. EXERCISE. Show that an irreducible homogeneous polynomial in $n+1$ variables over a field k describes an integral scheme of dimension $n-1$ in $\mathbb{P}_k^n$.

You should just note how short the proof of Theorem 12.3.3 is, and skip ahead to §12.3.5 until you have used the theorem enough that you very much appreciate its value.

12.3.4.* *Proof of Krull's Principal Ideal Theorem 12.3.3.* We dispose of the last sentence first: if $\mathfrak{p}$ is codimension 0, then it is an associated prime (Exercise 6.6.M), and hence any element of $\mathfrak{p}$ (including f) is a zerodivisor (Proposition 6.6.13).

We next wish to show that codim $\mathfrak{p} \leq 1$. We reduce to the case of a local ring: replacing A by $A_\mathfrak{p}$ (and $\mathfrak{p}$ by $\mathfrak{p}A_\mathfrak{p}$) does not change the problem, so we may assume that $\mathfrak{p}$ is a maximal ideal.

Suppose otherwise that $\mathfrak{q}$ is a strictly smaller prime ideal of A (so $f \notin \mathfrak{q}$ by the minimality of $\mathfrak{p}$ among prime ideals containing f). We wish to show that $\mathfrak{q}$ has codimension 0.

The key idea of the proof is the following clever construction.

12.3.B. EXERCISE. Let $\phi : A \to A_\mathfrak{q}$ be the localization, and define $\mathfrak{q}^{(n)} := \phi^{-1}(\mathfrak{q}^n A_\mathfrak{q})$. (This is called the nth **symbolic power** of $\mathfrak{q}$, but we will not use this terminology.) Note that $\mathfrak{q}^{(1)} \supset \mathfrak{q}^{(2)} \supset \cdots$ is a descending sequence of ideals of A. If $ag \in \mathfrak{q}^{(n)}$ and $g \notin \mathfrak{q}$, show that $a \in \mathfrak{q}^{(n)}$. (Translation: $\mathfrak{q}^{(n)}$ is $\mathfrak{q}$-*primary*, although we won't use this terminology either.)

Now $\sqrt{(f)} \subset A$ is the intersection of the minimal prime ideals containing f (Exercise 3.4.F), which is precisely the maximal ideal $\mathfrak{p}$. Thus in the local ring $A/(f)$, the nilradical ideal is the maximal ideal, so by Proposition 12.1.9 (on dimension 0 Noetherian local rings), any descending chain of ideals of $A/(f)$ eventually stabilizes. In particular, the image of $\mathfrak{q}^{(1)} \supset \mathfrak{q}^{(2)} \supset \cdots$ in $A/(f)$ eventually stabilizes, so

(12.3.4.1) $\mathfrak{q}^{(n)} \equiv \mathfrak{q}^{(n+1)}$ (mod f)   for $n \geq n_0$ sufficiently large.

We next claim that in fact $\mathfrak{q}^{(n)} = \mathfrak{q}^{(n+1)} + f\mathfrak{q}^{(n)}$ for $n \geq n_0$. Reason: Suppose $b_n \in \mathfrak{q}^{(n)}$. Then by (12.3.4.1), $b_n = b_{n+1} + af$ for $b_{n+1} \in \mathfrak{q}^{(n+1)} \subset \mathfrak{q}^{(n)}$ and $a \in A$. Hence $af \in \mathfrak{q}^{(n)}$; but $f \notin \mathfrak{q}$. By Exercise 12.3.B, $a \in \mathfrak{q}^{(n)}$, so $\mathfrak{q}^{(n)} = \mathfrak{q}^{(n+1)} + f\mathfrak{q}^{(n)}$, as claimed.

Thus by Nakayama's Lemma 8.2.9 (using A, $M = \mathfrak{q}^{(n)}/\mathfrak{q}^{(n+1)}$, $I = (f)$), $\mathfrak{q}^{(n)} = \mathfrak{q}^{(n+1)}$, from which $\mathfrak{q}^n A_\mathfrak{q} = \mathfrak{q}^{n+1} A_\mathfrak{q}$. By Nakayama's Lemma 8.2.9 (using $A_\mathfrak{q}$, $M = \mathfrak{q}^n A_\mathfrak{q}$, $I = \mathfrak{q}A_\mathfrak{q}$), $\mathfrak{q}^n A_\mathfrak{q} = 0$, so by Proposition 12.1.9, $A_\mathfrak{q}$ has dimension 0, so $\mathfrak{q}$ has codimension 0 in A, as desired.   □

12.3.5. We now use Krull's Principal Ideal Theorem to prove some geometric facts. We start with a fun fact in preparation for Exercise 12.3.D(a), which will also be very useful in other circumstances.

12.3.C. IMPORTANT EXERCISE (TO BE USED REPEATEDLY). Suppose k is a field (possibly finite!) and $n \in \mathbb{Z}^+$. Let $\{p_1, \ldots, p_m\} \in \mathbb{P}_k^n$ be a collection of points, not necessarily closed. Show that there exists a nonempty hypersurface (of some unspecified degree) not containing any of the

p_i. Hint: Show this by induction on m. To get from m to $m+1$: take a hypersurface not vanishing on $p_1, \ldots, p_m$. If it doesn't vanish on p_{m+1}, we are done. Otherwise, call this hypersurface f_{m+1}. Do something similar with $m+1$ replaced by i for each $1 \leq i \leq m$. Then consider $\sum_i f_1 \cdots \hat{f_i} \cdots f_{m+1}$.

12.3.D. IMPORTANT EXERCISE (TO BE USED REPEATEDLY). This is a cool argument.

(a) *(Hypersurfaces meet everything of dimension at least 1 in projective space, unlike in affine space.)* Suppose X is a closed subset of $\mathbb{P}_k^n$ of dimension at least 1, and H is a nonempty hypersurface in $\mathbb{P}_k^n$. Show that H meets X. (Hint: Note that the affine cone over H contains the origin in $\mathbb{A}_k^{n+1}$. Apply Krull's Principal Ideal Theorem 12.3.3 to the cone over X.)

(b) Suppose $X \hookrightarrow \mathbb{P}_k^n$ is a closed subset of dimension r. Show that any codimension r linear space meets X. Hint: Refine your argument in (a). (Exercise 12.3.G generalizes this to show that any two things in projective space that you would expect to meet for dimensional reasons do in fact meet.)

(c) Show further that there is an intersection of $r+1$ nonempty hypersurfaces missing X. (The key step: There is a hypersurface of sufficiently high degree that doesn't contain any generic point of X, by Exercise 12.3.C.) If k is infinite, show that there is a codimension $r+1$ *linear subspace* missing X. (The key step: Show that there is a hyperplane not containing any generic point of a component of X.)

(d) If k is an infinite field, show that there is an intersection of r hyperplanes meeting X in a finite number of points. (We will see in Exercise 13.4.C that if $k = \overline{k}$, for "most" choices of these r hyperplanes, this intersection is reduced, and in Exercise 18.6.M that the number of points is the "degree" of X. But first of course we must define "degree.")

The proof of prime avoidance (the following exercise) is similar to that of Exercise 12.3.D(c), and you may wish to solve it first. We will find it useful shortly.

12.3.E. EXERCISE. Prove Proposition 12.2.13 (prime avoidance). Hint: By induction on n. Don't look in the literature—you might find a much longer argument.

12.3.F. EXERCISE (BOUND ON CODIMENSION OF INTERSECTIONS IN $\mathbb{A}_k^N$). Let $k = \overline{k}$ be an algebraically closed field. Suppose X and Y are irreducible closed subvarieties (possibly singular) of codimension m and n respectively in $\mathbb{A}_k^d$. Show that every component of $X \cap Y$ has codimension at most $m+n$ in $\mathbb{A}_k^d$ as follows. Show that the diagonal $\mathbb{A}_k^d \cong \Delta \subset \mathbb{A}_k^d \times_k \mathbb{A}_k^d$ is a regular embedding of codimension d. (You will quickly guess the d equations for Δ.) Exercise 11.2.M identifies the intersection of X and Y in $\mathbb{A}^d$ with the intersection of $X \times Y$ with Δ in $\mathbb{A}^d \times_k \mathbb{A}_k^d$. Then show that locally, $X \cap Y$ is cut out in $X \times Y$ by d equations. Use Krull's Principal Ideal Theorem 12.3.3. You will also need Exercise 12.2.F. (The hypothesis of irreducibility can be replaced by pure dimensionality by appropriately generalizing Exercise 12.2.F. The hypothesis $k = \overline{k}$ can be removed by suitably invoking Exercise 12.1.H. See Exercise 13.2.N for a generalization.)

12.3.G. FUN EXERCISE (GENERALIZING EXERCISE 12.3.D(B)). Suppose X and Y are pure dimensional closed subvarieties of $\mathbb{P}^n$ of codimensions d and e respectively, and $d + e \leq n$. Show that X and Y intersect. Hint: Apply Exercise 12.3.F to the affine cones of X and Y. Recall the argument you used in Exercise 12.3.D(a) or (b).

12.3.H. USEFUL EXERCISE. Suppose f is an element of a Noetherian ring A, contained in no codimension 0 or 1 prime ideal. Show that f is invertible. (Hint: If a function vanishes nowhere, it is invertible, by Exercise 4.3.F(b).)

12.3.6. A useful characterization of unique factorization domains.

We can use Krull's Principal Ideal Theorem to prove one of the four useful criteria for unique factorization domains, promised in §5.4.5.

12.3.7. Proposition — *Suppose that A is a Noetherian integral domain. Then A is a unique factorization domain if and only if all codimension 1 prime ideals are principal.*

This contains Lemma 12.1.7 and (in some sense) its converse.

Proof. One direction is Lemma 12.1.7: if A is a unique factorization domain, then all codimension 1 prime ideals are principal. So we assume conversely that all codimension 1 prime ideals of A are principal, and we will show that A is a unique factorization domain. Now each codimension 1 prime ideal, being principal, is generated by a prime element, which is unique up to multiplication by a unit. All prime elements of an integral domain are irreducible, by the usual argument. (Suppose $p \in A$ is prime, and $p = ab$, with $a, b \in A \setminus \{0\}$. Then by primality of the ideal (p), either a or b is in (p), say, without loss of generality $a = kp$. Then $p = kpb$, from which $1 = bk$, from which b is unit. Thus p is irreducible.) Conversely, if $a \in A$ is irreducible, then by Krull's Principal Ideal Theorem 12.3.3, $V(a)$ contains some codimension 1 point $[(p)]$, so $a \in (p)$, say, $a = kp$. But by irreducibility of a, as p is not a unit, k must be a unit, from which $(a) = (p)$, so a is prime. Thus we have shown that the prime elements are the irreducible elements (and, incidentally, that up to multiplicative units, they correspond to the codimension 1 prime ideals).

Finally, A is a unique factorization domain by the usual argument (e.g., [DF, §8.3]): Each non-unit nonzero element can be factored into a finite number of irreducibles, and the uniqueness of such a factorization can be shown using the fact that those irreducibles are also prime. $\square$

12.3.8. Using Krull's Principal Ideal Theorem inductively. The following result will allow us to use Krull's Principal Ideal Theorem inductively.

12.3.9. Lemma — *Suppose $\mathfrak{p}_0 \subsetneq \mathfrak{p}_1 \subsetneq \cdots \subsetneq \mathfrak{p}_n$ is a strict chain of prime ideals in a Noetherian ring A, and $f \in \mathfrak{p}_n \setminus \mathfrak{p}_0$. Then there exists a strict chain of prime ideals*

$$\mathfrak{p}_0 = \mathfrak{q}_0 \subsetneq \mathfrak{q}_1 \subsetneq \cdots \subsetneq \mathfrak{q}_{n-1} \subsetneq \mathfrak{q}_n = \mathfrak{p}_n$$

in A, with $f \in \mathfrak{q}_1 \setminus \mathfrak{q}_0$.

You should feel compelled to immediately sketch a picture of this lemma. We are given Spec A, and a nested sequence of irreducible closed subsets, as well as a "hypersurface" containing the smallest closed subset. We are going to create a new sequence of irreducible closed subsets, with the same start and finish, which are all (after the first) contained in the hypersurface.

Proof. Our plan is to create the ideals as follows:

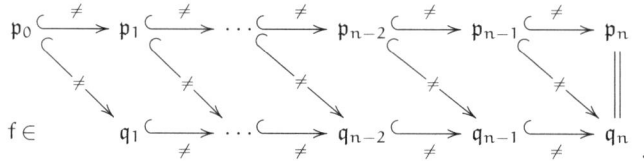

Unsurprisingly, we do this by descending induction. We prove the following. Suppose $\mathfrak{p}_{j-2} \subset \mathfrak{p}_{j-1} \subset \mathfrak{p}_j$ with $f \in \mathfrak{p}_j \setminus \mathfrak{p}_{j-2}$. We seek a prime ideal $\mathfrak{q}_{j-1}$ containing f, with $\mathfrak{p}_{j-2} \subsetneq \mathfrak{q}_{j-1} \subsetneq \mathfrak{p}_j$. Take $\mathfrak{q}_{j-1}$ to be a prime containing $(f, \mathfrak{p}_{j-2})$, and minimal among those contained in $\mathfrak{p}_j$. Now $\mathfrak{q}_{j-1}$ is codimension 1 in $A/\mathfrak{p}_{j-2}$ by Krull's Principal Ideal Theorem, and $\mathfrak{p}_j$ is codimension at least 2 (we have a chain $\mathfrak{p}_{j-2} \subset \mathfrak{p}_{j-1} \subset \mathfrak{p}_j$ of length 2). Thus $\mathfrak{q}_{j-1} \subsetneq \mathfrak{p}_j$. $\square$

12.3.I. EXERCISE. Suppose $(A, \mathfrak{m})$ is a Noetherian local ring, and $f \in \mathfrak{m}$. Show that $\dim A/(f) \geq \dim A - 1$.

The following generalization of Krull's Principal Ideal Theorem looks like it might follow by induction from Krull, but it is more subtle.

12.3.10. Krull's Height Theorem — *Suppose $X = \text{Spec } A$, where A is Noetherian, and Z is an irreducible component of $V(f_1, \ldots, f_r)$, where $f_1, \ldots, f_r \in A$. Then the codimension of Z is at most r.*

(As mentioned in §12.1.4, *height* is another common word for *codimension*.)

12.3.J. EXERCISE. Prove Krull's Height Theorem 12.3.10 by reducing it to the case of a Noetherian local ring $(A, \mathfrak{m})$, where $Z = V(\mathfrak{m})$, and then using Lemma 12.3.9 for the $f_1, \ldots, f_r$ in order.

12.3.K. IMPORTANT EXERCISE. Suppose $(A, \mathfrak{m})$ is a Noetherian local ring.

(a) *(Noetherian local rings have finite dimension, promised in Remark 12.1.11.)* Use Krull's Height Theorem 12.3.10 to prove that if there are $g_1, \ldots, g_\ell$ such that $V(g_1, \ldots, g_\ell) = \{[\mathfrak{m}]\}$, then $\dim A \leq \ell$. Hence show that A has finite dimension. (For comparison, Noetherian rings in general may have infinite dimension; see Exercise 12.1.M.)

(b) Let $d = \dim A$. Show that there exist $g_1, \ldots, g_d \in A$ such that $V(g_1, \ldots, g_d) = \{[\mathfrak{m}]\}$.

12.3.11. *Definition.* The geometric translation of the above exercise is the following. Given a d-dimensional "germ of a reasonable space" around a point p, p can be cut out set-theoretically by d equations, and we always need at least d equations. These d elements of A are called a **system of parameters** at p. Essentially equivalently, if $(A, \mathfrak{m})$ is a Noetherian local ring of dimension d, then $f_1, \ldots, f_d$ is a **system of parameters** for $(A, \mathfrak{m})$ if $\mathfrak{m} = \sqrt{(f_1, \ldots, f_d)}$.

The following result will be central in the next chapter.

12.3.12. Theorem — *For any Noetherian local ring $(A, \mathfrak{m}, k)$, $\dim A \leq \dim_k \mathfrak{m}/\mathfrak{m}^2$.*

Proof. Now $\mathfrak{m}$ is finitely generated, so $\mathfrak{m}/\mathfrak{m}^2$ is a finitely generated $(A/\mathfrak{m} = k)$-module, hence finite-dimensional. Say $\dim_k \mathfrak{m}/\mathfrak{m}^2 = n$. Choose a basis of $\mathfrak{m}/\mathfrak{m}^2$, and lift it to elements $f_1, \ldots, f_n$ of $\mathfrak{m}$. By Nakayama's Lemma (version 4, Exercise 8.2.H), $(f_1, \ldots, f_n) = \mathfrak{m}$. By Exercise 12.3.K, $\dim A \leq n$. $\qquad\square$

We conclude with two starred sections.

12.3.13.* Pathologies of the notion of "codimension". We can use Krull's Principal Ideal Theorem to produce the example of pathology in the notion of codimension promised earlier in this chapter. Let $A = k[x]_{(x)}[t]$. In other words, elements of A are polynomials in t, whose coefficients are quotients of polynomials in x, where no factors of x appear in the denominator. (Warning: A is not $k[x, t]_{(x)}$.) Clearly, A is an integral domain, so $xt - 1$ is not a zerodivisor. You can verify that $A/(xt - 1) \cong k[x]_{(x)}[1/x] \cong k(x)$— "in $k[x]_{(x)}$, we may divide by everything but x, and now we are allowed to divide by x as well"—so $A/(xt - 1)$ is a field. Thus $(xt - 1)$ is not just prime but also maximal. By Krull's Principal Ideal Theorem, $(xt - 1)$ is codimension 1. Thus $(0) \subset (xt - 1)$ is a maximal chain. However, A has dimension at least 2: $(0) \subset (t) \subset (x, t)$ is a chain of prime ideals of length 2. (In fact, A has dimension precisely 2, although we don't need this fact in order to observe the pathology.) Thus we have a codimension 1 prime in a dimension 2 ring whose quotient is dimension 0, not 1. Here is a picture of this partially ordered set of ideals.

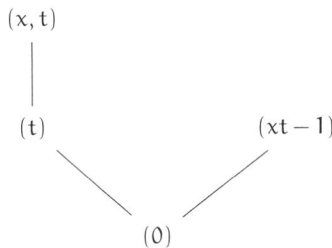

$$(x, t)$$
$$|$$
$$(t) \qquad\qquad (xt - 1)$$
$$\diagdown \qquad \diagup$$
$$(0)$$

This example comes from geometry, and it is enlightening to draw a picture; see Figure 12.3. $\operatorname{Spec} k[x]_{(x)}$ corresponds to a "germ" of $\mathbb{A}^1_k$ near the origin, and $\operatorname{Spec} k[x]_{(x)}[t]$ corresponds to "this $\times$ the affine line." You may be able to see from the picture some motivation for this pathology— $V(xt - 1)$ doesn't meet $V(x)$, so it can't have any specialization on $V(x)$, and there is nowhere else for $V(xt - 1)$ to specialize. It is disturbing that this misbehavior turns up even in a relatively benign-looking ring.

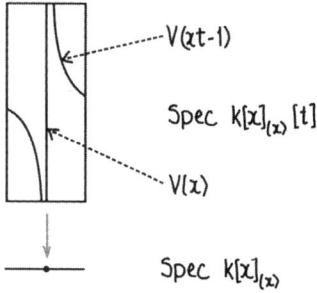

V(xt-1)

Spec $k[x]_{(x)}[t]$

V(x)

Spec $k[x]_{(x)}$

Figure 12.3 *Dimension and codimension behave oddly on the surface* Spec $k[x]_{(x)}[t]$.

12.3.14.* Lines on hypersurfaces, part 2. (Part 1 was §12.2.18.) We now give a geometric application of Krull's Principal Ideal Theorem 12.3.3, applied through Exercise 12.3.D(a). Throughout, we work over an algebraically closed field $\bar{k}$.

12.3.L. EXERCISE.

(a) Suppose

$$f(x_0, \ldots, x_n) = f_d(x_1, \ldots, x_n) + x_0 f_{d-1}(x_1, \ldots, x_n) + \cdots$$
$$+ x_0^{d-1} f_1(x_1, \ldots, x_n)$$

is a homogeneous degree d polynomial (so $\deg f_i = i$) cutting out a hypersurface X in $\mathbb{P}^n$ containing $p := [1, 0, \ldots, 0]$. Show that there is a line through p contained in X if and only if $f_1 = f_2 = \cdots = f_d = 0$ has a common zero in $\mathbb{P}^{n-1} = \operatorname{Proj} \bar{k}[x_1, \ldots, x_n]$. (Hint: Given a common zero $[a_1, \ldots, a_n] \in \mathbb{P}^{n-1}$, show that the line joining p to $[0, a_1, \ldots, a_n]$ is contained in X.)

(b) If $d \le n - 1$, show that through any point $p \in X$, there is a line contained in X. Hint: Exercise 12.3.D(a).

(c) If $d \ge n$, show that for "most hypersurfaces" X of degree d in $\mathbb{P}^n$ (for all hypersurfaces whose corresponding point in the parameter space $\mathbb{P}^{\binom{n+d}{d}-1}$—cf. Remark 4.5.3 and Exercise 9.3.J—lies in some nonempty Zariski-open subset), "most points $p \in X$" (all points in a nonempty dense Zariski-open subset of X) have no lines in X passing through them. (Hint: First show that there is a single p in a single X contained in no line. Chevalley's Theorem 8.4.2 may help.)

12.3.15. Remark. A projective (or proper) $\bar{k}$-variety X is **uniruled** if every point $p \in X$ is contained in some $\mathbb{P}^1 \subset X$. (We won't use this word beyond this remark.) Part (b) shows that all hypersurfaces of degree at most $n - 1$ are uniruled. One can show (using methods beyond what we know now; see [Ko1, Cor. IV.1.11]) that if char $\bar{k} = 0$, then every smooth hypersurface of degree at least $n + 1$ in $\mathbb{P}^n_{\bar{k}}$ is *not* uniruled (thus making the open set in (c) explicit). Furthermore, smooth hypersurfaces of degree n *are* uniruled, but covered by conics rather than lines. Thus there is a strong difference in how hypersurfaces behave depending on how the degree relates to $n + 1$. This is true in many other ways as well. Smooth hypersurfaces of degree less than $n + 1$ are examples of *Fano* varieties; smooth hypersurfaces of degree $n + 1$ are examples of *Calabi-Yau* varieties (with the possible exception of $n = 1$, depending on the definition); and smooth hypersurfaces of degree greater than $n + 1$ are examples of *general type* varieties. We define these terms in §21.5.5.

12.4 Dimensions of Fibers of Morphisms of Varieties

In this section, we show that the dimensions of fibers of morphisms of varieties behave in a way you might expect from our geometric intuition. The reason we have waited until now to discuss this is because we will use Theorem 12.2.9 (for varieties, codimension is the difference of dimensions).

Before we begin, let's make sure we are on the same page with respect to our intuition. Elimination theory (Theorem 8.4.10) tells us that the projection $\pi \colon \mathbb{P}^n_A \to \operatorname{Spec} A$ is closed. We can interpret this as follows. A closed subset X of $\mathbb{P}^n_A$ is cut out by a bunch of homogeneous equations in $n + 1$ variables (over A). The image of X is the subset of $\operatorname{Spec} A$ where these equations have a common nontrivial solution. If we try hard enough, we can describe this by saying that the *existence* of a nontrivial solution (or the existence of a preimage of a point under $\pi \colon X \to \operatorname{Spec} A$) is an "upper semicontinuous" fact. More generally, your intuition might tell you that the locus where a number of homogeneous polynomials in $n + 1$ variables over A have a solution space (in $\mathbb{P}^n_A$) of dimension at least d should be a closed subset of Spec A. (As a special case, consider linear equations. The condition for m linear equations in $n + 1$ variables to have a solution space of dimension at least $d + 1$ is a closed condition on the coefficients—do you see why, using linear algebra?) This intuition will be correct, and will use properness in a fundamental way (Theorem 12.4.3(b)). We will also make sense of upper semicontinuity in fiber dimension on the *source* (Theorem 12.4.3(a)). A useful example to think through is the map from the xy-plane to the xz-plane (Spec k[x, y] → Spec k[x, z]), given by $(x, z) \mapsto (x, xy)$. (This example also came up in §8.4.1.)

We begin our substantive discussion with an inequality that holds more generally in the locally Noetherian setting.

12.4.A. KEY EXERCISE (CODIMENSION BEHAVES AS YOU MIGHT EXPECT FOR A MOR-PHISM, OR "FIBER DIMENSIONS CAN NEVER BE LOWER THAN EXPECTED"). Suppose $\pi \colon X \to Y$ is a morphism of locally Noetherian schemes, and $p \in X$ and $q \in Y$ are points such that $q = \pi(p)$. Show that

$$\operatorname{codim}_X p \le \operatorname{codim}_Y q + \operatorname{codim}_{\pi^{-1}(q)} p$$

(see Figure 12.4). (This is similar to Exercise 12.1.K, but it is not the same!) Hint: Take a system of parameters (Definition 12.3.11) for q "in Y," and a system of parameters for p "in $\pi^{-1}(q)$," and use them to find $\operatorname{codim}_Y q + \operatorname{codim}_{\pi^{-1}(q)} p$ elements of $\mathscr{O}_{X,p}$ cutting out $p = \{[m]\}$ in Spec $\mathscr{O}_{X,p}$. Use Exercise 12.3.K.

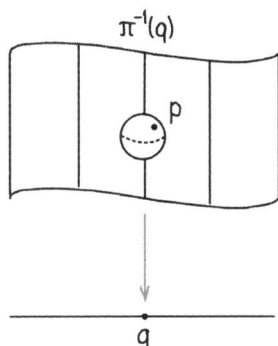

Figure 12.4 *Exercise 12.4.A: The codimension of a point in the total space is bounded by the sum of the codimension of the point in the fiber and the codimension of the image in the target.*

Does Exercise 12.4.A agree with your geometric intuition? You should be able to come up with enlightening examples where equality holds, and where equality fails. We will see that equality always holds for sufficiently nice—flat—morphisms; see Proposition 24.5.6.

We now show that the inequality of Exercise 12.4.A is actually an equality over "most of Y."

12.4.1. Theorem — *Suppose $\pi \colon X \to Y$ is a finite type dominant morphism of integral schemes such that the finitely generated field extension $K(X)/K(Y)$ has transcendence degree r. Then there exists a nonempty open subset $U \subset Y$ such that for all $q \in U$, the fiber over q is nonempty and has pure dimension r.*

An interesting fact: We will use the Noether Normalization Lemma 12.2.4 in the proof, even though the statement doesn't involve varieties! If the generality of the statement bothers you, we have the following special case for varieties.

12.4.2. Corollary — *Suppose $\pi \colon X \to Y$ is a (necessarily finite type) morphism of irreducible k-varieties, with $\dim X = m$ and $\dim Y = n$. Then there exists a nonempty open subset $U \subset Y$ such that for all $q \in U$, the fiber over q has pure dimension $m - n$, or is empty.*

(By convention, the empty set technically has pure dimension r—every irreducible component of the empty set has dimension r. This makes the statement of Corollary 12.4.2 a bit cleaner.)

12.4.B. EXERCISE. Verify that Theorem 12.4.1 implies Corollary 12.4.2.

Proof of Theorem 12.4.1. We begin with two quick reductions. (i) By shrinking Y if necessary, we may assume that Y is affine, say Spec B. (ii) We may also assume that X is affine, say Spec A. (Reason: Cover X with a finite number of affine open subsets $X_1, \ldots, X_a$, and take the intersection of the U's for each of the $\pi|_{X_i}$.)

In order to motivate the rest of the argument, we describe our goal. We will produce a nonempty distinguished open subset U of Spec B so that $\pi^{-1}(U) \to U$ factors through $\mathbb{A}^r_U$ via a finite surjective morphism:

(12.4.2.1)

$$X = \quad \text{Spec } A \xleftarrow{\quad \text{open emb.} \quad} \pi^{-1}(U)$$

$$\pi \downarrow \qquad \qquad \downarrow \text{finite surj.}$$

$$\mathbb{A}^r_U$$

$$\downarrow$$

$$Y = \quad \text{Spec } B \xleftarrow{\quad \text{open emb.} \quad} U$$

12.4.C. EXERCISE. Show that this suffices to prove Theorem 12.4.1. Hint: Suppose $p \in U$ is a point, with residue field κ. Show that any irreducible variety mapping finitely to $\mathbb{A}^r_\kappa$ has dimension at most r. This will show that the dimension of any component of $\pi^{-1}(p)$ has dimension at most r. To show equality, apply the Going-Down Theorem 12.2.12 appropriately to $\text{Spec } A \to \text{Spec } B[x_1, \ldots, x_r]$.

So we now work to build (12.4.2.1). We begin by noting that we have inclusions of B into both A and $K(B)$, and from both A and $K(B)$ into $K(A)$. The maps from A and $K(B)$ into $K(A)$ both factor through $A \otimes_B K(B)$ (whose Spec is the generic fiber of π), so the maps from both A and $K(B)$ to $A \otimes_B K(B)$ must be inclusions.

(12.4.2.2)

Clearly, $K(A) \otimes_B K(B) = K(A)$ (as $A \otimes_B K(B)$ can be interpreted as taking A and inverting the elements that are nonzero elements of B), and $A \otimes_B K(B)$ is a finitely generated algebra over the field $K(B)$.

By Noether normalization 12.2.4, we can find elements $t_1, \ldots, t_r \in A \otimes_B K(B)$, algebraically independent over $K(B)$, such that $A \otimes_B K(B)$ is integral over $K(B)[t_1, \ldots, t_r]$.

Now, we can think of the elements $t_i \in A \otimes_B K(B)$ as fractions, with numerators in A and (nonzero) denominators in B. If f is the product of the denominators appearing for each t_i, then by replacing B by B_f (replacing $\text{Spec } B$ by its distinguished open subset $D(f)$), we may assume that the t_i are all in A. Thus (after sloppily renaming B_f as B, and A_f as A) we can trim and extend (12.4.2.2) to the following.

Now A is finitely generated over B, and hence over $B[t_1, \ldots, t_r]$. But we cannot yet be sure that A is *finite* over $B[t_1, \ldots, t_r]$, so we are not done. We will have to localize B further.

Suppose A is generated over B by $u_1, \ldots, u_q$. Our Noether normalization argument implies that each u_i satisfies some monic equation $f_i(u_i) = 0$, where $f_i \in K(B)[t_1, \ldots, t_r][t]$ (i.e., monic considered as a polynomial in t with coefficients in $K(B)[t_1, \ldots, t_r]$). The coefficients of f_i (considered as polynomials in $t_1, \ldots, t_r, t$) are a priori fractions in B, but by multiplying by all those denominators, we can assume each $f_i \in B[t_1, \ldots, t_r][t]$, at the cost of losing monicity of the f_i (this time considered as polynomials in t). Let $b \in B$ be the product of the leading coefficients (considered as polynomials in t) of all the f_i. If $U = D(b)$ (the locus where b is invertible), then, over U, the f_i (can be taken to) have leading coefficient 1, so the u_i (in A_b) are integral over $B_b[t_1, \ldots, t_r]$. Thus $\operatorname{Spec} A_b \to \operatorname{Spec} B_b[t_1, \ldots, t_r]$ is finite and surjective (the latter by the Lying Over Theorem 8.2.5).

We have now constructed (12.4.2.1), as desired.   □

There are a couple of things worth pointing out about the proof. First, although the result was originally motivated by classical varieties over a field k, the proof uses the theory of varieties over *another* field, the function field $K(B)$. This is an example of how the introduction of generic points to algebraic geometry is useful even for considering more "classical" questions.

Second, the idea of the main part of the argument is that we have a result over the generic point ($\operatorname{Spec} A \otimes_B K(B)$ is finite and surjective over affine space over $K(B)$), and we want to "spread it out" to an open neighborhood of the generic point of $\operatorname{Spec} B$. We do this by realizing that "finitely many denominators" appear when correctly describing the problem, and inverting those denominators. This "spreading out" idea is a recurring theme.

12.4.D. EXERCISE (USEFUL CRITERION FOR IRREDUCIBILITY). Suppose $\pi\colon X \to Y$ is a proper morphism to an irreducible variety, and all the fibers of π are nonempty, and irreducible of the same dimension. Show that X is irreducible.

12.4.3. Theorem (upper semicontinuity of fiber dimension) — *Suppose $\pi\colon X \to Y$ is a morphism of finite type k-schemes.*

(a) *(upper semicontinuity on the source) The dimension of the fiber of π at $p \in X$ (the dimension of the largest component of $\pi^{-1}(\pi(p))$ containing p) is an upper semicontinuous function in p (i.e., on X).*

(b) *(upper semicontinuity on the target) If furthermore π is closed (e.g., if π is proper), then the dimension of the fiber of π over $q \in Y$ is an upper semicontinuous function in q (i.e., on Y).*

You should be able to immediately construct a counterexample to part (b) if the closedness hypothesis is dropped. (We also remark that Theorem 12.4.3(b) for projective morphisms is done, in a simple way, in Exercise 18.1.B.)

Proof. (a) Let F_n be the subset of X consisting of points where the fiber dimension is at least n. We wish to show that F_n is a closed subset for all n. We argue by induction on $\dim Y$. The base case $\dim Y = 0$ is trivial. So we fix Y, and assume the result for all smaller-dimensional targets.

12.4.E. EXERCISE. Show that it suffices to prove the result when X and Y are integral, and π is dominant.

Let $r = \dim X - \dim Y$ be the "relative dimension" of π. If $n \leq r$, then $F_n = X$ by Exercise 12.4.A (combined with Theorem 12.2.9, that codimension is the difference of dimensions for varieties).

If $n > r$, then let $U \subset Y$ be the dense open subset of Theorem 12.4.1, where "the fiber dimension is exactly r." Then F_n does not meet the preimage of U. By replacing Y with $Y \setminus U$ (and X by $X \setminus \pi^{-1}(U)$), we are done by the inductive hypothesis.

12.4.F. EASY EXERCISE. Prove (b) (using (a)).   □

12.4.4. Proposition ("generically finite implies generally finite"; cf. Exercise 10.3.G) — *Suppose $\pi\colon X \to Y$ is a generically finite morphism of irreducible k-varieties of dimension n. Then there is a dense open subset $V \subset Y$ above which π is finite.*

(If you wish, you can later relax the irreducibility hypothesis to simply requiring X and Y to be of pure dimension n.)

Proof. As in the proof of Theorem 12.4.1, we may assume that Y is affine, and that π is dominant. Suppose that $X = \cup_{i=1}^{n} U_i$, where the U_i are affine open subschemes of X. By Exercise 10.3.G, there are dense open subsets $V_i \subset Y$ over which $\pi|_{U_i}$ is finite. By replacing Y by an affine open subset of $\cap V_i$, we may assume that $\pi|_{U_i}$ is finite.

12.4.G. EXERCISE. Show that π is closed. Hint: You will just use that $\pi|_{U_i}$ is closed, and that there are a finite number of U_i.

Then $X \setminus U_1$ is a closed subset, so $\pi(X \setminus U_1)$ is closed.

12.4.H. EXERCISE. Show that this closed subset is not all of Y.

Define $V := Y \setminus \pi(X \setminus U_1)$. Then π is finite above V: it is the restriction of the finite morphism $\pi|_{U_1} : U_1 \to Y$ to the open subset V of the target Y. □

Chapter 13

Regularity and Smoothness

One natural notion we expect to see for geometric spaces is the notion of when an object is "smooth." In algebraic geometry, this notion, called *regularity*, is easy to define (Definition 13.2.1) but a bit subtle in practice. This will lead us to a different related notion of when *a variety is smooth* (Definition 13.2.4).

This chapter has many moving parts, of which §13.1–13.6 are the important ones. In §13.1, the *Zariski tangent space* is motivated and defined. In §13.2, we define *regularity* and *smoothness over a field*, the central topics of this chapter, and discuss some of their important properties. In §13.3, we give a number of important examples, mostly in the form of exercises. In §13.4, we discuss *Bertini's Theorem*, a fundamental classical result. In §13.5, we give many characterizations of *discrete valuation rings*, which play a central role in algebraic geometry. Having seen clues that "smoothness" is a "relative" notion rather than an "absolute" one, in §13.6 we define *smooth morphisms* (and in particular, *étale morphisms*), and give some of their properties. (We will revisit this definition in §21.2.28, once we know more.)

The remaining sections are less central. In §13.7, we discuss the valuative criteria for separatedness and properness. In §13.8, we mention some more sophisticated facts about regular local rings (the local rings at regular points). In §13.9, we prove the Artin-Rees Lemma, because it will have been invoked in §13.5 (and will be used later as well).

13.1 The Zariski Tangent Space

We begin by defining the tangent space of a scheme at a point. It behaves like the tangent space you know and love at "smooth" points, but also makes sense at other points. In other words, geometric intuition at the "smooth" points guides the definition, and then the definition guides the algebra at all points, which in turn lets us refine our geometric intuition.

The definition is short but surprising. The main difficulty is convincing yourself that it deserves to be called the tangent space. This is tricky to explain, because we want to show that it agrees with our intuition, but our intuition is worse than we realize. So you should just accept this definition for now, and later convince yourself that it is reasonable.

13.1.1. *Definition.* The **Zariski cotangent space** of a local ring $(A, \mathfrak{m})$ is defined to be $\mathfrak{m}/\mathfrak{m}^2$; it is a vector space over the residue field $A/\mathfrak{m}$. The dual vector space is the **Zariski tangent space**. If X is a scheme, the **Zariski cotangent space** $T^\vee_{X,p}$ at a point $p \in X$ is defined to be the Zariski cotangent space of the local ring $\mathcal{O}_{X,p}$, and similarly for the **Zariski tangent space** $T_{X,p}$. (Note that for *any* maximal ideal $\mathfrak{m}$ of *any* ring A, we have canonically $\mathfrak{m}/\mathfrak{m}^2 = \mathfrak{m}A_\mathfrak{m}/\mathfrak{m}^2 A_\mathfrak{m}$—make sure you see why.) Elements of the Zariski cotangent space are called **cotangent vectors** or **differentials**; elements of the tangent space are called **tangent vectors**.

The cotangent space is more algebraically natural than the tangent space, in that the definition is shorter. There is a moral reason for this: the cotangent space is more naturally determined in terms of functions on a space, and we are very much thinking about schemes in terms of "functions on them." This will come up later.

Here are two plausibility arguments that this is a reasonable definition. Hopefully one will catch your fancy.

In differential geometry, the tangent space at a point is sometimes defined as the vector space of derivations at that point. A derivation at a point p of a manifold is an $\mathbb{R}$-linear operation that takes in functions f near p (i.e., elements of $\mathcal{O}_p$), and outputs elements $f'(p)$ of $\mathbb{R}$, and satisfies the

Leibniz rule

$$(fg)' = f'g + g'f.$$

(We will later define derivations in a more general setting, §21.2.17.) A derivation is the same as a map $\mathfrak{m} \to \mathbb{R}$, where $\mathfrak{m}$ is the maximal ideal of $\mathscr{O}_p$. (The map $\mathscr{O}_p \to \mathbb{R}$ extends this, via the map $\mathscr{O}_p \to \mathfrak{m}$ given by $f - f(p)$.) But $\mathfrak{m}^2$ maps to 0, as if $f(p) = g(p) = 0$, then

$$(fg)'(p) = f'(p)g(p) + g'(p)f(p) = 0.$$

Thus a derivation induces a map $\mathfrak{m}/\mathfrak{m}^2 \to \mathbb{R}$, i.e., an element of $(\mathfrak{m}/\mathfrak{m}^2)^\vee$.

13.1.A. EXERCISE. Check that this is reversible, i.e., that any linear map $\mathfrak{m}/\mathfrak{m}^2 \to \mathbb{R}$ gives a derivation. In other words, verify that the Leibniz rule holds.

Here is a second, vaguer, motivation that this definition is plausible for the cotangent space of the origin of $\mathbb{A}^n$. (I prefer this one, as it is more primitive and elementary.) Functions on $\mathbb{A}^n$ should restrict to a linear function on the tangent space. What (linear) function does $x^2 + xy + x + y$ restrict to "near the origin"? You will naturally answer: $x + y$. Thus we "pick off the linear terms." Hence $\mathfrak{m}/\mathfrak{m}^2$ are the linear functionals on the tangent space, so $\mathfrak{m}/\mathfrak{m}^2$ is the cotangent space. In particular, you should picture functions vanishing at a point (i.e., lying in $\mathfrak{m}$) as giving functions on the tangent space in this obvious way.

13.1.2. *Old-fashioned example.* Computing the Zariski tangent space is actually quite hands-on, because you can compute it just as you did when you learned multivariable calculus. In $\mathbb{A}^3$, we have a curve cut out by $x + y + z^2 + xyz = 0$ and $x - 2y + z + x^2y^2z^3 = 0$. (You can use Krull's Principal Ideal Theorem 12.3.3 to check that this is a curve, but it is not important to do so.) What is the tangent line near the origin? (Is it even smooth there?) Answer: The first surface looks like $x + y = 0$ and the second surface looks like $x - 2y + z = 0$. The curve has tangent line cut out by $x + y = 0$ and $x - 2y + z = 0$. It is smooth (in the traditional sense). In multivariable calculus, the students work hard to get the answer, because we aren't allowed to tell them to just pick out the linear terms.

Let's make explicit the fact that we are using. If A is a ring, $\mathfrak{m}$ is a maximal ideal, and $f \in \mathfrak{m}$ is a function vanishing at the point $[\mathfrak{m}] \in \operatorname{Spec} A$, then the Zariski tangent space of $\operatorname{Spec} A/(f)$ at $\mathfrak{m}$ is cut out in the Zariski tangent space of $\operatorname{Spec} A$ (at $\mathfrak{m}$) by the single linear equation $f \pmod{\mathfrak{m}^2}$. The next exercise will force you to think this through.

13.1.B. IMPORTANT EXERCISE ("KRULL'S PRINCIPAL IDEAL THEOREM FOR TANGENT SPACES"—BUT MUCH EASIER THAN KRULL'S PRINCIPAL IDEAL THEOREM 12.3.3!). Suppose A is a ring, and $\mathfrak{m}$ a maximal ideal. If $f \in \mathfrak{m}$, show that the Zariski tangent space of A/f is cut out in the Zariski tangent space of A by $f \pmod{\mathfrak{m}^2}$. (Note: We can quotient by f and localize at $\mathfrak{m}$ in either order, as quotient and localization commute, (4.3.7.1).) Hence the dimension of the Zariski tangent space of $\operatorname{Spec} A/(f)$ at $[\mathfrak{m}]$ is the dimension of the Zariski tangent space of $\operatorname{Spec} A$ at $[\mathfrak{m}]$, or one less. (That last sentence should be suitably interpreted if the dimension is infinite, although it is less interesting in this case.)

Here is another example to see this principle in action, extending Example 13.1.2: $x + y + z^2 = 0$ and $x + y + x^2 + y^4 + z^5 = 0$ cut out a curve, which obviously passes through the origin. If I asked my multivariable calculus students to calculate the tangent line to the curve at the origin, they would do calculations which would boil down (without their realizing it) to picking off the linear terms. They would end up with the equations $x + y = 0$ and $x + y = 0$, which cut out a plane, not a line. They would be disturbed, and I would explain that this is because the curve isn't smooth at a point, and their techniques don't work. We, on the other hand, bravely declare that the tangent space is cut out by $x + y = 0$, and (will soon) use this pathology as *definition* of what makes a point singular (or nonregular). (Intuitively, the curve near the origin is very close to lying in the plane $x + y = 0$.) Notice: The cotangent space jumped up in dimension from what it was "supposed to be," not down. (This is not a coincidence: see Theorem 12.3.12.)

13.1.C. EXERCISE. Suppose Y and Z are closed subschemes of X, both containing the point $p \in X$.

(a) Show that $T_{Z,p}$ is naturally a sub-$\kappa(p)$-vector space of $T_{X,p}$.
(b) Show that $T_{Y \cap Z,p} = T_{Y,p} \cap T_{Z,p}$. (On the left, $\cap$ means scheme-theoretic intersection. On the right, $\cap$ means intersection of subspaces of the vector space $T_{X,p}$.)
(c) Show that $T_{Y \cup Z,p}$ contains the span of $T_{Y,p}$ and $T_{Z,p}$, where $\cup$, as usual in this context, is scheme-theoretic union.
(d) Show that $T_{Y \cup Z,p}$ can be strictly larger than the span of $T_{Y,p}$ and $T_{Z,p}$. (Hint: Figure 4.5.)

Here is a pleasant consequence of the notion of Zariski tangent space.

13.1.3. Problem. Consider the ring $A = k[x, y, z]/(xy - z^2)$. Show that (x, z) is not a principal ideal.

As $\dim A = 2$ (why?), and $A/(x, z) \cong k[y]$ has dimension 1, we see that this ideal is codimension 1 (as codimension is the difference of dimensions for irreducible varieties, Theorem 12.2.9). Our geometric picture is that $\operatorname{Spec} A$ is a cone (we can diagonalize the quadric as $xy - z^2 = ((x+y)/2)^2 - ((x-y)/2)^2 - z^2$, at least if $\operatorname{char} k \neq 2$—see Exercise 5.4.J), and that (x, z) is a line on the cone. (See Figure 13.1 for a sketch.) This suggests that we look at the cone point.

13.1.4. *Solution.* Let $\mathfrak{m} = (x, y, z)$ be the maximal ideal corresponding to the origin. Then $\operatorname{Spec} A$ has Zariski tangent space of dimension 3 at the origin, and $\operatorname{Spec} A/(x, z)$ has Zariski tangent space of dimension 1 at the origin. But $\operatorname{Spec} A/(f)$ must have Zariski tangent space of dimension at least 2 at the origin by Exercise 13.1.B. (As an added bonus, we have shown that A is not a unique factorization domain, by Lemma 12.1.7.))

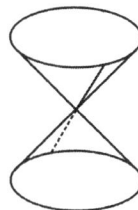

Figure 13.1 $V(x, z) \subset \operatorname{Spec} k[x, y, z]/(xy - z^2)$ *is a line on a cone.*

13.1.5.* *Remark.* Another approach to solving the problem, not requiring the definition of the Zariski tangent space, is to use the fact that the ring is graded (where x, y, and z each have degree 1), and the ideal (x, z) is a homogeneous ideal. (You may enjoy thinking this through.) The advantage of using the tangent space is that it applies to more general situations where there is no grading. For example, (a) (x, z) is not a principal ideal of $k[x, y, z]/(xy - z^2 - z^3)$. As a different example, (b) (x, z) is not a principal ideal of the local ring $(k[x, y, z]/(xy - z^2))_{(x,y,z)}$ (the "germ of the cone"). However, we remark that the graded case is still very useful. The construction of replacing a filtered ring by its "associated graded" ring can turn more general rings into graded rings (and can be used to turn example (a) into the graded case). The construction of completion can turn local rings into graded local rings (and can be used to turn example (b) into, essentially, the graded case). Filtered rings will come up in §13.9, the associated graded construction will implicitly come up in our discussions of the blow-up in §22.3, and many aspects of completions will be described in Chapter 28.

13.1.D. EXERCISE. Show that $(x, z) \subset k[w, x, y, z]/(wz - xy)$ is a codimension 1 ideal that is not principal, using the method of Solution 13.1.4. (See Figure 13.2 for the projectivization of this situation—a line on a smooth quadric surface.) This example was promised just after Exercise 5.4.D. An improvement is given in Exercise 15.5.N: this closed subset $x = z = 0$ is not even *set-theoretically* cut out by one equation.

13.1.E. EXERCISE. Let $A = k[w, x, y, z]/(wz - xy)$. Show that $\operatorname{Spec} A$ is not factorial. (Exercise 5.4.L shows that A is not a unique factorization domain, but this is not enough—why is the localization of A at the prime (w, x, y, z) not factorial? One possibility is to do this "directly," by trying to imitate the

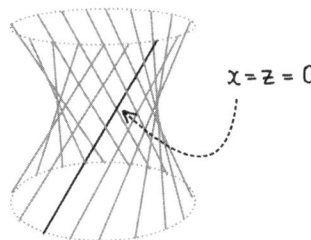

$x = z = 0$

Figure 13.2 *The line* $V(x, z)$ *on the smooth quadric surface* $V(wz - xy) \subset \mathbb{P}^3$.

solution to Exercise 5.4.L, but this is hard. Instead, use the intermediate result that, in a unique factorization domain, any codimension 1 prime is principal, Lemma 12.1.7, and considering Exercise 13.1.D.) As A is integrally closed if char $k \neq 2$ (Exercise 5.4.I(c)), this yields an example of a scheme that is normal but not factorial, as promised in Exercise 5.4.F. A slight generalization will be given in 22.4.N.

13.1.F. LESS IMPORTANT EXERCISE ("HIGHER-ORDER DATA"). (This exercise is fun, but won't be used.)

(a) In Exercise 3.7.B, you computed the equations cutting out the (union of the) three coordinate axes of $\mathbb{A}^3_k$. (Call this scheme X.) Your ideal should have had three generators. Show that the ideal cannot be generated by fewer than three elements. (Hint: Working modulo $\mathfrak{m} = (x, y, z)$ won't give any useful information, so work modulo a higher power of $\mathfrak{m}$.)

(b) Show that the coordinate axes in $\mathbb{A}^3_k$ are not regularly embedded in $\mathbb{A}^3_k$. (This was promised at the end of §9.5.)

13.1.6. *Morphisms and tangent spaces.* Suppose $\pi\colon X \to Y$, and $\pi(p) = q$. Then if we were in the category of differentiable manifolds, we would expect a tangent map, from the tangent space of p to the tangent space at q. Indeed that is the case; we have a map of stalks $\mathcal{O}_{Y,q} \to \mathcal{O}_{X,p}$, which sends the maximal ideal of the former $\mathfrak{n}$ to the maximal ideal of the latter $\mathfrak{m}$ (we checked that this is a "local morphism" when we briefly discussed locally ringed spaces; see §7.3.1). Thus $\mathfrak{n}^2$ maps to $\mathfrak{m}^2$, from which we see that $\mathfrak{n}/\mathfrak{n}^2$ maps to $\mathfrak{m}/\mathfrak{m}^2$. If $(\mathcal{O}_{X,p}, \mathfrak{m})$ and $(\mathcal{O}_{Y,q}, \mathfrak{n})$ have the same residue field κ, so that $\mathfrak{n}/\mathfrak{n}^2 \to \mathfrak{m}/\mathfrak{m}^2$ is a linear map of κ-vector spaces, then we have a natural map $(\mathfrak{m}/\mathfrak{m}^2)^\vee \to (\mathfrak{n}/\mathfrak{n}^2)^\vee$. This is the map from the tangent space of p to the tangent space at q that we sought. (Aside: Note that the *cotangent* map *always* exists, without requiring p and q to have the same residue field— a sign that cotangent spaces are more natural than tangent spaces in algebraic geometry.)

13.1.G. EXERCISE. Suppose X is a k-scheme. Describe a natural bijection from Mor_k $(\mathrm{Spec}\, k[\epsilon]/(\epsilon^2), X)$ to the data of a point p with residue field k (necessarily a closed point) and a tangent vector at p. (This is important, for example, in deformation theory. We won't use it, so you can treat it as a curiosity.)

13.1.H. EXERCISE (FUN ARITHMETIC PRACTICE). Find the dimension of the Zariski tangent space at the point $[(2, 2i)]$ of $\mathbb{Z}[2i] \cong \mathbb{Z}[x]/(x^2 + 4)$. Find the dimension of the Zariski tangent space at the point $[(2, x)]$ of $\mathbb{Z}[\sqrt{-2}] \cong \mathbb{Z}[x]/(x^2 + 2)$. (If you prefer geometric versions of the same examples, replace $\mathbb{Z}$ by $\mathbb{C}$, and 2 by y: consider $\mathbb{C}[x, y]/(x^2 + y^2)$ and $\mathbb{C}[x, y]/(x^2 + y)$.)

13.1.7. Important interpretation in the finite type case: The Jacobian corank.

13.1.I. IMPORTANT EXERCISE (THE JACOBIAN COMPUTES THE ZARISKI COTANGENT SPACE AT A RATIONAL POINT). Suppose X is a finite type k-scheme. Then locally it is of the form $\mathrm{Spec}\, k[x_1, \ldots, x_n]/(f_1, \ldots, f_r)$. Show that the Zariski cotangent space at a k-valued point (a closed point with residue field k) is given by the cokernel of the "Jacobian map" $k^r \to k^n$ given by the **Jacobian matrix**

(13.1.7.1)
$$J = \begin{pmatrix} \frac{\partial f_1}{\partial x_1}(p) & \cdots & \frac{\partial f_r}{\partial x_1}(p) \\ \vdots & \ddots & \vdots \\ \frac{\partial f_1}{\partial x_n}(p) & \cdots & \frac{\partial f_r}{\partial x_n}(p) \end{pmatrix}.$$

(This makes precise our example of a curve in $\mathbb{A}^3$ cut out by a couple of equations, where we picked off the linear terms; see Example 13.1.2.) You might be alarmed: what does $\frac{\partial f}{\partial x_1}$ *mean*? Do you need deltas and epsilons? No! Just define derivatives formally, e.g.,

$$\frac{\partial}{\partial x_1}(x_1^2 + x_1 x_2 + x_2^2) = 2x_1 + x_2.$$

Hint: Do this first when p is the origin, and consider linear terms, just as in Example 13.1.2 and Exercise 13.1.B. For the general case, "translate p to the origin."

13.1.8. *Remark.* This result can be extended to closed points of X whose residue field is separable over k (and, in particular, to *all* closed points if char k = 0 or if k is finite); see Remark 21.2.34. The fact that we wait until Chapter 21 to show this does not mean that it needs to be so complicated; a more elementary proof is possible.

13.1.9. *Warning.* It is more common in mathematics (but not universal) to define the Jacobian matrix as the transpose of (13.1.7.1).

13.1.J. EXERCISE (THE CORANK OF THE JACOBIAN IS INDEPENDENT OF THE PRESENTATION). Suppose A is a finitely generated k-algebra, generated by $x_1, \ldots, x_n$, with ideal of relations I generated by $f_1, \ldots, f_r$. Let p be a point of Spec A.

(a) Suppose $g \in I$. Show that appending the column of partials of g to the Jacobian matrix (13.1.7.1) does not change the **corank** (the dimension of the cokernel) at p. Hence show that the corank of the Jacobian matrix at p does not depend on the choice of generators of I.

(b) Suppose $q(x_1, \ldots, x_n) \in k[x_1, \ldots, x_n]$. Let h be the polynomial $y - q(x_1, \ldots, x_n) \in k[x_1, \ldots, x_n, y]$. Show that the Jacobian matrix of $(f_1, \ldots, f_r, h)$ with respect to the variables $(x_1, \ldots, x_n, y)$ has the same corank at p as the Jacobian matrix of $(f_1, \ldots, f_r)$ with respect to $(x_1, \ldots, x_n)$. Hence show that the corank of the Jacobian matrix at p is independent of the choice of generators for A.

Thus every finite type k-scheme X comes with an intrinsic "**Jacobian corank function**" from (the underlying set of) X to $\mathbb{Z}^{\geq 0}$, whose value at k-valued points p of X is $\dim_k T_{X,p}$. We temporarily call this function $JC : X \to \mathbb{Z}^{\geq 0}$, although we will later discover that it is really the rank of the coherent sheaf of differentials $\Omega_{X/k}$ (Exercise 21.2.E).

13.1.K. EASY EXERCISE. Show that JC is an upper semicontinuous function.

13.1.L. EXERCISE. Show that JC is preserved by field extension of k. More precisely, if X is a finite type k-scheme, and ℓ/k is a field extension, and under the base change map $X \times_k \ell \to X$, $p \mapsto q$, show that $JC(p) = JC(q)$.

13.1.M. EXERCISE. Prove that for any $p \in X$, $JC(p) \geq \dim_X p$. (Necessarily, X is finite type over k, but p need not be a closed point of X.) The dimension of X at p, $\dim_X p$, was defined in Exercise 12.2.K. Hint: Deal first with the case $k = \bar{k}$. Here use the closed points as leverage. Then prove the result for general k by showing that both sides of the inequality are preserved by field extension (Exercises 13.1.L and 12.2.M).

13.1.N. EXERCISE (COMPUTING THE JACOBIAN CORANK FUNCTION FOR PROJECTIVE K-SCHEMES). Suppose $X \subset \mathbb{P}^n_k$ is cut out by homogeneous polynomials $f_i \in k[x_0, \ldots, x_n]$ $(1 \leq i \leq r)$. Show that the Jacobian corank function of X is 1 less than the corank of the "projective Jacobian matrix":

(13.1.9.1)
$$JC(p) = \operatorname{corank} \begin{pmatrix} \frac{\partial f_1}{\partial x_0}(p) & \cdots & \frac{\partial f_r}{\partial x_0}(p) \\ \vdots & \ddots & \vdots \\ \frac{\partial f_1}{\partial x_n}(p) & \cdots & \frac{\partial f_r}{\partial x_n}(p) \end{pmatrix} - 1.$$

You will have to make sense of the right side, as the projective Jacobian matrix has entries that are not functions; what does the corank even mean?

13.2 Regularity, and Smoothness over a Field

The key idea in the definition of regularity is contained in Theorem 12.3.12 (the dimension of the Zariski tangent space is at least the dimension of the local ring).

13.2.1. *Definition.* If equality holds in Theorem 12.3.12, we say that A is a **regular local ring**. In particular, *regular local rings are Noetherian by definition.* (If a Noetherian ring A is regular at all of its prime ideals, i.e., if $A_\mathfrak{p}$ is a regular local ring for all prime ideals $\mathfrak{p}$ of A, then A is said to be a **regular ring**. We basically won't use this terminology. Fact 13.8.2 reassuringly implies that regular local rings are indeed regular rings.) A locally Noetherian scheme X is **regular** at a point p if the local ring $\mathscr{O}_{X,p}$ is regular. (The word **nonsingular** is often used as well, notably in the case when X is finite type over a field, but for the sake of consistency we will use "regular" throughout.) It is **singular** at the point otherwise, and we say that the point is a **singularity**. (The word **nonregular** is also used, but for the sake of consistency, we will use "singular," despite the fact that this choice is *inconsistent* with our choice of "regular" over "nonsingular." A possible motivation for the inconsistency: "regular local ring" and "singularity" are both standard terminology, so at some point we are forced to make a choice.) A scheme is **regular** (or **nonsingular**) if it is regular at all points. It is **singular** otherwise (i.e., if it is singular at *at least one* point—if it has a singularity).

13.2.A. EXERCISE. Show that a dimension 0 Noetherian local ring is regular if and only if it is a field.

You will hopefully gradually become convinced that this is the right notion of "smoothness" of schemes. Remarkably, Krull introduced the notion of a regular local ring for purely algebraic reasons, some time before Zariski realized that it was a fundamental notion in geometry in 1947.

13.2.B. EXERCISE. Suppose $(A, \mathfrak{m})$ is a regular local ring of dimension $n > 0$, and $f \in A$. Show that $A/(f)$ is a regular local ring of dimension $n - 1$ if and only if $f \in \mathfrak{m} \setminus \mathfrak{m}^2$. (Hint: Krull's Principal Ideal Theorem for tangent spaces, Exercise 13.1.B.)

13.2.C. EXERCISE (THE SLICING CRITERION FOR REGULARITY). Suppose X is a Noetherian scheme (such as a variety), D is an effective Cartier divisor on X (Definition 9.5.1), and $p \in D$. Show that if p is a regular point of D then p is a regular point of X. (Hint: Krull's Principal Ideal Theorem for tangent spaces, Exercise 13.1.B.)

13.2.D. EASY EXERCISE (PRACTICE WITH THE CONCEPT). Suppose $(A, \mathfrak{m})$ is a regular local ring of dimension d, and $f_1, \ldots, f_d \in \mathfrak{m}$ are such that their images in the vector space $\mathfrak{m}/\mathfrak{m}^2$ form a basis. Show that they are a system of parameters (Definition 12.3.11) for $(A, \mathfrak{m})$. (They are "particularly good" parameters.)

13.2.2. The Jacobian criterion for regularity.
A finite type k-scheme is locally of the form $\operatorname{Spec} k[x_1, \ldots, x_n]/(f_1, \ldots, f_r)$. The Jacobian criterion for regularity (Exercise 13.2.E) gives a hands-on method for checking for singularity at closed points, using the equations $f_1, \ldots, f_r$, if $k = \overline{k}$.

13.2.E. IMPORTANT EXERCISE (THE JACOBIAN CRITERION FOR REGULARITY—EASY, GIVEN EXERCISE 13.1.I). Suppose $X = \operatorname{Spec} k[x_1, \ldots, x_n]/(f_1, \ldots, f_r)$ has pure dimension d. Show that a k-valued point $p \in X$ is regular if and only if the Jacobian corank function at p is d.

13.2.F. EASY EXERCISE. Suppose $k = \overline{k}$. Show that the singular *closed* points of the hypersurface $f(x_1, \ldots, x_n) = 0$ in $\mathbb{A}^n_k$ are given by the equations

$$f = \frac{\partial f}{\partial x_1} = \cdots = \frac{\partial f}{\partial x_n} = 0.$$

(Translation: The singular points of $f = 0$ are where the gradient of f vanishes. This is not shocking.)

13.2.3. Smoothness over a field.
There seem to be two serious drawbacks with the Jacobian criterion. For finite type schemes over $\overline{k}$, the criterion gives a necessary condition for regularity, but it is not obviously sufficient, as we need

to check regularity at non-closed points as well. We can prove sufficiency by working hard to show Fact 13.8.2, which implies that the non-closed points must be regular as well. A second failing is that the criterion requires k to be algebraically closed. These problems suggest that old-fashioned ideas of using derivatives and Jacobians are ill-suited to the fancy modern notion of regularity. But that is wrong—the fault is with the concept of regularity. There is a better notion of *smoothness over a field*. Better yet, this idea generalizes to the notion of a smooth morphism of schemes (to be discussed in §13.6, and again in Chapter 24.8), which behaves well in all possible ways (including in the sense of §8.1). This is another sign that some properties we think of as of objects ("absolute notions") should really be thought of as properties of morphisms ("relative notions"). We know enough to imperfectly (but correctly) define what it means for a scheme to be k-**smooth**, or **smooth over** k.

13.2.4. *Definition.* A k-scheme is k-**smooth of dimension** d, or **smooth of dimension** d **over** k, if it is of pure dimension d, and there exists a cover by affine open sets $\operatorname{Spec} k[x_1, \ldots, x_n]/(f_1, \ldots, f_r)$ where the Jacobian matrix has corank d at *all* points (not just *closed* points). (In particular, smooth schemes are implicitly locally of finite type.) A k-scheme is **smooth** over k if it is smooth of some dimension. The k is often omitted when it is clear from context. By the theory of the Jacobian corank function developed in §13.1.7, for any pure dimension d variety, this can be checked on *any* affine open cover.

13.2.G. EXERCISE (FIRST EXAMPLES).

(a) Show that $\mathbb{A}_k^n$ is smooth for any n and k. For which characteristics is the curve $y^2 z = x^3 - xz^2$ in $\mathbb{P}_k^2$ smooth (cf. Exercise 13.3.C)?

(b) Suppose $f \in k[x_1, \ldots, x_n]$ is a polynomial such that the system of equations

$$f = \frac{\partial f}{\partial x_1} = \cdots = \frac{\partial f}{\partial x_n} = 0$$

has no solutions in $\overline{k}$. Show that the hypersurface $f = 0$ in $\mathbb{A}_k^n$ is smooth. (Compare this to Exercise 13.2.F, which has the additional hypothesis $k = \overline{k}$.)

13.2.H. EXERCISE (SMOOTHNESS OF VARIETIES IS "INDEPENDENT OF EXTENSION OF BASE FIELD"). Suppose X is a locally finite type k-scheme, and $k \subset \ell$ is a field extension. Show that X is smooth over k if and only if $X \times_{\operatorname{Spec} k} \operatorname{Spec} \ell$ is smooth over ℓ.

The next exercise shows that we need only check closed points, thereby making a connection to classical geometry.

13.2.I. EXERCISE. Show that if the Jacobian matrix for $X = \operatorname{Spec} k[x_1, \ldots, x_n]/(f_1, \ldots f_r)$ has corank d at all *closed* points, then it has corank d at *all* points. (Hint: The locus where the Jacobian matrix has corank d can be described in terms of the vanishing and nonvanishing of determinants of certain explicit matrices.)

13.2.5. *In defense of regularity.* Having made a spirited case for smoothness, we should be clear that regularity is still very useful. For example, it is the only concept that makes sense in mixed characteristic. In particular, $\mathbb{Z}$ is regular at its points (it is a regular ring), and more generally, discrete valuation rings are incredibly useful examples of regular rings.

13.2.6. Regularity vs. smoothness.

13.2.J. EXERCISE. Suppose X is a locally finite type scheme of pure dimension d over an algebraically closed field $k = \overline{k}$. Show that X is regular at its closed points if and only if it is smooth. (We will soon learn that for locally finite type $\overline{k}$-schemes, regularity at closed points is the same as regularity everywhere, Theorem 13.8.3.) Hint to show regularity implies smoothness: Use the Jacobian criterion to show that the corank of the Jacobian is d at the closed points of X. Then use Exercise 13.2.I.

More generally, if k is perfect (e.g., if char $k = 0$ or k is a finite field), then smoothness is the same as regularity at closed points (see Exercise 13.2.R). More generally still, we will later prove the following fact. We mention it now because it will make a number of statements cleaner long before we finally prove it. (There will be no circularity.)

13.2.7. Smoothness-Regularity Comparison Theorem —

(a) *If k is perfect, every regular finite type k-scheme is smooth over k.*
(b) *Every smooth k-scheme is regular (with no hypotheses on perfection).*

Part (a) will be proved in Exercise 13.2.R. Part (b) will be proved in §24.8.9. The fact that Theorem 13.2.7(b) will be proved so far in the future does not mean that we truly need to wait that long. But we deliberately postpone the proof until we have machinery that will do much of the work for us. If you wish for some insight right away, here is the outline of the argument for (b), which you can even try to implement after reading this chapter. We will soon show (in Exercise 13.3.P) that $\mathbb{A}_k^n$ is regular for all fields k. Once we know the definition of étale morphism, we will realize that every smooth variety locally admits an étale morphism to $\mathbb{A}_k^n$ (Observation 13.6.4). You can then show that if $\pi \colon X \to Y$ is an étale morphism of locally Noetherian schemes, and $p \in X$, then p is regular if $\pi(p)$ is regular. This will be shown in Exercise 24.8.E, but you could reasonably do this after reading the definition of étaleness.

13.2.8. *Caution: Regularity does not imply smoothness.* If k is not perfect, then regularity does *not* imply smoothness, as demonstrated by the following example. Let $k = \mathbb{F}_p(u)$, and consider the hypersurface $X = \operatorname{Spec} k[x]/(x^p - u)$. Now $k[x]/(x^p - u)$ is a field, hence regular. But if $f(x) = x^p - u$, then $f(u^{1/p}) = \frac{df}{dx}(u^{1/p}) = 0$, so the Jacobian criterion fails—X is not smooth *over* k. (Never forget that smoothness requires a choice of field—it is a "relative" notion, and we will later define smoothness over an arbitrary scheme, in §13.6.)

13.2.9. In case the previous example is too "small" to be enlightening (because the scheme in question *is* smooth over a *different* field, namely, $k[x]/(x^p - u)$), here is another. Let $k = \mathbb{F}_p(u)$ as before, with $p > 2$, and consider the curve $\operatorname{Spec} k[x, y]/(y^2 - x^p + u)$. Then the closed point $(y, x^p - u)$ is regular but not smooth.

Thus you should not use "regular" and "smooth" interchangeably.

13.2.10. Regular local rings are integral domains.

You might expect from geometric intuition that a scheme is "locally irreducible" at a "smooth" point. Put algebraically:

13.2.11. Theorem — *Suppose $(A, \mathfrak{m}, k)$ is a regular local ring of dimension n. Then A is an integral domain.*

Before proving it (in §13.2.14), we give some consequences.

13.2.K. EXERCISE. Suppose p is a regular point of a locally Noetherian scheme X. Show that only one irreducible component of X passes through p.

13.2.L. EASY EXERCISE. Show that a nonempty regular Noetherian scheme is irreducible if and only if it is connected.

13.2.M. IMPORTANT EXERCISE (REGULAR SCHEMES IN REGULAR SCHEMES ARE REGULAR EMBEDDINGS).

(a) Suppose $(A, \mathfrak{m}, k)$ is a regular local ring of dimension n, and $I \subset A$ is an ideal of A cutting out a regular local ring of dimension d. Let $r = n - d$. Show that $\operatorname{Spec} A/I$ is a regular embedding in $\operatorname{Spec} A$. Hint: Show that there are elements $f_1, \ldots, f_r$ of I giving a basis of the k-vector space $I/(I \cap \mathfrak{m}^2)$. Show that the quotient of A by both $(f_1, \ldots, f_r)$ and I yields dimension d regular local rings. By Exercise 12.1.D, a surjection of integral domains of the same dimension must be an isomorphism.

(b) Suppose $\pi\colon X\to Y$ is a closed embedding of regular schemes. Show that π is a regular embedding.

Exercise 13.2.M has the following striking geometric consequence.

13.2.N. EXERCISE (GENERALIZING EXERCISE 12.3.F).

(a) Suppose W is a smooth variety of pure dimension d over an algebraically closed field k, and X and Y are pure dimensional subvarieties (possibly singular) of W codimension m and n, respectively. Show that every component of $X\cap Y$ has codimension at most $m+n$ in W, as follows. Show that the diagonal $W\cong\Delta\subset W\times_k W$ is a regular embedding of codimension d. Then follow the rest of the hint to Exercise 12.3.F. (Why is there a hypothesis of smoothness rather than of regularity on W?)
(b) Remove the requirement from part (a) that k be algebraically closed.

13.2.12. *Remark.* The following example shows that the smoothness hypotheses in Exercise 13.2.N cannot be (completely) dropped. Let $W=\operatorname{Spec} k[w,x,y,z]/(wz-xy)$ be the cone over the smooth quadric surface, which is an integral threefold. Let X be the surface $w=x=0$ and Y the surface $y=z=0$; both lie in W. Then $X\cap Y$ is just the origin, so we have two codimension 1 subvarieties meeting in a codimension 3 subvariety. (It is no coincidence that X and Y are the affine cones over two lines in the same ruling; see Exercise 9.3.L. This example will arise again in Exercise 22.4.N.)

13.2.13. Proposition — *Smooth k-schemes are reduced.*

This argument is surprisingly serpentine!

Proof. Suppose X is a smooth k-scheme, and let $X_{\overline{k}}:=X\times_{\operatorname{Spec} k}\operatorname{Spec}\overline{k}$ as usual. Then $X_{\overline{k}}$ is smooth (Exercise 13.2.H), so $X_{\overline{k}}$ is regular at its closed points (Exercise 13.2.J), so the local rings of $X_{\overline{k}}$ at closed points are integral domains (Theorem 13.2.11) and thus reduced, so $X_{\overline{k}}$ is reduced at its closed points and hence reduced (Exercise 5.2.E or 6.6.N), so X is reduced (do you see why?). $\square$

13.2.O. EXERCISE. Suppose p is a closed point of a smooth k-scheme X of dimension n. Show that $\mathscr{O}_{X,p}$ is an integral domain (necessarily of dimension n).

13.2.14. *Proof of Theorem 13.2.11 (following [Liu, Prop. 2.11]).* We prove the result by induction on n. The case $n=0$ is clear (dimension 0 regular local rings are fields, Exercise 13.2.A). So assume $n>0$, and that the result holds for smaller dimension.

Suppose $\mathfrak{p}$ is prime, and $\dim A/\mathfrak{p}=n$. We wish to show that $\mathfrak{p}=(0)$. Now $A/\mathfrak{p}$ is a Noetherian local ring of dimension n, and its Zariski cotangent space is a quotient of that of A cut out by the "equations of $\mathfrak{p}$" (Exercise 13.1.B). But $\dim \mathfrak{m}/\mathfrak{m}^2=n$, so $\mathfrak{p}\subset\mathfrak{m}^2$. Hence $A/\mathfrak{p}$ is a regular local ring of dimension n.

Choose any $f\in\mathfrak{m}\setminus\mathfrak{m}^2$ (so $f\notin\mathfrak{p}$). By Exercise 12.3.I (applied to both $A/(f)$ and $A/(\mathfrak{p},f)$), $\dim A/(f)\geq n-1$, and $\dim A/(f,\mathfrak{p})\geq n-1$, but both rings have Zariski tangent space of dimension $n-1$, so both are regular local rings of dimension $n-1$, and by the inductive hypothesis both are integral domains. The only way one integral domain can be the quotient of another integral domain of the same dimension is if the quotient is an isomorphism (Exercise 12.1.D). Thus $(\mathfrak{p},f)=(f)$, from which $\mathfrak{p}\subset(f)$, from which $\mathfrak{p}\subset f\mathfrak{p}$, from which $\mathfrak{p}=f\mathfrak{p}$. Then by Nakayama's Lemma 8.2.9, $\mathfrak{p}=(0)$, as desired. $\square$

13.2.15.** Checking regularity of k-schemes at closed points by base changing to $\overline{k}$.
(We revisit these ideas using a different approach in 21.2.Y, so you should read this only if you are particularly curious.) The Jacobian criterion is a great criterion for checking regularity of finite type k-schemes at k-valued points. The following result extends its applicability to more general closed points.

Suppose X is a finite type k-scheme of pure dimension n, and $p \in X$ is a closed point with residue field k'. By the Nullstellensatz 3.2.6, k'/k is a finite extension of fields. Suppose that it is separable. Define $\pi: X_{\overline{k}} := X \times_k \overline{k} \to X$ by base change from $\operatorname{Spec} \overline{k} \to \operatorname{Spec} k$.

13.2.P. EXERCISE.

(a) Suppose $f(x) \in k[x]$ is a separable polynomial (i.e., f has distinct roots in $\overline{k}$), and irreducible, so $k'' := k[x]/(f(x))$ is a field extension of k. Show that $k'' \otimes_k \overline{k}$ is, as a ring, $\overline{k} \times \cdots \times \overline{k}$, where there are $\deg f = \deg k''/k$ factors.

(b) Show that $\pi^{-1}(p)$ consists of $\deg(k'/k)$ *reduced* points.

13.2.Q. EXERCISE. Suppose p is a closed point of X, with residue field k' that is separable over k of degree d. Show that $X_{\overline{k}}$ is regular at all the preimages $p_1, \ldots, p_d$ of p if and only if X is regular at p as follows.

(a) Reduce to the case $X = \operatorname{Spec} A$.

(b) Let $m \subset A$ be the maximal ideal corresponding to p. By tensoring the exact sequence

$$0 \longrightarrow m \longrightarrow A \longrightarrow k' \longrightarrow 0$$

with $\overline{k}$ (field extensions preserve exactness of sequences of vector spaces), we have the exact sequence

$$0 \longrightarrow m \otimes_k \overline{k} \longrightarrow A \otimes_k \overline{k} \longrightarrow k' \otimes_k \overline{k} \longrightarrow 0.$$

Show that $m \otimes_k \overline{k} \subset A \otimes_k \overline{k}$ is the ideal corresponding to the pullback of p to $\operatorname{Spec} A \otimes_k \overline{k}$. Verify that $(m \otimes_k \overline{k})^2 = m^2 \otimes_k \overline{k}$.

(c) By tensoring the short exact sequence of k-vector spaces

$$0 \longrightarrow m^2 \longrightarrow m \longrightarrow m/m^2 \longrightarrow 0$$

with $\overline{k}$, show that

$$\sum_{i=1}^{d} \dim_{\overline{k}} T_{X_{\overline{k}}, p_i} = d \dim_{k'} T_{X,p}.$$

(d) Use Exercise 12.1.H(a) and the inequalities $\dim_{\overline{k}} T_{X_{\overline{k}}, p_i} \geq \dim X_{\overline{k}}$, and $\dim_{k'} T_{X,p} \geq \dim X$ (Theorem 12.3.12) to conclude.

13.2.16. *Remark.* In fact, regularity at a single p_i is enough to conclude regularity at p. You can show this by following up on Exercise 13.2.Q; first deal with the case when k'/k is Galois, and obtain some transitive group action of $\operatorname{Gal}(k'/k)$ on $\{p_1, \ldots, p_d\}$. Another approach is given in Exercise 21.2.Y.

13.2.R. EXERCISE. Suppose k is perfect, and X is a finite type k-scheme. Show that X is smooth if and only if X is regular at all closed points. In particular, Theorem 13.2.7(a) holds. Hint: Let $X_{\overline{k}} := X \times_k \overline{k}$. Explain why X is smooth if and only if $X_{\overline{k}}$ is smooth if and only if $X_{\overline{k}}$ is regular at all closed points if and only if X is regular at all closed points. See Exercise 13.2.Q for the last step.

13.3 Examples

We now explore regularity and smoothness in practice through a series of examples and exercises.

13.3.1. Geometric examples.

13.3.A. EASY EXERCISE. Suppose k is a field. Show that $\mathbb{A}^1_k$ and $\mathbb{A}^2_{\overline{k}}$ are regular, by directly checking the regularity of all points. Show that $\mathbb{P}^1_k$ and $\mathbb{P}^2_{\overline{k}}$ are regular.

The generalization to arbitrary dimension is harder, so we leave it to Exercise 13.3.P. By contrast, it is basically immediate that $\mathbb{A}_k^n$ and $\mathbb{P}_k^n$ are smooth.

13.3.B. EXERCISE (THE JACOBIAN CRITERION FOR REGULARITY FOR PROJECTIVE HYPERSURFACES).

(a) Show that the non-smooth points of the hypersurface $f = 0$ in $\mathbb{P}_k^n$ correspond to the locus

$$f = \frac{\partial f}{\partial x_0} = \cdots = \frac{\partial f}{\partial x_n} = 0.$$

If the degree of the hypersurface is not divisible by char k (e.g., if char $k = 0$), show that it suffices to check $\frac{\partial f}{\partial x_0} = \cdots = \frac{\partial f}{\partial x_n} = 0$. Hint: Show that $(\deg f)f = \sum_i x_i \frac{\partial f}{\partial x_i}$.

(b) If furthermore $k = \overline{k}$, Show that the singular *closed* points are given by the same locus. (In fact, this gives all the singular points, not just the singular closed points cf. §13.2.3. We won't use this, so we won't prove it.)

13.3.C. EXERCISE. Suppose char $k \neq 2$. Show that $y^2 z = x^3 - xz^2$ in $\mathbb{P}_k^2$ is an irreducible smooth curve. (Eisenstein's Criterion gives one way of showing irreducibility. Warning: We didn't specify char $k \neq 3$, so be careful when using the Jacobian criterion.)

13.3.D. EXERCISE. Suppose char $k \neq 2$. Show that a quadric hypersurface in $\mathbb{P}^n$ is smooth if and only if it is maximal rank (defined in Exercise 5.4.J).

13.3.E. EXERCISE. Suppose $k = \overline{k}$ has characteristic 0. Show that there exists a regular (projective) plane curve of degree d. Hint: Try a "Fermat curve" $x^d + y^d + z^d = 0$. (Feel free to weaken the hypotheses. Bertini's Theorem 13.4.2 will give another means of showing existence.)

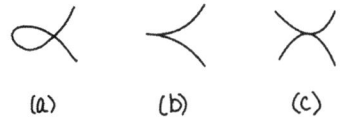

(a)   (b)   (c)

Figure 13.3 *Plane curve singularities.*

13.3.F. EXERCISE (SEE FIGURE 13.3). Assume $k = \overline{k}$ and char $k = 0$ to avoid distractions. Find the singular closed points (which will be the non-smooth points) of the following plane curves.

(a) $y^2 = x^2 + x^3$. This is an example of a *node*, which we saw earlier in Exercise 7.5.H.
(b) $y^2 = x^3$. This is called a *cusp*; we met it earlier in Exercise 10.7.F.
(c) $y^2 = x^4$. This is called a *tacnode*; we met it earlier in Exercise 10.7.G.

13.3.2. *Remark.* There are different possible definitions of node, cusp, and tacnode. One reasonable definition in the case $k = \overline{k}$ is the following. Suppose C is pure dimension 1, and $p \in C$ is a closed point. Let $(A, \mathfrak{m})$ be the local ring $\mathscr{O}_{C,p}$, so $A/\mathfrak{m} = k$. We say p is a **node** of C if $\dim_k \mathfrak{m}/\mathfrak{m}^2 = 2$, the kernel of

$$\operatorname{Sym}^2(\mathfrak{m}/\mathfrak{m}^2) \longrightarrow \mathfrak{m}^2/\mathfrak{m}^3$$

has dimension 1, and the kernel is not spanned by the square of an element in $\mathfrak{m}/\mathfrak{m}^2$. We won't need this definition, so you needn't memorize it. But can you figure out why it is reasonable? Can you make a reasonable guess for a definition of a cusp? Alternate definitions of node, cusp, and more will be given in §28.2.

13.3.G. EXERCISE. Show that the twisted cubic $\operatorname{Proj} k[w, x, y, z]/(wz - xy, wy - x^2, xz - y^2)$ is smooth. (You could do this using the fact that it is isomorphic to $\mathbb{P}^1$. But do this with the explicit equations, for the sake of practice. The twisted cubic was defined in Exercise 9.3.A.)

13.3.3. *Tangent planes and tangent lines.*
Suppose a scheme $X \subset \mathbb{A}^n$ is cut out by equations $f_1, \ldots, f_r$, and X is smooth of (pure) dimension d at the k-valued point $a = (a_1, \ldots, a_n)$. Then the **tangent d-plane to** X **at** a (sometimes denoted by

T_aX) is given by the r equations

$$\left(\frac{\partial f_i}{\partial x_1}(a)\right)(x_1 - a_1) + \cdots + \left(\frac{\partial f_i}{\partial x_n}(a)\right)(x_n - a_n) = 0,$$

where (as in (13.1.7.1)) $\frac{\partial f_i}{\partial x_1}(a)$ is the evaluation of $\frac{\partial f_i}{\partial x_1}$ at $a = (a_1, \ldots, a_n)$.

13.3.H. EXERCISE. Why is this independent of the *choice* of defining equations f_1, ..., f_r of X?

The Jacobian criterion for regularity (Exercise 13.2.E) ensures that these r equations indeed cut out a d-plane. (If $d = 1$, this is called the **tangent line**.) This can be readily shown to be notion of tangent plane that we see in multivariable calculus, but note that here this is the *definition*, and thus we don't have to worry about δ's and ϵ's. Instead we will have to just be careful that it behaves the way we want it to.

13.3.I. EXERCISE. Compute the tangent line to the curve of Exercise 13.3.F(b) at $(1,1)$.

13.3.J. EXERCISE. Compute the (projective) tangent line to the twisted cubic curve of Exercise 13.3.G at $[1,1,1,1]$.

13.3.K. EXERCISE. Suppose $X \subset \mathbb{P}^n_k$ (k, as usual, a field) is cut out by homogeneous equations $f_1, \ldots, f_r$, and $p \in X$ is a k-valued point that is smooth of (pure) dimension d. Define the (projective) tangent d-plane to X at p. (Definition 9.3.3 gives the definition of a d-plane in $\mathbb{P}^n_k$, but you shouldn't need to refer to it.)

13.3.4. *Side remark to help you think cleanly.* We would want the definition of tangent k-plane to be natural in the sense that for any automorphism σ of $\mathbb{A}^n_k$ (or, in the case of the previous Exercise, $\mathbb{P}^n_k$), $\sigma(T_pX) = T_{\sigma(p)}\sigma(X)$. You could verify this by hand, but you can also say this in a cleaner way, by interpreting the equations cutting out the tangent space in a coordinate free manner. Informally speaking, we are using the canonical identification of n-space with the tangent space to n-space at p, and using the fact that the Jacobian "linear transformation" cuts out T_pX in $T_p\mathbb{A}^n$ in a way independent of the choice of coordinates on $\mathbb{A}^n$ or our defining equations of X. Your solution to Exercise 13.3.H will help you start to think in this way.

13.3.L. EXERCISE. Suppose $X \subset \mathbb{P}^n_k$ is a degree d hypersurface cut out by $f = 0$, and L is a line not contained in X. Exercise 9.3.D (a case of Bézout's Theorem) showed that X and L meet at d points, counted "with multiplicity." Suppose L meets X "with multiplicity at least 2" at a k-valued point $p \in L \cap X$, and that p is a regular point of X. Show that L is contained in the tangent plane to X at p. (Do you have a picture of this in your mind?)

13.3.5. Arithmetic examples.

13.3.M. EASY EXERCISE. Show that $\operatorname{Spec} \mathbb{Z}$ is a regular curve.

13.3.N. EXERCISE. (This tricky exercise is for those who know about the primes of the Gaussian integers $\mathbb{Z}[i]$.) There are several ways of showing that $\mathbb{Z}[i]$ is dimension 1. (For example: (i) It is a principal ideal domain; (ii) it is the normalization of $\mathbb{Z}$ in the field extension $\mathbb{Q}(i)/\mathbb{Q}$; (iii) using Krull's Principal Ideal Theorem 12.3.3 and the fact that $\dim \mathbb{Z}[x] = 2$ by Exercise 12.1.I.) Show that $\operatorname{Spec} \mathbb{Z}[i]$ is a regular curve. (There are several ways to proceed. You could use Exercise 13.1.B. As an example to work through first, consider the prime $(1 + i)$, which is cut out by the equation $1 + x$ in $\operatorname{Spec} \mathbb{Z}[x]/(x^2 + 1)$.) We will later (§13.5.9) have a simpler approach once we discuss discrete valuation rings.

13.3.O. EXERCISE. Show that $[(5, 5i)]$ is the unique singular point of $\operatorname{Spec} \mathbb{Z}[5i]$. (Hint: $\mathbb{Z}[i]_5 \cong \mathbb{Z}[5i]_5$. Use the previous exercise.)

13.3.6. Back to geometry: $\mathbb{A}_k^n$ is regular.

The key step to showing that $\mathbb{A}_k^n$ is regular (where k is a field) is the following.

13.3.7. Proposition — *Suppose* $(B, \mathfrak{n}, k)$ *is a regular local ring of dimension* d. *Let* $\phi: B \to B[x]$. *Suppose* $\mathfrak{p}$ *is a prime ideal of* $A := B[x]$ *such that* $\mathfrak{n}B[x] \subset \mathfrak{p}$. *Then* $A_\mathfrak{p}$ *is a regular local ring.*

Proof. Geometrically: we have a morphism $\pi: X = \operatorname{Spec} B[x] \to Y = \operatorname{Spec} B$, and $\pi([\mathfrak{p}]) = [\mathfrak{n}]$. The fiber $\pi^{-1}([\mathfrak{n}])$ is $\operatorname{Spec}(B[x]/\mathfrak{n}B[x]) = \operatorname{Spec} k[x]$. Thus either (i) $\overline{[\mathfrak{p}]}$ is the fiber $\mathbb{A}_k^1$ above $[\mathfrak{n}]$, or (ii) $\overline{[\mathfrak{p}]}$ is a closed point of the fiber $\mathbb{A}_k^1$. (See Figure 13.4.)

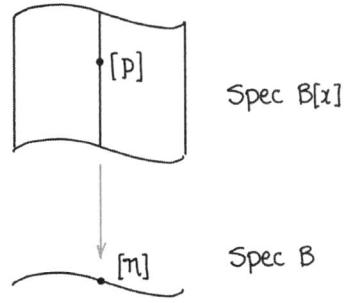

Figure 13.4 *The geometry behind the proof of Proposition 13.3.7.*

Before considering these two cases, we make two remarks. As $(B, \mathfrak{n})$ is a regular local ring of dimension d, $\mathfrak{n}$ is generated by d elements of B, say, $f_1, \ldots, f_d$ (Exercise 13.2.D), and there is a chain of prime ideals

(13.3.7.1) $$0 = \mathfrak{q}_0 \subset \mathfrak{q}_1 \subset \cdots \subset \mathfrak{q}_d = \mathfrak{n}.$$

Case (i): $\overline{[\mathfrak{p}]}$ *is the fiber* $\mathbb{A}_k^1$. In this case, $\mathfrak{p} = \mathfrak{n}B[x]$. Thus by Krull's Height Theorem 12.3.10, the codimension of $\mathfrak{p} = \mathfrak{n}B[x]$ is at most d, because $\mathfrak{p}$ is generated by the d elements $f_1, \ldots, f_d$. But $\mathfrak{p}$ has codimension at least d: given the chain (13.3.7.1) ending with $\mathfrak{n}$, we have a corresponding chain of prime ideals of B[x] ending with $\mathfrak{n}B[x]$. Hence the codimension of $\mathfrak{p}$ is precisely d, and $\mathfrak{p}A_\mathfrak{p}$ is generated by $f_1, \ldots, f_d$, implying that it is a regular local ring.

Case (ii): $\overline{[\mathfrak{p}]}$ *is a closed point of the fiber* $\mathbb{A}_k^1$. The closed point $\mathfrak{p}$ of k[x] corresponds to some monic irreducible polynomial $g(x) \in k[x]$. Arbitrarily lift the coefficients of g to B; we sloppily denote the resulting polynomial in B[x] by $g(x)$ as well. Then $\mathfrak{p} = (f_1, \ldots, f_d, g)$, so by Krull's Height Theorem 12.3.10, the codimension of $\mathfrak{p}$ is at most $d + 1$. But $\mathfrak{p}$ has codimension at least $d + 1$: given the chain (13.3.7.1) ending with $\mathfrak{n}$, we have a corresponding chain in B[x] ending with $\mathfrak{n}B[x]$, and we can extend it by appending $\mathfrak{p}$. Hence the codimension of $\mathfrak{p}$ is precisely $d + 1$, and $\mathfrak{p}A_\mathfrak{p}$ is generated by $f_1, \ldots, f_d, g$, implying that it is a regular local ring. $\square$

13.3.P. EXERCISE. Use Proposition 13.3.7 to show that if X is a regular (locally Noetherian) scheme, then so is $X \times \mathbb{A}^1$. In particular, show that $\mathbb{A}_k^n$ is regular. (Hint: Think about fibers.)

13.4   Bertini's Theorem

We now discuss Bertini's Theorem, a fundamental classical result.

13.4.1. *Definition: Dual projective space.* The **dual** (or **dual projective space**) to $\mathbb{P}_k^n$ (with projective coordinates $x_0, \ldots, x_n$) is informally the space of hyperplanes in $\mathbb{P}_k^n$. Somewhat more precisely, it is a projective space $\mathbb{P}_k^n$ with projective coordinates $a_0, \ldots, a_n$ (which we denote by $\mathbb{P}_k^{n\vee}$ in the futile hope of preventing confusion), along with the data of the "incidence variety" or "incidence correspondence" $I \subset \mathbb{P}^n \times \mathbb{P}^{n\vee}$ cut out by the equation $a_0 x_0 + \cdots + a_n x_n = 0$. (This is often called the **universal hyperplane**.) Note that the k-valued points of $\mathbb{P}^{n\vee}$ indeed correspond to hyperplanes in $\mathbb{P}^n$ defined over k, and this is also clearly a duality relation (there is a symmetry in the definition between the x-variables and the a-variables). This definition is concrete enough to use in practice, and extends over an arbitrary base (notably $\operatorname{Spec} \mathbb{Z}$). (But if you have a delicate and refined sensibility, you may want to come up with a coordinate-free definition.)

13.4.2. Bertini's Theorem — *Suppose* X *is a smooth closed subvariety of* $\mathbb{P}_k^n$ *of (pure) dimension* d. *Then there is a nonempty (= dense) open subset* U *of dual projective space* $\mathbb{P}_k^{n\vee}$ *such that for every point* $\mathfrak{p} = [H] \in U$, *the scheme* $H \cap X$ *is smooth over* $\kappa(\mathfrak{p})$ *of (pure) dimension* $d - 1$.

(You should perhaps think first of the case where $\kappa(p) = k$, i.e., H is a hyperplane cut out by a linear form with coefficients in k.) Any theorem of this flavor is often called a "Bertini Theorem." One example is the Kleiman–Bertini Theorem 21.6.10, which was *not* proved jointly by Kleiman and Bertini.

As an application of Bertini's Theorem 13.4.2, a general degree $d > 0$ hypersurface in $\mathbb{P}_k^n$ intersects X in a smooth subvariety of codimension 1 in X: replace $X \hookrightarrow \mathbb{P}^n$ with the composition

$$X \longrightarrow \mathbb{P}^n \overset{\nu_d}{\hookrightarrow} \mathbb{P}^N,$$

where ν_d is the dth Veronese embedding (9.3.8). Here "general" has its usual meaning in algebraic geometry; see §10.3.7. (See Exercise 13.4.B for a generalization.) A useful consequence of this, taking $X = \mathbb{P}^n$: we immediately see that there exists a smooth degree d hypersurface in $\mathbb{P}^n$ (over any infinite field k), without any messing around with specific equations (as in the special case Exercise 13.3.E).

Exercise 21.6.I gives a useful improvement of Bertini's Theorem in characteristic 0 (see Exercise 21.6.J).

Proof. The central idea of the proof is quite naive. We will describe the hyperplanes that are "bad," and show that they form a closed subset of dimension at most $n - 1$ of $\mathbb{P}_k^{n\vee}$, and hence that the complement is a dense open subset. Somewhat more precisely, we will define a projective variety $Z \subset X \times \mathbb{P}_k^{n\vee}$ that can informally be described as:

(13.4.2.1) $Z = \{(p \in X, H \subset \mathbb{P}_k^n) \ : \ p \in H,$

and either p is a singular point of $H \cap X$, or $X \subset H\}$.

We will see that the projection $\pi \colon Z \to X$ has fibers at closed points that are projective spaces of dimension $n - 1 - \dim X$, and hence that $\dim Z \leq n - 1$. Thus the image of Z in $\mathbb{P}_k^{n\vee}$ will be a closed subset (Theorem 8.4.10), of dimension of at most $n - 1$, so its complement will be open and nonempty. We now put this strategy into action.

13.4.3. We first define Z more precisely, in terms of equations on $\mathbb{P}^n \times \mathbb{P}^{n\vee}$, where the coordinates on $\mathbb{P}^n$ are $x_0, \ldots x_n$, and the dual coordinates on $\mathbb{P}^{n\vee}$ are $a_0, \ldots, a_n$. Suppose X is cut out by $f_1, \ldots, f_r$. Then we take these equations as the first of the defining equations of Z. (So far we have defined the subscheme $X \times \mathbb{P}^{n\vee}$.) We also add the equation $a_0 x_0 + \cdots + a_n x_n = 0$. (So far we have described the subscheme of $\mathbb{P}^n \times \mathbb{P}^{n\vee}$ corresponding to points (p, H), where $p \in X$ and $p \in H$.) Note that the projective Jacobian matrix (13.1.9.1)

$$\begin{pmatrix} \frac{\partial f_1}{\partial x_0}(p) & \cdots & \frac{\partial f_r}{\partial x_0}(p) \\ \vdots & \ddots & \vdots \\ \frac{\partial f_1}{\partial x_n}(p) & \cdots & \frac{\partial f_r}{\partial x_n}(p) \end{pmatrix}$$

has corank precisely $d + 1$ at all points of X by the projective Jacobian corank criterion (Exercise 13.1.N). We then require that the projective Jacobian matrix with a new column

$$\begin{pmatrix} a_0 \\ \vdots \\ a_n \end{pmatrix}$$

appended has corank $\geq d + 1$ (hence $= d + 1$). This is precisely the notion that we wished to capture—do you see why? This is cut out by equations (the determinants of certain minors). (Remark 13.4.4 works through an example, which may help clarify how this works.)

We next show that $\dim Z \leq n - 1$. For each closed point $p \in X$, let W_p be the locus of hyperplanes containing p, such that $H \cap X$ is singular at p, or else contains all of X; what is the dimension of W_p? Suppose $\dim X = d$. Then the restrictions on the hyperplanes in the definition of W_p correspond to $d + 1$ linear conditions. (Do you see why?) This means that W_p is a codimension $d + 1$, or dimension $n - d - 1$, projective space. Thus the fiber of $\pi\colon Z \to X$ over each closed point has pure dimension $n - d - 1$. By Key Exercise 12.4.A, this implies that $\dim Z \leq n - 1$. (If you wish, you can use Exercise 12.4.D to show that $\dim Z = n - 1$, and you can later show that Z is a projective bundle over X, once you know what a projective bundle is, Definition 17.2.4. But we don't need this for the proof.) $\qquad\square$

13.4.4. Remark. Here is an example that may help convince you that the algebra of paragraph 13.4.3 is describing the geometry we desire. Consider the plane conic $x_0^2 - x_1^2 - x_2^2 = 0$ over a field of characteristic not 2, which you might picture as the circle $x^2 + y^2 = 1$ from the real picture in the chart U_0. Consider the point $[1, 1, 0]$, corresponding to $(1, 0)$ on the circle. We expect the tangent line in the affine plane to be $x = 1$, which corresponds to $x_0 - x_1 = 0$. Let's see what the algebra gives us. The projective Jacobian matrix is

$$\begin{pmatrix} 2x_0 \\ -2x_1 \\ -2x_2 \end{pmatrix} = \begin{pmatrix} 2 \\ -2 \\ 0 \end{pmatrix},$$

which indeed has rank 1 as expected. Our recipe asks that the matrix

$$\begin{pmatrix} 2 & a_0 \\ -2 & a_1 \\ 0 & a_2 \end{pmatrix}$$

have rank 1 (i.e., $a_0 = -a_1$ and $a_2 = 0$), and also that $a_0 x_0 + a_1 x_1 + a_2 x_2 = 0$, which (you should check) is precisely what we wanted.

13.4.A. EASY EXERCISE.

(a) Prove Bertini's Theorem with the weaker hypothesis that X has finitely many non-smooth points.
(b) Prove Bertini's Theorem with the weaker hypothesis that $X \to \mathbb{P}_k^n$ is a locally closed embedding (but with X still smooth).

13.4.B. EXERCISE (MANY-HYPERSURFACE BERTINI). Suppose X is a smooth subvariety of $\mathbb{P}_k^n$ of (pure) dimension $d \geq r$. Suppose $e_1, \ldots, e_r$ are positive integers, so

(13.4.4.1) $$\mathbb{P}^{\binom{e_1+n}{n}-1} \times \cdots \times \mathbb{P}^{\binom{e_r+n}{n}-1}$$

parametrizes r-tuples of hypersurfaces $(H_1, \ldots, H_r)$ of degrees $e_1, \ldots, e_r$, respectively. Show that there is a dense open subset U of (13.4.4.1) such that for every point $p = [(H_1, \ldots, H_r)] \in U$, $H_1 \cap \cdots \cap H_r \cap X$ is smooth over $\kappa(p)$ of (pure) dimension $d - r$. Informally: for a general choice of hypersurfaces $H_1, \ldots, H_r$ of degrees $e_1, \ldots, e_r$ respectively, $H_1 \cap \cdots \cap H_r \cap X$ is smooth of (pure) dimension $d - r$. (Hint: Don't apply Bertini's Theorem 13.4.2. Instead, try to give the "same" proof.) The lesson of this exercise isn't that you should memorize this specific result. It is that you can prove any result of this flavor that you need.

13.4.C. EXERCISE. Suppose that X is a projective variety of dimension d in $\mathbb{P}^n$, with a dense open subset that is smooth. (This turns out to be automatic if the base field is perfect, by Theorem 21.6.1 on generic smoothness of varieties.) Show that the intersection of X with d general hyperplanes consists of a finite number of reduced points. More precisely: if $\mathbb{P}^{n\vee}$ is the dual projective space, then there is a nonempty Zariski-open subset $U \subset (\mathbb{P}^{n\vee})^d$ such that for each closed point $(H_1, \ldots, H_d)$ of U, the scheme-theoretic intersection $H_1 \cap \cdots \cap H_d \cap X$ consists of a finite number of reduced points. (The number of such points, counted correctly, is called the *degree* of the variety; see Exercise 18.6.M.)

13.4.5. Dual varieties. Our discussion of Bertini gives us an excuse to mention a classical construction. The image of Z (see (13.4.2.1)) in $\mathbb{P}^{n\vee}$ is called the **dual variety** of X. As $\dim Z = n - 1$, we "expect" the dual of X to be a hypersurface of $\mathbb{P}^{n\vee}$. It is a nonobvious fact that this in fact is a duality: the dual of the dual of X is X itself (see [H2, Thm. 15.24]). The following exercise will give you some sense of the dual variety.

13.4.D. EXERCISE. Show that the dual of a hyperplane in $\mathbb{P}^n$ is the corresponding point of the dual space $\mathbb{P}^{n\vee}$. In this way, the duality between $\mathbb{P}^n$ and $\mathbb{P}^{n\vee}$ is a special case of duality between projective varieties.

13.4.E. EXERCISE. Suppose $C \subset \mathbb{P}^2$ is a smooth conic over an algebraically closed field of characteristic not 2. Show that the dual variety to C is also a smooth conic. Thus, for example, through a general point in the plane (if $k = \bar{k}$), there are two tangents to C. (The points on a line in the dual plane correspond to those lines through a point of the original plane.)

13.4.F.* EXERCISE (THERE IS ONE SMOOTH CONIC TANGENT TO FIVE GENERAL LINES, AND GENERALIZATIONS). Continuing the notation of the previous problem, show that the number of smooth conics C containing i generally chosen points and tangent to $5-i$ generally chosen lines is $1, 2, 4, 4, 2, 1$, respectively, for $i = 0, 1, 2, 3, 4, 5$. You might interpret the symmetry of the sequence in terms of the duality between the conic and the dual conic. This fact was likely known in the paleolithic era.

13.5 Discrete Valuation Rings, and Algebraic Hartogs's Lemma

The case of (co)dimension 1 is important, because if you understand how prime ideals behave that are separated by dimension 1, then you can use induction to prove facts in arbitrary dimension. This is one reason why Krull's Principal Ideal Theorem 12.3.3 is so useful.

A dimension 1 regular local ring can be thought of as a "germ of a smooth curve" (see Figure 13.5). Two examples to keep in mind are $k[x]_{(x)} = \{f(x)/g(x) : x \not| g(x)\}$ and $\mathbb{Z}_{(5)} = \{a/b : 5 \not| b\}$. The first example is "geometric" and the second is "arithmetic," but hopefully it is clear that they have something fundamental in common.

Figure 13.5 A germ of a curve.

The purpose of this section is to give a long series of equivalent definitions of these rings. Before beginning, we quickly sketch these seven definitions. There are a number of ways a Noetherian local ring can be "nice." It can be regular, or a principal ideal domain, or a unique factorization domain, or normal. In dimension 1, these are the same. Also equivalent are nice properties of ideals: if m is principal; or if *all* ideals are either powers of the maximal ideal, or 0. Finally, the ring can have a *discrete valuation*, a measure of "size" of elements that behaves particularly well.

13.5.1. Theorem — *Suppose* $(A, \mathfrak{m})$ *is a Noetherian local ring of dimension* 1. *Then the following are equivalent.*

(a) $(A, \mathfrak{m})$ *is regular.*
(b) $\mathfrak{m}$ *is principal.*

Proof. Here is why (a) implies (b). If A is regular, then $\mathfrak{m}/\mathfrak{m}^2$ is one-dimensional. Choose any element $t \in \mathfrak{m} \setminus \mathfrak{m}^2$. Then t generates $\mathfrak{m}/\mathfrak{m}^2$, so generates m by Nakayama's Lemma 8.2.H (m is finitely generated by the Noetherian hypothesis). We call such an element a **uniformizer**.

Conversely, if m is generated by one element t over A, then $\mathfrak{m}/\mathfrak{m}^2$ is generated by one element t over $A/\mathfrak{m} = k$. Since $\dim_k \mathfrak{m}/\mathfrak{m}^2 \geq 1$ by Theorem 12.3.12, we have $\dim_k \mathfrak{m}/\mathfrak{m}^2 = 1$, so $(A, \mathfrak{m})$ is regular. □

We will soon use a useful fact, which is geometrically motivated, and is a special case of an important result, the Artin-Rees Lemma 13.9.3. We will prove it in §13.9.

13.5.2. Proposition — *If $(A, \mathfrak{m})$ is a Noetherian local ring, then $\cap_i \mathfrak{m}^i = 0$.*

13.5.3. The geometric intuition for this is that any function that is analytically zero at a point (vanishes to all orders) actually vanishes in an open neighborhood of that point. (Exercise 13.9.B will make this precise.) The geometric intuition also suggests an example showing that Noetherianness is necessary: consider e^{-1/x^2} in the ring of germs of C^∞-functions on $\mathbb{R}$ at the origin. (In particular, this implies that the ring of C^∞-functions on $\mathbb{R}$, localized at the origin, is not Noetherian!)

It is tempting to argue that

$$(13.5.3.1) \qquad\qquad \mathfrak{m}(\cap_i \mathfrak{m}^i) = \cap_i \mathfrak{m}^i,$$

and then to use Nakayama's Lemma 8.2.9 to argue that $\cap_i \mathfrak{m}^i = 0$. Unfortunately, it is not obvious that this first equality is true: product does not commute with infinite descending intersections in general. (Aside: Product also doesn't commute with finite intersections in general, as, for example, in $k[x, y, z]/(xz - yz)$, $(z)((x) \cap (y)) \neq (xz) \cap (yz)$.) We will establish Proposition 13.5.2 in Exercise 13.9.A(b). (We could do it directly right now without too much effort.)

13.5.4. Theorem — *Suppose $(A, \mathfrak{m})$ is a Noetherian local ring of dimension 1. Then (a) and (b) are equivalent to:*

(c) A is an integral domain, and all ideals are of the form $\mathfrak{m}^n$ (for $n \geq 0$) or (0).

Proof. Assume (a): suppose $(A, \mathfrak{m}, k)$ is a regular local ring of dimension 1. Then I claim that $\mathfrak{m}^n \neq \mathfrak{m}^{n+1}$ for any n. Otherwise, by Nakayama's Lemma, $\mathfrak{m}^n = 0$, from which $t^n = 0$. But A is an integral domain (by Theorem 13.2.11), so $t = 0$, from which $A = A/\mathfrak{m}$ is a field, which doesn't have dimension 1, contradiction.

I next claim that $\mathfrak{m}^n/\mathfrak{m}^{n+1}$ is dimension 1. Reason: $\mathfrak{m}^n = (t^n)$. So $\mathfrak{m}^n$ is generated as an A-module by one element, and $\mathfrak{m}^n/(\mathfrak{m}\mathfrak{m}^n)$ is generated as an $(A/\mathfrak{m} = k)$-module by one element (nonzero by the previous paragraph), so it is a one-dimensional vector space.

So we have a chain of ideals $A \supset \mathfrak{m} \supset \mathfrak{m}^2 \supset \mathfrak{m}^3 \supset \cdots$ with $\cap \mathfrak{m}^i = (0)$ (Proposition 13.5.2). We want to say that there is no room for any ideal besides these, because each pair is "separated by dimension 1," and there is "no room at the end." Proof: Suppose $I \subset A$ is an ideal. If $I \neq (0)$, then there is some n such that $I \subset \mathfrak{m}^n$ but $I \not\subset \mathfrak{m}^{n+1}$. Choose some $u \in I \setminus \mathfrak{m}^{n+1}$. Then $(u) \subset I$. But u generates $\mathfrak{m}^n/\mathfrak{m}^{n+1}$, hence by Nakayama's Lemma it generates $\mathfrak{m}^n$, so we have $\mathfrak{m}^n \subset I \subset \mathfrak{m}^n$, so we are done: (c) holds.

We now show that (c) implies (a). Assume (a) does not hold: suppose we have a dimension 1 Noetherian local integral domain that is not regular, so $\mathfrak{m}/\mathfrak{m}^2$ has dimension at least 2. Choose any $u \in \mathfrak{m} - \mathfrak{m}^2$. Then $(u, \mathfrak{m}^2)$ is an ideal, but $\mathfrak{m}^2 \subsetneq (u, \mathfrak{m}^2) \subsetneq \mathfrak{m}$. $\square$

13.5.A. EASY EXERCISE. Suppose $(A, \mathfrak{m})$ is a Noetherian dimension 1 local ring. Show that (a)–(c) above are equivalent to:

(d) A is a principal ideal domain.

13.5.5. Discrete valuation rings. We next define the notion of a discrete valuation ring. Suppose K is a field. A **discrete valuation** on K is a surjective homomorphism $v: K^\times \to \mathbb{Z}$ (in particular, $v(xy) = v(x) + v(y)$) satisfying

$$v(x + y) \geq \min(v(x), v(y))$$

except if $x + y = 0$ (in which case the left side is undefined). (Such a valuation is called *non-Archimedean*, although we will not use that term.) It is often convenient to say $v(0) = \infty$. More generally, a **valuation** is a surjective homomorphism $v: K^\times \to G$ to a totally ordered abelian group G, although this isn't so important to us. The symbol "val" will denote valuations.

Here are three key examples.

(i) *(the 5-adic valuation)* $K = \mathbb{Q}$, $v(r)$ is the "power of 5 appearing in r," e.g., $v(35/2) = 1$, $v(27/125) = -3$.
(ii) $K = k(x)$, $v(f)$ is the "power of x appearing in f."
(iii) $K = k(x)$, $v(f)$ is the negative of the degree. This is really the same as (ii), with x replaced by $1/x$.

Then $0 \cup \{x \in K^\times : v(x) \geq 0\}$ is a ring, which we denote by $\mathscr{O}_v$. It is called the **valuation ring** of v. (Not every valuation is discrete. Consider the ring of *Puiseux series* over a field k, $K = \cup_{n \geq 1} k((x^{1/n}))$, with $v: K^\times \to \mathbb{Q}$ given by $v(x^q) = q$.)

13.5.B. EXERCISE. Describe the valuation rings in the three examples (i)–(iii) above. (You will notice that they are familiar-looking dimension 1 Noetherian local rings. What a coincidence!)

13.5.C. EXERCISE. Show that $\{0\} \cup \{x \in K^\times : v(x) > 0\}$ is the unique maximal ideal of the valuation ring. (Hint: Show that everything in the complement is invertible.) Thus the valuation ring is a local ring.

An integral domain A is called a **discrete valuation ring** (or **DVR**) if there exists a discrete valuation v on its fraction field $K = K(A)$ for which $\mathscr{O}_v = A$. Similarly, A is a **valuation ring** if there exists a valuation v on K for which $\mathscr{O}_v = A$.

Now if A is a regular local ring of dimension 1, and t is a uniformizer (a generator of $\mathfrak{m}$ as an ideal, or, equivalently, of $\mathfrak{m}/\mathfrak{m}^2$ as a k-vector space), then any nonzero element r of A lies in some $\mathfrak{m}^n \setminus \mathfrak{m}^{n+1}$, so $r = t^n u$, where u is invertible (as t^n generates $\mathfrak{m}^n$ by Nakayama, and so does r), so $K(A) = A_t = A[1/t]$. So *any element of* $K(A)^\times$ *can be written uniquely as* ut^n, where u is invertible and $n \in \mathbb{Z}$. Thus we can define a valuation v by $v(ut^n) = n$.

13.5.D. EXERCISE. Show that v is a discrete valuation.

13.5.E. EXERCISE. Conversely, suppose $(A, \mathfrak{m})$ is a discrete valuation ring. Show that $(A, \mathfrak{m})$ is a regular local ring of dimension 1. (Hint: Show that the ideals are all of the form (0) or $I_n = \{r \in A : v(r) \geq n\}$, and (0) and I_1 are the only prime ideals. Thus we have Noetherianness, and dimension 1. Show that I_1/I_2 is generated by the image of any element of $I_1 - I_2$.)

Hence we have proved:

13.5.6. Theorem — *An integral domain A is a Noetherian local ring of dimension 1 satisfying (a)–(d) if and only if*

(e) A is a discrete valuation ring.

13.5.F. EXERCISE. Show that there is only one discrete valuation on a discrete valuation ring.

13.5.7. *Definition.* Thus any regular local ring of dimension 1 comes with a unique valuation on its fraction field. If the valuation of an element is $n > 0$, we say that the element has a **zero of order** n. If the valuation is $-n < 0$, we say that the element has a **pole of order** n. We will come back to this shortly, after dealing with (f) and (g).

13.5.8. Theorem — *Suppose $(A, \mathfrak{m})$ is a Noetherian local ring of dimension 1. Then (a)–(e) are equivalent to:*

(f) A is a unique factorization domain, and
(g) A is an integral domain, integrally closed (in its fraction field).

Proof. (a)–(e) clearly imply (f), because we have the following stupid unique factorization: each nonzero element of A can be written uniquely as ut^n, where $n \in \mathbb{Z}^{\geq 0}$ and u is invertible.

Now (f) implies (g), because unique factorization domains are integrally closed in their fraction fields (Exercise 5.4.F).

It remains to check that (g) implies (a)–(e). We will show that (g) implies (b).

Suppose $(A, \mathfrak{m})$ is a Noetherian local integral domain of dimension 1, integrally closed in its fraction field $K(A)$. Choose any nonzero $r \in \mathfrak{m}$. Then $S = A/(r)$ is a Noetherian local ring of dimension 0—its only prime is the image of $\mathfrak{m}$, which we denote by $\mathfrak{n}$ to avoid confusion. Then $\mathfrak{n}^N = 0$, where N is sufficiently large (Proposition 12.1.9). Hence there is some $\mathfrak{n}$ such that $\mathfrak{n}^n = 0$ but $\mathfrak{n}^{n-1} \neq 0$.

Now comes the crux of the argument. Thus in A, $\mathfrak{m}^n \subset (r)$ but $\mathfrak{m}^{n-1} \not\subset (r)$. Choose $s \in \mathfrak{m}^{n-1} - (r)$. Consider $s/r \in K(A)$. As $s \notin (r)$, $s/r \notin A$, so as A is integrally closed, s/r is not integral over A.

Now (inside $K(A)$) $\frac{s}{r}\mathfrak{m} \not\subset \mathfrak{m}$ (or else $\frac{s}{r}\mathfrak{m} \subset \mathfrak{m}$ would imply that $\mathfrak{m}$ is a faithful $A[\frac{s}{r}]$-module, contradicting Exercise 8.2.I). But $s\mathfrak{m} \subset \mathfrak{m}^n \subset r A$, so $\frac{s}{r}\mathfrak{m} \subset A$. Thus $\frac{s}{r}\mathfrak{m} = A$, from which $\mathfrak{m} = \frac{r}{s}A$, so $\mathfrak{m}$ is principal. $\qquad\square$

13.5.9. *Example: Spec $\mathbb{Z}[i]$ is regular.* We now have a simpler explanation of why Spec $\mathbb{Z}[i]$ is regular (Exercise 13.3.N). First, $\mathbb{Z}[i]$ is a unique factorization domain (§5.4.5(0)), hence normal (§5.4.6). It is dimension 1, so its closed (codimension 1) points are regular by Theorem 13.5.8. Its generic point is also regular, as $\mathbb{Z}[i]$ is an integral domain.

13.5.10. *Example: "Curves are regular almost everywhere."* By Theorem 13.5.8 and Exercise 10.7.O, every one-dimensional variety is regular away from finitely many closed points.

13.5.11. *Aside: Dedekind domains.* A **Dedekind domain** is a Noetherian integral domain of dimension at most 1 that is normal (integrally closed in its fraction field). The localization of a Dedekind domain at any prime but (0) (i.e., a codimension 1 prime) is hence a discrete valuation ring. This is an important notion, but we won't use it much. Rings of integers of number fields are examples; see §10.7.1. In particular, if n is a square-free integer congruent to 3 (mod 4), then $\mathbb{Z}[\sqrt{n}]$ is a Dedekind domain, by Exercise 5.4.I(a). If you wish you can prove *unique factorization of ideals in a Dedekind domain*: any nonzero ideal in a Dedekind domain can be uniquely factored into prime ideals. (Hint: The Jordan–Hölder package, §6.5.) By the same argument as in §13.5.9, Dedekind domains are regular.

13.5.12. Poles and zeroes of functions on Noetherian schemes. We now generalize the sense in which $(x-2)^2/(x-3)^4$ on $\mathbb{A}^1_{\mathbb{C}}$ has a double zero at $x = 2$ and a quadruple pole at $x = 3$.

13.5.13. *Definition: Regular in codimension 1 (R_1).* We say that a Noetherian ring A is **regular in codimension** 1 (or R_1 for short) if its localization at all codimension 1 primes are regular local rings. We say a locally Noetherian scheme A is **regular in codimension** 1 (or R_1) if it is regular at its codimension 1 points. By Theorem 13.5.8, Noetherian normal schemes are regular in codimension 1.

13.5.14. *Definition: Valuation at an R_1 point.* Suppose X is a locally Noetherian scheme. Then for any regular codimension 1 point p (i.e., any point p where $\mathcal{O}_{X,p}$ is a regular local ring of dimension 1), we have a discrete valuation val_p. If f is any nonzero element of the fraction field of $\mathcal{O}_{X,p}$ (e.g., if X is integral, and f is a nonzero element of the function field of X), then if $\mathrm{val}_p(f) > 0$, we say that f has a **zero of order** $\mathrm{val}_p(f)$ **at** p, and if $\mathrm{val}_p(f) < 0$, we say that f has a **pole of order** $- \mathrm{val}_p(f)$ **at** p. (We are not yet allowed to discuss order of vanishing at a point that is not regular and codimension 1. One can make a definition, but it doesn't behave as well as it does when have you have a discrete valuation.) Can you see why $75/34$ has a double zero at 5, and a single pole at 2? Can you describe the zeros and poles of $x^3(x+y)/(x^2+y)^3$ on $\mathbb{A}^2_k$?

13.5.G. EXERCISE (FINITENESS OF ZEROS AND POLES ON NOETHERIAN SCHEMES). Suppose X is an integral Noetherian scheme, and $f \in K(X)^\times$ is a nonzero element of its function field. Show that f has a finite number of zeros and poles. (Hint: Reduce to $X = \mathrm{Spec}\, A$. If $f = f_1/f_2$, where $f_i \in A$, prove the result for f_i.)

13.5.15. *Remark.* A (Noetherian) scheme can be singular in codimension 2 and still be normal. For example, you have shown that the cone $x^2 + y^2 = z^2$ in $\mathbb{A}^3$ in characteristic not 2 is normal (Exercise 5.4.I(b)), but it is singular at the origin (the Zariski tangent space is visibly three-dimensional).

But singularities of normal schemes are not so bad in some ways: we will very shortly (§13.5.18) discuss Algebraic Hartogs's Lemma 13.5.19 for Noetherian normal schemes, which states that you can extend functions over codimension 2 sets.

13.5.16. *Remark.* We know that for Noetherian rings we have implications

$$\text{unique factorization domain} \Longrightarrow \text{integrally closed} \Longrightarrow \text{regular in codimension 1.}$$

Hence for locally Noetherian schemes, we have similar implications:

$$\text{factorial} \Longrightarrow \text{normal} \Longrightarrow \text{regular in codimension 1.}$$

Here are two examples to show you that these inclusions are strict.

13.5.H. EXERCISE (THE **pinched plane**). Let A be the subring $k[x^3, x^2, xy, y]$ of $k[x, y]$. (Informally, we allow all polynomials that don't include a nonzero multiple of the monomial x.) Show that $\operatorname{Spec} k[x, y] \to \operatorname{Spec} A$ is a normalization. Show that A is not integrally closed. Show that $\operatorname{Spec} A$ is regular in codimension 1. (Hint for the last part: Show it is dimension 2, and when you throw out "the origin" you get something regular, by inverting x^2 and y respectively, and considering A_{x^2} and A_y.)

13.5.17. *Example (the quadric surface once again).* Suppose k is algebraically closed of characteristic not 2. Then $k[w, x, y, z]/(wz - xy)$ is integrally closed, but not a unique factorization domain; see Exercise 5.4.L (and Exercise 13.1.E).

13.5.18. Algebraic Hartogs's Lemma for Noetherian normal schemes.

Hartogs's Lemma in several complex variables states (informally) that a holomorphic function defined away from a codimension 2 set can be extended over that. We now describe an algebraic analog, for Noetherian normal schemes. (It may also be profitably compared to the Second Riemann Extension Theorem.) We will use this repeatedly and relentlessly when connecting line bundles and divisors in §15.4.

13.5.19. Algebraic Hartogs's Lemma — *Suppose A is an integrally closed Noetherian integral domain. Then*

$$A = \cap_{\mathfrak{p} \text{ codimension } 1} A_{\mathfrak{p}}.$$

The equality takes place in $K(A)$; recall that any localization of an integral domain A is naturally a subset of $K(A)$ (Exercise 1.2.C). Warning: Few people call this Algebraic Hartogs's Lemma. I call it this because it parallels the statement in complex geometry.

One might say that if $f \in K(A)$ does not lie in $A_{\mathfrak{p}}$, where $\mathfrak{p}$ has codimension 1, then f has a pole at $[\mathfrak{p}]$, and if $f \in K(A)$ lies in $\mathfrak{p}A_{\mathfrak{p}}$, where $\mathfrak{p}$ has codimension 1, then f has a zero at $[\mathfrak{p}]$.

13.5.20. It is worth *geometrically* interpreting Algebraic Hartogs's Lemma as saying that *a rational function on a Noetherian normal scheme with no poles is in fact regular* (an element of A). Informally: *"Noetherian normal schemes have the Hartogs property."*

One can state Algebraic Hartogs's Lemma more generally in the case that $\operatorname{Spec} A$ is a Noetherian normal scheme, meaning that A is a product of Noetherian normal integral domains; the reader may wish to do so.

Another generalization (and something closer to the "right" statement) is that if A is a subring of a field K, then the integral closure of A in K is the intersection of all valuation rings of K containing A; see [AtM, Cor. 5.22] for explanation and proof.

13.5.21.* *Proof of Algebraic Hartogs's Lemma 13.5.19.*
The key clever construction in the proof is the following. If A is an integral domain, and $r \in K(A)$, then $r \in A$ if and only if $r \in A_{\mathfrak{p}}$ for all primes $\mathfrak{p}$ of A, i.e., if r is regular at all points of Spec A. For this reason, it makes sense to ask at what points $r \in K(A)$ is not regular. Our "ideal of denominators" construction (from the proof Proposition 5.4.2) shows that it is a closed subset: if $I = \{a \in A : ar \in A\}$, then the points $[\mathfrak{p}]$ of $V(I) = \operatorname{Supp} A/I$ are precisely the points where r is not regular. We get a more refined understanding of this closed subset from an easy-to-define module $M^{a/b}$.

13.5.22. Suppose $r = a/b \notin A$, where $a, b \in A$. Let $\overline{a}$ be the image of a in $A/(b)$. The fact that $r \notin A$ means that $\overline{a} \neq 0$ in $A/(b)$. Let $M^{a/b}$ be the A-submodule of $A/(b)$ generated by $\overline{a}$. We have constructed $M^{a/b}$ so that $M_{\mathfrak{q}}^{a/b} = 0$ if and only if $a/b \in A_{\mathfrak{q}}$. (Translation: a is not a multiple of b in $A_{\mathfrak{p}}$ if and only if $\overline{a} \neq 0$ in $A_{\mathfrak{p}}/(b)$.)

We wish to show Algebraic Hartogs's Lemma, which says that if A is integrally closed, then it suffices to instead check only the height 1 prime ideals. We begin with an approximation of this fact, which works without the "integrally closed" hypothesis.

13.5.23. Lemma — *Suppose A is a Noetherian integral domain, and $r \in K(A)$. Then $r \in A$ if and only if for every prime $\mathfrak{p} \in \operatorname{Ass}(A/(g))$ for every nonzero $g \in A$, we have $r \in A_{\mathfrak{p}}$.*

Proof. Clearly, if $r \in A$, then $r \in A_{\mathfrak{p}}$ for all $\mathfrak{p}$. So assume that $r = a/b \notin A$. We will find such a $\mathfrak{p}$ so that $r \notin A_{\mathfrak{p}}$ as described in the statement of the lemma. We take $\mathfrak{p}$ to be any associated prime ideal of the nonzero module $M^{a/b}$ constructed in §13.5.22. As $M^{a/b}$ is a submodule of $A/(b)$, $\mathfrak{p}$ is an associated prime of $A/(b)$ as well (Exercise 6.6.G(a)). $\square$

13.5.24. Proposition — *If A is a Noetherian integrally closed integral domain, then*

(i) *A is regular in codimension 1 (Definition 13.5.13), and*
(ii) *for any $b \in A$, the A-module $A/(b)$ has no embedded points.*

Proof. (i) Integral closure is preserved by localization (Exercise 5.4.A), so for each codimension 1 prime $\mathfrak{p}$, $A_{\mathfrak{p}}$ is integrally closed. Thus by Theorem 13.5.8, $A_{\mathfrak{p}}$ is regular.

(ii) Let $\mathfrak{p}$ be an associated prime of $A/(b)$. We wish to show that $\mathfrak{p}$ is codimension 1 in A. The truth of the statement is preserved by replacing A by its localization $A_{\mathfrak{p}}$, so it suffices to prove the following: "Suppose $(A, \mathfrak{m})$ is a Noetherian local ring, an integrally closed integral domain, and $b \in \mathfrak{m} \setminus \{0\}$. Suppose $\mathfrak{m} \in \operatorname{Ass}(A/(b))$. Then $\mathfrak{m}$ is principal (and thus codimension at most 1)."

We now prove this statement. Choose $\overline{a} \in A/(b)$ such that $\operatorname{Ann}(\overline{a}) = \mathfrak{m}$. Then $\overline{a} \neq 0$. Lift $\overline{a}$ to $a \in A$. Then $a/b \in K(A) \setminus A$, and $(a/b)\mathfrak{m} \subset A$.

We show by contradiction that $(a/b)\mathfrak{m} \not\subset \mathfrak{m}$. Assume otherwise. Recall Exercise 8.2.I: "Suppose A is a subring of K, $r \in K$, and N is a faithful $A[r]$-module that is finitely generated as an A-module. Then r is integral over A." We apply this here with $r = a/b$ and $N = \mathfrak{m}$. Then a/b is integral over A, implying (from integral closure of A) that $a/b \in A$, contradiction.

Hence $(a/b)\mathfrak{m} = A$, from which $\mathfrak{m} = (b/a)A$, from which $\mathfrak{m}$ is principal, as desired. $\square$

13.5.25. Theorem — *Suppose A is a Noetherian integral domain, regular in codimension 1. Then (a) A is integrally closed if and only if (b) $A = \cap_{\mathfrak{p} \text{ codimension } 1} A_{\mathfrak{p}}$.*

This implies Algebraic Hartogs's Lemma 13.5.19.

Proof. (a) implies (b). We apply Lemma 13.5.23. From Proposition 13.5.24(ii), the only primes we need to check are codimension 1.

(b) implies (a). The $A_{\mathfrak{p}}$ are dimension 1 Noetherian domains that are regular, and hence integrally closed by Theorem 13.5.8. Suppose $r \in K(A)$ is integral over A. Then $r \in K(A)$ is integral over $A_{\mathfrak{p}}$ for all codimension 1 $\mathfrak{p}$, so $r \in A_{\mathfrak{p}}$, so $r \in \cap_{\mathfrak{p} \text{ codimension } 1} A_{\mathfrak{p}} = A$. $\square$

We conclude with a fact that we will later (§26.3), along with Proposition 13.5.24, recognize as Serre's $R_1 + S_2$ criterion for normality.

13.5.26. Proposition — *Suppose* A *is a Noetherian integral domain satisfying conditions (i) and (ii) of Proposition 13.5.24. Then* A *is integrally closed.*

Proof. Now $A = \cap_\mathfrak{p} A_\mathfrak{p}$ as $\mathfrak{p}$ runs over the primes described in Lemma 13.5.23. From condition (ii) of Proposition 13.5.24, these are all codimension 1 primes. The result then follows from Theorem 13.5.25, using condition (i) of Proposition 13.5.24. □

13.5.27. *Consequence of Algebraic Hartogs's Lemma 13.5.19 we will use.*

13.5.I. USEFUL EXERCISE. If f is a nonzero rational function on a locally Noetherian normal scheme, and f has no poles, show that f is regular.

13.6 Smooth (and Étale) Morphisms: First Definition

In §13.2.3, we defined smoothness over a field. We now know enough to define *smooth morphisms* (and, as a special case, *étale morphisms*) in general. Our definition will be imperfect in a number of ways. For example, it will look a little surprising. As another sign, it will not be obvious that it actually generalizes smoothness over a field. A third flaw is that because it is of the form "there exist open covers satisfying some property," it is poorly designed to be used to show that some morphism is *not* smooth. But it has the major advantage that we can give the definition right now, and the basic properties of smooth morphisms will be straightforward to show.

13.6.1. *Differential geometric motivation (see Figure 24.6).* The notion of a smooth morphism is motivated by the following idea from differential geometry (see, e.g., [Tu] or [Var, §1.1–1.2] for the differential geometry). For the purposes of this discussion, we say a map $\pi\colon X \to Y$ of manifolds (or real analytic spaces, or any variant thereof) is *smooth of relative dimension n* if locally (on X) it looks like $Y' \times \mathbb{R}^n \to Y'$. (In differential geometry, this is usually called a *submersion*.) Translation: It is "locally on the source a smooth fibration." (If you have not heard the word fibration before, don't worry. Informally speaking, a *fibration* in some "category of geometric objects" is a map $\pi\colon X \to Y$ that locally on the source is a product, $V \times W \to V$; in some sort of category of manifolds, a fibration is "locally on the source a smooth fibration" if the W in the local description can be taken to be $\mathbb{R}^a$ for some a.)

In particular, "smooth of relative dimension 0" is the same as the important notion of "local isomorphism" ("isomorphism locally on the source"—not quite the same as "covering space"). Carrying this idea too naively into algebraic geometry leads to the "wrong" definition (see Exercise 13.6.F). We give a better definition now, and some equivalent definitions in §24.8.

13.6.2. *Definition.* A morphism $\pi\colon X \to Y$ is **smooth of relative dimension** n if there exist open covers $\{U_i\}$ of X and $\{V_i\}$ of Y, with $\pi(U_i) \subset V_i$, such that for every i we have a commutative diagram

where ρ is induced by the obvious map of rings in the opposite direction, and W is an open subscheme of $\operatorname{Spec} B[x_1, \ldots, x_{n+r}]/(f_1, \ldots, f_r)$, such that the determinant of the Jacobian matrix of the f_j's with respect to the *first* r x_i's,

$$(13.6.2.1) \qquad\qquad \det\left(\frac{\partial f_j}{\partial x_i}\right)_{i,j \le r} ,$$

is an invertible (= nowhere zero) function on W.

Étale means *smooth of relative dimension* 0.

13.6.3. *Quick observations.* Be sure to notice that if B is a field, then we have just the right number of equations (r of them) to hope to cut out something of dimension n in $\mathbb{A}_k^{n+r}$. More generally, intuitively and imprecisely, we have just the "right number of equations" to cut out W in $\mathbb{A}_B^{n+r}$, with no redundant ones to spare.

From the definition, smooth morphisms are locally of finite presentation (hence locally of finite type). Also, $\mathbb{A}_B^n \to \operatorname{Spec} B$ is immediately seen to be smooth of relative dimension n. From the definition, the locus on X where π is smooth of relative dimension n is open. In particular, the locus where π is étale is open.

As easy examples, open embeddings are étale, and the projection $\mathbb{A}^n \times Y \to Y$ is smooth of relative dimension n. You may already have some sense of what makes this definition imperfect. For example, how do you know a morphism is smooth if it isn't given to you in a form where the "right" variables x_i and the "right" equations are clear?

13.6.A. MOTIVATING EXERCISE (FOR THOSE WITH DIFFERENTIAL-GEOMETRIC BACKGROUND). Show how Definition 13.6.2 gives the definition in differential geometry, as described in §13.6.1. (Your argument might use the implicit function theorem, which "doesn't work" in algebraic geometry; see Exercise 13.6.F.) Show that for complex varieties, analytification turns étale morphisms into "local biholomorphisms."

13.6.B. EXERCISE. Show that the notion of smoothness of relative dimension n (and, in particular, étaleness) is local on both the source and the target.

We can thus make sense of the phrase "$\pi \colon X \to Y$ is smooth of relative dimension n at $p \in X$": it means there is an open neighborhood U of p such that $\pi|_U$ is smooth of relative dimension n.

The phrase **smooth morphism** (without reference to relative dimension n) often informally means "smooth morphism of some relative dimension n," but sometimes can mean "smooth of some relative dimension in the neighborhood of every point."

13.6.C. EASY EXERCISE. Show that the notion of smoothness of relative dimension n (and hence the notion of étaleness) is preserved by base change.

13.6.D. EXERCISE. Suppose $\pi \colon X \to Y$ is smooth of relative dimension m and $\rho \colon Y \to Z$ is smooth of relative dimension n. Show that $\rho \circ \pi$ is smooth of relative dimension $m + n$. (In particular, the composition of étale morphisms is étale. Furthermore, this implies that smooth morphisms are closed under product, by Exercise 8.1.A.)

Exercises 13.6.B–13.6.D imply that smooth morphisms (and étale morphisms) form a "reasonable" class in the sense of §8.1.

13.6.4. *Observation.* Suppose $\pi \colon X \to Y$ is smooth of relative dimension n. Then locally on X, π can be described as an étale cover of $\mathbb{A}_Y^n$. More precisely, for every $p \in X$, there is an open neighborhood U of p, such that $\pi|_U$ can be factored into

$$U \xrightarrow{\ \alpha\ } \mathbb{A}_Y^n \xrightarrow{\ \beta\ } Y,$$

where α is étale and β is the obvious projection.

13.6.E. EXERCISE. Show that $\operatorname{Spec} k \to \operatorname{Spec} k[\epsilon]/(\epsilon^2)$ is not an étale morphism. (Does your mental picture of étale morphisms already capture this fact?) Possible approach: If it were étale, show that it would induce an "isomorphism on tangent spaces."

13.6.5. More sophisticated approaches to Exercise 13.6.E may lead you to consider the following exercises: Classify the étale maps to $\operatorname{Spec} k$. Classify the étale maps to $\operatorname{Spec} \overline{k}[\epsilon]/(\epsilon^2)$. If $A = \overline{k}[x_1, \ldots, x_n]/I$ is such that $\operatorname{Spec} A$ has only one point (an example of an Artinian local scheme, Definition 6.5.9, and A is an Artinian local ring, §12.1.8), classify the étale maps to $\operatorname{Spec} A$. These questions will lead you in different important directions, to deformation theory, Witt vectors, and more.

13.6.F. EXERCISE. Suppose k is a field of characteristic not 2. Let $Y = \operatorname{Spec} k[t]$, and $X = \operatorname{Spec} k[u, 1/u]$. Show that the morphism $\pi \colon X \to Y$ induced by $t \mapsto u^2$ is étale. (Sketch π!) Show that there is no nonempty open subset U of X on which π is an isomorphism. (This shows that étale cannot be defined as "local isomorphism," i.e., "isomorphism locally on the source." In particular, the naive extension of the differential intuition of §13.6.1 does not give the right definition of étaleness in algebraic geometry.)

We now show that our new Definition 13.6.2 specializes to Definition 13.2.4 when the target is a field.

13.6.6. Theorem — *Suppose X is a k-scheme. Then the following are equivalent.*

(i) X *is smooth of relative dimension n over* $\operatorname{Spec} k$ *(Definition 13.6.2).*
(ii) X *has pure dimension n, and is smooth over k (in the sense of Definition 13.2.4).*

This allows us to use the phrases "smooth over k" and "smooth over $\operatorname{Spec} k$" interchangeably, as we would expect. More generally, **"smooth over a ring A"** means "smooth over $\operatorname{Spec} A$."

Proof. (i) implies (ii). Suppose X is smooth of relative dimension n over $\operatorname{Spec} k$ (Definition 13.6.2). We will show that X has pure dimension n, and is smooth over k (Definition 13.2.4). Our goal is a Zariski-local statement, so we may assume that X is an open subset

$$W \subset \operatorname{Spec} k[x_1, \dots, x_{n+r}]/(f_1, \dots, f_r),$$

as described in Definition 13.6.2.

The central issue is showing that X has "pure dimension n," as the latter half of the statement would then follow because the corank of the Jacobian matrix is then n.

Let $X_{\overline{k}} := X \times_k \overline{k}$. By Exercise 12.1.H, $X_{\overline{k}}$ is pure dimension n if and only if X has pure dimension n. Let p be a closed point of the (finite type scheme) $X_{\overline{k}}$. Then $X_{\overline{k}}$ has dimension at least n at p, by Krull's Height Theorem 12.3.10, combined with Theorem 12.2.9 (codimension is the difference of dimensions for varieties). But the Zariski tangent space at a closed point of a finite type scheme over $\overline{k}$ is cut out by the Jacobian matrix (Exercise 13.1.I), and thus has dimension exactly n (by the Jacobian condition in the definition of k-smooth). As the dimension of the Zariski tangent space bounds the dimension of the scheme (Theorem 12.3.12), $X_{\overline{k}}$ has pure dimension n as desired.

(ii) implies (i). It suffices to prove the result in a neighborhood of any given closed point p of X. (Here we use that any union of open neighborhoods of all the closed points is all of X, by Exercise 5.3.F.) We thus consider the following problem. Suppose we are given $X := \operatorname{Spec} k[x_1, \dots, x_{n+r}]/(f_1, \dots, f_N)$, and a point $p \in X$, and an open neighborhood U of p in X such that U is pure dimension n, and the Jacobian matrix of the f_j's with respect to the x_i's has corank n everywhere on U. We wish to show that the structure morphism $\pi : X \to \operatorname{Spec} k$ is smooth of relative dimension n at p (Definition 13.6.2). By permuting the f_i (and shrinking U), we may assume that the Jacobian matrix of the *first* f_1, ..., f_r already has corank n (i.e., rank r). Then $Y := \operatorname{Spec} k[x_1, \dots, x_{n+r}]/(f_1, \dots, f_r)$ is smooth at p of relative dimension n over $\operatorname{Spec} k$ (Definition 13.6.2), so the local ring $\mathcal{O}_{Y,p}$ is an integral domain by Exercise 13.2.O, of pure dimension n by the proof that (i) implies (ii) above.

For the same reasons, $\mathcal{O}_{X,p}$ is also an integral domain of pure dimension n. We have a surjection $\mathcal{O}_{Y,p} \twoheadrightarrow \mathcal{O}_{Y,p}/(f_{r+1}, \dots, f_N) = \mathcal{O}_{X,p}$ of integral domains of the same dimension, so by Easy Exercise 12.1.D, $\mathcal{O}_{Y,p} = \mathcal{O}_{X,p}$, i.e., $f_{r+1} = \dots = f_N = 0$ in $\mathcal{O}_{Y,p}$. Thus, near p, X is isomorphic to Y, so X is indeed smooth over k of relative dimension n (Definition 13.6.2), as desired. $\square$

13.7* Valuative Criteria for Separatedness and Properness

(The only reason this section is placed here is that we need the theory of discrete valuation rings.)

In reasonable circumstances, it is possible to verify separatedness by checking only maps from spectra of valuation rings. There are four reasons you might like this (even if you never use it). First, it gives useful intuition for what separated morphisms look like. Second, given that we understand schemes by maps to them (the Yoneda philosophy), we might expect to understand morphisms by mapping certain maps of schemes to them, and this is how you can interpret the diagram appearing in the valuative criterion. And the third concrete reason is that one of the two directions in the statement is much easier (a special case of the Reduced-to-Separated Theorem 11.3.2; see Exercise 13.7.A), and this is the direction we will repeatedly use. Finally, the criterion is very useful!

Similarly, there is a valuative criterion for properness.

In this section, we will meet the valuative criteria, but aside from outlining the proof of one result (the DVR version of the valuative criterion of separatedness), we will not give proofs, and satisfy ourselves with references. There are two reasons for this controversial decision. First, the proofs require the development of some commutative algebra involving valuation rings that we will not otherwise need. Second, we will not use these results in any substantial or necessary way later in this book.

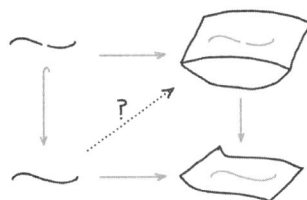

Figure 13.6 *The valuative criterion for separatedness.*

We begin with a valuative criterion for separatedness that applies in a case that will suffice for the interests of most people, that of finite type morphisms of Noetherian schemes. We will then give a more general version for more general readers.

13.7.1. Theorem (valuative criterion for separatedness, DVR version; see Figure 13.6) — *Suppose $\pi\colon X \to Y$ is a morphism of finite type of locally Noetherian schemes. Then π is separated if and only if the following condition holds: for any discrete valuation ring A, and any diagram of the form*

(13.7.1.1)
$$\begin{array}{ccc} \operatorname{Spec} K(A) & \longrightarrow & X \\ \text{\scriptsize open emb.} \downarrow & & \downarrow \pi \\ \operatorname{Spec} A & \longrightarrow & Y \end{array}$$

(where the vertical morphism on the left corresponds to the inclusion $A \hookrightarrow K(A)$), there is at most one morphism $\operatorname{Spec} A \to X$ such that the diagram

(13.7.1.2)
$$\begin{array}{ccc} \operatorname{Spec} K(A) & \longrightarrow & X \\ \text{\scriptsize open emb.} \downarrow & {\scriptstyle \leq 1}\nearrow & \downarrow \pi \\ \operatorname{Spec} A & \longrightarrow & Y \end{array}$$

commutes.

The idea behind the proof is explained in §13.7.3. We can show one direction right away, in the next exercise.

13.7.A. EXERCISE (THE EASY DIRECTION). Use the Reduced-to-Separated Theorem 11.3.2 to prove one direction of the theorem: that if π is separated, then the valuative criterion holds.

13.7.B. EXERCISE. Suppose X is an integral Noetherian separated curve. If $p \in X$ is a regular closed point, then $\mathcal{O}_{X,p}$ is a discrete valuation ring, so each regular point yields a discrete valuation on $K(X)$. Use the previous exercise to show that distinct points yield distinct discrete valuations.

Here is the intuition behind the valuative criterion (see Figure 13.6). We think of Spec of a discrete valuation ring A as a "germ of a curve," and $\operatorname{Spec} K(A)$ as the "germ minus the origin" (even though it is just a point!). Then the valuative criterion says that if we have a map from a germ of a curve to Y, and have a lift of the map away from the origin to X, then there is at most one way to lift the map from the entire germ. In the case where Y is the spectrum of a field, you can think of this as saying that limits of one-parameter families are unique (if they exist).

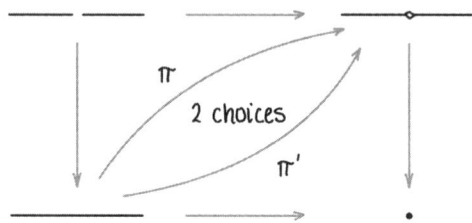

For example, this captures the idea of what is wrong with the map of the line with the doubled origin over k (Figure 13.7): we take Spec A to be the germ of the affine line at the origin, and consider the map of the germ minus the origin to the line with doubled origin. Then we have two choices for how the map can extend over the origin.

Figure 13.7 *The line with the doubled origin fails the valuative criterion for separatedness. (You may notice Figure 11.4 embedded in this diagram.)*

13.7.C. EXERCISE. Make this precise: show that map of the line with doubled origin over k to Spec k fails the valuative criterion for separatedness. (Earlier arguments were given in Exercises 11.2.I and 11.3.C.)

13.7.2. ** *Remark for experts: Moduli spaces and the valuative criterion for separatedness.* If $Y = $ Spec k, and X is a (fine) moduli space (a term we won't define here) of some type of object, then the question of the separatedness of X (over Spec k) has a natural interpretation: given a family of your objects parametrized by a "punctured discrete valuation ring," is there always at most one way of extending it over the closed point?

13.7.3. *Idea behind the proof of Theorem 13.7.1 (the valuative criterion for separatedness, DVR version) in the case of varieties.* (One direction was done in Exercise 13.7.A.) If π is *not* separated, our goal is to produce a diagram (13.7.1.1) that can be completed to (13.7.1.2) in more than one way. If π is not separated, then $\delta: X \to X \times_Y X$ is a locally closed embedding that is not a closed embedding.

13.7.D. EXERCISE. Show that you can find points p not in the diagonal Δ of $X \times_Y X$ and q in Δ such that $p \in \bar{q}$, and there are no points "between p and q" (no points r distinct from p and q with $p \in \bar{r}$ and $r \in \bar{q}$). (Exercise 8.4.C may shed some light.)

Let Q be the scheme obtained by giving the induced reduced subscheme structure to $\bar{q}$. Let $B = \mathscr{O}_{Q,p}$ be the local ring of Q at p.

13.7.E. EXERCISE. Show that B is a Noetherian local integral domain of dimension 1.

If B were regular, then we would be done: composing the inclusion morphism $Q \to X \times_Y X$ with the two projections induces the same morphism $q \to X$ (i.e., Spec $\kappa(q) \to X$) but different extensions to B precisely because p is not in the diagonal. To complete the proof, one shows that the normalization of B is Noetherian; this is where need the hypothesis that we are working with varieties, to invoke Exercise 10.7.M. Then localizing at any prime above p (there is one by the Lying Over Theorem 8.2.5) yields the desired discrete valuation ring A.

For an actual proof, see [Stacks, tag 0207] or [Gr-EGA, II.7.2.3].

With a more powerful invocation of commutative algebra, we can prove a valuative criterion with much less restrictive hypotheses.

13.7.4. Theorem (valuative criterion for separatedness) — *Suppose $\pi: X \to Y$ is a quasisepa- rated morphism. Then π is separated if and only if for any valuation ring A, and any diagram of the form (13.7.1.1), there is at most one morphism Spec $A \to X$ such that the diagram (13.7.1.2) commutes.*

Because I have already failed to completely prove the DVR version, I feel no urge to prove this harder fact. The proof of one direction, that π separated implies that the criterion holds, follows from an argument identical to the one in Exercise 13.7.A. For a complete proof, see [Stacks, tags 01KY and 01KZ] or [Gr-EGA, II.7.2.3].

13.7.5. Valuative criteria for (universal closedness and) properness.

There is a valuative criterion for properness too. It is philosophically useful, and sometimes directly useful, although we won't need it. It naturally comes from the valuative criterion for separatedness combined with a valuative criterion for universal closedness.

13.7.6. Theorem (valuative criterion for universal closedness and properness, DVR version) — *Suppose $\pi\colon X \to Y$ is a morphism of finite type of locally Noetherian schemes. Then π is universally closed (resp., proper) if and only if for any discrete valuation ring A, and any diagram (13.7.1.1), there is at least one (resp., exactly one) morphism $\mathrm{Spec}\,A \to X$ such that the diagram (13.7.1.2) commutes.*

See [Gr-EGA, II.7.3.8], [Ha1, Thm. II.4.7], or [Gr-EGA', I.5.5] for proofs. A comparison with Theorem 13.7.1 will convince you that these three criteria naturally form a family.

In the case where Y is a field, you can think of the valuative criterion of properness as saying that limits of one-parameter families in proper varieties always exist, and are unique. This is a useful intuition for the notion of properness.

13.7.F. EASY EXERCISE. Use the valuative criterion for properness to prove that $\mathbb{P}^n_A \to \mathrm{Spec}\,A$ is proper if A is Noetherian. (Don't be fooled: Because this requires the valuative criterion, this is a difficult way to prove a fact that we already showed in Theorem 11.4.5.)

13.7.G. EXERCISE (CF. EXERCISE 13.7.B). Suppose X is an irreducible regular (Noetherian) curve, proper over a field k (respectively, over $\mathbb{Z}$). Describe a bijection between the discrete valuations on $K(X)$ for which the elements of k (respectively $\mathbb{Z}$) have valuation 0 and the closed points of X.

13.7.7. *Remarks for experts.* There is a moduli-theoretic interpretation similar to that for separatedness (Remark 13.7.2): X is proper if and only if there is always precisely one way of filling in a family over a punctured spectrum of a discrete valuation ring.

13.7.8. Finally, here is a fancier version of the valuative criterion for universal closedness and properness.

13.7.9. Theorem (valuative criterion for universal closedness and properness) —
Suppose $\pi\colon X \to Y$ is a quasiseparated, finite type (hence quasicompact) morphism. Then π is universally closed (resp., proper) if and only if the following condition holds. For any valuation ring A and any diagram of the form (13.7.1.1), there is at least one (resp., exactly one) morphism $\mathrm{Spec}\,A \to X$ such that the diagram (13.7.1.2) commutes.

Clearly the valuative criterion for properness is a consequence of the valuative criterion for separatedness (Theorem 13.7.4) and the valuative criterion for universal closedness. For proofs, see [Stacks, tag 01KF] or [Gr-EGA, II.7.3.8].

13.7.10. On the importance of valuation rings in general. Although we have only discussed discrete valuation rings in depth, general valuation rings should not be considered an afterthought. Serre makes the case to Grothendieck in a letter, [GrS, p. 125]:

Tu es bien sévère pour les Valuations! Je persiste pourtant à les garder, pour plusieurs raisons: d'abord une pratique: n rédacteurs ont sué dessus, il n'y a rien à reprocher au fourbi, on ne doit pas le vider sans raisons très sérieuses (et tu n'en as pas). . . . Même un noethérien impénitent a besoin des valuations discrètes, et de leurs extension; en fait, Tate, Dwork, et tous les gen p-adiques te diront qu'on ne peut se limiter au cas discret et que le cas de rang 1 est indispensable; à ce moment, les méthodes noethériennes deviennent un vrai carcan, et on comprend beaucoup mieux en rédigeant le cas général que uniquement le cas de rang 1. . . . Il n'y a pas lieu d'en faire un plat, bien sûr, et c'est pourquoi j'avias vivement combattu le plan Weil initial qui en faisait le théorème central de l'Algèbre Commutative, mais d'autre part il faut le garder.

You are very harsh on Valuations! I persist nonetheless in keeping them, for several reasons, of which the first is practical: n people have sweated over them, there is nothing wrong with the result, and it should not be thrown out without very serious reasons (which you do not have). . . . Even an unrepentant Noetherian needs discrete valuations and their extensions; in fact, Tate, Dwork and all the p-adic people will tell you that one cannot restrict oneself to the discrete case and the rank 1 case is indispensable; Noetherian methods then become a burden, and one understands much better if one considers the general case and not only the rank 1 case. . . . It is not worth making a mountain out of it, of course, which is why I energetically fought Weil's original plan to make it the central theorem of Commutative Algebra, but on the other hand it must be kept.

13.8* More Sophisticated Facts about Regular Local Rings

Regular local rings have essentially every good property you could want, but some of them require hard work. We now discuss a few fancier facts that may help you sleep well at night.

13.8.1. Localizations of regular local rings are regular local rings.

13.8.2. Fact ([E, Cor. 19.14], [Mat2, Thm. 19.3]) — *If $(A, \mathfrak{m})$ is a regular local ring, then any localization of A at a prime is also a regular local ring.*

(We will not need this, and hence will not prove it.) This major theorem was an open problem in commutative algebra for a long time until settled by Serre and Auslander–Buchsbaum using homological methods.

Hence to check if $\operatorname{Spec} A$ is regular (A Noetherian), it suffices to check at closed points (at maximal ideals). Assuming Fact 13.8.2 (and using Exercise 5.1.E), you can check regularity of a Noetherian scheme by checking at closed points.

We will prove two important cases of Fact 13.8.2. The first you can do right now.

13.8.A. EXERCISE. Suppose X is a Noetherian dimension 1 scheme that is regular at its closed points. Show that X is reduced. Hence show (without invoking Fact 13.8.2) that X is regular.

The second important case will be proved in §21.6.2:

13.8.3. Theorem — *If X is a finite type scheme over a perfect field k that is regular at its closed points, then X is regular.*

More generally, Exercise 21.6.B will show that Fact 13.8.2 holds if A is the localization of a finite type algebra over a perfect field.

13.8.B. EXERCISE (GENERALIZING EXERCISE 13.3.E). Suppose k is an algebraically closed field of characteristic 0. Assuming Theorem 13.8.3, show that there exists a regular hypersurface of every positive degree d in $\mathbb{P}^n$. (As in Exercise 13.3.E, feel free to weaken the hypotheses.)

13.8.4. Regular local rings are unique factorization domains, integrally closed, and Cohen–Macaulay.

13.8.5. Fact (Auslander–Buchsbaum Theorem) — *Regular local rings are unique factorization domains.*

(This is a hard theorem, so we will not prove it, and will therefore not use it. For a proof, see [E, Thm. 19.19], [Mat2, Thm. 20.3], or [Gr-EGA, IV$_4$.21.11.1], or [Mu7, §III.7] in the special case of varieties.) Thus regular schemes are factorial, and hence normal by Exercise 5.4.F.

13.8.6. *Remark: Factoriality is weaker than regularity.* The implication "regular implies factorial" is strict. Here is an example showing this. Suppose k is an algebraically closed field of characteristic not 2. Let $A = k[x_1, \ldots, x_n]/(x_1^2 + \cdots + x_n^2)$ (cf. Exercise 5.4.I). Note that $\operatorname{Spec} A$ is clearly singular at the origin. In Exercise 15.5.R, we will show that A is a unique factorization domain when $n \geq 5$, so Spec A is factorial. In particular, $A_{(x_1,\ldots,x_n)}$ is a Noetherian local ring that is a unique factorization domain, but not regular. (More generally, it is a consequence of Grothendieck's proof of a conjecture of P. Samuel that a Noetherian local ring that is a complete intersection—in particular, a hypersurface—that is factorial in codimension at most 3 must be factorial, [SGA2, Exp. XI, Cor. 3.14]. For a shorter and somewhat more elementary proof, see [CL].)

13.8.7. *Regular local rings are integrally closed.* The Auslander–Buchsbaum Theorem (Fact 13.8.5) implies that regular local rings are integrally closed (by Exercise 5.4.F). We will prove this (without appealing to Fact 13.8.5) in §26.3.5.

13.8.8. *Regular local rings are Cohen–Macaulay.* In §26.2.5, we will show that regular local rings are Cohen–Macaulay (a notion defined in Chapter 26).

13.9* Filtered Rings and Modules, and the Artin-Rees Lemma

We conclude Chapter 13 by discussing the Artin-Rees Lemma 13.9.3, which was used to prove Proposition 13.5.2. The Artin-Rees Lemma generalizes the intuition behind Proposition 13.5.2, that any function that is analytically zero at a point actually vanishes in an open neighborhood of that point (§13.5.3). Because we will use it later, and because it is useful to recognize it in other contexts, we discuss it in some detail.

13.9.1. *Definitions.* Suppose I is an ideal of a ring A. A descending filtration of an A-module M

$$(13.9.1.1) \qquad\qquad M = M_0 \supset M_1 \supset M_2 \supset \cdots$$

is called an I-**filtration** if $I^d M_n \subset M_{n+d}$ for all $d, n \geq 0$. An example is the I-**adic filtration** where $M_k = I^k M$. We say an I-filtration is I-**stable** if for some s and all $d \geq 0$, $I^d M_s = M_{d+s}$. For example, the I-adic filtration is I-stable.

Let $A(I)_\bullet$ be the graded ring $\oplus_{n\geq 0} I^n$ (with the convention $I^0 = A$). This is called the **Rees algebra** of the ideal I in A, although we will not need this terminology. Define $M(I)_\bullet := \oplus M_n$. It is naturally a *graded* module over $A(I)_\bullet$.

13.9.2. Proposition — *If A is Noetherian, M is a finitely generated A-module, and (13.9.1.1) is an I-filtration, then $M(I)_\bullet$ is a finitely generated $A(I)_\bullet$-module if and only if the filtration (13.9.1.1) is I-stable.*

Proof. Note that $A(I)_\bullet$ is Noetherian (by Exercise 4.5.G(b), as A is Noetherian, and I is a finitely generated A-module).

Assume first that $M(I)_\bullet$ is finitely generated over the Noetherian ring $A(I)_\bullet$, and hence is Noetherian. Consider the increasing chain of $A(I)_\bullet$-submodules whose kth element L_k is

$$M \oplus M_1 \oplus M_2 \oplus \cdots \oplus M_k \oplus IM_k \oplus I^2 M_k \oplus \cdots$$

(which agrees with $M(I)_\bullet$ up until M_k, and then "I-stabilizes"). This chain must stabilize by Noetherianness. But $\cup L_k = M(I)_\bullet$, so for some $s \in \mathbb{Z}^{\geq 0}$, $L_s = M(I)_\bullet$, so $I^d M_s = M_{s+d}$ for all $d \geq 0$— (13.9.1.1) is I-stable.

For the other direction, assume that $M_{d+s} = I^d M_s$ for a fixed s and all $d \geq 0$. Then $M(I)_\bullet$ is generated over $A(I)_\bullet$ by $M \oplus M_1 \oplus \cdots \oplus M_s$. But each M_j is finitely generated, so $M(I)_\bullet$ is indeed a finitely generated $A(I)_\bullet$-module. $\qquad\square$

13.9.3. Artin-Rees Lemma — *Suppose A is a Noetherian ring, and (13.9.1.1) is an I-stable filtration of a finitely generated A-module M. Suppose that $L \subset M$ is a submodule, and let $L_n := L \cap M_n$. Then*

$$L = L_0 \supset L_1 \supset L_2 \supset \cdots$$

is an I-stable filtration of L.

Proof. Note that $L_\bullet$ is an I-filtration, as $IL_n \subset IL \cap IM_n \subset L \cap M_{n+1} = L_{n+1}$. Also, $L(I)_\bullet$ is an $A(I)_\bullet$-submodule of the finitely generated $A(I)_\bullet$-module $M(I)_\bullet$, and hence finitely generated by Exercise 3.6.Y (as $A(I)_\bullet$ is Noetherian, see the proof of Proposition 13.9.2). $\qquad\square$

An important special case is the following.

13.9.4. Corollary — *Suppose $I \subset A$ is an ideal of a Noetherian ring, and M is a finitely generated A-module, and L is a submodule. Then for some integer s, $I^d(L \cap I^s M) = L \cap I^{d+s} M$ for all $d \geq 0$.*

Warning: It need not be true that $I^d L = L \cap I^d M$ for all d. (Can you think of a counterexample to this statement?)

Proof. Apply the Artin-Rees Lemma 13.9.3 to the filtration $M_n = I^n M$. $\qquad\square$

13.9.A. EXERCISE (KRULL INTERSECTION THEOREM).

(a) Suppose I is an ideal of a Noetherian ring A, and M is a finitely generated A-module. Show that there is some $a \equiv 1 \pmod{I}$ such that $a \cap_{j=1}^\infty I^j M = 0$. Hint: Apply the Artin-Rees Lemma 13.9.3 with $L = \cap_{j=1}^\infty I^j M$ and $M_n = I^n M$. Show that $L = IL$, and apply the first version of Nakayama (Lemma 8.2.8).

(b) Show that if A is a Noetherian integral domain or a Noetherian local ring, and I is a proper ideal, then $\cap_{j=1}^\infty I^j = 0$. In particular, you will have proved Proposition 13.5.2: if $(A, \mathfrak{m})$ is a Noetherian local ring, then $\cap_i \mathfrak{m}^i = 0$.

13.9.B. EXERCISE. Make the following precise, and prove it (thereby justifying the intuition in §13.5.3): if X is a locally Noetherian scheme, and f is a function on X that is analytically zero at a point $p \in X$, then f vanishes in a (Zariski-)open neighborhood of p.

PART V

Quasicoherent Sheaves on Schemes, and Their Uses

V

Chapter 14

More on Quasicoherent and Coherent Sheaves

In Chapter 6, we discussed how, *algebraically*, quasicoherent and coherent sheaves generalized modules over a ring. In this chapter, we explore how, *geometrically*, quasicoherent and coherent sheaves generalize vector bundles. Informally, a vector bundle V on a geometric space X (such as a manifold) is a family of vector spaces continuously parametrized by points of X. In other words, for each point p of X, there is a vector space, and these vector spaces are glued into a space V so that, as p varies, the vector space above p varies continuously. Nontrivial examples to keep in mind are the tangent bundle to a differentiable manifold, and the Möbius strip over a circle (interpreted as a line bundle). We will make this somewhat more precise in §14.1, but if you have not seen this idea before, you shouldn't be concerned; the main thing we will use this notion for is motivation.

14.0.1. *(Locally) free sheaves.* A **free sheaf** on a ringed space X is an $\mathcal{O}_X$-module isomorphic to $\mathcal{O}_X^{\oplus I}$, where the sum is over some index set I. The **rank** of the free sheaf is the cardinality of I. (This agrees with its rank as an $\mathcal{O}$-module at a point; see Definition 4.3.8.) A **locally free sheaf** on a ringed space X is an $\mathcal{O}_X$-module locally isomorphic to a free sheaf. This corresponds to the notion of a vector bundle (§14.1). Quasicoherent sheaves form a convenient abelian category containing the locally free sheaves (i.e., vector bundles) that is much smaller than the category of $\mathcal{O}$-modules. Quasicoherent sheaves generalize locally free sheaves in much the way that modules generalize free modules. Coherent sheaves are, roughly speaking, a finite rank version of quasicoherent sheaves, which form a well-behaved abelian category containing finite rank locally free sheaves (or, equivalently, finite rank vector bundles). Just as the notion of free modules leads us to the notion of modules in general, and finitely generated modules, the notion of free sheaves will lead us inevitably to the notion of quasicoherent sheaves and coherent sheaves. (There is a slight fib in comparing finitely generated modules to coherent sheaves; see §6.4.)

14.1 Vector Bundles "=" Locally Free Sheaves

We recall somewhat more precisely the notion of vector bundles on manifolds. Arithmetically minded readers shouldn't tune out: for example, fractional ideals of the ring of integers in a number field (defined in §10.7.1) turn out to be an example of a "line bundle on a smooth curve" (Exercise 14.1.K).

A **rank n vector bundle on a manifold** M is a map $\pi \colon V \to M$ with the structure of an n-dimensional real vector space on $\pi^{-1}(p)$ for each point $p \in M$, such that for every $p \in M$, there is an open neighborhood U and a homeomorphism

$$\phi \colon U \times \mathbb{R}^n \to \pi^{-1}(U)$$

over U (so that the diagram

(14.1.0.1)

$$
\begin{array}{ccc}
U \times \mathbb{R}^n & \xrightarrow{\ \phi\ } & \pi^{-1}(U) \\
& \searrow{\scriptstyle \text{projection to first factor}} \quad \swarrow{\scriptstyle \pi|_{\pi^{-1}(U)}} & \\
& U &
\end{array}
$$

commutes) that is an isomorphism of vector spaces over each $q \in U$. An isomorphism (14.1.0.1) is called a **trivialization over** U. If $U = M$, we say that V is a **trivial bundle**.

We call n the **rank** of the vector bundle. A rank 1 vector bundle is called a **line bundle**. (It can also be convenient to be agnostic about the rank of the vector bundle, so it can have different ranks on different connected components. It is also sometimes convenient to consider infinite-rank vector bundles.)

14.1.1. Transition functions. Given trivializations over U_1 and U_2, over their intersection, the two trivializations must be related by an element T_{12} of GL_n with entries consisting of functions on $U_1 \cap U_2$. (There is an issue of convention as to whether T_{12} should be replaced by T_{12}^{-1}. Our choice: We want this to be a matrix multiplying on the left, sending U_1-coordinates to U_2-coordinates. This translates to: $T_{ij} = \phi_j \circ \phi_i^{-1}$, where ϕ_i and ϕ_j are the trivializations for U_i and U_j, respectively) If $\{U_i\}$ is a cover of M, and we are given trivializations over each U_i, then the $\{T_{ij}\}$ must satisfy the **cocycle condition**:

$$(14.1.1.1) \qquad T_{jk}|_{U_i \cap U_j \cap U_k} \circ T_{ij}|_{U_i \cap U_j \cap U_k} = T_{ik}|_{U_i \cap U_j \cap U_K}.$$

(This implies $T_{ij} = T_{ji}^{-1}$.) The data of the T_{ij} are called **transition functions** (or **transition matrices**) for the trivialization.

This is reversible: given the data of a cover $\{U_i\}$ and transition functions T_{ij}, we can recover the vector bundle (up to unique isomorphism) by "gluing together the various $U_i \times \mathbb{R}^n$ along $U_i \cap U_j$ using T_{ij}."

14.1.2. The sheaf of sections. Fix a rank n vector bundle $V \to M$. The sheaf of sections $\mathscr{F}$ of V (Exercise 2.2.G) is an $\mathscr{O}_M$-module—given any open set U, we can multiply a section over U by a function on U and get another section.

Moreover, given a trivialization over U, the sections over U are naturally identified with n-tuples of functions of U.

$$U \times \mathbb{R}^n$$
$$\pi \downarrow\uparrow \quad n\text{-tuple of functions}$$
$$U$$

Thus, given a trivialization, over each open set U_i, we have an isomorphism $\mathscr{F}|_{U_i} \xleftrightarrow{\quad} \mathscr{O}_{U_i}^{\oplus n}$. As mentioned in §14.0.1, we say that such an $\mathscr{F}$ is a **locally free sheaf of rank** n.

14.1.3. Transition functions for the sheaf of sections. Suppose we have a vector bundle on M, along with a trivialization over an open cover U_i. Suppose we have a section of the vector bundle over M. (This discussion will apply with M replaced by any open subset.) Then over each U_i, the section corresponds to an n-tuple of functions over U_i, say $\vec{s}^i$.

14.1.A. EXERCISE. Show that over $U_i \cap U_j$, the vector-valued function $\vec{s}^i$ is related to $\vec{s}^j$ by the (same) transition functions: $T_{ij}\vec{s}^i = \vec{s}^j$. (Don't do this too quickly—make sure your i's and j's are on the correct side.)

Given a locally free sheaf $\mathscr{F}$ with rank n, and a trivializing open neighborhood of $\mathscr{F}$ (an open cover $\{U_i\}$ such that over each U_i, $\mathscr{F}|_{U_i} \cong \mathscr{O}_{U_i}^{\oplus n}$ as $\mathscr{O}$-modules), we have transition functions $T_{ij} \in GL_n(\mathscr{O}(U_i \cap U_j))$ satisfying the cocycle condition (14.1.1.1). Thus the data of a locally free sheaf of rank n is *equivalent* to the data of a vector bundle of rank n. This change of perspective is useful, and is similar to an earlier change of perspective when we introduced ringed spaces: understanding spaces is the same as understanding (sheaves of) functions on the spaces, and understanding vector bundles (a type of "space over M") is the same as understanding (sheaves of) sections.

14.1.4. *Definition.* A rank 1 locally free sheaf is called an **invertible sheaf**. (Unimportant aside: "Invertible sheaf" is a heinous term for something that is essentially a line bundle. The motivation is that if X is a locally ringed space, and $\mathscr{F}$ and $\mathscr{G}$ are $\mathscr{O}_X$-modules with $\mathscr{F} \otimes_{\mathscr{O}_X} \mathscr{G} \cong \mathscr{O}_X$, then $\mathscr{F}$ and $\mathscr{G}$ are invertible sheaves [MO33489]. Thus in the monoid of $\mathscr{O}_X$-modules under tensor product, invertible sheaves are the invertible elements. We will never use this fact. People often informally use the phrase "line bundle" when they mean "invertible sheaf." The phrase *line sheaf* has been proposed but has not caught on.)

14.1.5. *Notation.* Recall (Definition 6.2.6) that given a section s of $\mathscr{F}$ and a section t of $\mathscr{G}$, we have a section $s \otimes t$ of $\mathscr{F} \otimes \mathscr{G}$. If $\mathscr{F}$ is an invertible sheaf, this section is often denoted by st.

14.1.6. Locally free sheaves more generally.

We can generalize the notion of locally free sheaves to schemes (or, more generally, ringed spaces) without change. A **locally free sheaf of rank** n **on a scheme** X is defined as an $\mathscr{O}_X$-module $\mathscr{F}$ that is locally a free sheaf of rank n. Precisely, there is an open cover $\{U_i\}$ of X such that for each U_i, $\mathscr{F}|_{U_i} \cong \mathscr{O}_{U_i}^{\oplus n}$. This open cover determines transition functions—the data of a cover $\{U_i\}$ of X, and functions $T_{ij} \in GL_n(\mathscr{O}(U_i \cap U_j))$ satisfying the cocycle condition (14.1.1.1)—which in turn determine the locally free sheaf. As before, given these data, we can find the sections over any open set U. Informally, they are sections of the free sheaves over each $U \cap U_i$ that agree on overlaps. More formally, for each i, they are

$$\vec{s}^i = \begin{pmatrix} s_1^i \\ \vdots \\ s_n^i \end{pmatrix} \in \Gamma(U \cap U_i, \mathscr{O}_X)^{\oplus n},$$

satisfying $T_{ij} \vec{s}^i = \vec{s}^j$ on $U \cap U_i \cap U_j$.

You should think of these as vector bundles, but just keep in mind that they are not the "same," just equivalent notions. We will later (Definition 17.1.5) define the "total space" of the vector bundle $V \to X$ (a scheme over X) in terms of the sheaf version of Spec. But the locally free sheaf perspective will prove to be more useful. As one example: The definition of a locally free sheaf is much shorter than that of a vector bundle.

As in our motivating discussion, it is sometimes convenient to let the rank vary among connected components, or to consider infinite rank locally free sheaves.

14.1.7. Pulling back vector bundles.

14.1.B. ESSENTIAL EXERCISE: PULLING BACK VECTOR BUNDLES. Suppose $\pi: X \to Y$ is a morphism of ringed spaces. Recall the definition of $\pi^* : Mod_{\mathscr{O}_Y} \to Mod_{\mathscr{O}_X}$, pulling back $\mathscr{O}$-modules. (Now is a good time to do Exercise 7.2.D, if you haven't done it before.) Suppose $n \in \mathbb{Z}^+$.

(a) Show that (canonically) $\pi^*(\mathscr{O}_Y^{\oplus n}) \cong \mathscr{O}_X^{\oplus n}$.
(b) If $\mathscr{G}$ is a locally free sheaf on Y of rank n, show that $\pi^*\mathscr{G}$ is a locally free sheaf on X of rank n.
(c) If $\mathscr{G}$ is a locally free sheaf on Y of rank n, and $\{U_i\}$ are trivializing neighborhoods for $\mathscr{G}$ on Y, with transition matrices T_{ij}, show that $\{\pi^{-1}U_i\}$ are trivializing neighborhoods for $\pi^*\mathscr{G}$ on X, with transition matrices π^*T_{ij}. (Do you see what that means?)

14.1.8. The problem with locally free sheaves (or, equivalently vector bundles), as opposed to vector spaces.

Recall that $\mathscr{O}_X$-modules form an abelian category: we can talk about kernels, cokernels, and so forth, and we can do homological algebra. Similarly, vector spaces form an abelian category. But locally free sheaves (i.e., vector bundles), along with reasonably natural maps between them (those that arise as maps of $\mathscr{O}_X$-modules), don't form an abelian category. As a motivating example in the category of manifolds, consider the map of the trivial line bundle on $\mathbb{R}$ (with coordinate t) to itself, corresponding to multiplying by the coordinate t. Then this map jumps rank, and if you try to define a kernel or cokernel you will get confused.

This problem is resolved by enlarging our notion of nice $\mathscr{O}_X$-modules in a natural way, to quasicoherent sheaves.

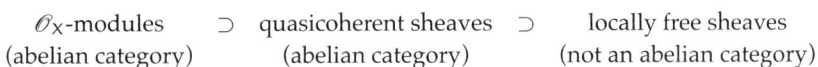

$\mathscr{O}_X$-modules	$\supset$	quasicoherent sheaves	$\supset$	locally free sheaves
(abelian category)		(abelian category)		(not an abelian category)

You can turn this into a *definition* of quasicoherent sheaves, equivalent to those we gave in §6.1. We want a notion that is local on X, of course. So we ask for the smallest abelian subcategory of $Mod_{\mathscr{O}_X}$ that is "local" and includes vector bundles. It turns out that the main obstruction to vector bundles' being an abelian category is the failure of cokernels of maps of locally free sheaves—as $\mathscr{O}_X$-modules—to be locally free; we could define quasicoherent sheaves to be those $\mathscr{O}_X$-modules that are locally cokernels, yielding a description that works more generally on ringed spaces, as described in Exercise 6.3.B. You may wish to later check that our definitions are equivalent to these.

Similarly, in the locally Noetherian setting, finite rank locally free sheaves will sit in a nice smaller abelian category, that of *coherent sheaves*.

quasicoherent sheaves	⊃	coherent sheaves	⊃	finite rank locally free sheaves
(abelian category)		(abelian category)		(not an abelian category)

14.1.9. Useful constructions on locally free sheaves: $\mathcal{H}om$, dual, ⊗.

We now give some useful constructions in the form of a series of exercises about locally free sheaves on a scheme. They are useful, important, and surprisingly nontrivial! Two hints: Exercises 14.1.C–14.1.I will apply for ringed spaces in general, so you shouldn't use special properties of schemes. Furthermore, they are all local on X, so you can reduce to the case where the locally free sheaves in question are actually free.

14.1.C. EXERCISE. Suppose $\mathscr{F}$ and $\mathscr{G}$ are locally free sheaves on X of rank m and n, respectively. Show that $\mathcal{H}om_{\mathscr{O}_X}(\mathscr{F},\mathscr{G})$ is a locally free sheaf of rank mn.

14.1.D. EXERCISE. If $\mathscr{E}$ is a locally free sheaf on X of (finite) rank n, Exercise 14.1.C implies that $\mathscr{E}^\vee := \mathcal{H}om(\mathscr{E},\mathscr{O}_X)$ is also a locally free sheaf of rank n. This is called the **dual** of $\mathscr{E}$ (cf. §2.3.4). Given transition functions for $\mathscr{E}$, describe transition functions for $\mathscr{E}^\vee$. (Note that if $\mathscr{E}$ is rank 1, i.e., invertible, the transition functions of the dual are the inverse of the transition functions of the original.) Show that $\mathscr{E} \cong \mathscr{E}^{\vee\vee}$. (Caution: Your argument showing that there is a canonical isomorphism $\mathscr{F} \overset{\sim}{\longleftrightarrow} (\mathscr{F}^\vee)^\vee$ better not also show that there is an isomorphism $\mathscr{F} \overset{\sim}{\longleftrightarrow} \mathscr{F}^\vee$! We will see an example in §15.1 of a locally free $\mathscr{F}$ that is not isomorphic to its dual: the invertible sheaf $\mathscr{O}(1)$ on $\mathbb{P}^n$.)

14.1.E. EXERCISE. If $\mathscr{F}$ and $\mathscr{G}$ are locally free sheaves, show that $\mathscr{F} \otimes \mathscr{G}$ is a locally free sheaf. (Here ⊗ is tensor product as $\mathscr{O}_X$-modules, defined in Exercise 2.6.K.) If $\mathscr{F}$ is an invertible sheaf, show that $\mathscr{F} \otimes \mathscr{F}^\vee \cong \mathscr{O}_X$.

14.1.F. EXERCISE. Recall that tensor products tend to be only right-exact in general (Exercise 1.2.H). Show that tensoring by a locally free sheaf is exact. More precisely, if $\mathscr{F}$ is a locally free sheaf, and $\mathscr{G}' \to \mathscr{G} \to \mathscr{G}''$ is an exact sequence of $\mathscr{O}_X$-modules, then so is $\mathscr{G}' \otimes \mathscr{F} \to \mathscr{G} \otimes \mathscr{F} \to \mathscr{G}'' \otimes \mathscr{F}$. (Possible hint: It may help to check exactness by checking exactness at stalks. Recall that the tensor product of stalks can be identified with the stalk of the tensor product, so, for example, there is a "natural" isomorphism $(\mathscr{G} \otimes_{\mathscr{O}_X} \mathscr{F})_p \overset{\sim}{\longleftrightarrow} \mathscr{G}_p \otimes_{\mathscr{O}_{X,p}} \mathscr{F}_p$, Exercise 2.6.K(b).)

14.1.G. EXERCISE. If $\mathscr{E}$ is a locally free sheaf of finite rank, and $\mathscr{F}$ and $\mathscr{G}$ are $\mathscr{O}_X$-modules, show that

$$\mathcal{H}om(\mathscr{F},\mathscr{G} \otimes \mathscr{E}) \cong \mathcal{H}om(\mathscr{F} \otimes \mathscr{E}^\vee,\mathscr{G}) \cong \mathcal{H}om(\mathscr{F},\mathscr{G}) \otimes \mathscr{E}.$$

In particular, if $\mathscr{D}$ is a locally free sheaf of finite rank, $\mathcal{H}om(\mathscr{D},\mathscr{G}) \cong \mathscr{D}^\vee \otimes \mathscr{G}$. (Possible hint: First consider the case where $\mathscr{E}$ is free.)

14.1.10. The Picard group.

14.1.H. EXERCISE AND IMPORTANT DEFINITION. Show that the invertible sheaves on X, up to isomorphism, form an abelian group under tensor product. This is called the **Picard group** of X, and is denoted by Pic X.

14.1.I. EASY EXERCISE (CF. EXERCISE 14.1.B). Show that Pic is a contravariant functor from the category of ringed spaces to the category *Ab* of abelian groups.

14.1.J.* EXERCISE (FOR THOSE WITH SUFFICIENT COMPLEX-ANALYTIC BACKGROUND). Recall the analytification functor (Exercises 7.3.P, 7.3.Q, and 11.2.K), that takes a complex finite type reduced scheme and produces a complex analytic space.

(a) If $\mathscr{L}$ is an invertible sheaf on a complex (algebraic) variety X, define (up to unique isomorphism) the corresponding invertible sheaf on the complex variety X_{an}.
(b) Show that the induced map $\text{Pic}\, X \to \text{Pic}\, X_{an}$ is a group homomorphism.
(c) Show that this construction is functorial, i.e., if $\pi\colon X \to Y$ is a morphism of complex varieties, the following diagram commutes:

$$
\begin{array}{ccc}
\text{Pic}\, Y & \xrightarrow{\pi^*} & \text{Pic}\, X \\
\downarrow & & \downarrow \\
\text{Pic}\, Y_{an} & \xrightarrow{\pi_{an}^*} & \text{Pic}\, X_{an}
\end{array}\quad,
$$

where the vertical maps are the ones you have defined.

14.1.K.* EXERCISE (FOR THOSE WITH SUFFICIENT ARITHMETIC BACKGROUND; SEE ALSO PROPOSITION 15.4.16 AND §15.5.5). Recall the definition of the ring of integers $\mathcal{O}_K$ in a number field K, Remark 10.7.1. A **fractional ideal** $\mathfrak{a}$ of $\mathcal{O}_K$ is a nonzero $\mathcal{O}_K$-submodule of K such that there is a nonzero $a \in \mathcal{O}_K$ such that $a\mathfrak{a} \subset \mathcal{O}_K$. Products of fractional ideals are defined analogously to products of ideals in a ring (defined in Exercise 3.4.C): $\mathfrak{a}\mathfrak{b}$ consists of (finite) $\mathcal{O}_K$-linear combinations of products of elements of $\mathfrak{a}$ and elements of $\mathfrak{b}$. Thus fractional ideals form an abelian semigroup under multiplication, with $\mathcal{O}_K$ as the identity. In fact, fractional ideals of $\mathcal{O}_K$ form a group.

(a) Explain how a fractional ideal on a ring of integers in a number field yields an invertible sheaf. (Although we won't need this, it is worth noting that a fractional ideal is the same as an invertible sheaf *with a trivialization at the generic point*.)
(b) A fractional ideal is **principal** if it is of the form $r\mathcal{O}_K$ for some $r \in K^\times$. Show that any two fractional ideals that differ by a principal fractional ideal yield the same invertible sheaf.
(c) Show that two fractional ideals that yield the same invertible sheaf differ by a principal ideal.
(d) The **class group** is defined to be the group of fractional ideals modulo the principal ideals (i.e., modulo $K^\times$). Give an isomorphism of the class group with the Picard group of $\mathcal{O}_K$.

(This discussion applies to any Dedekind domain. See Exercise 15.5.O for a follow-up.)

14.2 Locally Free Sheaves on Schemes in Particular

We now discuss aspects of locally free sheaves more specific to schemes. Based on your intuition for line bundles on manifolds, you might hope that every point has a "small" open neighborhood on which all invertible sheaves (or locally free sheaves) are trivial. Sadly, this is not the case. We will eventually see (§19.11.1) that for the curve $y^2 - x^3 - x = 0$ in $\mathbb{A}_\mathbb{C}^2$, every nonempty open set has nontrivial invertible sheaves. (This will use the fact that it is an open subset of an *elliptic curve*.) This is a persistent problem in algebraic geometry: the Zariski topology is too blunt.

14.2.A. EASY EXERCISE. Prove that locally free sheaves on a scheme are quasicoherent.

Similarly, finite rank locally free sheaves are always finite type, and if $\mathcal{O}_X$ is coherent, finite rank locally free sheaves on X are coherent. (Remember: If $\mathcal{O}_X$ is not coherent, then coherence is a pretty useless notion on X.)

14.2.B. UNIMPORTANT EXERCISE. Use the example of Exercise 6.1.B(b) to show that not every quasicoherent sheaf is locally free.

14.2.C. EXERCISE. Show that every (finite) rank n vector bundle on $\mathbb{A}^1_k$ is trivial of rank n. Hint: Finitely generated modules over a principal ideal domain are finite direct sums of cyclic modules, as mentioned in Remark 6.4.4. (Be careful: Before invoking this, you need to be sure that the module in question is finitely generated!) See the aside in §15.4.12 for the difficult generalization to $\mathbb{A}^n_k$.

14.2.1. *Definition.* We define **rational (and regular) sections of a locally free sheaf** on a scheme X just as we did rational (and regular) functions (see, for example, §6.6 and §7.5).

14.2.D. EXERCISE. Suppose s is a section of a locally free sheaf $\mathscr{F}$ on a scheme X. Define the notion of the **closed subscheme cut out by** $s = 0$, denoted (for obvious reasons) by $V(s)$. Be sure to check that your definition is independent of choices! (This exercise gives a new solution to Exercise 4.5.S.) Hint: Given a trivialization over an open set U, s corresponds to a number of functions $f_1, \ldots$ on U; on U, take the scheme cut out by these functions. Alternate hint that avoids coordinates: Figure out how to define it as the largest closed subscheme on which s restricts to 0.

14.2.E. EASY EXERCISE. Show that locally free sheaves on locally Noetherian normal schemes satisfy "Algebraic Hartogs's Lemma": sections defined away from a set of codimension at least 2 extend over that set. (Algebraic Hartogs's Lemma for Noetherian normal schemes is Theorem 13.5.19.)

14.2.F. EASY EXERCISE. Suppose s is a nonzero rational section of an invertible sheaf on a locally Noetherian normal scheme. Show that if s has no poles, then s is regular. (Hint: Exercise 13.5.I.)

14.2.2. Locally free sheaves in short exact sequences.

14.2.G. EXERCISE (POSSIBLE HELP FOR LATER PROBLEMS).

(a) Suppose

$$(14.2.2.1) \qquad 0 \longrightarrow \mathscr{F}' \longrightarrow \mathscr{F} \longrightarrow \mathscr{F}'' \longrightarrow 0$$

is a short exact sequence of quasicoherent sheaves on X. Suppose $U = \operatorname{Spec} A$ is an affine open set where $\mathscr{F}'$, $\mathscr{F}''$ are free, say, $\mathscr{F}'|_{\operatorname{Spec} A} = \widetilde{A}^{\oplus a}$, $\mathscr{F}''|_{\operatorname{Spec} A} = \widetilde{A}^{\oplus b}$. (Here a and b are assumed to be finite for convenience, but this is not necessary, so feel free to generalize to the infinite rank case.) Show that $\mathscr{F}$ is also free on $\operatorname{Spec} A$, and that $0 \to \mathscr{F}' \to \mathscr{F} \to \mathscr{F}'' \to 0$ can be interpreted as coming from the tautological exact sequence $0 \to A^{\oplus a} \to A^{\oplus(a+b)} \to A^{\oplus b} \to 0$. (As a consequence, given an exact sequence of quasicoherent sheaves (14.2.2.1) where $\mathscr{F}'$ and $\mathscr{F}''$ are locally free, $\mathscr{F}$ must also be locally free.)

(b) In the finite rank case, show that, given an open covering by trivializing affine open sets (of the form described in (a)), the transition functions (really, matrices) of $\mathscr{F}$ may be interpreted as block upper triangular matrices, where the top left $a \times a$ blocks are transition functions for $\mathscr{F}'$, and the bottom $b \times b$ blocks are transition functions for $\mathscr{F}''$.

14.2.H. EXERCISE. Suppose (14.2.2.1) is a short exact sequence of quasicoherent sheaves on X. By Exercise 14.2.G(a), if $\mathscr{F}'$ and $\mathscr{F}''$ are locally free, then $\mathscr{F}$ is too.

(a) If $\mathscr{F}$ and $\mathscr{F}''$ are locally free *of finite rank*, show that $\mathscr{F}'$ is too. Hint: Reduce to the case $X = \operatorname{Spec} A$ and $\mathscr{F}$ and $\mathscr{F}''$ free. Interpret the map $\phi \colon \mathscr{F} \to \mathscr{F}''$ as an $n \times m$ matrix M with values in A, with m the rank of $\mathscr{F}$ and n the rank of $\mathscr{F}''$. For each point p of X, show that there exist n columns $\{c_1, \ldots, c_n\}$ of M that are linearly independent at p and hence near p (as linear independence is given by nonvanishing of the appropriate $n \times n$ determinant). Thus X can be covered by distinguished open subsets in bijection with the choices of n columns of M. Restricting to one subset and renaming columns, reduce to the case where the determinant of the first n columns of M is invertible. Then change coordinates on $A^{\oplus m} = \mathscr{F}(\operatorname{Spec} A)$ so that M with respect to the new coordinates is the identity matrix in the first n columns, and 0 thereafter. Finally, in this case interpret $\mathscr{F}'$ as $\widetilde{A^{\oplus(m-n)}}$.

(b) If $\mathscr{F}'$ and $\mathscr{F}$ are both locally free, show that $\mathscr{F}''$ need not be. Hint: Consider (14.2.3.1), which we can interpret as the closed subscheme exact sequence (9.1.2.1) for a point on $\mathbb{A}^1$. (If an injection of locally free sheaves $\alpha\colon\mathscr{F}'\hookrightarrow\mathscr{F}$ is such that the cokernel $\mathscr{F}''$ is also locally free, people sometimes say that α is an *injection of vector bundles*, not just of locally free sheaves. Implicitly, they are considering a category of vector bundles that is different from the category of locally free sheaves; the cokernel of a map of vector bundles is then always a vector bundle. See §14.2.4 for further discussion.)

14.2.I. EXERCISE.

(a) Suppose $0\to\mathscr{F}'\to\mathscr{F}\to\mathscr{F}''\to 0$ is a short exact sequence of finite rank locally free sheaves. Show that
$$0\longrightarrow\mathscr{F}''^{\vee}\longrightarrow\mathscr{F}^{\vee}\longrightarrow\mathscr{F}'^{\vee}\longrightarrow 0$$
(obtained by applying $\mathscr{H}om(\cdot,\mathscr{O}_X)$ to the original sequence) is also a short exact sequence of finite rank locally free sheaves.

(b)* Suppose $\cdots\mathscr{F}_2\to\mathscr{F}_1\to\mathscr{F}_0\to 0$ is an exact sequence of finite rank locally free sheaves. Show that the induced sequence
$$0\longrightarrow\mathscr{F}_0^{\vee}\longrightarrow\mathscr{F}_1^{\vee}\longrightarrow\mathscr{F}_2^{\vee}\longrightarrow\cdots$$
is also an exact sequence of finite rank locally free sheaves.

14.2.3. Vector bundle maps.

Maps of vector bundles (in the more conventional sense) are more restrictive than morphisms of quasicoherent sheaves that happen to be locally free sheaves. A locally free subsheaf of a locally free sheaf does not always yield a "subvector bundle" of a vector bundle. The archetypal example is the exact sequence of quasicoherent sheaves on $\mathbb{A}^1_k=\operatorname{Spec}k[t]$ corresponding to the following exact sequence of $k[t]$-modules:

(14.2.3.1) $$0\longrightarrow tk[t]\longrightarrow k[t]\longrightarrow k\longrightarrow 0.$$

The locally free sheaf $\widetilde{tk[t]}$ is a subsheaf of $\widetilde{k[t]}$, but it does not correspond to a "subvector bundle"; the cokernel is not a vector bundle. (This example was mentioned in §14.1.8.)

Because maps of locally free sheaves that are actually "maps of vector bundles" have some particularly good and useful behavior, it seems worthwhile to give them a name. There is no accepted name, so we will just choose one.

14.2.4. *Definition.* Let X be a scheme (or, indeed, more generally, a ringed space), and let $\mathscr{E}$ and $\mathscr{F}$ be two finite rank locally free sheaves on X, of rank $a+b$ and $a+c$, respectively. We say that $\phi\colon\mathscr{E}\to\mathscr{F}$ is a **rank a map of vector bundles** if for every point $p\in X$, there is an open neighborhood U of p with a commutative diagram

$$
\begin{array}{ccc}
\mathscr{E}|_U & \xrightarrow{\ \ \phi\ \ } & \mathscr{F}|_U \\
\uparrow{\scriptstyle\sim} & & \uparrow{\scriptstyle\sim} \\
\mathscr{O}_U^{\oplus(a+b)} & \longrightarrow\ \mathscr{O}_U^{\oplus a}\hookrightarrow & \mathscr{O}_U^{\oplus(a+c)},
\end{array}
$$

where the map $\mathscr{O}_U^{\oplus(a+b)}\to\mathscr{O}_U^{\oplus a}$ is projection to the first a summands, and the map $\mathscr{O}_U^{\oplus a}\hookrightarrow\mathscr{O}_U^{\oplus(a+c)}$ is inclusion as the first a summands. We say that ϕ is a **map of vector bundles** if near every point $p\in X$, ϕ is a rank a map of vector bundles for some a. If $b=0$, we say ϕ is an **injection of vector bundles** (as in Exercise 14.2.H(b)); we use this terminology sparingly.

14.2.5. *Important but straightforward observations.* (i) The notion "rank a map of vector bundles" commutes with any base change $Y\to X$. (ii) The quasicoherent sheaves $\ker\phi$, $\operatorname{im}\phi$, and coker ϕ are locally free (of finite rank b, a, and c, respectively), and their construction commutes with any base change $Y \to X$.

$$
\begin{array}{ccccccccc}
0 & \longrightarrow & \ker \phi & \longrightarrow & \mathscr{E} & \stackrel{\phi}{\longrightarrow} & \mathscr{F} & \longrightarrow & \operatorname{coker} \phi & \longrightarrow & 0 \ , \\
& & \uparrow \wr & & \uparrow \wr & & \uparrow \wr & & \uparrow \wr & & \\
0 & \longrightarrow & \mathcal{O}_U^{\oplus b} & \longrightarrow & \mathcal{O}_U^{\oplus(a+b)} & \longrightarrow & \mathcal{O}_U^{\oplus(a+c)} & \longrightarrow & \mathcal{O}_U^{\oplus c} & \longrightarrow & 0
\end{array}
$$

(iii) If $\phi\colon \mathscr{E} \to \mathscr{F}$ is a rank a map of vector bundles, then at any point $p \in X$, the rank of $\phi|_p \colon \mathscr{E}|_p \to \mathscr{F}|_p$ is a.

In the next exercises we give three criteria for when a morphism ϕ of finite rank locally free sheaves is a map of vector bundles.

14.2.J. EXERCISE (USEFUL CHARACTERIZATION OF "MAP OF VECTOR BUNDLES"). Suppose $\phi\colon \mathscr{E} \to \mathscr{F}$ is a morphism of finite rank locally free sheaves, where $\mathscr{F}$ has rank $a + c$. Show that ϕ is a rank a map of vector bundles if and only if coker ϕ is locally free of rank c. (In particular, every surjection of finite rank locally free sheaves is a map of vector bundles.)

14.2.K. EXERCISE. Suppose $\phi\colon \mathscr{E} \to \mathscr{F}$ is a morphism of finite rank locally free sheaves, and $p \in X$. Show that the natural map $(\ker \phi)|_p \to \ker(\phi|_p)$ is surjective if and only if ϕ is a map of vector bundles in some neighborhood of p.

14.2.6.* Tensor algebra constructions.

(These important ideas require more work, so you may wish to leave them for later.) For the next exercises, recall the following. If M is an A-module, then the **tensor algebra** $T^\bullet(M)$ is a noncommutative algebra, graded by $\mathbb{Z}^{\geq 0}$, defined as follows. $T^0(M) = A$, $T^n(M) = M \otimes_A \cdots \otimes_A M$ (where n terms appear in the product), and multiplication is what you expect.

The **symmetric algebra** $\mathrm{Sym}^\bullet M$ is a symmetric algebra, graded by $\mathbb{Z}^{\geq 0}$, defined as the quotient of $T^\bullet(M)$ by the (two-sided) ideal generated by all elements of the form $x \otimes y - y \otimes x$ for all $x, y \in M$. Thus $\mathrm{Sym}^n M$ is the quotient of $M \otimes \cdots \otimes M$ by the relations of the form $m_1 \otimes \cdots \otimes m_n - m_1' \otimes \cdots \otimes m_n'$, where $(m_1', \ldots, m_n')$ is a rearrangement of $(m_1, \ldots, m_n)$.

The **exterior algebra** $\wedge^\bullet M$ is defined to be the quotient of $T^\bullet M$ by the (two-sided) ideal generated by all elements of the form $x \otimes x$ for all $x \in M$. Expanding $(a + b) \otimes (a + b)$, we see that $a \otimes b = -b \otimes a$ in $\wedge^2 M$. This implies that if 2 is invertible in A (e.g., if A is a field of characteristic not 2), $\wedge^n M$ is the quotient of $M \otimes \cdots \otimes M$ by the relations of the form $m_1 \otimes \cdots \otimes m_n - (-1)^{\mathrm{sgn}(\sigma)} m_{\sigma(1)} \otimes \cdots \otimes m_{\sigma(n)}$, where σ is a permutation of $\{1, \ldots, n\}$. The exterior algebra is a "skew-commutative" A-algebra.

Better: both Sym and $\wedge$ can be defined by universal properties. For example, any map of A-modules $M^{\otimes n} \to N$ that is symmetric in the n entries factors uniquely through $\mathrm{Sym}_A^n(M)$.

It is most correct to write $T_A^\bullet(M)$, $\mathrm{Sym}_A^\bullet(M)$, and $\wedge_A^\bullet(M)$, but the "base ring" A is usually omitted for convenience.

14.2.L. EXERCISE. If $\mathscr{F}$ is locally free of rank m, show that $T^n \mathscr{F}$, $\mathrm{Sym}^n \mathscr{F}$, and $\wedge^n \mathscr{F}$ are locally free, and find their ranks. We note that in this case, $\wedge^{\mathrm{rank}\, \mathscr{F}} \mathscr{F}$ is denoted by det $\mathscr{F}$, and is called the **determinant (line) bundle** or (both better and worse) the **determinant locally free sheaf**.

(Remark: These constructions can be defined for $\mathcal{O}$-modules on an arbitrary ringed space.)

You can also define the sheaf of noncommutative algebras $T^\bullet \mathscr{F}$, the sheaf of commutative algebras $\mathrm{Sym}^\bullet \mathscr{F}$, and the sheaf of skew-commutative algebras $\wedge^\bullet \mathscr{F}$.

14.2.M. EXERCISE. Suppose $0 \to \mathscr{F}' \to \mathscr{F} \to \mathscr{F}'' \to 0$ is an exact sequence of locally free sheaves. Show that for any r, there is a filtration of $\mathrm{Sym}^r \mathscr{F}$,

$$\mathrm{Sym}^r \mathscr{F} = \mathscr{G}^0 \supset \mathscr{G}^1 \supset \cdots \supset \mathscr{G}^r \supset \mathscr{G}^{r+1} = 0,$$

with subquotients

$$\mathscr{G}^p/\mathscr{G}^{p+1} \cong (\text{Sym}^p\,\mathscr{F}') \otimes (\text{Sym}^{r-p}\,\mathscr{F}'').$$

(Here are two different possible hints for this and Exercise 14.2.N: (1) Interpret the transition matrices for $\mathscr{F}$ as block upper triangular, with two blocks, where one diagonal block gives the transition matrices for $\mathscr{F}'$, and the other gives the transition matrices for $\mathscr{F}''$ (cf. Exercise 14.2.2.1(b)). Then appropriately interpret the transition matrices for $\text{Sym}^r\,\mathscr{F}$ as block upper triangular, with $r+1$ blocks. (2) It suffices to consider a small enough affine open set $\text{Spec}\,A$, where $\mathscr{F}'$, $\mathscr{F}$, $\mathscr{F}''$ are free, and to show that your construction behaves well with respect to localization at an element $f \in A$. In such an open set, the sequence is $0 \to A^{\oplus p} \to A^{\oplus(p+q)} \to A^{\oplus q} \to 0$ by the Exercise 14.2.G. Let $e_1,\dots,e_p$ be the standard basis of $A^{\oplus p}$, and $f_1,\dots,f_q$ be the the standard basis of $A^{\oplus q}$. Let $e_1',\dots,e_p'$ be denote by the images of $e_1,\dots,e_p$ in $A^{\oplus(p+q)}$. Let $f_1',\dots,f_q'$ be any lifts of $f_1,\dots,f_q$ to $A^{\oplus(p+q)}$. Note that f_i' is well-defined modulo $e_1',\dots,e_p'$. Note that

$$\text{Sym}^r\,\mathscr{F}|_{\text{Spec}\,A} \cong \oplus_{i=0}^r \text{Sym}^i\,\mathscr{F}'|_{\text{Spec}\,A} \otimes_{\mathscr{O}_{\text{Spec}\,A}} \text{Sym}^{r-i}\,\mathscr{F}''|_{\text{Spec}\,A}.$$

Show that $\mathscr{G}^p := \oplus_{i=p}^r \text{Sym}^i\,\mathscr{F}'|_{\text{Spec}\,A} \otimes_{\mathscr{O}_{\text{Spec}\,A}} \text{Sym}^{r-i}\,\mathscr{F}''|_{\text{Spec}\,A}$ gives a well-defined (locally free) subsheaf that is independent of the choices made, e.g., of the bases $e_1,\dots,e_p$ and $f_1,\dots,f_q$, and the lifts $f_1',\dots,f_q'$.)

14.2.N. USEFUL EXERCISE. Suppose $0 \to \mathscr{F}' \to \mathscr{F} \to \mathscr{F}'' \to 0$ is an exact sequence of locally free sheaves. Show that for any r, there is a filtration of $\wedge^r\mathscr{F}$,

$$\wedge^r\mathscr{F} = \mathscr{G}^0 \supset \mathscr{G}^1 \supset \cdots \supset \mathscr{G}^r \supset \mathscr{G}^{r+1} = 0,$$

with subquotients

$$\mathscr{G}^p/\mathscr{G}^{p+1} \cong (\wedge^p\mathscr{F}') \otimes (\wedge^{r-p}\mathscr{F}'')$$

for each p. In particular, if the sheaves have finite rank, then $\det\mathscr{F} = (\det\mathscr{F}') \otimes (\det\mathscr{F}'')$.

14.2.O. EXERCISE. Suppose $\mathscr{F}$ is locally free of rank n. Describe a map $\wedge^r\mathscr{F} \times \wedge^{n-r}\mathscr{F} \to \wedge^n\mathscr{F}$ that induces an isomorphism $\wedge^r\mathscr{F} \xrightarrow{\sim} (\wedge^{n-r}\mathscr{F})^\vee \otimes \wedge^n\mathscr{F}$. This is called a **perfect pairing of vector bundles**. (If you know about perfect pairings of vector spaces, do you see why this is a generalization?) You might use this later in showing duality of Hodge numbers of regular varieties over algebraically closed fields, Exercise 21.5.L.

14.2.P. EXERCISE (DETERMINANT LINE BUNDLES BEHAVE WELL IN EXACT SEQUENCES). Suppose $0 \to \mathscr{F}_1 \to \cdots \to \mathscr{F}_n \to 0$ is an exact sequence of finite rank locally free sheaves on X. Show that "the alternating product of determinant bundles is trivial":

$$\det(\mathscr{F}_1) \otimes \det(\mathscr{F}_2)^\vee \otimes \det(\mathscr{F}_3) \otimes \det(\mathscr{F}_4)^\vee \otimes \cdots \otimes \det(\mathscr{F}_n)^{(-1)^n} \cong \mathscr{O}_X.$$

(Hint: Break the exact sequence into short exact sequences. Use Exercise 14.2.H(a) to show that they are short exact sequences of *finite rank locally free sheaves*. Then use Exercise 14.2.N.)

14.2.Q. EXERCISE. Suppose $\mathscr{F}$ is a quasicoherent sheaf. Define the quasicoherent sheaves $T^n\,\mathscr{F}$, $\text{Sym}^n\,\mathscr{F}$, and $\wedge^n\,\mathscr{F}$. (One possibility: Describe them on each affine open set, and use the characterization of Important Exercise 6.2.D.)

14.3 More Pleasant Properties of Finite Type and Coherent Sheaves

We now describe a number of important facts about quasicoherent sheaves when we have some finiteness conditions.

14.3.A. EXERCISE ($\mathscr{H}om$ BEHAVES REASONABLY IF THE SOURCE IS FINITELY PRESENTED; CF. EXERCISE 1.5.H).

(a) ("$\mathscr{H}om(fpr, qcoh)$ is qcoh") Suppose $\mathscr{F}$ is a finitely presented (quasicoherent) sheaf on X, and $\mathscr{G}$ is a quasicoherent sheaf on X. Show that $\mathscr{H}om(\mathscr{F},\mathscr{G})$ is a quasicoherent sheaf. Hint: Take a "finite presentation" of $\mathscr{F}$ locally: on $U \subset X$, $\mathscr{O}^{\oplus m} \to \mathscr{O}^{\oplus n} \to \mathscr{F} \to 0$. Apply the left-exact functor (from $\mathscr{O}_X$-modules to $\mathscr{O}_X$-modules, §2.6.4) $\mathcal{H}om_{\mathscr{O}_X}(\cdot, \mathscr{G})$. (Aside: For an example of quasicoherent sheaves $\mathscr{F}$ and $\mathscr{G}$ on a scheme X such that $\mathcal{H}om(\mathscr{F}, \mathscr{G})$ is not quasicoherent, let A be a discrete valuation ring with uniformizer t, let $X = \operatorname{Spec} A$, and let $\mathscr{F} = \widetilde{M}$ and $\mathscr{G} = \widetilde{N}$ with $M = \oplus_{i=1}^{\infty} A$ and $N = A$. Then $M_t = \oplus_{i=1}^{\infty} A_t$, and, of course, $N_t = A_t$. Consider the homomorphism $\phi \colon M_t \to N_t$ sending 1 in the ith factor of M_t to $1/t^i$. Then ϕ is not the localization of any element of $\operatorname{Hom}_A(M, N)$.)

(b) ("$\mathcal{H}om(fpr, coh)$ is coh") If, further, $\mathscr{G}$ is coherent, show that $\mathcal{H}om(\mathscr{F}, \mathscr{G})$ is also coherent.

(c) Suppose $\mathscr{F}$ is a finitely presented sheaf on X, and $\mathscr{G}$ is a quasicoherent sheaf on X. Show that $\mathcal{H}om(\mathscr{F}, \cdot)$ gives a left-exact covariant functor $QCoh_X \to QCoh_X$ and $Coh_X \to Coh_X$, and that $\mathcal{H}om(\cdot, \mathscr{G})$ gives a left-exact contravariant functor from finitely presented sheaves on X to $QCoh_X$ (cf. Exercise 2.6.I).

14.3.1. *Duals of coherent sheaves.* From Exercise 14.3.A(b), if $\mathscr{F}$ is coherent, its dual $\mathscr{F}^{\vee} := \mathcal{H}om(\mathscr{F}, \mathscr{O}_X)$ (see §2.3.4) is too. This generalizes the notion of duals of vector bundles in Exercise 14.1.D. Your argument there generalizes to show that there is always a natural "double dual" morphism $\mathscr{F} \to (\mathscr{F}^{\vee})^{\vee}$. Unlike in the vector bundle case, this is not always an isomorphism. (For an example, let $\mathscr{F}$ be the coherent sheaf associated to $k[t]/(t)$ on $\mathbb{A}^1 = \operatorname{Spec} k[t]$, and show that $\mathscr{F}^{\vee} = 0$.) Coherent sheaves for which the "double dual" morphism is an isomorphism are called **reflexive sheaves**, but we won't use this notion. The canonical map $\mathscr{F} \otimes \mathscr{F}^{\vee} \to \mathscr{O}_X$ is called the *trace* map—can you see why?

14.3.B. EXERCISE. Suppose

$$(14.3.1.1) \qquad 0 \longrightarrow \mathscr{F} \longrightarrow \mathscr{G} \longrightarrow \mathscr{H} \longrightarrow 0$$

is an exact sequence of quasicoherent sheaves on a scheme X, where $\mathscr{H}$ is a locally free quasicoherent sheaf, and suppose $\mathscr{E}$ is a quasicoherent sheaf. By left-exactness of $\mathcal{H}om$ (Exercise 2.6.I),

$$0 \longrightarrow \mathcal{H}om(\mathscr{H}, \mathscr{E}) \longrightarrow \mathcal{H}om(\mathscr{G}, \mathscr{E}) \longrightarrow \mathcal{H}om(\mathscr{F}, \mathscr{E}) \longrightarrow 0$$

is exact except possibly on the right. Show that it is also exact on the right. (Hint: This is local, so you can assume that X is affine, say, $\operatorname{Spec} A$, and $\mathscr{H} = \widetilde{A^{\oplus I}}$, so (14.3.1.1) can be written as $0 \to M \to N \to A^{\oplus I} \to 0$. Show that this exact sequence splits, so we can write $N = M \oplus A^{\oplus I}$ in a way that respects the exact sequence.) In particular, if $\mathscr{F}, \mathscr{G}$, and $\mathscr{H}$ are coherent, and $\mathscr{H}$ is locally free, then we have an exact sequence of coherent sheaves

$$0 \longrightarrow \mathscr{H}^{\vee} \longrightarrow \mathscr{G}^{\vee} \longrightarrow \mathscr{F}^{\vee} \longrightarrow 0.$$

14.3.C. EXERCISE ("Supp($FT\mathscr{F}$) IS CLOSED"). Suppose $\mathscr{F}$ is a sheaf of abelian groups. (a) Show that the support of a finite type quasicoherent sheaf on a scheme X is a closed subset. (Hint: Exercise 6.6.B(b).) (b) Show that the support of a quasicoherent sheaf need not be closed. (Hint: If $A = \mathbb{C}[t]$, then $\mathbb{C}[t]/(t-a)$ is an A-module supported at a. Consider $\oplus_{a \in \mathbb{C}} \mathbb{C}[t]/(t-a)$. Be careful: this example won't work if $\oplus$ is replaced by $\prod$.)

We next come to a geometric interpretation of Nakayama's Lemma, which is why Nakayama's Lemma should be considered a geometric fact (with an algebraic proof). Informally: if some sections of a finite type quasicoherent sheaf generate the sheaf "at p" (the fiber at p), then they generate "near p" (in an actual open neighborhood of p).

14.3.D. USEFUL EXERCISE: GEOMETRIC NAKAYAMA (FOR FINITE TYPE SHEAVES, GENERATORS AT p ARE GENERATORS NEAR p). Suppose X is a scheme, and $\mathscr{F}$ is a finite type quasicoherent sheaf. Show that if $U \subset X$ is an open neighborhood of $p \in X$ and $a_1, \ldots, a_n \in \mathscr{F}(U)$

so that their images $a_1|_p, \ldots, a_n|_p$ generate the fiber $\mathscr{F}|_p$ (defined as $\mathscr{F}_p \otimes \kappa(p)$, §4.3.8), then there is an affine open neighborhood $p \in \operatorname{Spec} A \subset U$ of p such that "$a_1|_{\operatorname{Spec} A}, \ldots, a_n|_{\operatorname{Spec} A}$ generate $\mathscr{F}|_{\operatorname{Spec} A}$" in the following senses:

(i) $a_1|_{\operatorname{Spec} A}, \ldots, a_n|_{\operatorname{Spec} A}$ generate $\mathscr{F}(\operatorname{Spec} A)$ as an A-module;
(ii) for any $q \in \operatorname{Spec} A$, $a_1, \ldots, a_n$ generate the stalk $\mathscr{F}_q$ as an $\mathscr{O}_{X,q}$-module (and hence for any $q \in \operatorname{Spec} A$, the fibers $a_1|_q, \ldots, a_n|_q$ generate the fiber $\mathscr{F}|_q$ as a $\kappa(q)$-vector space).

In particular, if $\mathscr{F}|_p = 0$, then there exists an open neighborhood V of p such that $\mathscr{F}|_V = 0$.

14.3.E. USEFUL EXERCISE (FOR FINITELY PRESENTED SHEAVES, LOCALLY FREE = FREE STALKS). Suppose $\mathscr{F}$ is a finitely presented sheaf on a scheme X. Show that if $\mathscr{F}_p$ is a free $\mathscr{O}_{X,p}$-module for some $p \in X$, then $\mathscr{F}$ is locally free in some open neighborhood of p. Hence $\mathscr{F}$ is locally free if and only if $\mathscr{F}_p$ is a free $\mathscr{O}_{X,p}$-module for all $p \in X$. Translation: For a finitely presented sheaf, local freeness is a stalk-local property, and free stalks imply local freeness nearby.

Extended hint: Find an open neighborhood U of p, and n elements of $\mathscr{F}(U)$ that generate $\mathscr{F}_p$. Using Geometric Nakayama, Exercise 14.3.D, show that the sections generate $\mathscr{F}_q$ for all q in some open neighborhood Y of p in U. Thus you have described a surjection $\mathscr{O}_Y^{\oplus n} \to \mathscr{F}|_Y$. Show that the kernel of this map is finite type (using Lemma 6.4.6, "finitely presented implies always finitely presented"), and hence has closed support (say, $Z \subset Y$), which does not contain p. Thus $\mathscr{O}_{Y \backslash Z}^{\oplus n} \to \mathscr{F}|_{Y \backslash Z}$ is an isomorphism.

This is enlightening in a number of ways. It shows that for finitely presented sheaves, local freeness is a stalk-local condition. Furthermore, on an integral scheme, any finitely presented sheaf $\mathscr{F}$ is automatically free over the generic point (do you see why?), so every finitely presented sheaf on an integral scheme is locally free over a dense open subset. And any finitely presented sheaf that is 0 at the generic point of an irreducible scheme is necessarily 0 on a dense open subset. The last two sentences show the utility of generic points; such statements would have been more mysterious in classical algebraic geometry. An example showing the necessity of finiteness hypotheses: the $\mathbb{Z}$-module $A \subset \mathbb{Q}$, where A consists of fractions with square-free denominator; see [Stacks, tag 065J].

14.3.F. EXERCISE. (Torsion-free and torsion sheaves were defined in §6.1.4.)

(a) Show that torsion-free coherent sheaves on a regular (hence implicitly locally Noetherian) curve are locally free.
(b) Show that torsion coherent sheaves on a quasicompact regular integral curve are supported at a finite number of closed points.
(c) Suppose $\mathscr{F}$ is a coherent sheaf on a quasicompact (for convenience) regular curve. Describe a canonical short exact sequence $0 \to \mathscr{F}_{\text{tors}} \to \mathscr{F} \to \mathscr{F}_{\text{lf}} \to 0$, where $\mathscr{F}_{\text{tor}}$ is a torsion sheaf, and $\mathscr{F}_{\text{lf}}$ is locally free.

To answer Exercise 14.3.F, use Useful Exercise 14.3.E (local freeness can be checked at stalks) to reduce to the discrete valuation ring case, and recall Remark 6.4.4, the structure theorem for finitely generated modules over a principal ideal domain A: any such module can be written as the direct sum of principal modules $A/(a)$. Hence:

14.3.2. Proposition — *If M is a finitely generated module over a principal ideal domain, then M is torsion-free if and only if M is free.*

(Exercise 24.1.B is closely related.)

Proposition 14.3.2 is false without the finite generation hypothesis: consider $M = K(A)$ for a suitably general ring A. It is also false if we give up the principal ideal domain hypothesis: consider $(x, y) \subset \mathbb{C}[x, y]$ and $(x, y) \subset \mathbb{C}[x, y]/(xy)$. (These examples require some verification.) These examples show that Exercise 14.3.F(a) is false if we give up the "dimension 1" or "regular" hypothesis.

14.3.3. The fiber and rank of a quasicoherent sheaf at a point.

We now apply the definitions of fiber and rank of an $\mathscr{O}_X$-module at a point $p \in X$ (§4.3.8) to a quasicoherent sheaf $\mathscr{F}$ on a scheme X. The *fiber* of $\mathscr{F}$ at p is the vector space

$$\mathscr{F}|_p := \mathscr{F}_p/\mathfrak{m}\mathscr{F}_p = \mathscr{F}_p \otimes_{\mathscr{O}_{X,p}} \kappa(p),$$

where $\mathfrak{m}$ is the maximal ideal of $\mathscr{O}_{X,p}$, and $\kappa(p)$ is as usual the residue field $\mathscr{O}_{X,p}/\mathfrak{m}$ at p. A section of $\mathscr{F}$ over an open set containing p can be said to take on a value at p, which is an element of this vector space. The *rank* of a quasicoherent sheaf $\mathscr{F}$ at a point p is $\dim_{\kappa(p)} \mathscr{F}_p/\mathfrak{m}\mathscr{F}_p$ (possibly infinite). More explicitly, on any affine set $\operatorname{Spec} A$ where $p = [\mathfrak{p}]$ and $\mathscr{F}(\operatorname{Spec} A) = M$, the rank is $\dim_{\kappa(A/\mathfrak{p})} M_\mathfrak{p}/\mathfrak{p}M_\mathfrak{p}$. Note that this definition of rank is consistent with the notion of rank of a locally free sheaf. In the locally free case, the rank is a (locally) constant function of the point. The converse is sometimes true; see Exercise 14.3.K below.

14.3.G. EXERCISE. Consider the coherent sheaf $\mathscr{F}$ on $\mathbb{A}^1_k = \operatorname{Spec} k[t]$ corresponding to the module $k[t]/(t)$. Find the rank of $\mathscr{F}$ at every point of $\mathbb{A}^1$. Don't forget the generic point!

14.3.H. EXERCISE. Show that at any point, $\operatorname{rank}(\mathscr{F} \oplus \mathscr{G}) = \operatorname{rank}(\mathscr{F}) + \operatorname{rank}(\mathscr{G})$ and $\operatorname{rank}(\mathscr{F} \otimes \mathscr{G}) = \operatorname{rank}\mathscr{F}\operatorname{rank}\mathscr{G}$. (Hint: Show that direct sums and tensor products commute with ring quotients and localizations, i.e., $(M \oplus N) \otimes_R (R/I) \cong M/IM \oplus N/IN$, $(M \otimes_R N) \otimes_R (R/I) \cong (M \otimes_R R/I) \otimes_{R/I} (N \otimes_R R/I) \cong M/IM \otimes_{R/I} N/IM$, etc.)

If X is irreducible, and $\mathscr{F}$ is a quasicoherent (usually coherent) sheaf on X, then rank $\mathscr{F}$ (with no mention of a point) by convention means at the generic point. (For example, a rank 0 quasicoherent sheaf on an integral scheme is a torsion quasicoherent sheaf; see Definition 6.1.5.)

14.3.I. EASY EXERCISE. Show that the rank of coherent sheaves on an integral scheme X is additive in exact sequences: if

$$0 \longrightarrow \mathscr{F} \longrightarrow \mathscr{G} \longrightarrow \mathscr{H} \longrightarrow 0$$

is an exact sequence of coherent sheaves, show that

$$\operatorname{rank}\mathscr{F} - \operatorname{rank}\mathscr{G} + \operatorname{rank}\mathscr{H} = 0.$$

Hint: Localization is exact. (Caution: Your argument will use the fact that the rank is at the generic point; the example

$$0 \longrightarrow \widetilde{k[t]} \xrightarrow{\times t} \widetilde{k[t]} \longrightarrow \widetilde{k[t]/(t)} \longrightarrow 0$$

on $\mathbb{A}^1_k$ shows that rank at a closed point is not additive in exact sequences.)

If $\mathscr{F}$ is a finite type quasicoherent sheaf, then the rank of $\mathscr{F}$ at p is finite, and by Nakayama's Lemma, the rank is the minimal number of generators of $M_\mathfrak{p}$ as an $A_\mathfrak{p}$-module.

14.3.J. IMPORTANT EXERCISE (RANK IS UPPER SEMICONTINUOUS FOR FINITE TYPE QUASICOHERENT SHEAVES). If $\mathscr{F}$ is a finite type quasicoherent sheaf on X, show that $\operatorname{rank}(\mathscr{F})$ is an upper semicontinuous function on X. Hint: Generators at a point p are generators nearby by Geometric Nakayama's Lemma, Exercise 14.3.D. (The example in Exercise 14.3.C shows the necessity of the finite type hypothesis.)

14.3.K. IMPORTANT EXERCISE (FOR FINITE TYPE QUASICOHERENT SHEAVES ON REDUCED SCHEMES, CONSTANT RANK = VECTOR BUNDLE).

(a) If X is reduced, $\mathscr{F}$ is a finite type quasicoherent sheaf on X, and the rank of $\mathscr{F}$ is constant, show that $\mathscr{F}$ is locally free.

Hint: Prove the result in an open neighborhood of an arbitrary point p as follows. Suppose $n = \operatorname{rank}_p \mathscr{F}$. Choose n generators of the fiber $\mathscr{F}|_p$ (a basis as a $\kappa(p)$-vector space). By Geometric Nakayama's Lemma 14.3.D, we can find an affine open neighborhood $\operatorname{Spec} A \subset X$ of p, with $\mathscr{F}|_{\operatorname{Spec} A} = \widetilde{M}$, so that the chosen generators $\mathscr{F}|_p$ lift to generators $m_1, \ldots, m_n$ of

M. Let $\phi\colon A^{\oplus n} \twoheadrightarrow M$ $((r_1,\ldots,r_n)\mapsto\sum r_i m_i)$ be the corresponding surjection. Suppose $\ker\phi\neq 0$—say $(r_1,\ldots,r_n)\in\ker\phi$, with $r_i\neq 0$. As X is reduced, there is some $\mathfrak{p}$ where $r_i\notin\mathfrak{p}$. Then r_i is invertible in $A_\mathfrak{p}$, so $M_\mathfrak{p}$ has fewer than n generators, contradicting the constancy of rank.

(b) Then use upper semicontinuity of rank (Exercise 14.3.J) to show that finite type quasicoherent sheaves on a reduced scheme are locally free on a dense open set.

(c) Show that part (a) can be false without the condition of X being reduced. (Hint: Consider $A = k[x]/(x^2)$, $M = k$.)

14.3.L. EXERCISE. Suppose X is reduced, and $\phi\colon \mathscr{E}\to\mathscr{F}$ is a morphism of finite rank locally free sheaves. Show that ϕ is a rank a map of vector bundles (Definition 14.2.4) if and only if $\phi|_\mathfrak{p}\colon\mathscr{E}|_\mathfrak{p}\to\mathscr{F}|_\mathfrak{p}$ is rank a for all points $p\in X$. (Hint: Use the previous exercise, and Exercise 14.2.J.)

You can use the notion of rank to help visualize finite type quasicoherent sheaves, or even quasicoherent sheaves. For example, finite type quasicoherent sheaves generalize finite rank vector bundles as follows: to each point there is an associated vector space, and although the ranks can jump, they fit together in families as well as one might hope. You might try to visualize the example of Example 14.3.G. Nonreducedness can fit into the picture as well—how would you picture the coherent sheaf on $\operatorname{Spec} k[\epsilon]/(\epsilon^2)$ corresponding to $k[\epsilon]/(\epsilon)$ (Exercise 14.3.K(c))? How about $k[\epsilon]/(\epsilon^2)\oplus k[\epsilon]/(\epsilon)$?

14.3.4. *Degree of a finite morphism at a point.* Suppose $\pi\colon X\to Y$ is a finite morphism. Then $\pi_*\mathscr{O}_X$ is a finite type (quasicoherent) sheaf on Y, and the rank of this sheaf at a point p is called the **degree** of the finite morphism at p. By Exercise 14.3.J, the degree of π is a upper semicontinuous function on Y. The degree can jump: consider the closed embedding of a point into a line corresponding to $k[t]\to k$ given by $t\mapsto 0$. It can also be constant in cases that you might initially find surprising—see Exercise 10.3.4, where the degree is always 2, but the 2 is obtained in a number of different ways.

14.3.M. EXERCISE. Suppose $\pi\colon X\to Y$ is a finite morphism. By unwinding the definition, verify that the degree of π at p is the dimension of the space of functions on the scheme-theoretic preimage of p, considered as a vector space over the residue field $\kappa(p)$. In particular, the degree is zero if and only if $\pi^{-1}(p)$ is empty.

14.4 Pushforwards of Quasicoherent Sheaves

We conclude the chapter by discussing pushforward and pullbacks of quasicoherent sheaves, their properties, and some applications.

Suppose $B\to A$ is a morphism of rings. Recall (from Exercise 1.4.E) that $(\cdot\otimes_B A, \cdot_B)$ is an adjoint pair between the categories of A-modules and B-modules: we have a bijection

$$\operatorname{Hom}_A(N\otimes_B A, M)\cong\operatorname{Hom}_B(N, M_B)$$

functorial in both arguments. These constructions behave well with respect to localization (in an appropriate sense), and hence work (often) in the category of quasicoherent sheaves on schemes (and, indeed, always in the category of $\mathscr{O}$-modules on ringed spaces—see Definition 14.5.6—although we won't particularly care). The easier construction $(M\mapsto M_B)$ will turn into our old friend pushforward. The other $(N\mapsto A\otimes_B N)$ will be a relative of pullback, whom I'm reluctant to call an "old friend."

We begin with the pushforwards, for which we have already done much of the work.

The main moral of this section is that in "reasonable" situations, the pushforward of a quasicoherent sheaf is quasicoherent, and that this can be understood in terms of one of the module constructions defined above. We begin with a motivating example:

14.4.A. EXERCISE. Let $\pi\colon \operatorname{Spec} A \to \operatorname{Spec} B$ be a morphism of affine schemes, and suppose M is an A-module, so $\widetilde{M}$ is a quasicoherent sheaf on $\operatorname{Spec} A$. Give an isomorphism $\pi_*\widetilde{M} \xrightarrow{\sim} \widetilde{M_B}$. (Hint: There is only one reasonable way to proceed: look at distinguished open sets.)

In particular, $\pi_*\widetilde{M}$ is quasicoherent. Perhaps more important, this implies that the pushforward of a quasicoherent sheaf under an affine morphism is also quasicoherent.

14.4.B. EXERCISE. If $\pi\colon X \to Y$ is an affine morphism, show that π_* is an exact functor $QCoh_X \to QCoh_Y$.

The following result generalizes the fact that the pushforward of a quasicoherent sheaf under an affine morphism is also quasicoherent.

14.4.C. ESSENTIAL EXERCISE (COROLLARY TO EXERCISE 6.2.G: "$\pi_*\colon QCoh_X \to QCoh_Y$ FOR QCQS π"). Suppose $\pi\colon X \to Y$ is a quasicompact quasiseparated morphism, and $\mathscr{F}$ is a quasicoherent sheaf on X. Show that $\pi_*\mathscr{F}$ is a quasicoherent sheaf on Y. (For counterexamples without the quasicompact or quasiseparated hypotheses, see [Stacks, tag 078C].)

Coherent sheaves do not always push forward to coherent sheaves. For example, consider the structure morphism $\pi\colon \mathbb{A}^1_k \to \operatorname{Spec} k$, corresponding to $k \to k[t]$. Then $\pi_*\mathscr{O}_{\mathbb{A}^1_k}$ is the quasicoherent sheaf corresponding to $k[t]$, which is not a finitely generated k-module. But in good situations, coherent sheaves do push forward. For example:

14.4.D. EXERCISE. Suppose $\pi\colon X \to Y$ is a finite morphism of locally Noetherian schemes. If $\mathscr{F}$ is a coherent sheaf on X, show that $\pi_*\mathscr{F}$ is a coherent sheaf. Hint: Show first that $\pi_*\mathscr{O}_X$ is finite type.

Once we define cohomology of quasicoherent sheaves, we will quickly prove that if $\mathscr{F}$ is a coherent sheaf on $\mathbb{P}^n_k$, then $\Gamma(\mathbb{P}^n_k, \mathscr{F})$ is a finite-dimensional k-module, and, more generally, if $\mathscr{F}$ is a coherent sheaf on $\operatorname{Proj} S_\bullet$, then $\Gamma(\operatorname{Proj} S_\bullet, \mathscr{F})$ is a coherent A-module (where $S_0 = A$). This is a special case of the fact that the "pushforwards of coherent sheaves by projective morphisms are also coherent sheaves" (Theorem 18.7.1(d)). (The notion of projective morphism, a relative version of $\operatorname{Proj} S_\bullet \to \operatorname{Spec} A$, will be defined in §17.3.)

More generally, given Noetherian hypotheses, pushforwards of coherent sheaves by proper morphisms are also coherent sheaves (Grothendieck's Coherence Theorem 18.9.1).

14.4.1. *Remark.* Vector bundles do not always pushforward to vector bundles, which is another reason to work in the larger category of (quasi)coherent sheaves. (Can you think of an example? Hint: $\operatorname{Spec} k[t]/(t) \to \operatorname{Spec} k[t]$.) But in particularly good circumstances they can, which is one of the purposes of the Cohomology and Base Change Theorem 25.1.6.

14.5 Pullbacks of Quasicoherent Sheaves: Three Different Perspectives

We next discuss the pullback of a quasicoherent sheaf: if $\pi\colon X \to Y$ is a morphism of schemes, π^* is a covariant functor $QCoh_Y \to QCoh_X$. (In fact, our discussion will apply readily to pulling back $\mathscr{O}$-modules on ringed spaces; see §14.5.5 and §14.5.6.) The notion of the pullback of a quasicoherent sheaf can be confusing on first (and second) glance. (For example, it is *not* the inverse image sheaf, although we will see that it is related to it.)

14.5.1. Here are three contexts in which you have seen the pullback, or can understand it quickly. It may be helpful to keep these in mind, to keep you anchored in the long discussion that follows. Suppose $\mathscr{G}$ is a quasicoherent sheaf on a scheme Y.

  (i) *(restriction to open subsets)* If $i\colon U \hookrightarrow Y$ is an open embedding, then $i^*\mathscr{G}$ is $\mathscr{G}|_U$, the restriction of $\mathscr{G}$ to U (Example 2.2.8).
 (ii) *(restriction to points)* If $i\colon p \hookrightarrow Y$ is the "inclusion" of a point p in Y (for example, the closed embedding of a closed point; see Exercise 7.3.L(b)), then $i^*\mathscr{G}$ is $\mathscr{G}|_p$, the fiber of $\mathscr{G}$ at p (Definition 4.3.8).

The similarity of the notation $\mathscr{G}|_U$ and $\mathscr{G}|_p$ is precisely because both are pullbacks. Pullbacks (especially to locally closed subschemes or generic points) are often called *restriction*. For this reason, if $\pi\colon X \to Y$ is some sort of inclusion (such as a locally closed embedding, or an "inclusion of a generic point"), then $\pi^*\mathscr{G}$ is often written as $\mathscr{G}|_X$ and called the **restriction of $\mathscr{G}$ to X**, when π can be interpreted as some type of "inclusion."

(iii) *(pulling back vector bundles)* Suppose $\mathscr{G}$ is a locally free sheaf on Y, and $\pi\colon X \to Y$ is any morphism. Then $\pi^*\mathscr{G}$ is the pullback of $\mathscr{G}$ as a vector bundle, as defined in §14.1.7. (This will be established in §14.5.6.)

14.5.2. *Strategy.* We will see three different ways of thinking about the pullback. Each has significant disadvantages, but together they give a good understanding.

(a) Because we are understanding quasicoherent sheaves in terms of affine open sets, and modules over the corresponding rings, we begin with an interpretation in this vein. This will be very useful for proving and understanding facts. The disadvantage is that it is annoying to make a definition out of this (when the target is not affine), because gluing arguments can be tedious.

(b) As we saw with fibered product, gluing arguments can be made simpler using universal properties, so our second "definition" will be by universal property. This is elegant, but has the disadvantage that it still needs a construction, and because it works in the larger category of $\mathscr{O}$-modules, it isn't clear from the universal property that it takes quasicoherent sheaves to quasicoherent sheaves. But if the target is affine, our construction of (a) is easily seen to satisfy the universal property. Furthermore, the universal property is "local on the target": if $\pi\colon X \to Y$ is any morphism, $i\colon U \hookrightarrow Y$ is an open embedding, and $\mathscr{G}$ is a quasicoherent sheaf on Y, then if $\pi^*\mathscr{G}$ exists, then its restriction to $\pi^{-1}(U)$ is (canonically identified with) $(\pi|_U)^*(\mathscr{G}|_U)$. Thus if the pullback exists in general (even as an $\mathscr{O}$-module), affine-locally on Y it looks like the construction of (a) (and thus is quasicoherent).

(c) The third definition is one that works on ringed spaces in general. It is short, and is easily seen to satisfy the universal property. It doesn't obviously take quasicoherent sheaves to quasicoherent sheaves (at least in the way that we have defined quasicoherent sheaves)— a priori it takes quasicoherent sheaves to $\mathscr{O}$-modules. But thanks to the discussion at the end of (b) above, which used (a), this shows that the pullback of a quasicoherent sheaf is indeed quasicoherent.

14.5.3. First attempt at describing the pullback, using affines. Suppose $\pi\colon X \to Y$ is a morphism of schemes, and $\mathscr{G}$ is a quasicoherent sheaf on Y. We want to define the pullback quasicoherent sheaf $\pi^*\mathscr{G}$ on X in terms of affine open sets on X and Y. Suppose $\operatorname{Spec} A \subset X$, $\operatorname{Spec} B \subset Y$ are affine open sets, with $\pi(\operatorname{Spec} A) \subset \operatorname{Spec} B$. Suppose $\mathscr{G}|_{\operatorname{Spec} B} \cong \widetilde{N}$. Perhaps motivated by the fact that pullback should relate to tensor product, we want

$$(14.5.3.1) \qquad \qquad \Gamma(\operatorname{Spec} A, \pi^*\mathscr{G}) = N \otimes_B A.$$

More precisely, we would like $\Gamma(\operatorname{Spec} A, \pi^*\mathscr{G})$ and $N \otimes_B A$ to be identified. This could mean that we use this to construct a definition of $\pi^*\mathscr{G}$, by "gluing all this information together" (and showing it is well-defined). Or it could mean that we define $\pi^*\mathscr{G}$ in some other way, and then find a natural identification (14.5.3.1). The first approach can be made to work (and §14.5.4 is the first step), but we will follow the second.

14.5.4. We begin this project by *fixing* an affine open subset $\operatorname{Spec} B \subset Y$. To avoid confusion, let $\phi = \pi|_{\pi^{-1}(\operatorname{Spec} B)}\colon \pi^{-1}(\operatorname{Spec} B) \to \operatorname{Spec} B$. We will define a quasicoherent sheaf on $\pi^{-1}(\operatorname{Spec} B)$ that will turn out to be $\phi^*(\mathscr{G}|_{\operatorname{Spec} B})$ (and will also be the restriction of $\pi^*\mathscr{G}$ to $\pi^{-1}(\operatorname{Spec} B)$).

If $\operatorname{Spec} A_f \subset \operatorname{Spec} A$ is a distinguished open set, then

$$\Gamma(\operatorname{Spec} A_f, \phi^*\mathscr{G}) = N \otimes_B A_f = (N \otimes_B A)_f = \Gamma(\operatorname{Spec} A, \phi^*\mathscr{G})_f,$$

where "=" means "canonically isomorphic." Define the restriction map $\Gamma(\operatorname{Spec} A, \phi^*\mathcal{G}) \to \Gamma(\operatorname{Spec} A_f, \phi^*\mathcal{G})$,

(14.5.4.1) $\Gamma(\operatorname{Spec} A, \phi^*\mathcal{G}) \longrightarrow \Gamma(\operatorname{Spec} A, \phi^*\mathcal{G}) \otimes_A A_f,$

by $\alpha \mapsto \alpha \otimes 1$ (of course). Thus $\phi^*\mathcal{G}$ is (or: extends to) a quasicoherent sheaf on $\pi^{-1}(\operatorname{Spec} B)$ (by Exercise 6.2.D).

We have now defined a quasicoherent sheaf on $\pi^{-1}(\operatorname{Spec} B)$, for every affine open subset $\operatorname{Spec} B \subset Y$. We want to show that this construction, as $\operatorname{Spec} B$ varies, glues into a single quasicoherent sheaf on X.

You are welcome to do this gluing appropriately, for example, using the distinguished affine base of Y (§6.2.1). This works, but can be confusing, so we take another approach.

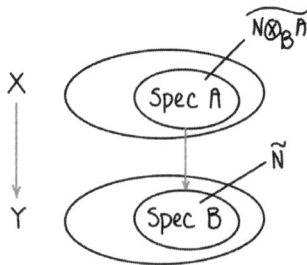

Figure 14.1 *The pullback of a quasi-coherent sheaf (module-theoretically).*

14.5.5. Universal property definition of pullback.

Suppose $\pi: X \to Y$ is a morphism of ringed spaces, and $\mathcal{G}$ is an $\mathcal{O}_Y$-module. (We are of course interested in the case where π is a morphism of schemes, and $\mathcal{G}$ is quasicoherent. Even once we specialize our discussion to schemes, much of our discussion will extend without change to this more general situation.) We "define" the pullback $\pi^*\mathcal{G}$ as an $\mathcal{O}_X$-module, using the following adjointness universal property: for any $\mathcal{O}_X$-module $\mathcal{F}$, there is a bijection $\operatorname{Hom}_{\mathcal{O}_X}(\pi^*\mathcal{G}, \mathcal{F}) \xleftarrow{\sim} \operatorname{Hom}_{\mathcal{O}_Y}(\mathcal{G}, \pi_*\mathcal{F})$, and these bijections are functorial in $\mathcal{F}$. By universal property nonsense, this determines $\pi^*\mathcal{G}$ up to unique isomorphism; we just need to make sure that it exists (which is why the word "define" is in quotes). Notice that we avoid worrying about when the pushforward of a quasicoherent sheaf $\mathcal{F}$ is quasicoherent by working in the larger category of $\mathcal{O}$-modules.

14.5.A. IMPORTANT EXERCISE. If Y is affine, say, $Y = \operatorname{Spec} B$, show that the construction of the quasicoherent sheaf in §14.5.4 satisfies this universal property of pullback of $\mathcal{G}$. Thus calling this sheaf $\pi^*\mathcal{G}$ is justified. Hint: Interpret both sides of the alleged bijection explicitly. The adjointness in the ring/module case should turn up.

14.5.B. IMPORTANT EXERCISE. Suppose $i: U \hookrightarrow X$ is an open embedding of ringed spaces, and $\mathcal{F}$ is an $\mathcal{O}_X$-module. Show that $\mathcal{F}|_U$ satisfies the universal property of $i^*\mathcal{F}$ (and thus deserves to be called $i^*\mathcal{F}$). In other words, for each $\mathcal{O}_U$-module $\mathcal{E}$, describe a bijection

$$\operatorname{Hom}_{\mathcal{O}_U}(\mathcal{F}|_U, \mathcal{E}) \xleftarrow{\sim} \operatorname{Hom}_{\mathcal{O}_X}(\mathcal{F}, i_*\mathcal{E}),$$

functorial in $\mathcal{E}$.

We next show that if $\pi^*\mathcal{G}$ satisfies the universal property (for the morphism $\pi: X \to Y$), then if $j: V \hookrightarrow Y$ is any open subset, and $i: U = \pi^{-1}(V) \hookrightarrow X$ (see (14.5.5.1)), then $(\pi^*\mathcal{G})|_U$ satisfies the universal property for $\pi|_U: U \to V$. Thus $(\pi^*\mathcal{G})|_U$ deserves to be called $\pi|_U^*(\mathcal{G}|_V)$. You will notice that we really need to work with $\mathcal{O}$-modules, not just with quasicoherent sheaves.

(14.5.5.1)

$$\pi^{-1}(V) == U \overset{i}{\hookrightarrow} X$$

$$\pi|_U \downarrow \qquad \qquad \downarrow \pi$$

$$V \overset{j}{\hookrightarrow} Y$$

If $\mathcal{F}'$ is an $\mathcal{O}_U$-module, we have a series of bijections (using Important Exercise 14.5.B and adjointness of pullback and pushforward):

$$\operatorname{Hom}_{\mathscr{O}_U}((\pi^*\mathscr{G})|_U, \mathscr{F}') \overset{\sim}{\longleftrightarrow} \operatorname{Hom}_{\mathscr{O}_U}(i^*(\pi^*\mathscr{G}), \mathscr{F}')$$

$$\overset{\sim}{\longleftrightarrow} \operatorname{Hom}_{\mathscr{O}_X}(\pi^*\mathscr{G}, i_*\mathscr{F}')$$

$$\overset{\sim}{\longleftrightarrow} \operatorname{Hom}_{\mathscr{O}_Y}(\mathscr{G}, \pi_* i_*\mathscr{F}')$$

$$\overset{\sim}{\longleftrightarrow} \operatorname{Hom}_{\mathscr{O}_Y}(\mathscr{G}, j_*(\pi|_U)_*\mathscr{F}')$$

$$\overset{\sim}{\longleftrightarrow} \operatorname{Hom}_{\mathscr{O}_V}(j^*\mathscr{G}, (\pi|_U)_*\mathscr{F}')$$

$$\overset{\sim}{\longleftrightarrow} \operatorname{Hom}_{\mathscr{O}_V}(\mathscr{G}|_V, (\pi|_U)_*\mathscr{F}').$$

We have thus described a bijection

$$\operatorname{Hom}_{\mathscr{O}_U}((\pi^*\mathscr{G})|_U, \mathscr{F}') \overset{\sim}{\longleftrightarrow} \operatorname{Hom}_{\mathscr{O}_V}(\mathscr{G}|_V, (\pi|_U)_*\mathscr{F}') ,$$

which is (by construction) functorial in $\mathscr{F}'$. Hence the discussion in the first paragraph of §14.5.5 is justified. For example, thanks to Important Exercise 14.5.A, the pullback π^* exists if Y is an open subset of an affine scheme.

At this point, we could show that the pullback exists, following the idea behind the construction of the fibered product: we would start with the definition when Y is affine, and "glue." We will instead take another route.

14.5.6. Third definition: Pullback of $\mathscr{O}$-modules via explicit construction. Suppose $\pi\colon X \to Y$ is a morphism of ringed spaces, and $\mathscr{G}$ is an $\mathscr{O}_Y$-module. Of course, our case of interest is: π is a morphism of schemes, and $\mathscr{G}$ is quasicoherent. Now $\pi^{-1}\mathscr{G}$ is a $\pi^{-1}\mathscr{O}_Y$-module. (Notice that we are using the ringed space $(X, \pi^{-1}\mathscr{O}_Y)$, not $(X, \mathscr{O}_X)$. The inverse image construction π^{-1} was discussed in §2.7.) Furthermore, $\mathscr{O}_X$ is also a $\pi^{-1}\mathscr{O}_Y$-module, via the map $\pi^{-1}\mathscr{O}_Y \to \mathscr{O}_X$ that is part of the data of the morphism π. Define the **pullback** of $\mathscr{G}$ by π as the $\mathscr{O}_X$-module

(14.5.6.1) $$\pi^*\mathscr{G} := \pi^{-1}\mathscr{G} \otimes_{\pi^{-1}\mathscr{O}_Y} \mathscr{O}_X.$$

It is immediate that pullback is a covariant functor $\pi^*\colon Mod_{\mathscr{O}_Y} \to Mod_{\mathscr{O}_X}$, and that $\pi^*\mathscr{O}_Y$ is canonically isomorphic to $\mathscr{O}_X$. We have now established §14.5.1(iii), that pullback of quasicoherent sheaves subsumes our older definition of pullback of locally free sheaves.

14.5.C. IMPORTANT EXERCISE. Show that this definition (14.5.6.1) of pullback satisfies the universal property (§14.5.5). Thus the pullback exists, at least as a functor $Mod_{\mathscr{O}_Y} \to Mod_{\mathscr{O}_X}$.

14.5.D. IMPORTANT EXERCISE (PULLBACK IS A COVARIANT FUNCTOR ON QUASICOHERENT SHEAVES). Show that if $\pi\colon X \to Y$ is a morphism of schemes, then π^* gives a covariant functor $QCoh_Y \to QCoh_X$. (You might use §14.5.4, Exercise 14.5.B, Exercise 14.5.A, and/or §14.5.5.)

If $\mathscr{G}$ is a quasicoherent sheaf on Y, note that we can interpret a global section of Y as a map $s\colon \mathscr{O}_Y \to \mathscr{G}$ (the section of $\mathscr{G}$ is the image of the section 1 of $\mathscr{O}_Y$). Note that a section of $\mathscr{G}$ is the data of a map $\mathscr{O}_Y \to \mathscr{G}$. By Exercise 14.5.D, if $s\colon \mathscr{O}_Y \to \mathscr{G}$ is a section of $\mathscr{G}$, then there is a natural section $\pi^*s\colon \mathscr{O}_X \to \pi^*\mathscr{G}$ of $\pi^*\mathscr{G}$. We call π^*s the **pullback of the section** s.

The following is immediate from the universal property, and Exercise 14.5.D. The "quasicompact and quasiseparated" hypotheses are solely to ensure that π_* indeed sends $QCoh_X$ to $QCoh_Y$ (Exercise 14.4.C).

14.5.7. Proposition — *Suppose $\pi\colon X \to Y$ is a quasicompact, quasiseparated morphism. Then $(\pi^*\colon QCoh_Y \to QCoh_X, \pi_*\colon QCoh_X \to QCoh_Y)$ are an adjoint pair: there is an isomorphism*

(14.5.7.1) $$\operatorname{Hom}_{\mathscr{O}_X}(\pi^*\mathscr{G}, \mathscr{F}) \overset{\sim}{\longleftrightarrow} \operatorname{Hom}_{\mathscr{O}_Y}(\mathscr{G}, \pi_*\mathscr{F}),$$

functorial in both $\mathscr{F} \in QCoh_X$ and $\mathscr{G} \in QCoh_Y$.

14.5.E. IMPORTANT EXERCISE (PULLBACK IS A right-exact FUNCTOR). Suppose $\pi: X \to Y$ is a morphism of schemes. Show that $\pi^*: QCoh_Y \to QCoh_X$ is a right-exact functor. Possible hint: $\otimes$ is right-exact.

For an important class of morphisms, called *flat morphisms*, we will see that pullback is an exact functor (Exercise 24.1.H).

14.5.F. EXERCISE (PULLBACK PRESERVES FINITE TYPE AND FINITELY PRESENTED QUASI-COHERENT SHEAVES). Suppose that $\pi: X \to Y$ is a morphism of schemes, and $\mathscr{G}$ is a quasicoherent sheaf on Y. If $\mathscr{G}$ is finite type (resp., finitely presented), show that $\pi^*\mathscr{G}$ is as well. Hint: Cover Y by open sets for which there are surjections $\mathscr{O}_U^{\oplus n} \to \mathscr{G}|_U \to 0$ (resp., $\mathscr{O}_U^{\oplus m} \to \mathscr{O}_U^{\oplus n} \to \mathscr{G}|_U \to 0$). Apply the the right-exact functor π^*. (Hence if X is locally Noetherian, and $\mathscr{G}$ is coherent, then so is $\pi^*\mathscr{G}$. It is not always true that the pullback of a coherent sheaf is coherent, and the interested reader can think of a counterexample.)

14.5.G. EASY EXERCISE. Suppose $\xi: W \to X$ and $\pi: X \to Y$ are morphisms of schemes, and $\mathscr{G}$ is a quasicoherent sheaf on Y. Show that there is a canonical isomorphism $\xi^*\pi^*\mathscr{G} \overset{\sim}{\longleftrightarrow} (\pi \circ \xi)^*\mathscr{G}$.

14.5.H. EXERCISE. Suppose $\pi: X \to Y$ is a morphism of schemes, with $\pi(p) = q$, and $\mathscr{G}$ is a quasicoherent sheaf on Y.

(a) (pullback on stalks) Show that pullback induces an isomorphism

(14.5.7.2)
$$(\pi^*\mathscr{G})_p \overset{\sim}{\longrightarrow} \mathscr{G}_q \otimes_{\mathscr{O}_{Y,q}} \mathscr{O}_{X,p}.$$

(b) (pullback on fibers) Show that pullback induces an isomorphism

$$(\pi^*\mathscr{G})|_p \overset{\sim}{\longrightarrow} \mathscr{G}|_q \otimes_{\kappa(q)} \kappa(p).$$

Possible hint: Given a ring map $B \to A$ with $[\mathfrak{m}] \mapsto [\mathfrak{n}]$, where $\mathfrak{m} \subset A$ and $\mathfrak{n} \subset B$ are maximal ideals, show that $(N \otimes_B A) \otimes_A (A/\mathfrak{m}) \cong (N \otimes_B (B/\mathfrak{n})) \otimes_{B/\mathfrak{n}} (A/\mathfrak{m})$ by showing both sides are isomorphic to $N \otimes_B (A/\mathfrak{m})$.

14.5.I. EXERCISE: PULLBACK PRESERVES TENSOR PRODUCT. Suppose $\pi: X \to Y$ is a morphism of schemes, and $\mathscr{G}$ and $\mathscr{G}'$ are quasicoherent sheaves on Y. Show that $\pi^*(\mathscr{G} \otimes_{\mathscr{O}_Y} \mathscr{G}') = \pi^*\mathscr{G} \otimes_{\mathscr{O}_X} \pi^*\mathscr{G}'$.

14.5.J. UNIMPORTANT EXERCISE. Verify that the following is an example showing that pullback is not left-exact. Consider the exact sequence of sheaves on $\mathbb{A}_k^1$, where $i: p \hookrightarrow \mathbb{A}_k^1$ is the origin:

$$0 \longrightarrow \mathscr{O}_{\mathbb{A}_k^1}(-p) \longrightarrow \mathscr{O}_{\mathbb{A}_k^1} \longrightarrow i_*\mathscr{O}|_p \longrightarrow 0.$$

(This is the closed subscheme exact sequence (9.1.2.1) for $p \in \mathbb{A}_k^1$, and corresponds to the exact sequence of $k[t]$-modules $0 \to tk[t] \to k[t] \to k \to 0$. Warning: Here $\mathscr{O}|_p$ is not the stalk $\mathscr{O}_p$; it is the structure sheaf of the scheme p.) Restrict to p.

14.5.K. EXERCISE (THE PUSH-PULL MAP FOR QUASICOHERENT SHEAVES). Suppose that

$$
\begin{array}{ccc}
X' & \overset{\psi'}{\longrightarrow} & X \\
{\scriptstyle \pi'}\downarrow & & \downarrow{\scriptstyle \pi} \\
Y' & \overset{\psi}{\longrightarrow} & Y
\end{array}
$$

is a commutative diagram of schemes, and π and π' are both quasicompact and quasiseparated (so pushforward by both sends quasicoherent sheaves to quasicoherent sheaves). Describe a natural map (functorial in $\mathscr{F} \in QCoh_X$) of quasicoherent sheaves on Y',

$$\psi^* \pi_* \mathscr{F} \longrightarrow \pi'_* (\psi')^* \mathscr{F},$$

following Definition 2.7.4 of the push-pull map for sheaves in general, as you may have done in Exercise 7.2.D(f) in the case of $\mathscr{O}$-modules. (See Exercise 18.7.B for a generalization to cohomology.)

14.5.8. By applying the above exercise in the special case where Y' is a point q of Y, we see that there is a natural map from the fiber of the pushforward to the sections over the fiber:

(14.5.8.1) $$\pi_* \mathscr{F} \otimes \kappa(q) \longrightarrow \Gamma(\pi^{-1}(q), \mathscr{F}|_{\pi^{-1}(q)}).$$

One might hope that (14.5.8.1) is an isomorphism; i.e., that $\pi_* \mathscr{F}$ "glues together" the fibers $\Gamma(\pi^{-1}(q), \mathscr{F}|_{\pi^{-1}(q)})$. This is too much to ask, but at least (14.5.8.1) gives a map. (In fact, under just the right circumstances, (14.5.8.1) is an isomorphism; see §25.1.)

14.5.L. IMPORTANT EXERCISE (THE PROJECTION FORMULA, TO BE GENERALIZED IN EXERCISE 18.7.E). Suppose $\pi \colon X \to Y$ is quasicompact and quasiseparated, and $\mathscr{F}$ and $\mathscr{G}$ are quasicoherent sheaves on X and Y respectively.

(a) Describe a natural morphism $(\pi_* \mathscr{F}) \otimes \mathscr{G} \to \pi_* (\mathscr{F} \otimes \pi^* \mathscr{G})$.
(b) If $\mathscr{G}$ is locally free, show that this natural morphism is an isomorphism. (Hint: What if $\mathscr{G}$ is free?)
(c) If π is affine, then show that this natural morphism is an isomorphism.

(Sadly, the phrase "projection formula" is used for a number of facts, not all related.)

14.5.9. Remark: Pulling back ideal sheaves. There is an important subtlety in pulling back quasicoherent *ideal* sheaves. Suppose $i \colon X \hookrightarrow Y$ is a closed embedding, and $\mu \colon Y' \to Y$ is an arbitrary morphism. Let $X' := X \times_Y Y'$. As "closed embeddings pull back to closed embeddings" (§10.2.2), the pulled back map $i' \colon X' \to Y'$ is a closed embedding.

$$
\begin{array}{ccc}
X' & \longrightarrow & X \\
{\scriptstyle i'}\downarrow & & \downarrow{\scriptstyle i} \\
Y' & \xrightarrow{\ \mu\ } & Y
\end{array}
$$

Now μ^* induces canonical isomorphisms $\mu^* \mathscr{O}_Y \overset{\sim}{\longleftrightarrow} \mathscr{O}_{Y'}$ and $\mu^*(i_* \mathscr{O}_X) \overset{\sim}{\longleftrightarrow} (i'_* \mathscr{O}_{X'})$. (Why is the latter isomorphism true? Hint: Check affine-locally.) But it is *not* necessarily true that $\mu^* \mathscr{I}_{X/Y} = \mathscr{I}_{X'/Y'}$. (Exercise 14.5.J yields an example.) This is because the application of μ^* to the closed subscheme exact sequence

$$0 \longrightarrow \mathscr{I}_{X/Y} \longrightarrow \mathscr{O}_Y \longrightarrow i_* \mathscr{O}_X \longrightarrow 0$$

yields something that is a priori only right-exact:

$$\mu^* \mathscr{I}_{X/Y} \longrightarrow \mathscr{O}_{Y'} \longrightarrow i'_* \mathscr{O}_{X'} \longrightarrow 0.$$

Thus, as $\mathscr{I}_{X'/Y'}$ is the kernel of $\mathscr{O}_{Y'} \to i'_* \mathscr{O}_{X'}$, we see that $\mathscr{I}_{X'/Y'}$ is the image of $\mu^* \mathscr{I}_{X/Y}$ in $\mathscr{O}_{Y'}$. We can also see this explicitly from Exercise 10.2.B: affine-locally, the ideal of the pullback is generated by the pullback of the ideal.

Note also that if μ is an exact functor (which we will later see as a consequence of flatness, in Exercise 24.1.H), then $\mu^* \mathscr{I}_{X/Y} \to \mathscr{I}_{X'/Y'}$ *is* an isomorphism.

14.6   The Quasicoherent Sheaf Corresponding to a Graded Module

We now describe quasicoherent sheaves on a projective A-scheme. Recall that a projective A-scheme is produced from the data of a $\mathbb{Z}^{\geq 0}$-graded ring $S_\bullet$, with $S_0 = A$, and S_+ is a finitely generated ideal (a "finitely generated graded ring over A," §4.5.6). The resulting scheme is denoted by $\operatorname{Proj} S_\bullet$.

Suppose $M_\bullet$ is a graded $S_\bullet$-module, *graded by* $\mathbb{Z}$. (While reading the next section, you may wonder why we don't grade by $\mathbb{Z}^{\geq 0}$. You will see that it doesn't matter. A $\mathbb{Z}$-grading will make things cleaner when we produce an $M_\bullet$ from a quasicoherent sheaf on $\operatorname{Proj} S_\bullet$.) We define the quasicoherent sheaf $\widetilde{M_\bullet}$ as follows. (I will avoid calling it $\widetilde{M}$, as this might cause confusion with the affine case; but $\widetilde{M_\bullet}$ is *not* graded in any way.) For each homogeneous f of *positive degree*, we define a quasicoherent sheaf $\widetilde{M_\bullet}(f)$ on the distinguished open $D(f) = \{p : f(p) \neq 0\}$ by $((M_\bullet)_f)_0$ — note that $((M_\bullet)_f)_0$ is an $((S_\bullet)_f)_0$-module, and recall that $D(f)$ is identified with $\operatorname{Spec}((S_\bullet)_f)_0$ (Exercise 4.5.I). As in (4.5.7.1), the subscript 0 means "the 0-graded piece." We have obvious isomorphisms of the restriction of $\widetilde{M_\bullet}(f)$ and $\widetilde{M_\bullet}(g)$ to $D(fg)$, satisfying the cocycle conditions. (Think through this yourself, to be sure you agree with the word "obvious"!) By Exercise 2.5.E, these sheaves glue together to a single sheaf $\widetilde{M_\bullet}$ on $\operatorname{Proj} S_\bullet$. We then discard the temporary notation $\widetilde{M_\bullet}(f)$.

The $\mathcal{O}$-module $\widetilde{M_\bullet}$ is clearly quasicoherent, because it is quasicoherent on each $D(f)$, and quasicoherence is local.

14.6.A. EXERCISE. Give an isomorphism between the stalk of $\widetilde{M_\bullet}$ at a point corresponding to homogeneous prime $\mathfrak{p} \subset S_\bullet$ and $((M_\bullet)_\mathfrak{p})_0$. Here $((M_\bullet)_\mathfrak{p})_0$ means "the degree 0 homogeneous fractions in $(M_\bullet)_\mathfrak{p}$," and $\mathfrak{p}$ should be a "relevant" homogeneous prime (i.e., $[\mathfrak{p}]$ is a point of $\operatorname{Proj} S_\bullet$). (Remark: You can use this exercise to give an alternate definition of $\widetilde{M_\bullet}$ in terms of "compatible stalks"; cf. Exercise 4.5.P.)

Given a map of graded modules $\phi: M_\bullet \to N_\bullet$, we get an induced map of sheaves $\widetilde{M_\bullet} \to \widetilde{N_\bullet}$. Explicitly, over $D(f)$, the map $M_\bullet \to N_\bullet$ induces $(M_\bullet)_f \to (N_\bullet)_f$, which induces $\phi_f: ((M_\bullet)_f)_0 \to ((N_\bullet)_f)_0$; and this behaves well with respect to restriction to smaller distinguished open sets, i.e., the following diagram commutes.

$$
\begin{array}{ccc}
((M_\bullet)_f)_0 & \xrightarrow{\ \phi_f\ } & ((N_\bullet)_f)_0 \\
\downarrow & & \downarrow \\
((M_\bullet)_{fg})_0 & \xrightarrow{\ \phi_{fg}\ } & ((N_\bullet)_{fg})_0.
\end{array}
$$

Thus $\sim$ is a functor from the category of graded $S_\bullet$-modules to the category of quasicoherent sheaves on $\operatorname{Proj} S_\bullet$.

14.6.B. EASY EXERCISE. Show that $\sim$ is an exact functor. (Hint: Everything in the construction is exact.)

14.6.C. EXERCISE. Show that if $M_\bullet$ and $M'_\bullet$ agree in high enough degrees, then $\widetilde{M_\bullet} \cong \widetilde{M'_\bullet}$. Then show that the map from graded $S_\bullet$-modules (up to isomorphism) to quasicoherent sheaves on $\operatorname{Proj} S_\bullet$ (up to isomorphism) is not a bijection. (Really: show this is not an equivalence of categories.)

14.6.D. EXERCISE. Describe a map of S_0-modules $M_0 \to \Gamma(\operatorname{Proj} S_\bullet, \widetilde{M_\bullet})$. (This foreshadows the "saturation map" of §15.7.5 that takes a graded module to its saturation; see Exercise 15.7.C.)

14.6.1. Exercise 14.6.C shows that $\sim$ isn't an isomorphism (or equivalence) of categories, but it is close. The relationship is akin to that between presheaves and sheaves, and the sheafification functor (see §15.7). One convenient way of understanding this is by starting with the category $f.g.Mod_{S_\bullet}$ of finitely generated $S_\bullet$-modules, and creating a new category $\mathscr{C}$ with the same objects, but *different* morphisms: the morphisms from $M_\bullet$ to $N_\bullet$ in this new category are

$$\text{Hom}_{\mathscr{C}}(M_\bullet, N_\bullet) := \lim_{m \to \infty} \left(\text{Hom}_{f.g.Mod_{S_\bullet}} (M_{\bullet \geq m}, N_{\bullet \geq m}) \right).$$

(Here $M_{\bullet \geq m} = \oplus_{n \geq m} M_\bullet$, as you might expect.) Then you can show that $\mathscr{C}$ is equivalent to the category of finite type quasicoherent sheaves on $\text{Proj}\, S_\bullet$, and this equivalence corresponds to the map $M_\bullet \mapsto \widetilde{M_\bullet}$. You are welcome to prove these facts, but we will not use them.

14.6.2. Example: Homogeneous ideals of $S_\bullet$ give closed subschemes of $\text{Proj}\, S_\bullet$. Recall that a homogeneous ideal $I_\bullet \subset S_\bullet$ yields a closed subscheme $\text{Proj}\, S_\bullet / I_\bullet \hookrightarrow \text{Proj}\, S_\bullet$. For example, suppose $S_\bullet = k[w, x, y, z]$, so $\text{Proj}\, S_\bullet \cong \mathbb{P}^3$. The ideal $I_\bullet = (wz - xy, x^2 - wy, y^2 - xz)$ yields our old friend, the twisted cubic (defined in Exercise 9.3.A)

14.6.E. EXERCISE. Show that if the functor $\sim$ is applied to the exact sequence of graded $S_\bullet$-modules

$$0 \longrightarrow I_\bullet \longrightarrow S_\bullet \longrightarrow S_\bullet / I_\bullet \longrightarrow 0,$$

we obtain the closed subscheme exact sequence (9.1.2.1) for $\text{Proj}\, S_\bullet / I_\bullet \hookrightarrow \text{Proj}\, S_\bullet$.

All closed subschemes of $\text{Proj}\, S_\bullet$ arise in this way (see Exercise 15.7.I).

14.6.3. Remark. If $M_\bullet$ is finitely generated (resp., finitely presented, coherent), then so is $\widetilde{M_\bullet}$, [Roh, (4.4.11)]. We will not need this fact. (Quick proof of the first two: if $M_\bullet$ is finitely generated, resp., finitely presented, so is the quasicoherent sheaf $\widetilde{M_\bullet}$ on the affine cone $\text{Spec}\, M_\bullet$. These two notions are preserved by pullback, and the standard open subsets of $\text{Proj}\, S_\bullet$ are closed subschemes of the affine cone $\text{Spec}\, S_\bullet$.)

Chapter 15

Line Bundles, Maps to Projective Space, and Divisors

When learning algebraic geometry for the first time, it is surprising to hear that line bundles (invertible sheaves) are of such vital importance. *i)* They are, of course, the simplest vector bundles. *ii)* We can often classify them, using "codimension 1" information. In particular, we can build them and manipulate them. *iii)* Knowing line bundles on a variety X helps us classify X, and understand maps to projective space, and to other varieties. *iv)* And this is only the beginning.

In this chapter, we next describe convenient and powerful ways of working with and classifying line bundles. We begin with a fundamental example, the line bundles $\mathscr{O}(m)$ on projective space, §15.1. We then discuss how line bundles and some sections give maps to projective space, and indeed how all maps to projective schemes arise in this way, §15.2. This is simultaneously elementary, subtle, and powerful. In §15.3, we describe a concretization of the idea that "one-parameter families to projective space have unique limits." We then come to the most challenging part of the chapter, in §15.4. We introduce Weil divisors, and use them to determine all the line bundles on X in a number of circumstances. Our hard work leads to many fun examples we can work out by hand, in §15.5. We then discuss a different means of easily describing some line bundles as sheaves of ideals that happen to be invertible (effective Cartier divisors, §15.6). We conclude with a brief section once again connecting quasicoherent sheaves on projective schemes to graded modules, §15.7.

15.1 Some Line Bundles on Projective Space

We now describe an important family of invertible sheaves on projective space over a field k.

As a warm-up, we begin with the invertible sheaf $\mathscr{O}_{\mathbb{P}^1_k}(1)$ on $\mathbb{P}^1_k = \operatorname{Proj} k[x_0, x_1]$. The subscript $\mathbb{P}^1_k$ refers to the space on which the sheaf lives, and is often omitted when it is clear from the context. We describe the invertible sheaf $\mathscr{O}(1)$ using transition functions. It is trivial on the usual affine open sets $U_0 = D(x_0) = \operatorname{Spec} k[x_{1/0}]$ and $U_1 = D(x_1) = \operatorname{Spec} k[x_{0/1}]$. (We continue to use the convention $x_{i/j}$ for describing coordinates on patches of projective space; see §4.4.9.) Thus the data of a section over U_0 is a polynomial in $x_{1/0}$. The transition function from U_0 to U_1 is multiplication by $x_{0/1} = x_{1/0}^{-1}$. The transition function from U_1 to U_0 is hence multiplication by $x_{1/0} = x_{0/1}^{-1}$.

This information is summarized below:

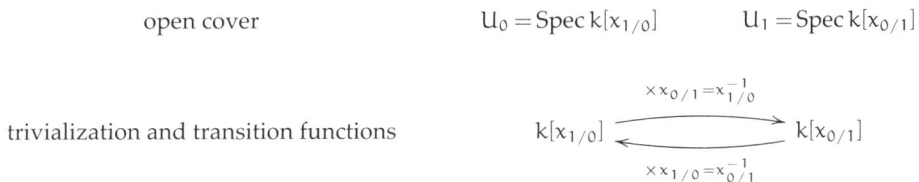

| open cover | $U_0 = \operatorname{Spec} k[x_{1/0}]$ | $U_1 = \operatorname{Spec} k[x_{0/1}]$ |

$$
\text{trivialization and transition functions} \qquad k[x_{1/0}] \underset{\times x_{1/0} = x_{0/1}^{-1}}{\overset{\times x_{0/1} = x_{1/0}^{-1}}{\rightleftarrows}} k[x_{0/1}]
$$

To test our understanding, let's compute the global sections of $\mathscr{O}(1)$. This will generalize our hands-on calculation that $\Gamma(\mathbb{P}^1_k, \mathscr{O}_{\mathbb{P}^1_k}) \cong k$ (Example 4.4.6). A global section is a polynomial $f(x_{1/0}) \in k[x_{1/0}]$ and a polynomial $g(x_{0/1}) \in k[x_{0/1}]$ such that $f(1/x_{0/1})x_{0/1} = g(x_{0/1})$. A little thought will show that f must be linear: $f(x_{1/0}) = ax_{1/0} + b$, and hence $g(x_{0/1}) = a + bx_{0/1}$. Thus

$$
\dim \Gamma(\mathbb{P}^1_k, \mathscr{O}(1)) = 2 \neq 1 = \dim \Gamma(\mathbb{P}^1_k, \mathscr{O}).
$$

Thus $\mathscr{O}(1)$ is not isomorphic to $\mathscr{O}$, and we have constructed our first (proved) example of a nontrivial line bundle!

We next define more generally $\mathscr{O}_{\mathbb{P}^1_k}(m)$ on $\mathbb{P}^1_k$. It is defined in the same way, except that the transition functions are the mth powers of those for $\mathscr{O}(1)$.

open cover	$U_0 = \operatorname{Spec} k[x_{1/0}]$	$U_1 = \operatorname{Spec} k[x_{0/1}]$

$$\text{trivialization and transition functions} \qquad k[x_{1/0}] \underset{\times x_{1/0}^m = x_{0/1}^{-m}}{\overset{\times x_{0/1}^m = x_{1/0}^{-m}}{\rightleftarrows}} k[x_{0/1}]$$

In particular, thanks to the explicit transition functions, we see that $\mathscr{O}(m) = \mathscr{O}(1)^{\otimes m}$ (with the obvious meaning if m is negative: $(\mathscr{O}(1)^{\otimes(-m)})^{\vee}$). Clearly, also $\mathscr{O}(m) \otimes \mathscr{O}(m') = \mathscr{O}(m + m')$.

15.1.A. IMPORTANT EXERCISE. Show that $\dim \Gamma(\mathbb{P}^1, \mathscr{O}(m)) = m + 1$ if $m \geq 0$, and 0 otherwise.

15.1.1. *Example.* Long ago (in Exercise 2.6.K(a)), we warned that sheafification was necessary when tensoring $\mathscr{O}_X$-modules: if $\mathscr{F}$ and $\mathscr{G}$ are two $\mathscr{O}_X$-modules on a ringed space, then it is not necessarily true that $\mathscr{F}(X) \otimes_{\mathscr{O}_X(X)} \mathscr{G}(X) \cong (\mathscr{F} \otimes \mathscr{G})(X)$. We now have an example: let $X = \mathbb{P}^1_k$, $\mathscr{F} = \mathscr{O}(1)$, $\mathscr{G} = \mathscr{O}(-1)$, and use the fact that $\mathscr{O}(-1)$ has no nonzero global sections.

15.1.B. EXERCISE. Show that if $m \neq m'$, then $\mathscr{O}(m) \not\cong \mathscr{O}(m')$. Hence conclude that we have an injection of groups $\mathbb{Z} \hookrightarrow \operatorname{Pic} \mathbb{P}^1_k$ given by $m \mapsto \mathscr{O}(m)$.

It is useful to identify the global sections of $\mathscr{O}(m)$ with the homogeneous polynomials of degree m in x_0 and x_1, i.e., with the degree m part of $k[x_0, x_1]$ (cf. §15.1.2 for the generalization to $\mathbb{P}^n$). Can you see this from your solution to Exercise 15.1.A? We will see that this identification is natural in many ways. For example, you can show that the definition of $\mathscr{O}(m)$ doesn't depend on a choice of affine cover, and this polynomial description is also independent of cover. (For this, see Example 4.5.13; you can later compare this to Exercise 25.1.N.) As an immediate check of the usefulness of this point of view, ask yourself: where does the section $x_0^3 - x_0 x_1^2$ of $\mathscr{O}(3)$ vanish? The section $x_0 + x_1$ of $\mathscr{O}(1)$ can be multiplied by the section x_0^2 of $\mathscr{O}(2)$ to get a section of $\mathscr{O}(3)$. Which one? Where does the rational section $x_0^4(x_1 + x_0)/x_1^7$ of $\mathscr{O}(-2)$ have zeros and poles, and to what order? (We saw the notions of zeros and poles in Definition 13.5.7, and will meet them again in §15.4, but you should intuitively be able to answer these questions already.)

We now define the invertible sheaf $\mathscr{O}_{\mathbb{P}^n_k}(m)$ on the projective space $\mathbb{P}^n_k$. On the usual affine open set $U_i = \operatorname{Spec} k[x_{0/i}, \ldots, x_{n/i}]/(x_{i/i} - 1) = \operatorname{Spec} A_i$, it is trivial, so sections (as an A_i-module) are isomorphic to A_i. The transition function from U_i to U_j is multiplication by $x_{i/j}^m = x_{j/i}^{-m}$.

$U_i = \operatorname{Spec} k[x_{0/i}, \ldots, x_{n/i}]/(x_{i/i} - 1)$	$U_j = \operatorname{Spec} k[x_{0/j}, \ldots, x_{n/j}]/(x_{j/j} - 1)$

$$k[x_{0/i}, \ldots, x_{n/i}]/(x_{i/i} - 1) \underset{\times x_{j/i}^m = x_{i/j}^{-m}}{\overset{\times x_{i/j}^m = x_{j/i}^{-m}}{\rightleftarrows}} k[x_{0/j}, \ldots, x_{n/j}]/(x_{j/j} - 1)$$

Note that these transition functions clearly satisfy the cocycle condition.

15.1.C. ESSENTIAL EXERCISE (CF. EXERCISE 9.3.J, THEOREM 18.1.2). Show that
$$\dim_k \Gamma(\mathbb{P}^n_k, \mathscr{O}_{\mathbb{P}^n_k}(m)) = \binom{n + m}{n}.$$

15.1.2. As in the case of $\mathbb{P}^1$, sections of $\mathscr{O}(m)$ on $\mathbb{P}^n_k$ are naturally identified with homogeneous degree m polynomials in our $n + 1$ variables. (The connection to homogeneous polynomials will be put on a more precise footing in the next exercise.) Thus $x + y + 2z$ is a section of $\mathscr{O}(1)$ on $\mathbb{P}^2$. It isn't a function, but we know where this section vanishes—precisely where $x + y + 2z = 0$.

Also, notice that for fixed n, $\binom{n+m}{n}$ is a polynomial in the variable m, of degree n, but only when $m \geq 0$ (or better: when $m \geq -n$). This should be telling you that this function "wants to be a polynomial," but won't succeed without assistance. We will later define $h^0(\mathbb{P}^n_k, \mathcal{O}(m)) :=$ $\dim_k \Gamma(\mathbb{P}^n_k, \mathcal{O}(m))$, and later still we will define higher cohomology groups, and we will define the *Euler characteristic* $\chi(\mathbb{P}^n_k, \mathcal{O}(m)) := \sum_{i=0}^{\infty} (-1)^i h^i(\mathbb{P}^n_k, \mathcal{O}(m))$ (cohomology will vanish in degree higher than n). We will discover the moral that the Euler characteristic is better behaved than h^0, and so we should now suspect (and later prove; see Theorem 18.1.2, Remark 18.3.5, and Exercise 18.3.C) that this polynomial is in fact the Euler characteristic, and the reason that it agrees with h^0 for $m \geq 0$ is that all the other cohomology groups vanish.

We finally note that we can define $\mathcal{O}(m)$ on $\mathbb{P}^n_A$ for any ring A: the above definition applies without change, and Exercise 15.1.C immediately generalizes to show that $\Gamma(\mathbb{P}^n_A, \mathcal{O}_{\mathbb{P}^n_A}(m))$ is a free A-module of rank $\binom{n+m}{n}$, with the same interpretation in terms of homogeneous polynomials.

15.1.D. EXERCISE (ENLIGHTENING EXAMPLE AND GOOD PRACTICE). Suppose $n > 0$. Let $\pi : \mathbb{A}^{n+1}_A \setminus V(x_0, \ldots, x_n) \to \mathbb{P}^n_A$ be the morphism $(x_0, \ldots, x_n) \mapsto [x_0, \ldots, x_n]$ (described in Exercise 7.3.F and Example 7.3.12). For any $m \in \mathbb{Z}$, describe an isomorphism

$$\pi^* \mathcal{O}_{\mathbb{P}^n_A}(m) \overset{\sim}{\longleftrightarrow} \mathcal{O}_{\mathbb{A}^{n+1}_A \setminus V(x_0, \ldots, x_n)}.$$

This induces a map $\pi^* : \Gamma(\mathbb{P}^n_A, \mathcal{O}_{\mathbb{P}^n_A}(m)) \to \Gamma(\mathbb{A}^{n+1}_A \setminus V(x_0, \ldots, x_n), \mathcal{O}_{\mathbb{A}^{n+1}_A})$. Show (as in §4.4.1) that the pullback

$$\alpha : A[x_0, \ldots, x_n] = \Gamma(\mathbb{A}^{n+1}_A, \mathcal{O}_{\mathbb{A}^{n+1}_A}) \longrightarrow \Gamma(\mathbb{A}^{n+1}_A \setminus V(x_0, \ldots, x_n), \mathcal{O}_{\mathbb{A}^{n+1}_A})$$

is an isomorphism, and explain how the composition

$$\alpha^{-1} \circ \pi^* : \Gamma(\mathbb{P}^n_A, \mathcal{O}_{\mathbb{P}^n_A}(m)) \longrightarrow A[x_0, \ldots, x_n]$$

interprets $\Gamma(\mathbb{P}^n_A, \mathcal{O}_{\mathbb{P}^n_A}(m))$ as the homogeneous degree m polynomials in $A[x_0, \ldots, x_n]$.

15.1.3. These are the only line bundles on $\mathbb{P}^n_k$.
Suppose that k is a field. We will see in §15.4.15 that these $\mathcal{O}(m)$ are the only invertible sheaves on $\mathbb{P}^n_k$. The next Exercise shows this when $n = 1$.

15.1.E. EXERCISE. Show that every invertible sheaf on $\mathbb{P}^1_k$ is of the form $\mathcal{O}(m)$ for some m. Hint: Use the classification of finitely generated modules over a principal ideal domain (Remark 6.4.4) to show that all invertible sheaves on $\mathbb{A}^1_k$ are trivial (a special case of Exercise 14.2.C). Reduce to determining possible transition functions between the two open subsets in the standard cover of $\mathbb{P}^1_k$.

Caution: There can exist invertible sheaves on $\mathbb{P}^1_A$ not of the form $\mathcal{O}(m)$. You may later be able to think of examples. (Hints to find an example for when you know more: What if Spec A is disconnected? Or if that leads to too silly an example, what if Spec A has nontrivial invertible sheaves?)

15.1.4. Line bundles on projective A-schemes.
Suppose $M_\bullet$ is a graded $S_\bullet$-module. Define the graded module $M(m)_\bullet$ by $M(m)_d := M_{m+d}$. Thus the quasicoherent sheaf $\widetilde{M(m)_\bullet}$ satisfies

$$\Gamma(D_+(f), \widetilde{M(m)_\bullet}) = ((M_\bullet)_f)_m,$$

where here the subscript means we take the mth graded piece. (These subscripts are admittedly confusing!)

15.1.F. EXERCISE. If $S_\bullet = A[x_0, \ldots, x_n]$, so $\operatorname{Proj} S_\bullet = \mathbb{P}^n_A$, show $\widetilde{S(m)_\bullet} \cong \mathcal{O}(m)$ using transition functions (cf. §15.1). (Recall from Exercise 15.1.D that the global sections of $\mathcal{O}(m)$ should be identified with the homogeneous degree m polynomials in $x_0, \ldots, x_n$. Can you see that in the context of this exercise?)

15.1.5. *Definition.* Motivated by this, if $S_\bullet$ is a graded ring generated in degree 1, we define $\mathcal{O}_{\operatorname{Proj} S_\bullet}(m)$ (or simply $\mathcal{O}(m)$, where $S_\bullet$ is implicit) by $\widetilde{S(m)}_\bullet$.

15.1.G. IMPORTANT EXERCISE. If $S_\bullet$ is generated in degree 1, show that $\mathcal{O}_{\operatorname{Proj} S_\bullet}(m)$ is an invertible sheaf.

If $S_\bullet$ is generated in degree 1, and $\mathscr{F}$ is a quasicoherent sheaf on $\operatorname{Proj} S_\bullet$, define $\mathscr{F}(m) := \mathscr{F} \otimes \mathcal{O}(m)$. This is often called **twisting** $\mathscr{F}$ by $\mathcal{O}(m)$, or **twisting** $\mathscr{F}$ by m. More generally, if $\mathscr{L}$ is an invertible sheaf, then $\mathscr{F} \otimes \mathscr{L}$ is often called **twisting** $\mathscr{F}$ by $\mathscr{L}$.

15.1.H. EXERCISE. If $S_\bullet$ is generated in degree 1, show that $\widetilde{M_\bullet}(m) \cong \widetilde{M(m)}_\bullet$. (Hereafter, we can be cavalier with the placement of the "dot" in such situations.)

15.1.I. EXERCISE. If $S_\bullet$ is generated in degree 1, show that $\mathcal{O}(m_1 + m_2) \cong \mathcal{O}(m_1) \otimes \mathcal{O}(m_2)$.

15.1.6.* *Unimportant remark.* Even if $S_\bullet$ is not generated in degree 1, then (if $S_\bullet$ is finitely generated) by Exercise 7.4.G, $S_{d\bullet}$ is generated in degree 1 for some d. In this case, we may define the invertible sheaves $\mathcal{O}(dm)$ for $m \in \mathbb{Z}$. This does *not* mean that we *can't* define $\mathcal{O}(1)$; this depends on $S_\bullet$. For example, if $S_\bullet$ is the polynomial ring $k[x, y]$ with the usual grading, except without linear terms (so $S_\bullet = k[x^2, xy, y^2, x^3, x^2y, xy^2, y^3]$), then $S_{2\bullet}$ and $S_{3\bullet}$ are both generated in degree 1, meaning that we may define $\mathcal{O}(2)$ and $\mathcal{O}(3)$. There is good reason to call their "difference" $\mathcal{O}(1)$. (This algebra may remind of you of the pinched plane, Exercise 13.5.H.)

15.2 Line Bundles and Maps to Projective Space

Recall that affine space $\mathbb{A}^n$ had a short description as a representable functor: maps (from X) to $\mathbb{A}^n$ are exactly the same data as a choice of n functions (on X), Exercise 7.6.E(a). In this section, we accidentally reinterpret projective space in the same way.

Suppose x_0, x_1, and x_2 are projective coordinates on $\mathbb{P}^2_k$. Do you see how

$$(15.2.0.1) \qquad\qquad [x_0^3, x_0^2 x_2, x_0 x_1 x_2 + x_1^3, x_2^3]$$

gives a morphism $\mathbb{P}^2_k \to \mathbb{P}^3_k$? The entries of (15.2.0.1) aren't functions, so we can't make sense of them individually. But for any "point of $\mathbb{P}^2_k$," their ratios can make sense nonetheless. For example, the point $[1, 1, 1]$ is sent to $[1, 1, 2, 1]$, and the same point $[2, 2, 2]$ is sent to $[8, 8, 16, 8]$, so there is no contradiction—the entries of (15.2.0.1) are all homogeneous of degree 3, so the quotient of one by another is a rational function. On the other hand,

$$(15.2.0.2) \qquad\qquad [x_0^3, x_0^2 x_2, x_0 x_1 x_2 + x_1^3, x_1^3]$$

doesn't give a morphism $\mathbb{P}^2_k \to \mathbb{P}^3_k$, because the point $[0, 0, 1]$ "doesn't know where to go." But you might be able to interpret (15.2.0.2) as giving a morphism $\mathbb{P}^2_k \setminus \{[0, 0, 1]\} \to \mathbb{P}^3_k$, or a rational map $\mathbb{P}^2_k \dashrightarrow \mathbb{P}^3_k$. This may even remind you of Exercise 7.3.O. We now make this precise, and state it much more generally. In particular, we won't need to work over a field.

15.2.A. EASY EXERCISE (A VITALLY IMPORTANT CONSTRUCTION). Suppose $s_0, \ldots, s_n$ are $n + 1$ global sections of an invertible sheaf $\mathscr{L}$ on a scheme X, with no common zero. Define a corresponding map to $\mathbb{P}^n$:

$$X \xrightarrow{\;[s_0, \ldots, s_n]\;} \mathbb{P}^n.$$

Hint: If U is an open subset on which $\mathscr{L}$ is trivial, choose a trivialization, then translate the s_i into functions using this trivialization, and use Exercise 7.3.O(a) to obtain a morphism $U \to \mathbb{P}^n$. Then show that all of these maps (for different U and different trivializations) "agree," using Exercise 7.3.O(b).

(In Theorem 15.2.2, we will see that this yields *all* maps to projective space.) Note that this exercise as written "works over $\mathbb{Z}$" (as *all* morphisms are "over" the final object in the category of schemes), although many readers will just work over a field k.

15.2.1. *Definitions.* If $\mathscr{L}$ is an invertible sheaf on X, then those points where all global sections of $\mathscr{L}$ vanish are called the **base points** of $\mathscr{L}$, and the set of base points is called the **base locus** of $\mathscr{L}$; it is a closed subset of X. (We can refine this to a closed subscheme: by taking the scheme-theoretic intersection of the vanishing loci of the sections of $\mathscr{L}$, we obtain the **scheme-theoretic base locus**.) The complement of the **base locus** is the **base-point-free locus**. If $\mathscr{L}$ has no base points, it is **base-point-free**. The base-point-free locus is an open subset of X.

More generally, if we have a bunch of global sections of $\mathscr{L}$ (perhaps not all of them), we can define the **base points** (and all the related notions, of the previous paragraph) of these sections. Thus, for example, Exercise 15.2.A states that if $s_0, \ldots, s_n$ are a base-point-free family of global sections of an invertible sheaf $\mathscr{L}$ on X, then they define a map to $\mathbb{P}^n$.

A **linear series** (or *linear system*) on a k-scheme X is a k-vector space V (usually finite-dimensional), an invertible sheaf $\mathscr{L}$, and a linear map $\lambda \colon V \to \Gamma(X, \mathscr{L})$. Such a linear series is often called "V," with the rest of the data left implicit. If the map λ is an isomorphism, it is called a **complete linear series**, and is often written $|\mathscr{L}|$. The language of base points (base-point-free, base locus, scheme-theoretic base locus, ...) readily applies to linear series (base-point-free linear series, ...).

As a reality check, Exercise 15.2.A states that an $(n+1)$-dimensional linear series on a k-scheme X, with choice of basis, with base-point-free locus U, defines a morphism $U \to \mathbb{P}^n_k$.

15.2.B. EASY EXERCISE. If $\mathscr{L}$ and $\mathscr{M}$ are base-point-free invertible sheaves, show that $\mathscr{L} \otimes \mathscr{M}$ is as well.

15.2.C. EXERCISE. Suppose we have a morphism $\pi \colon X \to \mathbb{P}^n_k$, corresponding to the base-point-free linear series $\Gamma(\mathbb{P}^n_k, \mathscr{O}(1)) \to \Gamma(X, \mathscr{L})$ (so $\mathscr{L} = \pi^* \mathscr{O}(1)$). If the scheme-theoretic image of X in $\mathbb{P}^n_k$ lies in a hyperplane, we say that the linear series (or X itself) is **degenerate** (and otherwise, **nondegenerate**). Show that a base-point-free linear series $V \to \Gamma(X, \mathscr{L})$ is nondegenerate if and only if the map $V \to \Gamma(X, \mathscr{L})$ is an inclusion. In particular, a complete linear series is always nondegenerate.

Theorem 15.2.2, the converse or completion to Exercise 15.2.A, will give one reason why line bundles are crucially important: They tell us about maps to projective space, and more generally, to quasiprojective A-schemes. Given that we have had a hard time naming any non-quasiprojective schemes, they tell us about maps to essentially all interesting schemes.

15.2.2. Important theorem — *For a fixed scheme X, morphisms $X \to \mathbb{P}^n$ are in bijection with the data $(\mathscr{L}, s_0, \ldots, s_n)$, where $\mathscr{L}$ is an invertible sheaf and $s_0, \ldots, s_n$ are sections of $\mathscr{L}$ with no common zeros (i.e., $(s_0, \ldots, s_n)$ is base-point-free), up to isomorphism of these data.*

(This works over $\mathbb{Z}$, or indeed any base.) Informally: morphisms to $\mathbb{P}^n$ correspond to $n+1$ sections of a line bundle, not all vanishing at any point, modulo global sections of $\mathscr{O}_X^*$, as multiplication by an invertible function gives an automorphism of $\mathscr{L}$. This is one of those important theorems in algebraic geometry that is easy to prove, but quite subtle in its effect on how one should think. It takes some time to properly digest. A "coordinate-free" version is given in Exercise 15.2.G.

15.2.3. Theorem 15.2.2 describes all morphisms to projective space, and hence by the Yoneda philosophy, this can be taken as the *definition* of projective space: It defines projective space up to unique isomorphism. *Projective space $\mathbb{P}^n$ (over $\mathbb{Z}$) is the moduli space of line bundles $\mathscr{L}$ along with $n+1$ sections of $\mathscr{L}$ with no common zeros.* (Can you give an analogous definition of projective space over X, denoted by $\mathbb{P}^n_X$?) Or, if you prefer, *maps $X \to \mathbb{P}^n_k$ of k-schemes correspond to $(n+1)$-dimensional base-point-free linear series on X, along with choice of basis.* (How would you rephrase this if k is replaced by another ring?)

Every time you see a map to projective space, you should immediately simultaneously keep in mind the invertible sheaf and sections (or, if you prefer, the linear series).

Maps to projective schemes can be described similarly. For example, if $Y \hookrightarrow \mathbb{P}_k^2$ is the curve $x_2^2 x_0 = x_1^3 - x_1 x_0^2$, then maps from a scheme X to Y are given by an invertible sheaf on X along with three sections s_0, s_1, s_2, with no common zeros, satisfying $s_2^2 s_0 - s_1^3 + s_1 s_0^2 = 0$.

15.2.4. The central idea of the proof of Theorem 15.2.2 is straightforward: we need to show a bijection, so we just figure out what the maps are from left to right, and from right to left, and show that they are indeed inverse. The key will be thinking about transition functions between the trivializing neighborhoods.

Here more precisely is the correspondence of Theorem 15.2.2. Any $n + 1$ sections of $\mathscr{L}$ with no common zeros determine a morphism to $\mathbb{P}^n$, by Exercise 15.2.A. Conversely, if you have a map to projective space $\pi \colon X \to \mathbb{P}^n$, then we have $n + 1$ sections of $\mathscr{O}_{\mathbb{P}^n}(1)$, corresponding to the hyperplane sections, $x_0, \ldots, x_n$. Then $\pi^* x_0, \ldots, \pi^* x_n$ are sections of $\pi^* \mathscr{O}_{\mathbb{P}^n}(1)$, and they have no common zero. (Reminder: It is helpful to think of pulling back invertible sheaves in terms of pulling back transition functions.)

So to prove Theorem 15.2.2, we just need to show that these two constructions compose to give the identity in either direction.

Proof of Important Theorem 15.2.2. Suppose we are given $n + 1$ sections $s_0, \ldots, s_n$ of an invertible sheaf $\mathscr{L}$, with no common zeros, that (via Exercise 15.2.A) induce a morphism $\pi \colon X \to \mathbb{P}^n$. For each s_i, we get a trivialization of $\mathscr{L}$ on the open set X_{s_i}, where s_i doesn't vanish. The transition functions for $\mathscr{L}$ are precisely s_i / s_j on $X_{s_i} \cap X_{s_j}$. As $\mathscr{O}(1)$ is trivial on the standard affine open sets $D_+(x_i)$ of $\mathbb{P}^n$, $\pi^*(\mathscr{O}(1))$ is trivial on $X_{s_i} = \pi^{-1}(D_+(x_i))$. Moreover, $s_i / s_j = \pi^*(x_i / x_j)$ (directly from the construction of π in Exercise 15.2.A). This gives an isomorphism $\mathscr{L} \xleftrightarrow{\sim} \pi^* \mathscr{O}(1)$—the two invertible sheaves have the same transition functions.

15.2.D. EXERCISE. Show that this isomorphism can be chosen so that

$$(\mathscr{L}, s_0, \ldots, s_n) \xleftrightarrow{\sim} (\pi^* \mathscr{O}(1), \pi^* x_0, \ldots, \pi^* x_n),$$

thereby completing one of the two implications of the theorem.

For the other direction, suppose we are given a map $\pi \colon X \to \mathbb{P}^n$. Let $s_i = \pi^* x_i \in \Gamma(X, \pi^*(\mathscr{O}(1)))$. As the x_i's have no common zeros on $\mathbb{P}^n$, the s_i's have no common zeros on X. The map $[s_0, \ldots, s_n]$ is the same as the map π. We see this as follows. The preimage of $D(x_i)$ (by the morphism $[s_0, \ldots, s_n]$) is $D(s_i) = D(\pi^* x_i) = \pi^{-1} D(x_i)$, so "the right open sets go to the right open sets." To show the two morphisms $D(s_i) \to D(x_i)$ (induced from $(s_1, \ldots, s_n)$ and π) are the same, we use the fact that maps to an affine scheme $D(x_i)$ are determined by their maps of global sections in the opposite direction (Essential Exercise 7.3.G). Both morphisms $D(s_i) \to D(x_i)$ correspond to the ring map $\pi^\sharp \colon x_{j/i} = x_j / x_i \mapsto s_j / s_i$. □

15.2.5. *Remark: Extending Theorem 15.2.2 to rational maps.* Suppose $s_0, \ldots, s_n$ are sections of an invertible sheaf $\mathscr{L}$ on a scheme X. Then Theorem 15.2.2 yields a morphism $X \setminus V(s_0, \ldots, s_n) \to \mathbb{P}^n$. In particular, if X is integral, and the s_i are not all 0, these data yield a rational map $X \dashrightarrow \mathbb{P}^n$.

15.2.6. Examples and applications.

15.2.7. *Example: The Veronese embedding is* $|\mathscr{O}_{\mathbb{P}_k^n}(d)|$. Consider the line bundle $\mathscr{O}_{\mathbb{P}_k^n}(d)$ on $\mathbb{P}_k^n$. We have checked that the sections of this line bundle form a vector space of dimension $\binom{n+d}{d}$, with a basis corresponding to homogeneous degree d polynomials in the projective coordinates for $\mathbb{P}_k^n$ (Exercises 15.1.C and 15.1.D). Also, they have no common zeros (as, for example, the subset of sections $x_0^d, x_1^d, \ldots, x_n^d$ have no common zeros). Thus the complete linear series is base-point-free, and determines a morphism $v_d \colon \mathbb{P}^n \to \mathbb{P}^{\binom{n+d}{d}-1}$. This is the Veronese embedding (Definition 9.3.8). For example, if $n = 2$ and $d = 2$, we get a map $\mathbb{P}_k^2 \to \mathbb{P}_k^5$.

In §9.3.8, we saw that this is a closed embedding. The following is a more general method of checking that maps to projective space are closed embeddings.

15.2.E. LESS IMPORTANT EXERCISE. Suppose $\pi\colon X \to \mathbb{P}^n_A$ corresponds to an invertible sheaf $\mathscr{L}$ on X, and sections $s_0, \ldots, s_n$. Show that π is a closed embedding if and only if

(i) each open set X_{s_i} is affine, and
(ii) for each i, the map of rings $A[y_0, \ldots, y_n] \to \Gamma(X_{s_i}, \mathscr{O})$ given by $y_j \mapsto s_j/s_i$ is surjective.

15.2.8. *Special case of Example 15.2.7.* Recall that the image of the Veronese embedding when $n = 1$ is called a *rational normal curve of degree* d (Exercise 9.3.I). Our map is $\mathbb{P}^1_k \to \mathbb{P}^d_k$, given by $[x, y] \mapsto [x^d, x^{d-1}y, \ldots, xy^{d-1}, y^d]$.

15.2.F. EXERCISE. Suppose we are given a map $\pi\colon \mathbb{P}^1_k \to \mathbb{P}^n_k$ where the corresponding invertible sheaf on $\mathbb{P}^1_k$ is $\mathscr{O}(d)$. (This can reasonably be called a *degree* d *map*; cf. Exercises 18.4.E and 18.6.H.) Show that if $d < n$, then the image is degenerate (defined in Exercise 15.2.C). Show that if $d = n$ and the image is nondegenerate, then the image is isomorphic (via an automorphism of projective space) to a rational normal curve.

15.2.G. EXERCISE.

(a) Suppose X is a k-scheme (although this statement can readily generalize). Show that a (finite-dimensional) base-point-free linear series V on X corresponding to $\mathscr{L}$ induces a morphism to projective space

$$\phi_V\colon X \longrightarrow \mathbb{P}V^\vee.$$

(This should be seen as a coordinate-free version of Theorem 15.2.2.)
(b) *Complete the sentence: "$\mathbb{P}V^\vee$ represents the functor ...." Explain the meaning of the statements $\Gamma(\mathbb{P}V, \mathscr{O}(1)) = V^\vee$ and $\Gamma(\mathbb{P}V, \mathscr{O}(2)) = \operatorname{Sym}^2 V^\vee$.

15.2.H. EXERCISE. Define the graded rings $R_\bullet = k[u, v, w]/(uw - v^2)$ and $S_\bullet = k[x, y]$ (with all variables having degree 1). By Exercise 9.3.H, we have an isomorphism $\operatorname{Proj} R_\bullet \xrightarrow{\sim} \operatorname{Proj} S_\bullet$ (via the Veronese embedding ν_2). Show that this isomorphism is not induced by a map of graded rings $S_\bullet \to R_\bullet$.

15.2.9. *Remark.* You may be able to show that, after "regrading," the isomorphism $\operatorname{Proj} R_\bullet \xleftrightarrow{\sim} \operatorname{Proj} S_\bullet$ *does* arise from a map of graded rings $(S_{2\bullet} \to R_\bullet)$. Exercise 19.11.B will give an example where it is not possible to remedy the lack of maps of graded rings by just regrading.

15.2.I. EXERCISE: AN EARLY LOOK AT INTERSECTION THEORY, RELATED TO BÉZOUT'S THEOREM. A classical definition of the degree of a curve in projective space is as follows: Intersect it with a "general" hyperplane, and count the number of points of intersection, with appropriate multiplicity. We interpret this in the case of $\pi\colon \mathbb{P}^1_k \to \mathbb{P}^n_k$. Show that there is a hyperplane H of $\mathbb{P}^n_k$ not containing $\pi(\mathbb{P}^1_k)$. Equivalently, $\pi^* H \in \Gamma(\mathbb{P}^1, \mathscr{O}_{\mathbb{P}^1}(d))$ (interpreted as the pullback of a section of $\mathscr{O}_{\mathbb{P}^n}(1)$) is not 0. Show that the number of zeros of $\pi^* H$ is precisely d. (You will have to define "appropriate multiplicity." What does it mean geometrically if π is a closed embedding, and $\pi^* H$ has a double zero? Aside: Can you make sense of this even if π is not a closed embedding?) Thus this classical notion of degree agrees with the notion of degree in Exercise 15.2.F. (See Exercise 9.3.D for another case of Bézout's Theorem. Here we intersect a degree d curve with a degree 1 hyperplane; there we intersect a degree 1 curve with a degree d hypersurface. Exercise 18.6.J will give a common generalization.)

15.2.10. *Example: The Segre embedding revisited.* The Segre embedding can also be interpreted in this way. This is a useful excuse to define some notation. Suppose $\mathscr{F}$ is a quasicoherent sheaf on a Z-scheme X, and $\mathscr{G}$ is a quasicoherent sheaf on a Z-scheme Y. Let π_X, π_Y be the projections from $X \times_Z Y$ to X and Y, respectively. Then $\mathscr{F} \boxtimes \mathscr{G}$ (pronounced "$\mathscr{F}$ box-times $\mathscr{G}$") is defined to be $\pi_X^* \mathscr{F} \otimes \pi_Y^* \mathscr{G}$. In particular, $\mathscr{O}_{\mathbb{P}^m \times \mathbb{P}^n}(a, b)$ is defined to be $\mathscr{O}_{\mathbb{P}^m}(a) \boxtimes \mathscr{O}_{\mathbb{P}^n}(b)$ (over any base Z). The Segre embedding $\mathbb{P}^m \times \mathbb{P}^n \to \mathbb{P}^{mn+m+n}$ corresponds to the complete linear series for the invertible sheaf $\mathscr{O}(1, 1)$.

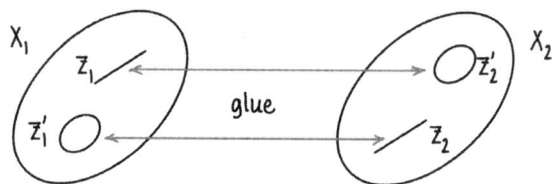

Figure 15.1 *Building a proper nonprojective variety.*

When we first saw the Segre embedding in §10.6, we saw (in different language) that this complete linear series is base-point-free. We also checked by hand (§10.6.1) that it is a closed embedding, essentially by Exercise 15.2.E.

Recall that if $\mathscr{L}$ and $\mathscr{M}$ are both base-point-free invertible sheaves on a scheme X, then $\mathscr{L} \otimes \mathscr{M}$ is also base-point-free (Exercise 15.2.B). We may interpret this fact using the Segre embedding (under reasonable hypotheses on X). If $\phi_{\mathscr{L}} : X \to \mathbb{P}^M$ is a morphism corresponding to a (base-point-free) linear series based on $\mathscr{L}$, and $\phi_{\mathscr{M}} : X \to \mathbb{P}^N$ is a morphism corresponding to a linear series on $\mathscr{M}$, then the Segre embedding yields a morphism $X \to \mathbb{P}^M \times \mathbb{P}^N \to \mathbb{P}^{(M+1)(N+1)-1}$, which corresponds to a base-point-free series of sections of $\mathscr{L} \otimes \mathscr{M}$.

15.2.11.** A proper nonprojective k-variety—and gluing schemes along closed subschemes.

(The rest of this section should be skipped on a first reading.) We conclude by using what we have developed to describe an example of a scheme that is proper but not projective (promised in Remark 11.4.6). We use a construction that looks so fundamental that you may be surprised to find that we won't use it in any meaningful way later.

Fix an algebraically closed field k. For $i = 1$ and 2, let $X_i \cong \mathbb{P}^3_k$, Z_i be a line in X_i, and Z_i' be a regular conic in X_i disjoint from Z_i (both Z_i and Z_i' isomorphic to $\mathbb{P}^1_k$). The construction of §15.2.12 will allow us to glue X_1 to X_2 so that Z_1 is identified with Z_2' and Z_1' is identified with Z_2; see Figure 15.1. (You will be able to make this precise after reading §15.2.12.) The result—call it X—is proper, by Exercise 15.2.M.

Then X is not projective. For if it were, then it would be embedded in projective space by some invertible sheaf $\mathscr{L}$. If X is embedded, then X_1 is too, so $\mathscr{L}$ must restrict to an invertible sheaf on X_1 of the form $\mathscr{O}_{X_1}(n_1)$, where $n_1 > 0$. You can check that the restriction of $\mathscr{L}$ to Z_1 is $\mathscr{O}_{Z_1}(n_1)$, and the restriction of $\mathscr{L}$ to Z_1' is $\mathscr{O}_{Z_1'}(2n_1)$. Symmetrically, the restriction of $\mathscr{L}$ to Z_2 is $\mathscr{O}_{Z_2}(n_2)$ for some $n_2 > 0$, and the restriction of $\mathscr{L}$ to Z_2' is $\mathscr{O}_{Z_2'}(2n_2)$. But after gluing, $Z_1 = Z_2'$, and $Z_1' = Z_2$, so we have $n_1 = 2n_2$ and $2n_1 = n_2$, which is impossible.

15.2.12. *Gluing two schemes together along isomorphic closed subschemes.*

It is straightforward to show that you can glue two schemes along isomorphic open subschemes. (More precisely, if X_1 and X_2 are schemes, with open subschemes U_1 and U_2 respectively, and an isomorphism $U_1 \xleftrightarrow{\sim} U_2$, you can make sense of gluing X_1 and X_2 along $U_1 \xleftrightarrow{\sim} U_2$. You should think this through.) You can similarly glue two schemes along isomorphic *closed* subschemes. We now make this precise. Suppose $Z_1 \hookrightarrow X_1$ and $Z_2 \hookrightarrow X_2$ are closed embeddings, and $\phi : Z_1 \xrightarrow{\sim} Z_2$ is an isomorphism. We will explain how to glue X_1 to X_2 along ϕ. The result will be called $X_1 \coprod_\phi X_2$.

15.2.13. *Motivating example.* Our motivating example is if $X_i = \operatorname{Spec} A_i$ and $Z_i = \operatorname{Spec} A_i/I_i$, and ϕ corresponds to $\phi^\sharp : A_2/I_2 \xrightarrow{\sim} A_1/I_1$. Then the result will be $\operatorname{Spec} R$, where R is the ring consisting of ordered pairs $(a_1, a_2) \in A_1 \times A_2$ that "agree via ϕ." More precisely, this is a fibered product of rings:

$$R := A_1 \times_{\phi^\sharp \,:\, A_1/I_1 \to A_2/I_2} A_2.$$

15.2.14. *The general construction, as a locally ringed space.* In our general situation, we might wish to cover X_1 and X_2 by open charts of this form. We would then have to worry about gluing

and choices, so to avoid this, we instead first construct $X_1 \coprod_\phi X_2$ as a locally ringed space. As a topological space, the definition is clear: we glue the underlying sets together along the underlying sets of $Z_1 \cong Z_2$, and topologize this glued set so that a subset of $X_1 \coprod_\phi X_2$ is open if and only if its restrictions to X_1 and X_2 are both open. For convenience, let Z be the image of Z_1 (or, equivalently, Z_2) in $X_1 \coprod_\phi X_2$. We next define the stalk of the structure sheaf at any point $p \in X_1 \coprod_\phi X_2$. If $p \in X_i \setminus Z = (X_1 \coprod_\phi X_2) \setminus X_{3-i}$ (hopefully the meaning of this is clear), we define the stalk as $\mathscr{O}_{X_i,p}$. If $p \in X_1 \cap X_2$, we define the stalk to consist of elements $(s_1, s_2) \in \mathscr{O}_{X_1,p} \times \mathscr{O}_{X_2,p}$ that agree in $\mathscr{O}_{Z_1,p} \cong \mathscr{O}_{Z_2,p}$. The meaning of everything in this paragraph will be clear to you if you can do the following.

15.2.J. EXERCISE. Define the structure sheaf of $\mathscr{O}_{X_1 \coprod_\phi X_2}$ in terms of compatible germs. (What should it mean for germs to be compatible? Hint: For $z \in Z$, suppose we have open subsets U_1 of X_1 and U_2 of X_2, with $U_1 \cap Z = U_2 \cap Z$, so U_1 and U_2 glue together to give an open subset U of $X_1 \coprod_\phi X_2$. Suppose we also have functions f_1 on U_1 and f_2 on U_2 that "agree on $U \cap Z$"—what does that mean? Then we declare that the germs of the "function on U obtained by gluing together f_1 and f_2" are compatible.) Show that the resulting ringed space is a locally ringed space.

We next want to show that the locally ringed space $X_1 \coprod_\phi X_2$ is a scheme. Clearly it is a scheme away from Z. We first verify a special case.

15.2.K. EXERCISE. Show that in Example 15.2.13 the construction of §15.2.14 indeed yields $\operatorname{Spec}(A_1 \times_{\phi^\sharp} A_2)$.

15.2.L. EXERCISE. In the general case, suppose $x \in Z$. Show that there is an affine open subset $\operatorname{Spec} A_i \subset X_i$ such that $Z \cap \operatorname{Spec} A_1 = Z \cap \operatorname{Spec} A_2$. Then use Exercise 15.2.J to show that $X_1 \coprod_\phi X_2$ is a scheme in an open neighborhood of x, and thus a scheme.

15.2.15. *Remarks.*

(a) As the notation suggests, this is a fibered coproduct in the category of schemes, and indeed in the category of locally ringed spaces. We won't need this fact, but you can prove it if you wish; it isn't hard. Unlike in the situation for products, fibered coproducts don't exist in general in the category of schemes. Miraculously (and for reasons that are specific to schemes), the resulting cofibered diagram is *also* a *fibered* (i.e., Cartesian) diagram. This has pleasant ramifications. For example, this construction "behaves well with respect to" (or "commutes with") flat base change.

(b) You might hope that if you have a single scheme X with two disjoint closed subschemes W' and W'', and an isomorphism $W' \overset{\sim}{\longrightarrow} W''$, then you should be able to glue X to itself along $W' \to W''$. This construction doesn't work, and indeed it may not be possible. You can still make sense of the quotient as an *algebraic space*, which we will not define here.

15.2.M. EXERCISE. We continue to use the notation X_i, ϕ, etc. Suppose we are working in the category of A-schemes.

(a) If X_1 and X_2 are universally closed, show that $X_1 \coprod_\phi X_2$ is as well.
(b) If X_1 and X_2 are separated, show that $X_1 \coprod_\phi X_2$ is as well.
(c) If X_1 and X_2 are finite type over a *Noetherian* ring A, show that $X_1 \coprod_\phi X_2$ is as well. (Hint: Reduce to the "affine" case of the Motivating Example 15.2.13. Choose generators $x_1, \ldots, x_n$ of A_1, and $y_1, \ldots, y_n$, such that x_i modulo I_1 agrees with y_i modulo I_2 via ϕ. Choose generators $g_1, \ldots, g_m$ of I_2—here use Noetherianness of A. Show that (x_i, y_i) and $(0, g_i)$ generate $R \subset A_1 \times A_2$, as follows. Suppose $(a_1, a_2) \in R$. Then there is some polynomial m such that $a_1 = m(x_1, \ldots, x_n)$. Hence $(a_1, a_2) - m((x_1, y_1), \ldots, (x_n, y_n)) = (0, a_2')$ for some $a_2' \in I_2$. Then a_2' can be written as $\sum_{i=1}^m \ell_i(y_1, \ldots, y_n)g_i$. But then $(0, a_2') = \sum_{i=1}^m \ell_i((x_1, y_1), \ldots, (x_n, y_n))(0, g_i)$.)

Thus if X_1 and X_2 are proper, so is $X_1 \coprod_\phi X_2$.

On the other hand, we have just seen (§15.2.11) that if X_1 and X_2 are *projective*, then $X_1 \coprod_\phi X_2$ needn't be projective.

15.2.16. *Fun with more general "gluing" (coproducts) of schemes.*
If you have read this far, you can see more interesting examples quite cheaply. Suppose more generally that we have a closed embedding $Z \hookrightarrow Y$, and only an affine morphism $Z \to X$. We can still "glue" X to Y along Z to get a scheme W, by following the same recipe.

15.2.N. EXTENDED EXERCISE. Given a closed embedding $\alpha : Z \hookrightarrow X$ and an affine morphism $\beta : Z \to Y$, show (following the strategy of §15.2.12) that there is a pushout diagram of schemes

(15.2.16.1)

$$
\begin{array}{ccc}
Z & \xrightarrow{\ \beta\ \text{affine}\ } & Y \\
{\scriptstyle \alpha\ \text{closed}}\big\downarrow\ \uparrow & & \big\downarrow {\scriptstyle \alpha'} \\
X & \xrightarrow{\ \beta'\ } & W.
\end{array}
$$

This diagram is necessarily unique up to unique isomorphism.

15.2.O. EXERCISE. Show that (15.2.16.1) is also a pullback diagram (i.e., a Cartesian diagram), and that α' is a closed embedding, and β' is an affine morphism.

15.2.17. *A scheme with no closed points.*
In the course of solving Extended Exercise 15.2.N, you will investigate the (much easier) affine case, which corresponds to the following diagram of rings.

(15.2.17.1)

$$
\begin{array}{ccc}
A/I & \xleftarrow{\ \beta^\sharp\ } & B \\
\big\uparrow & & \big\uparrow \\
A & \longleftarrow & A \times_{A/I} B.
\end{array}
$$

Thus the ring corresponding to W can be interpreted as

$$A \times_{A/I} B = \{(a, b) \in (A, B) : a \equiv \alpha^\sharp(b) \pmod{I}\}.$$

We can use this construction to see some interesting behavior. For example, we now describe a scheme X with no closed points. It will have countably many points $\{p_1, p_2, \dots\}$, and its open sets will be of the form $\{p_1, \dots, p_n\}$. (Clearly, this topological space has no closed points.) Each open set will be affine: $\{p_1, \dots, p_n\} = \operatorname{Spec} A_n$. We inductively build A_n as follows. Let $A_1 = k(x_1, x_2, \dots)$ (a countably generated transcendental extension of a field k). We inductively build A_{n+1} with the following special case of (15.2.17.1):

$$
\begin{array}{ccc}
k(x_n, x_{n+1}, \dots) & \longleftarrow & k(x_{n+1}, x_{n+2}, \dots)[x_n]_{x_n} \\
\big\uparrow & & \big\uparrow \\
A_n & \longleftarrow & A_{n+1}.
\end{array}
$$

The top row is just a discrete valuation ring mapping into its fraction field K. The left column is a complicated local ring whose residue field is that same field K.

This may be easier to understand when focusing on $\{p_1, p_2, p_3\}$. Let $F = k(x_3, x_4, \dots)$. Then A_1 is a field, A_2 is the discrete valuation ring $F(x_2)[x_1]_{x_1}$, and A_3 is the subring of A_2 whose residue modulo x_1 is in the discrete valuation ring $F[x_2]_{x_2}$. (The ring A_3 is an example of a *rank 2 valuation ring*; can you see why it deserves this name?)

15.2.P. EXERCISE. Use the above to describe a scheme $\cup_{n \geq 1} \operatorname{Spec} A_n$ with no closed points.

15.3   The Curve-to-Projective Extension Theorem

We now use the main theorem of the previous section, Theorem 15.2.2, to prove something useful and concrete. (In fact, we could have proved this far earlier—with a little cleverness you can replace the invocation of Theorem 15.2.2 by Exercise 7.3.M, and prove this as soon as you know about discrete valuation rings.)

As motivation, we consider a variation of the example at the start of §15.2: (15.2.0.2) was a failed attempt at a morphism $\mathbb{P}_k^2 \to \mathbb{P}_k^3$. There was a reason the example used $\mathbb{P}_k^2$ rather than $\mathbb{P}_k^1$ as the domain. For example, at first it appears as though the rational map $\mathbb{P}_k^1 \dashrightarrow \mathbb{P}_k^3$ given by

$$[x^7, x^3y^4, x^4y^3 + 4x^5y^2, x^5y^2 + x^7]$$

doesn't give a morphism $\mathbb{P}_k^1 \to \mathbb{P}_k^3$, because it is not defined at $[x, y] = [0, 1]$. But we can extend the map over $[0, 1]$, using

(15.3.0.1) $[x^7, x^3y^4, x^4y^3 + 4x^5y^2, x^5y^2 + x^7] = [x^4, y^4, xy^3 + 4x^2y^2, x^2y^2 + x^4]$

—we can just divide through by x^3 to make the morphism make sense. Similarly, it appears as though $[x^2, x^{-2}y^4, x^{-1}y^3 + 4y^2, y^2 + x^2]$ also is not defined at $[x, y] = [0, 1]$, but you may see what to do (and why it can be extended to the same map as (15.3.0.1)). This will generalize to the following.

15.3.1. The Curve-to-Projective Extension Theorem — *Suppose C is a pure dimension 1 Noetherian scheme over an affine base $S = \operatorname{Spec} A$, and $p \in C$ is a regular closed point of it. Suppose Y is a projective S-scheme. Then any morphism $C \setminus \{p\} \to Y$ (of S-schemes) extends to all of C.*

To preempt any confusion: C is "absolute pure dimension 1," not relative pure dimension 1.

In practice, we will use this theorem when $S = \operatorname{Spec} k$, and C is a k-variety. The only reason we assume S is affine is because we won't know the meaning of "projective S-scheme" until we know what a projective morphism is (§17.3). But the proof below extends immediately to general S once we know the meaning of the statement.

Note that if such an extension exists, then it is unique: the nonreduced locus of C is a closed subset (Exercise 6.6.N). Hence by replacing C by an open neighborhood of p that is reduced, we can use the Reduced-to-Separated Theorem 11.3.2, that maps from reduced schemes to separated schemes are determined by their behavior on a dense open set. Alternatively, maps to a separated scheme can be extended over an effective Cartier divisor in at most one way (Exercise 11.3.G).

The following exercise shows that the hypotheses are necessary.

15.3.A. EXERCISE. In each of the following cases, prove that the morphism $C \setminus \{p\} \to Y$ cannot be extended to a morphism $C \to Y$.

(a) *Projectivity of Y is necessary.* Suppose $C = \mathbb{A}_k^1$, $p = 0$, $Y = \mathbb{A}_k^1$, and $C \setminus \{p\} \to Y$ is given by "$t \mapsto 1/t$."

(b) *One-dimensionality of C is necessary.* Suppose $C = \mathbb{A}_k^2$, $p = (0, 0)$, $Y = \mathbb{P}_k^1$, and $C \setminus \{p\} \to Y$ is given by $(x, y) \mapsto [x, y]$.

(c) *Non-singularity of C is necessary.* Suppose $C = \operatorname{Spec} k[x, y]/(y^2 - x^3)$, $p = 0$, $Y = \mathbb{P}_k^1$, and $C \setminus \{p\} \to Y$ is given by $(x, y) \mapsto [x, y]$.

We remark that by combining this (easy) theorem with the (hard) valuative criterion for properness (Theorem 13.7.6), one obtains a proof of the properness of projective space bypassing the (tricky but not hard) Fundamental Theorem of Elimination Theory 8.4.10 (see Exercise 13.7.F). Fancier remark: The valuative criterion of properness can be used to show that Theorem 15.3.1 remains true if Y is only required to be proper, but it requires some thought.

15.3.2. *Central idea of proof.* The central idea of the proof may be summarized as "clear denominators," and is illustrated by the following motivating example. Suppose you have a morphism from $\mathbb{A}^1 \setminus \{0\}$ to projective space, and you wanted to extend it to $\mathbb{A}^1$. Suppose the map was given by $t \mapsto [t^4 + t^{-3}, t^{-2} + 4t]$. Then of course you would "clear the denominators," and replace the map by $t \mapsto [t^7 + 1, t + 4t^4]$. Similarly, if the map was given by $t \mapsto [t^2 + t^3, t^2 + t^4]$, you would divide by t^2, to obtain the map $t \mapsto [1 + t, 1 + t^2]$.

15.3.3. *Proof.* Our plan is to maneuver ourselves into the situation where we can apply the idea of §15.3.2. We begin with some quick reductions. The nonreduced locus of C is closed (Exercise 6.6.N) and doesn't contain p, so by replacing C by an appropriate open neighborhood of p, we may assume that C is reduced and affine.

We next reduce to the case where $Y = \mathbb{P}^n_A$. Choose a closed embedding $Y \to \mathbb{P}^n_A$. If the result holds for $\mathbb{P}^n$, and we have a morphism $C \to \mathbb{P}^n$ with $C \setminus \{p\}$ mapping to Y, then C must map to Y as well. Reason: We can reduce to the case where the source is an affine open subset, and the target is $\mathbb{A}^n_A \subset \mathbb{P}^n_A$ (and hence affine). Then the functions vanishing on $Y \cap \mathbb{A}^n_A$ pull back to functions that vanish at the generic points of the irreducible components of C and hence vanish everywhere on C (using reducedness of C), i.e., C maps to Y.

Choose a uniformizer $t \in \mathfrak{m} \setminus \mathfrak{m}^2$ in the local ring of C at p. As t is a function in some neighborhood of p, we may assume that t is a function on C (by replacing C by this neighborhood of p). Then V(t) contains p (as $t \in \mathfrak{m}$), but V(t) does not contain the component of C containing p (as $t \notin \mathfrak{m}^2$), so by replacing C by an affine open neighborhood of p in $(C \setminus V(t)) \cup \{p\}$—note that $(C \setminus V(t)) \cup \{p\} \subset C$ is open because C (almost) has the cofinite topology—we may assume that p is the only zero of the function t (and, of course, t vanishes at p with multiplicity 1).

We have a map $C \setminus \{p\} \to \mathbb{P}^n_A$, which by Theorem 15.2.2 corresponds to a line bundle $\mathscr{L}$ on $C \setminus \{p\}$ and $n+1$ sections of it with no common zeros in $C \setminus \{p\}$. Let U be a nonempty open set of $C \setminus \{p\}$ on which $\mathscr{L} \cong \mathcal{O}$. Then by replacing C by $U \cup \{p\}$, we interpret the map to $\mathbb{P}^n$ as $n+1$ rational *functions* $f_0, \ldots, f_n$, defined away away from p, with no common zeros away from p. Let $N = \min_i(\mathrm{val}_p f_i)$. Then $t^{-N}f_0, \ldots, t^{-N}f_n$ are $n+1$ functions with no common zeros. Thus they determine a morphism $C \to \mathbb{P}^n_A$ extending $C \setminus \{p\} \to \mathbb{P}^n_A$ as desired. $\square$

15.3.B. EXERCISE (USEFUL PRACTICE). Suppose X is a Noetherian k-scheme, and Z is an irreducible codimension 1 subvariety whose generic point is a regular point of X (so the local ring $\mathcal{O}_{X,Z}$ is a discrete valuation ring). Suppose $\pi: X \dashrightarrow Y$ is a rational map to a projective k-scheme. Show that the domain of definition of the rational map includes a dense open subset of Z. In other words, rational maps from Noetherian k-schemes to projective k-schemes can be extended over regular codimension 1 sets. (We have already seen this principle in action—see Exercise 7.5.J on the Cremona transformation.)

15.3.4.* **Extended example: Tangents to regular plane curves are limits of secants.**
We now discuss an extended example that may give insight into many different things. When learning calculus, we're sometimes taught that tangent lines to plane curves are limits of secants. We can now see this in terms of the Curve-to-Projective Extension Theorem 15.3.1, and deltas and epsilons are turned into easy algebra. This discussion also foreshadows the importance of the diagonal morphism in understanding differentials.

Let k be a field, which you should initially take to be algebraically closed. (You can later think about how this makes sense over arbitrary fields.) Suppose $C \subset \mathbb{A}^2_k$ is a plane curve, given by the equation $f(x, y) = 0$. Fix a point $p = (x_0, y_0)$ of C. Given a point $q = (x_1, y_1)$ of C, with $x_1 \neq x_0$, the slope of the line joining them is $m = (y_1 - y_0)/(x_1 - x_0)$, and the equation of the line joining them is

$$(y - y_0) = \frac{y_1 - y_0}{x_1 - x_0}(x - x_0).$$

More generally (even if we have $x_1 = x_0$, but still with $y_1 \neq y_0$), the line joining them has "slope" $[x_1 - x_0, y_1 - y_0] \in \mathbb{P}^1_k$, and the equation of the line is $(x_1 - x_0)(y - y_0) = (y_1 - y_0)(x - x_0)$. We thus have a map

(15.3.4.1) $C \setminus p \longrightarrow \mathbb{P}^1_k$

sending q to the "slope" of the secant line $\overline{pq}$. The Curve-to-Projective Extension Theorem 11.1.1 states that this will extend over p, where p is a smooth point of C. Let's work out the algebra. Keep in mind that now $x_0, y_0 \in k$, as p is a fixed point.

Define $F(x_1, y_1) := f(x_1, y_1) - f(x_0, y_0) \in k[x_1, y_1]$.

15.3.C. EXERCISE. Show that $F \in (x_1 - x_0, y_1 - y_0)$.

Hence we can write $F = g(x_1, y_1)(x_1 - x_0) + h(x_1, y_1)(y_1 - y_0)$ for some (nonunique) choice of $g, h \in k[x_1, y_1]$. Choose such a g and h.

15.3.D. EXERCISE. Show that if $p = (x_0, y_0)$ is a smooth point of C with nonvertical tangent vector, then $h(x_0, y_0) \neq 0$.

15.3.E. EXERCISE. Show that if $p = (x_0, y_0)$ is a smooth point of C, then the slope of the tangent line at p is $[-h(x_0, y_0), g(x_0, y_0)] \in \mathbb{P}_k^1$, and the equation of the tangent line is

$$g(x_0, y_0)(x - x_0) + h(x_0, y_0)(y - y_0) = 0.$$

15.4 Hard but Important: Line Bundles and Weil Divisors

The notion of Weil divisors gives a great way of understanding and classifying line bundles. Before we get started, you should be warned: this is one of those topics in algebraic geometry that is hard to digest—learning it changes the way in which you think about line bundles. But once you become comfortable with the imperfect dictionary to divisors, it becomes second nature.

In this section, we consider only *Noetherian normal schemes*. We do this because we will use finite decomposition into irreducible components (Exercise 5.3.B), and Algebraic Hartogs's Lemma 13.5.19. Some of what we discuss will apply in more general circumstances, and the expert is invited to consider generalizations by judiciously weakening hypotheses in various statements.

15.4.1. *Definition.* Define a **Weil divisor** as a formal $\mathbb{Z}$-linear combination of codimension 1 irreducible closed subsets of X. In other words, a Weil divisor is defined to be an object of the form

$$\sum_{Y \subset X \text{ codimension } 1} n_Y [Y],$$

where the n_Y are integers, all but a finite number of which are zero. Weil divisors obviously form an abelian group, denoted by Weil X. For example, if X is a curve, the Weil divisors are linear combinations of closed points.

15.4.2. We say that [Y] is an **irreducible** (Weil) divisor. A Weil divisor $D = \sum n_Y [Y]$ is said to be **effective** if $n_Y \geq 0$ for all Y. In this case we say $D \geq 0$, and by $D_1 \geq D_2$ we mean $D_1 - D_2 \geq 0$. The **support** of a Weil divisor D, denoted Supp D, is the subset $\cup_{n_Y \neq 0} Y$. If $U \subset X$ is an open set, we define the **restriction map** Weil $X \to$ Weil U by $\sum n_Y [Y] \mapsto \sum_{Y \cap U \neq \varnothing} n_Y [Y \cap U]$.

15.4.A. EASY EXERCISE. Show that the irreducible divisors on $\mathbb{P}_k^n$ correspond to irreducible homogeneous polynomials in $k[x_0, \ldots, x_n]$, up to multiplication by nonzero scalars $k^\times$.

15.4.3. Suppose that $\mathscr{L}$ is an invertible sheaf, and s a rational section not vanishing everywhere on any irreducible component of X. (Rational sections are given by a section over a dense open subset of X, with the obvious equivalence, §14.2.1.)

We can make sense of the **valuation of** s **along an irreducible Weil divisor** Y (denoted by $\text{val}_Y(s)$) as follows. Take any open set U containing the generic point of Y where $\mathscr{L}$ is trivializable, along with any trivialization over U; under this trivialization, s is a nonzero rational function on U, which thus has a valuation (§13.5.14). Any two such trivializations differ by an invertible function (transition functions are invertible), so this valuation is well-defined. Note that $\text{val}_Y(s) = 0$ for all but finitely many Y, by Exercise 13.5.G.

Thus s determines a Weil divisor

$$\operatorname{div}(s) := \sum_Y \operatorname{val}_Y(s)[Y].$$

The summation runs over all irreducible divisors Y of X. We call $\operatorname{div}(s)$ the **divisor of zeros and poles** of the rational section s (cf. Definition 13.5.7).

15.4.4. The important group of "line bundles with rational sections". Now consider the set $\{(\mathscr{L}, s)\}$ of pairs of line bundles $\mathscr{L}$ with rational sections s of $\mathscr{L}$, not the zero section on any irreducible component of X, *up to isomorphism*. (An isomorphism $(\mathscr{L}, s) \cong (\mathscr{L}', s')$ means an isomorphism of line bundles under which s is sent to s'.) This set (after taking quotient by isomorphism) forms an abelian group under tensor product $\otimes$, with identity $(\mathscr{O}_X, 1)$. (Tricky question: What is the inverse of $(\mathscr{L}, s)$ in this group?)

It is important to notice that if t is an invertible function on X, then multiplication by t gives an isomorphism

$$(\mathscr{L}, s) \xleftrightarrow{\sim} (\mathscr{L}, st).$$

Similarly, $(\mathscr{L}, s)/(\mathscr{L}, u) = (\mathscr{O}, s/u)$. Here s/u is a rational *function*.

The map div yields a group homomorphism

(15.4.4.1) $\operatorname{div} \colon \{(\mathscr{L}, s)\}/\text{isomorphism} \longrightarrow \operatorname{Weil} X.$

15.4.B. EASIER EXERCISE.

(a) *(divisors of rational functions)* Verify that on $\mathbb{A}^1_k$, $\operatorname{div}(x^3/(x+1)) = 3[(x)] - [(x+1)]$ ("$= 3[0] - [-1]$").

(b) *(divisor of rational sections of a nontrivial invertible sheaf)* On $\mathbb{P}^1_k$, there is a rational section of $\mathscr{O}(1)$ "corresponding to" $x^2/(x+y)$. Figure out what this means, and calculate $\operatorname{div}(x^2/(x+y))$.

The homomorphism (15.4.4.1) will be the key to determining all the line bundles on many X. (Note that any invertible sheaf will have such a rational section. For each irreducible component, take a nonempty open set not meeting any other irreducible component; then shrink it so that $\mathscr{L}$ is trivial; choose a trivialization; then take the union of all these open sets, and choose the section on this union corresponding to 1 under the trivialization.) We will see that in reasonable situations, this map div will be injective, and often an isomorphism. Thus by forgetting the rational section (i.e., taking an appropriate quotient), we will have described the Picard group of all line bundles. Let's put this strategy into action.

15.4.5. Proposition — *If X is normal and Noetherian then the map* div *is injective.*

Proof. Suppose $\operatorname{div}(\mathscr{L}, s) = 0$. Then s has no poles. By Exercise 14.2.F, s is a regular section. We now show that the morphism $\times s \colon \mathscr{O}_X \to \mathscr{L}$ is in fact an isomorphism; this will prove the proposition, as it will give an isomorphism $(\mathscr{O}_X, 1) \xleftrightarrow{\sim} (\mathscr{L}, s)$.

It suffices to show that $\times s$ is an isomorphism on an open subset U of X where $\mathscr{L}$ is trivial, as X is covered by trivializing open neighborhoods of $\mathscr{L}$ (as $\mathscr{L}$ is locally trivial). Choose an isomorphism $i \colon \mathscr{L}|_U \xrightarrow{\sim} \mathscr{O}_U$. Composing $\times s$ with i yields a map $\times s' \colon \mathscr{O}_U \to \mathscr{O}_U$ that is multiplication by a rational function $s' = i(s)$ that has no zeros and no poles. The rational function s' is regular because it has no poles (Exercise 13.5.I), and 1/s' is regular for the same reason. Thus s' is an invertible function on U, so $\times s'$ is an isomorphism. Hence $\times s$ is an isomorphism over U. $\square$

Motivated by this, we try to find an inverse to div, or at least to determine the image of div.

15.4.6. Important definition: $\mathscr{O}_X(D)$. *Assume now that X is normal*—this will be a standing assumption for the rest of this section.

Assume also that X is irreducible (purely to avoid making (15.4.6.1) look uglier—but feel free to relax this; see Exercise 15.4.C). In case it helps: recall that a normal scheme is a disjoint union of irreducible normal schemes (Exercise 5.4.B)—so this assumption truly is harmless.

Suppose D is a Weil divisor. Define the sheaf $\mathscr{O}_X(D)$ by

(15.4.6.1) $\Gamma(U, \mathscr{O}_X(D)) := \{t \in K(X)^\times : \operatorname{div}|_U t + D|_U \geq 0\} \cup \{0\}.$

Here $D|_U$ is the restriction of D to U (defined in §15.4.2), and $\operatorname{div}|_U t$ means the divisor of t considered as a rational function on U (i.e., consider just the irreducible divisors of U). The subscript X in $\mathscr{O}_X(D)$ is omitted when it is clear from context. The sections of $\mathscr{O}_X(D)$ over U are the rational functions on U that have poles and zeros "constrained by D": a positive coefficient in D allows a pole of that order; a negative coefficient demands a zero of that order. Away from the support of D, this is (isomorphic to) the structure sheaf (by Algebraic Hartogs's Lemma 13.5.19).

15.4.7. *Important remark.* It will be crucial to note that $\mathscr{O}_X(D)$ comes along with a canonical nonzero "rational section" corresponding to $1 \in K(X)^\times$. (It is a rational section in the sense that it is a section over a dense open set, namely the complement of Supp D.)

15.4.C. LESS IMPORTANT EXERCISE. Generalize the definition of $\mathscr{O}_X(D)$ to the case when X is not necessarily irreducible. (This is just a question of language. Once you have done this, feel free to drop this hypothesis in the rest of this section.)

15.4.D. EASY EXERCISE. Verify that $\mathscr{O}_X(D)$ is a quasicoherent sheaf. (Hint: The "distinguished affine" criterion for quasicoherence of Exercise 6.2.D.)

In good situations, $\mathscr{O}_X(D)$ is an invertible sheaf. For example, let $X = \mathbb{A}^1_k$. Consider

$$\mathscr{O}_X\left(-2[(x)] + [(x-1)] + [(x-2)]\right),$$

often written $\mathscr{O}(-2[0] + [1] + [2])$ for convenience. Then $3x^3/(x-1)$ is a global section; it has the required two zeros at $x = 0$ (and even one to spare), and takes advantage of the allowed pole at $x = 1$, and doesn't have a pole at $x = 2$, even though one is allowed. (Unimportant aside: The statement remains true in characteristic 2, although the explanation requires editing.)

15.4.E. EASY EXERCISE. (This is a consequence of later discussion as well, but you should be able to do this by hand.)

(a) Show that any global section of $\mathscr{O}_{\mathbb{A}^1_k}(-2[(x)] + [(x-1)] + [(x-2)])$ is a $k[x]$-multiple of $x^2/((x-1)(x-2))$.
(b) Extend the argument of (a) to give an isomorphism

$$\mathscr{O}_{\mathbb{A}^1_k}\left(-2[(x)] + [(x-1)] + [(x-2)]\right) \xrightarrow{\sim} \mathscr{O}_{\mathbb{A}^1_k}.$$

The next exercise is seemingly small, but is absolutely crucial to lessen confusion as the complications add up in the next few pages.

15.4.F. EXERCISE. We continue the example of Easy Exercise 15.4.E, letting $k = \mathbb{C}$ for concreteness. Let $X = \mathbb{A}^1_{\mathbb{C}}$, and $D = -2[(x)] + [(x-1)] + [(x-2)]$.

(a) Consider the section of $\mathscr{O}_X(D)$ corresponding to $x^2(x-3)/((x-1)(x-2)) \in K(X)$ (under the correspondence of (15.4.6.1)). This section vanishes at precisely one point. What point is it?
(b) Consider the section of $\mathscr{O}_X(D)$ corresponding to $x^2(x-1) \in K(X)$. This section vanishes at precisely two points. What are those points, and to what orders does it vanish at those points?
(c) Consider the *rational section* of $\mathscr{O}_X(D)$ corresponding to $1 \in K(X)$ (never forget, via the correspondence of (15.4.6.1)). Where are its zeros and poles? What are their orders? What is the most common wrong answer to this problem?

We next show that in good circumstances, $\mathscr{O}_X(D)$ is an invertible sheaf. (In fact the $\mathscr{O}_X(D)$ construction can occasionally be useful even if $\mathscr{O}_X(D)$ is *not* an invertible sheaf, but this won't concern us here. An example of an $\mathscr{O}_X(D)$ that is not an invertible sheaf is given in Exercise 15.4.K.)

15.4.G. HARD BUT IMPORTANT EXERCISE. Suppose $\mathscr{L}$ is an invertible sheaf, and s is a nonzero rational section of $\mathscr{L}$.

(a) Describe an isomorphism $\mathcal{O}(\operatorname{div} s) \overset{\sim}{\longleftrightarrow} \mathscr{L}$. (You will use the normality hypothesis!) Hint: Show that those open subsets U for which $\mathcal{O}(\operatorname{div} s)|_U \cong \mathcal{O}_U$ form a base for the Zariski topology. For each such U, define $\phi_U \colon \mathcal{O}(\operatorname{div} s)(U) \to \mathscr{L}(U)$ sending a rational function t (with zeros and poles "constrained by div s") to st. Show that ϕ_U is an isomorphism (with the obvious inverse map, division by s). Argue that this map induces an isomorphism of sheaves $\phi \colon \mathcal{O}(\operatorname{div} s) \overset{\sim}{\longrightarrow} \mathscr{L}$.

(b) Let σ be the map from K(X) to the rational sections of $\mathscr{L}$, where σ(t) is the rational section of $\mathcal{O}_X(\operatorname{div} s) \cong \mathscr{L}$ defined via (15.4.6.1) (as in Remark 15.4.7). Show that σ sends the canonical section "1" of $\mathcal{O}_X(D)$ to s. More concisely: σ(1) = s. (Hint: The map in part (a) sends 1 to s.)

In conclusion, we can identify which Weil divisors D are in the image of div (15.4.4.1): they are the D for which $\mathcal{O}_X(D)$ is a line bundle, and then we can construct the image—$(\mathcal{O}(D), \sigma(1))$ the unique $(\mathscr{L}, s)$ (up to isomorphism) such that div$(\mathscr{L}, s) = D$.

15.4.H. EXERCISE (THE EXAMPLE OF §15.1). Suppose $X = \mathbb{P}_k^n$, $\mathscr{L} = \mathcal{O}(1)$, s is the section of $\mathcal{O}(1)$ corresponding to x_0, and D = div s. Verify that $\mathcal{O}(mD) \cong \mathcal{O}(m)$, and the canonical rational section of $\mathcal{O}(mD)$ is precisely s^m. (Watch out for possible confusion: 1 has no pole along $x_0 = 0$, but σ(1) = s^m *does* have a zero if m > 0.) For this reason, $\mathcal{O}(1)$ is sometimes called the **hyperplane class** in Pic X. (Of course, x_0 can be replaced by any linear form.)

15.4.I. EXERCISE (THE PICARD GROUP OF PROJECTIVE SPACE). You have identified Weil$(\mathbb{P}_k^n)$ in Exercise 15.4.A. Use this to deduce that the only line bundles on $\mathbb{P}_k^n$ are isomorphic to $\mathcal{O}(m)$ for some m, so Pic $\mathbb{P}_k^n \cong \mathbb{Z}$.

15.4.8. Definition. If D is a Weil divisor on (Noetherian normal irreducible) X such that D = div f for some rational *function* f, we say that D is **principal**. Principal divisors clearly form a subgroup of Weil X; denote this group by Prin X. Note that div induces a group homomorphism $K(X)^\times \to$ Prin X. If X can be covered with open sets U_i such that on U_i, D is principal, we say that D is **locally principal**. (Caution: We used the phrase "locally principal" in a slightly different way, when referring to closed subschemes.) Locally principal divisors form a subgroup of Weil X, which we denote by LocPrin X. (This notation is not standard, and we only use it in (15.4.11.1).)

15.4.9. Important observation: "$\mathcal{O}(principal) \cong \mathcal{O}$, $\mathcal{O}(locally\ principal)$ = line bundle". As a consequence of Exercise 15.4.G(a) (taking $\mathscr{L} = \mathcal{O}$), if D is principal (and X is normal, our standing hypothesis), then (i) $\mathcal{O}(D) \cong \mathcal{O}$. (Diagram (15.4.11.1) will imply that the converse holds: if $\mathcal{O}(D) \cong \mathcal{O}$, then D is principal.) Thus (ii) if D is *locally* principal, $\mathcal{O}_X(D)$ is *locally* isomorphic to $\mathcal{O}_X$—$\mathcal{O}_X(D)$ is an invertible sheaf.

15.4.J. IMPORTANT EXERCISE. Suppose $\mathcal{O}_X(D)$ is an invertible sheaf. Show the converse to Important Observation 15.4.9: show that D is locally principal.

15.4.10. Remark. In definition (15.4.6.1), it may seem cleaner to consider those s such that div s $\geq$ D$|_U$. The reason for the convention comes from our desire that div σ(1) = D. (Taking the "opposite" convention would yield the dual bundle, in the case where D is locally principal.)

15.4.K. LESS IMPORTANT EXERCISE: A WEIL DIVISOR THAT IS NOT LOCALLY PRINCIPAL. Let $X = \operatorname{Spec} k[x, y, z]/(xy - z^2)$, a cone, and let D be the line z = x = 0 (see Figure 13.1).

(a) Show that D is not a locally principal divisor. (Hint: Consider the stalk at the origin. Use the Zariski tangent space; see Problem 13.1.3.) In particular, $\mathcal{O}_X(D)$ is not an invertible sheaf. (Caution: We earlier saw that it was not a locally principal closed subscheme in Problem 13.1.3. It is unfortunate that "locally principal" has two different but very similar meanings.)

(b) Show that $\mathrm{div}(x) = 2D$. This corresponds to the fact that the plane $x = 0$ is tangent to the cone X along D. Hence $2D$ *is* a locally principal divisor.

15.4.L. IMPORTANT EXERCISE. If X is Noetherian and factorial, show that for any Weil divisor D, $\mathscr{O}(D)$ is an invertible sheaf. (Hint: For motivation, consider the case where D is irreducible, say, $D = [Y]$, and cover X by open sets so that on each open set U there is a function whose divisor is $[Y \cap U]$. One open set will be $X - Y$. Next, we find an open set U containing an arbitrary $p \in Y$, and a function on U. As $\mathscr{O}_{X,p}$ is a unique factorization domain, the prime corresponding to Y is codimension 1 and hence principal by Lemma 12.1.7. Let f be a generator of this prime ideal, interpreted as an element of $K(X)$. It is regular at p, it has a finite number of zeros and poles, and through p, $[Y]$ is the "only zero" (the only component of the divisor of zeros). Let U be X minus all the other zeros and poles.)

15.4.11. The class group. We can now get a handle on the Picard group of a normal Noetherian scheme. Define the **class group** of X, $\mathrm{Cl}\, X$, by $\mathrm{Weil}\, X / \mathrm{Prin}\, X$. By taking the quotient of the inclusion (15.4.4.1) by $\mathrm{Prin}\, X$, we have the inclusion $\mathrm{Pic}\, X \hookrightarrow \mathrm{Cl}\, X$. This is summarized in the convenient and enlightening diagram.

(15.4.11.1)

$$
\begin{array}{ccccc}
& (\mathscr{O}(D),\sigma(1)) \mapsfrom D & & & \\
& \overleftarrow{\hspace{2cm}} & & \stackrel{= \text{ if } X}{\hookrightarrow} & \\
\{(\mathscr{L},s)\}/\text{iso.} & \xrightarrow[\text{div}]{\sim} & \mathrm{LocPrin}\, X & \underset{\text{factorial}}{} & \mathrm{Weil}\, X \\
\downarrow & & \downarrow\, /\,\mathrm{Prin}\, X & & \downarrow\, /\,\mathrm{Prin}\, X \\
\mathrm{Pic}\, X = \{\mathscr{L}\}/\text{iso.} & \xrightarrow{\sim} & \mathrm{LocPrin}\, X/\,\mathrm{Prin}\, X & \stackrel{= \text{ if } X}{\underset{\text{factorial}}{\hookrightarrow}} & \mathrm{Cl}\, X \\
& \overleftarrow{\hspace{2cm}} & & & \\
& \mathscr{O}(D) \mapsfrom D & & &
\end{array}
$$

This diagram is very important, and dense with meaning, it takes time to digest.

In particular, if A is a unique factorization domain, then all Weil divisors on $\mathrm{Spec}\, A$ *are principal by Lemma 12.1.7, so* $\mathrm{Cl}(\mathrm{Spec}\, A) = 0$*, and hence* $\mathrm{Pic}(\mathrm{Spec}\, A) = 0$.

15.4.12. As $k[x_1, \ldots, x_n]$ has unique factorization, $\mathrm{Cl}(\mathbb{A}_k^n) = 0$, so $\boxed{\mathrm{Pic}(\mathbb{A}_k^n) = 0}$. Geometers might find this believable—"$\mathbb{C}^n$ is a contractible complex manifold, and hence should have no nontrivial line bundles"—even if some caution is in order, as the kinds of line bundles being considered are entirely different: holomorphic vs. topological or C^∞. (Aside: For this reason, you might expect that $\mathbb{A}_k^n$ also has no nontrivial vector bundles. This is the Quillen–Suslin Theorem, formerly known as Serre's Conjecture, part of Quillen's work leading to his 1978 Fields Medal. The case $n = 1$ was Exercise 14.2.C. For a short proof by Vaserstein, see [Lan, p. 850].)

Removing a closed subset of X of codimension greater than 1 doesn't change the class group, as it doesn't change the Weil divisor group or the principal divisors. (Warning: It *can* affect the Picard group; see Exercise 15.5.M.)

Removing an irreducible closed subset of pure codimension 1 changes the Weil divisor group in a controllable way. Suppose Z is an *irreducible* codimension 1 subset (an irreducible divisor) of X. Then we clearly have an exact sequence:

$$0 \longrightarrow \mathbb{Z} \xrightarrow{1 \mapsto [Z]} \mathrm{Weil}\, X \longrightarrow \mathrm{Weil}(X - Z) \longrightarrow 0.$$

When we take the quotient by the subgroup of principal divisors, taking into account the fact that we may lose exactness on the left, we get an **excision exact sequence for class groups**:

(15.4.12.1)
$$\mathbb{Z} \xrightarrow{1 \mapsto [Z]} \mathrm{Cl}\, X \longrightarrow \mathrm{Cl}(X - Z) \longrightarrow 0.$$

(Do you see why?)

15.4.13. For example, if U is an open subscheme of $X = \mathbb{A}^n$, $\mathrm{Pic}\, U = \{0\}$.

15.4.14. As another application, let $X = \mathbb{P}^n_k$, and Z be the hyperplane $x_0 = 0$. We have

$$\mathbb{Z} \longrightarrow \mathrm{Cl}(\mathbb{P}^n_k) \longrightarrow \mathrm{Cl}(\mathbb{A}^n_k) \longrightarrow 0,$$

from which $\mathrm{Cl}(\mathbb{P}^n_k)$ is generated by the class $[Z]$, and $\mathrm{Pic}(\mathbb{P}^n_k)$ is a subgroup of this.

15.4.15. By Exercise 15.4.H, $[Z] \mapsto \mathscr{O}(1)$, and as $\mathscr{O}(m)$ is nontrivial for $m \neq 0$ (Exercise 15.1.B), $[Z]$ is not torsion in $\mathrm{Cl}\,\mathbb{P}^n_k$. Hence $\mathrm{Pic}(\mathbb{P}^n_k) \hookrightarrow \mathrm{Cl}(\mathbb{P}^n_k)$ is an isomorphism, and $\boxed{\mathrm{Pic}(\mathbb{P}^n_k) \cong \mathbb{Z}}$, with generator $\mathscr{O}(1)$. The **degree** of an invertible sheaf on $\mathbb{P}^n$ is defined using this: define $\deg \mathscr{O}(d)$ to be d. (You will have already proved that $\mathrm{Pic}(\mathbb{P}^n_k) \cong \mathbb{Z}$ if you did Exercise 15.4.I; but we will use the strategy here to great effect in §15.5.4.)

We have gotten good mileage from the fact that the Picard group of the spectrum of a unique factorization domain is trivial. More generally, Exercise 15.4.L gives us:

15.4.16. Proposition — *If X is Noetherian and factorial, then for any Weil divisor D, $\mathscr{O}(D)$ is invertible, and hence the map $\mathrm{Pic}\,X \to \mathrm{Cl}\,X$ is an isomorphism.*

This can be used to make the connection to the class group in number theory precise; see Exercise 14.1.K; see also §15.5.5.

15.4.17. Mild but important generalization: Twisting line bundles by divisors. The above constructions can be extended, with $\mathscr{O}_X$ replaced by an arbitrary invertible sheaf, as follows. Let $\mathscr{L}$ be an invertible sheaf on a normal Noetherian scheme X. Then define $\mathscr{L}(D)$ by $\mathscr{O}_X(D) \otimes \mathscr{L}$. If D is locally principal, then $\mathscr{L}(D)$ is a line bundle. Notice that in this case there are two different ways of interpreting sections of $\mathscr{L}(D)$ over an open set, each with different advantages: as a section of the new line bundle $\mathscr{L}(D)$, and as rational sections of $\mathscr{L}$ with constraints on poles and zeros given by the divisor D.

15.4.M. EASY EXERCISE.

(a) Assume for convenience that X is irreducible. Show that sections of $\mathscr{L}(D)$ can be interpreted as rational sections of $\mathscr{L}$ with zeros and poles constrained by D, just as in (15.4.6.1):

$$\Gamma(U, \mathscr{L}(D)) := \{t \text{ nonzero rational section of } \mathscr{L} \; : \; \mathrm{div}|_U t + D|_U \geq 0\} \cup \{0\}.$$

(b) Suppose D_1 and D_2 are locally principal. Show that

$$(\mathscr{O}(D_1))(D_2) \cong \mathscr{O}(D_1 + D_2).$$

15.4.18. A variation of the Qcqs Lemma. The Qcqs Lemma 6.2.9, proved in Exercise 6.2.G, has the following generalization.

15.4.N. IMPORTANT EXERCISE (TO BE USED REPEATEDLY). Suppose X is a quasicompact quasiseparated scheme, $\mathscr{L}$ is an invertible sheaf on X with section s, and $\mathscr{F}$ is a quasicoherent sheaf on X. Generalizing Definition 6.2.8, let X_s be the open subset of X where s doesn't vanish. We interpret s as a degree 1 element of the graded ring $R(\mathscr{L})_\bullet := \oplus_{n \geq 0} \Gamma(X, \mathscr{L}^{\otimes n})$. Note that $\oplus_{n \geq 0} \Gamma(X, \mathscr{F} \otimes_{\mathscr{O}_X} \mathscr{L}^{\otimes n})$ is a graded $R(\mathscr{L})_\bullet$-module.

(a) Describe a natural map

$$\left(\left(\oplus_{n \geq 0} \Gamma(X, \mathscr{F} \otimes_{\mathscr{O}_X} \mathscr{L}^{\otimes n}) \right)_s \right)_0 \longrightarrow \Gamma(X_s, \mathscr{F}).$$

(Possible hint: For quasicoherent sheaves, "tensor product has no need to be sheafified when restricted to affine subschemes," Exercise 6.2.F.)

(b) Show that this map is an isomorphism. (Hint: Show this map is an isomorphism in the affine case.)

Translation: Any section of $\mathscr{F}$ over X_s can be extended to a section over X after multiplying by some some appropriate power of s. And if we have two such extensions, they become equal after multiplying by another appropriate power of s.

15.5 The Payoff: Many Fun Examples

15.5.1. Fun examples: Projective transformations.

The fact that $\operatorname{Pic} \mathbb{P}_k^n$ is $\mathbb{Z}$ has many wonderful and cheap consequences.

15.5.A. EXERCISE (AUTOMORPHISMS OF PROJECTIVE SPACE). Show that all the automorphisms of projective space $\mathbb{P}_k^n$ (fixing k) correspond to $(n+1) \times (n+1)$ invertible matrices over k, modulo scalars (also known as $\operatorname{PGL}_{n+1}(k)$). (Hint: Suppose $\pi \colon \mathbb{P}_k^n \to \mathbb{P}_k^n$ is an automorphism. Show that there exists an isomorphism $\pi^* \mathscr{O}(1) \xrightarrow{\sim} \mathscr{O}(1)$. which then induces an issomorphism $\pi^* \colon \Gamma(\mathbb{P}^n, \mathscr{O}(1)) \to \Gamma(\mathbb{P}^n, \mathscr{O}(1)))$.

15.5.2. Automorphisms of projective space are often called **projective transformations**. Because of Exercise 15.5.A, in their incarnation of matrices modulo scalars, projective transformations are also called **projective changes of coordinates**. Exercise 15.5.A will be useful later, especially for the case $n = 1$. In this case, these automorphisms are called **fractional linear transformations**. (For experts: Why was Exercise 15.5.A not stated over an arbitrary base ring A? Where does the argument go wrong in that case? For what rings A does the result still work?)

15.5.B. EXERCISE. Show that $\operatorname{Aut}(\mathbb{P}_k^1)$ is strictly 3-transitive on k-valued points, i.e., given two triplets (p_1, p_2, p_3) and (q_1, q_2, q_3) each of distinct k-valued points of $\mathbb{P}^1$, there is precisely one automorphism of $\mathbb{P}^1$ sending p_i to q_i $(i = 1, 2, 3)$.

15.5.C. EXERCISE. Solve these problems over an arbitrary field k.

(a) Find a linear fractional transformation $f(t) \in \operatorname{PGL}(2)$ that has order precisely 3 in $\operatorname{PGL}(2)$.
(b) Show that any two order 3 elements of $\operatorname{PGL}(2)$ are conjugate. (Possible hint: Use transitivity.)

15.5.D. EXERCISE. Suppose $p_0, \ldots, p_{n+1}$ are $n+2$ distinct k-valued points of $\mathbb{P}_k^n$, no $n+1$ of which lie on a hyperplane. Show that there is a unique projective transformation taking p_i $(0 \le i \le n)$ to $[0, \ldots, 0, 1, 0, \ldots, 0]$ (where the 1 is in the ith position), and taking p_{n+1} to $[1, \ldots, 1]$.

15.5.E. FUN EXERCISE. Suppose X is a quasiprojective k-scheme, and $\pi \colon \mathbb{P}_k^n \to X$ is any morphism (over k). Show that either the image of π has dimension n, or π contracts $\mathbb{P}_k^n$ to a point. In particular, there are no nonconstant maps from projective space to a smaller-dimensional quasi-projective variety. Hint: Show that it suffices to assume k is algebraically closed, and, in particular, infinite. If $X \subset \mathbb{P}^N$, define d by $\pi^* \mathscr{O}_{\mathbb{P}^N}(1) \cong \mathscr{O}_{\mathbb{P}_k^n}(d)$. Try to show that $d = 0$. To do that, show that if $m \le n$ then m nonempty hypersurfaces in $\mathbb{P}^n$ have nonempty intersection. For this, use the fact that any nonempty hypersurface in $\mathbb{P}_k^n$ has nonempty intersection with any subscheme of dimension at least 1 (Exercise 12.3.D(a)).

15.5.F.* EXERCISE (FOR THOSE WHO HAVE READ §7.6.4, THE DOUBLE-STARRED SECTION ON GROUP SCHEMES). Explain how GL_n acts (nontrivially!) on $\mathbb{P}^{n-1}$ (over $\mathbb{Z}$, or over a field of your choice). (The group scheme GL_n was defined in Exercise 7.6.N. The *action* of a group scheme appeared earlier in Exercise 7.6.S(a).) Hint: This is much more easily done with the language of functors, §7.6, using our functorial description of projective space (§15.2.3), than with our old description of projective space in terms of patches. (A generalization to the Grassmannian will be given in Exercise 16.4.L.)

15.5.3. *Remark.* Over an algebraically closed field $\overline{k}$, GL_n acts transitively on the closed points of $\mathbb{P}_{\overline{k}}^n$, and the stabilizer of the point $[1, 0, \cdots, 0]$ consists of the subgroup P of matrices with 0's in the first column below the first row. (Side remark: The point is better written as a column vector, so the GL_n-action can be interpreted as matrix multiplication in the usual way.) This suggests that $\mathbb{P}_k^{n-1}$ is the quotient GL_n / P. This is largely true; but we first would have to make sense of the notion of group quotient.

15.5.4. More fun examples.

We can now actually calculate some Picard and class groups. First, a useful observation: Notice that you can restrict invertible sheaves on Y to any subscheme X, and this can be a handy way of checking that an invertible sheaf is not trivial. Effective Cartier divisors (§9.5.1) sometimes restrict too: if you have an effective Cartier divisor on Y, then it restricts to a closed subscheme on X, locally cut out scheme-theoretically by one equation. If you are fortunate and this equation doesn't vanish on any associated point of X (§6.6.2), then you get an effective Cartier divisor on X. You can check that the restriction of effective Cartier divisors corresponds to restriction of invertible sheaves (in the sense of Exercise 14.1.I).

15.5.G. EXERCISE: A TORSION PICARD GROUP. Suppose that Y is a hypersurface in $\mathbb{P}_k^n$ corresponding to an irreducible degree d polynomial. Show that $\operatorname{Pic}(\mathbb{P}_k^n - Y) \cong \mathbb{Z}/(d)$. (For differential geometers: this is related to the fact that $\pi_1(\mathbb{P}_k^n - Y) \cong \mathbb{Z}/(d)$.) Hint: (15.4.12.1).

The next two exercises explore consequences of Exercise 15.5.G, and provide us with some examples promised in Exercise 5.4.N.

15.5.H. EXERCISE (GENERALIZING EXERCISE 5.4.N). Keeping the same notation, assume $d > 1$ (so $\operatorname{Pic}(\mathbb{P}^n - Y) \neq 0$), and let $H_0, \ldots, H_n$ be the $n+1$ coordinate hyperplanes on $\mathbb{P}^n$. Show that $\mathbb{P}^n \setminus Y$ is affine, and $\mathbb{P}^n - Y - H_i$ is a distinguished open subset of it. Show that the $\mathbb{P}^n - Y - H_i$ form an open cover of $\mathbb{P}^n - Y$. Show that $\operatorname{Pic}(\mathbb{P}^n - Y - H_i) = 0$. Show that each $\mathbb{P}^n - Y - H_i$ is the Spec of a unique factorization domain, but $\mathbb{P}^n - Y$ is not. Thus the property of being a unique factorization domain is not an affine-local property—it satisfies only one of the two hypotheses of the Affine Communication Lemma 5.3.2.

15.5.I. EXERCISE. Keeping the same notation as that in the previous exercise, show that on $\mathbb{P}^n - Y$, H_i (restricted to this open set) is an effective Cartier divisor that is not cut out by a single equation. (Hint: Otherwise it would give a trivial element of the class group.)

15.5.J. EXERCISE. Show that $A := \mathbb{R}[x, y]/(x^2 + y^2 - 1)$ is not a unique factorization domain, but $A \otimes_{\mathbb{R}} \mathbb{C}$ is. Hint: Exercise 15.5.H. (Non-hint: Doesn't $(1-y)(1+y) = x^2$ suggest that $A \otimes_{\mathbb{R}} \mathbb{C}$ *isn't* a unique factorization domain?)

15.5.K. EXERCISE: PICARD GROUP OF $\mathbb{P}^1 \times \mathbb{P}^1$. Consider

$$X = \mathbb{P}_k^1 \times_k \mathbb{P}_k^1 \cong \operatorname{Proj} k[w, x, y, z]/(wz - xy),$$

a smooth quadric surface (see Figure 9.2, and Example 10.6.2). Show that $\operatorname{Pic} X \cong \mathbb{Z} \oplus \mathbb{Z}$ as follows: Show that if $L = \{\infty\} \times_k \mathbb{P}^1 \subset X$ and $M = \mathbb{P}^1 \times_k \{\infty\} \subset X$, then $X - L - M \cong \mathbb{A}^2$. This will give you a surjection $\mathbb{Z} \oplus \mathbb{Z} \twoheadrightarrow \operatorname{Cl} X$. Show that $\mathcal{O}(L)$ restricts to $\mathcal{O}$ on L and $\mathcal{O}(1)$ on M. Show that $\mathcal{O}(M)$ restricts to $\mathcal{O}$ on M and $\mathcal{O}(1)$ on L. (This exercise takes some time, but is enlightening.)

15.5.L. EXERCISE. Show that irreducible smooth projective surfaces (over k) can be birational but not isomorphic. Hint: Show $\mathbb{P}^2$ is not isomorphic to $\mathbb{P}^1 \times \mathbb{P}^1$ using the Picard group. (Aside: We will see in Exercise 20.2.D that the Picard group of the "blown-up plane" is $\mathbb{Z}^2$, but in Exercise 20.2.E we will see that the blown-up plane is not isomorphic to $\mathbb{P}^1 \times \mathbb{P}^1$, using a little more information in the Picard group.)

This is unlike the case for curves: birational irreducible smooth projective curves (over k) must be isomorphic, as we will see in Theorem 16.3.3. Nonetheless, any two surfaces are related in a simple way: if X and X' are projective, regular, and birational, then X can be sequentially blown up at judiciously chosen points, and X' can too, such that the two results are isomorphic (see [Ha1, Thm. V.5.5]; blowing up will be discussed in Chapter 22).

15.5.M. EXERCISE: PICARD GROUP OF THE CONE. Let $X = \operatorname{Spec} k[x, y, z]/(xy - z^2)$, a cone, where char $k \neq 2$. (The characteristic hypothesis is not necessary for the result, but is included so you can use Exercise 5.4.H to show normality of X.) Show that $\operatorname{Pic} X = 0$, and $\operatorname{Cl} X \cong \mathbb{Z}/2$. Hint:

Show that the class of $Z = \{x = z = 0\}$ (the "affine cone over a line") generates Cl X by showing that its complement $D(x)$ is isomorphic to an open subset of $\mathbb{A}^2_k$. Show that $2[Z] = \operatorname{div}(x)$ and hence principal, and that Z is not principal, Exercise 15.4.K. (Remark: You know enough to show that $X \setminus \{(0, 0, 0)\}$ is factorial. So although the class group is insensitive to removing loci of codimension greater than 1, §15.4.12, this is not true of the Picard group.)

A Weil divisor (on a normal scheme) with a nonzero multiple corresponding to a line bundle is called $\mathbb{Q}$-**Cartier**. (We won't use this terminology beyond the next exercise.) Exercise 15.5.M gives an example of a Weil divisor that does not correspond to a line bundle, but is nonetheless $\mathbb{Q}$-Cartier. We now give an example of a Weil divisor that is *not* $\mathbb{Q}$-Cartier.

15.5.N. EXERCISE (A NON-$\mathbb{Q}$-CARTIER DIVISOR). On the cone over the smooth quadric surface $X = \operatorname{Spec} k[w, x, y, z]/(wz - xy)$, let Z be the Weil divisor cut out by $w = x = 0$. Exercise 13.1.D showed that Z is not cut out scheme-theoretically by a single equation. Show more: that if $n \neq 0$, then $n[Z]$ is not locally principal. Hence show that Z is not even cut out locally *set-theoretically* by a single equation. Hint: Show that the complement of an effective Cartier divisor on an affine scheme is also affine, using Proposition 8.3.4. Then if some multiple of Z were locally principal, then the closed subscheme of the complement of Z cut out by $y = z = 0$ would be affine—any closed subscheme of an affine scheme is affine. But this is the scheme $y = z = 0$ (also known as the wx-plane) minus the point $w = x = 0$, which we have seen is non-affine, §4.4.1.

15.5.O.* EXERCISE (FOR THOSE WITH SUFFICIENT ARITHMETIC BACKGROUND). Identify the (ideal) class group of the ring of integers $\mathcal{O}_K$ in a number field K, as defined in Exercise 14.1.K, with the class group of $\operatorname{Spec} \mathcal{O}_K$, as defined in this section. In particular, you will recover the common description of the class group as formal sums of prime ideals, modulo an equivalence relation coming from principal fractional ideals.

15.5.5. More on class groups and unique factorization.
As mentioned in §5.4.5, there are few commonly used means of checking that a ring is a unique factorization domain. The next exercise is one of them, and it is useful. For example, it implies the classical fact that for rings of integers in number fields, the class group is the obstruction to unique factorization (see Exercise 14.1.K and Proposition 15.4.16).

15.5.P. EXERCISE. Suppose that A is a Noetherian integral domain. Show that A is a unique factorization domain if and only if A is integrally closed and Cl Spec $A = 0$. (One direction is easy: we have already shown that unique factorization domains are integrally closed in their fraction fields. Also, Lemma 12.1.7 shows that all codimension 1 prime ideals of a unique factorization domain are principal, so that implies that Cl Spec $A = 0$. It remains to show that if A is integrally closed and Cl Spec $A = 0$, then all codimension 1 prime ideals are principal, as this characterizes unique factorization domains (Proposition 12.3.7). Algebraic Hartogs's Lemma 13.5.19, may arise in your argument.) This is the third important characterization of unique factorization domains promised in §5.4.5.

My final favorite method of checking that a ring is a unique factorization domain (§5.4.5) is Nagata's Lemma. It is also the least useful.

15.5.Q.** EXERCISE (NAGATA'S LEMMA). Suppose A is a Noetherian domain, $x \in A$ an element such that (x) is prime and $A_x = A[1/x]$ is a unique factorization domain. Then A is a unique factorization domain. (Hint: Exercise 15.5.P. Use the short exact sequence

$$\mathbb{Z}[(x)] \longrightarrow \operatorname{Cl} \operatorname{Spec} A \longrightarrow \operatorname{Cl} \operatorname{Spec} A_x \longrightarrow 0$$

(15.4.12.1) to show that Cl Spec $A = 0$. Prove that $A[1/x]$ is integrally closed, then show that A is integrally closed as follows. Suppose $T^n + a_{n-1} T^{n-1} + \cdots + a_0 = 0$, where $a_i \in A$, and $T \in K(A)$. Then by integral closedness of A_x, we have that $T = r/x^m$, where if $m > 0$, then $r \notin (x)$. Then we quickly get a contradiction if $m > 0$.)

This leads to a fun algebra fact promised in Remark 13.8.6. Suppose k is an algebraically closed field of characteristic not 2. Let $A = k[x_1, \ldots, x_n]/(x_1^2 + \cdots + x_m^2)$, where $m \leq n$. When $m \leq 2$, we get some special behavior. (If $m = 0$, we get affine space; if $m = 1$, we get a nonreduced scheme; if $m = 2$, we get a reducible scheme that is the union of two affine spaces.)

If $m \geq 3$, we have verified that Spec A is normal, in Exercise 5.4.I(b). In fact, if $m \geq 3$, then A is a unique factorization domain *unless* $m = 3$ *or* $m = 4$ (Exercise 5.4.L; see also Exercise 13.1.E). For the case $m = 3$:

$$A = k[x, y, z, w_1, \ldots, w_{n-3}]/(x^2 + y^2 - z^2)$$

is not a unique factorization domain, as it is has nonzero class group (by essentially the same argument as for Exercise 15.5.M).

The failure at 4 comes from the geometry of the quadric surface: we have checked that in Spec $k[w, x, y, z]/(wz - xy)$, there is a codimension 1 irreducible subset—the cone over a line in a ruling—that is not principal.

15.5.R. EXERCISE (THE CASE M $\geq$ 5). Suppose that k is algebraically closed of characteristic not 2. Show that if $\ell \geq 3$, then

$$A = k[a, b, x_1, \ldots, x_n]/(ab - x_1^2 - \cdots - x_\ell^2)$$

is a unique factorization domain, by using Nagata's Lemma with $x = a$.

15.6 Effective Cartier Divisors "=" Invertible Ideal Sheaves

We now give a different means of describing invertible sheaves on a scheme. One advantage of this over Weil divisors is that it can give line bundles on everywhere nonreduced schemes (such a scheme can't be regular at any codimension 1 prime).

Suppose $D \hookrightarrow X$ is a closed subscheme such that the corresponding ideal sheaf $\mathscr{I}$ is an invertible sheaf. Then $\mathscr{I}$ is locally trivial; suppose U is a trivializing affine open set Spec A. Then the closed subscheme exact sequence (9.1.2.1)

$$0 \longrightarrow \mathscr{I} \longrightarrow \mathscr{O}_X \longrightarrow \mathscr{O}_D \longrightarrow 0$$

corresponds to

$$0 \longrightarrow I \longrightarrow A \longrightarrow A/I \longrightarrow 0$$

with $I \cong A$ as A-modules. Thus I is generated by a single element, say a, and this exact sequence starts as

$$0 \longrightarrow A \xrightarrow{\times a} A.$$

As multiplication by a is injective, a is not a zerodivisor. We conclude that D is locally cut out by a single equation that is not a zerodivisor. This was the definition of *effective Cartier divisor* given in §9.5.1. This argument is clearly reversible, so we have a quick new definition of effective Cartier divisor (an ideal sheaf $\mathscr{I}$ that is an invertible sheaf—or, equivalently, the corresponding closed subscheme).

15.6.A. EASY EXERCISE. Show that a is unique up to multiplication by an invertible function.

In the case where X is locally Noetherian, we can use the language of associated points (§6.6.2), so we can restate this definition as: D is locally cut out by a single equation, not vanishing at any associated point of X.

We now define an invertible sheaf corresponding to D. The seemingly obvious definition would be to take $\mathscr{I}_D$, but instead we do the "opposite."

15.6.1. *Definition: Invertible sheaves corresponding to effective Cartier divisors.* If D is an effective Cartier divisor, define the invertible sheaf $\mathscr{O}(D)$ to be $\mathscr{I}_D^\vee$. (The reason for the dual is

Exercise 15.6.C.) Define $\mathscr{O}(nD)$ to be $\mathscr{O}(D)^{\otimes n}$. In particular, define $\mathscr{O}(-D)$ to be $\mathscr{O}(D)^{\vee}$, i.e., $\mathscr{O}(-D) = \mathscr{I}_D$.

15.6.B. EXERCISE. If X is normal, check that this agrees with our earlier definition of $\mathscr{O}(D)$, Important Definition 15.4.6. (In Definition 15.4.6, D is a Weil divisor, but in this exercise, D is an effective Cartier divisor. So you will have to define the Weil divisor corresponding to an effective Cartier divisor on a normal Noetherian scheme X.)

Because the ideal sheaf $\mathscr{I}_D$ is $\mathscr{O}(-D)$, the closed subscheme exact sequence (9.1.2.1) may be written as

$$0 \longrightarrow \mathscr{O}(-D) \longrightarrow \mathscr{O} \longrightarrow \mathscr{O}_D \longrightarrow 0.$$

The invertible sheaf $\mathscr{O}(D)$ has a canonical section s_D: tensoring $0 \to \mathscr{I} \to \mathscr{O}$ with $\mathscr{I}^{\vee}$ gives us $\mathscr{O} \to \mathscr{I}^{\vee}$. (Easy unimportant fact: Instead of tensoring $\mathscr{I} \to \mathscr{O}$ with $\mathscr{I}^{\vee}$, we could have dualized $\mathscr{I} \to \mathscr{O}$, and we would get the same section.)

15.6.C. IMPORTANT AND SURPRISINGLY TRICKY EXERCISE. Recall that a section of a locally free sheaf on X cuts out a closed subscheme of X (Exercise 14.2.D). Show that the section s_D cuts out D. (Compare this to Remark 15.4.10.)

This construction is reversible:

15.6.D. EXERCISE. Suppose $\mathscr{L}$ is an invertible sheaf, and s is a section that is locally not a zero-divisor. (Make sense of this! In particular, if X is locally Noetherian, this means "s does not vanish at an associated point of X"; see §6.6.2.) Show that $s = 0$ cuts out an effective Cartier divisor D, and $\mathscr{O}(D) \cong \mathscr{L}$.

15.6.E. EXERCISE. Suppose $\mathscr{I}$ and $\mathscr{J}$ are invertible ideal sheaves (hence corresponding to effective Cartier divisors, say D and D', respectively). Show that $\mathscr{I}\mathscr{J}$ is an invertible ideal sheaf. (We define the **product of two quasicoherent ideal sheaves** $\mathscr{I}\mathscr{J}$ as you might expect: on each affine, we take the product of the two corresponding ideals, as defined in Exercise 3.4.C. To make sure this is well-defined, we need only check that if A is a ring, and $f \in A$, and $I, J \subset A$ are two ideals, then $(IJ)_f = I_f J_f$ in A_f.) We define the corresponding Cartier divisor to be $D + D'$. Verify that $\mathscr{O}(D + D') \cong \mathscr{O}(D) \otimes \mathscr{O}(D')$.

We thus have an important correspondence between *effective Cartier divisors* (closed subschemes whose ideal sheaves are invertible, or equivalently locally cut out by one non-zerodivisor, or, in the locally Noetherian case, locally cut out by one equation not vanishing at an associated point) and *ordered pairs* $(\mathscr{L}, s)$ where $\mathscr{L}$ is an invertible sheaf, and s is a section that is not locally a zerodivisor (or in the locally Noetherian case, not vanishing at an associated point). The effective Cartier divisors form an abelian semigroup. We have a map of semigroups, from effective Cartier divisors to invertible sheaves with sections not locally zerodivisors (and hence also to the Picard group of invertible sheaves).

15.6.2. *Normal (line) bundles to effective Cartier divisors.* If D is an effective Cartier divisor on a scheme X, then the restriction of the invertible sheaf $\mathscr{O}(D)$ to D itself is often called the **normal (line) bundle** to D (although perhaps it should be called the normal invertible sheaf). We denote it by $\mathscr{N}_{D/X} := \mathscr{O}_X(D)|_D$. The motivation for this language is that if X and D are smooth complex varieties, the normal bundle in this algebro-geometric sense agrees with the normal bundle in the differential-geometric sense. This interpretation breaks down in more "singular" situations, but the intuition is useful even then. We will define the normal bundle (and normal sheaf) in a more general setting in §21.2.15.

15.6.3. *Unimportant remark.* We get lots of invertible sheaves, by taking differences of two effective Cartier divisors. In fact, we "usually get them all"—it is very hard to describe an invertible sheaf on a finite type k-scheme that is not describable in such a way. For example, there are none if the scheme is regular or even factorial (basically by Proposition 15.4.16 for factoriality; and regular schemes are factorial by the Auslander–Buchsbaum Theorem 13.8.5). Exercise 16.2.F will imply that there are none if the scheme is projective. It holds in all other reasonable circumstances; see [Gr-EGA, IV$_4$.21.3.4]. However, it does not always hold; the first and best example is due to Kleiman; see [Kl3].

15.6.4.** **Cartier divisors.** There is a related notion of *Cartier divisor*. This notion is considered essential by many, so I will try to explain why we are not discussing it (except briefly here). Cartier divisors are indeed important, but arguably not at the level of discussion we are engaging in. Using Cartier divisors involves technical difficulties, in return for suprisingly little simplification of perspective and results. (The expert is welcome to develop a parallel discussion in the language of Cartier divisors in the margins to test this out.) For completeness, here is a definition. On a scheme X, we define an $\mathscr{O}_X$-module $\mathscr{M}_X$, whose sections on an affine open set $\operatorname{Spec} A$ are the (elements of the) total fraction ring of A (§6.6.36), with the "obvious" restriction maps. If X is integral, $\mathscr{M}_X$ is the constant sheaf corresponding to the function field $K(X)$. We have an induced map $\mathscr{O}_X \hookrightarrow \mathscr{M}_X$. Let $\mathscr{M}_X^*$ be the sheaf of groups corresponding (on the level of sections) to invertible elements, inducing $\mathscr{O}_X^* \hookrightarrow \mathscr{M}_X^*$. Then the **Cartier divisors** of X are defined as $\Gamma(X, \mathscr{M}_X^*/\mathscr{O}_X^*)$ (with its structure of an abelian group). Many seeming advantages of Cartier divisors actually come from other structures. Cartier divisors easily pull back under morphisms, but so do line bundles (or equivalently, invertible sheaves). Effective Cartier divisors are also very pleasant and simple. Cartier divisors easily yield invertible sheaves, but it is not *always* true that every invertible sheaf comes from a Cartier divisor, [Kl3].

15.7 The Graded Module Corresponding to a Quasicoherent Sheaf

(This section answers some fundamental questions, but it is surprisingly tricky. You may wish to skip this section, or at least the proofs, on first reading, unless you have a particular need for them.)

Throughout this section, $S_\bullet$ is a finitely generated graded algebra *generated in degree* 1, so, in particular, we have the invertible sheaf $\mathscr{O}(m)$ for all m by Exercise 15.1.G. Also, throughout, $M_\bullet$ is a graded $S_\bullet$-module, and $\mathscr{F}$ is a quasicoherent sheaf on $\operatorname{Proj} S_\bullet$.

We know how to get quasicoherent sheaves on $\operatorname{Proj} S_\bullet$ from graded $S_\bullet$-modules. We will now see that we can get them all in this way. We will define a functor $\Gamma_\bullet$ from (the category of) quasicoherent sheaves on $\operatorname{Proj} S_\bullet$ to (the category of) graded $S_\bullet$-modules that will attempt to reverse the $\sim$ construction. They are not quite inverses, as $\sim$ can turn two different graded modules into the same quasicoherent sheaf (see, for example, Exercise 14.6.C). But we will see a natural isomorphism $\widetilde{\Gamma_\bullet(\mathscr{F})} \overset{\sim}{\longleftrightarrow} \mathscr{F}$. In fact, $\Gamma_\bullet(\widetilde{M_\bullet})$ is a better ("saturated") version of $M_\bullet$, and there is a saturation functor $M_\bullet \to \Gamma_\bullet(\widetilde{M_\bullet})$ that is akin to groupification and sheafification—it is adjoint to the forgetful functor from saturated graded modules to graded modules. And thus we come to the fundamental relationship between $\sim$ and $\Gamma_\bullet$: they are an adjoint pair.

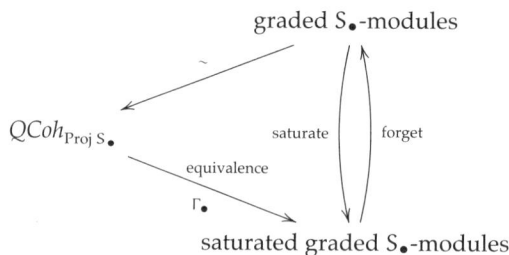

We now make some of this precise, but as little as possible to move forward. In particular, we will show that every quasicoherent sheaf on a projective A-scheme arises from a graded module (Corollary 15.7.4), and we will see that every closed subscheme of $\operatorname{Proj} S_\bullet$ arises from a homogeneous ideal $I_\bullet \subset S_\bullet$ (Exercise 15.7.I).

15.7.1. Definition of $\Gamma_\bullet$. When you do Essential Exercise 15.1.C (on global sections of $\mathscr{O}_{\mathbb{P}^n_k}(m)$), you will suspect that, in good situations,

$$M_m \cong \Gamma(\operatorname{Proj} S_\bullet, \widetilde{M}(m)_\bullet).$$

Motivated by this, we define

$$\Gamma_m(\mathscr{F}) := \Gamma(\operatorname{Proj} S_\bullet, \mathscr{F}(m)).$$

15.7.A. EXERCISE. Describe a morphism of S_0-modules $M_m \to \Gamma(\operatorname{Proj} S_\bullet, \widetilde{M(m)_\bullet})$, extending the $m = 0$ case of Exercise 14.6.D.

15.7.B. EXERCISE. Show that $\Gamma_\bullet(\mathscr{F})$ is a graded $S_\bullet$-module. (Hint: Consider $S_m \to \Gamma(\operatorname{Proj} S_\bullet, \mathscr{O}(m))$.)

15.7.C. EXERCISE. Show that the map $M_\bullet \to \Gamma_\bullet(\widetilde{M_\bullet})$ arising from the previous two exercises is a map of $S_\bullet$-modules. We call this the **saturation map**.

15.7.D. EXERCISE. Show that the saturation map need not be injective, nor need it be surjective. (Hint: $S_\bullet = k[x]$, $M_\bullet = k[x]/(x^2)$ or $M_\bullet = xk[x]$.)

15.7.E. EXERCISE. Show that $\Gamma_\bullet$ is a functor from $QCoh_{\operatorname{Proj} S_\bullet}$ to the category of graded $S_\bullet$-modules. In other words, if $\mathscr{F} \to \mathscr{G}$ is a morphism of quasicoherent sheaves on $\operatorname{Proj} S_\bullet$, describe the natural map $\Gamma_\bullet \mathscr{F} \to \Gamma_\bullet \mathscr{G}$, and show that such maps respect the identity and composition. Thus the saturation map can be better called the **saturation functor**.

15.7.2. *The special case* $M_\bullet = S_\bullet$. We have a saturation map $S_\bullet \to \Gamma_\bullet \widetilde{S_\bullet}$, which is a map of $S_\bullet$-modules. But $\Gamma_\bullet \widetilde{S_\bullet}$ has the structure of a graded ring (basically because we can multiply sections of $\mathscr{O}(m_1)$ by sections of $\mathscr{O}(m_2)$ to get sections of $\mathscr{O}(m_1 + m_2)$; see Exercise 15.1.I).

15.7.F. EXERCISE.

(a) Show that the saturation map is in fact a ring morphism.
(b) Show that the saturation map is an isomorphism when $S_\bullet$ is a polynomial ring (with the usual grading), in at least one variable.
(c) Show that the saturation map is *not* an isomorphism for $S_\bullet = k[x, y]/(xy)$, where both variables have degree 1.

15.7.3.* The reverse map. Now that we have defined the saturation map $M_\bullet \to \Gamma_\bullet \widetilde{M_\bullet}$, we will describe a map $\widetilde{\Gamma_\bullet \mathscr{F}} \to \mathscr{F}$. While subtler to define, it will have the advantage of being an isomorphism.

15.7.G. EXERCISE. Define the natural map $\widetilde{\Gamma_\bullet \mathscr{F}} \to \mathscr{F}$ as follows. First describe the map on sections over $D_+(f)$. Note that sections of the left side are of the form z/f^n, where $z \in \Gamma_{n \deg f}(\mathscr{F})$, and $z/f^n = z'/f^{n'}$ if there is some N with $f^N(f^{n'}z - f^n z') = 0$. Sections on the right are implicitly described in Exercise 15.4.N. Show that your map behaves well on overlaps $D_+(f) \cap D_+(g) = D_+(fg)$.

15.7.H. TRICKY EXERCISE. Show that the natural map $\widetilde{\Gamma_\bullet \mathscr{F}} \to \mathscr{F}$ is an isomorphism, by showing that it is an isomorphism of sections over $D_+(f)$ for any f. First show surjectivity, using Exercise 15.4.N to show that any section of $\mathscr{F}$ over $D_+(f)$ is of the form z/f^n, where $z \in \Gamma_{n \deg f}(\mathscr{F})$. Then verify that it is injective.

15.7.4. Corollary — *Every quasicoherent sheaf on a projective A-scheme arises from the $\sim$ construction.*

15.7.I. EXERCISE (CONVERSE TO EXERCISE 9.3.B). Show that each closed subscheme of $\operatorname{Proj} S_\bullet$ arises from a homogeneous ideal $I_\bullet \subset S_\bullet$. (Hint: Suppose Z is a closed subscheme of $\operatorname{Proj} S_\bullet$. Consider the exact sequence $0 \to \mathscr{I}_Z \to \mathscr{O}_{\operatorname{Proj} S_\bullet} \to \mathscr{O}_Z \to 0$. Apply $\Gamma_\bullet$, and then $\sim$. Be careful: $\Gamma_\bullet$ is left-exact, but not necessarily exact.) This fulfills promises made in Exercises 9.3.B and 14.6.E.

For the first time, we see that every closed subscheme of a projective scheme is cut out by homogeneous equations. This is the analog of the fact that every closed subscheme of an affine scheme is cut out by equations. It is disturbing that it is so hard to prove this fact. (For comparison, this was easy on the level of the Zariski topology—see Exercise 4.5.K(c).)

15.7.J. ** EXERCISE ($\Gamma_\bullet$ AND $\sim$ ARE ADJOINT FUNCTORS). Describe a natural bijection $\mathrm{Hom}(M_\bullet, \Gamma_\bullet \mathscr{F}) \cong \mathrm{Hom}(\widetilde{M_\bullet}, \mathscr{F})$, as follows.

(a) Show that maps $M_\bullet \to \Gamma_\bullet \mathscr{F}$ are the "same" as maps $((M_\bullet)_f)_0 \to ((\Gamma_\bullet \mathscr{F})_f)_0$ as f varies through S_+, which are "compatible" as f varies, i.e., if $D_+(g) \subset D_+(f)$, there is a commutative diagram

$$
\begin{array}{ccc}
((M_\bullet)_f)_0 & \longrightarrow & ((\Gamma_\bullet \mathscr{F})_f)_0 \\
\downarrow & & \downarrow \\
((M_\bullet)_g)_0 & \longrightarrow & ((\Gamma_\bullet \mathscr{F})_g)_0.
\end{array}
$$

More precisely, give a bijection between $\mathrm{Hom}(M_\bullet, \Gamma_\bullet \mathscr{F})$ and the set of compatible maps

$$
\left(\mathrm{Hom}((M_\bullet)_f)_0 \to ((\Gamma_\bullet \mathscr{F})_f)_0) \right)_{f \in S_+}.
$$

(b) Describe a bijection between the set of compatible maps $(\mathrm{Hom}((M_\bullet)_f)_0 \to ((\Gamma_\bullet \mathscr{F})_f)_0)_{f \in S_+}$ and the set of compatible maps $\Gamma(D_+(f), \widetilde{M_\bullet}) \to \Gamma(D_+(f), \mathscr{F})$.

15.7.5. ** **Saturated $S_\bullet$-modules.** We end with a remark: Different graded $S_\bullet$-modules give the same quasicoherent sheaf on $\mathrm{Proj}\, S_\bullet$, but the results of this section show that there is a "best" (= saturated) graded module for each quasicoherent sheaf, and there is a map from each graded module to its "best" version, $M_\bullet \to \Gamma_\bullet \widetilde{M_\bullet}$. A module for which this is an isomorphism (a "best" module) is called **saturated**. We won't use this term later.

This "saturation" map $M_\bullet \to \Gamma_\bullet \widetilde{M_\bullet}$ is analogous to the sheafification map, taking presheaves to sheaves. For example, the saturation of the saturation equals the saturation.

See §14.6.1 for another description of the category of saturated $S_\bullet$-modules.

Chapter 16

Maps to Projective Space, and Properties of Line Bundles

Line bundles on a variety X are important in part because they can tell a lot about maps from X to other varieties, because they tell us about maps from X to projective space (§15.2). The crucial concept of "ampleness" of a line bundle should be seen as some sort of "positivity" (§16.2), which will help us know when the line bundle has lots of sections and might even give a map to projective space. The fundamental facts of ampleness will take some work, but will be powerfully useful from then on. For example, in §16.3, we will see many applications to curves.

16.1   Globally Generated Quasicoherent Sheaves

We begin with a result letting us "get at" any finite type quasicoherent sheaf on a projective scheme in terms of line bundles. Throughout this section, $S_\bullet$ will be a finitely generated graded ring over A, generated in degree 1. We will prove the following result.

16.1.1. Theorem — *Any finite type quasicoherent sheaf $\mathscr{F}$ on $\operatorname{Proj} S_\bullet$ can be presented in the form*

$$\oplus_{finite} \mathscr{O}(-m) \longrightarrow \mathscr{F} \longrightarrow 0.$$

Because we can work with the line bundles $\mathscr{O}(-m)$ in a hands-on way, this result will give us great control over *all* coherent sheaves (and, in particular, vector bundles) on $\operatorname{Proj} S_\bullet$. As just a first example, it will allow us to show that every coherent sheaf on a projective k-scheme has a finite-dimensional space of global sections (Corollary 18.1.4). (This fact will grow up to be the fact that the higher pushforwards of coherent sheaves under proper morphisms are also coherent; see Theorem 18.7.1(d) and Grothendieck's Coherence Theorem 18.9.1.)

Rather than proceeding directly to a proof, we use this as an excuse to introduce notions that are useful in wider circumstances (*global generation, base-point-freeness, ampleness*), and their interrelationships. But first we use it as an excuse to mention an important classical result.

16.1.2. *The Hilbert Syzygy Theorem.*
Given any coherent sheaf $\mathscr{F}$ on $\mathbb{P}^n_k$, Theorem 16.1.1 gives the existence of a surjection $\phi \colon \oplus_{\text{finite}} \mathscr{O}(-m) \to \mathscr{F} \to 0$. The kernel of the surjection is also coherent, so iterating this construction, we can construct an infinite resolution of $\mathscr{F}$ by a direct sum of line bundles:

$$\cdots \oplus_{\text{finite}} \mathscr{O}(m_{2,j}) \longrightarrow \oplus_{\text{finite}} \mathscr{O}(m_{1,j}) \longrightarrow \oplus_{\text{finite}} \mathscr{O}(m_{0,j}) \longrightarrow \mathscr{F} \longrightarrow 0.$$

16.1.3. The Hilbert Syzygy Theorem — *There exists a* finite *resolution, of length at most* $n+1$.

See the comments after Theorem 3.6.17 for the original history of this result. We won't use the theorem, but a proof will be given in the optional section §24.4.

16.1.4. Globally generated sheaves. Suppose X is a scheme, and $\mathscr{F}$ is an $\mathscr{O}$-module. The most important definition of this section is the following: $\mathscr{F}$ is **globally generated** (or **generated by global sections**) if it admits a surjection from a free sheaf on X:

$$\mathscr{O}^{\oplus I} \longrightarrow \mathscr{F} .$$

Here I is some index set. The global sections in question are the images of the $|I|$ sections corresponding to 1 in the various summands of $\mathscr{O}_X^{\oplus I}$; those images generate the stalks of $\mathscr{F}$. We say $\mathscr{F}$ is **finitely globally generated** (or **generated by a finite number of global sections**) if the index set I can be taken to be finite.

More definitions in more detail: We say that $\mathscr{F}$ is **globally generated at a point** p (or sometimes **generated by global sections at** p) if we can find $\phi: \mathscr{O}^{\oplus I} \to \mathscr{F}$ that is surjective on stalks at p,

$$\mathscr{O}_p^{\oplus I} \xrightarrow{\phi_p} \mathscr{F}_p.$$

(It would be more precise to say that the stalk of $\mathscr{F}$ at p is generated by global sections of $\mathscr{F}$.) The key insight is that global generation at p means that every germ at p is a linear combination (over the local ring $\mathscr{O}_{X,p}$) of germs of global sections.

Note that $\mathscr{F}$ is *globally generated* if it is globally generated at *all* points p. (Reason: Exercise 2.4.D showed that isomorphisms can be checked on the level of stalks. An easier version of the same argument shows that surjectivity can also be checked on the level of stalks.) Notice that we can take a single index set for all of X, by taking the union of all the index sets for each p.

16.1.A. EASY EXERCISE. An invertible sheaf $\mathscr{L}$ on X is globally generated if and only if for any point $p \in X$, there is a global section of $\mathscr{L}$ not vanishing at p, i.e., $\mathscr{L}$ is base-point-free (Definition 15.2.1).

16.1.B. EASY EXERCISE (REALITY CHECK). Show that every quasicoherent sheaf on every *affine* scheme is globally generated. Show that every finite type quasicoherent sheaf on every affine scheme is generated by a finite number of global sections. (Hint for both: For any A-module M, there is a surjection onto M from a free A-module.)

Clearly, if $\mathscr{F}$ and $\mathscr{G}$ are globally generated, then so is $\mathscr{F} \oplus \mathscr{G}$ (and similarly for finitely globally generated, generated at a point p, etc.). Similarly for tensor product:

16.1.C. EASY EXERCISE (GLOBALLY GENERATED $\otimes$ GLOBALLY GENERATED IS GLOBALLY GENERATED). Show that if quasicoherent sheaves $\mathscr{F}$ and $\mathscr{G}$ are globally generated at a point p, then so is $\mathscr{F} \otimes \mathscr{G}$. (If $\mathscr{F}$ and $\mathscr{G}$ are invertible sheaves, this is Exercise 15.2.B.)

16.1.D. EASY BUT IMPORTANT EXERCISE. Suppose $\mathscr{F}$ is a finite type quasicoherent sheaf on X.

(a) Show that $\mathscr{F}$ is globally generated at p if and only if "the fiber of $\mathscr{F}$ is generated by global sections at p," i.e., the global sections span the fiber $\mathscr{F}_p/\mathfrak{m}\mathscr{F}_p$ as a vector space, where $\mathfrak{m}$ is the maximal ideal of $\mathscr{O}_{X,p}$. (Hint: Geometric Nakayama, Exercise 14.3.D.)
(b) Show that if $\mathscr{F}$ is globally generated at p, then "$\mathscr{F}$ is globally generated near p": there is an open neighborhood U of p such that $\mathscr{F}$ is globally generated at every point of U.
(c) Suppose further that X is a quasicompact scheme. Show that if $\mathscr{F}$ is globally generated at all closed points of X, then $\mathscr{F}$ is globally generated at all points of X. (Note that nonempty quasicompact schemes *have* closed points, Exercise 5.1.E.)

16.1.E. EASY EXERCISE. If $\mathscr{F}$ is a finite type quasicoherent sheaf on X, and X is quasicompact, show that $\mathscr{F}$ is globally generated if and only if it is generated by a *finite number* of global sections.

We are now able to state a celebrated result of Serre.

16.1.5. Serre's Theorem A — *Suppose $S_\bullet$ is generated in degree 1, and finitely generated over $A = S_0$. Let $\mathscr{F}$ be any finite type quasicoherent sheaf on $\operatorname{Proj} S_\bullet$. Then there exists some $\mathfrak{m}_0$ such that for all $\mathfrak{m} \geq \mathfrak{m}_0$, $\mathscr{F}(\mathfrak{m})$ is finitely globally generated.*

We could now prove Serre's Theorem A directly, but will continue to use it as an excuse to introduce more ideas; it will be a consequence of Theorem 16.2.2.

16.1.6. *Proof of Theorem 16.1.1 assuming Serre's Theorem A (Theorem 16.1.5).* Suppose we have n global sections $s_1, \ldots, s_n$ of $\mathscr{F}(\mathfrak{m})$ that generate $\mathscr{F}(\mathfrak{m})$. This gives a map

$$\mathscr{O}_{\operatorname{Proj} S_\bullet}^{\oplus n} \longrightarrow \mathscr{F}(\mathfrak{m})$$

given by $(f_1, \ldots, f_n) \mapsto f_1 s_1 + \cdots + f_n s_n$ on any open set. Because these global sections generate $\mathscr{F}(m)$, this is a surjection. Tensoring with $\mathscr{O}(-m)$ (which is exact, as tensoring with any locally free sheaf is exact, Exercise 14.1.F) gives the desired result. □

16.2   Ample and Very Ample Line Bundles

We now introduce ampleness, a central "positivity" property of line bundles that will prove important for a number of disparate reasons—geometric, algebraic, topological, computational, and more.

Suppose $\pi\colon X \to \operatorname{Spec} A$ is a morphism, and $\mathscr{L}$ is an invertible sheaf on X. (The case when A is a field is the one of most immediate interest.) We say that $\mathscr{L}$ is **very ample over** A or π-**very ample**, or **relatively very ample** if $X \cong \operatorname{Proj} S_\bullet$, where $S_\bullet$ is a finitely generated graded ring over A generated in degree 1 (Definition 4.5.6), and $\mathscr{L} \cong \mathscr{O}_{\operatorname{Proj} S_\bullet}(1)$. One often just says **very ample** if the structure morphism is clear from the context. Note that the existence of a very ample line bundle implies that X is projective.

16.2.A. EASY BUT IMPORTANT EXERCISE (EQUIVALENT DEFINITION OF VERY AMPLE OVER A). Suppose $\pi\colon X \to \operatorname{Spec} A$ is proper, and $\mathscr{L}$ is an invertible sheaf on X. Show that $\mathscr{L}$ is very ample if and only if there exist a finite number of global sections $s_0, \ldots s_n$ of $\mathscr{L}$, with no common zeros, such that the morphism

$$[s_0, \ldots, s_n]\colon X \longrightarrow \mathbb{P}_A^n$$

is a closed embedding.

16.2.B. EASY EXERCISE (VERY AMPLE IMPLIES BASE-POINT-FREE). Show that a very ample invertible sheaf $\mathscr{L}$ on a proper A-scheme must be base-point-free. In other words, show that for any point p on the scheme, there is a section of $\mathscr{L}$ not vanishing there.

16.2.C. EXERCISE (VERY AMPLE $\otimes$ BASE-POINT-FREE IS VERY AMPLE, HENCE VERY AMPLE $\otimes$ VERY AMPLE IS VERY AMPLE). Suppose $\mathscr{L}$ and $\mathscr{M}$ are invertible sheaves on a proper A-scheme X, $\mathscr{L}$ is very ample over A, and $\mathscr{M}$ is base-point-free. Show that $\mathscr{L} \otimes \mathscr{M}$ is very ample. (Hint: $\mathscr{L}$ gives a closed embedding $X \hookrightarrow \mathbb{P}^m$, and $\mathscr{M}$ gives a morphism $X \to \mathbb{P}^n$. Show that the product map $X \to \mathbb{P}^m \times \mathbb{P}^n$ is a closed embedding, using the Cancellation Theorem 11.1.1 for closed embeddings on $X \to \mathbb{P}^m \times \mathbb{P}^n \to \mathbb{P}^m$. Finally, consider the composition $X \hookrightarrow \mathbb{P}^m \times \mathbb{P}^n \hookrightarrow \mathbb{P}^{mn+m+n}$, where the last closed embedding is the Segre embedding.)

16.2.D. EXERCISE (VERY AMPLE $\boxtimes$ VERY AMPLE IS VERY AMPLE; CF. EXAMPLE 15.2.10). Suppose X and Y are proper A-schemes, and $\mathscr{L}$ (resp., $\mathscr{M}$) is a very ample invertible sheaf on X (resp., Y). If $\pi_X\colon X \times_A Y \to X$ and $\pi_Y\colon X \times_A Y \to Y$ are the usual projections, show that $\pi_X^* \mathscr{L} \otimes \pi_Y^* \mathscr{M}$ (also known as $\mathscr{L} \boxtimes \mathscr{M}$; see §15.2.10) is very ample on $X \times_A Y$.

16.2.1. *Definition.* We say an invertible sheaf $\mathscr{L}$ on a proper A-scheme X is **ample over** A, or π-**ample** (where $\pi\colon X \to \operatorname{Spec} A$ is the structure morphism), or **relatively ample**, if one of the following equivalent conditions holds.

16.2.2. Theorem — *Suppose $\pi\colon X \to \operatorname{Spec} A$ is proper, and $\mathscr{L}$ is an invertible sheaf on X. Then following are equivalent.*

Projective geometry:

(a) *For some $N > 0$, $\mathscr{L}^{\otimes N}$ is very ample over A.*
(a′) *For all $n \gg 0$, $\mathscr{L}^{\otimes n}$ is very ample over A.*

Global generation:

(b) *For all finite type quasicoherent sheaves $\mathscr{F}$, there is an n_0 such that for $n \geq n_0$, $\mathscr{F} \otimes \mathscr{L}^{\otimes n}$ is globally generated.*

Topology:

(c) As f runs over all the global sections of $\mathscr{L}^{\otimes n}$ (over all $n > 0$), the open subsets $X_f = \{p \in X : f(p) \neq 0\}$ form a base for the topology of X.

(c′) As f runs over the global sections of $\mathscr{L}^{\otimes n}$ (over all $n > 0$), those open subsets X_f that are affine form a base for the topology of X.

(c″) As f runs over the global sections of $\mathscr{L}^{\otimes n}$ (over all $n > 0$), those open subsets X_f that are affine cover X.

(Variants of Theorem 16.2.2 in the "absolute" and "relative" settings will be given in Theorems 16.2.6 and 17.3.7, respectively.)

Note that (b) and (c)–(c″) make no reference to the structure morphism π. We will later (Theorem 18.8.1) meet a cohomological criterion for ampleness (due to Serre). The Kodaira Embedding Theorem (see, for example, [GH1, p. 181]) also gives a criterion for ampleness in the complex category: if X is a compact complex manifold, then an invertible sheaf $\mathscr{L}$ on X is ample if and only if it admits a Hermitian metric with positive curvature everywhere.

The different flavors of these conditions give some indication that ampleness is better behaved than very ampleness in a number of ways. We mention without proof another property ("ampleness is an open condition in families"): if $\pi : X \to T$ is a finitely presented proper morphism, and $\mathscr{L}$ is an invertible sheaf on X, then those points on T where $\mathscr{L}$ is ample on the fiber form an open subset of T. Furthermore, on this open subset, $\mathscr{L}$ is relatively ample over the base. We won't use these facts (proved in [Gr-EGA, IV_3.9.6.4]), but they are good to know.

Before getting to the proof of Theorem 16.2.2, we give some sample applications. We begin by noting the fact that (a) implies (b) gives Serre's Theorem A (Theorem 16.1.5).

16.2.E. IMPORTANT EXERCISE. Suppose $\mathscr{L}$ and $\mathscr{M}$ are invertible sheaves on a proper A-scheme X, and $\mathscr{L}$ is ample. Show that $\mathscr{L}^{\otimes n} \otimes \mathscr{M}$ is very ample for $n \gg 0$. (Hint: Use both (a) and (b) of Theorem 16.2.2, and Exercise 16.2.C.)

16.2.F. IMPORTANT EXERCISE. Show that every line bundle on a projective A-scheme X is the difference of two very ample line bundles. More precisely, for any invertible sheaf $\mathscr{L}$ on X, we can find two very ample invertible sheaves $\mathscr{M}$ and $\mathscr{N}$ such that $\mathscr{L} \cong \mathscr{M} \otimes \mathscr{N}^{\vee}$. (Hint: Use the previous Exercise.)

16.2.G. IMPORTANT EXERCISE (USED REPEATEDLY). Suppose $\pi : X \to Y$ is a finite morphism of proper A-schemes, and $\mathscr{L}$ is an ample line bundle on Y. Show that $\pi^* \mathscr{L}$ is ample on X. Hint: Use the criterion of Theorem 16.2.2(c″). (Remark for those who have read about ampleness in the absolute setting in §16.2.5: The result applies in that situation, i.e., with "proper A-schemes" changed to "schemes," without change. The only additional thing to note is that ampleness of $\mathscr{L}$ on Y implies that Y is quasicompact from the definition, and separated from Theorem 16.2.6(d). A relative version of this result appears in §17.3.6. It can be generalized even further, with "π finite" replaced by "π quasiaffine" — to be defined in §17.3.11 — see [Gr-EGA, II.5.1.12].)

16.2.H. EXERCISE (AMPLE $\otimes$ AMPLE IS AMPLE, AMPLE $\otimes$ BASE-POINT-FREE IS AMPLE). Suppose $\mathscr{L}$ and $\mathscr{M}$ are invertible sheaves on a proper A-scheme X, and $\mathscr{L}$ is ample. Show that if $\mathscr{M}$ is ample or base-point-free, then $\mathscr{L} \otimes \mathscr{M}$ is ample.

16.2.I. LESS IMPORTANT EXERCISE (AMPLE $\boxtimes$ AMPLE IS AMPLE). Solve Exercise 16.2.D with "very ample" replaced by "ample."

16.2.J. EXERCISE (CF. EXERCISE 12.3.C). Suppose X is a projective k-scheme, $\mathscr{L}$ is an ample line bundle on X, $\mathscr{M}$ is a line bundle on X, and $q_1, \ldots, q_m$ are points of X (not necessarily closed). Show that for every $N \gg 0$, there is a section of $\mathscr{L}^{\otimes N} \otimes \mathscr{M}$ that does not vanish on any of the q_i. (Hint: If q_1 is not in the closure of any of the other q_i, use the criterion of Theorem 16.2.2(b) with a judiciously chosen $\mathscr{F}$ to find a section of $\mathscr{L}^{\otimes N} \otimes \mathscr{M}$ vanishing at all of the q_i except nonvanishing at q_1.) This exercise is interesting mainly when k is a finite field. You might use it to solve starred

Exercise 18.4.L (by way of Exercise 16.2.L), and it will be used in the proof of Proposition 20.1.6 (by way of Exercise 16.2.K).

16.2.K. EXERCISE. Suppose X is a projective k-scheme, $\mathscr{M}$ is a line bundle on X, and $q_1, \ldots, q_m$ are points of X (not necessarily closed). Show that there are effective Cartier divisors D_1 and D_2 on X not containing $q_1, \ldots, q_m$ such that $\mathscr{M} \cong \mathscr{O}(D_1 - D_2)$. Hint: Exercise 16.2.J.

16.2.L.* EXERCISE. Suppose C is a reduced projective projective curve over a field k, $\mathscr{L}$ is a line bundle on C, and $q_1, \ldots, q_m$ are closed points of C. Using Exercise 16.2.J, show that we can write $\mathscr{L} \cong \mathscr{O}(\sum n_j p_j)$, where the p_j are regular points distinct from the q_i. Hint: Use that the nonregular points of C form a finite set (§13.5.10). Write $\mathscr{L}$ as the difference of two very ample line bundles.

16.2.3. *Proof of Theorem 16.2.2 in the case X is Noetherian.* Noetherian hypotheses are used at only one point in the proof, and we explain how to remove them.

We work roughly from the easiest to the hardest implications. Obviously, *(a') implies (a)*, and *(c') implies both (c) and (c")*.

We now show that *(c) implies (c')*. Suppose we have a point p in an open subset U of X. We seek an *affine* X_f containing p and contained in U. By shrinking U, we may assume that U is affine. From (c), U contains some X_f containing p. But this X_f is affine, as it is the complement of the vanishing locus of a section of a line bundle on an affine scheme (Exercise 8.3.E), so (c') holds. Note for future reference that the equivalence of (c) and (c') did not require the hypothesis of properness.

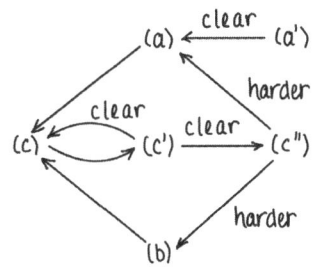

Figure 16.1 *Structure of the proof of Theorem 16.2.2, up to the last step.*

We next show that *(a) implies (c)*. We embed X in projective space by some power of $\mathscr{L}$. Given a closed subset $Z \subset X$, and a point p of the complement $X \setminus Z$, we seek a section of some $\mathscr{L}^{\otimes N}$ that vanishes on Z and not on p. The existence of such a section follows from the fact that $V(I(Z)) = Z$ (Exercise 4.5.K(c)): there is some (homogeneous) element of $I(Z)$ that does not vanish on p. (This also shows that (a) implies (c') as well, but we won't need it.)

We next show that *(b) implies (c)*. Suppose we have a point p in an open subset U of X. We seek a section of $\mathscr{L}^{\otimes N}$ that doesn't vanish at p, but vanishes on $X \setminus U$. Let $\mathscr{I}$ be the sheaf of ideals of functions vanishing on $X \setminus U$ (the quasicoherent sheaf of ideals cutting out $X \setminus U$, with reduced structure). As X is Noetherian, $\mathscr{I}$ is finite type, so

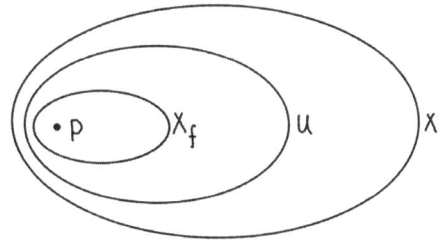

Figure 16.2 *Visual aid for the proof of Theorem 16.2.2.*

by (b), $\mathscr{I} \otimes \mathscr{L}^{\otimes N}$ is generated by global sections for some N, so there is some global section of it not vanishing at p. (*Noetherian note:* This is the only part of the argument where we use Noetherian hypotheses. They can be removed as follows. Show that for a quasicompact quasiseparated scheme, every ideal sheaf is generated by its finite type subideal sheaves. Indeed, any quasicoherent sheaf on a quasicompact quasiseparated scheme is the union of its finite type quasicoherent subsheaves. This can be proved by hand, and is a suitable but hard exercise; see, for example, [Stacks, tag 01PJ]. One of these finite type ideal sheaves doesn't vanish at p; use this as $\mathscr{I}$ instead.)

We now have to start working harder.

We next show that *(c") implies (b)*.

We wish to show that $\mathscr{F} \otimes \mathscr{L}^{\otimes n}$ is globally generated for $n \gg 0$. We first show that (c") implies that for some $N > 0$, $\mathscr{L}^{\otimes N}$ is globally generated, as follows. Use quasicompactness of X to select a finite number of affine $X_{f_1}, \ldots, X_{f_n}$ that cover X, where $f_i \in \Gamma(X, \mathscr{L}^{\otimes N_i})$. By replacing f_i with

$f_i^{\otimes(\prod_j N_j)/N_i}$, we may assume that they are all sections of the *same* power $\mathscr{L}^{\otimes N}$ of $\mathscr{L}$ (taking $N = \prod_j N_j$). Then $\mathscr{L}^{\otimes N}$ is generated by these global sections.

We next show that it suffices to show that for all finite type quasicoherent sheaves $\mathscr{F}$, $\mathscr{F} \otimes \mathscr{L}^{\otimes mN}$ is globally generated for $m \gg 0$. For if we knew this, we could apply it to $\mathscr{F}$, $\mathscr{F} \otimes \mathscr{L}, \ldots, \mathscr{F} \otimes \mathscr{L}^{\otimes(N-1)}$ (a finite number of times), and the result would follow. For this reason, we can replace $\mathscr{L}$ by $\mathscr{L}^{\otimes N}$. In other words, to show that (c″) implies (b), we may also assume the additional hypothesis that $\mathscr{L}$ is globally generated.

For each closed point p, choose an affine open neighborhood of the form X_f, using (c″). Then $\mathscr{F}|_{X_f}$ is generated by a finite number of global sections (Easy Exercise 16.1.B—a finite type quasicoherent sheaf on Spec A corresponds to a finitely generated A-module). By Exercise 15.4.N, each of these generators can be expressed as a quotient of a section (over X) of $\mathscr{F} \otimes \mathscr{L}^{\otimes M(p)}$ by $f^{M(p)}$. (Note: We can take a single $M(p)$ for each p.) Then $\mathscr{F} \otimes \mathscr{L}^{\otimes M(p)}$ is globally generated at p by a finite number of global sections. By Exercise 16.1.D(b), $\mathscr{F} \otimes \mathscr{L}^{\otimes M(p)}$ is globally generated at all points in some open neighborhood U_p of p. As $\mathscr{L}$ is also globally generated, this implies that $\mathscr{F} \otimes \mathscr{L}^{\otimes M'}$ is globally generated at all points of U_p for $M' \geq M(p)$ (cf. Easy Exercise 16.1.C). From quasicompactness of X, a finite number of these U_p cover X, so we are done (by taking the maximum of these $M(p)$).

Our penultimate step is to show that *(c″) implies (a)*. We assume (c″). Our goal is to find sections of some $\mathscr{L}^{\otimes N}$ that embed X into projective space. Choose a cover of (quasicompact) X by n affine open subsets $X_{a_1}, \ldots, X_{a_n}$, where $a_1, \ldots, a_n$ are all sections of powers of $\mathscr{L}$. By replacing each section with a suitable power, we may assume that they are all sections of the *same* power of $\mathscr{L}$, say, $\mathscr{L}^{\otimes N}$. Say $X_{a_i} = \operatorname{Spec} A_i$, where (using that π is finite type) $A_i = \operatorname{Spec} A[a_{i1}, \ldots, a_{ij_i}]/I_i$. By Exercise 15.4.N, each a_{ij} is of the form $s_{ij}/a_i^{m_{ij}}$, where $s_{ij} \in \Gamma(X, \mathscr{L}^{\otimes Nm_{ij}})$ (for some m_{ij}). Let $m = \max_{i,j} m_{ij}$. Then for each i, j, $a_{ij} = (s_{ij} a_i^{m-m_{ij}})/a_i^m$. For convenience, let $b_i = a_i^m$, and $b_{ij} = s_{ij} a_i^{m-m_{ij}}$; these are all global sections of $\mathscr{L}^{\otimes mN}$. Now consider the linear series generated by the b_i and b_{ij}. As the $D(b_i) = X_{a_i}$ cover X, this linear series is base-point-free, and hence (by Exercise 15.2.A) gives a morphism to $\mathbb{P}^Q$ (where $Q = \#b_i + \#b_{ij} - 1$). Let $x_1, \ldots, x_n, \ldots, x_{ij}, \ldots$ be the projective coordinates on $\mathbb{P}^Q$, so $f^* x_i = b_i$, and $f^* x_{ij} = b_{ij}$. Then the morphism of affine schemes $X_{a_i} \to D(x_i)$ is a closed embedding, as the associated maps of rings is a surjection (the generator a_{ij} of A_i is the image of x_{ij}/x_i).

At this point, we note for future reference that we have shown the following. If $X \to \operatorname{Spec} A$ is finite type, and $\mathscr{L}$ satisfies (c″), then X is a locally closed embedding into a projective A-scheme. (We did not use separatedness.) We conclude our proof that (c″) implies (a) by using Exercise 11.4.C ("image of proper is proper") to show that the image of this locally closed embedding into a projective A-scheme is in fact closed, so X is a projective A-scheme. (Do you see why a locally closed embedding with closed image is in fact a closed embedding?)

Finally, we note that *(a) and (b) together imply (a′)*: if $\mathscr{L}^{\otimes N}$ is very ample (from (a)), and $\mathscr{L}^{\otimes n}$ is base-point-free for $n \geq n_0$ (from (b)), then $\mathscr{L}^{\otimes n}$ is very ample for $n \geq n_0 + N$ by Exercise 16.2.C. $\qquad \square$

16.2.4. ** *Semiample line bundles.* Just as an invertible sheaf is ample if some tensor power of it is very ample, an invertible sheaf $\mathscr{L}$ is said to be **semiample** if some tensor power of it is base-point-free. (Translation: $\mathscr{L}$ is ample if some power gives a closed embedding into projective space, and $\mathscr{L}$ is semiample if some power gives just a *morphism* to projective space.) We won't use this notion.

16.2.5. * **Ampleness in the absolute setting.** (We will not use this section in any serious way later.) Note that global generation is already an absolute notion, i.e., is defined for a quasicoherent sheaf on a scheme, with no reference to any morphism. An examination of the proof of Theorem 16.2.2 shows that ampleness may similarly be interpreted in an absolute setting. We make this precise. Suppose $\mathscr{L}$ is an invertible sheaf on a *quasicompact* scheme X. We say that $\mathscr{L}$ is **ample** if as f runs over the sections of $\mathscr{L}^{\otimes n}$ ($n > 0$), the open subsets $X_f = \{p \in X : f(p) \neq 0\}$ form a base for the topology of X. (We emphasize that quasicompactness in X is part of the condition of ampleness of $\mathscr{L}$.) For example, (i) if X is an affine scheme, every invertible sheaf is ample, and (ii) if X is a projective A-scheme, $\mathscr{O}(1)$ is ample.

16.2.M. EASY EXERCISE (PROPERTIES OF ABSOLUTE AMPLENESS).

(a) Fix a positive integer n. Show that $\mathscr{L}$ is ample if and only if $\mathscr{L}^{\otimes n}$ is ample.
(b) Show that if $Z \hookrightarrow X$ is a closed embedding, and $\mathscr{L}$ is ample on X, then $\mathscr{L}|_Z$ is ample on Z.

The following result will give you some sense of how ampleness behaves. We will not use it, and hence omit the proof (which is given in [Stacks, tag 01Q3]). However, many parts of the proof are identical to (or generalize) the corresponding arguments in Theorem 16.2.2. The labeling of the statements parallels the labelling of the statements in Theorem 16.2.2.

16.2.6. Theorem (cf. Theorem 16.2.2) — *Suppose $\mathscr{L}$ is an invertible sheaf on a quasicompact scheme X. Then the following are equivalent.*

(b) *X is quasiseparated, and for every finite type quasicoherent sheaf $\mathscr{F}$, there is an n_0 such that for $n \geq n_0$, $\mathscr{F} \otimes \mathscr{L}^{\otimes n}$ is globally generated.*

(c) *As f runs over the global sections of $\mathscr{L}^{\otimes n}$ ($n > 0$), the open subsets $X_f = \{p \in X : f(p) \neq 0\}$ form a base for the topology of X (i.e., $\mathscr{L}$ is ample).*

(c') *As f runs over the global sections of $\mathscr{L}^{\otimes n}$ ($n > 0$), those open subsets X_f that are affine form a base for the topology of X.*

(c'') *As f runs over the global sections of $\mathscr{L}^{\otimes n}$ ($n > 0$), those open subsets X_f that are affine cover X.*

(d) *Let $S_\bullet$ be the graded ring $\oplus_{n \geq 0} \Gamma(X, \mathscr{L}^{\otimes n})$. Then the open sets X_f (with $f \in S_+$ homogeneous of positive degree) cover X, and the associated map $X \to \operatorname{Proj} S_\bullet$ is an open embedding. (Warning: $S_\bullet$ need not be finitely generated, and $\operatorname{Proj} S_\bullet$ is not necessarily finite type.)*

Part (d) implies that X is separated (and thus quasiseparated).

16.2.7.* Transporting global generation, base-point-freeness, and ampleness to the relative situation.

These notions can be "relativized." We could do this right now, but we wait until §17.3.5, when we will have defined the notion of a projective morphism, and thus a "relatively very ample" line bundle.

16.3 Applications to Curves

We now apply what we have learned to curves. (The only reason this discussion comes so late is because we need Exercise 16.2.G, which implies that if $X \to Y$ is finite, and Y is projective over k, then X is projective over k too.) Because the word "curve" can have different meanings, we will try to be pedantic about hypotheses in statements of theorems. (In common usage, a curve often means a variety over a field k of pure dimension 1.)

16.3.1. Theorem (every integral curve has a birational model that is regular and projective) — *If C is an integral finite type one-dimensional k-scheme, then there exists a regular projective k-variety C' (of dimension 1) birational to C.*

Proof. We can assume C is affine. By the Noether Normalization Lemma 12.2.4 , we can find some $x \in K(C) \setminus k$ with $K(C)/k(x)$ a finite extension of fields. By identifying a standard open subset of $\mathbb{P}_k^1$ with $\operatorname{Spec} k[x]$, and taking the normalization of $\mathbb{P}_k^1$ in the function field $K(C)$ (defined in Exercise 10.7.I), we obtain a finite morphism $C' \to \mathbb{P}_k^1$, where C' is a curve (finiteness by Exercise 10.7.M, $\dim C' = \dim \mathbb{P}_k^1$ by Exercise 12.1.G), and regular (it

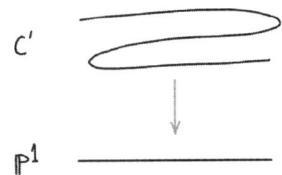

Figure 16.3 *Constructing a projective regular model of a curve C over k via a finite cover of $\mathbb{P}^1$.*

is reduced, hence regular at the generic point, and regular at the closed points by the main theorem on discrete valuation rings in §13.5). Also, C' is birational to C as they have isomorphic function fields (Exercise 7.5.E).

Finally, $C' \to \mathbb{P}^1_k$ is finite, and $\mathbb{P}^1_k \to \operatorname{Spec} k$ is projective, so by Exercise 16.2.G, $C' \to \operatorname{Spec} k$ is projective. $\qquad \square$

16.3.2. Theorem — *If C is an irreducible, regular, one-dimensional k-variety, then there is an open embedding $C \hookrightarrow C'$ into some projective regular curve C' (over k).*

Proof. We first prove the result in the case where C is affine. As in the proof of Theorem 16.3.1, we have the diagram

$$
\begin{array}{ccc}
C & & C', \\
\pi \,\Big\downarrow \text{finite} & & \Big\downarrow \pi' \text{ finite} \\
\mathbb{A}^1_k & \hookrightarrow & \mathbb{P}^1_k
\end{array}
$$

where π' is the normalization of $\mathbb{P}^1_k$ in the function field extension $K(C)/K(\mathbb{A}^1_k)$, and C' is projective. By the universal property of normalization in a function field extension, the normalization of $\mathbb{A}^1_k$ in the function field extension $K(C)/K(\mathbb{A}^1_k)$ is C itself, as C is already normal. Thus $(\pi')^{-1}(\mathbb{A}^1_k) = C$, yielding the desired open embedding $C \to C'$.

We next consider the case of general C. Let C_1 be any nonempty affine open subset of C. By the discussion in the previous paragraph, we have a regular projective compactification $\widetilde{C}_1$. The Curve-to-Projective Extension Theorem 15.3.1 (applied successively to the finite number of points $C \setminus C_1$) implies that the morphism $C_1 \hookrightarrow \widetilde{C}_1$ extends to a birational morphism $C \to \widetilde{C}_1$. Because points of a *separated* regular curve are determined by their valuation (Exercise 13.7.B), this is an inclusion of sets. Because the topology on curves is the "cofinite topology," it expresses C as an open subset of $\widetilde{C}_1$. But we are not done yet!

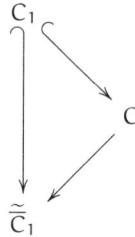

$$
\begin{array}{c}
C_1 \\
\Big\downarrow \qquad \searrow \\
\qquad C \\
\swarrow \\
\widetilde{C}_1
\end{array}
$$

We must show that $C \to \widetilde{C}_1$ is an *open embedding of schemes*. We show it is an open embedding near a point $p \in C$ as follows. Let C_2 be an affine open neighborhood of p in C. We repeat the construction we used on C_1, to obtain the following diagram, with open embeddings marked.

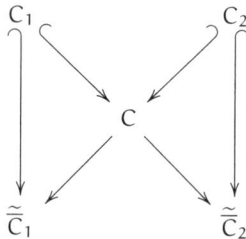

$$
\begin{array}{ccc}
C_1 & & C_2 \\
\Big\downarrow \searrow & & \swarrow \Big\downarrow \\
& C & \\
\swarrow & & \searrow \\
\widetilde{C}_1 & & \widetilde{C}_2
\end{array}
$$

By the Curve-to-Projective Extension Theorem 15.3.1, the map $C_1 \to \widetilde{C}_2$ extends to $\pi_{12} \colon \widetilde{C}_1 \to \widetilde{C}_2$, and we similarly have a morphism $\pi_{21} \colon \widetilde{C}_2 \to \widetilde{C}_1$, extending $C_2 \to \widetilde{C}_1$. The composition $\pi_{21} \circ \pi_{12}$ is the identity morphism (as it is the identity rational map; see Theorem 11.3.2). The same is true for $\pi_{12} \circ \pi_{21}$, so π_{12} and π_{21} are isomorphisms. The enhanced diagram

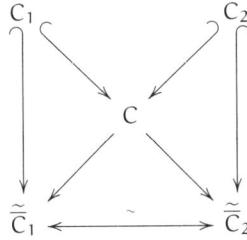

commutes (by Theorem 11.3.2 again, implying that morphisms of reduced separated schemes are determined by their behavior on dense open sets). But $C_2 \to \widetilde{\overline{C}}_1$ is an open embedding (in particular, at p), so $C \to \widetilde{\overline{C}}_1$ is an open embedding at p as well. $\square$

16.3.A. EXERCISE. Show that all regular proper curves over k are projective.

16.3.3. Theorem (various categories of curves are the same) — *The following categories are equivalent:*

  (i) *integral regular projective one-dimensional k-varieties, and surjective k-morphisms;*
 (ii) *integral regular projective one-dimensional k-varieties, and dominant k-morphisms;*
(iii) *integral regular projective one-dimensional k-varieties, and dominant rational maps over k;*
 (iv) *integral one-dimensional k-varieties, and dominant rational maps over k; and*
  (v) *the opposite category (defined in §1.1.16) of finitely generated fields of transcendence degree 1 over k, and k-homomorphisms.*

All morphisms and maps in the following discussion are assumed to be defined over k.

(Aside: The interested reader can tweak the proof below to show the following variation of the theorem. In (i)–(iv), consider only geometrically irreducible curves, and in (v), consider only fields K such that $K \cap k^s = k$ in $\overline{K}$. This variation allows us to exclude "weird" curves we may not want to consider. For example, if $k = \mathbb{R}$, then we are allowing curves such as $\mathbb{P}^1_\mathbb{C}$ that are not geometrically irreducible, as $\mathbb{P}^1_\mathbb{C} \times_\mathbb{R} \operatorname{Spec} \mathbb{C} \cong \mathbb{P}^1_\mathbb{C} \coprod \mathbb{P}^1_\mathbb{C}$.)

Proof. Every surjective morphism is a dominant morphism, and every dominant morphism is a dominant rational map, and each integral regular projective curve is a quasiprojective curve, so we have shown (informally speaking) how to get from (i) to (ii) to (iii) to (iv). To get from (iv) to (i), suppose we have a dominant rational map $C_1 \dashrightarrow C_2$ of integral curves. Replace C_1 by a dense open set so the rational map is a morphism $C_1 \to C_2$. This induces a map of normalizations $\widetilde{C}_1 \to \widetilde{C}_2$ of regular irreducible curves. Let $\widetilde{\overline{C}}_i$ be a regular projective compactification of $\widetilde{C}_i$ (for $i = 1, 2$), as in Theorem 16.3.2. Then the morphism $\widetilde{C}_1 \to \widetilde{C}_2$ extends to a morphism $\widetilde{\overline{C}}_1 \to \widetilde{\overline{C}}_2$ by the Curve-to-Projective Extension Theorem 15.3.1. This morphism is surjective (do you see why?), so we have produced a morphism in category (i).

16.3.B. EXERCISE. Put the above pieces together to describe equivalences of categories (i) through (iv).

It remains to connect (v). This is essentially the content of Exercise 7.5.E; details are left to the reader. $\square$

Theorem 16.3.3 has a number of implications. For example, each quasiprojective reduced curve is birational to precisely one projective regular curve. Here is another interesting consequence.

16.3.4. Important notion: Degree of a projective morphism from a curve to a regular curve.

You might already have a reasonable sense that a map of compact Riemann surfaces has a well-behaved degree, that the number of preimages of a point of C' is constant, so long as the preimages are counted with appropriate multiplicity. For example, if f locally looks like $z \mapsto z^m = y$, then near

$y = 0$ and $z = 0$ (but not *at* $z = 0$), each point has precisely $\mathfrak{m}$ preimages, but as y goes to 0, the $\mathfrak{m}$ preimages coalesce. Enlightening Example 10.3.4 showed this phenomenon in a more complicated context.

We now show the algebraic version of this fact. Suppose $\pi \colon C \to C'$ is a surjective (or, equivalently, dominant) map of regular projective curves. We will show that π has a well-behaved degree, in a sense that we will now make precise.

First we show that π is finite. Theorem 18.1.6 (finite = projective + finite fibers) implies this, but we haven't proved it yet. So, instead, we show the finiteness of π as follows. Let C'' be the normalization of C' in the function field of C. Then we have an isomorphism $K(C) \cong K(C'')$ (really, equality) which leads to birational maps $C \dashrightarrow C''$, which extend to morphisms as both C and C'' are regular and projective (by the Curve-to-Projective Extension Theorem 15.3.1). Thus this yields an isomorphism of C and C''. But $C'' \to C'$ is a finite morphism by the finiteness of normalization (Exercise 10.7.M).

16.3.5. Proposition — *Suppose that $\pi \colon C \to C'$ is a finite morphism, where C' is a regular Noetherian scheme of pure dimension 1, and C is a Noetherian scheme of pure dimension 1 with no embedded points (the most important case: C is reduced). Then $\pi_* \mathscr{O}_C$ is locally free of finite rank.*

The "no embedded points" hypothesis is the same as requiring that every associated point of C map to a generic point of (some component of) C'. Notice that if C is reduced, then there are no restrictions on how bad the singularities of C can be!

We will prove Proposition 16.3.5 in §16.3.11, after showing how useful it is. The regularity hypothesis on C' is necessary: the normalization of a nodal curve (Figure 8.4) is an example where most points have one preimage, and one point (the "node") has two. (We will later see, in Remark 24.3.8 and §24.3.12, that what matters in the hypotheses of Proposition 16.3.5 is that the morphism is finite and *flat*.)

16.3.6. *Definition.* If C' is irreducible, the rank of this locally free sheaf is the **degree** of π.

16.3.C. EXERCISE. Recall that the degree of a rational map from one integral curve to another is defined as the degree of the function field extension (Definition 12.2.2). Show that (with the notation of Proposition 16.3.5) if C and C' are integral, the degree of π as a rational map is the same as the rank of $\pi_* \mathscr{O}_C$.

16.3.7. *Remark for those with complex-analytic background (algebraic degree = analytic degree).* If $C \to C'$ is a finite map of regular complex algebraic curves, Proposition 16.3.5 establishes that the algebraic degree as defined above is the same as the analytic degree (counting preimages, with multiplicity).

16.3.D. EXERCISE. We continue the notation and hypotheses of Proposition 16.3.5. Suppose p is a closed point of C'. The scheme-theoretic preimage $\pi^{-1}(p)$ of p is a dimension 0 scheme over k.

(a) Suppose C' is finite type over a field k, and n is the dimension of the k-vector space of global sections of the structure sheaf of $\pi^{-1}(p)$. Show that $n = (\deg \pi)(\deg p)$. (The degree of a point was defined in §5.3.8.)

(b) Suppose that C is regular, and $\pi^{-1}(p) = \{p_1, \ldots, p_m\}$. Suppose t is a uniformizer of the discrete valuation ring $\mathscr{O}_{C',p}$. Show that

$$\deg \pi = \sum_{i=1}^{m} (\operatorname{val}_{p_i} \pi^* t) \deg(\kappa(p_i)/\kappa(p)),$$

where $\deg(\kappa(p_i)/\kappa(p))$ denotes the degree of the field extension of the residue fields. (Can you extend (a) to remove the hypotheses of working over a field? If you are a number theorist, can you recognize (b) in terms of splitting prime ideals in extensions of rings of integers in number fields?)

16.3.E. EXERCISE. Suppose that C is an irreducible regular projective curve over k, and s is a nonzero rational function on C. Show that the number of zeros of s (counted with appropriate multiplicity) equals the number of poles. Hint: Recognize this as the degree of a morphism $s \colon C \to \mathbb{P}^1_k$. Use Exercise 16.3.D(b) for $0 \in \mathbb{P}^1$ and $\infty \in \mathbb{P}^1$. (In the complex category, this is an important consequence of the Residue Theorem.)

16.3.8. ** *Important definition we could make now (although we wait until §18.4.2).* At this point, we can easily make the important definition of the degree of a line bundle on a smooth projective curve. Feel free to think this through now, although we wait until §18.4.2 to make a (seemingly different) definition, and show it is equivalent to this in Important Easy Exercise 18.4.C. Suppose C is an irreducible regular projective curve over k, and $\mathscr{L}$ is a line bundle on C. Then $\mathscr{L}$ has a nonzero rational section s (do you see why?). If s and t are any two nonzero rational sections, then by Exercise 16.3.E, for the the rational function s/t, the number of zeros equals the number of poles (counted correctly). Thus the number of zeros minus the number of poles (counted correctly) of a nonzero rational section s is independent of the choice of s. This integer could be taken to be the definition of the *degree of the line bundle* $\mathscr{L}$ on the curve C. (Because we are not taking this as our "official definition," we do not write this phrase in bold.)

16.3.9. *Remark.* In Exercise 18.4.E, we will see that the number of zeros (and poles) in the previous exercise is the degree of the line bundle giving the map to $\mathbb{P}^1$ (via Important Theorem 15.2.2).

16.3.10. *Revisiting Example 10.3.4.* Proposition 16.3.5 and Exercise 16.3.D make precise what general behavior we observed in Example 10.3.4. Suppose C′ is irreducible, and that d is the rank of this allegedly locally free sheaf. Then the fiber over any point of C′ with residue field K is the Spec of an algebra of dimension d over K. This means that the number of points in the fiber, counted with appropriate multiplicity, is always d.

As a motivating example, we revisit Example 10.3.4, the map $\mathbb{Q}[y] \to \mathbb{Q}[x]$ given by $x \mapsto y^2$, the projection of the parabola $x = y^2$ to the x-axis. We observed the following.

(i) The fiber over $x = 1$ is $\mathbb{Q}[y]/(y^2 - 1)$, so we get 2 points.
(ii) The fiber over $x = 0$ is $\mathbb{Q}[y]/(y^2)$ — we get one point, with multiplicity 2, arising because of the nonreducedness.
(iii) The fiber over $x = -1$ is $\mathbb{Q}[y]/(y^2 + 1) \cong \mathbb{Q}(i)$ — we get one point, with multiplicity 2, arising because of the field extension.
(iv) Finally, the fiber over the generic point $\operatorname{Spec}\mathbb{Q}(x)$ is $\operatorname{Spec}\mathbb{Q}(y)$, which is one point, with multiplicity 2, arising again because of the field extension (as $\mathbb{Q}(y)/\mathbb{Q}(x)$ is a degree 2 extension).

We thus see three sorts of behaviors ((iii) and (iv) are really the same). Note that even if you only work with algebraically closed fields, you will still be forced to deal with this third type of behavior, because residue fields at generic points are usually not algebraically closed (witness case (iv) above).

16.3.11. *Proof of Proposition 16.3.5.* The key idea, useful in other circumstances, is to reduce to a fact about discrete valuation rings.

The question is local on the target, so we may assume that C′ is affine. Now C′ is normal, so by Exercise 5.4.B, its connected components are irreducible. Replacing C′ by each of its irreducible components, we may also assume C′ is integral.

By Exercise 14.3.K, if the rank of the finite type quasicoherent sheaf $\pi_* \mathscr{O}_C$ is constant, then (as C′ is reduced) $\pi_* \mathscr{O}_C$ is locally free. We will show this by showing the rank at any closed point p of C′ is the same as the rank at the generic point. (This is another example of the usefulness of the generic point.)

Suppose $C' = \operatorname{Spec} A'$, where A′ is an integral domain, and $p = [\mathfrak{m}]$. As π is an affine morphism, C is an affine scheme as well; say, $C = \operatorname{Spec} A$.

We wish to show that (i) $\dim_{A'/m}(A/m)$ (the rank of $\pi_*\mathcal{O}_C$ at p) equals (ii) $\dim_{K(A')}(A'^\times)^{-1}A$ (the rank of $\pi_*\mathcal{O}_C$ at the generic point). In other words, we take A (considered as an A'-module), and (i) quotient by m, and (ii) invert all nonzero elements of A', and in each case compute the result's dimension over the appropriate field.

Both (i) and (ii) factor through localizing at m, so it suffices to show that A_m is a finite rank free A'_m-module, of rank d, say, as the answers to both (i) and (ii) will then be d.

Now A'_m is a discrete valuation ring; let t be its uniformizer. We can assume that $t \in A'$ (as otherwise, we replace A' by a suitable localization, at the "denominators" of t). Then A_m is a finitely generated A'_m-module, and hence by Remark 6.4.4 is a finite sum of principal modules, of the form A'_m or $A'_m/(t^n)$ (for various n). We wish to show that there are no summands of the latter type. But if there were, then t (interpreted as an element of A_m) would be a zerodivisor of A_m, and thus (interpreted as an element of A) a zerodivisor of A. But then by Proposition 6.6.13, there is an associated point of C in $\pi^{-1}(p)$, contradicting the hypotheses that C has no embedded points. $\square$

16.4* The Grassmannian as a Moduli Space

(In this section and others, the symbol k is used both for a field, and for a dimension. For this we apologize, and hope it does not cause too much confusion.)

We first defined projective space inelegantly in §4.4.9, and in §15.2.3 we gave a clean (if perhaps surprising) functorial definition. Similarly, in §7.7, we gave a preliminary description of the Grassmannian. We are now in a position to give a better definition. Before proceeding, you should first go back and read the doubled-starred §10.1.7, if you have not done so already.

We describe the "Grassmannian (contravariant) functor" (which we also denote by $G(k,n)$), then show that it is representable (§7.6.2). The construction works over an arbitrary base scheme, so we work over the final object $\text{Spec }\mathbb{Z}$. (You should think through what to change if you wish to work with, for example, complex schemes.) The functor is defined as follows. To a scheme B, we associate the set of *locally free rank k quotients of the rank n free sheaf,*

$$(16.4.0.1) \qquad \mathcal{O}_B^{\oplus n} \longrightarrow \mathcal{Q},$$

up to isomorphism. An isomorphism of two such quotients $\phi: \mathcal{O}_B^{\oplus n} \to \mathcal{Q} \to 0$ and $\phi': \mathcal{O}_B^{\oplus n} \to \mathcal{Q}' \to 0$ is an isomorphism $\sigma: \mathcal{Q} \xrightarrow{\sim} \mathcal{Q}'$ such that the diagram

$$\begin{array}{ccc} \mathcal{O}^{\oplus n} & \xrightarrow{\phi} & \mathcal{Q} \\ & \searrow{\phi'} & \downarrow{\sigma} \\ & & \mathcal{Q}' \end{array}$$

commutes. By Exercise 14.2.H(a), $\ker \phi$ is locally free of rank $n-k$. (Thus, if you prefer, you can extend (16.4.0.1), and instead consider the functor to take B to short exact sequences

$$(16.4.0.2) \qquad 0 \longrightarrow \mathcal{S} \longrightarrow \mathcal{O}^{\oplus n} \longrightarrow \mathcal{Q} \longrightarrow 0$$

of locally free sheaves over B, of ranks $n-k$, n, and k, respectively.)

It may surprise you that we are considering rank k *quotients* of a rank n sheaf, not rank k *subobjects*, given that the Grassmannian should parametrize k-dimensional subspace of an n-dimensional space. This is done for several reasons. One is that the kernel of a surjective map of locally free sheaves must be locally free, while the cokernel of an injective map of locally free sheaves need not be locally free (Exercise 14.2.H(a) and (b), respectively). (If you prefer, rather than surjections of finite rank locally free sheaves, you can equivalently consider injections of finite rank locally free sheaves that are "injections of vector bundles" in the sense of Definition 14.2.4.) Another reason: We will see in §25.3.3 that the geometric incarnation of this problem indeed translates to this. We can already see a key example here: if $k=1$, our definition yields one-dimensional quotients $\mathscr{O}^{\oplus n} \to \mathscr{L} \to 0$. But this is precisely the data of n sections of $\mathscr{L}$, with no common zeros, which by Theorem 15.2.2 (the functorial description of projective space) corresponds precisely to maps to $\mathbb{P}^{n-1}$, so the $k=1$ case parametrizes what we want.

We now show that the Grassmannian (contravariant) functor is representable for given n and k.

So far we have described what the Grassmannian functor does on *objects* B; to fully describe a functor, we need to say what it does to *morphisms* $B_1 \to B_2$. Can you see what the definition must be?

16.4.A. EXERCISE. Show that the Grassmannian functor is a Zariski sheaf (§10.1.8).

Hence by Key Exercise 10.1.H, to show that the Grassmannian functor is representable, we need only cover it with open subfunctors that are representable.

Throughout the rest of this section, a k-**subset** is a subset of $\{1, \dots, n\}$ of size k.

16.4.B. EXERCISE.

(a) Suppose I is a k-subset. Make the following statement precise: there is an open subfunctor $G(k,n)_I$ of $G(k,n)$ where the k sections of $\mathscr{Q}$ corresponding to I (of the n sections of $\mathscr{Q}$ coming from the surjection $\phi\colon \mathscr{O}^{\oplus n} \to \mathscr{Q}$) are linearly independent. Hint: In a trivializing open neighborhood of $\mathscr{Q}$, where we can choose an isomorphism $\mathscr{Q} \xrightarrow{\sim} \mathscr{O}^{\oplus k}$, ϕ can be interpreted as a $k \times n$ matrix M, and this locus is where the determinant of the $k \times k$ matrix consisting of the I columns of M is a unit. Show that this locus behaves well under transitions between trivializations.

(b) Show that these open subfunctors $G(k,n)_I$ cover the functor $G(k,n)$ (as I runs through the k-subsets).

Hence by Exercise 10.1.H, to show $G(k,n)$ is representable, we need only show that $G(k,n)_I$ is representable for arbitrary I. After renaming the summands of $\mathscr{O}^{\oplus n}$, without loss of generality we may assume $I = \{1, \dots, k\}$.

16.4.C. EXERCISE. Show that $G(k,n)_{\{1\dots,k\}}$ is represented by $\mathbb{A}^{k(n-k)}$ as follows. (You will have to make this precise.) Given a surjection $\phi\colon \mathscr{O}^{\oplus n} \to \mathscr{Q}$, let $\phi_i\colon \mathscr{O} \to \mathscr{Q}$ be the map from the ith summand of $\mathscr{O}^{\oplus n}$. (Really, ϕ_i is just a section of $\mathscr{Q}$.) For the open subfunctor $G(k,n)_I$, show that

$$\phi_1 \oplus \cdots \oplus \phi_k\colon \mathscr{O}^{\oplus k} \longrightarrow \mathscr{Q}$$

is an isomorphism. For a scheme B, the bijection $G(k,n)_I(B) \leftrightarrow \mathrm{Hom}(B, \mathbb{A}^{k(n-k)})$ is given as follows. Given an element $\phi \in G(k,n)_I(B)$, for $j \in \{k+1, \dots, n\}$, $\phi_j = a_{1j}\phi_1 + a_{2j}\phi_2 + \cdots + a_{kj}\phi_k$, where a_{ij} are functions on B. But $k(n-k)$ functions on B is the same as a map to $\mathbb{A}^{k(n-k)}$ (Exercise 7.6.E). Conversely, given $k(n-k)$ functions a_{ij} ($1 \le i \le k < j \le n$), define a surjection $\phi\colon \mathscr{O}^{\oplus n} \to \mathscr{O}^{\oplus k}$ as follows: $(\phi_1 \dots, \phi_k)$ is the identity, and $\phi_j = a_{1j}\phi_1 + a_{2j}\phi_2 + \cdots + a_{kj}\phi_k$ for $j > k$.

You have now shown that $G(k,n)$ is representable, by covering it with $\binom{n}{k}$ copies of $\mathbb{A}^{k(n-k)}$. (You might wish to relate this to the description you gave in §7.7.) In particular, the Grassmannian over a field is smooth, and irreducible of dimension $k(n-k)$. (The Grassmannian over any base is smooth over that base, because $\mathbb{A}^{k(n-k)}_B \to B$ is smooth; see §13.6.2.)

16.4.1. *The universal exact sequence over the Grassmannian.* Note that we have a tautological exact sequence

$$0 \longrightarrow \mathscr{S} \longrightarrow \mathscr{O}^{\oplus n} \longrightarrow \mathscr{Q} \longrightarrow 0.$$

The bundle $\mathscr{Q}$ is called the **tautological quotient bundle** on the Grassmannian, and $\mathscr{S}$ is called the **tautological subbundle**. (Exercise 1.2.Z(b) may give some inkling as to why this might be called the "universal exact sequence.")

16.4.2. The Plücker embedding.

By applying $\wedge^k$ to a surjection $\phi: \mathscr{O}^{\oplus n} \to \mathscr{Q}$ (over an arbitrary base B), we get a surjection

$$\wedge^k \phi: \mathscr{O}^{\oplus \binom{n}{k}} \longrightarrow \det \mathscr{Q}$$

(Exercise 14.2.N). But a surjection from a rank N free sheaf to a line bundle is the same as a map to $\mathbb{P}^{N-1}$ (Theorem 15.2.2).

16.4.D. EXERCISE. Use this to describe a map $P: G(k, n) \to \mathbb{P}^{\binom{n}{k}-1}$. (This is just a tautology: a natural transformation of functors induces a map of the representing schemes. This is Yoneda's Lemma, although if you didn't do Exercise 1.2.Z, you may wish to do this exercise by hand. But once you do, you may as well go back to prove Yoneda's Lemma and do Exercise 1.2.Z, because the argument is just the same!)

16.4.E. EXERCISE. The projective coordinate x_I on $\mathbb{P}^{\binom{n}{k}-1}$ corresponding to the Ith factor of $\mathscr{O}^{\oplus \binom{n}{k}}$ may be interpreted as the determinant of the map $\phi_I: \mathscr{O}^{\oplus k} \to \mathscr{Q}$, where the $\mathscr{O}^{\oplus k}$ consists of the summands of $\mathscr{O}^{\oplus n}$ corresponding to I. Make this precise.

16.4.F. EXERCISE. Show that the standard open set U_I of $\mathbb{P}^{\binom{n}{k}-1}$ corresponding to k-subset I (i.e., where the corresponding coordinate x_I doesn't vanish) pulls back to the open subscheme $G(k, n)_I \subset G(k, n)$. Denote this map by $P_I: G(k, n)_I \to U_I$.

16.4.G. EXERCISE. Show that P_I is a closed embedding as follows. We may deal with the case $I = \{1, \ldots, k\}$. Note that $G(k, n)_I$ is affine—you described it Spec $\mathbb{Z}[a_{ij}]_{1 \le i \le k < j \le n}$ in Exercise 16.4.C. Also, U_I is affine, with coordinates $x_{I'/I}$, as I' varies over the other k-subsets. (The case $k = 1$ is the useful notation $x_{i/j}$ we used to name coordinates on charts for projective space, §4.4.9.) You want to show that the map

$$P_I^\sharp: \mathbb{Z}[x_{I'/I}]_{I' \subset \{1,\ldots,n\}, |I'|=k}/(x_{I/I} - 1) \longrightarrow \mathbb{Z}[a_{ij}]_{1 \le i \le k < j \le n}$$

is a surjection. By interpreting the map $\phi: \mathscr{O}^{\oplus n} \to \mathscr{O}^{\oplus k}$ as a $k \times n$ matrix M whose left k columns are the identity matrix and whose remaining entries are a_{ij} ($1 \le i \le k < j \le n$), interpret $P_I^\sharp$ as taking $x_{I'/I}$ to the determinant of the $k \times k$ submatrix corresponding to the columns in I'. For each (i, j) (with $1 \le i \le k < j \le n$), find some I' so that $x_{I'/I} \mapsto \pm a_{ij}$. (Let $I' = \{1, \ldots, i-1, i+1, \ldots, k, j\}$.)

Hence $G(k, n) \to \mathbb{P}^{\binom{n}{k}-1}$ is a closed embedding, so $G(k, n)$ is projective over $\mathbb{Z}$.

16.4.H. UNIMPORTANT EXERCISE. As an entertaining geometric consequence: if V is a vector space over a field, show that the "pure tensors in $\wedge^k V$ are pure in exactly one way": more precisely, if $v_1 \wedge \cdots \wedge v_k = w_1 \wedge \cdots \wedge w_k \ne 0$ in $\wedge^k V$, show that there is a $k \times k$ matrix of determinant 1 relating the v_i to the w_i.

16.4.3. *The Plücker equations.*

The equations of $G(k, n) \to \mathbb{P}^{\binom{n}{k}-1}$ are particularly nice. There are quadratic relations among the $k \times k$ minors of a $k \times (n-k)$ matrix, called the Plücker equations. By our construction, they are equations satisfied by $G(k, n)$. It turns out that these equations cut out $G(k, n)$, and in fact generate the homogeneous ideal of $G(k, n)$, but this takes more work (see [F1, §9.1, Lem. 1]). We explore this in one example.

16.4.I. EXERCISE. Suppose v_1, v_2, v_3, and v_4 are four vectors in a two-dimensional vector space V over some field. Show that

$$(v_1 \wedge v_2)(v_3 \wedge v_4) - (v_1 \wedge v_3)(v_2 \wedge v_4) + (v_1 \wedge v_4)(v_2 \wedge v_3) = 0.$$

The tricky part of this exercise is figuring out how to interpret multiplication (of Plücker coordinates) here.

16.4.J. EXERCISE. Note that the Plücker embedding embeds the Grassmannian $G(2,4)$ into $\mathbb{P}^5$.

(a) Show that $G(2,4)$ is cut out by the quadratic equation

$$x_{12}x_{34} - x_{13}x_{24} + x_{14}x_{23} = 0.$$

(Hint: Use Exercise 16.4.I to show that the quadratic vanishes on $G(2,4)$. But that isn't enough.)

(b) Show that every smooth quadric hypersurface in $\mathbb{P}^5$ over an algebraically closed field $\overline{k}$ of characteristic not 2 is isomorphic to the Grassmannian (over $\overline{k}$). (For comparison, every smooth quadric hypersurface in $\mathbb{P}^1_k$ is two points; every smooth quadric hypersurface in $\mathbb{P}^2_k$ is isomorphic to $\mathbb{P}^1_k$, §7.5.10; and every smooth quadric hypersurface in $\mathbb{P}^3_k$ is isomorphic to $\mathbb{P}^1_k \times_{\overline{k}} \mathbb{P}^1_k$, Example 10.6.2.)

16.4.K. EXERCISE (RULINGS ON QUADRICS IN $\mathbb{P}^5$). Suppose $k = \overline{k}$. From Remark 9.3.10, we expect two three-dimensional families of planes in $G(2,4)$ (interpreted as a hypersurface in $\mathbb{P}^5$ via the Plücker embedding; see Exercise 16.4.I). One of them may be described as follows: for each point $p \in \mathbb{P}^3_k$, we have a two-dimensional family of lines through p; this is a plane in $G(2,4)$. There is a three-dimensional family of planes corresponding to the choice of p. This is one of the two rulings. What is the other one? Prove (as rigorously as you can manage, given what you know) that these are both rulings.

16.4.4. *Further discussion.*

16.4.L. EXERCISE. Show that the group scheme GL_n acts on the Grassmannian $G(k,n)$. (The *action* of a group scheme appeared earlier in Exercise 7.6.S(a).) Hint: This is much more easily done with the language of functors, §7.6, than with the description of the Grassmannian in terms of patches, §7.7. (Exercise 15.5.F was the special case of projective space.)

16.4.M.** EXERCISE (GRASSMANNIAN BUNDLES). Suppose $\mathscr{F}$ is a rank n locally free sheaf on a scheme X. Define the **Grassmannian bundle** $G(k, \mathscr{F})$ over X. Intuitively, if $\mathscr{F}$ is a varying family of n-dimensional vector spaces over X, $G(k, \mathscr{F})$ should parametrize k-dimensional quotients of the fibers. You may want to define the functor first, and then show that it is representable. Your construction will behave well under base change.

16.4.5. *(Partial) flag varieties.* The discussion here extends without change to partial flag varieties (§7.7.1), and the interested reader should think this through.

Chapter 17

Projective Morphisms, and Relative Versions of Spec and Proj

In this chapter, we will use universal properties to define two useful constructions, $Spec$ of a sheaf of algebras $\mathscr{R}$ and $Proj$ of a sheaf of graded algebras $\mathscr{S}_\bullet$ on a scheme X. These will generalize (globalize) our constructions of both Spec of A-algebras and Proj of graded A-algebras. We will see that affine morphisms are precisely those of the form $Spec\,\mathscr{R} \to X$, and so we will *define* projective morphisms to be those of the form $Proj\,\mathscr{S}_\bullet \to X$.

In both cases, our plan is to make a notion we know well over a ring work more generally over a scheme. The main issue is how to glue the constructions over each affine open subset together. The slick way we will proceed is by giving a universal property, then showing that the affine construction satisfies this universal property, then showing that the universal property behaves well with respect to open subsets, then using the idea that lets us glue together the fibered product (or normalization) to do all the hard gluing work. The most annoying part of this plan is finding the right universal property, especially in the $Proj$ case.

17.1 Relative Spec of a (Quasicoherent) Sheaf of Algebras

Given an A-algebra, R, we can take its Spec to get an affine scheme over Spec A: Spec R → Spec A. We will now see a universal property description of a globalization of that notation. Consider an arbitrary scheme X, and a quasicoherent sheaf of algebras $\mathscr{R}$ on it. We will define how to take Spec of this sheaf of algebras, and we will get a scheme $Spec\,\mathscr{R} \to X$ that is "affine over X," i.e., the structure morphism is an affine morphism. You can think of this in two ways.

17.1.1. First, and most concretely, for any affine open set $\operatorname{Spec} A \subset X$, $\Gamma(\operatorname{Spec} A, \mathscr{R})$ is some A-algebra; call it R. Then above Spec A, $Spec\,\mathscr{R}$ will be Spec R.

17.1.2. Second, it will satisfy a universal property. We could define the A-scheme Spec R by the fact that morphisms to Spec R (from an A-scheme W, over Spec A) correspond to maps of A-algebras $R \to \Gamma(W, \mathscr{O}_W)$ (this is our old friend Exercise 7.3.G). The universal property for $\beta: Spec\,\mathscr{R} \to X$ generalizes this. Given a morphism $\mu: W \to X$, the X-morphisms $W \to Spec\,\mathscr{R}$ are in functorial (in W) bijection with morphisms α of sheaves of algebras making

$$(17.1.2.1)$$

commute. Here the map $\mathscr{O}_X \to \mu_* \mathscr{O}_W$ is that coming from the map of ringed spaces, and the map $\mathscr{O}_X \to \mathscr{R}$ comes from the $\mathscr{O}_X$-algebra structure on $\mathscr{R}$. (For experts: It needn't be true that $\mu_* \mathscr{O}_W$ is quasicoherent, but that doesn't matter. Non-experts should completely ignore this parenthetical comment.)

By universal property nonsense, these data determine $\beta: Spec\,\mathscr{R} \to X$ up to unique isomorphism, assuming that it exists.

Fancy translation: In the category of X-schemes, $\beta: Spec\,\mathscr{R} \to X$ represents the functor

$$(\mu: W \to X) \longmapsto \{(\alpha: \mathscr{R} \to \mu_* \mathscr{O}_W)\}.$$

17.1.A. EXERCISE. Show that if X is affine, say, Spec A, and $\mathscr{R} = \widetilde{R}$, where R is an A-algebra, then Spec R → Spec A satisfies this universal property. (Hint: Exercise 7.3.G.)

17.1.3. Proposition — *Suppose* $\beta\colon Spec\,\mathscr{R} \to X$ *satisfies the universal property for* $(X, \mathscr{R})$, *and* $i\colon U \hookrightarrow X$ *is an open subscheme. Then* $\beta|_U\colon Spec\,\mathscr{R} \times_X U = (Spec\,\mathscr{R})|_U \to U$ *satisfies the universal property for* $(U, \mathscr{R}|_U)$.

Proof. Suppose W is a U-scheme, with structure morphism $\nu\colon W \to U$. A U-morphism $W \to Spec\,\mathscr{R} \times_X U$ is the same as an X-morphism $W \to Spec\,\mathscr{R}$ (where we already know by assumption that $\mu = i \circ \nu\colon W \to X$ factors through U).

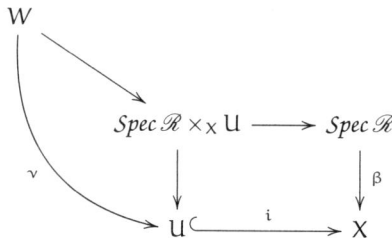

$$
\begin{array}{ccc}
W & & \\
& \searrow & \\
& Spec\,\mathscr{R} \times_X U \longrightarrow Spec\,\mathscr{R} & \\
\nu & \downarrow \qquad\qquad \downarrow \beta & \\
& U \overset{i}{\hookrightarrow} X &
\end{array}
$$

By the universal property of $Spec\,\mathscr{R}$, this is the same information as a map $\mathscr{R} \to \mu_*\mathscr{O}_W$, which by adjointness of (μ^*, μ_*) (§14.5.5) is the same as $\mu^*\mathscr{R} \to \mathscr{O}_W$, which (from $\nu^*(\mathscr{R}|_U) = \nu^* i^* \mathscr{R} = \mu^*\mathscr{R}$) is $\nu^*(\mathscr{R}|_U) \to \mathscr{O}_W$. By adjointness of (ν^*, ν_*), this is the same as $\mathscr{R}|_U \to \nu_*\mathscr{O}_W$, verifying the universal property. $\qquad\square$

Combining Exercise 17.1.A and Proposition 17.1.3, we have shown the existence of $Spec\,\mathscr{R}$ in the case that Y is an open subscheme of an affine scheme, and $\mathscr{R}$ is the restriction to Y of a quasicoherent sheaf of algebras on the affine scheme.

17.1.B. EXERCISE. Show the existence of $Spec\,\mathscr{R}$ in general, following the philosophy of our construction of the fibered product, normalization, and so forth.

17.1.4. We make some quick observations. First, $Spec\,\mathscr{R}$ can be "computed affine-locally on X." We also have an isomorphism $\phi\colon \mathscr{R} \overset{\sim}{\longrightarrow} \beta_*\mathscr{O}_{Spec\,\mathscr{R}}$.

17.1.C. EXERCISE. Given an X-morphism γ,

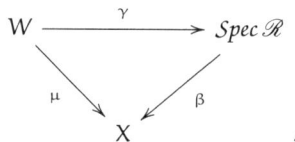

$$
\begin{array}{ccc}
W & \overset{\gamma}{\longrightarrow} & Spec\,\mathscr{R} \\
& \mu \searrow \quad \swarrow \beta & \\
& X &
\end{array}\quad ,
$$

show that the morphism α (in (17.1.2.1)) is the composition

$$
\mathscr{R} \overset{\phi}{\longrightarrow} \beta_*\mathscr{O}_{Spec\,\mathscr{R}} \longrightarrow \beta_*\gamma_*\mathscr{O}_W = \mu_*\mathscr{O}_W.
$$

The *Spec* construction gives an important way to understand affine morphisms. Note that $Spec\,\mathscr{R} \to X$ is an affine morphism. The "converse" is also true:

17.1.D. EXERCISE. Show that if $\mu\colon Z \to X$ is an affine morphism, then we have a natural isomorphism $Z \overset{\sim}{\hookrightarrow} Spec\,\mu_*\mathscr{O}_Z$ of X-schemes.

Hence we can recover any affine morphism using the *Spec* construction. More precisely, a morphism is affine if and only if it is of the form $\mu\colon Spec\,\mathscr{R} \to X$ constructed above.

17.1.E. EXERCISE. Suppose $\mu\colon Spec\,\mathscr{R} \to X$ is such a morphism. Show that the category of quasicoherent sheaves on $Spec\,\mathscr{R}$ is equivalent to the category of quasicoherent sheaves on X with the structure of $\mathscr{R}$-modules (quasicoherent $\mathscr{R}$-modules on X). Show that the category of finite type quasicoherent sheaves on $Spec\,\mathscr{R}$ is equivalent to the category of quasicoherent sheaves on X with the structure of "finite type" $\mathscr{R}$-modules (finite type quasicoherent $\mathscr{R}$-modules on X).

This is useful if X is quite simple but $\mathscr{Spec}\,\mathscr{R}$ is complicated. We will use this when $X \cong \mathbb{P}^1$, and $\mathscr{Spec}\,\mathscr{R}$ is a more complicated curve, and more generally when proving Serre duality in Chapter 29.

17.1.F. EXERCISE (THE FORMATION OF $\mathscr{Spec}$ "COMMUTES WITH BASE CHANGE"). Suppose $\mu\colon Z \to X$ is any morphism, and $\mathscr{R}$ is a quasicoherent sheaf of algebras on X. Show that there is a natural isomorphism $Z \times_X \mathscr{Spec}\,\mathscr{R} \overset{\sim}{\longleftarrow} \mathscr{Spec}\,\mu^*\mathscr{R}$. (If you think about it, you will see that this generalizes the statement and possibly proof of Proposition 17.1.3.)

17.1.5. Definition. An important example of the $\mathscr{Spec}$ construction is the **total space of a finite rank locally free sheaf** $\mathscr{F}$, which we define to be $\mathscr{Spec}(\mathrm{Sym}^\bullet\,\mathscr{F}^\vee)$.

17.1.G. EXERCISE. Suppose $\mathscr{F}$ is a locally free sheaf of rank n. Show that the total space of $\mathscr{F}$ is a rank n *vector bundle*, i.e., that given any point $p \in X$, there is an open neighborhood $p \in U \subset X$ such that

$$\mathscr{Spec}\left(\mathrm{Sym}^\bullet\,\mathscr{F}^\vee|_U\right) \cong \mathbb{A}_U^n.$$

Show that $\mathscr{F}$ is isomorphic to the sheaf of sections of the total space $\mathscr{Spec}(\mathrm{Sym}^\bullet\,\mathscr{F}^\vee)$. (Possible hint: Use transition functions.) For this reason, the total space is also called the **vector bundle associated to a locally free sheaf** $\mathscr{F}$. (Caution: Some authors, e.g., [Stacks, tag 01M2], call $\mathscr{Spec}(\mathrm{Sym}^\bullet\,\mathscr{F})$, the *dual* of this vector bundle, the vector bundle associated to $\mathscr{F}$.)

In particular, if $\mathscr{F} = \mathscr{O}_X^{\oplus n}$, then $\mathscr{Spec}(\mathrm{Sym}^\bullet\,\mathscr{F}^\vee)$ is $\mathbb{A}_X^n$.

(Aside: You may notice that the construction $\mathscr{Spec}\,\mathrm{Sym}^\bullet$ can be applied to any coherent sheaf $\mathscr{F}$ (without dualizing, i.e., $\mathscr{Spec}(\mathrm{Sym}^\bullet\,\mathscr{F})$). This is sometimes called the **abelian cone** associated to $\mathscr{F}$. This concept can be useful, but we won't need it.)

17.1.H. EXERCISE (THE TAUTOLOGICAL BUNDLE ON $\mathbb{P}^n$ IS $\mathscr{O}(-1)$). Suppose k is a field. Define the subset $X \subset \mathbb{A}_k^{n+1} \times \mathbb{P}_k^n$ corresponding to "points of $\mathbb{A}_k^{n+1}$ on the corresponding line of $\mathbb{P}_k^n$," so that the fiber of the map $\pi\colon X \to \mathbb{P}^n$ corresponding to a point $l = [x_0, \cdots, x_n]$ is the line in $\mathbb{A}_k^{n+1}$ corresponding to l, i.e., the scalar multiples of $(x_0, \ldots, x_n)$. Show that $\pi\colon X \to \mathbb{P}_k^n$ is (the line bundle corresponding to) the invertible sheaf $\mathscr{O}(-1)$. (Possible hint: Work first over the usual affine open sets of $\mathbb{P}_k^n$, and figure out transition functions.) For this reason, $\mathscr{O}(-1)$ is often called the **tautological bundle** of $\mathbb{P}_k^n$ (even over an arbitrary base, not just a field). (Side Remark: The projection $X \to \mathbb{A}_k^{n+1}$ is the blow-up of $\mathbb{A}_k^{n+1}$ at the "origin"; see Exercise 10.3.F.)

17.1.6. To read later: The weird functor $\pi^{!?}$, an occasional right-adjoint to pushforward. Because the next topic is used at two different places (Tricky Exercise 18.8.C and §29.3) that you may or may not read, there is no other natural place to put this discussion. But you should not read this now, and instead wait until you need it, as otherwise it will be completely unmotivated.

We now define a *right adjoint* $\pi^{!?}$ to the pushforward π_*, when π is a finite morphism. (Caution: The notation $\pi^{!?}$ is nonstandard, and is introduced only for the purposes of the arguments we will give.) This is surprising, as we usually think of π_* as a *right-adjoint* (to the pullback π^*), not a left adjoint. This can be seen as roughly analogous to the surprising occasional *left-adjoint* to the pullback: extension by zero (§23.4.7).

17.1.7. Definition. The definition is simple but strange. Suppose $\pi\colon X \to Y$ is a finite morphism of locally Noetherian schemes. If $\mathscr{G}$ is a quasicoherent sheaf on Y, then $\mathscr{Hom}_Y(\pi_*\mathscr{O}_X, \mathscr{G})$ is also a quasicoherent sheaf on Y (as $\pi_*\mathscr{O}_X$ is finitely presented, Exercise 14.3.A(b)). It furthermore has the structure of a $(\pi_*\mathscr{O}_X)$-module, which means that it corresponds to a quasicoherent sheaf on X (by Exercise 17.1.E, which shows that this correspondence is functorial in $\mathscr{G}$). Let $\pi^{!?}\mathscr{G}$ be this quasicoherent sheaf. Then $\pi^{!?}$ gives a covariant functor $QCoh_Y \to QCoh_X$, and also $Coh_Y \to Coh_X$.

17.1.I. EXERCISE. Describe a natural isomorphism of quasicoherent sheaves on Y

$$\pi_*\pi^{!?}\mathscr{G} \overset{\sim}{\longrightarrow} \mathscr{Hom}_Y(\pi_*\mathscr{O}_X, \mathscr{G}),$$

functorial in $\mathscr{G}$. Equivalently, describe an isomorphism of functors $QCoh_Y \to QCoh_Y$:

$$\pi_* \pi^{!?}(\cdot) \xrightarrow{\;\sim\;} \mathscr{H}om_Y(\pi_* \mathscr{O}_X, \cdot).$$

17.1.8. To show that $\pi^{!?}$ is right-adjoint to π_*, we study the affine case, or, equivalently, the ring-theoretic version. Suppose $B \to A$ is a ring morphism, M is an A-module, and N is a B-module. Note that $\mathrm{Hom}_B(A, N)$ naturally has the structure of an A-module. Also, M naturally carries the structure of a B-module; when we wish to emphasize its structure as a B-module, we sometimes call it M_B (see Exercise 1.4.E).

Consider the (very strange) map

(17.1.8.1) $$\mathrm{Hom}_A(M, \mathrm{Hom}_B(A, N)) \longrightarrow \mathrm{Hom}_B(M_B, N)$$

defined as follows. Given $m \in M$, and an element ϕ of $\mathrm{Hom}_A(M, \mathrm{Hom}_B(A, N))$, send m to $\phi_m(1)$.

17.1.J. EXERCISE.

(a) Show that (17.1.8.1) is a homomorphism of B-modules.
(b) Show that (17.1.8.1) is a bijection. Thus $(M \mapsto M_B, N \mapsto \mathrm{Hom}_B(A, N))$ is an adjoint pair $Mod_A \leftrightarrow Mod_B$.

17.1.K. EXERCISE. Suppose $\pi \colon X \to Y$ is a finite morphism of locally Noetherian schemes.

(a) Explain how the functor $Mod_B \to Mod_A$ given by $N \mapsto \mathrm{Hom}_B(A, N)$ in the previous exercise is the affine-local description of $\pi^{!?}$ in Definition 17.1.7.
(b) Show that $(\pi_*, \pi^{!?})$ is an adjoint pair between $QCoh_X$ and $QCoh_Y$.
(c) Show that $(\pi_*, \pi^{!?})$ is an adjoint pair between Coh_X and Coh_Y.
(d) If $\mathscr{F} \in QCoh_X$ and $\mathscr{G} \in QCoh_Y$, describe a natural isomorphism

(17.1.8.2) $$\pi_* \mathscr{H}om_X(\mathscr{F}, \pi^{!?}\mathscr{G}) \xrightarrow{\;\sim\;} \mathscr{H}om_Y(\pi_* \mathscr{F}, \mathscr{G})$$

(functorial in $\mathscr{F}$ and $\mathscr{G}$), which affine-locally is the isomorphism (17.1.8.1) described in Exercise 17.1.J.

17.2 Relative Proj of a (Quasicoherent) Sheaf of Graded Algebras

In parallel with the relative version $\mathcal{S}pec$ of Spec, we define a relative version of Proj, denoted by $\mathcal{P}roj$ (called "relative Proj" or "sheaf Proj"), of a quasicoherent graded sheaf of algebras (satisfying some hypotheses) on a scheme X. We have already done the case where the base X is affine, in §4.5.7, using the regular Proj construction over a ring A. The elegant way to proceed would be to state the right universal property, and then use this cleverly to glue together the constructions over each affine, just as we did in the constructions of fibered product, normalization, and $\mathcal{S}pec$. But because graded rings and graded modules make everything confusing, we do not do this. Instead we guiltily take a more pedestrian approach. (The universal property can be made to work; see [Stacks, tag 01O0]. But I recommend against it—a universal property should make your life easier, not harder.)

17.2.A. EXERCISE (Proj COMMUTES WITH AFFINE BASE CHANGE). Suppose $A \to B$ is map of rings, and $S_\bullet$ is a $\mathbb{Z}^{\geq 0}$-graded ring over A.

(a) Give a canonical isomorphism

(17.2.0.1) $$\alpha \colon \mathrm{Proj}_B(S_\bullet \otimes_A B) \xrightarrow{\;\sim\;} (\mathrm{Proj}_A S_\bullet) \times_{\mathrm{Spec}\, A} \mathrm{Spec}\, B.$$

(The **subscripts on the** Proj are new notation. They are redundant, and just remind us that we are thinking of $\mathrm{Proj}_A S_\bullet$ as an A-scheme, for example.)

(b) (easy) Suppose X is a projective A-scheme (§4.5.10). Show that $X \times_{\mathrm{Spec}\, A} \mathrm{Spec}\, B$ is a projective B-scheme.

(c) Suppose $S_\bullet$ is generated in degree 1, so $\mathscr{O}_{\mathrm{Proj}_A S_\bullet}(1)$ is an invertible sheaf (Exercise 15.1.G). Clearly, $S_\bullet \otimes_A B$ is generated in degree 1 as a B-algebra. Describe an isomorphism

$$\mathscr{O}_{\mathrm{Proj}_B (S_\bullet \otimes_A B)}(1) \xleftrightarrow{\ \sim\ } \alpha^* \gamma^* \mathscr{O}_{\mathrm{Proj}_A S_\bullet}(1),$$

where γ is the top morphism in the Cartesian diagram

$$
\begin{array}{ccc}
(\mathrm{Proj}_A S_\bullet) \times_{\mathrm{Spec}\, A} \mathrm{Spec}\, B & \xrightarrow{\ \gamma\ } & \mathrm{Proj}_A S_\bullet \\
\downarrow & & \downarrow \\
\mathrm{Spec}\, B & \longrightarrow & \mathrm{Spec}\, A \ .
\end{array}
$$

Possible hint: Transition functions.

We now give a general means of constructing schemes over X (from [Stacks, tag 01LH]), if we know what they should be over any affine open set, and how these behave under open embeddings of one affine open set into another.

17.2.B. EXERCISE. Suppose we are given a scheme X, and the following data:

(i) For each affine open subset $U \subset X$, we are given some morphism $\pi_U : Z_U \to U$ (a "scheme over U").

(ii) For each (open) inclusion of affine open subsets $V \subset U \subset X$, we are given an open embedding $\rho_V^U : Z_V \hookrightarrow Z_U$.

Assume this data satisfies:

(a) for each $V \subset U \subset X$, ρ_V^U induces an isomorphism $Z_V \xrightarrow{\sim} \pi_U^{-1}(V)$ of schemes over V, and

(b) whenever $W \subset V \subset U \subset X$ are three nested affine open subsets, $\rho_W^U = \rho_V^U \circ \rho_W^V$.

Show that there exists an X-scheme $\pi : Z \to X$, and isomorphisms $i_U : \pi^{-1}(U) \xrightarrow{\sim} Z_U$ over each affine open set U, such that for nested affine open sets $V \subset U$, ρ_V^U agrees with the composition

$$Z_V \xrightarrow{\ i_V^{-1}\ } \pi^{-1}(V) \hookrightarrow \pi^{-1}(U) \xrightarrow{\ i_U\ } Z_U.$$

Hint (cf. Exercise 4.4.A): Construct Z first as a set, then as a topological space, then as a scheme. (Your construction will be independent of choices. Your solution will work in more general situations, for example, when the category of schemes is replaced by ringed spaces, and when the affine open subsets are replaced by any base of the topology.)

17.2.C. IMPORTANT EXERCISE AND DEFINITION (relative $\mathcal{P}roj$). Suppose $\mathscr{S}_\bullet = \oplus_{n \geq 0} \mathscr{S}_n$ is a quasicoherent sheaf of $\mathbb{Z}^{\geq 0}$-graded algebras on a scheme X. Over each affine open subset $\mathrm{Spec}\, A \cong U \subset X$, we have a U-scheme $\mathrm{Proj}_A \mathscr{S}_\bullet(U) \to U$. Show that these can be glued together to form an X-scheme, which we call $\mathcal{P}roj_X \mathscr{S}_\bullet$; we have a "structure morphism" $\beta : \mathcal{P}roj_X \mathscr{S}_\bullet \to X$. (The structure morphism β is part of the definition.) Hint/Warning: This problem would be easier if Exercise 17.2.B required considering only *distinguished* affine inclusions. But that would make Exercise 17.2.B notably harder. Instead, use Exercise 17.2.A.

By the construction of Exercise 17.2.B, the preimage over any affine open set can be computed using the original Proj construction. (You may enjoy going back and giving constructions of X^{red}, the normalization of X, and $\mathcal{S}pec$ of a quasicoherent sheaf of $\mathscr{O}$-algebras using this idea. But there is a moral price to be paid by giving up the universal property.)

17.2.1. Ongoing (reasonable) hypotheses on $\mathscr{S}_\bullet$: "Finite generation in degree 1". The Proj construction is most useful when applied to an A-algebra $S_\bullet$ satisfying some reasonable hypotheses (§4.5.6), notably when $S_\bullet$ is a finitely generated $\mathbb{Z}^{\geq 0}$-graded A-algebra, and ideally if it is generated in degree 1. For this reason, in the rest of the book, we will enforce these assumptions on $\mathscr{S}_\bullet$, once we make sense of them for quasicoherent sheaves of algebras. (If you later need to relax these hypotheses—for example, to keep the finite generation hypothesis but remove the "generation in degree 1" hypothesis—it will not be too difficult.) Precisely, **we now always require that (i) $\mathscr{S}_0 = \mathscr{O}_X$, (ii) $\mathscr{S}_\bullet$ is "generated in degree 1," and (iii) $\mathscr{S}_1$ is finite type.** The cleanest way to make condition (ii) precise is to require the natural map

$$\mathrm{Sym}^\bullet_{\mathscr{O}_X} \mathscr{S}_1 \longrightarrow \mathscr{S}_\bullet$$

to be surjective. Because the $\mathrm{Sym}^\bullet$ construction may be computed affine-locally (§14.2.6), we can check generation in degree 1 on any affine cover. The motivation for these hypotheses is the following construction.

17.2.D. IMPORTANT EXERCISE: $\mathscr{O}(1)$ ON $\mathcal{P}roj\,\mathscr{S}_\bullet$. If $\mathscr{S}_\bullet$ is finitely generated in degree 1 (Hypotheses 17.2.1), construct an invertible sheaf $\mathscr{O}_{\mathcal{P}roj_X\,\mathscr{S}_\bullet}(1)$ on $\mathcal{P}roj_X\,\mathscr{S}_\bullet$ that "restricts to $\mathscr{O}_{\mathrm{Proj}_A\,\mathscr{S}_\bullet(\mathrm{Spec}\,A)}(1)$ over each affine open subset $\mathrm{Spec}\,A \subset X$."

17.2.E. EXERCISE ("$\mathcal{P}roj$ COMMUTES WITH BASE CHANGE"). Suppose $\mathscr{S}_\bullet$ is a quasicoherent sheaf of $\mathbb{Z}^{\geq 0}$-graded algebras on X. Let $\rho: Z \to X$ be any morphism. Give a natural isomorphism

$$\left(\mathcal{P}roj\,\rho^*\mathscr{S}_\bullet, \mathscr{O}_{\mathcal{P}roj\,\rho^*\mathscr{S}_\bullet}(1) \right) \longleftrightarrow \left(Z \times_X \mathcal{P}roj\,\mathscr{S}_\bullet, \psi^*\mathscr{O}_{\mathcal{P}roj\,\mathscr{S}_\bullet}(1) \right),$$

where ψ is the "top" morphism in the base change diagram

$$
\begin{array}{ccc}
Z \times_X \mathcal{P}roj\,\mathscr{S}_\bullet & \xrightarrow{\ \psi\ } & \mathcal{P}roj\,\mathscr{S}_\bullet \\
\downarrow & & \downarrow{\scriptstyle\beta} \\
Z & \xrightarrow[\ \rho\ ]{} & X.
\end{array}
$$

17.2.2. *Definition: π-very ample.* Suppose $\pi: X \to Y$ is proper. If $\mathscr{L}$ is an invertible sheaf on X, then we say that $\mathscr{L}$ is **very ample (with respect to π)**, **relatively very ample**, or (awkwardly) π-**very ample** if we can write $X = \mathcal{P}roj_Y\,\mathscr{S}_\bullet$ with $\mathscr{L} \cong \mathscr{O}(1)$, where $\mathscr{S}_\bullet$ is a quasicoherent sheaf of algebras on Y satisfying Hypotheses 17.2.1 ("finite generation in degree 1"). (The notion of very ampleness can be extended to more general situations; see, for example, [Stacks, tag 01VM]. But this is of interest only to people with esoteric tastes.)

17.2.F. EXERCISE. Suppose $\mathscr{S}_\bullet$ is finitely generated in degree 1 (Hypotheses 17.2.1). Describe a map of graded quasicoherent sheaves $\phi: \mathscr{S}_\bullet \to \oplus_{n=0}^\infty \beta_*\mathscr{O}(n)$ (β is the structure morphism; see Exercise 17.2.C). Hint: Exercise 15.7.C.

17.2.G. EXERCISE. Suppose $\mathscr{L}$ is an invertible sheaf on X, and $\mathscr{S}_\bullet$ is a quasicoherent sheaf of graded algebras on X generated in degree 1 (Hypotheses 17.2.1). Define $\mathscr{S}'_\bullet = \oplus_{n=0}^\infty \left(\mathscr{S}_n \otimes \mathscr{L}^{\otimes n} \right)$. Then $\mathscr{S}'_\bullet$ has a natural algebra structure inherited from $\mathscr{S}_\bullet$; describe it. Give a natural isomorphism of "X-schemes with line bundles"

$$\left(\mathcal{P}roj\,\mathscr{S}'_\bullet, \mathscr{O}_{\mathcal{P}roj\,\mathscr{S}'_\bullet}(1) \right) \longleftrightarrow \left(\mathcal{P}roj\,\mathscr{S}_\bullet, \mathscr{O}_{\mathcal{P}roj\,\mathscr{S}_\bullet}(1) \otimes \beta^*\mathscr{L} \right),$$

where $\beta: \mathcal{P}roj\,\mathscr{S}_\bullet \to X$ is the structure morphism. In other words, informally speaking, the $\mathcal{P}roj$ is the same, but the $\mathscr{O}(1)$ is twisted by $\mathscr{L}$.

17.2.3. Projectivization of vector bundles.

17.2.4. *Definition.* If $\mathscr{F}$ is a *finite rank locally free sheaf* on X, then $\mathcal{P}roj(\mathrm{Sym}^\bullet\,\mathscr{F}^\vee)$ is called its **projectivization**, and is denoted by $\mathbb{P}\mathscr{F}$. (The reason for the dual is the same as for $\mathcal{S}pec(\mathrm{Sym}^\bullet\,\mathscr{F}^\vee)$ in Definition 17.1.5.) You can check that this construction behaves well with respect to base change.

Define $\mathbb{P}_X^n := \mathbb{P}(\mathscr{O}_X^{\oplus(n+1)})$. (Then $\mathbb{P}_{\mathrm{Spec}\,A}^n$ agrees with our earlier definition of $\mathbb{P}_A^n$—cf. Exercise 4.5.Q—and $\mathbb{P}_X^n$ agrees with our earlier usage; see, for example, the proof of Theorem 11.4.5.) If $\mathscr{F}$ is locally free of rank $n+1$, then $\mathbb{P}\mathscr{F}$ is a **projective bundle** or $\mathbb{P}^n$-**bundle** over X. By Exercise 17.2.G, if $\mathscr{F}$ is a finite rank locally free sheaf on X, there is a canonical isomorphism $\mathbb{P}\mathscr{F} \xrightarrow{\sim} \mathbb{P}(\mathscr{L} \otimes \mathscr{F})$.

More generally, if $\mathscr{F}$ is a finite type quasicoherent sheaf on X, then one might define similarly its projectivization $\mathit{Proj}(\mathrm{Sym}^\bullet \mathscr{F}^\vee)$. Be careful, though. For example, if $\mathscr{G}$ is a torsion sheaf on an integral scheme, then $\mathscr{F}^\vee = 0$, so, with this definition, $\mathbb{P}\mathscr{F} = \varnothing$. So this isn't a great notion.

17.2.5. *Huge notation caution.* There is violent disagreement on whether $\mathbb{P}\mathscr{F}$ should be defined as $\mathrm{Sym}^\bullet \mathscr{F}$ or $\mathrm{Sym}^\bullet \mathscr{F}^\vee$, parallel to the disagreement as to whether the projectivization of a vector space parametrizes one-dimensional subspaces or quotients (cf. Exercise 17.2.I). Hence it is safest to avoid the notation $\mathbb{P}\mathscr{F}$, or at least to state at the outset which convention you are following. *In this book, we will try to avoid the notation $\mathbb{P}\mathscr{F}$, and will use it only when $\mathscr{F}$ is finite rank locally free ("a vector bundle"), and we will mean it as defined here, $\mathit{Proj}\,(\mathrm{Sym}^\bullet \mathscr{F}^\vee)$.*

17.2.6. *Example: Ruled surfaces.* If C is a regular curve and $\mathscr{F}$ is locally free of rank 2, then $\mathbb{P}\mathscr{F}$ is called a **ruled surface** over C. If C is further isomorphic to $\mathbb{P}^1$, then $\mathbb{P}\mathscr{F}$ is called a **Hirzebruch surface**. All vector bundles on $\mathbb{P}^1$ split as a direct sum of line bundles (see §18.5.5 for a proof), so each Hirzebruch surface is of the form $\mathbb{P}(\mathscr{O}(n_1) \oplus \mathscr{O}(n_2))$. By Exercise 17.2.G, this depends only on $n_2 - n_1$. The Hirzebruch surface $\mathbb{P}(\mathscr{O} \oplus \mathscr{O}(n))$ $(n \geq 0)$ is often denoted by $\mathbb{F}_n$. We will discuss the Hirzebruch surfaces in greater length in §20.2.10. We will see that the $\mathbb{F}_n$ are all distinct in Exercise 20.2.Q.

17.2.H. EXERCISE. If $\mathscr{S}_\bullet$ is finitely generated in degree 1 (Hypotheses 17.2.1), describe a canonical closed embedding

$$\mathit{Proj}\,\mathscr{S}_\bullet \overset{i}{\hookrightarrow} \mathit{Proj}_X \mathrm{Sym}^\bullet \mathscr{S}_1$$

with β to X

and an isomorphism $\mathscr{O}_{\mathit{Proj}\,\mathscr{S}_\bullet}(1) \xrightarrow{\sim} i^* \mathscr{O}_{\mathit{Proj}_X \mathrm{Sym}^\bullet \mathscr{S}_1}(1)$ arising from the surjection

$$\mathrm{Sym}^\bullet \mathscr{S}_1 \twoheadrightarrow \mathscr{S}_\bullet.$$

In particular, if $\mathscr{S}_1$ is locally free, then $\mathit{Proj}\,\mathrm{Sym}^\bullet \mathscr{S}_1 = \mathbb{P}\mathscr{S}_1^\vee$, so we have embedded $\mathit{Proj}\,\mathscr{S}_\bullet$ in a projective bundle on X.

17.2.I. EXERCISE. Suppose $\mathscr{F}$ is a locally free sheaf of rank $n+1$ on X. Exhibit a bijection between the set of sections $s\colon X \to \mathbb{P}\mathscr{F}$ of $\mathbb{P}\mathscr{F} \to X$ and the set of surjective homomorphisms $\mathscr{F} \to \mathscr{L} \to 0$ of $\mathscr{F}$ onto invertible sheaves on X. This functorial description of $\mathbb{P}\mathscr{F}$ in some sense generalizes the functorial description of projective space in §15.2.3.

17.2.7. *Remark (the relative version of the projective and affine cone).* There is a natural morphism from $\mathit{Spec}\,\mathscr{S}_\bullet$ minus the zero-section to $\mathit{Proj}\,\mathscr{S}_\bullet$ (cf. Exercise 9.3.N). Just as $\mathrm{Proj}\,S_\bullet[T]$ contains a closed subscheme identified with $\mathrm{Proj}\,S_\bullet$ whose complement can be identified with $\mathrm{Spec}\,S_\bullet$ (Exercise 9.3.O), $\mathit{Proj}\,\mathscr{S}_\bullet[T]$ contains a closed subscheme identified with $\mathit{Proj}\,\mathscr{S}_\bullet$ whose complement can be identified with $\mathit{Spec}\,\mathscr{S}_\bullet$. You are welcome to think this through.

17.3   Projective Morphisms

In §17.1, we reinterpreted affine morphisms: $X \to Y$ is an affine morphism if there is an isomorphism $X \xrightarrow{\sim} \mathit{Spec}\,\mathscr{B}$ of Y-schemes for some quasicoherent sheaf of algebras $\mathscr{B}$ on Y. We will *define* the notion of a projective morphism similarly.

374 Chapter 17 Projective Morphisms, and Relative Versions of Spec and Proj

You might think that because projectivity is such a classical notion, there should be some obvious definition that is reasonably behaved. But this is not the case, and there are many possible variant definitions of projective (see [Stacks, tag 01W8]). All are imperfect, including the accepted definition we give here (see the warnings in §17.3.4).

17.3.1. *Definition.* A morphism $\pi\colon X \to Y$ is **projective** if there is an isomorphism

$$X \xrightarrow{\;\sim\;} \mathcal{P}roj\,\mathscr{S}_\bullet$$

for a quasicoherent sheaf of algebras $\mathscr{S}_\bullet$ on Y satisfying "finite generation in degree 1" (Hypotheses 17.2.1). We say X is a **projective Y-scheme**, or X is **projective over** Y. Using Exercise 7.4.D, this generalizes the notion of a projective A-scheme.

17.3.A. EXERCISE.

(a) *(useful characterization of projective morphisms)* Suppose $\pi\colon X \to Y$ is a morphism. Show that π is projective if and only if there exist a finite type quasicoherent sheaf $\mathscr{S}_1$ on Y, and a closed embedding $i\colon X \hookrightarrow \mathcal{P}roj_Y \operatorname{Sym}^\bullet \mathscr{S}_1$ (over Y, i.e., commuting with the maps to Y). Hint: Exercise 17.2.H.

(b) *(useful characterization of projective morphisms, with line bundle)* Suppose $\mathscr{L}$ is an invertible sheaf on X, and $\pi\colon X \to Y$ is a morphism. Show that π is projective, with $\mathcal{O}(1) \cong \mathscr{L}$, if and only if there exist a finite type quasicoherent sheaf $\mathscr{S}_1$ on Y, a closed embedding $i\colon X \hookrightarrow \mathcal{P}roj_Y \operatorname{Sym}^\bullet \mathscr{S}_1$ (over Y, i.e., commuting with the maps to Y), and an isomorphism $i^* \mathcal{O}_{\mathcal{P}roj_Y \operatorname{Sym}^\bullet \mathscr{S}_1}(1) \xleftrightarrow{\;\sim\;} \mathscr{L}$.

(c) Suppose, furthermore, that Y admits an ample line bundle in the sense of §16.2.5, as is the case whenever Y is projective, affine, or, more generally, quasi-projective. Show that π is projective if and only if there exists a closed embedding $X \to \mathbb{P}^n_Y$ (over Y) for some n. This is the definition of projective morphism given in [Ha1, p. 103]. (If you want to avoid the starred section §16.2.5, you can assume that Y is projective over $\operatorname{Spec} A$ and use the definition of ample from §16.2.1. You will then have dealt with the important case where Y is projective, but missed out on other potentially interesting cases, such as when Y is affine or otherwise quasiprojective (but not proper).) Hint: The harder direction is the forward implication. Use the finite type quasicoherent sheaf $\mathscr{S}_1$ from (a). Tensor $\mathscr{S}_1$ with a high enough power of $\mathscr{M}$ so that it is finitely globally generated (Theorem 16.2.6, or Theorem 16.2.2 in the proper setting), to obtain a surjection

$$\mathcal{O}_Y^{\oplus (n+1)} \longrightarrow \mathscr{S}_1 \otimes \mathscr{M}^{\otimes N}.$$

Then use Exercise 17.2.G.

17.3.2. Properties of projective morphisms.
The property of a morphism being projective is clearly preserved by base change, as the $\mathcal{P}roj$ construction behaves well with respect to base change (Exercise 17.2.E). Also, projective morphisms are proper: properness is local on the target (Theorem 11.4.4(b)), and we saw earlier that projective A-schemes are proper over A (Theorem 11.4.5). In particular (by definition of properness), projective morphisms are separated, finite type, and universally closed.

17.3.B. IMPORTANT EXERCISE: FINITE MORPHISMS ARE PROJECTIVE (CF. EXERCISE 8.3.I). Show that finite morphisms are projective as follows. Suppose $Z \to X$ is finite, so that $Z \cong \operatorname{Spec} \mathscr{B}$, where $\mathscr{B}$ is a finite type quasicoherent sheaf of algebras on X. Describe a sheaf of graded algebras $\mathscr{S}_\bullet$ where $\mathscr{S}_0 \cong \mathcal{O}_X$ and $\mathscr{S}_n \cong \mathscr{B}$ for $n > 0$. Describe an X-isomorphism $Z \xleftrightarrow{\;\sim\;} \mathcal{P}roj\,\mathscr{S}_\bullet$.

In particular, closed embeddings are projective. We have the sequence of implications for morphisms

$$\text{closed embedding} \Longrightarrow \text{finite} \Longrightarrow \text{projective} \Longrightarrow \text{proper}.$$

17.3.3. We have shown that finite morphisms are projective (Exercise 17.3.B), and have finite fibers (Exercise 8.3.J). We will show the converse in Theorem 18.1.6, and state the extension to proper morphisms immediately after (although we won't prove it until Theorem 28.5.2).

17.3.4. *Warnings about projective morphisms.*
First, notice that $\mathscr{O}(1)$, an important part of the concept of *Proj*, is not mentioned in the definition. (I would prefer that it be part of the definition, but this isn't accepted practice.)

Second, [Ha1, p. 103] defines projective morphism differently; we follow the more general definition of Grothendieck. Although these definitions are the same in most circumstances (Exercise 17.3.A(c)), you should be cautious when reading the literature. For example, finite morphisms are not always projective in the sense of [Ha1], while they *are* projective with this definition (Exercise 17.3.B).

Third, projective morphisms are *not* reasonable in the sense of §8.1! Although projective morphisms are preserved by base change (§17.3.2), the property of being projective is *not* local on the target (see §24.7.7 for an example). (With the additional data of the invertible sheaf $\mathscr{O}(1)$, and in Noetherian circumstances, it is; see §17.3.8.) Furthermore, we will show that projective morphisms are preserved by composition only when the final target is affine or Noetherian (Exercise 17.3.I).

But many of the consequences of reasonableness still hold, as the next few exercises show. We first prove that the class of projective morphisms, despite its unreasonable nature, still satisfies the Cancellation Theorem 11.1.1.

17.3.C. EXERCISE (THE CANCELLATION THEOREM FOR PROJECTIVE MORPHISMS).

(a) Suppose $\pi: X \to Y$ is a closed embedding, and $\rho: Y \to Z$ is projective. Show that $\rho \circ \pi$ is projective.
(b) Show that a morphism $\rho: Y \to Z$ has projective diagonal if and only if ρ is separated.
(c) Prove the Cancellation Theorem for Projective Morphisms: Suppose $\pi: X \to Y$ is a morphism, $\rho: Y \to Z$ is a separated morphism (i.e., by (b), has projective diagonal), and $\rho \circ \pi$ is projective. Show that π is projective. (Possible hint: Use the diagram in the hint to Exercise 11.1.A.)

17.3.D. EXERCISE. Show that a morphism (over Spec k) from a projective k-scheme to a separated k-scheme is always projective. (Hint: The previous exercise.)

17.3.E. EXERCISE. Suppose $\pi: X \to Y$ and $\pi': X' \to Y'$ are projective morphisms of S-schemes. Show that $\pi \times \pi': X \times_S X' \to Y \times_S Y'$ is also projective. Hint: Because of the unreasonable nature of projective morphisms, you cannot invoke Exercise 8.1.A. Use Exercise 10.6.D instead.

17.3.5.* Global generation and (very) ampleness in the relative setting.
We extend the discussion of §16.1 to the relative setting, in order to give ourselves the language of relatively base-point-freeness. We won't use this discussion, so on a first reading you should skip it. But these ideas come up repeatedly in the research literature.

Suppose $\pi: X \to Y$ is a quasicompact quasiseparated morphism. If $\mathscr{F}$ is a quasicoherent sheaf on X, we say that $\mathscr{F}$ is **relatively globally generated** or **globally generated with respect to** π if the natural map of quasicoherent sheaves $\pi^*\pi_*\mathscr{F} \to \mathscr{F}$ is surjective. Quasicompactness and quasiseparatedness are needed ensure that $\pi_*\mathscr{F}$ is a quasicoherent sheaf, Exercise 14.4.C. But these hypotheses are not very restrictive. Global generation is most useful only in the quasicompact setting, and most people won't be bothered by quasiseparated hypotheses. (Unimportant aside: These hypotheses can be relaxed considerably. If $\pi: X \to Y$ is a morphism of *locally ringed spaces*—not necessarily schemes—with no other hypotheses, and $\mathscr{F}$ is a quasicoherent sheaf on X, then we say that $\mathscr{F}$ is **relatively globally generated** or **globally generated with respect to** π if the natural map $\pi^*\pi_*\mathscr{F} \to \mathscr{F}$ of $\mathscr{O}_X$-modules is surjective.)

Thanks to our hypotheses, as the natural map $\pi^*\pi_*\mathcal{F} \to \mathcal{F}$ is a morphism of quasicoherent sheaves, the condition of being relatively globally generated is affine-local on Y.

Suppose now that $\mathcal{L}$ is a locally free sheaf on X, and $\pi\colon X \to Y$ is a morphism. We say that $\mathcal{L}$ is **relatively base-point-free** or **base-point-free with respect to** π if it is relatively globally generated.

17.3.F. EXERCISE. Suppose $\mathcal{L}$ is a relatively base-point-free line bundle on X, $\pi\colon X \to Y$ is a quasicompact separated morphism, and $\pi_*\mathcal{L}$ is finite type on Y. (We will later show in Grothendieck's Coherence Theorem 18.9.1 that this latter statement is true if π is proper and Y is Noetherian. This is much easier if π is projective; see Theorem 18.7.1. We could work hard and prove it now, but it isn't worth the trouble.) Describe a canonical morphism $\psi\colon X \to \mathcal{P}roj_Y \operatorname{Sym}^\bullet(\pi_*\mathcal{L})$. (Possible hint: This generalizes the fact that base-point-free line bundles give maps to projective space, so generalize that argument; see §15.2.)

We say that $\mathcal{L}$ is **relatively ample** or **π-ample** or **relatively ample with respect to** π if for every affine open subset $\operatorname{Spec} B$ of Y, $\mathcal{L}|_{\pi^{-1}(\operatorname{Spec} B)}$ is ample on $\pi^{-1}(\operatorname{Spec} B)$ over B, or, equivalently (by §16.2.5), $\mathcal{L}|_{\pi^{-1}(\operatorname{Spec} B)}$ is (absolutely) ample on $\pi^{-1}(\operatorname{Spec} B)$. By the discussion in §16.2.5, if $\mathcal{L}$ is ample then π is necessarily quasicompact, and (by Theorem 16.2.6) separated; if π is affine, then all invertible sheaves are ample; and if π is projective, then the corresponding $\mathcal{O}(1)$ is ample. By Exercise 16.2.M, $\mathcal{L}$ is π-ample if and only if $\mathcal{L}^{\otimes n}$ is π-ample, and if $Z \hookrightarrow X$ is a closed embedding, then $\mathcal{L}|_Z$ is ample over Y.

From Theorem 16.2.6(d), we have a natural open embedding $X \hookrightarrow \mathcal{P}roj_Y \oplus\pi_*\mathcal{L}^{\otimes d}$. (Do you see what this map is? Also, be careful: $\oplus\pi_*\mathcal{L}^{\otimes d}$ need not be a finitely generated graded sheaf of algebras, so we are using the $\mathcal{P}roj$ construction where one of the usual hypotheses doesn't hold.)

The notions of relative global generation and relative ampleness are most useful in the proper setting, because of Theorem 16.2.2.

17.3.6. Many statements of §16.1 carry over without change. For example, we have the following. Suppose $\pi\colon X \to Y$ is proper, $\mathcal{F}$ and $\mathcal{G}$ are quasicoherent sheaves on X, and $\mathcal{L}$ and $\mathcal{M}$ are invertible sheaves on X. If π is affine, then $\mathcal{F}$ is relatively globally generated (from Easy Exercise 16.1.B). If $\mathcal{F}$ and $\mathcal{G}$ are relatively globally generated, so is $\mathcal{F} \otimes \mathcal{G}$ (Easy Exercise 16.1.C). If $\mathcal{L}$ is π-very ample (Definition 17.2.2), then it is π-base-point-free (Easy Exercise 16.2.B). If $\mathcal{L}$ is π-very ample, and $\mathcal{M}$ is π-base-point-free (if, for example, it is π-very ample), then $\mathcal{L} \otimes \mathcal{M}$ is π-very ample (Exercise 16.2.C). Exercise 16.2.G extends immediately to show that if

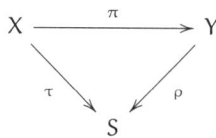

$$X \xrightarrow{\ \pi\ } Y$$
$$\tau \searrow \quad \swarrow \rho$$
$$S$$

is a finite morphism of S-schemes, and if $\mathcal{L}$ is a ρ-ample invertible sheaf on Y, then $\pi^*\mathcal{L}$ is τ-ample.

By the nature of the statements, some of the statements of §16.1 require quasicompactness hypotheses on Y, or other patches. For example:

17.3.7. Theorem — *Suppose $\pi\colon X \to Y$ is proper, $\mathcal{L}$ is an invertible sheaf on X, and Y is quasicompact. The following are equivalent.*

(a) *For some $N > 0$, $\mathcal{L}^{\otimes N}$ is π-very ample.*

(a') *For all $n \gg 0$, $\mathcal{L}^{\otimes n}$ is π-very ample.*

(b) *For all finite type quasicoherent sheaves $\mathcal{F}$, there is an n_0 such that for $n \geq n_0$, $\mathcal{F} \otimes \mathcal{L}^{\otimes n}$ is relatively globally generated.*

(c) *The invertible sheaf $\mathcal{L}$ is π-ample.*

17.3.G. EXERCISE. Prove Theorem 17.3.7 using Theorem 16.2.2. (Unimportant remark: The proof given of Theorem 16.2.2 used Noetherian hypotheses, but as stated there, they can be removed.)

After doing the above exercise, it will be clear how to adjust the statement of Theorem 17.3.7 if you need to remove the quasicompactness assumption on Y.

17.3.H. EXERCISE (USEFUL EQUIVALENT DEFINITION OF VERY AMPLENESS). Suppose $\pi: X \to Y$ is a proper morphism, Y is locally Noetherian (hence X is too, as π is finite type), and $\mathscr{L}$ is an invertible sheaf on X. Suppose that you know that in this situation $\pi_*\mathscr{L}$ is finite type. (We will later show this, as described in Exercise 17.3.F. This is where Noetherian hypotheses are used.) Show that $\mathscr{L}$ is π-very ample if and only if (i) $\mathscr{L}$ is relatively base-point-free, and (ii) the canonical morphism $\psi: X \to \mathscr{P}\!\mathit{roj}_Y \operatorname{Sym}^{\bullet}(\pi_*\mathscr{L})$ of Exercise 17.3.F is a closed embedding. Conclude that the notion of relative very ampleness is affine-local on Y (it may be checked on *any* affine cover Y), if π is proper.

17.3.8. As a consequence, Theorem 17.3.7 implies the notion of relative ampleness is affine-local on Y (if π is proper and Y is locally Noetherian).

17.3.I. IMPORTANT CHALLENGING EXERCISE (THE COMPOSITION OF PROJECTIVE MORPHISMS IS USUALLY PROJECTIVE). Suppose $\pi: X \to Y$ and $\rho: Y \to Z$ are projective morphisms, and Z is affine or Noetherian. Show that $\rho \circ \pi$ is projective. Hint: The criterion for projectivity given in Exercise 17.3.A(b) will be useful. (i) Deal first with the case where Z is affine. Build the following commutative diagram, thereby finding a closed embedding $X \hookrightarrow \mathbb{P}\mathscr{F}^{\oplus n}$ over Z. (See §17.2.3 for the definition of $\mathbb{P}\mathscr{V}$, including a caution about duals.) In this diagram, all inclusions are closed embeddings, and all script fonts refer to finite type quasicoherent sheaves.

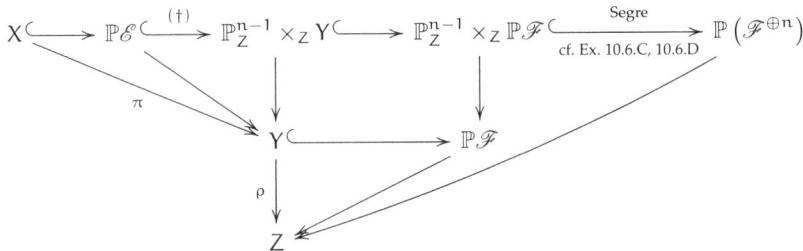

Construct the closed embedding (†) as follows. Suppose $\mathscr{M}$ is the very ample line bundle on Y over Z. Then $\mathscr{M}$ is ample, and so by Theorem 16.2.2, for $m \gg 0$, $\mathscr{E} \otimes \mathscr{M}^{\otimes m}$ is generated by a finite number of global sections. Suppose $\mathscr{O}_Y^{\oplus n} \to \mathscr{E} \otimes \mathscr{M}^{\otimes m}$ is the corresponding surjection. This induces a closed embedding $\mathbb{P}(\mathscr{E} \otimes \mathscr{M}^{\otimes m}) \hookrightarrow \mathbb{P}_Y^{n-1}$. But $\mathbb{P}(\mathscr{E} \otimes \mathscr{M}^{\otimes m}) \cong \mathbb{P}\mathscr{E}$ (Exercise 17.2.G), and $\mathbb{P}_Y^{n-1} = \mathbb{P}_Z^{n-1} \times_Z Y$. (ii) Unwind this diagram to show that (for Z affine) if $\mathscr{L}$ is π-very ample and $\mathscr{M}$ is ρ-very ample, then for $m \gg 0$, $\mathscr{L} \otimes \pi^*\mathscr{M}^{\otimes m}$ is $(\rho \circ \pi)$-very ample. Then deal with the general case by covering Z with a finite number of affine open subsets, using Exercise 17.3.H.

17.3.9. ** **Ample vector bundles.** The notion of an **ample vector bundle** is useful in some parts of the literature, so we define it, although we won't use the notion. A locally free sheaf $\mathscr{E}$ on a proper A-scheme X is **ample** if $\mathscr{O}_{\mathscr{P}\!\mathit{roj}_X \operatorname{Sym}^{\bullet} \mathscr{E}/X}(1)$ is an ample invertible sheaf. In particular, using Exercise 17.2.G, you can verify that an invertible sheaf is ample as a locally free sheaf (this definition) if and only if it is ample as an invertible sheaf (Definition 16.2.1), preventing a notational crisis. (The proper hypotheses can be relaxed; it is included only because Definition 16.2.1 of ampleness is only for proper schemes.)

17.3.10. ** **Quasiprojective morphisms.** In analogy with projective and quasiprojective A-schemes (§4.5.10), one may define quasiprojective morphisms. *If Y is quasicompact*, we say that $\pi: X \to Y$ is a **quasiprojective morphism** if π can be expressed as a quasicompact open embedding into a scheme projective over Y. This is not a great notion, and we will not use it. (The general definition of quasiprojective morphism is slightly delicate—see [Gr-EGA, II.5.3]—and we won't need it.)

17.3.11.** Quasiaffine morphisms.

Because we have introduced quasiprojective morphisms, we briefly introduce quasiaffine morphisms (and quasiaffine schemes), as some readers may have cause to use them. Many of these ideas could have been introduced long before, but because we will never use them, we deal with them all at once.

A scheme X is **quasiaffine** if it admits a quasicompact open embedding into an affine scheme. This implies that X is quasicompact and separated. If X is Noetherian (the most relevant case for most people), then any open embedding is of course automatically quasicompact.

17.3.J. EXERCISE. Show that X is quasiaffine if and only if the canonical map $X \to \operatorname{Spec} \Gamma(X, \mathscr{O}_X)$ (defined in Exercise 7.3.G and the paragraph following it) is a quasicompact open embedding. Thus a quasiaffine scheme comes with a *canonical* quasicompact open embedding into an affine scheme. Hint: Let $A = \Gamma(X, \mathscr{O}_X)$ for convenience. Suppose $X \hookrightarrow \operatorname{Spec} R$ is a quasicompact open embedding. We wish to show that $X \hookrightarrow \operatorname{Spec} A$ is a quasicompact open embedding. Factor $X \hookrightarrow \operatorname{Spec} R$ through $X \to \operatorname{Spec} A \to \operatorname{Spec} R$. Show that $X \to \operatorname{Spec} A$ is an open embedding in an open neighborhood of any chosen point $p \in X$, as follows. Choose $r \in R$ such that $p \in D(r) \subset X$. Notice that if $X_r = \{q \in X : r(q) \neq 0\}$, then $\Gamma(X_r, \mathscr{O}_X) = \Gamma(X, \mathscr{O}_X)_r$ by Exercise 15.4.N, using the fact that X is quasicompact and quasiseparated. Use this to show that the map $X_r \to \operatorname{Spec} A_r$ is an isomorphism.

It is not hard to show that X is quasiaffine if and only if $\mathscr{O}_X$ is ample, but we won't use this fact.

A morphism $\pi \colon X \to Y$ is **quasiaffine** if the inverse image of every affine open subset of Y is a quasiaffine scheme. By Exercise 17.3.J, this is equivalent to π being quasicompact and separated, and the natural map $X \to \mathcal{S}pec\, \pi_* \mathscr{O}_X$ being a quasicompact open embedding. This implies that the notion of quasiaffineness is local on the target (may be checked on an open cover), and also affine-local on the target (one may choose an affine cover, and check that the preimages of these open sets are quasiaffine). Quasiaffine morphisms are preserved by base change: if a morphism $X \hookrightarrow Z$ over Y is a quasicompact open embedding into a scheme affine over Y, then for any $W \to Y$, $X \times_Y W \hookrightarrow Z \times_Y W$ is a quasicompact open embedding into a scheme affine over W. (Interestingly, Exercise 17.3.J is *not* the right tool to use to show this base change property.)

One may readily check that quasiaffine morphisms are preserved by composition, [Stacks, tag 01SN] (so quasiaffine morphisms form a "reasonable class"). Thus quasicompact locally closed embeddings are quasiaffine. If X is affine, then $X \to Y$ is quasiaffine if and only if it is quasicompact (as the preimage of any affine open subset of Y is an open subset of an affine scheme, namely, X). In particular, from the Cancellation Theorem 11.1.1 for quasicompact morphisms, any morphism from an affine scheme to a quasiseparated scheme is quasiaffine.

Chapter 18

Čech Cohomology of Quasicoherent Sheaves

This topic is surprisingly simple and elegant. You may think cohomology must be complicated, and that this is why it appears so late in the book. But you will see that we need very little background. After defining schemes, we could have immediately defined quasicoherent sheaves, and then defined cohomology, and verified that it had many useful properties.

18.1 (Desired) Properties of Cohomology

Rather than immediately defining cohomology of quasicoherent sheaves, we first discuss why we care, and what properties it should have.

As $\Gamma(X, \cdot)$ is a left-exact functor, if $0 \to \mathscr{F} \to \mathscr{G} \to \mathscr{H} \to 0$ is a short exact sequence of quasicoherent sheaves on a scheme X, then

$$0 \longrightarrow \mathscr{F}(X) \longrightarrow \mathscr{G}(X) \longrightarrow \mathscr{H}(X)$$

is exact. We dream that this sequence continues to the right, giving a long exact sequence. More explicitly, there should be some covariant functors H^i ($i \geq 0$) from quasicoherent sheaves on X to groups such that H^0 is the global section functor Γ, and so that there is a "long exact sequence in cohomology."

(18.1.0.1)
$$0 \longrightarrow H^0(X, \mathscr{F}) \longrightarrow H^0(X, \mathscr{G}) \longrightarrow H^0(X, \mathscr{H})$$
$$\longrightarrow H^1(X, \mathscr{F}) \longrightarrow H^1(X, \mathscr{G}) \longrightarrow H^1(X, \mathscr{H}) \longrightarrow \cdots$$

(In general, whenever we see a left-exact or right-exact functor, we should hope for this, and in good cases our dreams will come true. The machinery behind this usually involves *derived functors*, which we will discuss in Chapter 23.)

Before defining cohomology groups of quasicoherent sheaves explicitly, we first describe their important properties, which are in some ways more important than the formal definition. The boxed properties will be the important ones.

Suppose X is a *separated and quasicompact* A-scheme. For each quasicoherent sheaf $\mathscr{F}$ on X, we will define A-modules $H^i(X, \mathscr{F})$. In particular, if $A = k$, they are k-vector spaces. In this case, we define $h^i(X, \mathscr{F}) = \dim_k H^i(X, \mathscr{F})$ (where k is left implicit on the left side).

(i) Each $\boxed{H^i \text{ is a covariant functor } QCoh_X \to Mod_A}$.

(ii) The functor H^0 is identified with the functor Γ: $\boxed{H^0(X, \mathscr{F}) = \Gamma(X, \mathscr{F}),}$ and the covariance of **(i)** for $i = 0$ is just the usual covariance for Γ ($\mathscr{F} \to \mathscr{G}$ induces $\Gamma(X, \mathscr{F}) \to \Gamma(X, \mathscr{G})$).

(iii) If $0 \to \mathscr{F} \to \mathscr{G} \to \mathscr{H} \to 0$ is a short exact sequence of quasicoherent sheaves on X, then we have a $\boxed{\text{long exact sequence}}$ (18.1.0.1). The maps $H^i(X, \mathscr{F}) \to H^i(X, \mathscr{G})$ come from covariance, and similarly for $H^i(X, \mathscr{G}) \to H^i(X, \mathscr{H})$. The *connecting homomorphisms* $H^i(X, \mathscr{H}) \to H^{i+1}(X, \mathscr{F})$ will have to be defined.

(iv) If $\pi \colon X \to Y$ is any morphism of quasicompact separated A-schemes, and $\mathscr{F}$ is a quasicoherent sheaf on X, then there is a natural morphism $\boxed{H^i(Y, \pi_*\mathscr{F}) \to H^i(X, \mathscr{F})}$ extending $\Gamma(Y, \pi_*\mathscr{F}) \to \Gamma(X, \mathscr{F})$. (Note that π is quasicompact and separated by the Cancellation Theorem 11.1.1 for quasicompact and separated morphisms, taking $Z = \operatorname{Spec} A$ in the statement of the Cancellation Theorem, so $\pi_*\mathscr{F}$ is indeed a quasicoherent sheaf by Exercise 14.4.C.)

We will later see this as part of a larger story, the *Leray spectral sequence* (Theorem 23.4.5). If $\mathscr{G}$ is a quasicoherent sheaf on Y, then setting $\mathscr{F} := \pi^*\mathscr{G}$ and using the adjunction map $\mathscr{G} \to \pi_*\pi^*\mathscr{G}$ and covariance (property **(i)**) gives a natural **pullback map** $H^i(Y, \mathscr{G}) \to H^i(X, \pi^*\mathscr{G})$ (via $H^i(Y, \mathscr{G}) \to H^i(Y, \pi_*\pi^*\mathscr{G}) \to H^i(X, \pi^*\mathscr{G})$) extending $\Gamma(Y, \mathscr{G}) \to \Gamma(X, \pi^*\mathscr{G})$. In this way, H^i is a "contravariant functor in the space."

(v) If $\pi: X \to Y$ is an affine morphism, and $\mathscr{F}$ is a quasicoherent sheaf on X, the natural map of **(iv)** is an isomorphism: $\boxed{H^i(Y, \pi_*\mathscr{F}) \xrightarrow{\sim} H^i(X, \mathscr{F}).}$ When π is a closed embedding and $Y = \mathbb{P}_A^N$, this isomorphism translates calculations on arbitrary projective A-schemes to calculations on $\mathbb{P}_A^N$.

(vi) (*affine cover cohomology vanishing*) If X can be covered by n affine open sets, then $\boxed{H^i(X, \mathscr{F}) = 0}$ for $i \geq n$ for all $\mathscr{F}$. In particular, on affine schemes, all higher ($i > 0$) quasicoherent cohomology groups vanish. The vanishing of H^1 in this case, along with the long exact sequence **(iii)**, implies that Γ is an exact functor for quasicoherent sheaves on affine schemes, something we already knew (Exercise 6.3.A).

18.1.1. *Remarks (not part of property* (vi)). It is also true that if $\dim X = n$, then $H^i(X, \mathscr{F}) = 0$ for all $i > n$ and for all $\mathscr{F}$ (**dimensional cohomology vanishing**). We will prove this for projective k-schemes (Theorem 18.2.6) and even quasiprojective k-schemes (Exercise 22.4.T). See §18.2.7 for discussion of the general case.

The converse to **(vi)** in the case when $n = 1$ is Serre's *cohomological criterion for affineness*; see §18.8.4.

Let's get back to our list.

(vii) The functor H^i behaves well under direct sums, and more generally under filtered colimits: $\boxed{H^i(X, \operatorname{colim} \mathscr{F}_j) = \operatorname{colim} H^i(X, \mathscr{F}_j).}$

(viii) We will also identify the cohomology of all $\mathscr{O}(m)$ on $\mathbb{P}_A^n$:

18.1.2. Theorem —

(a) $H^0(\mathbb{P}_A^n, \mathscr{O}_{\mathbb{P}_A^n}(m))$ *is a free A-module of rank* $\binom{n+m}{m}$ *if* $m \geq 0$.

(b) $H^n(\mathbb{P}_A^n, \mathscr{O}_{\mathbb{P}_A^n}(m))$ *is a free A-module of rank* $\binom{-m-1}{-n-m-1}$ *if* $m \leq -n-1$.

(c) $H^i(\mathbb{P}_A^n, \mathscr{O}_{\mathbb{P}_A^n}(m)) = 0$ *otherwise.*

We have already shown the first statement in Essential Exercise 15.1.C (see also §15.1.2). We saw there (and in Exercise 15.1.D) that it is advantageous to interpret $H^0(\mathbb{P}_A^n, \mathscr{O}_{\mathbb{P}_A^n}(m))$ as the degree m homogeneous polynomials in $A[x_0, \ldots, x_n]$. Similarly, we will find that $H^n(\mathbb{P}_A^n, \mathscr{O}_{\mathbb{P}_A^n}(m))$ can be interpreted as the homogeneous degree $-m$ terms of $(x_0 x_1 \cdots x_n)^{-1} A[x_0^{-1}, \ldots, x_n^{-1}]$ (see Remark 18.3.5).

Theorem 18.1.2 has a number of features that will be the first appearances of facts that we will prove later.

- The cohomology of these bundles vanish in degree above n (**(vi)** above, "affine cover cohomology vanishing").
- These cohomology groups are always *finitely generated* A-modules. This will be true for all coherent sheaves on projective A-schemes (Theorem 18.1.3(i)), and indeed (with more work) on proper A-schemes (Grothendieck's Coherence Theorem 18.9.1).
- The top cohomology group vanishes for $m > -n-1$. We will later see this as an example of *Kodaira vanishing*; see §21.5.7.

If A is a field k:

- The top cohomology group is one-dimensional for $m = -n-1$. This is the first appearance of the *canonical* or *dualizing sheaf*.

- There is a natural perfect pairing

$$H^i(\mathbb{P}^n, \mathscr{O}(m)) \times H^{n-i}(\mathbb{P}^n, \mathscr{O}(-n-1-m)) \longrightarrow H^n(\mathbb{P}^n, \mathscr{O}(-n-1)).$$

  This is the first appearance of *Serre duality*. (For the case $n = 1$, see Example 18.5.4.)
- The alternating sum $\sum (-1)^i h^i(\mathbb{P}^n, \mathscr{O}(m))$ is a polynomial in m. This is the first example of a *Hilbert polynomial*.

Before proving these facts, let's first use them to prove interesting things, as motivation.

By Theorem 16.1.1, for any coherent sheaf $\mathscr{F}$ on $\mathbb{P}^n_A$ we can find a surjection $\mathscr{O}(m)^{\oplus j} \to \mathscr{F}$, which yields the exact sequence

(18.1.2.1) $$0 \longrightarrow \mathscr{G} \longrightarrow \mathscr{O}(m)^{\oplus j} \longrightarrow \mathscr{F} \longrightarrow 0$$

for some coherent sheaf $\mathscr{G}$. We can use this to prove the following.

18.1.3. Theorem — *(i) For any coherent sheaf $\mathscr{F}$ on a projective A-scheme X where A is Noetherian, $H^i(X, \mathscr{F})$ is a coherent (finitely generated) A-module.*
(ii) (Serre vanishing) Furthermore, for $m \gg 0$, $H^i(X, \mathscr{F}(m)) = 0$ for all $i > 0$.

(A slightly fancier version of Serre vanishing will be given in Theorem 18.7.F.)

Proof. Because cohomology of a closed subscheme can be computed on the ambient space ((v) above), we may immediately reduce to the case $X = \mathbb{P}^n_A$.

(i) Consider the long exact sequence:

$$0 \longrightarrow H^0(\mathbb{P}^n_A, \mathscr{G}) \longrightarrow H^0(\mathbb{P}^n_A, \mathscr{O}(m)^{\oplus j}) \longrightarrow H^0(\mathbb{P}^n_A, \mathscr{F}) \longrightarrow$$

$$H^1(\mathbb{P}^n_A, \mathscr{G}) \longrightarrow H^1(\mathbb{P}^n_A, \mathscr{O}(m)^{\oplus j}) \longrightarrow H^1(\mathbb{P}^n_A, \mathscr{F}) \longrightarrow \cdots$$

$$\cdots \longrightarrow H^{n-1}(\mathbb{P}^n_A, \mathscr{G}) \longrightarrow H^{n-1}(\mathbb{P}^n_A, \mathscr{O}(m)^{\oplus j}) \longrightarrow H^{n-1}(\mathbb{P}^n_A, \mathscr{F}) \longrightarrow$$

$$H^n(\mathbb{P}^n_A, \mathscr{G}) \longrightarrow H^n(\mathbb{P}^n_A, \mathscr{O}(m)^{\oplus j}) \longrightarrow H^n(\mathbb{P}^n_A, \mathscr{F}) \longrightarrow 0$$

The exact sequence ends here because $\mathbb{P}^n_A$ is covered by $n + 1$ affine open sets ((vi) above, "affine cover cohomology vanishing"). Then $H^n(\mathbb{P}^n_A, \mathscr{O}(m)^{\oplus j})$ is finitely generated by Theorem 18.1.2(b) and (vii), hence $H^n(\mathbb{P}^n_A, \mathscr{F})$ is finitely generated for *all* coherent sheaves $\mathscr{F}$. Hence, in particular, $H^n(\mathbb{P}^n_A, \mathscr{G})$ is finitely generated. As $H^{n-1}(\mathbb{P}^n_A, \mathscr{O}(m)^{\oplus j})$ is finitely generated, and $H^n(\mathbb{P}^n_A, \mathscr{G})$ is too, we have that $H^{n-1}(\mathbb{P}^n_A, \mathscr{F})$ is finitely generated for all coherent sheaves $\mathscr{F}$. We continue inductively downward.

(ii) Twist (18.1.2.1) by $\mathscr{O}(N)$ for $N \gg 0$. Then

$$H^n(\mathbb{P}^n_A, \mathscr{O}(m + N)^{\oplus j}) = H^n(\mathbb{P}^n_A, \mathscr{O}(m + N))^{\oplus j} = 0$$

(by (vii) above), so $H^n(\mathbb{P}^n_A, \mathscr{F}(N)) = 0$. Translation: For any coherent sheaf, its top cohomology vanishes once you twist by $\mathscr{O}(N)$ for N sufficiently large. Hence this is true for $\mathscr{G}$ as well. Hence from the long exact sequence, $H^{n-1}(\mathbb{P}^n_A, \mathscr{F}(N)) = 0$ for $N \gg 0$. As in (i), we induct downward, until we get that $H^1(\mathbb{P}^n_A, \mathscr{F}(N)) = 0$. (The induction stops here, as it is *not* true that $H^0(\mathbb{P}^n_A, \mathscr{O}(m + N)^{\oplus j}) = 0$ for large N—quite the opposite.) $\square$

In particular, we have proved the following, which we would have cared about even before we knew about cohomology.

382 Chapter 18 Čech Cohomology of Quasicoherent Sheaves

18.1.4. Corollary — *Any projective k-scheme has a finite-dimensional space of functions. More generally, if A is Noetherian and $\mathscr{F}$ is a coherent sheaf on a projective A-scheme, then $H^0(X, \mathscr{F})$ is a coherent A-module.*

(We will generalize this in Theorem 18.7.1.) I want to emphasize how remarkable this proof is. It is a question about global sections, i.e., H^0, which we think of as the most down-to-earth cohomology group, yet the proof is by downward induction for H^n, starting with n large.

Corollary 18.1.4 is true more generally for proper k-schemes, not just projective k-schemes (Grothendieck's Coherence Theorem 18.9.1).

Here are some important consequences. They can also be shown directly, without the use of cohomology, but with much more elbow grease.

18.1.A. CRUCIAL EXERCISE (PUSHFORWARDS OF COHERENT SHEAVES BY PROJECTIVE MORPHISMS ARE COHERENT). Suppose $\pi: X \to Y$ is a projective morphism of locally Noetherian schemes. Show that the pushforward of a coherent sheaf on X is a coherent sheaf on Y. (See Grothendieck's Coherence Theorems 18.7.1 and 18.9.1 for generalizations.)

Finite morphisms are affine (from the definition) and projective (Exercise 17.3.B). We can now show that this is a characterization of finiteness.

18.1.5. Corollary (projective + affine = finite) — *Suppose Y is locally Noetherian. Then a morphism $\pi: X \to Y$ is projective and affine if and only if π is finite.*

We will see in Exercise 18.9.A that the projective hypotheses can be relaxed to proper.

Proof. We already know that finite morphisms are affine (by definition) and projective (Exercise 17.3.B), so we show the converse. Suppose π is projective and affine. By Exercise 18.1.A, $\pi_* \mathscr{O}_X$ is coherent and hence finite type. $\qquad\square$

The following result was promised in §17.3.3, and has a number of useful consequences.

18.1.6. Theorem (projective + finite fibers = finite) — *Suppose $\pi: X \to Y$ is a morphism, with Y Noetherian. Then π is projective and has finite fibers if and only if it is finite. Equivalently, π is projective and quasifinite if and only if it is finite.*

(Recall that quasifinite = finite fibers + finite type. But projective includes finite type.) It is true more generally that (with Noetherian hypotheses) proper + finite fibers = finite; see Theorem 28.5.2.

Proof. We have already shown that finite morphisms are projective (Exercise 17.3.B) and have finite fibers (Exercise 8.3.J). So we now assume that π is projective and has finite fibers, and show that π is finite. We show π is finite near a point $q \in Y$. Fix an affine open neighborhood $\operatorname{Spec} A$ of q in Y. For notational convenience, we rename $Y = \operatorname{Spec} A$ and $X = \pi^{-1}(\operatorname{Spec} A)$. Projectivity of π means we can factor π as

$$
\begin{array}{ccc}
X & \xrightarrow{\text{cl.emb.}} & \mathbb{P}^n_A \\
& \searrow{\scriptstyle \pi} \quad \downarrow{\scriptstyle \rho} & \\
& \operatorname{Spec} A & == Y.
\end{array}
$$

Choose a hypersurface H_q in $\mathbb{P}^n_{\kappa(q)}$ missing $\pi^{-1}(q) \subset X$ (Exercise 12.3.C). Choose any lift of its equation to A, to obtain a hypersurface H in $\mathbb{P}^n_A$ also missing $\pi^{-1}(q)$. See Figure 18.1.) Now $H \cap X$ is closed (in $\mathbb{P}^n_A$, so by the Fundamental Theorem of Elimination Theory 8.4.10, $H' := \pi(H \cap X)$ is closed in $\operatorname{Spec} A$. Also, H' doesn't contain q, so $U := \operatorname{Spec} A \setminus H'$ is an open subset of $Y = \operatorname{Spec} A$ containing q.

Finally, $\pi^{-1}(U) \to U$ is a projective morphism. It is also affine, because it is the composition of two affine morphisms

$$\pi^{-1}(U) \xrightarrow{\text{cl.emb.}} \mathbb{P}_U^n \setminus H \xrightarrow{\rho} U$$

(the second of which is affine by Exercise 4.5.I). We are then done by Corollary 18.1.5. $\qquad\square$

A similar trick was used in the proof of the Rigidity Lemma 11.4.12.

18.1.B. EXERCISE (UPPER SEMICONTINUITY OF FIBER DIMENSION ON THE TARGET, FOR PROJECTIVE MORPHISMS). Use a similar argument as in Theorem 18.1.6 to prove *upper semicontinuity of fiber dimension of projective morphisms*: Suppose $\pi : X \to Y$ is a projective morphism where Y is locally Noetherian (or, more generally, $\mathscr{O}_Y$ is coherent over itself). Show that $\{q \in Y : \dim \pi^{-1}(q) > k\}$ is a Zariski-closed subset of Y. In other words, the dimension of the fiber "jumps over Zariski-closed subsets" of the target. (You can interpret the case $k = -1$ as the fact that projective morphisms are closed, which is basically the Fundamental Theorem of Elimination Theory 8.4.10; cf. §17.3.2.) This exercise is rather important for having a sense of how projective morphisms behave. (The case of varieties was done earlier, in Theorem 12.4.3(b). This approach is much simpler.)

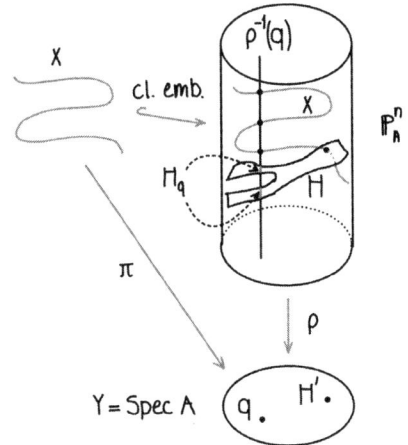

Figure 18.1 *The key construction in the proof of Theorem 18.1.6.*

The final exercise of the section is on a different theme.

18.1.C. EXERCISE. Suppose $0 \to \mathscr{F} \to \mathscr{G} \to \mathscr{H} \to 0$ is an exact sequence of quasicoherent sheaves on X, projective over A, with $\mathscr{F}$ coherent. Show that for $n \gg 0$,

$$0 \longrightarrow H^0(X, \mathscr{F}(n)) \longrightarrow H^0(X, \mathscr{G}(n)) \longrightarrow H^0(X, \mathscr{H}(n)) \longrightarrow 0$$

is also exact. (Hint: For $n \gg 0$, $H^1(X, \mathscr{F}(n)) = 0$.)

18.2 Definitions and Proofs of Key Properties

This section could be read much later; the facts we will use are all stated in the previous section. However, the arguments are not complicated, so you want to read this right away. As you read this, you should go back and check off all the facts in the previous section, to assure yourself that you understand everything promised.

18.2.1. Čech cohomology. Čech cohomology in general settings is defined using a limit over finer and finer covers of a space. In our algebro-geometric setting, the situation is much cleaner, and we can use a single cover.

Suppose X is quasicompact and separated, which is true, for example, if X is quasiprojective over A. In particular, X may be covered by a finite number of affine open sets, and the intersection of any two affine open sets is also an affine open set (by separatedness, Proposition 11.2.13). We will use quasicompactness and separatedness only in order to ensure these two nice properties.

Suppose $\mathscr{F}$ is a quasicoherent sheaf, and $\mathscr{U} = \{U_i\}_{i=1}^n$ is a *finite* collection of affine open sets covering X. For $I \subset \{1, \ldots, n\}$ define $U_I = \cap_{i \in I} U_i$, which is affine by the separated hypothesis. (Here is a strong analogy for those who have seen cohomology in other contexts: cover a topological space X with a finite number of open sets U_i, such that all intersections $\cap_{i \in I} U_i$ are contractible.)

Consider the **Čech complex**

$$(18.2.1.1) \qquad 0 \longrightarrow \prod_{\substack{|I| = 1 \\ I \subset \{1, \ldots, n\}}} \mathscr{F}(U_I) \longrightarrow \cdots \longrightarrow$$

$$\prod_{\substack{|I| = i \\ I \subset \{1, \ldots, n\}}} \mathscr{F}(U_I) \longrightarrow \prod_{\substack{|I| = i + 1 \\ I \subset \{1, \ldots, n\}}} \mathscr{F}(U_I) \longrightarrow \cdots .$$

The maps are defined as follows. The map from $\mathscr{F}(U_I) \to \mathscr{F}(U_J)$ is 0 unless $I \subset J$, i.e., $J = I \cup \{j\}$. If j is the kth element of J, then the map is $(-1)^{k-1}$ times the restriction map res_{U_I, U_J}.

18.2.A. EASY EXERCISE (FOR THOSE WHO HAVEN'T SEEN ANYTHING LIKE THE ČECH COMPLEX BEFORE). Show that the Čech complex is indeed a complex, i.e., that the composition of two consecutive arrows is 0.

Define $H^i_{\mathscr{U}}(X, \mathscr{F})$ to be the ith cohomology group of the complex (18.2.1.1). (The indexing starts with $i = 0$.) Note that if X is an A-scheme, then $H^i_{\mathscr{U}}(X, \mathscr{F})$ is an A-module. We have almost succeeded in defining the Čech cohomology group H^i, except our definition a priori depends on a choice of a cover $\mathscr{U}$. Note that $H^i_{\mathscr{U}}(X, \cdot)$ is clearly a covariant functor $QCoh_X \to Mod_A$.

18.2.B. EASY EXERCISE. Identify $H^0_{\mathscr{U}}(X, \mathscr{F})$ with $\Gamma(X, \mathscr{F})$. (Hint: Use the sheaf axioms for $\mathscr{F}$.)

18.2.C. EXERCISE. Suppose

$$(18.2.1.2) \qquad 0 \longrightarrow \mathscr{F} \longrightarrow \mathscr{G} \longrightarrow \mathscr{H} \longrightarrow 0$$

is a short exact sequence of sheaves of abelian groups on a topological space, and $\mathscr{U}$ is a finite open cover such that on any intersection U_I of open subsets in $\mathscr{U}$, the map $\Gamma(U_I, \mathscr{G}) \to \Gamma(U_I, \mathscr{H})$ is surjective. Show that we get a "long exact sequence of cohomology for $H^i_{\mathscr{U}}$" (where we take the same definition of $H^i_{\mathscr{U}}$). In our situation, where X is a quasicompact separated A-scheme, and (18.2.1.2) is a short exact sequence of quasicoherent sheaves on X, show that we get a long exact sequence for the A-modules $H^i_{\mathscr{U}}$.

In the proof of Theorems 18.2.4 and 18.7.1, we will make use of the fact that your construction of the connecting homomorphism will "commute with localization of A." More precisely, we will need the following.

18.2.D. EXERCISE. Suppose we are given a short exact sequence (18.2.1.2) of quasicoherent sheaves on a quasicompact separated A-scheme $\pi\colon X \to \operatorname{Spec} A$, a cover $\mathscr{U}$ of X by affine open sets, and some $f \in A$. The restriction of the sets of $\mathscr{U}$ to X_f yields an affine open cover $\mathscr{U}'$ of $X_f = \pi^{-1}(D(f))$. Identify the long exact sequence associated to (18.2.1.2) using $H^i_{\mathscr{U}}$, localized at f, with the long exact sequence associated to the restriction of (18.2.1.2) to X_f, using the affine open cover $\mathscr{U}'$. (First check that the maps given by covariance, such as $H^i_{\mathscr{U}}(\mathscr{F}) \to H^i_{\mathscr{U}}(\mathscr{G})$, "commute with localization," and then check that the connecting homomorphisms do as well.)

18.2.2. Theorem/definition — *Our standing assumption is that X is quasicompact and separated. $H^i_{\mathscr{U}}(X, \mathscr{F})$ is independent of the choice of (finite) cover $\{U_i\}$. More precisely, for any two covers $\{U_i\} \subset \{V_i\}$, the maps $H^i_{\{V_i\}}(X, \mathscr{F}) \to H^i_{\{U_i\}}(X, \mathscr{F})$ induced by the natural map of Čech complexes (18.2.1.1) are isomorphisms. Define the* **Čech cohomology group** $H^i(X, \mathscr{F})$ *to be this group.*

If you are unsure of what the "natural map of Čech complexes" is, by (18.2.3.1) it should become clear.

18.2.3. For experts: Maps of complexes inducing isomorphisms on cohomology groups are called **quasiisomorphisms**. We are actually getting a finer invariant than cohomology out of this construction; we are getting an element of the *derived category of A-modules*.

Proof. We need only prove the result when $|\{V_i\}| = |\{U_i\}| + 1$. We will show that if $\{U_i\}_{1 \le i \le n}$ is a cover of X, and U_0 is any other affine open set, then the map $H^i_{\{U_i\}_{0 \le i \le n}}(X, \mathscr{F}) \to H^i_{\{U_i\}_{1 \le i \le n}}(X, \mathscr{F})$ is an isomorphism. Consider the exact sequence of complexes

(18.2.3.1)

Throughout, $I \subset \{0, \dots, n\}$. The bottom two rows are Čech complexes with respect to two covers, and the map between them induces the desired map on cohomology. We get a long exact sequence of cohomology from this short exact sequence of complexes (Theorem 1.5.8). Thus we wish to show that the top row is exact and thus has vanishing cohomology. (Note that $U_0 \cap U_j$ is affine by our separatedness hypothesis, Proposition 11.2.13.) But the ith cohomology of the top row is precisely $H^{i-1}_{\{U_j \cap U_0\}_{j>0}}(U_0, \mathscr{F})$ except at the start, where we get 0 (because the complex starts off $0 \to \mathscr{F}(U_0) \to \prod_{j=1}^n \mathscr{F}(U_0 \cap U_j)$). So it suffices to show that higher Čech groups of affine schemes are 0. Hence we are done by the following result. □

18.2.4. Theorem — *The higher Čech cohomology $H^i_{\mathscr{U}}(X, \mathscr{F})$ of an affine A-scheme X vanishes (for any affine cover $\mathscr{U}$, $i > 0$, and quasicoherent $\mathscr{F}$).*

This is another "partition of unity" argument, in the spirit of the proof of Theorem 4.1.2.

Proof. We want to show that the "extended" complex

(18.2.4.1) $$0 \longrightarrow \mathscr{F}(X) \longrightarrow \prod_{|I|=1} \mathscr{F}(U_I) \longrightarrow \prod_{|I|=2} \mathscr{F}(U_I) \longrightarrow \cdots$$

(where the global sections $\mathscr{F}(X)$ have been appended to the start) has no cohomology, i.e., is exact. (Incidentally, this extended complex is called the **Čech resolution** of $\mathscr{F}(X)$, with respect to this choice of cover $\mathscr{U}$.) We do this with a trick.

Suppose first that some U_i, say U_0, is X. We deal with this case by sleight of hand. The complex (18.2.4.1) is the middle row of the following short exact sequence of complexes.

(18.2.4.2)

The top row is the same as the bottom row, slid over by 1. The corresponding long exact sequence of cohomology shows that the central row has vanishing cohomology. (You should show that the "connecting homomorphism" on cohomology is indeed an isomorphism.) This might remind you of the *mapping cone* construction (Exercise 1.6.E). (Here is a trick to accomplish the goal of this paragraph from a different point of view. Let id be the identity map between complex (18.2.4.1) and itself, and let 0 be the zero map between the complex (18.2.4.1) and itself. Describe an explicit homotopy between id and 0; cf. §1.5.7. The homotopy will use the "same idea" as (18.2.3.1). Then the identity map on homology equals the zero map on homology, Exercise 1.5.E, so the homology must be zero. This idea can be enlightening because it is used in other circumstances.)

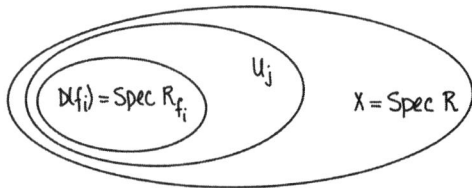

Figure 18.2 *The final step in the proof of Theorem 18.2.2.*

Finally, we prove the general case by another sleight of hand. Say $X = \operatorname{Spec} R$. We wish to show that the complex of A-modules (18.2.4.1) is exact. It is also a complex of R-modules, so we wish to show that the complex of R-modules (18.2.4.1) is exact. To show that it is exact, it suffices to show that for a cover of $\operatorname{Spec} R$ by distinguished open sets $D(f_i)$ ($1 \leq i \leq r$) (i.e., $(f_1, \ldots, f_r) = 1$ in R) the complex is exact. (Translation: Exactness of a sequence of sheaves may be checked locally.) We choose a cover so that each $D(f_i)$ is contained in one of our affine open subsets U_j (see Figure 18.2). The complex (18.2.4.1) localized at f_i is the Čech complex for the affine cover of $D(f_i)$ by $\{U_k \cap D(f_i)\}$. We want to show that this complex (for our fixed i and j) is exact. But this is the case dealt with in the previous paragraph—one of the affine open sets, $U_j \cap D(f_i)$, equals $D(f_i)$. □

We have now proved properties **(i)**–**(iii)** of the previous section. The "affine cover cohomology vanishing" property **(vi)** is also straightforward: if X is covered by n affine open sets, use these as the cover $\mathscr{U}$, and notice that the Čech complex ends by the nth step.

18.2.E. EXERCISE (PROPERTY (V)). Suppose $\pi \colon X \to Y$ is an affine morphism, and Y is a quasicompact and separated A-scheme (and hence X is, too, as affine morphisms are both quasicompact and separated). If $\mathscr{F}$ is a quasicoherent sheaf on X, describe a natural isomorphism $H^i(Y, \pi_*\mathscr{F}) \xleftrightarrow{\sim} H^i(X, \mathscr{F})$. (Hint: If $\mathscr{U}$ is an affine cover of Y, "$\pi^{-1}(\mathscr{U})$" is an affine cover of X. Use these covers to compute the cohomology of $\mathscr{F}$.)

18.2.F. EXERCISE (PROPERTY (IV)). Suppose $\pi \colon X \to Y$ is any quasicompact separated morphism, $\mathscr{F}$ is a quasicoherent sheaf on X, and Y is a quasicompact separated A-scheme. The hypotheses on π ensure that $\pi_*\mathscr{F}$ is a quasicoherent sheaf on Y. Describe a natural morphism $H^i(Y, \pi_*\mathscr{F}) \to H^i(X, \mathscr{F})$ extending $\Gamma(Y, \pi_*\mathscr{F}) \to \Gamma(X, \mathscr{F})$. (Aside: This morphism is an isomorphism for $i = 0$, but need not be an isomorphism for higher i: consider $i = 1$, $X = \mathbb{P}^1_k$, $\mathscr{F} = \mathscr{O}(-2)$, and let Y be a point $\operatorname{Spec} k$.)

18.2.G. EXERCISE. Prove Property **(vii)** of the previous section, that "cohomology commutes with filtered colimits." (This can be done by hand. Hint: In the category of modules over a ring, taking the colimit over a filtered sets is an exact functor, §1.5.14.)

We have now proved all of the properties of the previous section, except for **(viii)**, which we will get to in §18.3.

18.2.5. Useful facts about cohomology for k-schemes.

18.2.H. EXERCISE (COHOMOLOGY AND CHANGE OF BASE FIELD). Suppose X is a quasicompact separated k-scheme, and $\mathscr{F}$ is a quasicoherent sheaf on X. Give an isomorphism

$$H^i(X, \mathscr{F}) \otimes_k K \xleftrightarrow{\sim} H^i(X \times_{\operatorname{Spec} k} \operatorname{Spec} K, \mathscr{F} \otimes_k K)$$

for all i, where K/k is any field extension. Here $\mathscr{F} \otimes_k K$ means the pullback of $\mathscr{F}$ to $X \times_{\text{Spec } k} \text{Spec K}$. Hence $h^i(X, \mathscr{F}) = h^i(X \times_{\text{Spec } k} \text{Spec K}, \mathscr{F} \otimes_k K)$. (This is useful for relating facts about k-schemes to facts about schemes over algebraically closed fields. Your proof might use vector spaces—i.e., linear algebra—in a fundamental way. If it doesn't, you may prove something more general, if $k \to K$ is replaced by a flat ring map $B \to A$. Recall that $B \to A$ is flat if $\otimes_B A$ is an exact functor $Mod_B \to Mod_A$. A hint for this harder exercise: The FHHF Theorem, Exercise 1.5.I. See Exercise 18.7.B(b) for the next generalization of this.)

18.2.I. EXERCISE (BASE-POINT-FREENESS IS INDEPENDENT OF EXTENSION OF BASE FIELD). Suppose X is a quasicompact separated k-scheme, $\mathscr{L}$ is an invertible sheaf on X, and K/k is a field extension. Show that $\mathscr{L}$ is base-point-free if and only if its pullback to $X \times_{\text{Spec } k} \text{Spec K}$ is base-point-free. (Hint: Exercise 18.2.H with $i = 0$ implies that a basis of sections of $\mathscr{L}$ over k becomes, after tensoring with K, a basis of sections of $\mathscr{L} \otimes_k K$.)

18.2.6. Theorem (dimensional cohomology vanishing for quasicoherent sheaves on projective k-schemes) — *Suppose X is a projective k-scheme, and $\mathscr{F}$ is a quasicoherent sheaf on X. Then $H^i(X, \mathscr{F}) = 0$ for $i > \dim X$.*

In other words, cohomology vanishes above the dimension of X.

Proof. Suppose $X \hookrightarrow \mathbb{P}^N$, and let $n = \dim X$. We show that X may be covered by $n + 1$ affine open sets. Exercise 12.3.D shows that there are $n + 1$ effective Cartier divisors on $\mathbb{P}^N$ such that their complements $U_0, \ldots, U_n$ cover X. Then U_i is affine, so $U_i \cap X$ is affine, and thus we have covered X with $n + 1$ affine open sets. $\square$

(It turns out that $n + 1$ affine open sets are always necessary. One way of proving this is by showing that the complement of a dense affine open subset is always pure codimension 1; see, for example, [RotV, Lem. 2.3].)

18.2.7.* Dimensional cohomology vanishing more generally. Using the theory of blowing up, Theorem 18.2.6 can be extended to quasiprojective k-schemes; see §22.4.15. Dimensional cohomology vanishing is even true in much greater generality. To state it, we need to define cohomology with the more general machinery of derived functors (Chapter 23). If X is a Noetherian topological space (§3.6.14) and $\mathscr{F}$ is any sheaf of abelian groups on X, we have $H^i(X, \mathscr{F}) = 0$ for all $i > \dim X$. (Grothendieck sketches his elegant proof in [GrS, pp. 29–30]; see [Ha1, Theorem III.2.7] for a more detailed explanation.) In particular, if X is a k-variety of dimension n, we *always* have dimensional cohomology vanishing, even for crazy varieties that can't be covered with $n + 1$ affine open subsets (see §22.4.15).

18.3   Cohomology of Line Bundles on Projective Space

We finally prove the last promised basic fact about cohomology, property **(viii)** of §18.1, Theorem 18.1.2, on the cohomology of line bundles on projective space. But first we discuss a seemingly different problem, of the cohomology of the structure sheaf of affine space minus the origin.

18.3.1. *The problem.* For any ring A, let $X = \mathbb{A}_A^{n+1} \setminus V(x_0, \ldots, x_n)$ for $n \geq 1$. Here the coordinates on $\mathbb{A}^{n+1}$ are $x_0, \ldots, x_n$, and the motivation for the indexing will soon become clear. Find $H^i(X, \mathscr{O}_X)$.

18.3.2. *The answer.* We will show that $H^i(X, \mathscr{O}_X) = 0$ unless $i = 0$ (in which case it will be $A[x_0, \ldots, x_n]$) or $i = n$. We will see that $H^n(X, \mathscr{O}_X)$ can be identified with those "Laurent polynomials" of the form $(x_0 x_1 \cdots x_n)^{-1} A[x_0^{-1}, x_1^{-1}, \ldots, x_n^{-1}]$. (This gives another proof that $\mathbb{A}_k^2 \setminus \{(0,0)\}$ is not affine; cf. §4.4.3.) (Cf. Serre's cohomological criterion for affineness, §18.8.4.)

There is a natural open cover of X by $\{U_i\}$ where $U_i = D(x_i)$, and we use the corresponding Čech complex.

We discuss the case $n = 2$, because all of the complications present themselves here. In this case the Čech complex is

$$(18.3.2.1) \quad 0 \longrightarrow A[x_0, x_1, x_2, x_0^{-1}] \times A[x_0, x_1, x_2, x_1^{-1}] \times A[x_0, x_1, x_2, x_2^{-1}] \longrightarrow$$

$$A[x_0, x_1, x_2, x_0^{-1}, x_1^{-1}] \times A[x_0, x_1, x_2, x_1^{-1}, x_2^{-1}] \times A[x_0, x_1, x_2, x_0^{-1}, x_2^{-1}]$$

$$\longrightarrow A[x_0, x_1, x_2, x_0^{-1}, x_1^{-1}, x_2^{-1}] \longrightarrow 0.$$

We extend (18.3.2.1) by replacing the "$0 \to$" on the left by "$0 \to A[x_0, x_1, x_2] \to$":

$$(18.3.2.2)$$

$$\overset{H^0}{} \qquad \overset{U_0 \, U_1 \, U_2}{} \qquad \overset{U_{012}}{}$$

$$0 \longrightarrow A[x_0, x_1, x_2] \longrightarrow \cdots \longrightarrow \cdots \longrightarrow A[x_0, x_1, x_2, x_0^{-1}, x_1^{-1}x_2^{-1}] \longrightarrow 0.$$

18.3.A. EXERCISE. Show that if (18.3.2.2) is exact, except that at U_{012} the cohomology / cokernel is $x_0^{-1}x_1^{-1}x_2^{-1} A[x_0^{-1}, x_1^{-1}, x_2^{-1}]$, then Answer 18.3.2 is correct for $n = 2$.

Because the maps in (18.3.2.2) preserve multidegree (degrees of each x_i independently), we can study exactness of (18.3.2.2) monomial by monomial.

The "3 negative exponents" case. Consider first the monomial $x_0^{a_0} x_1^{a_1} x_2^{a_2}$, where the exponents a_i are all negative. Then (18.3.2.2) in this multidegree is:

$$0 \longrightarrow 0_{H^0} \longrightarrow 0_0 \times 0_1 \times 0_2 \longrightarrow 0_{01} \times 0_{12} \times 0_{02} \longrightarrow A_{012} \longrightarrow 0.$$

Here the subscripts serve only to remind us which "Čech" terms the factors correspond to. For example, A_{012} corresponds to the coefficient of $x_0^{a_0} x_1^{a_1} x_2^{a_2}$ in $A[x_0, x_1, x_2, x_0^{-1}, x_1^{-1}, x_2^{-1}]$. Clearly, this complex only has (co)homology at the U_{012} spot, as desired.

The "2 negative exponents" case. Consider next the case where *two* of the exponents, say, a_0 and a_1, are negative. Then the complex in this multidegree is

$$0 \longrightarrow 0_{H^0} \longrightarrow 0_0 \times 0_1 \times 0_2 \longrightarrow A_{01} \times 0_{12} \times 0_{02} \longrightarrow A_{012} \longrightarrow 0,$$

which is clearly exact.

The "1 negative exponent" case. We next consider the case where *one* of the exponents, say, a_0, is negative. Then the complex in this multidegree is

$$0 \longrightarrow 0_{H^0} \longrightarrow A_0 \times 0_1 \times 0_2 \longrightarrow A_{01} \times 0_{12} \times A_{02} \longrightarrow A_{012} \longrightarrow 0.$$

With a little thought (paying attention to the signs on the arrows $A \to A$), you will see that it is exact. (The subscripts, by reminding us of the subscripts in the original Čech complex, remind us what signs to take in the maps.)

The "0 negative exponent" case. Finally, consider the case where *none* of the exponents are negative. Then the complex in this multidegree is

$$(18.3.2.3) \quad 0 \longrightarrow A_{H^0} \longrightarrow A_0 \times A_1 \times A_2 \longrightarrow A_{01} \times A_{12} \times A_{02} \longrightarrow A_{012} \longrightarrow 0.$$

We wish to show that this is exact. We use the same strategy as we used at (18.2.4.2). We write (18.3.2.3) as the middle of a short exact sequence of complexes:

(18.3.2.4)

$$
\begin{array}{ccccccccc}
0 & \longrightarrow & 0 & \longrightarrow & A_2 & \longrightarrow & A_{02} \times A_{12} & \longrightarrow & A_{012} & \longrightarrow & 0 \\
& & \downarrow & & \downarrow & & \downarrow & & \downarrow & & \downarrow \\
0 & \longrightarrow & A_{H^0} & \longrightarrow & A_0 \times A_1 \times A_2 & \longrightarrow & A_{01} \times A_{12} \times A_{02} & \longrightarrow & A_{012} & \longrightarrow & 0 \\
& & \downarrow & & \downarrow & & \downarrow & & \downarrow & & \downarrow \\
0 & \longrightarrow & A_{H^0} & \longrightarrow & A_0 \times A_1 & \longrightarrow & A_{01} & \longrightarrow & 0 & \longrightarrow & 0 \;.
\end{array}
$$

Thus we get a long exact sequence in cohomology (Theorem 1.5.8). But the top and bottom rows are exact (basically from the "1-negative" case), i.e., cohomology-free, so the middle row must be exact too. (Alternatively, as with (18.2.4.2), let id be the identity map between complex (18.3.2.3) and itself. Describe an explicit homotopy between id and the zero map. Then the identity map on homology equals the zero map on homology, Exercise 1.5.E, which means the homology is zero.)

18.3.B. EXERCISE. Prove that Answer 18.3.2 is true for all $n \geq 1$. (I could of course just have given you the proof for general n, but seeing the argument in action may be enlightening. In particular, your argument may be much shorter. For example, the "2-negative" case could be done in the same way as the "1-negative" case, so you will not need $n+1$ separate cases if you set things up carefully.)

18.3.3. *Remark.* This argument is basically the proof that the reduced homology of the boundary of a simplex S (known in some circles as a "sphere") is 0, unless S is the empty set, in which case it is one-dimensional. The "empty set" case corresponds to the "3-negative" case.

We are now ready to prove Theorem 18.1.2.

18.3.4. *Proof of Theorem 18.1.2.* We use the standard cover $U_0 = D(x_0), \ldots, U_n = D(x_n)$ of $\mathbb{P}_A^n$.

We come to the central trick (Exercise 18.3.C): the Čech complex for $\mathscr{O}(m)$ is the degree m part of the Čech complex for $H^i(\mathbb{A}_A^{n+1}, \mathscr{O}_X)$! (You can stare at the $n=2$ case (18.3.2.1) if it helps.)

18.3.C. EXERCISE (CF. REMARK 18.3.5). If $I \subset \{0, \ldots, n\}$, then give an isomorphism (of A-modules) of $\Gamma(U_I, \mathscr{O}(m))$ with the homogeneous degree m Laurent monomials (in $x_0, \ldots, x_n$, with coefficients in A) where each x_i for $i \notin I$ appears with nonnegative degree. Your construction should be such that the restriction map $\Gamma(U_I, \mathscr{O}(m)) \to \Gamma(U_J, \mathscr{O}(m))$ ($I \subset J$) corresponds to the natural inclusion: a Laurent polynomial in $\Gamma(U_I, \mathscr{O}(m))$ maps to the *same* Laurent polynomial in $\Gamma(U_J, \mathscr{O}(m))$.

18.3.D. EXERCISE. Prove Theorem 18.1.2. □

18.3.5. *Remark.* Exercises 15.1.C and 15.1.D interpreted $H^0(\mathbb{P}_A^n, \mathscr{O}_{\mathbb{P}_A^n}(m))$ as the homogeneous degree m polynomials in $x_0, \ldots, x_n$ (with A-coefficients). We can now state this in a better way. Let

$$\pi : \mathbb{A}_A^{n+1} \setminus V(x_0, \ldots, x_n) \longrightarrow \mathbb{P}_A^n$$

be the morphism of Exercise 7.3.F and Example 7.3.12. You can show that the induced map

$$\pi^* : H^n(\mathbb{P}_A^n, \mathscr{O}(m)) \longrightarrow H^n(\mathbb{A}_A^{n+1} \setminus V(x_0, \ldots, x_n), \mathscr{O})$$

is an inclusion, and "identify the image."

18.4 Riemann–Roch, and Arithmetic Genus

We have seen some powerful uses of Čech cohomology, to prove things about spaces of global sections, and to prove Serre vanishing. We will now see some classical constructions come out very quickly and cheaply.

In this section, we will work over a field k. Suppose $\mathscr{F}$ is a coherent sheaf on a projective k-scheme X. Recall the notation (§18.1) $h^i(X, \mathscr{F}) := \dim_k H^i(X, \mathscr{F})$. By Theorem 18.1.3, $h^i(X, \mathscr{F})$ is finite. (The arguments in this section will extend without change to proper X once we have this finiteness for proper morphisms, by Grothendieck's Coherence Theorem 18.9.1.) Define the **Euler characteristic** of $\mathscr{F}$ by

$$\chi(X, \mathscr{F}) := \sum_{i=0}^{\dim X} (-1)^i h^i(X, \mathscr{F}).$$

We will see repeatedly here and later that Euler characteristics behave better than individual cohomology groups. As one sign, notice that for fixed n, and $m \geq 0$,

$$h^0(\mathbb{P}^n_k, \mathscr{O}(m)) = \binom{n+m}{m} = \frac{(m+1)(m+2)\cdots(m+n)}{n!}.$$

Notice that the expression on the right is a polynomial in m of degree n. (For later reference, notice also that the leading term is $m^n/n!$.) But it is not true that

$$h^0(\mathbb{P}^n_k, \mathscr{O}(m)) = \frac{(m+1)(m+2)\cdots(m+n)}{n!}$$

for *all* m—it breaks down for $m \leq -n-1$. Still, you can check (using Theorem 18.1.2) that

$$\chi(\mathbb{P}^n_k, \mathscr{O}(m)) = \frac{(m+1)(m+2)\cdots(m+n)}{n!}.$$

So one lesson is this: If one cohomology group (usually the top or bottom) behaves well in a certain range, and then messes up, likely it is because (i) it is actually the Euler characteristic which behaves well *always*, and (ii) the other cohomology groups vanish in that certain range.

In fact, we will see that it is often hard to calculate cohomology groups (even h^0), but it can be easier calculating Euler characteristics. So one important way of getting a hold of cohomology groups is by computing the Euler characteristic, and then showing that all the *other* cohomology groups vanish. Hence the ubiquity and importance of *vanishing theorems*. (A vanishing theorem usually states that a certain cohomology group vanishes under certain conditions.) We will see this in action when discussing curves. (One of the first applications will be (19.2.5.1).)

The following exercise shows another way in which the Euler characteristic behaves well: it is *additive in exact sequences*.

18.4.A. EXERCISE. Show that if $0 \to \mathscr{F} \to \mathscr{G} \to \mathscr{H} \to 0$ is an exact sequence of coherent sheaves on a projective k-scheme X, then $\chi(X, \mathscr{G}) = \chi(X, \mathscr{F}) + \chi(X, \mathscr{H})$. (Hint: Consider the long exact sequence in cohomology.) More generally, if

$$0 \longrightarrow \mathscr{F}_1 \longrightarrow \cdots \longrightarrow \mathscr{F}_n \longrightarrow 0$$

is an exact sequence of coherent sheaves, show that

$$\sum_{i=1}^{n} (-1)^i \chi(X, \mathscr{F}_i) = 0.$$

(This exercise both generalizes the "exact" case of Exercise 1.5.B—consider the case where $X = \operatorname{Spec} k$—and uses it in the proof.)

18.4.1. The Riemann–Roch Theorem for line bundles on projective curves.

18.4.B. ESSENTIAL EXERCISE: THE RIEMANN–ROCH THEOREM FOR LINE BUNDLES ON A PROJECTIVE CURVE (FIRST VERSION). Suppose $D = \sum_{p \in C^{\mathrm{reg}}} a_p [p]$ is a divisor on a projective (pure dimension 1) curve C over a field k (where $a_p \in \mathbb{Z}$, and all but finitely many a_p are 0). Here C^{reg} are the dimension 0 regular points of C. Define the **degree of** D by

$$\deg D = \sum a_p \deg p.$$

(The degree of a point p was defined in §5.3.8, as the degree of the field extension of the residue field over k.) Show that

$$\chi(C, \mathscr{O}_C(D)) = \deg D + \chi(C, \mathscr{O}_C)$$

by induction on $\sum |a_p|$ (where $D = \sum a_p[p]$ as above). Hint: To show that $\chi(C, \mathscr{O}_C(D)) = \deg p + \chi(C, \mathscr{O}_C(D - p))$, tensor the closed subscheme exact sequence (9.1.2.1)

$$0 \longrightarrow \mathscr{O}_C(-p) \longrightarrow \mathscr{O}_C \longrightarrow \mathscr{O}|_p \longrightarrow 0$$

by $\mathscr{O}_C(D)$, and use additivity of Euler characteristics in exact sequences (Exercise 18.4.A).

As every invertible sheaf $\mathscr{L}$ on a *regular* curve is of the form $\mathscr{O}_C(D)$ for some D (see §15.4), this exercise is very powerful.

18.4.2. *Important definition: Degree of a line bundle on a projective curve (generalized in §18.4.7).* Suppose $\mathscr{L}$ is an invertible sheaf on a projective curve C over k. (We make no assumption on the regularity, or even reducedness, of C!) Define the **degree of** $\mathscr{L}$ (denoted by $\deg \mathscr{L}$ or $\deg_C \mathscr{L}$) by

$$\deg \mathscr{L} := \chi(C, \mathscr{L}) - \chi(C, \mathscr{O}_C).$$

Thus we have a degree map $\deg : \operatorname{Pic} C \to \mathbb{Z}$. The set of degree d line bundles on C is denoted by $\operatorname{Pic}^d C := \deg^{-1}(d)$.

18.4.C. IMPORTANT EASY EXERCISE: DEGREE OF A DIVISOR = DEGREE OF THE CORRESPONDING LINE BUNDLE (JUSTIFYING REMARK 16.3.8). Suppose $\mathscr{L}$ is an invertible sheaf on a projective (pure dimension 1) curve C over k, and let s be a nonzero rational section of $\mathscr{L}$ on C, which is regular at all singular points of C. Let D be the divisor of zeros and poles of s:

$$D := \sum_{p \in C} v_p(s)[p].$$

Show that $\deg \mathscr{L} = \deg D$. In particular, the degree of a line bundle on a regular curve can be computed by counting zeros and poles of *any* rational section not vanishing on a component of C.

The next few exercises give useful applications.

18.4.D. EXERCISE. Give a new solution to Exercise 16.3.E (a nonzero rational function on a projective curve has the same number of zeros and poles, counted appropriately) using the ideas above.

18.4.E. EXERCISE. Suppose s_1 and s_2 are two sections of a degree d line bundle $\mathscr{L}$ on an irreducible regular projective curve C, with no common zeros. Then s_1 and s_2 determine a morphism $\pi : C \to \mathbb{P}^1_k$. Show that π is a finite morphism, and that the degree of π is d. (Translation: A two-dimensional base-point-free degree d linear series on C defines a degree d finite cover of $\mathbb{P}^1$.)

18.4.F. EXERCISE (TO BE EXTENDED IN EXERCISE 18.4.M). If $\mathscr{L}$ and $\mathscr{M}$ are two line bundles on a regular projective curve C, show that $\deg \mathscr{L} \otimes \mathscr{M} = \deg \mathscr{L} + \deg \mathscr{M}$. (Hint: Choose nonzero rational sections of $\mathscr{L}$ and $\mathscr{M}$.)

18.4.3. *Remark: The "degree" map* $\operatorname{Pic} \to \mathbb{Z}$ *is a homomorphism.* Thus for a regular projective curve C (over a field k), the degree map $\deg : \operatorname{Pic} \to \mathbb{Z}$ is a homomorphism of abelian groups. (This result will be extended to all proper curves over k in Exercise 18.4.M.)

18.4.G. EXERCISE. Suppose $\pi : C \to C'$ is a degree d morphism of integral projective regular curves, and $\mathscr{L}$ is an invertible sheaf on C'. Show that $\deg_C \pi^* \mathscr{L} = d \deg_{C'} \mathscr{L}$. Hint: Compute

$\deg_{C'} \mathscr{L}$ using any nonzero rational section s of $\mathscr{L}$, and compute $\deg_C \pi^* \mathscr{L}$ using the rational section $\pi^* s$ of $\pi^* \mathscr{L}$. Note that zeros pull back to zeros, and poles pull back to poles. Depending on your approach, Exercise 16.3.D might help.

18.4.H. [**] EXERCISE (COMPLEX-ANALYTIC INTERPRETATION OF DEGREE; ONLY FOR THOSE WITH SUFFICIENT ANALYTIC BACKGROUND). Suppose X is a connected regular projective complex curve. Show that the degree map is the composition of group homomorphisms

$$\operatorname{Pic} X \longrightarrow \operatorname{Pic} X_{an} \xrightarrow{c_1} H^2(X_{an}, \mathbb{Z}) \xrightarrow{\cap [X_{an}]} H_0(X_{an}, \mathbb{Z}) \cong \mathbb{Z}.$$

Hint: Show it for a generator $\mathscr{O}(p)$ of the group $\operatorname{Pic} X$, using explicit transition functions. (The first map was discussed in Exercise 14.1.J. The second map takes a line bundle to its first Chern class, and can be interpreted as follows. The transition functions for a line bundle yield a Čech 1-cycle for $\mathscr{O}_{X_{an}}^*$; this yields a map $\operatorname{Pic} X_{an} \to H^1(X_{an}, \mathscr{O}_{X_{an}}^*)$. Combining this with the map $H^1(X_{an}, \mathscr{O}_{X_{an}}^*) \to H^2(X_{an}, \mathbb{Z})$ from the long exact sequence in cohomology corresponding to the exponential exact sequence (2.4.9.1) yields the first Chern class map.)

18.4.4. Arithmetic genus.

Motivated by geometry (Miracle 18.4.5 below), we define the **arithmetic genus** of a dimension n scheme X as $p_a(X) := (-1)^n (\chi(X, \mathscr{O}_X) - 1)$. For integral projective curves over an algebraically closed field, $h^0(X, \mathscr{O}_X) = 1$ (§11.4.7), so $p_a(X) = h^1(X, \mathscr{O}_X)$. (In higher dimension, this is a less natural notion.)

We can restate the Riemann–Roch formula for curves (Exercise 18.4.B) as:

$$(18.4.4.1) \qquad h^0(C, \mathscr{L}) - h^1(C, \mathscr{L}) = \deg \mathscr{L} - p_a(C) + 1.$$

This is the most common formulation of the Riemann–Roch formula.

18.4.5. Miracle. If C is a regular irreducible projective complex curve, then the corresponding complex-analytic object, a compact *Riemann surface*, has a notion called the *genus* g, which is (informally speaking) the number of holes (see Figure 18.3). Miraculously, $g = p_a$ in this case (see Exercise 21.4.I), and for this reason, we will often write g for p_a when discussing regular (projective irreducible) curves, over any field. We will discuss genus further in §18.6.7, when we will be able to compute it in many interesting cases. (Warning: The arithmetic genus of $\mathbb{P}^1_{\mathbb{C}}$ as an $\mathbb{R}$-variety is -1!)

18.4.6. Degree of a coherent sheaf (e.g., degree of a vector bundle).

Suppose C is an irreducible reduced curve (possibly singular), over a field k. If $\mathscr{F}$ is a coherent sheaf on C, recall (from §14.3.3) that the *rank* of $\mathscr{F}$, denoted by rank $\mathscr{F}$, is its rank at the generic point of C.

18.4.7. *Definition.* If C is *projective*, define the **degree of** $\mathscr{F}$ by

$$(18.4.7.1) \qquad \deg \mathscr{F} = \chi(C, \mathscr{F}) - (\operatorname{rank} \mathscr{F}) \cdot \chi(C, \mathscr{O}_C).$$

The statement (18.4.7.1) is often called **Riemann–Roch for coherent sheaves** (or vector bundles) on a projective curve.

If $\mathscr{F}$ is an invertible sheaf (or if, more generally, the rank is the same on each irreducible component), we can drop the irreducibility hypothesis.

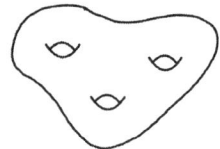

Figure 18.3 *A genus 3 Riemann surface.*

18.4.I. EASY EXERCISE. Show that degree (as a function of coherent sheaves on a fixed curve C) is additive in exact sequences.

18.4.J. EXERCISE. If $\mathscr{F}$ is a torsion sheaf on C, show that $\deg \mathscr{F} \geq 0$, with equality if and only if $\mathscr{F} = 0$.

18.4.K. EXERCISE. If $k = \bar{k}$, and $\mathscr{F}$ is a torsion sheaf on C, show that $\deg \mathscr{F} = \ell(\mathscr{F})$. (The length $\ell(\mathscr{F})$ of $\mathscr{F}$ was defined in §6.5.8.) What goes wrong if $k \neq \bar{k}$?

18.4.L.* EXERCISE. Suppose C is a projective curve over a field k, $\mathscr{L}$ is a line bundle on C, and $\mathscr{F}$ is a coherent sheaf on C. Using the following steps, show that $\chi(C, \mathscr{L} \otimes \mathscr{F}) - \chi(C, \mathscr{F})$ is the sum over the irreducible components C_i of C of the degree of $\mathscr{L}$ on C_i^{red} (the reduction was defined in §9.4.10) times the length of $\mathscr{F}$ at the generic point η_i of C_i (the length of $\mathscr{F}_{\eta_i}$ as an $\mathscr{O}_{\eta_i}$-module—length was defined in Definition 6.5.2). This could reasonably be called **Riemann–Roch for nonreduced curves**.

(a) First reduce to the case where $\mathscr{F}$ is scheme-theoretically supported on C^{red} (you defined the scheme-theoretic support of a finite type quasicoherent sheaf in Exercise 9.1.N), by showing that both sides of the alleged equality are additive in short exact sequences, and using the filtration
$$0 = \mathscr{I}^r \mathscr{F} \subset \mathscr{I}^{r-1} \mathscr{F} \subset \cdots \subset \mathscr{I} \mathscr{F} \subset \mathscr{F}$$
of $\mathscr{F}$, where $\mathscr{I}$ is the ideal sheaf cutting out C^{red} in C. Thus we need only consider the case where C is reduced.

(b) In this "reduced" case, by Exercise 16.2.L, we can write $\mathscr{L} \cong \mathscr{O}(\sum n_j p_j)$, where the p_j are regular points distinct from the associated points of $\mathscr{F}$. By an appropriate induction, prove what remains to be proved.

18.4.M.* EXERCISE. If $\mathscr{L}$ and $\mathscr{M}$ are line bundles on a projective curve C over a field k, show that $\deg_C(\mathscr{L} \otimes \mathscr{M}) = \deg_C(\mathscr{L}) + \deg_C(\mathscr{M})$. Thus the degree map $\deg : \operatorname{Pic} C \to \mathbb{Z}$ is a homomorphism of abelian groups, extending Remark 18.4.3.

18.4.8. In fact, all proper curves over k are projective (Remark 19.2.I), so "projective" can be replaced by "proper" in Exercises 18.4.L and 18.4.M. In this guise, we will use Exercise 18.4.L when discussing intersection theory in Chapter 20.

18.4.9. *Two properties of ample bundles on curves.* For Exercises 18.4.N and 18.4.O, either assume C is regular (so you can invoke Remark 18.4.3), or use starred Exercise 18.4.M.

18.4.N. EXERCISE. If C is a projective curve, and $\mathscr{L}$ is an ample line bundle on C, show that $\deg \mathscr{L} > 0$. (Hint: Show it if $\mathscr{L}$ is *very* ample.)

18.4.O. EXERCISE. Suppose $\mathscr{L}$ is a base-point-free invertible sheaf on a proper variety X, and hence induces some morphism $\phi : X \to \mathbb{P}^n$. Then $\mathscr{L}$ is ample if and only if ϕ is finite. (Hint: If ϕ is finite, use Exercise 16.2.G. If ϕ is not finite, show that there is a curve C contracted by ϕ, using Theorem 18.1.6. Show that $\mathscr{L}$ has degree 0 on C.)

18.4.10.* Extended example: The universal plane conic has no rational sections.
We use the theory of the degree to get an interesting consequence. We work over a fixed field k. We consider the following diagram.

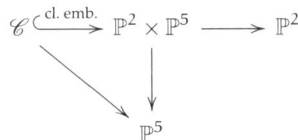

$$\mathscr{C} \overset{\mathrm{cl.\ emb.}}{\hookrightarrow} \mathbb{P}^2 \times \mathbb{P}^5 \longrightarrow \mathbb{P}^2$$
$$\searrow \qquad \downarrow$$
$$\mathbb{P}^5$$

If the $\mathbb{P}^2$ has projective coordinates x_0, x_1, x_2, then $\mathbb{P}^5$ has coordinates $a_{00}, a_{01}, a_{11}, a_{02}, a_{12}, a_{22}$, and $\mathscr{C}$ is cut out by the single equation
$$a_{00} x_0^2 + a_{01} x_0 x_1 + \cdots + a_{22} x_2^2 = 0.$$

We interpret $\mathbb{P}^5$ as the parameter space of conics (in $\mathbb{P}^2$), and $\mathscr{C}$ as the universal conic over $\mathbb{P}^5$ (parametrizing a conic C along with a point $p \in C$), which comes with a canonical projection $\mathscr{C} \to \mathbb{P}^2$.

18.4.P. EXERCISE. By interpreting $\mathscr{C}$ as a $\mathbb{P}^4$-bundle over $\mathbb{P}^2$ (with two irreducible divisors whose complement is an $\mathbb{A}^4$-bundle over $\mathbb{A}^2$), show that $\mathscr{C}$ is a smooth sixfold, and that $\operatorname{Pic}\mathscr{C} \cong \mathbb{Z} \times \mathbb{Z}$.

18.4.Q. EXERCISE. Fix a line $\ell \subset \mathbb{P}^2$ and a k-valued point $q \in \mathbb{P}^2$. Let D_ℓ be the divisor on $\mathscr{C}$ corresponding to (C, p), with p lying on ℓ. Let D_q be the divisor on $\mathscr{C}$ corresponding to (C, p), with $q \in C$. Using your description of $\mathscr{C}$ as a $\mathbb{P}^4$-bundle over $\mathbb{P}^2$, show that D_ℓ and D_q generate $\operatorname{Pic}\mathscr{C}$.

18.4.R. EXERCISE. Suppose K is a fiber of $\mathscr{C} \to \mathbb{P}^5$ over a point $r \in \mathbb{P}^5$—i.e., a conic in $\mathbb{P}^2$ over the field $\kappa(r)$. Suppose further that K is integral, and that neither q nor ℓ is contained in K. (This hypothesis is unnecessary, but simplifies the problem.) Show that $\deg_K \mathscr{O}(D_q) = 0$ and $\deg_K \mathscr{O}(D_\ell) = 2$. Hence show that if $\mathscr{L}$ is any invertible sheaf on $\mathscr{C}$, then $\deg_K(\mathscr{L})$ is even.

18.4.S. EXERCISE. Show that there is no rational section to the projection $\pi \colon \mathscr{C} \to \mathbb{P}^5$. Hint: If there were a regular section over an open subset U of $\mathbb{P}^5$, it would be a divisor on $\pi^{-1}(U)$; let D be its closure in $\mathscr{C}$. Show that D meets any fiber of π over U in multiplicity 1. Use Exercise 18.4.R to obtain a contradiction.

We can restate Exercise 18.4.S in the following dramatic way: There is no way to write down three rational functions X_0, X_1, X_2 in $a_{00}, \ldots, a_{22}$ such that

$$a_{00}X_0^2 + a_{01}X_0X_1 + \cdots + a_{22}X_2^2 = 0$$

without $X_0 = X_1 = X_2 = 0$.

The question of a rational point on a conic is one of arithmetic. (Think: $x^2 + y^2 = z^2$.) Our solution was topological. The unification of topology and arithmetic in this example is the beginning of a long and fruitful story in algebraic geometry.

18.4.11.* Numerical equivalence, the Néron–Severi group, nef line bundles, and the nef and ample cones.

Suppose X is a proper k-variety, and $\mathscr{L}$ is an invertible sheaf on X. If $i \colon C \hookrightarrow X$ is a one-dimensional closed reduced subscheme of X, define the degree of $\mathscr{L}$ on C by $\deg_C \mathscr{L} := \deg_C i^* \mathscr{L}$. If $\deg_C \mathscr{L} = 0$ for all C, we say that $\mathscr{L}$ is **numerically trivial**.

18.4.T. EXERCISE.

(a) Show that $\mathscr{L}$ is numerically trivial if and only if $\deg_C \mathscr{L} = 0$ for all *integral* curves C in X.
(b) Show that if $\pi \colon X \to Y$ is a proper morphism, and $\mathscr{L}$ is a numerically trivial invertible sheaf on Y, then $\pi^* \mathscr{L}$ is numerically trivial on X.
(c) Show that $\mathscr{L}$ is numerically trivial if and only if $\mathscr{L}$ is numerically trivial on each of the irreducible components of X.
(d) Show that if $\mathscr{L}$ and $\mathscr{L}'$ are numerically trivial, then $\mathscr{L} \otimes \mathscr{L}'$ and $\mathscr{L}^\vee$ are both numerically trivial.

18.4.12. *Numerical equivalence.* By part (d), the numerically trivial invertible sheaves form a subgroup of $\operatorname{Pic} X$, denoted by $\operatorname{Pic}^\tau X$. Two lines bundles equivalent modulo the subgroup of numerically trivial line bundles are called **numerically equivalent**. A property of invertible sheaves stable under numerical equivalence is said to be a *numerical property*. We will see that "nefness" and ampleness are numerical properties (Definition 18.4.13 and Remark 20.4.2, respectively).

We will later define the *Néron–Severi group* $\operatorname{NS}(X)$ of X as $\operatorname{Pic} X$ modulo algebraic equivalence (Exercise 24.7.5). (We will define algebraic equivalence once we have discussed flatness.) The highly nontrivial **Néron–Severi Theorem** (or **Theorem of the Base**) states that $\operatorname{NS}(X)$ is a finitely generated group. (See [ChJLO] for a thorough modern discussion, and [GH1, p. 462] for a simpler proof over $\mathbb{C}$.) The group $\operatorname{Pic} X / \operatorname{Pic}^\tau X$ is denoted by $N^1(X)$. We will see (in §24.7.5) that it is a quotient of $\operatorname{NS}(X)$, so it is also finitely generated. As the group $N^1(X)$ is clearly abelian and torsion-free, it is a finite free $\mathbb{Z}$-module (by the classification of finitely generated modules over a principal ideal domain; see §0.3). The rank of $N^1(X)$ is called the **Picard number**, and is denoted by $\rho(X)$ (although we won't have need of this notion, except in our discussion of the Hodge Index Theorem in §20.2.11). For example, $\rho(\mathbb{P}^n) = 1$ and $\rho((\mathbb{P}^1)^n) = n$. We define $N^1_{\mathbb{Q}}(X) := N^1(X) \otimes_{\mathbb{Z}} \mathbb{Q}$ (so $\rho(X) = \dim_{\mathbb{Q}} N^1_{\mathbb{Q}}(X)$), and call the elements of this group $\mathbb{Q}$-**line bundles**, for lack of any common term in the literature.

18.4.U. EXERCISE (FINITENESS OF PICARD NUMBER IN THE COMPLEX CASE, ONLY FOR THOSE WITH SUFFICIENT BACKGROUND). Show (without the Néron–Severi Theorem) that if X is a complex proper variety, then $\rho(X)$ is finite, by interpreting it as a subquotient of $H^2(X, \mathbb{Z})$. Hint: Show that the image of $(\mathscr{L}, C)$ under the map $H^2(X, \mathbb{Z}) \times H_2(X, \mathbb{Z}) \to H_0(X, \mathbb{Z}) \to \mathbb{Z}$ is $\deg_C \mathscr{L}$. Hint: Figure out how to reduce to the case where C is a smooth projective curve, then use Exercise 18.4.H.

18.4.13. *Definition.* We say that an invertible sheaf $\mathscr{L}$ is **numerically effective**, or **nef** if for all such C, $\deg_C \mathscr{L} \geq 0$. Clearly nefness is a numerical property.

18.4.V. EASY EXERCISE (CF. EXERCISE 18.4.T).

(a) Show that $\mathscr{L}$ is nef if and only if $\deg_C \mathscr{L} \geq 0$ for all *integral* curves C in X.
(b) Show that if $\pi: X \to Y$ is a proper morphism, and $\mathscr{L}$ is a nef invertible sheaf on Y, then $\pi^* \mathscr{L}$ is nef on X. (Hint: Exercise 18.4.G will be needed.)
(c) Show that $\mathscr{L}$ is nef if and only if $\mathscr{L}$ is nef on each of the irreducible components of X.
(d) Show that if $\mathscr{L}$ and $\mathscr{L}'$ are nef, then $\mathscr{L} \otimes \mathscr{L}'$ is nef. Thus the nef elements of $\operatorname{Pic} X$ form an abelian semigroup.
(e) Show that ample invertible sheaves are nef.
(f) Suppose $n \in \mathbb{Z}^{>0}$. Show that $\mathscr{L}$ is nef if and only if $\mathscr{L}^{\otimes n}$ is nef.

18.4.W. EXERCISE. Define what it means for a $\mathbb{Q}$-line bundle to be nef. Show that the nef $\mathbb{Q}$-line bundles form a closed cone in $N^1_{\mathbb{Q}}(X)$. This is called the **nef cone**.

18.4.X. EXERCISE. Describe the nef cones of $\mathbb{P}^2_k$ and $\mathbb{P}^1_k \times_k \mathbb{P}^1_k$. (Notice in the latter case that the two boundaries of the cone correspond to linear series contracting one of the $\mathbb{P}^1$'s. This is true in general: informally speaking, linear series corresponding to the boundaries of the cone give interesting contractions. Another example will be given in Exercise 20.2.F.)

It is a surprising fact that whether an invertible sheaf $\mathscr{L}$ on X is ample depends only on its class in $N^1_{\mathbb{Q}}(X)$, i.e., on how it intersects the curves in X. Because of this (as for any $n \in \mathbb{Z}^{\geq 0}$, $\mathscr{L}$ is ample if and only if $\mathscr{L}^{\otimes n}$ is ample; see Theorem 16.2.2), it makes sense to define when a $\mathbb{Q}$-line bundle is ample. Then by Exercise 16.2.H, the ample divisors form a cone in $N^1_{\mathbb{Q}}(X)$, necessarily contained in the nef cone by Exercise 18.4.V(e). It turns out that if X is projective, the ample divisors are precisely the interior of the nef cone. The new facts in this paragraph are a consequence of Kleiman's criterion for ampleness, Theorem 20.4.6.

18.5   A First Glimpse of Serre Duality

A common version of Riemann–Roch involves Serre duality, which, unlike Riemann–Roch, is hard to prove.

18.5.1. Theorem (Serre duality for smooth projective varieties) — *Suppose X is a geometrically irreducible smooth projective k-variety, of dimension n. Then there is an invertible sheaf ω_X (or simply ω) on X such that*

$$h^i(X, \mathscr{F}) = h^{n-i}(X, \omega_X \otimes \mathscr{F}^\vee)$$

for all $i \in \mathbb{Z}$ and all finite rank locally free sheaves $\mathscr{F}$.

The invertible sheaf ω_X is called the *canonical sheaf* or *dualizing sheaf*, which will be formally defined in §29.1.5. We will see in Chapter 29 that Theorem 18.5.1 is a consequence of a perfect pairing

$$(18.5.1.1) \qquad H^i(X, \mathscr{F}) \times H^{n-i}(X, \omega_X \otimes \mathscr{F}^\vee) \longrightarrow H^n(X, \omega_X) \cong k,$$

and that smoothness can be relaxed somewhat.

18.5.2. *Further miracle: The sheaf of algebraic volume forms is Serre-dualizing.* The invertible sheaf ω_X turns out to be the determinant $\mathscr{K}_X$ of the cotangent bundle $\Omega_{X/k}$ of X (see §21.5.3); we will define the cotangent bundle in Chapter 21. We make the connection $\omega_X = \mathscr{K}_X$ in §29.4; see Desideratum 29.1.1.

18.5.3. *Back to Riemann–Roch.* For the purposes of restating Riemann–Roch for a curve C, it suffices to note that $h^1(C, \mathscr{L}) = h^0(C, \omega_C \otimes \mathscr{L}^\vee)$. Then the Riemann–Roch formula can be rewritten as
$$h^0(C, \mathscr{L}) - h^0(C, \omega_C \otimes \mathscr{L}^\vee) = \deg \mathscr{L} - p_a(C) + 1.$$

18.5.A. EXERCISE (ASSUMING SERRE DUALITY). Suppose C is a geometrically integral smooth projective curve over k.

(a) Show that $h^0(C, \omega_C)$ is the genus g of C. (The genus g, in the guise of the arithmetic genus p_a, was defined in §18.4.4–18.4.5.)
(b) Show that $\deg \omega_C = 2g - 2$. (Hint: Riemann–Roch for $\mathscr{L} = \omega_C$.)

18.5.4. *Example.* If $C = \mathbb{P}^1_k$, Exercise 18.5.A implies that $\omega_C \cong \mathscr{O}(-2)$. Moreover, we also have a natural perfect pairing (cf. (18.5.1.1))

$$H^0(\mathbb{P}^1, \mathscr{O}(n)) \times H^1(\mathbb{P}^1, \mathscr{O}(-2-n)) \longrightarrow k \cong H^1(\mathbb{P}^1, \mathscr{O}(-2)).$$

We can interpret this pairing as follows. If $n < 0$, both factors on the left are 0, so we assume $n > 0$. Then $H^0(\mathbb{P}^1, \mathscr{O}(n))$ corresponds to homogeneous degree n polynomials in x and y, and $H^1(\mathbb{P}^1, \mathscr{O}(-2-n))$ corresponds to homogeneous degree $-2 - n$ Laurent polynomials in x and y so that the degrees of x and y are both at most -1 (see Remark 18.3.5). You can quickly check that the dimension of both vector spaces are $n + 1$. The pairing is given as follows: Multiply the polynomial by the Laurent polynomial, to obtain a Laurent polynomial of degree -2. Read off the coefficient of $x^{-1}y^{-1}$. (This works more generally for $\mathbb{P}^n_A$; see the discussion after the statement of Theorem 18.1.2.)

18.5.B. EXERCISE (ASSUMING SERRE DUALITY): AMPLE DIVISORS ON A CONNECTED SMOOTH PROJECTIVE VARIETY ARE CONNECTED. Suppose X is a connected smooth projective $\bar{k}$-variety of dimension at least 2, and D is an effective ample divisor. Show that D is connected. Hint: Suppose $D = V(s)$, where s is a section of an ample invertible sheaf $\mathscr{L}$. Then $V(s^n) = V(s)$ (as vanishing *sets*, not vanishing *schemes*) for all $n > 0$, so we may replace $\mathscr{L}$ with a high power of our choosing. Use the long exact sequence for $0 \to \mathscr{O}_X(-nD) \to \mathscr{O}_X \to \mathscr{O}_{V(s^n)} \to 0$ to show that for $n \gg 0$, $h^0(X, \mathscr{O}_{V(s^n)}) = 1$. (This may remind you of §11.4.7.)

Once we know that Serre duality holds for Cohen–Macaulay projective schemes (§29.3), this result will automatically extend to these schemes. (A related result is Exercise 18.6.U, which doesn't use Serre duality.) On the other hand, the result is false if X is the union of two 2-planes in $\mathbb{P}^4$ meeting at a point (why?), so this will imply that this X is not Cohen–Macaulay. (We will show this in another way in Counterexample 26.2.2.)

18.5.5.* **Classification of vector bundles on $\mathbb{P}^1_k$.**
As promised in Example 17.2.6, we classify the vector bundles on $\mathbb{P}^1_k$. We discuss this in §18.5 because you might use Serre duality at one step (although you might not).

18.5.6. Theorem — *If $\mathscr{E}$ is a rank r vector bundle on $\mathbb{P}^1_k$, then $\mathscr{E} \cong \mathscr{O}(a_1) \oplus \cdots \oplus \mathscr{O}(a_r)$, for a unique nondecreasing sequence of integers $a_1, \ldots, a_r$.*

This result was proved independently many times (see, for example, Dedekind and Weber's article [DW, §22]), and is a special case of a theorem of Grothendieck, [Gr2]. It is sometimes called Grothendieck's Theorem, because Grothendieck doesn't have enough theorems named after him.

18.5.7. For $\mathbb{P}^n_k$ more generally, the case $r = 1$ was shown in §15.4.15, but the statement is false for $r > 1$ and $n > 1$. A counterexample is given in Exercise 21.3.H, which shows that it is not even possible to *filter* the rank 2 bundle $\Omega_{\mathbb{P}^2}$ on $\mathbb{P}^2$ by line bundles. One true generalization is a theorem of Horrocks, which states that a finite rank locally free sheaf $\mathscr{E}$ on $\mathbb{P}^n_k$ splits precisely when "all the middle cohomology of all of its twists is zero"—when $H^i(\mathbb{P}^n_k, \mathscr{E}(m)) = 0$ for $0 < i < n$ and all $m \in \mathbb{Z}$. This has the surprising consequence that if $n \geq 2$, then a finite rank locally free sheaf on $\mathbb{P}^n_k$ splits if and only if its restriction to some previously chosen 2-plane ($\mathbb{P}^2_k$) splits—all of the additional complication turns up already in dimension 2. See [OSS, §2.3] for more.

Proof. Note that the classification makes no reference to cohomology, but the proof uses cohomology in an essential way. It is possible to prove Theorem 18.5.6 with no cohomological machinery (see, for example, [HM] or [GW, p. 314-5]), or with more cohomological machinery (see, for example, [Ha1, Ex. V.2.6]).

18.5.C. EXERCISE. Suppose that $\mathscr{E} \cong \mathscr{O}(a_1) \oplus \cdots \oplus \mathscr{O}(a_r)$, with $a_1 \leq \cdots \leq a_r$. Show that the a_i can be determined from the numbers $\dim_k \operatorname{Hom}(\mathscr{O}(m), \mathscr{E})$, as m ranges over the integers. Use this to show the uniqueness part of Theorem 18.5.6.

We now begin the proof, by induction on r. Fix a rank r locally free sheaf $\mathscr{E}$. The case $r = 1$ was established in §15.4.15, so we assume $r > 1$.

18.5.D. EXERCISE. Show that for $m \ll 0$, $\operatorname{Hom}(\mathscr{O}(m), \mathscr{E}) \neq 0$, and for $m \gg 0$, $\operatorname{Hom}(\mathscr{O}(m), \mathscr{E}) = 0$. Hint: Show that $\mathscr{H}om(\mathscr{O}(m), \mathscr{E}) = \mathscr{E}(-m)$ using Exercise 14.1.G. Use Serre's Theorem A (Theorem 16.1.5) for the first part. Feel free to use Serre duality and Serre vanishing for the second part. But you may prefer to come up with an argument without Serre duality, to avoid invoking something we have not yet proved.

Thus there is some a_r for which $\dim \operatorname{Hom}(\mathscr{O}(a_r), \mathscr{E}) > 0$, but for which $\operatorname{Hom}(\mathscr{O}(m), \mathscr{E}) = 0$ for all $m > a_r$. Choose a nonzero map $\phi \colon \mathscr{O}(a_r) \to \mathscr{E}$.

18.5.E. EXERCISE. Show that ϕ is an injection. (Hint: $\mathscr{O}(a_r)$ is torsion-free, and thus the kernel is torsion-free.)

18.5.F. EXERCISE. Let $\mathscr{F}$ be the cokernel of ϕ. Show that $\mathscr{F}$ is locally free. Hint: Exercise 14.3.F(c) gives an exact sequence $0 \to \mathscr{F}_{\mathrm{tors}} \to \mathscr{F} \to \mathscr{F}_{\mathrm{lf}} \to 0$, where $\mathscr{F}_{\mathrm{lf}}$ is locally free. Let $\mathscr{L}$ be the kernel of the surjection $\mathscr{E} \to \mathscr{F}_{\mathrm{lf}}$. Show that $\mathscr{L}$ is locally free, and thus is isomorphic to $\mathscr{O}(N)$ for some N. Show that there is a nonzero map $\phi' \colon \mathscr{O}(a_r) \to \mathscr{O}(N)$. Show (using the same idea as in the previous exercise) that this nonzero map ϕ' must be an injection, so $a_r \leq N$. Show that $N \leq a_r$ because there is a nonzero map $\mathscr{O}(N) \to \mathscr{E}$ (recall how a_r was chosen). Show that ϕ' is an isomorphism, and thus that $\mathscr{F} = \mathscr{F}_{\mathrm{lf}}$.

By our inductive hypothesis, we have $\mathscr{F} = \mathscr{O}(a_1) \oplus \cdots \oplus \mathscr{O}(a_{r-1})$, where $a_1 \leq \cdots \leq a_{r-1}$, so we have a short exact sequence

$$(18.5.7.1) \qquad 0 \longrightarrow \mathscr{O}(a_r) \longrightarrow \mathscr{E} \longrightarrow \mathscr{O}(a_1) \oplus \cdots \oplus \mathscr{O}(a_{r-1}) \longrightarrow 0.$$

We next show that $a_r \geq a_i$ for $i < r$. We are motivated by the fact that for a quasicoherent sheaf $\mathscr{G}$ on $\mathbb{P}^1$,

$$\operatorname{Hom}_{\mathbb{P}^1}(\mathscr{O}(a_r + 1), \mathscr{G}) = H^0(\mathbb{P}^1, \mathscr{G}(-a_r - 1))$$

(Exercise 14.1.G). Tensor (18.5.7.1) with $\mathscr{O}(-a_r - 1)$ (preserving exactness, by Exercise 14.1.F), and take the long exact sequence in cohomology. Part of the long exact sequence is

$$H^0(\mathbb{P}^1, \mathscr{E}(-a_r - 1)) \longrightarrow H^0(\mathbb{P}^1, \mathscr{F}(-a_r - 1)) \longrightarrow H^1(\mathbb{P}^1, \mathscr{O}(-1)).$$

Notice that $H^0(\mathbb{P}^1, \mathscr{E}(-a_r - 1)) = 0$ (as $\mathrm{Hom}(\mathscr{O}(a_r + 1), \mathscr{E}) = 0$, by the definition of a_r), and $H^1(\mathbb{P}^1, \mathscr{O}(-1)) = 0$ (Theorem 18.1.2(b)). Thus

$$
\begin{aligned}
0 &= H^0(\mathbb{P}^1, \mathscr{F}(-a_r - 1)) \\
&= H^0(\mathbb{P}^1, \oplus_i \mathscr{O}(a_i - a_r - 1)) \\
&= \oplus_i H^0(\mathbb{P}^1, \mathscr{O}(a_i - a_r - 1)).
\end{aligned}
$$

Hence $a_i - a_r - 1 < 0$ (by Exercise 15.1.A), so $a_r \geq a_i$, as desired.

Finally, we wish to show that exact sequence (18.5.7.1) expresses $\mathscr{E}$ as a direct sum (of the subsheaf and quotient sheaf). For simplicity, we focus on the case $r = 2$, and return to the general case in Exercise 18.5.I.

18.5.G. EXERCISE. Show that the transition functions for the vector bundle, in appropriate coordinates, are given by

$$
\begin{pmatrix} t^{-a_2} & \alpha(t) \\ 0 & t^{-a_1} \end{pmatrix},
$$

where $\alpha(t)$ is a Laurent polynomial in t. Hint/reminder: Recall that all vector bundles on $\mathbb{A}^1_k$ are trivial, Exercise 14.2.C. Transition matrices for extensions of one vector bundle (with known transition matrices) by another were discussed in Exercise 14.2.G.

18.5.H. EXERCISE. Implicit in the above 2×2 matrix is a choice of a basis of a rank 2 free module M over the ring $k[t]$ corresponding to one of the standard affine open subsets $\mathrm{Spec}\, k[t]$, and a rank 2 free module M' over the ring $k[1/t]$ over the other standard open subset. Show that by an appropriate "upper-triangular" change of basis of M, you can arrange for $\alpha(t)$ to have no monomials of degree $\geq -a_2$. Show that by an appropriate "upper-triangular" change of basis of M', you can arrange for $\alpha(t)$ to have no monomials of degree $\leq -a_1$. Thus by choosing bases of M and M' appropriately, we can take $\alpha(t) = 0$. Show that this implies that $\mathscr{E} = \mathscr{O}(a_1) \oplus \mathscr{O}(a_2)$.

18.5.I. EXERCISE. Finish the proof of Theorem 18.5.6 for arbitrary r. Hint: There are no new ideas beyond the case $r = 2$. $\square$

18.6 Hilbert Functions, Hilbert Polynomials, and Genus

If $\mathscr{F}$ is a coherent sheaf on a projective k-scheme $X \subset \mathbb{P}^n$, define the **Hilbert function of** $\mathscr{F}$ by

$$
h_{\mathscr{F}}(m) := h^0(X, \mathscr{F}(m)).
$$

Note that the Hilbert function depends on the projective embedding of X (by way of its line bundle), although this dependency is suppressed in the notation. The **Hilbert function of** X is the Hilbert function of the structure sheaf. The ancients were aware that the Hilbert function is "eventually polynomial," i.e., for large enough m, it agrees with some polynomial. This polynomial contains lots of interesting geometric information, as we will soon see. In modern language, we expect that this "eventual polynomiality" arises because the Euler characteristic should be a polynomial, and that for $m \gg 0$, the higher cohomology vanishes. This is indeed the case, as we now verify.

18.6.1. Theorem — *If $\mathscr{F}$ is a coherent sheaf on a projective k-scheme $X \hookrightarrow \mathbb{P}^n_k$, $\chi(X, \mathscr{F}(m))$ is a polynomial of degree equal to* $\dim \mathrm{Supp}\, \mathscr{F}$*. Hence by Serre vanishing (Theorem 18.1.3 (ii)), for $m \gg 0$, $h^0(X, \mathscr{F}(m))$ is a polynomial $p_{\mathscr{F}}(m)$ of degree $\dim \mathrm{Supp}\, \mathscr{F}$. In particular, for $m \gg 0$, $h^0(X, \mathscr{O}_X(m))$ is polynomial with degree equal to $\dim X$.*

18.6.2. Definition. The polynomial $p_{\mathscr{F}}(m)$ defined in Theorem 18.6.1 is called the **Hilbert polynomial**. If $X \subset \mathbb{P}^n$ is a projective k-scheme, define $p_X(m) := p_{\mathscr{O}_X}(m)$.

18.6.3. In Theorem 18.6.1, $\mathscr{O}_X(m)$ is the restriction or pullback of $\mathscr{O}_{\mathbb{P}^n_k}(m)$. Both the degree of the 0 polynomial and the dimension of the empty set are defined to be -1. In particular, the only coherent sheaf with Hilbert polynomial 0 is the zero-sheaf.

This argument uses the notion of associated points of a coherent sheaf on a locally Noetherian scheme, §6.6.2. (The resolution given by the Hilbert Syzygy Theorem 16.1.3 can give a shorter proof; but we won't prove it until §24.4.)

Proof. Define $p_{\mathscr{F}}(m) = \chi(X, \mathscr{F}(m))$. We will show that $p_{\mathscr{F}}(m)$ is a polynomial of the desired degree.

We first use Exercise 18.2.H to reduce to the case where k is algebraically closed, and, in particular, infinite. (This is one of those cases where even if you are concerned with potentially arithmetic questions over some non-algebraically closed field like $\mathbb{F}_p$, you are forced to consider the "geometric" situation where the base field is algebraically closed.)

The coherent sheaf $\mathscr{F}$ has a finite number of associated points (§6.6.17). We show a useful fact that we will use again.

18.6.A. EXERCISE. Suppose X is a projective k-scheme with k infinite, and $\mathscr{F}$ is a coherent sheaf on X. Show that if $\mathscr{L}$ is a very ample invertible sheaf on X, then there is an effective Cartier divisor D on X with $\mathscr{L} \cong \mathscr{O}(D)$, where D does not contain the associated points of $\mathscr{F}$. (Hint: Show that given any finite set of points of $\mathbb{P}^n_k$, there is a hyperplane not containing any of them. This is a variant of the key step in Exercise 12.3.D(c).)

Thus there is a hyperplane $x = 0$ ($x \in \Gamma(X, \mathscr{O}(1))$) missing this finite number of points. (This is where we use the infinitude of k.)

Then the map $\mathscr{F}(-1) \xrightarrow{\times x} \mathscr{F}$ is injective. (On any affine open subset, $\mathscr{F}$ corresponds to a module, and x is not a zerodivisor on that module, as it doesn't vanish at any associated point of that module; see Proposition 6.6.13.) Thus we have a short exact sequence

(18.6.3.1) $$0 \longrightarrow \mathscr{F}(-1) \longrightarrow \mathscr{F} \longrightarrow \mathscr{G} \longrightarrow 0,$$

where $\mathscr{G}$ is a coherent sheaf.

Now $\operatorname{Supp} \mathscr{G} = (\operatorname{Supp} \mathscr{F}) \cap V(x)$. (Reason: Exercise 6.6.D gives the affine/module version of this statement.) Hence $V(x)$ meets all positive-dimensional components of $\operatorname{Supp} \mathscr{F}$ (Exercise 12.3.D(a)), so $\dim \operatorname{Supp} \mathscr{G} = \dim \operatorname{Supp} \mathscr{F} - 1$ by Krull's Principal Ideal Theorem 12.3.3 unless $\mathscr{F} = 0$ (in which case we already know the result, so assume this is not the case).

Twisting (18.6.3.1) by $\mathscr{O}(m)$ yields

$$0 \longrightarrow \mathscr{F}(m-1) \longrightarrow \mathscr{F}(m) \longrightarrow \mathscr{G}(m) \longrightarrow 0.$$

Euler characteristics are additive in exact sequences (Exercise 18.4.A), from which $p_{\mathscr{F}}(m) - p_{\mathscr{F}}(m-1) = p_{\mathscr{G}}(m)$. Now $p_{\mathscr{G}}(m)$ is a polynomial of degree $\dim \operatorname{Supp} \mathscr{F} - 1$.

The result is then a consequence from the following elementary fact about polynomials in one variable.

18.6.B. EXERCISE. Suppose f and g are functions on the integers, $f(m+1) - f(m) = g(m)$ for all m, and $g(m)$ is a polynomial of degree $d \geq 0$. Show that f is a polynomial of degree $d+1$. □

Example 1. The Hilbert polynomial of projective space is $p_{\mathbb{P}^n}(m) = \binom{m+n}{m}$, where we interpret this as the polynomial $(m+1)\cdots(m+n)/n!$.

Example 2. Suppose H is a degree d hypersurface in $\mathbb{P}^n$, with $i: H \hookrightarrow \mathbb{P}^n$ the closed embedding. Then from the closed subscheme exact sequence

$$0 \longrightarrow \mathscr{O}_{\mathbb{P}^n}(-d) \longrightarrow \mathscr{O}_{\mathbb{P}^n} \longrightarrow i_* \mathscr{O}_H \longrightarrow 0,$$

we have

(18.6.3.2) $$p_H(m) = p_{\mathbb{P}^n}(m) - p_{\mathbb{P}^n}(m - d) = \binom{m+n}{n} - \binom{m+n-d}{n}.$$

(Implicit in this argument is the fact that $(i_* \mathscr{O}_H) \otimes \mathscr{O}_{\mathbb{P}^n}(m) \cong i_*(\mathscr{O}_H \otimes i^* \mathscr{O}_{\mathbb{P}^n}(m))$. This follows from the projection formula, Exercise 14.5.L(b). You can also show this directly with transition functions.)

18.6.C. EXERCISE. Show that the twisted cubic ($\mathbb{P}^1$ embedded in $\mathbb{P}^3$ by the complete linear series $|\mathscr{O}_{\mathbb{P}^1}(3)|$) has Hilbert polynomial $3m + 1$. (The twisted cubic was defined in Exercise 9.3.A.)

18.6.D. EXERCISE. More generally, find the Hilbert polynomial for the dth Veronese embedding of $\mathbb{P}^n$ (i.e., the closed embedding of $\mathbb{P}^n$ in a bigger projective space by way of the line bundle $\mathscr{O}(d)$, §9.3.6).

18.6.E. EXERCISE (TO BE USED SEVERAL TIMES IN CHAPTER 19). Suppose

$$X \overset{i}{\hookrightarrow} Y \hookrightarrow \mathbb{P}^n_k$$

is a sequence of closed embeddings.

(a) Show that $p_X(m) \leq p_Y(m)$ for $m \gg 0$. Hint: Let $\mathscr{I}_{X/Y}$ be the ideal sheaf of X in Y. Consider the exact sequence

$$0 \longrightarrow \mathscr{I}_{X/Y}(m) \longrightarrow \mathscr{O}_Y(m) \longrightarrow i_* \mathscr{O}_X(m) \longrightarrow 0.$$

(b) If $p_X(m) = p_Y(m)$ for $m \gg 0$, show that $X = Y$. Hint: Show that if the Hilbert polynomial of a coherent sheaf $\mathscr{F}$ is 0, then $\mathscr{F} = 0$. (Handy trick: For $m \gg 0$, $\mathscr{F}(m)$ is generated by global sections.) Apply this to $\mathscr{F} = \mathscr{I}_{X/Y}$.

From the Hilbert polynomial, we can extract many invariants, of which two are particularly important. The first is the *degree*, and the second is the arithmetic genus (§18.6.7). The **degree of a projective k-scheme of dimension** n is defined to be the leading coefficient of the Hilbert polynomial (the coefficient of m^n) times $n!$.

Using the examples above, we see that the degree of $\mathbb{P}^n$ in itself is 1. The degree of the twisted cubic is 3.

18.6.F. EXERCISE. Show that the degree is always an integer. Hint: By induction, show that any polynomial in m of degree k taking on only integer values must have as the coefficient of m^k an integral multiple of $1/k!$. Hint for this: If $f(x)$ takes on only integral values and is of degree k, then $f(x + 1) - f(x)$ takes on only integral values and is of degree $k - 1$.

18.6.G. EXERCISE. Show that the degree of a degree d hypersurface (Definition 9.3.2) is d (preventing a notational crisis).

18.6.H. EXERCISE (DEGREE = DEGREE). Suppose a regular curve C is embedded in projective space via an invertible sheaf of degree d (as defined in §18.4.6). In other words, this line bundle determines a closed embedding. Show that the degree of C under this embedding is d, preventing another notational crisis. (A similar notational crisis was averted in Exercise 18.4.E.) Hint: Riemann–Roch, Exercise 18.4.B. (The hypothesis of regularity is not necessary. It is included only to make this exercise easier.)

18.6.I. EXERCISE. Show that the degree of the dth Veronese embedding of $\mathbb{P}^n$ is d^n.

18.6.J. EXERCISE (BÉZOUT'S THEOREM, GENERALIZING EXERCISES 9.3.D AND 15.2.I). Suppose $X \subset \mathbb{P}^n_k$ is a projective scheme of dimension at least 1, and H is a hypersurface not containing any associated points of X. (For example, if X is reduced and thus has no embedded points, we are just requiring H not to contain any irreducible components of X.) Show that $\deg(H \cap X) = (\deg H)(\deg X)$.

18.6.4. *Bézout's Theorem for plane curves.* As a consequence, we have *Bézout's Theorem for plane curves*: if C and D are plane curves of degrees m and n respectively, with no common components, then C and D meet at mn points, counted with appropriate multiplicity. (To apply Exercise 18.6.J, you need to know that plane curves have no embedded points. You can do this either by using Exercise 6.6.F, or by assuming that one of the curves is reduced.) The multiplicity of intersection is just the degree of the intersection as a k-scheme.

18.6.K. EXERCISE. Suppose C is a degree 1 curve in $\mathbb{P}^3_k$ (or, more precisely, a degree 1 pure-one-dimensional closed subscheme of $\mathbb{P}^3_k$, with no embedded points). Show that C is a line. Hint: Reduce to the case $k = \overline{k}$. Suppose p and q are distinct closed points on C. Use Bézout's Theorem (Exercise 18.6.J) to show that any hyperplane containing p and q must contain C, and thus that $C \subset \overline{pq}$.

18.6.L. FUN EXERCISE. Let k be a field, which we assume to be algebraically closed for convenience (although you are free to remove this hypothesis if you wish). Suppose C is a degree d integral curve in $\mathbb{P}^N_k$ with $N \geq d$. Show that C is contained in a linear $\mathbb{P}^d_k \subset \mathbb{P}^N_k$. In particular, if $N > d$, C is degenerate (defined in Exercise 15.2.C). (Exercise 18.6.K is not quite a special case of this problem, but the hint may still be helpful.)

18.6.M. EXERCISE (ANOTHER FORM OF BÉZOUT'S THEOREM). Classically, the degree of a complex projective variety of dimension n was defined as follows. We slice the variety with n generally chosen hyperplanes. Then the intersection will be a finite number of reduced points, by Exercise 13.4.C (a consequence of Bertini's Theorem 13.4.2). The degree is this number of points. Use Bézout's Theorem to make sense of this in a way that agrees with our definition of degree.

Thus the classical definition of the degree, which involved making a choice and then showing that the result is independent of choice, has been replaced by making a cohomological definition involving Euler characteristics. This should remind you of how we got around to "correctly" understanding the degree of a line bundle. It was traditionally defined as the degree of a divisor of any nonzero rational section (Important Exercise 18.4.C), and we found a better definition in terms of Euler characteristics (§18.4.6).

18.6.5.** *Aside: Connection to the topological definition of degree (for those with sufficient background).* Another definition of degree of a dimension d *complex* projective variety $X \subset \mathbb{P}^n_{\mathbb{C}}$ is as the number d such that $[X]$ is d times the "positive" generator of $H_{2d}(\mathbb{P}^n_{\mathbb{C}}, \mathbb{Z})$. You can show this by induction on d as follows. Suppose X is a complex projective variety of dimension d. For a generally chosen hyperplane, $H \cap X$ is a complex projective variety of (complex) dimension $d - 1$. (Do you see why?) Show that $[H \cap X] = c_1(\mathcal{O}(1)) \cup [X]$ in $H_{2(d-1)}(X, \mathbb{Z})$, by suitably generalizing the solution to Exercise 18.4.H. (A further generalization is given in §20.1.9.) For this reason, $c_1(\mathcal{O}(1))$ is often called the "hyperplane class."

18.6.6. *Revisiting an earlier example.* We revisit the enlightening example of §10.3.4 and §16.3.10: let $k = \mathbb{Q}$, and consider the parabola $x = y^2$. We intersect it with the four lines, $x = 1$, $x = 0$, $x = -1$, and $x = 2$, and see that we get 2 each time (counted using the same convention as that used the last time we saw this example).

If we intersect it with $y = 2$, we only get one point—but that's because this isn't a projective curve, and we really should be doing this intersection on $\mathbb{P}^2_k$, and, in this case, the conic meets the line in two points, one of which is "at ∞."

18.6.N. EXERCISE. Show that the degree of the d-fold Veronese embedding of $\mathbb{P}^n$ is d^n in a different way from Exercise 18.6.I, as follows. Let $v_d : \mathbb{P}^n \to \mathbb{P}^N$ be the Veronese embedding. To find the degree of the image, we intersect it with n hyperplanes in $\mathbb{P}^N$ (scheme-theoretically), and find the number of intersection points (counted with multiplicity). But the pullback of a hyperplane in $\mathbb{P}^N$ to $\mathbb{P}^n$ is a degree d hypersurface. Perform this intersection in $\mathbb{P}^n$, and use Bézout's theorem (Exercise 18.6.J).

18.6.O. EXERCISE (DEGREE IS ADDITIVE FOR UNIONS). Suppose X and Y are two d-dimensional closed subschemes of $\mathbb{P}^n_k$, with no d-dimensional irreducible components in common. Show that $\deg X \cup Y = \deg X + \deg Y$.

18.6.7. Arithmetic genus, again.

There is another central piece of information residing in the Hilbert polynomial. Notice that $p_X(0) = \chi(X, \mathscr{O}_X)$ is an *intrinsic* invariant of the scheme X, independent of the projective embedding. It has the same information as the arithmetic genus $(-1)^{\dim X}(\chi(X, \mathscr{O}_X) - 1)$ (§18.4.4).

Imagine how amazing this must have seemed to the ancients: they defined the Hilbert function by counting how many "functions of various degrees" there are; then they noticed that when the degree gets large, it agrees with a polynomial; and then when they plugged 0 into the polynomial—extrapolating backward, to where the Hilbert function and Hilbert polynomials didn't agree—they found a magic invariant! Furthermore, in the case when X is a complex curve, this invariant was basically the topological genus!

We can now see a large family of curves over an algebraically closed field that is provably not $\mathbb{P}^1$! Note that the Hilbert polynomial of $\mathbb{P}^1$ is $(m + 1)/1 = m + 1$, so $\chi(\mathscr{O}_{\mathbb{P}^1}) = 1$. Suppose C is a degree d curve in $\mathbb{P}^2$. Then the Hilbert polynomial of C is

$$p_{\mathbb{P}^2}(m) - p_{\mathbb{P}^2}(m - d) = (m + 1)(m + 2)/2 - (m - d + 1)(m - d + 2)/2.$$

Plugging in $m = 0$ gives us $-(d^2 - 3d)/2$. Thus when $d > 2$, we have a curve that cannot be isomorphic to $\mathbb{P}^1$! (And it is not hard to show that there exists a *regular* degree d curve, Exercise 13.3.E.)

Now from $0 \to \mathscr{O}_{\mathbb{P}^2}(-d) \to \mathscr{O}_{\mathbb{P}^2} \to \mathscr{O}_C \to 0$, using $h^1(\mathbb{P}^2, \mathscr{O}_{\mathbb{P}^2}(-d)) = 0$, we have that $h^0(C, \mathscr{O}_C) = 1$. As $h^0 - h^1 = \chi$, we have

$$(18.6.7.1) \qquad\qquad h^1(C, \mathscr{O}_C) = (d - 1)(d - 2)/2.$$

We now revisit an interesting question we first saw in §7.5.11. If k is an algebraically closed field, is every finitely generated transcendence degree 1 extension of k isomorphic to $k(x)$? In that section, we found ad hoc (but admittedly beautiful) examples showing that the answer is "no." But we now have a better answer. The question initially looks like an algebraic question, but we now recognize it as a fundamentally geometric one. There is an integer-valued cohomological invariant of such field extensions that has good geometric meaning: the genus.

Equation (18.6.7.1) yields examples of curves of genus $0, 1, 3, 6, \dots$ (corresponding to degree 1 or 2, 3, 4, 5, $\dots$). This begs some questions, such as: Are there curves of other genera? (We will see soon, in §19.5.5, that the answer is yes.) Are there other genus 0 curves? (Not if k is algebraically closed, but sometimes yes otherwise—consider $x^2 + y^2 + z^2 = 0$ in $\mathbb{P}^2_{\mathbb{R}}$, which has no $\mathbb{R}$-points and hence is not isomorphic to $\mathbb{P}^1_{\mathbb{R}}$; we will discuss this more in §19.3.) Do we have all the curves of genus 3? (Almost all, but not quite. We will see more in §19.7.) Do we have all the curves of genus 6? (We are missing "most of them," as will be suggested by §19.8.4.)

18.6.8. *Caution.* The Euler characteristic of the structure sheaf is an incomplete invariant. We will now see that it doesn't always distinguish between isomorphism classes of irreducible smooth projective varieties.

18.6.P. EXERCISE. If a hypersurface $X \subset \mathbb{P}^n_k$ has degree $\leq n$, show that $\chi(X, \mathscr{O}_X) = 1$.

Thus, for example, if $X = \mathbb{P}^1_k \times \mathbb{P}^1_k$, then $\chi(X, \mathscr{O}_X) = 1$, as X can be described as a quadric surface in $\mathbb{P}^3_k$. Then $\mathbb{P}^2_k$ and $\mathbb{P}^1_k \times \mathbb{P}^1_k$ both have structure sheaf Euler characteristic 1 (see Theorem 18.1.2 for the case of $\mathbb{P}^2_k$), but are not isomorphic—$\operatorname{Pic} \mathbb{P}^2_k \cong \mathbb{Z}$ (§15.4.15) while $\operatorname{Pic}(\mathbb{P}^1_k \times \mathbb{P}^1_k) \cong \mathbb{Z} \oplus \mathbb{Z}$ (Exercise 15.5.K).

18.6.9. Complete intersections in $\mathbb{P}^n_k$. Recall that a codimension r complete intersection in $\mathbb{P}^n_k$ is the complete intersection of r hypersurfaces in $\mathbb{P}^n_k$ (§9.5.8), which by Krull's Height Theorem 12.3.10 has pure codimension r.

18.6.Q. EXERCISE. Suppose X is a dimension m complete intersection in $\mathbb{P}^N_k$. Show that $h^i(X, \mathscr{O}_X(n)) = 0$ for $0 < i < m$.

18.6.R. EXERCISE. Find the genus of the complete intersection of two quadric surfaces in $\mathbb{P}^3_k$.

18.6.S. EXERCISE. More generally, find the genus of the complete intersection of a degree m surface with a degree n surface in $\mathbb{P}^3_k$. (If $m = 2$ and $n = 3$, you should get genus 4. We will see in §19.8 that in some sense most genus 4 curves arise in this way. Note that Bertini's Theorem 13.4.2 ensures that there *are* regular curves of this form.)

18.6.T. EXERCISE. Show that the rational normal curve of degree d in $\mathbb{P}^d$ is *not* a complete intersection if $d > 2$. (Hint: If it *were* the complete intersection of $d - 1$ hypersurfaces, what would the degree of the hypersurfaces be? Why could none of the degrees be 1?)

18.6.U. EXERCISE (POSITIVE-DIMENSIONAL COMPLETE INTERSECTIONS IN $\mathbb{P}^n_k$ ARE CONNECTED; CF. EXERCISE 18.5.B). Show that complete intersections in $\mathbb{P}^n_k$ of *positive* dimension are connected. Hint: Show that $h^0(X, \mathscr{O}_X) = 1$. (This argument will even show that they are geometrically connected, defined in §10.5.1, as h^0 is preserved by field extension; see Exercise 18.2.H.)

Notice that this statement does not require X to be reduced, so, for example, this implies that that if X is the scheme cut out by $x^2 = 0$ in $\mathbb{P}^2_k$, then $h^0(X, \mathscr{O}_X) = 1$—there are no nonzero nilpotent functions, even though there are local functions that are nonzero and nilpotent! This example came up earlier in §5.2.3 and §11.4.7.

18.6.V. EXERCISE. Show that the union of two planes in $\mathbb{P}^4$ meeting at a point is not a complete intersection. Hint: It is connected, but you can slice with another hyperplane and get something not connected (see Exercise 18.6.U).

18.7 Higher Pushforward (or Direct Image) Sheaves

Cohomology groups were defined for $X \to \operatorname{Spec} A$, where the structure morphism is quasicompact and separated; for any quasicoherent $\mathscr{F}$ on X, we defined $H^i(X, \mathscr{F})$. We will now define a "relative" version of this notion, for quasicompact and separated morphisms $\pi: X \to Y$: for any quasicoherent $\mathscr{F}$ on X, we will define $R^i\pi_*\mathscr{F}$, a quasicoherent sheaf on Y. (Now would be a good time to do Exercise 1.5.I, the FHHF Theorem, if you haven't done it before.)

We have many motivations for doing this. In no particular order:

(1) It "globalizes" what we did before with cohomology.
(2) If $0 \to \mathscr{F} \to \mathscr{G} \to \mathscr{H} \to 0$ is a short exact sequence of quasicoherent sheaves on X, then we know that $0 \to \pi_*\mathscr{F} \to \pi_*\mathscr{G} \to \pi_*\mathscr{H}$ is exact, and higher pushforwards will extend this to a long exact sequence.
(3) We will later see that this will show how cohomology groups vary in families, especially in "nice" situations. Intuitively, if we have a nice family of varieties, and a family of sheaves on them, we could hope that the cohomology varies nicely in families, and in fact, in "nice" situations, this is true. (As always, "nice" usually means "flat," whatever that means. We will see that Euler characteristics are locally constant in proper flat families in §24.7, and the Cohomology and Base Change Theorem 25.1.6 will show that in particularly good situations, dimensions of cohomology groups are constant.)

All of the important properties of cohomology described in §18.1 will carry over to this more general situation. Best of all, there will be no extra work required.

In the notation $R^i\pi_*\mathscr{F}$ for higher pushforward sheaves, the "R" stands for "right derived functor," and corresponds to the fact that we get a long exact sequence in cohomology extending to the right (from the 0th terms). In Chapter 23, we will see that in good circumstances, if we have a left-exact functor, there is a long exact sequence going off to the right, in terms of right derived functors. Similarly, if we have a right-exact functor (e.g., if M is an A-module, then $\otimes_A M$ is a right-exact functor from the category of A-modules to itself), there may be a long exact sequence going off to the left, in terms of left derived functors.

Suppose $\pi\colon X \to Y$, and $\mathscr{F}$ is a quasicoherent sheaf on X. For each $\operatorname{Spec} A \subset Y$, we have A-modules $H^i(\pi^{-1}(\operatorname{Spec} A), \mathscr{F})$. We now show that these patch together to form a quasicoherent sheaf, in the sense of §6.2.3. We need check only one fact: that this behaves well with respect to taking distinguished open sets. In other words, we must check that for each $f \in A$, the natural map

$$H^i(\pi^{-1}(\operatorname{Spec} A), \mathscr{F}) \longrightarrow H^i(\pi^{-1}(\operatorname{Spec} A_f), \mathscr{F})$$

(induced by the map of spaces in the opposite direction—H^i is contravariant in the space) is precisely the localization $\otimes_A A_f$. But this can be verified easily. Let $\{U_i\}$ be an affine open cover of $\pi^{-1}(\operatorname{Spec} A)$. We can compute $H^i(\pi^{-1}(\operatorname{Spec} A), \mathscr{F})$ using the Čech complex (18.2.1.1). But this induces a cover of $\pi^{-1}(\operatorname{Spec} A_f)$ in a natural way: if $U_i = \operatorname{Spec} A_i$ is an affine open subset of $\pi^{-1}(\operatorname{Spec} A)$, we define $U_i' := \operatorname{Spec}(A_i)_{\pi^\sharp f}$. The resulting Čech complex for $\operatorname{Spec} A_f$ is the localization of the Čech complex for $\operatorname{Spec} A$. As taking cohomology of a complex commutes with localization (as discussed in the FHHF Theorem, Exercise 1.5.I), we have defined a quasicoherent sheaf on Y by the characterization of quasicoherent sheaves in §6.2.3.

Define the i**th higher pushforward sheaf** or the i**th higher direct image sheaf** $R^i\pi_*\mathscr{F}$ to be this quasicoherent sheaf.

Caution: Although when $U \subset Y$ is an affine open subset, $H^0(U, R^\bullet\pi_*\mathscr{F}) = H^\bullet(\pi^{-1}(U), \mathscr{F})$, this is not true for open subsets $U \subset Y$ in general. You might later try to construct an example of this behavior.

18.7.1. Theorem — *Suppose $\pi\colon X \to Y$ is a quasicompact separated morphism of schemes. Then:*

(a) *$R^i\pi_*$ is a covariant functor $\operatorname{QCoh}_X \to \operatorname{QCoh}_Y$.*

(b) *We can identify $R^0\pi_*$ with π_*. (More precisely: we have a natural isomorphism of functors.)*

(c) *(the **long exact sequence of higher pushforward sheaves**) A short exact sequence $0 \to \mathscr{F} \to \mathscr{G} \to \mathscr{H} \to 0$ of sheaves on X induces a long exact sequence*

$$(18.7.1.1) \qquad 0 \longrightarrow R^0\pi_*\mathscr{F} \longrightarrow R^0\pi_*\mathscr{G} \longrightarrow R^0\pi_*\mathscr{H} \longrightarrow$$
$$R^1\pi_*\mathscr{F} \longrightarrow R^1\pi_*\mathscr{G} \longrightarrow R^1\pi_*\mathscr{H} \longrightarrow \cdots$$

of sheaves on Y.

(d) *(projective pushforwards of coherent are coherent: Grothendieck's Coherence Theorem for projective morphisms) If π is a projective morphism, Y is locally Noetherian, and $\mathscr{F}$ is a coherent sheaf on X, then for all i, $R^i\pi_*\mathscr{F}$ is a coherent sheaf on Y.*

18.7.2. *Unimportant remark.* If X and Y are Noetherian, the hypotheses "quasicompact and separated" on π can be removed; see Unimportant Remark 23.5.8 for the "separated" hypothesis.

18.7.3. *Proof of Theorem 18.7.1.* We first show covariance: if $\mathscr{F} \to \mathscr{G}$ is a morphism of quasicoherent sheaves on X, we define a map $R^i\pi_*\mathscr{F} \to R^i\pi_*\mathscr{G}$. (It will be clear we will have shown that $R^i\pi_*$ is a functor.) It suffices to define this map on the "distinguished affine base" of Y (Definition 6.2.1). Thus it suffices to show the following: if X' is a quasicompact separated A-scheme, and $\mathscr{F} \to \mathscr{G}$ is a morphism of quasicoherent sheaves on X, then the map $H^i(X', \mathscr{F}) \to H^i(X', \mathscr{G})$ constructed in §18.2 (property **(i)** of §18.1) "commutes with localization at $f \in A$." But this was shown in Exercise 18.2.D.

In a similar way, we construct the connecting homomorphism $R^i \pi_* \mathcal{H} \to R^{i+1} \pi_* \mathcal{F}$ in the long exact sequence (18.7.1.1), by showing that the construction in the case where $Y = \operatorname{Spec} A$ "commutes with localization at $f \in A$." Again, this was shown in Exercise 18.2.D.

It suffices to check all other parts of this statement on affine open subsets of Y, so they all follow from the analogous statements in Čech cohomology (§18.1). □

The following result is handy, and essentially immediate from our definition.

18.7.A. EASY EXERCISE. Show that if π is affine, then for $i > 0$, $R^i \pi_* \mathcal{F} = 0$.

18.7.4. *How higher pushforwards behave with respect to base change.*

18.7.B. EXERCISE (HIGHER PUSHFORWARDS AND BASE CHANGE).

(a) *(the push-pull map for cohomology)* Suppose $\psi \colon Y' \to Y$ is any morphism, and $\pi \colon X \to Y$ is quasicompact and separated. Suppose $\mathcal{F}$ is a quasicoherent sheaf on X. Let

(18.7.4.1)
$$
\begin{array}{ccc}
X' & \xrightarrow{\psi'} & X \\
{\scriptstyle \pi'} \downarrow & & \downarrow {\scriptstyle \pi} \\
Y' & \xrightarrow{\psi} & Y
\end{array}
$$

be a Cartesian diagram. Describe a natural morphism

$$
\psi^* (R^i \pi_* \mathcal{F}) \longrightarrow R^i \pi'_* (\psi')^* \mathcal{F}
$$

of quasicoherent sheaves on Y'. (Hint: The FHHF Theorem, Exercise 1.5.I.) Show that the $i = 0$ case is the push-pull map of Exercise 14.5.K. So this might reasonably be called the **push-pull map for cohomology**.

(b) *(cohomology commutes with affine flat base change)* If $\psi \colon Y' \to Y$ is an affine morphism, and for a cover $\operatorname{Spec} A_i$ of Y, where $\psi^{-1}(\operatorname{Spec} A_i) = \operatorname{Spec} B_i$, B_i is a *flat* A_i-algebra (§1.5.13: $\otimes_{A_i} B_i$ is exact), and the diagram in (a) is a Cartesian diagram, show that the push-pull map of (a) is an isomorphism. (Exercise 18.2.H was a special case of this exercise. You can likely generalize this to non-affine morphisms—and thus show that cohomology commutes with flat base change (Theorem 24.1.9)—but we wait until Chapter 24 to discuss flatness at length.)

18.7.C. EXERCISE (CF. EXERCISE 14.5.K). Prove Exercise 18.7.B(a) *without* the hypothesis that (18.7.4.1) is a Cartesian diagram, but adding the requirement that π' quasicompact and separated (just so our definition of $R^i \pi'_*$ applies). In the course of the proof, you will see a map arising in the Leray spectral sequence (Theorem 23.4.5). (Hint: *Use* Exercise 18.7.B(a).)

A useful special case of Exercise 18.7.B(a) is the following.

18.7.D. EXERCISE. If $q \in Y$, describe a natural morphism $(R^i \pi_* \mathcal{F}) \otimes \kappa(q) \to H^i(\pi^{-1}(q), \mathcal{F}|_{\pi^{-1}(q)})$. (See §14.5.8 for the case $i = 0$.) Be sure you understand the meaning of the left side! Hint: The FHHF Theorem, Exercise 1.5.I.

Thus the fiber of the pushforward may not be the cohomology of the fiber, but at least it always maps to it. We will later see that in good situations this map is an isomorphism, and thus the higher pushforward sheaf indeed "patches together" the cohomology on fibers (the Cohomology and Base Change Theorem 25.1.6).

18.7.E. EXERCISE (THE PROJECTION FORMULA, GENERALIZING EXERCISE 14.5.L). Suppose $\pi \colon X \to Y$ is quasicompact and separated, and $\mathcal{F}$ and $\mathcal{G}$ are quasicoherent sheaves on

X and Y respectively.

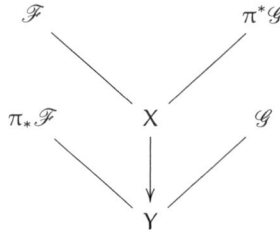

(a) Describe a natural morphism

$$(R^i\pi_*\mathscr{F}) \otimes \mathscr{G} \longrightarrow R^i\pi_*(\mathscr{F} \otimes \pi^*\mathscr{G}).$$

(Hint: The FHHF Theorem, Exercise 1.5.I.)

(b) If $\mathscr{G}$ is locally free, show that this natural morphism is an isomorphism.

The following fact uses the same trick as Theorem 18.1.6 and Exercise 18.1.B.

18.7.5. Theorem (relative dimensional cohomology vanishing) — *If $\pi: X \to Y$ is a projective morphism and Y is locally Noetherian, then the higher pushforwards vanish in degree higher than the maximum dimension of the fibers.*

This is false without the projective hypothesis (see Example 18.7.6 below). In particular, you might hope that just as dimensional cohomology vanishing generalized from projective varieties to quasiprojective varieties or more general settings (§18.2.7), relative dimensional cohomology vanishing would generalize from projective morphisms to quasiprojective morphisms, but this is not the case.

18.7.6. *Example.* Consider the open embedding $\pi: \mathbb{A}_k^n - \{0\} \to \mathbb{A}_k^n$ for $n > 1$. From §18.3.1–18.3.2, $R^{n-1}\pi_*\mathscr{O}_{\mathbb{A}_k^n - \{0\}} \neq 0$.

Proof of Theorem 18.7.5. Let $\mathfrak{m}$ be the maximum dimension of all the fibers.

The question is local on Y, so we will show that the result holds near a point p of Y. We may assume that Y is affine, and hence that $X \hookrightarrow \mathbb{P}_Y^n$.

Let k be the residue field at p. Then $\pi^{-1}(p)$ is a projective k-scheme of dimension at most $\mathfrak{m}$. By Exercise 12.3.D we can find affine open sets $D(f_1), \ldots, D(f_{m+1})$ that cover $\pi^{-1}(p)$. In other words, the intersection of the $V(f_i)$ does not intersect $\pi^{-1}(p)$.

If $Y = \operatorname{Spec} A$ and $p = [\mathfrak{p}]$ (so $k = A_\mathfrak{p}/\mathfrak{p}A_\mathfrak{p}$), then arbitrarily lift each f_i from an element of $k[x_0, \ldots, x_n]$ to an element f_i' of $A_\mathfrak{p}[x_0, \ldots, x_n]$. Let F be the product of the denominators of the f_i'; note that $F \notin \mathfrak{p}$, i.e., $p = [\mathfrak{p}] \in D(F)$. Then $f_i' \in A_F[x_0, \ldots, x_n]$. The intersection of their zero loci $\cap_i V(f_i') \subset \mathbb{P}_{A_F}^n$ is a closed subscheme of $\mathbb{P}_{A_F}^n$. Intersect it with X to get another closed subscheme of $\mathbb{P}_{A_F}^n$. Take its image under π; as projective morphisms are closed, we get a closed subset of $D(F) = \operatorname{Spec} A_F$. But this closed subset does not include p; hence we can find an affine open neighborhood $\operatorname{Spec} B$ of p in Y missing the image. But if f_i'' are the restrictions of f_i' to $B[x_0, \ldots, x_n]$, then $D(f_i'')$ cover $\pi^{-1}(\operatorname{Spec} B)$; in other words, $\pi^{-1}(\operatorname{Spec} B)$ is covered by $\mathfrak{m} + 1$ affine open sets, so by affine cover cohomology vanishing (Property **(vi)** of §18.1), its cohomology vanishes in degree at least $\mathfrak{m} + 1$. But the higher direct image sheaf is computed using these cohomology groups, hence the higher pushforward sheaf $R^i\pi_*\mathscr{F}$ vanishes on $\operatorname{Spec} B$ too. $\square$

18.7.F. EXERCISE (RELATIVE SERRE VANISHING; CF. THEOREM 18.1.3(II)). Suppose $\pi: X \to Y$ is a proper morphism of Noetherian schemes, and $\mathscr{L}$ is a π-ample invertible sheaf on X. Show that for any coherent sheaf $\mathscr{F}$ on X, for $m \gg 0$, $R^i\pi_*(\mathscr{F} \otimes \mathscr{L}^{\otimes m}) = 0$ for all $i > 0$.

18.8*   Serre's Characterizations of Ampleness and Affineness

Theorem 16.2.2 gave a number of characterizations of ampleness, in terms of projective geometry, global generation, and the Zariski topology. Here is another characterization, this time cohomological, under Noetherian hypotheses. Because (somewhat surprisingly) we won't use this result, this section is starred.

18.8.1. Theorem (Serre's cohomological criterion for ampleness) — *Suppose A is a Noetherian ring, X is a proper A-scheme, and $\mathscr{L}$ is an invertible sheaf on X. Then the following are equivalent.*

(a–c) The invertible sheaf $\mathscr{L}$ is ample on X (over A).

  (e) For all coherent sheaves $\mathscr{F}$ on X, there is an n_0 such that for $n \geq n_0$, $H^i(X, \mathscr{F} \otimes \mathscr{L}^{\otimes n}) = 0$ for all $i > 0$.

The label (a–c) is intended to reflect the statement of Theorem 16.2.2. We avoid the label (d) because it appeared in Theorem 16.2.6. (Aside: The "properness" assumption cannot be removed; see Example 18.7.6.) Before getting to the proof, we motivate this result by giving some applications. (As a warm-up, you can give a second solution to Exercise 16.2.G in the Noetherian case, using the affineness of π to show that $H^i(X, \mathscr{F} \otimes (\pi^*\mathscr{L})^{\otimes m}) = H^i(Y, (\pi_*\mathscr{F}) \otimes \mathscr{L}^{\otimes m})$.)

18.8.A. EXERCISE. Suppose X is a proper A-scheme (A Noetherian), and $\mathscr{L}$ is an invertible sheaf on X. Show that $\mathscr{L}$ is ample on X if and only if $\mathscr{L}|_{X^{red}}$ is ample on X^{red}. Hint: For the "only if" direction, use Exercise 16.2.G. For the "if" direction, let $\mathscr{I}$ be the ideal sheaf cutting out the closed subscheme X^{red} in X. Filter $\mathscr{F}$ by powers of $\mathscr{I}$:

$$0 = \mathscr{I}^r\mathscr{F} \subset \mathscr{I}^{r-1}\mathscr{F} \subset \cdots \subset \mathscr{I}\mathscr{F} \subset \mathscr{F}.$$

(Essentially the same filtration appeared in Exercise 18.4.L, for similar reasons.) Show that each quotient $\mathscr{I}^n\mathscr{F}/\mathscr{I}^{n-1}\mathscr{F}$, twisted by a high enough power of $\mathscr{L}$, has no higher cohomology. Use descending induction on n to show each part $\mathscr{I}^n\mathscr{F}$ of the filtration (and hence, in particular, $\mathscr{F}$) has this property as well.

18.8.B. EXERCISE. Suppose X is a proper A-scheme (A Noetherian), and $\mathscr{L}$ is an invertible sheaf on X. Show that $\mathscr{L}$ is ample on X if and only if $\mathscr{L}$ is ample on each component. Hint: Follow the outline of the solution to the previous exercise, taking instead $\mathscr{I}$ as the ideal sheaf of one component. Perhaps first reduce to the case where $X = X^{red}$.

18.8.C. TRICKY EXERCISE. Suppose C is a proper *reduced* curve (over a field k), and $\mathscr{L}$ is a line bundle on C, such that $\nu^*\mathscr{L}$ is an ample line bundle on the normalization $\tilde{C}$ (where the normalization as usual is $\nu : \tilde{C} \to C$). Show that $\mathscr{L}$ is ample on C, as follows. First read §17.1.6 to understand the morphism $\nu^{!?} : QCoh_C \to QCoh_{\tilde{C}}$. Then for every quasicoherent sheaf $\mathscr{F}$ on C, describe an exact sequence of quasicoherent sheaves on C:

$$0 \longrightarrow \mathscr{T} \longrightarrow \nu_*\nu^{!?}\mathscr{F} \longrightarrow \mathscr{F} \longrightarrow \mathscr{Q} \longrightarrow 0,$$

where $\mathscr{T}$ and $\mathscr{Q}$ are both supported on a set of dimension 0 (a finite set of points). Then show that for $n \gg 0$, $H^i(C, \mathscr{F} \otimes \mathscr{L}^{\otimes n}) = 0$ for $i > 0$. You will use the projection formula (Exercise 18.7.E(b)) to show that $H^i(C, (\nu_*\nu^{!?}\mathscr{F}) \otimes \mathscr{L}^{\otimes n}) = 0$ for $i > 0$ and $n \gg 0$.

18.8.2. *Very ample versus ample.* The previous exercises don't work with "ample" replaced by "very ample," which shows again how the notion of ampleness is better behaved than very ampleness.

18.8.3. *Proof of Theorem 18.8.1.* For (a–c) implies (e), use the fact that $\mathscr{L}^{\otimes N}$ is very ample for some N (Theorem 16.2.2(a)), and apply Serre vanishing (Theorem 18.1.3(ii)) to $\mathscr{F}$, $\mathscr{F} \otimes \mathscr{L}, \ldots,$ and $\mathscr{F} \otimes \mathscr{L}^{\otimes(N-1)}$.

So we now assume (e), and show that $\mathscr{L}$ is ample by criterion (b) of Theorem 16.2.2: we will show that for any coherent sheaf $\mathscr{F}$ on X, $\mathscr{F} \otimes \mathscr{L}^{\otimes n}$ is globally generated for $n \gg 0$.

We begin with a special case: we will show that $\mathscr{L}^{\otimes n}$ is globally generated (i.e., base-point-free) for $n \gg 0$. To do this, it suffices to show that every closed point p has an open neighborhood U so that there exists some N_p so that for $n \geq N_p$, $\mathscr{L}^{\otimes n}$ is globally generated for all points of U_p. (Reason: By quasicompactness, every closed subset of X contains a closed point, by Exercise 5.1.E. So as p varies over the closed points of X, these U_p cover X. By quasicompactness again, we can cover X by a finite number of these U_p. Let N be the maximum of the corresponding N_p. Then for $n \geq N$, $\mathscr{L}^{\otimes n}$ is globally generated in each of these U_p, and hence on all of X.)

Let p be a closed point of X. For all n, $\mathscr{I}_p \otimes \mathscr{L}^{\otimes n}$ is coherent (by our Noetherian hypotheses). (Here $\mathscr{I}_p$ is the ideal sheaf of p.) By (e), there exists some n_0 so that for $n \geq n_0$, $H^1(X, \mathscr{I}_p \otimes \mathscr{L}^{\otimes n}) = 0$. By the long exact sequence arising from the closed subscheme exact sequence

$$0 \longrightarrow \mathscr{I}_p \otimes \mathscr{L}^{\otimes n} \longrightarrow \mathscr{L}^{\otimes n} \longrightarrow \mathscr{L}^{\otimes n}|_p \longrightarrow 0,$$

we have that $\mathscr{L}^{\otimes n}$ is globally generated at p for $n \geq n_0$. By Exercise 16.1.D(b), there is an open neighborhood V_0 of p such that $\mathscr{L}^{\otimes n_0}$ is globally generated at all points of V_0. Thus $\mathscr{L}^{\otimes kn_0}$ is globally generated at all points of V_0 for all positive integers k (using Easy Exercise 16.1.C). For each $i \in \{1, \ldots, n_0 - 1\}$, there is an open neighborhood V_i of p such that $\mathscr{L}^{\otimes(n_0+i)}$ is globally generated at all points of V_i (again by Exercise 16.1.D(b)). We may take each V_i to be contained in V_0. By Easy Exercise 16.1.C, $\mathscr{L}^{\otimes(kn_0+n_0+i)}$ is globally generated at every point of V_i (as this is the case for $\mathscr{L}^{\otimes kn_0}$ and $\mathscr{L}^{\otimes(n_0+i)}$). Thus in the open neighborhood $U_p := \cap_{i=0}^{n-1} V_i$, $\mathscr{L}^{\otimes n}$ is globally generated for $n \geq N_p := 2n_0$.

We have now shown that there exists some N such that for $n \geq N$, $\mathscr{L}^{\otimes n}$ is globally generated. Now suppose $\mathscr{F}$ is a coherent sheaf. To conclude the proof, we will show that $\mathscr{F} \otimes \mathscr{L}^{\otimes n}$ is globally generated for $n \gg 0$. This argument has a similar flavor to what we have done so far, so we give it as an exercise.

18.8.D. EXERCISE. Suppose p is a closed point of X.

(a) Show that for $n \gg 0$, $\mathscr{F} \otimes \mathscr{L}^{\otimes n}$ is globally generated at p.

(b) Show that there exists an open neighborhood U_p of p such that for $n \gg 0$, $\mathscr{F} \otimes \mathscr{L}^{\otimes n}$ is globally generated at every point of U_p. Caution: While it is true that by Exercise 16.1.D(b), for each $n \gg 0$, there is some open neighborhood V_n of p such that $\mathscr{F} \otimes \mathscr{L}^{\otimes n}$ is globally generated there, it need not be true that

(18.8.3.1) $$\cap_{n \gg 0} V_n$$

is an open set. You may need to use the fact that $\mathscr{L}^{\otimes n}$ is globally generated for $n \geq N$ to replace (18.8.3.1) by a finite intersection.

18.8.E. EXERCISE. Conclude the proof of Theorem 18.8.1 by showing that $\mathscr{F} \otimes \mathscr{L}^{\otimes n}$ is globally generated for $n \gg 0$. □

18.8.4. Serre's cohomological characterization of affineness. Serre's characterization of affineness is similar to Theorem 18.8.1.

18.8.5. Theorem (Serre's cohomological characterization of affineness) — *Suppose X is a quasicompact separated scheme. Then the following are equivalent.*

(a) *The scheme X is affine.*

(b) *For every quasicoherent sheaf $\mathscr{F}$ on X, $H^i(X, \mathscr{F}) = 0$ for all $i > 0$.*

(c) *For every quasicoherent sheaf $\mathscr{F}$ on X, $H^1(X, \mathscr{F}) = 0$.*

Proof. Clearly (a) implies (b) implies (c) (the former from "affine cover cohomology vanishing," §18.1 **(vi)**), so we show that (c) implies (a).

Let $A = \Gamma(X, \mathscr{O}_X)$, so we have a canonical map $\pi : X \to \operatorname{Spec} A$ (defined in Exercise 7.3.G and the paragraph following it), and it suffices to prove that π is an affine morphism. In other words, we wish to find $f_i \in A$ such that (i) the f_i generate A (so $\cup D(f_i) = \operatorname{Spec} A$), and also (ii) X_{f_i} is affine (where $X_{f_i} = \{p \in X : f_i(p) \neq 0\}$, Definition 6.2.8).

We first produce affine X_{f_i} covering X. For any point $p \in X$, let U be an affine neighborhood of $p \in X$, and let $\mathscr{I}$ be the ideal sheaf of $X \setminus U$ (with induced reduced subscheme structure). Let q be any closed point of U (it needn't be a closed point of X!) in the closure of p.

18.8.F. EXERCISE. Show that $\mathscr{I} \to \mathscr{I}|_{\overline{q}}$ is a surjection of quasicoherent sheaves, and use the hypothesis (c) to show that it is surjective on global sections. Show that there exists $f \in A$, such that $X_f \subset U$ is affine, and contains p.

By quasicompactness of X, we may cover X with a finite number of such X_f, say $f \in \{f_1, \ldots, f_n\}$. In order to show that $X \to \operatorname{Spec} A$ is an affine morphism, we need that $D(f_1) \cup \cdots \cup D(f_n) = \operatorname{Spec} A$, i.e., $(f_1, \ldots, f_n) = A$.

18.8.G. EXERCISE. Prove this, by showing the map of quasicoherent sheaves on X

$$\cdot(f_1, \ldots, f_n) : \mathscr{O}_X^{\oplus n} \longrightarrow \mathscr{O}_X$$

is a surjection, and then taking global sections and invoking (c). $\qquad\qquad\square$

18.8.H. EXERCISE. With the same hypotheses as in Theorem 18.8.5 above, show that (a)–(c) are equivalent to:

(d) For every quasicoherent sheaf of ideals $\mathscr{I}$ on X, $H^1(X, \mathscr{I}) = 0$.

18.8.6. Serre proved an analogous result in complex analytic geometry: Stein spaces are also characterized by the vanishing of cohomology of coherent sheaves.

18.9*   From Projective to Proper Hypotheses: Chow's Lemma and Grothendieck's Coherence Theorem

The main proofs in this section are double-starred because the results are not absolutely necessary in the rest of our discussions, and may not be worth reading right now. But just knowing the statement of Grothendieck's Coherence Theorem 18.9.1 (generalizing Theorem 18.7.1(d)) will allow you to immediately translate many of our arguments about projective schemes and morphisms to proper schemes and morphisms, and Chow's Lemma is a multipurpose tool to extend results from the projective situation to the proper situation in general.

So you have four choices: (i) Read the proof of Grothendieck's Coherence Theorem; (ii) skim the proof and convince yourself that you *could* understand it if you wanted to; (iii) skip the proof and assume it is true; or (iv) just retreat to projective hypotheses whenever Grothendieck's Coherence Theorem is invoked. Life is full of compromises, and I won't judge you.

18.9.1. Grothendieck's Coherence Theorem — *Suppose $\pi : X \to Y$ is a proper morphism of locally Noetherian schemes. Then for any coherent sheaf $\mathscr{F}$ on X, $R^i \pi_* \mathscr{F}$ is coherent on Y.*

The case $i = 0$ has already been mentioned a number of times.

18.9.A. EXERCISE. Recall that finite morphisms are affine (by definition) and proper. Use Theorem 18.9.1 to show that if $\pi : X \to Y$ is proper and affine and Y is Noetherian, then π is finite. (Hint: Mimic the proof of the weaker result where proper is replaced by projective, Corollary 18.1.5.)

The proof of Theorem 18.9.1 requires two sophisticated facts. The first is the Leray spectral sequence (Theorem 23.4.5, which applies in this situation because of Exercise 23.5.F). Suppose $\pi : X \to Y$ and $\rho : Y \to Z$ are quasicompact separated morphisms. Then for any quasicoherent sheaf

$\mathscr{F}$ on X, there is a spectral sequence with the E_2 term given by $R^p\rho_*(R^q\pi_*\mathscr{F})$ converging to $R^{p+q}(\rho\circ\pi)_*\mathscr{F}$. Because this would be a reasonable (but hard) exercise in the case we need it (where Z is affine), we will feel comfortable using it. But because we will later prove it, we won't prove it now.

We will also need Chow's Lemma.

18.9.2. Chow's Lemma (proper = projective minus birational projective) — *Suppose* $\pi\colon X\to\operatorname{Spec}A$ *is a proper morphism, and A is Noetherian. Then there exists* $\mu\colon X'\to X$ *which is surjective and projective, such that* $\pi\circ\mu$ *is also projective, and such that* μ *is an isomorphism on a dense open subset of X:*

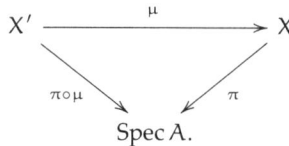

$$X' \xrightarrow{\ \mu\ } X$$
$$\pi\circ\mu \searrow \qquad \swarrow \pi$$
$$\operatorname{Spec}A.$$

In particular, if X is a proper k-variety, it admits a projective birational morphism from a projective k-variety.

Many generalizations of results from projective to proper situations go through Chow's Lemma. We will prove this version, and state other versions of Chow's Lemma, in §18.9.4. Assuming these two facts, we now prove Grothendieck's Coherence Theorem 18.9.1 in a series of exercises.

18.9.3.** *Proof of Grothendieck's Coherence Theorem 18.9.1, assuming Chow's Lemma 18.9.2.* The question is local on Y, so we may assume Y is affine, say, $Y=\operatorname{Spec}A$. We work by Noetherian induction on $\operatorname{Supp}\mathscr{F}$, with the base case when $\operatorname{Supp}\mathscr{F}=\varnothing$, i.e., $\mathscr{F}=0$, which is obvious. So fix $\mathscr{F}$, and assume the result is known for all coherent sheaves with strictly smaller support.

18.9.B. EXERCISE. Show that we may assume that $\operatorname{Supp}\mathscr{F}=X$. (Hint: Replace X by the scheme-theoretic support of $\mathscr{F}$, defined in Exercise 9.1.N.)

We now invoke Chow's Lemma to construct a projective morphism $\mu\colon X'\to X$ that is an isomorphism on a dense open subset U of X (so $X\setminus U$ is a closed subset of X, strictly smaller than X), and such that $\pi\circ\mu\colon X'\to\operatorname{Spec}A$ is projective.

Then $\mathscr{G}=\mu^*\mathscr{F}$ is a coherent sheaf on X', $\mu_*\mathscr{G}$ is a coherent sheaf on X (by the projective case, Theorem 18.7.1(d)), and the adjunction map $\mathscr{F}\to\mu_*\mathscr{G}=\mu_*\mu^*\mathscr{F}$ is an isomorphism on U. The kernel $\mathscr{E}$ and cokernel $\mathscr{H}$ are coherent sheaves on X that have strictly smaller support:

$$0\longrightarrow\mathscr{E}\longrightarrow\mathscr{F}\longrightarrow\mu_*\mathscr{G}\longrightarrow\mathscr{H}\longrightarrow0.$$

18.9.C. EXERCISE. By the inductive hypothesis, the higher pushforwards of $\mathscr{E}$ and $\mathscr{H}$ are coherent. Show that if all the higher pushforwards of $\mu_*\mathscr{G}$ are coherent, then the higher pushforwards of $\mathscr{F}$ are coherent.

So we are reduced to showing that the higher pushforwards of $\mu_*\mathscr{G}$ are coherent for any coherent $\mathscr{G}$ on X'.

The Leray spectral sequence (Theorem 23.4.5) for $X'\xrightarrow{\mu}X\xrightarrow{\pi}\operatorname{Spec}A$ has E_2 page given by $R^p\pi_*(R^q\mu_*\mathscr{G})$ converging to $R^{p+q}(\pi\circ\mu)_*\mathscr{G}$. Now $R^q\mu_*\mathscr{G}$ is coherent by Theorem 18.7.1(d). Furthermore, as μ is an isomorphism on a dense open subset U of X, $R^q\mu_*\mathscr{G}$ is zero on U, and is thus supported on the complement of U, whose support is strictly smaller than X. Hence by our inductive hypothesis, $R^p\pi_*(R^q\mu_*\mathscr{G})$ is coherent for all p, and all $q\geq1$. The only possibly noncoherent sheaves on the E_2 page are in the row $q=0$—precisely the sheaves we are interested in. Also, by Theorem 18.7.1(d) applied to $\pi\circ\mu$, $R^{p+q}(\pi\circ\mu)_*\mathscr{G}$ is coherent.

18.9.D. EXERCISE. Show that $E_n^{p,q}$ is always coherent for any $n \geq 2$, $q > 0$. Show that $E_n^{p,0}$ is coherent for a given $n \geq 2$ if and only if $E_?^{p,0}$ is coherent. Show that $E_\infty^{p,q}$ is coherent, and hence that $E_2^{p,0}$ is coherent, thereby completing the proof of Grothendieck's Coherence Theorem 18.9.1.

$\square$

18.9.4.** Proof of Chow's Lemma 18.9.2.

We use the properness hypothesis on π through each of its three constituent parts: finite type, separated, universally closed. The parts using separatedness are particularly tricky.

As X is Noetherian, it has finitely many irreducible components. Cover X with affine open sets $U_1, \ldots, U_n$. We may assume that each U_i meets each irreducible component. (If some U_i does not meet an irreducible component Z, then take any affine open subset Z' of $Z \setminus \overline{X - Z}$, and replace U_i by $U_i \cup Z'$.) Then $U := \cap_i U_i$ is a dense open subset of X. As each U_i is finite type over A, we can choose a closed embedding $U_i \subset \mathbb{A}_A^{n_i}$. Let $\overline{U}_i$ be the (scheme-theoretic) closure of U_i in $\mathbb{P}_A^{n_i}$.

Now we have the diagonal morphism $U \to X \times_A \prod \overline{U}_i$ (where the product is over $\mathrm{Spec}\, A$), which is a locally closed embedding (the composition of the closed embedding $U \hookrightarrow U^{n+1}$ with the open embedding $U^{n+1} \hookrightarrow X \times_A \prod \overline{U}_i$). Let X' be the scheme-theoretic closure of U in $X \times_A \prod \overline{U}_i$. Let μ be the composed morphism $X' \to X \times_A \prod \overline{U}_i \to X$, so we have a diagram

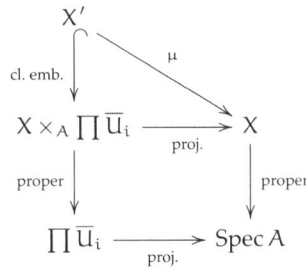

(where the square is Cartesian). The morphism μ is projective (as it is the composition of two projective morphisms and X is quasicompact, Exercise 17.3.I). We will conclude the argument by showing that $\mu^{-1}(U) = U$ (or, more precisely, μ is an isomorphism above U), and that $X' \to \prod \overline{U}_i$ is a closed embedding (from which the composition

$$X \longrightarrow \prod \overline{U}_i \longrightarrow \mathrm{Spec}\, A$$

is projective).

18.9.E. EXERCISE. Suppose $T_0, \ldots, T_n$ are *separated* schemes over A with isomorphic open sets, which we sloppily call V in each case. Then V is a locally closed subscheme of $T_0 \times_A \cdots \times_A T_n$. Let $\overline{V}$ be the closure of this locally closed subscheme. Show that

$$\begin{aligned}
V &\cong \overline{V} \cap (V \times_A T_1 \times_A \cdots \times_A T_n) \\
&= \overline{V} \cap (T_0 \times_A V \times_A T_2 \times_A \cdots \times_A T_n) \\
&= \cdots \\
&= \overline{V} \cap (T_0 \times_A \cdots \times_A T_{n-1} \times_A V).
\end{aligned}$$

(Hint for the first isomorphism: The graph of the morphism $V \to T_1 \times_A \cdots \times_A T_n$ is a closed embedding, as $T_1 \times_A \cdots \times_A T_n$ is separated over A, by Proposition 11.2.6. Thus the scheme-theoretic closure of V in $V \times_A T_1 \times_A \cdots \times_A T_n$ is V itself. Finally, the scheme-theoretic closure can be computed locally, essentially by Theorem 9.4.4.)

18.9.F. EXERCISE. Using (the idea behind) the previous exercise, show that $\mu^{-1}(U) = U$.

It remains to show that $X' \to \prod \overline{U}_i$ is a closed embedding. Now $X' \to \prod \overline{U}_i$ is closed (it is the composition of two closed maps), so it suffices to show that $X' \to \prod \overline{U}_i$ is a locally closed embedding.

18.9.G. EXERCISE. Let A_i be the closure of U in

$$B_i := X \times_A \overline{U}_1 \times_A \cdots \times_A U_i \times_A \cdots \times_A \overline{U}_n$$

(only the ith term is missing the bar), and let C_i be the closure of U in

$$D_i := \overline{U}_1 \times_A \cdots \times_A U_i \times_A \cdots \times_A \overline{U}_n.$$

Show that there is an isomorphism $A_i \xrightarrow{\sim} C_i$ induced by the projection $B_i \to D_i$. Hint: Note that the section $D_i \to B_i$ of the projection $B_i \to D_i$, given informally by $(t_1, \ldots, t_n) \mapsto (t_i, t_1, \ldots, t_n)$, is a closed embedding, as it can be interpreted as the graph of a morphism to a separated A-scheme (Proposition 11.2.6). So U can be interpreted as a locally closed subscheme of D_i, which in turn can be interpreted as a closed subscheme of B_i. Thus the closure of U in D_i may be identified with its closure in B_i.

As the U_i cover X, the $\mu^{-1}(U_i)$ cover $\overline{X}$. But $\mu^{-1}(U_i) = A_i$ (closure can be be computed locally—the closure of U in B_i is the intersection of B_i with the closure $\overline{X}$ of U in $X \times_A \overline{U}_1 \times_A \cdots \times_A \overline{U}_n$).

Hence over each U_i, we get a closed embedding of $A_i \hookrightarrow D_i$, and thus $X' \to \prod \overline{U}_i$ is a locally closed embedding, as desired. $\qquad\square$

18.9.5. Other versions of Chow's Lemma. We won't use these versions, but their proofs are similar to what we have already shown.

18.9.6. *Remark.* Notice first that if X is reduced (resp., irreducible, integral), then X' can be taken to be reduced (resp., irreducible, integral) as well.

18.9.H. EXERCISE. By suitably crossing out lines in the proof above, weaken the hypothesis "π is proper" to "π finite type and separated," at the expense of weakening the conclusion "$\pi \circ \mu$ is projective" to "$\pi \circ \mu$ is quasiprojective."

18.9.7. *Remark.* The target $\operatorname{Spec} A$ can be generalized to a scheme S that is (i) Noetherian, or (ii) separated and quasicompact with finitely many irreducible components. This can be combined with Remark 18.9.6 and Exercise 18.9.H. See [Gr-EGA, II.5.6.1] for a proof. See also [GW, Thm. 13.100] for a version that is slightly more general.

Chapter 19

Application: Curves

We now use what we have developed to study something explicit—curves. *Throughout this chapter, we will assume that all curves are projective, geometrically integral, regular curves over a field* k. We will sometimes add the hypothesis that k is algebraically closed. Most people are happy with working over algebraically closed fields, and those people should ignore the adverb "geometrically."

We certainly do not need the massive machinery we have developed in order to understand curves, but with the perspective we have gained, the development is quite clean. The key ingredients we will need are as follows. We use a criterion for a morphism to be a closed embedding, which we prove in §19.1. We use the "black box" of Serre duality (to be proved in Chapter 29). In §19.2, we use this background to observe a very few useful facts, which we will use repeatedly. Finally, in the course of applying them to understand curves of various genera, we develop the theory of hyperelliptic curves in a hands-on way (§19.5), in particular, proving a special case of the Riemann–Hurwitz formula.

If you are jumping into this chapter without reading much beforehand, you should skip §19.1, taking Theorem 19.1.1 as a black box. Depending on your background, you may want to skip §19.2 as well, taking the "crucial tools" as a black box.

19.1 A Criterion for a Morphism to Be a Closed Embedding

We will repeatedly use a criterion for when a morphism is a closed embedding, which is not special to curves. Before stating it, we recall some facts about closed embeddings. Suppose $\pi\colon X \to Y$ is a closed embedding. Then π is projective, and it is injective on points. This is not enough to ensure that it is a closed embedding, as the example of the normalization of the cusp shows (Figure 10.5). Another example is the following.

19.1.A. EXERCISE (FROBENIUS; CF. §8.3.Q). Show that $\pi\colon \mathbb{A}^1_{\mathbb{F}_p} \to \mathbb{A}^1_{\mathbb{F}_p}$ given by $x \mapsto x^p$ is a bijection on points, and induces an isomorphism of residue fields on closed points, yet is not a closed embedding.

The additional information you need to ensure that a morphism π is a closed embedding is that the tangent map is injective at all closed points.

19.1.B. EXERCISE. Show (directly, not invoking Theorem 19.1.1) that in the two examples described above (the normalization of a cusp and the Frobenius morphism), the tangent map is *not* injective at all closed points.

19.1.1. Theorem — *Suppose* $k = \overline{k}$, *and* $\pi\colon X \to Y$ *is a projective morphism of finite type* k*-schemes that is injective on closed points and injective on tangent vectors at closed points. Then* π *is a closed embedding.*

The converse is straightforward, so you should think of this result as "if and only if."

(Remark: In this case of finite type schemes over an algebraically closed field, "injective on tangent vectors at closed points" is equivalent to "unramified"; see Exercise 21.7.E. We will define unramified in §21.7.)

The example $\operatorname{Spec} \mathbb{C} \to \operatorname{Spec} \mathbb{R}$ shows that we need the hypothesis that k is algebraically closed in Theorem 19.1.1. Those allergic to algebraically closed fields should still pay attention, as we will use this to prove things about curves over k where k is *not* necessarily algebraically closed (see also Exercises 10.2.K and 19.1.F).

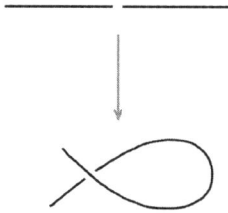

Figure 19.1 *(variant of Figure 8.4) The projectivity hypothesis in Theorem 19.1.1 cannot be dropped.*

We need the hypothesis that the morphism be projective, as shown by the example of Figure 19.1. It is the normalization of the node, except we erase one of the preimages of the node. We map $\mathbb{A}^1$ to the plane, so that its image is a curve with one node. We then consider the morphism we get by discarding one of the preimages of the node. Then this morphism is an injection on points, and is also injective on tangent vectors, but it is not a closed embedding. (In the world of differential geometry, this fails to be an embedding because the map doesn't give a homeomorphism onto its image.)

Theorem 19.1.1 appears to be fundamentally a statement about varieties, but it isn't. We will reduce it to the following result.

19.1.2. Theorem — *Suppose $\pi\colon X \to Y$ is a finite morphism of Noetherian schemes whose degree at every point of Y (§14.3.4) is 0 or 1. Then π is a closed embedding.*

19.1.C. EXERCISE. Suppose $\pi\colon X \to Y$ is a finite morphism whose degree at every point of Y is 0 or 1. Show that π is injective on points (easy). If $p \in X$ is any point, show that π induces an isomorphism of residue fields $\kappa(\pi(p)) \xrightarrow{\sim} \kappa(p)$. Show that π induces an injection of tangent spaces. Thus key hypotheses of Theorem 19.1.1 are implicitly in the hypotheses of Theorem 19.1.2.

19.1.3. *Reduction of Theorem 19.1.1 to Theorem 19.1.2.* The property of being a closed embedding is local on the target, so we may assume that Y is affine, say Spec B.

I next claim that π has finite fibers, not just finite fibers above closed points: the fiber dimension for projective morphisms is upper semicontinuous (Exercise 18.1.B, or Theorem 12.4.3(b)), so the locus where the fiber dimension is at least 1 is a closed subset, so if it is nonempty, it must contain a closed point of Y. Thus the fiber over any point is a dimension 0 finite type scheme over that point, hence a finite set.

Hence π is a projective morphism with finite fibers, thus finite by Theorem 18.1.6.

19.1.D. EXERCISE. Show that the degree of π is at most 1 at all closed points of Y. (Hint 1: You will use the "injectivity on tangent vectors" hypothesis here. Hint 2: What are the finite morphisms to Spec $\overline{k}$?)

But the degree of a finite morphism is upper semicontinuous, (§14.3.4), hence is at most 1 at all points (do you see why?). We have thus shown that Theorem 19.1.1 would follow from Theorem 19.1.2.

19.1.4. *Proof of Theorem 19.1.2.* The problem is local on Y, so we may assume Y is affine, say, $Y = \operatorname{Spec} B$. Thus X is affine too, say, Spec A, and π corresponds to a ring morphism $B \to A$. We wish to show that this is a surjection of rings, or (equivalently) of B-modules. Let K be the cokernel of this morphism of B-modules:

$$(19.1.4.1) \qquad B \longrightarrow A \longrightarrow K \longrightarrow 0.$$

We wish to show that $K = 0$. It suffices to show that for any maximal ideal $\mathfrak{n}$ of B, $K_\mathfrak{n} = 0$. (Do you remember why?) Localizing (19.1.4.1) at $\mathfrak{n}$, we obtain the exact sequence

$$B_\mathfrak{n} \longrightarrow A_\mathfrak{n} \longrightarrow K_\mathfrak{n} \longrightarrow 0.$$

Applying the right-exact functor $\otimes_{B_\mathfrak{n}} (B_\mathfrak{n}/\mathfrak{n}B_\mathfrak{n})$, we obtain

$$(19.1.4.2) \qquad B_\mathfrak{n}/\mathfrak{n}B_\mathfrak{n} \xrightarrow{\;\alpha\;} A_\mathfrak{n}/\mathfrak{n}A_\mathfrak{n} \longrightarrow K_\mathfrak{n}/\mathfrak{n}K_\mathfrak{n} \longrightarrow 0.$$

Now Spec $A_\mathfrak{n}/\mathfrak{n}A_\mathfrak{n}$ is the scheme-theoretic preimage of $[\mathfrak{n}] \in \operatorname{Spec} B$, so, by hypothesis, it is either empty, or the map α is an isomorphism (in which case we will see that the map α will turn out to be the map of residue fields). In the first case, $A_n/nA_n = 0$, from which $K_n/nK_n = 0$. In the second case, $K_n/nK_n = 0$ as well. Applying Nakayama's Lemma 8.2.9 (noting that A is a finitely generated B-module, hence K is too, hence K_n is a finitely generated B_n-module), $K_n = 0$ as desired. □

Exercise 10.2.K can be used to extend Theorem 19.1.1's reach to general fields k, not necessarily algebraically closed. The next exercise is an example of how.

19.1.E. EXERCISE. Use Theorem 19.1.1 to show that the dth Veronese embedding from $\mathbb{P}_k^n$, corresponding to the complete linear series $|\mathcal{O}_{\mathbb{P}_k^n}(d)|$, is a closed embedding. Do the same for the Segre embedding from $\mathbb{P}_k^m \times_{\operatorname{Spec} k} \mathbb{P}_k^n$. (This is just for practice for using this criterion. It is a weaker result than what we had before; we have earlier checked both of these statements over an arbitrary base ring in Remark 9.3.8 and §10.6, respectively, and we are now checking it only over fields. However, see Exercise 19.1.F below.)

19.1.F. LESS IMPORTANT EXERCISE. Using the ideas from this section, prove that the dth Veronese embedding from $\mathbb{P}_\mathbb{Z}^n$ (over the integers!) is a closed embedding. (Again, we have done this before. This exercise is simply to show that these methods can easily extend to work more generally.)

19.2   A Series of Crucial Tools

We are now ready to start understanding curves in a hands-on way. We will repeatedly make use of the following series of crucial remarks, and it will be important to have them at the tip of your tongue.

19.2.1. In what follows, C *will be a projective, geometrically regular, geometrically integral curve over a field* k, *and* $\mathcal{L}$ *is an invertible sheaf on* C.

19.2.2. Reminder: Serre duality. Serre duality (Theorem 18.5.1) on a geometrically irreducible regular genus g curve C over k involves an invertible sheaf ω_C (of degree $2g - 2$, with g linearly independent sections, Exercise 18.5.A), such that for any finite rank locally free sheaf $\mathcal{F}$ on C, $h^i(C, \mathcal{F}) = h^{1-i}(C, \omega_C \otimes \mathcal{F}^\vee)$ for $i = 0, 1$. (Better: there is a duality between the two cohomology groups.)

19.2.3. Negative degree line bundles have no nonzero section. If $\deg \mathcal{L} < 0$, then $h^0(C, \mathcal{L}) = 0$. Reason: $\deg \mathcal{L}$ is the number of zeros minus the number of poles (suitably counted) of any rational section (Important Exercise 18.4.C). If there is a regular section (i.e., with no poles), then this is necessarily nonnegative. Refining this argument yields the following.

19.2.4. Degree 0 line bundles, and recognizing when they are trivial. If $\deg \mathcal{L} = 0$, then $h^0(C, \mathcal{L}) = 0$ or 1; and if $h^0(C, \mathcal{L}) = 1$ then $\mathcal{L} \cong \mathcal{O}_C$. Reason: If there is a nonzero section s, it has no poles, and hence no zeros, because $\deg \mathcal{L} = 0$. Then $\operatorname{div} s = 0$, so $\mathcal{L} \cong \mathcal{O}_C(\operatorname{div} s) = \mathcal{O}_C$. (Recall how this works; cf. Important Exercise 15.4.G: s gives a trivialization for the invertible sheaf. We have a natural bijection for any open set $\Gamma(U, \mathcal{L}) \xleftrightarrow{\sim} \Gamma(U, \mathcal{O}_U)$, where the map from left to right is $s' \mapsto s'/s$, and the map from right to left is $f \mapsto sf$.) Conversely, for a geometrically integral projective variety, $h^0(\mathcal{O}) = 1$. (§11.4.7 shows this for k algebraically closed—this is where geometric integrality is used—and Exercise 18.2.H shows that cohomology commutes with base field extension.)

Serre duality turns these statements about line bundles of degree at most 0 into statements about line bundles of degree at least $2g - 2$, as follows.

19.2.5. We know $h^0(C, \mathcal{L})$ if the degree is sufficiently high. If $\deg \mathcal{L} > 2g - 2$, then

(19.2.5.1)
$$\boxed{h^0(C, \mathcal{L}) = \deg \mathcal{L} - g + 1.}$$

So we know $h^0(C, \mathscr{L})$ if $\deg \mathscr{L} \gg 0$. (*This is important—remember this!*) Reason: $h^1(C, \mathscr{L}) = h^0(C, \omega_C \otimes \mathscr{L}^\vee)$; but $\omega_C \otimes \mathscr{L}^\vee$ has negative degree (as ω_C has degree $2g - 2$), and thus this invertible sheaf has no sections. The result then follows from the Riemann–Roch Theorem in the guise of (18.4.4.1).

19.2.A. USEFUL EXERCISE (RECOGNIZING ω_C AMONG DEGREE $2g - 2$ LINE BUNDLES). Suppose $\mathscr{L}$ is a degree $2g - 2$ invertible sheaf. Show that it has $g - 1$ or g sections, and it has g sections if and only if $\mathscr{L} \cong \omega_C$.

19.2.B. EXERCISE. Suppose C is a curve of genus $g > 1$, over a field k that is not necessarily algebraically closed. Show that C has a closed point of degree at most $2g - 2$ over the base field. (For comparison: if $g = 1$, for any n, there is a genus 1 curve over $\mathbb{Q}$ with no point of degree less than n, [MO286390]!)

19.2.6. Twisting $\mathscr{L}$ by a (degree 1) point changes h^0 by at most 1. Suppose p is any closed point of degree 1 (i.e., the residue field of p is k). Then $h^0(C, \mathscr{L}) - h^0(C, \mathscr{L}(-p)) = 0$ or 1. (The twist of $\mathscr{L}$ by a divisor, such as $\mathscr{L}(-p)$, was defined in §15.4.17.) Reason: Consider $0 \to \mathscr{O}_C(-p) \to \mathscr{O}_C \to \mathscr{O}|_p \to 0$; tensor with $\mathscr{L}$ (this is exact as $\mathscr{L}$ is locally free) to get

$$0 \longrightarrow \mathscr{L}(-p) \longrightarrow \mathscr{L} \longrightarrow \mathscr{L}|_p \longrightarrow 0.$$

Then $h^0(C, \mathscr{L}|_p) = 1$, so as the long exact sequence of cohomology starts off

$$0 \longrightarrow H^0(C, \mathscr{L}(-p)) \longrightarrow H^0(C, \mathscr{L}) \longrightarrow H^0(C, \mathscr{L}|_p),$$

we are done.

19.2.7. Criterion for $\mathscr{L}$ to be base-point-free. Suppose for this remark that k is algebraically closed, so *all* closed points have degree 1 over k. Then if $h^0(C, \mathscr{L}) - h^0(C, \mathscr{L}(-p)) = 1$ for *all* closed points p, then $\mathscr{L}$ is base-point-free, and hence induces a morphism from C to projective space (Theorem 15.2.2). Reason: Given any p, our equality shows that there exists a section of $\mathscr{L}$ that does not vanish at p—so, by definition, p is not a base point of $\mathscr{L}$.

19.2.8. Next, suppose p and q are distinct (closed) points of degree 1. Then $h^0(C, \mathscr{L}) - h^0(C, \mathscr{L}(-p - q)) = 0$, 1, or 2 (by repeating the argument of Remark 19.2.6 twice). If $h^0(C, \mathscr{L}) - h^0(C, \mathscr{L}(-p - q)) = 2$, then necessarily

(19.2.8.1) $\quad h^0(C, \mathscr{L}) = h^0(C, \mathscr{L}(-p)) + 1 = h^0(C, \mathscr{L}(-q)) + 1 = h^0(C, \mathscr{L}(-p - q)) + 2.$

Then the linear series $\mathscr{L}$ separates points p and q, i.e., the corresponding map f to projective space satisfies $f(p) \neq f(q)$. Reason: There is a hyperplane of projective space passing through $f(q)$ but not passing through $f(p)$, or, equivalently, there is a section of $\mathscr{L}$ vanishing at q but not vanishing at p. This is because of the last equality in (19.2.8.1).

19.2.9. By the same argument as above, if p is a (closed) point of degree 1, then $h^0(C, \mathscr{L}) - h^0(C, \mathscr{L}(-2p)) = 0$, 1, or 2. I claim that if this is 2, then the map corresponding to $\mathscr{L}$ (which is already seen to be base-point-free at p from the above) separates the tangent vectors at p. To show this, we need to show that the cotangent map is *surjective*. To show surjectivity onto a one-dimensional vector space, we just need to show that the map is nonzero. So it suffices to find a function on the target vanishing at the image of p that pulls back to a function that vanishes at p to order 1 but not 2. In other words, we want a section of $\mathscr{L}$ vanishing at p to order 1 but not 2. But that is the content of the statement $h^0(C, \mathscr{L}(-p)) - h^0(C, \mathscr{L}(-2p)) = 1$.

19.2.10. Criterion for $\mathscr{L}$ to be very ample. Combining some of our previous comments: suppose C is a curve over an *algebraically closed* field k, and $\mathscr{L}$ is an invertible sheaf such that for *all* closed points p and q, *not necessarily distinct*, $h^0(C, \mathscr{L}) - h^0(C, \mathscr{L}(-p - q)) = 2$; then $\mathscr{L}$ gives a *closed embedding into projective space*, as it separates points and tangent vectors, by Theorem 19.1.1.

19.2.C. EXERCISE. Suppose that k is algebraically closed, so the previous remark applies. Show that $C \setminus \{p\}$ is affine. Hint: Show that if $j \gg 0$, then $\mathcal{O}(jp)$ is base-point-free and has at least two linearly independent sections, one of which has divisor jp. Use these two sections to map to $\mathbb{P}^1$ so that the set-theoretic preimage of ∞ is p. Argue that the map is finite, and that $C \setminus \{p\}$ is the preimage of $\mathbb{A}^1$. (A trivial variation of this argument shows that $C \setminus \{p_1, \ldots, p_n\}$ is affine if $n > 0$.)

19.2.11. Powerful conclusion: Numerical criteria for $\mathscr{L}$ to be base-point-free and very ample. We can combine much of the above discussion to give the following useful fact. If k is algebraically closed, then $\deg \mathscr{L} \geq 2g$ implies that $\mathscr{L}$ is base-point-free (and hence determines a morphism to projective space). Also, $\deg \mathscr{L} \geq 2g + 1$ implies that for any basis of sections $s_0, \ldots, s_{\deg \mathscr{L} - g}$ of $\mathscr{L}$, the morphism given by $[s_0, \ldots, s_{\deg \mathscr{L} - g}]$ is in fact a closed embedding into $\mathbb{P}^{\deg \mathscr{L} - g}$ (so $\mathscr{L}$ is very ample). Remember this!

19.2.D. EXERCISE (ON A PROJECTIVE, REGULAR, INTEGRAL CURVE OVER $\overline{k}$, AMPLE = POSITIVE DEGREE). Show that an invertible sheaf $\mathscr{L}$ on a projective, regular, integral curve over $\overline{k}$ is ample if and only if $\deg \mathscr{L} > 0$.

(This can be extended to curves over general fields using Exercise 19.2.E below.) Thus there is a blunt purely numerical criterion for ampleness of line bundles on curves. This generalizes to projective varieties of higher dimension; this is called *the Nakai–Moishezon criterion for ampleness*, Theorem 20.4.1.

19.2.E. EXERCISE (EXTENSION TO NON-ALGEBRAICALLY CLOSED FIELDS). Show that the statements in §19.2.11 hold even without the hypothesis that k is algebraically closed. (Hint: To show one of the facts about some curve C and line bundle $\mathscr{L}$, consider instead $C \times_{\operatorname{Spec} k} \operatorname{Spec} \overline{k}$. Then show that if the pullback of $\mathscr{L}$ here has sections giving you one of the two desired properties, then there are sections downstairs with the same properties. You may want to use facts that we have used, such as the fact that base-point-freeness is independent of any extension of the base field, Exercise 18.2.I, or that the property of an affine morphism over k being a closed embedding holds if and only if it does after an extension of k, Exercise 10.2.K.)

19.2.F. EXERCISE (EXTENDING EXERCISE 19.2.D). Suppose $\mathscr{L}$ is an invertible sheaf on a projective, geometrically regular, geometrically integral curve C (over a field k). Show that $\mathscr{L}$ is ample if and only if $\deg \mathscr{L} > 0$. Hint: Reduce to the case where k is algebraically closed, with the help of Exercise 10.2.K.

19.2.G. EXERCISE. Suppose C is a curve (in the sense of §19.2.1, as usual in this section), and $p \in C$ is a closed point. Show that $C \setminus \{p\}$ is affine. Hint: Exercise 19.2.F.

We are now ready to take these facts and go to the races, in §19.3.

19.2.12.* Remark: Proper curves are projective.
(In this final part of §19.2, we no longer assume our curves are smooth, or even regular or reduced.) Our discussion here, along with Serre's cohomological criterion for ampleness (§18.8), is the key ingredient for showing that every proper curve over a field is projective.

19.2.H. EXERCISE. Show that every *reduced* proper curve C is projective. Hint: Choose a regular point on each irreducible component of C, and let $\mathscr{L}$ be the corresponding invertible sheaf. We hope that $\mathscr{L}$ will be ample. Recall that a line bundle on C is ample if its pullback to the normalization of C is ample (Tricky Exercise 18.8.C). So show that the pullback to the normalization is ample.

19.2.I. EXERCISE. Show that every proper (not necessarily reduced) curve C is projective, using Exercise 18.8.A, after first finding an invertible sheaf on C that you will show to be ample.

19.2.J. EXERCISE. Show that a line bundle on a projective curve is ample if and only if it has positive degree on each component.

19.3 Curves of Genus 0

We are now ready to (in some form) answer the question: What are the curves of genus 0?

In §7.5.10, we saw a genus 0 curve (over a field k) that was *not* isomorphic to $\mathbb{P}^1$: $x^2 + y^2 + z^2 = 0$ in $\mathbb{P}^2_{\mathbb{R}}$. (It has genus 0 by (18.6.7.1).) We have already observed that this curve is *not* isomorphic to $\mathbb{P}^1_{\mathbb{R}}$, because it doesn't have an $\mathbb{R}$-valued point. On the other hand, we haven't seen a genus 0 curve over an algebraically closed field with this property. This is no coincidence: the lack of an existence of a k-valued point is the only obstruction to a genus 0 curve being $\mathbb{P}^1$.

19.3.1. Proposition — *Suppose C is genus 0, and C has a k-valued (degree 1) point. Then $C \cong \mathbb{P}^1_k$.*

Thus we see that all genus 0 (integral, regular, projective) curves over an algebraically closed field are isomorphic to $\mathbb{P}^1$.

Proof. Let p be the point, and consider $\mathscr{L} = \mathscr{O}_C(p)$. Then $\deg \mathscr{L} = 1$, so we can apply what we know above: first, $h^0(C, \mathscr{L}) = 2$ (Remark 19.2.5), and second, these two sections give a closed embedding into $\mathbb{P}^1_k$ (Remark 19.2.11, as extended in Exercise 19.2.E). But the only closed embedding of a curve into the integral curve $\mathbb{P}^1_k$ is an isomorphism! □

As a bonus, Proposition 19.3.1 implies that $x^2 + y^2 + z^2 = 0$ in $\mathbb{P}^2_{\mathbb{R}}$ has no *line bundles* of degree 1 over $\mathbb{R}$; otherwise, we could just apply the above argument to the corresponding line bundle. This example shows us that over a non-algebraically closed field, there can be genus 0 curves that are not isomorphic to $\mathbb{P}^1_k$. The next result lets us get our hands on them as well.

19.3.2. Claim — *All genus 0 curves can be described as conics in $\mathbb{P}^2_k$.*

Proof. Any genus 0 curve has a degree -2 line bundle—the canonical sheaf ω_C. Thus any genus 0 curve has a degree 2 line bundle: $\mathscr{L} = \omega_C^\vee$. We apply Exercise 19.2.E: $\deg \mathscr{L} = 2 \geq 2g + 1$, so this line bundle gives a closed embedding $i: C \hookrightarrow \mathbb{P}^2$. Exercise 18.6.H ensures that the image is degree 2 as a closed subscheme, but we haven't quite shown that the curve can be cut out by one homogeneous polynomial of degree 2. (This is more subtle than you might think—you can try to make your own argument before reading on.) We have the isomorphism $\mathrm{Sym}^2 H^0(\mathbb{P}^2, \mathscr{O}(1)) \xrightarrow{\sim} H^0(\mathbb{P}^2, \mathscr{O}(2))$ (both sides are the homogeneous quadratic polynomials in three variables). Consider the map $\mathrm{Sym}^2 H^0(C, \mathscr{L}) \to H^0(C, \mathscr{L}^{\otimes 2})$, which (by (19.2.5.1)) is a map from a vector space of dimension 6 to a vector space of dimension 5. This map has nontrivial kernel, so $i(C)$ is scheme-theoretically contained in $V(f)$ for some homogeneous conic f.

19.3.A. EXERCISE. Use Exercise 18.6.E to show that $i(C) = V(f)$ scheme-theoretically, completing the proof. □

19.3.B. EXERCISE. Suppose C is a genus 0 curve (projective, geometrically integral, and regular). Show that C has a point of degree at most 2. (The degree of a point was defined in §5.3.8.)

The geometric means of finding Pythagorean triples presented in §7.5.9 looked quite different, but was really the same. There was a genus 0 curve C (a plane conic) with a k-valued point p, and we proved that it was isomorphic to $\mathbb{P}^1_k$. The line bundle used to show the isomorphism wasn't the degree 1 line bundle $\mathscr{O}_C(p)$; it was the degree 1 line bundle $\mathscr{O}_{\mathbb{P}^2}(1)|_C \otimes \mathscr{O}_C(-p)$.

19.4 Classical Geometry Arising from Curves of Positive Genus

We will use the following proposition and corollary later, and we take this as an excuse to revisit some very classical geometry from a modern standpoint.

19.4.1. Proposition — *Recall our standing assumptions for this chapter (§19.2.1), that C is a projective, geometrically regular, geometrically integral curve over a field k. Suppose C is not isomorphic to $\mathbb{P}^1_k$ (with no assumptions on the genus of C), and $\mathscr{L}$ is an invertible sheaf of degree 1. Then $h^0(C, \mathscr{L}) < 2$.*

Proof. We could prove this similarly to Proposition 19.3.1 (and the reader may wish to do so). We instead prove it "by hand" because Proposition 19.3.1 requires Serre Duality, which we will not prove until Chapter 29 (and which is hard), and we wish to emphasize that we do not need Serre Duality for this result. Also, some readers may read this argument without ever reading about Serre Duality.

Assume the statement is false, and let s_1 and s_2 be two (independent) sections. As the degree of the divisor of zeros of s_i is the degree of $\mathscr{L}$, each vanishes at a single point p_i (to order 1).

Now $p_1 \neq p_2$. (Otherwise, if $p_1 = p_2$, then s_1/s_2 has no poles or zeros, and hence is an invertible function by Algebraic Hartogs's Lemma—see §13.5.20—and a constant function using §11.4.7 after base change to $\overline{k}$. This contradicts the linear independence of s_1 and s_2.)

Thus s_1 and s_2 generate a base-point-free linear series, so we have an induced morphism $C \to \mathbb{P}^1$. This is a finite degree 1 morphism from one regular curve to another (Exercise 18.4.E), which hence induces a degree 1 extension of function fields, i.e., an isomorphism of function fields, which means that the curves are isomorphic (cf. Theorem 16.3.3). But we assumed that C is not isomorphic to $\mathbb{P}^1_k$, so we have a contradiction. $\qquad\square$

19.4.2. Corollary — *If C is a projective, geometrically regular, geometrically integral curve over k, not isomorphic to $\mathbb{P}^1_k$, and p and q are degree 1 points, then $\mathscr{O}_C(p) \cong \mathscr{O}_C(q)$ if and only if $p = q$.*

19.4.A. EXERCISE. Prove this. (Possibly useful: Show/use that $\mathscr{O}_C(p)$ has a global section s with $\mathrm{div}(s) = p$.)

19.4.B. EXERCISE. Show that if k is algebraically closed, then C has genus 0 if and only if all degree 0 line bundles are trivial.

19.4.C. EXERCISE. Suppose C is a regular plane curve of degree $e > 2$, and D_1 and D_2 are two plane curves of the same degree d not containing C. By Bézout's Theorem for plane curves (§18.6.4), D_i meets C at de points, counted "correctly." Suppose D_1 and D_2 meet C at $de - 1$ of the "same points," plus one more. Show that the remaining points are the same as well. More precisely, suppose there is a divisor E on C of degree $de - 1$, and degree 1 (k-valued) points p_1 and p_2 such that $D_i \cap C = E + p_i$ (as divisors on C). Show that $p_1 = p_2$. (The case $d = e = 3$ is Chasles's Theorem, and is the first case of the Cayley–Bacharach Theorem; see [EGH].)

As an entertaining application of Exercise 19.4.C, we can prove two classical results. For convenience, in this discussion we will assume k is algebraically closed, although this assumption can be easily removed.

19.4.3. Pappus's Theorem (Pappus of Alexandria) — *(See Figure 19.2.) Suppose ℓ and $\mathfrak{m}$ are distinct lines in the plane, and α, β, and γ are distinct points on ℓ, and α', β', and γ' are distinct points on $\mathfrak{m}$, and all six points are distinct from $\ell \cap \mathfrak{m}$. Then points $x = \overline{\alpha\beta'} \cap \overline{\alpha'\beta}$, $y = \overline{\alpha\gamma} \cap \overline{\alpha\gamma'}$, and $z = \overline{\beta\gamma'} \cap \overline{\beta'\gamma}$ are collinear.*

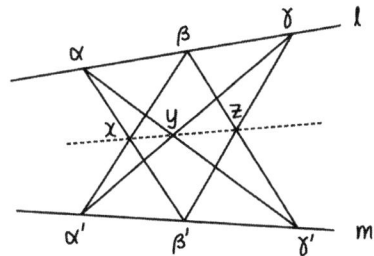

Figure 19.2 *Pappus's Theorem.*

Pascal's "Mystical Hexagon" Theorem was discovered by Pascal at age 16. (What were you doing at age 16?)

19.4.4. Pascal's "Mystical Hexagon" Theorem — *(See Figure 19.3.) If a hexagon $\alpha\gamma'\beta\alpha'\gamma\beta'$ is inscribed in a smooth conic K, and opposite pairs of sides are extended until they meet, the three intersection points $x = \overline{\alpha\beta'} \cap \overline{\alpha'\beta}$, $y = \overline{\alpha'\gamma} \cap \overline{\alpha\gamma'}$, and $z = \overline{\beta\gamma'} \cap \overline{\beta'\gamma}$ are collinear.*

Pappus's Theorem can be seen as a degeneration of Pascal's Theorem: the conic degenerates into the union of two lines, and the six points degenerate so that α, β, and γ are on one, and α', β', and γ' are on the other. We thus prove Pascal's Theorem, and you should check that the proof readily applies to proving Pappus's Theorem.

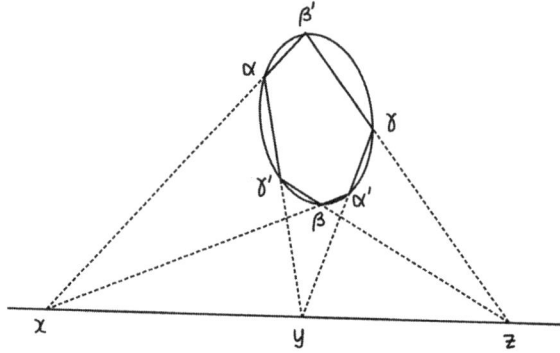

Figure 19.3 *Pascal's "Mystical Hexagon" Theorem.*

19.4.D. EXERCISE. Suppose α, β, γ, α', β' are five points in $\mathbb{P}^2_{\mathbb{R}}$, no three on a line. Show that there is a unique conic K passing through the five points. Show that K is regular. Explain how to construct the tangent to K at α using only a straightedge. (Hint for the last part: Apply Pascal's Theorem, taking $\gamma' = \alpha$. You may need to first figure out why you can apply Pascal's Theorem in this degenerate case.)

In particular, given an ellipse $K \subset \mathbb{R}^2$ and a point $\alpha \in K$, you will now be able to construct the tangent to K at α using only a straightedge.

19.4.5. *Proof of Pascal's Theorem 19.4.4.* We wish to show that the line $\overline{xy}$ meets $\overline{\beta\gamma'}$ and $\overline{\beta'\gamma}$ at the same point. Let C be the curve that is the union of K and the line $\overline{xy}$. (Warning, cf. §19.2.1: For one time in this chapter, we are not assuming C to be regular!)

Let D_1 be the union of the three lines $\overline{\alpha\beta'}$, $\overline{\beta\gamma'}$, and $\overline{\gamma\alpha'}$, and let D_2 be the union of the three lines $\overline{\alpha'\beta}$, $\overline{\beta'\gamma}$, and $\overline{\gamma'\alpha}$. Note that D_1 meets C at the nine points α, β, γ, α', β', γ', x, y, and $\overline{xy} \cap \overline{\beta\gamma'}$, and D_2 meets C at the same nine points, except $\overline{xy} \cap \overline{\beta\gamma'}$ is replaced by $\overline{xy} \cap \overline{\beta'\gamma}$. Thus if we knew Corollary 19.4.2 so that it applied to our C (which is singular), then we would be done (cf. Exercise 19.4.C).

So we extend Corollary 19.4.2 to our situation. So to do this, we extend Proposition 19.4.1 to our situation. We have a plane cubic C that is the union of a line L and a conic K, and two points $p = \overline{xy} \cap \overline{\beta\gamma'}$ and $q = \overline{xy} \cap \overline{\beta'\gamma}$, with $\mathscr{O}(p) \cong \mathscr{O}(q)$. (Reason: Both are the restriction of $\mathscr{O}_{\mathbb{P}^2}(3)$ to C, twisted by $-(\alpha + \beta + \gamma + \alpha' + \beta' + \gamma' + x + y)$.) Call this invertible sheaf $\mathscr{L}$. Suppose $p \neq q$. Then the two sections of $\mathscr{L}$ with zeros at p and q give a base-point-free linear series, and thus a morphism $\pi \colon C \to \mathbb{P}^1$, with $\pi^{-1}(0) = p$ and $\pi^{-1}(\infty) = q$.

By the argument in the proof of Proposition 19.4.1, π gives an isomorphism of L with $\mathbb{P}^1$. As π is proper, $\pi(K)$ is closed, and as K is irreducible, $\pi(K)$ is irreducible. As $\pi(K)$ does not contain 0 (or ∞), it can't be all of $\mathbb{P}^1$. Hence $\pi(K)$ is a point.

The conic K meets the line L in two points with multiplicity (by Bézout's Theorem for plane curves, §18.6.4, or by simple algebra). If $K \cap L$ consists of two points a and b, we have a contradiction: K can't be contracted to both $\pi(a)$ and $\pi(b)$. But what happens if K meets L at one point, i.e., if the conic is tangent to the line?

19.4.E. EXERCISE. Finish the proof of Pascal's Theorem by dealing with this case. Hint: Nilpotents will come to the rescue. The intuition is as follows: $K \cap L$ is a subscheme of L of length 2 (the length of a scheme was defined in §6.5.8), but the scheme-theoretic image $\pi(K)$ can be shown to be a reduced closed point. □

19.4.6. *Remark.* The key motivating fact that makes our argument work, Proposition 19.4.1, is centrally about curves not of genus 0, yet all the curves involved in Pappus's Theorem and Pascal's Theorem have genus 0. The insight to keep in mind is that union of curves of genus 0 need not have genus 0. In our case, it mattered that cubic curves have genus 1, even if they are a union of $\mathbb{P}^1$'s.

19.5   Hyperelliptic Curves

We next discuss an important class of curves, the hyperelliptic curves. In this section, *we assume* k *is algebraically closed of characteristic not 2*. (These hypotheses can be relaxed, at some cost.)

A (projective regular integral) genus g curve C is **hyperelliptic** if it admits a double cover of (i.e., degree 2, necessarily finite, morphism to) $\mathbb{P}^1_k$. (See §16.3.4 for a discussion of degrees of maps from curves to curves.) For convenience, when we say C is hyperelliptic, we will implicitly have in mind a *choice* of double cover $\pi\colon C \to \mathbb{P}^1$. (We will later see that if $g \geq 2$, then there is at most one such double cover, Proposition 19.5.7, so this is not a huge assumption.) The map π is called the **hyperelliptic map**.

By Exercise 16.3.D, the preimage of any closed point p of $\mathbb{P}^1$ consists of either one point or two points. If $\pi^{-1}(p)$ is a single point, we say p is a **branch point**, and $\pi^{-1}(p)$ is a **ramification point** of π. (The notion of ramification will be defined more generally in §21.7.)

19.5.1. Theorem (hyperelliptic Riemann–Hurwitz formula) — *Suppose* $k = \overline{k}$ *and* char $k \neq 2$, $\pi\colon C \to \mathbb{P}^1_k$ *is a double cover by a projective regular irreducible genus* g *curve over* k. *Then* π *has* $2g + 2$ *branch points.*

This is a special case of the Riemann–Hurwitz formula, which we will state and prove in §21.4. You may have already heard about genus 1 complex curves double covering $\mathbb{P}^1$, branched over four points.

To prove Theorem 19.5.1, we first prove the following.

19.5.2. Proposition — *Assume* char $k \neq 2$ *and* $k = \overline{k}$. *Given* r *distinct points* $p_1, \ldots, p_r \in \mathbb{P}^1$, *there is precisely one double cover branched at precisely these points if* r *is even, and none if* r *is odd.*

Proof. Pick points 0 and ∞ of $\mathbb{P}^1$ distinct from the r branch points. All r branch points are in $\mathbb{P}^1 - \infty = \mathbb{A}^1 = \operatorname{Spec} k[x]$. Suppose we have a double cover of $\mathbb{A}^1$, $C' \to \mathbb{A}^1$, where x is the coordinate on $\mathbb{A}^1$. This induces a quadratic field extension K over $k(x)$. As char $k \neq 2$, this extension is Galois. Let $\sigma\colon K \to K$ be the Galois involution. Let y be a nonzero element of K such that $\sigma(y) = -y$, so 1 and y form a basis for K over the field $k(x)$, and are eigenvectors of σ. Now $\sigma(y^2) = y^2$, so $y^2 \in k(x)$. We can replace y by an appropriate $k(x)$-multiple so that y^2 is a polynomial, with no repeated factors, and monic. (This is where we use the hypothesis that k is algebraically closed, to get leading coefficient 1.)

Thus $y^2 = x^N + a_{N-1}x^{N-1} + \cdots + a_0$, where the polynomial on the right (call it f(x)) has no repeated roots. The Jacobian criterion for regularity (in the guise of Exercise 13.2.F) implies that this curve C'_0 in $\mathbb{A}^2 = \operatorname{Spec} k[x, y]$ is regular. Then C'_0 is normal and has the same function field as C'. Thus C'_0 and C' are both normalizations of $\mathbb{A}^1$ in the finite extension of fields generated by y, and hence are isomorphic. Thus we have identified C' in terms of an explicit equation.

The branch points correspond to those values of x for which there is exactly one value of y, i.e., the roots of f(x). In particular, $N = r$, and

$$f(x) = (x - p_1) \cdots (x - p_r),$$

where the p_i are interpreted as elements of $\overline{k}$.

Having mastered the situation over $\mathbb{A}^1$, we return to the situation over $\mathbb{P}^1$. We will examine the branched cover over the affine open set $\mathbb{P}^1 \setminus \{0\} = \operatorname{Spec} k[u]$, where $u = 1/x$. The previous argument applied to $\operatorname{Spec} k[u]$ rather than $\operatorname{Spec} k[x]$ shows that any such double cover must be of the form

$$C'' = \operatorname{Spec} k[Z, u]/\left(Z^2 - (u - 1/p_1) \cdots (u - 1/p_r)\right)$$

$$= \operatorname{Spec} k[Z, u]/\left(\left((-1)^r \prod p_i\right) Z^2 - u^r f(1/u)\right)$$

$$\longrightarrow \operatorname{Spec} k[u] = \mathbb{A}^1.$$

So if there is a double cover over all of $\mathbb{P}^1$, it must be obtained by gluing C'' to C', "over" the gluing of $\operatorname{Spec} k[x]$ to $\operatorname{Spec} k[u]$ to obtain $\mathbb{P}^1$.

Thus in $K(C)$, we must have

$$z^2 = u^r f(1/u) = f(x)/x^r = y^2/x^r$$

(where z is obtained from Z by multiplying by a square root of $(-1)^r \prod p_i$), from which $z^2 = y^2/x^r$.

If r is even, considering $K(C)$ as generated by y and x, there are two possible values of z: $z = \pm y/x^{r/2}$. After renaming z by $-z$ if necessary, there is a single way of gluing these two patches together (we choose the positive square root).

If r is odd, the result follows from Exercise 19.5.A below.

19.5.A. EXERCISE. Suppose char $k \neq 2$. Show that x does not have a square root in the field $k(x)[y]/(y^2 - f(x))$, where f is a polynomial with nonzero roots $p_1, \ldots, p_r$. (Possible hint: Why is $\sqrt{3} \notin \mathbb{Q}(\sqrt{2})$?) Explain how this proves Proposition 19.5.2 in the case where r is odd. $\square$

For future reference, we collect here our explicit (two-affine) description of the hyperelliptic cover $C \to \mathbb{P}^1$.

(19.5.2.1)

$$\begin{array}{ccc}
\operatorname{Spec} k[x,y]/(y^2 - f(x)) & \xrightarrow[\substack{y = z/u^{r/2}}]{\substack{z = y/x^{r/2}}} & \operatorname{Spec} k[u,z]/(z^2 - u^r f(1/u)) \\
\downarrow & & \downarrow \\
\operatorname{Spec} k[x] & \xrightarrow[\substack{x = 1/u}]{\substack{u = 1/x}} & \operatorname{Spec} k[u]
\end{array}$$

19.5.3. If k is not algebraically closed. If k is not algebraically closed (but of characteristic not 2), the above argument shows that if we have a double cover of $\mathbb{A}^1$, then it is of the form $y^2 = af(x)$, where f is monic (and $a \neq 0$). Furthermore, if a and a' differ (multiplicatively) by an element of $(k^\times)^2$, then $y^2 = af(x)$ is isomorphic to $y^2 = a'f(x)$. You may be able to use this to show that (assuming that $k^\times \neq (k^\times)^2$) a double cover is *not* determined by its branch points. Moreover, this failure is classified by $k^\times/(k^\times)^2$. Thus we have lots of curves that are not isomorphic over k, but become isomorphic over $\overline{k}$. These are often called *twists* of each other.

(In particular, once we define elliptic curves, you will be able to show that there exist two elliptic curves over $\mathbb{Q}$ with the same j-invariant that are not isomorphic; see Exercise 19.10.H.)

19.5.4. Back to proving the hyperelliptic Riemann–Hurwitz formula, Theorem 19.5.1. Our explicit description of the unique double cover of $\mathbb{P}^1$ branched over r different points will allow us to compute the genus, thereby completing the proof of Theorem 19.5.1.

We continue the notation (19.5.2.1) of the proof of Proposition 19.5.2. Suppose $\mathbb{P}^1$ has affine cover by $\operatorname{Spec} k[x]$ and $\operatorname{Spec} k[u]$, with $u = 1/x$, as usual. Suppose $C \to \mathbb{P}^1$ is a double cover, given by $y^2 = f(x)$ over $\operatorname{Spec} k[x]$, where f has degree r, and $z^2 = u^r f(1/u)$. Then C has an affine open cover by $\operatorname{Spec} k[x,y]/(y^2 - f(x))$ and $\operatorname{Spec} k[u,z]/(z^2 - u^r f(1/u))$. The corresponding Čech complex for $\mathscr{O}_C$ is

$$0 \longrightarrow k[x,y]/(y^2 - f(x)) \times k[u,z]/(z^2 - u^r f(1/u)) \xrightarrow{\quad d \quad}$$

$$\left(k[x,y]/(y^2 - f(x)) \right)_x \longrightarrow 0.$$

The second (and last) nonzero part of the complex has basis consisting of monomials $x^n y^\epsilon$, where $n \in \mathbb{Z}$ and $\epsilon = 0$ or 1. To compute the genus $g = h^1(C, \mathscr{O}_C)$, we must compute coker d. We can use the first factor $k[x,y]/(y^2 - f(x))$ to hit the monomials $x^n y^\epsilon$, where $n \in \mathbb{Z}^{\geq 0}$ and $\epsilon = 0$ or 1. The image of the second factor is generated by elements of the form $u^m z^\epsilon$, where $m \geq 0$ and $\epsilon = 0$ or 1. But $u^m z^\epsilon = x^{-m}(y/x^{r/2})^\epsilon$. By inspection, the cokernel has basis generated by monomials $x^{-1}y$, $x^{-2}y, \ldots, x^{-r/2+1}y$, and thus has dimension $r/2 - 1$. Hence $g = r/2 - 1$, from which Theorem 19.5.1 follows. $\square$

19.5.5. Curves of every genus. As a consequence of the hyperelliptic Riemann–Hurwitz formula (Theorem 19.5.1), we see that there are curves of every genus $g \geq 0$ over an algebraically closed field of characteristic not 2: to get a curve of genus g, consider the branched cover branched over

$2g + 2$ distinct points. The unique genus 0 curve is of this form, and we will see in §19.6.2 that every genus 2 curve is of this form. We will soon see that every genus 1 curve (reminder: over an algebraically closed field!) is too (§19.9.5). But it is too much to hope that all curves are of this form, and we will soon see (§19.7.4) that there are genus 3 curves that are *not* hyperelliptic, and we will even get heuristic evidence that "most" genus 3 curves are not hyperelliptic. We will later give vague evidence (that can be made precise) that "most" genus g curves are not hyperelliptic if $g > 2$ (§19.8.4).

19.5.B. EXERCISE. Verify that a curve C of genus at least 1 admits a degree 2 cover of $\mathbb{P}^1$ if and only if it admits a degree 2 invertible sheaf $\mathscr{L}$ with $h^0(C, \mathscr{L}) = 2$. Possibly in the course of doing this, verify that if C is a curve of genus at least 1, and C has a degree 2 invertible sheaf $\mathscr{L}$ with at least 2 (linearly independent) sections, then $\mathscr{L}$ has precisely two sections, and that this $\mathscr{L}$ is base-point-free and gives a hyperelliptic map.

19.5.6. Proposition — *Assume $g \geq 2$. If $\mathscr{L}$ corresponds to a hyperelliptic cover $C \to \mathbb{P}^1$, then $\mathscr{L}^{\otimes(g-1)} \cong \omega_C$.*

Proof. Compose the hyperelliptic map with the $(g-1)$th Veronese embedding

$$(19.5.6.1) \qquad C \xrightarrow{\;|\mathscr{L}|\;} \mathbb{P}^1 \xrightarrow{\;\nu_{g-1} = |\mathscr{O}_{\mathbb{P}^1}(g-1)|\;} \mathbb{P}^{g-1}.$$

The composition corresponds to $\mathscr{L}^{\otimes(g-1)}$. This invertible sheaf has degree $2g - 2$ (by Exercise 18.4.G).

We next show that the pullback $H^0(\mathbb{P}^{g-1}, \mathscr{O}(1)) \to H^0(C, \mathscr{L}^{\otimes(g-1)})$ is injective, i.e., any nonzero $s \in H^0(\mathbb{P}^{g-1}, \mathscr{O}(1))$ pulls back to a nonzero element of $H^0(C, \mathscr{L}^{\otimes(g-1)})$, i.e., the image of C in $\mathbb{P}^{g-1}$ is nondegenerate. This is because the image of C in $\mathbb{P}^{g-1}$ is the image of $\mathbb{P}^1$ by ν_{g-1} (a rational normal curve; see §9.3.6); but the image of $\mathbb{P}^1$ under the complete linear system $|\mathscr{O}_{\mathbb{P}^1}(g-1)|$ is nondegenerate (Exercise 15.2.C).

Thus $\mathscr{L}^{\otimes(g-1)}$ has at least g sections. But by Exercise 19.2.A, the only invertible sheaf of degree $2g - 2$ with (at least) g sections is the canonical sheaf. $\square$

As an added bonus, we see that the composition of (19.5.6.1) is the *complete* linear series $|\mathscr{L}^{\otimes(g-1)}|$—all sections of $\mathscr{L}^{\otimes(g-1)}$ come up in this way.

19.5.7. Proposition (a genus ≥ 2 curve can be hyperelliptic in only one way) — *Any curve C of genus at least 2 admits at most one double cover of $\mathbb{P}^1$. More precisely, if $\mathscr{L}$ and $\mathscr{M}$ are two degree 2 line bundles yielding maps $C \to \mathbb{P}^1$, then $\mathscr{L} \cong \mathscr{M}$.*

Proof. If C is hyperelliptic, then we can recover the hyperelliptic map by considering the canonical linear series given by ω_C (the **canonical map**, which we will use again repeatedly in the next few sections): it is a double cover of a degree $g - 1$ rational normal curve (by the previous proposition), which is isomorphic to $\mathbb{P}^1$. This double cover is the hyperelliptic cover (also by the proof of the previous proposition). Thus the hyperelliptic cover $C \to \mathbb{P}^1$ is determined by the canonical map (so there can be at most one). This map $C \to \mathbb{P}^1$ must be induced by both $\mathscr{L}$ and $\mathscr{M}$, from which we have $\mathscr{L} \cong \mathscr{M}$ (using Theorem 15.2.2, relating maps to projective space and line bundles). $\square$

19.5.C. EXERCISE. Suppose C is a hyperelliptic curve of genus at least 2. Show that the hyperelliptic involution commutes with any other automorphism g of C. (Hint: What happens when you conjugate the involution by g?)

19.5.8. *The "space of hyperelliptic curves".* Thanks to Proposition 19.5.7, we can now classify hyperelliptic curves of genus at least 2. Hyperelliptic curves of genus $g \geq 2$ correspond to precisely $2g + 2$ distinct points on $\mathbb{P}^1$ modulo S_{2g+2}, and modulo automorphisms of $\mathbb{P}^1$. Thus "the space of hyperelliptic curves" has dimension

$$2g + 2 - \dim \mathrm{Aut}\, \mathbb{P}^1 = 2g - 1.$$

This is not a well-defined statement, because we haven't rigorously defined "the space of hyper-elliptic curves"—it is an example of a *moduli space*. For now, take this as a plausibility statement. It is also plausible that this space is irreducible and reduced—it is the image of something irreducible and reduced.

19.6   Curves of Genus 2

19.6.1. The reason for leaving genus 1 for later. It might make most sense to jump to genus 1 at this point, but the theory of elliptic curves is especially rich and subtle, so we will leave it for §19.9.

In general, curves have quite different behaviors (topologically, arithmetically, geometrically) depending on whether $g = 0$, $g = 1$, or $g \geq 2$. This trichotomy extends to varieties of higher dimension. We already have some inkling of it in the case of curves. Arithmetically, genus 0 curves can have lots and lots of rational points, genus 1 curves can have lots of rational points, and by Faltings's Theorem (Mordell's Conjecture) any curve of genus at least 2 has at most finitely many rational points. (Thus even before the proof of important cases of the Taniyama–Shimura conjecture [Wi, TWi], and hence Fermat's Last Theorem, we knew that $x^n + y^n = z^n$ in $\mathbb{P}^2$ has at most finitely many rational solutions for $n \geq 4$, as such curves have genus $\binom{n-1}{2} > 1$; see (18.6.7.1).) In the language of differential geometry, Riemann surfaces of genus 0 are positively curved, Riemann surfaces of genus 1 are flat, and Riemann surfaces of genus at least 2 are negatively curved. It is a fact that curves of genus at least 2 have finite automorphism groups (see §21.4.9), while curves of genus 1 have one-dimensional automorphism groups (see Question 19.10.6), and the unique curve of genus 0 over an algebraically closed field has a three-dimensional automorphism group (see Exercises 15.5.A and 15.5.B). (See Exercise 21.5.F for more on this issue.)

19.6.2. Back to curves of genus 2.
Over an algebraically closed field, we saw in §19.3 that there is only one genus 0 curve. In §19.5 we established that there are hyperelliptic curves of genus 2. How can we get a hold of curves of genus 2? For example, are they all hyperelliptic? "How many" are there? We now tackle these questions.

Fix a curve C of genus $g = 2$. Then ω_C is degree $2g - 2 = 2$, and has two sections (Exercise 19.2.A). By Exercise 19.5.B, ω_C is base-point-free, and thus induces a double cover $C \to \mathbb{P}^1$ (unique up to automorphisms of $\mathbb{P}^1$, which we studied in §15.5.2). Conversely, any double cover $C \to \mathbb{P}^1$ arises from a degree 2 invertible sheaf with at least two sections. Hence if $g(C) = 2$, this invertible sheaf must be the canonical sheaf, just because it is a degree $2g - 2 = 2$ line bundle with $g = 2$ sections (or, alternatively, by the easiest case of Proposition 19.5.6).

Hence we have a natural bijection between genus 2 curves and genus 2 double covers of $\mathbb{P}^1$ (up to automorphisms of $\mathbb{P}^1$). If the characteristic is not 2, the hyperelliptic Riemann–Hurwitz formula (Theorem 19.5.1) shows that the double cover is branched over $2g + 2 = 6$ geometric points. In particular, we have a "three-dimensional space of genus 2 curves." This is not rigorous, but we can certainly show that there are an infinite number of genus 2 curves. Precisely:

19.6.A. EXERCISE. Fix an algebraically closed field k of characteristic not 2. Show that there are an infinite number of (pairwise) nonisomorphic genus 2 curves over k.

19.6.B. EXERCISE. Show that every genus 2 curve (over any field of characteristic not 2) has finite automorphism group.

19.7   Curves of Genus 3

19.7.1. The canonical embedding.
We begin with some general discussion about genus $g > 1$, then specialize to genus 3.

19.7.A. IMPORTANT EXERCISE. Assume $k = \bar{k}$ (purely to avoid distraction—feel free to remove this hypothesis). Suppose C is a genus g curve, with $g > 1$. Show that if C is not hyperelliptic, then the canonical invertible sheaf gives a closed embedding $C \hookrightarrow \mathbb{P}^{g-1}$. (In the hyperelliptic case, we have already seen that the canonical sheaf gives us a double cover of a rational normal curve.) Hint: Recall that $h^0(C, \omega_C) = g$ (Exercise 19.2.A). Use our "closed embedding test" of §19.1, in the form of §19.2.10. Recall that if $h^0(C, \mathscr{O}(p + q)) = 2$, then C is hyperelliptic (Exercise 19.5.B).

This exercise gives a second proof that all genus 2 curves are hyperelliptic.

19.7.2. *Definition.* Such a curve embedded in $\mathbb{P}^{g-1}$ by the canonical sheaf (also known as the canonical line bundle) is called a **canonical curve**, and this closed embedding is called the **canonical embedding** of C.

19.7.3. Back to genus 3.

Suppose now that C is a curve of genus 3. Then ω_C has degree $2g - 2 = 4$, and has $g = 3$ sections. By Exercise 19.7.A, if (and only if) C is not hyperelliptic, the canonical map describes C as a degree 4 curve in $\mathbb{P}^2$.

Conversely, any quartic plane curve is canonically embedded. Reason: The curve has genus 3 (see (18.6.7.1)), and is mapped by an invertible sheaf of degree 4 with three sections. But by Exercise 19.2.A, the only invertible sheaf of degree $2g - 2$ with g sections is ω_C.

In particular, each nonhyperelliptic genus 3 curve can be described as a quartic plane curve in precisely one way (up to automorphisms of $\mathbb{P}^2$, which were described in Exercise 15.5.A). We have shown a bijection between nonhyperelliptic genus 3 curves and plane quartic curves up to projective linear transformations (§15.5.2).

19.7.4. *Remark.* In particular, as there exist regular plane quartics (Exercise 13.3.E), there exist nonhyperelliptic genus 3 curves.

19.7.B. EXERCISE (CF. §19.5.8). Give a heuristic (non-rigorous) argument that the nonhyperelliptic curves of genus 3 form a family of dimension 6. (Hint: Count the dimension of the family of regular quartic plane curves, and quotient by $\operatorname{Aut}\mathbb{P}^2 = \operatorname{PGL}(3)$.)

The genus 3 curves thus seem to come in two families: the hyperelliptic curves (a family of dimension 5), and the nonhyperelliptic curves (a family of dimension 6). This is misleading—they actually come in a single family of dimension 6. In fact, hyperelliptic curves are naturally limits of nonhyperelliptic curves. We can write down an explicit family. (This explanation necessarily requires some hand-waving, as it involves topics we haven't seen yet.) Suppose we have a hyperelliptic curve branched over $2g + 2 = 8$ points of $\mathbb{P}^1$. Choose an isomorphism of $\mathbb{P}^1$ with a conic in $\mathbb{P}^2$. There is a regular quartic plane curve meeting the conic at precisely those eight points. (This requires a short argument using Bertini's Theorem 13.4.2, which we omit.) Then if f is the equation of the conic, and g is the equation of the quartic, then $f^2 + t^2 g$ is a family of quartics that are smooth for most t (smoothness is an open condition in t, as we will see in Theorem 21.6.6 on generic smoothness). The $t = 0$ case is a double conic. Then it is a fact that if you normalize the total space of the family, the central fiber (above $t = 0$) turns into our hyperelliptic curve. Thus we have expressed our hyperelliptic curve as a limit of nonhyperelliptic curves.

19.7.C. UNIMPORTANT EXERCISE. A (projective) curve (over a field k) admitting a degree 3 cover of $\mathbb{P}^1$ is called **trigonal**. Show that every nonhyperelliptic genus 3 complex curve is trigonal, by taking the quartic model in $\mathbb{P}^2$, and projecting to $\mathbb{P}^1$ from any point on the curve. Do this by choosing coordinates on $\mathbb{P}^2$ so that p is at $[0, 0, 1]$. (The same idea, applied to cubics rather than quartics, will be used in §19.9.9.)

19.7.5.* A genus 3 curve with no nontrivial automorphisms.
We have seen that a (smooth projective integral) curve of genus at most 2 always has nontrivial automorphisms. It turns out that there are genus 3 curves with no nontrivial automorphisms.

19.7.D. EXERCISE. Suppose $C' \subset \mathbb{P}^2$ is a smooth plane quartic curve (over any field k). Show that there is bijection between automorphisms of C' and automorphisms of $\mathbb{P}^2$ preserving C'.

Thus to find a genus 3 curve with no nontrivial automorphisms, we need only find a smooth quartic plane curve C' such that the only automorphism of $\mathbb{P}^2$ fixing C' as a set must be the identity. Your intuition may (correctly) tell you that most quartics have this property. But exhibiting a specific C' (with proof) requires rolling up our sleeves and getting to work. Poonen gives automorphism free curves over any field in [P1]; an example in characteristic 0 is

$$y^3 z - 3yz^3 = 3x^4 - 4x^3 z + z^4.$$

19.7.E. EXERCISE. Suppose C is a smooth projective curve with no nontrivial automorphisms. Show that no two distinct open subsets of C are isomorphic.

19.7.6. *Genus 3 curves with nontrivial automorphisms.*
Certainly genus 3 curves can have nontrivial automorphisms: witness hyperelliptic curves. More impressive is the Klein quartic

$$x^3 y + y^3 z + z^3 x = 0$$

in $\mathbb{P}^2_{\mathbb{C}}$, which has 168 automorphisms. (Can you find them all?) In fact, the automorphism group of the Klein quartic is the unique finite simple group of order 168 (the second-smallest nonabelian finite simple group). In §21.4.9, we will see that no genus 3 curve can have more automorphisms (assuming the hard fact that curves of genus greater than 1 have finite automorphism groups).

19.8 Curves of Genus 4 and 5

We first consider nonhyperelliptic curves C of genus 4. Note that $\deg \omega_C = 6$ and $h^0(C, \omega_C) = 4$, so the canonical map expresses C as a sextic curve in $\mathbb{P}^3$. Let $i : C \hookrightarrow \mathbb{P}^3$ be the canonical (closed) embedding. We shall see that all such C are complete intersections of quadric surfaces and cubic surfaces, and conversely all regular complete intersections of quadric surfaces and cubic surfaces are genus 4 nonhyperelliptic curves, canonically embedded.

By (19.2.5.1) (Riemann–Roch and Serre duality),

$$h^0(C, i^* \mathscr{O}_{\mathbb{P}^3}(2)) = h^0(C, \omega_C^{\otimes 2}) = \deg \omega_C^{\otimes 2} - g + 1 = 12 - 4 + 1 = 9.$$

On the other hand, $h^0(\mathbb{P}^3, \mathscr{O}_{\mathbb{P}^3}(2)) = \binom{3+2}{2} = 10$ (Exercise 15.1.C). Thus the restriction map

$$H^0(\mathbb{P}^3, \mathscr{O}_{\mathbb{P}^3}(2)) \longrightarrow H^0(C, i^* \mathscr{O}_{\mathbb{P}^3}(2))$$

must have a nontrivial kernel, so there is at least one quadric surface Q in $\mathbb{P}^3$ that contains our curve C. Now quadric surfaces are either double planes, or the union of two planes, or cones, or regular quadrics. (They correspond to quadric forms of rank 1, 2, 3, and 4, respectively. The rank of a quadratic form was defined in Exercise 5.4.J.) But C can't lie in a plane, so Q must be a cone or regular. In particular, Q is irreducible.

Now C can't lie on *two* (distinct) such quadrics, say, Q and Q'. Otherwise, as Q and Q' have no common components (they are irreducible and not the same!), $Q \cap Q'$ is a curve (not necessarily reduced or irreducible). By Bézout's Theorem (Exercise 18.6.J), $Q \cap Q'$ is a curve of degree 4. Thus our curve C, being of degree 6, cannot be contained in $Q \cap Q'$. (If you don't see why directly, Exercise 18.6.E might help.)

19.8.1. We next consider cubic surfaces. By (19.2.5.1) again,

$$h^0(C, \omega_C^{\otimes 3}) = \deg \omega_C^{\otimes 3} - g + 1 = 18 - 4 + 1 = 15.$$

Now $\mathrm{Sym}^3\,\Gamma(C, \omega_C)$ has dimension $\binom{3+3}{3} = 20$. Thus C lies on at least a five-dimensional vector space of cubics. Now a four-dimensional subspace comes from multiplying the quadratic form Q by a linear form $(?w + ?x + ?y + ?z)$. Hence there is still one cubic K whose underlying form is not divisible by the quadric form Q (i.e., K doesn't contain Q.) Then K and Q share no component, so $K \cap Q$ is a complete intersection containing C as a closed subscheme. Now $K \cap Q$ and C are both degree 6 (the former by Bézout's Theorem, Exercise 18.6.J, and the latter because C is embedded by a degree 6 line bundle, Exercise 18.6.H). Also, $K \cap Q$ and C both have arithmetic genus 4 (the former by Exercise 18.6.S). These two invariants determine the (linear) Hilbert polynomial, so $K \cap Q$ and C have the same Hilbert polynomial. Hence $C = K \cap Q$ by Exercise 18.6.E.

We now show the converse, and that any regular complete intersection C of a quadric surface with a cubic surface is a canonically embedded genus 4 curve. By Exercise 18.6.S, such a complete intersection has genus 4.

19.8.A. EXERCISE. Show that $\mathscr{O}_C(1)$ has at least four sections. (Translation: C doesn't lie in a hyperplane.)

The only degree $2g - 2$ invertible sheaf with (at least) g sections is the canonical sheaf (Exercise 19.2.A), so $\mathscr{O}_C(1) \cong \omega_C$, and C is indeed canonically embedded.

19.8.B. EXERCISE. Give a heuristic argument suggesting that the nonhyperelliptic curves of genus 4 "form a family of dimension 9."

On to genus 5!

19.8.C. EXERCISE. Suppose C is a nonhyperelliptic genus 5 curve. Show that the canonical curve is degree 8 in $\mathbb{P}^4$. Show that it lies on a three-dimensional vector space of quadric surfaces (i.e., it lies on three linearly independent quadrics). Show that a regular complete intersection of three quadric surfaces is a canonical(ly embedded) genus 5 curve.

Unfortunately, not all canonical genus 5 curves are the complete intersection of three quadric surfaces in $\mathbb{P}^4$. But in the same sense that most genus 3 curves can be described as plane quartics, most canonical genus 5 curves are complete intersections of three quadric surfaces, and most genus 5 curves are nonhyperelliptic. The correct way to say this is that there is a dense Zariski-open locus in the moduli space of genus 5 curves consisting of nonhyperelliptic curves whose canonical embedding is cut out by three quadric surfaces.

(Those nonhyperelliptic genus 5 canonical curves not cut out by a three-dimensional vector space of quadric surfaces are precisely the trigonal curves; see Exercise 19.7.C. The triplets of points mapping to the same point of $\mathbb{P}^1$ under the trigonal map turn out to lie on a line in the canonical embedding. Any quadric vanishing along those three points must vanish along the line—basically, any quadratic polynomial with three zeros must be the zero polynomial.)

19.8.D. EXERCISE. Assuming the discussion above, count complete intersections of three quadric surfaces to give a heuristic argument suggesting that the curves of genus 5 "form a family of dimension 12."

19.8.2. We have now understood curves of genus 3 through 5 by thinking of canonical curves as complete intersections. Sadly, our luck has run out.

19.8.E. EXERCISE. Show that if $C \subset \mathbb{P}^{g-1}$ is a canonical curve of genus $g \geq 6$, then C is *not* a complete intersection. (Hint: Bézout's Theorem, Exercise 18.6.J.)

19.8.3. There is still something beautiful going on. For example, for most genus 6 curves over $k = \bar{k}$, the canonical embedding (degree 10 in $\mathbb{P}^5$) arises in the following way. Start with the complete intersection of $G(2, 5)$ (dimension 6; do you see why?) under its Plücker embedding in $\mathbb{P}^9$. Intersect it with a linear space of codimension 4, yielding a smooth surface in $\mathbb{P}^5$. This surface is isomorphic to $\mathbb{P}^2$ blown up at four points. (It is a *del Pezzo surface*, to be defined in Exercise 21.5.H.

This particular surface appears in the table in Remark 27.3.7.) Slice (intersect) the surface with a quadric hypersurface to obtain a genus 6 curve.

19.8.4. *Some discussion on curves of general genus.* However, we still have some data. If $\mathcal{M}_g$ is this ill-defined "moduli space of genus g curves," we have heuristics to find its dimension for low g. In genus 0, over an algebraically closed field, there is only one genus 0 curve (Proposition 19.3.1), so it appears that dim $\mathcal{M}_0 = 0$. In genus 1, over an algebraically closed field, we will soon see that the elliptic curves are classified by the j-invariant (Exercise 19.10.G), so it appears that dim $\mathcal{M}_1 = 1$. We have also informally computed dim $\mathcal{M}_2 = 3$, dim $\mathcal{M}_3 = 6$, dim $\mathcal{M}_4 = 9$, dim $\mathcal{M}_5 = 12$. What is the pattern? In fact, in some strong sense it was known by Riemann that dim $\mathcal{M}_g = 3g - 3$ for $g > 1$. What goes wrong in genus 0 and genus 1? As a clue, recall our insight when discussing Hilbert functions (§18.6) that whenever some function is "eventually polynomial," we should assume that it "wants to be polynomial," and there is some better function (usually an Euler characteristic) that *is* polynomial, and that cohomology-vanishing ensures that the original function and the better function "eventually agree." Making sense of this in the case of $\mathcal{M}_g$ is far beyond the scope of our current discussion, so we will content ourselves by observing the following facts. *Every* regular curve of genus greater than 1 has a finite number of automorphisms—a zero-dimensional automorphism group. *Every* regular curve of genus 1 has a one-dimensional automorphism group (see Question 19.10.6). And the only regular curve of genus 0 has a three-dimensional automorphism group (Exercise 15.5.B). (See Aside 21.5.12 for more discussion.) So notice that for all $g \geq 0$,

$$\dim \mathcal{M}_g - \dim \operatorname{Aut} C_g = 3g - 3,$$

where $\operatorname{Aut} C_g$ means the automorphism group of any curve of genus g.

In fact, in the language of stacks (or orbifolds), it makes sense to say that the dimension of the moduli space of (smooth projective geometrically irreducible) genus 0 curves is -3, and the dimension of the moduli space of genus 1 curves is 0.

19.9 Curves of Genus 1

Finally, we come to the very rich case of curves of genus 1. We will present the theory by discussing interesting things about line bundles of steadily increasing degree. Although it is not particularly essential to us now, recall that the degree d line bundles on C are denoted by $\operatorname{Pic}^d C$ (Definition 18.4.2).

19.9.1. Line bundles of degree 0 ($\operatorname{Pic}^0 C$).
Suppose C is a genus 1 curve. Then $\deg \omega_C = 2g - 2 = 0$ and $h^0(C, \omega_C) = g = 1$ (by Exercise 19.2.A). But the only degree 0 invertible sheaf with a section is the structure sheaf (§19.2.4), so we conclude that $\omega_C \cong \mathcal{O}_C$. (If you know that complex genus 1 curves are of the form $\mathbb{C}$ modulo a lattice, and Miracle 18.5.2 that the sheaf of differentials is the canonical sheaf, then you might not be surprised.)

We move on to line bundles of higher degree. Next, note that if $\deg \mathscr{L} > 0$, then Riemann–Roch and Serre duality (19.2.5.1) give

$$h^0(C, \mathscr{L}) = \deg \mathscr{L} - g + 1 = \deg \mathscr{L}.$$

19.9.2. Line bundles of degree 1 ($\operatorname{Pic}^1(C)$).
Each degree 1 (k-valued) point q determines a line bundle $\mathcal{O}_C(q)$, and two distinct points determine two distinct line bundles (as a degree 1 line bundle has only one section, up to scalar multiples). Conversely, any degree 1 line bundle $\mathscr{L}$ is of the form $\mathcal{O}_C(q)$ (as $\mathscr{L}$ has a section—then just take its divisor of zeros).

Thus we have a canonical bijection between degree 1 line bundles and degree 1 (closed) points. (If k is algebraically closed, as all closed points have residue field k, this means that we have a canonical bijection between degree 1 line bundles and closed points.)

Define an **elliptic curve** to be a genus 1 curve E with a choice of k-valued point p. The choice of this point should always be considered part of the definition of an elliptic curve—"elliptic curve" is not a synonym for "genus 1 curve." (Note: A genus 1 curve need not have any k-valued points at all! For example, the hypereliptic genus 1 curve obtained (as in §19.5.2) from $y^2 = -x^4 - 1$ in $\mathbb{A}^2_{\mathbb{Q}}$ has no $\mathbb{R}$-points, and hence no $\mathbb{Q}$-points. Of course, if $k = \bar{k}$, then any closed point is k-valued, by the Nullstellensatz 3.2.5.) We will often denote elliptic curves by E rather than C.

If (E, p) is an elliptic curve, then there is a canonical bijection between the set of degree 0 invertible sheaves (up to isomorphism) and the set of degree 1 points of E: twist the degree 1 line bundles by $\mathscr{O}(-p)$. Explicitly, the bijection is given by

$$\mathscr{L} \longmapsto \mathrm{div}(\mathscr{L}(p))$$

$$\mathscr{O}(q-p) \longmapsfrom q \quad .$$

But the degree 0 invertible sheaves form a group (under tensor product), so we have proved:

19.9.3. Proposition (the group law on the degree 1 points of an elliptic curve) — *The above bijection defines an abelian group structure on the degree 1 points of an elliptic curve, where p is the identity.*

From now on, we will identify closed points of E with degree 0 invertible sheaves on E without comment.

For those familiar with the complex analytic picture, this is not surprising: E is isomorphic to the complex numbers modulo a lattice: $E \cong \mathbb{C}/\Lambda$. (We haven't shown this, of course.)

Proposition 19.9.3 is currently just a bijection of sets. Given that E has a much richer structure (it has a generic point, and the structure of a variety), this is a sign that there should be a way of defining some *scheme* $\mathrm{Pic}^0(E)$, and that this should be an isomorphism of schemes. We will soon show (Theorem 19.10.4) that this group structure on the degree 1 points of E comes from a group variety structure on E.

19.9.4. *Aside: The Mordell–Weil Theorem, group, and rank.* This is a good excuse to mention the *Mordell–Weil Theorem*: for any elliptic curve E over $\mathbb{Q}$, the $\mathbb{Q}$-points of E form a *finitely generated* abelian group, often called the *Mordell–Weil group*. By the classification of finitely generated abelian groups (a special case of the classification of finitely generated modules over a principal ideal domain, Remark 6.4.4), the $\mathbb{Q}$-points are a direct sum of a torsion part, and of a free $\mathbb{Z}$-module. The rank of the $\mathbb{Z}$-module is called the *Mordell–Weil rank*.

19.9.5. Line bundles of degree 2 ($\mathrm{Pic}^2(E)$); see Figure 19.4.

Note that $\mathscr{O}_E(2p)$ has two sections, so E admits a double cover of $\mathbb{P}^1$ (Exercise 19.5.B). One of the ramification points is p: one of the sections of $\mathscr{O}_E(2p)$ vanishes at p to order 2, so there is a point of $\mathbb{P}^1$ whose preimage consists of p (with multiplicity 2). Assume now that $k = \bar{k}$ and char $k \neq 2$, so we can use the hyperelliptic Riemann–Hurwitz formula (Theorem 19.5.1), which implies that E has four ramification points (p and three others). Conversely, given four points in $\mathbb{P}^1$, there exists a unique

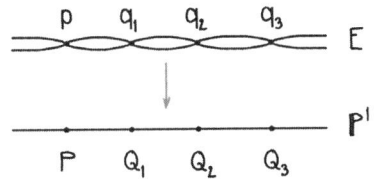

Figure 19.4 *Elliptic curves yield double covers of the projective line, branched at four points (§19.9.5).*

double cover branched at those four points (Proposition 19.5.2). Thus elliptic curves correspond to four distinct points in $\mathbb{P}^1$, where one is marked P, up to automorphisms of $\mathbb{P}^1$. Equivalently, by placing P at ∞, elliptic curves correspond to three points in $\mathbb{A}^1$, up to affine maps $x \mapsto ax + b$.

19.9.A. EXERCISE. Show that the other three ramification points are precisely the (non-identity) 2-torsion points in the group law. (Hint: If one of the points is q, show that $\mathscr{O}(2q) \cong \mathscr{O}(2p)$, but $\mathscr{O}(q)$ is not congruent to $\mathscr{O}(p)$.)

Thus (if char $k \neq 2$ and $k = \bar{k}$) every elliptic curve has precisely four 2-torsion points. If you are familiar with the complex picture $E \cong \mathbb{C}/\Lambda$, this isn't surprising.

19.9.6. *Follow-up remark.* An elliptic curve with *full level* n *structure* is an elliptic curve with an isomorphism of its n-torsion points with $(\mathbb{Z}/n)^2$. (This notion has problems if n is divisible by char k.) Thus an elliptic curve with *full level 2 structure* is the same thing as an elliptic curve with an ordering of the three other branch points in its degree 2 cover description. Thus (if $k = \bar{k}$) these objects are parametrized by the λ-line, which we discuss below.

Follow-up to the follow-up. There is a notion of moduli spaces of elliptic curves with full level n structure. Such moduli spaces are smooth curves (where this is interpreted appropriately—they are stacks), and have smooth compactifications. A *weight k level n modular form* is a section of $\omega_M^{\otimes k}$ where ω_M is the canonical sheaf of this moduli space ("modular curve").

19.9.7. *The cross-ratio and the j-invariant.* (We maintain the assumption that k is algebraically closed and characteristic not 2.) If the three other branch points are temporarily labeled $Q_1, Q_2, Q_3 \in \mathbb{P}^1$, there is a unique automorphism of $\mathbb{P}^1$ taking P, Q_1, Q_2 to $\infty, 0, 1$, respectively (as Aut $\mathbb{P}^1$ is 3-transitive, Exercise 15.5.B). Suppose that Q_3 is taken to some number λ under this map, where necessarily $\lambda \neq 0, 1, \infty$.

The value λ is called the **cross-ratio** of the four points (P, Q_1, Q_2, Q_3) of $\mathbb{P}^1$ (invented by Pappus of Alexandria, [MO152008]).

19.9.B. EXERCISE. Show that isomorphism classes of four ordered distinct (closed) points on $\mathbb{P}^1$ (over an algebraically closed field k), up to projective equivalence (automorphisms of $\mathbb{P}^1$), are classified by the cross-ratio.

19.9.C. EXERCISE. Show that the cross-ratio of (a, b, c, d) is the same as the cross-ratio of (b, a, d, c). The group S_4 acts on 4-tuples of points on $\mathbb{P}^1$, and hence on their cross-ratios. Show that the action of S_4 on the cross-ratio is trivial when restricted to the "Klein 4-group" $V_4 \subset S_4$.

We have not defined the notion of *moduli space*, but the previous exercise illustrates the fact that $\mathbb{P}^1 \setminus \{0, 1, \infty\}$ (the image of the cross-ratio map) is the moduli space for four ordered distinct points of $\mathbb{P}^1$ up to projective equivalence.

Notice:

- If we had instead sent P, Q_2, Q_1 to $(\infty, 0, 1)$, then Q_3 would have been sent to $1 - \lambda$.
- If we had instead sent P, Q_1, Q_3 to $(\infty, 0, 1)$, then Q_2 would have been sent to $1/\lambda$.
- If we had instead sent P, Q_3, Q_1 to $(\infty, 0, 1)$, then Q_2 would have been sent to $1 - 1/\lambda = (\lambda - 1)/\lambda$.
- If we had instead sent P, Q_2, Q_3 to $(\infty, 0, 1)$, then Q_1 would have been sent to $1/(1 - \lambda)$.
- If we had instead sent P, Q_3, Q_2 to $(\infty, 0, 1)$, then Q_1 would have been sent to $1 - 1/(1 - \lambda) = \lambda/(\lambda - 1)$.

Thus these six values (which correspond to S_3, the symmetric group permuting Q_1, Q_2, and Q_3) yield the same elliptic curve, and this elliptic curve will (upon choosing an ordering of the other three branch points) yield one of these six values.

This is fairly satisfactory already. To check if two elliptic curves (E, p), (E', p') over $k = \bar{k}$ are isomorphic, we write both as double covers of $\mathbb{P}^1$ ramified at p and p', respectively, then order the remaining branch points, then compute their respective λ's (say λ and λ', respectively), and see if they are related by one of the six expressions above:

(19.9.7.1) $$\lambda' = \lambda, \ 1/\lambda, \ 1 - \lambda, \ 1/(1 - \lambda), \ (\lambda - 1)/\lambda, \text{ or } \lambda/(\lambda - 1).$$

It would be far more convenient if, instead of a "six-valued invariant" λ, there were a single invariant (let's call it j), such that $j(\lambda) = j(\lambda')$ if and only if one of the equalities of (19.9.7.1) holds. This j-function should presumably be algebraic, so it would give a map j from the λ-line $\mathbb{A}^1 \setminus \{0, 1\}$ to the j-line $\mathbb{A}^1$. By the Curve-to-Projective Extension Theorem 15.3.1, this would extend to a morphism $j \colon \mathbb{P}^1 \to \mathbb{P}^1$. By Exercise 16.3.D, because this is (for most λ) a 6-to-1 map, the degree of this cover is 6 (or more correctly, at least 6).

We can make this dream more precise as follows. The elliptic curves over k correspond to k-valued points of $\mathbb{P}^1 \setminus \{0, 1, \infty\}$, modulo the action of S_3 on λ given above. Consider the subfield K of $k(\lambda)$ fixed by S_3. Then the extension $k(\lambda)/K$ is necessarily Galois (see, for example, [DF, §14.2, Thm. 9]), and degree 6. We are hoping that this subfield is of the form $k(j)$, and if so, we would obtain the j-map $\mathbb{P}^1 \to \mathbb{P}^1$ as described above. One could show that K is finitely generated over k, and then invoke Lüroth's Theorem, which we will soon prove in Example 21.4.7; but we won't need this.

Instead, we will just hunt for such a j. Note that λ should satisfy a sextic polynomial over $k(j)$ (or, more precisely, given what we know right now, a polynomial of degree *at least* 6), as for each j-invariant, there are six values of λ in general.

As you are undoubtedly aware, there *is* such a j-invariant. Here is the formula for the j-invariant that everyone uses:

$$(19.9.7.2) \qquad\qquad j = 2^8 \frac{(\lambda^2 - \lambda + 1)^3}{\lambda^2 (\lambda - 1)^2}.$$

You can readily check that $j(\lambda) = j(1/\lambda) = j(1 - \lambda) = \cdots$, and that as j (in lowest terms) has a degree 6 numerator and degree < 6 denominator, j indeed determines a degree 6 map from $\mathbb{P}^1$ (with coordinate λ) to $\mathbb{P}^1$ (with coordinate j). But this complicated-looking formula begs the question: where did this formula come from? How did someone think of it? We will largely answer this, but we will ignore the 2^8. This, as you might imagine, arises from characteristic 2 issues, and in order to invoke the results of §19.5, we have been assuming char $k \neq 2$. (To see why the 2^8 is forced upon us by characteristic 2, see [De, p. 64]. From a different, complex-analytic, point of view, the 2^8 comes from the fact that the q-expansion of j begins $j = q^{-1} + 744 + \cdots$: the q^{-1} term has coefficient 1; see [Se5, p. 90, Rem. 2]. These seemingly different explanations are related by the theory of the Tate curve; see [Si2, Thm. V.3.1].)

Rather than using the formula handed to us, let's try to guess what j is. We won't expect to get the same formula as (19.9.7.2), but our answer should differ by an automorphism of the j-line ($\mathbb{P}^1$)—we will get $j' = (aj + b)/(cj + d)$ for some a, b, c, d (Exercise 15.5.A).

We are looking for some $j'(\lambda)$ such that $j'(\lambda) = j'(1/\lambda) = \cdots$. Hence we want some expression in λ that is invariant under this S_3-action. A first possibility would be to take the product of the six numbers:

$$\lambda \cdot (1 - \lambda) \cdot \frac{1}{\lambda} \cdot \frac{\lambda - 1}{\lambda} \cdot \frac{1}{1 - \lambda} \cdot \frac{\lambda}{\lambda - 1}.$$

This is silly, as the product is obviously 1.

A better idea is to add them all together:

$$\lambda + (1 - \lambda) + \frac{1}{\lambda} + \frac{\lambda - 1}{\lambda} + \frac{1}{1 - \lambda} + \frac{\lambda}{\lambda - 1}.$$

This also doesn't work, as they add to 3—the six terms come in pairs adding to 1.

(Another reason you might realize this can't work: If you look at the sum, you will realize that you will get something of the form "degree at most 3" divided by "degree at most 2." Then if $j' = p(\lambda)/q(\lambda)$, then λ is a root of a cubic over j. But we said that λ should satisfy a sextic over j'. The only way we avoid a contradiction is if $j' \in k$.)

But you will undoubtedly have another idea immediately. One good idea is to take the second symmetric function in the six roots. An equivalent one that is easier to work out by hand is to add up the squares of the six terms. Even before doing the calculation, we can see that this will work: it will clearly produce a fraction whose numerator and denominator have degree at most 6, and it is not constant, as when λ is some fixed small number (say $1/2$), the sum of squares is some small real number, while when λ is a large real number, the sum of squares will have to be some large real number (different from the value when $\lambda = 1/2$).

When you add up the squares by hand (which is not hard), you will get

$$j' = \frac{2\lambda^6 - 6\lambda^5 + 9\lambda^4 - 8\lambda^3 + 9\lambda^2 - 6\lambda + 2}{\lambda^2(\lambda - 1)^2}.$$

Indeed $k(j) \cong k(j')$: you can check (again by hand) that

$$2j/2^8 = \frac{2\lambda^6 - 6\lambda^5 + 12\lambda^4 - 14\lambda^3 + 12\lambda^2 - 6\lambda + 2}{\lambda^2(\lambda - 1)^2}.$$

Thus $2j/2^8 - j' = 3$.

19.9.8. Line bundles of degree 3 ($\mathrm{Pic}^3(E)$).

In the discussion of degree 2 line bundles in §19.9.5, we assumed char $k \neq 2$ and $k = \bar{k}$, in order to invoke the Riemann–Hurwitz formula. In this section, we will start with no assumptions, and add them as we need them. In this way, you will see what partial results hold with weaker assumptions.

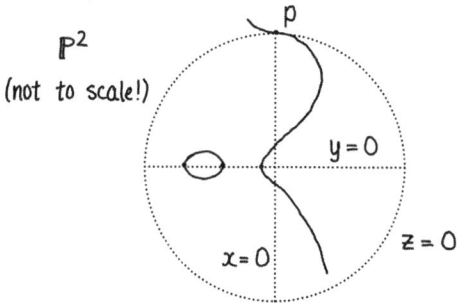

P²
(not to scale!)

Figure 19.5 *Finding the Weierstrass cubic incarnation of an elliptic curve.*

Consider the degree 3 invertible sheaf $\mathscr{O}_E(3p)$. By Riemann–Roch (19.2.5.1), $h^0(E, \mathscr{O}_E(3p)) = \deg(3p) - g + 1 = 3$. As $\deg \mathscr{O}_E(3p) > 2g$, this gives a closed embedding (Remark 19.2.11 and Exercise 19.2.E). Thus we have a closed embedding $E \hookrightarrow \mathbb{P}^2_k$ as a cubic curve. Moreover, there is a line in $\mathbb{P}^2_k$ meeting E at point p with multiplicity 3, corresponding to the section of $\mathscr{O}(3p)$ vanishing precisely at p with multiplicity 3. (A line in the plane meeting a smooth curve with multiplicity at least 2 is a *tangent line*; see Definition 13.3.3. A line in the plane meeting a smooth curve with multiplicity at least 3 is said to be a **flex line**, and that point is a **flex point** or **flex** of the curve.) (See Figure 19.5.)

Choose projective coordinates on $\mathbb{P}^2_k$ so that p maps to $[0, 1, 0]$, and the flex line is the line at infinity $z = 0$. Then the cubic is of the following form:

$$\begin{aligned}
?\,x^3 &+ 0\,x^2y + 0\,xy^2 + 0\,y^3 \\
&+ ?\,x^2z + ?\,xyz + ?\,y^2z \\
&+ ?\,xz^2 + ?\,yz^2 \\
&+ ?\,z^3.
\end{aligned} = 0$$

The coefficient of x^3 is not 0 (or else this cubic is divisible by z). Dividing the entire equation by this coefficient, we can assume that the coefficient of x^3 is 1. The coefficient of y^2z is not 0 either (or else this cubic is singular at $x = z = 0$). We can scale z (i.e., replace z by a suitable multiple) so that the coefficient of y^2z is -1. If the characteristic of k is not 2, then we can then replace y by $y + ?\,x + ?\,z$ so that the coefficients of xyz and yz^2 are 0, and if the characteristic of k is not 3, we can replace x by $x + ?z$ so that the coefficient of x^2z is also 0. In conclusion, if char $k \neq 2, 3$, the elliptic curve may be written as

(19.9.8.1) $$y^2z = x^3 + axz^2 + bz^3.$$

This is called the **Weierstrass normal form** of the curve.

19.9.9. *From the Weierstrass cubic to the double cover.* We see the hyperelliptic (double cover) description of the curve by setting $z = 1$, or, more precisely, by working in the distinguished open set $z \neq 0$ and using inhomogeneous coordinates. Here is the geometric explanation of why the double cover description is visible in the cubic description. Project the cubic from $p = [0, 1, 0]$ (see Figure 19.6). This is a map $E \setminus p \to \mathbb{P}^1$ (given by $[x, y, z] \dashrightarrow [x, z]$), and is basically the vertical projection of the cubic to the x-axis. (Figure 19.6 may help you visualize this.) By the Curve-to-Projective Extension Theorem 15.3.1, the morphism $E - p \to \mathbb{P}^1$ extends over p. If $\mathscr{L} := \mathscr{O}(3p)$ is the line bundle giving the morphism $E \hookrightarrow \mathbb{P}^2$ (via Theorem 15.2.2 describing maps to projective space in terms of line bundles), then the two sections x and z giving the map to $\mathbb{P}^1$ vanish at p to order 1 and 3, respectively. To "resolve" the rational map into an honest morphism, we interpret x and z as sections of $\mathscr{L}(-p) = \mathscr{O}(2p)$, and now they

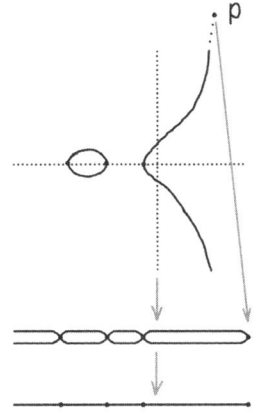

Figure 19.6 *From the Weierstrass cubic to the double cover interpretation of an elliptic curve (cf. Figure 19.5).*

generate a base-point-free linear series, and thus a morphism to $\mathbb{P}^1$. (You may be able to interpret this as implementing the proof of Theorem 15.2.2 in a different language.) A similar idea, applied to quartics rather than cubics, was used in Exercise 19.7.C.

As a consequence, with a little sweat, we can compute the j-invariant as a function of a and b:

$$j(a, b) = \frac{2^8 3^3 a^3}{4a^3 + 27b^2}.$$

19.9.D. EXERCISE. Show that the flexes of the cubic are the 3-torsion points in the group E. ("Flex" was defined in §19.8: it is a point where the tangent line meets the curve with multiplicity at least 3 at that point. In fact, if k is algebraically closed and char $k \neq 3$, there are nine of them. This won't be surprising if you are familiar with the complex story, $E \cong \mathbb{C}/\Lambda$.)

19.9.10. The group law, geometrically.
The group law has a beautiful classical description in terms of the Weierstrass form. Consider Figure 19.7. In the Weierstrass coordinates, the origin p is the only point of E meeting the line at infinity $(z = 0)$; in fact, the line at infinity corresponds to the tautological section of $\mathscr{O}(3p)$. If a line meets E at three points p_1, p_2, p_3, then

$$\mathscr{O}(p_1 + p_2 + p_3) \cong \mathscr{O}(3p),$$

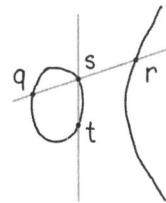

Figure 19.7 *The group law on the elliptic curve, geometrically.*

from which (in the group law) $p_1 + p_2 + p_3 = 0$.

Hence to find the inverse of a point s, we consider the intersection of E with the line $\overline{sp}$; $-s$ is the third point of intersection. To find the sum of two points q and r, we consider the intersection of E with the line $\overline{qr}$, and call the third point s. We then compute $-s$ by connecting s to p, obtaining $q + r$.

We could give this description of a group law on a cubic curve in Weierstrass normal form to anyone familiar with the notion of projective space, and the notion of a group, but we would then have to prove that the construction we are giving indeed defines a group. In particular, we would have to prove associativity, which is not a priori clear. But in this case we have already established that the degree 1 points form a group, by giving a bijection to $\mathrm{Pic}^0 E$, and we are merely interpreting the group law on $\mathrm{Pic}^0 E$.

Note that this description works even in characteristic 2 and 3; we do not need the cubic to be in Weierstrass normal form, and we need only that $\mathscr{O}(3p)$ gives a closed embedding into $\mathbb{P}^2$.

19.9.11. Degree 4 line bundles. You have probably forgotten that we began by studying line bundles degree by degree. The story doesn't stop in degree 3. In the same way that we showed that a canonically embedded nonhyperelliptic curve of genus 4 is the complete intersection in $\mathbb{P}_k^3$ of a quadric surface and a cubic surface (§19.8), we can show the following.

19.9.E. EXERCISE. Show that the complete linear series for $\mathscr{O}(4p)$ embeds E in $\mathbb{P}^3$ as the complete intersection of two quadric surfaces. Hint: Show the image of E is contained in at least two linearly independent quadric surfaces. Show that neither can be reducible, so they share no components. Use Bézout's Theorem (Exercise 18.6.J) and Exercise 18.6.E, as in §19.8.1.

The beautiful structure doesn't stop with degree 4, but it gets more complicated. For example, the degree 5 embedding is not a complete intersection (of hypersurfaces), but is the complete intersection of $G(2, 5)$ under its Plücker embedding with five hyperplanes (or, perhaps better, a codimension 5 linear space). In seemingly different terminology, its equations are 4×4 Pfaffians of a general 5×5 skew-symmetric matrix of linear forms, although I won't say what this means. (This construction may remind you of a similar construction for genus 6 curves in §19.8.3. This is no coincidence.)

19.10   Elliptic Curves Are Group Varieties

> Ever since 1949, I considered the construction of an algebraic theory of the Picard variety as the task of greatest urgency in abstract algebraic geometry.
>
> — A. Weil [Weil, p. 537]

We initially described the group law on the degree 1 points of an algebraic curve in a rather abstract way. From that definition, it was not clear that over $\mathbb{C}$ the group operations (addition, inverse) are continuous. But the explicit description in terms of the Weierstrass cubic makes this clear. In fact, we can observe even more: addition and inverse are algebraic in general. Better yet, elliptic curves are group varieties. (Thus they are abelian varieties; see Definition 11.4.11.)

19.10.1. (This is a clue that $\operatorname{Pic}^0(E)$ really wants to be a scheme, and not just a group. Once the notion of "moduli space of line bundles on a variety" is made precise, this can be shown; see, for example, [FGIKNV, Part 5].)

We begin with the "group inverse" morphism, as a warm-up.

19.10.2. Proposition — *If* char $k \neq 2, 3$, *there is a morphism of* k-*varieties* $E \to E$ *sending a (degree 1) point to its inverse, and this construction behaves well under field extension of* k.

In other words, the "inverse map" in the group law actually arises from a morphism of schemes—it isn't just a set map. (You are welcome to think through the two remaining characteristics, and to see that essentially the same proof applies. But the proof of Theorem 19.10.4 will give you a better sense of how to proceed.)

Proof. Consider the map (the hyperelliptic involution) $y \mapsto -y$ of the Weierstrass normal form.   □

The algebraic description of addition would be a big mess if we were to write it down. We will be able to show algebraicity by a trick—not by writing it down explicitly, but by thinking through how we *could* write it down explicitly. The main part of the trick is the following proposition. We give it in some generality just because it can be useful, but you may prefer to assume that $k = \bar{k}$ and C is a regular cubic. If C is a curve, let C^{reg} be the regular points of C, or, equivalently, the normal points (Theorem 13.5.8), which form an open subset of C (Exercise 10.7.O).

19.10.3. Proposition — *Suppose* $C \subset \mathbb{P}^2_k$ *is a geometrically integral cubic curve (so, in particular, C contains no lines). There is a unique morphism* $t \colon C^{reg} \times C^{reg} \to C^{reg}$ *such that*

(a) *if p and q are distinct regular k-valued points of C, then $t(p, q)$ is obtained by intersecting the line $\overline{pq}$ with C, and taking the third "residual" point of intersection with C. More precisely, $\overline{pq}$ will meet C at three points with multiplicity (Exercise 9.3.D), including p and q; $t(p, q)$ is the third point; and*

(b) *this property remains true after extension to $\overline{k}$.*

Furthermore, if p is a k-valued point of C^{reg}, then $t(p, p)$ is where the tangent line ℓ to C at p meets C again. More precisely, ℓ will meet C at three points with multiplicity, which includes p with multiplicity 2; $t(p, p)$ is the third point.

We will need property (b) because C may have few enough k-valued points (perhaps none!) that the morphism t cannot be determined by its behavior on them. In the course of the proof, we will see that (b) can be extended to "this property remains true after any field extension of k."

Proof. We first show (in this paragraph) that if p and q are distinct regular points, then the third point r of intersection of $\overline{pq}$ with C is also regular. If $r = p$ or $r = q$, we are done. Otherwise, the cubic obtained by restricting C to $\overline{pq}$ has three distinct (hence reduced, i.e., multiplicity 1) roots, p, q, and r. Thus $C \cap \overline{pq}$ is regular at r, so r is a regular point of C by the slicing criterion for regularity, Exercise 13.2.C.

We now assume that $k = \overline{k}$, and leave the general case to the end. Fix p, q, and r, where $p \neq q$, and r is the "third" point of intersection of $\overline{pq}$ with C. We will describe a morphism $t_{p,q}$ in an open neighborhood of $(p, q) \in C^{reg} \times C^{reg}$. By Exercise 11.3.B, showing that morphisms of varieties over $\overline{k}$ are determined by their behavior on closed ($\overline{k}$-valued) points, these morphisms glue together (uniquely) to give a morphism t, completing the proof in the case $k = \overline{k}$.

Choose projective coordinates on $\mathbb{P}^2$ in such a way that $U_0 \cong \operatorname{Spec} k[x_1, x_2]$ contains p, q, and r, and the line $\overline{pq}$ is not "vertical." More precisely, in $\operatorname{Spec} k[x_1, x_2]$, say $p = (p_1, p_2)$ (in terms of "classical coordinates"—more pedantically, $p = [(x_1 - p_1, x_2 - p_2)]$), $q = (q_1, q_2)$, $r = (r_1, r_2)$, and $p_1 \neq q_1$. In these coordinates, the curve C is cut out by some cubic, which we also sloppily denote by C: $C(x_1, x_2) = 0$.

Now suppose $P = (P_1, P_2)$ and $Q = (Q_1, Q_2)$ are two points of $C \cap U_0$ (not necessarily our p and q). We attempt to compute the third point of intersection of $\overline{PQ}$ with C, in a way that works on an open subset of $C \times C$ that includes (p, q). To do this explicitly requires ugly high school algebra, but because we know how it looks, we will be able to avoid dealing with any details!

The line $\overline{PQ}$ is given by $x_2 = mx_1 + b$, where $m = \frac{P_2 - Q_2}{P_1 - Q_1}$ and $b = P_2 - mP_1$ are both rational functions of P and Q. Then m and b are defined for all P and Q such that $P_1 \neq Q_1$ (and hence for an open neighborhood of (p, q), as $p_1 \neq q_1$, and as $P_1 \neq Q_1$ is an open condition).

Now we solve for $C \cap \overline{PQ}$, by substituting $x_2 = mx_1 + b$ into C, to get $C(x_1, mx_1 + b)$. This is a cubic in x_1, say,

$$\gamma(x_1) = Ax_1^3 + Bx_1^2 + Cx_1 + D = 0.$$

The coefficients of γ are rational functions of P_1, P_2, Q_1, and Q_2. The cubic γ has three roots (with multiplicity) so long as $A \neq 0$, which is a Zariski-open condition on m and b, and hence a Zariski-open condition on P_1, P_2, Q_1, Q_2. As $P, Q \in C \cap \overline{PQ} \cap U_0$, P_1 and Q_1 are two of the roots of $\gamma(x_1) = 0$. Since the line $\overline{PQ}$ contains p q, and r, the third root of $\gamma(x_1) = 0$ is R_1. The sum of the roots of $\gamma(x_1) = 0$ is $-B/A$ (by Viète's formula), so the third root of γ is $R_1 := -B/A - P_1 - Q_1$. Thus if we take $R_2 = mR_1 + b$, we have found the third point of intersection of $\overline{PQ}$ with C (which happily lies in U_0). We have thus described a morphism from the open subset of $(C^{reg} \cap U_0) \times (C^{reg} \cap U_0)$, containing (p, q), that does what we want. (Precisely, the open subset is defined by $A \neq 0$, which can be explicitly unwound.) We have thus completed the proof of Proposition 19.10.3 (except for the last paragraph) for $k = \overline{k}$. (Those who believe they are interested only in algebraically closed fields can skip ahead.)

We extend this to Proposition 19.10.3 for every field k except $\mathbb{F}_2$. Suppose $U_0^{[x_1,x_2]} =$ $\operatorname{Spec} k[x_1, x_2]$ is any affine open subset of $\mathbb{P}_k^2$, along with choice of coordinates. (The awkward superscript "$[x_1, x_2]$" is there to emphasize that the particular coordinates are used in the construction.) Then the construction above gives a morphism *defined over* k from an open subset of $(C^{\text{reg}} \cap U_0^{[x_1,x_2]}) \times (C^{\text{reg}} \cap U_0^{[x_1,x_2]})$ (note that all of the hypothetical algebra was done over k) that sends P and Q to the third point of intersection of $\overline{PQ}$ with C. Note that this construction commutes with any field extension, as the construction is insensitive to the field we are working over. Thus after base change to the algebraic closure, the map also has the property that it takes as input two points, and spits out the third point of intersection of the line with the cubic. Furthermore, all of these maps (as $U_0^{[x_1,x_2]}$ varies over all complements U_0 of lines "with k-coefficients," and choices of coordinates on U_0) can be glued together: they agree on their pairwise overlaps (as after base change to $\overline{k}$ they are the same, by our previous discussion, and two maps that are the same after base change to $\overline{k}$ were the same to begin with by Exercise 10.2.J), and this is what is required to glue them together (Exercise 7.2.A).

We can geometrically interpret the open subset $(C^{\text{reg}} \cap U_0^{[x_1,x_2]}) \times (C^{\text{reg}} \cap U_0^{[x_1,x_2]})$ by examining the construction: it is defined in the locus $\{P = (P_1, P_2), Q = (Q_1, Q_2)\}$, where (i) $P_1 \neq Q_1$, and (ii) the third point of intersection R of $\overline{PQ}$ with C also lies in U_0.

So which points (P, Q) of $C^{\text{reg}} \times C^{\text{reg}}$ are missed? Most important: we have omitted the case where $P = Q$.

19.10.A. EXERCISE. Use the algebra of Extended Example 15.3.4 to deal with the case where $P = Q$. (This will take some time and care. Ideally you should avoid doing any messy algebra.)

We now deal with the remaining cases where $P \neq Q$. Condition (i) isn't important; if (P, Q) satisfies (ii) but not (i), we can swap the roles of x_1 and x_2, and (P, Q) will then satisfy (i). The only way (P, Q) cannot be covered by one of these open sets is if there is *no* U_0 (a complement of a line defined over k) that includes P, Q, and R.

19.10.B. EXERCISE. Use $|k| > 2$ to show that there is a linear form on $\mathbb{P}^2$ with coefficients in k that misses P, Q, and R. (This is sadly *not* true if $k = \mathbb{F}_2$—do you see why?)

19.10.C. EXERCISE. Prove the last statement of Proposition 19.10.3.

19.10.D.** UNIMPORTANT EXERCISE. Complete the proof by dealing with the case $k = \mathbb{F}_2$. Hint: First produce the morphism t over $\mathbb{F}_4$. The goal is then to show that this t is really "defined over" $\mathbb{F}_2$ ("descends to" $\mathbb{F}_2$). The morphism t is initially described locally by considering the complement of a line defined over $\mathbb{F}_4$ (and then letting the line vary). Instead, look at the map by looking at the complement of a line and its "conjugate." The complement of the line and its conjugate is an affine $\mathbb{F}_2$-variety. The partially defined map t on this affine variety is a priori defined over $\mathbb{F}_4$, and is preserved by conjugation. Show that this partially defined map is "really" defined over $\mathbb{F}_2$. (If you figure out what all of this means, you will have an important initial insight into the theory of "descent.") ☐

We can now use this to define the group variety structure on E.

19.10.4. Theorem — *Suppose* (E, p) *is an elliptic curve (a regular genus 1 curve over k, with a k-valued point p). Take the Weierstrass embedding of E in $\mathbb{P}_k^2$, via the complete linear series $|\mathcal{O}_E(3p)|$. Define the k-morphism* e: $\operatorname{Spec} k \to E$ *by sending* $\operatorname{Spec} k$ *to p. Define the k-morphism* i: $E \to E$ *via* $q \mapsto t(p, q)$, *or, more precisely, as the composition*

$$E \xrightarrow{(\text{id}, e)} E \times E \xrightarrow{t} E.$$

Define the k-morphism m: $E \times E \to E$ *via* $(q, r) \mapsto t(p, t(q, r))$. *Then* (E, e, i, m) *is a group variety over k.*

By the construction of t, all of these morphisms "commute with arbitrary base extension."

Proof. We need to check that various pairs of morphisms described in axioms (i)–(iii) of §7.6.4 are equal. For example, in axiom (iii), we need to show that $m \circ (i, id) = m \circ (id, i)$; all of the axioms are clearly of this sort.

Assume first that $k = \bar{k}$. Then each of these pairs of morphisms agree as maps of $\bar{k}$-points: Pic E is a group, and under the bijection between Pic E and E of Proposition 19.9.3, the group operations translate into the maps described in the statement of Theorem 19.10.4 by the discussion of §19.9.10.

But morphisms of $\bar{k}$-varieties are determined by their maps on the level of $\bar{k}$-points (Exercise 11.3.B), so each of these pairs of morphisms are the same.

For general k, we note that, from the $\bar{k}$ case, these morphisms agree after base change to the algebraic closure. Then by Exercise 10.2.J, they must agree to begin with. $\square$

19.10.5. *Features of this construction.* The most common derivation of the properties of an elliptic curve is to describe the elliptic curve as a cubic, and then to describe addition using the explicit construction with lines. Then one has to work to prove that the multiplication described is associative.

Instead, we started with something that was patently a group (the degree 0 line bundles). We interpreted the maps used in the definition of the group (addition and inverse) geometrically using our cubic interpretation of elliptic curves. This allowed us to see that these maps were algebraic.

As a bonus, we see that, in some vague sense, the Picard group of an elliptic curve wants to be an algebraic variety.

19.10.E. EXERCISE. Suppose p and q are k-points of a genus 1 curve E. Show that there is an automorphism of E sending p to q.

19.10.F. EXERCISE. Suppose (E, p) is an elliptic curve over an algebraically closed field k of characteristic not 2 or 3. Show that the automorphism group of (E, p) is isomorphic to $\mathbb{Z}/2$, $\mathbb{Z}/4$, or $\mathbb{Z}/6$. (An automorphism of an elliptic curve (E, p) over $k = \bar{k}$ is an automorphism of E fixing p scheme-theoretically, or, equivalently, fixing the k-valued point p by Exercise 11.3.B.) Hint: Reduce to the question of automorphisms of $\mathbb{P}^1$ fixing a point ∞ and a set of distinct three points $\{p_1, p_2, p_3\} \in \mathbb{P}^1 \setminus \{\infty\}$. (The algebraic closure of k is not essential, so feel free to remove this hypothesis, using Exercise 10.2.J. What happens if the characteristic is 3?)

19.10.G. EXERCISE. Explain why genus 1 curves over an algebraically closed field of characteristic not 2 are classified (up to isomorphism) by j-invariant. (Caution: Note that the problem says "genus 1," not "elliptic.")

19.10.H. EXERCISE. Give (with proof) two elliptic curves over $\mathbb{Q}$ with the same j-invariant that are not isomorphic. (Hint: §19.5.3.)

19.10.6. *Vague question.* What are the possible automorphism groups of a genus 1 curve over an algebraically closed k of characteristic not 2? You should be able to convince yourself that the group has "dimension 1."

19.10.I. IMPORTANT EXERCISE: A DEGENERATE ELLIPTIC CURVE. Suppose char $k \neq 2$. Consider the arithmetic genus 1 curve $C \subset \mathbb{P}^2_k$ given by $y^2z = x^3 + x^2z$, with the point $p = [0, 1, 0]$. Emulate the above argument to show that $C \setminus \{[0, 0, 1]\}$ is a group variety (where the operation is that given by tensoring line bundles). Show that it is isomorphic to $\mathbb{G}_m$ (the multiplicative group scheme Spec $k[t, t^{-1}]$; see Exercise 7.6.E) with coordinate $t = (y - x)/(y + x)$, by showing an isomorphism of schemes, and showing that multiplication and inverse in both group varieties agree under this isomorphism.

19.10.J. EXERCISE: AN EVEN MORE DEGENERATE ELLIPTIC CURVE. Consider the arithmetic genus 1 curve $C \subset \mathbb{P}^2_k$ given by $y^2z = x^3$, with the point $p = [0, 1, 0]$. Emulate the above argument to show that $C \setminus \{[0,0,1]\}$ is a group variety (where the operation is that given by tensoring line bundles). Show that it is isomorphic to $\mathbb{A}^1$ (with additive group structure) with coordinate $t = x/y$, by showing an isomorphism of schemes, and showing that multiplication/addition and inverse in both group varieties agree under this isomorphism.

19.10.7.* Toward Poncelet's Porism.

These ideas lead to a beautiful classical fact, Poncelet's Porism. (A *porism* is an archaic name for a type of mathematical result.)

19.10.8. Poncelet's Porism — *Suppose C and D are two ellipses in $\mathbb{R}^2$, with C containing D. Choose any point p_0 on C. Choose one of the two tangents ℓ_1 from p_0 to D. Then ℓ_1 meets C at two points in total: p_0 and another point p_1. From p_1, there are two tangents to D, ℓ_1 and another line ℓ_2. The line ℓ_2 meets C at some other point p_2. Continue this to get a sequence of points $p_0, p_1, p_2, \ldots$. Suppose this sequence starting with p_0 is periodic, i.e., $p_0 = p_n$ for some n. Then it is periodic with any starting point $p \in C$ (see Figure 19.8).*

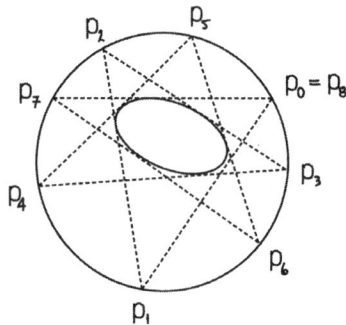

Figure 19.8 *Poncelet's Porism 19.10.8.*

It is possible to prove Poncelet's Porism in an elementary manner, but a proof involving elliptic curves is quite beautiful, and gives connections to more sophisticated ideas. Rather than proving Poncelet's Porism, we discuss some related facts.

19.10.K. EXERCISE. Suppose E is a smooth degree 3 projective plane curve over an algebraically closed field, $q, r \in E$, and n is a positive integer. Let $F \colon E \to E$ be the morphism $x \mapsto t(r, t(q, x))$. Suppose that $F^n(p) = p$ for some closed point $p \in E$. Show that $F^n(x) = x$ for all points $x \in E$.

You should feel the urge to improve this result. (Can you show that if $F^n(p) = p$ for some $p \in E$, then F^n is the identity morphism on E? Can you extend this to the case where the base field is not algebraically closed?)

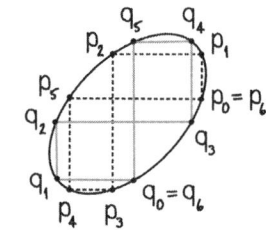

Figure 19.9 *Exercise 19.10.L (with 17 replaced by 3).*

19.10.L. EXERCISE. Suppose E is an ellipse in $\mathbb{R}^2$. For each point $(x, y) \in E$, there is another point $G(x, y) := (x, y')$ (possibly the same), with the same first coordinate, corresponding to the "other" intersection of the vertical line through (x, y) with E. Similarly, there is another point $H(x, y) := (x', y)$, with the same second coordinate. Let $F = G \circ H$. Show that if $F^{17}(p) = p$ for some point $p \in E$, then $F^{17}(x) = x$ for all points $x \in E$. (See Figure 19.9, where 17 is replaced by 3.) Hint: This might be best done not using any fancy methods from algebraic geometry.

Again, you should wish to improve this. To what extent is this dependent on the real numbers? What if the ellipse was instead a different conic section? (And what does this have to do with elliptic curves? Why is the conic called E?!)

19.10.M. ** EXERCISE. Let k be a field, algebraically closed purely for convenience. Suppose Q is a smooth quadric surface in $\mathbb{P}^3_k$, and K is a cone (a rank 3 quadric) in $\mathbb{P}^3$, such that $E = Q \cap K$ is a smooth curve.

(a) Show that E has genus 1.
(b) If ℓ is a line on Q, show that E meets ℓ at two points (with multiplicity).
(c) Fix one of the rulings (family of lines) of Q. Show that there exists a morphism $G \colon E \to E$ that takes a closed point p of E, and sends it to the other point of E on the line ℓ in the ruling containing p.

(d) Define $H: E \to E$ similarly, using the other ruling of Q. Fix a positive integer n. Let $F = G \circ H$. Show that if $F^n(p) = p$ for one closed point $p \in E$, then $F^n(x) = x$ for all $x \in E$.

We now connect Exercise 19.10.M to Poncelet's Porism 19.10.8. Let v be the cone point of K. Then $v \notin Q$, as otherwise v would be a singular point of $Q \cap K$ (do you see why?). It is a fact that projection from v gives a morphism $Q \to \mathbb{P}_k^2$ that is branched over a smooth conic D. (You may be able to make this precise, and prove it, after reading Chapter 21.) The lines on Q project to tangent lines to D in $\mathbb{P}_k^2$. The projection from v contracts $K \setminus \{v\}$ to a conic C. You may be able to interpret the statement of Poncelet's Porism 19.10.8 (for this C and D) in terms of Exercise 19.10.M. (Caution: The field is not algebraically closed in Poncelet's Porism 19.10.8, and *is* algebraically closed in Exercise 19.10.M.) You may even be able to take an arbitrary C and D as in Poncelet's Porism 19.10.8, and reverse-engineer K and Q as in Exercise 19.10.M, thereby proving Poncelet's Porism (and discovering why there is an elliptic curve hidden in its statement).

19.11 Counterexamples and Pathologies Using Elliptic Curves

We now give some fun counterexamples using our understanding of elliptic curves. The extra juice elliptic curves give us comes from the fact that elliptic curves are the simplest varieties with "continuous Picard groups."

19.11.1. A scheme that is factorial, yet no affine open neighborhood of any point is Spec of a unique factorization domain.

Suppose E is an elliptic curve over $\mathbb{C}$ (or some other uncountable algebraically closed field). Consider $p \in E$. The local ring $\mathscr{O}_{E,p}$ is a discrete valuation ring and hence a unique factorization domain (Theorem 13.5.8). Then any affine open neighborhood of E is of the form $E \setminus q_1 \setminus \cdots \setminus q_n$ for some points $q_1, \ldots, q_n$. I claim that its Picard group is nontrivial. Recall the exact sequence:

$$\mathbb{Z}^{\oplus n} \xrightarrow{\;(a_1,\ldots,a_n) \mapsto a_1 q_1 + \cdots + a_n q_n\;} \operatorname{Pic} E \longrightarrow \operatorname{Pic}(E \setminus q_1 \setminus \cdots \setminus q_n) \longrightarrow 0.$$

But the group on the left is countable, and the group in the middle is uncountable, so the group on the right, $\operatorname{Pic}(E \setminus q_1 \setminus \cdots \setminus q_n)$, is nonzero. We have shown that every nonempty open subset of E has nonzero line bundles, as promised in Remark 14.2.

As $E \setminus q_1 \setminus \cdots \setminus q_n$ is affine, by Exercise 15.5.P the corresponding ring is not a unique factorization domain. To summarize: complex elliptic curves are factorial, but no affine open subset has a ring that has unique factorization, as promised in §5.4.4.

19.11.A. EXERCISE.
The above argument shows that over an uncountable field, $\operatorname{Pic} E$ is not a finitely generated group. Show that even over the countable field $\overline{\mathbb{Q}}$, $\operatorname{Pic} E$ is not a finitely generated group, as follows. If the elliptic curve E is generated by $q_1, \ldots, q_n$, then there is a finite extension of fields K of $\mathbb{Q}$ over which all q_i are defined—the compositum of the residue fields of the q_i. (Careful—do you see what it means for the q_i to be "defined" over a smaller field than $\overline{\mathbb{Q}}$?) Show that any point in the subgroup of E generated by the q_i must also be defined over K. Show that E has a point not defined over K. Use this to show that $\operatorname{Pic} E$ is not finitely generated. (The same argument works with $\overline{\mathbb{Q}}$ replaced by $\overline{\mathbb{F}}_p$.)

19.11.2. *Remark.*
In contrast to the above discussion, over $\mathbb{Q}$, the Mordell–Weil Theorem states that $\operatorname{Pic} E$ *is* finitely generated (see Aside 19.9.4).

19.11.3.** A complex surface with infinitely many algebraic structures (for those with sufficient complex geometric background).

As remarked in §7.3.16, a complex manifold may have many algebraic structures. The following example of M. Kim gives an example with *infinitely* many algebraic structures. Suppose E is an elliptic curve over $\mathbb{C}$ with origin p, and let $\mathscr{L}$ be a nontrivial line bundle on $E - p$. Then $\mathscr{L}$ is *analytically* trivial because $E - p$ is a Stein space, so the analytification of the total space is independent of the choice of $\mathscr{L}$. However, you know enough to show (with work) that there are infinitely many pairwise (algebraically) nonisomorphic $\mathscr{L}$, and also that their total spaces are likewise pairwise (algebraically) nonisomorphic. This gives a complex surface with infinitely many algebraic structures (indeed a continuum of them). See [MO68421] for more details.

19.11.4. Counterexamples using a non-torsion point.

We next give a number of counterexamples using the existence of a non-torsion point of a complex elliptic curve. We first show the existence of such a point.

19.11.5. We have a "multiplication by n" map $[n]: E \to E$, which sends p to np. If $n = 0$, the map $[n]$ has degree 0 (even though $[n]$ isn't dominant, degree is still defined; see Definition 12.2.2). If $n = 1$, the map $[n]$ has degree 1. Given the complex picture of a torus, you might not be surprised that the degree of $[n]$ is n^2. If $n = 2$, we have almost shown that it has degree 4, as we have checked that there are precisely four points q such that $2p = 2q$ (Exercise 19.9.A). All that really shows (using Exercise 16.3.D(b)) is that the degree is at least 4. (We could check by hand that the degree is 4 is we really wanted to.)

19.11.6. Proposition — *Suppose E is an elliptic curve over a field k of characteristic not 2. For each $n > 0$, the "multiplication by n" morphism $[n]$ has positive degree, so there are only a finite number of n-torsion points.*

Proof. We may assume $k = \overline{k}$, as the degree of a map of curves is independent of field extension.

We prove the result by induction; it is true for $n = 1$ and $n = 2$.

If n is odd ($2k + 1$, say), then assume otherwise that $nq = 0$ for all closed points q. Let r be a nontrivial 2-torsion point, so $2r = 0$. But $nr = 0$ as well, so $r = (n - 2k)r = 0$, contradicting $r \neq 0$.

If n is even, then $[n] = [2] \circ [n/2]$ (degree is multiplicative under composition of rational maps, §12.2.2), and by our inductive hypothesis both $[2]$ and $[n/2]$ have positive degree. $\qquad\square$

In particular, the total number of torsion points on E is countable. If k is an uncountable field, then E has an uncountable number of closed points (consider an open subset of the curve as $y^2 = x^3 + ax + b$; there are uncountably many choices for x, and each of them has one or two choices for y). Thus we have the following.

19.11.7. Corollary — *If E is a curve over an uncountable algebraically closed field of characteristic not 2 (e.g., $\mathbb{C}$), then E has a non-torsion point.*

Proof. For each n, there are only finitely many n-torsion points. Thus there are (at most) countably many torsion points. The curve E has uncountably many closed points. (One argument for this: Take a double cover $\pi: E \to \mathbb{P}^1$. Then $\mathbb{P}^1$ has uncountably many closed points, and π is surjective on closed points.) $\qquad\square$

19.11.8. *Remark.* This argument clearly breaks down over countable fields. In fact, over $\overline{\mathbb{F}}_p$, all points of an elliptic curve E are torsion. (Any given point is defined over some finite field $\mathbb{F}_{p^r}$. The points defined over $\mathbb{F}_{p^r}$ form a subgroup of E, using the explicit geometric construction of the group law, and there are finite number of points over $\mathbb{F}_{p^r}$—certainly no more than the number of $\mathbb{F}_{p^r}$-points of $\mathbb{P}^2$.) But over $\overline{\mathbb{Q}}$, there are elliptic curves with non-torsion points. Even better, there are examples over $\mathbb{Q}$: $[2, 1, 8]$ is a $\mathbb{Q}$-point of the elliptic curve $y^2 z = x^3 + 4xz^2 - z^3$ that is not torsion. The proof would carry us too far afield, but one method is to use the Nagell–Lutz Theorem (see, for example, [Si1, Cor. 7.2]).

We now use the existence of a non-torsion point to create some interesting pathologies.

19.11.9. A map of projective varieties not arising from a map of graded rings, even after regrading.

19.11.B. EXERCISE. Suppose $E \hookrightarrow \mathbb{P}^2_{\mathbb{C}}$ is a smooth complex plane cubic (hence genus 1), yielding a graded ring $S_\bullet$. Let $t: E \to E$ be translation by a non-torsion point. Show that t does not correspond to a map of graded rings $S_\bullet \to S_\bullet$, even after regrading (cf. Remark 15.2.9). (If you wish, you can show the following. If $u: E \to E$ is a translation, show that u corresponds to some map of regraded rings $S_{n\bullet} \to S_{n\bullet}$ if and only if u is translation by a torsion point.)

19.11.10. An affine open subset of an affine scheme that is not a distinguished open set.

We can use this to construct an example of an affine scheme X and an affine open subset Y that is not distinguished in X. Let $X = E - p$, which is affine (see Exercise 19.2.C, or, better, note that the linear series $\mathcal{O}(3p)$ sends E to $\mathbb{P}^2$ in such a way that the "line at infinity" meets E only at p; then $E - p$ has a closed embedding into the affine scheme $\mathbb{A}^2$).

Let q be another point on E so that $q - p$ is non-torsion. Then $E - p - q$ is affine (Exercise 19.2.C). Assume that it is distinguished. Then there is a function f on $E - p$ that vanishes on q (to some positive order d). Thus f is a rational function on E that vanishes at q to order d, and (as the total number of zeros minus poles of f is 0) has a pole at p of order d. But then $d(p - q) = 0$ in $\mathrm{Pic}^0 E$, contradicting our assumption that $p - q$ is non-torsion.

In particular, $E - p$ is an affine scheme, and q is locally cut out by one equation, but it is not globally cut out *even set-theoretically* by one equation. This was promised in Exercise 8.3.E and §9.5.1.

19.11.11. A proper (nonprojective) surface with no nontrivial line bundles.

We next use a non-torsion point p on an elliptic curve E to construct a proper nonprojective surface with no nontrivial line bundles. Let X_1 be $\mathbb{P}^2$, and Z_1 a smooth cubic in X_1, identified with E. Let $Z_2 \subset X_2$ be exactly the same. Glue X_1 to X_2 along the isomorphism $Z_1 \overset{\sim}{\longleftrightarrow} Z_2$ given not by the identity but by translation by p. (This gluing construction was described in §15.2.12.) Call the result X. By Exercise 15.2.M (and the sentence thereafter), X is proper.

19.11.C. EXERCISE. Show that every line bundle on X is trivial. Hint: Suppose $\mathscr{L}$ is an invertible sheaf on X. Then $\mathscr{L}|_{X_i}$ is $\mathcal{O}(d_i)$ for some d_i (for $i = 1, 2$). The restriction of $\mathscr{L}|_{X_1}$ to Z_1 must agree with the restriction of $\mathscr{L}|_{X_2}$ to Z_2. Use this to show that $d_1 = d_2 = 0$. Explain why gluing two trivial bundles (on X_1 and X_2, respectively) together in this way yields a trivial bundle on X.

19.11.D. EXERCISE. Show that X is not projective.

See Exercise 20.2.G for a somewhat simpler example of a proper nonprojective surface.

19.11.12. A Picard group that has no chance of being a scheme.

We informally observed that the Picard group of an elliptic curve "wants to be" a scheme (see §19.10.5). This is true of projective (and even proper) varieties in general (see [FGIKNV, Ch. 9]). On the other hand, if we work over $\mathbb{C}$, the affine scheme $E - p - q$ (in the language of §19.11.10 above) has a Picard group that can be interpreted as $\mathbb{C}$ modulo a lattice modulo a non-torsion point (e.g., $\mathbb{C}/\langle 1, i, \pi \rangle$). This has no reasonable interpretation as a manifold, let alone a variety. So the fact that the Picard group of proper varieties turns out to be a scheme should be seen as quite remarkable.

19.11.13. A variety whose ring of functions is not finitely generated.

We next show an example of a complex variety whose ring of global sections is not finitely generated. (An example over $\mathbb{Q}$ can be constructed in the same way using the curve of Remark 19.11.8. If you wish, you can show that these examples are even quasiaffine.) This is related to Hilbert's fourteenth problem (see [Mu5, §3]).

19.11.E. PRELIMINARY EXERCISE. Suppose X is a scheme, and L is the total space of a line bundle corresponding to invertible sheaf $\mathscr{L}$, so $L = \mathit{Spec} \oplus_{n \geq 0} (\mathscr{L}^\vee)^{\otimes n}$. (This construction first appeared in Definition 17.1.5.) Show that if X is quasicompact, then $H^0(L, \mathcal{O}_L) = \oplus_n H^0(X, (\mathscr{L}^\vee)^{\otimes n})$. (Possible hint: Choose a trivializing cover for $\mathscr{L}$. Rhetorical question: Can you figure out the more general statement if $\mathscr{L}$ is a rank r locally free sheaf?)

Let E be an elliptic curve over some ground field k, $\mathscr{N}$ a degree 0 non-torsion invertible sheaf on E, and $\mathscr{P}$ a positive-degree invertible sheaf on E. Then $H^0(E, \mathscr{N}^m \otimes \mathscr{P}^n)$ is nonzero if and only if either (i) $n > 0$, or (ii) $m = n = 0$ (in which case the sections are elements of k).

19.11.F. EASY EXERCISE. Show that the ring $R = \oplus_{m,n \geq 0} H^0(E, \mathscr{N}^m \otimes \mathscr{P}^n)$ is not finitely generated.

19.11.G. EXERCISE. Let X be the total space of the vector bundle associated to $(\mathscr{N} \oplus \mathscr{P})^\vee$ over E. Show that the ring of global sections of X is R, and hence is not finitely generated. (Hint: Interpret X as a line bundle over a line bundle over E.)

19.11.H. EXERCISE. Show that X (as in the above exercise) is a variety (and, in particular, Noetherian) whose ring of global sections is not Noetherian.

Chapter 20*

Application: A Glimpse of Intersection Theory

This section is intended to show some of the interesting things we can do with what we have built so far. But it can be skipped on a first reading, by people in a hurry.

The only reason this chapter appears after Chapter 19 is because we will use Exercise 19.2.F in the double-starred proof of the Nakai–Moishezon criterion for ampleness (§20.4.5).

Throughout this chapter, X will be a k-variety. In most applications, X will be projective.

20.1 Intersecting n Line Bundles with an n-Dimensional Variety

The central tool in this chapter is the following.

20.1.1. Definition: Intersection product, or intersection number $(\mathscr{L}_1 \cdots \mathscr{L}_n)$, $(\mathscr{L}_1 \cdots \mathscr{L}_n \cdot \mathscr{F})$, $(\mathscr{L}_1 \cdots \mathscr{L}_n \cdot Y)$.

Suppose $\mathscr{F}$ is a coherent sheaf on X with proper support (automatic if X is proper) *of dimension at most n*, and $\mathscr{L}_1, \ldots, \mathscr{L}_n$ are invertible sheaves on X. Let $(\mathscr{L}_1 \cdot \mathscr{L}_2 \cdots \mathscr{L}_n \cdot \mathscr{F})$ be the signed sum over the 2^n subsets of $\{1, \ldots, n\}$:

$$(20.1.1.1) \qquad \sum_{\{i_1, \ldots, i_m\} \subset \{1, \ldots, n\}} (-1)^m \chi(X, \mathscr{L}_{i_1}^{\vee} \otimes \cdots \otimes \mathscr{L}_{i_m}^{\vee} \otimes \mathscr{F}).$$

We call this the **intersection of** $\mathscr{L}_1, \ldots, \mathscr{L}_n$ **with** $\mathscr{F}$. (*Never forget* that whenever we write $(\mathscr{L}_1 \cdots \mathscr{L}_n \cdot \mathscr{F})$, we are implicitly assuming that $\dim \operatorname{Supp} \mathscr{F} \leq n$.)

The case we will find most useful is if $\mathscr{F}$ is the structure sheaf of a closed subscheme Y (of dimension at most n). In this case, we may write it $(\mathscr{L}_1 \cdot \mathscr{L}_2 \cdots \mathscr{L}_n \cdot Y)$, and call it **the intersection of** $\mathscr{L}_1, \ldots, \mathscr{L}_n$ **with Y**.

If Y = X, we may write it $(\mathscr{L}_1 \cdot \mathscr{L}_2 \cdots \mathscr{L}_n)$, and call it **the intersection of** $\mathscr{L}_1, \ldots, \mathscr{L}_n$ (with the "X" left implicit).

If the $\mathscr{L}_i$ are all the same, say $\mathscr{L}$, one often writes $(\mathscr{L}^n \cdot \mathscr{F})$ or $(\mathscr{L}^n \cdot Y)$ or $(\mathscr{L}^n)$. (Be careful with this confusing notation: $\mathscr{L}^n$ does not mean $\mathscr{L}^{\otimes n}$.)

In some circumstances the convention is to omit the parentheses.

We will prove many things about the intersection product in this chapter. One fact will be left until we study flatness (Exercise 24.7.D): that it is "deformation-invariant"—that it is constant in "nice" families.

20.1.A. EXERCISE (REALITY CHECK). Show that if $\mathscr{L}_1 \cong \mathscr{O}_X$ then $(\mathscr{L}_1 \cdot \mathscr{L}_2 \cdots \mathscr{L}_n \cdot \mathscr{F}) = 0$.

20.1.B. EXERCISE. Show that the intersection product (20.1.1.1) is preserved by field extension of k.

The following exercise suggests that the intersection product might be interesting, as it "interpolates" between two useful notions: the degree of a line bundle on a curve and Bézout's Theorem.

20.1.C. EXERCISE.

(a) If X is a projective curve, and $\mathscr{L}$ is an invertible sheaf on X, show that $(\mathscr{L} \cdot X) = \deg_X \mathscr{L}$.

(b) Suppose k is an infinite field, $X = \mathbb{P}^N$, and Y is a dimension n subvariety of X. If $H_1, \ldots, H_n$ are generally chosen hypersurfaces of degrees $d_1, \ldots, d_n$, respectively (so $\dim(H_1 \cap \cdots \cap H_n \cap Y) = 0$ by Exercise 12.3.D(d)), then by an appropriate version of Bézout's Theorem (see perhaps

Exercise 13.4.C or Exercise 18.6.J),

$$\deg(H_1 \cap \cdots \cap H_n \cap Y) = d_1 \cdots d_n \deg(Y).$$

Show that

$$(\mathscr{O}_X(H_1) \cdots \mathscr{O}_X(H_n) \cdot Y) = d_1 \cdots d_n \deg(Y).$$

We now describe some of the properties of the intersection product. In the course of proving Exercise 20.1.C(b) you may in effect solve the following exercise.

20.1.D. EXERCISE. Suppose D is an effective Cartier divisor on X that restricts to an effective Cartier divisor $D|_Y = D \cap Y$ on Y (i.e., remains locally not a zerodivisor on Y). Show that

$$(\mathscr{L}_1 \cdots \mathscr{L}_{n-1} \cdot \mathscr{O}(D) \cdot Y) = (\mathscr{L}_1 \cdots \mathscr{L}_{n-1} \cdot D|_Y).$$

More generally, if D is an effective Cartier divisor on X that does not contain any associated point of $\mathscr{F}$, show that

$$(\mathscr{L}_1 \cdots \mathscr{L}_{n-1} \cdot \mathscr{O}(D) \cdot \mathscr{F}) = (\mathscr{L}_1 \cdots \mathscr{L}_{n-1} \cdot \mathscr{F}|_D).$$

(A similar idea came up in the proof that the Hilbert polynomial is actually polynomial; see the discussion around (18.6.3.1).)

20.1.2. *Definition.* For this reason, if D is an effective Cartier divisor, in the symbol for the intersection product, we often write D instead of $\mathscr{O}(D)$. We interchangeably think of intersecting effective Cartier divisors rather than line bundles. For example, we will discuss the special case of intersection theory on a surface in §20.2, and when we intersect two curves C and D, we will write the intersection as $(C \cdot D)$ or even $C \cdot D$.

20.1.3. Proposition — *Assume X is projective. For fixed $\mathscr{F}$, the intersection product $(\mathscr{L}_1 \cdots \mathscr{L}_n \cdot \mathscr{F})$ is a symmetric multilinear function of the $\mathscr{L}_1, \ldots, \mathscr{L}_n$.*

Proposition 20.1.3 is actually true with "projective" replaced by "proper"; see [Kl1, Prop. 2] or [Ko1, Prop. VI.2.7]. Unlike most extensions to the proper case, this is not just an application of Chow's Lemma. It involves a different approach, involving a beautiful trick called *dévissage*.

20.1.4. Important remark: Additive notation for line bundles. There is a standard, useful, but confusing convention suggested by the multilinearity of the intersection product: we sometimes write tensor product of invertible sheaves *additively*. This is convenient for working with multilinearity. (Some people try to avoid confusion by using divisors rather than line bundles, as we add divisors when we "multiply" the corresponding line bundles. This is psychologically helpful, but may add more confusion, as one then has to worry about the whether and why and how and when line bundles correspond to divisors.) We will use this, for example, in Exercises 20.2.B–20.2.G, §20.2.11, and §20.4.

20.1.5. *Proof of Proposition 20.1.3.* Symmetry is clear. By Exercise 20.1.B, we may assume that k is infinite (e.g., algebraically closed). We now prove the result by induction on n.

20.1.E. EXERCISE (BASE CASE). Prove the result when $n = 1$. Hint: Exercise 18.4.M.

We now assume the result for smaller values of n. We use a trick. We wish to show that (for arbitrary $\mathscr{L}_1, \mathscr{L}_1', \mathscr{L}_2, \ldots, \mathscr{L}_n$),

(20.1.5.1) $$(\mathscr{L}_1 \cdot \mathscr{L}_2 \cdots \mathscr{L}_n \cdot \mathscr{F}) + (\mathscr{L}_1' \cdot \mathscr{L}_2 \cdots \mathscr{L}_n \cdot \mathscr{F}) - ((\mathscr{L}_1 \otimes \mathscr{L}_1') \cdot \mathscr{L}_2 \cdots \mathscr{L}_n \cdot \mathscr{F})$$

is 0.

20.1.F. EXERCISE. Rewrite (20.1.5.1) as

(20.1.5.2) $$(\mathscr{L}_1 \cdot \mathscr{L}_1' \cdot \mathscr{L}_2 \cdots \mathscr{L}_n \cdot \mathscr{F}).$$

(There are now $n + 1$ line bundles appearing in the product, but this does not contradict the definition of the intersection product, as $\dim \operatorname{Supp} \mathscr{F} \leq n < n + 1$.)

20.1.G. EXERCISE. Use the inductive hypothesis to show that (20.1.5.1) is 0 if $\mathscr{L}_n \cong \mathscr{O}(D)$ for D an effective Cartier divisor missing the associated points of $\mathscr{F}$.

In particular, if $\mathscr{L}_n$ is very ample, then (20.1.5.1) is 0, as Exercise 18.6.A shows that there exists a section of $\mathscr{L}_n$ nonvanishing at the associated points of $\mathscr{F}$. (Caution: Exercise 18.6.A requres the field to be infinite. How would you edit this to work if k were finite? Hint: Show a variant of Exercise 18.6.A, that there exists a section of $\mathscr{L}_n^{\otimes N}$ nonvanishing at the associated points of $\mathscr{F}$ for some $N \gg 0$. Hint for this: Compare Exercise 12.3.D(c) and Exercise 12.3.D(d).)

By the symmetry of its incarnation as (20.1.5.2), expression (20.1.5.1) vanishes if $\mathscr{L}_1$ is very ample. We will use this observation repeatedly in the final equation display of the proof.

By Exercise 16.2.F, *any* invertible sheaf on X may be written "as the difference of two very amples," so write $\mathscr{L}_1 = \mathscr{A} \otimes \mathscr{B}^\vee$ and $\mathscr{L}_1' = \mathscr{A}' \otimes (\mathscr{B}')^\vee$, where $\mathscr{A}, \mathscr{B}, \mathscr{A}'$, and $\mathscr{B}'$ are very ample. Then

$$\begin{aligned}
((\mathscr{L}_1 \otimes \mathscr{L}_1') \cdot \mathscr{L}_2 \cdots \mathscr{L}_n \cdot \mathscr{F}) &= ((\mathscr{A} \otimes \mathscr{A}' \otimes (\mathscr{B} \otimes \mathscr{B}')^\vee) \cdot \mathscr{L}_2 \cdots \mathscr{L}_n \cdot \mathscr{F}) \\
&= ((\mathscr{A} \otimes \mathscr{A}') \cdot \mathscr{L}_2 \cdots \mathscr{L}_n \cdot \mathscr{F}) - ((\mathscr{B} \otimes \mathscr{B}') \cdot \mathscr{L}_2 \cdots \mathscr{L}_n \cdot \mathscr{F}) \\
&= (\mathscr{A} \cdot \mathscr{L}_2 \cdots \mathscr{L}_n \cdot \mathscr{F}) + (\mathscr{A}' \cdot \mathscr{L}_2 \cdots \mathscr{L}_n \cdot \mathscr{F}) \\
&\quad - (\mathscr{B} \cdot \mathscr{L}_2 \cdots \mathscr{L}_n \cdot \mathscr{F}) - (\mathscr{B}' \cdot \mathscr{L}_2 \cdots \mathscr{L}_n \cdot \mathscr{F}) \\
&= ((\mathscr{A} \otimes \mathscr{B}^\vee) \cdot \mathscr{L}_2 \cdots \mathscr{L}_n \cdot \mathscr{F}) + ((\mathscr{A}' \otimes (\mathscr{B}')^\vee) \cdot \mathscr{L}_2 \cdots \mathscr{L}_n \cdot \mathscr{F}) \\
&= (\mathscr{L}_1 \cdot \mathscr{L}_2 \cdots \mathscr{L}_n \cdot \mathscr{F}) + (\mathscr{L}_1' \cdot \mathscr{L}_2 \cdots \mathscr{L}_n \cdot \mathscr{F}),
\end{aligned}$$

as desired. $\qquad \square$

We have an added bonus arising from the proof.

20.1.H. EXERCISE. Suppose X is projective. Show that if $\dim \operatorname{Supp} \mathscr{F} < n+1$, and $\mathscr{L}_1, \mathscr{L}_1', \mathscr{L}_2, \ldots, \mathscr{L}_n$ are invertible sheaves on X, then (20.1.5.2) vanishes. In other words, the intersection product of $n+1$ invertible sheaves with a coherent sheaf $\mathscr{F}$ vanishes if the $\dim \operatorname{Supp} \mathscr{F} < n+1$. (In fact, the result holds with "projective" replaced by "proper," as the results it relies on hold in this greater generality.)

20.1.6. Proposition — *Suppose X is projective. The intersection product depends only on the numerical equivalence classes of the $\mathscr{L}_i$.*

(Numerical equivalence was defined in §18.4.12—two line bundles are linearly equivalent if their degrees are the same on any curve.) Again, the result remains true with "projective" replaced by "proper." But we use the fact that every line bundle is the difference of two very ample line bundles in both the proof of Proposition 20.1.3 and in the proof of Proposition 20.1.6 itself. For the proof of the proposition in the proper case, see [FGIKNV, Prop. B.20].

Proof. Suppose $\mathscr{L}_1$ is numerically equivalent to $\mathscr{L}_1'$, and $\mathscr{L}_2, \ldots, \mathscr{L}_n$, and $\mathscr{F}$ are arbitrary. We wish to show that $(\mathscr{L}_1 \cdot \mathscr{L}_2 \cdots \mathscr{L}_n \cdot \mathscr{F}) = (\mathscr{L}_1' \cdot \mathscr{L}_2 \cdots \mathscr{L}_n \cdot \mathscr{F})$. We proceed by induction on n. The case $n=1$ follows from Exercise 18.4.L. We assume that $n > 1$, and assume the result for "smaller n." By Exercise 16.2.K, we may write $\mathscr{L}_n \cong \mathscr{O}(D_1 - D_2)$, where D_1 and D_2 are effective Cartier divisors missing the associated points of $\mathscr{F}$. By multilinearity of the intersection product, it suffices to prove the result in the case when $\mathscr{L}_n = \mathscr{O}(D)$, where D is an effective Cartier divisor missing the associated points of $\mathscr{F}$. Then

$$\begin{aligned}
(\mathscr{L}_1 \cdot \mathscr{L}_2 \cdots \mathscr{L}_n \cdot \mathscr{F}) &= (\mathscr{L}_1 \cdot \mathscr{L}_2 \cdots \mathscr{L}_{n-1} \cdot \mathscr{F}|_D) & \text{(Exercise 20.1.D)} \\
&= (\mathscr{L}_1' \cdot \mathscr{L}_2 \cdots \mathscr{L}_{n-1} \cdot \mathscr{F}|_D) & \text{(inductive hypothesis)} \\
&= (\mathscr{L}_1' \cdot \mathscr{L}_2 \cdots \mathscr{L}_n \cdot \mathscr{F}) & \text{(Exercise 20.1.D)}.
\end{aligned}$$

$\qquad \square$

20.1.7. *Asymptotic Riemann–Roch.*

If Y is a projective (= proper; see Remark 19.2.I) curve, $\chi(Y, \mathscr{L}^{\otimes m}) = m \deg_Y \mathscr{L} + \chi(Y, \mathscr{O}_Y)$ is a linear polynomial in m (see Exercises 18.4.L and 18.4.M), whose leading term is an intersection product. This generalizes as follows.

20.1.I. EXERCISE (ASYMPTOTIC RIEMANN–ROCH). Suppose X is projective. (As usual, the result will remain true with "projective" replaced by "proper.") Suppose $\mathscr{F}$ is a coherent sheaf on X with $\dim \operatorname{Supp} \mathscr{F} \leq n$, and $\mathscr{L}$ is a line bundle on X. Show that $q(m) := \chi(X, \mathscr{L}^{\otimes m} \otimes \mathscr{F})$ is a polynomial in m of degree at most n. Show that the coefficient of m^n in $q(m)$ (the "leading term") is $(\mathscr{L}^n \cdot \mathscr{F})/n!$. Hint: Exercise 20.1.H implies that

$$(\underbrace{\mathscr{L} \cdot \mathscr{L} \cdots \mathscr{L}}_{n+1 \text{ times}} \cdot (\mathscr{L}^{\otimes i} \otimes \mathscr{F})) = 0,$$

which in turn (using (20.1.1.1)) implies that

$$\sum_{j=0}^{n+1} (-1)^j \binom{n+1}{j} q(m+j) = 0$$

for all m. Your argument may resemble the proof of polynomiality of the Hilbert polynomial, Theorem 18.6.1, so you may find further hints there. Exercise 18.6.B, in particular, might help.

We know all the coefficients of this polynomial if X is a curve, by Riemann–Roch (see (18.4.7.1)), or basically by definition. We will know/interpret all the coefficients if X is a regular projective surface and $\mathscr{F}$ is an invertible sheaf when we prove Riemann–Roch for surfaces (Exercise 20.2.B(b)). To understand the general case, we need the theory of Chern classes. The result is the Hirzebruch–Riemann–Roch Theorem, which can be further generalized to the celebrated Grothendieck–Riemann–Roch Theorem (see [F2, §15.2]).

20.1.J. EXERCISE (THE PROJECTION FORMULA). Suppose $\pi: X_1 \to X_2$ is a (projective) morphism of integral projective varieties of the same dimension n, and $\mathscr{L}_1, \ldots, \mathscr{L}_n$ are invertible sheaves on X_2. Show that

$$(\pi^* \mathscr{L}_1 \cdots \pi^* \mathscr{L}_n) = \deg(X_1/X_2)(\mathscr{L}_1 \cdots \mathscr{L}_n).$$

(The first intersection is on X_1, and the second is on X_2.) Hint: Let $d = \deg \pi$, and assume $d > 0$. (Deal with the case where π is not dominant separately, so $d = 0$ by convention, using Chevalley's Theorem 8.4.2.) Argue that by the multilinearity of the intersection product, it suffices to deal with the case where the $\mathscr{L}_i$ are very ample. Then choose sections of each $\mathscr{L}_i$, all of whose intersections lie in an open subset U where π has "genuine degree d." To find U: first use Exercise 10.3.G to find a dense open subset $U' \subset X_2$ over which π is finite. Then use Useful Exercise 14.3.E to show that there exists a dense open subset $U \subset U'$ on which $\pi_* \mathscr{O}$ is a locally free sheaf of rank d. In the "flatness" language of Chapter 24, you are showing that there is a dense open subset U of X_2 over which π is finite and flat (and hence has "constant degree"; see Remark 24.3.8). (As usual, the result holds with "projective" replaced with "proper"; see [Ko1, Prop. VI.2.11].)

20.1.8. *Remark: A more general projection formula.* Suppose $\pi: X_1 \to X_2$ is a proper morphism of proper varieties, and $\mathscr{F}$ is a coherent sheaf on X_1 with $\dim \operatorname{Supp} \mathscr{F} \leq n$ (so $\dim \operatorname{Supp} \pi_* \mathscr{F} \leq n$, using Exercise 12.2.C). Suppose also that $\mathscr{L}_1, \ldots, \mathscr{L}_n$ are invertible sheaves on X_2. Then

$$(\pi^* \mathscr{L}_1 \cdots \pi^* \mathscr{L}_n \cdot \mathscr{F}) = (\mathscr{L}_1 \cdots \mathscr{L}_n \cdot \pi_* \mathscr{F}).$$

This is also called the **projection formula** (and generalizes, in a nonobvious way, Exercise 20.1.J). Proofs are given in [FGIKNV, B.15] and [Ko1, Prop. VI.2.11]. Because we won't use this version of the projection formula, we do not give a proof here.

20.1.K. EXERCISE (INTERSECTING WITH AMPLE LINE BUNDLES). Suppose X is a projective k-variety, and $\mathscr{L}$ is an ample line bundle on X. Show that for any subvariety Y of X of dimension

n, $(\mathscr{L}^n \cdot Y) > 0$. (Hint: Use Proposition 20.1.3 and Theorem 16.2.2 to reduce to the case where $\mathscr{L}$ is very ample. Then show that $(\mathscr{L}^n \cdot Y) = \deg Y$ in the embedding into projective space induced by the linear series $|\mathscr{L}|$.)

The Nakai–Moishezon criterion (Theorem 20.4.1) states that this characterizes ampleness.

20.1.9.** *Cohomological interpretation in the complex projective case, generalizing Exercise 18.4.H and §18.6.5.* If $k = \mathbb{C}$, we can interpret $(\mathscr{L}_1 \cdots \mathscr{L}_n \cdot Y)$ as the degree of

$$(20.1.9.1) \qquad c_1((\mathscr{L}_1)_{\mathrm{an}}) \cup \cdots \cup c_1((\mathscr{L}_n)_{\mathrm{an}}) \cap [Y_{\mathrm{an}}]$$

in $H_0(Y_{\mathrm{an}}, \mathbb{Z})$. (Recall $c_1((\mathscr{L}_i)_{\mathrm{an}}) \in H^2(X_{\mathrm{an}}, \mathbb{Z})$, as discussed in Exercise 18.4.H.) One way of proving this is to use multilinearity of both the intersection product and (20.1.9.1) to reduce to the case where the $\mathscr{L}_n$ is very ample, so $\mathscr{L}_n \cong \mathcal{O}(D)$, where D restricts to an effective Cartier divisor E on Y. Then show that if $\mathscr{L}$ is an analytic line bundle on Y_{an} with nonzero section E_{an}, then $c_1(\mathscr{L}) \cap [Y_{\mathrm{an}}] = [E_{\mathrm{an}}]$. Finally, use induction on n and Exercise 20.1.D.

20.2 Intersection Theory on a Surface

We now apply the general machinery of §20.1 to the case of a regular projective surface X. (What matters is that X is Noetherian and factorial, so $\operatorname{Pic} X \to \operatorname{Cl} X$ is an isomorphism, Proposition 15.4.16. Recall that regular schemes are factorial by the Auslander–Buchsbaum Theorem 13.8.5.)

20.2.A. EXERCISE / DEFINITION. Suppose C and D are effective divisors (i.e., curves) on X. (Does this mean "Weil divisor" or "Cartier divisor"?)

(a) Show that

$$(20.2.0.1) \qquad \deg_C \mathcal{O}_X(D)|_C$$

$$(20.2.0.2) \qquad = (\mathcal{O}(C) \cdot \mathcal{O}(D) \cdot X)$$

$$(20.2.0.3) \qquad = \deg_D \mathcal{O}_X(C)|_D.$$

We call this the **intersection number** of C and D, and denote it by $C \cdot D$.

(b) If C does not contain any associated point of D (so, in particular, C and D have no components in common), show that

$$(20.2.0.4) \qquad C \cdot D = h^0(C \cap D, \mathcal{O}_{C \cap D}),$$

where $C \cap D$ is the scheme-theoretic intersection of C and D on X. (When speaking of associated points of D, we by definition must be considering D as an effective Cartier divisor.)

20.2.1. *Important aside.* The hypothesis in Exercise 20.2.A(b), that C does not contain any associated point of D, is a red herring. In fact, D can never have any embedded points, as we will see in §26.2.5 when discussing Cohen–Macaulay rings. (This should help motivate you to learn about Cohen–Macaulayness.) The case of $\mathbb{A}^2$ (and hence $\mathbb{P}^2$) can be done by hand (Exercise 6.6.F).

Thus the hypothesis in Exercise 20.2.A(b) can be replaced by the more simple "C and D have no common components."

20.2.2. *Remark.* In particular, if C and D have no irreducible component in common, then $C \cdot D \geq 0$. This means that we cannot hope to understand intersection numbers using moving lemmas as in topology—if we have some irreducible curve E with $E^2 < 0$, we cannot "move" E to something (call it F) "equivalent" to E, to compute $E \cdot E$ as $E \cdot F$.

20.2.3. *Remark.* The degree of the normal bundle to C in X (defined in §15.6.2) is the **self-intersection** $C \cdot C$ of C, as both are $\deg_C \mathcal{O}_X(C)|_C$.

20.2.4. *Advantages and disadvantages.* We thus have four descriptions of the intersection number (20.2.0.1)–(20.2.0.4), each with advantages and disadvantages. The Euler characteristic description

(20.2.0.2) is remarkably useful (for example, in the exercises below), but the geometry is obscured. The definition $\deg_C \mathscr{O}_X(D)|_C$, (20.2.0.1), is not obviously symmetric in C and D. The definition $h^0(C \cap D, \mathscr{O}_{C \cap D})$, (20.2.0.4), is clearly local—to each point of $C \cap D$, we have a vector space. For example, we know that in $\mathbb{A}_k^2$, $y - x^2 = 0$ meets the x-axis with multiplicity 2, because h^0 of the scheme-theoretic intersection $(k[x, y]/(y - x^2, y))$ has dimension 2. (This h^0 is also the *length* of the dimension 0 scheme, by Exercise 6.6.Y.)

By Proposition 20.1.3, the intersection number induces a bilinear "intersection form"

(20.2.4.1) $$\operatorname{Pic} X \times \operatorname{Pic} X \longrightarrow \mathbb{Z}.$$

By Asymptotic Riemann–Roch (Exercise 20.1.I), $\chi(X, \mathscr{O}(nD))$ is a quadratic polynomial in n.

20.2.5. You can verify that Exercise 20.2.A recovers Bézout's Theorem for plane curves (see Exercise 18.6.J), using $\chi(\mathbb{P}^2, \mathscr{O}(n)) = (n+2)(n+1)/2$ (from Theorem 18.1.2), and the fact that effective Cartier divisors on $\mathbb{P}_k^2$ have no embedded points (Exercise 6.6.F).

Before getting to a number of interesting explicit examples, we derive a couple of fundamental theoretical facts.

20.2.B. EXERCISE. Assuming Serre duality (Theorem 18.5.1) for a smooth projective surface X, prove the following. (We are mixing divisor and invertible sheaf notation, as described in Remark 20.1.4, so be careful. Here K_X is a divisor corresponding to ω_X.)

(a) (*sometimes called the adjunction formula*) $C \cdot (K_X + C) = 2p_a(C) - 2$ for any curve $C \subset X$. Hint: Compute $C \cdot (-K_X - C)$ instead. (See Exercise 21.5.B and §29.4 for other versions of the adjunction formula.)

(b) (*Riemann–Roch for surfaces*)

$$\chi(X, \mathscr{O}_X(D)) = D \cdot (D - K_X)/2 + \chi(X, \mathscr{O}_X)$$

for any Weil divisor D (cf. Riemann–Roch for curves, Exercise 18.4.B).

20.2.6. Two explicit examples: $\mathbb{P}^1 \times \mathbb{P}^1$ and $\operatorname{Bl}_p \mathbb{P}^2$.

20.2.C. EXERCISE: $X = \mathbb{P}^1 \times \mathbb{P}^1$. Recall from Exercise 15.5.K that $\operatorname{Pic}(\mathbb{P}^1 \times \mathbb{P}^1) = \mathbb{Z}\ell \times \mathbb{Z}m$, where ℓ is the curve $\mathbb{P}^1 \times \{0\}$ and m is the curve $\{0\} \times \mathbb{P}^1$. Show that the intersection form (20.2.4.1) is given by $\ell \cdot \ell = m \cdot m = 0$, $\ell \cdot m = 1$. (Hint: You can compute the cohomology groups of line bundles on $\mathbb{P}^1 \times \mathbb{P}^1$ using Exercise 20.2.A(b).) What is the class of the diagonal in $\mathbb{P}^1 \times \mathbb{P}^1$ in terms of these generators?

20.2.D. EXERCISE: THE BLOWN UP PROJECTIVE PLANE. (You needn't have read Chapter 22 to do this exercise.) Let $X = \operatorname{Bl}_p \mathbb{P}^2$ be the blow-up of $\mathbb{P}_k^2$ at a k-valued point (the origin, say) p—see Exercise 10.3.F, which describes the blow-up of $\mathbb{A}_k^2$, and "compactify." Interpret $\operatorname{Pic} X$ as generated (as an abelian group) by ℓ and e, where ℓ is a line not passing through the origin, and e is the exceptional divisor. Show that the intersection form (20.2.4.1) is given by $\ell \cdot \ell = 1$, $e \cdot e = -1$, and $\ell \cdot e = 0$. Hence show that $\operatorname{Pic} X \cong \mathbb{Z}\ell \times \mathbb{Z}e$ (as promised in the aside in Exercise 15.5.L). In particular, the exceptional divisor has negative self-intersection. (This exercise will be generalized in §22.4.13.)

20.2.7. *Hint.* Here is a possible hint to get the intersection form in Exercise 20.2.D. The scheme-theoretic preimage in $\operatorname{Bl}_p \mathbb{P}^2$ of a line through the origin is the scheme-theoretic union of the exceptional divisor e and the "proper transform" m of the line through the origin. Show that $\ell = e + m$ in $\operatorname{Pic}(\operatorname{Bl}_p \mathbb{P}^2)$ (writing the Picard group operation additively; cf. Remark 20.1.4). Show that $\ell \cdot m = e \cdot m = 1$ and $m \cdot m = 0$.

20.2.8. *Definition: (-1)-curve.* Notice that the exceptional divisor e has self-intersection -1. We will see more generally in Exercise 22.4.O that this is the case for all exceptional divisors of blow-ups of smooth surfaces (over k) blown up at a k-valued point. We give such curves a name. If X is a surface over k, and $C \subset X$ is a curve in X consisting of smooth points, with $C \cong \mathbb{P}_k^1$, and $C \cdot C = -1$, we say that C is a (-1)-**curve**. (We can restate "$C \cdot C = -1$" as "C has normal bundle $\mathscr{O}(-1)$.")

20.2.E. EXERCISE. Show that the blown up projective plane $\mathrm{Bl}_p\,\mathbb{P}^2$ in Exercise 20.2.D is not isomorphic to $\mathbb{P}^1 \times \mathbb{P}^1$, perhaps considering their (isomorphic) Picard groups, and identifying which classes are effective (represented by effective divisors). (This is an example of a pair of smooth projective birational surfaces that have isomorphic Picard groups, but that are not isomorphic. This exercise shows that $\mathbb{F}_0$ is not isomorphic to $\mathbb{F}_1$. The method and result are generalized in Exercise 20.2.Q.) A simpler approach but without the same generalizations: Show that the self-intersections of all the elements of the Picard group give different subsets of $\mathbb{Z}$ in the two cases.

20.2.F. EXERCISE (CF. EXERCISE 18.4.X). Show that the nef cone (Exercise 18.4.W) of $\mathrm{Bl}_p\,\mathbb{P}^2$ is generated by ℓ and m. Hint: Show that ℓ and m are nef. By intersecting line bundles with the *curves* e and m, show that nothing outside the cone spanned by ℓ and m is nef. (Side Remark: Note that as in Exercise 18.4.X, linear series corresponding to the boundaries of the cone give "interesting contractions.")

20.2.G. EXERCISE: ANOTHER PROPER NONPROJECTIVE SURFACE. Show the existence of a proper nonprojective surface over a field as follows, paralleling the construction of a proper nonprojective threefold in §15.2.11. Take two copies of the blown up projective plane $\mathrm{Bl}_p\,\mathbb{P}^2$, gluing ℓ on the first to e on the second, and e on the first to ℓ on the second. Hint: Show that if $\mathscr{L}$ is a line bundle having positive degree on each effective curve, then $\mathscr{L} \cdot \ell > \mathscr{L} \cdot e$, using $\ell = e + m$ from Hint 20.2.7. (See §19.11.11 for another example of a proper nonprojective surface.)

20.2.9. Fibrations.

Suppose $\pi\colon X \to B$ is a morphism from a regular projective surface to a regular curve, not contracting any irreducible components of X, and $b \in B$ is a closed point. Let $F = \pi^{-1}(b)$. Then $\mathscr{O}_X(F) = \pi^*\mathscr{O}_B(b)$, which is isomorphic to $\mathscr{O}$ on F. Thus $F \cdot F = \deg_F \mathscr{O}_X(F) = 0$: "the self-intersection of a fiber is 0." The same argument works without X being regular, so long as you phrase it properly: $(\pi^*\mathscr{O}_X(b))^2 = 0$.

20.2.H. EXERCISE. Suppose E is an elliptic curve, with origin p. On $E \times E$, let Δ be the diagonal. By considering the "difference" map $E \times E \to E$, for which $\pi^{-1}(p) = \Delta$, show that $\Delta^2 = 0$. Show that $N^1_{\mathbb{Q}}(E \times E)$ (defined in §18.4.12) has rank at least 3. Show that in general for schemes X and Y, $\mathrm{Pic}\,X \times \mathrm{Pic}\,Y \to \mathrm{Pic}(X \times Y)$ (defined by $\boxtimes$, i.e., pulling back and tensoring) need not be isomorphism; the case of $X = Y = \mathbb{P}^1$ is misleading.

Remark: $\dim_{\mathbb{Q}} N^1_{\mathbb{Q}}(E \times E)$ is always 3 or 4. It is 4 if there is a nontrivial endomorphism from E to itself (i.e., not just multiplication $[n]$ by some n, §19.11.5, followed by a translation; the additional class comes from the graph of this endomorphism. (See [Mu3, §21, App. III] for an introduction to the tools needed to show this.)

Our next goal is to describe the self-intersection of a curve on a ruled surface (Exercise 20.2.J). To set this up, we have a useful preliminary result.

20.2.I. EXERCISE (THE NORMAL BUNDLE TO A SECTION OF THE PROJECTIVIZATION OF A RANK 2 VECTOR BUNDLE). Suppose X is a scheme, and $\mathscr{V}$ is a rank 2 locally free sheaf on X. By Exercises 17.2.I and 14.2.H(a), the short exact sequences

$$(20.2.9.1) \qquad 0 \longrightarrow \mathscr{S} \longrightarrow \mathscr{V} \longrightarrow \mathscr{Q} \longrightarrow 0$$

on X, where $\mathscr{S}$ and $\mathscr{Q}$ have rank 1, correspond to the sections $\sigma\colon X \to \mathbb{P}\mathscr{V}$ to the projection $\mathbb{P}\mathscr{V} \to X$. (See §17.2.3 for the definition of $\mathbb{P}\mathscr{V}$, including a caution about duals.) Show that the normal bundle (§15.6.2) to $\sigma(X)$ in $\mathbb{P}\mathscr{V}$ is $\mathscr{Q} \otimes \mathscr{S}^\vee$. (A generalization is stated in §21.3.10.) Hint: (i) For simplicity, it is convenient to assume $\mathscr{S} = \mathscr{O}_X$, by replacing $\mathscr{V}$ by $\mathscr{V} \otimes \mathscr{S}^\vee$, as the statement of the problem respects tensoring by an invertible sheaf (see Exercise 17.2.G). (ii) Assume now (*with* loss of generality) that $\mathscr{Q} \cong \mathscr{O}_X$, and the exact sequence splits. Then describe the section as $\sigma\colon X \to \mathbb{P}^1 \times X$, with X mapping to the 0 section. Describe an isomorphism of $\mathscr{O}_X$ with the normal bundle to $\sigma(X) \to \mathbb{P}^1 \times X$. (Do *not* just say that the normal bundle "is trivial.") (iii) Now consider the case where $\mathscr{Q}$ is general. Choose trivializing open neighborhoods U_i of $\mathscr{Q}$, and let g_{ij} be the transition function for $\mathscr{Q}$. On the overlap between two trivializing open neighborhoods $U_i \cap U_j$, determine how your two isomorphisms of $\mathscr{O}_X$ with $N_{\sigma(X)/\mathbb{P}^1_X}$ from (ii) (one for U_i, one for U_j) are related. In particular, show that they differ by g_{ij}.

20.2.J. EXERCISE (SELF-INTERSECTIONS OF SECTIONS OF RULED SURFACES). Suppose C is a regular curve, and $\mathscr{V}$ is a rank 2 locally free sheaf on C. Then $\mathbb{P}\mathscr{V}$ is a ruled surface (Example 17.2.6). Fix a section σ of $\mathbb{P}\mathscr{V}$ corresponding to a filtration (20.2.9.1). Show that $\sigma(C) \cdot \sigma(C) = \deg_C(\mathscr{Q} \otimes \mathscr{S}^\vee)$.

20.2.10. The Hirzebruch surfaces $\mathbb{F}_n = \mathbb{P}(\mathscr{O}_{\mathbb{P}^1} \oplus \mathscr{O}_{\mathbb{P}^1}(n))$.

Recall the definition of the Hirzebruch surface $\mathbb{F}_n = \mathbb{P}(\mathscr{O}_{\mathbb{P}^1} \oplus \mathscr{O}_{\mathbb{P}^1}(n))$ in Example 17.2.6. It is a $\mathbb{P}^1$-bundle over $\mathbb{P}^1$; let $\pi: \mathbb{F}_n \to \mathbb{P}^1$ be the structure morphism. Using Exercise 20.2.J, corresponding to

$$0 \longrightarrow \mathscr{O}(n) \longrightarrow \mathscr{O} \oplus \mathscr{O}(n) \longrightarrow \mathscr{O} \longrightarrow 0,$$

we have a section of π of self-intersection $-n$; call it $E \subset \mathbb{F}_n$. Similarly, corresponding to

$$0 \longrightarrow \mathscr{O} \longrightarrow \mathscr{O} \oplus \mathscr{O}(n) \longrightarrow \mathscr{O}(n) \longrightarrow 0,$$

we have a section $C \subset \mathbb{F}_n$ of self-intersection n. Let p be any k-valued point of $\mathbb{P}^1$, and let $F = \pi^{-1}(p)$.

20.2.K. EXERCISE. Show that the line bundle $\mathscr{O}(F)$ is independent of the choice of p.

20.2.L. EXERCISE. Show that $\operatorname{Pic} \mathbb{F}_n$ is generated by E and F. In the course of doing this, you might develop "local charts" for $\mathbb{F}_n$, which could help you solve later exercises.

20.2.M. EXERCISE. Compute the intersection matrix on $\operatorname{Pic} \mathbb{F}_n$. Show that E and F are independent, and thus $\operatorname{Pic} \mathbb{F}_n \cong \mathbb{Z}E \oplus \mathbb{Z}F$. Calculate C in terms of E and F.

20.2.N. EXERCISE. Show how to identify $\mathbb{F}_n \setminus E$, along with the structure map π, with the total space of the line bundle $\mathscr{O}(n)$ on $\mathbb{P}^1$, with C as the 0-section. Similarly show how to identify $\mathbb{F}_n \setminus C$ with the total space of the line bundle $\mathscr{O}(-n)$ on $\mathbb{P}^1$, with E as the 0-section.

20.2.O. EXERCISE. Show that $h^0(\mathbb{F}_n, \mathscr{O}_{\mathbb{F}_n}(C)) > 1$. Hint: As $\mathscr{O}_{\mathbb{F}_n}(C)$ has a section—namely the "canonical one" whose divisor is C. Use Exercise 20.2.M to find another section.

20.2.P. EXERCISE. Show that every effective curve on $\mathbb{F}_n$ is a nonnegative linear combination of E and F. (Conversely, it is clear that for every nonnegative a and b, $\mathscr{O}(aE + bF)$ has a section, corresponding to the effective curve "$aE + bF$." The extension of this to $N^1_{\mathbb{Q}}$ (which was defined in §18.4.12) is called the **effective cone**, and this notion, extended to proper varieties more generally, can be very useful. This exercise shows that E and F generate the effective cone of $\mathbb{F}_n$.) Hint: Show that because "F moves," any effective curve must intersect F nonnegatively, and similarly, because "C moves" (Exercise 20.2.O), any effective curve must intersect C nonnegatively. If $\mathscr{O}(aE + bF)$ has a section corresponding to an effective curve D, what does this say about a and b?

20.2.Q. EXERCISE. By comparing effective cones, and the intersection pairing, show that the $\mathbb{F}_n$ are pairwise nonisomorphic. (This result was promised in Example 17.2.6. Exercise 20.2.E is a special case.)

Exercise 20.2.Q is difficult to do otherwise, and foreshadows the fact that nef and effective cones are useful tools in classifying and understanding varieties in general. In particular, they are central to the minimal model program.

20.2.R. EXERCISE. If $n = 0$, show that there are no curves on $\mathbb{F}_n$ of negative self-intersection. If $n \geq 0$, show that E is the unique curve on $\mathbb{F}_n$ with self-intersection $-n$, and there are no reduced curves on $\mathbb{F}_n$ of smaller self-intersection. This again gives another (related) means of showing that the $\mathbb{F}_n$ are pairwise nonisomorphic.

20.2.S. EXERCISE. Show that the nef cone of $\mathbb{F}_n$ is generated by C and F. (We will soon see that by Kleiman's criterion for ampleness, Theorem 20.4.7, the ample cone is the interior of this cone, so we have now identified the ample line bundles on $\mathbb{F}_n$.)

20.2.T. EXERCISE. We have seen earlier (Exercises 20.2.F and 18.4.X) that the boundary of the nef cone gives "interesting contractions." Show that the map given by the linear series corresponding to $\mathcal{O}(F)$ is the projection of $\mathbb{F}_n$ onto $\mathbb{P}^1$, and the map given by the linear series corresponding to $\mathcal{O}(C)$ sends $\mathbb{F}_n$ to the projective cone over the degree n rational normal curve.

After this series of exercises, you may wish to revisit Exercises 20.2.C–20.2.F, and interpret them as special cases: $\mathbb{F}_0 \cong \mathbb{P}^1 \times \mathbb{P}^1$ and $\mathbb{F}_1 \cong \mathrm{Bl}_p \mathbb{P}^2$.

20.2.11. The Hodge Index Theorem.

We use what we have learned to prove a celebrated result, the Hodge Index Theorem. The "index" in the name comes from the following notion.

20.2.12. *Index of a real quadratic form.* By arguments similar to the classification of quadratic forms over algebraically closed fields (Exercise 5.4.J), one can show that, given a symmetric bilinear form on a *real* vector space V of (finite) dimension n, there are integers a_+, a_0, a_- such that then given any orthogonal basis $v_1, \ldots, v_n$, the number of $v_i \cdot v_i$ that are positive (resp., zero, negative) is a_+ (resp., a_0, a_-). The ordered pair (a_+, a_-) is often called the **index**, and is sometimes called the **signature**. (This result is sometimes called *Sylvester's law of inertia.* You can prove it yourself as follows. Given any basis $(v_1, \ldots, v_n)$ for V, we can construct the $n \times n$ intersection matrix $(v_i \cdot v_j)_{1 \le i, j \le n}$. It is symmetric, so by the spectral theorem, it has real eigenvalues. If you change basis, how do the eigenvalues change? Why are the number of positive eigenvalues preserved?)

20.2.13. The Hodge Index Theorem — *Suppose X is an irreducible smooth projective surface (over a field k) with $\mathscr{L}$, $\mathscr{H} \in \mathrm{Pic}\, X$, with $\mathscr{H} \cdot \mathscr{H} > 0$ and $\mathscr{L} \cdot \mathscr{H} = 0$. Then (a) $\mathscr{L} \cdot \mathscr{L} \le 0$, and (b) equality holds if and only if $\mathscr{L}$ is numerically trivial.*

20.2.U. MOTIVATING EXERCISE. If $\rho(X)$ is finite, show that the Hodge Index Theorem is equivalent to the statement that $N^1_\mathbb{R}(X) := N^1_\mathbb{Q}(X) \otimes_\mathbb{Q} \mathbb{R}$, with its "intersection form," has index $(1, \rho(X) - 1)$. (We defined $N^1(X)$ and $\rho(X)$ in §18.4.12.) Don't forget that X is projective, and thus there is an ample line bundle on X! You should think of the positive "term" as coming from the ample class; the "rest" is negative. We won't use the fact that $\rho(X)$ is finite, because we haven't proved it, but it will motivate us.

20.2.14. *Proof of the Hodge Index Theorem 20.2.13.* Motivated by Exercise 20.2.U, we first prove the result under the additional assumption that $\mathscr{H}$ is ample.

20.2.15. We begin with the following fact. If $\mathscr{H}$ is an ample invertible sheaf, $\mathscr{M}$ is any invertible sheaf, and $\mathscr{M} \cdot \mathscr{H} > \omega_X \cdot \mathscr{H}$, then $h^2(X, \mathscr{M}) = 0$. (Translation: For all invertible sheaves "more positive than ω_X," there is no top cohomology; cf. §19.2.5, the corresponding statement in dimension 1.) Reason: By Serre duality (Theorem 18.5.1), we wish to show that $h^0(X, \omega_X \otimes \mathscr{M}^\vee) = 0$. If $\omega_X \otimes \mathscr{M}^\vee$ had a nonzero section, then its (effective nonzero) divisor would intersect $\mathscr{H}$ nonnegatively (Exercise 20.1.K), so (using additive notation; see Remark 20.1.4) $(\omega_X - \mathscr{M}) \cdot \mathscr{H} \ge 0$, contradicting $\mathscr{M} \cdot \mathscr{H} > \omega_X \cdot \mathscr{H}$.

20.2.16. We next show that if $\mathscr{H}$ is ample, and $\mathscr{L}$ is any invertible sheaf with $\mathscr{L} \cdot \mathscr{L} > 0$ and $\mathscr{L} \cdot \mathscr{H} > 0$, then for $n \gg 0$, $\mathscr{L}^{\otimes n}$ has a section. Here is why. For $n \gg 0$,

$$\mathscr{L}^{\otimes n} \cdot \mathscr{H} = n\mathscr{L} \cdot \mathscr{H} > \omega_X \cdot \mathscr{H},$$

so $h^2(X, \mathscr{L}^{\otimes n}) = 0$ by §20.2.15 (taking $\mathscr{M} = \mathscr{L}^{\otimes n}$). Then for $n \gg 0$, by Riemann–Roch for surfaces (Exercise 20.2.B(b)),

$$h^0(X, \mathscr{L}^{\otimes n}) - h^1(X, \mathscr{L}^{\otimes n}) = \chi(X, \mathscr{L}^{\otimes n})$$
$$= (n^2/2)\mathscr{L} \cdot \mathscr{L} - (n/2)\mathscr{L} \cdot \omega_X + \chi(X, \mathscr{O}_X).$$

As $\mathscr{L} \cdot \mathscr{L} > 0$, the right side is positive for $n \gg 0$, so

$$h^0(X, \mathscr{L}^{\otimes n}) \geq h^0(X, \mathscr{L}^{\otimes n}) - h^1(X, \mathscr{L}^{\otimes n}) > 0,$$

as desired.

20.2.17. We are now ready to prove the Hodge Index Theorem for ample $\mathscr{H}$. Assume first that $\mathscr{L} \cdot \mathscr{L} > 0$ (and $\mathscr{L} \cdot \mathscr{H} = 0$). Then for $n \gg 0$, $\mathscr{H}' := \mathscr{L} \otimes \mathscr{H}^{\otimes n}$ is ample (indeed, very ample) by Exercise 16.2.E. Then $\mathscr{L} \cdot \mathscr{H}' = \mathscr{L} \cdot \mathscr{L} > 0$, so by §20.2.16, $\mathscr{L}^{\otimes n}$ is effective, which implies that $\mathscr{L} \cdot \mathscr{H} = (\mathscr{L}^{\otimes n} \cdot \mathscr{H})/n > 0$ (using Exercise 20.1.K again), contradicting our hypothesis.

20.2.18. Assume finally that $\mathscr{L} \cdot \mathscr{H} = 0$, and (i) $\mathscr{L} \cdot \mathscr{L} = 0$, but that (ii) $\mathscr{L}$ is not numerically trivial. By (ii), we can find an invertible sheaf $\mathscr{Q}$ such that $\mathscr{Q} \cdot \mathscr{L} \neq 0$. Then we can find an invertible sheaf $\mathscr{R}$ such that $\mathscr{R} \cdot \mathscr{L} \neq 0$ and $\mathscr{R} \cdot \mathscr{H} = 0$: take

$$\mathscr{R} = (\mathscr{H} \cdot \mathscr{H})\mathscr{Q} - (\mathscr{Q} \cdot \mathscr{H})\mathscr{H}.$$

Then take $\mathscr{L}' = \mathscr{L}^{\otimes n} \otimes \mathscr{R}$, so $\mathscr{L}' \cdot \mathscr{H} = 0$, but because $\mathscr{R} \cdot \mathscr{L} \neq 0$, we can find some $n \in \mathbb{Z}$ so that

$$\mathscr{L}' \cdot \mathscr{L}' = 2n\mathscr{L} \cdot \mathscr{R} + \mathscr{R} \cdot \mathscr{R} > 0.$$

Then the argument of §20.2.17 applies, with $\mathscr{L}'$ in the place of $\mathscr{L}$. We have thus proved "the Hodge Index Theorem for ample line bundles."

20.2.19. It remains to prove the Hodge Index Theorem without the assumption that $\mathscr{H}$ is ample. So now assume we have $\mathscr{H}$ and $\mathscr{L}$ as in the statement of the theorem. Let $\mathscr{A} \in \operatorname{Pic} X$ be an ample line bundle on X. (If you would like to complete the proof from here on your own, the following hint should suffice: Try to apply the Hodge Index Theorem for ample line bundles three times, where the ample line bundle is $\mathscr{A}$, of course, and the "other" line bundle is sequentially $\mathscr{L}$, $\mathscr{H}$, and (20.2.19.1).)

If $\mathscr{A} \cdot \mathscr{L} = 0$, then we are done without even using $\mathscr{H}$, by the Hodge Index Theorem for ample line bundles (using $\mathscr{A}$ and $\mathscr{L}$). So now assume that $\mathscr{A} \cdot \mathscr{L} \neq 0$. Similarly, $\mathscr{A} \cdot \mathscr{H} \neq 0$ (as otherwise appplying the Hodge Index Theorem for ample line bundles using $\mathscr{A}$ and $\mathscr{H}$ would give $\mathscr{H} \cdot \mathscr{H} \leq 0$, contradicting the hypothesis $\mathscr{H} \cdot \mathscr{H} > 0$). Take

$$(20.2.19.1) \qquad \mathscr{M} = (\mathscr{A} \cdot \mathscr{L})\mathscr{H} - (\mathscr{A} \cdot \mathscr{H})\mathscr{L}.$$

Note that $\mathscr{A} \cdot \mathscr{M} = 0$. By the Hodge Index Theorem for ample line bundles (applied this time to $\mathscr{A}$ and $\mathscr{M}$), we have that $\mathscr{M} \cdot \mathscr{M} \leq 0$, i.e.,

$$(\mathscr{A} \cdot \mathscr{L})^2(\mathscr{H} \cdot \mathscr{H}) + (\mathscr{A} \cdot \mathscr{H})^2(\mathscr{L} \cdot \mathscr{L}) \leq 0.$$

But $(\mathscr{A} \cdot \mathscr{L})^2$, $\mathscr{H} \cdot \mathscr{H}$, and $(\mathscr{A} \cdot \mathscr{H})^2$ are all positive, so $\mathscr{L} \cdot \mathscr{L} < 0$ as desired. $\square$

20.2.20. *Generalizations and variations.* The hypotheses can be weakened considerably. We used smoothness only because we need Serre duality, with an invertible sheaf ω_X. We will see in §29.4 that we need less than smoothness for this. But we can do much better, and a proof where X is merely required to be geometrically irreducible and proper requires only a minor modification of our argument; see [FGIKNV, Thm. B.27].

20.3 The Grothendieck Group of Coherent Sheaves, and an Algebraic Version of Homology

The construction of the intersection product (20.1.1.1) may leave you hungry for something more, especially in light of the cohomological interpretation of §20.1.9. You may want some sort of homology-like theory that is a repository for cycles of different dimensions, on which (Chern classes of) line bundles can act. We can actually do this easily, given what we know.

20.3.1. *Definition.* If X is a k-variety, we define the **Grothendieck group of coherent sheaves**, which we denote by $K(X)$ (and which is often denoted by $K_0(X)$), as the abelian group generated by symbols of the form $[\mathscr{F}]$, where $\mathscr{F}$ is a coherent sheaf on X, subject to the relations that $[\mathscr{F}] = [\mathscr{F}']$ if $\mathscr{F} \cong \mathscr{F}'$ and $[\mathscr{F}'] + [\mathscr{F}''] = [\mathscr{F}]$ if there is a short exact sequence

$$0 \longrightarrow \mathscr{F}' \longrightarrow \mathscr{F} \longrightarrow \mathscr{F}'' \longrightarrow 0.$$

(Caution: The notation $K(X)$ is now used for two concepts, the Grothendieck group and the function field. The meaning of the symbol should be clear from the context.) By construction, the Grothendieck group is the universal construction of an operator on the category of coherent sheaves that "behaves well in exact sequences." For example, if X is proper, then:

(i) We have a map $\chi \colon Coh_X \to \mathbb{Z}$, as the Euler characteristic of a coherent sheaf is finite by Theorem 18.1.3(i) in the projective case, and, more generally, by Grothendieck's Coherence Theorem 18.9.1 in the proper case. This map descends to $\chi \colon K(X) \to \mathbb{Z}$ by Exercise 18.4.A (which extends without change to the proper case).

(ii) If X is integral, then the rank function

$$\mathrm{rank} \colon Coh_X \longrightarrow \mathbb{Z}$$

descends to a function $\mathrm{rank} \colon K(X) \to \mathbb{Z}$. (The argument of Exercise 14.3.I applies.)

20.3.2. *Definition and nonstandard notation.* The Grothendieck group is filtered by dimension: let $K(X)^{\leq d}$ be the subgroup of $K(X)$ generated by coherent sheaves supported in dimension at most d. (This is called the **dimensional filtration** or **coniveau filtration on** $K(X)$, although we won't need these names.) Let $A'_d(X)$ be the dth graded piece of $K(X)$, i.e., $A'_d(X) := K(X)^{\leq d}/K(X)^{\leq d-1}$.

If $\mathscr{L}$ is an invertible sheaf on X, define $\mathscr{L} \cdot \colon K(X) \to K(X)$ by $\mathscr{L} \cdot [\mathscr{F}] := [\mathscr{F}] - [\mathscr{L}^\vee \otimes \mathscr{F}]$. (Do you see why this operator is well-defined?)

20.3.A. EXERCISE. If X is projective and k is infinite, show that $\mathscr{L} \cdot$ sends $K(X)^{\leq d}$ to $K(X)^{\leq d-1}$. (Hint: For each fixed $\mathscr{F}$ supported on a subset of dimension at most d, write $\mathscr{L}$ as a difference of two very ample invertible sheaves, and choose sections of those two, missing the associated points of $\mathscr{F}$.)

20.3.3. *Remark.* The previous exercise holds true without k being infinite or X being proper; see [Ko1, Prop. VI.2.5].

20.3.4. For the rest of this section, we assume X is projective and k is infinite. (But in light of Remark 20.3.3, these hypotheses can be removed.) By Exercise 20.3.A, $\mathscr{L} \cdot$ descends to a map $A'_d(X) \to A'_{d-1}(X)$; we denote this operator by $c_1(\mathscr{L}) \cap$. (It is the action of the first Chern class.)

20.3.B. EXERCISE ("C_1 IS ADDITIVE"). Show that $c_1(\mathscr{L} \otimes \mathscr{L}') \cap = (c_1(\mathscr{L}) \cap) + (c_1(\mathscr{L}') \cap)$. Hint: Show that $((\mathscr{L} \otimes \mathscr{L}') \cdot) - (\mathscr{L} \cdot) - (\mathscr{L}' \cdot) = -(\mathscr{L} \cdot) \circ (\mathscr{L}' \cdot)$, and thus sends $K(X)^{\leq d}$ to $K(X)^{\leq d-2}$. (This will remind you of the trick in the proof of Proposition 20.1.3, and indeed is the motivation for that trick. Caution: The action of $\mathrm{Pic}(X)$ is not additive on $K(X)$; it only becomes additive once we pass to the associated graded ring $A'_\bullet(X)$.)

If $\mathscr{F}$ has support of dimension at most n, and $\mathscr{L}_1, \ldots, \mathscr{L}_n$ are n invertible sheaves, we can now reinterpret the intersection product $(\mathscr{L}_1 \cdot \mathscr{L}_2 \cdots \mathscr{L}_n \cdot \mathscr{F})$ as

$$c_1(\mathscr{L}_1) \cap \cdots \cap c_1(\mathscr{L}_n) \cap [\mathscr{F}],$$

where $[\mathscr{F}]$ is interpreted as lying in either $K(X)$ or $A'_n(X)$. You can now go back and read §20.1, and reprove all the results with this starting point, sometimes obtaining interesting generalizations. We have now promoted the intersection product from an integer to a class in $A'_0(X)$, which may have much more information. (This is true even if X is a curve of positive genus over $\mathbb{C}$, using the theory developed in [F2].)

These $A'_d(X)$ behave like homology groups in a number of ways. If $Y \subset X$ is a closed subvariety of pure dimension d, we have a class $[Y] := [\mathscr{O}_Y] \in A'_d(X)$. The groups $A'_d(X)$ have appropriate functorial properties. For example, if $\pi: X_1 \to X_2$ is a proper morphism (even if the X_i are not themselves proper), we have a map $\pi_*: K(X_1) \to K(X_2)$ given by

(20.3.4.1) $$\pi_*[\mathscr{F}] = [\pi_*\mathscr{F}] - [R^1\pi_*\mathscr{F}] + \cdots$$

(using the long exact sequence for higher pushforwards, Theorem 18.7.1 (c)) that descends to a map $A'_d(X_1) \to A'_d(X_2)$, where the "later terms" on the right side of (20.3.4.1) disappear, so $\pi_*[\mathscr{F}] = [\pi_*\mathscr{F}]$. This pushforward interacts well with the first Chern class of line bundles, yielding the projection formula of Remark 20.1.8.

If π is instead a *flat morphism*, soon to be discussed at length in Chapter 24, then π^* is exact (Exercise 24.1.H), so we have a map $\pi^*: K(X_2) \to K(X_1)$. If the "relative dimension" of this map is r (to be properly defined in Definition 24.5.7), this yields a map $\pi^*: A'_d(X_2) \to A'_{d+r}(X_1)$. This interacts well with first Chern classes and proper pushforwards.

If $k = \mathbb{C}$ and X is proper, there is a map $A'_d(X) \to H_{2d}(X, \mathbb{Q})$, which behaves as you might hope (for example, in its interaction with Chern classes of line bundles). If X is *not* proper (but $k = \mathbb{C}$), then the map is to Borel-Moore homology rather than usual homology.

Our groups $A'_d(X)$ are a good approximation of the theory of Chow groups $A_d(X)$, as developed in [F2]. In fact, there is a surjective map

(20.3.4.2) $$A_d(X) \longrightarrow A'_d(X)$$

[F2, Examp. 15.1.5], and this map is an isomorphism once tensored with $\mathbb{Q}$ [F2, Thms. 18.2 and 18.3]. This is the beginning of a long and rich story in algebraic geometry.

20.3.5. *Side remark.* The surjection map (20.3.4.2) need not be an isomorphism; see [SGA6, Exp. XIV, §4.5–4.7] (although its kernel must be torsion, as described above). As a tantalizing example, if X is the group E_8, the kernel has 2-torsion, 3-torsion, and 5-torsion; see [KN, DZ].

20.4**   The Nakai–Moishezon and Kleiman Criteria for Ampleness

Exercise 20.1.K stated that if X is projective k-variety, and $\mathscr{L}$ is an ample line bundle on X, then for any subvariety Y of X of dimension n, $(\mathscr{L}^n \cdot Y) > 0$. The Nakai–Moishezon criterion states that this is a characterization:

20.4.1. Theorem (Nakai–Moishezon criterion for ampleness) — *If $\mathscr{L}$ is an invertible sheaf on a projective k-scheme X, and for every irreducible subvariety Y of X, $(\mathscr{L}^{\dim Y} \cdot Y) > 0$, then $\mathscr{L}$ is ample.*

20.4.2. *Remarks.* We note that X need only be proper for this result to hold ([Kl1, Thm. III.1.1]).

Before proving the Nakai–Moishezon criterion, we point out some consequences related to our discussion of numerical equivalence in §18.4.11. By Proposition 20.1.6, $(\mathscr{L}^{\dim Y} \cdot Y)$ depends only on the numerical equivalence class of $\mathscr{L}$, so ampleness is a numerical property. As a result, the notion of ampleness makes sense on $N^1_{\mathbb{Q}}(X)$. As the tensor product of two ample invertible sheaves is ample (Exercise 16.2.H), the ample $\mathbb{Q}$-line bundles in $N^1_{\mathbb{Q}}(X)$ form a cone, called the **ample cone** of X.

20.4.3. Proposition — *If X is a projective k-scheme, the ample cone is open.*

20.4.4. In the rest of this section, we often use additive notation for the tensor product of invertible sheaves, as described in Remark 20.1.4. This is because we want to deal with intersections on the

$\mathbb{Q}$-vector space $N^1_{\mathbb{Q}}(X)$. For example, by $((a\mathscr{L}_1 + b\mathscr{L}'_1) \cdot \mathscr{L}_2 \cdots \mathscr{L}_n \cdot \mathscr{F})$ (where $a, b \in \mathbb{Q}$), we mean $a(\mathscr{L}_1 \cdot \mathscr{L}_2 \cdots \mathscr{L}_n \cdot \mathscr{F}) + b(\mathscr{L}'_1 \cdot \mathscr{L}_2 \cdots \mathscr{L}_n \cdot \mathscr{F})$.

Proof. We will need the fact that $N^1_{\mathbb{Q}}(X)$ is a finite-dimensional vector space (§18.4.12), which unfortunately we have not explained except over $\mathbb{C}$ (Exercise 18.4.U).

Suppose $\mathscr{A}$ is an ample invertible sheaf on X. We will describe a small open neighborhood of $[\mathscr{A}]$ in $N^1_{\mathbb{Q}}(X)$ consisting of ample $\mathbb{Q}$-line bundles. Choose invertible sheaves $\mathscr{L}_1, \dots, \mathscr{L}_n$ on X whose classes form a basis of $N^1_{\mathbb{Q}}(X)$. By Exercise 16.2.E, there is some m such that $\mathscr{A}^{\otimes m} \otimes \mathscr{L}_i$ and $\mathscr{A}^{\otimes m} \otimes \mathscr{L}^{\vee}_i$ are both very ample for all n. Thus (in the additive notation of §20.4.4), $\mathscr{A} + \frac{1}{m}\mathscr{L}_i$ and $\mathscr{A} - \frac{1}{m}\mathscr{L}_i$ are both ample. As the ample $\mathbb{Q}$-line bundles form a cone, it follows that $\mathscr{A} + \epsilon_1 \mathscr{L}_1 + \cdots + \epsilon_n \mathscr{L}_n$ is ample for $\sum_i |\epsilon_i| \le 1/m$. $\qquad\square$

20.4.5. *Proof of the Nakai–Moishezon criterion, Theorem 20.4.1.* We prove the Nakai–Moishezon criterion in several steps.

20.4.A. UNIMPORTANT EXERCISE. Prove the case where $\dim X = 0$.

Step 1: Initial reductions. Suppose $\mathscr{L}$ satisfies the hypotheses of the theorem; we wish to show that $\mathscr{L}$ is ample. By Exercises 18.8.A and 18.8.B, we may assume that X is integral. Moreover, we can work by induction on dimension, so we can assume that $\mathscr{L}$ is ample on any closed subvariety. The base case is dimension 1, which was done in Exercise 19.2.H.

Step 2: Sufficiently high powers of $\mathscr{L}$ have sections. We show that $H^0(X, \mathscr{L}^{\otimes m}) \ne 0$ for $m \gg 0$.

Our plan is as follows. By Asymptotic Riemann–Roch (Exercise 20.1.I), $\chi(X, \mathscr{L}^{\otimes m}) = m^n(\mathscr{L}^n)/n! + \cdots$ grows (as a function of m) without bound. A plausible means of attack is to show that $h^i(X, \mathscr{L}^{\otimes m}) = 0$ for $i > 0$ and $m \gg 0$. We won't do that, but will do something similar.

By Exercise 16.2.C, $\mathscr{L}$ is the difference of two very ample line bundles, say $\mathscr{L} \cong \mathscr{A} \otimes \mathscr{B}^{-1}$ with $\mathscr{A} = \mathscr{O}(A)$, and $\mathscr{B} = \mathscr{O}(B)$. From $0 \to \mathscr{O}(-A) \to \mathscr{O} \to \mathscr{O}_A \to 0$, we have

$$(20.4.5.1) \qquad 0 \longrightarrow \mathscr{L}^{\otimes m}(-B) \longrightarrow \mathscr{L}^{\otimes(m+1)} \longrightarrow \mathscr{L}^{\otimes(m+1)}|_A \longrightarrow 0.$$

From $0 \to \mathscr{O}(-B) \to \mathscr{O} \to \mathscr{O}_B \to 0$, we have

$$(20.4.5.2) \qquad 0 \longrightarrow \mathscr{L}^{\otimes m}(-B) \longrightarrow \mathscr{L}^{\otimes m} \longrightarrow \mathscr{L}^{\otimes m}|_B \longrightarrow 0.$$

Choose m large enough so that both $\mathscr{L}^{\otimes(m+1)}|_A$ and $\mathscr{L}^{\otimes m}|_B$ have vanishing higher cohomology (i.e., $h^{>0} = 0$ for both; use the inductive hypothesis that $\mathscr{L}$ is ample on all proper closed subvarieties, and Serre vanishing, Theorem 18.1.3(ii)). This implies that for $i \ge 2$,

$$H^i(X, \mathscr{L}^{\otimes m}) \cong H^i(X, \mathscr{L}^{\otimes m}(-B)) \quad \text{(long exact sequence for (20.4.5.2))}$$
$$\cong H^i(X, \mathscr{L}^{\otimes m+1}) \quad \text{(long exact sequence for (20.4.5.1))},$$

so the higher cohomology stabilizes (is constant) for large m. From

$$\chi(X, \mathscr{L}^{\otimes m}) = h^0(X, \mathscr{L}^{\otimes m}) - h^1(X, \mathscr{L}^{\otimes m}) + \text{constant},$$

$H^0(\mathscr{L}^{\otimes m}) \ne 0$ for $m \gg 0$, completing Step 2.

So by replacing $\mathscr{L}$ by a suitably large multiple (ampleness is independent of taking tensor powers, Theorem 16.2.2), we may assume $\mathscr{L}$ has a section, with corresponding effective divisor D. We now use D as a crutch.

Step 3: $\mathscr{L}^{\otimes m}$ is globally generated for $m \gg 0$.

As D is effective, $\mathscr{L}^{\otimes m}$ is globally generated on the complement of D: we have a section nonvanishing on that big open set. Thus any base locus must be contained in D. Consider the short exact sequence

$$(20.4.5.3) \qquad 0 \longrightarrow \mathscr{L}^{\otimes(m-1)} \longrightarrow \mathscr{L}^{\otimes m} \longrightarrow \mathscr{L}^{\otimes m}|_D \longrightarrow 0.$$

Now $\mathscr{L}|_D$ is ample by our inductive hypothesis. Choose m so large that $H^1(X, \mathscr{L}^{\otimes m}|_D) = 0$ (Serre vanishing, Theorem 18.1.3(ii)). From the exact sequence associated to (20.4.5.3),

$$\phi_m \colon H^1(X, \mathscr{L}^{\otimes(m-1)}) \longrightarrow H^1(X, \mathscr{L}^{\otimes m})$$

is surjective for $m \gg 0$. Using the fact that the $H^1(X, \mathscr{L}^{\otimes m})$ are finite-dimensional vector spaces, as m grows, $H^1(X, \mathscr{L}^{\otimes m})$ must eventually stabilize, so the ϕ_m are isomorphisms for $m \gg 0$.

Thus for large m, from the long exact sequence in cohomology for (20.4.5.3), $H^0(X, \mathscr{L}^{\otimes m}) \to H^0(X, \mathscr{L}^{\otimes m}|_D)$ is surjective for $m \gg 0$. But $H^0(X, \mathscr{L}^{\otimes m}|_D)$ has no base points by our inductive hypothesis (applied to D), i.e., for any point p of D there is a section of $\mathscr{L}^{\otimes m}|_D$ not vanishing at p, so $H^0(X, \mathscr{L}^{\otimes m})$ has no base points on D either, completing Step 3.

Step 4. Thus $\mathscr{L}$ is a base-point-free line bundle with positive degree on each curve (by hypothesis of Theorem 20.4.1), so by Exercise 18.4.O we are done. $\qquad\square$

The following result is the key to proving Kleiman's criterion for ampleness, Theorem 20.4.7.

20.4.6. Kleiman's Theorem — *Suppose X is a projective k-scheme. If $\mathscr{L}$ is a nef invertible sheaf on X, then $(\mathscr{L}^k \cdot V) \geq 0$ for every irreducible subvariety $V \subset X$ of dimension k.*

As usual, this extends to the proper case; see [Kl1, Thm. IV.2.1]. And, as usual, we postpone the proof until after we appreciate the consequences.

20.4.B. EXERCISE.

(a) *(limit of amples is nef)* If $\mathscr{L}$ and $\mathscr{H}$ are any two invertible sheaves such that $\mathscr{L} + \epsilon\mathscr{H}$ is ample for all sufficiently small $\epsilon > 0$, show that $\mathscr{L}$ is nef. (Hint: $\lim_{\epsilon \to 0}$. This doesn't require Kleiman's Theorem.)

(b) *(nef + ample = ample)* Suppose X is a projective k-scheme, $\mathscr{H}$ is ample, and $\mathscr{L}$ is nef. Show that $\mathscr{L} + \epsilon\mathscr{H}$ is ample for all $\epsilon \in \mathbb{Q}^{\geq 0}$. (Hint: Use the Nakai–Moishezon criterion; show that $((\mathscr{L} + \epsilon\mathscr{H})^k \cdot V) > 0$. This may help you appreciate the additive notation.)

20.4.7. Theorem (Kleiman's criterion for ampleness) — *Suppose X is a projective k-scheme.*

(a) *The nef cone is the closure of the ample cone.*
(b) *The ample cone is the interior of the nef cone.*

Side remark: Of course, (a) is false if "projective" is relaxed to "proper" (as the ample cone of a proper nonprojective variety is empty, but 0 is nef so the nef cone is nonempty). However, part (b) is true if X is proper and factorial (see [Kl1, Thm. IV.2.2] for a proof and a more general statement). Hence if X is smooth, proper, and nonprojective, then the interior of the nef cone is empty.

20.4.C. EXERCISE. In [Kl1, p. 326, Ex. 2], Kleiman describes Mumford's example of a non-ample line bundle on a smooth projective surface that meets every curve positively. Doesn't this contradict Theorem 20.4.7? Did Mumford and Kleiman make a mistake?

Proof. (a) Ample invertible sheaves are nef (Exercise 18.4.V(e)), and the nef cone is closed (Exercise 18.4.W), so the closure of the ample cone is contained in the cone. Conversely, each nef element of $N^1_{\mathbb{Q}}(X)$ is the limit of ample classes by Exercise 20.4.B(a), so the nef cone is contained in the closure of the ample cone.

(b) As the ample cone is open (Proposition 20.4.3), the ample cone is contained in the interior of the nef cone. Conversely, suppose $\mathscr{L}$ is in the interior of the nef cone, and $\mathscr{H}$ is any ample class. Then $\mathscr{L} - \epsilon\mathscr{H}$ is nef for all small enough positive ϵ. Then by Exercise 20.4.B(b), $\mathscr{L} = (\mathscr{L} - \epsilon\mathscr{H}) + \epsilon\mathscr{H}$ is ample. $\qquad\square$

Suitably motivated, we prove Kleiman's Theorem 20.4.6.

Proof. We may immediately reduce to the case where X is irreducible and reduced. We work by induction on $n := \dim X$. The base case $n = 1$ is obvious. So we assume that $(\mathscr{L}^{\dim V} \cdot V) \geq 0$ for all irreducible V not equal to X. We need only show that $(\mathscr{L}^n \cdot X) \geq 0$.

Fix some very ample invertible sheaf $\mathscr{H}$ on X.

20.4.D. EXERCISE. Show that $(\mathscr{L}^k \cdot \mathscr{H}^{n-k} \cdot X) \geq 0$ for all $k < n$. (Hint: Use the inductive hypothesis.)

Consider $P(t) := ((\mathscr{L} + t\mathscr{H})^n \cdot X) \in \mathbb{Q}[t]$. We wish to show that $P(0) \geq 0$. Assume otherwise that $P(0) < 0$. Now for $t \gg 0$, $\mathscr{L} + t\mathscr{H}$ is ample, so $P(t)$ is positive for large t. Thus $P(t)$ has positive real roots. Let t_0 be the largest positive real root of t. (In fact, there is only one positive root, as Exercise 20.4.D shows that all the nonconstant coefficients of $P(t)$ are nonnegative.)

20.4.E. EXERCISE. Show that for (rational) $t > t_0$, $\mathscr{L} + t\mathscr{H}$ is ample. Hint: Use the Nakai–Moishezon criterion; if $V \neq X$ is an irreducible subvariety, show that $((\mathscr{L} + t\mathscr{H})^{\dim V} \cdot V) > 0$ by expanding $(\mathscr{L} + t\mathscr{H})^{\dim V}$.

Let $Q(t) := (\mathscr{L} \cdot (\mathscr{L} + t\mathscr{H})^{n-1} \cdot X)$ and $R(t) := (t\mathscr{H} \cdot (\mathscr{L} + t\mathscr{H})^{n-1} \cdot X)$, so $P(t) = Q(t) + R(t)$.

20.4.F. EXERCISE. Show that $Q(t) \geq 0$ for all rational $t > t_0$. Hint (which you will have to make sense of): $(\mathscr{L} + t\mathscr{H})$ is ample by Exercise 20.4.E, so for N sufficiently large, $N(\mathscr{L} + t\mathscr{H})$ is very ample. Use the idea of the proof of Proposition 20.1.6 to intersect X with $n - 1$ divisors in the class of $N(\mathscr{L} + t\mathscr{H})$ so that "$((N(\mathscr{L} + t\mathscr{H}))^{n-1} \cdot X)$ is an effective curve C." Then $(\mathscr{L} \cdot C) \geq 0$ as $\mathscr{L}$ is nef. Thus $Q(t_0) \geq 0$.

20.4.G. EXERCISE. Show that $R(t_0) > 0$. (Hint: Exercise 20.4.D.)

Thus $P(t_0) > 0$ as desired. $\qquad\square$

Chapter 21

Differentials

Differentials are an intuitive geometric notion, and we are going to figure out the right description of them algebraically. The algebraic manifestation is somewhat nonintuitive, so it is helpful to understand differentials first in terms of geometry. Also, although the algebraic statements are odd, none of the proofs are hard or long. You will notice that this topic could have been done as soon as we knew about morphisms and quasicoherent sheaves. We have usually introduced new ideas through a number of examples, but in this case we will spend a fair amount of time discussing theory, and only then get to examples.

21.1 Motivation and Game Plan

Suppose X is a "smooth" k-variety. We would like to define a tangent bundle. We will see that the right way to do this will easily apply in much more general circumstances.

- We will see that the cotangent sheaf is more "natural" for schemes than tangent bundle / sheaf. This is similar to the fact that the Zariski *cotangent space* is more natural than the *tangent space* (i.e., if A is a ring and m is a maximal ideal, then m/m^2 is "more natural" than $(m/m^2)^\vee$), as we have repeatedly discussed since §13.1. In both cases this is because we are understanding "spaces" via their (sheaf of) functions on them, which is somehow dual to the geometric pictures you have of spaces in your mind. So we will define the cotangent sheaf first. An element of the (co)tangent space will be called a **(co)tangent vector**.
- Our undergraduate intuition will continue to work—we can calculate using the Jacobian matrix, the answers will be the same in cases where our undergraduate calculations apply, and we will even have the advantage of avoiding δ's and ϵ's.
- Our construction will automatically apply for general X, even if X is not "smooth" (or even at all nice, e.g., finite type). The cotangent sheaf will not necessarily be locally free, but it will still be a quasicoherent sheaf.
- Better yet, this construction will naturally work "relatively." For *any* $\pi\colon X \to Y$, we will define $\Omega_\pi = \Omega_{X/Y}$, a quasicoherent sheaf on X, the sheaf of *relative differentials*. The fiber of this sheaf at a point will be the cotangent vectors of the fiber of the map. This will specialize to the earlier case by taking $Y = \operatorname{Spec} k$. The idea is that this glues together the cotangent sheaves of the fibers of the family. Figure 21.1 is a sketch of the relative tangent space of a map $X \to Y$ at a point $p \in X$— it is the tangent to the fiber. (The tangent space is easier to draw than the cotangent space!) An element of the relative (co)tangent space is called a **relative (co)tangent vector**.

Thus the central concept of this chapter is the cotangent sheaf $\Omega_\pi = \Omega_{X/Y}$ for a morphism $\pi\colon X \to Y$ of schemes. A good picture to have in your mind is the following. If $\pi\colon X \to Y$ is a submersion of manifolds (a map inducing a surjection on tangent spaces), you might hope that the tangent spaces to the fibers at each point $p \in X$ might fit together to form a vector bundle. This is the relative tangent bundle (of π), and its dual is $\Omega_{X/Y}$ (see Figure 21.1). Even if you are not geometrically minded, you will find this useful. (For an arithmetic example, see Exercise 21.2.H.)

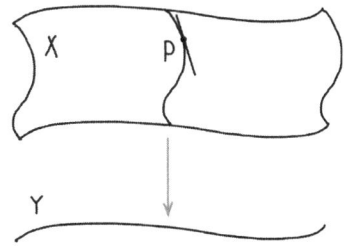

Figure 21.1 *The relative tangent space of a morphism* $X \to Y$ *at a point* p.

21.2   Definitions and First Properties

21.2.1. The affine case: Three definitions.

We first study the affine case. Suppose A is a B-algebra, so we have a morphism of rings $\phi\colon B \to A$ and a morphism of schemes $\operatorname{Spec} A \to \operatorname{Spec} B$. I will define an A-module $\Omega_{A/B}$ in three ways. This is called the **module of relative differentials** or the **module of Kähler differentials**. The module of differentials will be defined to be this module, as well as a map $d\colon A \to \Omega_{A/B}$ satisfying three properties. (Caution: Although d sends an A-module to an A-module, it is not in general A-linear. A priori we take it as a homomorphism of abelian groups, but we will momentarily make it a homomorphism of B-modules, Exercise 21.2.A.)

  (i)   *additivity:* $da + da' = d(a + a')$.
 (ii)   *Leibniz:* $d(aa') = a\,da' + a'\,da$.
(iii)   *triviality on pullbacks:* $db = 0$ for $b \in \phi(B)$.

These properties will not be surprising if you have seen differentials in any other context.

21.2.A. TRIVIAL EXERCISE. Show that d is B-linear.

21.2.B. EXERCISE. Prove the quotient rule: if $a' = as$, then $s^2\,da = s\,da' - a'\,ds$.

21.2.C. EXERCISE. State and prove the chain rule for $d(f(g))$, where f is a polynomial with B-coefficients, and $g \in A$. (As motivation, think of the case $B = k$. So, for example, $da^n = na^{n-1}\,da$, and, more generally, if f is a polynomial in one variable, $df(a) = f'(a)\,da$, where f' is defined formally: if $f = \sum c_i x^i$ then $f' = \sum c_i i x^{i-1}$.)

We will now see three definitions of the module of Kähler differentials, which will soon "sheafify" to the sheaf of relative differentials. The first definition is a concrete hands-on definition. The second is by universal property. And the third will globalize well, and will allow us to define $\Omega_{X/Y}$ conveniently in general.

21.2.2. First definition of differentials: Explicit description. We define $\Omega_{A/B}$ to be finite A-linear combinations of symbols "da" for $a \in A$, subject to the three rules (i)–(iii) above. For example, take $A = k[x, y]$, $B = k$. Then a sample differential is $3x^2\,dy + 4\,dx \in \Omega_{A/B}$. We have identities such as $d(3xy^2) = 3y^2\,dx + 6xy\,dy$.

21.2.3. Key fact. Note that if A is generated over B (as an algebra) by $x_i \in A$ (where i lies in some index set, possibly infinite), subject to some relations r_j (where j lies in some index set, and each is a polynomial in the x_i), then the A-module $\Omega_{A/B}$ is generated by the dx_i, subject to the relations $dr_j = 0$. In short, we needn't take every single element of A; we can take a generating set. And we needn't take every single relation among these generating elements; we can take generators of the relations.

21.2.D. EXERCISE. Verify Key Fact 21.2.3. (If you wish, use the affine conormal exact sequence, Theorem 21.2.12, to verify it; different people prefer to work through the theory in different orders. Just take care not to make circular arguments.)

In particular:

21.2.4. Proposition — *If A is a finitely generated B-algebra, then $\Omega_{A/B}$ is a finite type (i.e., finitely generated) A-module. If A is a finitely presented B-algebra, then $\Omega_{A/B}$ is a finitely presented A-module.*

Recall (§8.3.13) that a ring A is *finitely presented* over another ring B if it can be expressed with finite number of generators and finite number of relations:

$$A = B[x_1, \ldots, x_n]/(r_1(x_1, \ldots, x_n), \ldots, r_j(x_1, \ldots, x_n)).$$

If A is Noetherian, then finitely presented is the same as finite type, as the "finite number of relations" comes for free, so most of you will not care.

Let's now see some examples. Among these examples are three particularly important building blocks for ring maps: adding free variables; localizing; and taking quotients. If we know how to deal with these, we know (at least in theory) how to deal with any ring map. (They were similarly useful in understanding the fibered product in practice, in §10.2.)

21.2.5. Example: Taking a quotient. If $A = B/I$, then $\Omega_{A/B} = 0$: $da = 0$ for all $a \in A$, as each such a is the image of an element of B. This should be believable; in this case, there are no "relative tangent vectors."

21.2.6. Example: Adding variables. If $A = B[x_1, \ldots, x_n]$, then $\Omega_{A/B} = A\,dx_1 \oplus \cdots \oplus A\,dx_n$. (Note that this argument applies even if we add an arbitrarily infinite number of indeterminates.) The intuitive geometry behind this makes the answer very reasonable. The cotangent bundle of affine n-space should indeed be free of rank n.

21.2.7. Explicit example: An affine plane curve. Consider the plane curve $y^2 = x^3 - x$ in $\mathbb{A}_k^2$, where the characteristic of k is not 2 (see Figure 19.7 for a picture). Let $A = k[x,y]/(y^2 - x^3 + x)$ and $B = k$. By Key Fact 21.2.3, the module of differentials $\Omega_{A/B}$ is generated by dx and dy, subject to the relation

$$2y\,dy = (3x^2 - 1)\,dx.$$

Thus in the locus where $y \neq 0$, dx is a generator (as dy can be expressed in terms of dx). We conclude that where $y \neq 0$, $\widetilde{\Omega}_{A/B}$ is isomorphic to the trivial line bundle (invertible sheaf). Similarly, in the locus where $3x^2 - 1 \neq 0$, dy is a generator. These two loci cover the entire curve, as solving $y = 0$ gives $x^3 - x = 0$, i.e., $x = 0$ or ± 1, and in each of these cases $3x^2 - 1 \neq 0$. We have shown that $\widetilde{\Omega}_{A/B}$ is an invertible sheaf.

We can interpret dx and dy geometrically. Where does the differential dx vanish? The previous paragraph shows that it doesn't vanish on the patch where $2y \neq 0$. On the patch where $3x^2 - 1 \neq 0$, where dy is a generator, $dx = (2y/(3x^2 - 1))dy$, from which we see that dx vanishes precisely where $y = 0$. You should find this believable from the picture. We have shown that $dx = 0$ precisely where the curve has a vertical tangent vector. Once we can pull back differentials (Exercise 21.2.K(a) or Theorem 21.2.26), we can interpret dx as the pullback of a differential on the x-axis to Spec A (pulling back along the projection to the x-axis). When we do that, using the fact that dx doesn't vanish on the x-axis, we can interpret the locus where $dx = 0$ as the locus where the projection map branches. (Can you compute where $dy = 0$, and interpret it geometrically?)

This discussion applies to plane curves more generally. Suppose $A = k[x,y]/f(x,y)$, where, for convenience, $k = \overline{k}$. Then the same argument as the one given above shows that $\widetilde{\Omega}_{A/k}$ is free of rank 1 on the open set $D(\partial f/\partial x)$, and also on $D(\partial f/\partial y)$. If Spec A is a regular curve, then these two sets cover all of Spec A. (Conversely, Exercise 13.2.F—basically the Jacobian criterion for regularity—gives regularity at the closed points. Furthermore, the curve must be reduced, or else as the nonreduced locus is closed, it would be nonreduced at a closed point, contradicting regularity. Finally, reducedness at a generic point is equivalent to regularity—a scheme whose underlying set is a point is reduced if and only if it is regular. Alternatively, we could invoke a big result, Fact 13.8.2, to get regularity at the generic point from regularity at the closed points.)

If, on the other hand, the plane curve is singular, then Ω is *not* locally free of rank 1. For example, consider the plane curve Spec A where $A = \mathbb{C}[x,y]/(y^2 - x^3)$, so

$$\Omega_{A/\mathbb{C}} = (A\,dx \oplus A\,dy)/(2y\,dy - 3x^2\,dx).$$

Then the fiber of $\Omega_{A/\mathbb{C}}$ over the origin (computed by setting $x = y = 0$) is rank 2, as it is generated by dx and dy, with no relation.

Implicit in the above discussion is the following exercise, showing that Ω can be computed using the Jacobian matrix.

21.2.E. IMPORTANT BUT EASY EXERCISE (JACOBIAN DESCRIPTION OF $\Omega_{A/B}$). Suppose $A = B[x_1, \ldots, x_n]/(f_1, \ldots, f_r)$. Then

$$\Omega_{A/B} = \left(\bigoplus_i A \, dx_i \right) / (A \, df_1 + \cdots + A \, df_r)$$

may be interpreted as the cokernel of the Jacobian matrix (13.1.7.1)

$$J \colon A^{\oplus r} \longrightarrow A^{\oplus n}.$$

(You might notice that if J has full rank everywhere, so A is smooth over B, then $\Omega_{A/B}$ is locally free; see Exercise 21.2.Q for more.)

The next result connects the cotangent module $\Omega_{A/B}$ to the cotangent space at a (rational) point.

21.2.F. EXERCISE (THE FIBER OF Ω AT A RATIONAL POINT IS THE COTANGENT SPACE). Suppose B is a finite type k-algebra, and $\mathfrak{m} \subset B$ is a maximal ideal with residue field k. Then there is an isomorphism of k-vector spaces $\delta \colon \mathfrak{m}/\mathfrak{m}^2 \xrightarrow{\sim} \Omega_{B/k} \otimes_B k$ (where the k on the right is a B-module via the isomorphism $k \xleftarrow{\sim} B/\mathfrak{m}$). Hint: Both are "computed" using the Jacobian matrix.

(Exercise 21.2.F is true even without the finite type hypothesis, although the hint doesn't apply in that case. You are welcome to prove this more general statement, using the universal property of $\Omega_{B/k}$. But we won't need it.)

Corollary 21.2.33 will give a quite different proof, and generalize it to the case where $B/\mathfrak{m}$ is a finite separable extension of k. It will allow us to extend the Jacobian description of the Zariski cotangent space to closed points whose residue field is a separable algebraic extension of the ground field, not just the ground field itself.

21.2.G. EXERCISE. Suppose X is a finite type scheme over an algebraically closed field. Show that the function from the closed points of X to $\mathbb{Z}^{\geq 0}$ given by $p \mapsto \dim T_p X$ is upper semicontinuous in the Zariski topology. (Be clear on what the Zariski topology is on the set of closed points; cf. §8.4.8.) Corollary 21.2.33 will immediately extend this to all fields of characteristic 0, and all finite fields.

21.2.8. Example: Localization. If T is a multiplicative subset of B, and $A = T^{-1}B$, then $\Omega_{A/B} = 0$. Reason: By the quotient rule (Exercise 21.2.B), if $a = b/t$, then $da = (t \, db - b \, dt)/t^2 = 0$. If $A = B_f$, this is intuitively believable; then $\operatorname{Spec} A$ is an open subset of $\operatorname{Spec} B$, so there should be no relative (co)tangent vectors.

21.2.H. IMPORTANT EXERCISE (FIELD EXTENSIONS). This notion of relative differentials is interesting even for finite extensions of fields. In other words, even when you map a reduced point to a reduced point, something interesting can happen with differentials.

(a) Suppose K/k is a separable algebraic extension. Show that $\Omega_{K/k} = 0$. Do not assume that K/k is a finite extension! (Hint: For any $\alpha \in K$, there is a polynomial $f(x)$ such that $f(\alpha) = 0$ and $f'(\alpha) \neq 0$.)

(b) Suppose k is a field of characteristic p, $K = k(t^p)$, $L = k(t)$. Compute $\Omega_{L/K}$ (where $K \hookrightarrow L$ is the "obvious" inclusion).

We now delve a little deeper, and discuss two useful and geometrically motivated exact sequences.

21.2.9. Theorem (cotangent exact sequence, affine version) — *Suppose $C \to B \to A$ are ring morphisms. Then there is a natural exact sequence of A-modules*

$$A \otimes_B \Omega_{B/C} \xrightarrow{\; a \otimes db \mapsto a \, db \;} \Omega_{A/C} \xrightarrow{\; da \mapsto da \;} \Omega_{A/B} \longrightarrow 0.$$

The proof will be quite straightforward algebraically, but the statement comes fundamentally
from geometry, and that is how best to remember
it. Figure 21.2 is a sketch of a map $\pi\colon X \to Y$. Here
X should be interpreted as Spec A, Y as Spec B, and
Spec C (we will soon call it Z) is a point. (If you would
like a picture with a higher-dimensional Spec C, just
"take the product of Figure 21.2 with a curve.") In the
Figure, Y is "smooth," and X is "smooth over Y"—
which means roughly that all the fibers are smooth.
Suppose p is a point of X. Then the tangent space
of the fiber of π at p is certainly a subspace of the
tangent space of the total space of X at p. The cokernel is naturally the pullback of the tangent space of
Y at $\pi(p)$. This short exact sequence for each p should
be part of a short exact sequence of "relative tangent
sheaves"

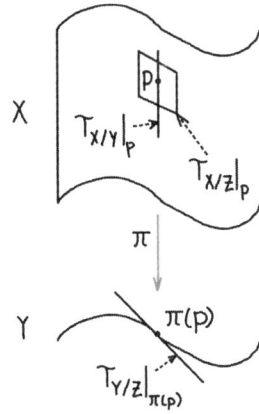

Figure 21.2 *The geometry behind the cotangent exact sequence.*

$$0 \longrightarrow \mathscr{T}_{X/Y} \longrightarrow \mathscr{T}_{X/Z}$$
$$\longrightarrow \pi^* \mathscr{T}_{Y/Z} \longrightarrow 0$$

on X. (We will formally define "relative tangent sheaf" in §21.2.18.) Dualizing this yields

$$0 \longrightarrow \pi^* \Omega_{Y/Z} \longrightarrow \Omega_{X/Z}$$
$$\longrightarrow \Omega_{X/Y} \longrightarrow 0.$$

This is precisely the statement of Theorem 21.2.9, except we also have left-exactness. This
discrepancy is because the statement of the theorem is more general; we will see in Exercise 21.2.S
that in the "smooth" case, we indeed have left-exactness.

21.2.10. *Intriguing remark.* As always, whenever you see something right-exact, you should suspect that there should be some sort of (co)homology theory so that this is the end of a long exact
sequence. This is indeed the case, and this exact sequence involves *André-Quillen homology* (see [E,
p. 386] for more). You should expect that the next term to the left should be the first homology
corresponding to A/B, and in particular shouldn't involve C. So if you already suspect that you
have exactness on the left in the case where A/B and B/C are "smooth" (whatever that means),
and the intuition of Figure 21.2 applies, then you should expect further that all that is necessary is
that A/B be "smooth," and that this would imply that the first André-Quillen homology should
be zero. Even though you wouldn't precisely know what all the words meant, you would be completely correct! You would also be developing a vague inkling about the *cotangent complex*. We will
see examples when left-exactness holds in a sufficiently "smooth" situation in Proposition 21.4.2
and Exercise 21.2.S. For a more general statement, see [Stacks, tag 06B6]. See also [Liu, Cor. 6.3.22]
and [Liu, Prop. 6.3.11] for clean discussions of the next two terms in the sequence: "cotangent exact
sequence," "conormal exact sequence,". . . .

21.2.11. *Proof of Theorem 21.2.9 (the cotangent exact sequence, affine version).* First, note that
surjectivity of $\Omega_{A/C} \to \Omega_{A/B}$ is clear, as this map is given by $da \mapsto da$ (where $a \in A$).

Next, the composition over the middle term is clearly 0, as this composition is given by $a \otimes db \mapsto adb \mapsto 0$.

Finally, we wish to identify $\Omega_{A/B}$ as the cokernel of $A \otimes_B \Omega_{B/C} \to \Omega_{A/C}$. Now $\Omega_{A/B}$ is exactly
the same as $\Omega_{A/C}$, except we have extra relations: $db = 0$ for $b \in B$. These are precisely the images
of $1 \otimes db$ on the left.   □

21.2.12. Theorem (conormal exact sequence, affine version) — *Suppose* B *is a* C-*algebra,* I *is an ideal of* B, *and* A = B/I. *Then there is a natural exact sequence of* A-*modules*

$$I/I^2 \xrightarrow{\quad \delta \quad} A \otimes_B \Omega_{B/C} \xrightarrow{\quad a \otimes db \mapsto a\, db \quad} \Omega_{A/C} \longrightarrow 0.$$

The map δ *is given by* $i \mapsto 1 \otimes di$, *or, more formally,* $1 \otimes d: B/I \otimes_B I \to B/I \otimes_B \Omega_{B/C}$.

(You will recognize the map $A \otimes_B \Omega_{B/C} \to \Omega_{A/C}$ from the cotangent exact sequence, Theorem 21.2.9. A special case of the map δ secretly appeared in Exercise 21.2.F.) The proof is algebraic, so the geometric discussion afterward may help clarify how you should really think of it.

Proof. We will identify the cokernel of $\delta: I/I^2 \to A \otimes_B \Omega_{B/C}$ with $\Omega_{A/C}$. Consider $A \otimes_B \Omega_{B/C}$. As an A-module, it is generated by db (where $b \in B$), subject to three types of relations: $dc = 0$ for $c \in \phi(C)$ (where $\phi: C \to B$ describes B as a C-algebra), additivity, and the Leibniz rule. Given any relation *in* B, d of that relation is 0.

Now $\Omega_{A/C}$ is defined similarly, except there are more relations *in* A; these are precisely the elements of $I \subset B$. Thus we obtain $\Omega_{A/C}$ by starting out with $A \otimes_B \Omega_{B/C}$, and adding the additional relations di, where $i \in I$. But this is precisely the image of δ! (Be sure that you see how the identification of the cokernel of δ with $\Omega_{A/C}$ is precisely via the map $a \otimes db \mapsto a\, db$.)   □

We now give a geometric interpretation of the conormal exact sequence, and in particular define conormal modules/sheaves/bundles.

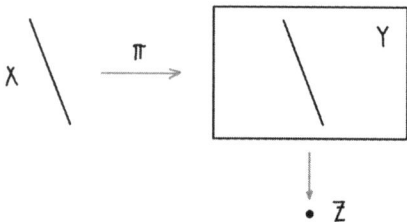

Figure 21.3 *The geometry behind the conormal exact sequence.*

As with the cotangent exact sequence (Theorem 21.2.9), the conormal exact sequence is fundamentally about geometry. To motivate it, consider the sketch of Figure 21.3. In the sketch, everything is "smooth," X is one-dimensional, Y is two-dimensional, j is the inclusion $j: X \hookrightarrow Y$, and Z is a point. Then at a point $p \in X$, the tangent space $\mathscr{T}_X|_p$ clearly injects into the tangent space of $j(p)$ in Y, and the cokernel is the normal vector space to X in Y at p. This should give an exact sequence of bundles on X:

$$0 \longrightarrow \mathscr{T}_X \longrightarrow j^* \mathscr{T}_Y \longrightarrow \mathscr{N}_{X/Y} \longrightarrow 0.$$

Dualizing this should give

$$0 \longrightarrow \mathscr{N}_{X/Y}^\vee \longrightarrow j^* \Omega_{Y/Z} \longrightarrow \Omega_{X/Z} \longrightarrow 0.$$

This is precisely what appears in the statement of the theorem, except (i) the exact sequence in algebraic geometry is not necessarily exact on the left, and (ii) we see I/I^2 instead of $\mathscr{N}_{\operatorname{Spec} A / \operatorname{Spec} B}^\vee$.

21.2.13. We resolve the first issue (i) by expecting that the sequence of Theorem 21.2.12 is exact on the left in appropriately "smooth" situations, and this is indeed the case; see Theorem 21.2.31 and Remark 21.2.32. (If you enjoyed Remark 21.2.10, you might correctly guess several things. The next term on the left should be the first André-Quillen homology of A/C, so we should only need that A/C is smooth, and B should be irrelevant. Also, if $A = B/I$, then we should expect that I/I^2 is the first André-Quillen homology of A/B.)

21.2.14. *Conormal modules and conormal sheaves.* We resolve the second issue (ii) by *declaring* I/I^2 to be the **conormal module**, and in Definition 21.2.15 we will define the obvious analog as the *conormal sheaf.*

Here is some geometric intuition as to why we might want to call (the sheaf associated to) I/I^2 the conormal sheaf, which will likely confuse you, but may offer some enlightenment. First, if Spec A is a closed point of Spec B, we expect the conormal space to be precisely the cotangent space.

And indeed if $A = B/\mathfrak{m}$, the Zariski cotangent space is $\mathfrak{m}/\mathfrak{m}^2$. (We made this subtle connection in §13.1.) In particular, at some point you will develop a sense of why the conormal (= cotangent) space to the origin in $\mathbb{A}_k^2 = \operatorname{Spec} k[x, y]$ is naturally the space of linear forms $\alpha x + \beta y$. But then consider the z-axis in $\operatorname{Spec} k[x, y, z] = \mathbb{A}_k^3$, cut out by $I = (x, y)$. Elements of I/I^2 may be written as $\alpha(z)x + \beta(z)y$, where $\alpha(z)$ and $\beta(z)$ are polynomial. This reasonably should be the conormal space to the z-axis: as z varies, the coefficients of x and y vary. More generally, the same idea suggests that the conormal module/sheaf to any coordinate k-plane inside n-space corresponds to I/I^2. Now consider a k-dimensional manifold X inside an n-dimensional manifold Y, with the classical topology. We can apply the same construction: if $\mathscr{I}$ is the ideal sheaf of X in Y, then $\mathscr{I}/\mathscr{I}^2$ can be identified with the conormal sheaf (essentially the conormal vector bundle), because, analytically locally, $X \hookrightarrow Y$ can be identified with $\mathbb{R}^k \hookrightarrow \mathbb{R}^n$. For this reason, you might hope that in algebraic geometry, if $\operatorname{Spec} A \hookrightarrow \operatorname{Spec} B$ is an inclusion of something "smooth" in something "smooth," I/I^2 should be the conormal module (or, after applying the functor $\sim$, the conormal sheaf). Motivated by this, we define the conormal module as I/I^2 *always*, and then notice that it has good properties (such as Theorem 21.2.12), but take care to learn what unexpected behavior it might have when we are not in the "smooth" situation, by working out examples such as that of §21.2.7.

21.2.15. *Definition.* Suppose $i\colon X \hookrightarrow Y$ is a closed embedding of schemes cut out by ideal sheaf $\mathscr{I}$. Define the **conormal sheaf for a closed embedding** by $\mathscr{I}/\mathscr{I}^2$, denoted by $\mathscr{N}_{X/Y}^\vee$. (The product of quasicoherent ideal sheaves was defined in Exercise 15.6.E.) Important: We interpret $\mathscr{N}_{X/Y}^\vee$ as a quasicoherent sheaf on X, **not** (just) Y—be sure you understand why we may do so. (Exercise 17.1.E is one reason, but it may confuse the issue more than help.)

Define the **normal sheaf** as its dual $\mathscr{N}_{X/Y} := \mathcal{H}om(\mathscr{N}_{X/Y}^\vee, \mathscr{O}_X)$. This is imperfect notation, because it suggests that the dual of $\mathscr{N}$ is always $\mathscr{N}^\vee$. This is not always true, as for A-modules, the natural morphism from a module to its double-dual is not always an isomorphism. (Modules for which this is true are called **reflexive**, but we won't use this notion.)

If the normal sheaf is locally free of finite rank, then we call the associated vector bundle the **normal bundle** of X in Y.

21.2.I. EASY EXERCISE. Define the **conormal sheaf** $\mathscr{N}_{X/Y}^\vee$ (and hence the normal sheaf $\mathscr{N}_{X/Y}$) *for a locally closed embedding* $i\colon X \hookrightarrow Y$ of schemes, a quasicoherent sheaf on X. (Make sure your definition is well-defined!)

In the good situation of a regular embedding, the conormal sheaf (and hence the normal sheaf) is locally free of finite rank (or, more informally, a vector bundle). In particular, the dual of $\mathscr{N}$ is indeed $\mathscr{N}^\vee$. As a warm-up we deal with the important codimension 1 case.

21.2.J. EXERCISE: NORMAL BUNDLES TO EFFECTIVE CARTIER DIVISORS. Suppose $D \subset X$ is an effective Cartier divisor (§9.5.1). Show that the conormal sheaf $\mathscr{N}_{D/X}^\vee$ is $\mathscr{O}(-D)|_D$ (and, in particular, is an invertible sheaf), and hence that the normal sheaf is $\mathscr{O}(D)|_D$. This justifies our earlier definition of normal bundles to effective Cartier divisors, in §15.6.2. It may be surprising that the normal sheaf should be locally free if $X \cong \mathbb{A}^2$ and D is the union of the two axes (and, more generally, if X is regular but D is singular), because you may be used to thinking that a "tubular neighborhood" is isomorphic to the normal bundle.

We now treat the general case.

21.2.16. Proposition —

(a) *If the ideal $I \subset B$ is generated by a regular sequence $x_1, \ldots, x_r$, then the map $\gamma\colon (B/I)^{\oplus r} \to I/I^2$ given by*

(21.2.16.1) $$(a_1, \ldots, a_r) \longmapsto a_1 x_1 + \cdots + a_r x_r$$

is an isomorphism, hence describing I/I^2 as a free module of rank r over B/I with basis $x_1, \ldots, x_r$.

(b) *The (co)normal sheaf of a codimension r regular embedding is locally free of rank r.*

(No Noetherian hypotheses are needed.) Exercise 24.6.D gives a different, short proof of (b) (and indeed (a) as well, if you pay attention) in the commonly used case where A contains a field.

Proof. (a) Clearly (21.2.16.1) is surjective, so we must prove injectivity. Suppose we have $b_1, \ldots, b_r \in B$ with $b_1 x_1 + \cdots + b_r x_r \in I^2$; we wish to show that $b_i \in I$ for all i. This is a consequence of the following two statements, (a_k) and (b_k) (both for $1 \le k \le r$).

(a_k): If we have $b_1, \ldots, b_k \in B$ with

(21.2.16.2) $$b_1 x_1 + \cdots + b_k x_k \in (x_1, \ldots, x_k)^2,$$

then $b_k \in (x_1, \ldots, x_k)$.

(b_k): If we have $b_1, \ldots, b_{k-1} \in B$ with

(21.2.16.3) $$b_1 x_1 + \cdots + b_{k-1} x_{k-1} \in (x_1, \ldots, x_k)^2,$$

then there are $d_1, \ldots, d_{k-1} \in B$ with

(21.2.16.4) $$(b_1 - d_1 x_k) x_1 + \cdots + (b_{k-1} - d_{k-1} x_k) x_{k-1} \in (x_1, \ldots, x_{k-1})^2.$$

We first show (a_k). Modulo $(x_1, \ldots, x_{k-1})$, (21.2.16.2) becomes $b_k x_k \in (x_k^2)$. Thus (as x_k is a non-zerodivisor modulo $(x_1, \ldots, x_{k-1})$, by definition of regular sequence), b_k is a multiple of x_k modulo $(x_1, \ldots, x_{k-1})$, so $b_k \in (x_1, \ldots, x_k)$, as desired.

We next show (b_k). Now (21.2.16.3) implies

$$b_1 x_1 + \cdots + b_{k-1} x_{k-1} = C + (c_1 x_1 + \cdots + c_k x_k) x_k$$

for some $c_i \in B$ and $C \in (x_1, \ldots, x_{k-1})^2$. Thus $0 \equiv (c_1 x_1 + \cdots + c_k x_k) x_k \pmod{(x_1, \ldots, x_{k-1})}$. As x_k is not a zerodivisor modulo $(x_1, \ldots, x_{k-1})$, we have

$$c_1 x_1 + \cdots + c_k x_k = d_1 x_1 + \cdots + d_{k-1} x_{k-1}$$

for some $d_1, \ldots, d_{k-1} \in B$. Thus

$$(b_1 - d_1 x_k) x_1 + \cdots + (b_{k-1} - d_{k-1} x_k) x_{k-1} = C \in (x_1, \ldots, x_{k-1})^2,$$

yielding (21.2.16.4), as desired.

(b) follows immediately from (a). $\qquad\square$

We will soon meet a related but harder fact, that if $\mathscr{I}$ is the ideal sheaf of a regular embedding of codimension r, then $\mathscr{I}^n / \mathscr{I}^{n+1}$ is locally free, because it is $\mathrm{Sym}^n(\mathscr{I}/\mathscr{I}^2)$ (Corollary 22.3.9, or Exercise 24.6.E).

21.2.17. Second definition: Universal property. Here is a second definition that is important philosophically, by universal property. Of course, it is a characterization rather than a definition: by universal property nonsense, it shows that if the module exists (with the d map), then it is unique up to unique isomorphism, and then one still has to construct it to make sure that it exists.

Suppose A is a B-algebra, and M is an A-module. A B-**linear derivation of** A **into** M is a map $d \colon A \to M$ of B-modules (*not necessarily a map of A-modules*) satisfying the Leibniz rule: $d(fg) = f\, dg + g\, df$. As an example, suppose $B = k$, and $A = k[x]$, and $M = A$. Then d/dx is a k-linear derivation. As a second example, if $B = k$, $A = k[x]$, and $M = k$, then $(d/dx)|_0$ (the operator "evaluate the derivative at 0") is a k-linear derivation.

A third example is $d \colon A \to \Omega_{A/B}$, and indeed $d \colon A \to \Omega_{A/B}$ is the *universal B-linear derivation of* A. Precisely, the map $d \colon A \to \Omega_{A/B}$ is defined by the following universal property: any other

B-linear derivation $d': A \to M$ factors uniquely through d:

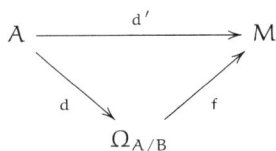

$$
\begin{array}{ccc}
A & \xrightarrow{\quad d' \quad} & M \\
& \searrow{\scriptstyle d} \quad \nearrow{\scriptstyle f} & \\
& \Omega_{A/B} &
\end{array} \quad .
$$

Here f is a map of A-modules. (Note again that d and d' are not necessarily maps of A-modules—they are only B-linear.) By universal property nonsense, if it exists, it is unique up to unique isomorphism. The map $d: A \to \Omega_{A/B}$ clearly satisfies this universal property, essentially by definition.

Depending on how your brain works, you may prefer using the first (constructive) or the second (universal property) definition to do the next two exercises.

21.2.K. EXERCISE.

(a) *(pullback of differentials)* If

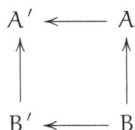

$$
\begin{array}{ccc}
A' & \longleftarrow & A \\
\uparrow & & \uparrow \\
B' & \longleftarrow & B
\end{array}
$$

is a commutative diagram, describe a natural homomorphism of A'-modules $A' \otimes_A \Omega_{A/B} \to \Omega_{A'/B'}$. An important special case is $B = B'$.

(b) *(differentials behave well with respect to base extension, affine case)* If furthermore the above diagram is a "tensor diagram" (i.e., $A' \cong B' \otimes_B A$, so the diagram is "co-Cartesian"), then show that $A' \otimes_A \Omega_{A/B} \to \Omega_{A'/B'}$ is an isomorphism. (Depending on how you proceed, this may be trickier than you expect.)

21.2.L. EXERCISE: LOCALIZATION (STRONGER FORM; CF. EXAMPLE 21.2.8). Suppose $\phi: B \to A$ is a map of rings, S is a multiplicative subset of A, and T is a multiplicative subset of B with $\phi(T) \subset S$, so we have the following commutative diagram.

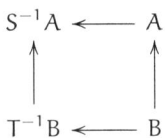

$$
\begin{array}{ccc}
S^{-1}A & \longleftarrow & A \\
\uparrow & & \uparrow \\
T^{-1}B & \longleftarrow & B
\end{array}
$$

Show that the pullback of differentials $S^{-1}\Omega_{A/B} \to \Omega_{S^{-1}A/T^{-1}B}$ of Exercise 21.2.K(a) is an isomorphism. (This should be believable from the intuitive picture of "relative tangent vectors.") An important case is when $T = \{1\}$.

21.2.M. EXERCISE (FIELD EXTENSIONS CONTINUED; CF. EXERCISE 21.2.H).

(a) Compute $\Omega_{k(t)/k}$. (Hint: §21.2.6 followed by Exercise 21.2.L.)

(b) If K/k is *separably generated by* $t_1, \ldots, t_n \in K$ (i.e., $t_1, \ldots, t_n$ form a transcendence basis, and $K/k(t_1, \ldots, t_n)$ is algebraic and separable, §10.5.18), show that $\Omega_{K/k}$ is a free K-module (i.e., vector space) with basis $dt_1, \ldots, dt_n$. Possible approach: Use the cotangent exact sequence (Theorem 21.2.9) for $k \hookrightarrow k(t_1, \ldots, t_n) \hookrightarrow K$ to show that the dt_i span $\Omega_{K/k}$ as a K-vector space. The tricky part is showing that the dt_i are linearly independent. Do this by showing that there exists a unique map $\Omega_{K/k} \to K$ sending dt_1 to 1, and dt_i to 0 for $i > 1$. Do this first for the case where $K/k(t_1, \ldots, t_n)$ is generated by one element; then for the case where it is finitely generated; then for the general case, by defining it on all finitely generated subextensions, and using uniqueness to show that they "all agree."

21.2.18. Third definition: Global. We now want to globalize this definition for an arbitrary morphism of schemes $\pi\colon X \to Y$. We could do this "affine by affine"; we just need to make sure that the above notion behaves well with respect to "change of affine sets." Thus a relative differential on X would be the data of, for every affine $U \subset X$, a differential of the form $\sum a_i \, db_i$, and on the intersection of two affine open sets $U \cap U'$, with representatives $\sum a_i \, db_i$ on U and $\sum a_i' \, db_i'$ on U', and equality on the overlap. Instead, we take a different approach. I will give the (seemingly unintuitive) definition, then tell you how to think about it, and then get back to the definition.

Let $\pi\colon X \to Y$ be any morphism of schemes. Recall that $\delta\colon X \to X \times_Y X$ is a locally closed embedding (Proposition 11.2.1(b)). *Define* the **relative cotangent sheaf** $\Omega_{X/Y}$ (or Ω_π) as the conormal sheaf $\mathscr{N}_{X/X \times_Y X}^\vee$ of the diagonal (see §21.2.14—and if $X \to Y$ is separated you needn't even worry about Exercise 21.2.I). (Now is also as good a time as any to define the **relative tangent sheaf** $\mathscr{T}_{X/Y}$ as the dual $\mathscr{H}om(\Omega_{X/Y}, \mathscr{O}_X)$ to the relative cotangent sheaf. If we are working in the category of k-schemes, then $\Omega_{X/k}$ and $\mathscr{T}_{X/k}$ are often called the **cotangent sheaf** and **tangent sheaf** of X, respectively.)

We now define d: $\mathscr{O}_X \to \Omega_{X/Y}$. Let $\mathrm{pr}_1\colon X \times_Y X \to X$ and $\mathrm{pr}_2\colon X \times_Y X \to X$ be the two projections. Then define d: $\mathscr{O}_X \to \Omega_{X/Y}$ on the open set U as follows:

$$df = \mathrm{pr}_2^* f - \mathrm{pr}_1^* f.$$

(How do you interpret $\mathrm{pr}_2^* f - \mathrm{pr}_1^* f$ as a section of $\Omega_{X/Y}$?)

Warning: d is not a morphism of quasicoherent sheaves on X, although it *is* $\mathscr{O}_Y$-linear in the only possible meaning of that phrase.

We will soon see that d is indeed a derivation of the sheaf $\mathscr{O}_X$ (in the only possible meaning of the phrase), and at the same time see that our new notion of differentials agrees with our old definition on affine open sets, and hence globalizes the definition. Note that for any open subset $U \subset X$, d induces a map

(21.2.18.1) $$\Gamma(U, \mathscr{O}_X) \longrightarrow \Gamma(U, \Omega_{X/Y}),$$

which we also call d, and interpret as "taking the derivative."

21.2.19. *Motivation.* Before connecting this to our other definitions, let me try to convince you that this is a reasonable definition to make. (This discussion is informal and nonrigorous.) Say, for example, that Y is a point, and X a manifold. Then the tangent bundle $T_{X \times X}$ on $X \times X$ is $\mathrm{pr}_1^* T_X \oplus \mathrm{pr}_2^* T_X$, where pr_1 and pr_2 are the projections from $X \times X$ onto its two factors. Restrict this to the diagonal Δ, and look at the normal bundle exact sequence:

$$0 \longrightarrow T_\Delta \longrightarrow T_{X \times X}|_\Delta \longrightarrow N_{\Delta/X} \longrightarrow 0.$$

Now the left morphism sends v to (v, v), so the cokernel can be interpreted as $(v, -v)$. Thus $N_{\Delta/X}$ is isomorphic to T_X. Thus we can turn this on its head: we know how to find the normal bundle (or, more precisely, the conormal sheaf), and we can use this to define the tangent bundle (or, more precisely, the cotangent sheaf).

21.2.20. *More motivation.* If you work through Extended Example 15.3.4, you may see many of the ideas here already faintly visible there, for example, $\mathrm{pr}_2^* f - \mathrm{pr}_1^* f$, and working in the ideal of the diagonal, modulo its square.

21.2.21. *Testing this out in the affine case.* Let's now see how this works for the special case $\operatorname{Spec} A \to \operatorname{Spec} B$. Then the diagonal morphism $\delta\colon \operatorname{Spec} A \hookrightarrow \operatorname{Spec} A \otimes_B A$ corresponds to the ideal I of $A \otimes_B A$ that is the kernel of the ring map

$$\alpha\colon \sum x_i \otimes y_i \longrightarrow \sum x_i y_i.$$

21.2.22. The ideal I of $A \otimes_B A$ is generated by the elements of the form $1 \otimes a - a \otimes 1$. Reason: If $\alpha(\sum x_i \otimes y_i) = 0$, i.e., $\sum x_i y_i = 0$, then

$$\sum x_i \otimes y_i = \sum (x_i \otimes y_i - x_i y_i \otimes 1) = \sum (x_i \otimes 1)(1 \otimes y_i - y_i \otimes 1).$$

The derivation is $d \colon A \to A \otimes_B A$, $a \mapsto (1 \otimes a - a \otimes 1)$ (an element of I, taken modulo I^2). (We shouldn't really call this "d" until we have verified that it agrees with our earlier definition, but we irresponsibly will anyway.)

Let's check that d is indeed a derivation (§21.2.17). Clearly d is B-linear, so we check the Leibniz rule:

$$d(aa') - a\, da' - a'\, da = 1 \otimes aa' - aa' \otimes 1 - a \otimes a' + aa' \otimes 1 - a' \otimes a + a'a \otimes 1$$

$$= -a \otimes a' - a' \otimes a + a'a \otimes 1 + 1 \otimes aa'$$

$$= (1 \otimes a - a \otimes 1)(1 \otimes a' - a' \otimes 1)$$

$$\in I^2.$$

Thus by the universal property of $\Omega_{A/B}$, we have a natural morphism $\Omega_{A/B} \to I/I^2$ of A-modules.

21.2.23. Theorem — *The natural morphism* $f \colon \Omega_{A/B} \to I/I^2$ *induced by the universal property of* $\Omega_{A/B}$ *is an isomorphism.*

Proof. We will show this as follows. (i) We will show that f is surjective, and (ii) we will describe $g \colon I/I^2 \to \Omega_{A/B}$ such that $g \circ f \colon \Omega_{A/B} \to \Omega_{A/B}$ is the identity (showing that f is injective).

(i) The map f sends da to $1 \otimes a - a \otimes 1$, and such elements generate I (§21.2.22), so f is surjective.

(ii) Consider the map $A \otimes_B A \to \Omega_{A/B}$ defined by $x \otimes y \mapsto x\, dy$. (This is a well-defined map, by the universal property of $\otimes$; see §1.2.5.) Define $g \colon I/I^2 \to \Omega_{A/B}$ as the restriction of this map to I. We need to check that this is well-defined, i.e., that elements of I^2 are sent to 0, i.e., we need that

$$\left(\sum x_i \otimes y_i\right)\left(\sum x_j' \otimes y_j'\right) = \sum_{i,j} x_i x_j' \otimes y_i y_j' \longmapsto 0$$

when $\sum_i x_i y_i = \sum x_j' y_j' = 0$. But by the Leibniz rule,

$$\sum_{i,j} x_i x_j'\, d(y_i y_j') = \sum_{i,j} x_i x_j' y_i\, dy_j' + \sum_{i,j} x_i x_j' y_j'\, dy_i$$

$$= \left(\sum_i x_i y_i\right)\left(\sum_j x_j'\, dy_j'\right) + \left(\sum_i x_i\, dy_i\right)\left(\sum_j x_j' y_j'\right)$$

$$= 0.$$

Then $g \circ f$ is indeed the identity, as

$$da \overset{f}{\longmapsto} 1 \otimes a - a \otimes 1 \overset{g}{\longmapsto} 1\, da - a\, d1 = da \;,$$

as desired. $\qquad\qquad\qquad\qquad\qquad\qquad\qquad\qquad\qquad\qquad\qquad\qquad\square$

21.2.N. EASY EXERCISE. Suppose $\pi \colon X \to Y$ is a morphism of schemes, with open subschemes $\operatorname{Spec} A \subset X$ and $\operatorname{Spec} B \subset Y$, with $\operatorname{Spec} A \subset \pi^{-1}(\operatorname{Spec} B)$. Identify $\Omega_{X/Y}|_{\operatorname{Spec} A}$ with $\widetilde{\Omega}_{A/B}$, and identify $d \colon \Gamma(\operatorname{Spec} A, \mathscr{O}_X) \to \Gamma(\operatorname{Spec} A, \Omega_{X/Y})$ with $d \colon A \to \Omega_{A/B}$. Thus the global construction indeed naturally "glues together" the affine construction.

We can now use our understanding of how Ω works on affine open sets to generalize previous statements to non-affine settings.

21.2.O. EXERCISE. If $U \subset X$ is an open subset, show that the map (21.2.18.1) is a derivation.

21.2.P. EXERCISE. Suppose $\pi\colon X \to Y$ is locally of finite type, and Y (and hence X) is locally Noetherian. Show that $\Omega_{X/Y}$ is a coherent sheaf on X. (Feel free to weaken the Noetherian hypotheses for weaker conclusions.)

21.2.Q. EXERCISE. Suppose $\pi\colon X \to Y$ is smooth of relative dimension n. Prove that $\Omega_{X/Y}$ is locally free of rank n. Hint: The Jacobian description of Ω, Exercise 21.2.E.

The cotangent exact sequence and the conormal exact sequence for schemes now directly follow.

21.2.24. Theorem —

(a) *(cotangent exact sequence) Suppose* $X \xrightarrow{\pi} Y \xrightarrow{\rho} Z$ *are morphisms of schemes. Then there is an exact sequence of quasicoherent sheaves on X,*

$$\pi^*\Omega_{Y/Z} \longrightarrow \Omega_{X/Z} \longrightarrow \Omega_{X/Y} \longrightarrow 0,$$

globalizing Theorem 21.2.9.

(b) *(conormal exact sequence) Suppose* $\rho\colon Y \to Z$ *is a morphism of schemes, and* $i\colon X \hookrightarrow Y$ *is a closed embedding, with conormal sheaf $\mathcal{N}^\vee_{X/Y}$. Then there is an exact sequence of sheaves on X;*

$$\mathcal{N}^\vee_{X/Y} \xrightarrow{\delta} i^*\Omega_{Y/Z} \longrightarrow \Omega_{X/Z} \longrightarrow 0,$$

globalizing Theorem 21.2.12.

21.2.R. EXERCISE. Prove Theorem 21.2.24. (What needs to be checked?)

You should expect these exact sequences to be left-exact as well in the presence of appropriate smoothness (see Remark 21.2.10 and §21.2.13). The following is an important special case for the cotangent exact sequence. (Theorem 21.2.31 will give a similar statement for the conormal exact sequence, for smooth varieties.)

21.2.S. EXERCISE (LEFT-EXACTNESS OF THE COTANGENT EXACT SEQUENCE WHEN π AND ρ ARE SMOOTH). Show that the cotangent exact sequence is exact on the left if π and ρ are smooth. Hint: The $(m+n) \times (m+n)$ matrix you used in Exercise 13.6.D is "block upper triangular."

21.2.25. *Pulling back relative differentials.* Not surprisingly, the sheaf of relative differentials pulls back, and it behaves well under base change.

21.2.26. Theorem (pullback of differentials) —

(a) *If*

$$
\begin{array}{ccc}
X' & \xrightarrow{\mu} & X \\
\downarrow & & \downarrow \\
Y' & \longrightarrow & Y
\end{array}
$$

is a commutative diagram of schemes, there is a natural homomorphism $\mu^*\Omega_{X/Y} \to \Omega_{X'/Y'}$ of quasicoherent sheaves on X'. An important special case is $Y = Y'$.

(b) *(Ω behaves well under base change) If furthermore the above diagram is a Cartesian square (so $X' \cong X \times_Y Y'$), then $\mu^*\Omega_{X/Y} \to \Omega_{X'/Y'}$ is an isomorphism.*

21.2.T. EXERCISE. Derive Theorem 21.2.26 from Exercise 21.2.K. (Why does the construction of Exercise 21.2.K(a) "glue well"?)

As a particular case of Theorem 21.2.26(b), the fiber of the sheaf of relative differentials is indeed the sheaf of differentials of the fiber. Thus the sheaf of differentials notion indeed "glues together" the differentials on each fiber.

21.2.U. EXERCISE. Suppose $\alpha\colon X \to Z$ and $\beta\colon Y \to Z$ are two morphisms, and let $\alpha'\colon X \times_Z Y \to Y$ and $\beta'\colon X \times_Z Y \to X$ be the morphisms induced by fibered product. Describe an isomorphism $\Omega_{X \times_Z Y/Z} \xleftarrow{\sim} (\beta')^* \Omega_{X/Z} \oplus (\alpha')^* \Omega_{Y/Z}$. (The right side is often written as $\Omega_{X/Z} \boxplus \Omega_{Y/Z}$.)

21.2.27. Important: Smoothness of varieties revisited.

Suppose k is a field. Since §13.2.3, we have used an awkward definition of k-smoothness, and we now make our definition of k-smoothness more robust.

21.2.28. *Redefinition.* A k-scheme X is k-**smooth of dimension** n or **smooth of dimension** n **over** k if it is locally of finite type, of pure dimension n, and $\Omega_{X/k}$ is locally free of rank n. The dimension n is often omitted, and one might (possibly) want to call something smooth if it is the (scheme-theoretic) disjoint union of things smooth of various dimensions.

21.2.V. EXERCISE. Show that Redefinition 21.2.28 of k-smoothness agrees with Definition 13.2.4. (Hint: Both can be described by the cokernel of the Jacobian.)

21.2.29. *Remark.* We will also revisit the definition of smooth morphism (Definition 13.6.2) before long, in Theorem 24.8.8.

21.2.W. ** EXERCISE (FOR THOSE WITH BACKGROUND IN COMPLEX GEOMETRY).** Suppose X is a complex algebraic variety. Show that the analytification X^{an} of X (defined in Exercise 7.3.P) is smooth (in the differential-geometric sense) if and only if X is smooth (in the algebro-geometric sense, over $\mathbb{C}$). In this case, show that complex dimension of the complex manifold X^{an} (half the real dimension) is $\dim X$. Hint: The Jacobian criterion applies in both settings.

21.2.30. Left-exactness of the conormal exact sequence for embeddings of smooth varieties.

As described in §21.2.13, we expect the conormal exact sequence to be exact on the left in appropriately "smooth" situations.

21.2.31. Theorem (conormal exact sequence for smooth varieties) — *Suppose* $i\colon X \hookrightarrow Y$ *is a closed embedding of smooth varieties over a field* k, *with conormal sheaf* $\mathcal{N}_{X/Y}^\vee$. *Then* $\mathcal{N}_{X/Y}^\vee$ *is locally free, and the conormal exact sequence (Theorem 21.2.24(b)) .is exact on the left:*

$$(21.2.31.1) \qquad 0 \longrightarrow \mathcal{N}_{X/Y}^\vee \xrightarrow{\delta} i^* \Omega_{Y/k} \longrightarrow \Omega_{X/k} \longrightarrow 0$$

is exact.

By dualizing, i.e., applying $\mathcal{H}om(\cdot, \mathscr{O}_X)$, we obtain the **normal exact sequence**

$$0 \longrightarrow \mathscr{T}_{X/k} \longrightarrow \mathscr{T}_{Y/k}|_X \longrightarrow \mathcal{N}_{X/Y} \longrightarrow 0$$

(exact by Exercise 14.2.I), which is geometrically more intuitive (see Figure 21.3 and the discussion after the proof of Theorem 21.2.12).

21.2.32. *Remark.* The conormal sequence is exact on the left in even more general circumstances. Essentially all that is required is appropriate smoothness of $i \circ g\colon X \to Z$; see [Stacks, tag 06B7], and [Stacks, tag 06BB] for related facts. In particular, the smoothness of Y is irrelevant! As a consequence, Important Corollary 21.2.33 below also holds for finite type Y in general.

Proof.

21.2.X. EXERCISE. Show that Theorem 21.2.31 is true if and only if it is true upon base change to the algebraic closure $\overline{k}$ of k. Hint: Show that construction of all the terms in (21.2.31.1) plays well with algebraic field extensions. For the case of differentials, see Theorem 21.2.26. For the case of the conormal sheaf, check that the definition plays well with algebraic field extensions.

We may thus assume that k is algebraically closed, and thus (by Exercise 13.2.J) closed points of smooth k-schemes are regular. Let $\mathscr{I}$ be the ideal sheaf of $i: X \hookrightarrow Y$. For any closed point $p \in X$, by Exercise 13.2.M(b), i is a regular embedding in an open neighborhood of p. Hence i is a regular embedding. (Do you see why?) Thus by Proposition 21.2.16(b), $\mathscr{I}/\mathscr{I}^2$ is locally free of rank r. By Exercise 6.6.G(a), the associated points of $\ker \delta$ are a subset of the associated points of $\mathscr{I}/\mathscr{I}^2$, so to show that $\ker \delta = 0$, it suffices to check this at the associated points of $\mathscr{I}/\mathscr{I}^2$, which are precisely the associated points of X (as $\mathscr{I}/\mathscr{I}^2$ is locally free).

By Proposition 13.2.13 (smooth implies reduced), X is reduced. Thus we need only check at the generic points of (the irreducible components of) X (as X has no embedded points; see §6.6.2 or Exercise 6.6.R). But this involves checking the left-exactness of a right-exact sequence of vector spaces (X reduced implies that the stalk at the generic point is a vector space), and as the dimension of the left (the codimension of X in Y by Proposition 21.2.16(b)) is precisely the difference of the other two, we are done. □

21.2.33. Important corollary — *Suppose Y is a smooth k-variety, and $q \in Y$ is a closed point whose residue field $\kappa(q)$ is separable over k. (This is automatic if $\operatorname{char} k = 0$ or if k is a finite field.) Then the conormal exact sequence for $q \hookrightarrow Y$ yields an isomorphism of the Zariski cotangent space of Y at q with the fiber of $\Omega_{Y/k}$ at q.*

Proof. Apply Theorem 21.2.31 with $X = q$. Note that q is indeed a smooth k-variety (using separability of $\kappa(q)/k$!), and that $\Omega_{q/k} = 0$. □

21.2.34. *Remark.* This result generalizes Exercise 21.2.F to separable closed points if Y is smooth, and as mentioned in Remark 21.2.32, even this smoothness condition on Y can be removed. As described after the statement of Exercise 21.2.F, this has a number of consequences. For example, it extends the Jacobian description of the Zariski tangent space to separable closed points.

21.2.Y. EXERCISE (CF. §13.2.15). Suppose p is a closed point of a k-variety X, with residue field $\kappa(p)$ that is separable over k. Define $\pi: X_{\overline{k}} := X \times_k \overline{k} \to X$ by base change from $\operatorname{Spec} \overline{k} \to \operatorname{Spec} k$. Suppose $q \in \pi^{-1}(p)$. Show that $X_{\overline{k}}$ is regular at q if and only if X is regular at p.

21.3 Examples

The examples below are organized by topic, not by difficulty.

21.3.1. The geometric genus of a curve. A smooth irreducible projective curve C (over a field k) has **geometric genus** $h^0(C, \Omega_{C/k})$. (This will be generalized to higher dimension in §21.5.3.) This is always finite, as $\Omega_{C/k}$ is coherent (Exercise 21.2.P), and coherent sheaves on projective k-schemes have finite-dimensional spaces of sections (Theorem 18.1.3(i)). Sadly, this isn't really a new invariant, by the following.

21.3.A. EASY EXERCISE. Assuming Miracle 18.5.2 (that $\Omega_{C/k}$ is Serre-dualizing), show that for a smooth irreducible projective curve C, the geometric genus equals the arithmetic genus, i.e., $h^0(C, \Omega_{C/k}) = h^1(C, \mathscr{O}_C)$.

Now let's compute some differentials!

21.3.2. The projective line. As an important first example, consider $\mathbb{P}^1_k$, with the usual projective coordinates x_0 and x_1. As usual, the first patch corresponds to $x_0 \neq 0$, and is of the form $\operatorname{Spec} k[x_{1/0}]$, where $x_{1/0} = x_1/x_0$. The second patch corresponds to $x_1 \neq 0$, and is of the form $\operatorname{Spec} k[x_{0/1}]$, where $x_{0/1} = x_0/x_1$.

Both patches are isomorphic to $\mathbb{A}_k^1$, and $\Omega_{\mathbb{A}_k^1} \cong \mathcal{O}_{\mathbb{A}_k^1}$. (More precisely, $\Omega_{k[x]/k} = k[x]\,dx$.) Thus $\Omega_{\mathbb{P}_k^1}$ is an invertible sheaf (a line bundle). The invertible sheaves on $\mathbb{P}_k^1$ are of the form $\mathcal{O}(m)$. So which invertible sheaf is $\Omega_{\mathbb{P}^1/k}$?

Let's take a section, $dx_{1/0}$ on the first patch. It has no zeros or poles there, so let's check what happens on the other patch. As $x_{1/0} = 1/x_{0/1}$, we have $dx_{1/0} = -(1/x_{0/1}^2)\,dx_{0/1}$. Thus this section has a double pole where $x_{0/1} = 0$. Hence $\Omega_{\mathbb{P}_k^1/k} \cong \mathcal{O}(-2)$. As a consequence, we have shown that the geometric genus of $\mathbb{P}_k^1$ is 0.

Note that the above argument works equally well if k were replaced by $\mathbb{Z}$: our theory of Weil divisors and line bundles of Chapter 15 applies ($\mathbb{P}_{\mathbb{Z}}^1$ is factorial), so the previous argument essentially without change shows that $\Omega_{\mathbb{P}_{\mathbb{Z}}^1/\mathbb{Z}} \cong \mathcal{O}(-2)$. And because Ω behaves well with respect to base change (Theorem 21.2.26(b)), and any scheme maps to $\operatorname{Spec}\mathbb{Z}$, this implies that $\Omega_{\mathbb{P}_B^1/B} \cong \mathcal{O}_{\mathbb{P}_B^1}(-2)$ for *any* base scheme B.

(Also, as suggested by §18.5.2, this shows that $\Omega_{\mathbb{P}^1/k}$ is the canonical sheaf for $\mathbb{P}_k^1$; see also Example 18.5.4. But given that we haven't yet proved Serre duality, this isn't so meaningful.)

Side remark: the fact that the degree of the tangent bundle is 2 is related to the "Hairy Ball Theorem" (the dimension 2 case of [Hat, Thm. 2.28]).

21.3.3. Hyperelliptic curves.

Throughout this discussion of hyperelliptic curves, we suppose that $k = \bar{k}$ and $\operatorname{char} k \neq 2$, so we may apply the discussion of §19.5. Consider a double cover $\pi\colon C \to \mathbb{P}_k^1$ by a regular projective curve C, branched over $2g + 2$ distinct points. We will use the explicit coordinate description of hyperelliptic curves of (19.5.2.1). In particular, π is unbranched at 0. By Theorem 19.5.1, C has genus g.

21.3.B. EXERCISE: DIFFERENTIALS ON HYPERELLIPTIC CURVES. What is the degree of the invertible sheaf $\Omega_{C/k}$? (Hint: Let x be a coordinate on one of the coordinate patches of $\mathbb{P}_k^1$. Consider $\pi^* dx$ on C, and count poles and zeros. Use the explicit coordinates of §19.5. You should find that $\pi^* dx$ has $2g + 2$ zeros and four poles, counted with multiplicity, for a total of $2g - 2$.) Doing this exercise will set you up well for the Riemann–Hurwitz formula, in §21.4.

21.3.C. EXERCISE. Show that $h^0(C, \Omega_{C/k}) = g$ (and hence that the geometric genus of C is g) as follows.

(a) Show that any regular differential ω on $\operatorname{Spec} k[x, y]/(y^2 - f(x))$ (i.e., an element of $\Omega_{(k[x,y]/(y^2-f(x)))/k}$) preserved by the involution $y \mapsto -y$ is pulled back from a differential on $\operatorname{Spec} k[x]$. (Hint: Make sense of the statement "ω/dx is a rational function on $\operatorname{Spec} k[x, y]/(y^2 - f(x))$ preserved by the involution $y \mapsto -y$" and show that ω/dx is a rational function in x.)

(b) Use (a) to show that any differential $\omega \in H^0(C, \Omega_{C/k})$ preserved by the involution $i\colon y \mapsto -y$ must be pulled back from $\mathbb{P}^1$ by π, and hence must be zero. Show that every differential $\omega \in H^0(C, \Omega_{C/k})$ satisfies $i^*\omega = -\omega$.

(c) Show that $\frac{dx}{y}$ is a (regular) differential on $\operatorname{Spec} k[x, y]/(y^2 - f(x))$. Show that for $0 \leq i < g$, $x^i(dx)/y$ extends to a global differential ω_i on C (i.e., with no poles).

(d) Show that the ω_i ($0 \leq i < g$) are linearly independent differentials, i.e., linearly independent in the vector space $H^0(C, \Omega_{C/k})$. (Hint: Let $\{p, q\} = \pi^{-1}(0)$. Show that the valuation of ω_i at both p and q is i. If $\omega := \sum_{j=s}^{g-1} a_j \omega_j$ is a nontrivial linear combination of the ω_i, with $a_j \in k$, and $a_s \neq 0$, show that the valuation of ω at p is s, and hence $\omega \neq 0$.)

(e) Show that the ω_i span the vector space of differentials $H^0(C, \Omega_{C/k})$. (Hint: if $\omega \in H^0(C, \Omega_{C/k})$ use (d) to show that there are unique a_i such that $\omega' := \omega - \sum_{i=0}^{g-1} a_i \omega_i$ vanishes at p to order $\geq g$. By (b), ω' also vanishes at q to the same order. Use Exercise 21.3.B to show that ω' must be zero.)

Hence $\Omega_{C/k}$ is an invertible sheaf of degree $2g - 2$ with g sections.

21.3.D.[*] EXERCISE (TOWARD SERRE DUALITY). (You may later see this as an example of Serre duality in action.)

(a) Show that $h^1(C, \Omega_{C/k}) = 1$. Interpret a generator of $H^1(C, \Omega_{C/k})$ as $x^{-1} dx$. (In particular, the pullback map $H^1(\mathbb{P}^1, \Omega_{\mathbb{P}^1/k}) \to H^1(C, \Omega_{C/k})$ is an isomorphism.)
(b) Describe a natural perfect pairing

$$H^0(C, \Omega_{C/k}) \times H^1(C, \mathscr{O}_C) \longrightarrow H^1(C, \Omega_{C/k}).$$

In terms of our explicit coordinates, you might interpret it as follows. Recall from the proof of the hyperelliptic Riemann–Hurwitz formula (Theorem 19.5.1) that $H^1(C, \mathscr{O}_C)$ can be interpreted as

$$\left\langle \frac{y}{x}, \frac{y}{x^2}, \ldots, \frac{y}{x^g} \right\rangle.$$

Then the pairing

$$\left\langle \frac{dx}{y}, x\frac{dx}{y}, \ldots, x^{g-1}\frac{dx}{y} \right\rangle \times \left\langle \frac{y}{x}, \frac{y}{x^2}, \ldots, \frac{y}{x^g} \right\rangle \longrightarrow \left\langle x^{-1} dx \right\rangle$$

is basically "multiply and read off the $x^{-1} dx$ term." Or in fancier terms: "multiply and take the residue." (You may want to compare this to Example 18.5.4.)

21.3.4. Discrete valuation rings. The following exercise is used in the proof of the Riemann–Hurwitz formula, §21.4.

21.3.E. EXERCISE. Suppose that the discrete valuation ring $(A, \mathfrak{m}, k)$ is a localization of a finitely generated k-algebra. Let t be a uniformizer of A. Show that the differentials are free of rank 1 and generated by dt, i.e., $\Omega_{A/k} = A\, dt$, as follows. (It is also possible to show this using the ideas from §21.5.)

(a) Show that $\Omega_{A/k}$ is a finitely generated A-module.
(b) Show that $\langle dt \rangle = \Omega_{A/k}$, i.e., that $\times dt \colon A \to \Omega_{A/k}$ is a surjection, as follows. Let π be the projection $A \to A/\mathfrak{m} = k$, so for $a \in A$, $a - \pi(a) \in \mathfrak{m}$. Define $\sigma(a) = (a - \pi(a))/t$. Show that $\Omega_{A/k} = \langle dt \rangle + \mathfrak{m}\Omega_{A/k}$, using the fact that for every $a \in A$,

$$da = \sigma(a)\, dt + t\, d\sigma(a).$$

Apply Nakayama's Lemma version 3 (Exercise 8.2.G) to $\langle dt \rangle \subset \Omega_{A/k}$. (This argument, with essentially no change, can be used to show that if $(A, \mathfrak{m}, k)$ is a localization of a finitely generated algebra over k, and $t_1, \ldots, t_n$ generate $\mathfrak{m}$, then $dt_1, \ldots, dt_n$ generate $\Omega_{A/k}$.)
(c) By part (b), $\Omega_{A/k}$ is a principal A-module. Show that $\times dt$ is an injection as follows. By the classification of finitely generated modules over discrete valuation rings (Remark 6.4.4), it suffices to show that $t^m\, dt \neq 0$ for all m. The surjection $A \to A/(t^N)$ induces a map $\Omega_{A/k} \to \Omega_{(A/(t^N))/k}$, so it suffices to show that $t^m\, dt$ is nonzero in $\Omega_{(A/(t^N))/k}$. Show that $A/(t^N) \cong k[t]/(t^N)$. The usual differentiation rule for polynomials gives a map $k[t]/(t^N) \to k[t]/(t^{N-1})$, which is a derivation of $k[t]/(t^N)$ over k, and $t^m\, dt$ will not map to 0 so long as N is sufficiently large. Put the pieces together and complete the proof. (An extension of these ideas can show that if $(A, \mathfrak{m}, k)$ is a localization of a finitely generated algebra over k that is a regular local ring, then $\Omega_{A/k}$ is free of rank $\dim A$.)

21.3.5. Projective space and the Euler exact sequence. We next examine the differentials of projective space $\mathbb{P}^n_k$, or, more generally, $\mathbb{P}^n_A$, where A is an arbitrary ring. As projective space is covered by affine open sets of the form $\mathbb{A}^n$, on which the differentials form a rank n locally free sheaf, $\Omega_{\mathbb{P}^n_A/A}$ is also a rank n locally free sheaf.

21.3.6. Theorem (the Euler exact sequence) — *The sheaf of differentials $\Omega_{\mathbb{P}^n_A/A}$ satisfies the following exact sequence:*

$$0 \longrightarrow \Omega_{\mathbb{P}^n_A/A} \xrightarrow{\ \alpha\ } \mathscr{O}_{\mathbb{P}^n_A}(-1)^{\oplus(n+1)} \xrightarrow{\ \phi\ } \mathscr{O}_{\mathbb{P}^n_A} \longrightarrow 0.$$

This is handy, because you can get a hold of $\Omega_{\mathbb{P}^n_A/A}$ in a concrete way. See Exercise 21.5.Q for an application. By dualizing this exact sequence, we have an exact sequence

$$0 \longrightarrow \mathscr{O}_{\mathbb{P}^n_A} \xrightarrow{\phi^\vee} \mathscr{O}_{\mathbb{P}^n_A}(1)^{\oplus(n+1)} \longrightarrow \mathscr{T}_{\mathbb{P}^n_A/A} \longrightarrow 0.$$

21.3.7.* *Proof of Theorem 21.3.6.* What is really going on in this proof is that we consider those differentials on $\mathbb{A}^{n+1}_A \setminus \{0\}$ that are pullbacks of differentials on $\mathbb{P}^n_A$. For a different explanation, in terms of the Koszul complex (briefly discussed in §24.4.1), see [E, §17.5].

We first describe a map $\phi\colon \mathscr{O}(-1)^{\oplus(n+1)} \to \mathscr{O}$, and later identify the kernel with $\Omega_{\mathbb{P}^n_A/A}$. The map is given by

$$\phi\colon (s_0, s_1, \ldots, s_n) \longmapsto x_0 s_0 + x_1 s_1 + \cdots + x_n s_n.$$

(Do you see why?) You should think of this as a "degree 1" map, as each x_i has degree 1.

21.3.8. Remark. The dual $\phi^\vee\colon \mathscr{O} \to \mathscr{O}(1)^{\oplus(n+1)}$ of ϕ gives a map $\mathbb{P}^n \to \mathbb{P}^n$ via Important Theorem 15.2.2 on maps to projective space. This map is the identity, and this is one way of describing ϕ in a "natural" (coordinate-free) manner.

21.3.F. EASY EXERCISE. Show that ϕ is surjective, by checking on the open set $D(x_i)$. (There is a one-line solution.)

Now we must identify the kernel of ϕ with the differentials, and we can do this on each $D(x_i)$, so long as we do it in a way that works simultaneously for each open set. So we consider the open set U_0, where $x_0 \neq 0$, and we have the usual coordinates $x_{j/0} = x_j/x_0$ $(1 \leq j \leq n)$. Given a differential

$$f_1(x_{1/0}, \ldots, x_{n/0})\, dx_{1/0} + \cdots + f_n(x_{1/0}, \ldots, x_{n/0})\, dx_{n/0},$$

we must produce $n+1$ sections of $\mathscr{O}(-1)$. As motivation, we just look at the first term, and pretend that the projective coordinates are actual coordinates.

$$f_1\, dx_{1/0} = f_1\, d(x_1/x_0)$$

$$= f_1\, \frac{x_0\, dx_1 - x_1\, dx_0}{x_0^2}$$

$$= -\frac{x_1}{x_0^2} f_1\, dx_0 + \frac{f_1}{x_0}\, dx_1.$$

Note that x_0 times the "coefficient of dx_0" plus x_1 times the "coefficient of dx_1" is 0, and also both coefficients are of homogeneous degree -1. Motivated by this, we define

$$(21.3.8.1) \qquad \alpha\colon f_1\, dx_{1/0} + \cdots + f_n\, dx_{n/0} \mapsto \left(-\frac{x_1}{x_0^2} f_1 - \cdots - \frac{x_n}{x_0^2} f_n, \frac{f_1}{x_0}, \frac{f_2}{x_0}, \ldots, \frac{f_n}{x_0} \right).$$

Note that over U_0, this indeed gives an injection of $\Omega_{\mathbb{P}^n_A/A}$ to $\mathscr{O}(-1)^{\oplus(n+1)}$ that surjects onto the kernel of $\mathscr{O}(-1)^{\oplus(n+1)} \to \mathscr{O}_X$ (if $(g_0, \ldots, g_n)$ is in the kernel, take $f_i = x_0 g_i$ for $i > 0$).

Let's make sure this construction, applied to two different coordinate patches (say, U_0 and U_1), gives the same answer. (This verification is best ignored on a first reading.) Note that

$$f_1\, dx_{1/0} + f_2\, dx_{2/0} + \cdots = f_1\, d\frac{1}{x_{0/1}} + f_2\, d\frac{x_{2/1}}{x_{0/1}} + \cdots$$

$$= -\frac{f_1}{x_{0/1}^2}\, dx_{0/1} + \frac{f_2}{x_{0/1}}\, dx_{2/1} - \frac{f_2 x_{2/1}}{x_{0/1}^2}\, dx_{0/1} + \cdots$$

$$= -\frac{f_1 + f_2 x_{2/1} + \cdots}{x_{0/1}^2}\, dx_{0/1} + \frac{f_2}{x_{0/1}}\, dx_{2/1} + \cdots.$$

Under this map, the $dx_{2/1}$ term goes to the second factor (where the factors are indexed 0 through n) in $\mathscr{O}(-1)^{\oplus(n+1)}$, and yields f_2/x_0, as desired (and similarly for $dx_{j/1}$ for $j > 2$). Also, the $dx_{0/1}$

term goes to the "zero" factor, and yields

$$-\left(\sum_{i=1}^{n} f_i(x_i/x_1)/(x_0/x_1)^2\right)/x_1 = -\sum_{i=1}^{n} f_i(x_i/x_1)x_i/x_0^2,$$

as desired. Finally, the "first" factor must be correct because the sum over i of x_i times the ith factor is 0.

21.3.G. EXERCISE. Finish the proof of Theorem 21.3.6, by verifying that this map, $\alpha: \Omega_{\mathbb{P}^n_A/A} \to \mathcal{O}_{\mathbb{P}^n_A}(-1)^{\oplus(n+1)}$, identifies $\Omega_{\mathbb{P}^n_A/A}$ with ker ϕ. □

21.3.H. EXERCISE (PROMISED IN § 18.5.7). Show that $h^1(\mathbb{P}^2_k, \Omega_{\mathbb{P}^2_k/k}) > 0$. Show that $\Omega_{\mathbb{P}^2_k/k}$ is not the direct sum of line bundles, and can't even be *filtered* into line bundles. Show that Theorem 18.5.6 (that all finite rank vector bundles on $\mathbb{P}^1$ split into line bundles) cannot be extended to rank 2 vector bundles on $\mathbb{P}^2_k$.

21.3.9. *Generalizations of the Euler exact sequence.* Generalizations of the Euler exact sequence are quite useful. We won't use them later, so no proofs will be given. First, the argument generalizes readily if Spec A is replaced by an arbitrary base scheme. The Euler exact sequence generalizes further in a number of ways. As a first step, suppose $\mathcal{V}$ is a rank $n+1$ locally free sheaf (or vector bundle) on a scheme X. Suppose $\pi: \mathbb{P}\mathcal{V} \to X$ is the corresponding projective bundle. (See §17.2.3 for the definition of $\mathbb{P}\mathcal{V}$, including a caution about duals.) Then $\Omega_{\mathbb{P}\mathcal{V}/X}$ sits in an Euler exact sequence:

$$0 \longrightarrow \Omega_{\mathbb{P}\mathcal{V}/X} \longrightarrow \mathcal{O}_{\mathbb{P}\mathcal{V}}(-1) \otimes \pi^* \mathcal{V}^\vee \longrightarrow \mathcal{O}_{\mathbb{P}\mathcal{V}} \longrightarrow 0.$$

It is not obvious that this is useful, but we have already implicitly seen it in the case of $\mathbb{P}^1$-bundles over curves, in Exercise 20.2.J, where the normal bundle to a section was identified in this way.

21.3.10.** *Generalization to the Grassmannian.* For another generalization, fix a base field k, and let $G(m, n+1)$ be the space of sub-vector spaces of dimension m in an $(n+1)$-dimensional vector space V (the Grassmannian, §16.4). Over $G(m, n+1)$ we have a short exact sequence of locally free sheaves,

$$0 \longrightarrow \mathcal{S} \longrightarrow \mathcal{O}_{G(m,n+1)} \otimes V^\vee \longrightarrow \mathcal{Q} \longrightarrow 0,$$

where $\mathcal{O}_{G(m,n+1)} \otimes V^\vee$ is the "trivial bundle whose fibers are $V^\vee$" (do you understand what that means?), and $\mathcal{S}$ is the "universal vector bundle" or "tautological vector (sub)bundle." Then there is a canonical isomorphism

(21.3.10.1) $$\Omega_{G(m,n+1)/k} \xleftrightarrow{\sim} \mathcal{H}om(\mathcal{Q}, \mathcal{S}).$$

21.3.I. EXERCISE. Recall that in the case of projective space, i.e., $m = 1$, $\mathcal{S} = \mathcal{O}(-1)$ (Exercise 17.1.H). Verify (21.3.10.1) in this case using the Euler exact sequence (Theorem 21.3.6).

21.3.J. EXERCISE. Prove (21.3.10.1), and explain how it generalizes Exercise 20.2.I. (The hint to Exercise 20.2.I may help.)

This Grassmannian fact generalizes further to Grassmannian bundles, and to flag varieties, and to flag bundles.

21.4 The Riemann–Hurwitz Formula

The Riemann–Hurwitz formula generalizes our calculation of the genus g of a double cover of $\mathbb{P}^1$ branched at $2g+2$ points, Theorem 19.5.1, to higher degree covers, and to higher genus target curves.

21.4.1. *Definition.* A finite morphism between integral schemes $X \to Y$ is said to be **separable** if it is dominant, and the induced extension of function fields $K(X)/K(Y)$ is a separable extension. Similarly, a generically finite morphism is **generically separable** if it is dominant, and the induced extension of function fields is a (finite) separable extension. Note that finite morphisms of integral schemes are automatically separable in characteristic 0.

21.4.2. Proposition — *If $\pi\colon X \to Y$ is a generically separable morphism of irreducible varieties of the same dimension n, and Y is smooth, then the cotangent exact sequence (Theorem 21.2.24) is exact on the left as well:*

(21.4.2.1)
$$0 \longrightarrow \pi^*\Omega_{Y/k} \overset{\phi}{\longrightarrow} \Omega_{X/k} \longrightarrow \Omega_{X/Y} \longrightarrow 0.$$

This is an example of left-exactness of the cotangent exact sequence in the presence of appropriate "smoothness"; see Remark 21.2.10.

Proof. We must check that ϕ is injective. Now $\Omega_{Y/k}$ is a rank n locally free sheaf on Y, so $\pi^*\Omega_{Y/k}$ is a rank n locally free sheaf on X. A locally free sheaf on an integral scheme (such as $\pi^*\Omega_{Y/k}$) is torsion-free (any nonzero section over any open set is nonzero at the generic point; see §6.1.4), so if a subsheaf of it (such as $\ker\phi$) is nonzero, it is nonzero at the generic point. Thus to show the injectivity of ϕ, we need only check that ϕ is an inclusion at the generic point. We thus tensor with $\mathcal{O}_\eta$, where η is the generic point of X. Tensoring with $\mathcal{O}_\eta$ is an exact functor (localization is exact, Exercise 1.5.G), and $\mathcal{O}_\eta \otimes \Omega_{X/Y} = 0$ (as $K(X)/K(Y)$ is a separable extension by hypothesis, and Ω for separable field extensions is 0 by Exercise 21.2.H(a)). Also, $\mathcal{O}_\eta \otimes \pi^*\Omega_{Y/k}$ and $\mathcal{O}_\eta \otimes \Omega_{X/k}$ are both n-dimensional $\mathcal{O}_\eta$-vector spaces (they are the stalks of rank n locally free sheaves at the generic point). Thus by considering

$$\mathcal{O}_\eta \otimes \pi^*\Omega_{Y/k} \longrightarrow \mathcal{O}_\eta \otimes \Omega_{X/k} \longrightarrow \mathcal{O}_\eta \otimes \Omega_{X/Y} \longrightarrow 0$$

(which is $\mathcal{O}_\eta^{\oplus n} \to \mathcal{O}_\eta^{\oplus n} \to 0 \to 0$) we see that $\mathcal{O}_\eta \otimes \pi^*\Omega_{Y/k} \to \mathcal{O}_\eta \otimes \Omega_{X/k}$ is injective, and thus that $\pi^*\Omega_{Y/k} \to \Omega_{X/k}$ is injective. $\qquad\square$

People not confined to characteristic 0 should note what goes wrong for non-separable morphisms. For example, suppose k is a field of characteristic p, and consider the map $\pi\colon \mathbb{A}_k^1 = \operatorname{Spec} k[t] \to \mathbb{A}_k^1 = \operatorname{Spec} k[u]$ given by $u = t^p$. Then Ω_π is the trivial invertible sheaf generated by dt. As another (similar but different) example, if $K = k(x)$ and $K' = k(x^p)$, then the inclusion $K' \hookrightarrow K$ induces $\pi\colon \operatorname{Spec} K[t] \to \operatorname{Spec} K'[t]$. Once again, Ω_π is an invertible sheaf, generated by dx (which in this case is pulled back from $\Omega_{K/K'}$ on $\operatorname{Spec} K$). In both of these cases, we have maps from one affine line to another, and there are nonzero relative tangent vectors.

21.4.A. EXERCISE. If X and Y are smooth varieties of dimension n, and $\pi\colon X \to Y$ is generically separable, show that the locus of non-smoothness is pure codimension 1, and has a natural interpretation as an effective divisor, as follows. Interpret ϕ as an $n \times n$ Jacobian matrix (13.1.7.1) in appropriate local coordinates, and hence interpret the locus where ϕ is not an isomorphism as (locally) the vanishing scheme of the determinant of an $n \times n$ matrix.

We call the locus of non-smoothness (in X) the **ramification locus**, and its image (in Y) the **branch locus**. Because of this exercise, we use the terms **ramification divisor** and (assuming the image of the ramification divisor is indeed a divisor) **branch divisor**.

Before getting down to our case of interest, dimension 1, we begin with something (literally) small but fun. Suppose $\pi\colon X \to Y$ is a surjective k-morphism from a smooth k-scheme that contracts a subset of codimension greater than 1. More precisely, suppose π is an isomorphism over an open subset of Y, from an open subset U of X whose complement has codimension greater than 1. Then by Exercise 21.4.A, Y *cannot* be smooth. (*Small resolutions*, to be defined in Exercise 22.4.N, are examples of such π. In particular, you can find an example there.)

Suppose now that X and Y are dimension 1. Then the ramification locus is a finite set (**ramification points**) of X, and the branch locus is a finite set (**branch points**) of Y. (Figure 21.4 shows a morphism with two ramification points and one branch point.) *Now assume that* $k = \overline{k}$. We examine $\Omega_{X/Y}$ near a point $p \in X$.

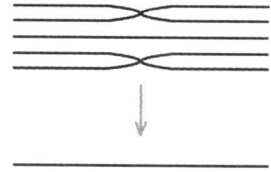

Figure 21.4 *An example where the branch divisor appears with multiplicity 2 (see Exercise 21.4.C).*

As motivation for what we will see, we note that in complex geometry, nonconstant maps from (complex) curves to curves may be written in appropriate local coordinates as $x \mapsto x^m = y$, from which we see that dy pulls back to $mx^{m-1}dx$, so $\Omega_{X/Y}$ locally looks like functions times dx modulo multiples of $mx^{m-1}dx$.

Consider now our map $\pi : X \to Y$, and fix $p \in X$, and $q = \pi(p)$. Because the construction of Ω behaves well under base change (Theorem 21.2.26(b)), we may replace Y with Spec of the local ring $\mathscr{O}_{Y,q}$ at q, i.e., we may assume $Y = \operatorname{Spec} B$, where B is a discrete valuation ring (as Y is a regular curve), with residue field $k = \overline{k}$ corresponding to q. By restricting to an affine open neighborhood of p in X, we assume that X is affine too. Similarly, as the construction of Ω behaves well with respect to localization (Exercise 21.2.L), we may replace X by $\operatorname{Spec} \mathscr{O}_{X,p}$, and thus assume $X = \operatorname{Spec} A$, where A is a discrete valuation ring, and π corresponds to $B \to A$, inducing an isomorphism of residue fields (with $k = \overline{k}$).

Suppose their uniformizers are s and t, respectively, with $t \mapsto us^n$, where u is an invertible element of A.

$$
\begin{array}{ccc}
X & A & us^n \\
\downarrow & \uparrow & \updownarrow \\
Y & B & t
\end{array}
$$

Recall that the differentials of a discrete valuation ring over k are generated by d of the uniformizer (Exercise 21.3.E). Then

$$dt = d(us^n) = uns^{n-1}\,ds + s^n\,du.$$

This differential on Spec A vanishes to order at least $n - 1$, and precisely $n - 1$ if n is not divisible by the characteristic. If the order is $n - 1$, we call it **tame** ramification, while if the order is greater, it is called **wild** ramification. We define one plus this order to be the the **ramification index** at this point of X.

21.4.B. EXERCISE. Show that the degree of $\Omega_{X/Y}$ at $p \in X$ is precisely the ramification order of π at p, minus 1. (The degree of a coherent sheaf on a curve was defined in §18.4.6. You will need to make sense of the length of a coherent sheaf at a point; see, for example, Exercise 6.5.M.)

21.4.C. EXERCISE: INTERPRETING THE RAMIFICATION DIVISOR IN TERMS OF NUMBER OF PREIMAGES. For this exercise, we take the base field k to be algebraically closed. Suppose all the ramification above $q \in Y$ is tame (which is always true in characteristic 0). Show that the degree of the branch divisor at q is $\deg \pi - |\pi^{-1}(q)|$. Thus the multiplicity of the branch divisor counts the extent to which the number of preimages is less than the degree (see Figure 21.4).

21.4.3. Theorem (the Riemann–Hurwitz formula) — *Suppose* $\pi : X \to Y$ *is a finite separable morphism of projective regular curves, of pure degree* n. *Let* R *be the ramification divisor. Then*

(21.4.3.1) $$2g(X) - 2 = n(2g(Y) - 2) + \deg R.$$

21.4.D. EXERCISE. Prove the Riemann–Hurwitz formula. Hint: Apply the fact that degree is additive in exact sequences (Exercise 18.4.I) to (21.4.2.1). Recall that degrees of line bundles pull back well under finite morphisms of integral projective curves, Exercise 18.4.G. A torsion coherent sheaf on a reduced curve (such as Ω_π) is supported in dimension 0 (Exercise 14.3.F(b)), so $\chi(\Omega_\pi) = h^0(\Omega_\pi)$. Show that the degree of R as a divisor is the same as its degree in the sense of h^0.

Here are some applications of the Riemann–Hurwitz formula.

21.4.4. *Example.* The degree of R is always even: any cover of a curve must be branched over an even number of points (counted with appropriate multiplicity).

21.4.E. EASY EXERCISE. Show that there is no nonconstant map from a smooth projective irreducible genus 2 curve to a smooth projective irreducible genus 3 curve. (Hint: deg R $\geq$ 0.)

21.4.5. *Definition.* If the only connected finite étale cover of a connected scheme X is an isomorphism, we say that X is **simply connected in the étale topology**. Do you see why "simply connected" is a reasonable name for the concept? We won't use this terminology beyond the following brief discussion.

21.4.6. *Example ("$\mathbb{P}^1_k$ is simply connected").* If $k = \bar{k}$, the only connected unbranched finite separable cover of $\mathbb{P}^1_k$ is the isomorphism, for the following reason. Suppose X is connected and $X \to \mathbb{P}^1_k$ has no ramification. Then X is a curve, and regular by Exercise 24.8.F(a). (Feel free to take this for granted, since this exercise is in the future. There will be no circularity.) Applying the Riemann–Hurwitz formula, using that the ramification divisor is 0, we have $2 - 2g_X = 2d$, with $d \geq 1$ and $g_X \geq 0$, from which $d = 1$ and $g_X = 0$.

21.4.F. EXERCISE ("IN CHARACTERISTIC 0, $\mathbb{A}^1_k$ IS SIMPLY CONNECTED"). Show that if $k = \bar{k}$ has characteristic 0, the only connected unbranched cover of $\mathbb{A}^1_k$ is itself. (Aside: In characteristic p, this needn't hold; Spec $k[x, y]/(y^p - y - x) \to$ Spec $k[x]$ is such a map, as you can show yourself. Once the theory of the algebraic fundamental group is developed, this translates to: "$\mathbb{A}^1$ is not simply connected in characteristic p." This cover is an example of an *Artin-Schreier cover*. Fun fact: The group $\mathbb{Z}/p$ acts on this cover via the map $y \mapsto y + 1$. This is an example of a *Galois* cover; you can check that the extension of function fields is Galois.)

21.4.G. UNIMPORTANT EXERCISE. Extend Example 21.4.6 and Exercise 21.4.F, by removing the $k = \bar{k}$ hypothesis, and changing "connected" to "geometrically connected."

21.4.7. *Example: Lüroth's Theorem.* Continuing the notation of Theorem 21.4.3, suppose $g(X) = 0$. Then from the Riemann–Hurwitz formula (21.4.3.1), $g(Y) = 0$. (Otherwise, if $g(Y)$ were at least 1, then the right side of the Riemann–Hurwitz formula would be nonnegative, and thus couldn't be -2, which is the left side.) Informally: the only maps from a genus 0 curve to a curve of positive genus are the constant maps. This has a nonobvious algebraic consequence, by our identification of covers of curves with field extensions (Theorem 16.3.3): all subfields of $k(x)$ containing k are of the form $k(y)$, where $y = f(x)$ for some $f \in k(x)$.

$$
\begin{array}{ccc}
k(x) & \mathbb{P}^1 & \\
\uparrow & \downarrow & \\
K(C) & C & = \mathbb{P}^1
\end{array}
$$

(It turns out that the hypotheses char $k = 0$ and $k = \bar{k}$ are not necessary.) This is Lüroth's Theorem.

21.4.H. EXERCISE. Use Lüroth's Theorem to give new geometric solutions to Exercises 7.5.K and 7.5.M. (These arguments will be less ad hoc, and more suitable for generalization, than the algebraic solutions suggested in the hints to those exercises.)

21.4.I.* EXERCISE (GEOMETRIC GENUS EQUALS TOPOLOGICAL GENUS). This exercise is intended for those with some complex background, who know that the Riemann–Hurwitz formula holds in the complex analytic category. Suppose C is an irreducible regular projective complex curve. Show that there is an algebraic nonconstant map $\pi\colon C \to \mathbb{P}^1_{\mathbb{C}}$. Describe the corresponding map of Riemann surfaces. Use the Riemann–Hurwitz formula to show that the algebraic notion of genus (as computed using the branched cover π) agrees with the topological notion of genus (using the same branched cover). (Recall that assuming Miracle 18.5.2—that the canonical bundle is Serre-dualizing—we know that the geometric genus equals the arithmetic genus, Exercise 21.3.A.)

21.4.J. UNIMPORTANT EXERCISE. Suppose $\pi: X \to Y$ is a dominant morphism of regular curves, and R is the ramification divisor of π. Show that $\Omega_X(-R) \cong \pi^*\Omega_Y$. (This exercise is geometrically pleasant, but we won't use it.) Hint: This says that we can interpret the invertible sheaf $\pi^*\Omega_Y$ over an open set U of X as precisely those differentials on U vanishing along the ramification divisor.

21.4.8. *Informal example: The degree of the discriminant of degree* d *polynomials in one variable.* You may be aware that there is a degree $2d - 2$ polynomial in the coefficients $a_d, \ldots, a_0$ of the degree d polynomial

$$f(x) = a_d x^d + a_{d-1} x^{d-1} + \cdots + a_0 = 0$$

that vanishes precisely when $f(x)$ has a multiple root. For example, when $d = 2$, the discriminant is $a_1^2 - 4a_0 a_2$. We can "compute" this degree $2d - 2$ using the Riemann–Hurwitz formula as follows. (You should try to make sense of the following informal and imprecise discussion.) We work over an algebraically closed field k of characteristic 0 for the sake of simplicity. If we take two general degree d polynomials, $g(x)$ and $h(x)$, the degree of the discriminant "should be" the number of $\lambda \in k$ for which $g(x) - \lambda h(x)$ has a double root. Consider the morphism $\mathbb{P}^1 \to \mathbb{P}^1$ given by $x \mapsto [g(x), h(x)]$. (Here we use "affine coordinates" on the source $\mathbb{P}^1$: by x we mean $[x, 1]$.) Then this morphism has degree d. A ramification point a mapping to the branch point $[\lambda, 1]$ in the target $\mathbb{P}^1$ corresponds to a being a double root of $g(x) - \lambda h(x)$. Thus the number of branch points should be the desired degree of the discriminant. By the Riemann–Hurwitz formula there are $2d - 2$ branch points (admittedly, with multiplicity). It is possible to turn this into a proof, and it is interesting to do so.

21.4.9.* Bounds on automorphism groups of curves.

It is a nontrivial fact that irreducible smooth projective curves of genus $g \geq 2$ have finite automorphism groups. (See [FK, §5.1] for a proof over $\mathbb{C}$. See [Schm] for the first proof in arbitrary characteristic, although better approaches are now available.) Granting this fact, we can show that in characteristic 0, the automorphism group has order at most $84(g - 1)$ (*Hurwitz's Automorphisms Theorem*), as follows.

Suppose C is an irreducible smooth projective curve over an algebraically closed field $k = \bar{k}$ of characteristic 0, of genus $g \geq 2$. Suppose that G is a *finite* group of automorphisms of C. We now show that $|G| \leq 84(g - 1)$. (The case where k is not algebraically closed is quickly dispatched by base-changing to $\bar{k}$.)

21.4.K. EXERCISE.

(a) Let C' be the smooth projective curve corresponding to the field extension $K(C)^G$ of k (via Theorem 16.3.3). ($K(C)^G$ means the G-invariants of $K(C)$.) Describe a morphism $\pi: C \to C'$ of degree $|G|$, as well as a faithful G-action on C that commutes with π.

(b) Show that above each branch point of π, the preimages are all ramified to the same order (as G acts transitively on them). Suppose there are n branch points and the ith one has ramification r_i (each $|G|/r_i$ times).

(c) Use the Riemann–Hurwitz formula to show that

$$(2g - 2) = |G| \left(2g(C') - 2 + \sum_{i=1}^{n} \frac{r_i - 1}{r_i} \right).$$

To maximize $|G|$, we wish to minimize

(21.4.9.1) $$2g(C') - 2 + \sum_{i=1}^{n} \frac{r_i - 1}{r_i}$$

subject to (21.4.9.1) being positive. Note that $1/42$ is possible: take $g(C') = 0$, $n = 3$, and $(r_1, r_2, r_3) = (2, 3, 7)$.

21.4.L. EXERCISE. Show that you can't do better than $1/42$ by considering the following cases separately:

(a) $g(C') > 1$,
(b) $g(C') = 1$,
(c) $g(C') = 0$ and $n \geq 5$,
(d) $g(C') = 0$ and $n = 4$, and
(e) $g(C') = 0$ and $n = 3$.

21.4.M. EXERCISE. Use the fact that (21.4.9.1) is at least $1/42$ to prove the result.

21.4.10. *Remark.* In positive characteristic, there can be many more automorphisms. For example [Stic], in characteristic p, if p^n is a prime power that is not 2, then the completion ("compactification") of the affine curve $y^{p^n} = x + x^{p^n+1}$ has genus $g = p^n(p^n - 1)/2$ and automorphism group of order $p^{3n}(p^{3n} + 1)(p^{2n} - 1)$.

21.5   Understanding Smooth Varieties Using Their Cotangent Bundles

In this section, we construct birational invariants of varieties over algebraically closed fields (such as the geometric genus), motivate the notion of an unramified morphism, show that varieties are "smooth almost everywhere," and get a first glimpse of Hodge theory.

21.5.1. The geometric genus, and other birational invariants from i-forms.

Suppose X is a projective scheme over k. Then $h^0(X, \Omega_{X/k})$ is an invariant of X. In fact it and related invariants are *birational invariants* if X is smooth, as shown in the following Exercise 21.5.A. We first define the **sheaf of (relative) i-forms** $\Omega^i_{X/Y} := \wedge^i \Omega_{X/Y}$. Sections of $\Omega^i_{X/Y}$ (over some open set) are called **(relative) i-forms** (over that open set).

21.5.2. *Joke.* Old MacDonald had a form; $e_i \wedge e_i = 0$.

21.5.A. EXERCISE ($H^0(X, \Omega^i_{X/k})$ ARE BIRATIONAL INVARIANTS). Suppose X and X' are birational irreducible smooth projective k-varieties. Show (for each i) that $H^0(X, \Omega^i_{X/k}) \cong H^0(X', \Omega^i_{X'/k})$. Hint: Fix a birational map $\phi : X \dashrightarrow X'$. By Exercise 15.3.B, the complement of the domain of definition U of ϕ is codimension at least 2. By pulling back i-forms from X' to U, we get a map $\phi^* : H^0(X', \Omega^i_{X'/k}) \to H^0(U, \Omega^i_{X/k})$. Use Algebraic Hartogs's Lemma 13.5.19 and the fact that Ω^i is locally free to show the map extends to a map $\phi^* : H^0(X', \Omega^i_{X'/k}) \to H^0(X, \Omega^i_{X/k})$. If $\psi : X' \dashrightarrow X$ is the inverse rational map, we similarly get a map $\psi^* : H^0(X, \Omega^i_{X/k}) \to H^0(X', \Omega^i_{X'/k})$. Show that ϕ^* and ψ^* are inverses by showing that each composition is the identity on a dense open subset of X or X'.

21.5.3. *The canonical bundle $\mathscr{K}_X$ and the geometric genus $p_g(X)$.* If X is a dimension n smooth k-variety, the invertible sheaf (or line bundle) $\det \Omega_{X/k} = \Omega^n_{X/k}$ (the sheaf of "algebraic volume forms") has particular importance, and is called the **canonical (invertible) sheaf**, or the **canonical (line) bundle**. It is denoted by $\mathscr{K}_X$ (or $\mathscr{K}_{X/k}$). As mentioned in §18.5.2, if X is projective, then $\mathscr{K}_X$ is the canonical sheaf ω_X appearing in the statement of Serre duality, something we will establish in §29.4 (see Desideratum 29.1.1).

21.5.B. EXERCISE (THE ADJUNCTION FORMULA FOR $\mathscr{K}_X$). Suppose X is a smooth variety, and Z is a smooth subvariety of X. Show that

$$\mathscr{K}_Z \cong \mathscr{K}_X|_Z \otimes \det \mathscr{N}_{Z/X}.$$

(Hint: Apply Exercise 14.2.P to Theorem 21.2.31.) In particular, by Exercise 21.2.J, if Z is codimension 1, then

$$\mathscr{K}_Z \cong (\mathscr{K}_X \otimes \mathscr{O}_X(Z))|_Z,$$

which is more commonly and compactly written as $\mathscr{K}_X(Z)|_Z$. The adjunction formula is often used inductively, for complete intersections; see Exercise 21.5.G for an example. (See Exercise 20.2.B(a) and §29.4 for other versions of the adjunction formula.)

21.5.4. *Definition.* If X is a projective (or even proper) smooth k-variety, the birational invariant $h^0(X, \mathscr{K}_X) = h^0(X, \Omega^n_{X/k})$ has particular importance. It is called the **geometric genus**, and is denoted by $p_g(X)$. We saw this in the case of curves in §21.3.1. If X is an irreducible variety that is *not* smooth or projective, the phrase geometric genus refers to $h^0(X', \mathscr{K}_{X'})$ for some smooth projective X' *birational* to X. (By Exercise 21.5.A, this is independent of the choice of X'.) For example, if X is an irreducible reduced projective curve over k, the geometric genus is the geometric genus of the normalization of X. (But in higher dimension, it is not clear if there exists such an X'. It is a nontrivial fact that this is true in characteristic 0, and it is not yet known in positive characteristic; see Remark 22.4.6.)

It is a miracle that for a complex curve the geometric genus is the same as the topological genus and the arithmetic genus. Exercise 18.5.A showed that the geometric and arithmetic genus are equal (assuming Miracle 18.5.2 that the canonical bundle is Serre-dualizing). We will connect the geometric genus to the topological genus in our discussion of the Riemann–Hurwitz formula soon (Exercise 21.4.I).

21.5.C. EXERCISE. Suppose Z is a regular degree d surface in $\mathbb{P}^3_k$. Compute the geometric genus $p_g(Z)$ of Z. Show that no regular quartic surface in $\mathbb{P}^3_k$ is rational (i.e., birational to $\mathbb{P}^2_k$, Definition 7.5.5). (Such quartic surfaces are examples of *K3 surfaces*; see Exercise 21.5.I.)

21.5.5. Important classes of varieties: Fano, Calabi-Yau, general type.
Suppose X is a smooth projective k-variety. Then X is said to be **Fano** if $\mathscr{K}_X^\vee$ is ample, and **Calabi-Yau** if $\mathscr{K}_X \cong \mathscr{O}_X$. (Caution: There are other definitions of Calabi-Yau.)

21.5.D. EXERCISE. For all $j \geq 0$, the j**th plurigenus** of a smooth projective k-variety is $h^0(X, \mathscr{K}_X^{\otimes j})$. Show that the jth plurigenus is a birational invariant.

(By contrast, $h^0(X, \mathscr{K}^{\otimes j})$ is not a birational invariant if $j < 0$. For example, you can show that $h^0(X, \mathscr{K}^\vee)$ differs for $X = \mathbb{P}^2_k$ and $X = \mathrm{Bl}_p \mathbb{P}^2_k$.)

The **Kodaira dimension** of X, denoted by $\kappa(X)$, tracks the rate of growth of the plurigenera. It is the smallest k such that $h^0(X, \mathscr{K}_X^{\otimes j})/j^k$ is bounded (as j varies through the positive integers), except that if all the plurigenera $h^0(X, \mathscr{K}_X^{\otimes j})$ are 0, we say $\kappa(X) = -1$. It is a nontrivial fact that the Kodaira dimension always exists, and that it is the maximum of the dimensions of images of the jth "pluricanonical rational maps." The latter fact implies that the Kodaira dimension is an integer between -1 and dim X inclusive. Exercise 21.5.D shows that the Kodaira dimension is a birational invariant. If $\kappa(X) = \dim X$, we say that X is of **general type**. (For more information on the Kodaira dimension, see [Ii, §10.5].)

21.5.E. EXERCISE. Show that if $\mathscr{K}_X$ is ample then X is of general type.

21.5.F. EXERCISE. Show that a smooth geometrically irreducible projective curve (over a field k) is Fano (resp., Calabi-Yau, general type) if its genus is 0 (resp., 1, greater than 1).

The "trichotomy" of Exercise 21.5.F is morally the reason that curves behave differently depending on which of the three classes they lie in (see §19.6.1). In some sense, important aspects of this trichotomy extend to higher dimension, which is part of the reason for making these definitions. We now explore this trichotomy in the case of complete intersections.

21.5.G. EXERCISE. Suppose X is a smooth complete intersection in $\mathbb{P}^N_k$ of hypersurfaces of degree $d_1, \ldots, d_n$, where k is algebraically closed.

(a) Show that $\mathscr{K}_X \cong \mathscr{O}(-N - 1 + d_1 + \cdots + d_n)|_X$.
(b) Show that if $-N - 1 + d_1 + \cdots + d_n$ is negative (resp., zero, positive) then X is Fano (resp., Calabi-Yau, general type).

(c) Find all possible values of N and $d_1, \ldots, d_n$ where X is Calabi-Yau of dimension at most 3. Notice how small this list is.

21.5.H. EASY EXERCISE. A smooth Fano surface is called a **del Pezzo surface** (part of the Enriques–Kodaira classification of surfaces). Show that the following surfaces are del Pezzo: (a) $\mathbb{P}^2_k$; (b) $\mathbb{P}^1_k \times \mathbb{P}^1_k$; and (c) smooth surfaces in $\mathbb{P}^3_k$ of degree at most 3. (This list has some redundancy! The remaining examples of del Pezzo surfaces are described in Remark 27.3.7. One of them appeared earlier in §19.8.3.)

21.5.I. EXERCISE. A **K3 surface** over a field k is a proper smooth geometrically connected Calabi-Yau surface X over k such that $H^1(X, \mathscr{O}_X) = 0$. (Weil has written that K3 surfaces were named in honor of Kummer, Kähler, and Kodaira, and the mountain K2, [BPV, p. 288].) Prove that the dimension 2 (smooth) Calabi-Yau complete intersections of Exercise 21.5.G(c) are all K3 surfaces.

21.5.6. *Tantalizing side remark.* If you compare the degrees of the hypersurfaces cutting out complete intersection K3 surfaces (in Exercise 21.5.G(c)), with the degrees of hypersurfaces cutting out complete intersection canonical curves (see §19.8.2 and Exercise 19.8.E), you will notice a remarkable coincidence. Of course, this is not a coincidence at all.

21.5.J. EXERCISE (NOT EVERYTHING FITS INTO THIS TRICHOTOMY). Suppose C is a smooth projective irreducible complex curve of genus greater than 1. Show that $C \times_{\mathbb{C}} \mathbb{P}^1_{\mathbb{C}}$ is neither Fano, nor Calabi-Yau, nor general type. Hint: Exercise 21.2.U.

21.5.7.* Kodaira vanishing.
The Kodaira vanishing theorem is an important tool in a number of different areas of algebraic geometry. We state it for the sake of culture, but do not prove it (and hence will not use it).

21.5.8. The Kodaira Vanishing Theorem — *Suppose k is a field of characteristic 0, and X is a smooth projective k-variety. Then for any ample invertible sheaf $\mathscr{L}$, $H^i(X, \mathscr{K}_X \otimes \mathscr{L}) = 0$ for $i > 0$.*

21.5.9. The restriction on characteristic is necessary; Raynaud gave an example where Kodaira vanishing fails in positive characteristic, [Ra]. The original proof is by proving the complex-analytic version, and then transferring it into a complex algebraic statement using Serre's GAGA Theorem [Se3]. The general characteristic 0 case can be reduced to $\mathbb{C}$—any reduction of this sort is often called (somewhat vaguely) an application of the *Lefschetz principle*. (For a precise formulation of the Lefschetz principle, with a short proof that applies in any characteristic, see [Ek]. See [FR] for a generalization. See [MO90551] for more context.) Raynaud later gave a dramatic algebraic proof of Kodaira vanishing *using* positive characteristic (!); see [DeI] or [Il].

21.5.K. EXERCISE. Prove the Kodaira Vanishing Theorem in the case $\dim X = 1$, and for $\mathbb{P}^n_k$. (Neither case requires the characteristic 0 hypothesis.)

21.5.10.* A first glimpse of Hodge theory.
The invariant $h^j(X, \Omega^i_{X/k})$ is called the **Hodge number** $h^{i,j}(X)$. By Exercise 21.5.A, $h^{i,0}$ are birational invariants. We will soon see (in Exercise 21.5.O) that this isn't true for all $h^{i,j}$.

21.5.L. EXERCISE. Suppose X is a smooth projective variety. Assuming Miracle 18.5.2 (that the canonical bundle is Serre-dualizing), show that Hodge numbers satisfy the symmetry $h^{p,q} = h^{n-p,n-q}$. (Exercise 14.2.O will be useful.)

21.5.M. EXERCISE (THE HODGE NUMBERS OF PROJECTIVE SPACE). Show that $h^{p,q}(\mathbb{P}^n_k) = 1$ if $0 \le p = q \le n$ and $h^{p,q}(\mathbb{P}^n_k) = 0$ otherwise. Hint: Use the Euler exact sequence (Theorem 21.3.6) and apply Exercise 14.2.N.

21.5.11. *Remark: The Hodge diamond.* Over $k = \mathbb{C}$, further miracles occur. If X is an irreducible smooth projective complex variety, then it turns out that there is a direct sum decomposition

$$(21.5.11.1) \qquad H^m(X, \mathbb{C}) = \oplus_{i+j=m} H^j(X, \Omega^i_{X/\mathbb{C}}),$$

from which $h^m(X, \mathbb{C}) = \sum_{i+j=m} h^{i,j}$, so the Hodge numbers (purely algebraic objects) yield the Betti numbers (a priori topological information). Moreover, complex conjugation interchanges $H^j(X, \Omega^i_{X/\mathbb{C}})$ with $H^i(X, \Omega^j_{X/\mathbb{C}})$, from which

$$(21.5.11.2) \qquad h^{i,j} = h^{j,i}.$$

(Aside: This additional symmetry holds in characteristic 0 in general, but can fail in positive characteristic; see, for example, [Ig, p. 966], [Mu1, §II], [Se4, Prop. 16].) This is the beginning of the vast and rich subject of Hodge theory (see [GH1, §0.6] for more, or [Vo] for much more).

If we write the Hodge numbers in a diamond, with $h^{i,j}$ the ith entry in the $(i+j)$th row, then the diamond has the two symmetries coming from Serre duality and complex conjugation. For example, the Hodge diamond of an irreducible smooth projective complex surface will be of the following form:

$$
\begin{array}{ccccc}
 & & 1 & & \\
 & q & & q & \\
p_g & & h^{1,1} & & p_g \\
 & q & & q & \\
 & & 1, & &
\end{array}
$$

where p_g is the geometric genus of the surface, and $q = h^{0,1} = h^{1,0} = h^{2,1} = h^{1,2}$ is called the **irregularity** of the surface. As another example, by Exercise 21.5.M, the Hodge diamond of $\mathbb{P}^n$ is all 0 except for 1's down the vertical axis of symmetry.

You won't need the unproved statements (21.5.11.1) or (21.5.11.2) to solve the following problems.

21.5.N. EXERCISE. Assuming Miracle 18.5.2 (that the canonical bundle is Serre-dualizing), show that the Hodge diamond of a smooth projective geometrically irreducible genus g curve over a field k is the following.

$$
\begin{array}{ccc}
 & 1 & \\
g & & g \\
 & 1 &
\end{array}
$$

21.5.O. EXERCISE. Show that the Hodge diamond of $\mathbb{P}^1_k \times \mathbb{P}^1_k$ is the following.

$$
\begin{array}{ccccc}
 & & 1 & & \\
 & 0 & & 0 & \\
0 & & 2 & & 0 \\
 & 0 & & 0 & \\
 & & 1 & &
\end{array}
$$

By comparing your answer to the Hodge diamond of $\mathbb{P}^2_k$ (Exercise 21.5.M), show that $h^{1,1}$ is not a birational invariant.

Notice that in both cases, $h^{1,1}$ is the Picard number ρ (defined in §18.4.12). In general, $\rho \leq h^{1,1}$ (see [GH1, §3.5, pp. 456–7] in characteristic 0).

21.5.12.* Aside: Infinitesimal deformations and automorphisms.
It is beyond the scope of this book to make this precise, but if X is a variety, $H^0(X, \mathscr{T}_X)$ parametrizes infinitesimal automorphisms of X, and $H^1(X, \mathscr{T}_X)$ parametrizes infinitesimal deformations. As an

example, if $X = \mathbb{P}^1$ (over a field), $\mathscr{T}_{\mathbb{P}^1} \cong \mathscr{O}(2)$ (§21.3.2), so $h^0(\mathbb{P}^1, \mathscr{T}_{\mathbb{P}^1}) = 3$, which is precisely the dimension of the automorphism group of $\mathbb{P}^1$ (Exercise 15.5.A).

21.5.P. EXERCISE. Compute $h^0(\mathbb{P}^n_k, \mathscr{T}_{\mathbb{P}^n_k})$ using the Euler exact sequence (Theorem 21.3.6). Compare this to the dimension of the automorphism group of $\mathbb{P}^n_k$ (Exercise 15.5.A).

21.5.Q. EXERCISE. Show that $H^1(\mathbb{P}^n_A, \mathscr{T}_{\mathbb{P}^n_A}) = 0$. (Thus projective space can't deform, and is "rigid.")

21.5.R. EXERCISE. Assuming Miracle 18.5.2 (that the canonical bundle is Serre-dualizing), compute $h^i(C, \mathscr{T}_C)$ for a genus g smooth projective geometrically irreducible curve over k, for $i = 0$ and 1. You should notice that $h^1(C, \mathscr{T}_C)$ for genus 0, 1, and $g > 1$ is 0, 1, and $3g - 3$, respectively; after doing this, reread §19.8.4.

21.6 Generic Smoothness, and Consequences

We can now verify something you may already have intuited. In positive characteristic, this is a hard theorem, in that it uses a result from commutative algebra that we have not proved.

21.6.1. Theorem (generic smoothness of varieties) — *If X is an integral finite type k-scheme over a perfect field* k *of dimension* n, *there is a dense open subset* U *of X such that* U *is smooth (over* k*) of dimension* n.

Hence, by Exercise 13.2.R combined with Fact 13.8.2 (localization of regular is regular), U is regular. Theorem 21.6.4 will generalize this to smooth *morphisms*, at the expense of restricting to characteristic 0.

See §13.2.9 for an example of what goes wrong if k is not perfect.

Proof. The $n = 0$ case is immediate, so we assume $n > 0$.

We will show that the rank at the generic point is n. Then by upper semicontinuity of the rank of a coherent sheaf (Exercise 14.3.J), it must be n in an open neighborhood of the generic point, and we are done.

We thus have to check that if K is the fraction field of a dimension n integral finite type k-scheme, i.e., (by Theorem 12.2.1) if K/k is a transcendence degree n extension, then $\Omega_{K/k}$ is an n-dimensional vector space. But every finitely generated extension of a perfect field is separably generated (Theorem 10.5.17), so the result follows by Exercise 21.2.M(b). □

21.6.2. Proof of Theorem 13.8.3, that the localization of a regular local ring is regular in the case of varieties over perfect fields. Theorem 21.6.1 allows us to prove Theorem 13.8.3 (an important case of Fact 13.8.2).

Suppose Y is a variety over a perfect field k that is regular at its closed points (so Y is smooth, by Exercise 13.2.R), and let η be a point of Y. We will show that Y is regular at η. Let $X = \bar{\eta}$. By Theorem 21.6.1, X contains a dense (= nonempty) open subset of smooth points. By shrinking Y by discarding the points of X outside that open subset, we may assume X is smooth.

Then Theorem 21.2.31 (exactness of the conormal exact sequence for smooth varieties) implies that $\mathscr{I}/\mathscr{I}^2$ is a locally free sheaf of rank $\operatorname{codim}_{X/Y} = \dim \mathscr{O}_{Y,\eta}$.

21.6.A. EXERCISE. Let m be the maximal ideal of $\mathscr{O}_{Y,\eta}$. Identify the stalk of $\mathscr{I}/\mathscr{I}^2$ at the generic point η of X with $\mathfrak{m}/\mathfrak{m}^2$. Conclude the proof of Theorem 13.8.3. □

21.6.B. TRICKY EXERCISE (LOCALIZATION OF REGULAR LOCAL RINGS OF VARIETIES ARE REGULAR, PROMISED JUST AFTER THEOREM 13.8.3). Suppose $(A, \mathfrak{m})$ is a regular local ring that is the localization of a finitely generated k-algebra, where k is perfect. Show that the localization of A at a prime is also a regular local ring.

21.6.3. Generic smoothness for morphisms, and the Kleiman–Bertini Theorem.

We will now discuss a number of important results that fall under the rubric of "generic smoothness for morphisms." All require working in characteristic 0.

21.6.4. Theorem (generic smoothness on the source) —

(a) Suppose $\pi: X \to Y$ is a dominant finite type morphism of integral Noetherian schemes, such that char $K(Y)$ $(= \text{char } K(X))$ is zero. Then there is a nonempty $(= \text{dense})$ open set $U \subset X$ such that $\pi|_U$ is smooth of relative dimension $\operatorname{trdeg} K(X)/K(Y)$.

(b) Let k be a field of characteristic 0, and let $\pi: X \to Y$ be a dominant morphism of integral k-varieties. Then there is a nonempty $(= \text{dense})$ open set $U \subset X$ such that $\pi|_U$ is smooth of relative dimension $\dim X - \dim Y$.

21.6.C. EXERCISE (PROMISED IN EXERCISE 12.2.D). Suppose $f_1, \ldots, f_n \in k[x_1, \ldots, x_n]$, where k is a field of characteristic 0. Show that $f_1, \ldots, f_n$ are algebraically dependent if and only if the determinant of the Jacobian matrix is identically zero. Hint: Why is this exercise given just after Theorem 21.6.4?

21.6.5. *Example.* Theorem 21.6.4 fails in positive characteristic: consider the purely inseparable extension $\mathbb{F}_p(t)/\mathbb{F}_p(t^p)$. The same problem can arise even over an *algebraically closed* field of characteristic p: consider $\mathbb{A}^1_k = \operatorname{Spec} k[t] \to \operatorname{Spec} k[u] = \mathbb{A}^1_k$, given by $u \mapsto t^p$.

Proof. Let $n = \operatorname{trdeg} K(X)/K(Y)$. We may replace Y by an affine open subset $\operatorname{Spec} B$, and then replace X by an affine open subset $\operatorname{Spec} A$, so π corresponds to the ring map $B \to A$. Choose n elements $x_1, \ldots, x_n \in A$ that form a transcendence basis for $K(A)/K(B)$. (Do you see why we may choose them to lie in A?) Choose additional elements $x_{n+1}, \ldots, x_N \in A$ so that $x_1, \ldots, x_N$ generate A as a B-algebra.

For $i = 0, \ldots, n$, let F_i be the subfield of $K(A)$ generated over $K(B)$ by $x_1, \ldots, x_i$. For example, $F_0 = K(B)$, $F_N = K(A)$, and $F_i/K(B)$ is a purely transcendental field extension for $i \leq n$. For $i = n+1, \ldots, N$, the minimal polynomial for $x_i \in K(A)$ over F_{i-1} (an element of $F_{i-1}[t]$) may be interpreted as $m_i(x_1, \ldots, x_{i-1}, t)$ for some $m_i \in K(B)[y_1, \ldots, y_{i-1}, t]$. By multiplying m_i by the products of the denominators of its coefficients, we obtain $M_i \in B[y_1, \ldots, y_{i-1}, t]$ so that $M_i(x_1, \ldots, x_{i-1}, t) \in F_{i-1}[t]$ is also a minimal polynomial for $x_i \in K(A)$ over F_{i-1}. Thus

(21.6.5.1) $$K(A) = K(B)(x_1, \ldots, x_n)[y_{n+1}, \ldots, y_N]/I,$$

where

$$I = (M_{n+1}(x_1, \ldots, x_n, y_{n+1}), M_{n+2}(x_1, \ldots, x_n, y_{n+1}, y_{n+2}), \ldots, M_N(x_1, \ldots, x_n, y_{n+1}, \ldots, y_N)).$$

Define $A' := B[y_1, \ldots, y_N]/(M_{n+1}(y_1, \ldots, y_{n+1}), \ldots, M_N(y_1, \ldots, y_N))$.

Let $X' := \operatorname{Spec} A'$. Let $Z_1, \ldots, Z_s$ be the irreducible components of X'. (There are finitely many by Noetherianity of B: Exercise 3.6.T and Proposition 3.6.15.) Precisely one of them, say, $Z = Z_1$, dominates $\operatorname{Spec} B$ (because $A' \otimes_B K(B) \cong K(A)$ by (21.6.5.1)). Let $U' \subset X'$ be the open subset $Z_1 \setminus Z_1 \cap (Z_2 \cup \cdots \cup Z_s)$.

At the generic point of U', the Jacobian matrix of $M_{n+1}(x_{n+1}), \ldots, M_N(x_N)$ with respect to $x_1, \ldots, x_N$ has corank n. (Here we use the separability of the minimal polynomials M_i.) Thus there is an open subset $U'' \subset U'$ where the Jacobian has corank n, so $U'' \to \operatorname{Spec} B$ is smooth of relative dimension n. Furthermore U'' is irreducible. Now U'' and X are birational, so by Proposition 7.5.6 there is an open subset $U \subset X$ that is isomorphic to an open subset of U''. This is the open subset we seek.

21.6.D. EXERCISE. Prove Theorem 21.6.4 (b) (using part (a)). □

If furthermore X is smooth, the situation is even better.

21.6.6. Theorem (generic smoothness on the target) — *Suppose* $\pi\colon X \to Y$ *is a morphism of* k-*varieties, where* char $k = 0$, *and* X *is smooth (over* k*). Then there is a dense open subset* U *of* Y *such that* $\pi|_{\pi^{-1}(U)}$ *is a smooth morphism.*

Note that $\pi^{-1}(U)$ may be empty! Indeed, if π is not dominant, we will have to take such a U.

To prove Theorem 21.6.6, we use a neat trick. Suppose $\pi\colon X \to Y$ is a morphism of schemes that are finite type over k, where char $k = 0$. Define

$$X_r = \left\{ p \in X \;:\; \mathrm{rank}\,\left(\pi^*\Omega_{Y/k} \to \Omega_{X/k}\right)\big|_p \leq r \right\}.$$

21.6.E. EXERCISE. Show that X_r is a constructible subset of X. Hint: If X and Y are both smooth, show that the rank condition implies that X_r is cut out by "determinantal equations," or use an appropriate form of upper semicontinuity.

By Chevalley's Theorem 8.4.2, $\pi(X_r)$ is constructible, so we can make sense of $\dim \pi(X_r)$.

21.6.7. Lemma — *If* $\pi\colon X \to Y$ *is a morphism of schemes that are finite type over* k, *where* char $k = 0$, *then* $\dim \pi(X_r) \leq r$.

Proof. In this proof, we make repeated use of the identification of Zariski cotangent spaces at closed points with the fibers of cotangent sheaves, using Corollary 21.2.33. We can replace X by an irreducible component of X_r, and Y by the closure of that component's image in Y (with reduced subscheme structure; see §9.4.9). The resulting map will have all of X contained in X_r, so we may as well assume $X = X_r$. This boils down to the following linear algebra observation: if a linear map $\rho\colon V_1 \to V_2$ has rank at most r, and V_i' is a quotient of V_i, with ρ sending V_1' to V_2', then the restriction of ρ to V_1' has rank at most that of ρ itself. Thus we have a dominant morphism $\pi\colon X \to Y$, and we wish to show that $\dim Y \leq r$. By generic smoothness of varieties (Theorem 21.6.1), Y has a dense open set that is smooth, so we may assume that Y is smooth. By generic smoothness on the source for $\pi\colon X \to Y$, there is a nonempty open subset $U \subset X$ such that $\pi\colon U \to Y$ is smooth. But then for any point $p \in U$, the cotangent map $\pi^*\Omega_Y|_{\pi(p)} \to \Omega_X|_p$ is injective (Exercise 21.2.S), and has rank at most r. Taking p to be a *closed* point, we have $\dim Y = \dim_{\pi(p)} Y \leq \dim \Omega_Y|_{\pi(p)} \leq r$. $\square$

There is not much left to do to prove the theorem.

21.6.8. *Proof of Theorem 21.6.6.* Reduce to the case where Y is smooth over k (by restricting to a smaller open set, using generic smoothness of Y, Theorem 21.6.4(b)). Say $n = \dim Y$. Now $\dim \overline{\pi(X_{n-1})} \leq n-1$ by Lemma 21.6.7 , so remove $\overline{\pi(X_{n-1})}$ from Y as well. Then the rank of $\pi^*\Omega_{Y/k} \to \Omega_{X/k}$ is at least n for each closed point of X. But as Y is regular of dimension n, we have that $\pi^*\Omega_{Y/k} \to \Omega_{X/k}$ is injective for every closed point of X, and hence $\pi^*\Omega_{Y/k} \to \Omega_{X/k}$ is an injective map of sheaves (do you see why?). Thus π is smooth by Exercise 24.8.G(b). $\square$

21.6.9.** The Kleiman–Bertini Theorem. The same idea of bounding the dimension of some "bad locus" can be used to prove the Kleiman–Bertini Theorem 21.6.10 (due to Kleiman), which is useful in (for example) enumerative geometry. Throughout this discussion, $k = \bar{k}$, although the definitions and results can be generalized. Suppose G is a group variety (over $k = \bar{k}$), and we have a G-action on a variety X. We say that the **action is transitive** if it is transitive on closed points. A better definition (which you can show is equivalent) is that the action morphism $G \times X \to X$ restricted to a fiber above any closed point of X is surjective. A variety X with a transitive G-action is said to be a **homogeneous space** for G. For example, G acts on itself by left-translation, and via this action G is a homogeneous space for itself.

21.6.F. EASY EXERCISE. Suppose G is a group variety over an algebraically closed field k of characteristic 0. Show that every homogeneous space X for G is smooth. (In particular, taking $X = G$, we see that G is smooth.) Hint: X has a dense open set U that is smooth by Theorem 21.6.4(b), and G acts transitively on the closed points of X, so we can cover X with translates of U.

21.6.10. The Kleiman–Bertini Theorem, [Kl2, Thm. 2] — *Suppose X is homogeneous space for a group variety G (over a field $k = \overline{k}$ of characteristic 0). Suppose $\alpha\colon Y \to X$ and $\beta\colon Z \to X$ are morphisms from smooth k-varieties Y and Z.*

(a) *Then there is a nonempty open subset $V \subset G$ such that for every $\sigma \in V(k)$, $Y \times_X Z$ defined by*

$$
\begin{array}{ccc}
Y \times_X Z & \longrightarrow & Z \\
\downarrow & & \downarrow{\scriptstyle\beta} \\
Y & \xrightarrow{\ \sigma \circ \alpha\ } & X
\end{array}
$$

(Y is "translated by σ") is smooth of dimension $\dim Y + \dim Z - \dim X$ (but possibly empty).

(b) *Furthermore, there is a nonempty open subset $V \subset G$ such that*

(21.6.10.1)
$$(G \times_k Y) \times_X Z \longrightarrow G$$

is a smooth morphism of relative dimension $\dim Y + \dim Z - \dim X$ over V.

The first time you hear this, you should think of the special case where $Y \to X$ and $Z \to X$ are locally closed embeddings (Y and Z are smooth subvarieties of X). In this case, the Kleiman–Bertini theorem says that the second subvariety will meet a "general translate" of the first "transversely." (There is no such thing as a smooth morphism of negative relative dimension. So what happens if $\dim Y + \dim Z - \dim X$ is negative?)

Proof. It is more pleasant to describe this proof "backward," by considering how we would prove it ourselves. We will use generic smoothness twice.

Clearly (b) implies (a), so we prove (b).

In order to show that the morphism (21.6.10.1) is smooth over a nonempty open set $V \subset G$, it would suffice to apply generic smoothness on the target (Theorem 21.6.6) to (21.6.10.1). Hence it suffices to show that $(G \times_k Y) \times_X Z$ is a smooth k-variety. Now Z is smooth over k, so it suffices to show that $(G \times_k Y) \times_X Z \to Z$ is a smooth morphism (as the composition of two smooth morphisms is smooth, Exercise 13.6.D). But this is obtained by base change from $G \times_k Y \to X$, so it suffices to show that this latter morphism is smooth (as smoothness is preserved by base change).

Now $G \times_k Y \to X$ is a G-equivariant morphism. (By **"G-equivariant,"** we mean that the G-action on both sides respects the morphism.) By generic smoothness of the target (Theorem 21.6.6), this is smooth over a dense open subset X. But then by transitivity of the G-action on X, this morphism is smooth everywhere.

21.6.G. EXERCISE. Refine the above argument to show the desired statement about relative dimension. ☐

21.6.H. EXERCISE (POOR MAN'S KLEIMAN–BERTINI). Prove Theorem 21.6.10(a) without the hypotheses on k (on algebraic closure or characteristic), and without the smoothness in the conclusion. Hint: This is a question about dimensions of fibers of morphisms, so you could have solved this after reading §12.4.

21.6.I. EXERCISE (IMPROVED CHARACTERISTIC 0 BERTINI). Suppose Z is a smooth k-variety, where $\operatorname{char} k = 0$ and $k = \overline{k}$. Let V be a finite-dimensional base-point-free linear series on Z, i.e., a finite vector space of sections of some invertible sheaf $\mathscr{L}$ on Z. Show that a general element of V, considered as a closed subscheme of Z, is regular. (More explicitly: each element $s \in V$ gives a closed subscheme of Z. Then for a general s, considered as a point of $\mathbb{P}V$, the corresponding closed subscheme is smooth over k.) Hint: Figure out what this has to do with the Kleiman–Bertini Theorem 21.6.10. Let $n = \dim V$, $G = GL(V)$, $X = \mathbb{P}V^{\vee}$, take Z in Kleiman–Bertini to be the Z of the problem, and let Y be the "universal hyperplane" over $\mathbb{P}V^{\vee}$ (the incidence variety $I \subset \mathbb{P}V \times \mathbb{P}V^{\vee}$ of Definition 13.4.1).

21.6.J. EASY EXERCISE. Interpret Bertini's Theorem 13.4.2 over a characteristic 0 field as a corollary of Exercise 21.6.I.

In characteristic 0, Exercise 21.6.I is a good improvement on Bertini's Theorem. For example, we don't need $\mathscr{L}$ to be very ample, or X to be projective. But unlike Bertini's Theorem, Exercise 21.6.I fails in positive characteristic, as demonstrated by the one-dimensional linear series $\{pQ : Q \in \mathbb{P}^1\}$. This is essentially Example 21.6.5. (Do you see why this does not contradict Bertini's Theorem 13.4.2?)

21.7 Unramified Morphisms

Suppose $\pi \colon X \to Y$ is a morphism of schemes. The support of the quasicoherent sheaf $\Omega_\pi = \Omega_{X/Y}$ is called the **ramification locus**, and the image of its support, $\pi(\operatorname{Supp} \Omega_{X/Y})$, is called the **branch locus**. If $\Omega_\pi = 0$, we say that π is **formally unramified**, and if π is also furthermore locally of finite type, we say π is **unramified**. (Caution: There is some lack of consensus in the definition of "unramified"; "locally of finite type" is sometimes replaced by "locally of finite presentation," which was the definition originally used in [Gr-EGA].)

21.7.A. EASY EXERCISE (EXAMPLES OF UNRAMIFIED MORPHISMS).

(a) Show that locally closed embeddings are unramified.
(b) Show that if S is a multiplicative subset of the ring B, then $\operatorname{Spec} S^{-1}B \to \operatorname{Spec} B$ is formally unramified. (Thus if η is the generic point of an integral scheme Y, then $\operatorname{Spec} \mathscr{O}_{Y,\eta} \to Y$ is formally unramified.)
(c) Show that finite separable field extensions (or, more correctly, the corresponding maps of schemes) are unramified.

21.7.B. EXERCISE (PRACTICE WITH THE CONCEPT).

(a) Show that the normalization of the node in Exercise 10.7.E (see Figure 8.4) is unramified.
(b) Show that the normalization of the cusp in Exercise 10.7.F (see Figure 10.5) is *not* unramified.

21.7.C. EASY EXERCISE (UNRAMIFIED MORPHISMS ARE A "REASONABLE" CLASS OF MORPHISMS). Show that the class of unramified morphisms is "reasonable" in the sense of §8.1.

21.7.D. EXERCISE (CHARACTERIZATIONS OF UNRAMIFIED MORPHISMS BY THEIR FIBERS). Suppose $\pi \colon X \to Y$ is locally of finite type.

(a) Show that π is unramified if and only if for each $q \in Y$, $\pi^{-1}(q)$ is the (scheme-theoretic) disjoint union of schemes of the form $\operatorname{Spec} K$, where K is a finite separable extension of $\kappa(q)$.
(b) Show that π is unramified if and only if for each *geometric* point $\bar{q}$, $\pi^{-1}(\bar{q}) := \bar{q} \times_Y X$ is the (scheme-theoretic) disjoint union of copies of $\bar{q}$.

21.7.E. EXERCISE (THE MEANING OF UNRAMIFIEDNESS IN "VERY GEOMETRIC" SETTINGS). Show that a morphism $\pi \colon X \to Y$ of finite type schemes over an algebraically closed field is unramified if and only if π is injective on tangent vectors at closed points. (This was mentioned in the remark immediately after Theorem 19.1.1.)

21.7.F. EXERCISE (FOR USE IN EXERCISE 21.7.G(b)).

(a) Suppose $\pi \colon X \to Y$ is a locally finite type morphism of locally Noetherian schemes. Show that π is unramified if and only if $\delta_\pi \colon X \to X \times_Y X$ is an open embedding. Hint: Show the following. If $\phi \colon X \to Z$ is a closed embedding of Noetherian schemes, and the ideal sheaf $\mathscr{I}$ of ϕ satisfies

$\mathscr{I} = \mathscr{I}^2$, then ϕ is also an open embedding. For that, show that if $(A, \mathfrak{m})$ is a Noetherian local ring, and I is a proper ideal of A satisfying $I = I^2$, then $I = 0$. For that, in turn, use Nakayama version 2 (Lemma 8.2.9). Also use the fact that Supp $\mathscr{I}$ is closed (using Exercise 14.3.C), so its complement is open.

(b)* Adapt your proof of (a) to drop the locally Noetherian hypothesis. Hint: Show that if π is locally of finite type, then δ_π is locally finitely presented.

21.7.G. EASY EXERCISE. Suppose $\pi: X \to Y$ and $\rho: Y \to Z$ are locally of finite type. Let $\tau = \rho \circ \pi$:

(21.0.1)

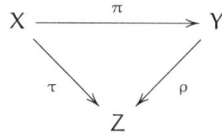

(a) Show that if τ is unramified, then so is π.
(b) Show that if ρ is unramified, and τ is smooth of relative dimension n (e.g., étale if $n = 0$), then π is smooth of relative dimension n.

(Does this agree with your geometric intuition?)

21.7.H. EXERCISE. Suppose $\pi: X \to Y$ is locally of finite type. Show that the locus in X where π is unramified is open.

21.7.I. UNIMPORTANT EXERCISE (FOR NUMBER THEORISTS). Suppose $\phi: (A, \mathfrak{m}) \to (B, \mathfrak{n})$ is a local homomorphism of local rings. In algebraic number theory, such a ring morphism is said to be *unramified* if $B/\phi(\mathfrak{m})B$ is a finite separable extension of $A/\mathfrak{m}$. Show that if ϕ is finite type, this agrees with our definition.

21.7.1. *Arithmetic side remark: The different and the discriminant.* If B is the ring of integers in a number field (§10.7.1), the **different ideal** of B is the annihilator of $\Omega_{B/\mathbb{Z}}$. It measures how "(un)ramified" Spec $B \to $ Spec $\mathbb{Z}$ is, and is a ring-theoretic version of the ramification locus. The **discriminant ideal** can be interpreted as the ideal of $\mathbb{Z}$ corresponding to effective divisor on Spec $\mathbb{Z}$ that is the "push forward" (not defined here, but defined as you might expect) of the divisor corresponding to the different. It is a ring-theoretic version of the branch locus. If B/A is an extension of rings of integers of number fields, the **relative different ideal** (of B) and **relative discriminant ideal** (of A) are defined similarly. (We won't use these ideas.)

Chapter 22*

Blowing Up

We next discuss an important construction in algebraic geometry, the blow-up of a scheme along a closed subscheme (cut out by a finite type ideal sheaf). The theory could have mostly been developed immediately after Chapter 17, but the interpretation in terms of the conormal cone/bundle/sheaf of many classical examples makes it natural to discuss blowing up after differentials.

We won't use blowing up much in later chapters, so feel free to skip this topic for now. But it is an important tool. For example, one can use it to resolve singularities, and, more generally, indeterminacy of rational maps. In particular, blow-ups can be used to relate birational varieties to each other.

We will start with a motivational example that will give you a picture of the construction in a particularly important (and the historically earliest) case, in §22.1. We will then see a formal definition, in terms of a universal property, §22.2. The definition won't immediately have a clear connection to the motivational example. We will deduce some consequences of the definition (assuming that the blow-up actually exists). We will then prove that the blow-up exists, by describing it quite explicitly, in §22.3. As a consequence, we will find that the blow-up morphism is projective, and we will deduce more consequences from this. In §22.4, we will do a number of explicit computations, to see various sorts of applications, and to see that many things can be computed by hand.

22.1 Motivating Example: Blowing Up the Origin in the Plane

We will generalize the following notion, which will correspond to "blowing up" the origin of the affine plane $\mathbb{A}_k^2$ (Exercise 10.3.F). Our discussion will be informal. Consider the subset of $\mathbb{A}^2 \times \mathbb{P}^1$ corresponding to the following. We interpret $\mathbb{P}^1$ as parametrizing the lines through the origin. Consider the subvariety

$$\mathrm{Bl}_{(0,0)} \mathbb{A}^2 := \left\{ (p \in \mathbb{A}^2, [\ell] \in \mathbb{P}^1) \; : \; p \in \ell) \right\},$$

which is the data of a point p in the plane, and a line ℓ containing both p and the origin. Algebraically: let x and y be coordinates on $\mathbb{A}^2$, and X and Y be projective coordinates on $\mathbb{P}^1$ ("corresponding" to x and y); we will consider the subset $\mathrm{Bl}_{(0,0)} \mathbb{A}^2$ of $\mathbb{A}^2 \times \mathbb{P}^1$ corresponding to $xY - yX = 0$. We have the useful diagram

$$
\begin{array}{ccccc}
\mathrm{Bl}_{(0,0)} \mathbb{A}^2 & \hookrightarrow & \mathbb{A}^2 \times \mathbb{P}^1 & \longrightarrow & \mathbb{P}^1 \\
& \searrow_{\beta} & \downarrow & & \\
& & \mathbb{A}^2 & &
\end{array}
\quad .
$$

You can verify that $\mathrm{Bl}_{(0,0)} \mathbb{A}^2$ is smooth over k (Definition 13.2.4 or 21.2.28) directly (you can now make the paragraph after Exercise 10.3.F precise), but here is an informal argument, using the projection $\mathrm{Bl}_{(0,0)} \mathbb{A}^2 \to \mathbb{P}^1$. The projective line $\mathbb{P}^1$ is smooth, and for each point $[\ell]$ in $\mathbb{P}^1$, we have a smooth choice of points on the line ℓ. Thus we are verifying smoothness by way of a fibration over $\mathbb{P}^1$.

We next consider the projection to $\mathbb{A}^2$, $\beta: \mathrm{Bl}_{(0,0)} \mathbb{A}^2 \to \mathbb{A}^2$. This is an isomorphism away from the origin. Loosely speaking, if p is not the origin, there is precisely one line containing p and the origin. On the other hand, if p *is* the origin, then there is a full $\mathbb{P}^1$ of lines containing p and the origin. Thus the preimage of $(0,0)$ is a curve, and hence a divisor (an effective Cartier divisor, as the blown-up surface is regular). This is called the *exceptional divisor* of the blow-up.

If we have some curve $C \subset \mathbb{A}^2$ singular at the origin, it can be potentially partially desingularized, using the blow-up, by taking the closure of $C \setminus \{(0,0)\}$ in $\mathrm{Bl}_{(0,0)} \mathbb{A}^2$. (A **desingularization** or a **resolution of singularities** of a variety X is a proper birational morphism $\widetilde{X} \to X$ from a regular scheme.) For example, consider the curve $y^2 = x^3 + x^2$, which is regular except for a node at the origin. We can take the preimage of the curve minus the origin, and take the closure of this locus in the blow-up, and we will obtain a regular curve; the two branches of the node downstairs are separated upstairs. (You can check this in Exercise 22.4.B once we have defined things properly. The result will be called the *proper transform* (or *strict transform*) of the curve.) We are interested in desingularizations for many reasons. Because we understand regular curves quite well, we could hope to understand other curves through their desingularizations. This philosophy holds true in higher dimension as well.

More generally, we can blow up $\mathbb{A}^n$ at the origin (or, more informally, "blow up the origin"), getting a subvariety of $\mathbb{A}^n \times \mathbb{P}^{n-1}$. Algebraically, if $x_1, \ldots, x_n$ are coordinates on $\mathbb{A}^n$, and $X_1, \ldots, X_n$ are projective coordinates on $\mathbb{P}^{n-1}$, then the blow-up $\mathrm{Bl}_{\vec{0}} \mathbb{A}^n$ is given by the equations $x_i X_j - x_j X_i = 0$. Once again, this is smooth: $\mathbb{P}^{n-1}$ is smooth, and for each point $[\ell] \in \mathbb{P}^{n-1}$, we have a smooth choice of $p \in \ell$.

We can extend this further, by blowing up $\mathbb{A}^{n+m}$ along a coordinate m-plane $\mathbb{A}^m$ by adding m more variables $x_{n+1}, \ldots, x_{n+m}$ to the previous example; we get a subset of $\mathbb{A}^{n+m} \times \mathbb{P}^{n-1}$.

Because, in complex geometry, submanifolds of manifolds locally "look like" coordinate m-planes in n-space, you might imagine that we could extend this to blowing up a regular subvariety of a regular variety. In the course of making this precise, we will accidentally generalize this notion greatly, defining the blow-up of any finite type quasicoherent sheaf of ideals in a scheme. In general, blowing up may not have such an intuitive description as in the case of blowing up something regular inside something regular—it can do great violence to the scheme—but even then, it is very useful.

Our description will depend only on the closed subscheme being blown up, and not on coordinates. That remedies a defect that was already present in the first example, of blowing up the plane at the origin. It is not obvious that if we picked different coordinates for the plane (preserving the origin as a closed subscheme) that we wouldn't have two different resulting blow-ups.

As is often the case, there are two ways of understanding the notion of blowing up, and each is useful in different circumstances. The first is by universal property, which lets you show some things without any work. The second is an explicit construction, which lets you get your hands dirty and compute things (and implies, for example, that the blow-up morphism is projective).

The motivating example here may seem like a very special case, but if you understand the blow-up of the origin in n-space well enough, you will understand blowing up in general.

22.2 Blowing Up, by Universal Property

We now define the blow-up by a universal property. The disadvantage of starting here is that this definition won't obviously be the same as (or even related to) the examples of §22.1.

Suppose $X \hookrightarrow Y$ is a closed subscheme corresponding to a finite type quasicoherent sheaf of ideals. (If Y is locally Noetherian, the "finite type" hypothesis is automatic, so Noetherian readers can ignore it.)

The blow-up of $X \hookrightarrow Y$ is a Cartesian diagram

(22.2.0.1)

$$
\begin{array}{ccc}
E_X Y & \longhookrightarrow & \mathrm{Bl}_X Y \\
\downarrow & & \downarrow{\scriptstyle \beta} \\
X & \longhookrightarrow & Y
\end{array}
$$

such that $E_X Y$ (the scheme-theoretic preimage of X by β) is an effective Cartier divisor (defined in §9.5.1) on $\mathrm{Bl}_X Y$, such that any other such Cartesian diagram

(22.2.0.2)

$$
\begin{array}{ccc}
D & \hookrightarrow & W \\
\downarrow & & \downarrow \\
X & \longrightarrow & Y,
\end{array}
$$

where D is an effective Cartier divisor on W, factors uniquely through it:

$$
\begin{array}{ccc}
D & \hookrightarrow & W \\
\downarrow & & \downarrow \\
E_X Y & \hookrightarrow & \mathrm{Bl}_X Y \\
\downarrow & & \downarrow \\
X & \hookrightarrow & Y.
\end{array}
$$

We call $\mathrm{Bl}_X Y$ the **blow-up** (of Y along X, or of Y with **center** X). (Other somewhat archaic terms for this are *monoidal transformation*, *σ-process*, *quadratic transformation*, and *dilation*.) We call $E_X Y$ the **exceptional divisor** of the blow-up. (Bl and β stand for "blow-up," and E stands for "exceptional.")

(Caution: In the minimal model program, "exceptional divisor" has a slightly different meaning; if $X \to Y$ is a birational morphism, then any Weil divisor on X whose image in Y is smaller dimension is called an **exceptional divisor** in this context.)

By a typical universal property argument, if the blow-up exists, it is unique up to unique isomorphism. (We can even recast this more explicitly in the language of Yoneda's Lemma: consider the category of diagrams of the form (22.2.0.2), where morphisms are diagrams of the form

$$
\begin{array}{ccc}
D & \hookrightarrow & W \\
& & \\
& D' & \hookrightarrow W' \\
& & \\
X & \hookrightarrow & Y.
\end{array}
$$

Then the blow-up is a final object in this category, if one exists.)

If $Z \hookrightarrow Y$ is any closed subscheme of Y, then the scheme-theoretic preimage $\beta^{-1}Z$ is called the **total transform** of Z. We will soon see that β is an isomorphism away from X (Observation 22.2.2). $\overline{\beta^{-1}(Z - X)}$ is called the **proper transform** or **strict transform** or **birational transform** of Z. (We will use the first terminology. We will also define it in a more general situation.) We will soon see (in the Blow-up Closure Lemma 22.2.7) that the proper transform is naturally isomorphic to $\mathrm{Bl}_{Z \cap X} Z$, where $Z \cap X$ is the scheme-theoretic intersection.

We will soon show that the blow-up always exists, and describe it explicitly. We first make a series of observations, *assuming that the blow-up exists*.

22.2.1. *Observation.* If X is the empty set, then $\mathrm{Bl}_X Y = Y$. More generally, if X is an effective Cartier divisor, then the blow-up is an isomorphism. (Reason: $\mathrm{id}_Y : Y \to Y$ satisfies the universal property.)

22.2.A. EXERCISE. If U is an open subset of Y, then $\mathrm{Bl}_{U \cap X} U \cong \beta^{-1}(U)$, where $\beta : \mathrm{Bl}_X Y \to Y$ is the blow-up.

Thus "we can compute the blow-up locally."

22.2.B. EXERCISE. Show that if Y_α is an open cover of Y (as α runs over some index set), and the blow-up of Y_α along $X \cap Y_\alpha$ exists, then the blow-up of Y along X exists.

22.2.2. *Observation.* Combining Observation 22.2.1 and Exercise 22.2.A, we see that the blow-up is an isomorphism away from the locus you are blowing up:

$$\beta|_{Bl_X Y \setminus E_X Y} \colon Bl_X Y \setminus E_X Y \to Y \setminus X$$

is an isomorphism.

22.2.3. *Observation.* If $X = Y$, then the blow-up is the empty set: the only map $W \to Y$ such that the pullback of X is an effective Cartier divisor is $\varnothing \hookrightarrow Y$. In this case we have "blown Y out of existence"!

22.2.C. EXERCISE (BLOW-UP PRESERVES IRREDUCIBILITY AND REDUCEDNESS). Show that if Y is irreducible, and X doesn't contain the generic point of Y, then $Bl_X Y$ is irreducible. Show that if Y is reduced, then $Bl_X Y$ is reduced.

22.2.4. Existence in a first nontrivial case: Blowing up a locally principal closed subscheme.

We next see why $Bl_X Y$ exists if $X \hookrightarrow Y$ is locally cut out by one equation. As the question is local on Y (Exercise 22.2.B), we reduce to the affine case $\operatorname{Spec} A/(t) \hookrightarrow \operatorname{Spec} A$. (A good example to think through is $A = k[x, y]/(xy)$ and $t = x$.) Let

$$I = \ker(A \to A_t) = \{ a \in A \ : \ t^n a = 0 \text{ for some } n > 0 \},$$

and let $\phi \colon A \to A/I$ be the projection.

22.2.D. EXERCISE. Show that $\phi(t)$ is not a zerodivisor in A/I.

22.2.E. EXERCISE. Show that $\beta \colon \operatorname{Spec} A/I \to \operatorname{Spec} A$ is the blow-up of $\operatorname{Spec} A$ along $\operatorname{Spec} A/t$. In other words, show that

$$
\begin{array}{ccc}
\operatorname{Spec} A/(t, I) & \longrightarrow & \operatorname{Spec} A/I \\
\downarrow & & \downarrow {\scriptstyle \beta} \\
\operatorname{Spec} A/t & \longrightarrow & \operatorname{Spec} A
\end{array}
$$

is a "blow-up diagram" (22.2.0.1). Hint: In checking the universal property, reduce to the case where W (in (22.2.0.2)) is affine. Then solve the resulting problem about rings. Depending on how you proceed, you might find Exercise 11.3.G, about the uniqueness of extension of maps over effective Cartier divisors, helpful.

22.2.F. EXERCISE. Show that $\operatorname{Spec} A/I$ is the scheme-theoretic closure of $D(t)$ in $\operatorname{Spec} A$.

Thus you might geometrically interpret $\operatorname{Spec} A/I \to \operatorname{Spec} A$ as "shaving off any fuzz supported in $V(t)$." In the Noetherian case, this can be interpreted as removing those associated points lying in $V(t)$. This is intended to be vague, and you should think about how to make it precise only if you want to.

22.2.5. The Blow-up Closure Lemma.

Suppose we have a Cartesian diagram

$$
\begin{array}{ccc}
W & \overset{\text{cl. emb.}}{\lhook\joinrel\longrightarrow} & Z \\
\downarrow & & \downarrow \\
X & \underset{\text{cl. emb.}}{\lhook\joinrel\longrightarrow} & Y,
\end{array}
$$

where the bottom closed embedding corresponds to a finite type ideal sheaf (and hence the upper closed embedding does too). The first time you read this, it may be helpful to consider only the special case where $Z \to Y$ is a closed embedding.

Then take the fibered product of this square by the blow-up $\beta\colon \mathrm{Bl}_X Y \to Y$, to obtain the fibered cube

(22.2.5.1)

The top square is the Cartesian diagram

(22.2.5.2)

22.2.G. EASY EXERCISE. Why is (22.2.5.2) a Cartesian diagram? Show the useful canonical isomorphisms

$$W \times_Y E_X Y \cong Z \times_Y E_X Y \cong W \times_Y \mathrm{Bl}_X Y.$$

22.2.6. The bottom closed embedding of (22.2.5.2) is locally cut out by one equation, and thus the same is true of the top closed embedding as well. However, the local equation on $Z \times_Y \mathrm{Bl}_X Y$ need not be a non-zerodivisor, and thus the top closed embedding is not necessarily an effective Cartier divisor.

Let $\overline{Z}$ be the scheme-theoretic closure of

$$(Z \times_Y \mathrm{Bl}_X Y) \setminus (W \times_Y \mathrm{Bl}_X Y)$$

in $Z \times_Y \mathrm{Bl}_X Y$. (As $W \times_Y \mathrm{Bl}_X Y$ is locally principal, we are in precisely the situation of §22.2.4, so the scheme-theoretic closure is not mysterious.) Note that in the special case where $Z \to Y$ is a closed embedding, $\overline{Z}$ is the proper transform, as defined in §22.2. For this reason, it is reasonable to call $\overline{Z}$ the **proper transform of** Z even if Z *isn't* a closed embedding. Similarly, it is reasonable to call $Z \times_Y \mathrm{Bl}_X Y$ the **total transform of** Z even if Z isn't a closed embedding.

Define $E_{\overline{Z}} \hookrightarrow \overline{Z}$ as the pullback of $E_X Y$ to $\overline{Z}$, i.e., by the Cartesian diagram

Note that $E_{\overline{Z}}$ is an effective Cartier divisor on $\overline{Z}$. (It is locally cut out by one equation, pulled back from a local equation of $E_X Y$ on $\mathrm{Bl}_X Y$. Can you see why this is locally not a zerodivisor?) It can be helpful to note that the top square of the diagram above is a blow-up square, by Exercises 22.2.E and 22.2.F (and the fact that blow-ups can be computed affine-locally).

22.2.7. Blow-up Closure Lemma — *Using the preceding notation: $(\mathrm{Bl}_W Z, E_W Z)$ is canonically isomorphic to $(\overline{Z}, E_{\overline{Z}})$. More precisely: if the blow-up $\mathrm{Bl}_X Y$ exists, then $(\overline{Z}, E_{\overline{Z}})$ is the blow-up of Z along W.*

This will be very useful. We make a few initial comments. The first three apply to the special case where $Z \to Y$ is a closed embedding, and the fourth comment basically tells us we shouldn't have concentrated on this special case.

(1) First, note that if $Z \to Y$ is a closed embedding, then the Blow-Up Closure Lemma states that the proper transform (as defined in §22.2) is the blow-up of Z along the scheme-theoretic intersection $W = X \cap Z$.

(2) In particular, the Blow-Up Closure Lemma lets you actually compute blow-ups, and we will do lots of examples soon. For example, suppose C is a plane curve, singular at a point p, and we want to blow up C at p. Then we could instead blow up the plane at p (which we have already described how to do, even if we haven't yet proved that it satisfies the universal property of blowing up), and then take the scheme-theoretic closure of $C \setminus \{p\}$ in the blow-up.

(3) More generally, if W is some nasty subscheme of Z that we wanted to blow up, and Z is a finite type k-scheme, then the same trick would work. We could work locally (Exercise 22.2.A), so we may assume that Z is affine. If W is cut out by r equations $f_1, \ldots, f_r \in \Gamma(Z, \mathcal{O}_Z)$, then complete the f's to a generating set $f_1, \ldots, f_n$ of $\Gamma(Z, \mathcal{O}_Z)$. This gives a closed embedding of Z into $\mathbb{A}^n$ such that W is the scheme-theoretic intersection of Z with a coordinate linear space $\mathbb{A}^{n-r}$.

(4) Most generally still, this reduces the existence of the blow-up to a specific special case. (If you prefer to work over a fixed field k, feel free to replace $\mathbb{Z}$ by k in this discussion.) Suppose that for each n, $\mathrm{Bl}_{U_{(x_1,\ldots,x_n)}} \mathrm{Spec}\, \mathbb{Z}[x_1, \ldots, x_n]$ exists. Then I claim that the blow-up always exists. Here's why. We may assume that Y is affine, say $\mathrm{Spec}\, B$, and $X = \mathrm{Spec}\, B/(f_1, \ldots, f_n)$. Then we have a morphism $Y \to \mathbb{A}^n_{\mathbb{Z}}$ given by $x_i \mapsto f_i$, such that X is the scheme-theoretic preimage of the origin. Hence by the blow-up closure lemma, $\mathrm{Bl}_X Y$ exists. (We won't follow this approach, however.)

22.2.H. EXERCISE. Prove the Blow-up Closure Lemma 22.2.7. Hint: Obviously, construct maps in both directions, using the universal property. Constructing the following diagram may or may not help. (You will recognize the bottom cube as (22.2.5.1).)

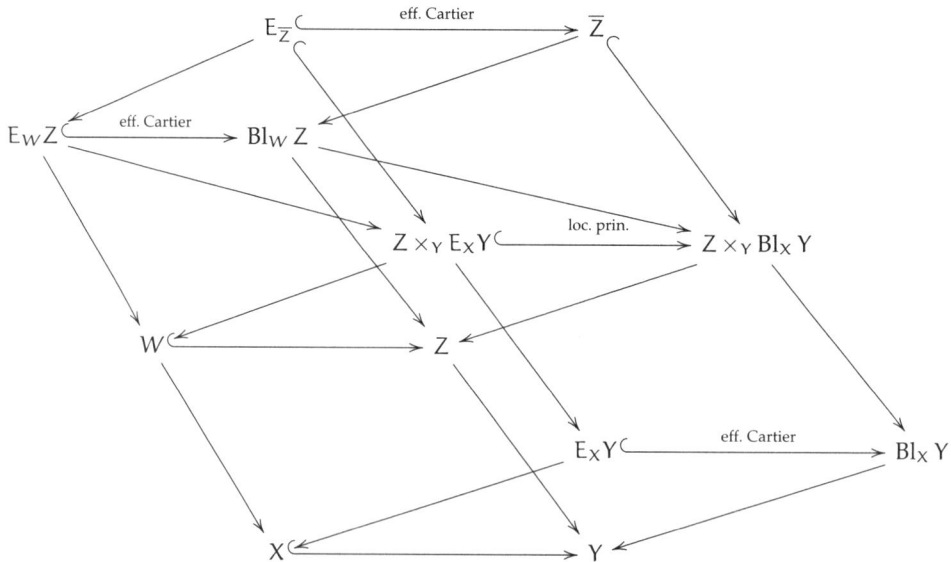

Hooked arrows indicate closed embeddings; and when morphisms are furthermore locally principal or even effective Cartier, they are so indicated. Exercise 11.3.G, on the uniqueness of extension of maps over effective Cartier divisors, may or may not help. Note that if $Z \to Y$ is actually a closed embedding, then so is $Z \times_Y \mathrm{Bl}_X Y \to \mathrm{Bl}_X Y$, and hence also $\overline{Z} \to \mathrm{Bl}_X Y$.

22.3 The Blow-up Exists, and Is Projective

22.3.1. It is now time to show that the blow-up always exists. We will see two arguments, which are enlightening in different ways. Both will imply that the blow-up morphism is projective, and hence quasicompact, proper, finite type, and separated. In particular, if $Y \to Z$ is quasicompact (resp., proper, finite type, separated), so is $\mathrm{Bl}_X Y \to Z$. (And if $Y \to Z$ is projective, and Z is quasicompact, then $\mathrm{Bl}_X Y \to Z$ is projective. See the solution to Exercise 17.3.I for the reason for this annoying extra hypothesis.) The blow-up of a k-variety is a k-variety (using the fact that reducedness is preserved, Exercise 22.2.C), and the blow-up of an irreducible k-variety is an irreducible k-variety (using the fact that irreducibility is preserved, also Exercise 22.2.C).

22.3.2. Theorem ($\mathcal{P}roj$ description of the blow-up) — *Suppose* $X \hookrightarrow Y$ *is a closed subscheme cut out by a finite type quasicoherent sheaf of ideals* $\mathcal{I} \hookrightarrow \mathcal{O}_Y$. *Then*

$$\mathcal{P}roj_Y \left(\mathcal{O}_Y \oplus \mathcal{I} \oplus \mathcal{I}^2 \oplus \mathcal{I}^3 \oplus \cdots \right) \longrightarrow Y$$

satisfies the universal property of blowing up.

(We made sense of products of ideal sheaves, and hence $\mathcal{I}^n$, in Exercise 15.6.E.)

We will prove Theorem 22.3.2 soon (§22.3.3), after seeing what it tells us. Because $\mathcal{I}$ is finite type, the graded sheaf of algebras has degree 1 piece that is finite type. The graded sheaf of algebras is also clearly generated in degree 1. Thus the sheaf of algebras satisfy Hypotheses 17.2.1 ("finite generation in degree 1").

But first, we should make sure that the preimage of X is indeed an effective Cartier divisor. We can work affine-locally (Exercise 22.2.A), so we may assume that $Y = \mathrm{Spec}\, B$, and X is cut out by the finitely generated ideal I. Then

$$\mathrm{Bl}_X Y = \mathcal{P}roj_B \left(B \oplus I \oplus I^2 \oplus \cdots \right).$$

We are slightly abusing notation by using the notation $\mathrm{Bl}_X Y$, as we haven't yet shown that this satisfies the universal property. Recall (§13.9.1) that the graded ring $B \oplus I \oplus \cdots$, denoted by $B(I)_\bullet$, is called the *Rees algebra* of the ideal I in B.

The preimage of X isn't just any effective Cartier divisor; it corresponds to the invertible sheaf $\mathcal{O}(1)$ on this $\mathcal{P}roj$. Indeed, $\mathcal{O}(1)$ corresponds to taking our graded ring, chopping off the bottom piece, and sliding all the graded pieces to the left by 1 (§15.1.4); it is the invertible sheaf corresponding to the graded module

$$IB(I)_\bullet = I \oplus I^2 \oplus I^3 \oplus \cdots$$

(where that first summand I has grading 0). But this can be interpreted as the scheme-theoretic preimage of X, which corresponds to the ideal I of B:

$$IB(I)_\bullet = I \left(B \oplus I \oplus I^2 \oplus \cdots \right) \lhook\joinrel\longrightarrow B \oplus I \oplus I^2 \oplus \cdots = B(I)_\bullet.$$

Thus the scheme-theoretic preimage of $X \hookrightarrow Y$ to

$$\mathcal{P}roj(\mathcal{O}_Y(\mathcal{I})_\bullet) = \mathcal{P}roj(\mathcal{O}_Y \oplus \mathcal{I} \oplus \mathcal{I}^2 \oplus \cdots),$$

corresponding to the ideal sheaf corresponding to $\mathcal{I}\mathcal{O}_Y(\mathcal{I})_\bullet = \mathcal{I} \oplus \mathcal{I}^2 \oplus \mathcal{I}^3 \oplus \cdots$, is an effective Cartier divisor in class $\mathcal{O}(1)$. Once we have verified that this construction indeed yields the blow-up, this divisor will be our exceptional divisor $E_X Y$.

Moreover, we see that the exceptional divisor can be described beautifully as a *Proj* over X:

(22.3.2.1) $E_X Y = \mathcal{P}roj_X \left(\mathcal{O}_Y / \mathcal{I} \oplus \mathcal{I}/\mathcal{I}^2 \oplus \mathcal{I}^2/\mathcal{I}^3 \oplus \cdots \right) = \mathcal{P}roj_X \left(\mathcal{O}_Y(\mathcal{I})_\bullet / (\mathcal{I}\mathcal{O}_Y(\mathcal{I})_\bullet) \right).$

We will later see that in good circumstances (if X is a regular embedding in Y), this is a projectivization of a vector bundle (the "projectivized normal bundle"); see Exercise 22.3.D(a).

22.3.3. Proof of Theorem 22.3.2. Reduce to the case of affine target Y = Spec B, with X corresponding to ideal I ⊂ B. Reduce to the case of affine source Spec R, with principal effective Cartier divisor V(t) (t a non-zerodivisor in R). (A principal effective Cartier divisor is locally cut out by a single non-zerodivisor.) Thus we have reduced to the case $\pi:$ Spec R → Spec B, corresponding to $\phi:$ B → R. Choose $x_1, \ldots, x_n$, with $(x_1, \ldots, x_n) = I$, and $(\phi(x_1), \ldots, \phi(x_n)) = (t)$. (Here we use the hypothesis that the scheme-theoretic pullback of V(I) is V(t), and our understanding of scheme-theoretic pullback from Exercise 10.2.B). Hence

(22.3.3.1) $(\phi(x_1)/t, \ldots, \phi(x_n)/t) = R.$

(Because t is a non-zerodivisor, $\phi(x_i)/t$ makes sense.) We will describe *one* map Spec R → Proj B(I)$_\bullet$ that will extend the map on the open set Spec R$_t$ → Spec B. It is then unique, by Exercise 11.3.G. Consider the ring morphism $\phi': B(I)_\bullet \to R$ sending elements X of degree d to $\phi(X)/t^d$. This induces a map π' from Spec R to the "affine cone" Spec B(I)$_\bullet$.

Let X_i be the degree 1 element of B(I)$_\bullet$ corresponding to x_i. Now ϕ' induces

$$\left((B(I)_\bullet)_{X_i} \right)_0 \to R_{\phi(x_i)/t},$$

and hence, geometrically, $D(\phi(x_i)/t) \to D(X_i)$ from a distinguished open subset of Spec R to a distinguished open subset of the affine cone Spec B(I)$_\bullet$. As x runs through the set $\{x_1, \ldots, x_n\}$, $D(\phi(x_i)/t)$ runs through a distinguished open cover of Spec R (from (22.3.3.1)). Thus the image of $\pi':$ Spec R → Spec B(I)$_\bullet$ misses the vanishing set of the irrelevant ideal $(X_1, \ldots, X_n) \subset$ B(I)$_\bullet$, so the map π' factors through Spec B(I)$_\bullet \setminus V(X_1, \ldots, X_n)$ and hence maps to Proj B(I)$_\bullet$ (everything commuting with morphisms to Spec B):

$$
\begin{array}{ccccc}
\text{Spec } B(I)_\bullet & \longleftarrow & \text{Spec } B(I)_\bullet \setminus V(X_1, \ldots, X_n) & \longrightarrow & \text{Proj } B(I)_\bullet \\
{\scriptstyle \pi'} \uparrow & \nearrow & \downarrow & \swarrow & \\
\text{Spec } R & \xrightarrow{\ \pi\ } & \text{Spec } B. &&
\end{array}
$$

We have constructed our desired morphism Spec R → Proj B(I)$_\bullet$ "lifting" π. □

Here are some applications and observations arising from this construction of the blow-up. First, we can verify that our initial motivational examples are indeed blow-ups. For example, blowing up $\mathbb{A}^2$ (with coordinates x and y) at the origin yields: B = k[x, y], I = (x, y), and Proj(B ⊕ I ⊕ $I^2 \oplus \cdots$) = Proj B[X, Y], where the elements of B have degree 0, and X and Y are degree 1 and "correspond to" x and y, respectively.

22.3.4. *Normal bundles to exceptional divisors.* The normal bundle to an effective Cartier divisor D is the (space associated to the) invertible sheaf $\mathcal{O}(D)|_D$, the invertible sheaf corresponding to the D on the total space, then restricted to D (Exercise 21.2.J). Thus in the case of the blow-up of a point in the plane, the exceptional divisor has normal bundle $\mathcal{O}(-1)$. (As an aside: Castelnuovo's Criterion, Theorem 28.6.1, states that conversely given a smooth surface containing $E \cong \mathbb{P}^1$ with normal bundle $\mathcal{O}(-1)$, E can be blown-down to a point on another smooth surface.) In the case of the blow-up of a regular subvariety of a regular variety, the blow-up turns out to be regular (see Theorem 22.3.10), the exceptional divisor is a projective bundle over X, and the normal bundle to the exceptional divisor restricts to $\mathcal{O}(-1)$ (Exercise 22.3.D).

22.3.A. HARDER BUT ENLIGHTENING EXERCISE. If $i: X \hookrightarrow \mathbb{P}^n$ is a projective scheme, identify the exceptional divisor of the blow-up of the affine cone over X (§9.3.11) at the origin with X itself, and show that its normal bundle (§22.3.4) is isomorphic to $\mathscr{O}_X(-1) := i^* \mathscr{O}_{\mathbb{P}^n}(-1)$. (In the case $X = \mathbb{P}^1$, we recover the blow-up of the plane at a point. In particular, we recover the important fact that the normal bundle to the exceptional divisor is $\mathscr{O}(-1)$.)

22.3.5. *The normal cone.* Motivated by (22.3.2.1), as well as Exercise 22.3.D below, we make the following definition. If X is a closed subscheme of Y cut out by $\mathscr{I}$, then the **normal cone** $N_X Y$ of X in Y is defined as

$$(22.3.5.1) \qquad N_X Y := \mathcal{S}pec_X \left(\mathscr{O}_Y / \mathscr{I} \oplus \mathscr{I}/\mathscr{I}^2 \oplus \mathscr{I}^2/\mathscr{I}^3 \oplus \cdots \right).$$

This can profitably be thought of as an algebro-geometric version of an infinitesimal "tubular neighborhood." But some cautions are in order. If Y is smooth, $N_X Y$ may not be smooth. (You can work out the example of $Y = \mathbb{A}_k^2$ and $X = V(xy)$.)

And even if X and Y are smooth, then although $N_X Y$ is smooth (as we will see shortly, Exercise 22.3.D), it doesn't "embed" in any way in Y. As a simple example, consider a smooth point $X = p$ of a non-rational curve $Y = C$. Then $N_X Y$ is rational and cannot admit an open embedding into Y.

22.3.6. *The tangent cone.* If X is a closed point p, then the normal cone is called the **tangent cone** to Y at p. The **projectivized tangent cone** is the exceptional divisor $E_X Y$ (the *Proj* of the same graded sheaf of algebras). Following §9.3.12, the tangent cone and the projectivized tangent cone can be put together in the projective completion of the tangent cone, which contains the tangent cone as an open subset, and the projectivized tangent cone as a complementary effective Cartier divisor.

In Exercise 22.3.D, we will see that at a regular point of Y, the tangent cone may be identified with the tangent space, and the normal cone may often be identified with the total space of the normal bundle.

22.3.B. EXERCISE. Suppose $Y = \operatorname{Spec} k[x, y]/(y^2 - x^2 - x^3)$ (the nodal cubic; see the bottom of Figure 8.4). Assume (to avoid distraction) that $\operatorname{char} k \neq 2$. Show that the tangent cone to Y at the origin is isomorphic to $\operatorname{Spec} k[x, y]/(y^2 - x^2)$. Thus, informally, the tangent cone "looks like" the original variety "infinitely magnified."

22.3.C. EXERCISE. Suppose $S_\bullet$ is a finitely generated graded algebra over a field k. Exercise 22.3.A gives an isomorphism of $\operatorname{Proj} S_\bullet$ with the exceptional divisor to the blow-up of $\operatorname{Spec} S_\bullet$ at the origin. Show that the tangent cone to $\operatorname{Spec} S_\bullet$ at the origin is isomorphic to $\operatorname{Spec} S_\bullet$ itself. (Your geometric intuition should lead you to find these facts believable.)

22.3.7. Blowing up regular embeddings.

The case of blow-ups of regular embeddings $X \subset Y$ is particularly pleasant. For example, the exceptional divisor is a projective bundle over X.

22.3.8. Theorem — *If $I \subset A$ is generated by a regular sequence $a_1, \ldots, a_d$, then the natural map $\operatorname{Sym}_A^n(I/I^2) \to I^n/I^{n+1}$ is an isomorphism.*

22.3.9. Corollary — *If a closed embedding $i : X \hookrightarrow Y$ is a regular embedding with ideal sheaf $\mathscr{I} \subset \mathscr{O}_Y$, then the natural map $\operatorname{Sym}^n(\mathscr{I}/\mathscr{I}^2) \to \mathscr{I}^n/\mathscr{I}^{n+1}$ is an isomorphism. Furthermore, in combination with Proposition 21.2.16, we see that $\mathscr{I}^n/\mathscr{I}^{n+1}$ is a locally free sheaf.*

Exercise 24.6.E gives a somewhat simpler proof of Theorem 22.3.8 (and Corollary 22.3.9) in the case where A is a k-algebra.

Before starting the proof of Theorem 22.3.8 in §22.3.11, we show its utility.

22.3.D. EXERCISE.

(a) Suppose $X \to Y$ is a regular embedding with ideal sheaf $\mathscr{I}$. Identify the total space (§17.1.5) of the normal sheaf (the "normal bundle") with the normal cone $N_X Y$ (22.3.5.1), and show that the exceptional divisor $E_X Y$ is a projective bundle (the "projectivized normal bundle") over X.

(b) Show that the normal bundle to $E_X Y$ in $Bl_X Y$ is $\mathcal{O}(-1)$ (for the projective bundle over X).

(c) Assume further that X is a reduced closed point p. Show that p is a regular point of Y. Identify the total space of the tangent space to p with the tangent cone to Y at p.

22.3.10. Theorem — *Suppose $X \hookrightarrow Y$ is a closed embedding of smooth varieties over k. Then $Bl_X Y$ is also smooth.*

Proof. We use the fact that smooth varieties are regular, the Smoothness-Regularity Comparison Theorem 13.2.7(b), whose proof we still have to complete.

We may assume that $k = \bar{k}$, because the statement of the result is preserved by base change to the algebraic closure. Reason: Smoothness can be checked after base change to $\bar{k}$, by Exercise 13.2.H. The blow-up construction commutes with base change to the algebraic closure: you can examine the construction, or wait until you can invoke the generalization in Exercise 24.1.P(a)

We need only check smoothness of $Bl_X Y$ at the points of $E_X Y$, by Observation 22.2.2. By Exercise 13.2.M(b), $X \hookrightarrow Y$ is a regular embedding. Then by Exercise 22.3.D(a), $E_X Y$ is a projective bundle over X, and thus smooth, and hence regular at its closed points. But $E_X Y$ is an effective Cartier divisor on $Bl_X Y$. By the slicing criterion for regularity (Exercise 13.2.C), it follows that $Bl_X Y$ is regular at the closed points of $E_X Y$, hence smooth at all points of $E_X Y$. □

22.3.11.* *Proving Theorem 22.3.8.*
The proof of Theorem 22.3.8 may reasonably be skipped on a first reading. We prove Theorem 22.3.8 following [F2, A.6.1], which in turn follows [Dav]. The proof will be completed in §22.3.13. To begin, let α be the map of graded rings

$$\alpha: (A/I)[X_1, \ldots, X_d] \longrightarrow \oplus_{n=0}^{\infty} I^n / I^{n+1},$$

which takes X_i to the image of a_i in I/I^2. Clearly, α is surjective.

22.3.E. EXERCISE. Show that Theorem 22.3.8 would follow from the statement that α is an isomorphism.

Because a_1 is a non-zerodivisor, we can interpret $A[a_2/a_1, \ldots, a_d/a_1]$ as a subring of the total fraction ring (defined in §6.6.35). In particular, as A is a subring of its total fraction ring, the map $A \to A[a_2/a_1, \ldots, a_d/a_1]$ is an injection. Define

$$\beta: A[T_2, \ldots, T_d] \longrightarrow A[a_2/a_1, \ldots, a_d/a_1]$$

by $T_i \mapsto a_i/a_1$. Clearly, the map β is surjective, and $L_i := a_1 T_i - a_i$ lies in $\ker(\beta)$.

22.3.12. Lemma — *The kernel of β is $(L_2, \ldots, L_d)$.*

Proof. We prove the result by induction on d. We consider first the base case $d = 2$. Suppose $F[T_2] \in \ker \beta$, so $F(a_2/a_1) = 0$. Then applying the algorithm for the Remainder Theorem, dividing $a_1^{\deg F} F(T_2)$ by $a_1 T_2 - a_2 = L_2$,

(22.3.12.1) $$a_1^{\deg F} F(T_2) = G(T_2)(a_1 T_2 - a_2) + R,$$

where $G(T_2) \in A[T_2]$, and $R \in A$ is the remainder. Substituting $T_2 = a_2/a_1$ (and using the fact that $A \to A[a_2/a_1]$ is injective), we have that $R = 0$. Then $(a_1 T_2 - a_2) G(T_2) \equiv 0 \pmod{(a_1^{\deg F})}$. Using the fact that a_2 is a non-zerodivisor modulo $(a_1^{\deg F})$ (as $a_1^{\deg F}, a_2$ is a regular sequence by Exercise 9.5.E), a short induction shows that the coefficients of $G(T_2)$ must all be divisible by $a_1^{\deg F}$. Thus $F(T_2)$ is divisible by $a_1 T_2 - a_2 = L_2$, so the case $d = 2$ is proved.

We now consider the general case $d > 2$, assuming the result for all smaller d. Let $A' = A[a_2/a_1]$. Then $a_1, a_3, a_4, \ldots, a_d$ is a regular sequence in A'. Reason: a_1 is a non-zerodivisor in

A'. $A'/(a_1) = (A[T_2]/(a_1T_2 - a_2))/(a_1) = A[T_2]/(a_1T_2 - a_2, a_1) = A[T_2]/(a_1, a_2)$. Then a_3 is a non-zerodivisor in this ring because it is a non-zerodivisor in $A/(a_1, a_2)$, a_4 is a non-zerodivisor in $A[T_2]/(a_1, a_2, a_3)$ because it is a non-zerodivisor in $A/(a_1, a_2, a_3)$, and so forth. Condition (ii) of the Definition 9.5.4 of regular sequence also holds, as

$$A'/(a_1, a_3, a_4, \ldots, a_d) = A[T_2]/(a_1, a_2, \ldots, a_d) \neq 0.$$

Consider the composition

$$A[T_2, \ldots, T_d] \to A'[T_3, \ldots, T_d] \longrightarrow A'[a_3/a_1, \ldots, a_d/a_1] = A[a_2/a_1, \ldots, a_d/a_1].$$

By the case $d = 2$, the kernel of the first map is L_2. By the inductive hypothesis, the kernel of the second map is $(L_3, \ldots, L_d)$. The result follows. □

22.3.13. *Proof of Theorem 22.3.8.* By Exercise 22.3.E, it suffices to prove that the surjection α is an isomorphism. Suppose $F \in \ker(\alpha)$; we wish to show that $F = 0$. We may assume that F is homogeneous, say of degree n. Consider the map $\alpha' : A[X_1, \ldots, X_d] \to \oplus_{n=0}^{\infty} I^n/I^{n+1}$ lifting α. Lift F to $A[X_1, \ldots, X_d]$, so $F \in \ker(\alpha')$. We wish to show that $F \in IA[X_1, \ldots, X_d]$. Suppose $F(a_1, \ldots, a_d) = x \in I^{n+1}$. Then we can write x as $F'(a_1, \ldots, a_d)$, where F' is a homogeneous polynomial of the same degree as F, with coefficients in I. Then by replacing F by $F - F'$, we are reduced to the following problem: suppose $F \in A[X_1, \ldots, X_d]$ is homgeneous of degree n, and $F(a_1, \ldots, a_d) = 0$, we wish to show that $F \in IA[X_1, \ldots, X_n]$. But if $F(a_1, \ldots, a_d) = 0$, then $F(1, a_2/a_1, \ldots, a_d/a_1) = 0$ in $A[a_2/a_1, \ldots, a_d/a_1]$. But Lemma 22.3.12 identifies the kernel of β, so

$$F(1, T_2, T_3, \ldots, T_d) \in (a_1T_2 - a_2, a_1T_3 - a_3, \ldots, a_1T_d - a_d).$$

Thus the coefficients of F are in $(a_1, \ldots, a_d) = I$, as desired. □

22.4 Examples and Computations

In this section we will work through a number of explicit of examples, to get a sense of how blow-ups behave, how they are useful, and how one can work with them explicitly. **Throughout we work over a field** k, **and we assume throughout that** char $k = 0$ **to avoid distraction.** The examples and exercises are loosely arranged by topic, but not in order of importance.

22.4.1. Example: Blowing up the plane along the origin.

Let's first blow up the plane $\mathbb{A}^2$ along the origin, and see that the result agrees with our discussion in §22.1. Let x and y be the coordinates on $\mathbb{A}^2$. The blow-up is $\operatorname{Proj} k[x, y, X, Y]$, where $xY - yX = 0$. (Here x and y have degree 0 and X and Y have degree 1.) This is naturally a closed subscheme of $\mathbb{A}^2 \times \mathbb{P}^1$, cut out (in terms of the projective coordinates X and Y on $\mathbb{P}^1$) by $xY - yX = 0$. We consider the two usual patches on $\mathbb{P}^1$: $[X, Y] = [s, 1]$ and $[1, t]$. The first patch yields $\operatorname{Spec} k[x, y, s]/(sy - x)$, and the second gives $\operatorname{Spec} k[x, y, t]/(y - xt)$. Notice that both are smooth: the first is $\operatorname{Spec} k[y, s] \cong \mathbb{A}^2$, and the second is $\operatorname{Spec} k[x, t] \cong \mathbb{A}^2$.

We now describe the exceptional divisor. We first consider the first (s) patch. The ideal is generated by (x, y), which in our ys-coordinates is $(ys, y) = (y)$, which is indeed principal. Thus on this patch the exceptional divisor is generated by y. Similarly, in the second patch, the exceptional divisor is cut out by x. (This can be a little confusing, but there is no contradiction!) This explicit description will be useful in working through some of the examples below.

22.4.A. EXERCISE. Let p be a k-valued point of $\mathbb{P}^2$. Exhibit an isomorphism between $\operatorname{Bl}_p \mathbb{P}^2$ and the Hirzebruch surface $\mathbb{F}_1 = \mathbb{P}_{\mathbb{P}^1}(\mathcal{O}_{\mathbb{P}^1} \oplus \mathcal{O}_{\mathbb{P}^1}(1))$ (Example 17.2.6). (The map $\operatorname{Bl}_p \mathbb{P}^2 \to \mathbb{P}^1$ informally corresponds to taking a point to the line connecting it to the origin. Do not be afraid: You can do this by explicitly working with coordinates.)

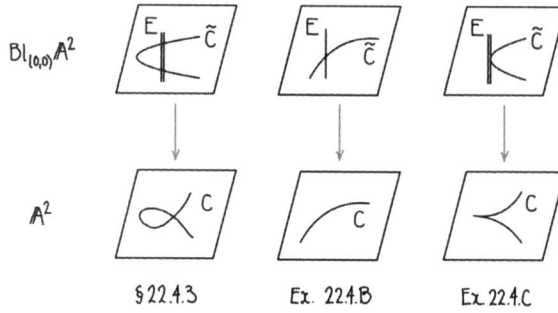

Figure 22.1 *Resolving curve singularities.*

22.4.2. Resolving singularities.

22.4.3. *The proper transform of a nodal curve (Figure 22.1).* (You may wish to flip to Figure 8.4 while thinking through this exercise.) Consider next the curve $y^2 = x^3 + x^2$ inside the plane $\mathbb{A}^2$ (the nodal cubic, see Exercise 22.3.B). Let's blow up the origin, and compute the total and proper transform of the curve. (By the Blow-up Closure Lemma 22.2.7, the latter is the blow-up of the nodal curve at the origin.) In the first patch, we get $y^2 - s^2 y^2 - s^3 y^3 = 0$. This factors: we get the exceptional divisor y with multiplicity 2, and the curve $1 - s^2 - s^3 y = 0$. You can easily check that the proper transform is regular. Also, notice that the proper transform $\tilde{C}$ meets the exceptional divisor at two points, $s = \pm 1$. This corresponds to the two tangent directions at the origin (as $s = x/y$). (You should do the second patch yourself.)

22.4.B. EXERCISE (FIGURE 22.1). Describe both the total and the proper transform of the curve C given by $y = x^2 - x$ in $\mathrm{Bl}_{(0,0)} \mathbb{A}^2$. Show that the proper transform of C is isomorphic to C. Interpret the intersection of the proper transform of C with the exceptional divisor E as the slope of C at the origin.

22.4.C. EXERCISE: BLOWING UP A CUSPIDAL PLANE CURVE (CF. EXERCISE 10.7.F). Describe the proper transform of the cuspidal curve C given by $y^2 = x^3$ in the blown-up plane $\mathrm{Bl}_{(0,0)} \mathbb{A}^2$. Show that it is regular. Show that the proper transform of C meets the exceptional divisor E at one point, and is tangent to E there.

The previous two exercises are the first in an important sequence of singularities, which we now discuss.

22.4.D. EXERCISE: RESOLVING A_n CURVE SINGULARITIES. Resolve the singularity $y^2 = x^{n+1}$ in $\mathbb{A}^2$, by first blowing up its singular point, then considering its proper transform and deciding what to do next.

22.4.4. *Toward a definition of A_n curve singularities.* You will notice that your solution to Exercise 22.4.D depends only on the "power series expansion" of the singularity at the origin, and not on the precise equation. For example, if you compare §22.4.3 with the $n = 1$ case of Exercise 22.4.D, you will see that they are "basically the same." We will make this precise in Definition 28.2.C.

22.4.E. EXERCISE (WARM-UP TO EXERCISE 22.4.F). Blow up the cone point $z^2 = x^2 + y^2$ (Figure 3.4) at the origin. Show that the resulting surface is regular. Show that the exceptional divisor is isomorphic to $\mathbb{P}^1$. (Remark: You can check that the normal bundle to this $\mathbb{P}^1$ is not $\mathscr{O}_{\mathbb{P}^1}(-1)$, as is the case when you blow up a point on a smooth surface; see §22.3.4; it is $\mathscr{O}_{\mathbb{P}^1}(-2)$. Doesn't this contradict Exercise 22.3.D(b)?)

22.4.F. EXERCISE (RESOLVING A_n SURFACE SINGULARITIES). Resolve the singularity $z^2 = y^2 + x^{n+1}$ in $\mathbb{A}^3$ by first blowing up its singular point, then considering its proper transform, and deciding what to do next. (A k-surface singularity analytically isomorphic to this is called an

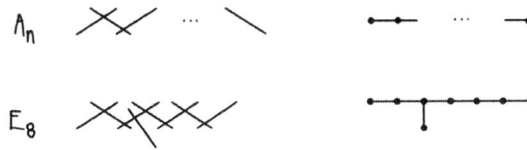

Figure 22.2 *The exceptional divisors for resolutions of some ADE surface singularities, and their corresponding dual graphs (see Remark 22.4.5).*

A_n **surface singularity**. For example, the cone shown in Figure 3.4 is an A_1 surface singularity. We make this precise in Exercise 28.2.C.) This exercise is a bit time-consuming, but is rewarding in that it shows that you can really resolve singularities by hand.

22.4.5. *Remark: ADE-surface singularities and Dynkin diagrams (see Figure 22.2).* A k-singularity analytically isomorphic to $z^2 = x^2 y + y^{n-1}$ (resp., $z^2 = x^3 + y^4$, $z^2 = x^3 + xy^3$, $z^2 = x^3 + y^5$) is called a D_n surface singularity (resp., E_6, E_7, E_8 surface singularity). We will make this precise in Exercise 28.2.C, and you will then be able to guess the definition of the corresponding curve singularity. If you (minimally) desingularize each of these surfaces by sequentially blowing up singular points as in Exercise 22.4.F, and look at the arrangement of exceptional divisors (the various exceptional divisors and how they meet), you will discover the corresponding Dynkin diagram. More precisely, if you create a graph where the vertices correspond to exceptional divisors, and two vertices are joined by an edge if the two divisors meet, you will find the underlying graph of the corresponding Dynkin diagram. This is the start of several very beautiful stories; see Remark 27.3.7 for a first glimpse of one of them.

22.4.6. *Remark: Resolution of singularities.* Hironaka's Theorem on resolution of singularities implies that this idea of trying to resolve singularities by blowing up singular loci in general can succeed in characteristic 0 (see [Hir], and [Ko2]). (The case of dimension 1 will be shown in §28.4.4, and the case of dimension 2 will be discussed in §28.6.4.) It is not known if an analogous statement is true in positive characteristic (except in dimension at most 3; see [CP]), but de Jong's Alteration Theorem [dJ] gives a result that is good enough for most applications. Rather than producing a birational proper map $\widetilde{X} \to X$ from something regular, it produces a proper map from something regular that is generically finite (and the corresponding extension of function fields is separable).

Here are some other exercises related to resolution of singularities.

22.4.G. EXERCISE. Blowing up a nonreduced subscheme of a regular scheme can give you something singular, as shown in this example. Describe the blow-up of the closed subscheme $V(y, x^2)$ in $\operatorname{Spec} k[x, y] = \mathbb{A}^2$. Show that you get an A_1 surface singularity.

22.4.H. EXERCISE. Desingularize the tacnode (see Exercise 10.7.G and Definition 28.2.2) $y^2 = x^4$, not in two steps (as in Exercise 22.4.D), but in a single step by blowing up (y, x^2).

22.4.I. EXERCISE (RESOLVING A SINGULARITY BY AN UNEXPECTED BLOW-UP). Suppose Y is the cone $x^2 + y^2 = z^2$, and X is the line cut out by $x = 0$, $y = z$ on Y. Show that $\operatorname{Bl}_X Y$ is regular. (In this case we are blowing up a codimension 1 locus that is not an effective Cartier divisor; see Problem 13.1.3. But it *is* an effective Cartier divisor away from the cone point, so you should expect your answer to be an isomorphism away from the cone point.)

22.4.7. *Multiplicity of a function at a regular point.* In order to pose Exercise 22.4.J, we introduce a useful concept. If f is a function on a locally Noetherian scheme X, its **multiplicity at a regular point** p is the largest m such that f lies in the mth power of the maximal ideal in the local ring $\mathscr{O}_{X,p}$. For example, if $f \neq 0$, $V(f)$ is singular at p if and only if $m > 1$. (Do you see why?)

22.4.J. EXERCISE. Show that the multiplicity of the exceptional divisor in the total transform of a subscheme Z of $\mathbb{A}^n$ when you blow up the origin is the smallest multiplicity (at the origin) of a defining equation of Z. (For example, in the case of the nodal and cuspidal curves above,

Example 22.4.3 and Exercise 22.4.C, respectively, the exceptional divisor appears with multiplicity 2.) Caution: You need to make this question precise before solving it.

22.4.8. Resolving rational maps.

22.4.K. EXERCISE (UNDERSTANDING THE BIRATIONAL MAP $\mathbb{P}^2 \dashrightarrow \mathbb{P}^1 \times \mathbb{P}^1$ VIA BLOW-UPS). Let p and q be two distinct k-points of $\mathbb{P}^2$, and let r be a k-point of $\mathbb{P}^1 \times \mathbb{P}^1$. Describe an isomorphism $\mathrm{Bl}_{\{p,q\}} \mathbb{P}^2 \xrightarrow{\sim} \mathrm{Bl}_r \mathbb{P}^1 \times \mathbb{P}^1$. (Possible hint: Consider lines ℓ through p and m through q; the choice of such a pair corresponds to a point of $\mathbb{P}^1 \times \mathbb{P}^1$. A point s of $\mathbb{P}^2$ not on line $\overline{pq}$ yields a pair of lines $(\overline{ps}, \overline{qs})$ of $\mathbb{P}^1 \times \mathbb{P}^1$. Conversely, a choice of lines (ℓ, m) such that neither ℓ nor m is line $\overline{pq}$ yields a point $s = \ell \cap m \in \mathbb{P}^2$. This describes a birational map $\mathbb{P}^2 \dashrightarrow \mathbb{P}^1 \times \mathbb{P}^1$. Exercise 22.4.A is related.)

Exercise 22.4.K is an example of the general phenomenon explored in the next two exercises.

22.4.L. HARDER BUT USEFUL EXERCISE (BLOW-UPS RESOLVE BASE LOCI OF RATIONAL MAPS TO PROJECTIVE SPACE). Suppose we have a scheme Y, an invertible sheaf $\mathscr{L}$, and a number of sections $s_0, \dots, s_n$ of $\mathscr{L}$ (a *linear series*, Definition 15.2.1). Then away from the closed subscheme X cut out by $s_0 = \dots = s_n = 0$ (the base locus of the linear series), these sections give a morphism to $\mathbb{P}^n$. Show that this morphism extends uniquely to a morphism $\mathrm{Bl}_X Y \to \mathbb{P}^n$, where this morphism corresponds to the invertible sheaf $(\beta^* \mathscr{L})(-E_X Y)$, where $\beta \colon \mathrm{Bl}_X Y \to Y$ is the blow-up morphism. In other words, "blowing up the scheme-theoretic base locus resolves this rational map." Hint: It suffices to consider an affine open subset of Y where $\mathscr{L}$ is trivial. Uniqueness might use Exercise 11.3.G.

22.4.9. *Remarks.* (i) Exercise 22.4.L immediately implies that blow-ups can be used to resolve rational maps to projective schemes $Y \dashrightarrow Z \hookrightarrow \mathbb{P}^n$.

(ii) The following interpretation is enlightening. The linear series on Y pulls back to a linear series on $\mathrm{Bl}_X Y$, and the base locus of the linear series on Y pulls back to the base locus on $\mathrm{Bl}_X Y$. The base locus on $\mathrm{Bl}_X Y$ is $E_X Y$, an effective Cartier divisor. Because $E_X Y$ is not just locally principal, but also locally a non-zerodivisor, it can be "divided out" from the $\beta^* s_i$ (yielding a section of $(\beta^* \mathscr{L})(-E_X Y)$), thereby removing the base locus, and leaving a base-point-free linear series. (In a sense that can be made precise through the universal property, this is the smallest "modification" of Y that can remove the base locus.) If X is already Cartier (as, for example, happens with any nontrivial linear series if Y is a regular pure-dimensional curve), then we can remove a base locus by just "dividing out X."

(iii) You may wish to revisit Exercise 19.7.C, and interpret it in terms of Exercise 22.4.L.

22.4.10. *Examples.* (i) The rational map $\mathbb{P}^n \dashrightarrow \mathbb{P}^{n-1}$ given by $[x_0, \dots, x_n] \dashrightarrow [x_1, \dots, x_n]$, defined away from $p = [1, 0, \dots, 0]$, is resolved by blowing up p. Then by the Blow-up Closure Lemma 22.2.7, if Y is any locally closed subscheme of $\mathbb{P}^n$, we can project to $\mathbb{P}^{n-1}$ once we blow up p in Y, and the invertible sheaf giving the map to $\mathbb{P}^{n-1}$ is (somewhat informally speaking) $\beta^*(\mathscr{O}_{\mathbb{P}^n}(1)) \otimes \mathscr{O}(-E_p Y)$.

(ii) Consider two general homogeneous cubic functions C_1 and C_2 in three variables, yielding two cubic curves in $\mathbb{P}^2$. They are smooth (by Bertini's Theorem 13.4.2), and meet in nine points $p_1, \dots, p_9$ (using our standing assumption that we work over an algebraically closed field). Then $[C_1, C_2]$ gives a rational map $\mathbb{P}^2 \dashrightarrow \mathbb{P}^1$. To resolve the rational map, we blow up $p_1, \dots, p_9$. The result is (generically) an *elliptic fibration* $\mathrm{Bl}_{\{p_1,\dots,p_9\}} \mathbb{P}^2 \to \mathbb{P}^1$. (This is by no means a complete argument or even a precise statement.)

(iii) Fix six general points $p_1, \dots, p_6$ in $\mathbb{P}^2$. There is a four-dimensional vector space of cubics vanishing at these points, and they vanish scheme-theoretically precisely at these points. This yields a rational map $\mathbb{P}^2 \dashrightarrow \mathbb{P}^3$, which is resolved by blowing up the six points. The resulting morphism turns out to be a closed embedding, and the image in $\mathbb{P}^3$ is a (smooth) cubic surface. This is the famous fact that the blow-up of the plane at six general points may be represented as a (smooth) cubic in $\mathbb{P}^3$. (Again, this argument is not intended to be complete.) See §27.3.5 for a more precise and more complete discussion.

In reasonable circumstances, Exercise 22.4.L has an interpretation in terms of graphs of rational maps.

22.4.M. EXERCISE. Suppose $s_0, \ldots, s_n$ are sections of an invertible sheaf $\mathscr{L}$ on an integral scheme X, not all 0. By Remark 15.2.5, these data give a rational map $\phi \colon X \dashrightarrow \mathbb{P}^n$. Give an isomorphism between the graph of ϕ (§11.3.4) and $\mathrm{Bl}_{V(s_0,\ldots,s_n)} X$. (Your argument will not require working over a field k; it will work in general.)

You may enjoy exploring the previous idea by working out how the Cremona transformation $\mathbb{P}^2 \dashrightarrow \mathbb{P}^2$ (Exercise 7.5.J) can be interpreted in terms of the graph of the rational map $[x, y, z] \dashrightarrow [1/x, 1/y, 1/z]$.

22.4.N.* EXERCISE. Resolve the rational map

$$\operatorname{Spec} k[w, x, y, z]/(wz - xy) \xrightarrow{\ \ \ \ \ [w,x]\ \ \ \ \ } \mathbb{P}^1$$

from the cone over the quadric surface to the projective line. Let X be the resulting variety, and $\pi \colon X \to \operatorname{Spec} k[w, x, y, z]/(wz - xy)$ the projection to the cone over the quadric surface. Show that π is an isomorphism away from the cone point, and that the preimage of the cone point is isomorphic to $\mathbb{P}^1$ (and thus has codimension 2, and therefore is different from the resolution obtained by simply blowing up the cone point). Possible hint: If Q is the quadric in $\mathbb{P}^3$ cut out by $wz - xy = 0$, then factor the rational map as $\operatorname{Spec} k[w, x, y, z]/(wz - xy) \setminus \{0\} \to Q$ (cf. Exercise 9.3.N), followed by the isomorphism $Q \xrightarrow{\sim} \mathbb{P}^1 \times \mathbb{P}^1$ (Example 10.6.2), followed by projection onto one of the factors.

This is an example of a small resolution. (A **small resolution** $X \to Y$ is a resolution where the locus of points of Y where the fiber has dimension r is of codimension greater than $2r$. We will not use this notion again in any essential way. Caution: This is the definition of "small resolution" used in intersection homology; in the minimal model program, a small resolution is a resolution that is an isomorphism in codimension 1. So pay attention to the context of what you are reading.) Notice that this resolution of the morphism involves blowing up the base locus $w = x = 0$, which is a cone over one of the lines on the quadric surface $wz = xy$. We are blowing up an effective Weil divisor, which is necessarily not Cartier as the blow-up is not an isomorphism. In Exercise 13.1.E, we saw that (w, x) was not principal, while here we see that (w, x) is not even locally principal. Essentially by Exercise 15.5.N, $V(w, x)$ cannot even be the support of a locally principal divisor.

22.4.11. *Remark: Nonisomorphic small resolutions.* If you instead resolved the map $[w, y]$, you would obtain a similar looking small resolution

$$\pi' \colon X' \longrightarrow \operatorname{Spec} k[w, x, y, z]/(wz - xy)$$

(it is an isomorphism away from the origin, and the fiber over the origin is $\mathbb{P}^1$). But it is different! More precisely, there is no morphism $X \to X'$ making the following the diagram commute.

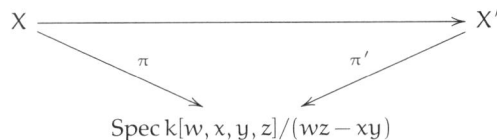

The birational map $X \dashrightarrow X'$ is called the **Atiyah flop**, [At1].

22.4.12. *Factorization of birational maps.* We end our discussion of resolution of rational maps by noting that just as Hironaka's Theorem states that one may resolve all singularities of varieties in characteristic 0 by a sequence of blow-ups along smooth subvarieties (§22.4.6), the *Weak Factorization Theorem* (first proved by Włodarczyk [Wł]) states that any two birational varieties X and Y in characteristic 0 may be related by blow-ups and "blow-downs" along smooth subvarieties. More precisely, there are varieties $X_0, \ldots, X_n$, $X_{01}, \ldots, X_{(n-1)n}$, with $X_0 = X$ and $X_n = Y$, with morphisms $X_{i(i+1)} \to X_i$ and $X_{i(i+1)} \to X_{i+1}$ ($0 \le i < n$) that are blow-ups of smooth subvarieties.

22.4.13. Blow-ups and line bundles.

22.4.O. EXERCISE (GENERALIZING EXERCISE 20.2.D). Suppose X is a regular projective surface over k, and p is a k-valued point. Let $\beta \colon \mathrm{Bl}_p X \to X$ be the blow-up morphism, and let $E = E_p X$ be the exceptional divisor. Consider the exact sequence

$$\mathbb{Z} \xrightarrow{\gamma \colon 1 \mapsto [E]} \mathrm{Pic}(\mathrm{Bl}_p X) \xrightarrow{\alpha} \mathrm{Pic}(\mathrm{Bl}_p X \setminus E) \longrightarrow 0$$

from (15.4.12.1). (Technically, (15.4.12.1) has Pic replaced by Cl. But because regular local rings are unique factorization domains by §13.8.5, Pic = Cl in this case by Proposition 15.4.16.) Note that $\mathrm{Bl}_p X \setminus E = X \setminus p$. Show that $\mathrm{Pic}(X \setminus p) = \mathrm{Pic}\, X$. Show that $\beta^* \colon \mathrm{Pic}\, X \to \mathrm{Pic}\, \mathrm{Bl}_p X$ gives a section to α. Use §22.3.4 to show that $\mathscr{O}_X(E)|_E \cong \mathscr{O}_E(-1)$ (so in the language of Chapter 20, E is a (-1)-curve, Definition 20.2.8), and from that show that γ is an injection. Conclude that $\mathrm{Pic}\, \mathrm{Bl}_p X \cong \mathrm{Pic}\, X \oplus \mathbb{Z}$. Describe how to find the intersection matrix on $N^1_{\mathbb{Q}}(\mathrm{Bl}_p X)$ from that of $N^1_{\mathbb{Q}}(X)$.

22.4.P. EXERCISE. Suppose D is an effective Cartier divisor (a curve) on X. Let $\mathrm{mult}_p D$ be the multiplicity of D at p (Exercise 22.4.J), and let D^{pr} be the proper transform of D. Show that $\beta^* D = D^{\mathrm{pr}} + (\mathrm{mult}_p D)E$ as effective Cartier divisors. More precisely, show that the product of the local equation for D^{pr} and the $(\mathrm{mult}_p D)$th power of the local equation for E is the local equation for $\beta^* D$, and hence that (i) $\beta^* D$ is an effective Cartier divisor, and (ii) $\beta^* \mathscr{O}_X(D) \cong \mathscr{O}_{\mathrm{Bl}_p X}(D^{\mathrm{pr}}) \otimes \mathscr{O}_{\mathrm{Bl}_p X}(E)^{\otimes(\mathrm{mult}_p D)}$. (A special case is the equation $\ell = e + m$ in Hint 20.2.7.)

22.4.14. *Change of the canonical line bundle under blow-ups.*
As motivation for how the canonical line bundle changes under blowing up, consider the blow-up $\beta \colon \mathrm{Bl}_{(0,0)} \mathbb{A}^2 \to \mathbb{A}^2$. Let $X = \mathrm{Bl}_{(0,0)} \mathbb{A}^2$ and $Y = \mathbb{A}^2$ for convenience. We use Exercise 21.4.A to relate $\beta^* \mathscr{K}_Y$ to $\mathscr{K}_X$.

We pick a generator for $\mathscr{K}_Y$ near $(0,0)$: $dx \wedge dy$. (This is, in fact, a generator for $\mathscr{K}_Y$ everywhere on $\mathbb{A}^2$, but for the sake of generalization, we point out that all that matters is that it is a generator at $(0,0)$, and hence *near* $(0,0)$ by geometric Nakayama, Exercise 14.3.D.) When we pull it back to X, we can interpret it as a section of $\mathscr{K}_X$, which will generate $\mathscr{K}_X$ away from the exceptional divisor E, but may contain E with some multiplicity μ. Recall that X can be interpreted as the data of a point in $\mathbb{A}^2$ as well as the choice of a line through the origin. We consider the open subset U where the line is not vertical, and thus can be written as $y = mx$. Here we have natural coordinates: $U = \mathrm{Spec}\, k[x, y, m]/(y - mx)$, which we can interpret as $\mathrm{Spec}\, k[x, m]$. The exceptional divisor E meets U, at $x = 0$ (in the coordinates on U), so we can calculate μ on this open set. Pulling back $dx \wedge dy$ to U, we get

$$dx \wedge dy = dx \wedge d(xm) = m(dx \wedge dx) + x(dx \wedge dm) = x(dx \wedge dm)$$

as $dx \wedge dx = 0$. Thus $\beta^*(dx \wedge dy)$ vanishes to order 1 along E.

22.4.Q. EXERCISE (CF. UNIMPORTANT EXERCISE 21.4.J). Explain how this determines a canonical isomorphism $\mathscr{K}_X \xleftrightarrow{\sim} (\beta^* \mathscr{K}_Y)(E)$.

22.4.R. EXERCISE. Repeat the above calculation in dimension n. Show that the exceptional divisor appears with multiplicity $(n-1)$.

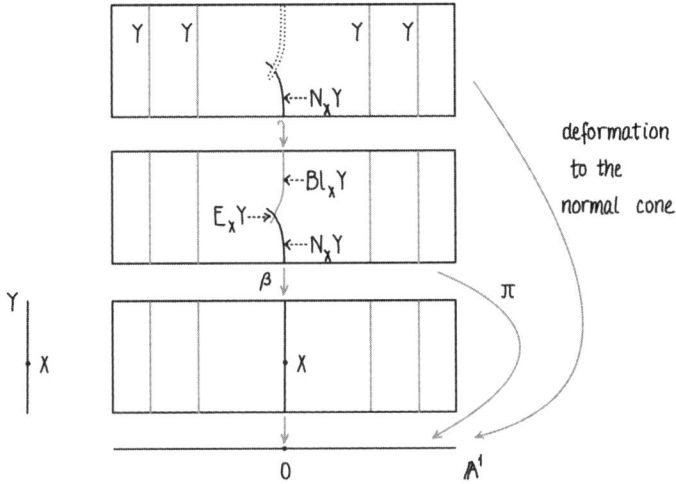

Figure 22.3 *Deformation to the normal cone (§22.4.17).*

22.4.S. EXERCISE. Suppose k is perfect.

(a) Suppose Y is a surface over k, and p is a regular k-valued point, and let $\beta \colon X \to Y$ be the blow-up of Y at p. Show that $\mathscr{K}_X \cong (\beta^* \mathscr{K}_Y)(E)$. Hint: To find a generator of $\mathscr{K}_X$ near p, choose generators $\overline{x}$ and $\overline{y}$ of $\mathfrak{m}/\mathfrak{m}^2$ (where $\mathfrak{m}$ is the maximal ideal of $\mathscr{O}_{Y,p}$), and lift them to elements of $\mathscr{O}_{X,p}$. Why does $dx \wedge dy$ generate $\mathscr{K}_X$ at p?

(b) Repeat part (a) in arbitrary dimension (following Exercise 22.4.R).

(c) Suppose Z is a smooth m-dimensional (closed) subvariety of a smooth n-dimensional variety Y, and let $\beta \colon X \to Y$ be the blow-up of Y along Z. Show that $\mathscr{K}_X \cong (\beta^* \mathscr{K}_Y)((n-m-1)E)$. (Recall from Theorem 22.3.10 that $X = \mathrm{Bl}_Z Y$ is smooth.)

22.4.15. Dimensional cohomology vanishing for quasiprojective schemes (promised in §18.2.7).

Using the theory of blowing up, Theorem 18.2.6 (dimensional cohomology vanishing for quasicoherent sheaves on projective k-schemes) can be extended to quasiprojective k-schemes. Suppose X is a quasiprojective k-variety of dimension n. We show that X may be covered by $n + 1$ affine open subsets. As X is quasiprojective, there is some projective variety Y with an open embedding $X \hookrightarrow Y$. By replacing Y with the closure of X in Y, we may assume that $\dim Y = n$. Put any subscheme structure Z on the complement of X in Y (for example, the reduced subscheme structure, §9.4.9). Let $Y' = \mathrm{Bl}_Z Y$. Then Y' is a projective variety (§22.3.1), which can be covered by $n + 1$ affine open subsets. The complement of X in Y' is an effective Cartier divisor ($E_Z Y$), so the restriction to X of each of these affine open subsets of Y is also affine, by Exercise 8.3.E.

22.4.16. *Remarks.* (i) You might then hope that *any* dimension n variety can be covered by $n + 1$ affine open subsets. This is not true. For each integer m, there is a threefold that requires at least m affine open sets to cover it; see [RotV, Ex. 4.9]. By the discussion above, this example is necessarily not quasiprojective. (ii) Here is a fact useful in invariant theory, which can be proved in the same way. Suppose $p_1, \dots, p_n$ are closed points on a quasiprojective k-variety X. Then there is an affine open subset of X containing all of them.

22.4.T. EXERCISE (DIMENSIONAL COHOMOLOGY VANISHING FOR QUASIPROJECTIVE VARIETIES). Suppose X is a quasiprojective k-scheme of dimension d. Show that for any quasicoherent sheaf $\mathscr{F}$ on X, $H^i(X, \mathscr{F}) = 0$ for $i > d$.

22.4.17. ** **Deformation to the normal cone (Figure 22.3).**
The following construction is key to the modern understanding of intersection theory in algebraic geometry, as developed by Fulton and MacPherson, [F2].

22.4.U. EXERCISE: DEFORMATION TO THE NORMAL CONE. Suppose Y is a k-variety, and $X \hookrightarrow Y$ is a closed subscheme.

(a) Show that the exceptional divisor of $\beta \colon \mathrm{Bl}_{X \times 0}(Y \times \mathbb{A}^1) \to Y \times \mathbb{A}^1$ is isomorphic to the projective completion of the normal cone to X in Y.

(b) Let $\pi \colon \mathrm{Bl}_{X \times 0}(Y \times \mathbb{A}^1) \to \mathbb{A}^1$ be the composition of β with the projection to $\mathbb{A}^1$. Show that $\pi^{-1}(0)$ is the scheme-theoretic union of $\mathrm{Bl}_X Y$ with the projective completion of the normal cone to X in Y, and the intersection of these two subschemes may be identified with $E_X Y$, which is a closed subscheme of $\mathrm{Bl}_X Y$ in the usual way (as the exceptional divisor of the blow-up $\mathrm{Bl}_X Y \to Y$), and a closed subscheme of the projective completion of the normal cone as described in Exercise 9.3.O.

The map

$$\mathrm{Bl}_{X \times 0}(Y \times \mathbb{A}^1) \setminus \mathrm{Bl}_X Y \longrightarrow \mathbb{A}^1$$

is called the **deformation to the normal cone** (short for *deformation of Y to the normal cone of X in Y*). Notice that the fiber above every k-point away from $0 \in \mathbb{A}^1$ is canonically isomorphic to Y, and the fiber over 0 is the normal cone. Because this family is "nice" (more precisely, *flat*, the topic of Chapter 24), we can prove things about general Y (near X) by way of this degeneration. (We will see in §24.3.10 that the deformation to the normal cone is a *flat* morphism, which is useful in intersection theory.)

PART VI

More Cohomological Tools

VI

Chapter 23

Derived Functors

> *Ça me semble extrêmement plaisant de ficher comme ça beaucoup de choses, pas drôles quand on les prend séparément, sous le grand chapeau des foncteurs dérivés.*
>
> *I find it very agreeable to stick all sorts of things, which are not much fun when taken individually, together under the heading of derived functors.*
>
> —A. Grothendieck, letter to J.-P. Serre, February 18, 1955 [GrS, p. 6]

In this chapter, we discuss derived functors, introduced by Grothendieck in his celebrated "Tôhoku article" [Gr1], and their applications to sheaves. For quasicoherent sheaves on quasicompact separated schemes, derived functor cohomology will agree with Čech cohomology (§23.5). Čech cohomology will suffice for most of our purposes, and is quite down-to-earth and computable, but derived functor cohomology is worth seeing. First, it will apply much more generally in algebraic geometry (e.g., étale cohomology) and elsewhere, although this is beyond the scope of this book. Second, it will easily provide us with some useful notions, such as the Ext functors and the Leray spectral sequence. But derived functors can be intimidating the first time you see them, so feel free to just skim the main results, and to return to them later.

23.1 The Tor Functors

We begin with a warm-up: the case of Tor. This is a hands-on example, but if you understand it well, you will understand derived functors in general. Tor will be useful to prove facts about flatness, which we will discuss in §24.2. Tor is short for "torsion" (see Remark 24.2.1).

If you have never seen this notion before, you may want to just remember its properties. But I will prove everything anyway—it is surprisingly easy.

The idea behind Tor is as follows. Whenever we see a right-exact functor, we always hope that it is the end of a long exact sequence. Informally, given a short exact sequence

$$(23.1.0.1) \qquad 0 \longrightarrow N' \longrightarrow N \longrightarrow N'' \longrightarrow 0,$$

we hope $M \otimes_A N' \to M \otimes_A N \to M \otimes_A N'' \to 0$ will extend to a long exact sequence

$$(23.1.0.2) \qquad \cdots \longrightarrow \mathrm{Tor}_i^A(M, N') \longrightarrow \mathrm{Tor}_i^A(M, N) \longrightarrow \mathrm{Tor}_i^A(M, N'') \longrightarrow \cdots$$

$$\longrightarrow \mathrm{Tor}_1^A(M, N') \longrightarrow \mathrm{Tor}_1^A(M, N) \longrightarrow \mathrm{Tor}_1^A(M, N'')$$

$$\longrightarrow M \otimes_A N' \longrightarrow M \otimes_A N \longrightarrow M \otimes_A N'' \longrightarrow 0.$$

More precisely, we are hoping for *covariant functors* $\mathrm{Tor}_i^A(M, \cdot)$ from A-modules to A-modules (functoriality giving 2/3 of the morphisms in (23.1.0.2)), with $\mathrm{Tor}_0^A(M, N) \equiv M \otimes_A N$, and natural *connecting homomorphisms*

$$\delta \colon \mathrm{Tor}_{i+1}^A(M, N'') \longrightarrow \mathrm{Tor}_i^A(M, N')$$

for every short exact sequence (23.1.0.1) giving the long exact sequence (23.1.0.2). ("Natural" means: given a morphism of short exact sequences, the natural square you would write down involving the δ-morphism must commute.)

It turns out to be not too hard to make this work, and this will also motivate derived functors. Let's now define $\operatorname{Tor}_i^A(M, N)$.

Take any resolution $\mathscr{R}$ of N by free modules:

$$\cdots \longrightarrow A^{\oplus n_2} \longrightarrow A^{\oplus n_1} \longrightarrow A^{\oplus n_0} \longrightarrow N \longrightarrow 0.$$

More precisely, build this resolution from right to left. Start by choosing generators of N as an A-module, giving us $A^{\oplus n_0} \to N \to 0$. Then choose generators of the kernel, and so on. Note that we are not requiring the n_i to be finite (although we could, if N is a finitely generated module and A is Noetherian). Truncate the resolution, by stripping off the last term N (replacing $\to N \to 0$ with $\to 0$). Then tensor with M (which does not preserve exactness). Note that $M \otimes (A^{\oplus n_i}) = M^{\oplus n_i}$, as tensoring with M commutes with arbitrary direct sums (Exercise 1.2.M). Let $\operatorname{Tor}_i^A(M, N)_{\mathscr{R}}$ be the homology of this complex at the ith stage ($i \geq 0$). The subscript $\mathscr{R}$ reminds us that our construction depends on the resolution, although we will soon see that it is independent of $\mathscr{R}$.

We make some quick observations.

- $\operatorname{Tor}_0^A(M, N)_{\mathscr{R}} \cong M \otimes_A N$, canonically. Reason: As tensoring is right-exact, and $A^{\oplus n_1} \to A^{\oplus n_0} \to N \to 0$ is exact, we have that $M^{\oplus n_1} \to M^{\oplus n_0} \to M \otimes_A N \to 0$ is exact, and hence that the homology of the truncated complex $M^{\oplus n_1} \to M^{\oplus n_0} \to 0$ is $M \otimes_A N$.
- If $M \otimes \cdot$ is exact (i.e., M is *flat*, §1.5.13), then $\operatorname{Tor}_i^A(M, N)_{\mathscr{R}} = 0$ for all $i > 0$. (This characterizes flatness; see Exercise 23.1.C.)

Now given two modules N and N' and resolutions $\mathscr{R}$ and $\mathscr{R}'$ of N and N', we can "lift" any morphism $N \to N'$ to a morphism of the two resolutions:

$$
\begin{array}{ccccccccccc}
\cdots & \longrightarrow & A^{\oplus n_i} & \longrightarrow & \cdots & \longrightarrow & A^{\oplus n_1} & \longrightarrow & A^{\oplus n_0} & \longrightarrow & N & \longrightarrow & 0 \\
 & & \downarrow & & & & \downarrow & & \downarrow & & \downarrow & & \\
\cdots & \longrightarrow & A^{\oplus n_i'} & \longrightarrow & \cdots & \longrightarrow & A^{\oplus n_1'} & \longrightarrow & A^{\oplus n_0'} & \longrightarrow & N' & \longrightarrow & 0
\end{array}
$$

We do this inductively on i. Here we use the freeness of $A^{\oplus n_i}$: if $a_1, \ldots, a_{n_i}$ are generators of $A^{\oplus n_i}$, to lift the map $b: A^{\oplus n_i} \to A^{\oplus n_{i-1}'}$ (the composition of the differential $A^{\oplus n_i} \to A^{\oplus n_{i-1}}$ with the previously constructed $A^{\oplus n_{i-1}} \to A^{\oplus n_{i-1}'}$) to $c: A^{\oplus n_i} \to A^{\oplus n_i'}$, we arbitrarily lift $b(a_j)$ from $A^{\oplus n_{i-1}'}$ to $A^{\oplus n_i'}$, and declare this to be $c(a_j)$. (Warning for people who care about such things: we are using the Axiom of Choice here.)

Denote the choice of lifts by $\mathscr{R} \to \mathscr{R}'$. Now truncate both complexes (remove column $N \to N'$) and tensor with M. Maps of complexes induce maps of homology (Exercise 1.5.D), so we have described maps (a priori depending on $\mathscr{R} \to \mathscr{R}'$)

$$\operatorname{Tor}_i^A(M, N)_{\mathscr{R}} \longrightarrow \operatorname{Tor}_i^A(M, N')_{\mathscr{R}'}.$$

Recall (§1.5.7) that we say two maps of complexes $f: C_\bullet \to C_\bullet'$ and $g: C_\bullet \to C_\bullet'$ are *homotopic* if there is a sequence of maps $w: C_i \to C_{i+1}'$ such that $f - g = dw + wd$. We saw that two homotopic maps give the same map on homology (Exercise 1.5.E).

23.1.A. CRUCIAL EXERCISE. Show that any two lifts $\mathscr{R} \to \mathscr{R}'$ are homotopic.

We now pull these observations together. (Be sure to digest these completely!)

(1) We get a map of A-modules $\operatorname{Tor}_i^A(M, N)_{\mathscr{R}} \to \operatorname{Tor}_i^A(M, N')_{\mathscr{R}'}$, independent of the lift $\mathscr{R} \to \mathscr{R}'$.

(2) Hence for any two resolutions $\mathscr{R}$ and $\mathscr{R}'$ of an A-module N, we get a *canonical* isomorphism $\operatorname{Tor}_i^A(M, N)_{\mathscr{R}} \xleftarrow{\sim} \operatorname{Tor}_i^A(M, N)_{\mathscr{R}'}$. Here's why. Choose lifts $\mathscr{R} \to \mathscr{R}'$ and $\mathscr{R}' \to \mathscr{R}$. The composition $\mathscr{R} \to \mathscr{R}' \to \mathscr{R}$ is homotopic to the identity (as it is a lift of the identity map $N \to N$). Thus if $f_{\mathscr{R} \to \mathscr{R}'}: \operatorname{Tor}_i^A(M, N)_{\mathscr{R}} \to \operatorname{Tor}_i^A(M, N)_{\mathscr{R}'}$ is the map induced by $\mathscr{R} \to \mathscr{R}'$, and

similarly $f_{\mathscr{R}' \to \mathscr{R}}$ is the map induced by $\mathscr{R}' \to \mathscr{R}$, then $f_{\mathscr{R}' \to \mathscr{R}} \circ f_{\mathscr{R} \to \mathscr{R}'}$ is the identity, and similarly $f_{\mathscr{R} \to \mathscr{R}'} \circ f_{\mathscr{R}' \to \mathscr{R}}$ is the identity.

(3) Hence $\mathrm{Tor}_i^A(M, \cdot)$ doesn't depend on the choice of resolution. It is a covariant functor $Mod_A \to Mod_A$.

23.1.1. *Remark.* Note that if N is a free module, then $\mathrm{Tor}_i^A(M, N) = 0$ for all M and all $i > 0$, as N has the trivial resolution $0 \to N \to N \to 0$ (it is "its own resolution").

Finally, we get long exact sequences:

23.1.2. Proposition — *For any short exact sequence* (23.1.0.1) *we get a long exact sequence of* Tor's (23.1.0.2).

Proof. Given a short exact sequence (23.1.0.1), choose resolutions of N' and N''. Then use these to get a resolution for N as follows.

(23.1.2.1)

The map $A^{\oplus(n'_{i+1}+n''_{i+1})} \to A^{\oplus(n'_i+n''_i)}$ is the composition $A^{\oplus n'_{i+1}} \to A^{\oplus n'_i} \hookrightarrow A^{\oplus(n'_i+n''_i)}$ along with a lift of $A^{\oplus n''_{i+1}} \to A^{\oplus n''_i}$ to $A^{\oplus(n'_i+n''_i)}$ ensuring that the middle row is a *complex*.

23.1.B. EXERCISE. Verify that it is possible to choose such a lift of $A^{\oplus n''_{i+1}} \to A^{\oplus n''_i}$ to $A^{\oplus(n'_i+n''_i)}$.

Hence the middle row of (23.1.2.1) is *exact* (not just a complex), using the long exact sequence in cohomology (Theorem 1.5.8), and the fact that the top and bottom rows are exact. Thus the middle row is a resolution, and (23.1.2.1) is a short exact sequence of resolutions. (This is sometimes called the *horseshoe construction*, as the filling in of the middle row looks like filling in the middle of a horseshoe.) It may be helpful to notice that the columns other than the "N-column" are all "direct sum exact sequences," and the horizontal maps in the middle row are "block upper triangular."

Then truncate (removing the right column $0 \to N' \to N \to N'' \to 0$), tensor with M (obtaining a short exact sequence of complexes), and take cohomology, yielding the desired long exact sequence. $\square$

23.1.C. EXERCISE (THE Tor_1 CRITERION FOR FLATNESS). Show that the following are equivalent conditions on an A-module M.

(i) M is flat.
(ii) $\mathrm{Tor}_i^A(M, N) = 0$ for all $i > 0$ and all A-modules N.
(iii) $\mathrm{Tor}_1^A(M, N) = 0$ for all A-modules N.

23.1.3. *Caution.* Given that free modules are immediately seen to be flat, you might think that Exercise 23.1.C implies Remark 23.1.1. This would follow if we knew that $\mathrm{Tor}_i^A(M,N) \cong \mathrm{Tor}_i^A(N,M)$, which is clear for $i=0$ (as $\otimes$ is symmetric), but we won't know this about Tor_i when $i>0$ until Exercise 23.3.A.

23.1.D. EXERCISE. Show that the connecting homomorphism δ constructed above is independent of all choices (of resolutions, etc.). Try to do this with as little annoyance as possible. (Possible hint: Given two sets of choices used to build (23.1.2.1), build a map—a three-dimensional diagram—from one version of (23.1.2.1) to the other version.)

23.1.E. UNIMPORTANT EXERCISE. Show that $\mathrm{Tor}_i^A(M, \cdot)$ is an *additive* functor (Definition 1.5.1). (We won't use this later, so feel free to skip it.)

We have thus established the foundations of Tor.

23.2 Derived Functors in General

23.2.1. Projective resolutions. We used very little about free modules in the above construction of Tor—in fact, we used only that free modules are **projective**, i.e., those modules P such that for any surjection $M \longrightarrow\!\!\!\!\!\rightarrow N$, it is possible to lift any morphism $P \to N$ to $P \to M$:

(23.2.1.1)

$$
\begin{array}{ccc}
& & P \\
& \overset{\text{exists}}{\swarrow} & \downarrow \\[-0.3em]
& & \\
M & \longrightarrow\!\!\!\!\!\rightarrow & N.
\end{array}
$$

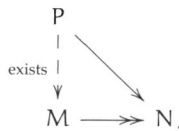

(As noted in §23.1, this uses the Axiom of Choice.) Equivalently, $\mathrm{Hom}(P, \cdot)$ is an exact functor (recall that $\mathrm{Hom}(Q, \cdot)$ is always left-exact for any Q). More generally, the same idea yields the definition of a **projective object in any abelian category**. Hence by following through our entire argument with projective modules replacing free modules throughout, (i) we can compute $\mathrm{Tor}_i^A(M,N)$ by taking any projective resolution of N, and (ii) $\mathrm{Tor}_i^A(M,N) = 0$ for any projective A-module N.

23.2.A. EXERCISE. Show that an object P is projective if and only if every short exact sequence $0 \to Q \to R \to P \to 0$ splits. Hence show that an A-module P is projective if and only if P is a direct summand of a free module.

23.2.B. EXERCISE. Show that projective modules are flat. Hint: Exercise 23.2.A. Be careful if you want to use Exercise 23.1.C; see Caution 23.1.3. (In fact, finitely generated projective modules over local rings are even free; see Remark 24.3.6.)

23.2.2. Definition: Derived functors.
The above description was low-tech, but immediately generalizes drastically. All we are using is that $M \otimes_A \cdot$ is a right-exact functor, and that for any A-module N, we can find a surjection $P \longrightarrow\!\!\!\!\!\rightarrow N$ from a projective module. In general, if F is *any* right-exact covariant functor from the category of A-modules to any abelian category, this construction will define a sequence of functors $L_i F$ such that $L_0 F = F$ and the $L_i F$'s give a long exact sequence. We can make this more general still. We say that an abelian category **has enough projectives** if for any object N there is a surjection onto it from a projective object. Then if F is any right-exact covariant functor from an abelian category with enough projectives to any abelian category, then we can define the **left derived functors** to F, denoted by $L_i F$ ($i \geq 0$). You should reread §23.1 and see that throughout we only use the fact that we have a projective resolution (repeatedly lifting maps as in (23.2.1.1)), as well as the fact that F sends products to products (a consequence of additivity of

the functor; see Remark 1.5.2), to show that F applied to (23.1.2.1) preserves the exactness of the columns.

23.2.C. EXERCISE. The notion of an **injective object** in an abelian category is dual to the notion of a projective object.

(a) State precisely the definition of an injective object.
(b) Define derived functors for (i) covariant left-exact functors (these are called **right derived functors**), (ii) contravariant left-exact functors (also called **right derived functors**), and (iii) contravariant right-exact functors (these are called **left derived functors**), making explicit the necessary assumptions of the category having enough injectives or projectives.

23.2.3. *Notation.* Recall from Definition 23.2.2 that if F is a right-exact functor, its (left) derived functors are denoted by $L_i F$ ($i \geq 0$, with $L_0 F = F$). Similarly, if F is a left-exact functor, its (right-) derived functors are denoted by $R^i F$. The i is a superscript, to indicate that the long exact sequence is "ascending in i."

23.2.4. The Ext functors.

23.2.D. EASY EXERCISE (AND DEFINITION): Ext FUNCTORS FOR A-MODULES, FIRST VERSION. As $\mathrm{Hom}(\cdot, N)$ is a contravariant left-exact functor in Mod_A, which has enough projectives, define $\mathrm{Ext}^i_A(M, N)$ as the ith right derived functor of $\mathrm{Hom}(\cdot, N)$, applied to M. State the corresponding long exact sequence for Ext-modules.

23.2.E. EASY EXERCISE (AND DEFINITION): Ext FUNCTORS FOR A-MODULES, SECOND VERSION. The category Mod_A has enough injectives (see §23.2.5). As $\mathrm{Hom}(M, \cdot)$ is a covariant left-exact functor in Mod_A, define $\mathrm{Ext}^i_A(M, N)$ as the ith right derived functor of $\mathrm{Hom}(M, \cdot)$, applied to N. State the corresponding long exact sequence for Ext-modules.

We seem to have a problem with the previous two exercises: we have defined $\mathrm{Ext}^i(M, N)$ twice, and we have two different long exact sequences! Fortunately, these two definitions of Ext agree (see Exercise 23.3.B), and two long exact sequences for Ext are better than one.

23.2.F. EASY EXERCISE. Use the definition of Ext in Exercise 23.2.D to show that if A is a Noetherian ring, and M and N are finitely generated A-modules, then $\mathrm{Ext}^i_A(M, N)$ is a finitely generated A-module.

Ext-functors (for sheaves) will play a key role in the proof of Serre duality; see §29.2.

23.2.5.* The category of A-modules has enough injectives. We will need the fact that Mod_A has enough injectives, but the details of the proof won't come up again, so feel free to skip this discussion.

23.2.G. EXERCISE. Suppose Q is an A-module, such that for every ideal $I \subset A$, every homomorphism $I \to Q$ extends to $A \to Q$. Show that Q is an injective A-module. Hint: Suppose $N \subset M$ is an inclusion of A-modules, and we are given $\beta \colon N \to Q$. We wish to show that β extends to $M \to Q$. Use Zorn's Lemma to show that among those A-modules N' with $N \subset N' \subset M$, such that β extends to N', there is a maximal one. If this N' is not M, give an extension of β to $N' + Am$, where $m \in M \setminus N'$, obtaining a contradiction.

23.2.H. EASY EXERCISE (USING ZORN'S LEMMA). Show that a $\mathbb{Z}$-module (i.e., abelian group) Q is injective if and only if it is **divisible** (i.e., for every $q \in Q$ and $n \in \mathbb{Z}^{\neq 0}$, there is $q' \in Q$ with $nq' = q$). Hence show that any quotient of an injective $\mathbb{Z}$-module is also injective.

23.2.I. EXERCISE. Show that the category of $\mathbb{Z}$-modules $Mod_{\mathbb{Z}} = Ab$ has enough injectives. (Hint: If M is a $\mathbb{Z}$-module, then write it as the quotient of a free $\mathbb{Z}$-module F by some K. Show that M is contained in the divisible group $(F \otimes_{\mathbb{Z}} \mathbb{Q})/K$.)

23.2.J. EXERCISE. Suppose Q is an injective $\mathbb{Z}$-module, and A is a ring. Show that $\text{Hom}_{\mathbb{Z}}(A, Q)$ is an injective A-module. Hint: First describe the A-module structure on $\text{Hom}_{\mathbb{Z}}(A, Q)$. You will only use the fact that $\mathbb{Z}$ is a ring, and that A is an algebra over that ring.

23.2.K. EXERCISE. Show that Mod_A has enough injectives. Hint: Suppose M is an A-module. By Exercise 23.2.I, we can find an inclusion *of $\mathbb{Z}$-modules* $M \hookrightarrow Q$ where Q is an injective $\mathbb{Z}$-*module*. Describe a sequence of inclusions of A-modules

$$M \hookrightarrow \text{Hom}_{\mathbb{Z}}(A, M) \hookrightarrow \text{Hom}_{\mathbb{Z}}(A, Q).$$

(The A-module structure on $\text{Hom}_{\mathbb{Z}}(A, M)$ is via the A-action on the left argument A, not via the A-action on the right argument M.) The right term is injective by the previous Exercise 23.2.J.

23.2.6.* Universal δ-functors.

(This discussion is best skipped on a first reading; you should move directly to §23.3. We won't use the notion of δ-functors until §29.2.2.) We now describe a more general variant of derived functors, as you may use them in the discussion of Serre duality in Chapter 29. The advantage of the notion of universal δ-functor is that we can apply it even in cases where $\mathscr{A}$ does not have enough injectives.

Abstracting key properties of derived functors, we define the data of a (cohomological) δ-functor, from an abelian category $\mathscr{A}$ to another abelian category $\mathscr{B}$. A δ-**functor** is a collection of additive functors $T^i : \mathscr{A} \to \mathscr{B}$ (where T^i is taken to be 0 if $i < 0$), along with morphisms $\delta^i : T^i(A'') \to T^{i+1}(A')$ for each short exact sequence

(23.2.6.1) $$0 \longrightarrow A' \longrightarrow A \longrightarrow A'' \longrightarrow 0$$

in $\mathscr{A}$, satisfying two properties:

(i) *(short exact sequences yield long exact sequences)* For each short exact sequence (23.2.6.1), the sequence

$$\cdots \longrightarrow T^{i-1}(A'') \xrightarrow{\delta^{i-1}} T^i(A') \longrightarrow T^i(A) \longrightarrow T^i(A'') \xrightarrow{\delta^i} T^{i+1}(A') \longrightarrow \cdots$$

(where the unlabeled maps come from the covariance of the T^i) is exact. In particular, T^0 is left-exact.

(ii) *(functoriality of (i))* For each morphism of short exact sequences in $\mathscr{A}$

$$\begin{array}{ccccccccc}
0 & \longrightarrow & a' & \longrightarrow & a & \longrightarrow & a'' & \longrightarrow & 0 \\
& & \downarrow & & \downarrow & & \downarrow & & \\
0 & \longrightarrow & A' & \longrightarrow & A & \longrightarrow & A'' & \longrightarrow & 0
\end{array}$$

(where the squares commute), the δ^i's give a commutative diagram

$$\begin{array}{ccc}
T^i(a'') & \xrightarrow{\delta^i} & T^{i+1}(a') \\
\downarrow & & \downarrow \\
T^i(A'') & \xrightarrow{\delta^i} & T^{i+1}(A')
\end{array}$$

(where the vertical arrows come from the covariance of T^i and T^{i+1}).

Derived functor cohomology is clearly an example of a δ-functor; Čech cohomology of sheaves on quasicompact separated schemes is another. (You can make these statements precise if you wish.)

23.2.7. *Remark/definition.* More precisely, we should call the above a **covariant δ-functor**. There is similarly a notion of a **contravariant δ-functor**, whose definition you may be able to guess.

23.2.L. EXERCISE. Figure out the right definition of **morphism of δ-functors** $\mathscr{A} \to \mathscr{B}$. (It should then be clear that the δ-functors from $\mathscr{A}$ to $\mathscr{B}$ form a category.)

23.2.8. *Definition.* A (cohomological) δ-functor (T^i, δ^i) is **universal** if for any other δ-functor (T'^i, δ'^i), and any natural transformation of functors $\alpha \colon T^0 \to T'^0$, there is a unique morphism of δ-functors $(T^i, \delta^i) \to (T'^i, \delta'^i)$ extending α. By universal property nonsense (and Exercise 23.2.L), given any covariant left-exact functor $F \colon \mathscr{A} \to \mathscr{B}$, there is at most one universal δ-functor (T^i, δ^i) extending F (i.e., with a natural isomorphism $T^0 \overset{\sim}{\longleftrightarrow} F$). The key fact about universal δ-functors is the following.

23.2.9. Theorem — *Suppose (T^i, δ^i) is a covariant δ-functor from $\mathscr{A}$ to $\mathscr{B}$, and for all $A \in \mathscr{A}$, there exists a monomorphism $A \to J$ with $T^i J = 0$ for all $i > 0$. Then (T^i, δ^i) is universal.*

23.2.M.** EXERCISE. Prove Theorem 23.2.9. Partial hint: Motivated by Corollary 23.2.11 below, follow our discussion of derived functors. Better hint (because this exercise is hard): Follow the hints in [Weib, Exercise 2.4.5], or follow the proof of [Lan, Ch. XX, Thm. 7.1].

23.2.10. *Remark.* An additive functor $F \colon \mathscr{A} \to \mathscr{B}$ is said to be **effaceable** if for every $A \in \mathscr{A}$, there is a monomorphism $A \to J$ with $F(J) = 0$. The hypotheses of Theorem 23.2.9 can be weakened to require only that T^i is effaceable for each $i > 0$, and you are welcome to prove that instead. (Indeed, [Weib, Exercise 2.4.5], [Lan, Ch. XX, Thm. 7.1], and the original source [Gr1, II.2.2.1] deal with this case.) We give the statement of Theorem 23.2.9 for simplicity, as we will only use this version.

23.2.11. Corollary — *If $\mathscr{A}$ has enough injectives, and F is a left-exact covariant functor $\mathscr{A} \to \mathscr{B}$, then the $R^i F$ (with the δ^i that accompany them) form a universal δ-functor.*

Proof. Each element of $\mathscr{A}$ admits a monomorphism into an injective element; this is just the definition of "enough injectives" (Exercise 23.2.C). Higher derived functors of an injective object I are always 0: just compute the higher derived functor by taking the injective resolution of I "by itself." □

23.3 Derived Functors and Spectral Sequences

A number of useful facts can be easily proved using spectral sequences. By doing these exercises, you will lose any fear of spectral sequence arguments in similar situations, as you will realize they are all the same.

Before you read this section, you should read §1.6 on spectral sequences.

23.3.1. Symmetry of Tor.

23.3.A. EXERCISE (SYMMETRY OF Tor). Show that there is an isomorphism $\operatorname{Tor}_i^A(M, N) \overset{\sim}{\longleftrightarrow} \operatorname{Tor}_i^A(N, M)$. (Hint: Take a free resolution of M and a free resolution of N. Take their "product" to somehow produce a double complex. Use both orientations of the obvious spectral sequence and see what you get.)

On a related note:

23.3.B. EXERCISE. Show that the two definitions of $\operatorname{Ext}^i(M, N)$ given in Exercises 23.2.D and 23.2.E agree.

23.3.2. Derived functors can be computed using acyclic resolutions. Suppose $F \colon \mathscr{A} \to \mathscr{B}$ is a right-exact additive functor of abelian categories, and that $\mathscr{A}$ has enough projectives. We say that $A \in \mathscr{A}$ is F-**acyclic** (or just **acyclic** if the F is clear from context) if $L_i F A = 0$ for $i > 0$. In Exercise 23.3.D, we will see that derived functors $L_i F B$ of an object B of $\mathscr{A}$ can be computed by using "acyclic resolutions." We set the stage with a useful construction.

23.3.3. *Building a "projective resolution" of an exact sequence.* Suppose $\cdots \to E_2 \to E_1 \to E_0 \to 0$ is an exact sequence in an abelian category with enough projectives. We explain how to inductively build a double complex of projectives

such that the rows and columns are all exact. Suppose you have built part of the complex, and are trying to build the $P_{m,n}$ term:

(23.3.3.1)

For reasons that will soon become clear, we assume that for any $b \in P_{m-1,n-1}$ whose image in both $P_{m-1,n-2}$ and $P_{m-2,n-1}$ is zero, there is a $c \in P_{m,n-1}$ whose image in $P_{m-1,n-1}$ is b, and whose image in $P_{m,n-2}$ is zero; and symmetrically there is a $c' \in P_{m-1,n}$ whose image in $P_{m-1,n-1}$ is b, and whose image in $P_{m-2,n}$ is zero.

Consider

$$K := \ker(P_{m,n-1} \oplus P_{m-1,n} \to P_{m,n-2} \oplus P_{m-1,n-1} \oplus P_{m-2,n}).$$

If K surjects onto both $\ker(P_{m-1,n} \to P_{m-2,n})$ and $\ker(P_{m,n-1} \to P_{m,n-2})$, then we could take any surjection from a projective object $P \twoheadrightarrow K$, then take $P_{m,n}$ to be P (with its map to $P_{m,n-1}$ and the negative of its map to $P_{m-1,n}$). With this choice, we would have ensured "horizontal exactness" at $P_{m-1,n}$, and "vertical exactness" at $P_{m,n-1}$, and commutativity of the square in (23.3.3.1).

We now verify our two desired surjections, by "diagram-chasing" (which we can do; see §1.5.5). Suppose $a \in \ker(P_{m-1,n} \to P_{m-2,n})$, and let b be its image in $P_{m-1,n-1}$. We wish to find $c \in \ker(P_{m,n-1} \to P_{m,n-2})$ so that c maps to b in $P_{m-1,n-1}$ and 0 in $P_{m,n-2}$. (Then $(-c, a) \in P_{m,n-1} \oplus P_{m-1,n}$ would be our desired element of K mapping to $a \in P_{m-1,n}$.) But such a c exits by our assumption! And, symmetrically, we get surjectivity $K \twoheadrightarrow \ker(P_{m,n-1} \to P_{m,n-2})$.

The final task is to ensure these assumptions hold for later stages in the building of our double complex. We need to ensure that for any $b' \in P_{m-1,n}$ that maps to $(0, 0) \in P_{m-1,n-1} \oplus P_{m-2,n}$, there is an element of $P_{m,n}$ mapping to $(0, b') \in P_{m,n-1} \oplus P_{m-1,n}$. And we have to ensure the analogous statement with the roles of m and n reversed. We take $Q' = \ker(P_{m-1,n} \to P_{m-1,n-1} \oplus P_{m-2,n})$, and $Q'' = \ker(P_{m,n-1} \to P_{m,n-2} \oplus P_{m-1,n-1})$, and take surjections $P' \twoheadrightarrow Q'$ and $P'' \twoheadrightarrow Q''$ from projective objects. Then take $P_{m,n}$ to be $P \oplus P' \oplus P''$, where the maps from P to $P_{m,n-1}$ and $P_{m-1,n}$ are as described above; the map P' to $P_{m-1,n}$ is the above-described map (via Q'); the map P' to $P_{m,n-1}$ is zero; and the opposite for P'' (the map to $P_{m-1,n}$ is zero, and the map to $P_{m,n-1}$ is as implied above, by way of Q''). The summand P' ensures our first desired assumption, and the summand P'' ensures our second.

23.3.C. EXERCISE. Verify that the above construction indeed gives a projective resolution of an exact sequence. Where did you use that the sequence $E_\bullet$ was exact?

Now let's apply this.

23.3.D. EXERCISE. Show that you can compute the derived functors of an objects B of $\mathscr{A}$ using **acyclic resolutions** (not just projective resolutions), i.e., by taking a resolution

$$(23.3.3.2) \qquad \cdots \longrightarrow A_2 \longrightarrow A_1 \longrightarrow A_0 \longrightarrow B \longrightarrow 0$$

by F-acyclic objects A_i, truncating, applying F, and taking homology. Hence $\mathrm{Tor}_i(M, N)$ can be computed with a flat resolution of M or N. Hint: As described above, build a double complex of projectives "on top of" the exact sequence (23.3.3.2). Remove the bottom row, and the rightmost nonzero column, and then apply F, to obtain a new double complex. Use a spectral sequence argument to show that (i) the double complex has homology equal to $L_i F(B)$, and (ii) the homology of the double complex agrees with the construction given in the statement of the exercise. If this is too confusing, read more about the Cartan–Eilenberg resolution below.

23.3.4. The Grothendieck composition-of-functors spectral sequence.

Suppose $\mathscr{A}$, $\mathscr{B}$, and $\mathscr{C}$ are abelian categories, $F: \mathscr{A} \to \mathscr{B}$ and $G: \mathscr{B} \to \mathscr{C}$ are a left-exact additive covariant functors, and $\mathscr{A}$ and $\mathscr{B}$ have enough injectives. Thus right derived functors of F, G, and $G \circ F$ exist. A reasonable question is: How are they related?

23.3.5. Theorem (Grothendieck composition-of-functors spectral sequence). — *Suppose* $F: \mathscr{A} \to \mathscr{B}$ *and* $G: \mathscr{B} \to \mathscr{C}$ *are left-exact additive covariant functors, and* $\mathscr{A}$ *and* $\mathscr{B}$ *have enough injectives. Suppose further that* F *sends injective elements of* $\mathscr{A}$ *to* G-*acyclic elements of* $\mathscr{B}$. *Then for each* $X \in \mathscr{A}$, *there is a spectral sequence with* $_\to E_2^{p,q} = R^q G(R^p F(X))$ *converging to* $R^\bullet(G \circ F)(X)$.

We will soon see the Leray spectral sequence as an application (Theorem 23.4.5).

There is more one might want to extract from the proof of Theorem 23.3.5. For example, although E_0 page of the spectral sequence will depend on some choices (of injective resolutions), the E_2 page will be independent of choice. For our applications, we won't need this refinement.

We will have to work to establish Theorem 23.3.5, so the proof is possibly best skipped on a first reading.

23.3.6.* Proving Theorem 23.3.5.

Before we give the proof (in §23.3.8), we begin with some preliminaries to motivate it. In order to discuss derived functors applied to X, we choose an injective resolution of X:

$$0 \longrightarrow X \longrightarrow I^0 \longrightarrow I^1 \longrightarrow \cdots .$$

To compute the derived functors $R^p F(X)$, we apply F to the injective resolution $I^\bullet$:

$$0 \longrightarrow F(I^0) \longrightarrow F(I^1) \longrightarrow F(I^2) \longrightarrow \cdots .$$

Note that $F(I^p)$ is G-acyclic, by hypothesis of Theorem 23.3.5. If we were to follow our nose, we might take simultaneous injective resolutions $I^{\bullet,\bullet}$ of the terms in the above complex $F(I^\bullet)$ (the "dual" of §23.3.3—note that only the columns are required to be exact), and apply G, and consider the resulting double complex:

(23.3.6.1)

$$
\begin{array}{ccccccccc}
& & \vdots & & \vdots & & \vdots & & \\
& & \uparrow & & \uparrow & & \uparrow & & \\
0 & \longrightarrow & G(I^{0,2}) & \longrightarrow & G(I^{1,2}) & \longrightarrow & G(I^{2,2}) & \longrightarrow & \cdots \\
& & \uparrow & & \uparrow & & \uparrow & & \\
0 & \longrightarrow & G(I^{0,1}) & \longrightarrow & G(I^{1,1}) & \longrightarrow & G(I^{2,1}) & \longrightarrow & \cdots \\
& & \uparrow & & \uparrow & & \uparrow & & \\
0 & \longrightarrow & G(I^{0,0}) & \longrightarrow & G(I^{1,0}) & \longrightarrow & G(I^{2,0}) & \longrightarrow & \cdots \\
& & \uparrow & & \uparrow & & \uparrow & & \\
& & 0 & & 0 & & 0 & &
\end{array}
$$

23.3.E. EXERCISE. Consider the spectral sequence with upward orientation, starting with (23.3.6.1) as page E_0. Show that $E_2^{p,q}$ is $R^p(G \circ F)(X)$ if $q = 0$, and 0 otherwise.

We now see half of the terms in the conclusion of Theorem 23.3.5; we are halfway there. To complete the proof, we would want to consider another spectral sequence, with rightward orientation, but we need to know more about (23.3.6.1); we will build it more carefully.

23.3.7. *Cartan–Eilenberg resolutions.*
Suppose $\cdots \to C^{p-1} \to C^p \to C^{p+1} \to \cdots$ is a complex in an abelian category $\mathscr{B}$. We will build an injective resolution of $C^\bullet$

(23.3.7.1)

$$
\begin{array}{ccccccccc}
& & \vdots & & \vdots & & \vdots & & \\
& & \uparrow & & \uparrow & & \uparrow & & \\
0 & \longrightarrow & I^{0,2} & \longrightarrow & I^{1,2} & \longrightarrow & I^{2,2} & \longrightarrow & \cdots \\
& & \uparrow & & \uparrow & & \uparrow & & \\
0 & \longrightarrow & I^{0,1} & \longrightarrow & I^{1,1} & \longrightarrow & I^{2,1} & \longrightarrow & \cdots \\
& & \uparrow & & \uparrow & & \uparrow & & \\
0 & \longrightarrow & I^{0,0} & \longrightarrow & I^{1,0} & \longrightarrow & I^{2,0} & \longrightarrow & \cdots \\
& & \uparrow & & \uparrow & & \uparrow & & \\
0 & \longrightarrow & C^0 & \longrightarrow & C^1 & \longrightarrow & C^2 & \longrightarrow & \cdots \\
& & \uparrow & & \uparrow & & \uparrow & & \\
& & 0 & & 0 & & 0 & &
\end{array}
$$

satisfying some further properties.

We first define some notation for functions on a complex.

- Let $Z^p(K^\bullet)$ be the kernel of the pth differential of a complex $K^\bullet$.
- Let $B^{p+1}(K^\bullet)$ be the image of the pth differential of a complex $K^\bullet$. (The superscript is chosen so that $B^{p+1}(K^\bullet) \subset K^{p+1}$.)
- As usual, let $H^p(K^\bullet)$ be the cohomology at the pth step of a complex $K^\bullet$.

For each p, we have complexes

(23.3.7.2)

$$
\begin{array}{ccccc}
\vdots & & \vdots & & \vdots \\
\uparrow & & \uparrow & & \uparrow \\
Z^p(I^{\bullet,1}) & & B^p(I^{\bullet,1}) & & H^p(I^{\bullet,1}) \\
\uparrow & & \uparrow & & \uparrow \\
Z^p(I^{\bullet,0}) & & B^p(I^{\bullet,0}) & & H^p(I^{\bullet,0}) \\
\uparrow & & \uparrow & & \uparrow \\
Z^p(C^\bullet) & & B^p(C^\bullet) & & H^p(C^\bullet) \\
\uparrow & & \uparrow & & \uparrow \\
0 & & 0 & & 0
\end{array} \qquad .
$$

We will construct (23.3.7.1) so that the three complexes (23.3.7.2) are all injective resolutions (of their first nonzero terms). We begin by choosing injective resolutions $B^{p,*}$ of $B^p(C^\bullet)$ and $H^{p,*}$ of $H^p(C^\bullet)$; these will eventually be the last two lines of (23.3.7.2).

23.3.F. EXERCISE. Describe an injective resolution $Z^{p,*}$ of $Z^p(C^\bullet)$ (the first line of (23.3.7.2)) making the following diagram a short exact sequence of complexes.

(23.3.7.3)

$$
\begin{array}{ccccccccc}
& & \vdots & & \vdots & & \vdots & & \\
& & \uparrow & & \uparrow & & \uparrow & & \\
0 & \longrightarrow & B^{p,1} & \longrightarrow & Z^{p,1} & \longrightarrow & H^{p,1} & \longrightarrow & 0 \\
& & \uparrow & & \uparrow & & \uparrow & & \\
0 & \longrightarrow & B^{p,0} & \longrightarrow & Z^{p,0} & \longrightarrow & H^{p,0} & \longrightarrow & 0 \\
& & \uparrow & & \uparrow & & \uparrow & & \\
0 & \longrightarrow & B^p(C^\bullet) & \longrightarrow & Z^p(C^\bullet) & \longrightarrow & H^p(C^\bullet) & \longrightarrow & 0 \\
& & \uparrow & & \uparrow & & \uparrow & & \\
& & 0 & & 0 & & 0 & &
\end{array}
$$

Hint: The "dual" problem was solved in (23.1.2.1), by a "horseshoe construction."

23.3.G. EXERCISE. Describe an injective resolution $I^{p,*}$ of C^p making the following diagram a short exact sequence of complexes.

(23.3.7.4)

(The hint for the previous problem applies again. We remark that the bottom nonzero rows of (23.3.7.3) and (23.3.7.4) appeared in (1.5.6.3).)

23.3.H. EXERCISE/DEFINITION. Build an injective resolution (23.3.7.1) of $C^\bullet$ such that $Z^{p,*} = Z^p(I^{\bullet,*})$, $B^{p,*} = B^p(I^{\bullet,*})$, $H^{p,*} = H^p(I^{\bullet,*})$, so the three complexes (23.3.7.2) are injective resolutions. This is called a **Cartan–Eilenberg resolution** of $C^\bullet$.

23.3.8. *Proof of the Grothendieck spectral sequence, Theorem 23.3.5.* We pick up where we left off before our digression on Cartan–Eilenberg resolutions. Choose an injective resolution $I^\bullet$ of X. Apply the functor F, then take a Cartan–Eilenberg resolution $I^{\bullet,\bullet}$ of $FI^\bullet$, and then apply G, to obtain (23.3.6.1).

Exercise 23.3.E describes what happens when we take (23.3.6.1) as E_0 in a spectral sequence with upward orientation. So we now consider the rightward orientation.

From our construction of the Cartan–Eilenberg resolution, we have injective resolutions (23.3.7.2), and short exact sequences

(23.3.8.1) $$0 \longrightarrow B^p(I^{\bullet,q}) \longrightarrow Z^p(I^{\bullet,q}) \longrightarrow H^p(I^{\bullet,q}) \longrightarrow 0,$$

(23.3.8.2) $$0 \longrightarrow Z^p(I^{\bullet,q}) \longrightarrow I^{p,q} \longrightarrow B^{p+1}(I^{\bullet,q}) \longrightarrow 0$$

of *injective* objects (from the columns of (23.3.7.3) and (23.3.7.4)). This means that both are *split* exact sequences (the central term can be expressed as a direct sum of the outer two terms), so upon application of G, both exact sequences remain exact.

Applying the left-exact functor G to

$$0 \longrightarrow Z^p(I^{\bullet,q}) \longrightarrow I^{p,q} \longrightarrow I^{p+1,q},$$

we find that $GZ^p(I^{\bullet,q}) = \ker(GI^{p,q} \to GI^{p+1,q})$. But this kernel is the *definition* of $Z^p(GI^{\bullet,q})$, so we have an induced isomorphism $GZ^p(I^{\bullet,q}) = Z^p(GI^{\bullet,q})$ ("G and Z^p commute"). From the exactness of (23.3.8.2), upon application of G, we see that $GB^{p+1}(I^{\bullet,q}) = B^{p+1}(GI^{\bullet,q})$ (both are $\text{coker}(GZ^p(I^{\bullet,q}) \to GI^{p,q})$). From the exactness of (23.3.8.1), upon application of G, we see that $GH^p(I^{\bullet,q}) = H^p(GI^{\bullet,q})$ (both are $\text{coker}(GB^p(I^{\bullet,q}) \to GZ^p(I^{\bullet,q}))$—so "$G$ and H^p commute").

We return to considering the rightward-oriented spectral sequence with (23.3.6.1) as E_0. Taking cohomology in the rightward direction, we find $E_1^{p,q} = H^p(GI^{\bullet,q}) = GH^p(I^{\bullet,q})$ (as G and H^p commute). Now $H^p(I^{\bullet,q})$ is an injective resolution of $(R^pF)(X)$ (the last resolution of (23.3.7.2)). Thus when we compute E_2 by using the vertical arrows, we find $E_2^{p,q} = R^qG(R^pF(X))$.

Now verify yourself that this (combined with Exercise 23.3.E) concludes the proof. $\qquad\square$

23.4   Derived Functor Cohomology of $\mathscr{O}$-Modules

We wish to apply the machinery of derived functors to define cohomology of quasicoherent sheaves on a scheme X. Rather than working in the category $QCoh_X$, for a number of reasons it is simpler to work in the larger category $Mod_{\mathscr{O}_X}$ (but see Unimportant Remark 23.5.7).

23.4.1. Theorem — *Suppose $(X, \mathscr{O}_X)$ is a ringed space. Then the category of $\mathscr{O}_X$-modules $Mod_{\mathscr{O}_X}$ has enough injectives.*

As a side benefit (of use to others more than us), taking $\mathscr{O}_X = \mathbb{Z}$, we see that the category of sheaves of abelian groups on a fixed topological space has enough injectives.

23.4.2. *Proof.* We prove Theorem 23.4.1 in a series of exercises, following Godement, [GrS, pp. 27–28]. Suppose $\mathscr{F}$ is an $\mathscr{O}_X$-module. We will exhibit an injection $\mathscr{F} \hookrightarrow \mathscr{Q}'$ into an injective $\mathscr{O}_X$-module. For each $p \in X$, choose an inclusion $\mathscr{F}_p \hookrightarrow Q_p$ into an injective $\mathscr{O}_{X,p}$-module (which is possible as the category of $\mathscr{O}_{X,p}$-modules has enough injectives, Exercise 23.2.K).

23.4.A. EXERCISE. Show that the skyscraper sheaf $\mathscr{Q}_p := i_{p,*} Q_p$, with module Q_p at point $p \in X$, is an injective $\mathscr{O}_X$-module. (You can cleverly do this by abstract nonsense, using Exercise 23.5.B, but it is just as quick to do it by hand.)

23.4.B. EASY EXERCISE. Show that the product (possibly infinite) of injective objects in an abelian category is also injective.

By the previous two exercises, $\mathscr{Q}' := \prod_{p \in X} \mathscr{Q}_p$ is an injective $\mathscr{O}_X$-module.

23.4.C. EASY EXERCISE. By considering stalks, show that the natural map $\mathscr{F} \to \mathscr{Q}'$ is an injection.

This completes the proof of Theorem 23.4.1.   □

We can now make a number of definitions.

23.4.3. Definition. If $(X, \mathscr{O}_X)$ is a ringed space, and $\mathscr{F}$ is an $\mathscr{O}_X$-module, define $H^i(X, \mathscr{F})$ as $R^i\Gamma(X, \mathscr{F})$. If furthermore $\pi\colon (X, \mathscr{O}_X) \to (Y, \mathscr{O}_Y)$ is a map of ringed spaces, we have **higher pushforwards** (or **derived pushforwards**) $R^i\pi_*\colon Mod_{\mathscr{O}_X} \to Mod_{\mathscr{O}_Y}$.

We have defined these notions earlier in special cases, for quasicoherent sheaves on quasicompact separated schemes (for H^i), or for quasicompact separated morphisms of schemes (for $R^i\pi_*$), in Chapter 18. We will soon (§23.5) show that these older definitions agree with Definition 23.4.3. Thus the derived functor definition applies much more generally than our Čech definition. But it is worthwhile to note that almost everything we use will come out of the Čech definition. A notable exception is the Leray spectral sequence, which we now discuss.

23.4.4. The Leray spectral sequence.

23.4.5. Theorem (Leray spectral sequence) — *Suppose $\pi\colon (X, \mathscr{O}_X) \to (Y, \mathscr{O}_Y)$ is a morphism of ringed spaces. For any $\mathscr{O}_X$-module $\mathscr{F}$, there is a spectral sequence with E_2 term given by $H^q(Y, R^p\pi_*\mathscr{F})$ abutting to $H^{p+q}(X, \mathscr{F})$.*

This is an immediate consequence of the Grothendieck composition-of-functors spectral sequence (Theorem 23.3.5) once we prove that the pushforward of an injective $\mathscr{O}$-module is an acyclic $\mathscr{O}$-module. We do this now.

23.4.6. *Definition.* We make an intermediate definition that is independently important. A sheaf $\mathscr{F}$ on a topological space is **flasque** (also sometimes called *flabby*) if all restriction maps are surjective, i.e., if $\mathrm{res}_{U \subset V}\colon \mathscr{F}(V) \to \mathscr{F}(U)$ is surjective for all $U \subset V$.

23.4.D. EXERCISE. Suppose $(X, \mathscr{O}_X)$ is a ringed space.

(a) Show that if

(23.4.6.1) $$0 \longrightarrow \mathscr{F}' \longrightarrow \mathscr{F} \longrightarrow \mathscr{F}'' \longrightarrow 0$$

is an exact sequence of $\mathscr{O}_X$-modules, and $\mathscr{F}'$ is flasque, then (23.4.6.1) is exact on sections over any open set U, i.e., $0 \longrightarrow \mathscr{F}'(U) \longrightarrow \mathscr{F}(U) \longrightarrow \mathscr{F}''(U) \longrightarrow 0$ is exact.

(b) Given an exact sequence (23.4.6.1), if $\mathscr{F}'$ is flasque, show that $\mathscr{F}$ is flasque if and only if $\mathscr{F}''$ is flasque.

23.4.E. EASY EXERCISE (PUSHFORWARD OF FLASQUES ARE FLASQUE).

(a) Suppose $\pi\colon X \to Y$ is a continuous map of topological spaces, and $\mathscr{F}$ is a flasque sheaf of sets on X. Show that $\pi_* \mathscr{F}$ is a flasque sheaf on Y.

(b) Suppose $\pi\colon (X, \mathscr{O}_X) \to (Y, \mathscr{O}_Y)$ is a morphism of ringed spaces, and $\mathscr{F}$ is a flasque $\mathscr{O}_X$-module. Show that $\pi_* \mathscr{F}$ is a flasque $\mathscr{O}_Y$-module.

23.4.7. *Extension by zero, an occasional* left adjoint *to the inverse image functor.* In order to prove that injective sheaves are flasque (Exercise 23.4.H), we introduce the "extension by zero" functor. In addition to always being a left adjoint, π^{-1} can sometimes be a right adjoint, when π is an inclusion of an open subset. Suppose $i\colon U \hookrightarrow Y$ is an inclusion of an open set into Y. Define the **extension of i by zero** $i_!\colon Mod_{\mathscr{O}_U} \to Mod_{\mathscr{O}_Y}$ as follows. Suppose $\mathscr{F}$ is an $\mathscr{O}_U$-module. For open $W \subset Y$, define $(i_!^{\mathrm{pre}} \mathscr{F})(W) = \mathscr{F}(W)$ if $W \subset U$, and 0 otherwise (with the obvious restriction maps). This is clearly a presheaf $\mathscr{O}_Y$-module. Define $i_!$ as $(i_!^{\mathrm{pre}})^{\mathrm{sh}}$. Note that $i_! \mathscr{F}$ is an $\mathscr{O}_Y$-module, and that this defines a functor. (The symbol "!" is read as "shriek," presumably because people shriek when they see it. Thus "$i_!$" is read as "i-lower-shriek.")

23.4.F. EXERCISE. Show that $i_!^{\mathrm{pre}} \mathscr{F}$ satisfies the identity axiom, but need not be a sheaf. (We won't need this, but it may give some insight into why this is called "extension by zero." Possible source for an example: continuous functions on $\mathbb{R}$.)

23.4.G. EXERCISE.

(a) For $q \in Y$, show that $(i_! \mathscr{F})_q = \mathscr{F}_q$ if $q \in U$, and 0 otherwise.

(b) Show that $i_!$ is an exact functor.

(c) If $\mathscr{G}$ is an $\mathscr{O}_Y$-module, describe an inclusion $i_! i^{-1} \mathscr{G} \hookrightarrow \mathscr{G}$. (Interesting remark we won't need: Let Z be the complement of U, and $j\colon Z \to Y$ the natural inclusion. Then there is a short exact sequence
$$0 \longrightarrow i_! i^{-1} \mathscr{G} \longrightarrow \mathscr{G} \longrightarrow j_* j^{-1} \mathscr{G} \longrightarrow 0.$$

This is best checked by describing the maps, then checking exactness at stalks.)

(d) Show that $(i_!, i^{-1})$ is an adjoint pair, so there is a natural bijection
$$\mathrm{Hom}_{\mathscr{O}_Y}(i_! \mathscr{F}, \mathscr{G}) \longleftrightarrow \mathrm{Hom}_{\mathscr{O}_U}(\mathscr{F}, \mathscr{G}|_U)$$

for any $\mathscr{O}_U$-module $\mathscr{F}$ and $\mathscr{O}_Y$-module $\mathscr{G}$. (In particular, the sections of $\mathscr{G}$ over U can be identified with $\mathrm{Hom}_{\mathscr{O}_Y}(i_! \mathscr{O}_U, \mathscr{G})$.)

We are ready to use the tool of the extension by zero functor.

23.4.H. EXERCISE (INJECTIVE SHEAVES ARE FLASQUE). Suppose $(X, \mathscr{O}_X)$ is a ringed space, and $\mathscr{I}$ is an injective $\mathscr{O}_X$-module. Show that $\mathscr{I}$ is flasque. Hint: If $U \subset V \subset X$, then describe an injection of $\mathscr{O}_X$-modules $0 \to (i_U)_! \mathscr{O}_U \to (i_V)_! \mathscr{O}_V$, where $i_U\colon U \hookrightarrow X$ and $i_V\colon V \hookrightarrow X$ are the obvious open embeddings. Apply the exact contravariant functor $\mathrm{Hom}(\cdot, \mathscr{I})$.

23.4.I. EXERCISE (FLASQUE IMPLIES Γ-ACYCLIC). Suppose $\mathscr{F}$ is a flasque $\mathscr{O}_X$-module. Show that $\mathscr{F}$ is Γ-acyclic (that $H^i(X, \mathscr{F}) = 0$ for $i > 0$, §23.3.2) as follows. As $Mod_{\mathscr{O}_X}$ has enough injectives, choose an inclusion of $\mathscr{F}$ into some injective $\mathscr{I}$, and call its cokernel $\mathscr{G}$:

$$0 \longrightarrow \mathscr{F} \longrightarrow \mathscr{I} \longrightarrow \mathscr{G} \longrightarrow 0.$$

Then $\mathscr{I}$ is flasque by Exercise 23.4.H, so $\mathscr{G}$ is flasque by Exercise 23.4.D(b). Take the long exact sequence in (derived functor) cohomology, and show that $H^1(X, \mathscr{F}) = 0$. Your argument works for *any* flasque sheaf $\mathscr{F}$, so $H^1(X, \mathscr{G}) = 0$ as well. Show that $H^2(X, \mathscr{F}) = 0$. Turn this into an induction.

Thus if $\pi \colon X \to Y$ is a morphism of ringed spaces, and $\mathscr{I}$ is an injective $\mathscr{O}_X$-module, then $\mathscr{I}$ is flasque (Exercise 23.4.H), so $\pi_* \mathscr{I}$ is flasque (Exercise 23.4.E(b)), so $\pi_* \mathscr{I}$ is acyclic for the functor Γ (Exercise 23.4.I), so this completes the proof of the Leray spectral sequence (Theorem 23.4.5). $\quad\square$

23.4.J. EXERCISE. Extend the Leray spectral sequence (Theorem 23.4.5) to deal with a composition of higher pushforwards for

$$(X, \mathscr{O}_X) \xrightarrow{\ \pi\ } (Y, \mathscr{O}_Y) \xrightarrow{\ \rho\ } (Z, \mathscr{O}_Z).$$

23.4.8.** The category of $\mathscr{O}_X$-modules on a ringed space need not have enough projectives. In contrast to Theorem 23.4.1, the category of $\mathscr{O}_X$-modules on a ringed space need not have enough projectives. For example, let X be $\mathbb{P}^1_k$ with the Zariski-topology (in fact, we will need to know very little about X—only that it is not an *Alexandrov space*), but take $\mathscr{O}_X$ to be the constant sheaf $\underline{\mathbb{Z}}$. We will see that $Mod_{\mathscr{O}_X}$—i.e., the category of sheaves of abelian groups on X—does not have enough projectives. If $Mod_{\mathscr{O}_X}$ had enough projectives, then there would be a surjection $\psi \colon P \to \underline{\mathbb{Z}}$ from a projective sheaf. Fix a closed point $q \in X$. We will show that the map on stalks $\psi_q \colon P_q \to \underline{\mathbb{Z}}_q$ is the zero map, contradicting the surjectivity of ψ. For each open subset U of X, denote by $\underline{\mathbb{Z}}_U$ the extension by zero from U to X of the constant sheaf with values in $\mathbb{Z}$ (§23.4.7—$\underline{\mathbb{Z}}_U(V) = \mathbb{Z}$ if $V \subset U$, and $\underline{\mathbb{Z}}_U(V) = 0$ otherwise). For each open neighborhood V of q, let W be a strictly smaller open neighborhood. Consider the surjection $\underline{\mathbb{Z}}_{X-q} \oplus \underline{\mathbb{Z}}_W \to \underline{\mathbb{Z}}$. By projectivity of P, the surjection ψ lifts to $P \to \underline{\mathbb{Z}}_{X-q} \oplus \underline{\mathbb{Z}}_W$. The map $P(V) \to \underline{\mathbb{Z}}(V)$ factors through $\underline{\mathbb{Z}}_{X-q}(V) \oplus \underline{\mathbb{Z}}_W(V) = 0$, and hence must be the zero map. Thus the map $\psi_q \colon P_q \to \underline{\mathbb{Z}}_q$ is zero as well (do you see why?), as desired.

23.5 Čech Cohomology and Derived Functor Cohomology Agree

We next prove that Čech cohomology and derived functor cohomology agree, where the former is defined.

23.5.1. Theorem — *Suppose X is a quasicompact separated scheme, and $\mathscr{F}$ is a quasicoherent sheaf. Then the Čech cohomology of $\mathscr{F}$ agrees with the derived functor cohomology of $\mathscr{F}$.*

This statement is not as precise as it should be. We would want to know that this isomorphism is functorial in $\mathscr{F}$, and that it respects long exact sequences (so the connecting homomorphism defined for Čech cohomology agrees with that for derived functor cohomology). There is also an important extension to higher pushforwards. We leave these issues for the end of this section, §23.5.5.

In case you are curious: so long as it is defined appropriately, Čech cohomology agrees with derived functor cohomology in a wide variety of circumstances outside of scheme theory (if the underlying topological space is paracompact), but not always (see [Gr1, §3.8] for a counterexample). Remark 23.5.6 is also related.

The central idea in the proof (albeit with a twist) is a spectral sequence argument in the same style as those of §23.3, and uses two "cohomology-vanishing" ingredients, one for each orientation of the spectral sequence.

(A) If $(X, \mathscr{O}_X)$ is a ringed space, $\mathscr{Q}$ is an injective $\mathscr{O}_X$-module, and $X = \cup_i U_i$ is a finite open cover, then $\mathscr{Q}$ has no ith Čech cohomology with respect to this cover for $i > 0$.

(B) If X is an affine scheme, and $\mathscr{F}$ is a quasicoherent sheaf on X, then $(R^i\Gamma)(\mathscr{F}) = 0$ for $i > 0$.

Translation: **(A)** says that building blocks of derived functor cohomology have no Čech cohomology, and **(B)** says that building blocks of Čech cohomology have no derived functor cohomology.

23.5.A. PRELIMINARY EXERCISE. Suppose $(X, \mathscr{O}_X)$ is a ringed space, $\mathscr{Q}$ is an injective $\mathscr{O}$-module, and $i : U \hookrightarrow X$ is an open subset. Show that $\mathscr{Q}|_U$ is injective on U. Hint: Use the fact that i^{-1} has an *exact* left adjoint $i_!$ (extension by zero); see §23.4.7, and the following diagrams.

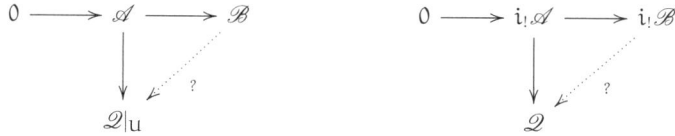

$$
\begin{array}{ccc}
0 \longrightarrow \mathscr{A} \longrightarrow \mathscr{B} \\
\downarrow \quad \swarrow \quad {}^? \\
\mathscr{Q}|_U
\end{array}
\qquad\qquad
\begin{array}{ccc}
0 \longrightarrow i_!\mathscr{A} \longrightarrow i_!\mathscr{B} \\
\downarrow \quad \swarrow \quad {}^? \\
\mathscr{Q}
\end{array}
$$

In the course of Exercise 23.5.A, you will have proved the following fact, which we shall use again in Exercise 29.4.A. (You can also use it to solve Exercise 23.4.A.)

23.5.B. EXERCISE (RIGHT ADJOINTS TO EXACT FUNCTORS PRESERVE INJECTIVES). Show that if (F, G) is an adjoint pair of additive functors between abelian categories, and F is exact, then G sends injective elements to injective elements.

23.5.2. *Proof of Theorem 23.5.1, assuming* **(A)** *and* **(B)**. As with the facts proved in §23.3, we take the only approach that is reasonable: we choose an injective resolution $0 \to \mathscr{F} \to \mathscr{Q}_\bullet$ of $\mathscr{F}$ and a Čech cover of X, mix these two types of information in a double complex, and toss it into our spectral sequence machine (§1.6). More precisely, choose a finite affine open cover $X = \cup_i U_i$ and an injective resolution

$$
0 \longrightarrow \mathscr{F} \longrightarrow \mathscr{Q}_0 \longrightarrow \mathscr{Q}_1 \longrightarrow \cdots .
$$

Consider the double complex

(23.5.2.1)

$$
\begin{array}{ccccccccc}
\vdots & & \vdots & & \vdots & & \vdots \\
\uparrow & & \uparrow & & \uparrow & & \uparrow \\
0 \to & \oplus_i \mathscr{Q}_2(U_i) & \to & \oplus_{i,j} \mathscr{Q}_2(U_{ij}) & \to & \oplus_{i,j,k} \mathscr{Q}_2(U_{ijk}) & \to & \cdots \\
\uparrow & & \uparrow & & \uparrow & & \uparrow \\
0 \to & \oplus_i \mathscr{Q}_1(U_i) & \to & \oplus_{i,j} \mathscr{Q}_1(U_{ij}) & \to & \oplus_{i,j,k} \mathscr{Q}_1(U_{ijk}) & \to & \cdots \\
\uparrow & & \uparrow & & \uparrow & & \uparrow \\
0 \to & \oplus_i \mathscr{Q}_0(U_i) & \to & \oplus_{i,j} \mathscr{Q}_0(U_{ij}) & \to & \oplus_{i,j,k} \mathscr{Q}_0(U_{ijk}) & \to & \cdots \\
\uparrow & & \uparrow & & \uparrow & & \uparrow \\
0 \to & 0 & \to & 0 & \to & 0 & \to & \cdots .
\end{array}
$$

We take this as the E_0 term in a spectral sequence. First, we use the rightward filtration. As higher Čech cohomology of injective $\mathscr{O}$-modules is 0 (assumption **(A)**), we get 0's everywhere except in "column 0," where we get $\mathscr{Q}_i(X)$ in row i:

$$
\begin{array}{ccccc}
\vdots & \vdots & \vdots & \vdots & \\
\uparrow & \uparrow & \uparrow & \uparrow & \\
0 & \mathscr{D}_2(X) & 0 & 0 & \cdots \\
\uparrow & \uparrow & \uparrow & \uparrow & \\
0 & \mathscr{D}_1(X) & 0 & 0 & \cdots \\
\uparrow & \uparrow & \uparrow & \uparrow & \\
0 & \mathscr{D}_0(X) & 0 & 0 & \cdots \\
\uparrow & \uparrow & \uparrow & \uparrow & \\
0 & 0 & 0 & 0 & .
\end{array}
$$

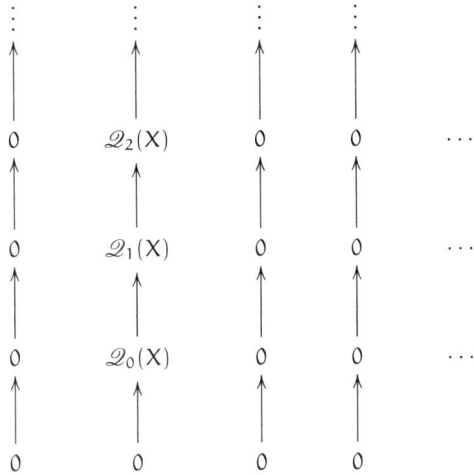

Then we take cohomology in the vertical direction, and we get derived functor cohomology of $\mathscr{F}$ on X on the E_2 page:

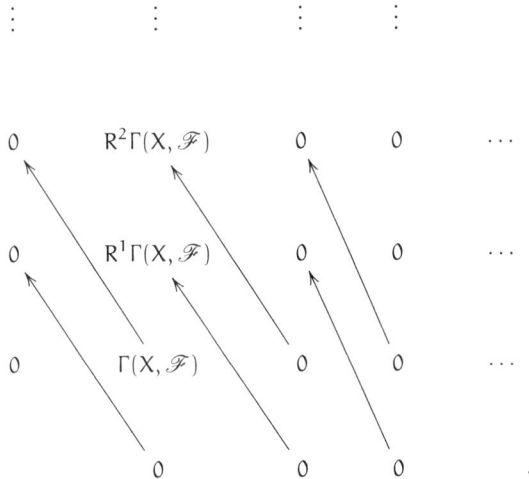

$$
\begin{array}{ccccc}
\vdots & \vdots & \vdots & \vdots & \\
0 & R^2\Gamma(X,\mathscr{F}) & 0 & 0 & \cdots \\
0 & R^1\Gamma(X,\mathscr{F}) & 0 & 0 & \cdots \\
0 & \Gamma(X,\mathscr{F}) & 0 & 0 & \cdots \\
& 0 & 0 & 0 & .
\end{array}
$$

We then start over on the E_0 page, and this time use the upward orientation (corresponding to choosing the upward arrow first). By Proposition 23.5.A, $\mathscr{D}_i|_{U_J}$ is injective on U_J, so we are computing the derived functor cohomology of $\mathscr{F}$ on U_J. Then the higher derived functor cohomology of $\mathscr{F}$ on U_J is 0 (assumption **(B)**), so all entries are 0 except possibly on row 0. Thus the E_1 term is:

(23.5.2.2)
$$
0 \longrightarrow 0 \longrightarrow 0 \longrightarrow 0 \longrightarrow \cdots
$$

$$
0 \longrightarrow 0 \longrightarrow 0 \longrightarrow 0 \longrightarrow \cdots
$$

$$
0 \longrightarrow \oplus_i \Gamma(U_i,\mathscr{F}) \longrightarrow \oplus_{i,j} \Gamma(U_{ij},\mathscr{F}) \longrightarrow \oplus_{i,j,k} \Gamma(U_{ijk},\mathscr{F}) \longrightarrow \cdots
$$

$$
0 \longrightarrow 0 \longrightarrow 0 \longrightarrow 0 \longrightarrow \cdots .
$$

Row 0 is precisely the Čech complex of $\mathscr{F}$, so the spectral sequence converges at the E_2 term, yielding the Čech cohomology. Since one orientation yields derived functor cohomology and one yields Čech cohomology, we are done. $\qquad\square$

So it remains to show **(A)** and **(B)**.

23.5.3. Ingredient (A): injectives have no Čech cohomology.

23.5.C. EXERCISE. Suppose $X = \cup_j U_j$ is a finite cover of X by open sets, and $\mathscr{F}$ is a flasque sheaf (Definition 23.4.6) on X. Show that the Čech complex for $\mathscr{F}$ with respect to $\cup_{j=1}^{n} U_j$ has no cohomology in positive degree, i.e., that it is exact except in degree 0 (where it has cohomology $\mathscr{F}(X)$, by the sheaf axioms). Hint: Use induction on n. Consider the short exact sequence of complexes (18.2.4.2) (see also (18.2.3.1)). The corresponding long exact sequence will immediately give the desired result for $i > 1$, and flasqueness will be used for $i = 1$.

Thus flasque sheaves have no Čech cohomology, so injective $\mathscr{O}$-modules, in particular (Exercise 23.4.H), have none. This is all we need for our algebro-geometric applications, but to show you how general this machinery is, we give an entertaining application.

23.5.D. UNIMPORTANT EXERCISE (PERVERSE PROOF OF INCLUSION-EXCLUSION THROUGH COHOMOLOGY OF SHEAVES). The inclusion-exclusion principle is (equivalent to) the following: suppose that X is a finite set, and U_i ($1 \leq i \leq n$) are finite sets covering X. As usual, define $U_I = \cap_{i \in I} U_i$ for $I \subset \{1, \ldots, n\}$. Then

$$|X| = \sum |U_i| - \sum_{|I|=2} |U_{|I|}| + \sum_{|I|=3} |U_{|I|}| - \sum_{|I|=4} |U_{|I|}| + \cdots .$$

Prove this by endowing X with the discrete topology, showing that the constant sheaf $\underline{\mathbb{Q}}$ is flasque, considering the Čech complex computing $H^i(X, \underline{\mathbb{Q}})$ using the cover $\{U_i\}$, and using Exercise 1.5.B.

23.5.4.* Ingredient (B): Quasicoherent sheaves on affine schemes have no derived functor cohomology.

We show the following statement by induction on k. Suppose X is an affine scheme, and $\mathscr{F}$ is a quasicoherent sheaf on X. Then $R^i \Gamma(X, \mathscr{F}) = 0$ for $0 < i \leq k$. The result is vacuously true for $k = 0$; so suppose we know the result for all $0 < k' < k$ (with X replaced by *any* affine scheme). Suppose $\alpha \in R^k \Gamma(X, \mathscr{F})$. We wish to show that $\alpha = 0$. Choose an injective resolution *by $\mathscr{O}_X$-modules,*

$$
\begin{array}{c}
\vdots \\
\uparrow {\scriptstyle d_1} \\
\mathscr{Q}_1 \\
\uparrow {\scriptstyle d_0} \\
\mathscr{Q}_0 \\
\uparrow \\
\mathscr{F} \\
\uparrow \\
0 \ .
\end{array}
$$

Then α has a representative α' in $\mathscr{Q}_k(X) = \Gamma(X, \mathscr{Q}_k)$, such that $d\alpha' = 0$. Because the injective resolution is exact, α' is locally a boundary. In other words, in the open neighborhood of any point $p \in X$, there is an open set V_p such that $\alpha'|_{V_p} = d\alpha''$ for some $\alpha'' \in \mathscr{Q}_{k-1}(V_p)$. (Be sure you see why this is true! Recall that taking cokernel of a map of sheaves requires sheafification; see Proposition 2.6.1.) By shrinking V_p if necessary, we can assume V_p is affine. By the quasicompactness of X, we can choose a finite number of the V_p's that cover X. Rename these as U_i, so we have an affine cover X. Consider the Čech cover of X with respect to *this* affine cover (*not* the affine cover you might have thought we would use—that of X by itself—but instead an affine cover tailored for our particular α). Consider the double complex (23.5.2.1), as the E_0 term in a spectral sequence.

First consider the rightward orientation. As in the argument in §23.5.2, the spectral sequence converges at E_2, where we get 0 everywhere, except that the derived functor cohomology appears in the 0th column.

Next, start over again, choosing the upward orientation. On the E_1 page, row 0 is the Čech complex, as in (23.5.2.2). All the rows between 1 and $k-1$ are 0 by our inductive hypothesis, but we don't yet know anything about the higher rows. Because we are interested in the kth derived functor, we focus on the kth antidiagonal ($E_\bullet^{p,k-p}$). The only possibly nonzero terms in this antidiagonal are $E_1^{k,0}$ and $E_1^{0,k}$. We look first at the term on the bottom row $E_1^{k,0} = \prod_{|I|=k} \Gamma(U_I, \mathscr{F})$, which is part of the Čech complex:

$$\cdots \longrightarrow \prod_{|I|=k-1} \Gamma(U_I, \mathscr{F}) \longrightarrow \prod_{|I|=k} \Gamma(U_I, \mathscr{F}) \longrightarrow \prod_{|I|=k+1} \Gamma(U_I, \mathscr{F}) \longrightarrow \cdots.$$

But we have already verified that the Čech cohomology of a quasicoherent sheaf on an affine scheme vanishes—this is the one spot where we use the quasicoherence of $\mathscr{F}$. Thus this term vanishes by the E_2 page (i.e., $E_i^{k,0} = 0$ for $i \geq 2$).

So the only term of interest in the kth antidiagonal of E_1 is $E_1^{0,k}$, which is the homology of

$$(23.5.4.1) \qquad \prod_i \mathscr{Q}_{k-1}(U_i) \longrightarrow \prod_i \mathscr{Q}_k(U_i) \longrightarrow \prod_i \mathscr{Q}_{k+1}(U_i),$$

which is $\prod_i R^k \Gamma(U_i, \mathscr{F})$ (using Preliminary Exercise 23.5.A, which stated that the $\mathscr{Q}_j|_{U_i}$ are injective on U_i, so they can be used to compute $R^k(\Gamma(U_i, \mathscr{F}))$). So $E_2^{0,k}$ is the homology of

$$0 \longrightarrow \prod_i R^k \Gamma(U_i, \mathscr{F}) \longrightarrow \prod_{i,j} R^k \Gamma(U_{ij}, \mathscr{F}),$$

and thereafter all differentials to and from the $E_\bullet^{0,k}$ terms will be 0, as the sources and targets of those arrows will be 0. Consider now our lift of α' of our original class $\alpha \in R^k \Gamma(X, \mathscr{F})$ to $\prod_i R^k \Gamma(U_i, \mathscr{F})$. Its image in the homology of (23.5.4.1) is *zero*—this was how we chose our cover U_i to begin with! Thus $\alpha = 0$, as desired, completing our proof. □

23.5.E.** EXERCISE. The proof is not quite complete. We have a class $\alpha \in R^k \Gamma(X, \mathscr{F})$, and we have interpreted $R^k \Gamma(X, \mathscr{F})$ as

$$\ker \left(\prod_i R^k \Gamma(U_i, \mathscr{F}) \longrightarrow \prod_{i,j} R^k \Gamma(U_{ij}, \mathscr{F}) \right).$$

We have two maps $R^k \Gamma(X, \mathscr{F}) \to R^k \Gamma(U_i, \mathscr{F})$, one coming from the natural restriction (under which we can see that the image of α is zero), and one coming from the actual spectral sequence machinery. Verify that they are the same map. (Possible hint: With the spectral sequence orientation used, the $E_\infty^{0,k}$ term is indeed the quotient of the homology of the double complex, so the map goes the right way.)

23.5.5.* Tying up loose ends.

23.5.F. IMPORTANT EXERCISE. State and prove the generalization of Theorem 23.5.1 to higher pushforwards $R^i \pi_*$, where $\pi \colon X \to Y$ is a quasicompact separated morphism of schemes.

23.5.G. EXERCISE. Show that the isomorphism of Theorem 23.5.1 is functorial in $\mathscr{F}$, i.e., given a morphism $\mathscr{F} \to \mathscr{G}$, the diagram

$$\begin{array}{ccc} H^i(X, \mathscr{F}) & \longleftrightarrow & R^i\Gamma(X, \mathscr{F}) \\ \downarrow & & \downarrow \\ H^i(X, \mathscr{G}) & \longleftrightarrow & R^i\Gamma(X, \mathscr{G}) \end{array}$$

commutes, where the horizontal arrows are the isomorphisms of Theorem 23.5.1, and the vertical arrows come from the functoriality of H^i and $R^i\Gamma$. (Hint: "Spectral sequences are functorial in E_0," which can be easily seen from the construction, although we haven't said it explicitly. See §1.6.7, which remarks on the functoriality of spectral sequences.)

23.5.H. EXERCISE. Show that the isomorphisms of Theorem 23.5.1 induce isomorphisms of long exact sequences.

23.5.6. *Remark.* If you wish, you can use the above argument to prove the following theorem of Leray. Suppose we have a sheaf of abelian groups $\mathscr{F}$ on a topological space X, and some covering $\{U_i\}$ of X such that the (derived functor) cohomology of $\mathscr{F}$ in positive degree vanishes on every finite intersection of the U_i. Then the cohomology of $\mathscr{F}$ can be calculated by the Čech cohomology of the cover $\{U_i\}$; there is no need to pass to the colimit of all covers, as is the case for Čech cohomology in general.

23.5.7. *Unimportant remark: Working in $QCoh_X$ rather than $Mod_{\mathscr{O}_X}$.* In our definition of derived functors of quasicoherent sheaves on X, we could have tried to work in the category of quasicoherent sheaves $QCoh_X$ itself, rather than in the larger category $Mod_{\mathscr{O}_X}$. There are several reasons why this would require more effort. It is not hard to show that $QCoh_X$ has enough injectives if X is Noetherian (see, for example, [Ha1, Exer. III.3.6(a)]). Because we don't have "extension by zero" (§23.4.7) in $QCoh$, the proofs that injective quasicoherent sheaves on an open set U restrict to injective quasicoherent sheaves on smaller open subsets V (the analog of Exercise 23.5.A) and that injective quasicoherent sheaves are flasque (the analog of Exercise 23.4.H) are harder. You can use this to show that H^i and $R^i\pi_*$ computed in $QCoh$ are the same as those computed in $Mod_{\mathscr{O}}$ (once you make the statements precise). It is true that injective elements of $QCoh_X$ (X Noetherian) are injective in $Mod_{\mathscr{O}_X}$, but this requires work (see [Mur, Prop. 68]).

It is true that $QCoh_X$ has enough injectives for *any* scheme X, but this is much harder; see [EE]. And as is clear from the previous paragraph, "enough injectives" is only the beginning of what we want.

23.5.8. *Unimportant remark.* Theorem 23.5.1 implies that if $\pi\colon X \to Y$ is quasicompact and separated, then $R^i\pi_*$ sends $QCoh_X$ to $QCoh_Y$ (by showing an isomorphism with Čech cohomology). If X and Y are Noetherian, the hypothesis "separated" can be relaxed to "quasiseparated," using the ideas of Unimportant Remark 23.5.7. But because morphisms from a Noetherian scheme are automatically quasiseparated (Easy Exercise 8.3.B(b)), we can remove the redundant "quasiseparated" assumption. This sounds exciting, but it is not nearly as useful as the separated case, because without Čech cohomology, it is hard to compute anything.

Chapter 24

Flatness

The concept of flatness is a riddle that comes out of algebra, but which technically is the answer to many prayers.

—D. Mumford [Mu7, III.10]

We return to the important concept of flatness. We could have discussed flatness at length as soon as we had discussed quasicoherent sheaves and morphisms. But it is an unexpected idea, and the algebra and geometry are not obviously connected, so we have left it for relatively late. Serre has stated that he introduced flatness purely for reasons of algebra in his landmark "GAGA" paper [Se3], and that it was Grothendieck who recognized its geometric significance. "Exact" seems more descriptive terminology than "flat" (so we would have "exact modules" and "exact morphisms"), but this has not caught on.

24.0.1. A flat morphism $\pi\colon X \to Y$ is the right notion of a "nice," or "nicely varying" family over Y. For example, if π is a projective flat family over a connected Noetherian base (translation: $\pi\colon X \to Y$ is a projective flat morphism, with Y connected and Noetherian), we will see that various numerical invariants of fibers are constant, including the dimension (§24.5.5), and numbers interpretable in terms of an Euler characteristic (see §24.7):

(a) the Hilbert polynomial (Corollary 24.7.2),
(b) the degree (in projective space) (Exercise 24.7.A(a)),
(b′) (as a special case of (b)) if π is finite, the degree of π (recovering and extending the fact that the degree of a projective map between regular curves is constant, §16.3.4; see Exercise 24.3.G and §24.3.12),
(c) the arithmetic genus (Exercise 24.7.A(b)),
(d) the degree of a line bundle if the fiber is a curve (Corollary 24.7.3), and
(e) intersections of divisors and line bundles (Exercise 24.7.D).

One might think that the right hypothesis might be smoothness, or more generally some sort of equisingularity, but we only need something weaker. And this is a good thing: branched covers are not fibrations in any traditional sense, yet they still behave well—the double cover $\mathbb{A}^1 \to \mathbb{A}^1$ given by $y \mapsto x^2$ has constant degree 2 (§10.3.4, revisited in §16.3.10). Another key example is that of a family of smooth curves degenerating to a nodal curve (Figure 24.1)—the topology of the (underlying analytic) curve changes, but the arithmetic genus remains constant. One can prove things about regular curves by first proving them about a

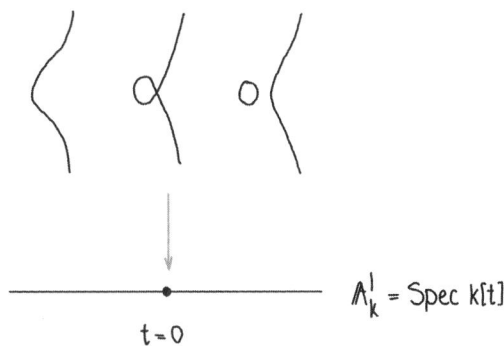

Figure 24.1 *A flat family of smooth curves degenerating to a nodal curve:* $y^2 = x^3 + x^2 + tx$.

nodal degeneration, and then showing that the result behaves well in flat families. Degeneration techniques such as this are ubiquitous in algebraic geometry.

Given the cohomological nature of the constancy of Euler characteristic result, you should not be surprised that the hypothesis needed (flatness) is cohomological in nature—it can be characterized by the vanishing of Tor (Exercise 23.1.C), which we use to great effect in §24.2.

But flatness is important for other reasons too. As a start: as this is the right notion of a "nice family," it allows us to correctly define the notion of moduli space. For example, the *Hilbert scheme* of $\mathbb{P}^n$ "parametrizes closed subschemes of $\mathbb{P}^n$." Maps from a scheme B to the Hilbert scheme correspond to (finitely presented) closed subschemes of $\mathbb{P}^n_B$ *flat* over B. By universal property nonsense, this defines the Hilbert scheme up to unique isomorphism (although we of course must show that it exists, which takes some effort—see Theorem 25.3.1 for more). The moduli space of smooth projective curves is defined by the universal property that maps to the moduli space correspond to projective flat (finitely presented) families whose geometric fibers are smooth curves. (Sadly, this moduli space does not exist.) On a related note, flatness is central in deformation theory: it is key to understanding how schemes (and other geometric objects, such as vector bundles) can deform (cf. §21.5.12). Finally, the notion of Galois descent generalizes to (faithfully) "flat descent," which allows us to "glue" in more exotic Grothendieck topologies in the same way we do in the Zariski topology (or more classical topologies); but this is beyond the scope of our current discussion.

Flatness has many aspects of different flavors, and it is easy to lose sight of the forest for the trees. Because the algebra of flatness seems so unrelated to the geometry, it can be nonintuitive. We will necessarily begin with algebraic foundations, but you should focus on the following points: methods of showing things are flat (both general criteria and explicit examples), and classification of flat modules over particular kinds of rings. You should try every exercise dealing with explicit examples such as these.

Here is an outline of the chapter, to help focus your attention.

- In §24.1, we discuss some of the easier facts, which are algebraic in nature.
- §24.2, §24.3, and §24.6 give ideal-theoretic criteria for flatness. §24.2 and §24.3 should be read together. The first uses Tor to understand flatness, and the second uses these insights to develop ideal-theoretic criteria for flatness. §24.6, on local criteria for flatness, is harder.
- §24.5 is relatively freestanding, and could be read immediately after §24.1. It deals with topological aspects of flatness, such as the fact that flat morphisms are open maps in good situations. (Recall that a map of topological spaces is an **open map** if the image of every open set is an open set.)
- In §24.7, we explain the fact that "the Euler characteristic of quasicoherent sheaves is constant in flat families" (with appropriate hypotheses), and its many happy consequences. This section is surprisingly easy given its utility.

You should focus on what flatness implies and how to "picture" it, but also on explicit criteria for flatness in different situations, such as for integral domains (Observation 24.1.2), principal ideal domains (Exercise 24.3.B), discrete valuation rings (Exercise 24.3.C), the dual numbers (Exercise 24.3.D), and local rings (Theorem 24.3.5).

> *Happy families are all alike; every unhappy family is unhappy in its own way.* (See, for example, Figure 24.2, and Exercises 24.1.G(d), 24.1.L, and 24.3.H.)
>
> —L. Tolstoy, Anna Karenina, [To, I. 1]

24.1   Easier Facts

Many facts about flatness are easy or immediate, although a number are tricky. As always, I will try to make clear which is which, to help you remember the easy facts and the key ideas of proofs of the harder facts. We will pick the low-hanging fruit first.

We recall the definition of a *flat A-module* (§1.5.13). If $M \in Mod_A$, $M \otimes_A \cdot$ is always right-exact (Exercise 1.2.H). We say that M **is a flat A-module** (or *flat over A* or *A-flat*) if $M \otimes_A \cdot$ is an exact functor. We say that a *ring morphism* $B \to A$ *is* **flat** if A is flat as a B-module. (In particular, the algebra structure of A is irrelevant.)

24.1.1. *Two key examples.*

(i) Free A-modules (even of infinite rank) are clearly flat. More generally, projective modules are flat (Exercise 23.2.B).

(ii) Localizations are flat: Suppose S is a multiplicative subset of B. Then $B \to S^{-1}B$ is a flat ring morphism because localization is exact (Exercise 1.5.G(a)).

24.1.A. EASY EXERCISE: FIRST EXAMPLES.

(a) *(trick question)* Classify flat modules over a field k.

(b) Show that $A[x_1, \ldots, x_n]$ is a flat A-module.

(c) Show that the ring morphism $\mathbb{Q}[x] \to \mathbb{Q}[y]$, with $x \mapsto y^2$, is flat. (This will help us understand Example 10.3.4 better; see §24.3.12.)

We make some quick but important observations.

24.1.2. *Important observation.* If x is a non-zerodivisor of A, and M is a flat A-module, then

$$M \xrightarrow{\times x} M$$

is injective. (Apply the exact functor $M \otimes_A \cdot$ to the exact sequence $0 \longrightarrow A \xrightarrow{\times x} A$.) In particular, *flat modules are torsion-free.* (Torsion-freeness was defined in §6.1.4.) This observation gives an easy way of recognizing when a module is *not* flat. We will use it many times.

24.1.B. EXERCISE: ANOTHER EXAMPLE. Show that a finitely generated module over a principal ideal domain is flat if and only if it is torsion-free if and only if it is free. Hint: Remark 6.4.4 classifies finitely generated modules over a principal ideal domain. (Exercise 24.3.B sheds more light on flatness over a principal ideal domain. Proposition 14.3.2 is also related.)

24.1.C. EXERCISE (FLATNESS IS PRESERVED BY CHANGE OF BASE RING). Show that if M is a flat B-module, and $B \to A$ is a homomorphism, then $M \otimes_B A$ is a flat A-module.

24.1.D. EXERCISE (FLATNESS IS PRESERVED BY COMPOSITION). Show that if A is a flat B-algebra, and M is A-flat, then M is also B-flat.

24.1.3. Proposition (flatness is a stalk/prime-local property) — *An A-module M is flat if and only if $M_\mathfrak{p}$ is a flat $A_\mathfrak{p}$-module for all prime ideals $\mathfrak{p} \subset A$.*

You are welcome to prove that it suffices to check only *maximal* ideals, but we don't need this extension.

Proof. Suppose first that M is a flat A-module. Given any exact sequence of $A_\mathfrak{p}$-modules

(24.1.3.1) $$0 \longrightarrow N' \longrightarrow N \longrightarrow N'' \longrightarrow 0,$$

we wish to show that

(24.1.3.2) $$0 \longrightarrow M_\mathfrak{p} \otimes_{A_\mathfrak{p}} N' \longrightarrow M_\mathfrak{p} \otimes_{A_\mathfrak{p}} N \longrightarrow M_\mathfrak{p} \otimes_{A_\mathfrak{p}} N'' \longrightarrow 0$$

is also exact. Now (24.1.3.1) is *also* an exact sequence of A-modules, so

(24.1.3.3) $$0 \longrightarrow M \otimes_A N' \longrightarrow M \otimes_A N \longrightarrow M \otimes_A N'' \longrightarrow 0$$

is exact (by A-flatness of M). But $M \otimes_A N$ is canonically isomorphic to $M \otimes_A A_\mathfrak{p} \otimes_{A_\mathfrak{p}} N = M_\mathfrak{p} \otimes_{A_\mathfrak{p}} N$ (and similarly for N', N''), so we can rewrite (24.1.3.3) as (24.1.3.2) (make sure you see why the maps in (24.1.3.3) and (24.1.3.2) are the same), as desired.

Assume next that $M_\mathfrak{p}$ is a flat $A_\mathfrak{p}$-module for all $\mathfrak{p}$. Given any short exact sequence (24.1.3.1), we seek to show that (24.1.3.3) is exact. By right-exactness of tensor product (Exercise 1.2.H), (24.1.3.3) is exact except possibly on the left. Let K be $\ker(M \otimes_A N' \to M \otimes_A N)$, so

$$(24.1.3.4) \qquad 0 \longrightarrow K \longrightarrow M \otimes_A N' \longrightarrow M \otimes_A N \longrightarrow M \otimes_A N'' \longrightarrow 0$$

is exact; we wish to show that $K = 0$. Given any $\mathfrak{p}$, localizing (24.1.3.1) at $\mathfrak{p}$ and tensoring with the exact $A_\mathfrak{p}$-module $M_\mathfrak{p}$ yields

$$(24.1.3.5) \qquad 0 \longrightarrow M_\mathfrak{p} \otimes_{A_\mathfrak{p}} N'_\mathfrak{p} \longrightarrow M_\mathfrak{p} \otimes_{A_\mathfrak{p}} N_\mathfrak{p} \longrightarrow M_\mathfrak{p} \otimes_{A_\mathfrak{p}} N''_\mathfrak{p} \longrightarrow 0.$$

Localizing (24.1.3.4) at $\mathfrak{p}$ and using the isomorphisms $M_\mathfrak{p} \otimes_{A_\mathfrak{p}} N_\mathfrak{p} \overset{\sim}{\longleftrightarrow} (M \otimes_A N)_\mathfrak{p}$, we obtain the exact sequence

$$0 \longrightarrow K_\mathfrak{p} \longrightarrow M_\mathfrak{p} \otimes_{A_\mathfrak{p}} N'_\mathfrak{p} \longrightarrow M_\mathfrak{p} \otimes_{A_\mathfrak{p}} N_\mathfrak{p} \longrightarrow M_\mathfrak{p} \otimes_{A_\mathfrak{p}} N''_\mathfrak{p} \longrightarrow 0,$$

which is the same as the exact sequence (24.1.3.5) except for the $K_\mathfrak{p}$. Hence $K_\mathfrak{p} = 0$ for all primes $\mathfrak{p} \subset A$, so $K = 0$ (by Exercise 4.1.F), as desired. □

24.1.4. Flatness for schemes.
Motivated by Proposition 24.1.3, the extension of the notion of flatness to schemes is straightforward.

24.1.5. *Definition: Flat quasicoherent sheaf.* We say that a quasicoherent sheaf $\mathscr{F}$ on a scheme X is **flat at** $p \in X$ if $\mathscr{F}_p$ is a flat $\mathscr{O}_{X,p}$-module. We say that a quasicoherent sheaf $\mathscr{F}$ on a scheme X is **flat** (over X) if it is flat at all $p \in X$. In light of Proposition 24.1.3, we can check this notion on an affine open cover of X.

24.1.6. *Definition: Flat morphism.* Similarly, we say that a morphism of schemes $\pi\colon X \to Y$ is **flat at** $p \in X$ if $\mathscr{O}_{X,p}$ is a flat $\mathscr{O}_{Y,\pi(p)}$-module. We say that a morphism of schemes $\pi\colon X \to Y$ is **flat** if it is flat at all $p \in X$. We can check flatness locally on the source and the target.

We can combine these two definitions into a single fancy definition.

24.1.7. *Definition: Flat quasicoherent sheaf over a base.* Suppose $\pi\colon X \to Y$ is a morphism of schemes, and $\mathscr{F}$ is a quasicoherent sheaf on X. We say that $\mathscr{F}$ **is flat (over Y) at** $p \in X$ if $\mathscr{F}_p$ is a flat $\mathscr{O}_{Y,\pi(p)}$-module. We say that $\mathscr{F}$ **is flat (over Y)** if it is flat at all $p \in X$.

Definitions 24.1.5 and 24.1.6 correspond to the cases $X = Y$ and $\mathscr{F} = \mathscr{O}_X$, respectively. (Definition 24.1.7 applies without change to the category of ringed spaces, but we won't use this.)

24.1.E. EASY EXERCISE (REALITY CHECK). Show that open embeddings are flat.

Our results about flatness over rings above carry over easily to schemes.

24.1.F. EXERCISE. Show that a map of rings $B \to A$ is flat if and only if the corresponding morphism of schemes $\operatorname{Spec} A \to \operatorname{Spec} B$ is flat. More generally, if $B \to A$ is a map of rings, and M is an A-module, show that M is B-flat if and only if $\widetilde{M}$ is flat over $\operatorname{Spec} B$. (Be careful: this requires more than merely invoking Proposition 24.1.3.) Hint: Working indirectly in both directions is a good way to proceed. "Suppose $M_\mathfrak{p}$ is not $B_\mathfrak{q}$-flat. Then there is an exact sequence of $B_\mathfrak{p}$-modules (hence also B-modules) $0 \to N' \to N \to N'' \to 0$ such that . . . ." "Suppose M is not B-flat. Then there is an exact sequence of B-modules . . . ."

Thus if $\pi\colon X \to Y$ is an affine morphism, and $\mathscr{F}$ is a quasicoherent sheaf on X, then $\mathscr{F}$ is flat over Y if and only if $\pi_* \mathscr{F}$ is flat over Y.

24.1.G. EASY EXERCISE (EXAMPLES AND REALITY CHECKS).

(a) If X is a scheme, and $p \in X$, show that the natural morphism $\mathrm{Spec}\, \mathscr{O}_{X,p} \to X$ is flat. (Hint: Localization is flat, §24.1.1.)
(b) Show that $\mathbb{A}^n_A \to \mathrm{Spec}\, A$ is flat.
(c) If $\mathscr{F}$ is a locally free sheaf on a scheme X, show that $\mathbb{P}\mathscr{F} \to X$ (Definition 17.2.4) is flat.
(d) Show that $\mathrm{Spec}\, k \to \mathrm{Spec}\, k[t]/(t^2)$ is not flat. (Draw a picture to try to see what is not "nice" about this morphism. Some more insight about flatness of the dual numbers will be given in the criterion of Exercise 24.3.D.)

24.1.H. EXERCISE. Show that if $\pi : X \to Y$ is a flat morphism, then $\pi^* : QCoh_Y \to QCoh_X$ is an exact functor.

24.1.I. EXERCISE (FLATNESS IS PRESERVED BY COMPOSITION). Suppose $\pi : X \to Y$ and $\mathscr{F}$ is a quasicoherent sheaf on X, flat over Y. Suppose also that $\psi : Y \to Z$ is a flat morphism. Show that $\mathscr{F}$ is flat over Z.

24.1.J. EXERCISE (FLATNESS IS PRESERVED BY BASE CHANGE). Suppose $\pi : X \to Y$ is a morphism, and $\mathscr{F}$ is a quasicoherent sheaf on X, flat over Y. If $\rho : Y' \to Y$ is any morphism, and $\rho' : X \times_Y Y' \to X$ is the induced morphism, show that $(\rho')^* \mathscr{F}$ is flat over Y'.

$$
\begin{array}{ccc}
(\rho')^*\mathscr{F} & & \mathscr{F} \\
| & & | \\
| & & | \\
X \times_Y Y' & \xrightarrow{\ \rho'\ } & X \\
\downarrow & & \downarrow{\scriptstyle\pi} \\
Y' & \xrightarrow{\ \rho\ } & Y
\end{array}
$$

In particular, using Exercise 24.1.A(a), if $Y = \mathrm{Spec}\, k$ (and X and Y' are *any* k-schemes, and $\mathscr{F}$ is *any* quasicoherent sheaf on X), then $(\rho')^* \mathscr{F}$ is flat over Y'. For example, $X \times_k Y'$ is *always* flat over Y'— "products over a field are flat over their factors." (Feel free to immediately generalize this further; for example, $\mathscr{F}$ can be a quasicoherent sheaf on a scheme Z over X, flat over Y.)

24.1.8. *Flat morphisms behave reasonably.* Since flatness is clearly local on the target from the definition, the previous two exercises show that the class of flat morphisms is "reasonable" in the sense of §8.1.

The following exercise is very useful for visualizing flatness and non-flatness (see, for example, Figure 24.2).

24.1.K. EXERCISE (FLAT MAPS SEND ASSOCIATED POINTS TO ASSOCIATED POINTS). Suppose $\pi : X \to Y$ is a flat morphism of locally Noetherian schemes. Show that any associated point of X must map to an associated point of Y. (Feel free to immediately generalize this to a coherent sheaf $\mathscr{F}$ on X, flat over Y, without π itself needing to be flat.) Hint: Suppose $\pi^\sharp : (B, \mathfrak{n}) \to (A, \mathfrak{m})$ is a local morphism of Noetherian local rings (i.e., $\pi^\sharp(\mathfrak{n}) \subset \mathfrak{m}$, §7.3.1). Suppose $\mathfrak{n}$ is not an associated prime of B. Show that there is an element $f \in \mathfrak{n}$ not in any associated prime of B (perhaps using prime avoidance, Proposition 12.2.13), and hence it is a non-zerodivisor. Show that $\pi^\sharp f \in \mathfrak{m}$ is a non-zerodivisor of A using Observation 24.1.2, and thus show that $\mathfrak{m}$ is not an associated prime of A.

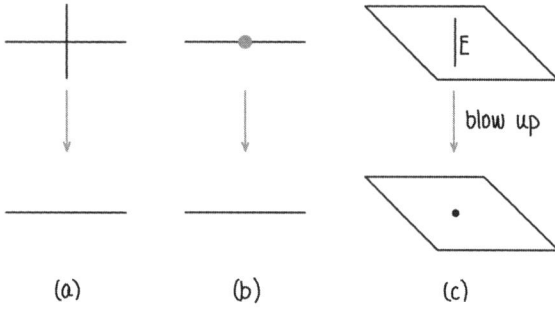

Figure 24.2 *Morphisms that are not flat (Exercise 24.1.L).*

24.1.L. EXERCISE. Use Exercise 24.1.K to show that the following morphisms are not flat (see Figure 24.2):

(a) $\operatorname{Spec} k[x, y]/(xy) \to \operatorname{Spec} k[x]$,
(b) $\operatorname{Spec} k[x, y]/(y^2, xy) \to \operatorname{Spec} k[x]$,
(c) $\operatorname{Bl}_{(0,0)} \mathbb{A}_k^2 \to \mathbb{A}_k^2$.

Hint for (c): First pull back to a line through the origin to obtain a something akin to (a). (This foreshadows the statement and proof of Proposition 24.5.6, which says that for flat morphisms "there is no jumping of fiber dimension.")

24.1.9. Theorem (cohomology commutes with flat base change) — *Suppose*

$$\begin{array}{ccc} X' & \xrightarrow{\rho'} & X \\ {\scriptstyle\pi'}\downarrow & & \downarrow{\scriptstyle\pi} \\ Y' & \xrightarrow{\rho} & Y \end{array}$$

is a Cartesian diagram, and π (and thus π') is quasicompact and separated (so higher pushforwards of quasicoherent sheaves by π and π' exist, as described in §18.7). Suppose also that ρ is flat, and $\mathscr{F}$ is a quasicoherent sheaf on X. Then the natural "push-pull" morphisms (Exercise 18.7.B(a)) $\rho^*(R^i\pi_*\mathscr{F}) \to R^i\pi'_*(\rho'^*\mathscr{F})$ are isomorphisms.

24.1.M. EXERCISE. Prove Theorem 24.1.9. Hint: Exercise 18.7.B(b) is the special case where ρ is affine. Extend it to the quasicompact and separated case using the same idea as the solution to Exercise 14.4.C (which used Exercise 6.2.G). Your proof of the case $i=0$ will only need a quasiseparated hypothesis in place of the separated hypothesis.

A useful special case is where Y' is the generic point of a reduced component of Y. In other words, in light of Exercise 24.1.G(a), the stalk of the higher pushforward of $\mathscr{F}$ at the generic point is the cohomology of $\mathscr{F}$ on the fiber over the generic point. This is a first example of something important: understanding cohomology of (quasicoherent sheaves on) fibers in terms of higher pushforwards. (We would certainly hope that higher pushforwards would tell us something about higher cohomology of fibers, but this is certainly not a priori clear!) In comparison to this result, which shows that cohomology of *any* quasicoherent sheaf commutes with *flat* base change, §24.7 and Chapter 25 deal with when and how cohomology of a *flat* quasicoherent sheaf commutes with *any* base change.

24.1.10. Pulling back closed subschemes (and ideal sheaves) by flat morphisms.
Closed subschemes pull back particularly well under flat morphisms, and this can be helpful to keep in mind. As pointed out in Remark 14.5.9, in the case of flat morphisms, pullback of ideal sheaves *as quasicoherent sheaves* agrees with pullback in terms of the pullback of the corresponding closed subschemes. In other words, closed subscheme exact sequences pull back (remain exact) under flat pullbacks. This is in fact not just a necessary condition for flatness; it is also sufficient, which can be shown using the ideal-theoretic criterion for flatness (Theorem 24.3.1). There is an analogous fact about pulling ideal sheaves of *flat* subschemes by *arbitrary* pullbacks; see §24.2.2.

24.1.N. EXERCISE.

(a) Suppose D is an effective Cartier divisor on Y and $\pi\colon X \to Y$ is a flat morphism. Show that the pullback of D to X (by π) is also an effective Cartier divisor.
(b) Use part (a) to show that under a flat morphism, regular embeddings pull back to regular embeddings.

24.1.O. EXERCISE.

(a) Suppose $\pi\colon X \to Y$ is a morphism, and $Z \hookrightarrow Y$ is a closed embedding cut out by an ideal sheaf $\mathscr{I} \subset \mathscr{O}_Y$. For this exercise only, for any ideal sheaf $\mathscr{J} \subset \mathscr{O}_Y$ corresponding to a closed subscheme $W \subset Y$, let $\pi^\flat \mathscr{J} \subset \mathscr{O}_X$ be the ideal sheaf corresponding to $W \times_Y X \subset X$ (so $\pi^* \mathscr{J} \to \pi^\flat \mathscr{J}$ is surjective but is not always an isomorphism, as discussed in Remark 14.5.9). Show that $(\pi^\flat \mathscr{I})^n = \pi^\flat (\mathscr{I}^n)$. (Products of quasicoherent ideal sheaves were defined in Exercise 15.6.E.)
(b) Suppose further that π is flat, $Y = \mathbb{A}_k^n$, and Z is the origin. Let $\mathscr{J} = \pi^* \mathscr{I}$ be the quasicoherent sheaf of algebras on X cutting out the pullback W of Z. Prove that the graded sheaf of algebras $\oplus_{n \geq 0} \mathscr{J}^n / \mathscr{J}^{n+1}$ (do you understand the multiplication?) is isomorphic to $\mathscr{O}_W[x_1, \ldots, x_n]$ (interpreted as a graded sheaf of algebras). (Hint: First show that $\mathscr{J}^n / \mathscr{J}^{n+1} \cong \mathrm{Sym}^n(\mathscr{J}/\mathscr{J}^2)$, Corollary 22.3.9. This approach will be revisited in Exercise 24.6.E.)

24.1.P. EXERCISE.

(a) Show that blowing up commutes with flat base change. If $\pi\colon X \to Y$ is any flat morphism, and $Z \hookrightarrow Y$ is any closed embedding, give a canonical isomorphism $(\mathrm{Bl}_Z Y) \times_Y X \longleftrightarrow \mathrm{Bl}_{Z \times_Y X} X$. (You can proceed by universal property, using Exercise 24.1.N(a), or by using the Proj construction of the blow-up and Exercise 24.1.O.)
(b) Give an example to show that blowing up does not commute with base change in general.

24.2 Flatness through Tor

We defined the Tor (bi-)functor in §23.1: $\mathrm{Tor}_i^A(M, N)$ is obtained by taking a free resolution of N, removing the N, tensoring it with M, and taking homology. Exercise 23.1.C characterized flatness in terms of Tor: M is A-flat if $\mathrm{Tor}_1^A(M, N) = 0$ for all N. In this section, we reap the easier benefits of this characterization, recalling key properties of Tor when needed. In §24.3, we work harder to extract more from Tor.

It is sometimes possible to compute Tor from its definition, as shown in the following exercise that we will use repeatedly.

24.2.A. EXERCISE. Define $(M:x)$ as $\{m \in M : xm = 0\} \subset M$—it consists of the elements of M annihilated by x. If x is not a zerodivisor, show that

$$\mathrm{Tor}_i^A(M, A/(x)) = \begin{cases} M/xM & \text{if } i = 0; \\ (M:x) & \text{if } i = 1; \\ 0 & \text{if } i > 1. \end{cases}$$

Hint: Use the resolution

$$0 \longrightarrow A \xrightarrow{\times x} A \longrightarrow A/(x) \longrightarrow 0$$

of $A/(x)$.

24.2.1. *Remark.* As a corollary of Exercise 24.2.A, we see again that flat modules over an integral domain are torsion-free (and, more generally, Observation 24.1.2). Also, Exercise 24.2.A gives the reason for the notation Tor—it is short for *torsion*.

24.2.B. EXERCISE. If B is A-flat, use the FHHF Theorem (Exercise 1.5.I(c)) to give an isomorphism $B \otimes_A \mathrm{Tor}_i^A(M, N) \xrightarrow{\sim} \mathrm{Tor}_i^B(B \otimes M, B \otimes N)$.

Recall that the Tor functor is symmetric in its entries (there is an isomorphism $\mathrm{Tor}_i^A(M, N) \xrightarrow{\sim} \mathrm{Tor}_i^A(N, M)$, Exercise 23.3.A). This gives us a quick but very useful result.

24.2.C. EASY EXERCISE. If $0 \to N' \to N \to N'' \to 0$ is an exact sequence of A-modules, and N'' is flat (e.g., free), show that $0 \to M \otimes_A N' \to M \otimes_A N \to M \otimes_A N'' \to 0$ is exact for *any* A-module M.

We would have cared about this result long before seeing Tor, so it gives some motivation for learning about Tor. (Unimportant side question: Can you prove this without Tor, using a diagram chase?)

24.2.D. EXERCISE (IMPORTANT CONSEQUENCE OF EXERCISE 24.2.C). Suppose $0 \to \mathscr{F}' \to \mathscr{F} \to \mathscr{F}'' \to 0$ is a short exact sequence of quasicoherent sheaves on a scheme Y, and $\mathscr{F}''$ is flat (e.g., locally free). Show that if $\pi \colon X \to Y$ is any morphism of schemes, the pulled back sequence $0 \to \pi^* \mathscr{F}' \to \pi^* \mathscr{F} \to \pi^* \mathscr{F}'' \to 0$ remains exact.

24.2.E. EXERCISE (CF. EXERCISE 14.2.H FOR THE ANALOGOUS FACTS ABOUT VECTOR BUNDLES). Suppose $0 \to M' \to M \to M'' \to 0$ is an exact sequence of A-modules.

(a) If M and M'' are both flat, show that M' is too. (Hint: Recall the long exact sequence for Tor, Proposition 23.1.2. Also, use that N is flat if and only if $\mathrm{Tor}_i(N, N') = 0$ for all $i > 0$ and all N', Exercise 23.1.C.)
(b) If M' and M'' are both flat, show that M is too. (Same hint.)
(c) If M' and M are both flat, show that M'' need not be flat.

24.2.F. EXERCISE. If $\cdots \to M_{-2} \to M_{-1} \to M_0 \to 0$ is an exact sequence of flat A-modules, show that it remains exact upon tensoring with any other A-module. (One possible approach: Break the exact sequence into short exact sequences.)

24.2.G. EASY EXERCISE. If $0 \to M_0 \to M_1 \to \cdots \to M_n \to 0$ is an exact sequence, and M_i is flat for $i > 0$, show that M_0 is flat too. (Hint: Break the exact sequence into short exact sequences.)

Exercises 24.2.F and 24.2.G will be useful.

24.2.2. Pulling back quasicoherent ideal sheaves of flat closed subschemes by arbitrary morphisms (promised in §24.1.10). Suppose

$$
\begin{array}{ccc}
W & \xrightarrow{\ \alpha\ } & X \\
\downarrow & & \downarrow \\
Y & \xrightarrow{\ \beta\ } & Z
\end{array}
$$

is a fibered product, and $V \hookrightarrow X$ is a closed subscheme. Then $Y \times_Z V$ is a closed subscheme of W (§10.2.2). There are two possible senses in which $\mathscr{I}_{V/X}$ can be "pulled back" to W: as a quasicoherent sheaf $\alpha^* \mathscr{I}_{V/X}$, and as the ideal of the "pulled back" closed subscheme $\mathscr{I}_{Y \times_Z V/W}$. As pointed out in Remark 14.5.9, these are not necessarily the same, but they *are* the same if β is flat. We now give another important case in which they are the same.

24.2.H. EXERCISE. If V is flat over Z (with no hypotheses on β), show that $\alpha^* \mathscr{I}_{V/X} \cong \mathscr{I}_{Y \times_Z V/W}$. Hint: Exercise 24.2.D.

24.2.3.* The Künneth formula for quasicoherent sheaves.

It is worthwhile to know that the Künneth formula holds for quasicoherent sheaves, and that the proof is reasonable. But we will not use it, so you can read this section only if you particularly feel like it.

24.2.I. EXERCISE. Suppose X is a separated scheme, and $j: \operatorname{Spec} A \hookrightarrow X$ is an open embedding. Show that $j_* \mathscr{O}_{\operatorname{Spec} A}$ is a quasicoherent sheaf on X, flat (over $\mathscr{O}_X$). (Partial hint: The Cancellation Theorem 11.1.1 for affine morphisms.)

Suppose further that $X = \cup_{i=1}^n U_i$ is a cover of X by a finite number of affine open sets, so X is quasicompact. (We still assume X separated.) Denote the cover by $\mathfrak{U} := \{U_i\}$. As usual with Čech covers, for $I \subset \{1, \ldots, n\}$, define $U_I = \cap_{i \in I} U_i$, and let $j^I : U_I \hookrightarrow X$ be the corresponding open embedding.

For the purposes of this discussion only, define the *Čech complex of sheaves* for the cover $\mathfrak{U}$ of X, denoted by $C_X^\bullet(\mathfrak{U})$, by

$$ 0 \longrightarrow \oplus_{|I|=1} j_*^I \mathscr{O}_{U_I} \longrightarrow \oplus_{|I|=2} j_*^I \mathscr{O}_{U_I} \longrightarrow \cdots \longrightarrow \oplus_{|I|=n} j_*^I \mathscr{O}_{U_I} \longrightarrow 0. $$

Define the *augmented Čech complex of sheaves for* $X = \{U_i\}$, denoted by $C_{X,aug}^\bullet(\mathscr{F})$, by prepending $\mathscr{O}_X = \oplus_{|I|=0} j_*^I \mathscr{O}_{U_I}$:

$$ 0 \longrightarrow \mathscr{O}_X \longrightarrow \oplus_{|I|=1} j_*^I \mathscr{O}_{U_I} \longrightarrow \oplus_{|I|=2} j_*^I \mathscr{O}_{U_I} \longrightarrow \cdots \longrightarrow \oplus_{|I|=n} j_*^I \mathscr{O}_{U_I} \longrightarrow 0. $$

(Be sure you understand the definitions of the maps in these complexes.)

24.2.J. EXERCISE. Show that the augmented Čech complex of sheaves $C_{X,aug}^\bullet(\mathfrak{U})$ is an exact sequence of flat quasicoherent sheaves on X. (Hint: Check on any affine open subset of X.)

Before proceeding further, we prove a useful homological statement. Recall that a finite exact sequence of flat modules remains exact upon tensoring with any other module (Exercise 24.2.F). Recall also that an exact sequence of modules remains exact upon tensoring with any flat module (a version of the FHHF theorem, Exercise 1.5.I(c)). The following statement generalizes both of these naturally.

24.2.K. EXERCISE. Suppose $C^\bullet$ is a finite exact sequence of A-modules, and $F^\bullet$ is a finite exact sequence of *flat* A-modules. Show that the total complex of the double complex $C^\bullet \otimes_A F^\bullet$ is also an exact sequence of A-modules. (If you wish, you might show that if $C^\bullet$ is merely a complex, not necessarily exact, then $C^\bullet \otimes_A F^\bullet$ has "the same cohomology" as $C^\bullet$, further extending the FHHF theorem. You may also wish to remove the finiteness assumptions irrelevant to your solution.)

24.2.L. EXERCISE. We return to the situation of Exercise 24.2.J. Suppose $\mathscr{F}$ is a quasicoherent sheaf (not necessarily flat) on X (which is quasicompact and separated). Show that $\mathscr{F} \otimes C_{X,aug}^\bullet(\mathfrak{U})$ is an exact sequence sequence of quasicoherent sheaves on X. (Equivalently, $\mathscr{F} \otimes C_X^\bullet(\mathfrak{U})$ is a complex, exact except at the first step, where it has kernel/cohomology sheaf canonically identified with $\mathscr{F}$.) Hint: Check on each affine open subset of X.

Following Serre, you might interpret Exercises 24.2.J and 24.2.L in terms of partitions of unity.

We use these pleasant exact sequences to prove a form of the Künneth formula. Suppose that X and Y are both quasicompact separated k-schemes (e.g., varieties over k). Name the projection maps $\pi_X : X \times_k Y \to X$ and $\pi_Y : X \times_k Y \to Y$. Let $\mathfrak{U}_X$ and $\mathfrak{U}_Y$ be finite covers of X and Y (respectively) by affine open sets.

24.2.M. EXERCISE. Show that $\pi_X^* C_{X,aug}^\bullet(\mathfrak{U}_X)$ is an exact complex of flat quasicoherent sheaves on $X \times_k Y$.

Define $C_X^\bullet \boxtimes C_Y^\bullet := \pi_X^* C_X^\bullet(\mathfrak{U}_X) \otimes \pi_Y^* C_Y^\bullet(\mathfrak{U}_Y)$, interpreted as the total complex associated to the double complex of the right side.

24.2.N. EXERCISE. Show that $C_X^\bullet \boxtimes C_Y^\bullet$ is a complex of flat quasicoherent sheaves on $X \times_k Y$, exact except at the first step, where the cohomology/kernel is canonically identified with $\mathscr{O}_{X \times Y}$. (You may find it helpful to prove a similarly statement for a similar defined $C_{X,aug}^\bullet \boxtimes C_{Y,aug}^\bullet$.)

24.2.O. EXERCISE. Suppose now that $\mathscr{F}$ is a quasicoherent sheaf on X, and $\mathscr{G}$ is a quasicoherent sheaf on Y. Show that $(\mathscr{F} \boxtimes \mathscr{G}) \otimes (C_X^\bullet \boxtimes C_Y^\bullet)$ is a complex of quasicoherent sheaves on $X \times_k Y$, exact except at the first step, where the cohomology/kernel is canonically identified with $\mathscr{F} \boxtimes \mathscr{G}$.

24.2.P. EXERCISE (THE KÜNNETH FORMULA FOR QUASICOHERENT SHEAVES). By suitably identifying $(\mathscr{F} \boxtimes \mathscr{G}) \otimes (C_X^\bullet \boxtimes C_Y^\bullet)$ with $(\mathscr{F} \otimes C_X^\bullet) \boxtimes (\mathscr{G} \otimes C_Y^\bullet)$, show that for all n,

$$H^n(X \times_k Y, \mathscr{F} \boxtimes \mathscr{G}) = \oplus_{i+j=n} H^i(X, \mathscr{F}) \otimes_k H^j(Y, \mathscr{G}).$$

(Hint: Don't use spectral sequences; work directly with the complexes. Show by direct calculation that if you have two complexes of k-modules, the cohomology of the total complex of their tensor product is the "direct sum of the tensor product of their cohomologies.")

24.2.4. *Remarks on hypotheses.* Why did we require quasicoherence? Where did we use the fact that we were working over a field k?

24.2.5. ** *Extensions.* Given quasicompact separated morphisms of schemes $\rho^X : X \to Z$ and $\rho^Y : Y \to Z$, the resulting $\rho^{X \times Y} : X \times_Z Y \to Z$, and quasicoherent sheaves $\mathscr{F}$ and X and $\mathscr{G}$ on Y, what is the relationship between $R^n \rho_*^{X \times Y} \mathscr{F} \boxtimes_Z \mathscr{G}$, $R^i \rho_*^X \mathscr{F}$, and $R^j \rho_*^Y \mathscr{G}$? By following through your proof, you may be able to extend the statement of Exercise 24.2.P considerably.

24.3 Ideal-Theoretic Criteria for Flatness

The following theorem will allow us to classify flat modules over a number of rings. It is a refined version of Exercise 23.1.C, that M is a flat A-module if and only if $\mathrm{Tor}_1^A(M, N) = 0$ for all A-modules N.

24.3.1. Theorem (ideal-theoretic criterion for flatness) — *The A-module M is flat if and only if $\mathrm{Tor}_1^A(M, A/I) = 0$ for every ideal I.*

24.3.2. *Remarks.* Before getting to the proof, we make some side remarks that may give some insight into how to think about flatness. Theorem 24.3.1 is profitably stated without the theory of Tor. It is equivalent to the statement that M is flat if and only if for all ideals $I \subset A$, $I \otimes_A M \to M$ is an injection, and you can reinterpret the proof in this guise. Perhaps better, M is flat if and only if $I \otimes_A M \to IM$ is an isomorphism for every ideal I.

Flatness is often informally described as "continuously varying fibers," and this can be made more precise as follows. An A-module M is flat if and only if it restricts nicely to closed subschemes of Spec A. More precisely, what we lose in this restriction, the submodule IM of elements which "vanish on Z," is easy to understand: it consists of formal linear combinations of elements $i \otimes m$, with no surprise relations among them—i.e., the tensor product $I \otimes_A M$. This is the content of the following exercise, in which you may use Theorem 24.3.1.

24.3.A. * **EXERCISE (THE EQUATIONAL CRITERION FOR FLATNESS).** Show that an A-module M is flat if and only if for every relation $\sum a_i m_i = 0$ with $a_i \in A$ and $m_i \in M$, there exist $m_j' \in M$ and $a_{ij} \in A$ such that $\sum_j a_{ij} m_j' = m_i$ for all i and $\sum_i a_i a_{ij} = 0$ in A for all j. (Translation: Whenever elements of M satisfy an A-linear relation, this is "because" of linear equations holding in A.)

24.3.3. *Unimportant remark.* In the statement of Theorem 24.3.1, it suffices to check only finitely generated ideals. This is essentially the content of the following statement, which you can prove if you wish: Show that an A-module M is flat if and only if for all *finitely generated ideals* I, the natural map $I \otimes_A M \to M$ is an injection. Hint: Use a counterexample for an ideal J that is not finitely generated to find another counterexample for an ideal I that *is* finitely generated.

24.3.4. *Proof of the ideal-theoretic criterion for flatness, Theorem 24.3.1.* By Exercise 23.1.C, we need only show that $\operatorname{Tor}_1^A(M, A/I) = 0$ for all I implies $\operatorname{Tor}_1^A(M, N) = 0$ for all A-modules N, and hence that M is flat.

We first prove that $\operatorname{Tor}_1^A(M, N) = 0$ for all *finitely generated* modules N, by induction on the number n of generators $a_1, \ldots, a_n$ of N. The base case (if $n = 1$, so $N \cong A/\operatorname{Ann}(a_1)$—our first use of the annihilator ideal in a long time) is our assumption. If $n > 1$, then $Aa_n \cong A/\operatorname{Ann}(a_n)$ is a submodule of N, and the quotient Q is generated by the images of $a_1, \ldots, a_{n-1}$, so the result follows by considering the Tor_1 portion of the Tor long exact sequence for

$$0 \longrightarrow A/\operatorname{Ann}(a_n) \longrightarrow N \longrightarrow Q \longrightarrow 0.$$

We deal with the case of general N by abstract nonsense. Notice that N is the union of its finitely generated submodules $\{N_\alpha\}$. In fancy language, this union is a filtered colimit—any two finitely generated submodules are contained in a finitely generated submodule (specifically, the submodule they generate). Filtered colimits of modules commute with cohomology (Exercise 1.5.M), so $\operatorname{Tor}_1(M, N)$ is the colimit over α of $\operatorname{Tor}_1(M, N_\alpha) = 0$, and is thus 0. $\qquad\square$

We now use Theorem 24.3.1 to get explicit characterizations of flat modules over three (types of) rings: principal ideal domains, dual numbers, and some local rings.

Recall Observation 24.1.2, that flatness implies torsion-free. The converse is true for principal ideal domains:

24.3.B. EXERCISE (FLAT = TORSION-FREE FOR A PID). Show that a module over a principal ideal domain is flat if and only if it is torsion-free.

24.3.C. EXERCISE (FLATNESS OVER A DVR). Suppose M is a module over a discrete valuation ring A with uniformizer t. Show that M is flat if and only if t is not a zerodivisor on M, i.e., (using the notation defined in Exercise 24.2.A) $(M : t) = 0$. (See Exercise 24.1.B for the case of finitely generated modules.) This yields a simple and very important geometric interpretation of flatness over a regular curve, which we discuss in §24.3.9.

24.3.D. EXERCISE (FLATNESS OVER THE DUAL NUMBERS). Show that M is flat over $k[t]/(t^2)$ if and only if the "multiplication by t" map $M/tM \to tM$ is an isomorphism. (This fact is important in deformation theory and elsewhere.) Hint: $k[t]/(t^2)$ has only three ideals.

24.3.5. Important theorem (flat = free = projective for finitely presented modules over local rings) — *Suppose $(A, \mathfrak{m})$ is a local ring (not necessarily Noetherian), and M is a finitely presented A-module. Then the following are equivalent.*

(a) *M is free.*
(b) *M is projective.*
(c) *M is flat.*
(d) *$\operatorname{Tor}_1^A(M, A/\mathfrak{m}) = 0$.*

24.3.6. *Remarks.* Warning: Modules over local rings can be flat without being free: $\mathbb{Q}$ is a flat $\mathbb{Z}_{(p)}$-algebra ($\mathbb{Z}_{(p)}$ is the localization of $\mathbb{Z}$ at p, not the p-adics), as all localizations are flat (§24.1.1), but it is not free (do you see why?).

Also, non-Noetherian people may be pleased to know that with a little work, "finitely presented" can be weakened to "finitely generated": use [Mat2, Thm. 7.10] in the proof below, where finite presentation comes up.

Finally, Proposition 24.4.5 is a close variant of Theorem 24.3.5 for graded rings.

Proof. For any ring, free modules are projective (§23.2.1), and projective modules are flat (Exercise 23.2.B), and flat modules M satisfy $\operatorname{Tor}_1^A(M, N) = 0$ for *any* N (Exercise 23.1.C), so we need only show that if $\operatorname{Tor}_1^A(M, A/\mathfrak{m}) = 0$, then M is free.

(At this point, you should see Nakayama coming from a mile away.) Now $M/\mathfrak{m}M$ is a finite-dimensional vector space over the field $A/\mathfrak{m}$. Choose a basis of $M/\mathfrak{m}M$, and lift it to elements $m_1, \ldots, m_n \in M$. Consider $A^{\oplus n} \to M$ given by $e_i \mapsto m_i$. We will show this is an isomorphism. It is surjective by Nakayama's Lemma (see Exercise 8.2.H): the image is all of M modulo the maximal ideal, hence is everything. As M is finitely presented, by Lemma 6.4.6 ("finitely presented implies always finitely presented"), the kernel K is finitely generated. Tensor $0 \to K \to A^{\oplus n} \to M \to 0$ with $A/\mathfrak{m}$. As M is flat, the result is still exact (Exercise 24.2.C):

$$0 \longrightarrow K/\mathfrak{m}K \longrightarrow (A/\mathfrak{m})^{\oplus n} \longrightarrow M/\mathfrak{m}M \longrightarrow 0.$$

But $(A/\mathfrak{m})^{\oplus n} \to M/\mathfrak{m}M$ is an isomorphism by construction, so $K/\mathfrak{m}K = 0$. As K is finitely generated, $K = 0$ by Nakayama's Lemma 8.2.9. $\square$

Here is an immediate and useful corollary—really just a geometric interpretation.

24.3.7. Corollary (flat = locally free for finitely presented sheaves) — *A finitely presented sheaf $\mathscr{F}$ on X is flat (over X) if and only if it is locally free.*

Proof. Local freeness of a finitely presented sheaf can be checked at the stalks, Exercise 14.3.E. $\square$

24.3.E. EXERCISE. Suppose $\pi \colon X \to Y$ is a finite flat morphism of locally Noetherian schemes, and $\mathscr{F}$ is a finite rank locally free sheaf on X.

(a) Show that $\pi_*\mathscr{F}$ is a finite rank locally free sheaf on Y.
(b) If Y is irreducible with generic point η, the degree of π above η is n, and $\mathscr{F}$ is locally free of rank r, show that $\pi_*\mathscr{F}$ is locally free of rank nr. (This may be trickier than it looks.)

24.3.8. *Remark.* In particular, taking $\mathscr{F} = \mathscr{O}_X$, we see that in the locally Noetherian setting, *finite flat morphisms have locally constant degree.*

24.3.F.* EXERCISE (INTERESTING VARIANT OF THEOREM 24.3.5, BUT UNIMPORTANT FOR US). Suppose A is a ring (not necessarily local), and M is a finitely presented A-module. Show that M is flat if and only if it is projective. Hint: Show that M is projective if and only if $M_\mathfrak{m}$ is free for every maximal ideal $\mathfrak{m}$. The harder direction of this implication uses the fact that $\mathrm{Hom}_{A_\mathfrak{m}}(M_\mathfrak{m}, N_\mathfrak{m}) = \mathrm{Hom}_A(M, N)_\mathfrak{m}$, which follows from Exercise 1.5.H. (Remark: There exist finitely *generated* flat modules that are not projective. They are necessarily not finitely presented. Example without proof: Let $A = \prod_{i=1}^\infty \mathbb{F}_2$, interpreted as functions $\mathbb{Z}^{\geq 0} \to \mathbb{Z}/2$, and let M be the module of functions modulo those of proper support, i.e., those vanishing at almost all points of $\mathbb{Z}^{\geq 0}$.)

24.3.G. EXERCISE. Prove the following useful criterion for flatness: Suppose $\pi \colon X \to Y$ is a finite morphism, and Y is reduced and locally Noetherian. Then π is flat if and only if $\pi_*\mathscr{O}_X$ is locally free, if and only if the rank of $\pi_*\mathscr{O}_X$ is locally constant ($\dim_{\kappa(q)}(\pi_*\mathscr{O}_X)_q \otimes \kappa(q)$ is a locally constant function of $q \in Y$). Partial hint: Exercise 14.3.K.

24.3.H. EXERCISE. Show that the normalization of the node (see Figure 8.4) is not flat. Hint: Use Exercise 24.3.G. (This exercise can be strengthened to show that nontrivial normalizations are *never* flat.)

24.3.I. EXERCISE. In $\mathbb{A}^4_k = \mathrm{Spec}\, k[w, x, y, z]$, let X be the union of the wx-plane with the yz-plane:

(24.3.8.1) $$X = \mathrm{Spec}\, k[w, x, y, z]/(wy, wz, xy, xz).$$

The projection $\mathbb{A}^4_k \to \mathbb{A}^2_k$ given by $k[a, b] \to k[w, x, y, z]$ with $a \mapsto w - y$, $b \mapsto x - z$ restricts to a morphism $X \to \mathbb{A}^2_k$. Show that this morphism is not flat.

24.3.9. Flat families over regular curves. Exercise 24.3.C gives an elegant geometric criterion for when morphisms to regular curves are flat.

24.3.J. EXERCISE (CRITERION FOR FLATNESS OVER A REGULAR CURVE). Suppose $\pi\colon X \to Y$ is a morphism from a locally Noetherian scheme to a regular (locally Noetherian) curve. (The local Noetherian hypothesis on X is so we can discuss its associated points.) Show that π is flat if and only if all associated points of X map to a generic point of Y. (This is a partial converse to Exercise 24.1.K, that flat maps always send associated points to associated points. As with Exercise 24.1.K, feel free to immediately generalize your argument to a coherent sheaf $\mathscr{F}$ on X.)

24.3.10. For example, a nonconstant map from an integral (locally Noetherian) scheme to a regular curve must be flat. (As another example, the deformation to the normal cone, discussed in the double-starred section §22.4.17 and depicted in Figure 22.3, is flat.) Exercise 24.3.H (and the comment after it) shows that the regular condition is necessary. The example of two planes meeting at a point in Exercise 24.3.I shows that the dimension 1 condition is necessary.

24.3.11.* *Unimportant remark: A valuative criterion for flatness.* Exercise 24.3.J shows that flatness over a regular curve is geometrically intuitive (and is "visualizable"). It gives a criterion for flatness in general: Suppose $\pi\colon X \to Y$ is finitely presented morphism. If π is flat, then for every morphism $Y' \to Y$ where Y' is the Spec of a discrete valuation ring, $\pi'\colon X \times_Y Y' \to Y'$ is flat, so no associated points of $X \times_Y Y'$ map to the closed point of Y'. If Y is reduced and locally Noetherian, then this is a sufficient condition; this can reasonably be called a *valuative criterion for flatness.* (Reducedness is necessary: consider Exercise 24.1.G(d).) This gives an excellent way to visualize flatness, which you should try to put into words (perhaps after learning about flat limits below). See [Gr-EGA, IV$_3$.11.8.1] for a proof (and an extension without Noetherian hypotheses).

24.3.12. *Revisiting the degree of a projective morphism from a curve to a regular curve.* As hinted after the statement of Proposition 16.3.5, we can now better understand why nonconstant projective morphisms from a curve to a regular curve have a well-defined degree, which can be determined by taking the preimage of any point (§16.3.4). (Example 10.3.4 was particularly enlightening.) This is because such maps are flat by Exercise 24.3.J, and then the degree is constant by Remark 24.3.8 (see also Exercise 24.3.G). Also, Exercise 24.3.G yields a new proof of Proposition 16.3.5.

24.3.13.* **Flat limits.** Here is an important consequence of Exercise 24.3.J, which we can informally state as: We can take flat limits over one-parameter families. More precisely: suppose A is a discrete valuation ring, and let 0 be the closed point of Spec A and η the generic point. Suppose X is a locally Noetherian scheme over A, and Y is a closed subscheme of $X|_\eta$. Let Y' be the scheme-theoretic closure of Y in X. Then $\pi\colon Y' \to \operatorname{Spec} A$ has no associated points over 0 by Exercise 9.4.D, so Y' is flat over Spec A by Exercise 24.3.J. Similarly, suppose Z is a one-dimensional Noetherian scheme, 0 is a regular point of Z, and $\pi\colon X \to Z$ is a morphism from a locally Noetherian scheme to Z. If Y is a closed subscheme of $\pi^{-1}(Z \setminus \{0\})$, flat over $Z \setminus \{0\}$, and Y' is the scheme-theoretic closure of Y in X, then Y' is flat over Z. In both cases, the closure $Y'|_0$ is often called the **flat limit** of Y. (Feel free to weaken the Noetherian hypotheses on X.)

24.3.K. EXERCISE. Suppose (with the language of the previous paragraph) that A is a discrete valuation ring, X is a locally Noetherian A-scheme, and Y is a closed subscheme of the generic fiber $X|_\eta$. Show that there is only one closed subscheme Y' of X such that $Y'|_\eta = Y$, and Y' is flat over A.

24.3.L. EXERCISE. Consider the family of pairs of points in $\mathbb{A}^2_{\mathbb{C}}$ with coordinates parametrized by $\mathbb{A}^1_{\mathbb{C}}$ with coordinate t, given by $\{(0,0), (3t, 4t)\}$.

(a) Find the flat limit as $t \to 0$. (It will be a length 2 subscheme. The main point of this exercise is for you to work through the algebra.)

(b) Find the flat limit "as $t \to \infty$."

24.3.M. EXERCISE. Suppose $C = V(f(x, y))$ is a smooth plane curve in $\mathbb{A}^2_C$, and p is a closed point of C, so $\mathscr{O}_{C,p}$ is a discrete valuation ring. Consider the family of pairs of points of C, parametrized by $q \in C \setminus \{p\}$, given by $\{p, q\}$. Explain why the flat limit over p (the "limit as $q \mapsto p$") can be identified with the tangent direction to C at p. Translation: The tangent line is indeed the limit of secants, without deltas and epsilons.

24.3.N. HARDER EXERCISE (AN EXPLICIT FLAT LIMIT). Let $X = \mathbb{A}^3 \times \mathbb{A}^1 \to Z = \mathbb{A}^1$ over a field k, where the coordinates on $\mathbb{A}^3$ are x, y, and z, and the coordinates on $\mathbb{A}^1$ are t. Define Y away from $t = 0$ as the union of the two lines $y = z = 0$ (the x-axis) and $x = z - t = 0$ (the y-axis translated by t). Find the flat limit at $t = 0$. (Hints: (i) It is *not* the union of the two axes, although it includes this union. The flat limit is nonreduced at the node, and the "fuzz" points out of the plane they are contained in. (ii) $(y, z)(x, z) \neq (xy, z)$. (iii) Once you have a candidate flat limit, be sure to check that it *is* the flat limit. (iv) If you get stuck, read Example 24.3.14 below.)

Consider a projective version of the previous example, where two lines in $\mathbb{P}^3$ degenerate to meet. The limit consists of two lines meeting at a node, with some nonreduced structure at the node. Before the two lines come together, their space of global sections is two-dimensional. When they come together, it is not immediately obvious that their flat limit also has two-dimensional space of global sections as well. The reduced version (the union of the two lines meeting at a point) has a one-dimensional space of global sections, but the effect of the nonreduced structure on the space of global sections may not be immediately clear. However, we will see that "cohomology groups can only jump up in flat limits," as a consequence (indeed, the main moral) of the Semicontinuity Theorem 25.1.1.

24.3.14.** Example of variation of cohomology groups in flat families.

We can use a variant of Exercise 24.3.N to see an example of a cohomology group actually jumping. We work over an algebraically closed field to avoid distractions. Before we get down to explicit algebra, here is the general idea. Consider a twisted cubic C in $\mathbb{P}^3$. A projection pr_p from a random point $p \in \mathbb{P}^3$ will take C to a nodal plane cubic curve. Picture this projection "dynamically," by choosing coordinates so p is at $[1, 0, 0, 0]$, and considering the map $\phi_t \colon [w, x, y, z] \mapsto [tw, x, y, z]$; ϕ_1 is the identity on $\mathbb{P}^3$, ϕ_t is an automorphism of $\mathbb{P}^3$ for $t \neq 0$, and ϕ_0 is the projection. The limit of $\phi_t(C)$ as $t \to 0$ will be a nodal cubic, with nonreduced structure at the node "analytically the same" as what we saw when two lines came together in Exercise 24.3.N. (The phrase "analytically the same" can be made precise once we define completions in §28.)

Let's now see this in practice. Rather than working directly with the twisted cubic, we use another example where we saw a similar picture. Consider the nodal (affine) plane cubic $y^2 = x^3 + x^2$. Its normalization (see Figure 8.4, Example (3) of §8.3.5, Exercise 10.7.E, ...) was obtained by adding an extra variable m corresponding to y/x (which can be interpreted as blowing up the origin; see §22.4.3). We use the variable m rather than t (used in §8.3.5) in order to reserve t for the parameter for the flat family.

We picture the nodal cubic C as lying in the xy-plane in $\mathbb{A}^3 = \operatorname{Spec} k[x, y, m]$, and the normalization $\tilde{C}$ projecting to it, with $m = y/x$. What are the equations for $\tilde{C}$? Clearly, they include the equations $y^2 = x^3 + x^2$ and $y = mx$, but these are not enough—the m-axis (i.e., $x = y = 0$) is also in $V(y^2 - x^3 - x^2, y - mx)$. A little thought (and the algebra we have seen earlier in this example) will make clear that we have a third equation $x = m^2 - 1$, which along with $y = mx$ implies $y^2 = x^2 + x^3$. *Now* we have enough equations: $k[x, y, m]/(x - (m^2 - 1), y - mx)$ is an integral domain, as it is clearly isomorphic to $k[m]$. Indeed, you should recognize this as the algebra appearing in Exercise 10.7.E, so we have described $\tilde{C}$ as

$$(24.3.14.1) \qquad \operatorname{Spec} k[x, y, m]/(x - (m^2 - 1), y - mx).$$

Next, we want to formalize our intuition of the dynamic projection to the xy-plane of $\tilde{C} \subset \mathbb{A}^3$. We picture it as follows. Given a point (x, y, z) at time 1, at time t we want it to be at (x, y, zt). At time $t = 1$, we "start with" $\tilde{C}$, and at time $t = 0$ we have (set-theoretically) C. Thus *at time $t \neq 0$, the* curve $\tilde{C}$ is sent to the curve

$$\operatorname{Spec} k[x, y, m]/(x - ((m/t)^2 - 1), y - (m/t)x) = \operatorname{Spec} k[x, y, m]/(m^2 - t^2(x+1), ty - mx).$$

The family over $\operatorname{Spec} k[t, t^{-1}]$ is thus

$$\operatorname{Spec} k[x, y, m, t, t^{-1}]/(m^2 - t^2(x+1), ty - mx) \longrightarrow \operatorname{Spec} k[t, t^{-1}].$$

We have inverted t because we are so far dealing only with nonzero t. For $t \neq 0$, this is certainly a "nice" family, and so surely flat.

24.3.O. EXERCISE. Check this, as painlessly as possible! Hint: By a change of coordinates, show that the family is constant "over $\operatorname{Spec} k[t, t^{-1}]$," and hence pulled back (in some way you must figure out) via $\operatorname{Spec} k[t, t^{-1}] \to \operatorname{Spec} k$ from

$$\operatorname{Spec} k[x, y, M]/(M^2 - (x+1), y - Mx) \longrightarrow \operatorname{Spec} k,$$

which is flat by Trick Question 24.1.A(a).

We now figure out the flat limit of this family over $t = 0$, in $\operatorname{Spec} k[x, y, m, t] \to \mathbb{A}^1 = \operatorname{Spec} k[t]$. We first hope that our flat family is given by the equations we have already written down:

$$\operatorname{Spec} k[x, y, m, t]/(m^2 - t^2(x+1), ty - mx).$$

But this is *not* flat over $\mathbb{A}^1 = \operatorname{Spec} k[t]$, as the fiber dimension jumps (§24.5.5): substituting $t = 0$ into the equations (obtaining the fiber over $0 \in \mathbb{A}^1$), we find $\operatorname{Spec} k[x, y, m]/(m^2, mx)$. This is set-theoretically the entire xy-plane ($m = 0$), which of course has dimension 2. Notice for later reference that this "false limit" is scheme-theoretically the xy-plane, *with some nonreduced structure along the y-axis*. (This may remind you of Figure 4.4.)

So we are missing at least one equation. One clue as to what equation is missing: The equation $y^2 = x^3 + x^2$ clearly holds for $t \neq 0$, and does *not* hold for our naive attempt at a limit scheme $m^2 = mx = 0$. In retrospect, we should not have discarded this original equation when writing (24.3.14.1). So we put this equation back in, and have a second hope for describing the flat family over $\mathbb{A}^1$:

$$\operatorname{Spec} k[x, y, m, t]/(m^2 - t^2(x+1), ty - mx, y^2 - x^3 - x^2) \longrightarrow \operatorname{Spec} k[t].$$

When $t = 0$, we get $\operatorname{Spec} k[x, y, m]/(m^2, mx, y^2 - x^3 - x^2)$, so it passes the "no-dimension-jumping" criterion for flatness (§24.5.5 again), which is necessary but not sufficient. But, unfortunately, it is not flat, and this not so easy to see.

24.3.P. EXERCISE. Show that in the ring

$$k[x, y, m, t]/(m^2 - t^2(x+1), ty - mx, y^2 - x^3 - x^2),$$

$my - tx(x+1) \neq 0$, but $t(my - tx(x+1)) = 0$.

This suggests what equation we need to add: $my - tx(x+1) = 0$. Let

$$A = k[x, y, m, t]/(ty - mx, m^2 - t^2(x+1), y^2 - x^2(x+1), my - tx(x+1))$$

for convenience. We next hope that $\operatorname{Spec} A \to \operatorname{Spec} k[t]$ is flat at $t = 0$. We can show this using Exercise 24.3.B, by showing that t is not a zerodivisor on A. We do this by giving a "normal form" for elements of A.

24.3.Q. EXERCISE. Show that each element of A can be written uniquely in the form $f(t)m + g(x, t)y + h(x, t)$, where f, g, and h are polynomials. Then show that t is not a zerodivisor on A, and conclude that $\operatorname{Spec} A \to \mathbb{A}^1$ is indeed flat.

In particular, the flat limit when $t = 0$ is given by

$$\operatorname{Spec} A/(t) = \operatorname{Spec} k[x, y, m]/(m^2, mx, my, y^2 - x^2 - x^3).$$

24.3.R. EXERCISE. Show that the flat limit is nonreduced, and the "nonreducedness has length 1 and supported at the origin." More precisely, if $X = \operatorname{Spec} A/(t)$, show that $\mathscr{I}_{X^{\mathrm{red}}}$ is a skyscraper sheaf, with value k, supported at the origin. Sketch this flat limit X.

24.3.15. Note that we have a nonzero global function on X, given by m, which is supported at the origin (i.e., 0 away from the origin).

We now use this example to get a projective example with interesting behavior. We take the projective completion of this example, to get a family of cubic curves in $\mathbb{P}^3$ degenerating to a nodal cubic C with a nonreduced point.

24.3.S. EXERCISE. Do this: Describe this family (in $\mathbb{P}^3 \times \mathbb{A}^1$) precisely.

Take the long exact sequence corresponding to

$$0 \longrightarrow \mathscr{I}_{C^{\mathrm{red}}} \longrightarrow \mathscr{O}_C \longrightarrow \mathscr{O}_{C^{\mathrm{red}}} \longrightarrow 0,$$

to get

$$0 \longrightarrow H^0(C, \mathscr{I}_{C^{\mathrm{red}}}) \xrightarrow{\alpha} H^0(C, \mathscr{O}_C) \longrightarrow H^0(C, \mathscr{O}_{C^{\mathrm{red}}})$$
$$\longrightarrow H^1(C, \mathscr{I}_{C^{\mathrm{red}}}) \longrightarrow H^1(C, \mathscr{O}_C) \longrightarrow H^1(C, \mathscr{O}_{C^{\mathrm{red}}})$$
$$\longrightarrow H^2(C, \mathscr{I}_{C^{\mathrm{red}}}).$$

We have $H^1(C, \mathscr{I}_{C^{\mathrm{red}}}) = H^2(C, \mathscr{I}_{C^{\mathrm{red}}}) = 0$ as $\mathscr{I}_{C^{\mathrm{red}}}$ is supported in dimension 0 (by dimensional cohomology vanishing, Theorem 18.2.6). Also, $H^i(C^{\mathrm{red}}, \mathscr{O}_{C^{\mathrm{red}}}) = H^i(C, \mathscr{O}_{C^{\mathrm{red}}})$ (property **(v)** of cohomology; see §18.1). The (reduced) nodal cubic C^{red} has $h^0(\mathscr{O}) = 1$ (§11.4.7) and $h^1(\mathscr{O}) = 1$ (cubic plane curves have genus 1, (18.6.7.1)). Also, $h^0(C, \mathscr{I}_{C^{\mathrm{red}}}) = 1$, as observed above. Finally, α is not 0, as there exists a nonzero function on C vanishing on C^{red} (§24.3.15—convince yourself that this function extends from the affine patch $\operatorname{Spec} A$ to the projective completion).

Using the long exact sequence, we conclude $h^0(C, \mathscr{O}_C) = 2$ and $h^1(C, \mathscr{O}_C) = 1$. Thus in this example we see that $(h^0(\mathscr{O}), h^1(\mathscr{O})) = (1, 0)$ for the general member of the family (twisted cubics are isomorphic to $\mathbb{P}^1$), and the special member (the flat limit) has $(h^0(\mathscr{O}), h^1(\mathscr{O})) = (2, 1)$. Notice that both cohomology groups have jumped, yet the Euler characteristic has remained the same. The first behavior, as stated after Exercise 24.3.N, is an example of the Semicontinuity Theorem 25.1.1. The second, constancy of Euler characteristics in flat families, is the subject of §24.7. (It is no coincidence that the example had a singular limit; see §25.1.3.)

24.4** Aside: The Koszul Complex and the Hilbert Syzygy Theorem

To show that the theory we have developed so far is already very useful, we use it to prove the Hilbert Syzygy Theorem 16.1.3. This in turn gives us an excuse to introduce an important tool in commutative algebra, the Koszul complex. We will not use the contents of this section, so you should read it only if you feel like it.

24.4.1. *The Koszul complex.* Suppose E is a rank r free A-module. Given an element $s \in E^\vee$, i.e., an A-linear map $s : E \to A$, the **Koszul complex** $K_\bullet(s)$ associated to s is the chain complex:

$$0 \longrightarrow \wedge^r E \xrightarrow{d_r} \wedge^{r-1} E \xrightarrow{d_{r-1}} \cdots \xrightarrow{d_2} \wedge^1 E \xrightarrow{d_1} A \longrightarrow 0.$$

The differential is defined by

$$d_k(e_1 \wedge \cdots \wedge e_k) = \sum_{i=1}^{k} (-1)^{i+1} s(e_i) \, e_1 \wedge \cdots \wedge \hat{e}_i \wedge \cdots \wedge e_k,$$

where $e_1, \dots, e_r \in E$ (and $\hat{e}_i$ as usual means "omit the term e_i"). It is not hard to see that this is a complex (and it also follows from Exercise 24.4.B). It is a little harder to see that it is well-defined.

For convenience, choose a basis of E, and let suppose $s: A^{\oplus r} \to A$ sends the ith basis element of $E = A^{\oplus r}$ to f_i. In this case we denote Koszul complex by $K_\bullet(f_1, \ldots, f_r)$.

24.4.A. EXERCISE. If $A = k[x_1, \ldots, x_r]$, show by hand that the Koszul complex $K_\bullet(x_1, \ldots, x_r)$ is exact, except at the last term, where the cohomology is $k = A/(x_1, \ldots, x_r)$. (One possible method: Watch what happens to each monomial, as in §18.3.)

Thus the Koszul complex yields a free resolution of $k[x_1, \cdots, x_r]/(x_1, \cdots, x_r)$. (Exercise 24.4.C will generalize this.)

24.4.B. EXERCISE (USEFUL INTERPRETATION OF THE KOSZUL COMPLEX). Show that the Koszul complex $K_\bullet(f_1, \ldots, f_r)$ is the complex

$$\otimes_{i=1}^{r} \left(0 \longrightarrow A \xrightarrow{\times f_i} A \longrightarrow 0 \right).$$

First figure out what this means!

24.4.C. EXERCISE. Show that if $f_1, \ldots, f_r$ is a regular sequence, then $K_\bullet(f_1, \ldots, f_r)$ is exact everywhere but at the last term, where the cohomology is $A/(f_1, \ldots, f_r)$. Hint: Use induction on r, and Exercise 24.4.B.

Hence if $f_1, \ldots, f_r$ is a regular sequence, the Koszul complex gives a free resolution of $A/(f_1, \ldots, f_r)$ of length r, called the **Koszul resolution** of $A/(f_1, \ldots, f_r)$.

24.4.2. *Remarks.* (a) The $n = 2$ case of the Koszul complex implicitly came up in the proof of Theorem 9.5.6—it is the total complex of the tiny double complex (9.5.6.1). (b) One can similarly take the Koszul complex of an A-module M: $K_\bullet(s, M) := K_\bullet(s) \otimes_A M$. (c) The cohomology of the Koszul complex is called **Koszul homology**. It measures the failure of a sequence to be regular.

24.4.3. Finitely generated modules over regular local rings have finite free resolutions.
For the next four exercises, suppose $(A, \mathfrak{m})$ is a (Noetherian) regular local ring of dimension n, and let $f_1, \ldots, f_n$ be generators of $\mathfrak{m}$.

24.4.D. EXERCISE. Show that for *any* A-module M, $\mathrm{Tor}_i^A(M, A/\mathfrak{m}) = 0$ for $i > n$. Hint: Take a suitable free resolution of $A/\mathfrak{m}$.

24.4.E. EXERCISE. If $\mathrm{Tor}_i^A(M, A/\mathfrak{m}) = 0$ for $i > b \geq 1$ and $\alpha: A^{\oplus a} \to M$ is a surjection, show that $\mathrm{Tor}_i^A(\ker \alpha, A/\mathfrak{m}) = 0$ for $i > b - 1$.

24.4.F. EXERCISE. Show that every finitely generated A-module M has a free resolution of length at most n. Hint: By Theorem 24.3.5, if M is a finitely generated A-module such that $\mathrm{Tor}_i^A(M, A/\mathfrak{m}) = 0$, then M is free.

24.4.G. EXERCISE. Give an example to show that this bound cannot be improved.

24.4.4. Proof of the Hilbert Syzygy Theorem 16.1.3.
We now prove the Hilbert Syzygy Theorem following the outline suggested by §24.4.3. Let $S_\bullet$ be the graded ring $k[x_0, \ldots, x_n]$, with the usual grading, where k is a field. Recall the graded module $S(e)_\bullet$ defined in §15.1.4 by $S(e)_d = S_{e+d}$, so $\widetilde{S(e)}_\bullet = \mathscr{O}_{\mathbb{P}_k^n}(e)$.

24.4.H. WARM-UP EXERCISE. Guided by the Koszul resolution of Exercise 24.4.A, describe an exact sequence on $\mathbb{P}_A^n$:

$$0 \longrightarrow \mathscr{O}(-n-1) \longrightarrow \cdots \longrightarrow \mathscr{O}(-j)^{\oplus \binom{n+1}{j}} \longrightarrow$$
$$\mathscr{O}(-1)^{\oplus(n+1)} \longrightarrow \mathscr{O} \longrightarrow 0.$$

Suppose $M_\bullet$ is finitely generated graded $S_\bullet$-module. Choose homogeneous elements $a_1, \ldots, a_r$ of degree $e_1, \ldots, e_r$ that generate $M_\bullet$, so the map

$$\times (a_1, \ldots, a_r) : \oplus_{i=1}^r S_\bullet \longrightarrow M_\bullet$$

is a surjection. We rewrite it as a degree-preserving map of graded modules:

$$\times (a_1, \ldots, a_r) : \oplus_{i=1}^r S(-e_i)_\bullet \longrightarrow\!\!\!\!\!\rightarrow M_\bullet.$$

24.4.5. Proposition (analog of Theorem 24.3.5) — *Suppose $M_\bullet$ is a finitely generated graded $S_\bullet$-module, and* $\operatorname{Tor}_1^{S_\bullet}(M_\bullet, k) = 0$. *(Here $k = S_\bullet/S_+$.) Then $M_\bullet \cong \oplus_{i=1}^n S(-d_i)_\bullet$ as graded modules for some choice of d_i.*

In other words, M is "free in the graded sense." To prove this, we start with a graded variant of Nakayama's Lemma.

24.4.I. EXERCISE (GRADED NAKAYAMA). Suppose $C_\bullet$ is a finitely generated $S_\bullet$-module such that $C_\bullet \otimes_{S_\bullet} k = 0$. Show that $C_\bullet = 0$. Hint: If $C_\bullet \neq 0$, look at the nonzero element of C of least degree. (You won't need to invoke the "usual" Nakayama's Lemma.)

24.4.6. *Proof of Proposition 24.4.5.* Suppose $\operatorname{Tor}_1^{S_\bullet}(M_\bullet, k) = 0$. Let $y_0, \ldots, y_r$ be homogeneous elements of $M_\bullet$ such that their images form a basis for $M_\bullet/(x_0, \ldots, x_n) = M_\bullet \otimes_{S_\bullet} k$. We have a map $\alpha : \oplus_{i=1}^r S(-\deg y_i)_\bullet \to M_\bullet$ that becomes an isomorphism upon tensoring with k. If $Q_\bullet = \operatorname{coker} \alpha$, then by applying $\otimes_{S_\bullet} k$ to

$$\oplus_{i=1}^r S(-\deg y_i)_\bullet \xrightarrow{\ \alpha\ } M_\bullet \longrightarrow Q_\bullet \longrightarrow 0$$

we see that $Q_\bullet \otimes_{S_\bullet} k = 0$ and hence $Q_\bullet = 0$ by Exercise 24.4.I, so α is surjective. Next, let $K_\bullet = \ker \alpha$. Applying $\otimes_{S_\bullet} k$ to

$$0 \longrightarrow K_\bullet \longrightarrow \oplus_{i=1}^r S(-\deg y_i)_\bullet \xrightarrow{\ \alpha\ } M_\bullet \longrightarrow 0$$

and using the hypothesis $\operatorname{Tor}_1^{S_\bullet}(M_\bullet, k) = 0$, we see that $K_\bullet \otimes_{S_\bullet} k = 0$ and thus $K_\bullet = 0$ as well, so α gives us our desired isomorphism. $\qquad\square$

We are now ready to completely follow the template of §24.4.3.

24.4.J. EXERCISE. Prove that every finite-generated graded $S_\bullet$-module $M_\bullet$ has a "graded free resolution" of length at most $n+1$:

$$0 \longrightarrow F_\bullet^{n+1} \longrightarrow \cdots \longrightarrow F_\bullet^1 \longrightarrow M_\bullet \longrightarrow 0,$$

where $F_\bullet^q = \oplus_{i=1}^{j_q} S(-a_{i,q})_\bullet$.

24.4.K. EXERCISE. Prove the Hilbert Syzygy Theorem 16.1.3.

24.5   Topological Implications of Flatness

We now discuss some topological aspects and consequences of flatness, which boil down to the Going-Down Theorem for flat morphisms (§24.5.3), which in turn comes from faithful flatness. Because dimension in algebraic geometry is a topological notion, we will show that dimensions of fibers behave well in flat families (§24.5.5).

24.5.1. Faithful flatness. The notion of faithful flatness is handy for many reasons, and we describe only a few. A B-module M is **faithfully flat** if for all complexes of B-modules

(24.5.1.1) $N' \longrightarrow N \longrightarrow N''$,

(24.5.1.1) is exact if and only if (24.5.1.1)$\otimes_B M$ is exact. A B-**algebra** A **is faithfully flat** if it is faithfully flat as a B-module.

24.5.A. EXERCISE. Show that a flat B-module M is faithfully flat if and only if for all B-modules N, $M \otimes_B N = 0$ implies that $N = 0$.

24.5.B. EXERCISE. Suppose M is a flat B-module. Show that the following are equivalent.

(a) M is faithfully flat;
(b) for all prime ideals $\mathfrak{p} \subset B$, $M \otimes_B \kappa(\mathfrak{p})$ is nonzero;
(c) for all maximal ideals $\mathfrak{m} \subset B$, $M \otimes_B \kappa(\mathfrak{m}) = M/\mathfrak{m}M$ is nonzero.

Suppose $\pi\colon X \to Y$ is a morphism of schemes. We say that π is **faithfully flat** if it is flat and surjective. (Unlike flatness, faithful flatness is not particularly useful for quasicoherent sheaves, so we do not define faithfully flat quasicoherent sheaves over a base.)

24.5.C. EXERCISE (CF. EXERCISE 24.5.B). Suppose $B \to A$ is a ring morphism. Show that A is faithfully flat over B if and only if $\operatorname{Spec} A \to \operatorname{Spec} B$ is faithfully flat.

24.5.2. *Faithfully flat morphisms behave "reasonably."* Faithfully flat morphisms behave "reasonably" in the sense of §8.1 (preserved by composition and base change, and local on the target) because both flatness and surjectivity have this property (§24.1.8 and Exercise 10.4.G, respectively).

24.5.D. EXERCISE. Suppose $\pi\colon \operatorname{Spec} A \to \operatorname{Spec} B$ is flat.

(a) Show that π is faithfully flat if and only if every *closed* point $q \in \operatorname{Spec} B$ is in the image of π. (Hint: Exercise 24.5.B(c).)
(b) Hence show that *every* flat (local) morphism of local rings (Definition 7.3.1) is faithfully flat. (Morphisms of local rings are assumed to be local, i.e., the maximal ideal pulls back to the maximal ideal.)

24.5.3. Going-Down for flat morphisms. A consequence of Exercise 24.5.D is the following useful result, whose statement makes no mention of faithful flatness. (The statement is not coincidentally reminiscent of the Going-Down Theorem for finite extensions of integrally closed domains, Theorem 12.2.12.)

24.5.E. EXERCISE (GOING-DOWN THEOREM FOR FLAT MORPHISMS).

(a) Suppose that $B \to A$ is a flat morphism of rings, corresponding to a map $\pi\colon \operatorname{Spec} A \to \operatorname{Spec} B$. Suppose $\mathfrak{q} \subset \mathfrak{q}'$ are prime ideals of B, and $\mathfrak{p}'$ is a prime ideal of A with $\pi([\mathfrak{p}']) = [\mathfrak{q}']$. Show that there exists a prime $\mathfrak{p} \subset \mathfrak{p}'$ of A with $\pi([\mathfrak{p}]) = [\mathfrak{q}]$. Hint: Show that $B_{\mathfrak{q}'} \to A_{\mathfrak{p}'}$ is a flat local ring homomorphism, and hence faithfully flat by the Exercise 24.5.D(b).
(b) Part (a) gives a geometric consequence of flatness. Draw a picture illustrating this.
(c) Recall the Going-Up Theorem, described in §8.2.4. State the Going-Down Theorem for flat morphisms in a way parallel to Exercise 8.2.F, and prove it.

24.5.F. EXERCISE. Suppose $\pi\colon X \to Y$ is an integral (e.g., finite) flat morphism, and Y has pure dimension n. Show that X has pure dimension n. (This generalizes Exercise 12.1.H(a).) Hint: π satisfies both Going-Up (see Exercise 8.2.F) and Going-Down (Exercise 24.5.E).

24.5.G. IMPORTANT EXERCISE: FLAT MORPHISMS ARE OPEN MAPS (IN REASONABLE SITUATIONS). Suppose $\pi\colon X \to Y$ is locally of finite type and flat, and Y (and hence X) is locally Noetherian. Show that π is an open map (i.e., sends open sets to open sets). Hint: Reduce to showing that $\pi(X)$ is open for all such π. Reduce to the case where X and Y are both affine. Use Chevalley's Theorem 8.4.2 to show that $\pi(X)$ is constructible. Use the Going-Down Theorem for flat morphisms, Exercise 24.5.E, to show that $\pi(X)$ is stable under generization. Conclude using Exercise 8.4.C.

24.5.4. *Remarks.*

(i) Of course, not all open morphisms are flat: witness $\operatorname{Spec} k[t]/(t) \to \operatorname{Spec} k[t]/(t^2)$.

(ii) Also, in quite reasonable circumstances, flat morphisms are *not* open: witness $\operatorname{Spec} k(t) \to \operatorname{Spec} k[t]$ (flat by Example 24.1.1(ii)).

24.5.H.* EXERCISE. Weaken the hypotheses of "locally of finite type" and "locally Noetherian" to just "locally finitely presented." Hint: As with the similar generalization in Exercise 10.3.I of Chevalley's Theorem 8.4.2, use the fact that any such morphism is "locally" pulled back from a Noetherian situation.

24.5.5. Dimensions of fibers are well-behaved for flat morphisms.

24.5.6. Proposition — *Suppose* $\pi: X \to Y$ *is a flat morphism of locally Noetherian schemes, with* $p \in X$ *and* $q \in Y$ *such that* $\pi(p) = q$. *Then*

$$\operatorname{codim}_X p = \operatorname{codim}_Y q + \operatorname{codim}_{\pi^{-1}(q)} p$$

(see Figure 12.4).

Informal translation: The dimension of the fibers is the difference of the dimensions of X and Y (at least locally). Compare this to Exercise 12.4.A, which stated that without the flatness hypothesis, we would only have inequality ($\leq$).

24.5.I. EXERCISE. Prove Proposition 24.5.6 as follows. As just mentioned, Exercise 12.4.A gives one inequality, so show the other. Given a chain of irreducible closed subsets in Y containing $\bar{q}$, and a chain of irreducible closed subsets in $\pi^{-1}(q) \subset X$ containing $\bar{p}$, construct a chain of irreducible closed subsets in X containing $\bar{p}$, using the Going-Down Theorem for flat morphisms (Exercise 24.5.E).

As a consequence of Proposition 24.5.6, if $\pi: X \to Y$ is a flat map of irreducible varieties, then the fibers of π all have pure dimension $\dim X - \dim Y$. (Warning: $\operatorname{Spec} k[t]/(t) \to \operatorname{Spec} k[t]/(t^2)$ does not exhibit dimensional jumping of fibers, is open, and sends associated points to associated points—cf. Exercise 24.1.K—but is not flat. If you prefer a reduced example, the normalization of the cuspidal plane cubic $\operatorname{Spec} k[t] \to \operatorname{Spec} k[x, y]/(y^2 - x^3)$, shown in Figure 10.5, also has these properties.) This leads us to the following useful definition.

24.5.7. *Definition.* If $\pi: X \to Y$ is a flat morphism that is locally of finite type, and all fibers of π have pure dimension n, we say that π is **flat of relative dimension** n. (In particular, when one says a morphism is flat of relative dimension n, the locally finite type hypotheses are implied. Remark 24.5.9 motivates this hypothesis.)

24.5.J. EXERCISE. Suppose $\pi: X \to Y$ is a flat morphism of finite type k-schemes, and Y is pure dimensional (so "codimension is the difference of dimensions"; cf. Theorem 12.2.9). Show that the following are equivalent.

(i) The scheme X has pure dimension $\dim Y + n$.

(ii) The morphism π is flat of relative dimension n.

24.5.K. EXERCISE. Show that the notion of a morphism being "flat of relative dimension n" is preserved by arbitrary base change. Hint to show that fiber dimension is preserved: Exercise 12.2.M.

24.5.L. EXERCISE. Suppose $\pi: X \to Y$ and $\rho: Y \to Z$ are flat of relative dimension m and n, respectively (hence locally of finite type). Show that $\rho \circ \pi$ is flat of relative dimension $m + n$.

24.5.8. *Remark.* For practice, you can now give a second solution to Exercise 12.2.F, that dimension is additive for products of varieties.

24.5.9. *Remark.* The reason for the "locally finite type" assumption in the definition of "flat of relative dimension n" is that we want this to be a "reasonable" class of morphisms. In particular, we want our notion of "flatness of relative dimension n" to be preserved by base change. Consider the Cartesian diagram

$$
\begin{array}{ccc}
\operatorname{Spec} k(x) \otimes_k k(y) & \longrightarrow & \operatorname{Spec} k(y) \\
{\scriptstyle \pi'}\downarrow & & \downarrow{\scriptstyle \pi} \\
\operatorname{Spec} k(x) & \longrightarrow & \operatorname{Spec} k.
\end{array}
$$

Both π and π' are trivially flat, because they are morphisms to spectra of fields (Exercise 24.1.A(a)). But the dimension of the fiber of π is 0, while (as described in Remark 12.2.17) the dimension of the fiber of π' is 1.

24.5.10. Morphisms to $\operatorname{Spec} k$ **are universally open.** We now deliver on a promise made in §10.5.6.

24.5.11. Theorem — *Every morphism $X \to \operatorname{Spec} k$ is universally open, i.e., remains open after any base change.*

* *Proof.* We wish to show that for any k-scheme Y, $X \times_k Y \to Y$ is an open map of topological spaces, that is, sends open sets to open sets. By considering an affine cover of Y, we see that we may reduce to the case when Y is affine, say, $\operatorname{Spec} S$. Then by considering an affine cover of X, we see that we may reduce to the case where X is affine, say, $\operatorname{Spec} R$. So we wish to show that $\pi\colon \operatorname{Spec} R \otimes_k S \to \operatorname{Spec} S$ is an open map.

Because distinguished affine open sets form a base of the Zariski topology of an affine scheme, it suffices to show that for all $f \in R \otimes_k S$, $\pi(D(f))$ is open. Write f as the finite sum $\sum_{i=1}^n r_i \otimes s_i$. Let R' be the subring of R generated by the r_i, so $R' \cong k[r_1, \ldots, r_n]/I$ for some (finitely generated) ideal I. To avoid confusion, we denote the element f when considered as an element of $R' \otimes_k S$ by f'. Consider the diagram

$$
\begin{array}{ccc}
\operatorname{Spec} R \otimes_k S & \xrightarrow{\quad\rho\quad} & \operatorname{Spec} R' \otimes_k S. \\
{\scriptstyle \pi}\searrow & & \swarrow{\scriptstyle \pi'} \\
& \operatorname{Spec} S &
\end{array}
$$

By Exercise 24.5.H, π' is an open map (as R' is finitely presented over k), so $\pi'(D(f'))$ is an open subset of $\operatorname{Spec} S$. We conclude by showing that $\pi(D(f)) = \pi'(D(f'))$. Notice that $\rho^{-1}(D(f')) = D(f)$ (as $\rho^\sharp(f') = f$), so we have the inclusion $\pi(D(f)) \subset \pi'(D(f'))$. To show the opposite inclusion, suppose $p \in \pi'(D(f'))$. Then the fiber of $\pi'|_{D(f')}$ over p is nonempty, i.e., the following ring is not the zero ring:

$$(R' \otimes_k S)_{f'} \otimes_S \kappa(p) = (R' \otimes_k \kappa(p))_{f'}.$$

Thus $R' \otimes_k \kappa(p)$ is not the zero ring, and f' is not nilpotent in this ring. But $R' \otimes_k \kappa(p)$ is a *subring* of $R \otimes_k \kappa(p)$ ($\cdot \otimes_k \kappa(p)$ is an exact functor; everything is flat over a field), so $R \otimes_k \kappa(p)$ is also not the zero ring, and f is not nilpotent in this ring. Hence $(R \otimes_k \kappa(p))_f \neq 0$, so the fiber of $\pi|_{D(f)}$ over p, $\operatorname{Spec}(R \otimes_k S)_f \otimes_S \kappa(p)$, is also nonempty. $\square$

24.5.12. Fancier flatness facts without proof.
We conclude with a couple of important facts, although we won't use them, so we will not prove them.

"Generic flatness" would be better called "general flatness" (cf. the discussion after Definition 3.6.11). Because we're not proving it, we may as well state it in some generality. For a slick proof of generic flatness under Noetherian hypotheses, see [Mu2, pp. 57–58].

24.5.13. Theorem (generic flatness, [Stacks, tag 052B]) — *If $\pi\colon X \to Y$ is a finite type morphism of schemes, Y is reduced, and $\mathscr{F}$ is a finite type quasicoherent sheaf on X, then there is a dense open subscheme $U \subset Y$ such that $\mathscr{F}|_{\pi^{-1}(U)}$ is of finite presentation over $\pi^{-1}(U)$, and flat over U.*

24.5.14. *Interpretation of the degree of a generically finite morphism.* We use this to interpret the degree of a rational map of varieties in terms of counting preimages, fulfilling a promise made in §12.2.2. Suppose $\pi\colon X \dashrightarrow Y$ is a generically finite rational map of k-varieties of degree d. For simplicity we assume that X and Y are irreducible. (Feel free to relax this.) Replace X by an open subset on which π is a morphism. By Proposition 12.4.4 ("generically finite implies generally finite"), there is a dense open subset V of Y over which π is finite. By Theorem 24.5.13, there is a dense open subset V' of V over which π is *finite and flat*. Then by Remark 24.3.8 ("finite flat morphisms have locally constant degree"), over this open subset V', π has "locally constant degree," which is necessarily d.

24.5.15. *The flat locus is open (flatness is an open condition).* Generic flatness can be used to show one last topological aspect of flatness: in reasonable circumstances, the locus where a quasicoherent sheaf is flat over a base is an open subset. More precisely:

24.5.16. Theorem (the flat locus is open) — *Suppose $\pi\colon X \to Y$ is a locally finite type morphism of locally Noetherian schemes, and $\mathscr{F}$ is a coherent sheaf on X.*

(a) *The locus of points of X at which $\mathscr{F}$ is Y-flat is an open subset of X.*
(b) *If π is closed (e.g., proper), then the locus of points of Y over which $\mathscr{F}$ is flat (i.e. those $q \in Y$ for which $\mathscr{F}$ is flat at all $p \in \pi^{-1}(q)$) is an open subset of Y.*

Part (b) follows immediately from part (a). Part (a) reduces to a nontrivial statement in commutative algebra; see, for example, [Mat2, Thm. 24.3] or [Gr-EGA, IV$_3$.11.1.1]. As is often the case, Noetherian hypotheses can be dropped in exchange for local finite presentation hypotheses on the morphism π; see [Gr-EGA, IV$_3$.11.3.1] or [Stacks, tag 00RC].

24.6 Local Criteria for Flatness

For a Noetherian local ring, there is a greatly improved version of the ideal-theoretic criterion of Theorem 24.3.1: we need check only *one* ideal—the maximal ideal.

24.6.1. Theorem (local criterion for flatness) — *Suppose $(B, \mathfrak{n}) \to (A, \mathfrak{m})$ is a local morphism of Noetherian local rings (§7.3.1), and that M is a finitely generated A-module. Then M is B-flat if and only if $\operatorname{Tor}_1^B(M, B/\mathfrak{n}) = 0$.*

This is a miracle: flatness over all of Spec B is determined by what happens over the closed point.

We have already seen Theorem 24.6.1 in a special case: if $(B, \mathfrak{n}) \to (A, \mathfrak{m})$ is an isomorphism, it is an immediate consequence of Theorem 24.3.5.

The following proof is due to Ogus, [OB]. To highlight what makes it work, we make a temporary and unneeded definition.

24.6.2. *Definition.* Suppose $F\colon \mathscr{B} \to \mathscr{A}$ is an additive functor of abelian categories. For the sake of this section only, we say F is **half-exact** if for every short exact sequence $0 \to M' \to M \to M'' \to 0$ in $\mathscr{B}$, $FM' \to FM \to FM''$ is exact in $\mathscr{A}$. Examples are left-exact functors, right-exact functors, and derived functors of either.

24.6.3. Theorem (Nakayama's Lemma for half-exact functors) — *Suppose $T\colon (B, \mathfrak{n}) \to (A, \mathfrak{m})$ is a local morphism of Noetherian local rings, and the additive functor $T\colon f.g.Mod_B \to f.g.Mod_A$ is B-linear and half-exact. If $T(B/\mathfrak{n}) = 0$ then $T(N) = 0$ for all finitely generated B-modules N (i.e., T is the zero functor).*

Proof. We will use Nakayama's Lemma for A, in the form that if $M \in f.g.Mod_A$, $a \in m$, and $\times a :$ $M \to M$ is a surjection, then $M = 0$. (By now you should see how this is one of many versions of Nakayama's Lemma from §8.2.7.)

Let $k = B/n$ for convenience. We wish to show that $T(M) = 0$ for all $M \in f.g.Mod_B$. If not, by Noetherianity of M, there is a maximal $M' \subset M$ with $T(M/M') \neq 0$. Chose such a maximal M'.

Let $N = M/M'$, so $T(N) \neq 0$, but $T(N'') = 0$ for every nontrivial quotient N'' of N. We thus wish to prove that if $N \in f.g.Mod_B$, and $T(N'') = 0$ for every proper quotient N'' of N, then $T(N) = 0$.

Let $I = Ann(N) = \{b \in B \ : \ bN = 0\}$. If $I = B$, then $N = 0$, so $T(N) = 0$. If $I = n$, then $N = (B/n)^{\oplus m}$ (N Noetherian implies m is finite), so $T(N) = T(B/n)^{\oplus m} = 0$. Otherwise, there exists $b \in n \setminus I$.

Consider the short exact sequence

$$0 \longrightarrow bN \longrightarrow N \longrightarrow N/bN \longrightarrow 0.$$

As $bN \neq 0$, N/bN is a nontrivial quotient of N, so $T(N/bN) = 0$. By the half-exactness of T, as $T(N) \neq 0$, we have that $T(bN) \neq 0$.

Next, consider the short exact sequence

$$0 \longrightarrow ker(N \xrightarrow{\times b} N) \longrightarrow N \xrightarrow{\times b} bN \longrightarrow 0.$$

As $T(bN) \neq 0$, bN is *not* a nontrivial quotient of N, so $ker(N \xrightarrow{\times b} N) = 0$. Thus we have a third short exact sequence

$$0 \longrightarrow N \xrightarrow{\times b} N \longrightarrow N/bN \longrightarrow 0,$$

from which

$$T(N) \xrightarrow{\times T(b)} T(N) \longrightarrow T(N/bN) = 0$$

is exact. (Here we are using the B-linearity of T to identify the map $T(N) \to T(N)$ as multiplication by $T(b) \in m$.) Then by Nakayama's Lemma for A, $T(N) = 0$, contradicting $T(N) \neq 0$. □

24.6.4. *Proof of Theorem 24.6.1.* Suppose M is finitely generated A-module. One direction is easy: if M is flat over B, then $Tor_1^B(M, B/n) = 0$ by the Tor criterion (Exercise 23.1.C).

For the harder direction, suppose now that $Tor_1^B(M, B/n) = 0$. Define $T : f.g.Mod_B \to f.g.Mod_A$ by $N \mapsto Tor_1^B(M, N)$. Then T is A-linear (do you see why?) and half-exact. By Nakayama's Lemma for half-exact functors, $T(B/n) = 0$ implies that $T(N) = 0$ for all finitely generated B-modules, and, in particular, all B/I, from which we see that M is flat over B by the ideal-theoretic criterion for flatness (Theorem 24.3.1). □

24.6.5. Slicing criteria for flatness.

A useful variant of the local criterion is the following. Suppose t is a non-zerodivisor of B in n (geometrically: an effective Cartier divisor on the target passing through the closed point). If M is flat over B, then t is not a zerodivisor of M (Observation 24.1.2). Also, M/tM is a flat B/tB-module (flatness commutes with base change, Exercise 24.1.J). The next result says that this is a characterization of flatness, at least when M is finitely generated, or somewhat more generally.

24.6.6. Theorem (slicing criterion for flatness on the target, see Figure 24.3) — *Suppose $(B, n) \to (A, m)$ is a local morphism of Noetherian local rings, M is a finitely generated A-module, and $t \in n$ is a non-zerodivisor on B. Then M is B-flat if and only if*

(i) *t is not a zerodivisor on M, and*

(ii) *M/tM is flat over B/(t).*

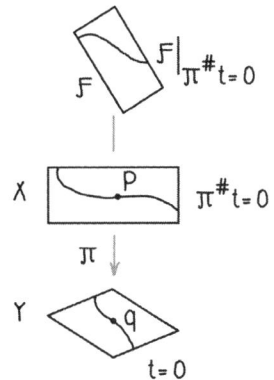

Figure 24.3 *The (necessary and sufficient) slicing criterion for flatness on the target, Theorem 24.6.6.*

(For slicing criteria for other properties, see Exercises 13.2.C, Theorem 24.6.9, and Theorem 26.2.3.)

To prove this, we show a lemma that we will also use to prove Theorem 24.6.14. (Notice that it has no hypotheses of local rings or finite generation.)

24.6.7. Lemma — *Suppose* $I \subset B$ *is an ideal, and* M *is a* B-*module. Suppose that* (i) $\mathrm{Tor}_1^B(B/I, M) = 0$, *and* (ii) M/IM *is flat over* B/I. *Then* $\mathrm{Tor}_1^B(B/J, M) = 0$ *for every ideal* J *containing* I.

Proof. The functor $\otimes_B B/J$ is the composition of the functor $\otimes_B B/I$ with the functor $\otimes_{B/I} B/J$. From the Grothendieck composition-of-functors spectral sequence (§23.3.5), there is a spectral sequence with E_2 given by $\mathrm{Tor}_q^{B/I}(B/J, \mathrm{Tor}_p^B(B/I, M))$ converging to $\mathrm{Tor}_{p+q}^B(B/J, M)$. The hypotheses of the theorem guarantee that

$$\mathrm{Tor}_0^{B/I}(B/J, \mathrm{Tor}_1^B(B/I, M)) = 0$$

by (i), and

$$\mathrm{Tor}_1^{B/I}(B/J, \mathrm{Tor}_0^B(B/I, M)) = \mathrm{Tor}_1^{B/I}(B/J, M/IM) = 0$$

by (ii) (using the ideal-theoretic criterion for flatness for B/I, Theorem 24.3.1). Thus $\mathrm{Tor}_1^B(B/J, M) = 0$, as desired. (This can also be proved readily by hand without the machinery of spectral sequences.) $\square$

24.6.8. *Proof of the slicing criterion for flatness on the target, Theorem 24.6.6.* Theorem 24.6.6 then follows by applying the previous lemma with $I = (t)$ and $J = \mathfrak{n}$, using $\mathrm{Tor}_1^B(B/tB, M) = \ker(M \xrightarrow{\times t} M)$ from Exercise 24.2.A, which is zero by hypothesis. $\square$

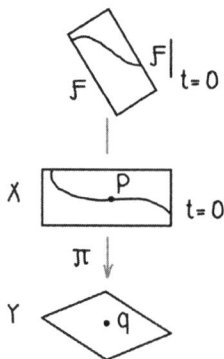

Figure 24.4 *The (sufficient) slicing criterion for flatness on the source, Theorem 24.6.9.*

24.6.A. EXERCISE. Use the slicing criterion to give a second solution to Exercise 24.3.I, on two planes in $\mathbb{P}^4$ meeting at a point. (This will use only the "easy direction" of the slicing criterion.)

Here is a similar slicing criterion for flatness, but in the *source*. Note that the criterion only goes one way.

24.6.9. Theorem (slicing criterion for flatness on the source; see Figure 24.4) — *Suppose* $(B, \mathfrak{n}) \to (A, \mathfrak{m})$ *is a local morphism of Noetherian local rings,* M *is a finitely generated* A-*module, and* $t \in \mathfrak{m} \subset A$ *is a non-zerodivisor on* $M/\mathfrak{n}M$. *If* M *is flat over* B, *then* M/tM *is flat over* B.

Proof. We have an exact sequence of A-modules (and hence B-modules),

$$M \xrightarrow{\times t} M \longrightarrow M/tM \longrightarrow 0.$$

Extend this to a flat resolution of B-modules (for example, by building it leftward using free B-modules):

(24.6.9.1) $$\cdots \longrightarrow M \xrightarrow{\times t} M \longrightarrow M/tM \longrightarrow 0.$$

We can compute $\mathrm{Tor}_1^B(B/tM, B/\mathfrak{n})$ using any flat resolution (Exercise 23.3.D), so we use (24.6.9.1). We apply $\otimes_B B/\mathfrak{n}$ to (24.6.9.1) and truncate the right term to get the complex whose cohomology will give us $\mathrm{Tor}_i^B(B/tM, B/\mathfrak{n})$:

$$\cdots \longrightarrow M/\mathfrak{n} \xrightarrow{\times t} M/\mathfrak{n} \longrightarrow 0.$$

But $\times t : M/\mathfrak{n} \to M/\mathfrak{n}$ has kernel zero by hypothesis, so the cohomology at the first step is zero, giving us that $\mathrm{Tor}_1^B(M/tM, B/\mathfrak{n}) = 0$. Then by the local criterion for flatness (Theorem 24.6.1), M/tN is flat over B as desired. $\square$

24.6.10.** *Relative effective Cartier divisors.* Theorem 24.6.9 has an immediate geometric interpretation: "Suppose $\pi\colon X \to Y$ is a morphism of Noetherian schemes, $\mathscr{F}$ is a coherent sheaf on X, and $Z \hookrightarrow X$ is a locally principal subscheme ...." In the special case where $\mathscr{F} = \mathscr{O}_X$, this leads to the notion of a **relative effective Cartier divisor**: a locally principal subscheme of X that is an effective Cartier divisor on all the fibers of π. Theorem 24.6.9 implies that if π is flat, then any relative effective Cartier divisor is also flat. (See Exercise 25.3.H for an important application.)

24.6.11. Over a field, regular embeddings are locally flat covers of the coordinate axes in affine space.

24.6.B. USEFUL EXERCISE. Let $(A, \mathfrak{m})$ be a Noetherian local ring, where A contains a field k (i.e., A is a k-algebra). Suppose $f_1, \ldots, f_n \in \mathfrak{m}$. These n functions on $\operatorname{Spec} A$ determine a morphism $\pi\colon \operatorname{Spec} A \to \mathbb{A}_k^n$. Show that π is flat if and only if $f_1, \ldots, f_n$ form a regular sequence. More generally, if M is a finitely generated A-module, show that M is flat over $\mathbb{A}_k^n$ if and only if $f_1, \ldots, f_n$ form an M-regular sequence. Hint: The slicing criterion for flatness on the target, Theorem 24.6.6.

Exercise 24.6.B makes a number of sophisticated facts surprisingly easy (so long as we are working over a field k).

24.6.C. EXERCISE. Use Exercise 24.6.B to give an immediate proof of Theorem 9.5.6, that an M-regular sequence in $\mathfrak{m}$ remains regular upon any reordering, so long as A is a k-algebra for some k. (The case of "mixed characteristic" is not covered.) Our original proof was notably harder (although more general).

24.6.D. EXERCISE. Use Exercise 24.6.B to reprove that the (co)normal sheaf of a codimension r regular embedding in a Noetherian k-scheme is locally free of rank r (Proposition 21.2.16(b) for k-schemes).

24.6.E. EXERCISE (CF. EXERCISE 24.1.O). Use Exercise 24.6.B to reprove that if $\mathscr{I}$ is the ideal sheaf of a codimension r regular embedding in a Noetherian k-schemes, then the natural map $\operatorname{Sym}^n(\mathscr{I}/\mathscr{I}^2) \to \mathscr{I}^n/\mathscr{I}^{n+1}$ is an isomorphism (Corollary 22.3.9 for k-schemes).

24.6.F.** EXERCISE. For this exercise, recall Exercise 24.4.A. If $(A, \mathfrak{m})$ is a Noetherian local ring containing a field k, and $f_1, \ldots, f_r$ is a regular sequence in $\mathfrak{m}$, and s sends the ith basis element of $E = A^{\oplus r}$ to f_i, use Exercise 24.6.B to show that the Koszul complex $K_\bullet(s)$ (§24.4.1) is exact except at the last term, where the cohomology is $A/(f_1, \ldots, f_r)$. (For a more general statement, see Exercise 24.4.C.)

24.6.12.** **Fibral flatness.** We conclude with a criterion for flatness that can be very useful, but which we will not need in the rest of the book. It applies when we have the following commutative diagram of locally Noetherian schemes, and where $\mathscr{F}$ is a coherent sheaf on X.

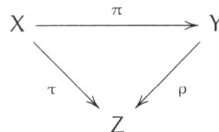

For each point $r \in Z$, let $\pi_r\colon X_r \to Y_r$ be the base-change of π.

24.6.13. Fibral flatness theorem, version one — *Suppose $\mathscr{F}$ is flat over Z, and for every point $r \in Z$, $\mathscr{F}|_{X_r}$ is flat over Y_r. Then $\mathscr{F}$ is flat over Y.*

24.6.G. EXERCISE. Show that to prove Theorem 24.6.13, it suffices to prove the following.

24.6.14. Fibral flatness theorem, commutative algebra version one — *Suppose*

(24.6.14.1)

$$(A, \mathfrak{m}) \longleftarrow (B, \mathfrak{n})$$
$$(R, \mathfrak{r})$$

is a commutative diagram of Noetherian local rings (sending maximal ideals to maximal ideals as usual), and M is a finitely generated A-module. Suppose M is flat over R, and M/\mathfrak{r}M is flat over B/\mathfrak{r}B. Then M is flat over B.

Proof. We will use Lemma 24.6.7. We wish to show that $\operatorname{Tor}_1^B(M, B/\mathfrak{n}) = 0$ (to invoke the local criterion for flatness, Theorem 24.6.1), and we know that $M/\mathfrak{r}M$ is flat over $B/\mathfrak{r}B$, so we wish to apply Lemma 24.6.7 for the module M, the ring B, and the ideals $J = \mathfrak{n}$ and $I = \mathfrak{r}B$. The only hypothesis of Lemma 24.6.7 remaining to show is that $\operatorname{Tor}_1^B(M, B/\mathfrak{r}B) = 0$.

Applying the right-exact functor $\cdot \otimes_B M$ to the surjection of B-modules $\mathfrak{r} \otimes_R B \twoheadrightarrow \mathfrak{r}B$ yields a surjection

$$\alpha : \mathfrak{r} \otimes_R M = \mathfrak{r} \otimes_R B \otimes_B M \longrightarrow (\mathfrak{r}B) \otimes_B M.$$

Consider the composition γ of maps of B-modules,

$$\mathfrak{r} \otimes_R M \xrightarrow{\alpha} (\mathfrak{r}B) \otimes_B M \xrightarrow{\beta} M.$$

Now α is surjective, and the composition $\beta \circ \alpha$ is injective (apply $\cdot \otimes_R M$ to $0 \to \mathfrak{r} \to R$, and use that M is R-flat). Thus α must be injective and thus an isomorphism, from which β must be injective.

Now apply $M \otimes_B \cdot$ to the short exact sequence

$$0 \longrightarrow \mathfrak{r}B \longrightarrow B \longrightarrow B/\mathfrak{r}B \longrightarrow 0.$$

Since B is flat over B, $\operatorname{Tor}_1^B(M, B) = 0$, so we have an exact sequence

$$0 = \operatorname{Tor}_1^B(M, B) \longrightarrow \operatorname{Tor}_1^B(M, B/\mathfrak{r}B) \longrightarrow$$
$$M \otimes_B \mathfrak{r}B \xrightarrow{\beta} M \otimes_B B \longrightarrow M \otimes_B B/\mathfrak{r}B \longrightarrow 0.$$

From the injectivity of β we have $\operatorname{Tor}_1^R(M, B/\mathfrak{r}B) = 0$, as desired. $\square$

Theorem 24.6.14 can be extended to the following (which can also lead to a geometric statement similar to Theorem 24.6.13).

24.6.15. Fibral flatness theorem, commutative algebra version two — *Suppose* (24.6.14.1) *is a commutative diagram of Noetherian local rings, and M is a **nonzero** finitely generated A-module. Then (a) is true if and only if (b) is true.*

(a) *(i) M is flat over R, and (ii) M/\mathfrak{r}M is flat over B/\mathfrak{r}B.*
(b) *(i) M is flat over B, and (ii) B is flat over R.*

24.6.H. EASY EXERCISE. Show that (b) implies (a).

Proof. Most of the result was shown in Theorem 24.6.14 and Exercise 24.6.H. It only remains to show that (a)(i), (a)(ii), and (b)(i) imply (b)(ii).

24.6.I. EXERCISE. Show that (a)(i), (a)(ii), and (b)(i) imply that M is faithfully flat over B.

We now show directly that B is flat over R. Let $N_\bullet$ be an exact sequence of R-modules. We wish to show that $B \otimes_R N_\bullet$ is exact. But by (a)(i), $M \otimes_R N_\bullet$ is exact, and $M \otimes_R N_\bullet = M \otimes_B (B \otimes_R N_\bullet)$ is exact. As M is faithfully flat over B, we have that $B \otimes_R N_\bullet$ is exact, as desired. $\square$

24.7 Flatness Implies Constant Euler Characteristic

We come to an important consequence of flatness promised in §24.0.1. We will see that this result implies many answers and examples to questions that we would have asked before we even knew about flatness.

24.7.1. Important theorem (Euler characteristic is constant in flat families) — *Suppose $\pi: X \to Y$ is a projective morphism of locally Noetherian schemes, and $\mathscr{F}$ is a coherent sheaf on X, flat over Y. Then $\chi(X_q, \mathscr{F}|_{X_q}) = \sum_{i \geq 0} (-1)^i h^i(X_q, \mathscr{F}|_{X_q})$ is a locally constant function of $q \in Y$ (where $X_q = \pi^{-1}(q)$).*

This is the first sign that "cohomology behaves well in flat families." (We will soon see a second: the Semicontinuity Theorem 25.1.1. A different proof of Theorem 24.7.1, giving an extension to the proper case, will be given in §25.2.5.) The Noetherian hypotheses are used to ensure that $\pi_* \mathscr{F}(m)$ is a coherent sheaf.

Theorem 24.7.1 gives a necessary condition for flatness. Converses (yielding a sufficient condition) are given in Exercise 24.7.C.

Proof. We make three quick reductions. (i) The question is local on the target Y, so we may reduce to case Y is affine, say, $Y = \operatorname{Spec} B$, so π factors through a closed embedding $X \hookrightarrow \mathbb{P}^n_B$ for some n. (ii) We may reduce to the case $X = \mathbb{P}^n_B$, by considering $\mathscr{F}$ as a sheaf on $\mathbb{P}^n_B$. (iii) We may reduce to showing that for $m \gg 0$, $h^0(X_q, \mathscr{F}(m)|_{X_q})$ is a locally constant function of $q \in Y$, and $h^i(X_q, \mathscr{F}(m)|_{X_q}) = 0$ for $i > 0$. (Serre vanishing ensures that $h^i(X_q, \mathscr{F}(m)|_{X_q}) = 0$ for $i > 0$ for $m \gg 0$, but a priori there could be a different m for each q, and we need m to be independent of q.)

Twist by $\mathscr{O}(m)$ for $m \gg 0$, so that all the higher cohomology $H^{i>0}(X, \mathscr{F}(m))$ of $\mathscr{F}$ vanishes (by Serre vanishing, Theorem 18.1.3(ii)). Now consider the Čech complex $\mathscr{C}^\bullet$ for $\mathscr{F}(m)$. Note that all the terms in the Čech complex $\mathscr{C}^\bullet$ are flat, because $\mathscr{F}$ is flat. (Do you see why?) As all higher cohomology groups (higher pushforwards) vanish, $\mathscr{C}^\bullet$ is exact except at the first term, where the cohomology is $\Gamma(\pi_* \mathscr{F}(m))$. We add the module $\Gamma(\pi_* \mathscr{F}(m))$ to the front of the complex, so it is once again exact:

$$(24.7.1.1) \qquad 0 \longrightarrow \Gamma(\pi_* \mathscr{F}(m)) \longrightarrow \mathscr{C}^1 \longrightarrow \mathscr{C}^2 \longrightarrow \cdots \longrightarrow \mathscr{C}^{n+1} \longrightarrow 0.$$

(We have done this trick of tacking on a module before, for example, in (18.2.4.1).) Thus, by Exercise 24.2.G, as we have an exact sequence in which all but the first terms are flat, the first term is flat as well. Thus $\pi_* \mathscr{F}(m)$ is a flat coherent sheaf on Y, and hence locally free (Corollary 24.3.7), and thus has locally constant rank.

Suppose $q \in Y$. We wish to show that the Hilbert function $h_{\mathscr{F}|_{X_q}}(m)$ is a locally constant function of q. To compute $h_{\mathscr{F}|_{X_q}}(m)$, we tensor the Čech resolution with $\kappa(q)$ and take cohomology. Now the extended Čech resolution (with $\Gamma(\pi_* \mathscr{F}(m))$ tacked on the front), (24.7.1.1), is an exact sequence of flat modules, and hence remains exact upon tensoring with $\kappa(q)$ (Exercise 24.2.F). Thus

$$\Gamma(\pi_* \mathscr{F}(m)) \otimes \kappa(q) \cong \Gamma(\pi_* \mathscr{F}(m)|_q),$$

so the Hilbert function $h_{\mathscr{F}|_{X_q}}(m)$ is the rank at q of a locally free sheaf, which is a locally constant function of $q \in Y$. $\qquad\square$

We now give some ridiculously useful consequences of Theorem 24.7.1.

24.7.2. Corollary — *Assume the same hypotheses and notation as in Theorem 24.7.1. Then the Hilbert polynomial of $\mathscr{F}|_{X_q}$ is locally constant as a function of $q \in Y$.*

24.7.A. CRUCIAL EXERCISE. Suppose $X \to Y$ is a projective flat morphism, where Y is connected. Show that the following functions of $q \in Y$ are constant: (a) the degree of the fiber, (b) the dimension of the fiber, (c) the arithmetic genus of the fiber.

24.7.B. EXERCISE. Use §24.3.9 and Exercise 24.7.A(a) to give another solution to Exercise 16.3.D(a) ("the degree of a finite morphism from a curve with no embedded points to an irreducible regular curve is constant").

24.7.C. UNIMPORTANT EXERCISE (CONVERSES TO THEOREM 24.7.1). (We won't use this exercise for anything.)

(a) Suppose A is a ring, and $S_\bullet$ is a finitely generated A-algebra that is flat over A. Show that $\operatorname{Proj} S_\bullet$ is flat over A.

(b) Suppose $\pi: X \to Y$ is a projective morphism of locally Noetherian schemes (which, as always, includes the data of an invertible sheaf $\mathscr{O}_X(1)$ on X), such that $\pi_* \mathscr{O}_X(m)$ is locally free for all $m \geq m_0$ for some m_0. Show that π is flat. Hint: Describe X as

$$\operatorname{Proj}\left(\mathscr{O}_Y \bigoplus (\oplus_{m \geq m_0} \pi_* \mathscr{O}_X(m))\right).$$

(c) More generally, suppose $\pi: X \to Y$ is a projective morphism of locally Noetherian schemes, and $\mathscr{F}$ is a coherent sheaf on X, such that $\pi_* \mathscr{F}(m)$ is locally free for all $m \geq m_0$ for some m_0. Show that $\mathscr{F}$ is flat over Y.

(d) Suppose $\pi: X \to Y$ is a projective morphism of locally Noetherian schemes, and $\mathscr{F}$ is a coherent sheaf on X, such that the Hilbert polynomial of $\mathscr{F}|_{X_q}$ is a locally constant function of $q \in Y$. If Y is reduced, show that $\mathscr{F}$ must be flat over Y. (Hint: Exercise 14.3.K shows that constant rank implies local freeness in particularly nice circumstances.)

Another consequence of Corollary 24.7.2 is something remarkably useful.

24.7.3. Corollary — *An invertible sheaf $\mathscr{L}$ on a flat projective family of curves X has locally constant degree on the fibers.*

(Recall that the definition of the degree of a line bundle on a projective curve requires no hypotheses on the curve such as regularity; see (18.4.7.1).)

Proof. An invertible sheaf $\mathscr{L}$ on a flat family of curves is always flat (as locally it is isomorphic to the structure sheaf). Hence $\chi(X_q, \mathscr{L}_q)$ is a constant function of q. By the definition of degree given in (18.4.7.1), $\deg(X_q, \mathscr{L}_q) = \chi(X_q, \mathscr{L}_q) - \chi(X_q, \mathscr{O}_{X_q})$. The result follows from the local constancy of $\chi(X_q, \mathscr{L}_q)$ and $\chi(X_q, \mathscr{O}_{X_q})$ (Theorem 24.7.1). $\qquad\square$

The following exercise is a serious generalization of Corollary 24.7.3.

24.7.D. * EXERCISE FOR THOSE WHO HAVE READ STARRED CHAPTER 20: INTERSECTION NUMBERS ARE LOCALLY CONSTANT IN FLAT FAMILIES. Suppose $\pi: X \to Y$ is a flat projective morphism to a connected scheme; $\mathscr{L}_1, \ldots, \mathscr{L}_n$ are line bundles on X; and $\mathscr{F}$ is a coherent sheaf on X, flat over Y, such that the support of $\mathscr{F}$ when restricted to any fiber of π has dimension at most n. If q is any point of Y, define (the temporary notation) $(\mathscr{L}_1 \cdot \mathscr{L}_2 \cdots \mathscr{L}_n \cdot \mathscr{F})_q$ to be the intersection on the fiber X_q of $\mathscr{L}_1, \ldots, \mathscr{L}_n$ with $\mathscr{F}|_{X_q}$ (Definition 20.1.1). Show that $(\mathscr{L}_1 \cdot \mathscr{L}_2 \cdots \mathscr{L}_n \cdot \mathscr{F})_q$ is independent of q. (The projective hypothesis can be relaxed to properness using §25.2.5.)

Corollary 24.7.3 motivates the following discussion.

24.7.4. *Definition.* Suppose $\mathscr{L}_1$ and $\mathscr{L}_2$ are line bundles on a k-variety X. We say that $\mathscr{L}_1$ and $\mathscr{L}_2$ are **algebraically equivalent** if there exist a connected (but not necessarily irreducible) k-variety Y with two k-valued points q_1 and q_2 and a line bundle $\mathscr{L}$ on $X \times Y$ such that the restriction of $\mathscr{L}$ to the fibers over q_1 and q_2 are isomorphic to $\mathscr{L}_1$ and $\mathscr{L}_2$, respectively.

24.7.E. * EXERCISE. Show that "algebraic equivalence" really is an equivalence relation. Show that the line bundles algebraically equivalent to $\mathscr{O}$ form a subgroup of $\operatorname{Pic} X$. This subgroup is denoted by $\operatorname{Pic}^0 X$. (You will use the notion of gluing two schemes together along isomorphic closed subschemes, §15.2.12.) Warning: Despite the unfortunately identical notation for degree 0 line bundles on curves (Definition 18.4.2), this is not the same concept.

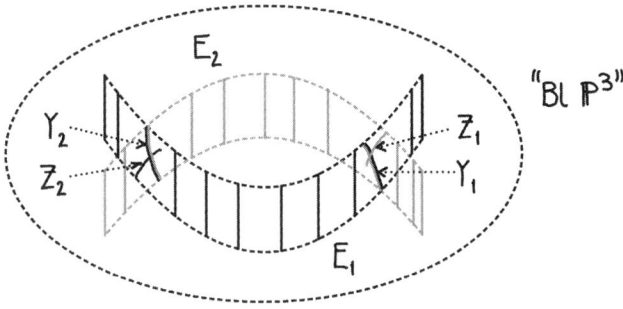

Figure 24.5 *Hironaka's example of a proper nonprojective smooth threefold.*

24.7.5. The **Néron–Severi group** is the group of line bundles $\mathrm{Pic}\, X$ modulo algebraic equivalence, $\mathrm{Pic}\, X/\mathrm{Pic}^0 X$. (This definition was promised in §18.4.12.) By Corollary 24.7.3, algebraic equivalence implies numerical equivalence (§18.4.12): $\mathrm{Pic}^0 X \subset \mathrm{Pic}^\tau X$. (Side remark: A line bundle $\mathscr{L}$ on a proper k-scheme X is numerically trivial if and only if there exists an integer $m \neq 0$ with $\mathscr{L}^{\otimes m}$ algebraically trivial. Thus $\mathrm{Pic}^\tau X/\mathrm{Pic}^0 X$ is torsion. See [SGA6, XIII, Thm. 4.6] for a proof, or [Laz, Cor. 1.4.38] for the projective case.)

24.7.6.* Hironaka's example of a proper nonprojective smooth threefold.

In §15.2.11, we produced a proper nonprojective variety, but it was singular. We can use Corollary 24.7.3 to give a *smooth* example, due to Hironaka. It is in in some sense "locally a blow-up of $\mathbb{P}^3$."

Inside $\mathbb{P}^3_k$, fix two conics C_1 and C_2, which meet in two (k-valued) points, p_1 and p_2. We construct a proper map $\pi\colon X \to \mathbb{P}^3_k$ as follows. For $i = 1$ and 2, let $U_i = \mathbb{P}^3 \setminus \{p_i\}$. For both i, consider the projective morphism $\pi_i\colon X_i \to U_i$ obtained by blowing up C_i, and then blowing up the proper transform of C_{3-i}. The two π_i glue together over $\mathbb{P}^3 \setminus \{p_1, p_2\}$ to give a morphism $\pi\colon X \to \mathbb{P}^3$. (Away from p_1 and p_2, C_1 and C_2 are disjoint, so blowing up one and then the other is the same as blowing up their union.)

Note that π is proper, as it is proper over U_1 and U_2, and the notion of properness is local on the target (Proposition 11.4.4(b)). As $\mathbb{P}^3_k$ is proper (over k, Theorem 8.4.10), and compositions of proper morphisms are proper (Proposition 11.4.4(c)), X is proper.

24.7.F. EXERCISE. Show that X is smooth. (Hint: Theorem 22.3.10.) Let E_i be the preimage of $C_i \setminus \{p_1, p_2\}$. Show that $\pi|_{E_i}\colon E_i \to C_i \setminus \{p_1, p_2\}$ is a $\mathbb{P}^1$-bundle (and flat).

24.7.G. EXERCISE. Let $\overline{E}_i$ be the closure of E_i in X. Show that $\overline{E}_i \to C_i$ is flat. (Hint: Exercise 24.3.J.)

24.7.H. EXERCISE. Show that $\pi^{-1}(p_i)$ is the union of two $\mathbb{P}^1$'s, say Y_i and Z_i, meeting at a point, such that $Y_i, Y_{3-i}, Z_{3-i} \in \overline{E}_i$ but $Z_i \notin \overline{E}_i$.

24.7.I. EXERCISE. Show that X is not projective, as follows. Suppose otherwise $\mathscr{L}$ is a very ample line bundle on X, so $\mathscr{L}$ has positive degree on every curve (including the Y_i and Z_i). Using flatness of $\overline{E}_i \to C_i$, and constancy of degree in flat families (Exercise 24.7.D), show that $\deg_{Y_i} \mathscr{L} = \deg_{Y_{3-i}} \mathscr{L} + \deg_{Z_{3-i}} \mathscr{L}$. Obtain a contradiction. (This argument will remind you of the argument of §15.2.11.)

24.7.7. *The notion of "projective morphism" is not local on the target.* Note that $\pi\colon X \to \mathbb{P}^3$ is not projective, as otherwise X would be projective (as the composition of projective morphisms is projective if the final target is quasicompact, Exercise 17.3.I). But away from each p_i, π *is* projective (as it is a composition of blow-ups, which are projective by construction, and the final target is quasicompact, so Exercise 17.3.I applies). Thus the notion of "projective morphism" is not local on the target. (A simpler but non-smooth example can be constructed using §15.2.11.)

24.8   Smooth and Étale Morphisms, and Flatness

We defined smooth and étale morphisms earlier (Definition 13.6.2), but we are now in a position to understand them much better. We will see that the notion of unramified morphism (§21.7) is a natural companion to them.

Our three algebro-geometric definitions won't be so obviously a natural triplet, but we will discuss the definitions given in [Gr-EGA] (§24.8.13), and in this context the three types of morphisms look very similar. (We briefly mention other approaches and definitions in §24.8.14.)

The three classes of morphisms we will discuss in this chapter are the analogs of the following types of maps of manifolds, in differential geometry (cf. §13.6.1).

- *Submersions* are maps inducing surjections of tangent spaces everywhere. They are useful in the notion of a fibration. (Perhaps a more relevant notion from differential geometry, allowing singularities, is: "locally on the source a smooth fibration.")
- *Isomorphisms locally on the source* (or *local isomorphisms*) are maps inducing isomorphisms of tangent spaces.
- *Immersions* are maps inducing injections of tangent spaces.

(Recall our warning from §9.2: "immersion" is often used in algebraic geometry with a different meaning.)

In order to better understand smooth and étale morphisms, we temporarily forget our earlier definitions, and consider some examples of things we want to be analogs of "local isomorphism" (or "locally on the source an isomorphism") and "locally on the source a smooth fibration," and see if they help us make good definitions.

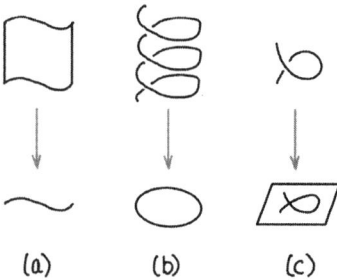

Figure **24.6** *Sketches of notions from differential geometry: (a) maps "locally on the source a smooth fibration," (b) local isomorphisms, and (c) immersions. (In algebraic geometry: (a) smooth morphisms, (b) étale morphisms, and (c) unramified morphisms.)*

24.8.1. "Local isomorphisms" (étale morphisms). Consider the parabola $x = y^2$ projecting to the x-axis, over the complex numbers. (This example has come up repeatedly, in one form or another, including in Exercise 13.6.F.) We might reasonably want this to be a local isomorphism (on the source) away from the origin. We might also want the notion of local isomorphism space to be an open condition: the locus where a morphism is a local isomorphism should be open on the source. This is true for the differential geometric definition. But then this morphism should be a local isomorphism over the generic point, and here we get a nontrivial residue field extension ($\mathbb{C}(y)/\mathbb{C}(y^2)$), not an isomorphism. Thus we are forced to consider (the Spec's of) certain finite extensions of fields to be "isomorphisms locally on the source"—in this case, even "covering spaces." (We will see in Exercise 24.8.D that we want precisely the finite separable extensions; you could have shown this earlier.)

Note also in this example there are no (nonempty) Zariski-open subsets $U \subset \operatorname{Spec} \mathbb{C}[x] \setminus \{0\}$ and $V \subset \operatorname{Spec} \mathbb{C}[x,y]/(x - y^2) \setminus \{(0,0)\}$ where the map sends U into V isomorphically (Exercise 13.6.F), so this is not a local isomorphism in a way you may have seen before. This leads to the notion of the étale topology, which is not even a topology in the usual sense, but a "Grothendieck topology" (§6.2.10). The étale topology is beyond the scope of this book.

24.8.2. Submersions (smooth morphisms).

24.8.3. *Fibers are smooth varieties.* As a first approximation of the algebro-geometric version of submersion, we will want the fibers to be smooth varieties (over the residue field). So the very first thing we need is to generalize the notion of "variety" over a base. It is reasonable to do this by having a locally finite type hypothesis. For somewhat subtle reasons, we will require the stronger condition that the morphism be locally of finite presentation. (If you really care, you can see where it comes up in our discussion. But, of course, there is no difference for Noetherian readers.)

The fibers are not just varieties; they should be *smooth* varieties (of dimension n, say). From the case of smoothness over a field, or from our intuition of what smooth varieties should look like, we expect the sheaf of differentials on the fibers to be locally free of rank n, or, even better, the sheaf of relative differentials to be locally free of rank n.

24.8.4. *Flatness.* At this point, our first approximation of "smooth morphism" is some version of "locally finitely presented, and fibers are smooth varieties." But that isn't quite enough. For example, a horrible map from a scheme X to a curve Y that maps a different regular variety to each point of Y (X is the infinite disjoint union of these) should not be considered a smooth fibration in any reasonable sense. Also, we might not want to consider $\operatorname{Spec} k \to \operatorname{Spec} k[\epsilon]/(\epsilon^2)$ to be a submersion; for example, this isn't surjective on tangent spaces, and, more generally, the picture "doesn't look like a fibration."

Both problems are failures of $\pi\colon X \to Y$ to be a nice, "'continuous" family. Whenever we are looking for some vague notion of "niceness" we know that "flatness" will be in the definition.

For comparison, note that "unramified" has no flatness hypothesis, and indeed we didn't expect it, as we would want the inclusion (closed embedding) of the origin into $\mathbb{A}^1$ to be unramified. But then weird things may be unramified. For example, if $X = \coprod_{z \in \mathbb{C}} \operatorname{Spec} \mathbb{C}$, then the morphism $X \to \mathbb{A}^1_{\mathbb{C}}$ sending the point corresponding to z to the point $z \in \mathbb{A}^1_{\mathbb{C}}$ is unramified. Such is life.

24.8.5. *Desired alternate definitions.* We might hope that a morphism $\pi\colon X \to Y$ is **smooth of relative dimension** n if and only if

  (i) π is locally of finite presentation,
 (ii) π is flat of relative dimension n, and
(iii) $\Omega_{X/Y}$ is locally free of rank n.

We might similarly hope that a morphism $\pi\colon X \to Y$ is **étale** if and only if

  (i) π is locally of finite presentation,
 (ii) π is flat, and
(iii) $\Omega_{X/Y} = 0$.

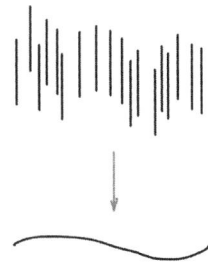

Figure 24.7 *We don't want this to be a "smooth morphism."*

We will shortly show (in Theorem 24.8.8 and Exercise 24.8.C, respectively) that these desired definitions are equivalent to Definition 13.6.2.

24.8.6. Different characterizations of smooth and étale morphisms.
The main result in this section is a description of equivalent characterizations of smooth morphisms, Theorem 24.8.8. We will state the theorem, and then give consequences, and then finally give a proof. As an important precursor:

24.8.7. Proposition — *Smooth morphisms are flat.*

Proof. We prove this first in the Noetherian setting. Suppose $\pi\colon X \to Y$ is a morphism of locally Noetherian schemes that is smooth of relative dimension n. We wish to show that it is flat at a point $p \in X$. By the definition of smooth morphisms (§13.6.2), we may replace X and Y with affine neighborhoods of p and $\pi(p)$, respectively, so that π is the morphism

(24.8.7.1) $$\operatorname{Spec} B[x_1, \ldots, x_{n+r}]/(f_1, \ldots, f_r) \longrightarrow \operatorname{Spec} B,$$

and p is a point where the Jacobian matrix is full rank.

24.8.A. EXERCISE. Show inductively, using the slicing criterion for flatness on the source (Theorem 24.6.9), that

$$\mathbb{A}_B^{n+r} \supset V(f_1) \supset V(f_1, f_2) \supset \cdots \supset V(f_1, \ldots, f_r) = X$$

are all smooth and flat over B at p, thereby proving Proposition 24.8.7 in the Noetherian case. (You will need the fact that smooth k-schemes are reduced, Proposition 13.2.13.)

(Noetherian readers can smugly skip to the end of the proof.) We prove the general case by a sneaky trick. We still write π locally as (24.8.7.1). By rearranging the x-variables, we may assume that the Jacobian of $f_1, \ldots, f_r$ by the first r variables $x_1, \ldots, x_r$ is invertible. Suppose

$$f_i(x_1, \ldots, x_{n+r}) = \sum_{a_1^i, \ldots, a_{n+r}^i} b_{a_1^i, \ldots, a_{n+r}^i}^i x_1^{a_1^i} \cdots x_{n+r}^{a_{n+r}^i},$$

where $b_{j,\ldots,k}^i \in B$. (These are finite sums!) Consider the polynomial algebra $P = \mathbb{Z}[c_{j,\ldots,k}^i]$, where the indices on the c-variables are precisely those that appear on the indices on the b-variables. Define $g_i \in P[x_1, \ldots, x_{n+r}]$ by

$$g_i(x_1, \ldots, x_{n+r}) = \sum_{a_1^i, \ldots, a_{n+r}^i} c_{a_1^i, \ldots, a_{n+r}^i}^i x_1^{a_1^i} \cdots x_{n+r}^{a_{n+r}^i},$$

Let $J \in P[x_1, \ldots, x_{n+r}]$ be the determinant of the Jacobian of $g_1, \ldots, g_r$ by $x_1, \ldots, x_r$. Finally, let $R := P_J$. We have a Cartesian diagram

$$
\begin{array}{ccc}
\operatorname{Spec} B[x_1, \ldots, x_{n+r}]/(f_1, \ldots, f_r) & \longrightarrow & \operatorname{Spec} R[x_1, \ldots, x_{n+r}]/(g_1, \ldots, g_r) \\
\pi \downarrow & & \downarrow \alpha \\
\operatorname{Spec} B & \longrightarrow & \operatorname{Spec} R
\end{array}
$$

sending $c_{j,\ldots,k}^i$ to $b_{j,\ldots,k}^i$, and g_i to f_i. Because J is invertible in R, α is smooth of relative dimension n (basically by definition). From the Noetherian case of the result already proved, α is flat. Because flatness is preserved by base change, we have that π is flat, as desired. □

We now come to the central result of this section.

24.8.8. Theorem — *Suppose $\pi\colon X \to Y$ is a morphism of schemes. Then the following are equivalent.*

(i) *The morphism π is smooth of relative dimension n (Definition 13.6.2).*
(ii) *The morphism π is locally finitely presented and flat of relative dimension n; and Ω_π is locally free of rank n (Desired Alternate Definition 24.8.5).*
(iii) *The morphism π is locally finitely presented and flat, and the fibers are smooth k-schemes of pure dimension n.*
(iv) *The morphism π is locally finitely presented and flat, and the geometric fibers are smooth k-schemes of pure dimension n.*

We first dispatch with the parts of this theorem we have basically already proved.

24.8.B. EXERCISE. Prove that (i) implies (ii) implies (iii). Show that (iii) is equivalent to (iv). Hints: Smooth implies flat by Proposition 24.8.7; the definition of smooth k-varieties in terms of differentials in Redefinition 21.2.28; and smoothness is independent of field extension by Exercise 13.2.H, respectively.

Before finishing the proof of Theorem 24.8.8 in §24.8.11, we motivate the result by giving a number of exercises and results assuming it.

24.8.C. EXERCISE. Show that a morphism π is étale (Definition 13.6.2) if and only if π is locally of finite presentation, is flat, and $\Omega_\pi = 0$ (Desired Alternate Definition 24.8.5).

24.8.D. EASY EXERCISE. Suppose $\pi: X \to Y$ is a morphism. Show that the following are equivalent.

(a) π is étale.
(b) π is smooth and unramified.
(c) π is locally finitely presented, flat, and unramified.
(d) π is locally finitely presented, flat, and for each $q \in Y$, $\pi^{-1}(q)$ is the disjoint union of schemes of the form $\operatorname{Spec} K$, where K is a finite separable extension of $\kappa(q)$.
(e) π is locally finitely presented, flat, and, for each *geometric* point $\overline{q}$ of Y, $\pi^{-1}(\overline{q})$ is the disjoint union of copies of $\overline{q}$.

24.8.E. IMPORTANT EXERCISE. If $\pi: X \to Y$ is étale, show that any preimage $p \in X$ of any regular point $q \in Y$ whose local ring has dimension n is also a regular point whose local ring has dimension n. Hint: Prove the result by induction on $\dim \mathscr{O}_{Y,q}$. "Slice" by an element of $\mathfrak{m}_{Y,q} \setminus \mathfrak{m}_{Y,q}^2$. Use the slicing criterion for regularity (Exercise 13.2.C).

24.8.9. *Proof of the Smoothness-Regularity Comparison Theorem 13.2.7(b) (every smooth k-scheme is regular).* By Observation 13.6.4, any dimension n smooth k-scheme X can locally be expressed as an étale cover of $\mathbb{A}_k^n$, which is regular by Exercise 13.3.P. Then by Exercise 24.8.E, X is regular. □

24.8.F. EXERCISE. Suppose $\pi: X \to Y$ is a morphism of pure dimensional k-varieties, and Y is smooth (over k). Use the cotangent exact sequence (Theorem 21.2.24) to show the following.

(a) Suppose that $\dim X = \dim Y = n$, and π is unramified. Show that X is smooth.
(b) Suppose $\dim X = m > \dim Y = n$, and the (scheme-theoretic) fibers of π over closed points are smooth of dimension $m - n$. Show that X is smooth.

24.8.G. EXERCISE. Suppose $\pi: X \to Y$ is a morphism (over k) of smooth pure dimensional k-varieties, where $\dim X = m$ and $\dim Y = n$. (The two parts are clearly essentially the same geometric situation.)

(a) Show that if the fibers of π are all smooth of dimension $m - n$, then π is smooth of relative dimension $m - n$.
(b) Show that if $\pi^* \Omega_Y \to \Omega_X$ is an "injection of vector bundles" (see Definition 14.2.4), or, equivalently, "T_π is surjective," then π is smooth of relative dimension $m - n$. Possible approach: For each point $q \in Y$, $\mathscr{O}_{Y,q}$ is a regular local ring. Work by induction on $\dim \mathscr{O}_{Y,q}$, and "slice" by an element of $\mathfrak{m}_{Y,q} \setminus \mathfrak{m}_{Y,q}^2$.

24.8.10. *Remark.* Suppose $\pi: X \to Y$ is locally finitely presented, $\rho: Y \to Z$ is étale, and $\tau = \rho \circ \pi$. Then π is smooth of dimension n (e.g., étale, taking $n = 0$) if and only if τ is. One direction is Exercise 13.6.D, and the other is Exercise 21.7.G(b). Similarly, π is unramified if and only if τ is (Exercises 21.7.C and 21.7.G(a)).

24.8.11. ** *Proof of Theorem 24.8.8.*
By Exercise 24.8.B, it remains only to show that (iii) implies (i). (This argument is reminiscent of the proof of Theorem 13.6.6.)

Fix a closed point $p \in X$. Let $q = \pi(p) \in Y$. We will show that there is an open neighborhood U_i of p and V_i of q of the form stated in Definition 13.6.2. It suffices to deal with the case that p is a closed point in $\pi^{-1}(q)$, because closed points are dense in finite type schemes (Exercise 5.3.F). (However, we cannot assume that q is a closed point of Y.)

As π is finitely presented, there are affine open neighborhoods $U \subset X$ of p and $V \subset Y$ of q, with $\pi(U) \subset V$, such that the morphism $\pi|_U \colon U \to V$ can be written as

$$U = \operatorname{Spec} B[x_1, \dots, x_N]/I \xrightarrow{\ \pi|_U\ } \operatorname{Spec} B = V,$$

where I is finitely generated. Choose generators $f_1, \dots, f_s$ of I. Consider the Cartesian diagram

$$
\begin{array}{ccc}
\overline{U} & \lhook\joinrel\longrightarrow & U \\
\| & & \| \\
\operatorname{Spec} \kappa(q)[x_1, \dots, x_N]/\overline{I} & \lhook\joinrel\longrightarrow & \operatorname{Spec} B[x_1, \dots, x_N]/I \\
\downarrow & & \downarrow{\scriptstyle \pi|_U} \\
\operatorname{Spec} \kappa(q) & \lhook\joinrel\longrightarrow & \operatorname{Spec} B \\
\| & & \| \\
q & \lhook\joinrel\longrightarrow & V
\end{array}
\qquad ,
$$

where $\overline{U}$ is the fiber of U over q, and $\overline{I}$ is the ideal cutting out $\overline{U}$ in $\mathbb{A}^N_{\kappa(q)}$.

Recall from §24.2.2 that for a closed subscheme flat over a base, the pullback of the defining ideal as a quasicoherent sheaf is the same as the ideal of the pulled back closed subscheme. Hence because $B[x_1, \dots, x_N]/I$ is flat over B, we have that $\overline{I} = I \otimes_B \kappa(q)$. Then $\overline{f}_1 := f_1 \otimes 1, \dots, \overline{f}_s := f_s \otimes 1$ generate $I \otimes_B \kappa(q) = \overline{I}$.

By hypothesis (iii), the fiber $\overline{U} \subset \mathbb{A}^N_{\kappa(q)}$ is smooth over $\kappa(q)$, so $\mathscr{O}_{\overline{U}, p}$ is an integral domain of dimension n (Exercise 13.2.O). Also by hypothesis (iii), the Jacobian matrix of $\overline{f}_1, \dots, \overline{f}_s$ (with respect to $x_1, \dots, x_N$) has corank n at p, i.e., the Jacobian matrix has rank $r := N - n$. So we can rename the $\overline{f}_i$ so that the Jacobian matrix of the *first* r of them, $\overline{f}_1, \dots, \overline{f}_r$, has corank n at p. Then $\overline{U}' := \operatorname{Spec} \kappa(q)[x_1, \dots, x_N]/(\overline{f}_1, \dots, \overline{f}_r)$ is smooth of relative dimension n over $\kappa(q)$ (in a neighborhood of p), so $\mathscr{O}_{\overline{U}', p}$ is an integral domain of dimension n (Exercise 13.2.O again). Then

$$\mathscr{O}_{\overline{U}', p} \longrightarrow\!\!\!\!\!\rightarrow \mathscr{O}_{\overline{U}', p}/(\overline{f}_{r+1}, \dots, \overline{f}_s) = \mathscr{O}_{\overline{U}, p}$$

is a surjection of integral domains of the same dimension n, and thus an isomorphism by Easy Exercise 13.2.M. Hence $\overline{U} = \overline{U}'$ near p in $\overline{U}$ (or, more precisely, there is a neighborhood of p in $\overline{U}$ that is also a neighborhood of p in $\overline{U}'$).

Let $U' \subset \operatorname{Spec} B[x_1, \dots, x_N]/(f_1, \dots, f_r)$ be the open subset where the Jacobian matrix of $f_1, \dots, f_r$ (with respect to $x_1, \dots, x_N$) has corank n (i.e., full rank r). Then $U' \to \operatorname{Spec} B$ is smooth of relative dimension n by definition. We will show that there is a neighborhood of p in U that is also a neighborhood of p in U', thereby completing the proof.

Let $R := B[x_1, \dots, x_n]/(f_1, \dots, f_r)$ and $J := (f_{r+1}, \dots, f_s) \subset R$ for convenience (see Figure 24.8). As J is a finitely generated ideal, it is supported on a closed subset of $\operatorname{Spec} R$ (Exercise 6.6.B(b)). We wish to show that p is not in the support of J in $\operatorname{Spec} R$. It suffices to show that $J \otimes_R \kappa(p) = 0$ by Nakayama's Lemma.

Because $U = \operatorname{Spec} R/J \subset X$ is smooth over B, R/J is flat over B (Proposition 24.8.7). Now $J \otimes_B \kappa(q)$ is the ideal cutting out $\overline{U}'$ in $\overline{U}$ (§24.2.2 again, using flatness of $U = \operatorname{Spec} R/J$), and $\overline{U} = \overline{U}'$ near p in $\overline{U}$. Hence $J \otimes_B \kappa(q) = 0$ near p (i.e., $\widetilde{J \otimes_B \kappa(q)} = 0$ in an open neighborhood of p), so $(J \otimes_B \kappa(q)) \otimes_R \kappa(p) = 0$. Then

$$J \otimes_R \kappa(p) = J \otimes_R (\kappa(p) \otimes_B \kappa(q)) = (J \otimes_B \kappa(q)) \otimes_R \kappa(p) = 0,$$

as desired. $\qquad\qquad\qquad\qquad\qquad\qquad\qquad\qquad\qquad\qquad\qquad\qquad\qquad\qquad\qquad\square$

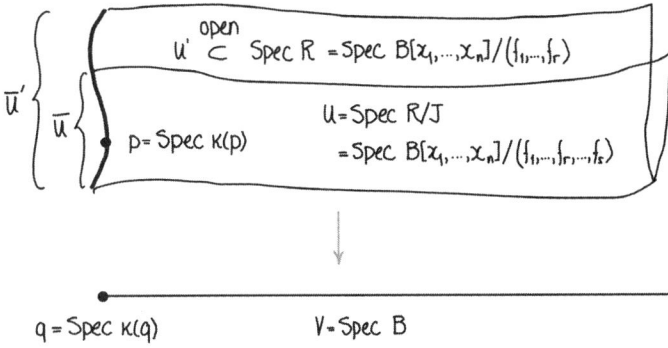

Figure 24.8 *Geometry behind part of the proof of Theorem 24.8.8.*

24.8.12.** Formally unramified, smooth, and étale.
[Gr-EGA] takes a different starting point for the definition of unramified, smooth, and étale morphisms. The definitions there make clear that these three definitions form a family.

The cost of these definitions are that they are perhaps less immediately motivated by geometry, and it is harder to show some basic properties. The benefit is that it is possible to show more (for example, left-exactness of the cotangent and conormal exact sequences, and good interpretations in terms of completions of local rings). But we simply introduce these ideas here, and do not explore them. See [BLR, §2.2] for an excellent discussion. (You should largely ignore what follows, unless you find later in life that you really care.)

24.8.13. *Definition.* We say that $\pi \colon X \to Y$ is **formally smooth** (resp., **formally étale, formally unramified**) if for all affine schemes Z, and every closed subscheme $Z_0 \subset Z$ defined by a nilpotent ideal, and every morphism $Z \to Y$, the canonical map $\operatorname{Hom}_Y(Z, X) \to \operatorname{Hom}_Y(Z_0, X)$ is surjective (resp., bijective, injective). This is summarized in the following diagram, which is reminiscent of the valuative criteria for separatedness and properness.

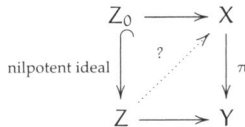

$$
\begin{array}{ccc}
Z_0 & \longrightarrow & X \\
{\scriptstyle\text{nilpotent ideal}}\Big\downarrow \quad {\scriptstyle ?} \nearrow & & \Big\downarrow {\scriptstyle \pi} \\
Z & \longrightarrow & Y
\end{array}
$$

(You can check that this is the same as the definition we would get by replacing "nilpotent" by "square-zero." This is sometimes an easier formulation to work with.)

Then [Gr-EGA] defines "smooth" as morphisms that are formally smooth and locally of finite presentation, and similarly for unramified and étale.

One can show that [Gr-EGA]'s definitions of formally unramified and smooth agree with the definitions we give. For "formally unramified" (where our definition given in §21.7 is not obviously the same as the definition of [Gr-EGA] given here), see [Gr-EGA, IV$_4$.17.2.1] or [Stacks, tag 00UO]. For "smooth," see [Gr-EGA, IV$_4$.17.5.1] or [Stacks, tag 00TN]. (Our characterization of étale as smooth of relative dimension 0 then agrees with [Gr-EGA]. Our definition of unramified as formally unramified plus locally of finite *type* disagrees with [Gr-EGA]; see §21.7.)

24.8.14. Other starting points. There are a number of quite different ways of defining smooth, étale, and unramified morphisms, and it is nontrivial to relate them. We have just described the approach of [Gr-EGA]. Another common approach is the characterization of smooth morphisms as locally finitely presented, flat, and with regular geometric fibers (see Theorem 24.8.8). Yet another definition is via a naive version of the cotangent complex; this is the approach taken by [Stacks, tag 00T2], and is less frightening than it sounds. Finally, the different characterizations of Exercise 24.8.D give alternate initial definitions of étaleness.

Chapter 25

Cohomology and Base Change Theorems

Cohomology and base change theorems are powerful results relating the higher pushforwards of coherent sheaves with the cohomology of those sheaves on fibers, and more generally explaining when higher pushforward commutes with base change. We first describe some of their statements, and immediate applications, in §25.1. We give their unexpectedly elegant proofs in §25.2. Then in §25.3 we use them to introduce a number of important ideas in moduli theory.

25.1 Statements and Applications

Higher pushforwards are easy to define, but it is hard to get a geometric sense of what they are, or how they behave. For example, given a morphism $\pi\colon X \to Y$, and a quasicoherent sheaf $\mathscr{F}$ on X, you might reasonably hope that the fibers of $R^i\pi_*\mathscr{F}$ are the cohomologies of $\mathscr{F}$ along the fibers. More precisely, given $\psi\colon q \to Y$ corresponding to the inclusion of a point (better: $\psi\colon \operatorname{Spec}\kappa(q) \to Y$), yielding the Cartesian diagram

(25.1.0.1)
$$
\begin{array}{ccc}
X_q & \xrightarrow{\;\psi'\;} & X \\
{\scriptstyle\pi'}\downarrow & & \downarrow{\scriptstyle\pi} \\
q & \xrightarrow{\;\psi\;} & Y,
\end{array}
$$

one might hope that the morphism

$$
\boxed{\phi_q^p\colon \psi^*(R^p\pi_*\mathscr{F}) \longrightarrow H^p(X_q, \mathscr{F}|_{X_q})}
$$

(the push-pull map in cohomology given in Exercise 18.7.C) is an isomorphism. (Note: $\mathscr{F}|_{X_q}$ and $(\psi')^*\mathscr{F}$ are symbols for the same thing. The first is often preferred, but we sometimes use the second because we will consider more general ψ and ψ'.) We could then picture $R^i\pi_*\mathscr{F}$ as somehow fitting together the cohomology groups of fibers into a coherent sheaf. (Warning: This is too much to hope for; see Exercise 25.1.A.)

It would also be wonderful if $h^p(X_q, (\psi')^*\mathscr{F})$ was constant, and ϕ_q^p put them together into a nice locally free sheaf (vector bundle) $R^p\pi_*\mathscr{F}$.

The base change $\psi\colon q \to Y$ should not be special. As long as we are dreaming, we may as well hope that in good circumstances, given a Cartesian diagram (18.7.4.1)

(25.1.0.2)
$$
\begin{array}{ccc}
W & \xrightarrow{\;\psi'\;} & X \\
{\scriptstyle\pi'}\downarrow & & \downarrow{\scriptstyle\pi} \\
Z & \xrightarrow{\;\psi\;} & Y,
\end{array}
$$

the natural push-pull morphism

(25.1.0.3)
$$
\boxed{\phi_Z^p\colon \psi^*(R^p\pi_*\mathscr{F}) \longrightarrow R^p\pi'_*(\psi')^*\mathscr{F}}
$$

of sheaves on Z (Exercise 18.7.B(a)) is an isomorphism. (In some cases, we can already address this question. For example, cohomology commutes with flat base change, Theorem 24.1.9, so the result holds if ψ is flat.)

567

We formalize our dreams into three nice properties that we might wish in this situation. We will see that they are closely related. Suppose $\mathscr{F}$ is a coherent sheaf on X, $\pi\colon X \to Y$ is proper, Y (hence X) is Noetherian, and $\mathscr{F}$ is flat over Y.

(a) Given a Cartesian square (25.1.0.1), is $\phi_q^p\colon R^p\pi_*\mathscr{F} \otimes \kappa(q) \to H^p(X_q, \mathscr{F}|_{X_q})$ an isomorphism?

(b) Given a Cartesian square (25.1.0.2), is $\phi_Z^p\colon \psi^*(R^p\pi_*\mathscr{F}) \to R^p\pi'_*(\psi')^*\mathscr{F}$ an isomorphism?

(c) Is $R^p\pi_*\mathscr{F}$ locally free?

We turn first to property (a). The dimension of the left side $R^p\pi_*\mathscr{F} \otimes \kappa(q)$ is an upper semicontinuous function of $q \in Y$ by upper semicontinuity of rank of finite type quasicoherent sheaves (Exercise 14.3.J). The Semicontinuity Theorem states that the dimension of the right is also upper semicontinuous. More formally:

25.1.1. Semicontinuity Theorem — *Suppose $\pi\colon X \to Y$ is a proper morphism of Noetherian schemes, and $\mathscr{F}$ is a coherent sheaf on X flat over Y. Then for each $p \geq 0$, the function $Y \to \mathbb{Z}$ given by $q \mapsto \dim_{\kappa(q)} H^p(X_q, \mathscr{F}|_{X_q})$ is an upper semicontinuous function of $q \in Y$.*

Translation: Ranks of cohomology groups are upper semicontinuous in proper flat families. (A proof will be given in §25.2.4.)

25.1.2. *Example.* You may already have seen an example of cohomology groups jumping, in §24.3.14. Here is a simpler example, albeit not of the structure sheaf. Let (E, p_0) be an elliptic curve over a field k, and consider the projection $\pi\colon E \times E \to E$ to the second factor. Let $\mathscr{L}$ be the invertible sheaf (line bundle) on $E \times E$ corresponding to the divisor that is the diagonal, minus the section of $p_0 \times E$ of π (where $p_0 \in E$). Then $\mathscr{L}|_{p_0}$ (i.e., $\mathscr{L}|_{E \times p_0}$) is trivial, but $\mathscr{L}|_p$ is nontrivial for any $p \neq p_0$ (as we showed in our study of genus 1 curves, in §19.9). Thus $h^0(E, \mathscr{L}|_p)$ is 0 in general, but jumps to 1 for $p = p_0$.

25.1.A. EXERCISE. Show that $\pi_*\mathscr{L} = 0$. Thus we cannot picture $\pi_*\mathscr{L}$ as "gluing together" h^0 of the fibers; in this example, cohomology does not commute with "base change" or "taking fibers."

25.1.3. *Side remark.* In characteristic 0, the cohomology of $\mathscr{O}$ doesn't jump in smooth families. Over $\mathbb{C}$, this is because Betti numbers are constant in connected families, and (21.5.11.1) (from Hodge theory) expresses the Betti constants h_{Betti}^k as sums (over $i + j = k$) of upper semicontinuous functions $h^j(\Omega^i)$, so the Hodge numbers $h^j(\Omega^i)$ must in fact be constant. The general characteristic 0 case can be reduced to $\mathbb{C}$ by an application of the Lefschetz principle (which also arose in §21.5.9). But ranks of cohomology groups of $\mathscr{O}$ for smooth families of varieties *can* jump in positive characteristic (see, for example, [MO70920]). Also, the example of §24.3.14 shows that the "smoothness" hypothesis cannot be removed.

25.1.4. Grauert's Theorem. If $R^p\pi_*\mathscr{F}$ is locally free (property (c)) and ϕ_q^p is an isomorphism (property (a)), then $h^p(X_q, \mathscr{F}|_{X_q})$ is clearly locally constant. The following is a partial converse.

25.1.5. Grauert's Theorem — *If $\pi\colon X \to Y$ is proper, Y is reduced and locally Noetherian, $\mathscr{F}$ is a coherent sheaf on X flat over Y, and $h^p(X_q, \mathscr{F}|_{X_q})$ is a locally constant function of $q \in Y$, then $R^p\pi_*\mathscr{F}$ is locally free, and ϕ_Z^p is an isomorphism for all $\psi\colon Z \to Y$.*

In other words, if cohomology groups of fibers have locally constant dimension (over a reduced base), then they can be fit together to form a vector bundle, and the fiber of the pushforward is identified with the cohomology of the fiber. Our dreams at the start of this chapter have come true!

(See §25.2.10 to remove Noetherian hypotheses.)

We further note that if Y is integral, π is proper, and $\mathscr{F}$ is a coherent sheaf on X flat over Y, then by the Semicontinuity Theorem 25.1.1 there is a dense open subset of Y on which $R^p\pi_*\mathscr{F}$ is locally free (and on which the fiber of the pth higher pushforward is the pth cohomology of the fiber).

The following statement is even more magical than Grauert's Theorem 25.1.5.

25.1.6. Cohomology and Base Change Theorem — *Suppose π is proper, Y is locally Noetherian, $\mathscr{F}$ is coherent over X and flat over Y, and ϕ_q^p is surjective. Then the following hold.*

(i) *There is an open neighborhood U of q such that for any $\psi\colon Z \to U$, ϕ_Z^p is an isomorphism. In particular, ϕ_q^p is an isomorphism.*

(ii) *Furthermore,*

    (a) *ϕ_q^{p-1} is surjective (hence an isomorphism by (i)) if and only if*

    (b) *$R^p\pi_*\mathscr{F}$ is locally free in some open neighborhood of q (or, equivalently, $(R^p\pi_*\mathscr{F})_q$ is a free $\mathscr{O}_{Y,q}$-module, Exercise 14.3.E).*

    (These then imply that $h^p(X_r, \mathscr{F}|_{X_r})$ is constant for r in an open neighborhood of q.)

(Proofs of Theorems 25.1.5 and 25.1.6 will be given in §25.2. Note in (ii) that if $p = 0$, ϕ_q^{p-1} is automatically surjective, as ϕ_q^{-1} is the zero map. See §25.2.10 to remove Noetherian hypotheses.)

This is amazing: the hypothesis that ϕ_q^p is surjective involves what happens only over points q of X, with *reduced* structure, yet it has implications over the (possibly nonreduced) scheme as a whole! This might remind you of the local criterion for flatness (Theorem 24.6.1), which indeed is the key technical ingredient of the proof.

Here are some consequences.

25.1.B. EXERCISE. Use Theorem 25.1.6 to give a second solution to Exercise 24.3.E. (This is a big weapon to bring to bear on this problem, but it is still enlightening; your original solution to Exercise 24.3.E may foreshadow the proof of the Cohomology and Base Change Theorem 25.1.6.)

25.1.C. EXERCISE. Suppose $\pi\colon X \to Y$ is proper, Y is locally Noetherian, and $\mathscr{F}$ is a coherent sheaf on X, flat over Y. Suppose further that $H^p(X_q, \mathscr{F}|_{X_q}) = 0$ for some $q \in Y$. Show that there is an open neighborhood U of q such that $(R^p\pi_*\mathscr{F})|_U = 0$.

25.1.D. EXERCISE. Suppose $\pi\colon X \to Y$ is proper, Y is locally Noetherian, and $\mathscr{F}$ is a coherent sheaf on X, flat over Y. Suppose further that $H^p(X_q, \mathscr{F}|_{X_q}) = 0$ for all $q \in Y$. Show that the $(p-1)$st cohomology commutes with arbitrary base change: ϕ_Z^{p-1} is an isomorphism for all $\psi\colon Z \to Y$.

25.1.E. EXERCISE. Suppose π is proper, Y is locally Noetherian, and $\mathscr{F}$ is a coherent sheaf on X flat over Y. Suppose further that $R^p\pi_*\mathscr{F} = 0$ for $p \geq p_0$. Show that $H^p(X_q, \mathscr{F}|_{X_q}) = 0$ for all $q \in Y$, $p \geq p_0$.

25.1.F. EXERCISE. Suppose π is proper, Y is locally Noetherian, and $\mathscr{F}$ is a coherent sheaf on X, flat over Y. Suppose further that Y is reduced. Show that there exists a dense open subset U of Y such that ϕ_Z^p is an isomorphism for all $\psi\colon Z \to U$ and all p. (Hint: Find suitable open neighborhoods of the generic points of Y. See Exercise 24.1.M and the paragraph following it.)

25.1.G. EXERCISE. Suppose $\pi\colon C \to S$ is a proper smooth morphism whose fibers are genus $g > 1$ curves (a "flat family of smooth genus g curves"). Suppose further that S is locally Noetherian.

(a) If $n > 1$, show that $\pi_*\Omega_{C/S}^{\otimes n}$ is a vector bundle of rank $(2n-1)(g-1)$, and that this construction commutes with any base change $S' \to S$ (with no Noetherian hypotheses on S'). More precisely, if

$$
\begin{array}{ccc}
C' & \longrightarrow & C \\
{\scriptstyle \pi'}\downarrow & & \downarrow{\scriptstyle \pi} \\
S' & \xrightarrow{\;\rho\;} & S
\end{array}
$$

    is a Cartesian diagram, show that $\rho^*\pi_*\Omega_{C/S}^{\otimes n} \cong \pi'_*\Omega_{C'/S'}^{\otimes n}$.

(b) If S is reduced, show that $\pi_*\Omega_{C/S}$ is a vector bundle of rank g, and that this construction commutes with any base change $S' \to S$ (with no hypotheses on S). *Remark:* In fact, the reducedness hypothesis can be removed. One bizarre way of doing this is by showing that there is a moduli space $\mathscr{M}_g$ of smooth genus $g > 1$ curves that is smooth (albeit a Deligne-Mumford stack, not a scheme), and then applying this argument to the universal curve over $\mathcal{M}_g$, from which all families of genus g curves are pulled back (cf. Exercise 25.3.A).

25.1.7. An important class of morphisms: Proper, $\mathscr{O}$-connected morphisms $\pi\colon X \to Y$ of locally Noetherian schemes.

If a morphism $\pi\colon X \to Y$ satisfies the property that the natural map $\mathscr{O}_Y \to \pi_*\mathscr{O}_X$ is an isomorphism, we say that π is $\mathscr{O}$-**connected**. Clearly the notion of $\mathscr{O}$-connectedness is local on the target, and preserved by composition.

25.1.H. EASY EXERCISE. Show that proper $\mathscr{O}$-connected morphisms of locally Noetherian schemes are surjective.

25.1.8. We will soon meet Zariski's Connectedness Lemma 28.4.1, which shows that proper, $\mathscr{O}$-connected morphisms of locally Noetherian schemes have connected fibers. In some sense, this class of morphisms is really the right class of morphisms capturing what we might want by "connected fibers"; this is the motivation for the terminology. The following result gives some evidence for this point of view, in the flat context.

25.1.I. IMPORTANT EXERCISE. Suppose $\pi\colon X \to Y$ is a proper *flat* morphism of locally Noetherian schemes, whose fibers satisfy $h^0(X_q, \mathscr{O}_{X_q}) = 1$. (Important remark: This is satisfied if π has geometrically connected and geometrically reduced fibers, by §11.4.7.) Show that π is $\mathscr{O}$-connected. Hint: Consider

$$(25.1.8.1) \qquad \mathscr{O}_Y \otimes \kappa(q) \longrightarrow (\pi_*\mathscr{O}_X) \otimes \kappa(q) \xrightarrow{\phi_q^0} H^0(X_q, \mathscr{O}_{X_q}) \xleftarrow{\;\sim\;} \kappa(q).$$

The composition is surjective, hence ϕ_q^0 is surjective, hence it is an isomorphism by the Cohomology and Base Change Theorem 25.1.6(i). Then by the Cohomology and Base Change Theorem 25.1.6(ii), $\pi_*\mathscr{O}_X$ is locally free, thus of rank 1. Perhaps use Nakayama's Lemma to show that a map of invertible sheaves $\mathscr{O}_Y \to \pi_*\mathscr{O}_X$ that is a surjection on fibers over points (with reduced structure) is necessarily an isomorphism of sheaves. Then extract "surjection on fibers" from (25.1.8.1).

25.1.9.* *Unimportant remark.* This class of proper, $\mathscr{O}$-connected morphisms is not preserved by arbitrary base change, and thus is not "reasonable" in the sense of §8.1. But you can show that they are preserved by *flat* base change, using the fact that cohomology commutes with flat base change, Theorem 24.1.9. Furthermore, the conditions of Exercise 25.1.I behave well under base change, and Noetherian hypotheses can be removed from the Cohomology and Base Change Theorem 25.1.6 (at the expense of finitely presented hypotheses; see §25.2.10), so the class of morphisms $\pi\colon X \to Y$ that are proper, finitely presented, and flat, with geometrically connected and geometrically reduced fibers, *is* "reasonable" (and useful).

25.1.10. We next address the following question. Suppose $\pi\colon X \to Y$ is a morphism of schemes. Given an invertible sheaf $\mathscr{L}$ on X, we ask when it is the pullback of an invertible sheaf $\mathscr{M}$ on Y. For this to be true, we certainly need that $\mathscr{L}$ is trivial on the fibers. We will see that if π is a proper $\mathscr{O}$-connected morphism of locally Noetherian schemes, then this often suffices. Given $\mathscr{L}$, we recover $\mathscr{M}$ as $\pi_*\mathscr{L}$; the fibers of $\mathscr{M}$ are one-dimensional, and glue together to form a line bundle. We now begin to make this precise.

25.1.J. EXERCISE. Suppose $\pi\colon X \to Y$ is a proper, $\mathscr{O}$-connected morphism of locally Noetherian schemes. Show that if $\mathscr{M}$ is any invertible sheaf on Y, then the natural morphism $\mathscr{M} \to \pi_*\pi^*\mathscr{M}$ is an isomorphism. In particular, we can recover $\mathscr{M}$ from $\pi^*\mathscr{M}$ by applying the pushforward π_*. (Your proof will work more generally when $\mathscr{M}$ is locally free of finite rank.)

25.1.11. Proposition — *Suppose $\pi\colon X \to Y$ is a **flat**, proper morphism of locally Noetherian schemes with geometrically connected and geometrically reduced fibers (hence $\mathscr{O}$-connected, by Exercise 25.1.I). Suppose also that Y is reduced, and $\mathscr{L}$ is an invertible sheaf on X that is trivial on the fibers of π (i.e., $\mathscr{L}|_{X_q}$ is a trivial* *invertible sheaf on* X_q *for all* $q \in Y$). *Then* $\pi_* \mathscr{L}$ *is an invertible sheaf on* Y (*call it* $\mathscr{M}$), *and the natural map* $\pi^* \mathscr{M} \to \mathscr{L}$ *is an isomorphism.*

Proof. By Grauert's Theorem 25.1.5, $\pi_* \mathscr{L}$ is locally free of rank 1 (again, call it $\mathscr{M}$), and the map $\mathscr{M} \otimes_{\mathscr{O}_Y} \kappa(q) \to H^0(X_q, \mathscr{L}|_{X_q})$ is an isomorphism. We have a natural map of invertible sheaves $\pi^* \mathscr{M} = \pi^* \pi_* \mathscr{L} \to \mathscr{L}$. To show that it is an isomorphism, we need only show that it is surjective. (Do you see why? If A is a ring, and $\phi \colon A \to A$ is a surjection of A-modules, why is ϕ an isomorphism?) For this, it suffices to show that it is surjective on the fibers of π. (Do you see why? Hint: If the cokernel of the map is not 0, then it is not 0 above some point of Y.) But this follows from the first line of the proof (using, for example, that $\mathscr{M} \cong \mathscr{O}$ in a neighborhood of q). $\qquad \square$

Proposition 25.1.11 has some pleasant consequences. For example, if you have two invertible sheaves $\mathscr{A}$ and $\mathscr{B}$ on X that are isomorphic on every fiber of π, then they differ by a pullback of an invertible sheaf on Y: just apply Proposition 25.1.11 to $\mathscr{A} \otimes \mathscr{B}^\vee$.

25.1.12. Projective bundles.

25.1.K. EXERCISE. Let X be a locally Noetherian scheme, and let $\mathrm{pr}_1 \colon X \times \mathbb{P}^n \to X$ be the projection onto the first factor. Suppose $\mathscr{L}$ is a line bundle on $X \times \mathbb{P}^n$, whose degree on every fiber of pr_1 is zero. Use the Cohomology and Base Change Theorem 25.1.6 to show that $(\mathrm{pr}_1)_* \mathscr{L}$ is an invertible sheaf on X. Show that the natural map $\mathrm{pr}_1^*((\mathrm{pr}_1)_* \mathscr{L}) \to \mathscr{L}$ is an isomorphism, as at the end of proof of Proposition 25.1.11.

Your argument will apply just as well to the situation where $\mathrm{pr}_1 \colon X \times \mathbb{P}^n \to X$ is replaced by a $\mathbb{P}^n$-bundle over X, $\mathrm{pr}_1 \colon Z \to X$; or by $\mathrm{pr}_1 \colon Z \to X$, which is a proper smooth morphism whose geometric fibers are integral curves of genus 0.

Furthermore, the locally Noetherian hypotheses can be removed; see §25.2.10.

25.1.L. EXERCISE. Let X be a connected Noetherian scheme. Show that $\mathrm{Pic}(X \times \mathbb{P}^n) \cong \mathrm{Pic}\, X \times \mathbb{Z}$. Hint: The map $\mathrm{Pic}\, X \times \mathrm{Pic}\, \mathbb{P}^n \to \mathrm{Pic}(X \times \mathbb{P}^n)$ is given by $(\mathscr{L}, \mathscr{O}(m)) \mapsto \mathrm{pr}_1^* \mathscr{L} \otimes \mathrm{pr}_2^* \mathscr{O}(m)$, where $\mathrm{pr}_1 \colon X \times \mathbb{P}^n \to X$ and $\mathrm{pr}_2 \colon X \times \mathbb{P}^n \to \mathbb{P}^n$ are the projections from $X \times \mathbb{P}^n$ to its factors. (The notation $\boxtimes$ is often used for this construction; see §15.2.10.)

A very similar argument will show that if Z is a $\mathbb{P}^n$-bundle over X, then $\mathrm{Pic}\, Z \cong \mathrm{Pic}\, X \times \mathbb{Z}$. You will undoubtedly also be able to figure out the right statement if X is not connected.

25.1.13. *Remark.* As mentioned in §19.10.1, the Picard group of a scheme often "wants to be a scheme." You may be able to make this precise in the case of $\mathrm{Pic}\, \mathbb{P}^n_{\mathbb{Z}}$. In this case, the *scheme* $\mathrm{Pic}\, \mathbb{P}^n_{\mathbb{Z}}$ is "$\mathbb{Z}$ copies of $\mathrm{Spec}\, \mathbb{Z}$," with the "obvious" group scheme structure. Can you figure out what functor it represents? Can you show that it represents this functor? This will require extending Exercise 25.1.L out of the Noetherian setting, using §25.2.10.

25.1.M. EXERCISE. Suppose $\pi \colon X \to Y$ is a projective flat morphism over a Noetherian integral scheme, all of whose geometric fibers are isomorphic to $\mathbb{P}^n$ (over the appropriate field). Show that π is a projective bundle if and only if there is an invertible sheaf $\mathscr{L}$ on X that restricts to $\mathscr{O}(1)$ on all the geometric fibers. (One direction is clear: if it is a projective bundle, then it has a $\mathscr{O}(1)$, which comes from the projectivization; see Exercise 17.2.D. In the other direction, the candidate vector bundle is $\pi_* \mathscr{L}$. Show that it is indeed a locally free sheaf of the desired rank. Show that its projectivization is indeed $\pi \colon X \to Y$.)

Caution: The map $\pi \colon \mathrm{Proj}\, \mathbb{R}[x, y, z]/(x^2 + y^2 + z^2) \to \mathrm{Spec}\, \mathbb{R}$ shows that not every projective flat morphism over a Noetherian integral scheme, all of whose geometric fibers are isomorphic to $\mathbb{P}^n$, is necessarily a $\mathbb{P}^n$-bundle. However, *Tsen's Theorem* implies that if the target is a *smooth curve over an algebraically closed field*, then the morphism *is* a $\mathbb{P}^n$-bundle (see [GS, Thm. 6.2.8]). Example 18.4.10 shows that "curve" cannot be replaced by "5-fold" in this statement—the "universal smooth plane conic" is not a $\mathbb{P}^1$-bundle over the parameter space $U \subset \mathbb{P}^5$ of smooth plane conics. If you wish, you can extend Example 18.4.10 to show that "curve" cannot even be replaced by "surface." (Just replace the $\mathbb{P}^5$ of all conics with a generally chosen $\mathbb{P}^2$ of conics—but then figure out what goes wrong if you try to replace it with a generally chosen $\mathbb{P}^1$ of conics.)

25.1.N. EXERCISE. Suppose $\pi\colon X \to Y$ is the projectivization of a finite rank locally free sheaf $\mathscr{F}$ over a locally Noetherian scheme (i.e., $X \cong \mathcal{P}roj(\operatorname{Sym}^\bullet \mathscr{F}^\vee))$ (§17.2.4). Recall from §17.2.4 that for any invertible sheaf $\mathscr{L}$ on Y, $X \cong \mathcal{P}roj(\operatorname{Sym}^\bullet(\mathscr{F}^\vee \otimes \mathscr{L}))$). Show that these are the only ways in which it is the projectivization of a vector bundle. (Hint: Recover $\mathscr{F}$ by pushing forward $\mathcal{O}(1)$.)

25.2 Proofs of Cohomology and Base Change Theorems

The key to proving the Semicontinuity Theorem 25.1.1, Grauert's Theorem 25.1.5, and the Cohomology and Base Change Theorem 25.1.6 is the following wonderful idea of Mumford (see [Mu3, p. 47 Lem. 1]). It turns questions of pushforwards (and how they behave under arbitrary base change) into something computable with vector bundles (hence questions of linear algebra). After stating it, we will interpret it.

25.2.1. Key theorem — *Suppose $\pi\colon X \to \operatorname{Spec} B$ is a proper morphism, and $\mathscr{F}$ is a coherent sheaf on X, flat over $\operatorname{Spec} B$, where B is Noetherian. Then there is a complex*

$$(25.2.1.1) \qquad \cdots \xrightarrow{\delta^{-2}} K^{-1} \xrightarrow{\delta^{-1}} K^0 \xrightarrow{\delta^0} K^1 \xrightarrow{\delta^1} \cdots \longrightarrow K^n \xrightarrow{\delta^n} 0$$

of finitely generated free B-modules and an isomorphism of functors

$$(25.2.1.2) \qquad H^p(X \times_B A, \mathscr{F} \otimes_B A) \overset{\sim}{\longleftrightarrow} H^p(K^\bullet \otimes_B A)$$

for all p, for all ring maps $B \to A$. (Here A needn't be Noetherian.)

The complex (25.2.1.1) is sometimes called the *Mumford complex*.

Because (25.2.1.1) is a complex of *free* B-modules, all of the information is contained in the maps, which are matrices with entries in B. This will turn questions about cohomology (and base change) into questions about linear algebra. For example, semicontinuity will turn into the fact that ranks of matrices (with functions as entries) drop on closed subsets (§12.2.16(ii)).

Although the complex (25.2.1.1) is infinite, by (25.2.1.2) it has no cohomology in negative degree, even after any ring extension $B \to A$ (as the left side of (25.2.1.2) is 0 for $p < 0$).

The idea behind the proof is as follows: take the Čech complex, produce a complex of finite rank free modules mapping to it "with the same cohomology" (a *quasiisomorphic complex*, §18.2.3). We first construct the complex so that (25.2.1.2) holds for $B = A$ in the next lemma, and then show the same complex works for general A, in Lemma 25.2.3 immediately thereafter.

25.2.2. Lemma — *Let B be a Noetherian ring. Suppose $C^\bullet$ is a complex of B-modules such that $H^i(C^\bullet)$ are finitely generated B-modules, and such that $C^p = 0$ for $p > n$. Then there exist a complex $K^\bullet$ of finite rank free B-modules such that $K^p = 0$ for $p > n$, and a homomorphism of complexes $\alpha\colon K^\bullet \to C^\bullet$ such that α induces isomorphisms $H^i(K^\bullet) \overset{\sim}{\longrightarrow} H^i(C^\bullet)$ for all i.*

Proof. We build this complex inductively. (This may remind you of §23.3.3.) Assume we have defined $(K^p, \alpha^p, \delta^p)$ for $p \geq m+1$ (as in (25.2.2.1)) such that the squares commute, and the top row is a complex, and α^q defines an isomorphism of cohomology $H^q(K^\bullet) \to H^q(C^\bullet)$ for $q \geq m+2$ and a surjection $\ker(\delta^{m+1}) \to H^{m+1}(C^\bullet)$, and the K^p are finite rank free B-modules. (Our base case is $m = p$: take $K^n = 0$ for $n > p$.)

$$(25.2.2.1)$$

$$
\begin{array}{ccccc}
& & K^{m+1} & \xrightarrow{\delta^{m+1}} & K^{m+2} & \xrightarrow{\delta^{m+2}} & \cdots \\
& & \downarrow{\scriptstyle \alpha^{m+1}} & & \downarrow{\scriptstyle \alpha^{m+2}} & & \\
\cdots \longrightarrow C^{m-1} \xrightarrow{\epsilon^{m-1}} C^m \xrightarrow{\epsilon^m} & C^{m+1} & \xrightarrow{\epsilon^{m+1}} & C^{m+2} & \longrightarrow & \cdots
\end{array}
$$

We construct $(K^m, \delta^m, \alpha^m)$. Choose generators of $H^m(C^\bullet)$, say, $c_1, \ldots, c_M$. Let

$$D^{m+1} := \ker\left(\ker(\delta^{m+1}) \xrightarrow{\alpha^{m+1}} H^{m+1}(C^\bullet)\right).$$

Choose generators of D^{m+1}, say, $d_1, \ldots, d_N$. (This is where we use the Noetherian hypotheses—to ensure this kernel D^{m+1} is finitely generated.) Let $K^m = B^{\oplus(M+N)}$. Define $\delta^m \colon K^m \to K^{m+1}$ by sending the last N generators to $d_1, \ldots, d_N$, and the first M generators to 0. Define α^m by sending the first M generators of $B^{\oplus(M+N)}$ to (lifts of) $c_1, \ldots, c_M$, and sending the last N generators to arbitrarily chosen lifts of the $\alpha^{m+1}(d_i)$ (as the $\alpha^{m+1}(d_i)$ are 0 in $H^{m+1}(C^\bullet)$, and thus lie in the image of ϵ^m), so the square (with upper left corner K^m) commutes. Then, by construction, we have completed our inductive step:

$$
\begin{array}{ccccccc}
K^m & \xrightarrow{\delta^m} & K^{m+1} & \xrightarrow{\delta^{m+1}} & K^{m+2} & \longrightarrow & \cdots \\
\downarrow{\scriptstyle \alpha^m} & & \downarrow{\scriptstyle \alpha^{m+1}} & & \downarrow{\scriptstyle \alpha^{m+2}} & & \\
\cdots \longrightarrow C^{m-1} \xrightarrow{\epsilon^{m-1}} & C^m & \xrightarrow{\epsilon^m} & C^{m+1} & \xrightarrow{\epsilon^{m+1}} & C^{m+2} & \longrightarrow \cdots .
\end{array}
$$
$\square$

25.2.3. Lemma — *Suppose $\alpha \colon K^\bullet \to C^\bullet$ is a morphism of complexes of **flat** B-modules, bounded on the right (i.e., $K^n = C^n = 0$ for $n \gg 0$), inducing isomorphisms of cohomology (a quasiisomorphism, §18.2.3). Then "this quasiisomorphism commutes with arbitrary change of base ring": for every B-algebra A, the maps $H^p(K^\bullet \otimes_B A) \to H^p(C^\bullet \otimes_B A)$ are isomorphisms.*

Proof. The mapping cone $M^\bullet$ of $\alpha \colon K^\bullet \to C^\bullet$ is exact by Exercise 1.6.E. Then $M^\bullet \otimes_B A$ is still exact, by Exercise 24.2.F. But $M^\bullet \otimes_B A$ is the mapping cone of

$$\alpha \otimes_B A \colon K^\bullet \otimes_B A \longrightarrow C^\bullet \otimes_B A,$$

so by Exercise 1.6.E, $\alpha \otimes_B A$ induces an isomorphism of cohomology (i.e., is a quasiisomorphism) too. $\square$

Proof of Key Theorem 25.2.1. Choose a finite affine covering of X. Take the Čech complex $C^\bullet$ for $\mathscr{F}$ with respect to this cover. By Grothendieck's Coherence Theorem 18.9.1 (which had Noetherian hypotheses), the cohomology of $\mathscr{F}$ is coherent. (Theorem 18.9.1 required serious work. If you need Theorem 25.2.1 only in the projective case, the analogous statement with projective hypotheses, Theorem 18.7.1(d), is much easier.) Apply Lemma 25.2.2 to get the nicer variant $K^\bullet$ of the same complex $C^\bullet$. By Lemma 25.2.3, if we tensor with A and take cohomology, we get the same answer whether we use $K^\bullet$ or $C^\bullet$. $\square$

We now use Theorem 25.2.1 to prove some of the fundamental results stated earlier: the Semicontinuity Theorem 25.1.1, Grauert's Theorem 25.1.5, and the Cohomology and Base Change Theorem 25.1.6. We will also (§25.2.5) give a new proof of Theorem 24.7.1, that Euler characteristics are locally constant in flat families (which applies more generally in proper situations).

25.2.4. Proof of the Semicontinuity Theorem 25.1.1. The result is local on Y, so we may assume Y is affine. Let $K^\bullet$ be a complex as in Key Theorem 25.2.1.

Then for $q \in Y$,

$$
\dim_{\kappa(q)} H^p(X_q, \mathscr{F}|_{X_q}) = \dim_{\kappa(q)} \ker(\delta^p \otimes_B \kappa(q)) - \dim_{\kappa(q)} \operatorname{im}(\delta^{p-1} \otimes_B \kappa(q))
$$
$$
= \dim_{\kappa(q)}(K^p \otimes_B \kappa(q)) - \dim_{\kappa(q)} \operatorname{im}(\delta^p \otimes_B \kappa(q))
$$
$$
(25.2.4.1) \qquad - \dim_{\kappa(q)} \operatorname{im}(\delta^{p-1} \otimes_B \kappa(q)).
$$

Now $\dim_{\kappa(q)} \operatorname{im}(\delta^p \otimes_B \kappa(q))$ is a lower semicontinuous function on Y. (Reason: The locus where the dimension is less than some number N is obtained by setting all $N \times N$ minors of the matrix $K^p \to K^{p+1}$ to 0; cf. §12.2.16(ii)).) The same is true for $\dim_{\kappa(q)} \operatorname{im}(\delta^{p-1} \otimes_B \kappa(q))$. The result follows. $\square$

25.2.5. A new proof (and extension to the proper case) of Theorem 24.7.1 that Euler characteristics of flat sheaves are locally constant.

If $K^\bullet$ were finite "on the left" as well—if $K^p = 0$ for $p \ll 0$ — then we would have a short proof of Theorem 24.7.1. By taking alternating sums (over p) of (25.2.4.1), we would have that

$$\chi(X_q, \mathscr{F}|_{X_q}) = \sum (-1)^p h^p(X_q, \mathscr{F}|_{X_q}) = \sum (-1)^p \operatorname{rank} K^p,$$

which is locally constant. The only problem is that the sums are infinite. We patch this problem by truncating the complex $K^\bullet$ below where there is cohomology. Define $J^\bullet$ by $J^p = K^p$ for $p \geq 0$, $J^p = 0$ for $p < -1$, and $J^{-1} := \ker(K^{-1} \to K^0)$. Then $J^\bullet$ is a complex in the obvious way, and the map of complexes $K^\bullet \to C^\bullet$ clearly factors through $J^\bullet$:

$$\begin{array}{ccccccccccc}
\cdots & \to & J^{-3} & \to & J^{-2} & \to & J^{-1} & \to & J^0 & \to & J^1 & \to & J^2 & \to & \cdots \\
& & \| & & \| & & \| & & \| & & \| & & \| & & \\
\cdots & \to & 0 & \to & 0 & \to & \ker(K^{-1}\to K^0) & \to & K^0 & \to & K^1 & \to & K^2 & \to & \cdots \\
& & \downarrow & & \downarrow & & \downarrow & & \downarrow & & \downarrow & & \downarrow & & \\
\cdots & \to & 0 & \to & 0 & \to & 0 & \to & C^0 & \to & C^1 & \to & C^2 & \to & \cdots
\end{array}$$

Clearly, $J^\bullet \to C^\bullet$ induces an isomorphism on cohomology (recall both have 0 cohomology for $p < 0$).

Now J^{-1} (the kernel of a map of coherent modules) is coherent. Consider the mapping cone $M^\bullet$ of $\beta \colon J^\bullet \to C^\bullet$:

$$0 \to J^{-1} \to C^{-1} \oplus J^0 \to C^0 \oplus J^1 \to \cdots \to C^{n-1} \oplus J^n \to C^n \to 0.$$

From Exercise 1.6.E, as $J^\bullet \to C^\bullet$ induces an isomorphism on cohomology, the mapping cone has no cohomology — it is exact. All terms in it are flat except possibly J^{-1} (the C^p are flat by assumption, and J^i is free for $i \neq -1$). Hence J^{-1} is flat too, by Exercise 24.2.G. But flat coherent sheaves are locally free (Corollary 24.3.7). Then Theorem 24.7.1 follows from

$$\chi(X_q, \mathscr{F}|_{X_q}) = \sum (-1)^p h^p(X_q, \mathscr{F}|_{X_q}) = \sum (-1)^p \operatorname{rank} J^p. \qquad \square$$

25.2.6. Proof of Grauert's Theorem 25.1.5 and the Cohomology and Base Change Theorem 25.1.6 (following E. Larson).

(⋆⋆ Experts: You might see in the proof that what makes pth cohomology commute with base change is that the complex of Key Theorem 25.2.1 can be "broken" into two complexes at the pth step. You might even want to interpret this in terms of the Čech complex as an object of the derived category of B-modules.)

Thanks to Theorem 25.2.1, Theorems 25.1.5 and 25.1.6 are now statements about a complex of free modules over a Noetherian ring.

25.2.7. Proof of Grauert's Theorem 25.1.5. By hypothesis, $h^p(X_q, \mathscr{F}|_{X_q})$ is a locally constant function of $q \in Y$. From (25.2.4.1), $h^p(X_q, \mathscr{F}|_{X_q}) = \operatorname{rank} K^p - \operatorname{rank} \operatorname{im}(\delta^p|_q) - \operatorname{rank} \operatorname{im}(\delta^{p-1}|_q)$. But $\operatorname{rank} K^p$ is constant, and $\operatorname{rank} \operatorname{im}(\delta^p|_q)$ and $\operatorname{rank} \operatorname{im}(\delta^{p-1}|_q)$ are lower semicontinuous, so in fact $\operatorname{rank} \operatorname{im}(\delta^p|_q)$ and $\operatorname{rank} \operatorname{im}(\delta^{p-1}|_q)$ must be locally constant. By Exercise 14.3.L, both δ^{p-1} and δ^p are maps of vector bundles (in the sense of Definition 14.2.4). Then by Observation 14.2.5(ii), $\operatorname{coker} \delta^{p-1}$ and $\operatorname{im} \delta^p$ are both locally free (of finite rank). In the short exact sequence

(25.2.7.1) $$0 \to H^p(K^\bullet) \to \operatorname{coker} \delta^{p-1} \to \operatorname{im} \delta^p \to 0$$

(Exercise 1.5.6.4, the "dual" definition of cohomology), both $\operatorname{coker} \delta^{p-1}$ and $\operatorname{im} \delta^p$ correspond to finite rank locally free sheaves. Thus $H^p(K^\bullet)$ does as well, by Exercise 14.2.H(a).

25.2.A. EXERCISE. Show (perhaps using (25.2.7.1)) that the construction of $H^p(K^\bullet)$ commutes with any base change, thereby completing the proof of Grauert's Theorem 25.1.5. $\qquad\square$

In order to prove the Cohomology and Base Change Theorem 25.1.6, we need a preliminary result.

25.2.8. Lemma — *Suppose*

$$
\begin{array}{ccccc}
K^{p-1} & \xrightarrow{\;\delta_K^{p-1}\;} & K^p & \xrightarrow{\;\delta_K^p\;} & K^{p+1} \\
\downarrow & & \downarrow & & \downarrow \\
J^{p-1} & \xrightarrow{\;\delta_J^{p-1}\;} & J^p & \xrightarrow{\;\delta_J^p\;} & J^{p+1}
\end{array}
$$

is a map of complexes, with the left vertical arrow surjective. Then $H^p(K^\bullet) \to H^p(J^\bullet)$ is surjective if and only if $\ker \delta_K^p \to \ker \delta_J^p$ *is surjective.*

Proof. The map $\operatorname{im} \delta_K^{p-1} \to \operatorname{im} \delta_J^{p-1}$ is surjective: any element α of $\operatorname{im} \delta_J^{p-1}$ lifts to J^{p-1}, then can lift to K^{p-1}, which then can map to K^p, which maps to α. Then apply the Snake Lemma 1.6.5 to

$$
\begin{array}{ccccccccc}
0 & \longrightarrow & \operatorname{im} \delta_K^{p-1} & \longrightarrow & \ker \delta_K^p & \longrightarrow & H^p(K^\bullet) & \longrightarrow & 0 \\
& & \downarrow & & \downarrow & & \downarrow & & \\
0 & \longrightarrow & \operatorname{im} \delta_J^{p-1} & \longrightarrow & \ker \delta_J^p & \longrightarrow & H^p(J^\bullet) & \longrightarrow & 0.
\end{array}
$$
$\qquad\square$

25.2.9. Proof of the Cohomology and Base Change Theorem 25.1.6. We focus on the complex near the pth step, and its restriction to the point $q \in X$:

$$
\begin{array}{ccccc}
K^{p-1} & \longrightarrow & K^p & \xrightarrow{\;\delta^p\;} & K^{p+1} \\
\downarrow & & \downarrow & & \downarrow \\
K^{p-1} \otimes_B \kappa(q) & \longrightarrow & K^p \otimes_B \kappa(q) & \xrightarrow{\;\delta^p \otimes \kappa(q)\;} & K^{p+1} \otimes_B \kappa(q)
\end{array} \qquad \text{free } B\text{-modules}
$$

- By hypothesis, $\phi_q^p : (R^p \pi_* \mathscr{F})|_q \to H^p(X_q, \mathscr{F}|_{X_q})$ is surjective.
- By Lemma 25.2.8, this is equivalent to $(\ker \delta^p) \otimes \kappa(q) \to \ker(\delta^p \otimes \kappa(q))$ being surjective.
- By Exercise 14.2.K, this is equivalent to δ^p being a map of vector bundles near q.
- This implies (by Observation 14.2.5(ii)) that $\ker \delta^p$ is finite rank locally free, and the construction of $\ker \delta^p$ commutes with any base change in a neighborhood of q.

Now $H^p(K^\bullet) = \operatorname{coker}(K^{p-1} \to \ker \delta^p)$, i.e., we have the exact sequence

$$
K^{p-1} \longrightarrow \ker \delta^p \longrightarrow H^p(K^\bullet) \longrightarrow 0.
$$

Thus $H^p(K^\bullet)$ commutes with any base change (as tensor product is right exact). This completes the proof of part (i) of the theorem.

For part (ii), consider again the map $K^{p-1} \to \ker \delta^p$ of finite rank locally free sheaves, whose cokernel is $H^p(K^\bullet)$. Now $H^p(K^\bullet)$ is locally free if and only if $K^{p-1} \to \ker \delta^p$ is a map of vector bundles, if and only if (since δ^p is a map of vector bundles) δ^{p-1} is a map of vector bundles, if and only if $H^{p-1}(K^\bullet) \to H^{p-1}(K^\bullet|_q)$ is surjective. This completes the proof of part (ii). $\qquad\square$

25.2.10.* Removing Noetherian conditions.

It can be helpful to have versions of the theorems of §25.1 without Noetherian conditions; important examples come from moduli theory, and will be discussed in the next section. Noetherian conditions can often be exchanged for finite presentation conditions. We begin with an extension of Exercise 10.3.H.

25.2.B. EXERCISE. Suppose $\pi\colon X \to \operatorname{Spec} B$ is a finitely presented morphism, and $\mathscr{F}$ is a finitely presented sheaf on X. Show that there exists a base change diagram of the form

(25.2.10.1)

$$
\begin{array}{ccc}
\mathscr{F} & & \mathscr{F}' \\
| & & | \\
| & & | \\
X & \xrightarrow{\ \sigma\ } & X' \\
{\scriptstyle\pi}\Big\downarrow & & \Big\downarrow{\scriptstyle\pi'} \\
\operatorname{Spec} B & \xrightarrow{\ \rho\ } & \operatorname{Spec}\mathbb{Z}[x_1,\ldots,x_N]/I
\end{array}
$$

where N is some integer, $I \subset \mathbb{Z}[x_1,\ldots,x_N]$, and π' is finitely presented (= finite type, as the target is Noetherian; see §8.3.13), and a finitely presented (= coherent) quasicoherent sheaf $\mathscr{F}'$ on X' with $\mathscr{F} \cong \sigma^*\mathscr{F}'$.

25.2.11. *Properties of π'.* (The ideal I appears in the statement of Exercise 25.2.B not because it is needed there, but to make the statement of this remark correct.) If π is proper, then diagram (25.2.10.1) can be constructed so that π' is also proper (using [Gr-EGA, IV$_3$.8.10.5]). Furthermore, if $\mathscr{F}$ is flat over Spec B, then (25.2.10.1) can be constructed so that $\mathscr{F}'$ is flat over $\operatorname{Spec}\mathbb{Z}[x_1,\ldots,x_N]/I$ (using [Gr-EGA, IV$_3$.11.2.6]). This requires significantly more work.

25.2.C. EXERCISE. *Assuming* the results stated in §25.2.11, prove the following results, with the "locally Noetherian" hypotheses removed, and "finite presentation" hypotheses added:

(a) the constancy of Euler characteristic in flat families (Theorem 24.7.1, extended to the proper case as in §25.2.5);
(b) the Semicontinuity Theorem 25.1.1;
(c) Grauert's Theorem 25.1.5 (you will have to show that $\mathbb{Z}[x_1,\ldots,x_N]/I$ in (25.2.10.1) can be taken to be reduced); and
(d) the Cohomology and Base Change Theorem 25.1.6.

25.2.12. *Necessity of finite presentation conditions.* The finite presentation conditions are necessary. There is a projective flat morphism to a connected target where the fiber dimension jumps. There is a finite flat morphism where the degree of the fiber is not locally constant. There is a projective flat morphism to a connected target where the fibers are curves, and the arithmetic genus is not constant. See [Stacks, tag 05LB] for the first example; the other two use the same idea.

25.3 Applying Cohomology and Base Change to Moduli Problems

The theory of moduli relies on ideas of cohomology and base change. We explore this by examining two special cases of one of the primordial moduli spaces, the Hilbert scheme: the Grassmannian, and the fact that degree d hypersurfaces in projective space are "parametrized" by another projective space (corresponding to degree d polynomials; see Remark 4.5.3).

As suggested in §24.0.1, the Hilbert functor $\operatorname{Hilb}_Y\mathbb{P}^n$ of $\mathbb{P}^n_Y$ parametrizes flat finitely presented closed subschemes of $\mathbb{P}^n_Y$, where Y is an arbitrary scheme. More precisely, it is a contravariant functor sending the Y-scheme X to the set of finitely presented closed subschemes of $X\times_Y\mathbb{P}^n_Y=\mathbb{P}^n_X$, flat over X (and sending morphisms $X_1\to X_2$ to pullbacks of flat families). An early achievement of Grothendieck was the construction of the Hilbert scheme, which can then be cleverly used to construct many other moduli spaces.

25.3.1. Theorem (Grothendieck) — $\text{Hilb}_{\mathbb{Z}} \, \mathbb{P}^n$ *is representable by a scheme locally of finite type.*

(Grothendieck's original argument is in [Gr4]. A readable construction is given in [Mu2], and in [FGIKNV, Ch. 5].)

We won't use the following exercise, but it is conceptually very important, and explains why "moduli spaces tautologically always have universal families" (with the "correct" definition of "moduli space"). The proof will remind you of Yoneda's Lemma. (You have already seen universal families of moduli spaces under other names, such as the universal hyperplane, Definition 13.4.1, or tautological bundles, §16.4.1.)

25.3.A. EXERCISE (THE UNIVERSAL FAMILY OVER THE HILBERT SCHEME). Assume Theorem 25.3.1. Show that there exists a closed subscheme Z of $\text{Hilb}_{\mathbb{Z}} \, \mathbb{P}^n \times \mathbb{P}^n_{\mathbb{Z}}$

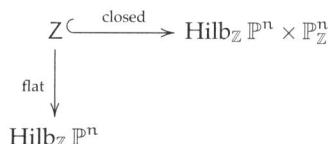

$$
\begin{array}{ccc}
Z & \overset{\text{closed}}{\hookrightarrow} & \text{Hilb}_{\mathbb{Z}} \, \mathbb{P}^n \times \mathbb{P}^n_{\mathbb{Z}} \\
{\scriptsize \text{flat}} \downarrow & & \\
\text{Hilb}_{\mathbb{Z}} \, \mathbb{P}^n & &
\end{array}
$$

(which we call the **universal family** over the Hilbert scheme) such that a flat family

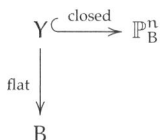

$$
\begin{array}{ccc}
Y & \overset{\text{closed}}{\hookrightarrow} & \mathbb{P}^n_B \\
{\scriptsize \text{flat}} \downarrow & & \\
B & &
\end{array}
$$

corresponds to a map $\pi \colon B \to \text{Hilb}_{\mathbb{Z}} \, \mathbb{P}^n$ if and only if Y is the pullback of Z by π, i.e., diagram

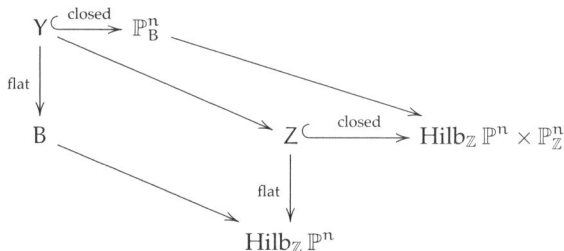

is a Cartesian diagram. Hint: The Hilbert functor applied to the Hilbert scheme by definition corresponds to the set $\text{Maps}(\text{Hilb}_{\mathbb{Z}} \, \mathbb{P}^n, \text{Hilb}_{\mathbb{Z}} \, \mathbb{P}^n)$. The universal family will correspond to the identity map.

25.3.B. EASY EXERCISE. Assuming Theorem 25.3.1, show that $\text{Hilb}_Y \, \mathbb{P}^n$ is representable, by showing that it is represented by $\text{Hilb}_{\mathbb{Z}} \, \mathbb{P}^n \times_{\mathbb{Z}} Y$. Thus the general case follows from the "universal" case of $Y = \mathbb{Z}$.

25.3.C. EXERCISE. Assuming Theorem 25.3.1, show that $\text{Hilb}_{\mathbb{Z}} \, \mathbb{P}^n$ is the disjoint union of schemes $\text{Hilb}_{\mathbb{Z}}^{p(m)} \, \mathbb{P}^n$, each one corresponding to finitely presented closed subschemes of $\mathbb{P}^n_{\mathbb{Z}}$ whose fibers have fixed Hilbert polynomial $p(m)$. Hint: Corollary 24.7.2.

Motivated by Exercise 25.3.C, what Grothendieck actually proved was the following.

25.3.2. Theorem (Grothendieck) — *Each $\text{Hilb}_{\mathbb{Z}}^{p(m)} \, \mathbb{P}^n$ is projective over $\mathbb{Z}$.*

25.3.D. EXERCISE. Prove that Theorem 25.3.2 implies Theorem 25.3.1.

In order to get some feeling for the Hilbert scheme, we discuss two important examples, without relying on Theorem 25.3.1.

25.3.3. The Grassmannian.

We have defined the Grassmannian $G(k, n)$ twice before, in §7.7 and §16.4. The second time involved showing the representability of a (contravariant) functor (from *Sheaves* to *Sets*), of rank k locally free quotient sheaves of a rank n free sheaf.

We now consider a parameter space for a more geometric problem. The space will again be $G(k, n)$, but because we won't immediately know this, we invent some temporary notation. Let $G'(k, n)$ be the contravariant functor (from *Schemes* to *Sets*) that assigns to a scheme B the set of *finitely presented* closed subschemes of $\mathbb{P}_B^{n-1}$, flat over B, whose fiber over any point $b \in B$ is a (linearly embedded) $\mathbb{P}_{\kappa(b)}^{k-1}$ in $\mathbb{P}_{\kappa(b)}^{n-1}$:

(25.3.3.1)

$$
\begin{array}{ccc}
X & \xrightarrow{\text{cl. subscheme}} & \mathbb{P}_B^{n-1} \\
{\scriptstyle\text{flat, f. pr.}}\downarrow & \swarrow{\scriptstyle\pi} & \\
B & &
\end{array}
$$

(This describes the map to *Sets*; you should think through how pullback makes this into a contravariant functor.)

25.3.4. Theorem — *The functor* $G'(k, n)$ *is represented by* $G(k, n)$.

Translation: There is a natural (i.e., functorial) bijection between diagrams of the form (25.3.3.1) (where the fibers are $\mathbb{P}^{k-1}$'s) and diagrams of the form (16.4.0.1) (the diagrams that $G(k, n)$ parametrizes, or represents).

Proof. One direction is notably easier. Suppose we are given a diagram of the form (16.4.0.1) over a scheme B,

(25.3.4.1)
$$ \mathcal{O}_B^{\oplus n} \longrightarrow \mathcal{Q}, $$

where $\mathcal{Q}$ is locally free of rank k. Applying $\mathcal{P}roj_B$ to the $\mathrm{Sym}^\bullet$ construction on both $\mathcal{O}_B^{\oplus n}$ and $\mathcal{Q}$, we obtain a closed embedding

(25.3.4.2)

$$
\begin{array}{ccccc}
\mathcal{P}roj_B\left(\mathrm{Sym}^\bullet \mathcal{Q}\right) & \xhookrightarrow{\hspace{3cm}} & \mathcal{P}roj_B\left(\mathrm{Sym}^\bullet \mathcal{O}_B^{\oplus n}\right) & & = \mathbb{P}^{n-1} \times B \\
& \searrow \qquad \swarrow & & &
\end{array}
$$
$$ B $$

(as, for example, in Exercise 17.2.H).

The fibers are linearly embedded $\mathbb{P}^{k-1}$'s (as base change, in this case to a point of B, commutes with the $\mathcal{P}roj$ construction, Exercise 17.2.E). Note that $\mathcal{P}roj(\mathrm{Sym}^\bullet \mathcal{Q})$ is flat and finitely presented over B, as it is a projective bundle. We have constructed a diagram of the form (25.3.3.1).

We now need to reverse this. The trick is to produce (25.3.4.1) from our geometric situation (25.3.3.1), and this is where cohomology and base change will be used.

Given a diagram of the form (25.3.3.1) (where the fibers are $\mathbb{P}^{k-1}$'s), consider the closed subscheme exact sequence for X:

$$ 0 \longrightarrow \mathcal{I}_X \longrightarrow \mathcal{O}_{\mathbb{P}_B^{n-1}} \longrightarrow \mathcal{O}_X \longrightarrow 0. $$

Tensor this with $\mathcal{O}_{\mathbb{P}_B^{n-1}}(1)$:

(25.3.4.3)
$$ 0 \longrightarrow \mathcal{I}_X(1) \longrightarrow \mathcal{O}_{\mathbb{P}_B^{n-1}}(1) \longrightarrow \mathcal{O}_X(1) \longrightarrow 0. $$

Note that $\mathcal{O}_X(1)$ restricted to each fiber of π is $\mathcal{O}(1)$ on $\mathbb{P}^{k-1}$ (over the residue field), for which all higher cohomology vanishes (§18.3).

25.3.E. EXERCISE. Show that $R^i\pi_*\mathscr{O}_X(1) = 0$ for $i > 0$, and $\pi_*\mathscr{O}_X(1)$ is locally free of rank k. Hint: Use the Cohomology and Base Change Theorem 25.1.6. Either use the non-Noetherian discussion of §25.2.10 (which we haven't proved), or else just assume B is locally Noetherian.

25.3.F. EXERCISE. Show that the long exact sequence obtained by applying π_* to (25.3.4.3) is just a short exact sequence of locally free sheaves

$$0 \longrightarrow \pi_*\mathscr{I}_X(1) \longrightarrow \pi_*\mathscr{O}_{\mathbb{P}^{n-1}_B}(1) \longrightarrow \pi_*\mathscr{O}_X(1) \longrightarrow 0$$

of ranks $n - k$, n, and k respectively, where the middle term is canonically identified with $\mathscr{O}_B^{\oplus n}$.

The surjection $\mathscr{O}_B^{\oplus n} \longrightarrow\!\!\!\!\!\rightarrow \pi_*\mathscr{O}_X(1)$ is precisely a diagram of the sort we wished to construct, (16.4.0.1).

25.3.G. EXERCISE. Close the loop, by using these two "inverse" constructions to show that $G(k, n)$ represents the functor $G'(k, n)$. $\qquad\square$

25.3.5. Hypersurfaces.

Ages ago (in Remark 4.5.3), we informally said that hypersurfaces of degree d in $\mathbb{P}^n$ are parametrized by a $\mathbb{P}^{\binom{n+d}{d}-1}$. We now make this precise. We work over a base $\mathbb{Z}$ for suitable generality. You are welcome to replace $\mathbb{Z}$ by a field of your choice, but by the same argument as in Easy Exercise 25.3.B, all other cases are obtained from this one by base change.

Define the contravariant functor $H_{d,n}\colon Sch \to Sets$ from schemes to sets as follows. To a scheme B, we associate the set of all closed subschemes $X \hookrightarrow \mathbb{P}^n_B$, flat and finitely presented over B, all of whose fibers are degree d hypersurfaces in $\mathbb{P}^n$ (over the appropriate residue field). To a morphism $B_1 \to B_2$, we obtain a map $H_{d,n}(B_2) \to H_{d,n}(B_1)$ by pullback.

25.3.6. Proposition — *The functor $H_{d,n}$ is represented by $\mathbb{P}^{\binom{n+d}{d}-1}$.*

As with the case of the Grassmannian, one direction is easy, and the other requires cohomology and base change.

25.3.H. EASY EXERCISE. Over $\mathbb{P}^{\binom{n+d}{d}-1}$, describe a closed subscheme $\mathscr{X} \hookrightarrow \mathbb{P}^n \times \mathbb{P}^{\binom{n+d}{d}-1}$ that will be the universal hypersurface (the universal family over the moduli space). Show that $\mathscr{X}$ is flat and finitely presented over $\mathbb{P}^{\binom{n+d}{d}-1}$. (For flatness, you can use the local criterion of flatness on the source, Theorem 24.6.9, but it is possible to deal with it easily by working by hand.)

Thus given any morphism $B \to \mathbb{P}^{\binom{n+d}{d}-1}$, by pullback, we have a degree d hypersurface X over B (an element of $H_{d,n}(B)$).

Our goal is to reverse this process: from a degree d hypersurface $\pi\colon X \to B$ over B (an element of $H_{d,n}(B)$), we want to describe a morphism $B \to \mathbb{P}^{\binom{n+d}{d}-1}$.

Consider the closed subscheme exact sequence for $X \hookrightarrow \mathbb{P}^n_B$, twisted by $\mathscr{O}_{\mathbb{P}^n_B}(d)$:

(25.3.6.1) $\qquad 0 \longrightarrow \mathscr{I}_X(d) \longrightarrow \mathscr{O}_{\mathbb{P}^n_B}(d) \longrightarrow \mathscr{O}_X(d) \longrightarrow 0.$

25.3.I. EXERCISE (CF. EXERCISE 25.3.E). Show that the higher pushforward (by π) of each term of (25.3.6.1) is 0, and that the long exact sequence of pushforwards of (25.3.6.1) is

$$0 \longrightarrow \pi_*\mathscr{I}_X(d) \longrightarrow \pi_*\mathscr{O}_{\mathbb{P}^n_B}(d) \longrightarrow \pi_*\mathscr{O}_X(d) \longrightarrow 0,$$

where the middle term is free of rank $\binom{n+d}{d}$, (whose summands can be identified with degree d monomials in the projective variables $x_1, \ldots, x_n$; see § 15.1.2), and the left term $\pi_*\mathscr{I}_X(d)$ is locally free of rank 1 (basically, a line bundle).

(It is helpful to interpret the middle term $\mathscr{O}_B^{\oplus\binom{n+d}{d}}$ as parametrizing homogeneous degree d polynomials in $n+1$ variables, and the rank 1 subsheaf of $\pi_*\mathscr{I}_X(d)$ as "the equation of X." This will motivate what comes next.)

Taking the dual of the injection $\pi_*\mathscr{I}_X(d) \hookrightarrow \mathscr{O}_B^{\oplus\binom{n+d}{d}}$, we have a surjection

$$\mathscr{O}_B^{\oplus\binom{n+d}{d}} \longrightarrow\!\!\!\!\!\longrightarrow \mathscr{L}$$

from a free sheaf onto an invertible sheaf $\mathscr{L} = (\pi_*\mathscr{I}_X(d))^\vee$, which (by the functorial description of projective space, Theorem 15.2.2) yields a morphism $B \to \mathbb{P}^{\binom{n+d}{d}-1}$.

25.3.J. EXERCISE. Close the loop: show that these two constructions are inverses, thereby proving Proposition 25.3.6.

25.3.7. *Remark.* The proof of the representability of the Hilbert scheme shares a number of features of our arguments about the Grassmannian and the parameter space of hypersurfaces.

Chapter 26

Depth and Cohen–Macaulayness

We now introduce the notion of depth. Depth is an algebraic rather than a geometric concept, so we concentrate on developing some geometric sense of what it means. Most important is the geometric bound on depth by dimension of (closures of) associated points (Theorem 26.1.3). A central tool to understanding depth is the Koszul complex (briefly discussed in §24.4.1), but we avoid this approach, as we can prove what we need directly.

When the depth of a local ring equals its dimension, the ring is said to be Cohen–Macaulay, and this is an important way in which schemes can be "nice." For example, regular local rings are Cohen–Macaulay (§26.2.5), as are regular embeddings in smooth varieties (Proposition 26.2.6). Cohen–Macaulayness will be key to the proof of Serre duality in Chapter 29, through the Miracle Flatness Theorem 26.2.11.

Another application of depth is Serre's $R_1 + S_2$ criterion for normality (Theorem 26.3.2), which we will use to prove that regular schemes are normal (§26.3.5) without having to show that regular local rings are factorial (the Auslander–Buchsbaum Theorem, Fact 13.8.5), and to prove that regular embeddings in smooth schemes are normal if they are regular in codimension 1 (§26.3.3).

26.1 Depth

26.1.1. *Informal introduction.* Given a point p on a scheme X, sometimes you can locally "slice" through the point—there is an effective Cartier divisor through p (in a neighborhood of p, say). You can do this if and only if the point is *not* an associated point of X. Sometimes you can slice again, if the first slice still doesn't have p as an associated point. The depth of (X at) the point p is how deep you can slice—how many slices you can make before you are blocked by the point becoming associated. Obviously you can't make more than $\dim X$ slices, since at each stage you lose a dimension, and after $\dim X$ slices, you will be left (set-theoretically) with just the point p (which is then an associated point). Thus the depth is bounded by the dimension. In figuring out the depth, you might think you would have to slice with care in order to maximize the number of slices you can take. You needn't worry; you can slice haphazardly, and will always end at the same step.

26.1.2. *Definition.* Suppose $(A, \mathfrak{m})$ is a Noetherian local ring, and M is a finitely generated A-module. The **depth of** M (denoted by $\operatorname{depth} M$) is the length of the longest M-regular sequence with elements in $\mathfrak{m}$. (The theory of regular sequences was discussed in §9.5.3.)

(More generally, if R is a Noetherian ring, $I \subset R$ is an ideal, and M is a finitely generated R-module, then the I-**depth of** M, denoted by $\operatorname{depth}_I M$, is the length of the longest M-regular sequence with elements in I. We won't need this.)

26.1.A. IMPORTANT EXERCISE. Suppose M is a nonzero finitely generated module over a Noetherian local ring $(A, \mathfrak{m})$. Show that $\operatorname{depth} M = 0$ if and only if every element of $\mathfrak{m}$ is a zerodivisor of M if and only if $\mathfrak{m}$ is an associated prime of M.

26.1.B. EXERCISE. Suppose M is a finitely generated module over a Noetherian local ring A. Show that $\operatorname{depth} M \leq \dim \operatorname{Supp} M$. In particular,

$$(26.1.2.1) \qquad\qquad \operatorname{depth} A \leq \dim A.$$

(We will improve this result in Theorem 26.1.7.)

At this point, it is hard to determine the depth of an A-module M. You can start trying to build an M-regular sequence by successively choosing $x_1, x_2, \ldots$, but how do you know you have made the right choices to find the *longest* one? The happy answer is that you can't go wrong; this is the content of the next result. (We then describe how to find the depth of a module M in practice, in §26.1.5.)

26.1.3. Theorem — *Suppose M is a finitely generated module over a Noetherian local ring $(A, \mathfrak{m})$. Then all maximal M-regular sequences contained in $\mathfrak{m}$ (M-regular sequences $x_1, \ldots, x_n$ in $\mathfrak{m}$ that cannot be extended to $x_1, \ldots, x_{n+1}$) have the same length. Thus the depth of M is the length of any maximal M-regular sequence.*

We prove Theorem 26.1.3 by giving a cohomological criterion, Theorem 26.1.4, for a regular sequence to be maximal. (You will then prove Theorem 26.1.3 in Exercise 26.1.D.) We will also use this criterion to give a better bound on the depth in Theorem 26.1.7. Theorem 26.1.3 is the key technical result of this chapter. An important moral is that depth should be understood as a "cohomological" property.

26.1.4. Theorem (cohomological criterion for existence of regular sequences) — *Suppose $(A, \mathfrak{m})$ is a Noetherian local ring, and M is a nonzero finitely generated A-module. The following are equivalent.*

(i) *For every finitely generated A-module N with $\operatorname{Supp} N = \{[\mathfrak{m}]\}$ (i.e., N has finite length, Exercise 6.6.X(a)), $\operatorname{Ext}_A^i(N, M) = 0$ for all $i < n$.*

(ii) *$\operatorname{Ext}_A^i(A/\mathfrak{m}, M) = 0$ for all $i < n$.*

(iii) *There exists an M-regular sequence in $\mathfrak{m}$ of length n.*

This result can be extended in various ways; see, for example, [Mat1, Thm. 28].

Proof. Clearly *(i) implies (ii)*.

26.1.C. EXERCISE. Prove that *(ii) implies (i)*. Hint: Use induction on the length of N, and the long exact sequence for $\operatorname{Ext}_A^i(\cdot, M)$.

Proof that (iii) implies (ii). The case $n = 0$ is vacuous. We inductively prove the result for all n. Suppose (iii) is satisfied, where $n \geq 1$, and assume that we know (ii) for "all smaller n." Choose a regular sequence $x_1, \ldots, x_n$ of length n. Then x_1 is a non-zerodivisor on M, so we have an exact sequence

$$(26.1.4.1) \qquad 0 \longrightarrow M \xrightarrow{\times x_1} M \longrightarrow M/x_1 M \longrightarrow 0.$$

Then $M/x_1 M$ has a regular sequence $x_2, \ldots, x_{n-1}$ of length $n - 1$, so by the inductive hypothesis, $\operatorname{Ext}_A^i(A/\mathfrak{m}, M/x_1 M) = 0$ for $i < n - 1$. Taking the Ext long exact sequence for $\operatorname{Ext}_A^i(A/\mathfrak{m}, \cdot)$ for (26.1.4.1), we find that

$$\operatorname{Ext}_A^i(A/\mathfrak{m}, M) \xrightarrow{\times x_1} \operatorname{Ext}_A^i(A/\mathfrak{m}, M)$$

is an injection for $i < n$. Now $\operatorname{Ext}_A^i(A/\mathfrak{m}, M)$ can be computed by taking an injective resolution of M, and applying $\operatorname{Hom}_A(A/\mathfrak{m}, \cdot)$. Hence as x_1 lies in $\mathfrak{m}$ (and thus annihilates $A/\mathfrak{m}$), multiplication by x_1 is the zero map. Thus (ii) holds for n as well.

Proof that (ii) implies (iii). The case $n = 0$ is vacuous.

We deal next with the case $n = 1$, by showing the contrapositive. Assume that there are no non-zerodivisors in $\mathfrak{m}$ on M, so by Exercise 26.1.A, $\mathfrak{m}$ is an associated prime of M. Thus (§6.6.7) we have an injection $A/\mathfrak{m} \hookrightarrow M$, yielding $\operatorname{Hom}_A(A/\mathfrak{m}, M) \neq 0$, as desired.

We now inductively prove the result for all $n > 1$. Suppose (ii) is satisfied, where $n \geq 2$, and assume that we know (iii) for "all smaller n." Then by the case $n = 1$, there exists a non-zerodivisor x_1 on M, so we have a short exact sequence (26.1.4.1). A portion of the Ext long exact sequence for

$\text{Ext}_A^i(A/\mathfrak{m}, \cdot)$ for (26.1.4.1) is

$$\text{Ext}_A^i(A/\mathfrak{m}, M) \longrightarrow \text{Ext}_A^i(A/\mathfrak{m}, M/x_1 M) \longrightarrow \text{Ext}_A^{i+1}(A/\mathfrak{m}, M).$$

By assumption, $\text{Ext}_A^i(A/\mathfrak{m}, M)$ and $\text{Ext}_A^{i+1}(A/\mathfrak{m}, M)$ are 0 for $i < n-1$, so $\text{Ext}_A^i(A/\mathfrak{m}, M/x_1 M) = 0$ for $i < n-1$, so by the inductive hypothesis, we have an $(M/x_1 M)$-regular sequence $x_2, \ldots, x_n$ of length $n-1$ in $\mathfrak{m}$. Adding x_1 to the front of this sequence, we are done. □

26.1.D. EXERCISE. Prove Theorem 26.1.3. Hint: By Theorem 26.1.4 (notably, the equivalence of (ii) and (iii)), you have control of how long an M-regular sequence in $\mathfrak{m}$ can be. Use the criterion, and the long exact sequence used in the proof of Theorem 26.1.4, to show that any M-regular sequence in $\mathfrak{m}$ can be extended to this length.

26.1.E. EXERCISE. Suppose M is a finitely generated module over a Noetherian local ring $(A, \mathfrak{m})$. If x is a non-zerodivisor in $\mathfrak{m}$, show that $\text{depth}(M/xM) = \text{depth } M - 1$. Hint: Theorem 26.1.3.

26.1.5. *Finding the depth of a module.* We can now compute the depth of M by successively finding non-zerodivisors, as follows. Is there a non-zerodivisor x on M in $\mathfrak{m}$?

(a) If not, then depth $M = 0$ (Exercise 26.1.A).
(b) If so, then choose any such x, and (using the previous exercise) repeat the process with M/xM.

The process must terminate by Exercise 26.1.B.

26.1.F. IMPORTANT EXERCISE. Suppose $(A, \mathfrak{m})$ is a dimension d regular local ring. Show that depth $A = d$. Hint: Exercise 13.2.B.

26.1.G. EXERCISE (CF. EXERCISE 24.3.I). Suppose $X = \text{Spec } R$, where

$$R = k[w, x, y, z]/(wy, wz, xy, xz),$$

the union of two coordinate two-planes in $\mathbb{A}_k^4$ meeting at the origin. Show that the depth of the local ring of X at the origin is 1. Hint: Show that $w - y$ is not a zerodivisor, and that $R/(w-y)$ has an embedded point at the origin.

26.1.6. Depth is bounded by the dimension of associated prime ideals.
Theorem 26.1.4 can be used to give an important improvement of the bound (26.1.2.1) on depth by the dimension:

26.1.7. Theorem — *The depth of a module M is at most the smallest* $\dim A/\mathfrak{p}$, *as $\mathfrak{p}$ runs over the associated prime ideals of M.*

(The example of two planes meeting at a point in Exercise 26.1.G shows that this bound is not sharp.) The key step in the proof of Theorem 26.1.7 is the following result of Ischebeck.

26.1.8. Lemma — *Suppose $(A, \mathfrak{m})$ is a Noetherian local ring, and M and N are nonzero finitely generated A-modules. Then $\text{Ext}_A^i(N, M) = 0$ for $i < \text{depth } M - \dim \text{Supp } N$.*

Proof. Consider the following statements.

$(general_r)$ Lemma 26.1.8 holds for $\dim \text{Supp } N \leq r$.
$(prime_r)$ Lemma 26.1.8 holds for $\dim \text{Supp } N \leq r$ and $N = A/\mathfrak{p}$ for some prime $\mathfrak{p}$.

Note that $(general_0)$ (and hence $(prime_0)$) is true, as in this case $\text{Supp } N = \{[\mathfrak{m}]\}$, and the result follows from Theorem 26.1.4 (from "(iii) implies (i)").

26.1.H. EXERCISE. Show that $(prime_r)$ implies $(general_r)$. Hint: For a short exact sequence $0 \to M' \to M \to M'' \to 0$, $\text{Supp } M = \text{Supp } M' \cup \text{Supp } M''$ (Exercise 6.6.B).

We conclude the proof by showing that *(general$_{r-1}$)* implies *(prime$_r$)* for $r \geq 1$. Fix a prime $\mathfrak{p} \subset A$ with $\dim A/\mathfrak{p} = r$. Since $\dim A/\mathfrak{p} > 0$, we can choose $x \in \mathfrak{m} \setminus \mathfrak{p}$. Consider the exact sequence

$$(26.1.8.1) \qquad 0 \longrightarrow A/\mathfrak{p} \xrightarrow{\times x} A/\mathfrak{p} \longrightarrow A/(\mathfrak{p} + (x)) \longrightarrow 0,$$

noting that $\dim \operatorname{Supp} A/(\mathfrak{p} + (x)) \leq r - 1$ (do you see why?). Then the Ext long exact sequence obtained by applying $\operatorname{Hom}_A(\cdot, M)$ to (26.1.8.1), along with the vanishing of $\operatorname{Ext}_A^i(A/(\mathfrak{p} + (x))), M)$ for $i < \operatorname{depth} M - r + 1$ (by *(general$_{r-1}$)*), implies that

$$\operatorname{Ext}_A^i(A/\mathfrak{p}, M) \xrightarrow{\times x} \operatorname{Ext}_A^i(A/\mathfrak{p}, M)$$

is an isomorphism for $i < \operatorname{depth} M - r$. But $\operatorname{Ext}_A^i(A/\mathfrak{p}, M)$ is a finitely generated A-module (Exercise 23.2.F), so by Nakayama's Lemma 8.2.8, $\operatorname{Ext}_A^i(A/\mathfrak{p}, M) = 0$ for $i < \operatorname{depth} M - r$. $\square$

26.1.I. EASY EXERCISE. Prove Theorem 26.1.7. Hint: If $\mathfrak{p} \in \operatorname{Ass}(M)$, then $\operatorname{Hom}_A(A/\mathfrak{p}, M) \neq 0$ (§6.6.7).

26.2 Cohen–Macaulay Rings and Schemes

26.2.1. *Definition.* A Noetherian local ring $(A, \mathfrak{m})$ is **Cohen–Macaulay** if $\operatorname{depth} A = \dim A$, i.e., if equality holds in (26.1.2.1). (One may define **Cohen–Macaulay module** similarly, but we won't need this concept.) A locally Noetherian scheme is **Cohen–Macaulay** if all of its local rings are Cohen–Macaulay. We won't use this language, but a ring A is a **Cohen–Macaulay ring** if all of its localizations $A_\mathfrak{p}$ at all primes $\mathfrak{p}$ are Cohen–Macaulay, or, equivalently, if $\operatorname{Spec} A$ is Cohen–Macaulay. (Fact 26.2.14 below, that Cohen–Maculayness of a local ring is preserved by localization, ensures that there is no contradiction between this definition and the definition of a Cohen–Macaulay *local* ring.)

26.2.A. EXERCISE. Show that every locally Noetherian scheme of dimension 0 is Cohen–Macaulay. Show that a locally Noetherian scheme of dimension 1 is Cohen–Macaulay if and only if it has no embedded points.

26.2.2. *(Counter)example.* Let X be the example of Exercise 26.1.G—two planes meeting at a point. By Exercise 26.1.G, X is not Cohen–Macaulay.

26.2.B. EXERCISE. Suppose A is a Cohen–Macaulay Noetherian local ring. Use Theorem 26.1.7 to show that $\operatorname{Spec} A$ is pure dimensional, and has no embedded points. (It is not true that Noetherian local rings of pure dimension having no embedded prime ideals are Cohen–Macaulay; see Example 26.2.2.)

26.2.3. Theorem (slicing criterion for Cohen–Macaulayness) — *Suppose $(A, \mathfrak{m})$ is a Noetherian local ring, and $x \in \mathfrak{m}$ is a non-zerodivisor. Then $(A, \mathfrak{m})$ is Cohen–Macaulay if and only if $(A/x, \mathfrak{m})$ is Cohen–Macaulay.*

Compare this to the slicing criteria for regularity and flatness (Exercise 13.2.C and Theorem 24.6.6, respectively).

26.2.4. *Geometric interpretation of the slicing criterion.* Suppose X is a locally Noetherian scheme, and D is an effective Cartier divisor. If X is Cohen–Macaulay, then so is D. If D is Cohen–Macaulay, then X is Cohen–Macaulay at the points of D.

26.2.C. EXERCISE. Prove Theorem 26.2.3, using Theorem 26.1.3, the fact that maximal regular sequences (in $\mathfrak{m}$) all have the same length. (Hint: Recall Exercise 12.3.I, that $\dim A/(f) \geq \dim A - 1$ for any $f \in A$.)

26.2.D. EXERCISE. Show that if $(A, \mathfrak{m})$ is Cohen–Macaulay, then a set of elements $x_1, \ldots, x_r \in \mathfrak{m}$ is a regular sequence (for A) if and only if $\dim A/(x_1, \ldots, x_r) = \dim A - r$.

26.2.5. By Exercise 26.1.F, regular local rings (Noetherian by definition) are Cohen–Macaulay. In particular, as smooth schemes over a field k are regular (Theorem 13.2.7(b), proved in §24.8.9), we see that smooth k-schemes are Cohen–Macaulay. (In combination with Exercise 26.2.B, this explains why effective Cartier divisors on smooth varieties have no embedded points, justifying a comment made in Aside 20.2.1.) Combining this with Exercise 26.2.D or §26.2.4, we have the following.

26.2.6. Proposition — *Regular embeddings in smooth k-schemes are Cohen–Macaulay.*

26.2.7. As a consequence of Proposition 26.2.6 and the fact that Cohen–Macaulay schemes have no embedded points (Exercise 26.2.B), we see that regular embeddings in smooth k-schemes (in $\mathbb{A}^n_k$ or $\mathbb{P}^n_k$, for example) have no embedded points, generalizing Exercise 6.6.F (the hypersurface in $\mathbb{A}^n_k$ case). This is not clear without the theory of Cohen–Macaulayness!

26.2.8. *Alternate definition of Cohen–Macaulayness.* The slicing criterion (Theorem 26.2.3) gives an enlightening alternative inductive definition of Cohen–Macaulayness in terms of effective Cartier divisors, in the spirit of the method of §26.1.5 for computing depth. Suppose as before that $(A, \mathfrak{m})$ is a Noetherian local ring.

(i)   If $\dim A = 0$, then A is Cohen–Macaulay (Exercise 26.2.A).

(iia) If $\dim A > 0$, and every element of $\mathfrak{m}$ is a zerodivisor, then A is *not* Cohen–Macaulay (by Exercise 26.1.A).

(iib) Otherwise, choose *any* non-zerodivisor x in $\mathfrak{m}$. Then A is Cohen–Macaulay if and only if $A/(x)$ (necessarily of dimension $\dim A - 1$ by Krull's Principal Ideal Theorem 12.3.3) is Cohen–Macaulay.

The following example could have been stated (but not proved) before we knew any algebraic geometry at all. (We work over $\mathbb{C}$ rather than over an arbitrary field only to ensure that the statement requires as little background as possible.)

26.2.E. FUN EXERCISE (MAX NOETHER'S AF + BG THEOREM). Suppose $f, g \in \mathbb{C}[x_0, x_1, x_2]$ are two homogeneous polynomials, cutting out two curves in $\mathbb{P}^2_{\mathbb{C}}$ that meet "transversely," i.e., at a finite number of reduced points. Suppose $h \in \mathbb{C}[x_0, x_1, x_2]$ is a homogeneous polynomial vanishing at these points. Show that $h \in (f, g)$. Hint: Show that the intersection of the affine cones $V(f)$ and $V(g)$ in $\mathbb{A}^3$ has no embedded points. (This problem is quite nontrivial to do without the theory developed in this chapter! As a sign that this is subtle: you can easily construct *three* quadratics $e, f, g \in \mathbb{C}[x_0, x_1, x_2]$ cutting out precisely the two points $[1, 0, 0]$ and $[0, 1, 0] \in \mathbb{P}^2_{\mathbb{C}}$, yet the line $z = 0$ is not in the ideal (e, f, g) for degree reasons.)

As additional motivation for Cohen–Macaulayness, the next exercise is seemingly basic, but we need Cohen–Macaulayness to prove it cleanly.

26.2.F. EXERCISE. Suppose $H_1, \ldots, H_n$ are n hypersurfaces in $\mathbb{P}^n_{\mathbb{C}}$ of degrees $d_1, \ldots, d_n$, respectively, and suppose their scheme-theoretic intersection Z is a finite number of points. Show that the sum (over the points of Z) of the lengths of Z at the points is $d_1 \cdots d_n$. Equivalently: Show that the length of Z is $d_1 \cdots d_n$. Informally: The hypersurfaces intersect at $d_1 \cdots d_n$ points, "counted correctly." Hint: Bézout's Theorem, in the form of Exercise 18.6.J.

26.2.9. Miracle flatness.

We conclude with a remarkably simple and useful criterion for flatness, which we shall use in the proof of Serre duality. The main content is the following algebraic result.

26.2.10. Miracle Flatness Theorem (algebraic version) — *Suppose $\phi \colon (B, \mathfrak{n}) \to (A, \mathfrak{m})$ is a (local) homomorphism of Noetherian local rings, such that A is Cohen–Macaulay, and B is regular, and*

$A/\mathfrak{n}A = A \otimes_B (B/\mathfrak{n})$ *(the ring corresponding to the fiber) has pure dimension* $\dim A - \dim B$. *Then* ϕ *is flat.*

Proof. We prove Theorem 26.2.10 by induction on $\dim B$. If $\dim B = 0$, then B is a field (Exercise 13.2.A), so the result is immediate, as everything is flat over a field (Exercise 24.1.A(a)). Assume next that $\dim B > 0$, and we have proved the result for all "B of smaller dimension." Choose $x \in \mathfrak{n} \setminus \mathfrak{n}^2$, so B/x is a regular local ring of dimension $\dim B - 1$ (Exercise 12.3.I). Then

$$\dim A/xA \leq \dim B/(x) + \dim A/\mathfrak{n}A \quad \text{(Key Exercise 12.4.A)}$$

$$= \dim B - 1 + \dim A/\mathfrak{n}A \quad \text{(Exercise 12.3.I)}$$

$$= \dim A - 1 \qquad\qquad \text{(by hypothesis of Theorem 26.2.10).}$$

By (Exercise 12.3.I), $\dim A/xA \geq \dim A - 1$, so we have $\dim A/xA = \dim B/(x) + \dim A/\mathfrak{n}A = \dim A - 1$. By Exercise 26.2.D, A/xA is a Cohen–Macaulay ring, and x is a non-zerodivisor on A. The inductive hypothesis then applies to $(B/(x), \mathfrak{n}) \to (A/xA, \mathfrak{m})$, so A/xA is flat over $B/(x)$. Then by the local slicing criterion for flatness (on the target, Theorem 24.6.6), $B \to A$ is flat, as desired.   □

26.2.11. Miracle Flatness Theorem — *Suppose* $\pi \colon X \to Y$ *is a morphism of pure dimensional finite type k-schemes, where X is Cohen–Macaulay, Y is regular, and the fibers of π have dimension* $\dim X - \dim Y$. *Then* π *is flat.*

26.2.G. EXERCISE. Prove the Miracle Flatness Theorem 26.2.11. (Do not forget that schemes usually have non-closed points!)

The geometric situation in the Miracle Flatness Theorem 26.2.11 is part of the following pretty package.

26.2.H. EXERCISE (THE MIRACLE FLATNESS PACKAGE). Suppose $\pi \colon X \to Y$ is a map of locally finite type k-schemes, where both X and Y are pure dimensional, and Y is regular. Show that if any two of the following hold, then the third does as well:

(i) π is flat of relative dimension $\dim X - \dim Y$.
(ii) X is Cohen–Macaulay.
(iii) Every fiber X_y is Cohen–Macaulay of pure dimension $\dim X - \dim Y$.

Hint: If $\phi \colon B \to A$ is a flat ring map, then ϕ sends non-zerodivisors to non-zerodivisors (Observation 24.1.2).

The statement of Exercise 26.2.H can be improved, at the expense of killing the symmetry. In the implication that (ii) and (iii) imply (i), the Cohen–Macaulay hypotheses in (iii) are not needed.

26.2.12. *Example.* As an example of Exercise 26.2.H in action, we consider the example of two planes meeting a point, continuing the notation of Exercise 26.1.G. Consider the morphism $\pi \colon X \to Y := \mathbb{A}^2$ given by $(w - y, x - z)$. Note that Y is regular (§13.3.6). But X is not Cohen–Macaulay (Exercise 26.1.G) and π is not flat (Exercise 24.3.I), so by Exercise 26.2.H, you can use either of these to prove the other.

26.2.13. Fancy properties of Cohen–Macaulayness.
We mention a few additional properties of Cohen–Macaulayness without proof. They are worth seeing, but we will not use them.

If A is a Noetherian ring that is the quotient of a Cohen–Macaulay ring, the locus of Cohen–Macaulay points in $\operatorname{Spec} A$ is open; see [Mat2, Ex. 24.2]. Related:

26.2.14. Fact (cf. Fact 13.8.5) — *Any localization of a Cohen–Macaulay Noetherian local ring at a prime ideal is also Cohen–Macaulay.*

(See [Stacks, tag 00NB], [E, Prop. 18.8], [Mat2, Thm. 17.3(iii)], or [Mat1, Thm. 30] for a direct proof.) In particular, for varieties, Cohen–Macaulayness may be checked at closed points

(Exercise 5.3.F). The fact that Cohen–Macaulayness is preserved by localization can be used to quickly show that Cohen–Macaulay local rings are catenary (see [E, Cor. 18.10], [Mat1, Thm. 31(ii)], or [Mat2, Thm. 17.9]). A ring A is Cohen–Macaulay if and only if $A[x]$ is Cohen–Macaulay ([E, Prop 18.9], [Mat2, Thm 17.7], [Mat1, Thm. 33]), if and only if $A[[x]]$ is [Mat2, p. 137]. A local ring is Cohen–Macaulay if and only if its completion is Cohen–Macaulay ([E, Prop. 18.8], [Mat2, Thm 17.5]).

For these and many other cheerful facts, and, in general, a thorough introduction to Cohen–Macaulayness, see [BruH].

26.3 Serre's $R_1 + S_2$ Criterion for Normality

The notion of depth yields a useful criterion for normality, due to Serre.

26.3.1. *Definition.* Suppose A is a Noetherian ring, and $k \in \mathbb{Z}^{\geq 0}$. We say A has **property** R_k (A is **regular in codimension** $\leq k$, or, more sloppily, A is **regular in codimension** k) if for every prime $\mathfrak{p} \subset A$ of codimension at most k, $A_\mathfrak{p}$ is regular. (We have repeatedly seen the importance of regularity in codimension 1.) Thus a Noetherian ring is regular and only if it has property R_k for all k.

We say A has **property** S_k if for every prime $\mathfrak{p} \subset A$, the local ring $A_\mathfrak{p}$ has depth at least $\min(k, \dim A_\mathfrak{p})$—"the local rings are Cohen–Macaulay up until codimension k, and have depth at least k thereafter." Thus a Noetherian ring is Cohen–Macaulay if and only if it has property S_k for all k.

26.3.A. EASY EXERCISE. Note that a Noetherian ring A trivially always has property S_0. Show that a Noetherian ring A has property R_0 if Spec A is "generically reduced": it is reduced at the generic point of each of its irreducible components. Show that a Noetherian ring A has property S_1 if and only if Spec A has no embedded points. (Possible hint: Exercise 26.1.A.)

26.3.B. EXERCISE (IMPORTANT ELEMENTARY INTERPRETATION OF S_2). Suppose A is a Noetherian integral domain. Show that A has property S_2 if and only if for all nonzero $g \in A$, $A/(g)$ has no embedded primes.

26.3.C. EXERCISE. Show that a Noetherian ring A has no nilpotents (Spec A is reduced) if and only if it has properties R_0 and S_1. (Hint: Show that a Noetherian scheme is reduced if and only if each irreducible component is generically reduced, and it has no embedded points.)

Incrementing the subscripts in Exercise 26.3.C yields Serre's criterion.

26.3.2. Theorem (Serre's criterion for normality) — *A Noetherian ring A is normal if and only if it has properties R_1 and S_2.*

(Recall from §5.4.1 that a Noetherian ring A is normal if its localizations $A_\mathfrak{p}$ at all primes $\mathfrak{p}$ are integrally closed domains.) Thus failure of normality can have two possible causes: it can be failure of R_1 (something we already knew, from the equivalence of (a) and (g) in Theorem 13.5.8), or it can be the more subtle failure of S_2. Examples of varieties satisfying R_1 but not S_2 are given in Example 26.3.4 (two planes meeting at a point) and Exercise 26.3.F (the pinched plane).

26.3.D. EASY EXERCISE. Prove Serre's criterion for normality (Theorem 26.3.2). Hint: Look closely at Proposition 13.5.24 and Proposition 13.5.26.

26.3.3. Applications. First, Serre's Criterion for normality implies that Cohen–Macaulay schemes are normal if and only if they are regular in codimension 1. Thus to check normality of hypersurfaces (or, more generally, regular embeddings) in $\mathbb{P}^n_k$ (or, more generally, in any smooth variety), it suffices to check that their singular locus has codimension greater than 1. (You should think through the details of why these statements are true.) In particular, this gives a new (more complicated) proof of Exercise 5.4.I(b) and (c).

588 Chapter 26 Depth and Cohen–Macaulayness

26.3.E. EXERCISE (PRACTICE WITH THE CONCEPT). Show that two-dimensional normal varieties are Cohen–Macaulay.

26.3.4. *Example: Two planes meeting at a point.* The variety X of Exercise 26.1.G (two planes meeting at a point) is not normal (why?), but it is regular away from the origin. This implies that X does not have property S_2 (and hence is not Cohen–Macaulay), without the algebraic manipulations of Exercise 26.1.G.

We already knew that this example was not Cohen–Macaulay, but the same idea can show that the pinched plane $\operatorname{Spec} k[x^3, x^2, xy, y]$ (appearing in Exercise 13.5.H) is not Cohen–Macaulay. Because of the "extrinsic" description of the ring, it is difficult to do this in another way.

26.3.F. EXERCISE: THE PINCHED PLANE IS NOT COHEN–MACAULAY. Let A be the subring $k[x^3, x^2, xy, y]$ of $k[x, y]$. Show that A is not Cohen–Macaulay. Hint: Exercise 13.5.H showed that A is not integrally closed.

26.3.5. *Regular local rings are integrally closed.* Serre's criterion for normality can be used to show that regular local rings are integrally closed without going through the difficult Fact 13.8.5 (the Auslander–Buchsbaum Theorem) that they are unique factorization domains. Regular local rings are Cohen–Macaulay (§26.2.5), and (by Exercise 26.3.C) regular in codimension 0 as they are integral domains (Theorem 13.2.11), so we need only show that regular local rings are regular in codimension 1. We can invoke a different difficult Fact 13.8.2 that localizations of regular local rings are again regular, but at least we have shown this for localizations of finitely generated algebras over a perfect field (see Theorem 13.8.3 and Exercise 21.6.B).

26.3.6. *Caution.* As is made clear by the following exercise, the condition S_2 is a condition on *all* prime ideals, not just those of codimension at most 2.

26.3.G. EXERCISE. Give an example of a variety satisfying R_1, and Cohen–Macaulay at all points of codimension at most 2, which is not normal.

26.3.7.** *Projective normality.* We will not use the notion of projective normality, but mention it because many readers may come across it, and it is easy to introduce. It is a property of a closed embedding of $\pi : X \hookrightarrow \mathbb{P}^n_k$. Let $\mathscr{L} = \pi^* \mathscr{O}(1)$ be the line bundle giving the closed embedding (§15.2), and for convenience let $R(\mathscr{L})_\bullet := \oplus_{m \geq 0} \Gamma(X, \mathscr{L}^{\otimes m})$ be the graded ring of "sections of all powers of $\mathscr{L}$." Then we say (Definition 1) $X \hookrightarrow \mathbb{P}^n_k$ is **projectively normal** if X is normal and for all $m \geq 0$, $H^0(\mathbb{P}^n_k, \mathscr{O}(m)) \to H^0(X, \mathscr{O}_X(m))$ is surjective. In other words, all sections of $\mathscr{L}^{\otimes m}$ for all $m \geq 0$ "come from" polynomials; there are no "new" sections not "arising" from $H^0(X, \mathscr{L})$. Equivalently (Definition 2), via $\mathscr{L} = \pi^* \mathscr{O}(1)$, $R(\mathscr{L})_\bullet$ is a quotient of the polynomial ring $k[x_0, \dots, x_n]$. You will quickly be able to show that this is equivalent to the statement (Definition 3) that (X is normal and) $H^1(\mathbb{P}^n_k, \mathscr{I}_X(m)) = 0$ for all $m \geq 0$, and also to the statement (Definition 4) that X is normal, $R(\mathscr{L})_\bullet$ is generated in degree 1, and $X \hookrightarrow \mathbb{P}^n_k$ is an embedding by a complete linear series (assuming it is nondegenerate).

26.3.H. EXERCISE. Show that complete intersections in $\mathbb{P}^n_k$ are projectively normal if and only if they are normal if and only if they are regular in codimension 1.

It is not too hard to show that $X \hookrightarrow \mathbb{P}^n_k$ is normal if and only if its affine cone is normal, [Ha1, Ex. II.5.14].

Chapter 27

The Twenty-Seven Lines on a Cubic Surface

> *Wake an algebraic geometer in the dead of night, whispering: "27". Chances are, he will respond: "lines on a cubic surface."*
>
> —R. Donagi and R. Smith [DS] (on page 27, of course)

Since the middle of the nineteenth century, geometers have been entranced by the fact that there are 27 lines on every smooth cubic surface, and by the remarkable structure of the configuration of the lines. Their discovery by Cayley and Salmon in 1849 has been called the beginning of modern algebraic geometry, [Do, p. 55].

The reason so many people are bewitched by this fact is because it requires some magic, and this magic connects to many other things, including fundamental ideas we have discussed, other beautiful classical constructions (such as Pascal's Mystical Hexagon Theorem 19.4.4, the fact that most smooth quartic plane curves have 28 bitangents, exceptional Lie groups, . . .), and many themes in modern algebraic geometry (deformation theory, intersection theory, enumerative geometry, arithmetic and diophantine questions, . . .). It will be particularly pleasant for us, as it takes advantage of many of the things we have learned.

You are now ready to be initiated into the secret fellowship of the 27 lines.

27.0.1. Theorem — *Every smooth cubic surface in $\mathbb{P}_{\overline{k}}^3$ contains exactly 27 lines.*

Theorem 27.0.1 is closely related to the following.

27.0.2. Theorem — *Every smooth cubic surface over $\overline{k}$ is isomorphic to $\mathbb{P}^2$ blown up at six points.*

There are many reasons why people consider these facts magical. First, there is the fact that there are *always* 27 lines. Unlike most questions in enumerative geometry, there are no weasel words such as "a general cubic surface" or "most cubic surfaces" or "counted correctly"—as in, "every monic degree d polynomial has d roots—counted correctly." And somehow (and we will see how) it is precisely the smoothness of the surface that makes it work.

Second, there is the magic that you *always* get the blow-up of the plane at six points (§27.3).

Third, there is the magical incidence structure of the 27 lines, which relates to E_6 in Lie theory. The Weyl group of E_6 is the symmetry group of the incidence structure (see Remark 27.2.5). In a natural way, the 27 lines form a basis of the 27-dimensional fundamental representation of E_6.

27.0.3. On the structure of this chapter. Throughout this chapter, X will be a smooth cubic surface over an algebraically closed field $\overline{k}$. In §27.1, we establish some preliminary facts. In §27.2, we prove Theorem 27.0.1, away from characteristic 3. In §27.3, we prove Theorem 27.0.2. We remark here that the only input that §27.3 needs from §27.2 is Exercise 27.2.J. This can be done directly by hand (see, for example, [Rei1, §7], [Be, Thm. IV.13], or [Sh, pp. 246–7]), and Theorem 27.0.2 readily implies Theorem 27.0.1, using Exercise 27.3.F. We would thus have another, shorter, proof of Theorem 27.0.1. The reasons for giving the argument of §27.2 are that it is natural given what we have done so far, it gives you some glimpse of some ideas used more broadly in the subject (the key idea is that a map from one moduli space to another is finite and flat), and it may help you further appreciate and digest the tools we have developed.

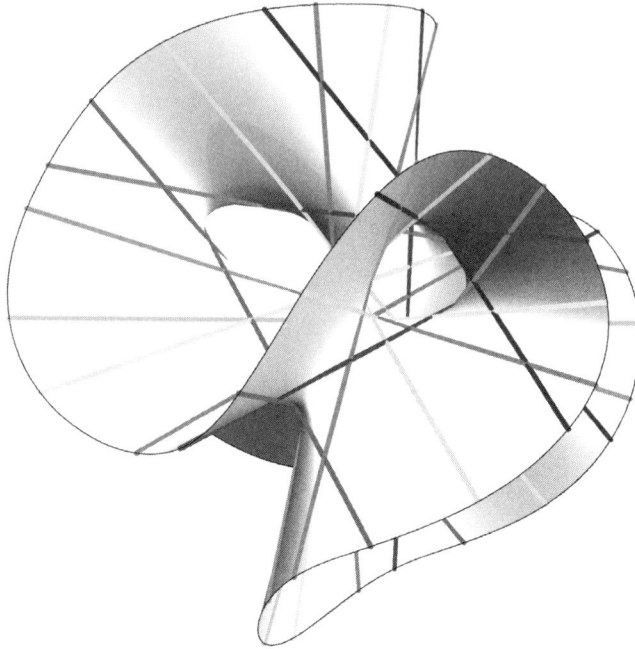

Figure 27.1 *The 27 lines on the Clebsch cubic surface (courtesy of Greg Egan)*

27.0.4. On the location of this chapter.

This topic is Chapter 27 for purely numerological reasons; it could certainly be read after Chapter 24. This is a feature rather than a bug: you needn't read the details of the later chapters in order to enjoy this one. In fact, you should probably read this many chapters earlier, taking key facts as "black boxes," to appreciate the value of seemingly technical definitions and theorems before you learn about them in detail. This topic is a fitting conclusion to an energetic introduction to algebraic geometry, even if you do not earlier cover everything needed to justify all details.

27.1 Preliminary Facts

By Essential Exercise 15.1.C, there is a 20-dimensional vector space of cubic forms in four variables, so the cubic surfaces in $\mathbb{P}^3$ are parametrized by $\mathbb{P}^{19}$. Let $\Delta \subset \mathbb{P}^{19}$ be the locus corresponding to singular cubics. Bertini's Theorem 13.4.2 shows that the locus of smooth cubics $\mathbb{P}^{19} \setminus \Delta$ is dense and open in the space $\mathbb{P}^{19}$ of all cubics.

27.1.A. EXERCISE. Show that any smooth cubic surface X is "anticanonically embedded"—it is embedded by the complete anticanonical linear series $\mathscr{K}_X^\vee$. Hint: The adjunction formula, Exercise 21.5.B.

27.1.B. EXERCISE. Suppose $X \subset \mathbb{P}^3_k$ is a smooth cubic surface. Suppose C is a curve on X. Show that C is a line if and only if C is a (-1)-curve (Definition 20.2.8). Hint: The adjunction formula again, perhaps in the guise of Exercise 20.2.B(a); and also Exercise 18.6.K.

It will be useful to find a *single* cubic surface over a *single* field with exactly 27 lines:

27.1.C. HANDS-ON EXERCISE. Show that the **Fermat cubic surface**

$$(27.1.0.1) \qquad\qquad x_0^3 + x_1^3 + x_2^3 + x_3^3 = 0$$

in $\mathbb{P}^3_{\mathbb{C}}$ has precisely 27 lines, each of the form

$$x_0 + \omega x_i = x_j + \omega' x_k = 0,$$

where $\{1, 2, 3\} = \{i, j, k\}$, $j < k$, and ω and ω' are cube roots of 1 (possibly the same). This will require some brute force. Possible hint: Up to a permutation of coordinates, show that every line in $\mathbb{P}^3$ can be written $x_0 = ax_2 + bx_3$, $x_1 = cx_2 + dx_3$. Show that this line is on (27.1.0.1) if and only if

$$\text{(27.1.0.2)} \qquad a^3 + c^3 + 1 = b^3 + d^3 + 1 = a^2 b + c^2 d = ab^2 + cd^2 = 0.$$

Show that if a, b, c, and d are all nonzero, then (27.1.0.2) has no solutions.

27.1.1.* The singular cubic surfaces give an irreducible divisor Δ in the parameter space $\mathbb{P}^{19}$ of all cubic surfaces. This discussion is starred not because it is hard, but because it is not needed in the rest of the chapter. Nonetheless, it is a pretty application of what we have learned, and it foreshadows key parts of the argument in §27.2.

Bertini's Theorem 13.4.2 shows that the locus of smooth cubics in the $\mathbb{P}^{19}$ of all cubics is dense and open, but the proof suggests more: that the complement Δ of this locus is of pure codimension 1. We now show that this is the case, and even that this "discriminant hypersurface" Δ is irreducible.

27.1.D. EXERCISE. (Hint for both: Recall the solution to Exercise 12.2.N.)

(a) Define the incidence correspondence $Y \subset \mathbb{P}^{19} \times \mathbb{P}^3$ corresponding to the data of a cubic surface X, along with a *singular* point $p \in X$. (This is part of the exercise! We need Y as a scheme, not just as a set.) Let μ be the projection $Y \to \mathbb{P}^{19}$.

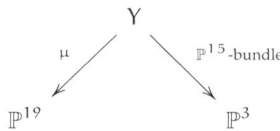

(b) Show that Y is an irreducible smooth variety of dimension 18, by describing it as a $\mathbb{P}^{15}$-bundle over $\mathbb{P}^3$.

27.1.E. EXERCISE. Show that there exists a cubic surface with a single singular point. Feel free to assume k is your favorite field; the main point is to familiarize yourself with geometric ideas, not the peculiarities of positive characteristic fields. (Hint if char $k \neq 2$: Exercise 13.3.C.)

27.1.F. EXERCISE. Show that $\mu(Y) = \Delta$ is a closed irreducible subset of codimension 1. Hint for the codimension 1 statement: Exercise 12.4.A or Theorem 12.4.1, in combination with Exercise 27.1.E.

Your argument will generalize with essentially no change to deal with degree d hypersurfaces in $\mathbb{P}^n$.

27.2   Every Smooth Cubic Surface (over $\bar{k}$) Contains 27 Lines

We are now ready to prove Theorem 27.0.1. Until Exercise 27.2.K, to avoid distraction, we assume char $\bar{k} = 0$. However, the following argument carries through without change if char $\bar{k} \neq 3$. The one required check—and the reason for the restriction on the characteristic—is that Exercise 27.1.C works with $\mathbb{C}$ replaced by any such $\bar{k}$.

27.2.A. EXERCISE. (Hint for both: Recall the solution to Exercise 12.2.N.)

(a) Define the incidence correspondence $Z \subset \mathbb{P}^{19} \times \mathbb{G}(1, 3)$ corresponding to the data of a line ℓ in $\mathbb{P}^3$ contained in a cubic surface X. (As in Exercise 27.1.D, this is part of the exercise.) Let π

be the projection $Z \to \mathbb{P}^{19}$.

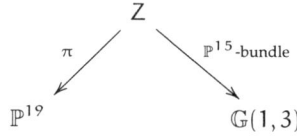

$$
\begin{array}{ccc}
 & Z & \\
\pi \swarrow & & \searrow \; \mathbb{P}^{15}\text{-bundle} \\
\mathbb{P}^{19} & & \mathbb{G}(1,3)
\end{array}
$$

(b) Show that Z is an irreducible smooth variety of dimension 19, by describing it as a $\mathbb{P}^{15}$-bundle over the Grassmannian $\mathbb{G}(1,3)$.

27.2.B. EXERCISE. Use the fact that there exists a cubic surface with a finite number of lines (Exercise 27.1.C), and the behavior of dimensions of fibers of morphisms (Exercise 12.4.A or Theorem 12.4.1), to show the following.

(a) Every cubic surface contains a line, i.e., π is surjective. (Hint: Show that π is projective.)
(b) "Most cubic surfaces have a finite number of lines": There is a dense open subset $U \subset \mathbb{P}^{19}$ such that the cubic surfaces parametrized by closed points of U have a positive finite number of lines. (Hint: Upper semicontinuity of fiber dimension, Theorem 12.4.3.)

The following fact is the key result in the proof of Theorem 27.0.1, and one of the main miracles of the 27 lines, which ensures that the lines stay distinct on a smooth surface. It states, informally, that two lines can't come together without damaging the surface. This is really a result in deformation theory: we are explicitly showing that a line in a smooth cubic surface has no first-order deformations.

27.2.1. Theorem — *If ℓ is a line in a regular cubic surface X, then $\{\ell \subset X\}$ is a reduced isolated point of the fiber of $\pi \colon Z \to \mathbb{P}^{19}$.*

Before proving Theorem 27.2.1, we use it to prove Theorem 27.0.1.

27.2.2. *Proof of Theorem 27.0.1, assuming Theorem 27.2.1.* Now π is a projective morphism, and by Theorem 27.2.1, over $\mathbb{P}^{19} \setminus \Delta$, π has relative dimension 0, and hence has finite fibers. Hence, by Theorem 18.1.6 (projective + finite fibers = finite), π is finite over $\mathbb{P}^{19} \setminus \Delta$.

Furthermore, Z and $\mathbb{P}^{19}$ are both smooth of dimension 19 (over $\mathbb{Z}$), and the fibers of $\pi \colon Z \to \mathbb{P}^{19}$ are (geometrically) reduced points above $\mathbb{P}^{19} \setminus \Delta$, again by Theorem 27.2.1. Then by Exercise 24.8.G, π is étale over $\mathbb{P}^{19} \setminus \Delta$ (and, in particular, flat).

Thus, over $\mathbb{P}^{19} \setminus \Delta$, π is a finite flat morphism, and so the fibers of π (again, away from Δ) always have the same number of points, "counted correctly" (Remark 24.3.8). But by Theorem 27.2.1, above each closed point of $\mathbb{P}^{19} \setminus \Delta$, each point of the fiber of π counts with multiplicity 1 (here using $k = \bar{k}$). Finally, by Exercise 27.1.C, the Fermat cubic surface gives an example of one regular cubic surface with precisely 27 lines, so (as $\mathbb{P}^{19} \setminus \Delta$ is connected) we are done. $\qquad \square$

We have actually shown that, away from Δ, $Z \to \mathbb{P}^{19}$ is a finite étale morphism of degree 27.

27.2.3.* *Proof of Theorem 27.2.1.* Choose projective coordinates so that the line ℓ is given, in a distinguished affine subset (with coordinates named x, y, z), by the z-axis. (We use affine coordinates to help visualize what we are doing, although this argument is better done in projective coordinates. On a second reading, you should translate this to a fully projective argument.)

27.2.C. EXERCISE. Consider the lines of the form $(x, y, z) = (a, b, 0) + t(a', b', 1)$ (where $(a, b, a', b') \in \mathbb{A}^4$ is fixed, and t varies in $\mathbb{A}^1$). Show that a, b, a', b' can be interpreted as the "usual" coordinates on one of the standard open subsets of the Grassmannian (see §7.7), with $[\ell]$ as the origin.

Having set up local coordinates on the moduli space, we can get down to business. Suppose $f(x, y, z) = 0$ is the (affine version) of the equation for the cubic surface X. Because X contains the z-axis ℓ, $f(x, y, z) \in (x, y)$. More generally, the line

(27.2.3.1) $(x, y, z) = (a, b, 0) + t(a', b', 1)$

lies in X precisely when $f(a + ta', b + tb', t)$ is 0 as a cubic polynomial in t. This is equivalent to four equations in a, a', b, and b', corresponding to the coefficients of t^3, t^2, t, and 1. This is better than just a set-theoretic statement:

27.2.D. EXERCISE. Verify that these four equations are local equations for the scheme-theoretic fiber (or scheme-theoretic preimage) $\pi^{-1}([X])$.

Now we come to the crux of the argument, where we use the regularity of X (along ℓ). We have a precise question in algebra. We are given a cubic surface X given by $f(x, y, z) = 0$, containing ℓ, and we know that X is regular (including "at ∞," i.e., in $\mathbb{P}^3$). To show that $[\ell] = V(a, a', b, b')$ is a reduced isolated point in the fiber of π, we work in the ring $\overline{k}[a, a', b, b']/(a, a', b, b')^2$, i.e., we impose the equations

(27.2.3.2) $a^2 = aa' = \cdots = (b')^2 = 0,$

and try to show that $a = a' = b = b' = 0$. (It is essential that you understand why we are setting $(a, a', b, b')^2 = 0$. You can also interpret this argument in terms of the

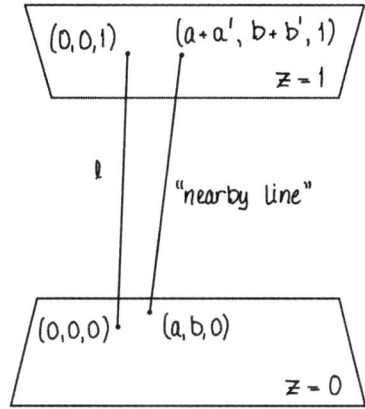

Figure 27.2 *Parameters for the space of "lines near ℓ," in terms of where they meet the $z = 0$ plane and the $z = 1$ plane.*

derivatives of the functions involved — which after all can be interpreted as forgetting higher-order information and remembering only linear terms in the relevant variables; cf. Exercise 13.1.I. See [Mu6, §8D] for a description of this calculation in terms of derivatives.)

Suppose $f(x, y, z) = c_{x^3} x^3 + c_{x^2 y} x^2 y + \cdots + c_1 1 = 0$, where $c_{x^3}, c_{x^2 y}, \cdots \in \overline{k}$. Because $\ell \subset X$, i.e., $f \in (x, y)$, we have $c_1 = c_z = c_{z^2} = c_{z^3} = 0$. We now substitute (27.2.3.1) into f, and then apply (27.2.3.2). Only the coefficients of f of monomials involving precisely one x or y survive:

$$c_x(a + a't) + c_{xz}(a + a't)t + c_{xz^2}(a + a't)t^2 + c_y(b + b't) + c_{yz}(b + b't)t + c_{yz^2}(b + b't)t^2$$

$$= (a + a't)(c_x + c_{xz}t + c_{xz^2}t^2) + (b + b't)(c_y + c_{yz}t + c_{yz^2}t^2)$$

is required to be 0 as a polynomial in t. (Recall that $c_x, \ldots, c_{yz^2}$ are fixed elements of $\overline{k}$.) Let $C_x(t) = c_x + c_{xz}t + c_{xz^2}t^2$ and $C_y(t) = c_y + c_{yz}t + c_{yz^2}t^2$ for convenience.

Now X is regular at $(0, 0, 0)$ precisely when c_x and c_y are not both zero (as $c_z = 0$). More generally, X is regular at $(0, 0, t_0)$ precisely if $c_x + c_{xz}t_0 + c_{xz^2}t_0^2 = C_x(t_0)$ and $c_y + c_{yz}t_0 + c_{yz^2}t_0^2 = C_y(t_0)$ are not both zero. You should be able to quickly check that X is regular at the point of ℓ "at ∞" precisely if c_{xz^2} and c_{yz^2} are not both zero. We summarize this as follows: X is regular at every point of ℓ precisely if the two quadratics $C_x(t)$ and $C_y(t)$ have no common roots, including "at ∞."

We now use this to force $a = a' = b = b' = 0$ using $(a + a't)C_x(t) + (b + b't)C_y(t) \equiv 0$.

We deal first with the special case where C_x and C_y have two distinct roots, both finite (i.e., c_{xz^2} and c_{yz^2} are nonzero). If t_0 and t_1 are the roots of $C_x(t)$, then substituting t_0 and t_1 into $(a + a't)C_x(t) + (b + b't)C_y(t)$, we obtain $b + b't_0 = 0$ and $b + b't_1 = 0$, from which $b = b' = 0$. Similarly, $a = a' = 0$.

27.2.E. EXERCISE. Deal with the remaining cases to conclude the proof of Theorem 27.2.1. (It is possible to do this quite cleverly. For example, you may be able to re-choose coordinates to ensure that C_x and C_y have finite roots.) □

27.2.4. The configuration of lines.

By the "configuration of lines" on a cubic surface, we mean the data of which pairs of the 27 lines intersect. We can readily work this out in the special case of the Fermat cubic surface (Exercise 27.1.C). (It can be more enlightening to use the description of X as a blow-up of $\mathbb{P}^2$; see Exercise 27.3.F.) We now show that the configuration is the "same" (interpreted appropriately) for all smooth cubic surfaces.

27.2.F. EXERCISE. Construct a degree 27! finite étale map $W \to \mathbb{P}^{19} \setminus \Delta$ that parametrizes a cubic surface along with an *ordered list* of 27 distinct lines. Hint: Let W' be the 27th fibered power of Z over $\mathbb{P}^{19} \setminus \Delta$, interpreted as parametrizing a cubic surface with an ordered list of 27 lines, not necessarily distinct. Let W be the subset corresponding to where the lines are distinct, and show that W is open and closed in W', and thus a union of connected components of W'. An alternative way of identifying W within W', in a way that might generalize to Exercise 27.2.H: make sense of the fact for each i, j, "$\ell_i \cdot \ell_j$" is a locally constant function on W', so the locus where these functions are positive for all $i \neq j$ is both open and closed; show that this is the W we seek.

We now make sense of the statement that the configuration of lines on the Fermat cubic surface (call it X_0) is the "same" as the configuration on some other smooth cubic surface (call it X_1). Lift the point $[X_0]$ to a point $w_0 \in W$. Let W'' be the connected component of W containing w_0.

27.2.G. EXERCISE. Show that the morphism $W'' \to \mathbb{P}^{19} \setminus \Delta$ is finite and étale (and surjective).

Choose a point $w_1 \in W''$ mapping to $[X_1]$. Because W parametrizes a "labeling" or ordering of the 27 lines on a surface, we now have chosen an identification of the lines on X_0 with those of X_1. Let the lines be $\ell_1, \ldots, \ell_{27}$ on X_0, and let the corresponding lines on X_1 be $m_1, \ldots, m_{27}$.

27.2.H. EXERCISE (USING STARRED EXERCISE 24.7.D). Show that $\ell_i \cdot \ell_j = m_i \cdot m_j$ for all i and j.

27.2.I. EXERCISE. Show that for each smooth cubic surface $X \subset \mathbb{P}^3_k$, each line on X meets exactly 10 other lines $\ell_1, \ell_1', \ldots, \ell_5, \ell_5'$ on X, where ℓ_i and ℓ_i' meet for each i, and no other pair of the lines meet.

27.2.J. EXERCISE. Show that every smooth cubic surface contains two disjoint lines ℓ and ℓ', such that there are precisely five other lines $\ell_1, \ldots, \ell_5$ meeting both ℓ and ℓ'. (In fact, for any two disjoint lines ℓ and ℓ' in a smooth cubic surface, there are precise five other lines meeting them; but this is strictly harder than what the problem asks.)

27.2.5. *Remark (cf. Remark 27.3.7): The Weyl group* $W(E_6)$. The symmetry group of the configuration of lines—i.e., the subgroup of the permutations of the 27 lines preserving the intersection data—magically turns out to be the Weyl group of E_6, a group of order 51840. (You know enough to at least verify that the size of the group is 51840, using the Fermat cubic surface of Exercise 27.1.C, but this takes some work.) It is no coincidence that the degree of W'' over $\mathbb{P}^{19} \setminus \Delta$ is 51840, and the Galois group of the Galois closure of $K(Z)/K(\mathbb{P}^{19} \setminus \Delta)$ is isomorphic to $W(E_6)$ (see [H1, III.3]).

27.2.K. ** **EXERCISE.** Prove Theorem 27.2.1 in arbitrary characteristic. Begin by figuring out the right statement of Exercise 27.2.A over $\mathbb{Z}$, and proving it. Then follow the argument given in this section, making changes when necessary.

27.2.6.* Fano varieties of lines, and Hilbert schemes.

In Exercises 27.2.A and 27.2.K, you constructed a moduli space of lines contained in a X, as a scheme. Your argument can be generalized to any $X \subset \mathbb{P}^N$. This construction is called the *Fano variety of lines* of X (not to be confused with the notion of a *Fano variety*, §21.5.5), and is an example of a Hilbert scheme.

27.3   Every Smooth Cubic Surface (over $\overline{k}$) is a Blown-Up Plane

We now prove Theorem 27.0.2. As stated in §27.0.3, this section is remarkably independent from the previous one; all we will need is Exercise 27.2.J, and it is possible to prove that in other ways.

Suppose X is a smooth cubic surface (over $\overline{k}$). Suppose ℓ is a line on X, and choose coordinates on the ambient $\mathbb{P}^3$ so that ℓ is cut out by x_0 and x_1. Projection from ℓ gives a rational map $\mathbb{P}^3 \dashrightarrow \mathbb{P}^1$ (given by $[x_0, x_1, x_2, x_3] \mapsto [x_0, x_1]$), which extends to a morphism on X. The reason is that this rational map is resolved by blowing up the closed subscheme $V(x_0, x_1)$ (Exercise 22.4.L). But (x_0, x_1) cuts out the effective Cartier divisor ℓ on X, and blowing up an effective Cartier divisor does not change X (Observation 22.2.1).

Now choose two disjoint lines ℓ and ℓ' as in Exercise 27.2.J, and consider the morphism $\rho \colon X \to \mathbb{P}^1 \times \mathbb{P}^1$, where the map to the first $\mathbb{P}^1$ is the projection from ℓ, and the map to the second $\mathbb{P}^1$ is the projection from ℓ'. The first $\mathbb{P}^1$ can then be identified with ℓ', and the second with ℓ.

27.3.A. EXERCISE. Show that the morphism ρ is birational. Hint: Given a general point of $(p, q) \in \ell' \times \ell$, we obtain a point of X as follows. The line pq in $\mathbb{P}^3$ meets the cubic X at three points by Bézout's Theorem 9.3.D: p, q, and some third point $x \in X$; send (p, q) to x. (This idea appeared earlier in the development of the group law on the cubic curve; see Proposition 19.10.3.) Given a general point $x \in X$, we obtain a point $(p, q) \in \ell' \times \ell$ by projecting from ℓ' and ℓ.

In particular, we have shown for the first time that every smooth cubic surface over $\overline{k}$ is rational.

27.3.B. EXERCISE: ρ CONTRACTS PRECISELY $\ell_1, \ldots, \ell_5$. Show that ρ is an isomorphism away from the ℓ_i mentioned in Exercise 27.2.J, and that each $\rho(\ell_i)$ is a point $p_i \in \ell' \times \ell$.

27.3.1. Proposition — *The morphism $\rho \colon X \to \mathbb{P}^1 \times \mathbb{P}^1$ is the blow-up of $\mathbb{P}^1 \times \mathbb{P}^1$ at the five p_i.*

We give two proofs. The first demonstrates the utility of a number of technical tools we have developed (or will develop in the future—there is one "forward reference" here). The second is a down-to-earth explicit "calculation." Different people prefer different approaches, but there are advantages in seeing them side by side. The first shows how to apply tools in quite general situations where we don't have explicit control over the geometry. The second gives us rather complete and direct control of the geometry in question, without needing formal functions.

27.3.2. *Proof 1 of Proposition 27.3.1.* By Castelnuovo's Criterion (Theorem 28.6.1), as the lines ℓ_i are (-1)-curves (Exercise 27.1.B), they can be contracted. More precisely, there is a morphism $\beta \colon X \to X'$ that is the blow-up of X' at five closed points $p_1', \ldots, p_5'$, such that ℓ_i is the exceptional divisor at p_i'. We wish to show that X' is $\mathbb{P}^1 \times \mathbb{P}^1$.

The morphism $\rho \colon X \to \mathbb{P}^1 \times \mathbb{P}^1$ yields a morphism $\rho' \colon X' \setminus \{p_1', \ldots p_5'\} \to \mathbb{P}^1 \times \mathbb{P}^1$. We now show that ρ' extends over p_i' for each i, sending p_i' to p_i. Choose an open neighborhood of $p_i \in \mathbb{P}^1 \times \mathbb{P}^1$ isomorphic to $\mathbb{A}^2$, with coordinates x and y. Then both x and y pull back to functions on a punctured open neighborhood of p_i' (i.e., there is some open neighborhood U of p_i' such that x and y are functions on $U \setminus \{p_i'\}$). By Algebraic Hartogs's Lemma 13.5.19, they extend over p_i', and this extension is unique as $\mathbb{P}^1 \times \mathbb{P}^1$ is separated—use the Reduced-to-Separated Theorem 11.3.2 if you really need to. Thus ρ' extends over p_i'. (Do you see why $\rho'(p_i') = p_i$?)

27.3.C. EXERCISE. Show that the birational morphism $\rho' \colon X' \to \mathbb{P}^1 \times \mathbb{P}^1$ is invertible. Hint: You can use Zariski's Main Theorem (in the guise of Exercise 28.5.E), but you needn't use something so powerful. Instead, note that the birational map ρ'^{-1} is a morphism away from $p_1, \ldots, p_5$. Use essentially the same argument as in the last paragraph to extend ρ'^{-1} over each p_i. $\qquad\square$

27.3.3. *Proof 2 of Proposition 27.3.1.* (I learned this slick version of a very classical argument from J. Kollár, who is famous for explaining complicated things in simple ways.) Choose coordinates on $\mathbb{P}^3$ so that ℓ is given by $x_0 = x_1 = 0$, and ℓ' is given by $x_2 = x_3 = 0$. Let the projective coordinates on ℓ be u_0 and u_1, and let the projective coordinates on ℓ' be v_0 and v_1 (so, for example, $\ell \hookrightarrow \mathbb{P}^3$ is given by $[u_0, u_1] \mapsto [0, 0, u_0, u_1]$). Consider the rational map

$$\alpha : \mathbb{P}^1_{u_0, u_1} \times \mathbb{P}^1_{v_0, v_1} \times \mathbb{P}^1_{s,t} \to \mathbb{P}^3$$

(where the subscripts indicate the names of the projective coordinates) given by

(27.3.3.1) $$([u_0, u_1], [v_0, v_1], [s, t]) \mapsto [sv_0, sv_1, tu_0, tu_1].$$

Because X contains $\ell = V(x_0, x_1)$ and $\ell' = V(x_2, x_3)$, the equation of $X \subset \mathbb{P}^3$ is of the form

$$F(x_0, x_1, x_2, x_3) = L_{02}x_0x_2 + L_{03}x_0x_3 + L_{12}x_1x_2 + L_{13}x_1x_3,$$

where the L_{ij} are linear forms in x_0, x_1, x_2, x_3. Thus

$$\alpha^* F = st\left(sG(u_0, u_1, v_0, v_1) + tH(u_0, u_1, v_0, v_1)\right).$$

From (27.3.3.1), G has degree 1 in the u_i and degree 2 in the v_i, and H has degree 2 in the u_i and degree 1 in the v_i.

27.3.D. EXERCISE. Interpret the morphism $X \to \mathbb{P}^1 \times \mathbb{P}^1$ as a rational map as pulling back $F = 0$ by α, then projecting to $\mathbb{P}^1_{u_0, u_1} \times \mathbb{P}^1_{v_0, v_1}$. Show that the five lines $\ell_1, \dots, \ell_5$ contracted (to $p_1, \dots, p_5$) by $X \to \mathbb{P}^1 \times \mathbb{P}^1$ must correspond set-theoretically to the locus $G = H = 0$ in some sense. Show that $G = H = 0$ set-theoretically cuts out the five points $p_1, \dots, p_5$ in $\mathbb{P}^1 \times \mathbb{P}^1$. Show that (in the intersection theory of $\mathbb{P}^1 \times \mathbb{P}^1$, Exercise 20.2.C) $V(G) \cdot V(H) = 5$. Show that near any of the five points $p_1, \dots, p_5$ of $\mathbb{P}^1 \times \mathbb{P}^1$, the effective Cartier divisors $V(G)$ and $V(H)$ must intersect trasnversely (with "multiplicity 1") — you'll have to make this precise. Show that the projection near each p_i must be the blow-up of $\mathbb{P}^1 \times \mathbb{P}^1$ at p_i. $\qquad\square$

27.3.4. *Conclusion of the proof of Theorem 27.0.2.* As a consequence we see that X is the blow-up of $\mathbb{P}^1 \times \mathbb{P}^1$ at five points. Because the blow-up of $\mathbb{P}^1 \times \mathbb{P}^1$ at one point is isomorphic to the blow-up of $\mathbb{P}^2$ at two points (Exercise 22.4.K), Theorem 27.0.2 follows. $\qquad\square$

27.3.5. Reversing the process. (This is a more precise version of 22.4.10(iii).) The process can be reversed: we can blow-up $\mathbb{P}^2$ at six points, and embed it in $\mathbb{P}^3$. We first explain why we can't blow up $\mathbb{P}^2$ at just any six points and hope to embed the result in $\mathbb{P}^3$. Because the cubic surface is embedded anticanonically (Exercise 27.1.A), we see that any curve C in X must satisfy $\deg_C \mathscr{K}_X < 0$.

27.3.E. EXERCISE. Suppose $\mathbb{P}^2$ is sequentially blown up at $p_1, \dots, p_6$, resulting in a smooth surface X.

(a) If p_i lies on the exceptional divisor of the blow-up at p_j ($i > j$), then show that there is a curve $C \subset X$ isomorphic to $\mathbb{P}^1$, with $\deg_C \mathscr{K}_X \geq 0$.

(b) If the p_i are distinct points on $\mathbb{P}^2$, and three of them are collinear, show that there is a curve $C \subset X$ isomorphic to $\mathbb{P}^1$, with $\deg_C \mathscr{K}_X \geq 0$.

(c) If the six p_i are distinct points on a smooth conic, show that there is a curve $C \subset X$ isomorphic to $\mathbb{P}^1$, with $\deg_C \mathscr{K}_X \geq 0$.

Thus the only chance we have of obtaining a smooth cubic surface by blowing up six points on $\mathbb{P}^2$ is by blowing up six distinct points, no three on a line and not all on a conic.

27.3.6. Proposition — *The anticanonical map of $\mathbb{P}^2$ blown up at six distinct points, no three on a line and not all on a conic, gives a closed embedding into $\mathbb{P}^3$, as a cubic surface.*

Because we won't use this, we only describe the main steps of the proof: first, count sections of the *anticanonical bundle* $\mathscr{K}_{\mathbb{P}^2}^\vee \cong \mathscr{O}_{\mathbb{P}^2}(3)$ (there is a four-dimensional vector space of cubics on $\mathbb{P}^2$

vanishing at the six points, and these correspond to sections of the anticanonical bundle of the blow-up via Exercise 22.4.S(a)). Then show that these sections separate points and tangent vectors of X, thus showing that the anticanonical linear series gives a closed embedding, Theorem 19.1.1. Judicious use of the Cremona transformation (Exercise 7.5.J) can reduce the amount of tedious case-checking in this step.

27.3.F. EXERCISE. Suppose X is the blow-up of $\mathbb{P}^2_{\bar{k}}$ at six distinct points $p_1, \ldots, p_6$, no three on a line and not all on a conic. Verify that the only (-1)-curves on X are the six exceptional divisors, the proper transforms of the 15 lines $p_i p_j$, and the proper transforms of the six conics through five of the six points, for a total of 27.

27.3.G. EXERCISE. Solve Exercises 27.2.I and 27.2.J again, this time using the description of X as a blow-up of $\mathbb{P}^2$.

27.3.7. *Remark.* If you blow up $4 \le n \le 8$ points on $\mathbb{P}^2$, with no three on a line and no six on a conic, then the symmetry group of the configuration of lines is a Weyl group, as shown in the following table.

n	4	5	6	7	8
	$W(A_4)$	$W(D_5)$	$W(E_6)$	$W(E_7)$	$W(E_8)$

(If you know about Dynkin diagrams, you may see the pattern, and may be able to interpret what happens for $n = 3$ and $n = 9$. The surface corresponding to $n = 4$ appeared in §19.8.3.) This generalizes part of Remark 27.2.5, and the rest of it can similarly be generalized.

27.3.H. EXERCISE. Show that the surfaces described in Remark 27.3.7 are del Pezzo surfaces (defined in Exercise 21.5.H). (You don't know need to know about Dynkin diagrams to do this. It turns out that every del Pezzo is of the sort described here or in Exercise 21.5.H.)

Chapter 28

Power Series and the Theorem on Formal Functions

Power series are a central tool in analytic geometry. Their analog in algebraic geometry, completion, is similarly useful. We will only touch on some aspects of the subject.

In §28.1, we deal with some algebraic preliminaries. In §28.2, we use completions to (finally) give a definition of various singularities, such as nodes. (We won't use these definitions in what follows.) In §28.3, we state the main technical result of the chapter, the Theorem on Formal Functions 28.3.2. The subsequent three sections give applications. In §28.4, we prove Zariski's Connectedness Lemma 28.4.1 and the Stein Factorization Theorem 28.4.3. In §28.5, we prove a commonly used version of (Grothendieck's version of) Zariski's Main Theorem 28.5.1; we rely on §28.4. In §28.6, we prove Castelnuovo's Criterion for contracting (-1)-curves, which we used in Chapter 27. The proof of Castelnuovo's Criterion also uses §28.4. Finally, in §28.7, we prove the Theorem on Formal Functions 28.3.2.

There are deliberately many small sections in this chapter, so you can see that they tend not to be as hard at they look, with the exception of the proof of the Theorem on Formal Functions itself, and possibly Theorem 28.1.6 relating completion to exactness and flatness.

28.1   Algebraic Preliminaries

Suppose A is a ring, and $J_1 \supset J_2 \supset \cdots$ is a decreasing sequence of ideals. The limit $\lim A/J_n$ is often denoted by $\hat{A}$, and is called the **completion** of A (for this sequence of ideals); the sequence of ideals is left implicit. The most important case of completion is if $J_n = I^n$, where I is an ideal of A. The limit $\lim A/I^n$ is called the I-**adic completion** of A, or **the completion of** A **along** I or **at** I.

28.1.1. *Example.* We define $k[[x_1, \ldots, x_n]]$ as the completion of $k[x_1, \ldots, x_n]$ at the maximal ideal $(x_1, \ldots, x_n)$. This is the ring of *formal power series* in n variables over k (Example 1.3.3).

The p-adic numbers (Example 1.3.4) are another example.

28.1.A. EXERCISE. Suppose that $J_1' \supset J_2' \supset \cdots$ is a decreasing series of ideals that is **cofinal** with J_n . In other words, for every J_n, there is some J_N' with $J_n \supset J_N'$, and for every J_n', there is some J_N with $J_n' \supset J_N$. Show that there is a canonical isomorphism $\lim A/J_n' \xrightarrow{\sim} \lim A/J_n$. Thus what matters is less the specific sequence of ideals than the "cofinal" equivalence class.

28.1.2. *Preliminary remarks.* We have an obvious morphism $A \to \hat{A}$.

If we put the discrete topology on A/J_n, then $\hat{A}$ naturally has the structure of a *topological ring* (a ring that is a topological space, where all the ring operations are continuous, or, equivalently, a ring object in the category of topological spaces). We then can have the notion of a *topological module* M over a topological ring A'—a module over the underlying ring A', with a topology, such that the action of A' on M is continuous. This is a useful point of view, but we won't use it.

28.1.B. EXERCISE. Suppose m is a maximal ideal of a ring A. Show that the completion of A at m is canonically isomorphic to the completion of A_m at m.

If (A, m) is a Noetherian local ring, then the natural map $A \to \hat{A}$ is an injection: anything in the kernel must lie in $\cap m^i$, which is 0 by the Krull Intersection Theorem (Exercise 13.9.A(b)). Thus "no information is lost by completing," just as analytic functions are (locally) determined by their power series expansion.

(In the case of completion at an ideal I, we say that A is I-**adically separated** if $A \to \hat{A}$ is an injection, and **complete with respect to** I if $A \to \hat{A}$ is an isomorphism, although we won't need these phrases. For example, the Krull Intersection Theorem implies that if I is a proper ideal of a Noetherian integral domain or a Noetherian local ring, then A is I-adically separated.)

28.1.C. EXERCISE. Suppose that $(A, \mathfrak{m})$ is a Noetherian local ring containing its residue field k (i.e., it is a k-algebra). Let $x_1, \ldots, x_n$ be elements of A whose images are a basis for $\mathfrak{m}/\mathfrak{m}^2$. Show that the map of k-algebras

(28.1.2.1)
$$k[[t_1, \ldots, t_n]] \longrightarrow \hat{A}$$

defined by $t_i \mapsto x_i$ is a surjection. (First explain why there *is* such a map!) As usual, for local rings, the completion is assumed to be at the maximal ideal.

Exercise 28.1.C is a special case of the *Cohen Structure Theorem*. (See [E, §7.4] for more on this topic.)

28.1.D. EXERCISE. Let X be a locally Noetherian scheme over k, let $p \in X$ be a rational (k-valued) point. Suppose $p \in X$ is regular, of codimension d (i.e., $\dim \mathcal{O}_{X,p} = d$). Describe an isomorphism $\hat{\mathcal{O}}_{X,p} \xleftarrow{\sim} k[[x_1, \ldots, x_d]]$ as topological rings. (Hint: As in Exercise 28.1.C, choose d elements of $\mathfrak{m}$ that restrict to a basis of $\mathfrak{m}/\mathfrak{m}^2$; these will be your $x_1, \ldots, x_d$. Show that the map (28.1.2.1) has no kernel. It may help to identify $\mathfrak{m}^n/\mathfrak{m}^{n+1}$ with $\mathrm{Sym}^n(\mathfrak{m}/\mathfrak{m}^2)$ using Theorem 22.3.8.)

The converse also holds: if $(A, \mathfrak{m}, k)$ is a Noetherian local ring that is a k-algebra, and $\hat{A} \cong k[[x_1, \ldots, x_d]]$, then A is a regular local ring of dimension d; see [AtM, Prop. 11.24] for the key step.

28.1.3. *Remark.* Suppose p is a smooth rational (k-valued) point of a k-variety of dimension d, and $f \in \mathcal{O}_{X,p}$ (f is a "local function"). By way of the isomorphism of Exercise 28.1.D, we can interpret f as an element of $k[[x_1, \ldots, x_d]]$. This should be interpreted as the "power series expansion of f at p," in the "local coordinates $x_1, \ldots, x_d$."

28.1.4. Some more geometric motivation.

Exercise 28.1.D may give some motivation for completion: "in the completion, regular schemes look like affine space." This is often stated in the suggestive language of "formally locally, regular schemes are isomorphic to affine space"; this will be made somewhat more precise in §28.3.1.

We now give a little more geometric motivation for completion, which we will not use later on. Recall from §13.6.1 that étale morphisms are designed to look like "local isomorphisms" in differential geometry. But Exercise 13.6.F showed that this metaphor badly fails in one important way. More precisely, let $Y = \mathrm{Spec}\, k[t]$, and $X = \mathrm{Spec}\, k[u, 1/u]$, and let $p \in X$ be given by $u = 1$, and $q \in Y$ be given by $t = 1$. Suppose $\mathrm{char}\, k \neq 2$. The morphism $\pi \colon X \to Y$ induced by $t \mapsto u^2$ is étale, and $\pi(p) = q$. But there is no open neighborhood of p that π maps isomorphically onto an open neighborhood of q. However, the following exercise shows that π induces an isomorphism of completions (an isomorphism of "formal neighborhoods").

28.1.E. EXERCISE. Continuing the notation of the previous paragraph, show that π induces an isomorphism of completions $\hat{\mathcal{O}}_{Y,q} \to \hat{\mathcal{O}}_{X,p}$.

28.1.F. EXERCISE. Suppose $Y = \mathrm{Spec}\, A$, and $X = \mathrm{Spec}\, A[t]/(f(t))$, and $\pi \colon X \to Y$ is the morphism induced by $A \to A[t]/(f(t))$. Suppose $p \in X$ is a closed point, $q \in Y$ is a closed point, $\pi(p) = q$, and π is étale at p and induces an isomorphism of residue fields. Show that π induces an isomorphism of completions $\hat{\mathcal{O}}_{Y,q} \to \hat{\mathcal{O}}_{X,p}$. (If you are not familiar with Hensel's Lemma, you might rediscover its central idea in the course of solving this exercise.)

With a little more care, you can show more generally that if $\pi \colon X \to Y$ is an étale morphism, $\pi(p) = q$, and π induces an isomorphism of residue fields at p, then π induces an isomorphism of completions $\hat{\mathcal{O}}_{Y,q} \to \hat{\mathcal{O}}_{X,p}$. (You may even wish to think about how to remove the hypothesis of isomorphism of residue fields.)

These ideas are close to the definition of "formal étaleness" discussed in §24.8.13.

You can interpret these results as the statement that "the implicit function theorem works formally locally," even though it doesn't work Zariski-locally. (The étale topology is somewhere between these two; it is partially designed so that the implicit function theorem in this sense always holds, essentially by fiat. But this is not the place to discuss the étale topology.)

Here are two more fun problems secretly about Hensel's Lemma.

28.1.G. EXERCISE. Show that $x^2 + y^2 = -2$ has solutions modulo 7^{100}.

28.1.H. EXERCISE. Suppose we have a plane curve $f(x, y) = 0$ in $\mathbb{A}^2_{\mathbb{Q}}$, such that $(0, a)$ is a smooth point whose tangent line is nonvertical. Show that there is precisely one power series solution $y = p(x)$ such that $f(x, p(x)) = 0$ (as a power series), and $p(0) = a$.

28.1.5. Completion and exactness.

We conclude this section with an interesting and useful statement.

28.1.6. Theorem — *Suppose A is a Noetherian ring, and $I \subset A$ is an ideal. For any A-module M, let $\hat{M} = \lim M/I^j M$ be the completion of M with respect to I.*

(a) *The completion $\hat{A}$ (of A with respect to I) is flat over A.*

(b) *If M is finitely generated, then the natural map $\hat{A} \otimes_A M \to \hat{M}$ is an isomorphism.*

(c) *If $0 \to M \to N \to P \to 0$ is a short exact sequence of finitely generated A-modules, then $0 \to \hat{M} \to \hat{N} \to \hat{P} \to 0$ is exact. (Thus completion preserves exact sequences of finitely generated modules, by Exercise 1.5.F.)*

We will use (a) in §28.2, and (b) in §28.7.

28.1.7. *Remark.* Before proving Theorem 28.1.6, we make some remarks. Parts (a) and (b) together clearly imply part (c), but we will use (c) to prove (a) and (b). Also, note a delicate distinction (which helps me remember the statement): if $0 \to M \to N \to P \to 0$ is an exact sequence of A-modules, *not necessarily finitely generated*, then

$$(28.1.7.1) \qquad 0 \longrightarrow \hat{A} \otimes_A M \longrightarrow \hat{A} \otimes_A N \longrightarrow \hat{A} \otimes_A P \longrightarrow 0$$

is *always* exact, but

$$(28.1.7.2) \qquad 0 \longrightarrow \hat{M} \longrightarrow \hat{N} \longrightarrow \hat{P} \longrightarrow 0$$

need *not* be exact—and when it *is* exact, it is often because the modules are finitely generated, and thus (28.1.7.2) is really (28.1.7.1).

Caution: Completion is not always exact. Consider the exact sequence of $k[t]$-modules

$$0 \longrightarrow \oplus_{n=1}^{\infty} k[t] \xrightarrow{\times (t, t^2, t^3, \dots)} \oplus_{n=1}^{\infty} k[t] \longrightarrow \oplus_{n=1}^{\infty} k[t]/(t^n) \longrightarrow 0.$$

After completion with respect to the ideal $((t, t, t, \dots))$, the sequence is no longer exact in the middle: $(t^2, t^3, t^4, \dots)$ maps to 0, but is not in the image of the completion of the previous term.

** *Proof.* The key step is to prove (c), which we do through a series of exercises. Suppose that $0 \to M \to N \to P \to 0$ is a short exact sequence of finitely generated A-modules.

28.1.I. EXERCISE. Show that $\hat{N} \to \hat{P}$ is surjective. Hint: Consider an element of $\hat{P}$ as a sequence $(p_j \in P/I^j P)_{j \geq 0}$, where the image of p_{j+1} is p_j; cf. Exercise 1.3.C. Build a preimage $(n_j \in N/I^j N)_{j \geq 0}$ by induction on j.

We now wish to identify $\ker(\hat{N} \to \hat{P})$ with $\hat{M}$.

28.1.J. EXERCISE. Show that for each $j \geq 0$,

(28.1.7.3) $$0 \longrightarrow M/(M \cap I^j N) \longrightarrow N/I^j N \longrightarrow P/I^j P \longrightarrow 0$$

is exact. (Possible hint: Show that $0 \to M \cap I^j N \to M \to N/I^j N \to P/I^j P \to 0$ is exact.)

The short exact sequences (28.1.7.3) form an inverse system as j varies. Its limit is left-exact (because limits always are), but it is also right-exact by Exercise 1.5.N, as the "transition maps on the left" $M/(M \cap I^{j+1} N) \to M/(M \cap I^j N)$ are clearly surjective. Thus

(28.1.7.4) $$0 \longrightarrow \lim M/(M \cap I^j N) \longrightarrow \hat{N} \longrightarrow \hat{P} \longrightarrow 0$$

is exact. To complete the proof of (c), it suffices (by Exercise 28.1.A) to show that the sequence of submodules $I^j M$ is cofinal with the sequence $M \cap I^j N$, so that the term $\lim M/(M \cap I^j N)$ on the left of (28.1.7.4) is naturally identified with $\hat{M}$.

28.1.K. EXERCISE. Prove this. Hint: Clearly, $I^j M \subset M \cap I^j N$. By Corollary 13.9.4 to the Artin–Rees Lemma 13.9.3, for some integer s, we have $M \cap I^{j+s} N = I^j (M \cap I^s N)$ for all $j \geq 0$, and clearly $I^j (M \cap I^s N) \subset I^j M$.

This completes the proof of part (c) of Theorem 28.1.6.
For part (b), present M as

(28.1.7.5) $$A^{\oplus m} \xrightarrow{\ \alpha\ } A^{\oplus n} \longrightarrow M \longrightarrow 0,$$

where α is an $m \times n$ matrix with coefficients in A. Completion is exact in this case by part (c), and clearly commutes with finite direct sums, so

$$\hat{A}^{\oplus m} \longrightarrow \hat{A}^{\oplus n} \longrightarrow \hat{M} \longrightarrow 0$$

is exact. Tensor product is right-exact, and commutes with direct sums (Exercise 1.2.M), so

$$\hat{A}^{\oplus m} \longrightarrow \hat{A}^{\oplus n} \longrightarrow \hat{A} \otimes_A M \longrightarrow 0$$

is exact as well. Notice that the maps from $\hat{A}^{\oplus m}$ to $\hat{A}^{\oplus n}$ in both right-exact sequences are the same; they are both α. Thus their cokernels are identified, and (b) follows.

Finally, to prove (a), we need to extend the ideal-theoretic criterion for flatness (Theorem 24.3.1) slightly. Recall (§24.3.2) that it is equivalent to the fact that an A-module M is flat if and only if for all ideals I, the natural map $I \otimes_A M \to M$ is an injection.

28.1.L. EXERCISE (STRONGER FORM OF THE IDEAL-THEORETIC CRITERION FOR FLATNESS). Show that an A-module M is flat if and only if for all *finitely generated ideals* I, the natural map $I \otimes_A M \to M$ is an injection. (Hint: If there is a counterexample for an ideal J that is not finitely generated, use it to find another counterexample for an ideal I that *is* finitely generated.)

By this criterion, to prove (a) it suffices to prove that the multiplication map $I \otimes_A \hat{A} \to \hat{A}$ is an injection for all finitely generated ideals I. But by part (b), this is the same showing that $\hat{I} \to \hat{A}$ is an injection; and this follows from part (c). $\qquad\square$

28.2   Types of Singularities

Singularities are often most easily defined in terms of completions.

28.2.1. *Definition.* Suppose X is a dimension 1 variety over $\overline{k}$, and $p \in X$ is a closed point. We say that X has a **node** at p if the completion of $\mathscr{O}_{X,p}$ at $\mathfrak{m}_{X,p}$ is isomorphic (as topological rings) to $\overline{k}[[x,y]]/(xy)$. (How does this compare to the definition of node in §13.3.2?)

28.2.A. EXERCISE. Suppose $k = \overline{k}$ and char $k \neq 2$. Show that the curve in $\mathbb{A}^2_k$ cut out by $y^2 = x^2 + x^3$, which we have been studying repeatedly since Figure 8.4, has a node at the origin.

28.2.B. EXERCISE. Suppose $k = \overline{k}$ and char $k \neq 2$, and we have $f(x,y) \in k[x,y]$.

(a) Show that $\operatorname{Spec} k[x,y]/(f(x,y))$ has a node at the origin if and only if f has no terms of degree 0 or 1, and the degree 2 terms are not a perfect square. (This generalizes Exercise 28.2.A.)

(b) Show that $\operatorname{Spec} k[x,y]/(f(x,y))$ has a node at the point (a,b) if and only if

$$\frac{\partial f}{\partial x} = \frac{\partial f}{\partial y} = 0 \quad \text{and} \quad \left(\frac{\partial^2 f}{\partial x^2}\right)\left(\frac{\partial^2 f}{\partial y^2}\right) \neq \left(\frac{\partial^2 f}{\partial x \partial y}\right)^2$$

at (a,b).

The definition of node outside the case of varieties over algebraically closed fields is more problematic, and we give some possible ways forward. For varieties over a non-algebraically closed field k, one can always base change to the closure $\overline{k}$. As an alternative approach, if p is a k-valued point of a variety over k (not necessarily algebraically closed), then we could take the same Definition 28.2.1; this might reasonably be called a *split node*, because the branches (or, more precisely, the tangent directions) are distinguished. Those singularities that are not split nodes, but which become nodes after base change to $\overline{k}$ (such as the origin in $\operatorname{Spec} \mathbb{R}[x,y]/(x^2+y^2)$, or $\operatorname{Spec} \mathbb{Q}[x,y]/(x^2-2y^2)$), might reasonably be called *non-split nodes*.

28.2.2. *Definition.* We may define other singularities similarly. To avoid complications, we do so over an algebraically closed field. Suppose X is a variety over $\overline{k}$, and p is a closed point of X, where char $k \neq 2,3$. We say that X has a **cusp** (resp., **tacnode, triple point**) at p if $\hat{\mathscr{O}}_{X,p}$ is isomorphic to the completion of $\overline{k}[x,y]/(y^2-x^3)$ (resp., $\overline{k}[x,y]/(y^2-x^4)$, $\overline{k}[x,y]/(y^3-x^3)$). (See Figure 13.3 for pictures of the first two.) We say that X has an **ordinary multiple point of multiplicity** m, or **ordinary m-fold point**, if $\hat{\mathscr{O}}_{X,p}$ is isomorphic to the completion of $\overline{k}[x,y]/(f(x,y))$, where f is a homogeneous polynomial of degree m with no repeated roots. (You can see quickly that an ordinary 2-fold point—or *ordinary double point*—is precisely a node, and an ordinary 3-fold point is a triple point.)

28.2.C. TRIVIAL EXERCISE. (For this exercise, work over an algebraically closed field for simplicity.) Define A_n curve singularity (see §22.4.4). Define A_n **surface singularity** (see Exercise 22.4.F). Define D_n, E_6, E_7, **and** E_8 **surface and curve singularities** (see Remark 22.4.5).

28.2.3. Using this definition. We now give an example of how this definition can be used.

28.2.D. EXERCISE. Suppose X is a $\overline{k}$-variety with a node at a closed point p. Show that the blow-up of X at p yields a morphism $\beta \colon \tilde{X} \to X$, where the exceptional divisor $\beta^{-1}(p)$ consists of two reduced smooth points. Hint: Use the fact that completion is flat (Theorem 28.1.6(a)), and that blowing up commutes with flat base change (Exercise 24.1.P(a)), to turn this into a calculation on the "formal model" of the node, $\operatorname{Spec} k[[x,y]]/(xy)$.

28.2.E. EXERCISE. Continuing the terminology of the previous exercise, describe an exact sequence

(28.2.3.1) $$0 \longrightarrow \mathscr{O}_X \longrightarrow \beta_* \mathscr{O}_{\tilde{X}} \longrightarrow \mathscr{O}_p \longrightarrow 0.$$

Hint: Faithful flatness.

28.2.F. EXERCISE. We continue the terminology of the previous two exercises. If X is a pure-dimensional reduced projective curve, show that $p_a(X) = p_a(\tilde{X}) + 1$.

Thus "resolving a node" of a curve reduces the arithmetic genus of the curve by 1. If you wish, you can readily show that "resolving a cusp" reduces the genus by 1, and "resolving a tacnode" reduces the genus by 2. In general, for each type of curve singularity, the contribution it makes to the genus—the difference between the genus of the curve and that of its "partial normalization at the singular point," when the curve is projective—is called the δ-**invariant**. Thus δ for a node or cusp is 1, and δ for a tacnode is 2.

28.2.G. EXERCISE. Show that for *any* singularity type (other than a smooth point), $\delta > 0$.

28.2.H. EXERCISE. Show that a reduced irreducible degree d plane curve can have at most $\binom{d-1}{2}$ singularities.

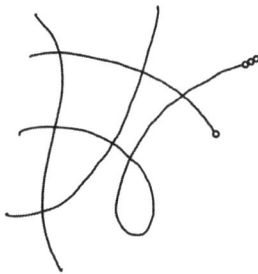

Figure 28.1 *This curve has genus 8 (count the "holes"!).*

28.2.I. EXERCISE. Here is an informal trick for speedily working out the genus of a nodal (projective reduced) curve when it comes up in examples. Assuming it is possible, draw a sketch on a piece of paper (i.e., the real plane) where nodes "look like nodes," and a curve of genus g is drawn with g holes, as in Figure 28.1. (In Figure 28.1, we see four irreducible components, of geometric genus 0, 0, 1, and 3, respectively. The first three components are smooth, while the fourth meets itself once. There are six additional nodes.) Then if the curve is connected, count the number of holes visible in the picture. Prove that this recipe works. (Feel free to figure out what to do if the curve is not connected.)

Figure 28.2 *These curves have arithmetic genus 6, 6, and 3, respectively.*

28.2.4. This can be quite useful. The sketches in Figure 28.2 can remind you that a degree d plane curve has arithmetic genus $\binom{d-1}{2}$, the curve on $\mathbb{P}^1 \times \mathbb{P}^1$ coming from a section of $\mathcal{O}(a, b)$ (§15.2.10) has genus $(a-1)(b-1)$, and a curve in class $2C + 2F$ on the Hirzebruch surface $\mathbb{F}_2$ (using the language of Exercise 20.2.M) has arithmetic genus 3. In each case, we are taking a suitable curve in the class, taking advantage of the fact that arithmetic genus is constant in flat families (Crucial Exercise 24.7.A).

28.2.5. Other definitions. If $\overline{k} = \mathbb{C}$, then definitions of this sort agree with the analytic definitions (see [Ar1, §1]). For example, a complex algebraic curve singularity is a node if and only if it is *analytically* isomorphic to an open neighborhood of $xy = 0$ in $\mathbb{C}^2$. There is also a notion of isomorphism "étale-locally," which we do not define here. Once again, this leads to the same definition of these types of singularities (see [Ar2, §2]).

28.3 The Theorem on Formal Functions

Suppose $\pi\colon X \to Y$ is proper morphism of locally Noetherian schemes, and $\mathcal{F}$ is a coherent sheaf on X, so $R^i\pi_*\mathcal{F}$ is a coherent sheaf (Grothendieck's Coherence Theorem 18.9.1). Fix a point $q \in Y$. We already have a sense that there is an imperfect relationship between the fiber of $R^i\pi_*\mathcal{F}$ at q and the cohomology of $\mathcal{F}$ "on the fiber X_q," and much of Chapter 25 was devoted to making this precise.

The Theorem on Formal Functions deals with this issue in a different way. Rather than comparing the fiber of $R^i\pi_*\mathcal{F}$ with the cohomology of $\mathcal{F}$ restricted to the fiber, it gives an isomorphism between information on "infinitesimal thickenings" of the fiber. Informally:

$$R^i\pi_*\mathscr{F} \longrightarrow H^i(X_q, \mathscr{F}|_{X_q})$$

.

$$\parallel$$

$$H^i(\text{0th order nbhd of } X_q, \mathscr{F})$$

$$\uparrow$$

$$H^i(\text{1st order nbhd of } X_q, \mathscr{F})$$

$$\uparrow$$

$$H^i(\text{2nd order nbhd of } X_q, \mathscr{F})$$

$$\uparrow$$

$$\vdots$$

$$\uparrow$$

completion of $R^i\pi_*\mathscr{F}$ at q $\xrightarrow{\text{Thm. 28.3.2: iso}}$ limit

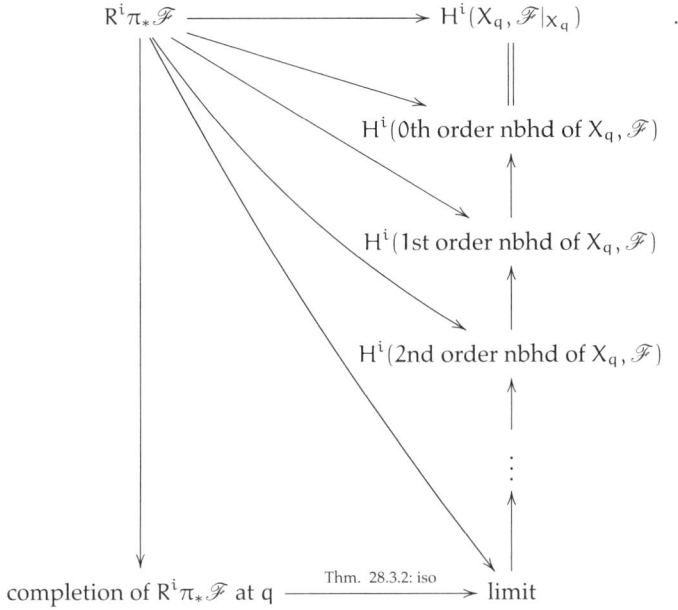

28.3.1. We now make this more precise. If Z is a closed subscheme of W cut out by ideal sheaf $\mathscr{I}$, we say that the closed subscheme cut out by $\mathscr{I}^{n+1}$ is the **nth order formal** (or **infinitesimal**) **neighborhood** (or **thickening**) of Z (in W). The phrase "formal neighborhood" without mention of an "order" refers to the information contained in all of these neighborhoods at once, often in the form of a limit of the sort we will soon describe. (This also leads us to the notion of *formal schemes*, an important notion we will nonetheless not need.)

We turn now to our situation of interest. Rather than the fiber of $R^i\pi_*\mathscr{F}$ at q, we consider its completion: we take the stalk at q, and complete it at the maximal ideal $\mathfrak{m}_q$. Rather than the cohomology of $\mathscr{F}$ along the fiber, we consider the cohomology of $\mathscr{F}$ when restricted to the nth formal neighborhood X_n of the fiber, and take the limit. The Theorem of Formal Functions says that these are canonically identified.

Before we state the Theorem on Formal Functions, we need to be precise about what these limits are, and how they are defined. To concentrate on the essential, we do this in the special case where q is a *closed* point of Y, and leave the (mild) extension to the general case to you (Exercise 28.3.A).

We deal first with $R^i\pi_*\mathscr{F}$. It is helpful to note that, by Exercise 28.1.B, we can compute the restriction to the nth formal neighborhood either by pulling back to Spec $\mathscr{O}_{Y,q}$ ("the germ of Y at q," i.e., tensoring with $\mathscr{O}_{Y,q}$), or in any affine open neighborhood of $q \in Y$. For convenience, we pick an affine open neighborhood Spec $B \subset Y$ of q. (Again, by Exercise 28.1.B, it won't matter which affine open subset we take.) Define $X_B := \pi^{-1}(\text{Spec } B)$ for convenience. Let $\mathfrak{m} \subset B$ be the maximal ideal corresponding to q. Then $((R^i\pi_*\mathscr{F})_q)\hat{\ }$ is canonically the completion of the coherent B-module $H^i(X_B, \mathscr{F}|_{X_B})$ at $\mathfrak{m}$:

$$\left((R^i\pi_*\mathscr{F})_q\right)\hat{\ } = \lim H^i(X_B, \mathscr{F}|_{X_B})/\mathfrak{m}^n H^i(X_B, \mathscr{F}|_{X_B}).$$

We turn next to the cohomology of $\mathscr{F}$ on thickenings of the fiber. We have closed embeddings $X_n \hookrightarrow X_{n+1}$, and thus maps of cohomology groups

$$H^i(X_{n+1}, \mathscr{F}|_{X_{n+1}}) \longrightarrow H^i(X_n, \mathscr{F}|_{X_n}).$$

We have base change maps

(28.3.1.1) $$H^i(X_B, \mathscr{F}|_{X_B})/\mathfrak{m}^n H^i(X_B, \mathscr{F}|_{X_B}) \longrightarrow H^i(X_n, \mathscr{F}|_{X_n})$$

(see (25.1.0.3)) such that the square

$$
\begin{array}{ccc}
H^i(X_B, \mathscr{F}|_{X_B})/\mathfrak{m}^{n+1} H^i(X_B, \mathscr{F}|_{X_B}) & \longrightarrow & H^i(X_{n+1}, \mathscr{F}|_{X_{n+1}}) \\
\downarrow & & \downarrow \\
H^i(X_B, \mathscr{F}|_{X_B})/\mathfrak{m}^n H^i(X_B, \mathscr{F}|_{X_B}) & \longrightarrow & H^i(X_n, \mathscr{F}|_{X_n})
\end{array}
$$

commutes. (Do you see why? Basically, this is again because they are base change maps.) Thus we have an induced map of limits:

(28.3.1.2) $$\widehat{((R^i \pi_* \mathscr{F})_q)} \longrightarrow \lim_{n \to \infty} H^i(X_n, \mathscr{F}|_{X_n}).$$

28.3.A. EXERCISE. Extend the previous discussion to the case where q is not a closed point of Y.

The Theorem on Formal Functions states that this is an isomorphism.

28.3.2. Theorem on Formal Functions — *Suppose* $\pi\colon X \to Y$ *is a proper morphism of locally Noetherian schemes,* $\mathscr{F}$ *is a coherent sheaf on* X, *and* $q \in Y$. *Then* (28.3.1.2) *is an isomorphism for all* $i \geq 0$.

Warning: The Theorem on Formal Functions does *not* imply anything about the maps "at finite level," i.e., about (28.3.1.1) for "finite n."

The proof of Theorem 28.3.2 is subtle, and is postponed to the double-starred §28.7. We first give four important applications to keep you motivated and excited: Zariski's Connectedness Lemma 28.4.1, Stein factorization 28.4.3, Zariski's Main Theorem 28.5.1, and Castelnuovo's Criterion 28.6.1 for contracting (-1)-curves on surfaces.

28.4   Zariski's Connectedness Lemma and Stein Factorization

We now state and prove Zariski's Connectedness Lemma, which was mentioned in §25.1.8.

28.4.1. Zariski's Connectedness Lemma — *If a proper morphism* $\pi\colon X \to Y$ *of locally Noetherian schemes is* $\mathcal{O}$-*connected (see §25.1.7), then* $\pi^{-1}(q)$ *is connected for every* $q \in Y$.

The proof requires the following result.

28.4.A. EASY EXERCISE (THE COMPLETION OF A LOCAL RING IS A LOCAL RING). Suppose $(A, \mathfrak{m})$ is a Noetherian local ring. Show that the completion of A along $\mathfrak{m}$, $\hat{A} := \lim A/\mathfrak{m}^n$, is a local ring, with maximal ideal $\mathfrak{m}\hat{A}$. Hint: Show that any element of $\hat{A}$ not in $\mathfrak{m}\hat{A}$ is invertible.

Proof. Assume otherwise that there is some $q \in Y$ such that $\pi^{-1}(q)$ is not connected, say, $\pi^{-1}(q) = X_1 \coprod X_2$ (where X_1 and X_2 are nonempty open subsets of the fiber; see Figure 28.3). Then the nth order formal neighborhood of $\pi^{-1}(q)$, having the same topological space, is also disconnected. We use the useful trick of idempotents (Remark 3.6.3). Let e_1 be the function 1 on X_1 and 0 on X_2, and let e_2 be the function 1 on X_2 and 0 on X_1. These functions make sense for any order formal neighborhood, and they have natural images in the inverse limit. By the isomorphism of the Theorem on Formal Functions 28.3.2 (applied to H^0),

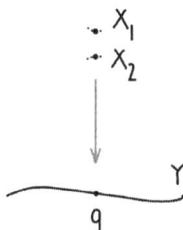

Figure 28.3 *"Formal neighborhoods" of X_1 and X_2 in the proof of Zariski's Connectedness Lemma 28.4.1.*

we get nonzero elements $e_1, e_2 \in \hat{\mathscr{O}}_{Y,q}$ with $e_1 + e_2 = 1$, $e_1 e_2 = 0$. Now, $\hat{\mathscr{O}}_{Y,q}$ is a local ring (Exercise 28.4.A). But $e_1 e_2 = 0$ implies that neither e_1 nor e_2 is invertible (or else the other one would be 0), so both e_1 and e_2 are in the maximal ideal, and hence $e_1 + e_2$ can't be 1. We thus have a contradiction. $\qquad\square$

28.4.2. Stein factorization. We next describe the construction known as "Stein factorization." We could have defined it long before, but now Zariski's Connectedness Lemma 28.4.1 will give us some impressive consequences.

28.4.3. Stein Factorization Theorem — *Any proper morphism $\pi\colon X \to Y$ of locally Noetherian schemes can be factored into $\beta \circ \alpha$, where $\alpha\colon X \to Y'$ is proper and $\mathscr{O}$-connected (hence has connected fibers, by Zariski's Connectedness Lemma 28.4.1), and $\beta\colon Y' \to Y$ is a finite morphism.*

(28.4.3.1)

$$
\begin{array}{ccc}
X & \xrightarrow{\;\;\overset{\mathscr{O}\text{-conn.}}{\alpha}\;\;} & Y' \\
 & {\scriptstyle\pi}\searrow & \downarrow{\scriptstyle\beta} \\
 & & Y
\end{array}
\quad\text{finite}
$$

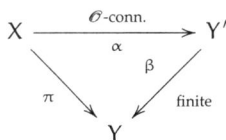

We note that by the Cancellation Theorem for projective morphisms (Exercise 17.3.C), if π happens to be projective, then so is α (as δ_β, being a closed embedding, is projective).

Although it is not in the statement of the theorem, the proof produces a *specific* factorization, which is called *the* **Stein factorization** of π. The picture to have in mind is that the Stein factorization (roughly) contracts the continuous parts of the fibers, in some canonical way; see Figure 28.4.

As usual, the Noetherian hypotheses can be removed (see [Stacks, tag 03H2]).

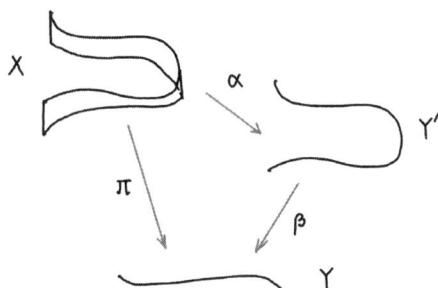

Proof of the Stein Factorization Theorem 28.4.3. By Grothendieck's Coherence Theorem 18.9.1, $\pi_* \mathscr{O}_X$ is a coherent sheaf (of algebras) on Y. Define $Y' := \operatorname{Spec} \pi_* \mathscr{O}_X$, so (as $\pi_* \mathscr{O}_X$ is finite type) the structure morphism $\beta\colon Y' \to Y$ is finite. We have a factorization

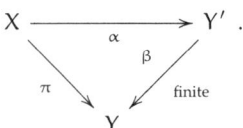

Figure 28.4 *Stein factorization.*

$$
\begin{array}{ccc}
X & \xrightarrow{\;\;\alpha\;\;} & Y' \\
 & {\scriptstyle\pi}\searrow & \downarrow{\scriptstyle\beta} \\
 & & Y
\end{array}
\quad\text{finite}
$$

By the Cancellation Theorem 11.1.1 for proper morphisms, α is proper. Finally, we show that α is $\mathscr{O}$-connected, i.e., that $\mathscr{O}_{Y'} \to \alpha_* \mathscr{O}_X$ is an isomorphism. Because β is an affine morphism (so β_* is an exact functor $Coh_{Y'} \to Coh_Y$), it suffices to check after applying the pushforward β_*. But $\beta_* \mathscr{O}_{Y'} \to \beta_* \alpha_* \mathscr{O}_X$ is the identity map $\pi_* \mathscr{O}_X \to \pi_* \mathscr{O}_X$ (here we use that $\beta_* \mathscr{O}_{Y'} = \pi_* \mathscr{O}_X$—see §17.1.4—and $\beta_* \alpha_* \mathscr{O}_X = \pi_* \mathscr{O}_X$). $\qquad\square$

28.4.B. EXERCISE. Suppose $\pi\colon X \to Y$ is a proper birational morphism of irreducible varieties. Suppose further that Y is normal. Show that π is $\mathscr{O}$-connected. (This applies, for example, to blow-ups of smooth varieties under smooth centers.) Hence, by Zariski's Connectedness Lemma 28.4.1, π has connected fibers. This is called **Zariski's Main Theorem** (the version for birational morphisms). Hint: Exercise 10.7.P.

28.4.C. EXERCISE. Suppose $\pi\colon X \to Y$ and $\pi'\colon X \to Y'$ are two proper birational morphisms of integral locally Noetherian schemes, each contracting the same *connected* set of X to a point (and an isomorphism elsewhere), and suppose Y and Y' are normal. Show that π and π' are the same (or, more precisely, that there is an isomorphism $i\colon Y \xrightarrow{\sim} Y'$ such that $\pi' = i \circ \pi$). Informal translation: If $X \to Y$ has connected fibers, then Y is determined by just knowing the locus contracted on X. Hint: Identify Y and Y' as topological spaces, and then identify π and π' as maps of topological spaces. Use Stein factorization to recover the structure sheaf as $\pi_* \mathcal{O}_X$.

28.4.D. EXERCISE. Show that the construction of the Stein factorization of a morphism $\pi\colon X \to Y$ of locally Noetherian schemes commutes with any flat base change $Z \to Y$ of locally Noetherian schemes.

28.4.4. Resolution of singularities of curves.

If C is a reduced projective curve over a field k, then we can resolve its singularities by normalization: $\nu\colon \tilde{C} \to C$. (See Remark 22.4.6 for some discussion of resolution of singularities, and Theorem 13.5.8 for the reason why normal curves are regular.) But as we have seen from examples, it requires luck and insight to figure out how to take an integral closure, and I often have neither. We now take the (inspired) guesswork out of desingularization by explaining how to desingularize by blowing up. The algorithm is simple: we find a singular point, then blow it up, then look for more singular points to repeat the process. Clearly, if there are no singular points to be found, then we are done; the problem is to show that this process is guaranteed to terminate. We do this by making use of an integer invariant: the arithmetic genus. We must show that (i) if $p_a(C) = p_a(\tilde{C})$, then ν is an isomorphism—C is already nonsingular; (ii) if $C' \to C$ is a birational morphism, then $p_a(C') \leq p_a(C)$ (so $p_a(\tilde{C})$ will be a lower bound for the arithmetic genus throughout the process); and (iii) blowing up a singular point decreases the arithmetic genus.

To set up these results, we consider the following situation. Suppose $\pi\colon C' \to C$ is a finite morphism (where C is as described in the previous paragraph, and C' is a reduced curve), that is an isomorphism away from a finite closed subset of C (hence π is birational). Then the pullback map $\mathcal{O}_C \to \pi_* \mathcal{O}_{C'}$ has no kernel, as C is reduced, and the map is an isomorphism on the generic points of the components of C. The cokernel $\mathcal{G}$ is supported on a finite set of closed points, and thus $H^i(C, \mathcal{G}) = 0$ for $i > 0$, and if $H^0(C, \mathcal{G}) = 0$, then $\mathcal{G} = 0$. From the exact sequence

$$0 \longrightarrow \mathcal{O}_C \longrightarrow \pi_* \mathcal{O}_{C'} \longrightarrow \mathcal{G} \longrightarrow 0,$$

and the fact that π is affine, $\chi(C', \mathcal{O}_{C'}) = \chi(C, \mathcal{O}_C) + \chi(C, \mathcal{G})$, from which $p_a(C) \geq p_a(C')$, with equality if and only if π is $\mathcal{O}$-connected (i.e., $\mathcal{G} = 0$).

28.4.E. EXERCISE. Show that if π (as above) is $\mathcal{O}$-connected, then π is an isomorphism. Hint: Show that affine $\mathcal{O}$-connected morphisms are isomorphisms.

To complete our strategy, it remains to show that if $\beta\colon C' \to C$ is the blow-up of C at a point p, and β is an isomorphism, then C is already nonsingular at that point. But if β is an isomorphism, then p is the exceptional divisor, and hence effective Cartier, and hence cut out locally by a single equation. But then $\mathfrak{m}_{C,p} \subset \mathcal{O}_{C,p}$ is principal, so $\mathcal{O}_{C,p}$ is a regular local ring of dimension 1.

(You should verify that we have completed our strategy. Can you extend this argument to the case where C is not projective?)

Resolution of singularities for surfaces will be briefly discussed in §28.6.4.

28.5 Zariski's Main Theorem

Zariski's Main Theorem is misnamed in all possible ways (although not horribly so). It is not a single *Theorem*, because there are many results that go by this name, and they often seem quite unrelated. What they have in common is that they are intellectual descendants of a particular result of Zariski. This result is not the *Main* Theorem of Zariski's prolific career; the name comes because it was the Main Theorem of a particular paper, [Z]. And finally, it is not *Zariski's* in any stronger sense than this; the modern versions are due to Grothendieck.

We have already seen one version of Zariski's Main Theorem: the "birational" form of Exercise 28.4.B. Our goal in this section is to prove a stronger statement that is close to the optimal version.

Before we start, we note that the fiber of a proper morphism over a point $q \in Y$ is a proper (hence finite type) scheme over the residue field $\kappa(q)$, and we say that a point is *isolated in its fiber* if it forms a component of the fiber of dimension 0. (Here component means both connected and irreducible component.)

28.5.1. Zariski's Main Theorem (Grothendieck's form) — *Suppose* $\pi \colon X \to Y$ *is a proper morphism of locally Noetherian schemes.*

(a) *The set of points of X that are isolated in their fiber forms an open subset $X_0 \subset X$.*
(b) *The morphism $\pi|_{X_0} \colon X_0 \to Y$ factors into an open embedding $X_0 \hookrightarrow Y'$ followed by a finite morphism $Y' \to Y$.*
(c) *The morphism $\pi \colon X \to Y$ factors through Y'.*

(28.5.1.1)

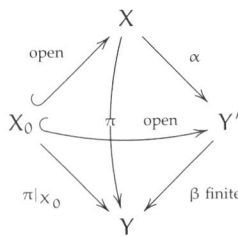

As with Stein factorization, the proof of the theorem yields a *specific* factorization. Indeed, $X \to Y' = \operatorname{Spec} \pi_* \mathscr{O}_X \to Y$ is the Stein factorization of $\pi \colon X \to Y$, giving (c) and part of (b) basically immediately.

If π is a morphism of *varieties*, we already know part (a), by upper semicontinuity of fiber dimension Theorem 12.4.3(a). For this reason, the proof in the case of varieties is easier, and (even if you are interested in the more general case) you are advised to read the proof restricting to this simpler case, to concentrate on the main ideas. But even in the case of varieties, we will need some of the ideas arising in the proof of (a), so you should read it until advised to skip ahead to (b).

Proof. (a) Take the Stein factorization (28.4.3.1) of π:

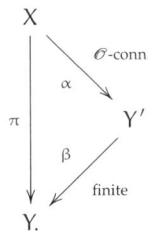

By the finiteness of β, a point of X is an isolated point in its fiber of π if and only if it is an isolated point in its fiber of α. Thus we may replace π by α. (Basically, we are reducing to the case where π is $\mathscr{O}$-connected, but we prefer to carefully call the morphism under consideration α so we can return to considering π in the proof of (b).)

Because fibers of proper $\mathscr{O}$-connected morphisms of locally Noetherian schemes are connected (Zariski's Connectedness Lemma 28.4.1), a point is isolated in its fiber of α if and only if it *is* its fiber.

If you are considering only the variety case, at this point you should jump to the proof of part (b).

To show that X_0 is open, we will identify it with the locus where $\Omega_\alpha = 0$ (the unramified locus of α—not of π!), which is open by Exercise 21.7.H.

28.5.A. EXERCISE. If $\Omega_\alpha|_p = 0$, show that p is isolated in its fiber of α (and hence of π).

We now show the other direction. Suppose p is isolated in its fiber, and let $r \in Y'$ be $\alpha(p)$. We will show that α induces an isomorphism $\alpha^\sharp \colon \mathscr{O}_{Y',r} \xrightarrow{\sim} \mathscr{O}_{X,p}$, so $\Omega_{\mathscr{O}_{X,p}/\mathscr{O}_{Y',r}} = 0$. As the construction of Ω behaves well with respect to localization on the source and target (Exercise 21.2.L), this implies that $(\Omega_\alpha)_p = 0$, so Ω_α is 0 at p.

The stalk $\mathscr{O}_{Y',r}$ is obtained by taking the colimit (of sections of $\mathscr{O}_{Y'}$) over all open subsets of Y' containing r. For any such open subset $V \subset Y'$, the condition $\alpha_* \mathscr{O}_X = \mathscr{O}_{Y'}$ gives an isomorphism $\Gamma(V, \mathscr{O}_{Y'}) \xleftarrow{\sim} \Gamma(\alpha^{-1}(V), \mathscr{O}_X)$ (compatible with inclusions). We will show that the system of open subsets $\alpha^{-1}(V)$ (as V varies through open neighborhoods of r), which each necessarily contain p, is cofinal with the system of open neighborhoods of p in X. To do this, we must show that for any open neighborhood U of p in X, there is an open neighborhood V of r in Y' such that $\alpha^{-1}(V) \subset U$. To do this, note that $X \setminus U$ is closed in X, so $\alpha(X \setminus U)$ is closed in Y' (as α is proper), so its complement $V := Y' \setminus \alpha(X \setminus U)$ is open. But $\alpha^{-1}(V) \subset U$, so we are done.

(b) *(set)* In our proof of (a), we have established that α gives a bijection between X_0 and an open subset of Y'.

(topology) We have basically also showed that α gives a homeomorphism: any open subset of Y' pulls back to an open subset of X_0 by the continuity of α; and if U is an open subset of X_0, then $\alpha(X \setminus U)$ is a closed subset of Y', so its complement (which is $\alpha(U)$, using that proper $\mathscr{O}$-connected morphisms are surjective, Exercise 25.1.H) is open.

(sheaf) Using this isomorphism of topological spaces, the condition of $\mathscr{O}$-connectedness of α (restricted to $\alpha(X_0)$) shows that α gives an isomorphism of ringed spaces (i.e., of schemes) $X_0 \xrightarrow{\sim} \alpha(X_0)$.

We have thus found our desired factorization. (Part (c) is immediate from our construction.) □

28.5.B. EXERCISE. Prove Zariski's Main Theorem with slightly different hypotheses: with Y affine, say $\operatorname{Spec} A$, and X a *quasiprojective* A-scheme. (Hint: by the Definition 4.5.10 of quasiprojective A-scheme, we can find an open embedding $X \hookrightarrow X'$ into a projective A-scheme. Apply Theorem 28.5.1 to $X' \to \operatorname{Spec} A$.)

As an application of Zariski's Main Theorem, we can finally prove a characterization of finite morphisms that we have mentioned a number of times.

28.5.2. Theorem — *Suppose $\pi\colon X \to Y$ is a morphism of locally Noetherian schemes. The following are equivalent.*

(a) *π is finite.*
(b) *π is affine and proper.*
(c) *π is proper and quasifinite.*

As usual, Noetherian hypotheses can be removed: [Gr-EGA, IV$_3$.8.11.1] shows that proper, quasifinite, locally finitely presented morphisms are finite.

Proof. Finite morphisms are affine by definition, and proper by Proposition 11.4.3, so (a) implies (b).

To show that (b) implies (c), we need only show that affine proper morphisms have finite fibers (as the "finite type" part of the definition of quasifiniteness is taken care of by properness). This was shown in Exercise 18.9.A.

Finally, we assume (c), that π is proper and quasifinite, and show (a), that π is finite. By Zariski's Main Theorem 28.5.1, we have a factorization (28.5.1.1).

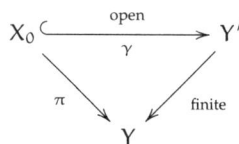

$$\begin{array}{ccc} X_0 & \xrightarrow[\gamma]{\text{open}} & Y' \\ & \searrow_{\pi} \quad \swarrow_{\text{finite}} & \\ & Y & \end{array}$$

Now $X = X_0$. (Do you see why? Hint: Exercise 8.4.D followed by Exercise 8.3.G.) By the Cancellation Theorem 11.1.1 for proper morphisms (using that $Y' \to Y$ is finite hence separated), γ is proper.

28.5.C. EXERCISE. Show that proper quasicompact open embeddings are closed embeddings. Hint: use Corollary 9.4.5 (for quasicompact morphisms, the closure of the image is the image of the closure) to show that the image is open and closed in the target.

Applying this to our situation, we see that $X \to Y$ is the composition of two finite morphisms $\gamma \colon X \hookrightarrow Y'$ and $Y' \to Y$, and is thus finite itself. $\square$

28.5.3. *Nagata's compactification theorem.* For the final two applications (Exercises 28.5.E and 28.5.F), we will need an annoying hypothesis, which is foreshadowed by the previous exercise. We say that a morphism $\pi \colon X \to Y$ of locally Noetherian schemes satisfies ($\dagger$) if for all affines U_i in some open cover of Y, the restriction $\pi|_{\pi^{-1}(U_i)} \colon \pi^{-1}(U_i) \to U_i$ factors as an open embedding into a scheme proper over U_i. A morphism satisfying ($\dagger$) is necessarily separated and finite type. It turns out that this is sufficient: **Nagata's Compactification Theorem** states that *every* separated finite type morphism of Noetherian schemes $\pi \colon X \to Y$ can be factored into a dense open embedding into a scheme proper over Y. (See [Lüt] for a proof. The Noetherian hypotheses can be replaced by the condition that Y is quasicompact and quasiseparated, see [Con].)

28.5.D. EXERCISE. Show that if $\pi \colon X \to Y$ is a morphism of varieties over k, and X is an open subset of a proper variety Z, then π satisfies ($\dagger$). (Hint: X is an open subscheme of its closure in $Z \times_k Y$.)

28.5.E. EXERCISE. Suppose $\pi \colon X \to Y$ is a quasifinite birational morphism of integral locally Noetherian schemes, and that Y is normal. Suppose further π satisfies ($\dagger$). Show that π is an open embedding. (Hence if furthermore π is a bijection, then π must be an isomorphism.) Hint: the condition of being an open embedding is local on the target.

28.5.F. EXERCISE. Suppose $\pi \colon X \to Y$ is a separated quasifinite morphism of Noetherian schemes.

(a) Assuming Nagata's Compactification Theorem (§28.5.3), show that π factors as an open embedding into a finite morphism.

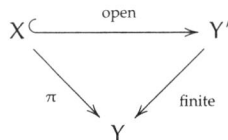

(b) Assuming instead that π satisfies ($\dagger$), show that π "locally" factors into an open embedding into a finite morphism, i.e., that there is a open cover U_i of Y such that over each U_i, there is such a factorization.

This justifies thinking of quasifinite morphisms as "open subsets of finite morphisms," as promised in §8.3.12.

28.5.4.** Other versions of Zariski's Main Theorem.
The Noetherian conditions in Theorem 28.5.1 can be relaxed (see [Stacks, tag 02LQ], [Gr-EGA, IV$_3$.8.12.13], or [GW, Thm. 12.73]): if $\pi \colon X \to Y$ is a separated morphism of finite type, and Y is quasicompact and quasiseparated, then the set of points X_0 isolated in their fiber is open in X, and

for every quasicompact open subset U of X_0, there is a factorization

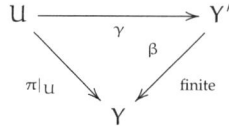

such that β is finite, $\gamma|_U$ is a quasicompact open embedding, and $\gamma^{-1}(\gamma(U)) = U$.

There are also topological and power series forms of Zariski's Main Theorem, see [Mu7, §III.9].

28.6 Castelnuovo's Criterion for Contracting (-1)-Curves

We showed in Exercise 22.4.O that if we blow up a regular surface at a (reduced) point, the exceptional divisor is a (-1)-curve (see Definition 20.2.8: it is isomorphic to $\mathbb{P}^1$, and has normal bundle $\mathscr{O}_C(-1)$). Castelnuovo's Criterion is the converse: if we have a quasiprojective surface containing a (-1)-curve, that surface is obtained by blowing up another surface at a reduced regular point. We say that we can *blow down* the (-1)-curve.

28.6.1. Theorem (Castelnuovo's Criterion) — *Let $C \subset X$ be a (-1)-curve on a smooth projective surface over k. Then there exists a birational morphism $\pi\colon X \to Y$ such that Y is a smooth projective surface, $\pi(C)$ is a k-valued point, π is the blow-up of Y at that point, and C is the exceptional divisor of the blow-up.*

By Exercise 28.4.C, there is only one way to "blow down C": this contraction is unique.

Proof. The proof is in three steps. *Step 1.* We construct $\pi\colon X \to Y$ that contracts C to a point q (and is otherwise an isomorphism). At this point we will know that Y is a projective variety. *Step 2.* Then we show that Y is smooth; this is the hard step, and requires the Theorem on Formal Functions 28.3.2. *Step 3.* Finally, we recognize π as the blow-up $\mathrm{Bl}_q Y \to Y$.

Step 1. We do this by finding an invertible sheaf $\mathscr{L}$ whose complete linear series will do the job. We start with a very ample invertible sheaf $\mathscr{H}$ on X such that $H^1(X, \mathscr{H}) = 0$. We can find such an invertible sheaf $\mathscr{H}$ by choosing any very ample invertible sheaf, and then taking a suitably large multiple, invoking Exercise 16.2.C (very ample $\otimes$ very ample = very ample) and Serre vanishing (Theorem 18.1.3(ii)). Let $d = \mathscr{H} \cdot C := \deg_C \mathscr{H}|_C$. We have $d > 0$ by the very ampleness of $\mathscr{H}$ (Exercise 20.1.K).

28.6.A. EXERCISE. Show that $H^1(X, \mathscr{H}(iC)) = 0$ for $0 \leq i \leq d$. Hint: Use induction on i, using

$$(28.6.1.1) \qquad 0 \longrightarrow \mathscr{H}((i-1)C) \longrightarrow \mathscr{H}(iC) \longrightarrow \mathscr{H}(iC)|_C \longrightarrow 0.$$

Hint: Note that $\mathscr{H}(iC)|_C$ is a line bundle on $C \cong \mathbb{P}^1$; which $\mathscr{O}(n)$ is it?

Define $\mathscr{L} := \mathscr{H}(dC)$. We will use the sections of $\mathscr{H}$ to obtain sections of $\mathscr{L}$, via the map $\phi\colon H^0(X, \mathscr{H}) \to H^0(X, \mathscr{H}(dC))$.

28.6.B. EXERCISE. Show that $\mathscr{L}$ is base-point-free. Hint: To show that $\mathscr{L}$ has no base points away from C, consider the image of ϕ. To show that $\mathscr{L}$ has no base points *on* C, use that, from (28.6.1.1), for $i = d$, we have that $H^0(X, \mathscr{L}) \to H^0(C, \mathscr{L}|_C)$ is surjective, and $\mathscr{L}|_C \cong \mathscr{O}_C$.

Thus the complete linear series $|\mathscr{L}|$ yields a morphism π' from X to some projective space $\mathbb{P}^N$.

28.6.C. EXERCISE. Show that π' precisely contracts C to a point q'. More explicitly, show that π' sends C to a point q', and $\pi'|_{X \setminus C}\colon X \setminus C \to \mathbb{P}^N \setminus \{q'\}$ is a closed embedding. Hint: To show that π' gives a closed embedding away from C, note that $|\mathscr{H}|$ separates points and tangent vectors, and argue that $|\mathscr{L}| = |\mathscr{H}(dC)|$ separates points and tangent vectors away from C. To show that C is contracted by π', use the fact that $\deg \mathscr{L}|_C = 0$.

Let Y' be the image of π', so we have a morphism $X \to Y'$ (which we also call π'). Let $\nu : Y \to Y'$ be the normalization of Y', so π' lifts to $\pi : X \to Y$ by the universal property of normalization (§10.7). (We are normalizing in order to use Exercise 28.4.B in Step 2.)

28.6.D. EASY EXERCISE. Show that $\nu^{-1}(q') \subset Y$ is a single point, which we call q, and that $\pi(C) =$ q. (Hint: The image of π is closed, and $\pi(C)$ is connected.)

28.6.E. EXERCISE. Let X be a smooth projective surface, containing a curve $C \subset X$ isomorphic to $\mathbb{P}^1_k$. If $C \cdot C := \deg_C \mathscr{O}(C)|_C < 0$, show that there is a morphism $\pi : X \to X'$ contracting C to a point, and leaving the rest of X unchanged. (This will not be used, and is included to give practice with the argument of Step 1.)

Step 2. We now show that Y is a smooth surface. As Y is normal, we have that $\pi_* \mathscr{O}_X \cong \mathscr{O}_Y$ ($\pi : X \to Y$ is $\mathscr{O}$-connected), by Exercise 28.4.B. We will show that Y is smooth at q by showing that $\hat{\mathscr{O}}_{Y,q} \cong k[[x, y]]$ as topological rings, as then $\dim_k(\mathfrak{m}/\mathfrak{m}^2) = 2$. (Do you see why this shows that Y is smooth, not just regular?) By the Theorem on Formal Functions 28.3.2, $\hat{\mathscr{O}}_{Y,q} \cong \lim_{n \to \infty} H^0(C'_n, \mathscr{O}_{C'_n})$, where C'_n is the closed subscheme of X that is the scheme-theoretic preimage of the nth order formal neighborhood of q.

We are not precisely sure what the nth order formal neighborhood is, but that is fine; this inverse system is cofinal with $\mathscr{O}(-nC) = \mathscr{I}^n_{C/X}$ (do you see why?), so we can take this inverse limit instead (by Exercise 28.1.A). It suffices to show that for all $n \geq 0$, $H^0(C_n, \mathscr{O}_{C_n}) \cong k[[x, y]]/(x, y)^n$, where C_n is defined by

$$0 \longrightarrow \mathscr{O}(-nC) \longrightarrow \mathscr{O}_X \longrightarrow \mathscr{O}_{C_n} \longrightarrow 0.$$

(Be careful: We do not know that $C_n = C'_n$, or even that $C_1 = C'_1$. For all we know, the fiber over q might be nonreduced!)

We do this by induction on n. As a k-vector space, this is easy, using

$$(28.6.1.2) \qquad 0 \longrightarrow \mathscr{I}^n/\mathscr{I}^{n+1} \longrightarrow \mathscr{O}_{C_{n+1}} \longrightarrow \mathscr{O}_{C_n} \longrightarrow 0$$

(the closed subscheme exact sequence (9.1.2.1) for C_n in C_{n+1}), using the canonical isomorphism $H^0(\mathbb{P}^1, \mathscr{O}_{\mathbb{P}^1}(n)) \xrightarrow{\sim} \mathrm{Sym}^n (H^0(\mathbb{P}^1, \mathscr{O}_{\mathbb{P}^1}(1)))$. The tricky thing is that we want this isomorphism as *rings*.

28.6.F. EXERCISE. Prove this. (This may remind you of how we found the ring of functions on the total space of an invertible sheaf in Exercise 19.11.E. This is no coincidence.) Hint: Interpret $\mathscr{I}^n/\mathscr{I}^{n+1}$ as $\mathrm{Sym}^n(\mathscr{I}/\mathscr{I}^2)$ (Corollary 22.3.9) and, in turn, as $\mathscr{O}_{\mathbb{P}^1}(n)$, and show that upon taking global sections, (28.6.1.2) remains exact. At this point you will have shown that some sort of associated graded ring of $H^0(C_n, \mathscr{O}_{C_n})$ is isomorphic to $k[[x, y]]/(x, y)^n$. Show that this is enough to establish the desired isomorphism of rings.

Hence we have smoothness, completing Step 2.

Step 3. We now must recognize π as the blow-up of Y at q. As C is an effective Cartier divisor on X, by the universal property of blowing up means that there is a unique morphism α making the following diagram commute:

(28.1.3)

$$\begin{array}{ccc} X & \xrightarrow{\quad \alpha \quad} & \mathrm{Bl}_q\, Y \\ & {}_{\pi}\searrow & \downarrow{}_{\beta} \\ & & Y \end{array}\qquad.$$

By the Cancellation Theorem for projective morphisms (Exercise 17.3.C), α is projective.

28.6.G. EXERCISE. Show that α has finite fibers.

As projective morphisms with finite fibers are finite (Theorem 18.1.6), α is a finite morphism. But then α is a birational finite morphism to a normal scheme, and thus an isomorphism by Exercise 10.7.P. □

So far we have only used the Theorem on Formal Functions 28.3.2 for H^0. The next exercise will give you a chance to see it in action for other cohomology groups.

28.6.H. EXERCISE.

(a) Suppose $\beta\colon X = \mathrm{Bl}_q\, Y \to Y$ is the blow-up of a smooth surface over k at a smooth k-valued point. Show that $R^1\beta_*\mathscr{O}_X = R^2\beta_*\mathscr{O}_X = 0$.
(b) Show that the natural maps $\beta^*\colon H^1(Y, \mathscr{O}_Y) \to H^1(X, \mathscr{O}_X)$ and $\beta^*\colon H^2(Y, \mathscr{O}_Y) \to H^2(X, \mathscr{O}_X)$ are isomorphisms.

28.6.2. Elementary transformations.

Recall the definition of the Hirzebruch surfaces $\mathbb{F}_n$ (over a field k) from Example 17.2.6. If $n > 0$, then on $\mathbb{F}_n$ we have a unique curve E of maximally negative self-intersection (Exercise 20.2.R). It has self-intersection $E \cdot E := \deg_E \mathscr{O}(E)|_E = -n$. It is a section of the projection $\pi\colon \mathbb{F}_n \to \mathbb{P}^1$. (If $n = 0$, so $\mathbb{F}_n \cong \mathbb{P}^1 \times \mathbb{P}^1$, E is usually defined as any "constant" section of $\pi\colon \mathbb{P}^1 \times \mathbb{P}^1 \to \mathbb{P}^1$—it is not unique.)

Suppose q is a k-valued point of $\mathbb{F}_n$. Let F be the fiber of π containing q.

28.6.I. EXERCISE. Show that $F \cdot F := \deg_F \mathscr{O}(F)|_F = 0$.

Let $\beta\colon \mathrm{Bl}_q\, \mathbb{F}_n \to \mathbb{F}_n$ be the blow-up at q, and let Z be the exceptional divisor (as the name "E" is taken). Let F' be the proper transform of F, and let E' be the proper transform of E.

28.6.J. EXERCISE. Show that $F' \cdot F' := \deg_{F'} \mathscr{O}(F')|_{F'} = -1$.

Thus by Castelnuovo's Criterion, we can blow down F', to obtain a new surface Y.

28.6.K. EXERCISE. Suppose $n > 0$. If $q \notin E$, show that $Y \cong \mathbb{F}_{n-1}$. If $q \in E$, show that $Y \cong \mathbb{F}_{n+1}$. Possible hint: If you knew that Y was a Hirzebruch surface $\mathbb{F}_m$, you could recognize it by the self-intersection of the unique curve of negative self-intersection if $m > 0$.

28.6.L. EXERCISE. Suppose $n = 0$. Show that $Y \cong \mathbb{F}_1$. Hint: Exercise 22.4.K.

This discussion can be generalized to $\mathbb{P}^1$-bundles over more general curves. The following is just a first step in the story.

28.6.M. EXERCISE. Suppose C is a smooth curve over k, $\pi\colon X \to C$ is a $\mathbb{P}^1$-bundle over C. If q is a k-valued point of X, then we can blow up q and blow down the proper transform of the fiber through q. Show that the result is another $\mathbb{P}^1$-bundle X' over C. (This is called an **elementary transformation** of the ruled surface.)

28.6.N. EXERCISE. Suppose $X = \mathscr{P}roj_C(\mathscr{L} \oplus \mathscr{M})$, and q lies on the section corresponding to

$$0 \longrightarrow \mathscr{L} \longrightarrow \mathscr{L} \oplus \mathscr{M} \longrightarrow \mathscr{M} \longrightarrow 0.$$

Show that $X' \cong \mathscr{P}roj_C(\mathscr{L}(\pi(q)) \oplus \mathscr{M}))$.

Can you speculate (or show) what happens if you are just told that $X \cong \mathscr{P}roj_C \mathscr{V}$, where $\mathscr{V}$ is a locally free sheaf of rank 2, and q lies on a section corresponding to (20.2.9.1)?

28.6.3. Minimal models of surfaces.

A surface over an algebraically closed field is called **minimal** if it has no (-1)-curves. If a surface X is not minimal, then we can choose a (-1)-curve, and blow it down. If the resulting surface is not minimal, we can again blow down a (-1)-curve, and so on. By the finiteness of the Picard number ρ (which we have admittedly not proved; see §18.4.12), as blowing down reduces ρ by 1 (by Exercise 22.4.O), this process must terminate. By this means we construct a *minimal model* of X—a minimal surface birational to X. If minimal models were unique, this would provide a means of classifying surfaces up to birationality: if X and X′ were smooth projective surfaces, to see if they were birational, we would find their minimal models, and see if they were isomorphic.

Sadly, minimal models are not unique. The example of $\mathbb{P}^1 \times \mathbb{P}^1$ and $\mathbb{P}^2$ show that there can be more than one minimal model in a birational equivalence class. (Why are these two surfaces minimal?) Furthermore, the isomorphism between the blow-up of $\mathbb{P}^2$ at two points with the blow-up of $\mathbb{P}^1 \times \mathbb{P}^1$ at one point (Exercise 22.4.K) shows that a smooth surface can have more than one minimal model.

Nonetheless, the failure of uniqueness is well understood: it was shown by Zariski that the only surfaces with more than one minimal model are those birational to $C \times \mathbb{P}^1$ for some curve C. This is part of the Enriques–Kodaira classification of surfaces. For more on this, see [Be], [Ba], or [BPV, §6].

28.6.4. Here are two more facts worth mentioning but not proving here. First, singular surfaces can be desingularized by normalizing and then repeatedly blowing up points; this explicit desingularization is analogous to the version for curves in §28.4.4, and is generalized by Hironaka's Theorem on resolution of singularities (Remark 22.4.6). Second, birational maps $X \dashrightarrow Y$ between smooth surfaces can be factored into a finite number of blow-ups followed by a finite number of blow-downs (cf. §22.4.12, on weak factorization). As a consequence, by Exercise 28.6.H, birational projective surfaces have the same $H^i(\mathcal{O})$ for $i = 0, 1, 2$. (Using facts stated in §21.5.10, you can show that, over $\mathbb{C}$, this implies that the entire Hodge diamond except for $h^{1,1}$ is a birational invariant of surfaces.)

28.7** Proof of the Theorem on Formal Functions 28.3.2

In this section, we will prove the following.

28.7.1. Theorem — *Suppose $X \to \operatorname{Spec} B$ is a proper morphism, with B Noetherian, and $\mathcal{F}$ is a coherent sheaf on X. Suppose I is an ideal of B, and $\hat{B}$ is the I-adic completion $\lim B/I^n$ of B. Let $X_n := X \times_B (B/I^{n+1})$ (the nth order formal neighborhood of the fiber X_0) and let $\mathcal{F}_n$ be the pullback of $\mathcal{F}$ to X_n. Then for all i, the natural map*

$$(28.7.1.1) \qquad H^i(X, \mathcal{F}) \otimes_B \hat{B} = \widehat{H^i(X, \mathcal{F})} \longrightarrow \lim H^i(X_n, \mathcal{F}_n)$$

is an isomorphism.

This proof is due to Serre, by way of B. Conrad.

28.7.A. EXERCISE. Prove that Theorem 28.7.1 implies Theorem 28.3.2. (This is essentially immediate if q is a closed point, and might require a little thought if q is not closed; cf. Exercise 28.3.A.)

On the right side of (28.7.1.1), we have morphisms $X_1 \hookrightarrow X_2 \hookrightarrow \cdots \hookrightarrow X$, from which we get restriction maps $H^i(X_1, \mathcal{F}_1) \leftarrow H^i(X_2, \mathcal{F}_2) \leftarrow \cdots \leftarrow H^i(X, \mathcal{F})$. But notice that the X_i are all supported on the same underlying set. It is helpful to keep in mind that the $\mathcal{F}_i$'s are all sheaves on the same topological space.

Our proof will use properness only through the fact that the pushforwards of coherent sheaves under proper morphisms are coherent (Grothendieck's Coherence Theorem 18.9.1), which requires hard work. You can retreat to the projective case with little loss.

Write $I^n \mathcal{F}$ for the coherent sheaf on X defined as you might expect: if $\mathscr{I}$ is the pullback of $\tilde{I}$ from $\operatorname{Spec} B$ to X, $I^n \mathcal{F}$ is the image of $\mathscr{I}^n \otimes \mathcal{F} \to \mathcal{F}$. Then $H^i(X_n, \mathcal{F}_n)$ can be interpreted as $H^i(X, \mathcal{F}/I^n \mathcal{F})$ (by §18.1 property **(v)**), so the terms in the limit on the right side of (28.7.1.1) all live on the single space X. From now on we work on X.

28.7.2. *The idea.* The map (28.7.1.1) in the statement of the theorem arises from the restriction $H^i(X, \mathscr{F}) \to H^i(X, \mathscr{F}/I^n\mathscr{F})$. We are led to consider the long exact sequence associated to

$$0 \longrightarrow I^n\mathscr{F} \longrightarrow \mathscr{F} \longrightarrow \mathscr{F}/I^n\mathscr{F} \longrightarrow 0,$$

and, in particular, the portion

(28.7.2.1)

$$H^i(X, I^n\mathscr{F}) \xrightarrow{a_n} H^i(X, \mathscr{F}) \xrightarrow{b_n} H^i(X, \mathscr{F}/I^n\mathscr{F}) \xrightarrow{c_n} H^{i+1}(X, I^n\mathscr{F}) \xrightarrow{d_n} H^{i+1}(X, \mathscr{F}).$$

The map

(28.7.2.2)

$$H^i(X, \mathscr{F})/\operatorname{im}(a_n) \xrightarrow{\beta_n} H^i(X, \mathscr{F}/I^n\mathscr{F})$$

resembles the map (28.7.1.1) in the statement of the theorem, and we should hope that the limit (as $n \to \infty$) of (28.7.2.2),

(28.7.2.3)

$$\lim_{n\to\infty} H^i(X, \mathscr{F})/\operatorname{im}(a_n) \xrightarrow{\beta} \lim_{n\to\infty} H^i(X, \mathscr{F}/I^n\mathscr{F}),$$

should be (28.7.1.1). Indeed, the right sides are the same: $\lim_{n\to\infty} H^i(X, \mathscr{F}/I^n\mathscr{F}) = \lim_{n\to\infty} H^i(X_n, \mathscr{F}_n)$.

28.7.3. *The plan.* Thus to prove the result, we wish to show two things.

(i) We have an identification

$$\lim_{n\to\infty} H^i(X, \mathscr{F})/\operatorname{im}(a_n) = \lim_{n\to\infty} H^i(X, \mathscr{F}) \otimes_B \hat{B}.$$

(ii) The map β of (28.7.2.3) is an isomorphism.

28.7.4. *Step (i).* By Theorem 28.1.6(b), we have the isomorphism

$$\lim_{n\to\infty} H^i(X, \mathscr{F}) \otimes_B \hat{B} \xrightarrow{\sim} \lim_{n\to\infty} H^i(X, \mathscr{F})/I^n H^i(X, \mathscr{F}).$$

So we wish to show that

$$\lim_{n\to\infty} H^i(X, \mathscr{F})/\operatorname{im}(a_n) = \lim_{n\to\infty} H^i(X, \mathscr{F})/I^n H^i(X, \mathscr{F}).$$

To do this, we show that $(\operatorname{im}(a_n))$ and $(I^n H^i(X, \mathscr{F}))$ are cofinal sequences of subgroups of $H^i(X, \mathscr{F})$ (Exercise 28.1.A).

First, we must show that, for any m, there is some $N(m)$ such that $I^{N(m)} H^i(X, \mathscr{F}) \subset \operatorname{im}(a_m)$. This is straightforward, as we can take $N(m) = m$: for any $x \in I^m$,

$$
\begin{array}{ccc}
H^i(X, \mathscr{F}) & \xrightarrow{\;\times x\;} & H^i(X, \mathscr{F}) \\
 \searrow{\scriptstyle \times x} & & \nearrow{\scriptstyle a_m} \\
& H^i(X, I^m\mathscr{F}) &
\end{array}
$$

commutes.

Second, we must show that for any m, there is some $N(m)$ such that $\operatorname{im}(a_{N(m)}) \subset I^m H^i(X, \mathscr{F})$. For this, we use a devious construction. Recall the Rees algebra $B(I)_\bullet = B \oplus I \oplus I^2 \oplus \cdots$ (§13.9.1). Consider the Cartesian diagram

(28.7.4.1)

$$
\begin{array}{ccc}
X & \xleftarrow{\text{affine}} Spec_X(\mathscr{O}_X(I)_\bullet) = X' \;, \\
{\scriptstyle \text{proper}} \downarrow & \qquad \downarrow {\scriptstyle \text{proper}} \\
\operatorname{Spec} B & \xleftarrow{\text{affine}} \operatorname{Spec} B(I)_\bullet
\end{array}
$$

where we define $X' := \mathcal{S}pec_X(\mathcal{O}_X(I)_\bullet)$ for convenience. The quasicoherent sheaf $\mathscr{F} \oplus I\mathscr{F} \oplus I^2\mathscr{F} \oplus \cdots$ on X is a finitely generated sheaf of modules over $\mathcal{O}_X(I)_\bullet$, and thus (by Exercise 17.1.E) corresponds to a coherent sheaf $\mathscr{F}'$ on X'. As $X' \to X$ is affine (and property §18.1(**v**)), we have

$$(28.7.4.2) \qquad H^i(X', \mathscr{F}') = H^i(X, \mathscr{F} \oplus I\mathscr{F} \oplus I^2\mathscr{F} \oplus \cdots) = H^i(X, \mathscr{F}) \oplus H^i(X, I\mathscr{F}) \oplus \cdots.$$

By Grothendieck's Coherence Theorem 18.9.1, $H^i(X', \mathscr{F}')$ is a finitely generated $B(I)_\bullet$-module. By Proposition 13.9.2, $IH^i(X, I^k\mathscr{F}) = H^i(X, I^{k+1}\mathscr{F})$ for $k \geq k_0$ for some k_0. Then

$$H^i(X, I^{k_0+m}\mathscr{F}) = I^m H^i(X, I^{k_0}\mathscr{F}) \subset I^m H^i(X, \mathscr{F}),$$

so we may take $N(m) = k_0 + m$, completing Step (i).

28.7.5. *Step (ii).* From (28.7.2.1) we have a short exact sequence

$$(28.7.5.1) \qquad 0 \longrightarrow H^i(X, \mathscr{F})/\operatorname{im}(a_n) \longrightarrow H^i(X, \mathscr{F}/I^n\mathscr{F}) \longrightarrow \ker(d_n) \longrightarrow 0.$$

Apply $\lim_{n\to\infty}$ to both sides. Limits are left-exact (§1.5.14), so the result is left-exact. Even better, the transition maps of the left term for our exact sequences

$$H^i(X, \mathscr{F})/\operatorname{im}(a_{n+1}) \longrightarrow H^i(X, \mathscr{F})/\operatorname{im}(a_n)$$

are clearly surjective, so by Exercise 1.5.N we have an *exact* sequence
(28.7.5.2)

$$0 \longrightarrow \lim H^i(X, \mathscr{F})/\operatorname{im}(a_n) \xrightarrow{\ \beta\ } \lim H^i(X, \mathscr{F}/I^n\mathscr{F}) \longrightarrow \lim \ker(d_n) \longrightarrow 0.$$

To complete step (ii), it remains to show that $\lim \ker(d_n) = 0$. We use the construction of X' in (28.7.4.1) again. Note first that we have maps $I \times \ker(d_n) \to \ker(d_{n+1})$, so $Q := \oplus_n \ker(d_n)$ is a module over $B(I)_\bullet$. Moreover, Q is a submodule of the $B(I)_\bullet$-module $\oplus_n H^i(X, I^{n+1}\mathscr{F})$, which we showed was finitely generated (over the Noetherian ring $B(I)_\bullet$); see (28.7.4.2). Thus Q is also a finitely generated graded $B(I)_\bullet$-module. Choose a finite set of homogeneous generators of Q, and let N be the largest degree of a generator. Also, since $\ker d_n = \operatorname{im} c_n$ (from the exactness of (28.7.2.1)), we see that $\ker d_n$ is annihilated by I^n, because $\mathscr{F}/I^n\mathscr{F}$ is. Thus the ideal $I^N \oplus I^{N+1} \oplus \cdots \subset B(I)_\bullet$ annihilates Q.

Consider the composition

$$I^r \otimes \ker d_n \longrightarrow \ker d_{n+r} \longrightarrow \ker d_n,$$

where the first map is from multiplication, and the second is the usual map induced by $H^i(X, I^{n+r}\mathscr{F}) \to H^i(X, I^n\mathscr{F})$. The first map is surjective for $r \geq 0$ and $n \geq N$ (as $B(I)_\bullet$ is generated in degree 1, and Q a finitely generated $B(I)_\bullet$-module generated in degrees at most N), and the composition is zero for $r \geq n$. Thus map $\ker d_{n+r} \to \ker d_n$ is zero for $r, n \geq N$. In particular, $\ker d_{N+r} \to \ker d_N$ is zero for $r \geq N$, so $\lim \ker(d_n) = 0$ as desired. $\qquad \square$

Chapter 29*

Proof of Serre Duality

Réfléchissant un peu à ton théorème de dualité, je m'aperçois que sa formulation générale est à peu près évidente, et d'ailleurs je viens de vérifier qu'elle se trouve implicitement (dans le cas de l'espace projectif) dans ton théorème donnant les $T^q(M)$ par des Ext. (J'ai bien l'impression, salaud, que tes §3 et 4 du Chap. 3 peuvent se faire aussi sans aucun calcul).

Thinking a bit about your duality theorem, I notice that its general form is almost obvious, and in fact I just checked that (for a projective space) it is implicitly contained in your theorem giving the $T^q(M)$ in terms of Exts. (I have the impression, you bastard, that §3 and 4 in your Chap. 3 could be done without any computation).

—A. Grothendieck, letter to J.-P. Serre, December 15, 1955 [GrS, p. 19]

We first met Serre duality in §18.5 (or even Theorem 18.1.2), and we have repeatedly seen how useful it is. We will prove Theorem 18.5.1 (Corollary 29.3.10, combined with Exercise 29.4.K; see Remark 29.4.10), as well as stronger versions, and we will be left with a desire to prove even more. We give three statements (*Serre duality for vector bundles; Serre duality for* Hom; and *Serre duality for* Ext), in two versions (*functorial* and *trace*). (These names are idiosyncratic and nonstandard.) We give several variants for a number of reasons. First, the easier statements will be easier to prove, and the hardest statements we won't be able to prove here. Second, they may help give you experience in knowing how to know what to hope for, and what to try to prove.

29.1   Desiderata

Throughout this chapter, we work over a field k; X will be a projective k-scheme, and $n = \dim X$. We will want a coherent sheaf $\boxed{\omega}$ (or, with more precision, $\boxed{\omega_X}$, or, even better, $\omega_{X/k}$) on X, the **canonical sheaf** or **dualizing sheaf**, which will play a role in the statements of duality. For the best statements, we will want a **trace morphism**

$$(29.1.0.1) \qquad\qquad \boxed{t \colon H^n(X, \omega_X) \to k.}$$

We will finally define ω and t in §29.1.5, but we first discuss what properties we wish them to have.

29.1.1. Desideratum: The determinant of the cotangent bundle is dualizing for smooth varieties. If X is smooth, we will want $\boxed{\omega_X = \mathscr{K}_X}$ in this case (recall $\mathscr{K}_X = \det \Omega_X$)—the miracle that the canonical bundle is Serre-dualizing (§18.5.2). In particular, ω_X is an invertible sheaf. This will be disturbingly harder to prove than the basic duality statements we show; we will only get to it later (Exercise 29.4.K). But we will prove more, for example, that ω_X is an invertible sheaf if X is a regular embedding in a smooth variety (Exercise 29.4.J).

29.1.2. Desideratum: Serre duality for vector bundles. The first version of duality, which (along with Desideratum 29.1.1) gives Theorem 18.5.1, is the following: If $\mathscr{F}$ is locally free of finite rank, then we have a functorial isomorphism

$$(29.1.2.1) \qquad\qquad H^i(X, \mathscr{F}) \cong H^{n-i}(X, \mathscr{F}^\vee \otimes \omega_X)^\vee.$$

More precisely, we want to construct a *particular* isomorphism (29.1.2.1), or, equivalently, a particular perfect pairing $H^i(X, \mathscr{F}) \times H^{n-i}(X, \mathscr{F}^\vee \otimes \omega_X) \to k$. This isomorphism will be *functorial* *in $\mathscr{F}$*, i.e., it gives a natural isomorphism of covariant functors

(29.1.2.2)
$$\boxed{H^i(X, \cdot) \xrightarrow{\sim} H^{n-i}(X, \cdot^\vee \otimes \omega_X)^\vee.}$$

We call this **functorial Serre duality for vector bundles**.

Better still, there should be a cup product in cohomology, which can be used to construct a map $H^i(X, \mathscr{F}) \times H^{n-i}(X, \mathscr{F}^\vee \otimes \omega_X) \to H^n(X, \omega_X)$. This should be functorial in $\mathscr{F}$ (in the sense that we get a natural transformation of functors $H^i(X, \cdot) \to H^{n-i}(X, \cdot^\vee \otimes \omega_X)^\vee \otimes H^n(X, \omega_X)$). Combined with the trace map (29.1.0.1), we get a map

(29.1.2.3)
$$\boxed{\begin{array}{c} H^i(X, \mathscr{F}) \times H^{n-i}(X, \mathscr{F}^\vee \otimes \omega_X) \longrightarrow H^n(X, \mathscr{F} \otimes \mathscr{F}^\vee \otimes \omega_X) \\ \\ \downarrow \\ \\ H^n(X, \omega_X) \xrightarrow{\ t\ } k, \end{array}}$$

which, if it is a perfect pairing, should yield (29.1.2.2). If the composition (29.1.2.3) is a perfect pairing, we say that X (with the additional data of (ω_X, t)) satisfies the **trace version of Serre duality for vector bundles**. (This was hinted at after the statement of Theorem 18.5.1.)

In fact, the trace version of Serre duality for vector bundles (and hence the functorial version) is true when X is Cohen–Macaulay, and, in particular, when X is smooth. We will prove the functorial version (see Corollary 29.3.10). We will give an indication of how the trace version can be proved in Remark 29.3.15.

29.1.3. Desideratum: Duality for more general X and $\mathscr{F}$. A weaker sort of duality will hold with weaker hypotheses. We will show that (without Cohen–Macaulay hypotheses) for any coherent sheaf $\mathscr{F}$ on a pure n-dimensional projective k-scheme X, there is a *functorial isomorphism* (functorial in $\mathscr{F}$)

(29.1.3.1)
$$\boxed{\operatorname{Hom}(\mathscr{F}, \omega_X) \xrightarrow{\sim} H^n(X, \mathscr{F})^\vee.}$$

We call this **functorial Serre duality for** Hom.

In parallel with Serre duality for vector bundles, we have a natural candidate for the perfect pairing:

(29.1.3.2)
$$\boxed{\operatorname{Hom}(\mathscr{F}, \omega_X) \times H^n(X, \mathscr{F}) \longrightarrow H^n(X, \omega_X) \xrightarrow{\ t\ } k,}$$

where the trace map t is some linear functional on $H^n(X, \omega_X)$. If this composition is a perfect pairing, we say that X (with the additional data of (ω_X, t)) satisfies the **trace version of Serre duality for** Hom. Unlike the trace version of Serre duality for vector bundles, we already know what the "cup product" map $\operatorname{Hom}(\mathscr{F}, \omega_X) \times H^n(X, \mathscr{F}) \to H^n(X, \omega_X)$ is: an element $[\sigma \colon \mathscr{F} \to \omega_X]$ of $\operatorname{Hom}(\mathscr{F}, \omega_X)$ that induces—by covariance of $H^n(X, \cdot)$; see §18.1—a map $H^n(X, \mathscr{F}) \to H^n(X, \omega_X)$. The resulting pairing is clearly functorial in $\mathscr{F}$, so the trace version of Serre duality for Hom implies the functorial version. Unlike the case of Serre duality for vector bundles, the functorial version of Serre duality for Hom also implies the trace version:

29.1.A. EXERCISE. Show that the functorial version of Serre duality for Hom implies the trace version. In other words, the trace map is already implicit in the functorial isomorphism $\operatorname{Hom}(\mathscr{F}, \omega_X) \xrightarrow{\sim} H^n(X, \mathscr{F})^\vee$. Hint: Consider the commuting diagram (coming from functoriality)

$$\begin{array}{ccc} \operatorname{Hom}(\mathscr{F}, \omega_X) & \longrightarrow & H^n(X, \mathscr{F})^\vee \\ \uparrow & & \uparrow \\ \operatorname{Hom}(\omega_X, \omega_X) & \longrightarrow & H^n(X, \omega_X)^\vee. \end{array}$$

29.1.4. Getting down to work. We begin by defining some of our terms.

29.1.5. *Definition.* Suppose X is a projective k-scheme of dimension n. A coherent sheaf $\omega = \omega_X = \omega_{X/k}$ along with a map $t\colon H^n(X, \omega_X) \to k$ is called **dualizing** if the natural map (cf. (29.1.3.1))

$$(29.1.5.1) \qquad \operatorname{Hom}(\mathscr{F}, \omega_X) \times H^n(X, \mathscr{F}) \longrightarrow H^n(X, \omega_X) \overset{t}{\longrightarrow} k$$

is a perfect pairing for all coherent sheaves $\mathscr{F}$ on X. We call ω_X the **canonical sheaf** and t the **trace map**. (The earlier discussion of ω_X and t was aspirational. This now is a definition.) If X has such (ω_X, t), we say that X satisfies **Serre duality for** Hom.

29.1.6. *Aside on notation.* The canonical sheaf as defined here is often called the **dualizing sheaf**. This is not terrible terminology, but it is perhaps best to reserve this phrase for sheaves satisfying the best version of Serre duality, the trace version for Ext (§29.1.13), which is beyond the scope of this book. Similarly, "dualizing" in Definition 29.1.5 might better be called "weakly dualizing" for the sake of precision.

There is an apparent collision of terminology between "canonical sheaf" ω_X here, and "canonical bundle" $\mathscr{K}_X = \det \Omega_X$ of §21.5.3 for smooth varieties. But the point of §29.4 is to show that these notions are the same where they both apply.

29.1.7. The following proposition justifies the use of the word "the" (as opposed to "a") in the phrase "the canonical sheaf."

29.1.8. Proposition — *If a canonical sheaf and trace (ω_X, t) exists for X, this data is unique up to unique isomorphism.*

Proof. Suppose we have two such (ω_X, t) and (ω'_X, t'). From the two morphisms

$$(29.1.8.1) \qquad \operatorname{Hom}(\mathscr{F}, \omega_X) \times H^n(X, \mathscr{F}) \longrightarrow H^n(X, \omega_X) \overset{t}{\longrightarrow} k,$$
$$\operatorname{Hom}(\mathscr{F}, \omega'_X) \times H^n(X, \mathscr{F}) \longrightarrow H^n(X, \omega'_X) \overset{t'}{\longrightarrow} k,$$

we get a natural bijection $\operatorname{Hom}(\mathscr{F}, \omega_X) \cong \operatorname{Hom}(\mathscr{F}, \omega'_X)$, which is functorial in $\mathscr{F}$. By the typical universal property argument (Exercise 1.2.Z), this induces a (unique) isomorphism $\omega_X \overset{\sim}{\longleftrightarrow} \omega'_X$. From (29.1.8.1), under this isomorphism, the two trace maps t and t' must be the same too. $\square$

We will prove the functorial and trace versions of Serre duality for Hom in Corollary 29.3.12. The special case of projective space will be a key ingredient. We prove this case now.

29.1.9. *Serre duality (for* Hom*) for projective space.* Define $\omega_{\mathbb{P}^n_k}$ (or just ω for convenience) as $\mathscr{O}_{\mathbb{P}^n_k}(-n-1)$. Let t be any isomorphism $H^n(\mathbb{P}^n_k, \omega_{\mathbb{P}^n_k}) \overset{\sim}{\longrightarrow} k$ (Theorem 18.1.2(b)). As the notation suggests, $(\omega_{\mathbb{P}^n_k}, t)$ will be dualizing for projective space $\mathbb{P}^n_k$.

29.1.B. EXERCISE. Suppose $\mathscr{F} = \mathscr{O}_{\mathbb{P}^n_k}(m)$. Show that the natural map (29.1.5.1) is a perfect pairing. (Hint: Do this by hand. See the discussion after Theorem 18.1.2.) Hence show that if $\mathscr{F}$ is a direct sum of line bundles on $\mathbb{P}^n_k$, the natural map (29.1.5.1) is a perfect pairing.

29.1.10. Proposition — *The functorial version (and hence the trace version by Exercise 29.1.A) of Serre duality for* Hom *holds for $\mathbb{P}^n_k$.*

Proof. We wish to show that the (functorial) natural map

$$\operatorname{Hom}(\mathscr{F}, \omega_{\mathbb{P}^n_k}) \overset{\sim}{\longrightarrow} H^n(\mathbb{P}^n_k, \mathscr{F})^{\vee},$$

coming from the cup pairing and the choice of isomorphism t, is an isomorphism (cf. (29.1.3.1)). Fix a coherent sheaf $\mathscr{F}$ on $\mathbb{P}^n_k$. By Theorem 16.1.1, we can present $\mathscr{F}$ as

$$(29.1.10.1) \qquad 0 \longrightarrow \mathscr{G} \longrightarrow \mathscr{E} \longrightarrow \mathscr{F} \longrightarrow 0,$$

where $\mathcal{E}$ is a finite direct sum of line bundles, and $\mathcal{G}$ is coherent. Applying the left-exact functor $\mathrm{Hom}(\cdot, \omega_{\mathbb{P}^n_k})$ to (29.1.10.1), we have the exact sequence

(29.1.10.2) $\qquad 0 \longrightarrow \mathrm{Hom}(\mathcal{F}, \omega_{\mathbb{P}^n_k}) \longrightarrow \mathrm{Hom}(\mathcal{E}, \omega_{\mathbb{P}^n_k}) \longrightarrow \mathrm{Hom}(\mathcal{G}, \omega_{\mathbb{P}^n_k}).$

Taking the long exact sequence in cohomology for (29.1.10.1) and dualizing, we have the exact sequence

(29.1.10.3) $\qquad 0 \longrightarrow \mathrm{H}^n(\mathbb{P}^n_k, \mathcal{F})^\vee \longrightarrow \mathrm{H}^n(\mathbb{P}^n_k, \mathcal{E})^\vee \longrightarrow \mathrm{H}^n(\mathbb{P}^n_k, \mathcal{G})^\vee.$

The (functorial) pairing (29.1.3.2) gives a map from (29.1.10.2) to (29.1.10.3):

(29.1.10.4)

$$
\begin{array}{ccccccccc}
0 & \longrightarrow & 0 & \longrightarrow & \mathrm{H}^n(\mathbb{P}^n_k, \mathcal{F})^\vee & \longrightarrow & \mathrm{H}^n(\mathbb{P}^n_k, \mathcal{E})^\vee & \longrightarrow & \mathrm{H}^n(\mathbb{P}^n_k, \mathcal{G})^\vee \\
& {\scriptstyle\alpha}\big\uparrow & & {\scriptstyle\beta}\big\uparrow & & {\scriptstyle\gamma}\big\uparrow & & {\scriptstyle\delta}\big\uparrow & & {\scriptstyle\epsilon}\big\uparrow \\
0 & \longrightarrow & 0 & \longrightarrow & \mathrm{Hom}(\mathcal{F}, \omega_{\mathbb{P}^n_k}) & \longrightarrow & \mathrm{Hom}(\mathcal{E}, \omega_{\mathbb{P}^n_k}) & \longrightarrow & \mathrm{Hom}(\mathcal{G}, \omega_{\mathbb{P}^n_k}).
\end{array}
$$

Maps α and β are obviously isomorphisms, and Exercise 29.1.B shows that δ is an isomorphism. Thus by the subtle version of the Five Lemma (Exercise 1.6.D, as β and δ are injective and α is surjective), γ is injective. This shows that the natural map $\mathrm{Hom}(\mathcal{F}', \omega_{\mathbb{P}^n_k}) \to \mathrm{H}^n(\mathbb{P}^n_k, \mathcal{F}')^\vee$ is injective for *all* coherent sheaves $\mathcal{F}'$, and, in particular, for $\mathcal{F}' = \mathcal{G}$. Thus ϵ is injective. Then by the dual of the subtle version of the Five Lemma (as β and δ are surjective, and ϵ is injective), γ is surjective. $\qquad\square$

29.1.11. *Mathematical puzzle.* Here is a puzzle to force you to confront a potentially confusing point. We will see that Desideratum 29.1.1 holds for $\mathbb{P}^1$, so $\omega_{\mathbb{P}^1} \cong \Omega_{\mathbb{P}^1}$. What then is the trace map $\mathrm{t}: \mathrm{H}^1(\mathbb{P}^1, \Omega_{\mathbb{P}^1}) \to k$? The Čech complex for $\mathrm{H}^1(\mathbb{P}^1, \Omega_{\mathbb{P}^1})$ (with the usual cover of $\mathbb{P}^1$) is given by

(29.1.11.1) $\qquad 0 \longrightarrow \Omega_{\mathbb{P}^1}(U_0) \times \Omega_{\mathbb{P}^1}(U_1) \xrightarrow{\ \alpha\ } \Omega_{\mathbb{P}^1}(U_0 \cap U_1) \longrightarrow 0.$

If $U_0 = \mathrm{Spec}\, k[x]$, and $U_0 \cap U_1 = \mathrm{Spec}\, k[x, 1/x]$, then the differentials on $U_0 \cap U_1$ are those of the form $f(x)dx$, where $f(x)$ is a Laurent polynomial (for example: $(x^{-3} + x^{-1} + 3 + 17x^4)\, dx$). To compute $\mathrm{H}^1(\mathbb{P}^1, \Omega_{\mathbb{P}^1})$, we need to find which such differentials on $U_0 \cap U_1$ are in the image of α in (29.1.11.1). Clearly, any term of the form $x^i dx$ (for $i \geq 0$) extends to a differential on U_0, and thus is in the image of α. A short calculation shows that any term of the form $x^i dx$ ($i < -1$) extends to a differential on U_1. Thus the cokernel of α can be described as the one-dimensional k-vector space generated by $x^{-1} dx$. We have an obvious isomorphism to k: take the coefficient of $x^{-1}dx$, which can be interpreted as "take the residue at 0." But there is another choice, which is equally good: take the residue at ∞—certainly there is no reason to privilege 0 over ∞ (or U_0 over U_1)! But these two residues are *not* the same—they add to 0 (as you can quickly calculate—you may also believe it because of the Residue Theorem in the theory of Riemann surfaces). So: which one is the trace?

29.1.12. Desideratum: A stronger version of duality, involving Ext.

Ta formule:

$$\mathrm{H}^{n-p}(X, \mathcal{F})' = \mathrm{Ext}^p_\mathcal{O}(X, \mathrm{F}, \Omega^n)$$

m'excite beaucoup, car je suis bien convaincu que c'est la bonne façon d'énoncer le théorème de dualité à la fois dans le cas analytique et dans le cas algébrique (le plus important—pour moi!).

I find your formula

$$H^{n-p}(X, \mathscr{F})' = \mathrm{Ext}_{\mathscr{O}}^{p}(X, F, \Omega^{n})$$

very exciting, as I am quite convinced that it is the right way to state the duality theorem in both the analytic case and the algebraic case (the more important one—for me!).

—J.-P. Serre, letter to A. Grothendieck, December 22, 1955 [GrS, p. 21]

The vector bundle and Hom versions of Serre duality have a common extension. For this, we need the notion of Ext for $\mathscr{O}$-modules. We will develop what we need in §29.2, but for now, it suffices to observe that if X is a ringed space, then the category of $\mathscr{O}_X$-modules has enough injectives (Theorem 23.4.1), so we can define $\mathrm{Ext}^i(\mathscr{F}, \cdot)$ as the derived functor of $\mathrm{Hom}(\mathscr{F}, \cdot)$ (like Ext-functors for A-modules, defined in §23.2.4).

If we have an isomorphism of functors

(29.1.12.1)

$$\boxed{\mathrm{Ext}^i(\cdot, \omega_X) \xrightarrow{\ \sim\ } H^{n-i}(X, \cdot)^{\vee},}$$

we say that X satisfies **functorial Serre duality for** Ext. (The case $i = 0$ is functorial Serre duality for Hom, or, by Exercise 29.1.A, the trace version.)

In Exercise 29.2.K, we will find that the functorial version of Serre duality for Ext (resp., the trace version) implies the functorial version of Serre duality for vector bundles (resp., the trace version). Thus to prove Theorem 18.5.1, it suffices to prove functorial Serre duality for Ext, and Desideratum 29.1.1 (see Remark 29.4.10). We will prove the functorial version of Serre duality for Ext when X is Cohen–Macaulay in Corollary 29.3.14.

29.1.13. *Trace version of Serre duality for* Ext. As with the previous versions, the functoriality of functorial Serre duality for Ext should come from somewhere. There is the cup product

$$\mathrm{Ext}^a(\mathscr{F}, \mathscr{G}) \times \mathrm{Ext}^b(\mathscr{E}, \mathscr{F}) \to \mathrm{Ext}^{a+b}(\mathscr{E}, \mathscr{F})$$

(the "Yoneda cup product"). If the composition

$$\boxed{\begin{array}{l} \mathrm{Ext}_X^i(\mathscr{F}, \omega_X) \times H^{n-i}(X, \mathscr{F}) = \mathrm{Ext}_X^i(\mathscr{F}, \omega_X) \times \mathrm{Ext}_X^{n-i}(\mathscr{O}_X, \mathscr{F}) \xrightarrow{\ \cup\ } \\ \mathrm{Ext}_X^n(\mathscr{O}_X, \omega_X) = H^n(X, \omega_X) \xrightarrow{\quad t \quad} k \end{array}}$$

is a perfect pairing for all i, we say that X satisfies the **trace version of Serre duality for** Ext. We will not be able to prove the trace version, as we will not define this cup product. But all pure dimensional proper Cohen–Macaulay k-schemes satisfy this strong form of duality, [Stacks, tag 0FVU].

29.1.14. Necessity of Cohen–Macaulay hypotheses. We remark that the Cohen–Macaulay hypotheses are necessary everywhere they are stated. The following example applies in all cases. Let X be the union of two 2-planes in $\mathbb{P}_k^4$, meeting at a point p. If there were a coherent canonical sheaf ω_X on X, then for $d \gg 0$, we would have $h^1(X, \mathscr{O}_X(-d)) = h^1(X, \omega_X(d))$ by Serre duality, which must be 0 by Serre vanishing (Theorem 18.1.3(ii)). We show that this is not the case.

29.1.C. EXERCISE. Let H be a hyperplane in $\mathbb{P}^4$ not passing through p; say, H is hyperplane $x_0 = 0$. Let Z be the intersection of dH (the divisor $x_0^d = 0$) with X. Show (cheaply) that for $d \geq 0$, $h^0(Z, \mathscr{O}_Z) > 1$. Use the exact sequence

$$0 \longrightarrow \mathscr{O}_X(-Z) \longrightarrow \mathscr{O}_X \longrightarrow \mathscr{O}_Z \longrightarrow 0$$

to show that $h^1(X, \mathscr{O}_X(-d)) > 0$.

29.2   Ext Groups and Ext Sheaves for $\mathcal{O}$-Modules

Recall that for any ringed space X, the category $Mod_{\mathcal{O}_X}$ has enough injectives (Theorem 23.4.1). Thus for any $\mathcal{O}_X$-module $\mathscr{F}$ on X, we may define

$$\mathrm{Ext}^i_X(\mathscr{F},\cdot)\colon Mod_{\mathcal{O}_X} \longrightarrow Mod_{\Gamma(X,\mathcal{O}_X)}$$

as the ith right derived functor of $\mathrm{Hom}_X(\mathscr{F},\cdot)$, and we have a corresponding long exact sequence for $\mathrm{Ext}^i_X(\mathscr{F},\cdot)$. We similarly define the sheaf version

$$\mathcal{E}xt^i_X(\mathscr{F},\cdot)\colon Mod_{\mathcal{O}_X} \longrightarrow Mod_{\mathcal{O}_X}$$

as the ith right derived functor of $\mathcal{H}om_X(\mathscr{F},\cdot)$. In both cases, the subscript X is often omitted when it is clear from the context, although this can be dangerous when more than one space is relevant to the discussion. (We saw Ext functors for A-modules in §23.2.4.)

    Warning: It is not clear (and in fact not true; see §23.4.8) that $Mod_{\mathcal{O}_X}$ has enough projectives, so we cannot define Ext^i as a derived functor in its left argument. Nonetheless, we will see that it behaves as though it *is* a derived functor—it is "computable by acyclics," and has a long exact sequence (Remark 29.2.1).

    Another warning: With this definition, it is not clear that if $\mathscr{F}$ and $\mathscr{G}$ are quasicoherent sheaves on a scheme, then the $\mathcal{E}xt^i(\mathscr{F},\mathscr{G})$ are quasicoherent, and indeed the aside in Exercise 14.3.A(a) points out this need not be true even for $i=0$. But Exercise 29.2.F will reassure you that all is well if $\mathscr{F}$ is coherent, and X is locally Noetherian.

    Exercise 23.5.A (an injective $\mathcal{O}_X$-module, when restricted to an open subset $U \subset X$, is injective on U) has a number of useful consequences.

29.2.A. EXERCISE. Suppose $\mathscr{I}$ is an injective $\mathcal{O}_X$-module. Show that $\mathcal{H}om_{\mathcal{O}_X}(\cdot,\mathscr{I})$ is an exact contravariant functor. (A related fact: $\mathrm{Hom}_{\mathcal{O}_X}(\cdot,\mathscr{I})$ is exact, by the definition of injectivity, Exercise 23.2.C(a).)

29.2.B. EXERCISE. Suppose X is a ringed space, $\mathscr{F}$ and $\mathscr{G}$ are $\mathcal{O}_X$-modules, and U is an open subset. Describe a canonical isomorphism $\mathcal{E}xt^i_X(\mathscr{F},\mathscr{G})|_U \overset{\sim}{\longleftrightarrow} \mathcal{E}xt^i_U(\mathscr{F}|_U,\mathscr{G}|_U)$. Hint: The restriction $Mod_{\mathcal{O}_X} \to Mod_{\mathcal{O}_U}$ is exact (cf. Exercises 2.7.D and 2.7.E); restriction preserves injectivity of $\mathcal{O}$-modules (Exercise 23.5.A); and the FHHF Theorem (Exercise 1.5.I).

29.2.C. EXERCISE. Suppose X is a ringed space, and $\mathscr{G}$ is an $\mathcal{O}_X$-module.

(a) Show that

$$\mathcal{E}xt^i(\mathcal{O}_X,\mathscr{G}) = \begin{cases} \mathscr{G} & \text{if } i=0\text{, and} \\ 0 & \text{otherwise.} \end{cases}$$

(b) Describe a canonical isomorphism $\mathrm{Ext}^i_X(\mathcal{O}_X,\mathscr{G}) \overset{\sim}{\longleftrightarrow} H^i(X,\mathscr{G})$.

29.2.D. EXERCISE. Use Exercises 29.2.B and 29.2.C(a) to show that if $\mathscr{E}$ is a locally free sheaf on X, then $\mathcal{E}xt^i(\mathscr{E},\mathscr{G}) = 0$ for $i > 0$.

    In the category of modules over a ring, we like projectives more than injectives, because free modules are easy to work with. It would be wonderful if locally free sheaves on schemes were always projective, but sadly this is not true. For if $\mathcal{O}_X$ were projective, then $\mathrm{Ext}^i(\mathcal{O}_X,\cdot)$ would be zero (as a functor) for $i > 0$, but by Exercise 29.2.C(b), it is the functor $H^i(X,\cdot)$, which is often nonzero. Nonetheless, we can still compute $\mathcal{E}xt$ using a locally free resolution, as shown by the following exercise.

29.2.E. IMPORTANT EXERCISE. Suppose X is a ringed space, and

$$(29.2.0.1) \qquad\qquad \cdots \longrightarrow \mathscr{E}_1 \longrightarrow \mathscr{E}_0 \longrightarrow \mathscr{F} \longrightarrow 0$$

is a resolution of an $\mathscr{O}_X$-module $\mathscr{F}$ by locally free sheaves. (Of course, we are most interested in the case where X is a scheme, and $\mathscr{F}$ is quasicoherent, or even coherent.) Let $\mathscr{E}_\bullet$ denote the truncation of (29.2.0.1), where $\mathscr{F}$ is removed. Describe an isomorphism $\mathcal{E}xt^i(\mathscr{F},\mathscr{G}) \overset{\sim}{\longleftrightarrow} H^i(\mathcal{H}om(\mathscr{E}_\bullet,\mathscr{G}))$. In other words, $\mathcal{E}xt^\bullet(\mathscr{F},\mathscr{G}))$ can be computed by taking a locally free resolution of $\mathscr{F}$, truncating, applying $\mathcal{H}om(\cdot,\mathscr{G})$, and taking homology. Hint: Choose an injective resolution

$$0 \longrightarrow \mathscr{G} \longrightarrow \mathscr{I}_0 \longrightarrow \mathscr{I}_1 \longrightarrow \cdots$$

and consider the spectral sequence whose E_0 term is

$$
\begin{array}{ccc}
\vdots & & \vdots \\
\uparrow & & \uparrow \\
\mathcal{H}om(\mathscr{E}_0,\mathscr{I}_1) & \longrightarrow & \mathcal{H}om(\mathscr{E}_1,\mathscr{I}_1) \longrightarrow \cdots \\
\uparrow & & \uparrow \\
\mathcal{H}om(\mathscr{E}_0,\mathscr{I}_0) & \longrightarrow & \mathcal{H}om(\mathscr{E}_1,\mathscr{I}_0) \longrightarrow \cdots .
\end{array}
$$

This result is important: to compute $\mathcal{E}xt$, we can compute it using finite rank locally free resolutions. You can work affine by affine (by Exercise 29.2.B), and on each affine you can use a free resolution of the left argument. As another consequence of Exercise 29.2.E:

29.2.F. EXERCISE.

(a) Suppose $\mathscr{F}$ is a coherent sheaf and $\mathscr{G}$ is a quasicoherent sheaf on a locally Noetherian scheme X. Show that $\mathcal{E}xt_X^i(\mathscr{F},\mathscr{G})$ is a quasicoherent sheaf on X. Hint: By Exercise 29.2.B, because $\mathcal{E}xt$ can be computed on an open cover, show that it suffices to show the result for X affine. But in that case, $\mathscr{F}$ has a *free* resolution. (You will discover that $\mathcal{E}xt_X^i(\mathscr{F},\mathscr{G})(\operatorname{Spec} A) = \operatorname{Ext}_A^i(\mathscr{F}(\operatorname{Spec} A), \mathscr{G}(\operatorname{Spec} A)).$)

(b) If, furthermore, $\mathscr{G}$ is coherent, show that $\mathcal{E}xt_X^i(\mathscr{F},\mathscr{G})$ is also coherent.

29.2.1. Remark. The statement "$\operatorname{Ext}^i(\mathscr{F},\mathscr{G})$ and $\mathcal{E}xt^i(\mathscr{F},\mathscr{G})$ *behave like a derived functor in the first argument*" is true in a number of ways. For example, we have a corresponding long exact sequence, as shown in the next exercise.

29.2.G. EXERCISE. Suppose $0 \to \mathscr{F}' \to \mathscr{F} \to \mathscr{F}'' \to 0$ is an exact sequence of $\mathscr{O}_X$-modules on a ringed space X. For any $\mathscr{O}_X$-module $\mathscr{G}$, describe a long exact sequence

$$0 \longrightarrow \operatorname{Hom}(\mathscr{F}'',\mathscr{G}) \longrightarrow \operatorname{Hom}(\mathscr{F},\mathscr{G}) \longrightarrow \operatorname{Hom}(\mathscr{F}',\mathscr{G})$$

$$\longrightarrow \operatorname{Ext}^1(\mathscr{F}'',\mathscr{G}) \longrightarrow \operatorname{Ext}^1(\mathscr{F},\mathscr{G}) \longrightarrow \operatorname{Ext}^1(\mathscr{F}',\mathscr{G}) \longrightarrow \cdots .$$

(Thus $\operatorname{Ext}^i(\cdot,\mathscr{G})$ forms a contravariant δ-functor, as defined in §23.2.7, despite not being defined as a derived functor.) Prove the analogous results for $\mathcal{E}xt$. Hint: Take an injective resolution $0 \to \mathscr{G} \to \mathscr{I}^0 \to \cdots$. Use the fact that if $\mathscr{I}$ is injective, then $\operatorname{Hom}(\cdot,\mathscr{I})$ is exact (the definition of injectivity, Exercise 23.2.C(a)). Hence get a short exact sequence of complexes

$$0 \longrightarrow \operatorname{Hom}(\mathscr{F}'',\mathscr{I}^\bullet) \longrightarrow \operatorname{Hom}(\mathscr{F},\mathscr{I}^\bullet) \longrightarrow \operatorname{Hom}(\mathscr{F}',\mathscr{I}^\bullet) \longrightarrow 0$$

and take the long exact sequence in cohomology.

29.2.2. Functorial Serre duality for Ext holds for projective space.

We now prove that functorial Serre duality for Ext holds for projective space. We will use the machinery of universal δ-functors (§23.2.6), so you may wish to either quickly skim that section, or else ignore this discussion.

29.2.H. EXERCISE. Show that $(\operatorname{Ext}_{\mathbb{P}_k^n}^i(\cdot, \omega_{\mathbb{P}_k^n}))$ is a (contravariant) universal δ-functor. Hint: Ext is not a derived functor in its first argument, so you can't use the "projective" version of Corollary 23.2.11. Instead, use Theorem 23.2.9, and the existence of a surjection $\mathcal{O}(m)^{\oplus N} \to \mathcal{F}$ for each $\mathcal{F}$, for some $m < 0$.

29.2.I. EXERCISE. Show that $(H^{n-i}(\mathbb{P}_k^n, \cdot)^\vee)$ is a universal δ-functor. (What are the δ-maps?) Hint: Try the same idea as in the previous exercise.

Proposition 29.1.10 gives an isomorphism of functors $\operatorname{Ext}_{\mathbb{P}_k^n}^0(\cdot, \omega_{\mathbb{P}_k^n}) \xrightarrow{\sim} H^n(\mathbb{P}_k^n, \cdot)^\vee$, so by the Definition 23.2.8 of universal δ-functor, we have an isomorphism of δ-functors $(\operatorname{Ext}_{\mathbb{P}_k^n}^i(\cdot, \omega_{\mathbb{P}_k^n})) \xrightarrow{\sim} (H^{n-i}(\mathbb{P}_k^n, \cdot)^\vee)$, thereby proving functorial Serre duality for Ext for $\mathbb{P}_k^n$.

29.2.3. Two useful exercises, and functorial Serre duality for vector bundles for projective space.

29.2.J. EXERCISE. Suppose X is a ringed space, $\mathcal{F}$ and $\mathcal{G}$ are $\mathcal{O}_X$-modules, and $\mathcal{E}$ is a locally free sheaf on X. Describe isomorphisms

$$\mathcal{E}xt^i(\mathcal{F} \otimes \mathcal{E}^\vee, \mathcal{G}) \xrightarrow{\sim} \mathcal{E}xt^i(\mathcal{F}, \mathcal{G} \otimes \mathcal{E}) \xrightarrow{\sim} \mathcal{E}xt^i(\mathcal{F}, \mathcal{G}) \otimes \mathcal{E}$$

and $\quad \operatorname{Ext}^i(\mathcal{F} \otimes \mathcal{E}^\vee, \mathcal{G}) \xrightarrow{\sim} \operatorname{Ext}^i(\mathcal{F}, \mathcal{G} \otimes \mathcal{E}).$

Hint: Show that if $\mathcal{I}$ is injective then $\mathcal{I} \otimes \mathcal{E}$ is injective.

29.2.K. EXERCISE. If $\mathcal{F}$ is a locally free sheaf on a scheme X, and $\mathcal{G}$ is any coherent sheaf on X, describe an isomorphism $\operatorname{Ext}^i(\mathcal{F}, \mathcal{G}) \xrightarrow{\sim} H^i(X, \mathcal{F}^\vee \otimes \mathcal{G})$ (functorial in $\mathcal{F}$). By taking $\mathcal{G}$ to be ω_X, show that functorial Serre duality for Ext implies functorial Serre duality for vector bundles. (You may wish to ponder the trace versions as well.) As a consequence, by §29.2.2, functorial Serre duality for vector bundles holds for projective space. Hint: Exercises 29.2.J and 29.2.C(b).

29.2.4. The local-to-global spectral sequence for Ext.

"Sheaf" $\mathcal{E}xt$ and "global" Ext are related by a spectral sequence. This is a direct application of the Grothendieck composition-of-functors spectral sequence (Theorem 23.3.5), once we show that $\mathcal{H}om(\mathcal{F}, \mathcal{I})$ is acyclic for the functor Γ.

29.2.L. EXERCISE. Suppose $\mathcal{I}$ is an injective $\mathcal{O}_X$-module. Show that $\mathcal{H}om(\mathcal{F}, \mathcal{I})$ is *flasque* (Definition 23.4.6), and thus Γ-acyclic by Exercise 23.4.I. Hint: Suppose $j \colon U \hookrightarrow V$ is an inclusion of open subsets. We wish to show that $\mathcal{H}om(\mathcal{F}, \mathcal{I})(V) \to \mathcal{H}om(\mathcal{F}, \mathcal{I})(U)$ is surjective. Note that $\mathcal{I}|_V$ is injective on V (Exercise 23.5.A). Apply the exact functor $\operatorname{Hom}_V(\cdot, \mathcal{I}|_V)$ to the inclusion $j_!(\mathcal{F}|_U) \hookrightarrow \mathcal{F}|_V$ of sheaves on V (§23.4.7).

29.2.M. EXERCISE (THE LOCAL-TO-GLOBAL SPECTRAL SEQUENCE FOR Ext). Suppose X is a ringed space, and $\mathcal{F}$ and $\mathcal{G}$ are $\mathcal{O}_X$-modules. Describe a spectral sequence with E_2-term $H^i(X, \mathcal{E}xt^j(\mathcal{F}, \mathcal{G}))$ abutting to $\operatorname{Ext}^{i+j}(\mathcal{F}, \mathcal{G})$. (Hint: Use the Grothendieck composition-of-functors spectral sequence, Theorem 23.3.5. Recall that $\operatorname{Hom}(\mathcal{F}, \cdot) = \Gamma(\mathcal{H}om(\mathcal{F}, \cdot))$, Exercise 2.3.C.)

29.3 Serre Duality for Projective k-Schemes

We now prove various versions of Serre duality for projective k-schemes, by leveraging Serre duality for projective space.

29.3.1. Recalling the weird functor $\pi^{!?}$.

The key construction is a *right adjoint* $\pi^{!?}$ to the pushforward π_*, when π is a finite morphism, discussed in §17.1.6. You should read this section if you have not done so already.

29.3.2. *Caution.* We have defined $\pi^{!?}$ only for quasicoherent sheaves on Y, not for all $\mathscr{O}_Y$-modules. If $\mathscr{G}$ is an $\mathscr{O}_X$-module (not necessarily quasicoherent), it is in general *not* clear how to make sense of this construction to define an $\mathscr{O}_Y$-module $\pi^{!?}\mathscr{G}$. (Try it and see!) However, in the special case where π is a closed embedding, we *can* make sense of $\pi^{!?}\mathscr{G}$, as described in the Exercise 29.4.A.

29.3.3. *Remark.* If π is finite and flat, which is the case most of interest to us, $\pi^{!?}$ agrees (in the only possible sense of the word) with $\pi^!$, one of Grothendieck's "six operations," and indeed this motivates our notation ("a strange variant of $\pi^!$"). But $\pi^!$ naturally lives in the world of derived categories, so we will not discuss it here. (For more on $\pi^!$, see [KS1, Ch. III].)

We now apply the machinery of $\pi^{!?}$ to Serre duality.

29.3.4. Projective n-dimensional schemes are finite covers of $\mathbb{P}^n$.

29.3.5. Proposition — *Suppose X is a projective k-scheme of dimension at most n.*

(a) *There exists a finite morphism $\pi\colon X \to \mathbb{P}^n$.*
(b) *If furthermore X is Cohen–Macaulay of pure dimension n, then π is flat.*

29.3.6. *Remark for experts.* We carefully allow n to be greater than dim X here, not because it is useful, but to point out that the proof works, and we are proving something strange if we don't have the right hypotheses. Proposition 29.3.5(a) is still interesting if dim X = n but X is not Cohen–Macaulay. What does Proposition 29.3.5(a) imply if dim X < n?

Proof. Part (b) follows from part (a) by the Miracle Flatness Theorem 26.2.11, so we will prove part (a).

Choose a closed embedding $j\colon X \hookrightarrow \mathbb{P}^N$. For simplicity of exposition, first assume that k is an infinite field. By Exercise 12.3.D(d), there is a linear space L of codimension n + 1 (1 less than the complementary dimension) disjoint from X. Projection from L yields a morphism $\pi\colon X \to \mathbb{P}^n$. The morphism π is affine ($\mathbb{P}^N \setminus L \to \mathbb{P}^n$ is affine, and the closed embedding $X \hookrightarrow \mathbb{P}^N \setminus L$ is, like all closed embeddings, affine) and projective (by the Cancellation Theorem for projective morphisms, Exercise 17.3.C; or more directly by Exercise 17.3.D), so π is finite (projective affine morphisms of locally Noetherian schemes are finite, Corollary 18.1.5).

29.3.A. EXERCISE. Prove Proposition 29.3.5(a), without the assumption that k is infinite. Hint: Using Exercise 12.3.D(c), show that there is some d such that there is an intersection of n + 1 degree d hypersurfaces missing X. Then apply the above argument to the dth Veronese embedding of $\mathbb{P}^N$ (§9.3.6). $\qquad\square$

29.3.7. *Remark.* More generally, the above argument shows that any projective k-scheme of dimension n (not necessarily pure dimensional) can be expressed as a finite cover of $\mathbb{P}^n$. This might be seen as a projective version of Noether normalization.

29.3.8. Serre duality on X via $\pi^{!?}$.

Suppose $\pi\colon X \to Y$ is a finite morphism of projective k-schemes of pure dimension n. (We will soon apply this in the case where $Y = \mathbb{P}_k^n$, but we may as well avoid distraction and needless specificity.)

29.3.9. Proposition — *Suppose π is flat. If functorial Serre duality for vector bundles holds for Y, with canonical sheaf ω_Y, then functorial Serre duality for vector bundles holds for X, with canonical sheaf $\pi^{!?}(\omega_Y)$.*

Proof. For each i, and each finite rank locally free sheaf $\mathscr{F}$ on X, we have isomorphisms (functorial in $\mathscr{F}$):

$$H^{n-i}(X, \mathscr{F}^\vee \otimes \pi^{!?}\omega_Y) \cong H^{n-i}(Y, \pi_*(\mathscr{F}^\vee \otimes \pi^{!?}\omega_Y)) \quad \text{(affineness of } \pi, \text{§18.1 (v))}$$

$$\cong H^{n-i}(Y, \pi_*(\mathscr{H}om_X(\mathscr{F}, \pi^{!?}\omega_Y))) \quad \text{(Exercise 14.1.G)}$$

$$\cong H^{n-i}(Y, \mathscr{H}om_Y(\pi_*\mathscr{F}, \omega_Y)) \quad \text{(equ. (17.1.8.2))}$$

$$\cong H^{n-i}(Y, (\pi_* \mathscr{F})^\vee \otimes \omega_Y) \quad \text{(Exercises 24.3.E and 14.1.G)}$$

$$\cong H^i(Y, \pi_* \mathscr{F})^\vee \quad \text{(Serre duality for vector bundles on Y)}$$

$$\cong H^i(X, \mathscr{F})^\vee \quad \text{(affineness of } \pi, \text{§18.1 (v))}.$$

(The hypothesis of flatness is used to show that $\pi_* \mathscr{F}$ is locally free.) $\qquad\square$

29.3.10. Corollary — *Functorial Serre duality for vector bundles holds for every Cohen–Macaulay pure dimensional projective k-scheme.*

Proof. Combine Proposition 29.3.9 with Proposition 29.3.5(b) and §29.2.2. $\qquad\square$

29.3.11. Proposition — *Suppose $\pi\colon X \to Y$ is a finite morphism of projective k-schemes. If Serre duality for Hom holds for Y with canonical sheaf (ω_Y, t_Y), then Serre duality for Hom holds for X with canonical sheaf $\pi^{!?} \omega_Y$.*

Recall from Exercise 29.1.A that for Serre duality for Hom, the trace version is equivalent to the functorial version. For this reason we just use the phrase "Serre duality for Hom."

Note that we have no flatness hypotheses on π (unlike the corresponding propositions for Serre duality for vector bundles and, soon, for Ext).

Proof. We have functorial isomorphisms

$$\operatorname{Hom}_X(\mathscr{F}, \pi^{!?}\omega_Y) \cong \operatorname{Hom}_Y(\pi_*\mathscr{F}, \omega_Y) \quad \text{(adjointness of } (\pi_*, \pi^{!?}))$$

$$\cong H^n(Y, \pi_*\mathscr{F})^\vee \quad \text{(Serre duality for Hom for Y)}$$

$$\cong H^n(X, \mathscr{F})^\vee \quad \text{(affineness of } \pi). \qquad\square$$

You can unwind the argument to obtain the trace map t_X for X in terms of the trace map t_Y for Y, as the composition

$$H^n(X, \pi^{!?}\omega_Y) \longrightarrow H^n(Y, \pi_*\pi^{!?}\omega_Y) \longrightarrow H^n(Y, \omega_Y) \xrightarrow{\ t_Y\ } k.$$

29.3.12. Corollary — *The functorial and trace versions of Serre duality for Hom hold for all pure dimensional projective k-schemes (with no Cohen–Macaulay hypotheses).*

Proof. Combine Proposition 29.3.11 with Proposition 29.3.5(a) and Proposition 29.1.10. $\qquad\square$

29.3.13. Proposition — *If X is a Cohen–Macaulay projective k-scheme of pure dimension n, and $\pi\colon X \to \mathbb{P}^n$ is a finite flat morphism (as in Proposition 29.3.5(b)), then functorial Serre duality for Ext holds for X with canonical sheaf $\omega_X := \pi^{!?}\omega_{\mathbb{P}^n}$.*

Unlike Propositions 29.3.9 and 29.3.11, we did not state this for general $\pi\colon X \to Y$; we will use the fact that the target is $\mathbb{P}^n$.

Proof. The argument will parallel that of §29.2.2: we will show an isomorphism of δ-functors

$$(29.3.13.1) \qquad \operatorname{Ext}^i_X(\mathscr{F}, \pi^{!?}\omega_{\mathbb{P}^n}) \xrightarrow{\ \sim\ } H^{n-i}(X, \mathscr{F})^\vee.$$

We already have an isomorphism for $i=0$ (Corollary 29.3.12), so it suffices to show that both sides of (29.3.13.1) are *universal* δ-functors (§23.2.6). We do this by verifying the criterion of Theorem 23.2.9. For any coherent sheaf $\mathscr{F}$ on X, we can find a surjection $\mathscr{O}(-m)^{\oplus N} \to \mathscr{F}$ for $m \gg 0$ (and some N), so it suffices to show that for $m \gg 0$,

$$\operatorname{Ext}^i_X(\mathscr{O}(-m), \pi^{!?}\omega_{\mathbb{P}^n}) = 0 \quad \text{and} \quad H^{n-i}(X, \mathscr{O}(-m))^\vee = 0$$

for $i > 0$ and $m \gg 0$. For the first, we have (for $i > 0$ and $m \gg 0$)

$$\operatorname{Ext}^i_X(\mathscr{O}(-m), \pi^{!?}\omega_{\mathbb{P}^n}) \cong H^i(X, (\pi^{!?}\omega_{\mathbb{P}^n})(m)) \quad \text{(Exercise 29.2.K)}$$

$$= 0 \quad \text{(Serre vanishing for X, Thm. 18.1.3(ii))}.$$

For the second, we have

$$H^{n-i}(X, \mathscr{O}_X(-m)) = H^{n-i}(X, \mathscr{O}_X \otimes \pi^* \mathscr{O}(-m))$$
$$\cong H^{n-i}(\mathbb{P}^n, (\pi_* \mathscr{O}_X) \otimes \mathscr{O}(-m))$$

by the projection formula (Exercise 18.7.E(b)). Now $\pi_* \mathscr{O}_X$ is locally free by Exercise 24.3.E, so by functorial Serre duality for vector bundles on $\mathbb{P}^n$, we have (for $m \gg 0$)

$$H^{n-i}(\mathbb{P}^n, (\pi_* \mathscr{O}_X) \otimes \mathscr{O}(-m)) \cong H^i(\mathbb{P}^n, (\pi_* \mathscr{O}_X)^\vee \otimes \mathscr{O}(m) \otimes \omega_{\mathbb{P}^n})^\vee$$
$$\cong H^i(\mathbb{P}^n, ((\pi_* \mathscr{O}_X)^\vee \otimes \omega_{\mathbb{P}^n}) \otimes \mathscr{O}(m))^\vee$$
$$= 0 \qquad \text{(Serre vanishing for } \mathbb{P}^n, \text{Thm. 18.1.3(ii))}. \qquad \square$$

29.3.14. Corollary — *Functorial Serre duality for* Ext *holds for all pure dimensional Cohen–Macaulay projective* k-*schemes.*

Proof. Combine Proposition 29.3.13 with Proposition 29.3.5(b) and the projective space case of §29.2.2. $\qquad \square$

29.3.15. ** *Remark: The trace version of Serre duality for vector bundles.* Here is an outline of how the above proof can be extended to prove the trace version of Serre duality for vector bundles on all pure dimensional Cohen–Macaulay projective k-schemes. Continuing the notation of Proposition 29.3.13, the Yoneda cup product (which we haven't defined) followed by the trace map gives a pairing

$$\operatorname{Ext}_X^i(\mathscr{F}, \omega_X) \times H^{n-i}(X, \mathscr{F}) \xrightarrow{\;\cup\;} H^n(X, \omega_X) \xrightarrow{\;t\;} k \ ,$$

which is functorial in $\mathscr{F}$. This induces a functorial map (29.3.13.1). One shows that this particular map is a map of δ-functors. Then by the above argument, this particular map is an isomorphism.

29.3.16. As described in §29.1.13, the situation is even better: pure-dimensional proper k-schemes satisfy the best sort of Serre duality, the trace version for Ext, [Stacks, tag 0FVU].

29.3.17. Conclusion. The proofs in this section are perhaps surprisingly easy. Once we defined $\pi^{!?}$, we only used the fact that it was right-adjoint to π_*. The price we pay is that we have very little idea of what the canonical sheaf looks like from the proof. For example, if X is smooth, it is not clear that ω_X is a line bundle! (We rectify this in the next section.)

You may also have a sense of how these ideas lead to a larger story: to define and prove duality more generally, we are trying to find a functor $\pi^!$ that is right-adjoint to π_* in as much generality as we can manage. This will lead us to work in a more general setting than coherent sheaves—the *derived category* of coherent sheaves (and beyond).

29.4   The Adjunction Formula for the ω_X, and $\omega_X = \mathscr{K}_X$

(In this final section, despite previous protestations about terminology in §29.1.6, we use the phrase "dualizing sheaf" for the canonical sheaf of Serre duality, because the entire point of this section is to show that the two uses of the phrase "canonical sheaf" refer to the same thing in the case of smooth projective varieties X: the sheaf ω_X arising in Sere duality for Hom, §29.1.5, and $\mathscr{K}_X = \det \Omega_X$.)

The dualizing sheaf behaves well with respect to slicing by effective Cartier divisors, and this will be useful for obtaining Desideratum 29.1.1, i.e., the miracle that $\mathscr{K} = \det \Omega$ is Serre-dualizing (§18.5.2). In order to show this, we give a description for the dualizing sheaf for a subvariety of a variety satisfying Serre duality for Ext.

As a warm-up, we extend the notion of $\pi^{!?}$ to $\mathscr{O}$-modules, in the special case where π is a closed embedding.

29.4.A. EXERCISE. Suppose $\pi\colon X \to Y$ is a *closed embedding* of schemes and $\mathscr{G}$ is an $\mathscr{O}_Y$-module. Explain why $\mathcal{H}om_Y(\pi_*\mathscr{O}_X, \mathscr{G})$ naturally has the structure of an $\mathscr{O}_X$-module. Hint: If $\mathscr{I}$ is the ideal sheaf of X, explain how $\mathcal{H}om_Y(\pi_*\mathscr{O}_X, \mathscr{G})$ (over some open subset $U \subset Y$) is annihilated by "functions vanishing on X" (elements of $\mathscr{I}(U)$).

29.4.1. Definition. Hence if $\pi\colon X \to Y$ is a closed embedding, we have defined a map $\pi^{!?}\colon Mod_{\mathscr{O}_Y} \to Mod_{\mathscr{O}_X}$ extending the map of Exercise 17.1.K(a). In this case, where π is a closed embedding, it is reasonable and correct to write $\pi^{!?}\mathscr{G} := \pi^{-1}\mathcal{H}om(\pi_*\mathscr{O}_X, \mathscr{G})$, i.e., to use the symbol π^{-1} in the definition.

29.4.B. EXERCISE. We remain in the situation of Exercise 29.4.A.

(a) Show that $(\pi_*, \pi^{!?})$ is an adjoint pair between $Mod_{\mathscr{O}_X}$ and $Mod_{\mathscr{O}_Y}$.
(b) Show that $\pi^{!?}$ sends injective $\mathscr{O}_Y$-modules to injective $\mathscr{O}_X$-modules. (Hint: π_* is exact; use Exercise 23.5.B.)

We now extend this discussion from $\mathcal{H}om$ to $\mathcal{E}xt$. Suppose $\pi\colon X \to Y$ is a closed embedding. Then for any $\mathscr{O}_Y$-module $\mathscr{F}$, $\mathcal{E}xt^i_Y(\pi_*\mathscr{O}_X, \mathscr{F})$ naturally has the structure of an $\mathscr{O}_X$-module. (Reason: We compute this by taking an injective resolution of $\mathscr{F}$ by $\mathscr{O}_Y$-modules, then truncating, then applying $\mathcal{H}om_Y(\pi_*\mathscr{O}_X, \cdot)$. But for any $\mathscr{O}_Y$-module $\mathscr{G}$, $\mathcal{H}om_Y(\pi_*\mathscr{O}_X, \mathscr{G})$ has the structure of an $\mathscr{O}_X$-module, indeed, $\pi^{!?}\mathscr{G}$.) To emphasize its structure as an $\mathscr{O}_X$-module, we write it as $\pi^{-1}\mathcal{E}xt^i_Y(\pi_*\mathscr{O}_X, \mathscr{F})$.

29.4.C. EXERCISE. Show that if Y (and hence X) is locally Noetherian, and $\mathscr{G}$ is a coherent sheaf on Y, then $\pi^{-1}\mathcal{E}xt^i_Y(\pi_*\mathscr{O}_X, \mathscr{G})$ is a coherent sheaf on X. (Hint: $\mathcal{E}xt^i_Y(\pi_*\mathscr{O}_X, \mathscr{G})$ is a coherent sheaf on Y, Exercise 29.2.F(b).)

29.4.2. The dualizing sheaf for a subvariety in terms of the dualizing sheaf of the ambient variety.

For the rest of this section, $\pi\colon X \to Y$ will be a closed embedding of pure dimensional projective k-schemes of dimension n and N, respectively, where Y satisfies functorial Serre duality for Ext. Let $r = N - n$ (the codimension of X in Y). By Corollary 29.3.12, X satisfies functorial Serre duality for Hom. We now identify ω_X in terms of ω_Y.

29.4.3. Theorem — *We have $\omega_X = \pi^{-1}\mathcal{E}xt^r_Y(\pi_*\mathscr{O}_X, \omega_Y)$.*

Before we prove Theorem 29.4.3, we explain what happened to the "earlier" $\mathcal{E}xt^i_Y(\pi_*\mathscr{O}_X, \omega_Y)$'s.

29.4.4. Proposition — *Suppose that $\pi\colon X \hookrightarrow Y$ is a closed embedding of pure dimensional projective k-schemes of dimension n and N, respectively, and Y satisfies functorial Serre duality for Ext. Then for all $i < r := N - n$, $\mathcal{E}xt^i_Y(\pi_*\mathscr{O}_X, \omega_Y) = 0$.*

Proof. As $\mathcal{E}xt^i_Y(\pi_*\mathscr{O}_X, \omega_Y)$ is coherent (Exercise 29.2.F), it suffices to show that $\mathcal{E}xt^i_Y(\pi_*\mathscr{O}_X, \omega_Y) \otimes \mathscr{O}(m)$ has no nonzero global sections for $m \gg 0$ (as for any coherent sheaf $\mathscr{G}$ on Y, $\mathscr{G}(m)$ is generated by global sections for $m \gg 0$ by Serre's Theorem A, Theorem 16.1.5). By Exercise 29.2.J,

$$H^j(Y, \mathcal{E}xt^i_Y(\pi_*\mathscr{O}_X, \omega_Y)(m)) = H^j(Y, \mathcal{E}xt^i_Y(\pi_*\mathscr{O}_X, \omega_Y(m))).$$

If $j > 0$, then for $m \gg 0$, by Serre vanishing, $H^j(Y, \mathcal{E}xt^i_Y(\pi_*\mathscr{O}_X, \omega_Y)(m)) = 0$. Thus by the local-to-global spectral sequence for Ext (Exercise 29.2.M),

$$H^0(Y, \mathcal{E}xt^i_Y(\pi_*\mathscr{O}_X, \omega_Y(m))) = Ext^i_Y(\pi_*\mathscr{O}_X, \omega_Y(m))).$$

By Exercise 29.2.J again, then functorial Serre duality for Ext on Y,

$$Ext^i_Y(\pi_*\mathscr{O}_X, \omega_Y(m))) = Ext^i_Y(\pi_*\mathscr{O}_X(-m), \omega_Y) = H^{N-i}(Y, \pi_*\mathscr{O}_X(-m)),$$

which is 0 if $N - i > n$, as the cohomology of a quasicoherent sheaf on a projective scheme vanishes in degree higher than the dimension of the sheaf's support (dimensional cohomology vanishing, Theorem 18.2.6). $\square$

The most difficult step in the proof of Theorem 29.4.3 is the following.

29.4.5. Lemma — *Suppose that* $\pi \colon X \to Y$ *is a closed embedding of codimension* r *of pure dimensional projective* k-*schemes, and* Y *satisfies functorial Serre duality for* Ext. *Then we have an isomorphism, functorial in* $\mathscr{F} \in Coh_X$:

$$(29.4.5.1) \qquad \operatorname{Hom}_X(\mathscr{F}, \pi^{-1} \mathcal{E}\mathit{xt}^r_Y(\pi_* \mathcal{O}_X, \omega_Y)) \xrightarrow{\ \sim\ } \operatorname{Ext}^r_Y(\pi_* \mathscr{F}, \omega_Y).$$

Proof. Choose an injective resolution

$$0 \longrightarrow \omega_Y \longrightarrow \mathscr{I}^0 \longrightarrow \mathscr{I}^1 \longrightarrow \mathscr{I}^2 \longrightarrow \cdots$$

of ω_Y. To compute the right side of (29.4.5.1), we drop the ω_Y from this resolution, and apply $\operatorname{Hom}_Y(\pi_* \mathscr{F}, \cdot)$, to obtain

$$0 \longrightarrow \operatorname{Hom}_Y(\pi_* \mathscr{F}, \mathscr{I}^0) \longrightarrow \operatorname{Hom}_Y(\pi_* \mathscr{F}, \mathscr{I}^1) \longrightarrow \operatorname{Hom}_Y(\pi_* \mathscr{F}, \mathscr{I}^2) \longrightarrow \cdots,$$

or (by adjointness of π_* and $\pi^{!?}$ for $\mathcal{O}$-modules when π is a closed embedding, Exercise 29.4.B(a)):

$$0 \longrightarrow \operatorname{Hom}_X(\mathscr{F}, \pi^{!?} \mathscr{I}^0) \longrightarrow \operatorname{Hom}_X(\mathscr{F}, \pi^{!?} \mathscr{I}^1) \longrightarrow \operatorname{Hom}_X(\mathscr{F}, \pi^{!?} \mathscr{I}^2) \longrightarrow \cdots.$$

Motivated by this, consider the complex

$$(29.4.5.2) \qquad 0 \longrightarrow \pi^{!?} \mathscr{I}^0 \longrightarrow \pi^{!?} \mathscr{I}^1 \longrightarrow \pi^{!?} \mathscr{I}^2 \longrightarrow \cdots.$$

Note first that (29.4.5.2) is indeed a complex, and second that $\pi^{!?} \mathscr{I}^i$ are injective $\mathcal{O}_X$-modules (by Exercise 29.4.B(b)).

29.4.D. EXERCISE. Show that the cohomology of (29.4.5.2) at the ith step is $\pi^{-1} \mathcal{E}\mathit{xt}^i_Y(\pi_* \mathcal{O}_X, \omega_Y)$.

Thus by Proposition 29.4.4, the complex (29.4.5.2) is exact before the rth step.

29.4.E. EXERCISE. Show that there exists a direct sum decomposition $\pi^{!?} \mathscr{I}^r = \mathscr{J} \oplus \mathscr{K}$, so that

$$0 \longrightarrow \pi^{!?} \mathscr{I}^0 \longrightarrow \pi^{!?} \mathscr{I}^1 \longrightarrow \cdots \longrightarrow \pi^{!?} \mathscr{I}^{r-1} \longrightarrow \mathscr{J} \longrightarrow 0$$

is exact. Hint: Work out the case $r=1$ first. Another hint: Notice that if

$$0 \longrightarrow \mathscr{I}^0 \xrightarrow{\ \alpha\ } \mathscr{F}$$

is exact and $\mathscr{I}^0$ is injective, then there is a map $\beta \colon \mathscr{F} \to \mathscr{I}^0$ "splitting" α, allowing us to write $\mathscr{F}$ as a direct sum $\mathscr{I}^0 \oplus \mathscr{J}^1$. If $\mathscr{F}$ is furthermore injective, then $\mathscr{J}^1$ is injective too. This is the beginning of an induction.

29.4.F. EXERCISE. (Keep the notation of the previous exercise.) Identify $\ker(\mathscr{K} \to \pi^{!?} \mathscr{I}^{r+1})$ with $\pi^{-1} \mathcal{E}\mathit{xt}^r_Y(\pi_* \mathcal{O}_X, \omega_Y)$.

29.4.G. EXERCISE. Put together the pieces above to complete the proof of Lemma 29.4.5. □

We are now ready to prove Theorem 29.4.3.

29.4.6. *Proof of Theorem 29.4.3.* Suppose $\mathscr{F}$ is a coherent sheaf on X. We wish to describe an isomorphism

$$\operatorname{Hom}_X(\mathscr{F}, \pi^{-1} \mathcal{E}\mathit{xt}^r_Y(\pi_* \mathcal{O}_X, \omega_Y)) \xrightarrow{\ \sim\ } H^n(X, \mathscr{F})^\vee,$$

functorial in $\mathscr{F}$. By Lemma 29.4.5, we have a functorial isomorphism

$$\operatorname{Hom}_X(\mathscr{F}, \pi^{-1} \mathcal{E}\mathit{xt}^r_Y(\pi_* \mathcal{O}_X, \omega_Y)) \xrightarrow{\ \sim\ } \operatorname{Ext}^r_Y(\pi_* \mathscr{F}, \omega_Y).$$

By functorial Serre duality for Ext on Y, we have a functorial isomorphism

$$\operatorname{Ext}^r_Y(\pi_* \mathscr{F}, \omega_Y) \xleftarrow{\sim} H^{N-r}(Y, \pi_* \mathscr{F})^\vee.$$

As $\mathscr{F}$ is a coherent sheaf on X, and $N - r = n$, the right side is precisely $H^n(X, \mathscr{F})^\vee$. $\qquad\square$

29.4.7. Applying Theorem 29.4.3.

We first apply Theorem 29.4.3 in the special case where X is an effective Cartier divisor on Y. We can compute the dualizing sheaf $\pi^{-1}\mathcal{E}xt^1_Y(\pi_*\mathcal{O}_X, \omega_Y)$ by computing $\mathcal{E}xt^1_Y(\pi_*\mathcal{O}_X, \omega_Y)$ using any locally free resolution (on Y) of $\pi_*\mathcal{O}_X$ (Exercise 29.2.E). But $\pi_*\mathcal{O}_X$ has a particularly simple resolution, the closed subscheme exact sequence (9.1.2.1) for X:

$$(29.4.7.1) \qquad 0 \longrightarrow \mathcal{O}_Y(-X) \longrightarrow \mathcal{O}_Y \longrightarrow \pi_*\mathcal{O}_X \longrightarrow 0.$$

We compute $\mathcal{E}xt^\bullet(\pi_*\mathcal{O}_X, \omega_Y)$ by truncating (removing the $\pi_*\mathcal{O}_X$), and applying $\mathcal{H}om(\cdot, \omega_Y)$: $\mathcal{E}xt^\bullet(\pi_*\mathcal{O}_X, \omega_Y)$ is the cohomology of

$$0 \longrightarrow \mathcal{H}om_Y(\mathcal{O}_Y, \omega_Y) \longrightarrow \mathcal{H}om(\mathcal{O}_Y(-X), \omega_Y) \longrightarrow 0,$$

i.e.,

$$0 \longrightarrow \omega_Y \longrightarrow \omega_Y \otimes \mathcal{O}_Y(X) \longrightarrow 0.$$

We immediately see that $\mathcal{E}xt^i(\pi_*\mathcal{O}_X, \omega_Y) = 0$ if $i \neq 0, 1$. Furthermore, $\mathcal{E}xt^0(\pi_*\mathcal{O}_X, \omega_Y) = 0$ by Proposition 29.4.4. (Unimportant aside: You can use this to show that ω_Y has no embedded points.)

We now consider $\omega_X = \operatorname{coker}(\omega_Y \to \omega_Y \otimes \mathcal{O}_Y(X))$. Tensoring (29.4.7.1) with the invertible sheaf $\mathcal{O}_Y(X)$, and then tensoring with ω_Y, yields

$$\omega_Y \longrightarrow \omega_Y \otimes \mathcal{O}_Y(X) \longrightarrow \omega_Y \otimes \pi_*(\mathcal{O}_Y(X)|_X) \longrightarrow 0.$$

The right term $\omega_Y \otimes \pi_*(\mathcal{O}_Y(X)|_X)$ is often (somewhat informally) written as $\omega_Y(X)|_X$. Thus

$$\omega_X = \operatorname{coker}(\omega_Y \to \omega_Y \otimes \mathcal{O}_Y(X)) = \omega_Y(X)|_X,$$

and this identification is *canonical*.

We have shown the following.

29.4.8. Proposition (the adjunction formula)

— *Suppose that Y is a Cohen–Macaulay projective scheme of pure dimension n (which satisfies functorial Serre duality for* Ext *by Corollary 29.3.14), and X is an effective Cartier divisor on Y. (Hence X is Cohen–Macaulay by §26.2.4, and thus satisfies functorial Serre duality for* Ext *by Corollary 29.3.14.) Then $\omega_X = \omega_Y(X)|_X$. In particular, if ω_Y is an invertible sheaf on Y, then ω_X is an invertible sheaf on X.*

As an immediate application, we have the following.

29.4.H. EXERCISE. Suppose X is a complete intersection in $\mathbb{P}^n$, of hypersurfaces of degrees $d_1, \dots, d_r$. (Note that X is Cohen–Macaulay by Proposition 26.2.6, and thus satisfies functorial Serre duality for Ext, by Corollary 29.3.14.) Show that $\omega_X \cong \mathcal{O}_X(-n - 1 + \sum d_i)$. If furthermore X is smooth, show that $\omega_X \cong \det \Omega_X$. (Hint for the last sentence: Use the adjunction formula for $\mathscr{K}$, Exercise 21.5.B).

But we can say more.

29.4.I. EXERCISE. Suppose $\pi: X \hookrightarrow Y$ is a codimension r regular embedding with normal sheaf $\mathscr{N}_{X/Y}$ (which is locally free, by Proposition 21.2.16(b)). Suppose $\mathscr{L}$ is an invertible sheaf on Y.

(a) Show that $\mathcal{E}xt^i_Y(\pi_*\mathcal{O}_X, \mathscr{L}) = 0$ if $i \neq r$.

(b) Describe a *canonical* isomorphism $\pi^{-1}\mathcal{E}xt^r_Y(\pi_*\mathcal{O}_X, \mathscr{L}) \xleftarrow{\sim} (\det \mathscr{N}_{X/Y}) \otimes_{\mathcal{O}_X} \mathscr{L}|_X$.

Hint for both parts: Deal with the case $\mathscr{L} = \mathcal{O}$ first, then notice that the question is local, if you have solved (b) carefully and correctly.

(Note that because Exercise 29.4.I has nothing explicitly to do with duality, we have no projectivity assumptions on Y; it is a completely local question.) From Exercise 29.4.I we deduce the following.

29.4.J. IMPORTANT EXERCISE. Suppose X is a codimension r regular embedding in $\mathbb{P}_k^n$. (Then X is Cohen–Macaulay by Proposition 26.2.6, and thus satisfies functorial Serre duality for Ext by Corollary 29.3.14.) Show that

$$\omega_X = \pi^{-1} \mathcal{E}xt_{\mathbb{P}^n}^r (\pi_* \mathscr{O}_X, \omega_{\mathbb{P}^n}) \cong (\det \mathscr{N}_{X/\mathbb{P}^n}) \otimes \omega_{\mathbb{P}^n}|_X.$$

In particular, ω_X is an invertible sheaf.

29.4.9. *Aside: Gorenstein singularities.* If a pure-dimensional Cohen–Maculay projective k-scheme X has $\omega_{X/k}$ an invertible sheaf, we say that X is *Gorenstein*. Given Proposition 29.4.8, you can show that this is a condition that can be checked (stalk-)locally at the closed points of a pure-dimensional projective k-scheme, similarly to the definition of Cohen–Macaulayness. Can you write down such a definition? The definition of Gorenstein singularity in correct generality certainly does not require the scheme in question to be projective. This aside will lead you to a correct if inelegant definition in terms of a "slicing criterion." All "local complete intersection" k-schemes are Gorenstein, and all Gorenstein k-schemes are Cohen–Macaulay by definition. For an excellent introduction to Gorenstein rings and schemes, see [BruH, §3].

29.4.K. IMPORTANT EXERCISE. Suppose X is a smooth pure codimension r subvariety of $\mathbb{P}_k^n$ (and hence a regular embedding, by Exercise 13.2.M(b)). Show that $\omega_X \cong \mathcal{K}_X$. Hint: Both sides satisfy adjunction (see Exercise 21.5.B for adjunction for Ω): they are isomorphic to $(\det \mathscr{N}_{X/Y}) \otimes (\omega_{\mathbb{P}^n}|_X) \cong (\det \mathscr{N}_{X/Y}) \otimes (\mathcal{K}_{\mathbb{P}^n}|_X)$.

29.4.10. *Remark.* As long promised, the version of Serre duality given in Theorem 18.5.1 and Further Miracle 18.5.2 now follows by combining Corollary 29.3.10 with Exercise 29.4.K.

> *Aux yeux de ces amateurs d'inquiétude et de perfection, un ouvrage n'est jamais achevé,— mot qui pour eux n'a aucun sens,—mais abandonné; et cet abandon, qui le livre aux flammes ou au public, (et qu'il soit l'effet de la lassitude ou de l'obligation de livrer), leur est une sorte d'accident, comparable à la rupture d'une réflexion …*
>
> To those who like to worry about understanding and to chase after perfection, a work is never completed—a word which for them has no meaning—but rather is abandoned. And this abandonment, which delivers the work to the flames or to the public (perhaps because of lassitude or of deadline), is to them an accident of sorts, akin to a knock on the door that ruptures their dream . . .
>
> —P. Valéry [Coh, pp. 7–8] (Tadashi Tokieda trans.)

> Well, I've been talking to you for two years and now I'm going to quit. In some ways I would like to apologize, and other ways not. I hope—in fact, I know—that two or three dozen of you have been able to follow everything with great excitement, and have had a good time with it. But I also know that "the powers of instruction are of very little efficacy except in those happy circumstances in which they are practically superfluous." So, for the two or three dozen who have understood everything, may I say I have done nothing but shown you the things. For the others, if I have made you hate the subject, I'm sorry. I never taught elementary physics before, and I apologize. I just hope that I haven't caused a serious trouble to you, and that you do not leave this exciting business. I hope that someone else can teach it to you in a way that doesn't give you indigestion, and that you will find someday that, after all, it isn't as horrible as it looks.
>
> —R. Feynman [FLS, Epilogue]

Bibliography

[Al]   P. Aluffi, *Algebra, Chapter 0*, Grad. Stud. in Math. **104**, Amer. Math. Soc., Providence, RI, 2009.

[Ar1]   M. Artin, "On the solutions of analytic equations," Invent. Math. **5** (1968), no. 4, 277–291.

[Ar2]   ――――, "Algebraic approximation of structures over complete local rings," Publ. Math. IHES **36** (1969), 23–58.

[Ar3]   ――――, *Algebra*, Prentice Hall, Inc., Englewood Cliffs, NJ, 1991.

[At1]   M. F. Atiyah, "On analytic surfaces with double points," Proc. Roy. Soc. London Ser. A **247** (1958), 237–244.

[At2]   ――――, "Mathematics in the 20th century," Amer. Math. Monthly **108** (2001), no. 7, 654–666.

[AtM]   M. F. Atiyah and I. G. Macdonald, *Introduction to Commutative Algebra*, Addison-Wesley Publ. Co., Reading, MA, 1969.

[Ba]   L. Bădescu, *Algebraic Surfaces*, V. Maşek trans., Universitext, Springer-Verlag, New York, 2001.

[BCDKT]   M. Baker, B. Conrad, S. Dasgupta, K. Kedlaya, and J. Teitelbaum, "p-adic geometry," D. Savitt and D. Thakur ed., Univ. Lect. Series **45**, Amer. Math. Soc., Providence, RI, 2008.

[BPV]   W. Barth, C. Peters, and A. van de Ven, *Compact Complex Surfaces*, Ergeb. Math. Grenz. **4**, Springer-Verlag, Berlin, 1984.

[Be]   A. Beauville, *Complex Algebraic Surfaces*, 2nd ed., R. Barlow trans., Cambridge U. P., Cambridge, 1996.

[BL]   C. Birkenhake and H. Lange, *Complex Abelian Varieties*, 2nd ed., Grund. Math. Wiss. **302**, Springer-Verlag, Berlin, 2004.

[BLR]   S. Bosch, W. Lütkebohmert, and M. Raynaud, *Néron Models*, Ergeb. Math. Grenz. **21**, Springer-Verlag, Berlin, 1990.

[Bo]   N. Bourbaki, *Éléments de mathématique, Fasc. XXX: Algèbre Commutative*, Actualités Scientifiques et Industrielles, no. 1308, Hermann, Paris, 1964.

[BGS]   S. Boissière, O. Gabber, and O. Serman, *Sur le produit de variétés localement factorielles ou* $\mathbb{Q}$-*factorielles*, arxiv:1104.1861v1, preprint 2011.

[BroP]   R. Brown and T. Porter, *Analogy, concepts and methodology, in mathematics*, UWB Math Preprint, May 26, 2006.

[BruH]   W. Bruns and J. Herzog, *Cohen–Macaulay Rings*, Cambridge U. P., Cambridge, 1993.

[CL]   F. Call and G. Lyubeznik, *A simple proof of Grothendieck's theorem on the parafactoriality of local rings*, in *Commutative algebra: syzygies, multiplicities, and birational algebra (South Hadley, MA, 1992)*, Contemp. Math. **159**, Amer. Math. Soc., Providence, RI, 1994, 15–18.

[Carr]   L. Carroll, *Alice's Adventures in Wonderland*, Macmillan, London, 1865.

[ChJLO]   R. Cheng, L. Ji, M. Larson, and N. Olander, *Theorem of the base*, in *Stacks Project Expository Collection (SPEC)*, 163–193, London Math. Soc. Lecture Note Ser. **480**, Cambridge U. P., Cambridge, 2022.

[Coh]   G. Cohen, *Essai d'explication du Cimetière Marin*, Librairie Gallimard, 1933.

[Con]   B. Conrad, "Deligne's notes on Nagata compactifications," J. Ramanujan Math. Soc. **22** (2007), no. 3, 205–257.

[CP]   V. Cossart and O. Piltant, "Resolution of singularities of arithmetical threefolds," J. Algebra **529** (2019), no. 7, 268–535.

[Dan]   V. I. Danilov, "Samuel's conjecture," Mat. Sb. (N.S.) **81 (123)** (1970), 132–144.

[Dav]   E. D. Davis, "Ideals of the principal class, R-sequences and a certain monoidal transformation," Pacific J. Math. **20** (1967), 197–205.

[DW]   R. Dedekind and H. Weber, *Theory of algebraic functions of one variable*, J. Stillwell trans. History of Math. **39**, Amer. Math. Soc., Providence, RI; London Math. Soc. London, 2012.

[De]   P. Deligne, *Courbes elliptiques: formulaire d'après Tate*, in *Modular Functions of One Variable IV (Proc. Internat. Summer School, Univ. Antwerp, Antwerp, 1972)*, 53–73, Lect. Notes in Math. **476**, Springer-Verlag, Berlin, 1975.

[DeI]   P. Deligne and L. Illusie, "Relèvements modulo p^2 et décomposition du complexe de de Rham," Invent. Math. **89** (1987), no. 2, 247–270.

[Di]   J. Dieudonné, "On regular sequences," Nagoya Math. J. **27** (1966), 355–356.

[Do]   I. Dolgachev, *Luigi Cremona and cubic surfaces*, in *Luigi Cremona (1830–1903)*, 55–70, Incontr. Studio, **36**, Instituto Lombardo di Scienze e Lettere, Milan, 2005.

[DS]   R. Donagi and R. C. Smith, "The structure of the Prym map," Acta Math. **146** (1981), no. 1–2, 25–102.

[DZ]   H. Duan and X. Zhao, *Schubert calculus and the integral cohomology of exceptional Lie groups*, arXiv:0711.2541v10, preprint 2007.

[DF]   D. Dummit and R. Foote, *Abstract Algebra*, 3rd ed., Wiley and Sons, Inc., Hoboken, NJ, 2004.

[E]   D. Eisenbud, *Commutative Algebra with a View to Algebraic Geometry*, Grad. Texts in Math. **150**, Springer-Verlag, New York, 1995.

[EGH]   D. Eisenbud, M. Green, and J. Harris, "Cayley–Bacharach theorems and conjectures," Bull. AMS **33** (1996), no. 3, 295–324.

[Ek]   P. Eklof, "Lefschetz's principle and local functors," Proc. Amer. Math. Soc. **37** (1973), no. 2, 1973.

[EE]   E. Enochs and S. Estrada, "Relative homological algebra in the category of quasi-coherent sheaves," Adv. Math. **194** (2005), 284–295.

[FGIKNV] B. Fantechi, L. Göttsche, L. Illusie, S. Kleiman, N. Nitsure and A. Vistoli, *Fundamental algebraic geometry: Grothendieck's FGA explained*, Mathematical Surveys and Monographs 123, Amer. Math. Soc., Providence, RI, 2005.

[FK]   H. M. Farkas and I. Kra, *Riemann Surfaces*, Grad. Texts in Math. **71**, Springer-Verlag, New York, 1980.

[FLS]   R. Feynman, R. Leighton, and M. Sands, *The Feynman Lectures on Physics, Volume III*, Addison Wesley, 1971.

[Fi]   F. Scott Fitzgerald, *The Crack-Up: A desolately frank document from one for whom the salt of life has lost its savor*, Esquire Magazine, February 1936, Esquire Inc., Chicago, IL.

[FR]   G. Frey and H.-G. Rück, "The strong Lefschetz principle in algebraic geometry," manuscripta math. **55** (1986), no. 3–4, 385–401.

[F1]   W. Fulton, *Young Tableaux: With Applications to Representation Theory and Geometry*, Cambridge U. P., Cambridge, 1997.

[F2]   ———, *Intersection Theory*, 2nd ed., Ergeb. Math. Grenz. **2**, Springer-Verlag, New York, 1998.

[Ge]   S. Germain, *Œuvres Philosophiques de Sophie Germain* (nov. éd.), Librarie de Firmin-Didot et Cie, Paris, 1896.

[GS]   P. Gille and T. Szamuely, *Central Simple Algebras and Galois Cohomology*, Cambridge U. P., Cambridge, 2006.

[GW]   U. Görtz and T. Wedhorn, *Algebraic Geometry I*, Vieweg + Teubner, Wiesbaden, 2010. (Appendices C, D, and E, collecting properties of schemes and morphisms and their relationships, are alone worth the price.)

[GR]   H. Grauert and R. Remmert, *Coherent Analytic Sheaves*, Grund. Math. Wiss. **265**, Springer-Verlag, Berlin, 1984.

[GH1]   P. Griffiths and J. Harris, *Principles of Algebraic Geometry*, Wiley-Interscience, New York, 1978.

[GH2]   ———, "On the Noether–Lefschetz theorem and some remarks on codimension two cycles," Math. Ann. **271** (1985), 31–51.

[Gr1]   A. Grothendieck, "Sur quelques points d'algèbre homologique," Tôhoku Math. J. (2) **9** (1957), 119–221.

[Gr2]   ———, "Sur la classification des fibrés holomorphes sur la sphère de Riemann," Amer. J. Math. **79** (1957), 121–138.

[Gr3]   ———, *The cohomology theory of abstract algebraic varieties*, in *Proc. Internat. Congress Math. (Edinburgh, 1958)*, Cambridge U. P., New York, 1960.

[Gr4]   ———, *Fondements de la géométrie algébrique*, in *Collected Bourbaki Talks*, Secrétariat Mathématique, Paris, 1962.

[Gr5]   _____, *Récoltes et Semailles*, unpublished.

[Gr-EGA]   _____(w. J. Dieudonné), *Éléments de Géométrie Algébrique*, Publ. Math. IHES, Part I: **4** (1960); Part II: **8** (1961); Part III: **11** (1961), 17 (1963); Part **IV**: 20 (1964), 24 (1965), 28 (1966), 32 (1967).

[Gr-EGA']   _____(w. J. Dieudonné), *Éléments de Géométrie Algébrique I: Le langage des schémas*, Grund. Math. Wiss. **166** (2nd ed.), Springer-Verlag, Berlin-New York, 1971.

[GrS]   A. Grothendieck and J.-P. Serre, *Grothendieck-Serre Correspondence*, P. Colmez and J.-P. Serre ed., C. Maclean trans., Amer. Math. Soc. and Soc. Math. de France, Paris, 2004.

[H1]   J. Harris, "Galois groups of enumerative problems," Duke Math. J. **46** (1979), no. 4, 685–724.

[H2]   _____, *Algebraic Geometry: A First Course*, Grad. Texts in Math. **133**, Springer-Verlag, New York, 1995.

[Ha1]   R. Hartshorne, *Algebraic Geometry*, Grad. Texts in Math. **52**, Springer-Verlag, New York-Heidelberg, 1977.

[Ha2]   _____, *Ample Subvarieties of Algebraic Varieties*, Lect. Notes in Math. **156**, Springer-Verlag, Berlin-New York, 1970.

[Hat]   A. Hatcher, *Algebraic Topology*, Cambridge U. P., Cambridge, 2002.

[HM]   M. Hazewinkel and C. Martin, "A short elementary proof of Grothendieck's theorem on algebraic vector bundles over the projective line," J. Pure Appl. Algebra **25** (1982), no. 2, 207–211.

[He]   R. C. Heitmann, "Examples of noncatenary rings," Trans. Amer. Math. Soc. **247** (1979), 125–136.

[Hil]   D. Hilbert, "Ueber die Theorie der algebraischen Formen," Math. Ann. **36** (1890), 473–534.

[Hir]   H. Hironaka, "Resolution of singularities of an algebraic variety over a field of characteristic zero. I, II," Ann. of Math. (2) **79** (1964), 109-203; ibid. (2) **79** (1964), 205-326.

[Ig]   J. Igusa, "On some problems in abstract algebraic geometry," Proc. Nat. Acad. Sci. **41** (1955), 964–967.

[Ii]   S. Iitaka, *Algebraic Geometry: An Introduction to Birational Geometry of Algebraic Varieties*, Grad. Texts in Math. **76**, Springer-Verlag, New York-Berlin, 1982.

[Il]   L. Illusie, *Frobenius and Hodge degeneration*, in *Introduction to Hodge Theory*, SMF/AMS Texts and Monographs **8**, 99–149, 2002.

[dJ]   A. J. de Jong, "Smoothness, semi-stability and alterations," IHES Publ. Math. **83** (1996), 51–93.

[KN]   S. Kaji and M. Nakagawa, *The Chow rings of the algebraic groups of type* E_6, E_7, *and* E_8, arXiv:0709.3702v3, preprint 2007.

[KS1]   M. Kashiwara and P. Schapira, *Sheaves on Manifolds*, Grund. Math. Wiss. **292**, Springer-Verlag, Berlin, 1994.

[KS2]   _____, *Categories and sheaves*, Grund. Math. Wiss. **332**, Springer-Verlag, Berlin, 2006.

[KPV]   K. Kedlaya, B. Poonen, and R. Vakil, *The William Lowell Putnam Mathematical Competition 1985–2000: Problems, Solutions, and Commentary*, Math. Ass. of Amer., Washington, DC, 2002.

[Kl1]   S. Kleiman, "Toward a numerical theory of ampleness," Ann. Math. (2) **84** (1966), no. 3, 293–344.

[Kl2]   _____, "The transversality of a general translate," Compositio Math. **28** (1974), no. 3, 287–297.

[Kl3]   _____, "Cartier divisors versus invertible sheaves," Comm. Alg. **28** (2000), no. 12, 5677–5678.

[Ko1]   J. Kollár, *Rational Curves on Algebraic Varieties*, Ergeb. Math. Grenz. **32**, Springer-Verlag, Berlin, 1996.

[Ko2]   _____, *Lectures on Resolutions of Singularities*, Annals of Math. Stud. **166**, Princeton U. P., Princeton, NJ, 2007.

[Lan]   S. Lang, *Algebra* (rev. 3rd ed.), Grad. Texts in Math. **211**, Springer-Verlag, New York, 2002.

[Laz]   R. Lazarsfeld, *Positivity in Algebraic Geometry I: Classical Setting: Line Bundles and Linear Series*, Ergeb. Math. Grenz. **48**, Springer-Verlag, Berlin, 2004.

[Lef]   S. Lefschetz, *L'analysis situs et la géométrie algébrique*, Gauthier-Villars, Paris, 1924.

[Lew]   R. Lewontin, *The Triple Helix: Gene, Organism, and Environment*, Harvard U. P., Cambridge MA, 2000.

[Lit]   J. E. Littlewood, *Littlewood's Miscellany*, Cambridge U. P., Cambridge, 1986.

[Liu]   Q. Liu, *Algebraic Geometry and Arithmetic Curves*, R. Erné trans., Oxford Grad. Texts in Math. **6**, Oxford U. P., Oxford, 2002.

[Lur]   J. Lurie, *Higher Algebra*, preprint, September 18, 2017.

[Lüt]	W. Lütkebohmert, "On compactification of schemes," Manuscripta Math. **80** (1), 95–111.
[Mac]	S. Mac Lane, *Categories for the Working Mathematician*, 2nd ed., Grad. Texts in Math. **5**, Springer-Verlag, New York, 1998.
[Mat1]	H. Matsumura, *Commutative Algebra*, 2nd ed., Math. Lecture Note Series **56**, Benjamin/Cummings Publ. Co., Inc., Reading, MA, 1980.
[Mat2]	_____, *Commutative Ring Theory*, M. Reid trans., 2nd ed, Cambridge Stud. in Adv. Math. **8**, Cambridge U. P., Cambridge, 1989.
[MO33489]	http://mathoverflow.net/questions/33489/
[MO68421]	Answers by G. Elencwajg and M. Kim to http://mathoverflow.net/questions/68421/
[MO70920]	Answers by T. Ekedahl to http://mathoverflow.net/questions/70920/
[MO90551]	Answer by M. Brandenburg to http://mathoverflow.net/questions/90551/
[MO129242]	Answer by L. Moret-Bailly to http://mathoverflow.net/questions/129242/
[MO152008]	Answer by J. Stillwell to http://mathoverflow.net/questions/152008/
[MO286390]	Answers to https://mathoverflow.net/questions/286390/
[Mc]	C. McLarty, *The Rising Sea: Grothendieck on simplicity and generality*, in J. J. Gray and K. H. Parshall eds., *Episodes in the History of Modern Algebra (1800–1950)*, Amer. Math. Soc., Providence, RI, 2007.
[Mu1]	D. Mumford, "Pathologies of modular algebraic surfaces," Amer. J. Math. **83** (no. 2), Apr. 1961, 339–342.
[Mu2]	_____, *Lectures on Curves on an Algebraic Surface*, Princeton U. P., Princeton NJ, 1966.
[Mu3]	_____, *Abelian Varieties*, TIFR Studies in Math. 5, Oxford U. P., London, 1970.
[Mu4]	_____, *Curves and Their Jacobians*, Univ. of Michigan Press, Ann Arbor, MI, 1975.
[Mu5]	_____, *Hilbert's fourteenth problem — The finite generation of subrings such as rings of invariants*, in *Mathematical Developments Arising from Hilbert Problems, Part 2*, F. Browder ed., Amer. Math. Soc., Providence RI, 1976.
[Mu6]	_____, *Algebraic Geometry I: Complex Projective Varieties*, Classics in Math., Springer-Verlag, Berlin, 1995.
[Mu7]	_____, *The Red Book of Varieties and Schemes* (2nd expanded ed.), Lect. Notes in Math. **1358**, Springer-Verlag, Berlin, 1999. (Caution: In order to keep readers on their toes, a number of typos were introduced between the first and second editions.)
[Mur]	D. Murfet, http://therisingsea.org/notes/ModulesOverAScheme.pdf
[N]	F. Nietzsche, *Twilight of the Idols, or, How to Philosophize with a Hammer*, 1889.
[OB]	A. Ogus and G. Bergman, "Nakayama's lemma for half-exact functors," Proc. Amer. Math. Soc. **31** (1972), 67–74.
[OSS]	C. Okonek, M. Schneider, and H. Spindler, *Vector Bundles on Complex Projective Spaces*, corrected reprint of the 1988 edition, Birkhäuser/Springer Basel AG, Basel, 2011.
[Pi]	P. Picasso, *Statement to Marius de Zayas*, 1923, in *Picasso Speaks*, The Arts, New York, May 1923, pp. 315-26.
[Po1]	H. Poincaré, "Analysis situs," J. de l'École Polytechnique (2) **1** (1895), 1–123.
[Po2]	_____, "Papers on topology: Analysis situs and its five supplements," J. Stillwell trans., Amer. Math. Soc., Providence, RI, 2010.
[P1]	B. Poonen, "Varieties without extra automorphisms, I: Curves," Math. Res. Lett. **7** (2000), no. 1, 67–76.
[P2]	_____, *Rational Points on Varieties*, Grad. Stud. Math., **186**, Amer. Math. Soc., Providence, RI, 2017.
[Ra]	M. Raynaud, *Contre-exemple au vanishing theorem en caractéristique* $p > 0$, in *C. P. Ramanujam — A Tribute*, 273–278, TIFR Studies in Math. **8**, Springer, Berlin-New York, 1978.
[RS]	M. Reed and B. Simon, *Methods of Modern Mathematical Physics I: Functional Analysis*, Academic Press, Inc., New York, 1980.
[Rei1]	M. Reid, *Undergraduate Algebraic Geometry*, London Math. Soc. Student Texts 12, Cambridge U. P., Cambridge, 1988.
[Rei2]	_____, *Undergraduate Commutative Algebra*, London Math. Soc. Student Texts 29, Cambridge U. P., Cambridge, 1995.

[Rem]   R. Remmert, *Local theory of complex spaces*, in *Several Complex Variables VII: Sheaf-Theoretical Methods in Complex Analysis*, Encl. of Math. Sci. **74**, Springer-Verlag, Berlin, 1994.

[Roh]   F. Rohrer, "Quasicoherent sheaves on toric schemes," Expo. Math. **32** (2014), 33–78.

[RotV]   M. Roth and R. Vakil, *The affine stratification number and the moduli space of curves*, CRM Proceedings and Lecture Notes **38**, Univ. de Montréal, 2004, 213–227.

[Sa]   P. Salmon, "Sulla fattorialità delle algebre graduate e degli anelli locali," Rend. Sem. Mat. Univ. Padova **41** (1968), 119–138.

[Schm]   H. L. Schmid, "Über die Automorphismen eines algebraischen Funktionenkörpers von Primzahlcharacteristik," J. Reine Angew. Math. **179** (1938), 5–14.

[Se1]   J.-P. Serre, "Faisceaux algébriques cohérents," Ann. of Math. (2) **61** (1955), 197–278.

[Se2]   _____, *Sur la dimension cohomologique des anneaux et des modules noethériens*, in *Proc. Intern. Symp. on Alg. Number Theory*, 176–189, Tokyo-Nikko, 1955.

[Se3]   _____, "Géométrie algébrique et géométrie analytique," Ann. de l'institut Fourier **6** (1956), 1–42.

[Se4]   _____, *Sur la topologie des variétés algébriques en caractéristique p*, in *Symposium Internacional de Topología Algebraica, International Symposium on Algebraic Topology*, 24–53, Univ. Nacional Autónoma de México and UNESCO, Mexico City, 1958.

[Se5]   _____, *A Course in Arithmetic*, Grad. Texts in Math. **7**, Springer-Verlag, New York, 1973.

[SGA2]   A. Grothendieck, *Séminaire de Géométrie Algébrique du Bois Marie 1962: Cohomologie locale des faisceaux cohérents et Théorèmes de Lefschetz locaux et globaux (SGA2)*, North-Holland Publ. Co., Amsterdam, 1968.

[SGA6]   P. Berthelot, A. Grothendieck, and L. Illusie eds., *Séminaire de Géométrie Algébrique du Bois Marie 1966–67: Théorie des intersections et théorème de Riemann Roch (SGA 6)*, Lect. Notes in Math. **225**, Springer-Verlag, Berlin-New York, 1971.

[Sh]   I. Shafarevich, *Basic Algebraic Geometry 1: Varieties in Projective Space*, 2nd ed., M. Reid trans., Springer-Verlag, Berlin, 1994.

[Si1]   J. Silverman, *The Arithmetic of Elliptic Curves*, 2nd ed., Grad. Texts in Math. **206**, Springer-Verlag, New York-Berlin, 2009.

[Si2]   _____, *Advanced Topics in the Arithmetic of Elliptic Curves*, Grad. Texts in Math. **151**, Springer-Verlag, New York-Berlin, 1994.

[Stacks]   The Stacks Project Authors, *Stacks Project*, http://stacks.math.columbia.edu. (To look up tags: http://stacks.math.columbia.edu/tags. To search: http://stacks.math.columbia.edu/search.)

[Stic]   H. Stichtenoth, "Über die Automorphismengruppe eines algebraischen Funktionenkörpers von Primzahlcharakteristik. I. Eine Abschätzung der Ordnung der Automorphismengruppe," Arch. Math. (Basel) **24** (1973), 527–544.

[Ta]   A. Tarski, *A decision method for elementary algebra and geometry*, RAND Corp., Santa Monica, CA, 1948.

[TWi]   R. Taylor and A. Wiles, "Ring-theoretic properties of certain Hecke algebras," Ann. of Math. (2) **141** (3) (1995), 553–572.

[To]   L. Tolstoy, *Anna Karenina*, http://www.gutenberg.org/files/1399/1399-h/1399-h.htm.

[Tu]   L. Tu, *An Introduction to Manifolds*, Universitext, Springer-Verlag, New York, 2011.

[Vak1]   R. Vakil, *The Rising Sea: Foundations of Algebraic Geometry*, preprint.

[Vak2]   _____, *Puzzling through exact sequences: A bedtime story with pictures*, https://www.3blue1brown.com/blog/exact-sequence-picturebook.

[van]   B. L. van der Waerden, *Algebra* vol. II, J. R. Schulenberger trans., Springer-Verlag, New York, 1991.

[Var]   V. S. Varadarajan, *Lie Groups, Lie Algebras, and Their Representations*, Grad. Texts in Math. **102**, Springer-Verlag, New York-Berlin, 1984.

[Vo]   C. Voisin, *Hodge Theory and Complex Algebraic Geometry I* and *II*, L. Schneps trans., Cambridge Stud. in Adv. Math. **76** and **77**, Cambridge U. P., Cambridge, 2007.

[Weib]   C. Weibel, *An Introduction to Homological Algebra*, Cambridge Stud. in Adv. Math. **38**, Cambridge U. P., Cambridge, 1994.

[Weil]   A. Weil, *Scientific Works — Collected Papers Vol. II (1951–1964)*, Springer-Verlag, New York-Heidelberg, 1979.

[Wey]   H. Weyl, *Philosophy of Mathematics and Natural Science*, Princeton U. P., Princeton, NJ, 2009.

[Wi]   A. Wiles, "Modular elliptic curves and Fermat's last theorem," Ann. of Math. (2) **141** (3) (1995), 443–551.

[Wł]   J. Włodarczyk, "Toroidal varieties and the weak factorization theorem," Invent. Math. **154** (2003), no. 2, 223–331.

[Z]   O. Zariski, "Foundations of a general theory of birational correspondences," Trans. Amer. Math. Soc. **53** (1943), 490–542.

Index

Page numbers in boldface indicate definitions.